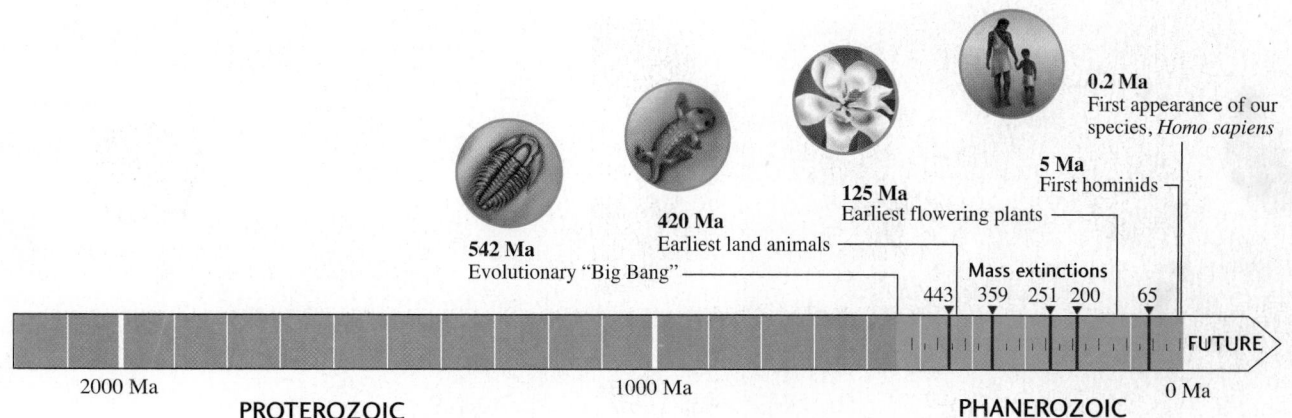

542 Ma
Evolutionary "Big Bang"

420 Ma
Earliest land animals

125 Ma
Earliest flowering plants

Mass extinctions

5 Ma
First hominids

0.2 Ma
First appearance of our
species, *Homo sapiens*

443 359 251 200 65

FUTURE

2000 Ma   1000 Ma   0 Ma

PROTEROZOIC   PHANEROZOIC

John Grotzinger
California Institute of Technology

Thomas H. Jordan
University of Southern California

W.H. Freeman and Company
A Macmillan Higher Education Company

UNDERSTANDING EARTH

7th Edition

Publisher: Jessica Fiorillo

Development Editor: Randi Blatt Rossignol

Acquisitions Editor: Bill Minick

Marketing Manager: Debbie Clare

Marketing Assistant: Samantha Zimbler

Media and Supplements Editor:
Amy Thorne

Associate Editor: Heidi Bamatter

Assistant Editor: Nicholas Ciani

Project Editor: Dennis Free, Aptara, Inc.

Photo Editor: Jennifer Atkins

Cover Design: Blake Logan

Interior Design: Patrice Sheridan

Illustrations: Precision Graphics

Illustration Coordinator: Janice Donnola

Production Manager: Paul Rohloff

Composition: Aptara, Inc.

Printing and Binding: RR Donnelley

Library of Congress Control Number: 2014930768

ISBN-13: 978-1-4641-3874-4
ISBN-10: 1-4641-3874-5

Printed in the United States of America

First printing

W. H. Freeman and Company
41 Madison Avenue
New York, NY 10010
Houndmills, Basingstoke RG21 6XS, England
www.whfreeman.com

We dedicate this book to Frank Press
and Ray Siever, pioneering educators
in the era of modern geology. This
book was possibly only because
they led the way.

MEET THE AUTHORS

John Grotzinger is a field geologist interested in the evolution of Earth's surface environments and biosphere. He also works on the early environmental evolution of Mars and on assessing its potential habitability. His research addresses the chemical development of the early oceans and atmosphere, the environmental context of early animal evolution, and the geologic factors that regulate sedimentary basins. His fieldwork has taken him to northwestern Canada, northern Siberia, southern Africa, the western United States, and via robot, to Mars. He received a B.S. in geoscience from Hobart College in 1979, an M.S. in geology from the University of Montana in 1981, and a Ph.D. in geology from Virginia Polytechnic Institute and State University in 1985. He spent three years as a research scientist at the Lamont-Doherty Geological Observatory before joining the MIT faculty in 1988. From 1979 to 1990, he was engaged in regional mapping for the Geological Survey of Canada. He currently works as the Chief Scientist for the Mars Curiosity Rover team, the first mission to assess the habitability of the ancient environments of another planet.

In 1998, Dr. Grotzinger was named the Waldemar Lindgren Distinguished Scholar at MIT, and in 2000, he became the Robert R. Shrock Professor of Earth and Planetary Sciences. In 2005, he moved from MIT to Caltech, where he is the Fletcher Jones Professor of Geology. He received the Presidential Young Investigator Award of the National Science Foundation in 1990, the Donath Medal of the Geological Society of America in 1992, the Charles Doolittle Walcott Medal of the National Academy of Sciences in 2007, and NASA's Outstanding Public Leadership Medal in 2013. He is a member of the American Academy of Arts and Sciences and the U.S. National Academy of Sciences.

Tom Jordan is a geophysicist interested in the composition, dynamics, and evolution of the solid Earth. He has conducted research into the nature of deep subduction, the formation of thickened keels beneath ancient continental cratons, and the question of mantle stratification. He has developed a number of seismological techniques for investigating Earth's interior that bear on geodynamic problems. He has also worked on modeling plate movements, measuring tectonic deformation, quantifying seafloor morphology, and characterizing large earthquakes. He received his Ph.D. in geophysics and applied mathematics at the California Institute of Technology (Caltech) in 1972 and taught at Princeton University and the Scripps Institution of Oceanography before joining the Massachusetts Institute of Technology (MIT) faculty as the Robert R. Shrock Professor of Earth and Planetary Sciences in 1984. He served as the head of MIT's Department of Earth, Atmospheric and Planetary Sciences for the decade 1988–1998. He moved from MIT to the University of Southern California (USC) in 2000, where he is University Professor and W. M. Keck Professor of Earth Sciences. He is currently the director of the Southern California Earthquake Center, where he coordinates an international research program in earthquake system science that involves over 600 scientists at more than 60 universities and research organizations.

Dr. Jordan received the Macelwane Medal of the American Geophysical Union in 1983, the Woollard Award of the Geological Society of America in 1998, and the Lehmann Medal of the American Geophysical Union in 2005. He is a member of the American Academy of Arts and Sciences, the U.S. National Academy of Sciences, and the American Philosophical Society.

BRIEF CONTENTS

Chapter 1
The Earth System   3

Chapter 2
Plate Tectonics: The Unifying Theory   27

Chapter 3
Earth Materials: Minerals and Rocks   57

Chapter 4
Igneous Rocks: Solids from Melts   91

Chapter 5
Sedimentation: Rocks Formed by
Surface Processes   115

Chapter 6
Metamorphism: Alteration of Rocks
by Temperature and Pressure   149

Chapter 7
Deformation: Modification of Rocks
by Folding and Fracturing   171

Chapter 8
Clocks in Rocks: Timing the
Geologic Record   195

Chapter 9
Early History of the Terrestrial
Planets   221

Chapter 10
History of the Continents   253

Chapter 11
Geobiology: Life Interacts
with Earth   283

Chapter 12
Volcanoes   313

Chapter 13
Earthquakes   347

Chapter 14
Exploring Earth's Interior   383

Chapter 15
The Climate System 407

Chapter 16
Weathering, Erosion, and Mass
Wasting: Interactions Between the
Climate and Plate Tectonic Systems 435

Chapter 17
The Hydrologic Cycle and Groundwater 466

Chapter 18
Stream Transport: From
Mountains to Oceans 499

Chapter 19
Winds and Deserts 529

Chapter 20
Coastlines and Ocean Basins 551

Chapter 21
Glaciers: The Work of Ice 587

Chapter 22
Landscape Development 617

Chapter 23
The Human Impact on Earth's
Environment 641

Appendix 1 Conversion Factors AP-1

Appendix 2 Numerical Data Pertaining
to Earth AP-2

Appendix 3 Chemical Reactions AP-3

Appendix 4 Properties of the Most Common
Minerals of Earth's Crust AP-5

Appendix 5 Practicing Geology Exercises:
Answers to Problems AP-9

Glossary GL-1

Index I-1

CONTENTS

Meet the Authors   vii

Preface   xix

Chapter 1

The Earth System   3

The Scientific Method   4

Geology as a Science   5

Earth's Shape and Surface   8

**Peeling the Onion: Discovery
of a Layered Earth**   9
Earth's Density 10
The Mantle and Core 10
The Crust 11
The Inner Core 11
*Chemical Composition of Earth's
Major Layers 12*

**Earth as a System of Interacting
Components**   13
The Climate System 15
The Plate Tectonic System 16
The Geodynamo 17
*Interactions Among Geosystems
Support Life 18*

An Overview of Geologic Time   18
*The Origin of Earth and Its Global
Geosystems 19*
The Evolution of Life 19

Welcome to Google Earth   21

Chapter 2

Plate Tectonics: The Unifying Theory   27

The Discovery of Plate Tectonics   28
Continental Drift 28
Seafloor Spreading 29
The Great Synthesis: 1963–1968 30

The Plates and Their Boundaries   31
Divergent Boundaries 36
Convergent Boundaries 36
Transform Faults 38
Combinations of Plate Boundaries 38

Rates and History of Plate Movements   39
*The Seafloor as a Magnetic
Tape Recorder 39*
Deep-Sea Drilling 41
*Measurements of Plate Movements
by Geodesy 41*

The Grand Reconstruction   42
Seafloor Isochrons 43
*Reconstructing the History
of Plate Movements 43*
The Breakup of Pangaea 44
*The Assembly of Pangaea
by Continental Drift 44*
Implications of the Grand Reconstruction 44

**Mantle Convection: The Engine of
Plate Tectonics**   45
*Where Do the Plate-Driving
Forces Originate? 45*

How Deep Does Plate Recycling Occur? 48
*What Is the Nature of Rising
   Convection Currents?* 49

**The Theory of Plate Tectonics and
   the Scientific Method**   52

Chapter 3

Earth Materials: Minerals and Rocks   57

What Are Minerals?   58

The Structure of Matter   59
The Structure of Atoms 59
Atomic Number and Atomic Mass 59
Chemical Reactions 60
Chemical Bonds 61

The Formation of Minerals   61
The Atomic Structure of Minerals 61
The Crystallization of Minerals 62
How Do Minerals Form? 63

Classes of Rock-Forming Minerals   64
Silicates 65
Carbonates 67
Oxides 67
Sulfides 67
Sulfates 68

Physical Properties of Minerals   69
Hardness 69
Cleavage 70
Fracture 71
Luster 71
Color 72
Density 73
Crystal Habit 73

What Are Rocks?   74
Properties of Rocks 74
Igneous Rocks 76
Sedimentary Rocks 77
Metamorphic Rocks 78

**The Rock Cycle: Interactions Between
   the Plate Tectonic and Climate Systems**   79

**Concentrations of Valuable
   Mineral Resources**   81
Hydrothermal Deposits 82
Igneous Deposits 83
Sedimentary Deposits 84

Chapter 4

Igneous Rocks: Solids from Melts   91

**How Do Igneous Rocks Differ
   from One Another?**   92
Texture 92
Chemical and Mineral Composition 94

How Do Magmas Form?   97
How Do Rocks Melt? 97
The Formation of Magma Chambers 99
Where Do Magmas Form? 99

Magmatic Differentiation   99
*Fractional Crystallization:
   Laboratory and Field Observations* 99
*Granite from Basalt: Complexities
   of Magmatic Differentiation* 101

Forms of Igneous Intrusions   102
Plutons 103
Sills and Dikes 103
Veins 105

Igneous Processes and Plate Tectonics   105
Spreading Centers as Magma Factories 106
Subduction Zones as Magma Factories 108
Mantle Plumes as Magma Factories 111

Chapter 5

Sedimentation: Rocks Formed
by Surface Processes   115

Surface Processes of the Rock Cycle   116
*Weathering and Erosion:
   The Source of Sediments* 117
*Transportation and Deposition: The
   Downhill Journey to Sedimentary Basins* 119
Oceans as Chemical Mixing Vats 122

**Sedimentary Basins: The Sinks
   for Sediments**   122
Rift Basins and Thermal Subsidence Basins 122
Flexural Basins 123

Sedimentary Environments   124
*Continental Sedimentary
   Environments* 124
Shoreline Sedimentary Environments 124
Marine Sedimentary Environments 124
*Siliciclastic versus Chemical and
   Biological Sedimentary Environments* 126

Sedimentary Structures   127
Cross-Bedding 127
Graded Bedding 127
Ripples 128
Bioturbation Structures 128
Bedding Sequences 129

**Burial and Diagenesis:
   From Sediment to Rock**   130
Burial 130
Diagenesis 130

**Classification of Siliciclastic Sediments
   and Sedimentary Rocks**   132
*Coarse-Grained Siliciclastics:
   Gravel and Conglomerate* 133

Medium-Grained Siliciclastics:
Sand and Sandstone 133
Fine-Grained Siliciclastics 134

Classification of Chemical and Biological
Sediments and Sedimentary Rocks 135
Carbonate Sediments and Rocks 136
Evaporite Sediments and Rocks:
Products of Evaporation 140
Other Biological and Chemical Sediments 142

Chapter 6

Metamorphism: Alteration of Rocks
by Temperature and Pressure 149

Causes of Metamorphism 150
The Role of Temperature 151
The Role of Pressure 152
The Role of Fluids 153

Types of Metamorphism 153
Regional Metamorphism 154
Contact Metamorphism 154
Seafloor Metamorphism 154
Other Types of Metamorphism 154

Metamorphic Textures 155
Foliation and Cleavage 155
Foliated Rocks 155
Granoblastic Rocks 157
Porphyroblasts 158

Regional Metamorphism
and Metamorphic Grade 159
Mineral Isograds: Mapping Zones
of Change 159
Metamorphic Grade and Parent
Rock Composition 160
Metamorphic Facies 161

Plate Tectonics and Metamorphism 162
Metamorphic Pressure-Temperature
Paths 162
Ocean-Continent Convergence 163
Continent-Continent Collision 165
Exhumation: A Link Between the
Plate Tectonic and Climate Systems 166

Chapter 7

Deformation: Modification of Rocks
by Folding and Fracturing 171

Plate Tectonic Forces 172
Mapping Geologic Structure 172
Measuring Strike and Dip 173
Geologic Maps 173
Geologic Cross Sections 175

How Rocks Deform 175
Brittle and Ductile Behavior of Rocks
in the Laboratory 175
Brittle and Ductile Behavior of Rocks
in Earth's Crust 176

Basic Deformation Structures 177
Faults 177
Folds 179
Circular Structures 182
Joints 183
Deformation Textures 183

Styles of Continental
Deformation 184
Tensional Tectonics 184
Compressive Tectonics 185
Shearing Tectonics 187

Unraveling Geologic History 188

Chapter 8

Clocks in Rocks: Timing
the Geologic Record 195

Reconstructing Geologic History from
the Stratigraphic Record 196
Principles of Stratigraphy 197
Fossils as Recorders of
Geologic Time 198
Unconformities: Gaps in the
Geologic Record 200
Cross-Cutting Relationships 201

The Geologic Time Scale:
Relative Ages 202
Intervals of Geologic Time 202
Interval Boundaries Mark
Mass Extinctions 202
Ages of Petroleum Source Rocks 206

Measuring Absolute Time
with Isotopic Clocks 207
Discovery of Radioactivity 208
Radioactive Isotopes: The Clocks
in Rocks 208
Isotopic Dating Methods 209

The Geologic Time Scale:
Absolute Ages 210
Eons: The Longest Intervals
of Geologic Time 210
Perspectives on Geologic Time 211

Recent Advances in Timing
the Earth System 212
Sequence Stratigraphy 212
Chemical Stratigraphy 213
Paleomagnetic Stratigraphy 213
Clocking the Climate System 213

Chapter 9
Early History of the Terrestrial Planets 221

Origin of the Solar System 222
The Nebular Hypothesis 222
The Sun Forms 223
The Planets Form 223
Small Bodies of the Solar System 224

Early Earth: Formation of a Layered Planet 225
Earth Heats Up and Melts 225
Differentiation of Earth's Core, Mantle, and Crust 226
Earth's Oceans and Atmosphere Form 227

Diversity of the Planets 228

What's in a Face? The Age and Complexion of Planetary Surfaces 230
The Man in the Moon: A Planetary Time Scale 230
Mercury: The Ancient Planet 231
Venus: The Volcanic Planet 233
Mars: The Red Planet 235
Earth: No Place Like Home 236

Mars Rocks! 237
Missions to Mars: Flybys, Orbiters, Landers, and Rovers 238
Mars Exploration Rovers: Spirit *and* Opportunity 240
Mars Science Laboratory (MSL): Curiosity 240
Recent Missions: Mars Reconnaissance Orbiter *and* Phoenix 243
Recent Discoveries: The Environmental Evolution of Mars 243

Exploring the Solar System and Beyond 245
Space Missions 245
The Cassini-Huygens Mission to Saturn 246
Other Solar Systems 248

Chapter 10
History of the Continents 253

The Structure of North America 254
The Stable Interior 254
The Appalachian Fold Belt 256
The Coastal Plain and Continental Shelf 256
The North American Cordillera 257

Tectonic Provinces Around the World 259
Types of Tectonic Provinces 260
Tectonic Ages 261
A Global Puzzle 262

How Continents Grow 262
Magmatic Addition 262
Accretion 263

How Continents Are Modified 266
Orogeny: Modification by Plate Collision 266
The Wilson Cycle 271
Epeirogeny: Modification by Vertical Movements 272

The Origins of Cratons 273

The Deep Structure of Continents 276
Cratonic Keels 276
Composition of the Keels 277
Age of the Keels 278

Chapter 11
Geobiology: Life Interacts with Earth 283

The Biosphere as a System 284
Ecosystems 284
Inputs: The Stuff Life Is Made Of 285
Processes and Outputs: How Organisms Live and Grow 286
Biogeochemical Cycles 288

Microorganisms: Nature's Tiny Chemists 288
Abundance and Diversity of Microorganisms 289
Microorganism-Mineral Interactions 292

Geobiologic Events in Earth's History 296
Origin of Life and the Oldest Fossils 296
Prebiotic Soup: The Original Experiment on the Origin of Life 297
The Oldest Fossils and Early Life 298
Origin of Earth's Oxygenated Atmosphere 299

Evolutionary Radiations and Mass Extinctions 300
Radiation of Life: The Cambrian Explosion 301
Tail of the Devil: The Demise of Dinosaurs 303
Global Warming Disaster: The Paleocene-Eocene Mass Extinction 304

Astrobiology: The Search for Extraterrestrial Life 306
Habitable Zones Around Stars 307
Habitable Environments on Mars 308

Chapter 12
Volcanoes 313

Volcanoes as Geosystems 314

Lavas and Other Volcanic Deposits 315
Types of Lava 315
Textures of Volcanic Rocks 319
Pyroclastic Deposits 319

Eruptive Styles and Landforms 320
Central Eruptions 321
Fissure Eruptions 325

**Interactions of Volcanoes with
Other Geosystems** 326
Volcanism and the Hydrosphere 327
Volcanism and the Atmosphere 328

The Global Pattern of Volcanism 329
Volcanism at Spreading Centers 330
Volcanism in Subduction Zones 330
*Intraplate Volcanism: The Mantle
Plume Hypothesis 331*

Volcanism and Human Affairs 334
Volcanic Hazards 334
Reducing the Risks of Volcanic Hazards 337
Natural Resources from Volcanoes 339

Chapter 13

Earthquakes 347

What Is an Earthquake? 348
The Elastic Rebound Theory 349
Fault Rupture During Earthquakes 351
Foreshocks and Aftershocks 351

How Do We Study Earthquakes? 353
Seismographs 354
Seismic Waves 355
Locating the Focus 357
Measuring the Size of an Earthquake 357
Determining Fault Mechanisms 361
*GPS Measurements and
"Silent" Earthquakes 362*

Earthquakes and Patterns of Faulting 362
*The Big Picture: Earthquakes
and Plate Tectonics 362*
Regional Fault Systems 364

Earthquake Hazards and Risks 365
How Earthquakes Cause Damage 366
Reducing Earthquake Risk 369

Can Earthquakes Be Predicted? 375
Long-Term Forecasting 375
Short-Term Prediction 375
Medium-Term Forecasting 376

Chapter 14

Exploring Earth's Interior 383

**Exploring Earth's Interior with
Seismic Waves** 384
Basic Types of Waves 384
Paths of Seismic Waves Through Earth 384
*Seismic Exploration of Near-Surface
Layering 387*

**Layering and Composition
of Earth's Interior** 388
The Crust 388
The Mantle 388
The Core-Mantle Boundary 390
The Core 390

Earth's Internal Temperature 390
Heat Flow Through Earth's Interior 390
Temperatures Inside Earth 392

**Visualizing Earth's Three-Dimensional
Structure** 394
Seismic Tomography 394
Earth's Gravitational Field 396

**Earth's Magnetic Field and
The Geodynamo** 396
The Dipole Field 396
Complexity of the Magnetic Field 396
Paleomagnetism 400
*The Magnetic Field and
the Biosphere 402*

Chapter 15

The Climate System 407

Components of the Climate System 408
The Atmosphere 408
The Hydrosphere 409
The Cryosphere 411
The Lithosphere 411
The Biosphere 412

The Greenhouse Effect 413
A Planet Without Greenhouse Gases 413
Earth's Greenhouse Atmosphere 414
*Balancing the Climate System
Through Feedbacks 414*
*Climate Models and
Their Limitations 415*

Climate Variation 416
Short-Term Regional Variations 417
*Long-Term Global Variations:
The Pleistocene Ice Ages 417*
*Long-Term Global Variations:
Paleozoic and Proterozoic Ice Ages 422*
*Variations During the Most Recent
Glacial Cycle 422*

The Carbon Cycle 423
*Geochemical Cycles and
How They Work 423*
The Cycling of Carbon 426
*Human Perturbations
of the Carbon Cycle 427*

**Twentieth-Century Warming:
Fingerprints of Anthropogenic
Global Change** 428

Chapter 16

Weathering, Erosion, and Mass Wasting: Interactions Between the Climate and Plate Tectonic Systems   435

Weathering, Erosion, Mass Wasting, and the Rock Cycle   436

Controls on Weathering   436
The Properties of Parent Rock   436
Climate: Rainfall and Temperature   437
The Presence or Absence of Soil   437
The Length of Exposure   437

Chemical Weathering   438
The Role of Water: Feldspar and Other Silicates   438
Carbon Dioxide, Weathering, and the Climate System   439
The Role of Oxygen: From Iron Silicates to Iron Oxides   441
Chemical Stability   442

Physical Weathering   443
How Do Rocks Break?   443
Interactions Between Weathering and Erosion   444

Soils: The Residue of Weathering   445
Soils as Geosystems   446
Paleosols: Working Backward from Soil to Climate   447

Mass Wasting   447
Slope Materials   449
Water Content   450
Slope Steepness   451
Triggers of Mass Movements   452

Classification of Mass Movements   453
Mass Movements of Rock   455
Mass Movements of Unconsolidated Material   456

Understanding the Origins of Mass Movements   460
Natural Causes of Mass Movements   461
Human Activities That Promote or Trigger Mass Movements   462

Chapter 17

The Hydrologic Cycle and Groundwater   469

The Geologic Cycling of Water   470
Flow and Reservoirs   470
How Much Water Is There?   470
The Hydrologic Cycle   470
How Much Water Can We Use?   471

Hydrology and Climate   472
Humidity, Rainfall, and Landscape   472
Droughts   473
The Hydrology of Runoff   475

The Hydrology of Groundwater   477
Porosity and Permeability   478
The Groundwater Table   480
Aquifers   482
Balancing Recharge and Discharge   482
The Speed of Groundwater Flows   485
Groundwater Resources and Their Management   485

Erosion by Groundwater   488

Water Quality   490
Contamination of the Water Supply   490
Reversing Contamination   491
Is the Water Drinkable?   491

Water Deep in the Crust   492
Hydrothermal Waters   492
Ancient Microorganisms in Deep Aquifers   494

Chapter 18

Stream Transport: From Mountains to Oceans   499

The Form of Streams   500
Stream Valleys   500
Channel Patterns   500
Stream Floodplains   503
Drainage Basins   503
Drainage Networks   505
Drainage Patterns and Geologic History   505

Where Do Channels Begin? How Running Water Erodes Soil and Rock   507
Abrasion   508
Chemical and Physical Weathering   508
The Undercutting Action of Waterfalls   508

How Currents Flow and Transport Sediment   509
Erosion and Sediment Transport   510
Sediment Bed Forms: Dunes and Ripples   511

Deltas: The Mouths of Rivers   513
Delta Sedimentation   513
The Growth of Deltas   513
Human Effects on Deltas   515
The Effects of Ocean Currents, Tides, and Plate Tectonic Processes   515

Streams as Geosystems   515
Discharge   516
Floods   518
Longitudinal Profiles   519
Lakes   522

Chapter 19

Winds and Deserts 529

Global Wind Patterns 530

Wind as a Transport Agent 531
Wind Strength 531
Particle Size 531
Surface Conditions 531
Materials Carried by Wind 531

Wind as an Agent of Erosion 534
Sandblasting 534
Deflation 534

Wind as a Depositional Agent 535
Where Sand Dunes Form 535
How Sand Dunes Form and Move 536
Dune Types 538
Dust Falls and Loess 538

The Desert Environment 540
Where Deserts Are Found 540
Desert Weathering and Erosion 542
Desert Sediments and Sedimentation 543
Desert Landscapes 544

Chapter 20

Coastlines and Ocean Basins 551

How Ocean Basins Differ from Continents 552

Coastal Processes 552
*Wave Motion: The Key to Shoreline
Dynamics* 553
The Surf Zone 554
Wave Refraction 555
Tides 556
*Hurricanes and Coastal
Storm Surges* 557

The Shaping of Shorelines 564
Beaches 564
*Erosion and Deposition at
Shorelines* 567
Effects of Sea Level Change 568

Continental Margins 570
The Continental Shelf 571
The Continental Slope and Rise 571
Submarine Canyons 571

Topography of the Deep Seafloor 572
*Probing the Seafloor from
Surface Ships* 572
Charting the Seafloor by Satellite 575
Profiles Across Two Oceans 575
Main Features of the Deep Seafloor 578

Ocean Sedimentation 578
Sedimentation on the Continental Shelf 579
Deep-Sea Sedimentation 579

Chapter 21

Glaciers: The Work of Ice 587

Ice as a Rock 588
Valley Glaciers 588
Continental Glaciers 590

How Glaciers Form 591
*Basic Ingredients: Freezing Cold and Lots
of Snow* 591
Glacial Growth: Accumulation 592
Glacial Shrinkage: Ablation 592
*The Glacial Budget: Accumulation
Minus Ablation* 592

How Glaciers Move 594
Mechanisms of Glacial Flow 594
Flow in Valley Glaciers 594
Antarctica in Motion 594

Glacial Landscapes 598
*Glacial Erosion and Erosional
Landforms* 599
*Glacial Sedimentation and Sedimentary
Landforms* 603
Permafrost 606

Glacial Cycles and Climate Change 606
The Wisconsin Glaciation 607
Glaciation and Sea Level Change 608
*The Geologic Record of Pleistocene
Glaciations* 609
The Geologic Record of Ancient Glaciations 609

Chapter 22

Landscape Development 617

Topography, Elevation, and Relief 618

**Landforms: Features Sculpted by Erosion
and Sedimentation** 621
Mountains and Hills 621
Plateaus 623
Stream Valleys 623
Structurally Controlled Ridges and Valleys 626
Structurally Controlled Cliffs 626

**How Interacting Geosystems Control
Landscapes** 626
*Feedback Between Climate and
Topography* 628
Feedback Between Uplift and Erosion 629

Models of Landscape Development 630
*Davis's Cycle: Uplift Is Followed
by Erosion* 632
*Penck's Model: Erosion Competes
with Uplift* 634
*Hack's Model: Erosion and Uplift Achieve
Equilibrium* 636

Chapter 23
The Human Impact on Earth's Environment 641

Civilization as a Global Geosystem 642
Natural Resources 642
Energy Resources 643
Rise of the Carbon Economy 644
Global Energy Consumption 645
Energy Resources for the Future 646
Carbon Flux from Energy Production 646

Fossil-Fuel Resources 646
How Do Oil and Gas Form? 647
Where Do We Find Oil and Gas? 647
Distribution of Oil Reserves 648
Oil Production and Consumption 649
When Will We Run Out of Oil? 650
Oil and the Environment 650
Natural Gas 651
Coal 653
Unconventional Hydrocarbon
  Resources 655

Alternative Energy Resources 655
Nuclear Energy 655
Biofuels 657
Solar Energy 658
Hydroelectric Energy 658
Wind Energy 658
Geothermal Energy 660

Global Change 660
Greenhouse Gases and Global Warming 660
Predictions of Future Global Warming 661
Consequences of Climate Change 662
Ocean Acidification 666
Loss of Biodiversity 666

Earth System Engineering and Management 667
Energy Policy 668
Use of Alternative Energy Resources 669
Engineering the Carbon Cycle 669
Stabilizing Carbon Emissions 669
Sustainable Development 670

Appendix 1 Conversion Factors AP-1

Appendix 2 Numerical Data Pertaining to Earth AP-2

Appendix 3 Chemical Reactions AP-3

Appendix 4 Properties of the Most Common Minerals of Earth's Crust AP-5

Appendix 5 Practicing Geology Exercises: Answers to Problems AP-9

Glossary G-1

Index I-1

PREFACE

■ Our Vision

Geology is everywhere in our daily lives. We are surrounded by materials and resources extracted from the Earth, from jewelry to the gasoline we use to fuel our cars, to the water we drink. Geological science routinely informs the decisions of public policy leaders in government, industry, and community organizations. Understanding our Earth has never been more important.

Because Earth science is so intertwined with our daily lives, our discipline evolves as the years go by, responding to the needs of what society compels us to understand. Decades ago most geologists worked in oil and mining companies, but today there is an exploding need for environmental specialists. As our world population grows we see the increased impact of hurricanes, tornadoes, and other environmental forces such as landslides. Even in the search for life on other planets we increasingly see the need for geological expertise in helping to reconstruct the environments on planets like Mars. There, geologists are exploring for traces of past life in rocks that are billions of years old, with robots that are hundreds of millions of miles away.

These diverse needs require a strong understanding of the basic concepts and principles of Earth science. Although the times change and the applications vary, understanding the basic composition of geologic materials, their origins, and how the planet acts as a system is imperative to understanding Earth. Everything from climate change, to the abundance of groundwater, to the frequency of large storms and volcanic eruptions, to the

location and cost of extracting rare elements from Earth is relevant. It is a simple fact that as the complexity of these challenges increases, the need for well-educated geologists to make wise decisions will increase as well. We bring that conviction to this book.

■ Content Updates and Revisions

Since the publication of the sixth edition, we have witnessed some major geologic events, seen new data on climate trends and global climate change, discovered new sources of natural resources and more modern methods of retrieving them, and have new policies that address how we impact and are impacted by geologic events. Some of the updated topics, as well as topics new to the seventh edition, are listed below:

Interactions among geosystems support life (Chapter 1)

Past climate changes (Chapter 2)

Ages of petroleum-source rocks (Chapter 8)

Current status and findings of Mars Mission (Chapters 9 and 11)

Iceland volcano, eruption clouds, and air traffic (Chapter 12)

Christchurch, New Zealand, earthquake of September 2010 (Chapter 13)

Tohoku, Japan, earthquake and tsunami of March 2011 (Chapter 13)

Earth Issues essay on the L'Aquila, Italy, earthquake and subsequent trial (Chapter 13)

Haiti earthquake of January 2010 (Chapter 13)

Current global seismicity maps (Chapter 13)

Land use policies, including the construction of nuclear power facilities (Chapter 13)

Earthquake and tsunami early warning systems (Chapter 13)

Current seismic tomography models of Earth's mantle (Chapter 14)

Twentieth-century warming (Chapter 15)

Recent drought in New Zealand (Chapter 17)

Discovery of Kenya aquifers and ancient aquifers (Chapter 17)

Hurricane Sandy (Chapter 20)

New IPCC data about the state of the East Antarctic ice sheet and changes in sea level (Chapter 21)

New trends in energy use (Chapter 23)

Hydraulic fracturing (fracking) as a method for extracting oil and gas (Chapters 5 and 23)

New IPCC scenarios for climate change (Chapter 23)

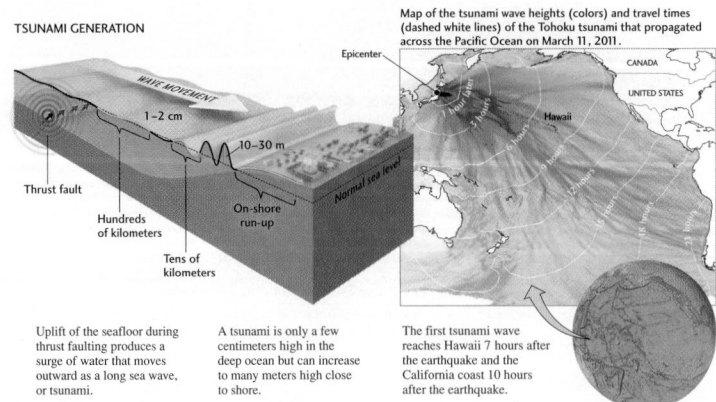

FIGURE 13.22 ▪ Earthquakes on megathrusts may generate tsunamis that can propagate across ocean basins. [Map by NOAA, Pacific Marine Environmental Laboratory.]

(a) Borehole is cased and surrounded by cement.

(b) Small holes are blasted through casing and cement.

(c) Surrounding rocks are hydraulically fractured by pumping water and sand into borehole at high pressure.

(d) Fracking generates small fissures that are kept open by the sand, allowing oil and gas to flow up the borehole to the wellhead.

FIGURE 23.15 ▪ Hydraulic fracturing or "fracking" is a technique for withdrawing oil and gas from shale and other tight formations by first pumping water and sand into a borehole at high pressures to create fractures through which the oil and gas can more readily flow. The boreholes are commonly drilled horizontally through nearly flat-lying shale formations.

▪ Emphasizing What Geologists Do

If you ask the question, "What do geologists do?" the answer will most likely be something about the study of rocks, volcanoes, or earthquakes. As with many sciences, a more complete understanding of the field of geology is obtained only through its study. It is up to us as instructors to teach our students that the price of gasoline depends partly on the work of geologists who study oil deposits; that geologists help to determine the safety of building locations; and that the water emerging from their faucets is brought to them with the help of geologists. The introductory geology course presents us with an extraordinary opportunity not only to share with students the beauty and power of geology, but also to cultivate greater appreciation for the work of all scientists and a better understanding of the world around us. Several features contribute to our effort to engage students in what geologists do.

Practicing Geology Exercises help students connect to important work currently under way in the field, making cutting-edge research and problem solving accessible to students at all levels. These exercises provide enough background for an informed discussion or activity based on the topic. Each essay includes detailed visualizations of the issue at hand, as well as a question that asks students to apply their knowledge independently. Practicing Geology exercises address questions such as:

- How Big Is Our Planet?
- What Happened in Baja? How Geologists Reconstruct Plate Movements
- Is It Worth Mining?
- How Do Valuable Metallic Ores Form?
- Organic-Rich Shales: Where Do We Look for Oil and Gas?
- How Do We Read Geologic History in Crystals?
- How Do We Use Geologic Maps to Find Oil?
- How Do Isotopes Tell Us the Ages of Earth Materials?
- How Do We Land a Spacecraft on Mars? Seven Minutes of Terror
- How Fast Are the Himalaya Rising and How Quickly Are They Eroding?
- How Do Geobiologists Find Evidence of Early Life in Rocks?
- Are the Siberian Traps a Smoking Gun of Mass Extinction?
- Can Earthquakes Be Controlled?
- The Principle of Isostasy: Why Are Oceans Deep and Mountains High?
- Where's the Missing Carbon?
- What Makes a Slope Too Unstable to Build On?
- How Much Water Can Our Well Produce?
- Can We Paddle Today? Using Streamgauge Data to Plan a Safe and Enjoyable River Trip
- Can We Predict the Extent of Desertification?
- Does Beach Restoration Work?
- Why Is Sea Level Rising?
- How Fast Do Streams Erode Bedrock?

PRACTICING GEOLOGY EXERCISE

How Fast Are the Himalaya Rising, and How Quickly Are They Eroding?

The Himalaya, the world's highest and most rugged mountains, are being raised by thrust faulting caused by the collision of India with Asia (see Figure 10.15). How rapidly are they rising, and how quickly are they being eroded away? The answers to these questions depend on accurate topographic mapping.

On February 6, 1800, Colonel William Lambton, of the 33rd Regiment of Foot of the British Army, received orders to begin the Great Trigonometrical Survey of India, the most ambitious scientific project of the nineteenth century. Over the next several decades, intrepid British explorers led by Lambton and his successor, George Everest, hauled bulky telescopes and heavy surveying equipment through the jungles of the Indian subcontinent, triangulating the positions of reference monuments established at high points in the terrain, from which they could accurately establish Earth's size and shape. Along the way, in 1852, the surveyors discovered that an obscure Himalayan peak, known on their maps only as "Peak XV," was the highest mountain on Earth. They promptly named it Mount Everest, in honor of their former boss. Its official Tibetan name, Chomolungma, means "Mother of the Universe."

On February 11, 2000, almost exactly 200 years after Lambton commenced his exploration, NASA launched another great survey, the Shuttle Radar Topography Mission (SRTM). The space shuttle *Endeavour* carried two large radar antennas into low Earth orbit, one in the cargo bay and the second mounted on a mast that could extend up to 60 m outward. Working together like a pair of eyes, these antennas mapped the height of the land

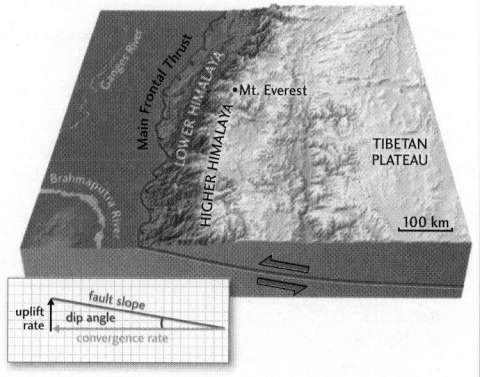

Cross section of the Himalaya, showing the approximate location of the thrust fault that is uplifting the mountains. The dip angle is about 10°.

Google Earth Projects. Satellite views of Earth are commonplace on news programs, on mapping Web sites, and in other aspects of popular media. Google Earth is by far the most widely used virtual globe browser. Taking advantage of student familiarity with these images and software, the Google Earth Projects guide students through focused explorations of key geologic locations. Balancing observation, core geologic concepts, geographic awareness, guided inquiry, and active learning, the students work through a series of questions aimed at producing a unique and insightful experience. After navigating to the appropriate location and checking their position with the image provided, students may answer the questions in a free-response format or within the text's accompanying learning system, GeologyPortal, which can automatically store and grade student responses.

Google Earth Project

Water, one of the most prolific weathering and transport agents on Earth, is constantly moving material from one location to another. Google Earth is an ideal tool for interpreting and appreciating this uniquely surficial process. Large rivers such as the Mississippi illustrate how efficiently river systems can gather sediment from mountainous regions of a continent (a source area) and transport it to the ocean, where deltas form (a sink area). What kinds of drainage and channel patterns do you find in the Mississippi drainage basin? How does the slope of the river channel change as one moves downstream? These questions and many more can be explored though the GE interface.

Image © USDA Farm Service Agency Image © 2009 TerraMetrics Data SIO, NOAA, U.S. Navy, GEBCO

This image shows the continental scale of the Mississippi River, from near its point of origin (Ft. Benton) to where it enters the Gulf of Mexico near New Orleans.

LOCATION   Missouri-Mississippi drainage basin, United States

GOAL   Understand source-to-sink transportation of sediment by river systems; observe meandering rivers with point bars, eroded outside banks, and oxbow lakes

LINKED   Figure 18.20

1. Type "Ft. Benton, Montana, United States" into the GE search window. Once you arrive there, zoom out to an eye altitude of 35 km. You will be looking down on the Missouri River, the longest tributary of the Mississippi River. Examine the stretch of river that flows from the southwest to the northeast through town and describe the channel pattern you see.

   a. distributary
   b. braided
   c. meandering
   d. artificially straightened

2. Using the cursor, determine the change in the elevation of the Missouri River channel over the 525 km between Ft. Benton, Montana, and Williston, North Dakota. Now compare that value with the change in the elevation of the Mississippi River channel over the same distance between Memphis, Tennessee, and Baton Rouge, Louisiana, to the south. Which relationship is most accurate?

   a. The slope of the Missouri River is steeper than that of the Mississippi River.
   b. The slope of the Mississippi River is steeper than that of the Missouri River.

FIGURE 10.7 ■ Image synthesized from satellite data of the Teton Range, Wyoming. The sharp eastern face of the mountain range, which has a vertical relief of more than 2000 m, is the result of normal faulting along the northeastern edge of the Basin and Range province. The view is from the northeast looking to the southwest. Grand Teton Mountain, near the center of the image, rises to an altitude of 4200 meters. [NASA/Goddard Space Flight Center, Landsat 7 Team.]

Field Sketches Introductory geology is widely known for being a particularly visual course. We are fortunate to display stunning vistas and spectacular natural phenomena in our courses and textbooks. A number of photos are accompanied by realistic field sketches, bridging the gap between what students see and what geologists see when they look at a geologic formation. The use of the field sketch style provides students with a sense of the hands-on way geologists work and enables them to develop a greater appreciation for the geologic structures they may see every day.

■ Multimedia and Supplements

The teaching and learning resources that accompany the seventh edition of *Understanding Earth* constitute a comprehensive and flexible ancillary package. They provide opportunities for active learning, student self-study, and automatic grading of homework, and they emphasize the visual aspects of the concepts presented in the text.

LAUNCHPAD UNITS make class prep a whole lot easier

At W. H. Freeman, we are committed to providing online instructional materials that meet the needs of instructors and students in powerful, yet simple ways—powerful enough to dramatically enhance teaching and learning, yet simple enough to use right away.

We've taken what we've learned from thousands of instructors and the hundreds of thousands of students to create a new generation of technology—featuring **LAUNCHPAD.** LaunchPad offers our acclaimed content curated and organized for easy assignability in a breakthrough user interface in which power and simplicity go hand in hand.

Combining a curated collection of video, tutorials, animations, projects, multimedia activities and exercises, and e-Book content, LaunchPad's interactive units give you a building block to use as-is, or as a starting point for your own learning units. An entire unit's worth of work can be assigned in seconds, drastically saving the amount of time it takes for you to have your course up and running.

■ **LearningCurve**
- Powerful adaptive quizzing, a game-like format, direct links to the e-Book, instant feedback, and the promise of better grades make using Learning-Curve a no-brainer.
- Customized quizzing tailored to each text adapts to students' responses and provides material at different difficulty levels and topics based on students' performance. Students love the simple yet powerful system, and instructors can access class reports to help refine lecture content.

■ **Interactive e-Book.** The **Interactive e-Book** is a complete online version of the textbook with easy access to rich multimedia resources that complete students' understanding.
- All text, graphics, tables, boxes, and end-of-chapter resources are included in the e-Book, and the e-Book provides instructors and students with powerful functionality to tailor their course resources to fit their needs.
- Quick, intuitive navigation to any section or subsection
- Full-text search, including the Glossary and Index
- Sticky-note feature allows users to place notes anywhere on the screen and choose the note color for easy categorization.
- "Top-note" feature allows users to place a prominent note at the top of the page to provide a more significant alert or reminder.
- Text highlighting in a variety of colors

■ **Video Exercises.** Students complete quizzes and do matching activities after viewing 2–5-minute Expeditions in Geology video tutorials shot around the world by Jerry Magloughlin of Colorado State University. Over two dozen videos are available, including
- *Natural Arches and Bridges.* Topics: Desert processes, erosion, stress. Filmed in the western United States and New Zealand.

- *Gneiss: The Lewisian Complex of Scotland.* Topics: Metamorphism, cratons, and crystalline basement. Filmed in the Outer Hebrides, Scotland.
- *Mount Vesuvius and the Plinian Eruption of 79* A.D. Topics: Stratovolcanoes, Plinian eruptions, and effects on humans. Filmed in Pompeii and on Mount Vesuvius, Italy.
- *Cinder Cones of Northern Arizona: Sunset and SP Craters.* Topics: Volcanism, cinder cones, and hot spots. Filmed in northern Arizona.

▪ **Image Map Activities.** These activities use figures from the text to assess key ideas, helping students to develop their visual literacy skills. Students must click the appropriate section(s) of the image and answer corresponding questions.

▪ **Animations, Flashcards, and other resources** highlight key concepts in introductory geology.

▪ **Assignments for Online Quizzing, Homework, and Self-Study.** Instructors can create and assign automatically graded homework and quizzes from the complete test bank, which is preloaded in LaunchPad. All quiz results feed directly into the instructor's gradebook.

▪ *Scientific American* **Newsfeed:** To demonstrate the continued process of science and the exciting new developments in the field, the *Scientific American* Newsfeed delivers regularly updated material from the well-known magazine. Articles, podcasts, news briefs, and videos on subjects related to geology are selected for inclusion by *Scientific American's* editors. The newsfeed provides several updates per week, and instructors can archive or assign the content they find most valuable.

▪ The **Gradebook** quickly and easily allows you to look up performance metrics for your whole class, for individual students, and for individual assignments. Having ready access to this information can help in both lecture prep and in making office hours more productive and efficient.

e-Books

The seventh edition of *Understanding Earth* is offered in two electronic versions: one is an interactive e-Book, available in the LaunchPad as described above, and the other is a PDF-based e-Book from CourseSmart. These options are provided to offer students and instructors flexibility in their use of course materials.

The CourseSmart e-Book offers the complete text in an easy-to-use, flexible format. Students can choose to view the CourseSmart e-Book online or to download it to a personal computer or a portable media player, such as an iPhone. To help students study and to mirror the experience of a printed textbook, the CourseSmart e-Book incorporates note taking, highlighting, and bookmark features.

Additional Resources for Instructors

The computerized test bank [ISBN 1-4641-7474-1] includes approximately 60 multiple-choice questions for each chapter (over 1300 questions in total) in an electronic format that allows instructors to edit, resequence, and add questions as they create tests.

Instructor's Resources

Images from the text, Image and Lecture PowerPoint presentations, clicker questions, and answers to end-of-chapter questions are available to instructors on the Book Companion Site at www.whfreeman.com/understandingearth7e.

Additional Resources for Students

Student Study Guide

The Student Study Guide includes tips on studying geology, chapter summaries, practice exams, and practice exercises.

Book Companion Site

The Book Companion Site, accessible at www.whfreeman.com/understandingearth7e, provides study tools aimed at helping students:

▪ Student Self-Quizzes, which can report to the instructor's gradebook
▪ Flashcards on key vocabulary

Lecture Tutorials in Introductory Geoscience, Second Edition (ISBN: 1-4641-0105-1)

Karen Kortz, Community College of Rhode Island
Jessica Smay, San Jose City College

A set of brief worksheets designed to be completed by students working alone or in groups, *Lecture Tutorials in Introductory Geoscience* engages students in the learning process and makes abstract concepts real. The tutorials are designed specifically to address misconceptions and difficult topics. Through the use of effective questioning, scaffolded learning, and a progression from simple to complex visuals, they help students construct correct scientific ideas and foster a meaningful and memorable learning experience. Research based on extensive classroom use shows that these Lecture Tutorials increase student learning more than lectures alone.

Research indicates that students learn more when they are actively engaged while learning. Lecture Tutorials are worksheets of carefully designed questions that require students to think about challenging subjects.

They are designed to be used after a brief lecture or introduction to the topic. Working in small groups, students are encouraged to "talk science," ask questions, and teach one another.

A geologist looking at terrain or a rock formation can often identify its structure and attempt to draw conclusions about its history; most introductory geoscience students cannot make the same connections. Lecture Tutorials use simplified images and questions to help students build a fundamental understanding of a concept, then move them into more complex interpretations of that concept. In the process, the activities create an environment in which students must confront their misconceptions. Those misconceptions were identified through literature searches of published misconceptions and through the classroom experience of the authors.

Lecture Tutorials scaffold student learning. Early questions are designed to introduce the students to the topic and help them consider what they do and do not know. The tutorial then focuses on underdeveloped or misunderstood concepts and slowly steps students through more difficult questions, helping them construct a new understanding. The final questions are higher-level questions, both scientifically and cognitively, that indicate whether students understand the material.

Lecture Tutorials do not need to stand alone in the classroom and can be used with other interactive teaching methods. They are designed to be used to complement lectures, laboratory exercises, textbook use, and online resources. While research shows they are most effective when used frequently in the lecture component of a course, instructors from around the country have successfully used them in laboratory settings or as homework.

Instructors can order *Lecture Tutorials in Introductory Geoscience* as a stand-alone item or packaged with a W. H. Freeman textbook.

- *Lecture Tutorials in Introductory Geoscience*
- *Lecture Tutorials in Introductory Geoscience and Understanding Earth,* Seventh Edition
- *Lecture Tutorials in Introductory Geoscience* and *The Essential Earth*

Students can also purchase *Lecture Tutorials in Introductory Geoscience* directly at www .whfreeman.com/ lecturetutorials.

■ Acknowledgments

It is a challenge both to geology instructors and to authors of geology textbooks to compress the many important aspects of geology into a single course and to inspire interest and enthusiasm in their students. To meet this challenge, we have called on the advice of many colleagues who teach in all kinds of college and university settings.

From the earliest planning stages of each edition of this book, we have relied on a consensus of views in designing an organization for the text and in choosing which topics to include. As we wrote and rewrote the chapters, we again relied on our colleagues to guide us in making the presentation pedagogically sound, accurate, accessible, and stimulating to students. To each one we are grateful.

Marianne Caldwell, *Hillsborough Community College*

Courtney Clamons, *Austin Community College*

Ellen Cowan, *Appalachian State University*

Meredith Denton-Hedrick, *Austin Community College*

Mark Feigensen, *Rutgers University*

Edward Garnero, *Arizona State University*

Richard Gibson, *Texas A&M University*

Bruce Herbert, *Texas A&M University*

Bernard Housen, *Western Washington University*

Qinhong Hu, *University of Texas at Arlington*

Maureen McCurdy Hillard, *Louisiana Tech University*

Daniel Kelley, *Louisiana State University*

Steppen Murphy, *Central Piedmont Community College*

Alycia Stigall, *Ohio University*

John Tacinelli, *Rochester Community and Technical College*

J. M. Wampler, *Georgia State University*

We remain grateful to the instructors who were involved or assisted in the planning or reviewing stages of the sixth edition:

Jake Armour, *University of North Carolina, Charlotte*

Emma Baer, *Shoreline Community College*

Graham B. Baird, *University of Northern Colorado*

Rob Benson, *Adams State College*

Barbara L. Brande, *University of Montevallo*

Denise Burchsted, *University of Connecticut*

Erik W. Burtis, *Northern Virginia Community College*

Chu-Yung Chen, *University of Illinois*

Geoffrey W. Cook, *University of Rhode Island*

Tim D. Cope, *DePauw University*

Michael Dalman, *Blinn College, Brenham*

Iver W. Duedall, *Florida Institute of Technology*

Stewart S. Farrar, *Eastern Kentucky University*

Mark D. Feigenson, *Rutgers University*

William Garcia, *University of North Carolina, Charlotte*

Michael D. Harrell, *Seattle Central Community College*

Elizabeth A. Johnson, *James Madison University*

Tamie J. Jovanelly, *Berry College*

David T. King, Jr., *Auburn University*

Steve Kluge, *State University of New York, Purchase*

Michael A. Kruge, *Montclair State University*

Steven Lee, *Jet Propulsion Laboratory*

Michael B. Leite, *Chadron State College*

Beth Lincoln, *Albion College*

Ryan Mathur, *Juniata College*

Stanley A. Mertzman, *Franklin & Marshall College*

James G. Mills, Jr., *DePauw University*

Sadredin C. Moosavi, *Tulane University*

Gregory Mountain, *Rutgers University*

Otto H. Muller, *Alfred University*

M. Susan Nagel, *University of Connecticut*

Heidi Natel, *U.S. Military Academy, West Point*

Jeffrey A. Nunn, *Louisiana State University*

Debajyoti Paul, *University of Texas, San Antonio*

Marilyn Velinsky Rands, *Lawrence Technological University*

Jason A. Rech, *Miami University*

Randye L. Rutberg, *Hunter College*

Anne Marie Ryan, *Dalhousie University*

Jane Selverstone, *University of New Mexico*

Steven C. Semken, *Arizona State University*

Eric Small, *University of Colorado, Boulder*

Neptune Srimal, *Florida International University*

Alexander K. Stewart, *St. Lawrence University*

Michael A. Stewart, *University of Illinois*

Gina Seegers Szablewski, *University of Wisconsin, Milwaukee*

Leif Tapanila, *Idaho State University*

Mike Tice, *Texas A&M University*

Others have worked with us more directly in writing and preparing the manuscript for publication. At our side always was our Science Editor at W. H. Freeman and Company: Randi Rossignol. Randi has worked with us for the past 12 years, and through four revisions of the text. It has been a pleasure for us to work with her every time! Bill Minick, Senior Editor, was very helpful in shepherding the process through to completion. Jen Griffes helped us with the simultaneous challenge of text revision and Mars Rover operations. Amy Thorne coordinated the media supplements, Blake Logan designed the text, Jennifer Atkins obtained many beautiful photographs, and Paul Rohloff managed the production of the text. Nick Ciani coordinated the transmittal process. We thank Dennis Free and Sheryl Rose from Aptara Corporation for creating the printed text from final manuscript. We are grateful to Jake Armour, University of North Carolina, Charlotte, who worked with us to create the Google Earth Projects. And our heartfelt thanks go to Emily Cooper, who created so many beautiful illustrations.

UNDERSTANDING EARTH

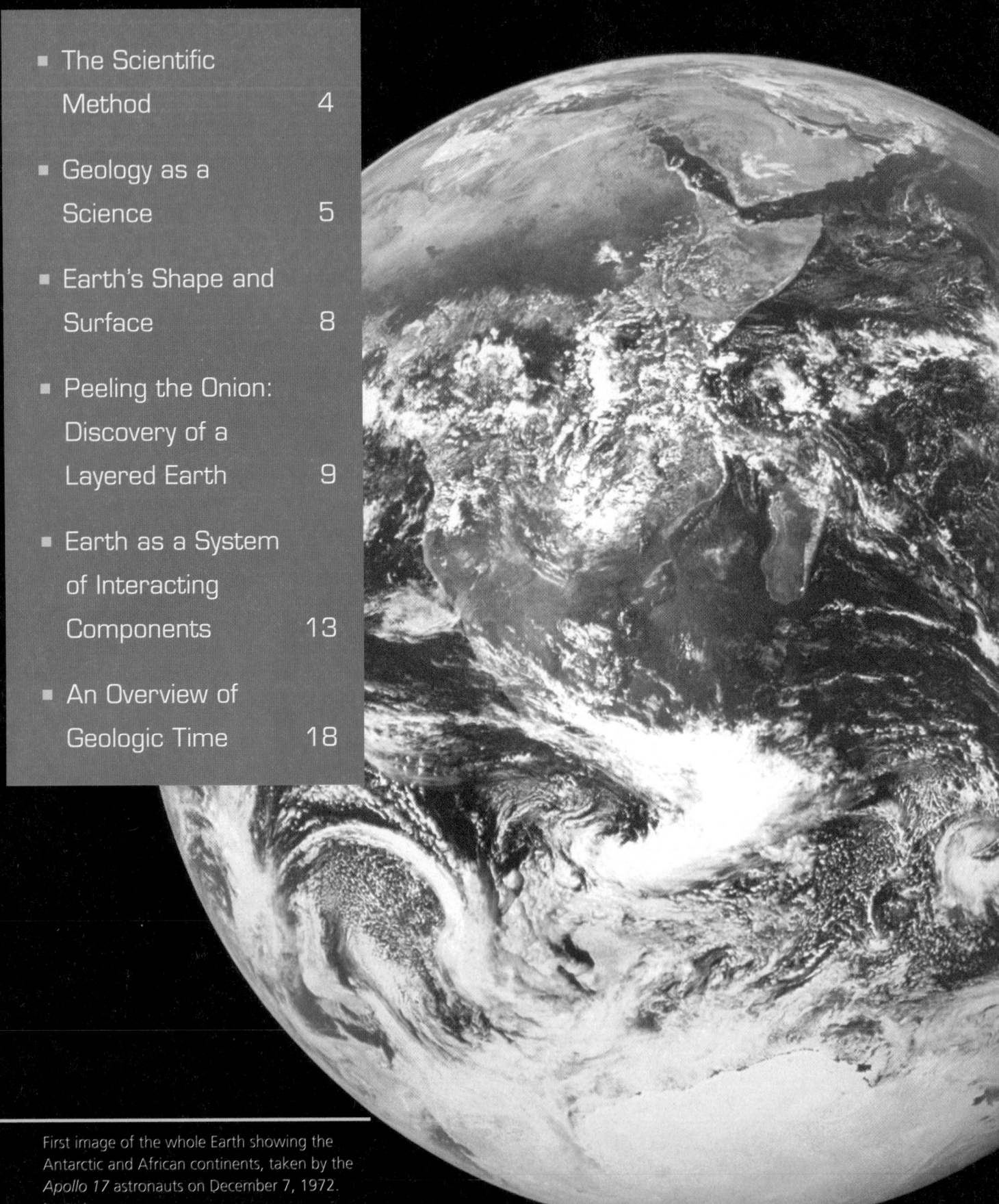

■ The Scientific
  Method   4

■ Geology as a
  Science   5

■ Earth's Shape and
  Surface   8

■ Peeling the Onion:
  Discovery of a
  Layered Earth   9

■ Earth as a System
  of Interacting
  Components   13

■ An Overview of
  Geologic Time   18

First image of the whole Earth showing the
Antarctic and African continents, taken by the
Apollo 17 astronauts on December 7, 1972.
[NASA.]

THE EARTH SYSTEM

1

EARTH IS A UNIQUE place, home to millions of organisms, including ourselves. No other planet we have yet discovered has the same delicate balance of conditions necessary to sustain life. Geology is the science that studies Earth: how it was born, how it evolved, how it works, and how we can help preserve its habitats for life. Geologists seek answers to many basic questions: Of what material is the planet composed? Why are there continents and oceans? How did the Himalaya, Alps, and Rocky Mountains rise to their great heights? Why are some regions subject to earthquakes and volcanic eruptions while others are not? How did Earth's surface environment and the life it contains evolve over billions of years? What changes are likely in the future? We think you will find the answers to such questions fascinating. Welcome to the science of geology!

We have organized the discussion of geology in this textbook around three basic concepts that will appear in almost every chapter: (1) Earth as a system of interacting components, (2) plate tectonics as a unifying theory of geology, and (3) changes in the Earth system through geologic time.

This chapter gives a broad picture of how geologists think. It starts with the scientific method, the observational approach to the physical universe on which all scientific inquiry is based. Throughout this textbook, you will see the scientific method in action as we describe how Earth scientists gather and interpret information about our planet. In this first chapter, we will illustrate how the scientific method has been applied to discover some of Earth's basic features—its shape and its internal layering.

To explain features that are millions or even billions of years old, Earth scientists look at what is happening on Earth today. We will introduce the study of our complex natural world as an *Earth* system involving many interrelated components. Some of these components, such as the atmosphere and oceans, are clearly visible above Earth's solid surface; others lie hidden deep within its interior. By observing the ways in which these components interact, scientists have built up an understanding of how the Earth system has changed through geologic time.

We will also introduce you to a geologist's view of time. You will think about time differently as you begin to comprehend the immense span of geologic history. Earth and the other planets in our solar system formed about 4.5 billion years ago. More than 3 billion years ago, living cells developed on Earth's surface, and life has been evolving ever since. Yet our human origins date back only a few million years—less than a tenth of a percent of Earth's existence. The decades of individual lives or even the thousands of years of recorded human history are inadequate to study Earth's long existence.

■ The Scientific Method

The term *geology* (from the Greek words for "Earth" and "knowledge") was coined by scientific philosophers more than 200 years ago to describe the study of rock formations and fossils. Through careful observations and reasoning, their successors developed the theories of biological evolution, continental drift, and plate tectonics—major topics of this textbook. Today, **geology** identifies the branch of Earth science that studies all aspects of the planet: its history, its composition and internal structure, and its surface features.

The goal of geology—and of science in general—is to explain the physical universe. Scientists believe that physical events have physical explanations, even if they may be beyond our present capacity to understand them. The **scientific method**, on which all scientists rely, is the general procedure for discovering how the universe works through systematic observations and experiments. Using the scientific method to make new discoveries and to confirm old ones is the process of *scientific research* (**Figure 1.1**).

When scientists propose a *hypothesis*—a tentative explanation based on data collected through observations and experiments—they present it to the community of scientists for criticism and repeated testing. A hypothesis is supported if it explains new data or predicts the outcome of new experiments. A hypothesis that is confirmed by other scientists gains credibility.

Here are four interesting scientific hypotheses we will encounter in this textbook:

- Earth is billions of years old.
- Coal is a rock formed from dead plants.
- Earthquakes are caused by the breaking of rocks along geologic faults.
- The burning of fossil fuels is causing global warming.

The first hypothesis agrees with the ages of thousands of ancient rocks as measured by precise laboratory techniques, and the next two hypotheses have also been confirmed by many independent observations. The fourth hypothesis has been more controversial, though so many new data

FIGURE 1.1 ■ Scientific research is the process of discovery and confirmation through observations of the real world. These geologists are researching soil samples near a lake in Minnesota. [USGS.]

support it that most scientists now accept it as true (see Chapters 15 and 23).

A coherent set of hypotheses that explains some aspect of nature constitutes a *theory*. Good theories are supported by substantial bodies of data and have survived repeated challenges. They usually obey *physical laws*, general principles about how the universe works that can be applied in almost every situation, such as Newton's law of gravity.

Some hypotheses and theories have been so extensively tested that all scientists accept them as true, at least to a good approximation. For instance, the theory that Earth is nearly spherical, which follows from Newton's law of gravity, is supported by so much experience and direct evidence (ask any astronaut) that we take it to be a fact. The longer a theory holds up to all scientific challenges, the more confidently it is held.

Yet theories can never be considered completely proved. The essence of science is that no explanation, no matter how believable or appealing, is closed to questioning. If convincing new evidence indicates that a theory is wrong, scientists will discard it or modify it to account for the data. A theory, like a hypothesis, must always be testable; any proposal about the universe that cannot be evaluated by observing the natural world should not be called a scientific theory.

For scientists engaged in research, the most interesting hypotheses are often the most controversial, rather than the most widely accepted. The hypothesis that fossil-fuel burning causes global warming has been widely debated. Because the long-term predictions of this hypothesis are so important, many Earth scientists are now vigorously testing it.

Knowledge based on many hypotheses and theories can be used to create a *scientific model*—a precise representation of how a natural process operates or how a natural system behaves. Scientists combine related ideas in a model to test the consistency of their knowledge and to make predictions. Like a good hypothesis or theory, a good model makes predictions that agree with observations.

A scientific model is often formulated as a computer program that simulates the behavior of a natural system through numerical calculations. The forecast of rain or sunshine you may see on TV tonight comes from a computer model of the weather. A computer can be programmed to simulate geologic phenomena that are too big to replicate in a laboratory or that operate over periods of time that are too long for humans to observe. For example, models used for predicting weather have been extended to predict climate changes decades into the future.

To encourage discussion of their ideas, scientists share those ideas and the data on which they are based. They present their findings at professional meetings, publish them in professional journals, and explain them in informal conversations with colleagues. Scientists learn from one another's work as well as from the discoveries of the past. Most of the great concepts of science, whether they emerge as a flash of insight or in the course of painstaking analysis, result from untold numbers of such interactions. Albert Einstein put it this way: "In science . . . the work of the individual is so bound up with that of his scientific predecessors and contemporaries that it appears almost as an impersonal product of his generation."

Because such free intellectual exchange can be subject to abuses, a code of ethics has evolved among scientists. Scientists must acknowledge the contributions of all others on whose work they have drawn. They must not falsify data, use the work of others without recognizing them, or be otherwise deceitful in their work. They must also accept responsibility for training the next generation of researchers and teachers. These principles are supported by the basic values of scientific cooperation, which a president of the National Academy of Sciences, Bruce Alberts, has aptly described as "honesty, generosity, a respect for evidence, openness to all ideas and opinions."

■ Geology as a Science

In the popular media, scientists are often portrayed as people who do experiments wearing white coats. That stereotype is not inappropriate: many scientific problems are best investigated in the laboratory. What forces keep atoms together? How do chemicals react with one another? Can viruses cause cancer? The phenomena that scientists observe to answer such questions are sufficiently small and happen quickly enough to be studied in the controlled environment of the laboratory.

The major questions of geology, however, involve processes that operate on much larger and longer scales. Controlled laboratory measurements yield critical data for testing geologic hypotheses and theories—the ages and properties of rocks, for instance—but they are usually insufficient to solve major geologic problems. Almost all of the great discoveries described in this textbook were made by observing Earth processes in their uncontrolled, natural environment.

For this reason, geology is an outdoor science with its own particular style and outlook. Geologists "go into the field" to observe nature directly (**Figure 1.2**). They learn how mountains were formed by climbing up steep slopes and examining the exposed rocks, and they deploy sensitive instruments to collect data on earthquakes, volcanic eruptions, and other activity within the solid Earth. They discover how ocean basins have evolved by sailing rough seas to map the ocean floor (**Figure 1.3**).

FIGURE 1.2 ■ Geology is principally an outdoor science. Here, Peter Gray welds one of the five Global Positioning System stations placed on the flanks of Mount St. Helens. The stations will monitor the changing shape of the land surface as molten rock moves upward within the volcano. [USGS/Lyn Topinka.]

(a)

FIGURE 1.3 ■ The research crew from the icebreaker Louis S. St-Laurent, lowers a corer that will gather mud and sediment from the ocean floor. [AP Photo/The Canadian Press, Jonathan Hayward.]

Geology is closely related to other areas of Earth science, including *oceanography,* the study of the oceans; *meteorology,* the study of the atmosphere; and *ecology,* which concerns the abundance and distribution of life. *Geophysics, geochemistry,* and *geobiology* are subfields of geology that apply the methods of physics, chemistry, and biology to geologic problems (**Figure 1.4**).

Geology is a *planetary science* that uses remote sensing devices, such as instruments mounted on Earth-orbiting spacecraft, to scan the entire globe (**Figure 1.5**). Geologists develop computer models that can analyze the huge quantities of data amassed by satellites to map the continents, chart the motions of the atmosphere and oceans, and monitor how our environment is changing.

A special aspect of geology is its ability to probe Earth's long history by reading what has been "written in stone." The **geologic record** is the information preserved in the rocks that have been formed at various times throughout Earth's history (**Figure 1.6**). Geologists decipher the geologic record by combining information from many kinds of work: examination of rocks in the field; careful mapping of their positions relative to older and younger rock formations; collection of representative samples; and determination of their ages using sensitive laboratory instruments (Figure 1.4b).

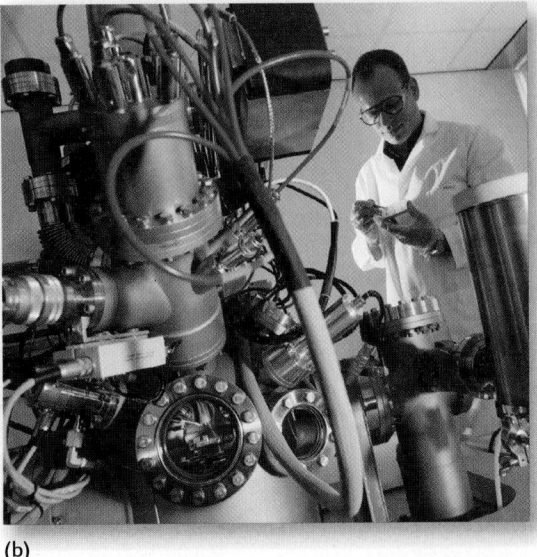

(b)

FIGURE 1.4 ■ A number of subfields contribute to the study of geology. (a) Geophysicists deploy instruments to measure the underground activity of a volcano. (b) A geochemist readies a rock sample for analysis by a mass spectrometer. (c) Geobiologists investigate underground life inside Spider Cave at Carlsbad Caverns, New Mexico. [(a) Hawaiian Volcano Observatory/USGS; (b) John McLean/Science Source; (c) AP Photo/Val Hildreth-Werker.]

(c)

FIGURE 1.5 ■ An astronaut checks out instrumentation for monitoring Earth's surface. [StockTrek/SuperStock.]

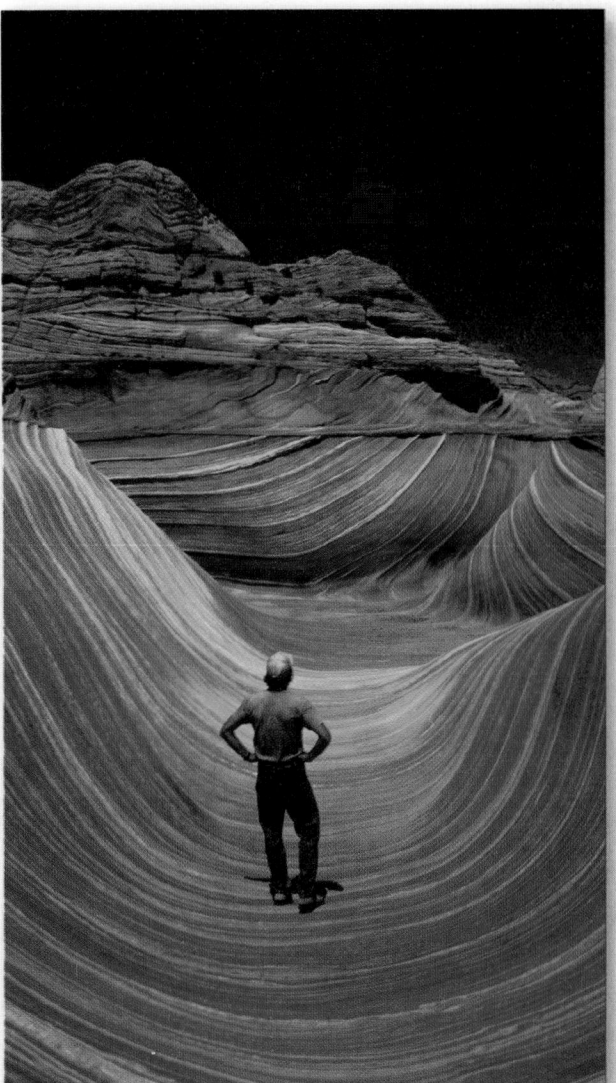

FIGURE 1.6 ■ The geologic record preserves evidence of Earth's long history. These multicolored layers of sand at Colorado National Monument were deposited more than 200 million years ago, when this part of the western United States was a vast Sahara-like desert. They were subsequently overlain by other rocks, welded by pressure into sandstone, uplifted by mountain-building events, and eroded by wind and water into today's stunning landforms. [Mark Newman/Lonely Planet Images/Getty Images, Inc.]

In *Annals of the Former World*, a compendium of colorful stories about geologists, the popular writer John McPhee offers his view of how geologists bring field and laboratory observations together to visualize the big picture:

They look at mud and see mountains, in mountains oceans, in oceans mountains to be. They go up to some rock and figure out a story, another rock, another story, and as the stories compile through time they connect—and long case histories are constructed and written from interpreted patterns of clues. This is detective work on a scale unimaginable to most detectives, with the notable exception of Sherlock Holmes.

The geologic record tells us that, for the most part, the processes we see in action on Earth today have worked in much the same way throughout the geologic past. This important concept is known as the **principle of uniformitarianism.** It was stated as a scientific hypothesis in the eighteenth century by a Scottish physician and geologist, James Hutton. In 1830, the British geologist Charles Lyell summarized the concept in a memorable line: "The present is the key to the past."

The principle of uniformitarianism does not mean that all geologic phenomena proceed at the same gradual pace. Some of the most important geologic processes happen as sudden events. A large meteorite that impacts Earth can gouge out a vast crater in a matter of seconds. A volcano can blow its top, and a fault can rupture the ground in an earthquake, almost as quickly. Other processes do occur much more slowly. Millions of years are required for continents to drift apart, for mountains to be raised and eroded, and for river systems to deposit thick layers of sediments. Geologic processes take place over a tremendous range of scales in both space and time (**Figure 1.7**).

Nor does the principle of uniformitarianism mean that we have to observe a geologic event to know that it is important in the current Earth system. Humans have not witnessed a large meteorite impact in recorded history, but we know these impacts have occurred many times in the geologic past and will certainly happen again. The same

Over millions of years, layers of sediments built up over the oldest rocks. The most recent layer—the top—is about 250 million years old.

About 50,000 years ago, the explosive impact of a meteorite (perhaps weighing 300,000 tons) created this 1.2-km-wide crater in just a few seconds.

(a)   The rocks at the bottom of the Grand Canyon are 1.7–2.0 billion years old.

(b)

FIGURE 1.7   ■   Some geologic processes take place over thousands of centuries, while others occur with dazzling speed. (a) The Grand Canyon, Arizona. (b) Meteor Crater, Arizona. [(a) John Wang/PhotoDisc/Getty Images; (b) John Sanford/Science Source.]

can be said of the vast volcanic outpourings that have covered areas bigger than Texas with lava and poisoned the global atmosphere with volcanic gases. The long history of Earth is punctuated by many such extreme, though infrequent, events that result in rapid changes in the Earth system. Geology is the study of *extreme events* as well as gradual change.

From Hutton's day onward, geologists have observed nature at work and used the principle of uniformitarianism to interpret features found in rock formations. This approach has been very successful. However, Hutton's principle is too confining for geologic science as it is now practiced. Modern geology must deal with the entire range of Earth's history, which began more than 4.5 billion years ago. As we will see in Chapter 9, the violent processes that shaped Earth's early history were distinctly different from

those that operate today. To understand that history, we will need some information about Earth's shape and surface, as well as its deep interior.

■ Earth's Shape and Surface

The scientific method has its roots in **geodesy**, a very old branch of Earth science that studies Earth's shape and surface. The concept that Earth is spherical rather than flat was advanced by Greek and Indian philosophers around the sixth century B.C., and it was the basis of Aristotle's theory of Earth put forward in his famous treatise, *Meteorologica,* published around 330 B.C. (the first Earth science

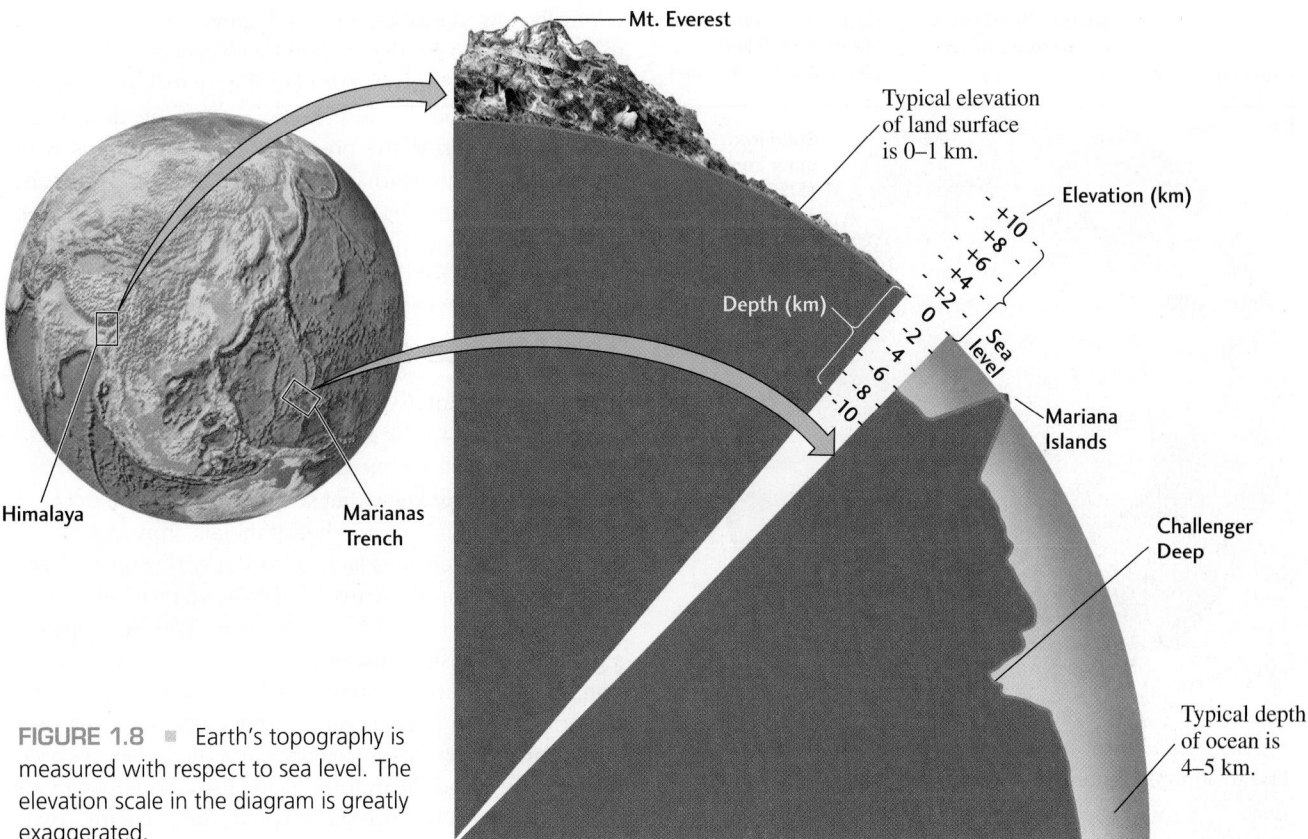

Mt. Everest

Typical elevation
of land surface
is 0–1 km.

Elevation (km)

Depth (km)

Sea
level

Mariana
Islands

Himalaya

Marianas
Trench

Challenger
Deep

Typical depth
of ocean is
4–5 km.

FIGURE 1.8 ▪ Earth's topography is measured with respect to sea level. The elevation scale in the diagram is greatly exaggerated.

textbook!). In the third century B.C., Eratosthenes used a clever experiment to measure Earth's radius, which turned out to be 6370 km (see the Practicing Geology exercise at the end of the chapter).

Much more precise measurements have shown that Earth is not a perfect sphere. Because of its daily rotation, it bulges out slightly at its equator and is slightly squashed at its poles. In addition, the smooth curvature of Earth's surface is broken by mountains and valleys and other ups and downs. This **topography** is measured with respect to *sea level,* a smooth surface set at the average level of ocean water that conforms closely to the squashed spherical shape expected for the rotating Earth. Many features of geologic significance stand out in Earth's topography (**Figure 1.8**). Its two largest features are continents, which have typical elevations of 0 to 1 km above sea level, and ocean basins, which have typical depths of 4 to 5 km below sea level. The elevation of Earth's surface varies by nearly 20 km from its highest point (Mount Everest in the Himalaya at 8850 m above sea level) to its lowest point (Challenger Deep in the Marianas Trench in the Pacific Ocean at 11,030 m below sea level). Although the Himalaya may loom large to us, their elevation is a small fraction of Earth's radius, only about one part in a thousand, which is why the globe looks like a smooth sphere when seen from outer space.

▪ Peeling the Onion: Discovery of a Layered Earth

Ancient thinkers divided the universe into two parts, the heavens above and Hades below. The sky was transparent and full of light, and they could directly observe its stars and track its wandering planets. But Earth's interior was dark and closed to human view. In some places, the ground quaked and erupted hot lava. Surely something terrible was going on down there!

So it remained until about a century ago, when geologists began to peer downward into Earth's interior, not with waves of light (which cannot penetrate rock), but with waves produced by earthquakes. An earthquake occurs when geologic forces cause brittle rocks to fracture, sending out vibrations like the cracking of ice on a river. These **seismic waves** (from the Greek word for earthquake, *seismos*), when recorded on sensitive instruments called *seismometers,* allow geologists to locate earthquakes and also to make pictures of Earth's inner workings, much as doctors use ultrasound and CAT scans to image the inside of your body. When the first networks of seismographs were installed around the world at the end of the nineteenth century, geologists began to discover that

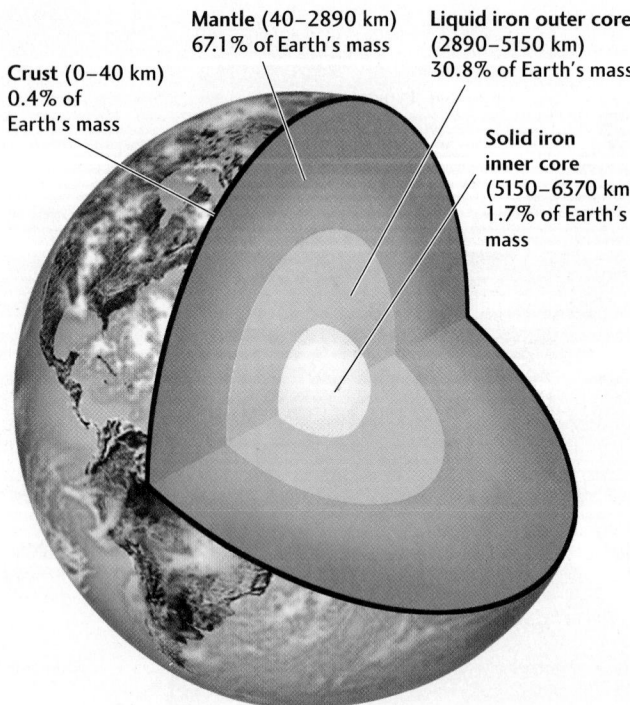

Crust (0–40 km)
0.4% of
Earth's mass

Mantle (40–2890 km)
67.1% of Earth's mass

Liquid iron outer core
(2890–5150 km)
30.8% of Earth's mass

Solid iron
inner core
(5150–6370 km)
1.7% of Earth's
mass

FIGURE 1.9 ■ Earth's major layers, showing their depths and their masses expressed as a percentage of Earth's total mass.

Earth's interior was divided into concentric layers of different compositions, separated by sharp, nearly spherical boundaries (**Figure 1.9**).

Earth's Density

Layering of Earth's deep interior was first proposed by the German physicist Emil Wiechert at the end of the nineteenth century, before much seismic data had become available. He wanted to understand why our planet is so heavy, or more precisely, so *dense.* The density of a substance is easy to calculate: just measure its mass on a scale and divide by its volume. A typical rock, such as the granite used for tombstones, has a density of about 2.7 grams per cubic centimeter (g/cm^3). Estimating the density of the entire planet is a little harder, but not much. Eratosthenes had shown how to measure Earth's volume in 250 B.C., and sometime around 1680, the great English scientist Isaac Newton figured out how to calculate its mass from the gravitational force that pulls objects to its surface. The details, which involved careful laboratory experiments to calibrate Newton's law of gravity, were worked out by another Englishman, Henry Cavendish. In 1798, he calculated Earth's average density to be about 5.5 g/cm^3, twice that of tombstone granite.

Wiechert was puzzled. He knew that a planet made entirely of common rocks could not have such a high density. Most common rocks, such as granite, contain a high proportion of silica (silicon plus oxygen; SiO_2) and have

relatively low densities, below 3 g/cm^3. Some iron-rich rocks brought to Earth's surface by volcanoes have densities as high as 3.5 g/cm^3, but no ordinary rock approached Cavendish's value. He also knew that, going downward into Earth's interior, the pressure on rock increases with the weight of the overlying mass. The pressure squeezes the rock into a smaller volume, making its density higher. But Wiechert found that even the effect of pressure was too small to account for the density Cavendish had calculated.

The Mantle and Core

In thinking about what lay beneath his feet, Wiechert turned outward to the solar system and, in particular, to meteorites, which are pieces of the solar system that have fallen to Earth. He knew that some meteorites are made of an *alloy* (a mixture) of two heavy metals, iron and nickel, and thus have densities as high as 8 g/cm^3 (**Figure 1.10**). He also knew that these two elements are relatively abundant throughout our solar system. So, in 1896, he proposed a grand hypothesis: sometime in Earth's past, most of the iron and nickel in its interior had dropped inward to its center under the force of gravity. This movement created a dense **core,** which was surrounded by a shell of silicate-rich rock that he called the **mantle** (using the German word for "coat"). With this hypothesis, he could come up with a two-layered Earth model that agreed with Cavendish's value for Earth's average density. He could also explain the existence of iron-nickel meteorites: they were chunks from the core of an Earthlike planet (or planets) that had broken apart, most likely by collision with other planets.

Wiechert got busy testing his hypothesis using seismic waves recorded by seismographs located around the globe (he designed one himself). The first results showed a shadowy inner mass that he took to be the core, but he had problems identifying some of the seismic waves. These waves come in two basic types: *compressional waves,* which expand and compress the material they move through as they travel through a solid, liquid, or gas; and *shear waves,* which move the material from side to side. Shear waves can propagate only through solids, which resist shearing, and not through fluids (liquids or gases) such as air and water, which have no resistance to this type of motion.

In 1906, a British seismologist, Robert Oldham, was able to sort out the paths traveled by these two types of seismic waves and show that shear waves did not propagate through the core. The core, at least in its outer part, was liquid! This finding turns out to be not too surprising. Iron melts at a lower temperature than silicates, which is why metallurgists can use containers made of ceramics (which are silicate materials) to hold molten iron. Earth's deep interior is hot enough to melt an iron-nickel alloy, but not silicate rock. Beno Gutenberg, one of Wiechert's students, confirmed Oldham's observations and, in 1914, determined that the depth of the *core-mantle boundary* was about 2890 km (see Figure 1.9).

(a)   (b)

FIGURE 1.10 ▪ Two common types of meteorites. (a) This stony meteorite, which is similar in composition to Earth's silicate mantle, has a density of about 3 g/cm³. (b) This iron-nickel meteorite, which is similar in composition to Earth's core, has a density of about 8 g/cm³. [John Grotzinger/Ramón Rivera-Moret/Harvard Mineralogical Museum.]

The Crust

Five years earlier, a Croatian scientist had detected another boundary at the relatively shallow depth of 40 km beneath the European continent. This boundary, named the *Mohorovičić discontinuity* (Moho for short) after its discoverer, separates a **crust** composed of low-density silicates, which are rich in aluminum and potassium, from the higher-density silicates of the mantle, which contain more magnesium and iron.

Like the core-mantle boundary, the Moho is a global feature. However, it was found to be substantially shallower beneath the oceans than beneath the continents. On average, the thickness of oceanic crust is only about 7 km, compared with almost 40 km for continental crust. Moreover, rocks in the oceanic crust contain more iron, and are therefore denser, than continental rocks. Because continental crust is thicker but less dense than oceanic crust, the continents ride higher by floating like buoyant rafts on the denser mantle (**Figure 1.11**), much as icebergs float on

the ocean. Continental buoyancy explains the most striking feature of Earth's surface topography: why the elevations shown in Figure 1.8 fall into two main groups, 0 to 1 km above sea level for much of the land surface and 4 to 5 km below sea level for much of the deep sea.

Shear waves travel well through the mantle and crust, so we know that both are solid rock. How can continents float on solid rock? Rock can be solid and strong over the short term (seconds to years), but weak over the long term (thousands to millions of years). The mantle below a depth of about 100 km has little strength, and over very long periods, it flows as it adjusts to support the weight of continents and mountains.

The Inner Core

Because the mantle is solid and the outer part of the core is liquid, the core-mantle boundary reflects seismic waves, just as a mirror reflects light waves. In 1936, Danish seismologist

Less dense continental crust floats on denser mantle.

Continental crust is less dense and thicker than oceanic crust and therefore rides higher.

0 (km)
10
20
30 — Mantle
40   (3.4 g/cm³)
50

Continental crust
(2.8 g/cm³)

Oceanic crust
(3.0 g/cm³)

Moho

Horizontal distance not to scale

FIGURE 1.11 ▪ Because crustal rocks are less dense than mantle rocks, Earth's crust floats on the mantle. Continental crust is thicker and has a lower density than oceanic crust, which causes it to ride higher, explaining the difference in elevation between continents and the deep seafloor.

Inge Lehmann discovered another sharp spherical boundary at the much greater depth of 5150 km, indicating a central mass with a higher density than the liquid core. Studies following her pioneering research showed that the inner core can transmit both shear waves and compressional waves. The **inner core** is therefore a solid metallic sphere suspended within the liquid **outer core**—a "planet within a planet." The radius of the inner core is 1220 km, about two-thirds the size of the Moon.

Geologists were puzzled by the existence of this "frozen" inner core. They knew that temperatures inside Earth should increase with depth. According to the best current estimates, Earth's temperature rises from about 3500°C at the core-mantle boundary to almost 5000°C at its center. If the inner core is hotter, how could it be solid while the outer core is molten? The mystery was eventually solved by laboratory experiments on iron-nickel alloys, which showed that the "freezing" was due to higher pressures, rather than lower temperatures, at Earth's center.

Chemical Composition of Earth's Major Layers

By the mid-twentieth century, geologists had discovered all of Earth's major layers—crust, mantle, outer core, and inner core—plus a number of more subtle features in its interior. They found, for example, that the mantle itself is layered into an *upper mantle* and a *lower mantle,* separated by a *transition zone* where the rock density increases in a series of steps. These density steps are not caused by changes in the rock's chemical composition, but rather by changes in the compactness of its constituent minerals due to the increasing pressure with depth. The two largest density jumps in the transition zone are located at depths of about 410 km and 660 km, but they are smaller than the density increases across the Moho and core-mantle boundaries, which *are* due to changes in chemical composition (**Figure 1.12**).

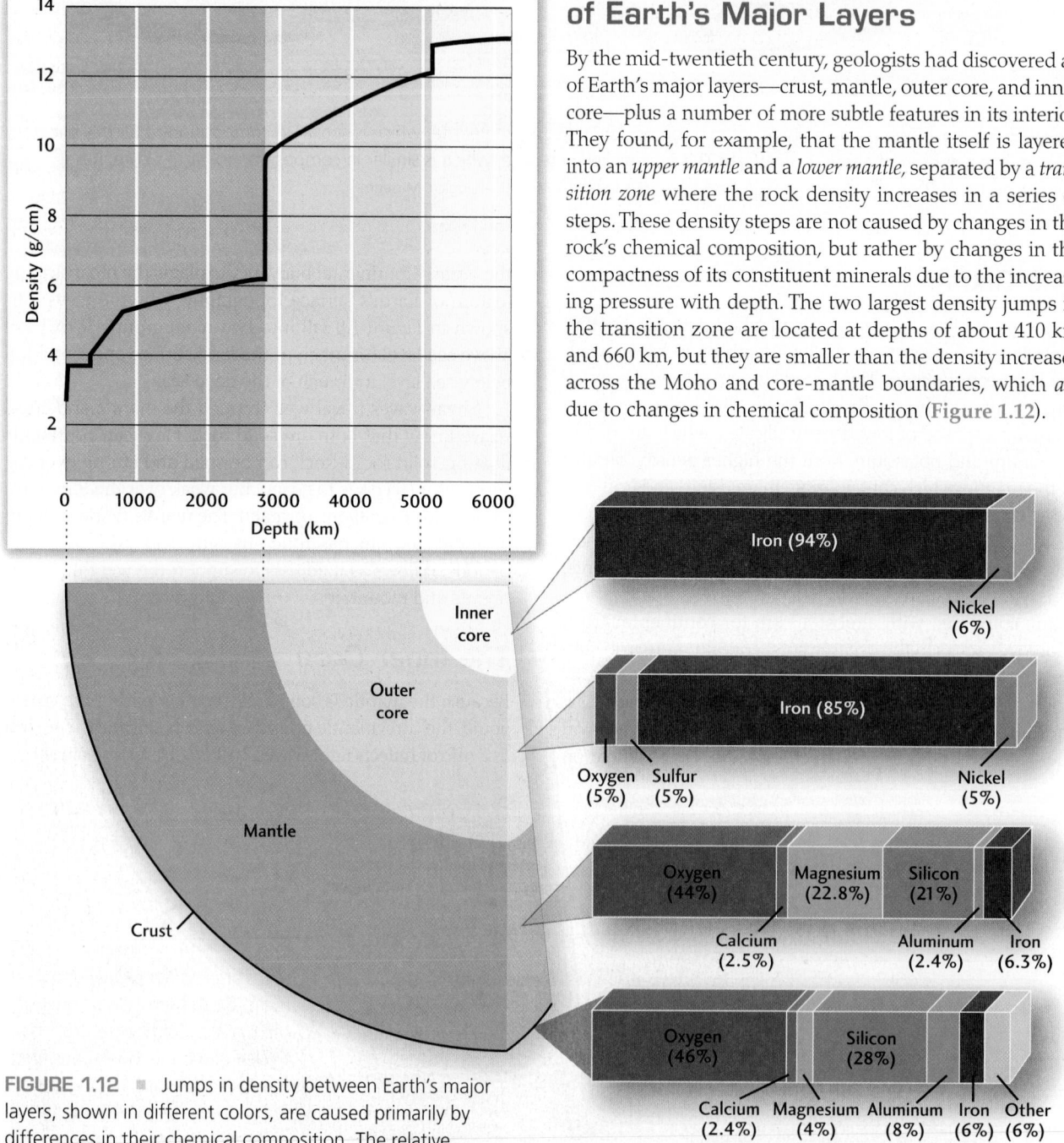

FIGURE 1.12 ▪ Jumps in density between Earth's major layers, shown in different colors, are caused primarily by differences in their chemical composition. The relative amounts of the main elements are depicted in the bars on the right.

Geologists were also able to show that Earth's outer core cannot be made of a pure iron-nickel alloy, because the densities of these metals are higher than the observed density of the outer core. About 10 percent of the outer core's mass must be made of lighter elements, such as oxygen and sulfur. On the other hand, the density of the solid inner core is slightly higher than that of the outer core and is consistent with a nearly pure iron-nickel alloy.

By bringing together many lines of evidence, geologists have put together a model of the composition of Earth and its various layers. In addition to the seismic data, that evidence includes the compositions of crustal and mantle rocks as well as the compositions of meteorites, thought to be samples of the cosmic material from which planets like Earth were originally made.

Only eight elements, out of more than a hundred, make up 99 percent of Earth's mass (see Figure 1.12). In fact, about 90 percent of Earth consists of only four elements: iron, oxygen, silicon, and magnesium. The first two are the most abundant elements, each accounting for nearly a third of the planet's overall mass, but they are distributed very differently. Iron, the densest of these common elements, is concentrated in the core, whereas oxygen, the least dense, is concentrated in the crust and mantle. The crust contains more silica than the mantle, and the core almost none. These relationships confirm Wiechert's hypothesis that the different compositions of Earth's layers are primarily the work of gravity. As you can see in Figure 1.12, the crustal rocks on which we stand are almost 50 percent oxygen!

■ Earth as a System of Interacting Components

Earth is a restless planet, continually changing through geologic activity such as earthquakes, volcanoes, and glaciation. This activity is powered by two heat engines: one internal, the other external (Figure 1.13). A *heat engine*—for example, the gasoline engine of an automobile—transforms heat into mechanical motion or work. Earth's *internal heat engine* is powered by the heat energy trapped in its deep interior during its violent origin and released inside the planet by radioactivity. This internal heat drives movement in the mantle and core, supplying the energy that melts rock, moves continents, and lifts up mountains. Earth's *external heat engine* is driven by solar energy: heat supplied to Earth's surface by the Sun. Heat from the Sun energizes the atmosphere and oceans and is responsible for Earth's climate and weather. Rain, wind, and ice erode mountains and shape the landscape, and the shape of the landscape, in turn, influences the climate.

All the parts of our planet and all their interactions, taken together, constitute the **Earth system.** Although Earth scientists have long thought in terms of natural systems, it was not until the late twentieth century that they had the tools to investigate how the Earth system actually works. Networks of instruments and Earth-orbiting satellites now collect information about the Earth system on a global scale, and computers are powerful enough to calculate the

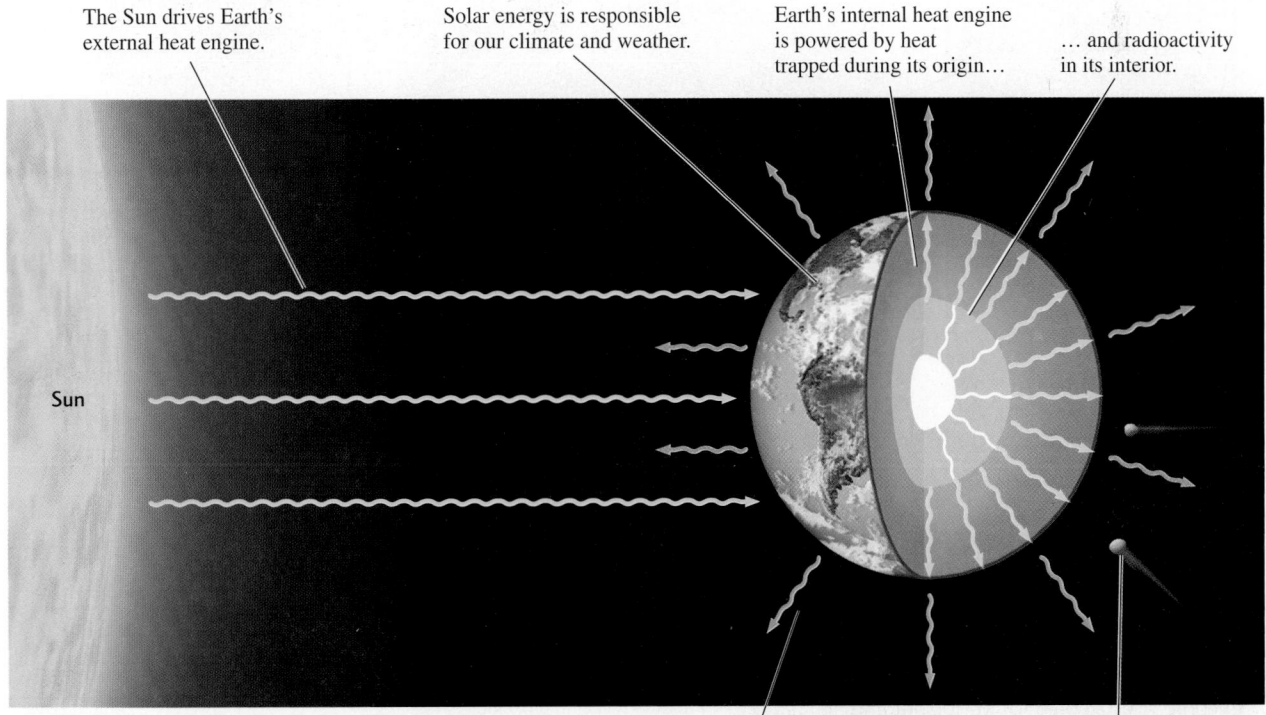

The Sun drives Earth's external heat engine.

Solar energy is responsible for our climate and weather.

Earth's internal heat engine is powered by heat trapped during its origin…

… and radioactivity in its interior.

Sun

Heat radiating from Earth balances solar input and heat from interior.

Meteors move mass from the cosmos to Earth.

FIGURE 1.13 ■ The Earth system is an open system that exchanges energy and mass with its surroundings.

mass and energy transfers within the system. The major components of the Earth system can be represented as a set of domains, or "spheres" (**Figure 1.14**). We have discussed some of these components already; we will define the others shortly.

We will talk about the Earth system throughout this textbook. Let's get started by looking at some of its basic features. The Earth system is an *open system* that exchanges energy and mass with its surroundings (see Figure 1.14). Radiant energy from the Sun energizes the weathering and erosion of Earth's surface, as well as the growth of plants, which feed almost all living things. Earth's climate is controlled by the balance between the solar energy coming into the Earth system and the heat energy Earth radiates back into space.

Early in the life of the solar system, collisions between Earth and other solid bodies were a very important process, growing the planet's mass and forming the Moon. These days, the exchange of mass between Earth and space is relatively small: on average, only about 40,000 tons of material—equivalent to a cube 24 m on a side—fall into Earth's atmosphere each year in the form of meteors and meteorites. Most meteors we see streaking across the sky are very small, perhaps a few grams in mass, although occasionally Earth encounters a larger chunk, with dangerous results (**Figure 1.15**).

Although we think of Earth as a single system, it is a challenge to study the whole thing all at once. Instead, we will focus our attention on the particular components of the Earth system (subsystems) that we are trying to

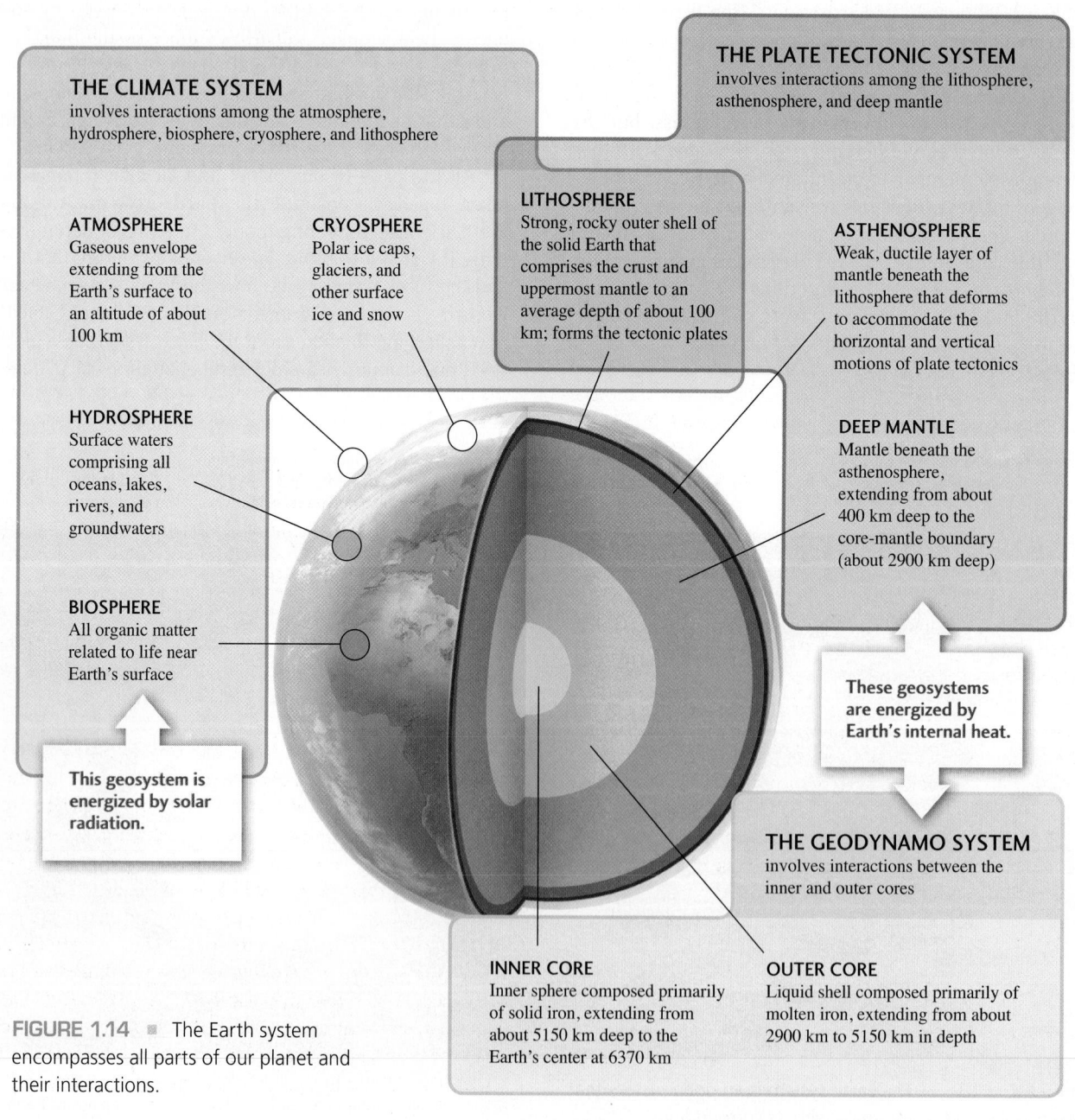

THE CLIMATE SYSTEM
involves interactions among the atmosphere, hydrosphere, biosphere, cryosphere, and lithosphere

THE PLATE TECTONIC SYSTEM
involves interactions among the lithosphere, asthenosphere, and deep mantle

ATMOSPHERE
Gaseous envelope extending from the Earth's surface to an altitude of about 100 km

CRYOSPHERE
Polar ice caps, glaciers, and other surface ice and snow

LITHOSPHERE
Strong, rocky outer shell of the solid Earth that comprises the crust and uppermost mantle to an average depth of about 100 km; forms the tectonic plates

ASTHENOSPHERE
Weak, ductile layer of mantle beneath the lithosphere that deforms to accommodate the horizontal and vertical motions of plate tectonics

HYDROSPHERE
Surface waters comprising all oceans, lakes, rivers, and groundwaters

DEEP MANTLE
Mantle beneath the asthenosphere, extending from about 400 km deep to the core-mantle boundary (about 2900 km deep)

BIOSPHERE
All organic matter related to life near Earth's surface

This geosystem is energized by solar radiation.

These geosystems are energized by Earth's internal heat.

THE GEODYNAMO SYSTEM
involves interactions between the inner and outer cores

INNER CORE
Inner sphere composed primarily of solid iron, extending from about 5150 km deep to the Earth's center at 6370 km

OUTER CORE
Liquid shell composed primarily of molten iron, extending from about 2900 km to 5150 km in depth

FIGURE 1.14 ■ The Earth system encompasses all parts of our planet and their interactions.

FIGURE 1.15 ▪ Explosion of the Chelyabinsk meteor above central Russia on Feb 15, 2013, released 20–30 times more energy than the Hiroshima atomic bomb, and its shock wave injured 1,500 people. This small asteroid had a diameter of about 20 meters and weighed about 11,000 tons, a reminder that Earth is an open system that continues to exchange both mass and energy with the solar system. [Camera Press/Ria Novosti//Redux.]

understand. For instance, in our discussion of global climate change, we will primarily consider interactions between the atmosphere and several other components that are driven by solar energy: the *hydrosphere* (Earth's surface waters and groundwaters), the *cryosphere* (Earth's ice caps, glaciers, and snowfields), and the *biosphere* (Earth's living organisms). Our coverage of how the continents are deformed to raise mountains will focus on interactions between the crust and the mantle that are driven by Earth's internal heat engine. Specialized subsystems that produce specific types of activity, such as climate change or mountain building, are called **geosystems.** The Earth system can be thought of as a collection of many open, interacting (and often overlapping) geosystems.

In this section, we will introduce three important geosystems that operate on a global scale: the climate system, the plate tectonic system, and the geodynamo. Later in this textbook, we will discuss a number of smaller geosystems, such as volcanoes that erupt hot lava (Chapter 12), hydrologic systems that give us our drinking water (Chapter 17), and petroleum reservoirs that produce oil and gas (Chapter 23).

The Climate System

Weather is the term we use to describe the temperature, precipitation, cloud cover, and winds observed at a particular location and time on Earth's surface. We all know how variable the weather can be—hot and rainy one day, cool and dry the next—depending on the movements of storm systems, warm and cold fronts, and other atmospheric disturbances. Because the atmosphere is so complex, even the best forecasters have a hard time predicting the weather more than 4 or 5 days in advance. However, we can guess in rough terms what our weather will be much further into the future, because weather is governed primarily by the changes in solar energy input on seasonal and daily cycles: summers are hot, winters cold; days are warmer, nights cooler. The **climate** produced by these weather cycles can be described by averaging temperatures and other variables over many years of observation. A complete description of climate also includes measures of how variable the weather has been, such as the highest and lowest temperatures ever recorded on a given day of the year.

The **climate system** includes all the Earth system components that determine climate on a global scale and how climate changes with time. In other words, the climate system involves not only the behavior of the atmosphere, but also its interactions with the hydrosphere, cryosphere, biosphere, and lithosphere (see Figure 1.15).

When the Sun warms Earth's surface, some of the heat is trapped by water vapor, carbon dioxide, and other gases in the atmosphere, much as heat is trapped by frosted glass in a greenhouse. This *greenhouse effect* explains why Earth has a climate that makes life possible. If its atmosphere contained no greenhouse gases, Earth's surface would be frozen solid! Therefore, greenhouse gases, particularly carbon dioxide, play an essential role in regulating climate. As we will learn in later chapters, the concentration of carbon dioxide in the atmosphere is a balance between the amount spewed out of Earth's interior in volcanic eruptions and the amount withdrawn during the weathering of silicate rocks. In this way, the behavior of the atmosphere is regulated by interactions with the lithosphere.

To understand these types of interactions, scientists build numerical models—virtual climate systems—on large computers, and they compare the results of their computer simulations with data from their observations. A particularly urgent problem to which these models are being applied is the global warming that is being caused by *anthropogenic* (human-generated) emissions of carbon dioxide and other greenhouse gases. Part of the public debate about global warming centers on the accuracy of predictions based on computer models. Skeptics argue that even the most sophisticated computer models are unreliable because they lack many features of the real Earth system. In Chapter 15, we will discuss some aspects of how the climate system works, and in Chapter 23, we will examine the practical problems posed by anthropogenic climate change.

The Plate Tectonic System

Some of Earth's most dramatic geologic events—volcanic eruptions and earthquakes, for example—result from interactions within Earth's interior. These phenomena are driven by Earth's internal heat, which is transferred upward through the circulation of material in Earth's mantle.

We have seen that Earth is zoned by chemistry: its crust, mantle, and core are chemically distinct layers. Earth is also zoned by *strength,* a property that measures how much an Earth material can resist being deformed. Material strength depends on both chemical composition (bricks are strong, soap bars are weak) and temperature (cold wax is strong, hot wax is weak).

In some ways, the outer part of the solid Earth behaves like a ball of hot wax. Cooling of the surface forms a strong outer shell, or **lithosphere** (from the Greek *lithos,* meaning "stone"), which encases a hot, weak **asthenosphere** (from the Greek *asthenes,* meaning "weak"). The lithosphere includes the crust and the top part of the mantle down to an average depth of about 100 km. The asthenosphere is the portion of mantle, perhaps 300 km thick, immediately below the lithosphere. When subjected to force, the lithosphere tends to behave like a nearly rigid and brittle shell, whereas the underlying asthenosphere flows like a moldable, or *ductile,* solid.

According to the remarkable theory of *plate tectonics,* the lithosphere is not a continuous shell; it is broken into about a dozen large plates that move over Earth's surface at rates of a few centimeters per year. Each lithospheric plate is a rigid unit that rides on the asthenosphere, which is also in motion. The lithosphere that forms a plate may vary from just a few kilometers thick in volcanically active areas to more than 200 km thick beneath the older, colder parts of continents. The discovery of plate tectonics in the 1960s led to the first unified theory that explained the worldwide distribution of earthquakes and volcanoes, continental drift, mountain building, and many other geologic phenomena. Chapter 2 describes the basic concepts of plate tectonics.

Why do the plates move across Earth's surface instead of locking up into a completely rigid shell? The forces that push and pull the plates come from the mantle. Driven by Earth's internal heat engine, hot mantle material rises at boundaries where plates separate, forming new lithosphere. The lithosphere cools and becomes more rigid as it moves away from these boundaries, eventually sinking back into the mantle under the pull of gravity at other boundaries where plates converge. This general process, in which hotter material rises and cooler material sinks, is called **convection** (**Figure 1.16**). Convection in the mantle can

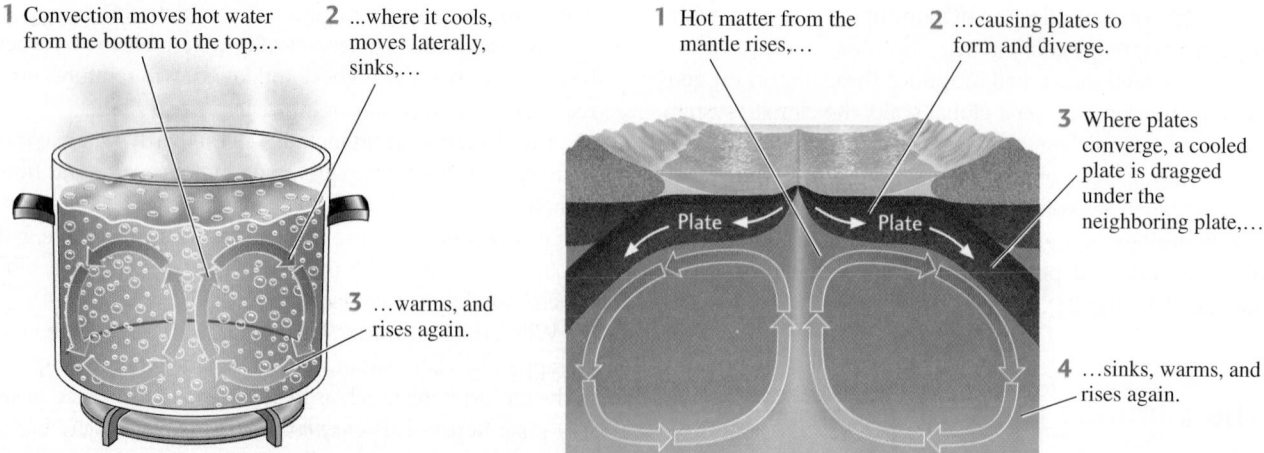

1 Convection moves hot water from the bottom to the top,...

2 ...where it cools, moves laterally, sinks,...

3 ...warms, and rises again.

1 Hot matter from the mantle rises,...

2 ...causing plates to form and diverge.

3 Where plates converge, a cooled plate is dragged under the neighboring plate,...

4 ...sinks, warms, and rises again.

Plate ← → Plate

FIGURE 1.16 ▪ Convection in Earth's mantle can be compared to the pattern of movement in a pot of boiling water. Both processes carry heat upward through the movement of matter.

be compared to the pattern of movement in a pot of boiling water. Both processes transfer energy by the movement of mass, but mantle convection is much slower because the solid mantle rocks are much more resistant to deformation than ordinary fluids such as water.

The convecting mantle and its overlying mosaic of lithospheric plates constitute the **plate tectonic system**. As with the climate system (which involves a wide range of convective processes in the atmosphere and oceans), scientists use computer simulations to study the plate tectonic system and test the agreement of their models against observations.

The Geodynamo

The third global geosystem involves interactions that produce a **magnetic field** deep inside Earth in its liquid outer core. This magnetic field reaches far into outer space, causing compass needles to point north and shielding the biosphere from harmful solar radiation. When rocks form, they become slightly magnetized by this magnetic field, so geologists can study how the field behaved in the past and use it to help them decipher the geologic record.

Earth rotates about an axis that goes through its north and south poles. Earth's magnetic field behaves as if a powerful bar magnet were located at Earth's center and inclined about 11° from this rotational axis. The magnetic force points into Earth at the north magnetic pole and outward from Earth at the south magnetic pole (**Figure 1.17**). At any place on Earth (except near the magnetic poles), a compass needle that is free to swing under the influence of the magnetic field will rotate into a position parallel to

the local line of force, approximately in the north-south direction.

Although a permanent magnet at Earth's center could explain the dipolar (two-pole) nature of the observed magnetic field, this hypothesis can be easily rejected. Laboratory experiments have demonstrated that the field of a permanent magnet is destroyed when the magnet is heated above about 500°C. We know that the temperatures in Earth's deep interior are much higher than that—thousands of degrees at its center—so, unless the magnetism were constantly regenerated, it could not be maintained.

Scientists theorize that heat flowing out of Earth's core causes convection that generates and maintains the magnetic field. Why is a magnetic field created by convection in the outer core, but not by convection in the mantle? First, the outer core is made primarily of iron, which is a very good electrical conductor, whereas the silicate rocks of the mantle are poor electrical conductors. Second, the convective flow is a million times more rapid in the liquid outer core than in the solid mantle. The rapid flow stirs up electric currents in the liquid iron-nickel alloy to produce the magnetic field. Thus, this **geodynamo** is more like an electromagnet than a bar magnet (see Figure 1.17).

For some 400 years, scientists have known that a compass needle points north because of Earth's magnetic field. Imagine how stunned they were half a century ago when they found geologic evidence that the direction of the magnetic force can be reversed. Over about half of geologic time, a compass needle would have pointed south! These *magnetic reversals* occur at irregular intervals ranging from tens of thousands to millions of years. The processes that cause them are not well understood, but computer models of the geodynamo show sporadic reversals occurring in

(a) Bar magnet   **(b) Electromagnetic**   **(c) Geodynamo**

FIGURE 1.17 ■ (a) A bar magnet creates a dipolar field with north and south poles. (b) A dipolar field can also be produced by electric currents flowing through a coil of metallic wire, as shown for this battery-powered electromagnet. (c) Earth's magnetic field, which is approximately dipolar above Earth's surface, is produced by electric currents flowing in the liquid-metal outer core, which are powered by convection.

FIGURE 1.18 ▪ Earth's magnetic field protects life by shielding Earth's surface from harmful solar radiation. This solar wind contains highly energetic charged particles ejected from the Sun, which distorts Earth's magnetic field lines, shown here in light blue. The distances in this picture are not to scale. [SOHO (ESA and NASA).]

the absence of any external factors—that is, purely through interactions within Earth's core. As we will see in the next chapter, magnetic reversals, which leave their imprint on the geologic record, have helped geologists figure out the movements of the lithospheric plates.

Interactions Among Geosystems Support Life

The natural environment—the habitat of life—is largely controlled by the climate system. The biosphere participates as an active component of this geosystem, regulating,

for example, the amount of carbon dioxide, methane, and other greenhouse gases in the atmosphere, which in turn determines the planet's surface temperature. As we shall see in Chapter 11, the evolution of the biosphere and atmosphere have gone hand-in-hand throughout the last 3.5 billion years of climate-system history.

Perhaps less obvious is the coupling of the natural environment to the other two global geosystems. Plate tectonics produces volcanoes that resupply the atmosphere and oceans with water and gases from Earth's deep interior, and it is responsible for the tectonic processes that raise mountains. The interactions of the atmosphere, hydrosphere, and cryosphere with the surface topography create a variety of habitats that enrich the biosphere and, through the erosion of rock and dissolution of minerals, provide life with essential nutrients.

Unlike the convective motions of plate tectonics, the swirling currents in Earth's outer core are too deep to deform the crust or alter its chemistry. However, the magnetic field produced by this geodynamo reaches outward into space far beyond Earth's atmosphere (see Figure 1.17). There it forms a barrier to highly energetic particles that stream outward from the Sun at speeds of more than 400 km/s—the *solar wind* (**Figure 1.18**). Without this shield, Earth's surface would be bombarded by harmful solar radiation, which would kill many forms of life that now prosper in its biosphere.

▪ An Overview of Geologic Time

So far, we have discussed Earth's size and shape, its internal layering and composition, and the operation of its three major geosystems. How did Earth get its layered structure

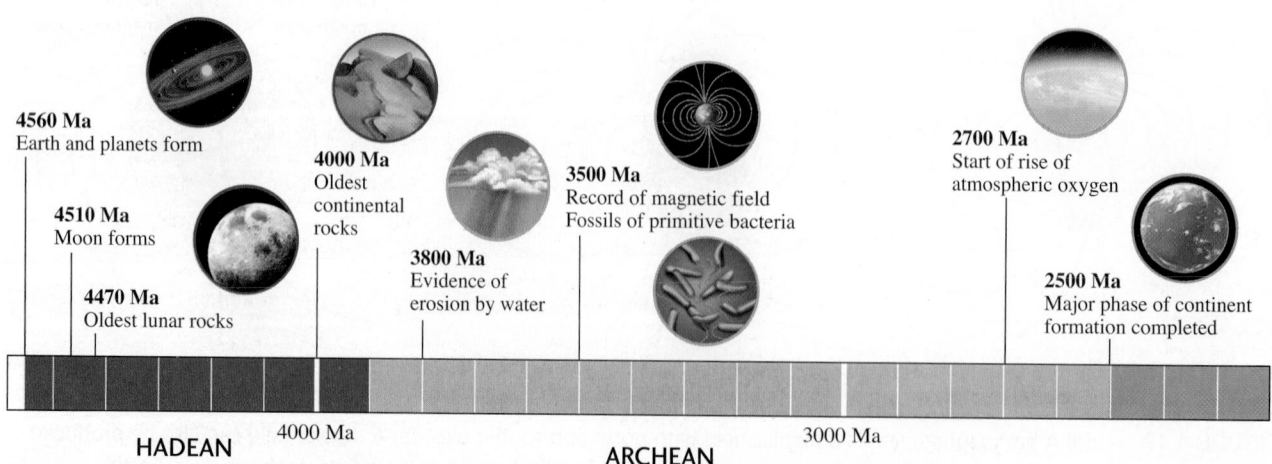

4560 Ma
Earth and planets form

4510 Ma
Moon forms

4470 Ma
Oldest lunar rocks

4000 Ma
Oldest continental rocks

3800 Ma
Evidence of erosion by water

3500 Ma
Record of magnetic field
Fossils of primitive bacteria

2700 Ma
Start of rise of atmospheric oxygen

2500 Ma
Major phase of continent formation completed

HADEAN   4000 Ma   ARCHEAN   3000 Ma

FIGURE 1.19 ▪ This geologic time line shows some of the major events observed in the geologic record, beginning with the formation of the planets. (Ma, million years ago.)

in the first place? How have the global geosystems evolved through geologic time? To begin to answer these questions, we present a brief overview of geologic time from the birth of the planet to the present. Later chapters will fill in the details.

Comprehending the immensity of geologic time is a challenge. John McPhee notes that geologists look into the "deep time" of Earth's early history (measured in billions of years), just as astronomers look into the "deep space" of the outer universe (measured in billions of light-years). Figure 1.19 presents the "arrow of geologic time," marked with some major events and transitions.

The Origin of Earth and Its Global Geosystems

Using evidence from meteorites, geologists have been able to show that Earth and the other planets of the solar system formed about 4.56 billion years ago through the rapid condensation of a dust cloud that circulated around the young Sun. This violent process, which involved the aggregation and collision of progressively larger clumps of matter, will be described in more detail in Chapter 9. Within just 100 million years (a relatively short time, geologically speaking), the Moon had formed and Earth's core had separated from its mantle. Exactly what happened during the next several hundred million years is hard to know. Very little of the rock record survived intense bombardment by the large meteorites that were constantly smashing into Earth. This early period of Earth's history is appropriately called the geologic "dark ages."

The oldest rocks now found on Earth's surface are over 4 billion years old. Rocks as ancient as 3.8 billion years show evidence of erosion by water, indicating the existence of a hydrosphere and the operation of a climate system not too different from that of the present. Rocks only slightly younger, 3.5 billion years old, record a magnetic field about as strong as the one we see today, showing that the geodynamo was operating by that time. By 2.5 billion years ago, enough low-density crust had collected at Earth's surface to form large continental masses. The geologic processes that then modified those continents were very similar to those we see operating today.

The Evolution of Life

Life also began very early in Earth's history, as we can tell from the study of **fossils**, traces of organisms preserved in the geologic record. Fossils of primitive bacteria have been found in rocks dated at 3.5 billion years ago. A key event was the evolution of organisms that release oxygen into the atmosphere and oceans. The buildup of oxygen in the atmosphere was under way by 2.7 billion years ago. Atmospheric oxygen concentrations probably rose to modern levels in a series of steps over a period as long as 2 billion years.

Life on early Earth was simple, consisting mostly of small, single-celled organisms that floated near the surface of the oceans or lived on the seafloor. Between 1 billion and 2 billion years ago, more complex life-forms such as algae and seaweeds evolved. The first animals appeared about 600 million years ago, evolving in a series of waves. In a period starting 542 million years ago and probably lasting less than 10 million years, eight entirely new branches of the animal kingdom were established, including the ancestors of nearly all animals inhabiting Earth today. It was during this evolutionary explosion, sometimes called biology's "Big Bang," that animals with shells first left their shelly fossils in the geologic record.

Although biological evolution is often viewed as a very slow process, it is punctuated by brief periods of rapid change. Spectacular examples are *mass extinctions*, during

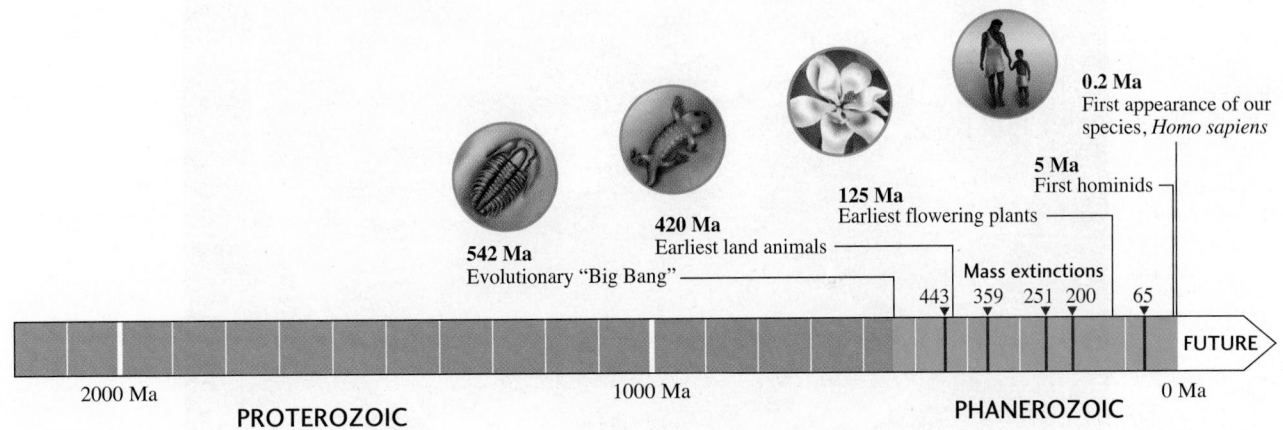

0.2 Ma
First appearance of our species, *Homo sapiens*

5 Ma
First hominids

125 Ma
Earliest flowering plants

420 Ma
Earliest land animals

542 Ma
Evolutionary "Big Bang"

Mass extinctions
443   359 251 200   65

FUTURE

2000 Ma   1000 Ma   0 Ma
PROTEROZOIC   PHANEROZOIC

which many kinds of organisms suddenly disappeared from the geologic record. Five of these huge turnovers are marked on the geologic time line in Figure 1.19. The most recent one was caused by a large meteorite impact 65 million years ago. The meteorite, not much larger than 10 km in diameter, caused the extinction of half of Earth's species, including all the dinosaurs.

The causes of the other mass extinctions are still being debated. In addition to meteorite impacts, scientists have proposed other kinds of extreme events, such as rapid climate changes brought on by glaciations and massive eruptions of volcanic material. The evidence is often ambiguous or inconsistent, however. For example, the largest mass extinction of all time took place about 251 million years ago, wiping out nearly 95 percent of all species. A meteorite impact has been proposed by some investigators as the cause, but the geologic record shows that ice sheets expanded and seawater chemistry changed at this time—a finding that is consistent with a major climate change. At the same time, an enormous volcanic eruption covered an area in Siberia almost half the size of the United States with 2 million to 3 million cubic kilometers of lava. This mass extinction has been dubbed "Murder on the Orient Express" because there are so many suspects!

Google Earth Project

Earth is a dynamic and complex system of interrelated components. A great many factors work to shape Earth's surface, and they are brought together by the overarching theory of plate tectonics. In our first exercise, we will use GE to explore the topographic extremes of our planet; we will use subsequent exercises in later chapters to explore the origin of those features. Let's start at the roof of the world: the Himalaya.

LOCATION   Topographic exploration from the Himalaya, in central Asia, to the Challenger Deep, off the southern coast of Guam in the Pacific Ocean

GOAL   Demonstrate the topographic variation of our planet and introduce the tools of Google Earth

LINKED   Figure 1.8

Data SIO, NOAA, U.S. Navy, NGA, GEBCO Image © 2009 TerraMetrics Data @ MIRC/JHA Image ©2009 DigitalGlobe

Mass extinctions reduce the number of species competing for space in the biosphere. By "thinning out the crowd," these extreme events can promote the evolution of new species. After the demise of the dinosaurs 65 million years ago, mammals became the dominant class of animals. The rapid evolution of mammals into species with bigger brains and more dexterity led to the emergence of humanlike species (*hominids*) about 5 million years ago and our own species, *Homo sapiens* (Latin for "knowing human"), about 200,000 years ago. As newcomers to the biosphere, we are just beginning to leave our mark on the geologic record. Indeed, our short history as a species can be appreciated by noting that it spans less than a line's width on the geologic time line (see Figure 1.19).

■ Welcome to Google Earth

Google Earth (GE) is a spatial dataset interface available through the Internet search engine Google that can be downloaded free. This interface uses aerial and satellite photographs at a variety of spatial resolutions overlaid on

1. Enter "Mt. Everest" into your GE search engine and use the cursor to find its highest point. What is its approximate elevation above sea level (*above mean sea level*, or amsl)? It may be helpful to tilt your frame of view to the north in order to pick out the highest point.
   a. 10,400 m amsl
   b. 7380 m amsl
   c. 8850 m amsl
   d. 9230 m amsl

2. Zoom out from Mt. Everest proper and take a look at the shape of the Himalaya as a whole (try an eye altitude of 4400 km). Which of the following descriptions best captures what you see?
   a. A triangular mountain range composed of a single high peak
   b. An east-west–oriented mountain range composed of dozens of high peaks along the southern rim of a high plateau
   c. A north-south–oriented mountain range composed of high peaks in the middle and lower peaks around the edges
   d. A circular mountain range closed around a central broad dome

3. From the Himalaya, move to one of the deepest places on Earth's surface by typing "Challenger Deep" into the search panel. GE should take you immediately out to sea off the coast of the Philippines. Use the GE "line" measurement tool to determine the approximate horizontal surface distance between the two locations. What is that distance?
   a. 6300 km
   b. 2200 km
   c. 185,000 km
   d. 75,500 km

4. Zoom out from Challenger Deep to an eye altitude of 4200 km. Notice the unique surface feature that links Challenger Deep to deep regions of the ocean here. How would you describe this large-scale feature?
   a. Challenger Deep is part of an undersea mountain range with a roughly north-south orientation.
   b. Challenger Deep is part of an arcuate deep-sea trench in the Pacific Ocean that trends almost east-west at this location.
   c. Challenger Deep is the deepest part of a broad, almost flat plain near the middle of the Pacific Ocean.
   d. Challenger Deep is at the top of an undersea volcano that rises above the Pacific Ocean floor.

Optional Challenge Question
5. Using the answer to question 1 and using your cursor to note the maximum depth of Challenger Deep below mean sea level, calculate the approximate total difference in elevation of the two locations. Which of the following numbers is closest to that difference?
   a. 14,000 m
   b. 20,000 m
   c. 18,000 m
   d. 26,000 m

digital elevation model datasets to give the images a three-dimensional quality. Since the data are geo-referenced in all three dimensions, they can be used to make measurements of distance with GE's "path" and "line" measurement tools. Elevation, latitude, and longitude are continuously tracked for any specific location of the cursor and are displayed at the bottom of the screen. GE also offers navigation tools in the upper right corner of the screen that allow you to zoom in and out as well as to alter the azimuth and aspect of your view.

One of the newest functions of GE is the ability to move backward in time at some locations by accessing archived spatial datasets. In the spirit of all Internet search engines, Google also provides a "Fly to" search window you can use to transport yourself to specific virtual locations. You can bookmark favorite locations as well as link locations to geo-referenced digital images taken at the same locations. Please make use of some or all of these tools while familiarizing yourself with the interface, and have fun doing it! For specific instruction on using Google Earth, go to *Google Earth Tutorial* at www.whfreeman.com/understandingearth7e.

SUMMARY

What is geology? Geology is the study of Earth—its history, its composition and internal structure, and its surface features.

How do geologists study Earth? Geologists, like other scientists, use the scientific method. They develop and test hypotheses, which are tentative explanations for natural phenomena based on observations and experiments. They share their data and test one another's hypotheses. A coherent set of hypotheses that have survived repeated challenges constitutes a theory. Hypotheses and theories can be combined into a scientific model that represents a natural system or process. Confidence grows in those hypotheses, theories, and models that withstand repeated tests and are able to predict the results of new observations or experiments.

What is Earth's shape? Earth's overall shape is a sphere with an average radius of 6370 km that bulges slightly at the equator and is slightly squashed at the poles due to the planet's rotation. Its topography varies by about 20 km from the highest point on its surface to the lowest point. Its elevations fall into two main groups: 0 to 1 km above sea level over much of the continents and 4 to 5 km below sea level for much of the ocean basins.

What are Earth's major layers? Earth's interior is divided into concentric layers of different compositions separated by sharp, nearly spherical boundaries. The outer layer is the crust, made up mainly of silicate rock, which varies in thickness from about 40 km in the case of continental crust to about 7 km for oceanic crust. Below the crust is the mantle, a thick shell of denser silicate rock that extends to the core-mantle boundary at a depth of about 2890 km. The core, which is composed primarily of iron and nickel, is divided into two layers: a liquid outer core and a solid inner core, separated by a boundary at a depth of 5150 km. Jumps in density between these layers are caused primarily by differences in their chemical composition.

How do we study Earth as a system of interacting components? When we try to understand a complex system such as the Earth system, we find that it is often easier to focus on its subsystems (which we call geosystems). This textbook focuses on three major global geosystems: the climate system, which involves interactions among the atmosphere, hydrosphere, cryosphere, biosphere, and lithosphere; the plate tectonic system, which involves interactions among Earth's solid components; and the geodynamo, which involves interactions within Earth's core. The climate system is driven by heat from the Sun, whereas the plate tectonic system and the geodynamo are driven by Earth's internal heat.

What are the basic elements of plate tectonics? The lithosphere is broken into about a dozen large plates. Driven by convection in the mantle, these plates move over Earth's surface at rates of a few centimeters per year. Each plate acts as a rigid unit riding on the ductile asthenosphere, which is also in motion. Hot mantle material rises at boundaries where plates form and separate, cooling and becoming more rigid as it moves away. Eventually, most of it sinks back into the mantle at boundaries where plates converge.

What are some major events in Earth's history? Earth formed 4.56 billion years ago. Rocks as old as 4.3 billion years have survived in Earth's crust. Liquid water existed on Earth's surface by 3.8 billion years ago. Rocks about 3.5 billion years old show evidence of a magnetic field, and the earliest evidence of life has been found in rocks of the same age. By 2.7 billion years ago, the oxygen content of the atmosphere was rising because of oxygen production by early organisms, and by 2.5 billion years ago, large continental masses had formed. Animals appeared suddenly about 600 million years ago, diversifying rapidly in a great evolutionary explosion. The subsequent evolution of life was marked by a series of extreme events that killed off many species, allowing new species to evolve. A dramatic example was the impact of a large meteorite 65 million years ago. Our species, *Homo sapiens*, first appeared about 200,000 years ago.

KEY TERMS AND CONCEPTS

asthenosphere (p. 16)

climate (p. 15)

climate system (p. 15)

convection (p. 16)

core (p. 10)

crust (p. 11)

Earth system (p. 13)

fossil (p. 19)

geodesy (p. 8)

geodynamo (p. 17)

geologic record (p. 6)

geology (p. 4)

geosystem (p. 15)

inner core (p. 12)

lithosphere (p. 16)

magnetic field (p. 17)

mantle (p. 10)

outer core (p. 12)

plate tectonic system (p. 17)

principle of
uniformitarianism (p. 7)

scientific method (p. 4)

seismic wave (p. 9)

topography (p. 9)

PRACTICING GEOLOGY EXERCISE

How Big Is Our Planet?

How Eratosthenes measured Earth's circumference.

How was it discovered that Earth is round with a circumference of 40,000 km? No one had looked down on Earth from space before the early 1960s, but its shape and size was understood long before that time. In 1492, Columbus set a westward course for India because he believed in a theory of geodesy that had been favored by Greek philosophers: *we live on a sphere.* His math was poor, however, so he badly underestimated Earth's circumference. Instead of a shortcut, he took the long way around, finding a New World instead of the Spice Islands! Had Columbus properly understood the ancient Greeks, he might not have made this fortuitous mistake, because they had accurately measured Earth's size more than 17 centuries earlier.

The credit for determining Earth's size goes to Eratosthenes, a Greek who was chief librarian at the

Great Library of Alexandria in Egypt. Sometime around 250 B.C., a traveler told him about an interesting observation. At noon on the first day of summer (June 21), a deep well in the city of Syene, about 800 km south of Alexandria, was completely lit up by sunlight because the Sun was directly overhead. Acting on a hunch, Eratosthenes did an experiment. He set up a vertical pole in his own city, and at high noon on the first day of summer, the pole cast a shadow.

Eratosthenes assumed that the Sun was very far away, so that the light rays falling on the two cities were parallel. Knowing that the Sun cast a shadow in Alexandria but was directly overhead at the same time in Syene, Eratosthenes could demonstrate with simple geometry that the ground surface must be curved. He knew that the most perfect curved surface is a sphere, so he hypothesized that Earth had a spherical shape (the Greeks admired geometric perfection). By measuring the length of the pole's shadow in Alexandria, he calculated that if vertical lines through the two cities could be extended to Earth's center, they would intersect at an angle of about 7°, which is about 1/50 of a full circle (360°). The distance between the two cities was known to be about 800 km in today's measurements. From these figures, Eratosthenes calculated a circumference for Earth that is very close to the modern value:

Earth's circumference = 50 × distance from Syene to
Alexandria = 50 × 800 km = 40,000 km

With this figure for Earth's circumference, it was a simple matter to calculate its radius. Eratosthenes knew that, for any circle, the circumference is equal to 2π (pi) times the radius, where π is about 3.14. Therefore, he divided his estimate of Earth's circumference by 2π to find its radius:

$$\text{radius} = \frac{\text{circumference}}{2\pi}$$

$$\frac{40,000 \text{ km}}{6.28} = 6370 \text{ km}$$

By these calculations, Eratosthenes arrived at a simple and elegant scientific model: *Earth is a sphere with a radius of about 6370 km.*

In this powerful demonstration of the scientific method, Eratosthenes made observations (the length of the shadow), formed a hypothesis (spherical shape), and applied some mathematical theory (spherical geometry) to propose a remarkably accurate model of Earth's physical form. His model correctly predicted other types of measurements, such as the distance at which a ship's tall mast would disappear over the horizon. Moreover, knowing Earth's shape and size allowed Greek astronomers to calculate the sizes of the Moon and Sun and the distances of these bodies from Earth. This story makes clear why well-designed experiments and good measurements are central to the scientific method: they give us new information about the natural world.

BONUS PROBLEM: The volume of a sphere is given by

$$\text{volume} = \frac{4\pi}{3} (\text{radius})^3$$

From this formula, calculate Earth's volume in cubic kilometers.

EXERCISES

1. Illustrate the differences between a hypothesis, a theory, and a model with some examples drawn from this chapter.

2. Give an example of how the model of Earth's spherical shape developed by Eratosthenes can be experimentally tested.

3. Give two reasons why Earth's shape is not a perfect sphere.

4. If you made a model of Earth that was 10 cm in radius, how high would Mount Everest rise above sea level?

5. It is thought that a large meteorite impact 65 million years ago caused the extinction of half of Earth's living species, including all the dinosaurs. Does this event disprove the principle of uniformitarianism? Explain your answer.

6. How does the chemical composition of Earth's crust differ from that of its mantle? From that of its core?

7. Explain how Earth's outer core can be liquid while the mantle is solid.

8. How do the terms *weather* and *climate* differ? Express the relationship between climate and weather using examples from your experience.

9. Earth's mantle is solid, but it undergoes convection as part of the plate tectonic system. Explain why these statements are not contradictory.

THOUGHT QUESTIONS

1. How does science differ from religion as a way to understand the world?

2. Imagine you are a tour guide on a journey from Earth's surface to its center. How would you describe the material that your tour group encounters on the way down? Why is the density of the material always increasing as you go deeper?

3. How does viewing Earth as a system of interacting components help us to understand our planet? Give an example of an interaction between two or more geosystems that could affect the geologic record.

4. In what general ways are the climate system, the plate tectonic system, and the geodynamo similar? In what ways are they different?

5. Not every planet has a geodynamo. Why not? If Earth did not have a magnetic field, what might be different about our planet?

6. Based on the material presented in this chapter, what can we say about how long ago the three major global geosystems began to operate?

7. If no theory can be completely proved, why do almost all geologists believe strongly in Darwin's theory of evolution?

MEDIA SUPPORT

 1-1 Animation: Earth's Major Layers

■ The Discovery of
Plate Tectonics   28

■ The Plates and
Their Boundaries   31

■ Rates and History of
Plate Movements   39

■ The Grand
Reconstruction   42

■ Mantle Convection:
The Engine of Plate
Tectonics   45

■ The Theory of Plate
Tectonics and the
Scientific Method   51

Mount Everest, Nepal, the highest mountain in
the world, as viewed from Kala Pattar.
[Michael C. Klesius/National Geographic/Getty Images.]

PLATE TECTONICS: THE UNIFYING THEORY

2

THE LITHOSPHERE—EARTH'S STRONG, rigid outer shell of rock—is broken into about a dozen plates, which slide past, converge with, or separate from each other as they move over the weaker, ductile asthenosphere. Plates are formed where they separate and recycled where they converge in a continuous process of creation and destruction. Continents, embedded in the lithosphere, drift along with the moving plates.

The theory of plate tectonics describes the movements of plates and the forces acting on them. It also explains volcanoes, earthquakes, and the distribution of mountain chains, rock assemblages, and structures on the seafloor—all of which result from events at plate boundaries. Plate tectonics provides a conceptual framework for a large part of this textbook and, indeed, for much of geology.

This chapter lays out the theory of plate tectonics and how it was discovered, describes plate movements today and in the geologic past, and examines how the forces that drive these movements arise from the mantle convection system.

■ The Discovery of Plate Tectonics

In the 1960s, a great revolution in thinking shook the world of geology. For almost 200 years, geologists had been developing various theories of *tectonics* (from the Greek *tekton,* meaning "builder")—the general term used to describe mountain building, volcanism, earthquakes, and other processes that construct geologic features on Earth's surface. It was not until the discovery of plate tectonics, however, that a single theory could satisfactorily explain the whole range of geologic processes. Physics had a comparable revolution at the beginning of the twentieth century, when the theory of relativity revised the physical laws that govern space, time, mass, and motion. Biology had a similar revolution in the middle of the twentieth century, when the discovery of DNA allowed biologists to explain how organisms transmit the information that controls their growth and functioning from generation to generation.

The basic ideas of plate tectonics were put together as a unified theory of geology about 50 years ago. The scientific synthesis that led to the theory of plate tectonics, however, really began much earlier in the twentieth century with the recognition of evidence for continental drift.

Continental Drift

Such changes in the superficial parts of the globe seemed to me unlikely to happen if the earth were solid to the center. I therefore imagined that the internal parts might be a fluid more dense, and of greater specific gravity than any of the solids we are acquainted with, which therefore might swim in or upon that fluid. Thus the surface of the earth would be a shell, capable of being broken and disordered by the violent movements of the fluid on which it rested.

(Benjamin Franklin, 1782, in a letter to French geologist Abbé J. L. Giraud-Soulavie)

The concept of **continental drift**—large-scale movements of continents—has been around for a long time. In the late sixteenth century and in the seventeenth century, European scientists noticed the jigsaw-puzzle fit of the coasts on both sides of the Atlantic Ocean, as if the Americas, Europe, and Africa had once been part of a single continent and had subsequently drifted apart. By the close of the nineteenth century, the Austrian geologist Eduard Suess had put together some of the pieces of the puzzle. He postulated that the present-day southern continents had once formed a single giant continent called *Gondwana* (or *Gondwanaland*). In 1915, Alfred Wegener, a German meteorologist who was recovering from wounds suffered in World War I, wrote a book on the breakup and drift of continents, in which he laid out the remarkable similarity of geologic features on opposite sides of the Atlantic (**Figure 2.1**). In the years that

FIGURE 2.1 ■ The jigsaw-puzzle fit of the continents bordering the Atlantic Ocean formed the basis of Alfred Wegener's theory of continental drift. In his book, titled *The Origin of Continents and Oceans,* Wegener cited as additional evidence the similarity of geologic features on opposite sides of the Atlantic. The matches between ancient crystalline rocks in adjacent regions of South America and Africa and of North America and Europe are shown here.
[Geographic fit from data of E. C. Bullard; geologic data from P. M. Hurley.]

followed, Wegener postulated a supercontinent, which he called **Pangaea** (Greek for "all lands"), that broke up into the continents as we know them today.

Although Wegener was correct in asserting that the continents had drifted apart, his hypotheses about how fast they were moving and what forces were pushing them across Earth's surface turned out to be wrong, as we will see, and those errors reduced his credibility among other scientists. After about a decade of spirited debate, physicists convinced geologists that Earth's outer layers were too rigid for continental drift to occur, and Wegener's ideas were rejected by all but a few geologists.

Wegener and other advocates of the drift hypothesis pointed not only to the geographic matching of geologic features but also to similarities in rock ages and trends in geologic structures on opposite sides of the Atlantic. They also offered arguments, accepted now as good evidence of drift, based on fossil and climate data. Identical 300-million-year-old fossils of the reptile *Mesosaurus,* for example, have been found in Africa and in South America, but nowhere else, suggesting that the two continents were

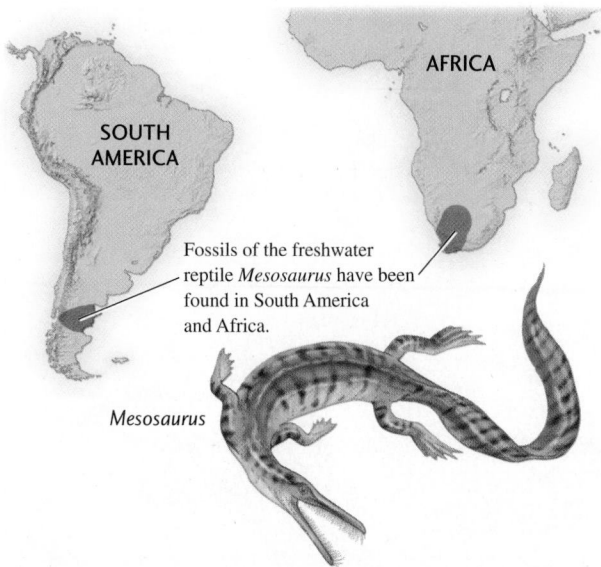

FIGURE 2.2 ■ Fossils of the freshwater reptile *Mesosaurus*, 300 million years old, are found in South America and Africa and nowhere else in the world. If *Mesosaurus* were able to swim across the South Atlantic Ocean, it should have been able to cross other oceans and should have spread more widely. The observation that it did not suggests that South America and Africa must have been joined about 300 million years ago. [After A. Hallam, "Continental Drift and the Fossil Record," *Scientific American* (November 1972): 57–66.]

joined when *Mesosaurus* lived (**Figure 2.2**). The animals and plants on the different continents showed similarities in their evolution until the postulated breakup time. After that, they followed different evolutionary paths because of their isolation and changing environments on the separating continents. In addition, rocks deposited by glaciers that existed 300 million years ago were found distributed

across South America, Africa, India, and Australia. If these southern continents had once been part of Gondwana near the South Pole, a single continental glacier could account for all of these glacial deposits.

Seafloor Spreading

The geologic evidence did not convince the skeptics, who maintained that continental drift was physically impossible. No one had yet come up with a plausible driving force that could have split Pangaea and moved the continents apart. Wegener, for example, thought the continents floated like boats across the solid oceanic crust, dragged along by the tidal forces of the Sun and Moon. His hypothesis was quickly rejected, however, because it could be shown that tidal forces are much too weak to move continents.

The breakthrough came when scientists realized that convection in Earth's mantle (described in Chapter 1) could push and pull continents apart, creating new oceanic crust through the process of **seafloor spreading.** In 1928, the British geologist Arthur Holmes proposed that convection currents "dragged the two halves of the original continent apart, with consequent mountain building in the front where the currents are descending, and the ocean floor development on the site of the gap, where the currents are ascending." Many still argued, however, that Earth's crust and mantle are rigid and immobile, and Holmes conceded that "purely speculative ideas of this kind, specially invented to match the requirements, can have no scientific value until they acquire support from independent evidence."

That evidence emerged from extensive explorations of the seafloor after World War II. Marine geologist Maurice "Doc" Ewing showed that the seafloor of the Atlantic Ocean is made of young basalt, not old granite, as some geologists had previously thought (**Figure 2.3**). Moreover, the mapping of an undersea mountain chain called the Mid-Atlantic Ridge led to the discovery of a deep crack-like valley, or *rift,* running down its crest (**Figure 2.4**). Two of the geologists who mapped this feature were Bruce Heezen and Marie Tharp, colleagues of Doc Ewing at Columbia University (**Figure 2.5**). "I thought it might be a rift valley," Tharp said years later. Heezen initially dismissed the idea as "girl talk," but they soon found that almost all

FIGURE 2.3 ■ This photo, taken in the summer of 1947, shows Maurice "Doc" Ewing (*center*) beaming as he looks at a piece of young basalt dredged from the depths of the Atlantic Ocean by the research vessel *Atlantis I.* On the near left is Frank Press, who initiated the series of geology textbooks that includes this one. [Lamont-Doherty Earth Observatory, Columbia University.]

FIGURE 2.4 ■ The North Atlantic seafloor, showing the cracklike rift valley running down the center of the Mid-Atlantic Ridge and the locations of earthquakes (black dots).

earthquakes in the Atlantic Ocean occurred near the rift, confirming Tharp's hunch. Because most earthquakes are generated by faulting, their results indicated that the rift was a tectonically active feature. Other mid-ocean ridges with similar rifts and earthquake activity were found in the Pacific and Indian oceans.

In the early 1960s, Harry Hess of Princeton University and Robert Dietz of the Scripps Institution of Oceanography proposed that Earth's crust separates along the rifts in mid-ocean ridges, and that new crust is formed by the upwelling of hot molten rock into these cracks. The new seafloor—actually the surface of newly created lithosphere—spreads laterally away from the rifts and is replaced by even newer crust in a continuing process of plate creation.

The Great Synthesis: 1963–1968

The seafloor spreading hypothesis put forward by Hess and Dietz explained how the continents could move apart through the creation of new lithosphere at mid-ocean ridges. But it raised another question: Could the seafloor and its underlying lithosphere be destroyed by recycling back into Earth's interior? If not, Earth's surface area would have to increase over time. For a while in the early 1960s, some physicists and geologists, including Heezen, actually believed in this idea of an expanding Earth. Other geologists recognized that the seafloor was indeed being recycled. They were convinced this was happening in several regions of intense volcanic and earthquake activity around the margins of the Pacific Ocean basin, known collectively as the Ring of Fire (**Figure 2.6**). The details of the process, however, remained unclear.

In 1965, the Canadian geologist J. Tuzo Wilson first described tectonics around the globe in terms of rigid plates moving over Earth's surface. He characterized three basic types of boundaries where plates move apart, come together, or slide past each other. Soon after, other scientists showed that almost all contemporary tectonic deformation—the process by which rocks are folded, faulted, sheared, or compressed by tectonic forces—is concentrated at these boundaries. They measured the rates and directions of crustal movements and demonstrated that these movements are mathematically consistent with a system of rigid plates moving over the planet's spherical surface.

FIGURE 2.5 ■ Marie Tharp and Bruce Heezen inspecting a map of the seafloor. Their discovery of tectonically active rifts on mid-ocean ridges provided important evidence for seafloor spreading. [Marie Tharp, www.marietharp.com.]

The basic elements of the new theory of **plate tectonics** were established by the end of 1968. By 1970, the evidence for plate tectonics had become so persuasive that almost all Earth scientists embraced the theory. Textbooks were revised, and specialists began to consider the implications of the new concept for their own fields.

■ The Plates and Their Boundaries

According to the theory of plate tectonics, the lithosphere is not a continuous shell, but is broken into a mosaic of rigid plates that move over Earth's surface (**Figure 2.7**). Each plate travels as a distinct unit, riding on the asthenosphere, which is also in motion. The largest is the Pacific Plate, which comprises much (though not all) of the Pacific Ocean basin. Some of the plates are named after the continents they include, but in no case is a plate identical with a continent. The North American Plate, for instance, extends from the Pacific coast of North America to the middle of the Atlantic Ocean, where it meets the Eurasian and African plates.

In addition to the 13 major plates, there are a number of smaller ones. An example is the Juan de Fuca Plate, a small piece of oceanic lithosphere trapped between the giant Pacific and North American plates just offshore of the northwestern United States. Others are continental fragments, such as the small Anatolian Plate, which includes much of Turkey.

To see plate tectonics in action, go to a plate boundary. Depending on which boundary you visit, you may find earthquakes, volcanoes, rising mountains, long, narrow rifts, folding, or faulting. Many geologic features develop through the interactions of plates at their boundaries.

There are three basic types of plate boundaries (**Figure 2.8**), all defined by the direction of movement of the plates relative to each other:

- At **divergent boundaries,** plates move apart and new lithosphere is created (plate area increases).

- At **convergent boundaries,** plates come together and one plate is recycled into the mantle (plate area decreases).

- At **transform faults,** plates slide horizontally past each other (plate area does not change).

Like many models of nature, these three plate boundary types are idealized. There are also "oblique" boundaries that combine divergence or convergence with some amount of transform faulting. Moreover, what actually goes on at a plate boundary depends on the type of lithosphere involved, because continental and oceanic lithosphere

FIGURE 2.6 ■ The Pacific Ring of Fire, with its active volcanoes (large red circles) and frequent earthquakes (small black dots), marks convergent plate boundaries where oceanic lithosphere is being recycled.

FIGURE 2.7 ▪ Earth's surface is a mosaic of 13 major plates, as well as a number of smaller plates, of rigid lithosphere that move slowly over the ductile asthenosphere. Only one of the smaller plates—the Juan de Fuca Plate, off the west coast of North America—is shown on this map. The arrows show the relative movement of two plates at a point on their boundary. The numbers next to the arrows give the relative plate velocities in millimeters per year. [Plate boundaries by Peter Bird, UCLA.]

63

47

Juan de
Fuca Plate

34

North
American
Plate

50

64

11

89

Cocos
Plate

Caribbean
Plate

12

118

50

Pacific Plate

138

73

Nazca
Plate

79

150

55

80

92

South
American
Plate

18

81

18

64

Antarctic Plate

18

18

Eurasian
Plate

22

24

African
Plate

27

31

35

34

50

DIVERGENT BOUNDARIES

(a) Oceanic Spreading Center

Rifting and spreading along a mid-ocean ridge create new oceanic lithosphere.

Mid-Atlantic Ridge

North American Plate

Eurasian Plate

(b) Continental Rift Zone

Rifting and spreading zones on continents are characterized by parallel rift valleys, volcanism, and earthquakes.

Great Rift Valley

African Plate

Somali Subplate

CONVERGENT BOUNDARIES

(c) Ocean–Ocean Convergence

Where oceanic lithosphere meets oceanic lithosphere, one plate is subducted under the other, and a deep-sea trench and a volcanic island arc are formed.

Mariana Islands

Marianas Trench

Philippine Plate

Pacific Plate

FIGURE 2.8 ▪ The interactions of lithospheric plates at their boundaries depend on the relative direction of plate movement and the type of lithosphere involved.

(d) Ocean–Continent Convergence

When oceanic lithosphere meets continental lithosphere, the oceanic lithosphere is subducted, and a volcanic mountain belt is formed at the continental margin.

Andes

Peru-Chile Trench

Nazca Plate

South American Plate

(e) Continent–Continent Convergence

Where two continents converge, the crust crumples and thickens, creating high mountains and a wide plateau.

Himalaya

Tibetan Plateau

Indian Plate

Eurasian Plate

TRANSFORM-FAULT BOUNDARIES

(f) Continental Transform Fault

At transform faults, plates slip horizontally past each other.

San Andreas Fault

Pacific Plate

North American Plate

(g) Mid-Ocean Ridge Transform Fault

Mid-ocean ridges are typically offset by transform faults.

Eurasian Plate

North American Plate

behave differently. The continental crust is made of rocks that are both lighter and weaker than those of either the oceanic crust or the mantle beneath the crust (see Figure 1.11). Later chapters will examine these differences in more detail; for now, you need to keep in mind only two of their consequences:

1. Because continental crust is lighter, it is not as easily recycled back into the mantle as oceanic crust.

2. Because continental crust is weaker, plate boundaries that involve continental crust tend to be more spread out and more complicated than those that involve oceanic crust.

Divergent Boundaries

Divergent boundaries are where the plates move apart. Divergent boundaries within ocean basins are narrow rifts that approximate the idealization of plate tectonics. Divergent boundaries within continents are usually more complicated and distributed over a wider area. This difference is illustrated in Figures 2.8a and 2.8b.

OCEANIC SPREADING CENTERS On the seafloor, the boundary between separating plates is marked by a **mid-ocean ridge,** an undersea mountain chain that exhibits earthquakes, volcanism, and rifting, all caused by the tensional (stretching) forces of mantle convection that are pulling the two plates apart. As the seafloor spreads, hot molten rock, called *magma,* wells up into the rifts to form new oceanic crust. Figure 2.8a shows what happens at one such **spreading center** on the Mid-Atlantic Ridge, where the North American and Eurasian plates are separating. (A more detailed map of the Mid-Atlantic Ridge is shown in Figure 2.4.) The island of Iceland exposes a segment of the otherwise submerged Mid-Atlantic Ridge, allowing geologists to view the processes of plate separation and seafloor spreading directly (**Figure 2.9**). The Mid-Atlantic Ridge continues in the Arctic Ocean north of Iceland and, to the south, it connects to a nearly globe-encircling system of mid-ocean ridges that wind through the Indian and Pacific oceans, ending along the west coast of North America. Seafloor spreading at these mid-ocean ridges has created the millions of square kilometers of oceanic crust that now form the floors of the world's oceans.

CONTINENTAL RIFTING Early stages of plate separation, such as the divergence that forms the Great Rift Valley of east Africa (see Figure 2.8b), can be found on some continents. These divergent boundaries are characterized by rift valleys, volcanism, and earthquakes distributed over a wider zone than is found at oceanic spreading centers. The Red Sea and the Gulf of California are rifts that are further along in the spreading process (**Figure 2.10**). In these cases, the continents have separated enough for new oceanic crust to form along the spreading axis, creating a deep basin into which the ocean has flooded.

FIGURE 2.9 ■ The Mid-Atlantic Ridge, a divergent plate boundary, rises above sea level in Iceland. This cracklike rift valley, filled with newly formed volcanic rock, indicates that plates are being pulled apart. [Ragnar Th Sigurdsson © ARCTIC IMAGES/Alamy.]

Sometimes continental rifting slows or stops before the continent actually splits apart. The Rhine Valley, along the border of Germany and France in western Europe, is a weakly active continental rift that may be this type of "failed" spreading center. Will the East African Rift continue to open, causing the Somali Subplate to split away from Africa completely and form a new ocean basin, as happened between Africa and the island of Madagascar? Or will the spreading slow and eventually stop, as appears to be happening in western Europe? Geologists don't know the answer.

Convergent Boundaries

Lithospheric plates cover the globe, so if they separate in one place, they must converge somewhere else, if Earth's surface area is to remain the same. (As far as we can tell, our planet is not expanding!) Where plates come together, they form convergent boundaries. The geologic processes that act during plate convergence make these boundaries more complex than the other boundary types.

OCEAN-OCEAN CONVERGENCE If the lithosphere of both converging plates is oceanic, one plate descends

(a)

Mediterranean Sea

Gulf of 'Aqaba

Gulf of Suez

Nile River

Red Sea

(b)

Mexico

Gulf of California

Baja California

Pacific Ocean

FIGURE 2.10 ■ Rifting of continental crust. (a) The Arabian Plate, on the right, is moving northeastward relative to the African Plate, on the left, opening the Red Sea (lower right). The Gulf of Suez is a failed rift that became inactive about 5 million years ago. North of the Red Sea, most of the plate motion is now taken up by rifting and transform faulting along the Gulf of 'Aqaba and its northward extension. (b) Baja California, on the Pacific Plate, is moving northwestward relative to the North American Plate, opening the Gulf of California between Baja and the Mexican mainland. [(a) Courtesy MDA Information Systems LLC; (b) Jeff Schmaltz, MODIS Rapid Response Team, NASA/GSFC.]

beneath the other in a process known as **subduction** (see Figure 2.8c). The lithosphere of the subducting plate sinks into the asthenosphere and is eventually recycled by the mantle convection system. This sinking produces a long, narrow deep-sea trench. In the Marianas Trench of the western Pacific, the ocean reaches its greatest depth, about 11 km—deeper than the height of Mount Everest.

As the cold slab of lithosphere descends deeper into Earth's interior, the pressure on it increases. Water trapped in the rocks is squeezed out and rises into the asthenosphere above the slab. This fluid causes the mantle material above it to melt. The resulting magma produces a chain of volcanoes, called an **island arc,** behind the trench. The subduction of the Pacific Plate has formed the volcanically active Aleutian Islands west of Alaska as well as the Mariana Islands and other island arcs in the western Pacific. The lithospheric slabs descending into the mantle cause earthquakes as deep as 690 km beneath some island arcs.

OCEAN-CONTINENT CONVERGENCE If one plate has a continental edge, it overrides the oceanic lithosphere

of the other plate because continental lithosphere is less dense and therefore less easily subducted (see Figure 2.8d). The submerged margin of the continent is crumpled by the convergence, deforming the continental crust and up-lifting rocks into a mountain belt roughly parallel to the deep-sea trench. The enormous compressive (squeezing) forces of convergence and subduction produce great earth-quakes along the subduction zone. Over time, materials are scraped off the descending slab and incorporated into the adjacent mountain belt, leaving geologists with a complex (and often confusing) record of the subduction process. As in the case of ocean-ocean convergence, the water carried downward by the subducting oceanic lithosphere causes mantle material to melt; the resulting magma rises and forms volcanoes in the mountain belt behind the trench.

The west coast of South America, where the South American Plate converges with the Nazca Plate, is a sub-duction zone of this type. A great chain of high mountains, the Andes, rises on the continental side of this convergent boundary, and a deep-sea trench lies just off the coast. The volcanoes here are active and deadly. One of them, Nevado del Ruiz in Colombia, killed 25,000 people when it erupted in 1985. Some of the world's largest earthquakes have been recorded along this boundary.

Another example is the Cascadia subduction zone, where the small Juan de Fuca Plate converges with the North American Plate off the coast of western North America. This convergent boundary gives rise to the dangerous volcanoes of the Cascade Range, such as Mount St. Helens, which had a major eruption in 1980 and a minor one in 2004. There is increasing worry that a great earthquake will occur in the Cascadia subduction zone and cause devastating damage along the coasts of Oregon, Washington, and British Columbia. Such an earthquake could cause a disastrous tsunami as large as the one generated by the great Tohoku earthquake of March 11, 2011, which occurred in a subduction zone off the northeastern coast of Honshu, Japan.

CONTINENT-CONTINENT CONVERGENCE Where two continents converge (see Figure 2.8e), the kind of subduction seen at other convergent boundaries cannot occur. The geologic consequences of such a continent-continent collision are impressive. The collision of the Indian and Eurasian plates, both with continents at their leading edges, provides the best example. The Eurasian Plate overrides the Indian Plate, but India and Asia remain afloat on the mantle. The collision creates a double thickness of crust, forming the highest mountain range in the world, the Himalaya, as well as the vast high Tibetan Plateau. Severe earthquakes occur in the crumpling crust of this and other continent-continent collision zones.

Many episodes of mountain building throughout Earth's history were caused by continent-continent collisions. The Appalachian Mountains, which run along the east coast of North America, were uplifted when North America, Eurasia, and Africa collided to form the supercontinent Pangaea about 300 million years ago.

Transform Faults

At boundaries where plates slide past each other, lithosphere is neither created nor destroyed. Such boundaries are transform faults: fractures along which the plates slip horizontally past each other (see Figure 2.8f, g).

The San Andreas fault in California, where the Pacific Plate slides past the North American Plate, is a prime example of a continental transform fault (see Figure 2.8f). Because the plates have been sliding past each other for millions of years, the rocks facing each other on the two sides of the fault are of different types and ages (**Figure 2.11**). Large earthquakes, such as the one that destroyed San Francisco in 1906, can occur on transform faults. There is much concern that, within the next several decades, a sudden rupture of the San Andreas fault, or on related faults near Los Angeles and San Francisco, will result in an extremely destructive earthquake.

Transform-fault boundaries are typically found along mid-ocean ridges where the continuity of a spreading zone is broken and the boundary is offset in a steplike pattern. An example can be found along the boundary between the African Plate and the South American Plate in the central Atlantic Ocean (see Figure 2.8g). Transform faults can also connect divergent plate boundaries with convergent plate boundaries and convergent boundaries with other convergent boundaries. Can you find examples of these types of transform-fault boundaries in Figure 2.7?

Combinations of Plate Boundaries

Each plate is bordered by some combination of divergent, convergent, and transform-fault boundaries. For example, the Nazca Plate is bounded on three sides by spreading

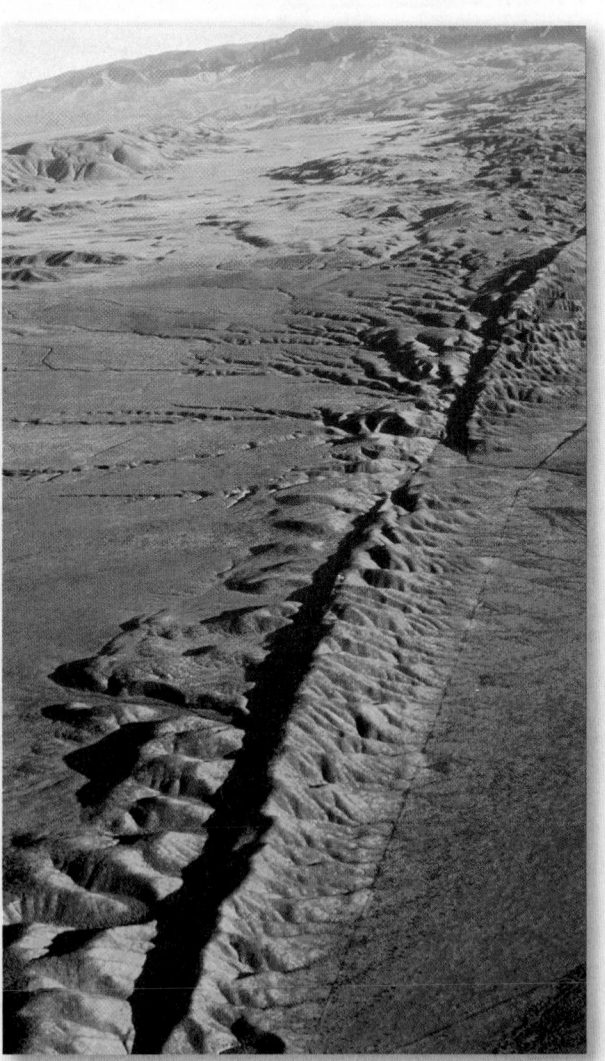

FIGURE 2.11 ■ A view southeast along the San Andreas fault in the Carrizo Plain of central California. The San Andreas is a transform fault, forming a portion of the sliding boundary between the Pacific Plate (*right*) and the North American Plate (*left*). [Kevin Schafer/Peter Arnold, inc./Alamy.]

centers, offset in a steplike pattern by transform faults, and on one side by the Peru-Chile subduction zone (see Figure 2.7). The North American Plate is bounded on the east by the Mid-Atlantic Ridge, a spreading center, and on the west by subduction zones and transform-fault boundaries.

◾ Rates and History of Plate Movements

How fast do plates move? Do some plates move faster than others, and if so, why? Is the velocity of plate movements today the same as it was in the geologic past? Geologists have developed ingenious methods to answer these questions and thereby to gain a better understanding of plate tectonics. In this section, we will examine three of these methods.

The Seafloor as a Magnetic Tape Recorder

During World War II, extremely sensitive magnetometers were developed to detect submarines by the magnetic fields emanating from their steel hulls. Geologists modified these instruments slightly and towed them behind research ships to measure the local magnetic field created by magnetized rocks on the seafloor. Steaming back and forth across the ocean, seagoing scientists discovered regular patterns in the strength of the local magnetic field that completely surprised them. In many areas, the intensity of the magnetic field alternated between high and low values in long, narrow parallel bands, called **magnetic anomalies,** that were almost perfectly symmetrical with respect to the crest of a mid-ocean ridge (**Figure 2.12**). The detection of these patterns was one of the great discoveries that confirmed the seafloor spreading hypothesis and led to the theory of plate tectonics. It also allowed geologists to trace plate movements far back in geologic time. To understand these advances, we need to look more closely at how rocks become magnetized.

THE ROCK RECORD OF MAGNETIC REVERSALS ON LAND Magnetic anomalies are evidence that Earth's magnetic field does not remain constant over time. At present, the north magnetic pole is closely aligned with the geographic north pole (see Figure 1.17), but small changes in the geodynamo can flip the orientation of the north and south magnetic poles by 180°, causing a magnetic reversal.

In the early 1960s, geologists discovered that a precise record of this peculiar behavior can be obtained from layered flows of volcanic lava. When iron-rich lavas cool, they become slightly but permanently magnetized in the direction of Earth's magnetic field. This phenomenon is called *thermoremanent magnetization* because the cold lava "remembers" the magnetization long after the magnetic field has changed.

In layered lava flows, such as those in a volcanic cone, the rocks at the top represent the most recent layer, while layers deeper in the cone are older. The age of each layer can be determined by various dating methods (described in Chapter 8). The direction of magnetization of rock samples from each layer then reveals the direction of Earth's magnetic field at the time when that layer cooled (Figure 2.12b). By repeating these measurements at hundreds of places around the world, geologists have worked out the **magnetic time scale** of the past 200 million years. Figure 2.12c shows this time scale for the most recent 5 million years. About half of all volcanic rocks studied have been found to be magnetized in a direction opposite to that of Earth's present magnetic field. Apparently, the field has flipped frequently over geologic time, so normal fields (same as now) and reversed fields (opposite to now) are equally likely. Major periods during which the field is normal or reversed are called *magnetic chrons* (from the Greek for "time"); they last about half a million years, on average, although the pattern of reversals becomes highly irregular as we move back in geologic time. Within the major chrons are short-lived reversals of the field, known as *magnetic subchrons*, which may last anywhere from several thousand to 200,000 years.

MAGNETIC ANOMALY PATTERNS ON THE SEAFLOOR The banded patterns of magnetism found on the seafloor puzzled scientists until 1963, when two Englishmen, F. J. Vine and D. H. Mathews—and, independently, two Canadians, L. Morley and A. Larochelle—made a startling proposal. Based on evidence of magnetic reversals that geologists had collected from lava flows on land, they reasoned that the bands of high and low magnetic intensity on the seafloor corresponded to bands of rock that were magnetized during ancient episodes of normal and reversed magnetism. That is, when a research ship was above rocks magnetized in the normal direction, it would record a locally stronger field, or a *positive magnetic anomaly*. When it was above rocks magnetized in the reversed direction, it would record a locally weaker field, or a *negative magnetic anomaly.*

The seafloor spreading hypothesis explained these observations: when the plates move apart at a mid-ocean ridge, magma rises from Earth's interior and flows into the rift, where it cools, solidifies, and becomes magnetized in the direction of Earth's magnetic field at the time. As the seafloor spreads away from the ridge, approximately half of the newly magnetized material moves to one side and half to the other, forming two symmetrical magnetized bands. Newer material fills the rift, continuing the process. In this

(a) **1** A ship towing a sensitive magnetometer…

2 …recorded alternating bands of high and low magnetism, which proved to be roughly symmetrical on both sides of the Mid-Atlantic Ridge.

Mid-Atlantic Ridge

High intensity

Low intensity

Iceland

Mid-Atlantic Ridge

(b) **3** Volcanic lavas also revealed magnetic anomalies. When iron-rich lava cools, it becomes magnetized in the direction of Earth's magnetic field.

Youngest

Normal

Reversed

Oldest

4 Lavas "remember" the magnetic field (thermoremanent magnetization).

5 Older, deeper layers preserve the direction of the magnetic field at earlier times.

6 By determining the ages of magnetic reversals at many volcanoes, scientists constructed a magnetic time scale.

(c) **7** The magnetic time scale was used to date the magnetic anomalies on the seafloor. The magnetic chrons of the last 5 million years, represented here as blue and brown bands, were named after prominent scientists who studied Earth's magnetic field: Gilbert, Gauss, Matuyama, and Brunhes.

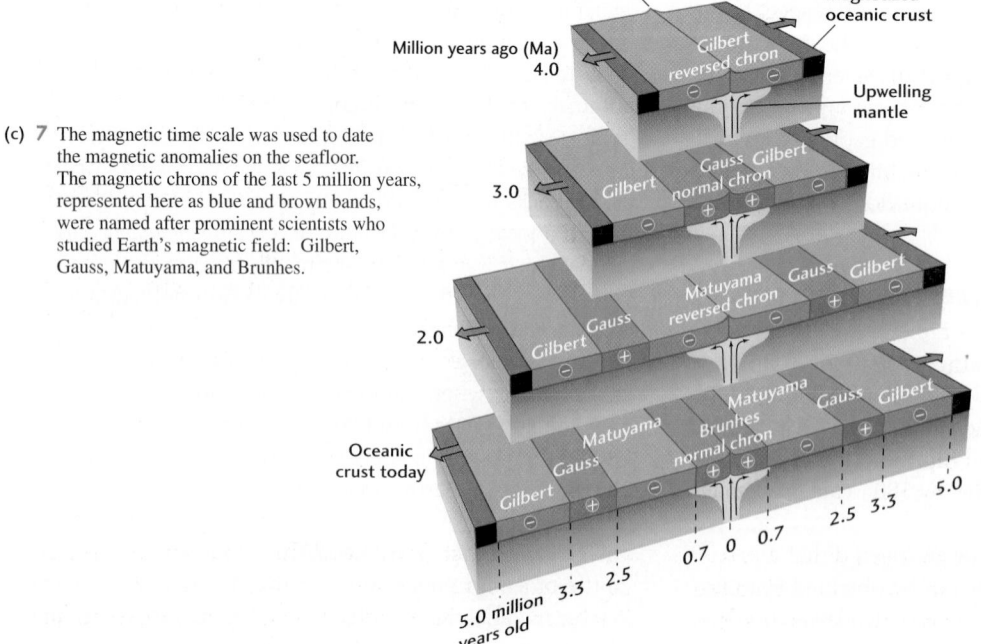

Mid-ocean ridge

Magnetized oceanic crust

Million years ago (Ma)
4.0

Gilbert reversed chron

Upwelling mantle

3.0

Gilbert Gauss Gilbert normal chron

2.0

Gilbert Gauss Matuyama reversed chron Gauss Gilbert

Oceanic crust today

Gilbert Gauss Matuyama Brunhes normal chron Matuyama Gauss Gilbert

5.0 million years old 3.3 2.5 0.7 0 0.7 2.5 3.3 5.0

FIGURE 2.12 ■ Magnetic anomalies allow geologists to measure the rate of seafloor spreading. (a) An oceanographic survey over the Mid-Atlantic Ridge just southwest of Iceland revealed a banded pattern of magnetic field intensity. (b) Geologists found and dated similar magnetic anomalies in volcanic lavas on land to construct a magnetic time scale. (c) That magnetic time scale was used to date the magnetic anomalies on the seafloor worldwide.

way, the seafloor acts like a magnetic tape recorder that encodes the history of the opening of the oceans.

INFERRING SEAFLOOR AGES AND RELATIVE PLATE VELOCITY By using the ages of magnetic reversals that had been worked out from magnetized lavas on land, geologists were able to assign ages to the bands of magnetized rocks on the seafloor. They could then calculate how fast the seafloor was spreading by using the formula *speed = distance ÷ time,* where distance is measured from the mid-ocean ridge axis and time equals seafloor age. For instance, the magnetic anomaly pattern in Figure 2.12c shows that the boundary between the Gauss normal chron and the Gilbert reversed chron, which was dated from lava flows at 3.3 million years of age, is located about 30 km away from the crest of the Mid-Atlantic Ridge just southwest of Iceland. Thus, seafloor spreading has moved the North American and Eurasian plates apart by about 60 km in 3.3 million years, giving a spreading rate of 18 km per million years or, equivalently, 18 mm/year. On a divergent plate boundary, the combination of the spreading rate and the spreading direction gives the **relative plate velocity,** the velocity at which one plate moves relative to the other.

The speed record for spreading can be found on the East Pacific Rise just south of the equator, where the Pacific and Nazca plates are separating at a rate of about 150 mm/year—much faster than the rate in the North Atlantic. A rough average spreading rate for mid-ocean ridges around the world is 50 mm/year. This is approximately the rate at which your fingernails grow—so, geologically speaking, the plates move very fast indeed!

Mapping magnetic anomalies on the seafloor has been an amazingly effective and very expedient method for working out the history of ocean basins. Simply by steaming a ship back and forth over the ocean surface, measuring the magnetic fields of the seafloor rocks, and correlating the pattern of magnetic anomalies with the magnetic time scale, geologists have been able to determine the ages of various regions of the seafloor by remote sensing—that is, without directly sampling the oceanic crust far below the surface. In effect, they have learned how to "replay the tape."

Deep-Sea Drilling

In 1968, a seafloor drilling program was launched as a joint project of several major U.S. oceanographic institutions and the National Science Foundation. Later, many nations joined the effort (**Figure 2.13**). Using hollow drills, scientists brought up cores containing sections of seafloor rock from many locations in the oceans.

Small particles falling through the ocean waters—dust from the atmosphere, organic material from marine plants and animals—begin to accumulate as seafloor sediments on new oceanic crust as soon as it forms.

FIGURE 2.13 ■ The deep-sea drilling vessel *JOIDES Resolution* is 143 m long and carries a drilling derrick 61 m high that is capable of drilling into the seafloor beneath the deepest ocean. Rock samples recovered from the seafloor have confirmed the ages of seafloor rocks deduced from magnetic anomalies. Such samples have also shed new light on the history of ocean basins and ancient climate conditions. [Courtesy Integrated Ocean Drilling Program/United States Implementing Organization (IODP/USIO).]

Therefore, the age of the oldest sediments in a core—those immediately on top of the crust—tells the geologist how old the crust is at that spot. The ages of sediments can be calculated from the fossil skeletons of tiny single-celled planktonic organisms that live at the surface of the open ocean and sink to the bottom when they die. Geologists found that the ages of the samples in the drill cores increased with distance from mid-ocean ridges, and that the age of the samples at any one place agreed almost perfectly with the age of the seafloor determined from magnetic reversal data. This agreement validated the magnetic time scale and provided strong evidence for seafloor spreading.

Measurements of Plate Movements by Geodesy

ASTRONOMICAL POSITIONING Astronomical positioning—measuring the positions of points on Earth's surface in relation to the fixed stars in the night sky—is a technique of **geodesy,** the ancient science of measuring the shape of

Earth and locating points on its surface. Surveyors have used astronomical positioning for centuries to determine geographic boundaries on land, and sailors have used it to locate their ships at sea. Four thousand years ago, Egyptian builders used astronomical positioning to aim the Great Pyramid due north.

Because of the high accuracy that would have been required to observe plate movements directly, geodetic techniques did not play a significant role in the discovery of plate tectonics. Geologists had to rely on evidence of seafloor spreading from the geologic record—the magnetic anomalies and ages of fossils described earlier. Beginning in the late 1970s, however, an astronomical positioning method was developed that used signals from distant "quasi-stellar radio sources" (quasars) recorded by huge radio telescopes. This method can measure intercontinental distances to an amazing accuracy of 1 mm. In 1986, a team of scientists using this method showed that the distance between radio telescopes in Europe (Sweden) and North America (Massachusetts) had increased 19 mm/year over a period of 5 years, very close to the rate predicted by geologic models of plate tectonics.

Today, the Great Pyramid of Egypt is not aimed directly north, but slightly east of north. Did the ancient Egyptian astronomers make a mistake in orienting the pyramid 40 centuries ago? Archaeologists think not. Over this period, Africa drifted enough to rotate the pyramid out of alignment with true north.

THE GLOBAL POSITIONING SYSTEM Doing geodesy with big radio telescopes is expensive and is not a practical means of investigating plate movements in remote areas of the world where no radio telescopes exist. Since the mid-1980s, geologists have used a constellation of 24 Earth-orbiting satellites, called the Global Positioning System (GPS), to make the same types of measurements with the same astounding accuracy. The satellite constellation serves as an outside frame of reference, just as the fixed stars and quasars do in astronomical positioning. The satellites emit high-frequency radio waves keyed to precise atomic clocks on board. Those signals can be picked up by inexpensive, portable radio receivers not much bigger than this textbook (**Figure 2.14**). These devices are similar to the GPS receivers used in automobiles and by hikers, though much more precise.

Geologists use GPS to measure plate movements on a regular basis at many locations around the globe. Changes in distance between land-based GPS receivers placed on different plates, recorded over several years, agree in both magnitude and direction with those calculated from magnetic anomalies on the seafloor, indicating that plate movements are remarkably steady over periods ranging from just a few years to millions of years, as demonstrated in the Practicing Geology Exercise at the end of this chapter.

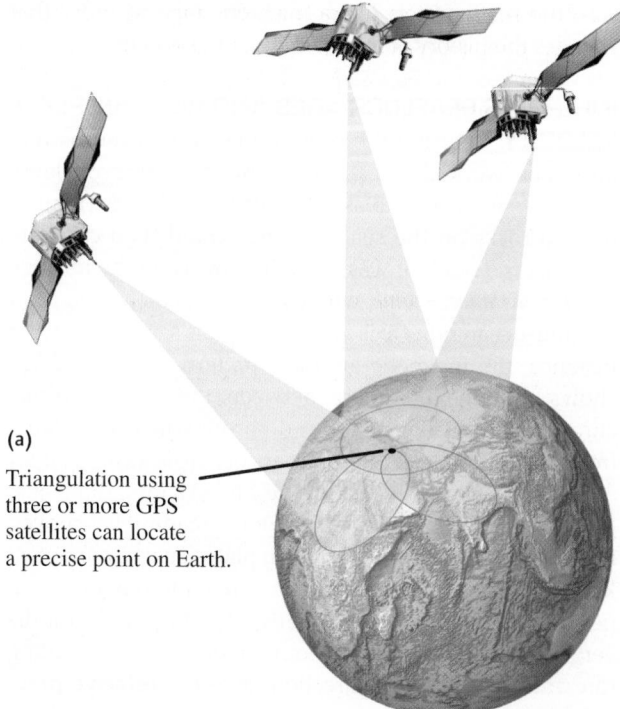

(a)
Triangulation using three or more GPS satellites can locate a precise point on Earth.

(b)

A GPS station

FIGURE 2.14 ■ The Global Positioning System allows geologists to monitor plate movements. (a) GPS satellites provide a fixed frame of reference outside Earth. (b) Small GPS receivers can be easily placed anywhere on Earth. Displacements of receiver locations over a period of years can be used to measure plate movements. [Photo courtesy of Southern California Earthquake Center.]

■ The Grand Reconstruction

The supercontinent Pangaea was the only major landmass that existed 250 million years ago. One of the great triumphs of modern geology was the reconstruction of the events that led to the assembly of Pangaea and to its later fragmentation into the continents we know today. Let's use what we have learned about plate tectonics to see how this feat was accomplished.

Seafloor Isochrons

The color map in **Figure 2.15** shows the ages of the rocks on the seafloor as determined from magnetic anomaly data and deep-sea drilling. Each colored band represents the span of time when the rocks within that band formed. Notice that the seafloor becomes progressively older on both sides of the mid-ocean ridges. The boundaries between bands are contours of equal seafloor age, or **isochrons.**

Isochrons tell us the time that has elapsed since the rocks were injected as magma into a spreading zone and, therefore, the amount of spreading that has occurred since they formed. For example, the distance from a ridge axis to a 100-million-year isochron (boundary between green and blue bands) indicates the extent of new seafloor created

over that time span. The more widely spaced isochrons of the eastern Pacific signify faster spreading rates there than in the Atlantic.

In 1990, after a 20-year search, geologists found the oldest oceanic rocks by drilling into the seafloor of the western Pacific. These rocks turned out to be about 200 million years old, only 4 percent of Earth's age. In contrast, the oldest known continental rocks are over 4 billion years in age. All existing seafloor is geologically young compared with the continents. Over a period of 100 million to 200 million years in some places, and only tens of millions of years in others, oceanic lithosphere forms by seafloor spreading, cools, and is recycled into the underlying mantle.

Reconstructing the History of Plate Movements

Earth's plates behave as rigid bodies. That is, the distances between three points on the same rigid plate—say, New

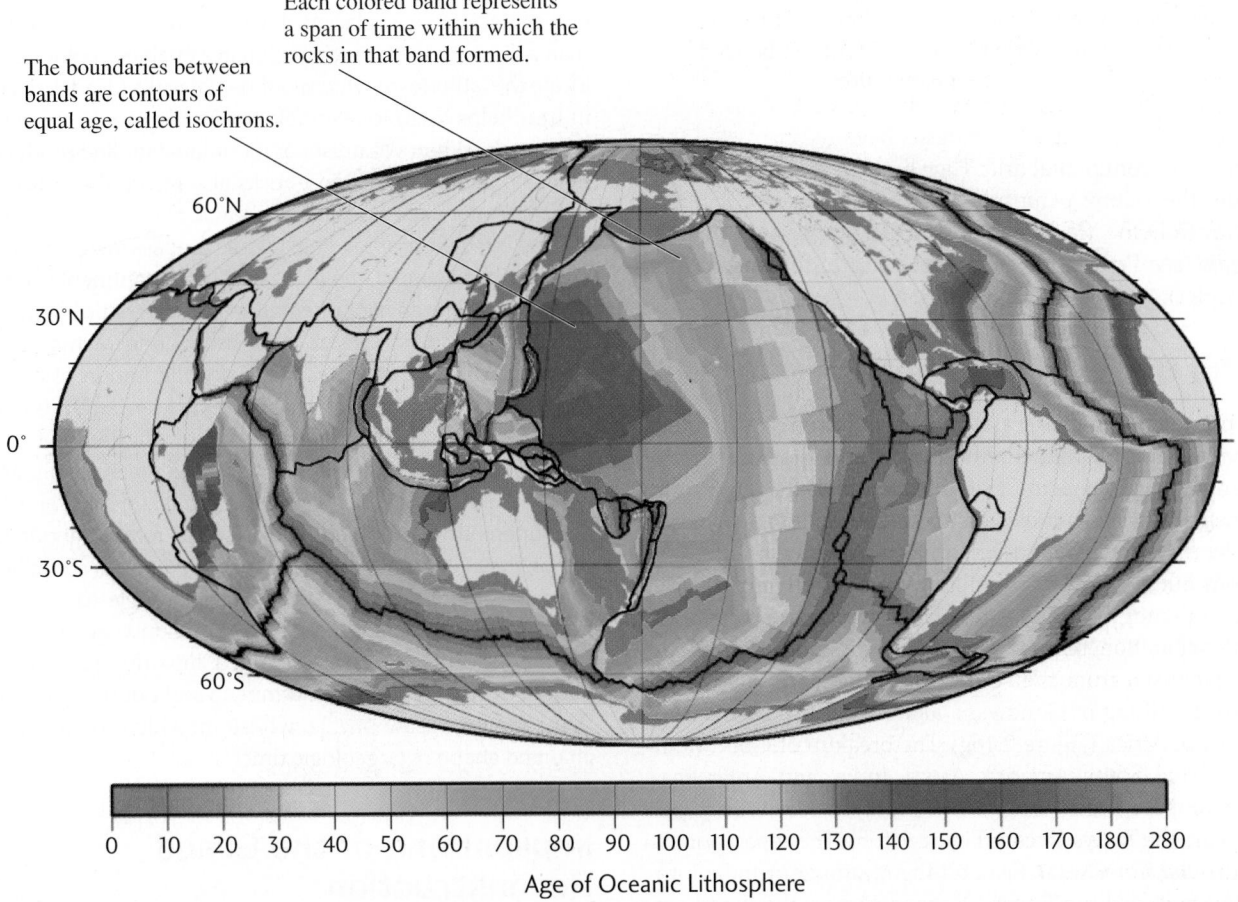

FIGURE 2.15 ■ This global isochron map shows the ages of rocks on the seafloor. The time scale at the bottom gives the age of the seafloor in millions of years since its creation at mid-ocean ridges. Light gray indicates land; dark gray indicates shallow water over continental shelves. Mid-ocean ridges, along which new seafloor is extruded, coincide with the youngest rocks (red).

[Courtesy of R. Dietmar Müller.]

York, Miami, and Bermuda on the North American Plate—do not change very much, no matter how far the plate moves. But the distance between, say, New York and Lisbon increases over time because those two cities are on two different plates that are separating along the Mid-Atlantic Ridge. The direction of movement of one plate in relation to another depends on two geometric principles that govern the behavior of rigid plates on a sphere:

- *Transform-fault boundaries indicate the directions of relative plate movement.* With few exceptions, no overlap, buckling, or separation occurs along typical transform-fault boundaries in the oceans. The two plates merely slide past each other without creating or destroying plate material. Therefore, the orientation of the fault measures the direction in which one plate is sliding with respect to the other (see Figure 2.8f, g).

- *Seafloor isochrons reveal the positions of divergent boundaries in earlier times.* Isochrons on the seafloor are roughly parallel and symmetrical to the ridge axis along which they were created (see Figure 2.15). Because each isochron was at the divergent boundary at an earlier time, isochrons that are of the same age but on opposite sides of a mid-ocean ridge can be brought together to show the positions of the plates, and the configuration of the continents embedded in them, as they were in that earlier time.

Using these principles, geologists have reconstructed the history of continental drift. They have shown, for example, how the skinny peninsula of Baja California was rifted away from the Mexican mainland during the last 5 million years (see Practicing Geology Exercise at the end of this chapter).

The Breakup of Pangaea

On a much grander scale, geologists have reconstructed the opening of the Atlantic Ocean and the breakup of Pangaea (**Figure 2.16**). Figure 2.16e shows the supercontinent Pangaea as it existed about 240 million years ago. It began to break apart when North America rifted away from Europe about 200 million years ago (Figure 2.16f). The opening of the North Atlantic was accompanied by the separation of the northern continents (referred to as Laurasia) from the southern continents (Gondwana) and the rifting of Gondwana along what is now the east coast of Africa (Figure 2.16g). The breakup of Gondwana separated South America, Africa, India, and Antarctica, creating the South Atlantic and Southern oceans and narrowing the Tethys Ocean (Figure 2.16h). The separation of Australia from Antarctica and the ramming of India into Eurasia closed the Tethys Ocean, giving us the world we see today (Figure 2.16i).

The plate movements have not ceased, of course, so the configuration of the continents will continue to evolve. A plausible scenario for the distribution of continents and plate boundaries 50 million years in the future is shown in Figure 2.16j.

The Assembly of Pangaea by Continental Drift

The isochron map in Figure 2.15 tells us that all of the seafloor on Earth's surface today has been created since the breakup of Pangaea. We know from the geologic record in older continental mountain belts, however, that plate tectonics had been operating for billions of years before this breakup. Evidently, seafloor spreading took place just as it does today, and there were previous episodes of continental drift and collision. Subduction into the mantle has destroyed the seafloor created in those earlier times, however, so we must rely on the older evidence preserved on continents to identify and chart the movements of ancient continents (*paleocontinents*).

Old mountain belts, such as the Appalachians of North America and the Urals, which separate Europe from Asia, help us locate ancient collisions of the paleocontinents. In many places, the rocks reveal ancient episodes of rifting and subduction. Rock types and fossils also indicate the distribution of ancient seas, glaciers, lowlands, mountains, and climates. Knowledge of ancient climates enables geologists to locate the latitudes at which continental rocks formed, which in turn helps them to assemble the jigsaw puzzle of paleocontinents. When volcanism or mountain building produces new continental rocks, these rocks also record the direction of Earth's magnetic field, just as oceanic crust does when it is created by seafloor spreading. Like a compass frozen in time, the thermoremanent magnetization of a continental fragment records its ancient orientation and magnetic latitude.

The left side of Figure 2.16 shows one of the latest efforts to depict the pre-Pangaean configuration of continents. It is truly impressive that modern science can recover the geography of this strange world of hundreds of millions of years ago. The evidence from rock types, fossils, and magnetization has allowed scientists to reconstruct an earlier supercontinent, called **Rodinia,** that formed about 1.1 billion years ago and began to break up about 750 million years ago (Figure 2.16a). They have been able to chart its fragments over the subsequent 500 million years as those fragments drifted and reassembled into the supercontinent Pangaea. Geologists continue to sort out the details of this complex jigsaw puzzle, whose individual pieces have changed shape over geologic time.

Implications of the Grand Reconstruction

Hardly any branch of geology remains untouched by this grand reconstruction of the continents. Economic geologists have used the former fit of the continents to find mineral and oil deposits by correlating the rock formations in which these resources exist on one continent with their predrift continuations on another continent. Paleontologists have rethought some aspects of evolution in light of continental drift. Geologists have broadened their focus from the geology of a particular region to a world-encompassing picture. The concept of plate tectonics provides a way to interpret, in global terms, such geologic processes as rock formation, mountain building, and climate change.

PAST CLIMATE CHANGES Over millions of years, movements of the tectonic plates have rearranged the continents and oceans, affecting the climate system in profound ways. In the present arrangement, the waters of the Southern Ocean are able to circulate all the way around Antarctica, forming a "circumpolar seaway" that isolates the continent from the warmer water and air of tropical latitudes. This isolation keeps the southern polar regions colder than they might otherwise be, maintaining a massive ice sheet across the entire Antarctic continent.

The situation was rather different 66 million years ago, as shown in panel (h) of Figure 2.16. Australia was still connected to Antarctica, allowing currents of warmer water to flow southward and heat the polar continent. Also at this time, the North and South American continents were separated, so that water could flow between the Atlantic and Pacific oceans. The circumpolar seaway did not form until Australia broke away from Antarctica around 40 million years ago. Somewhat later, only about 5 million years ago, subduction in the eastern Pacific Ocean formed the isthmus of Panama, connecting North and South America and isolating the Atlantic from the Pacific.

These changes, combined with the collision of India with Asia, which formed the high plateau of Tibet (see Figure 2.16g), cooled the entire planet enough to create ice sheets of Antarctica in the southern hemisphere and Greenland in the northern hemisphere. The resulting modification of the climate system is thought to have initiated oscillations of climate between very cold periods (ice ages, described in Chapter 21) and somewhat warmer periods, such as the one we now enjoy.

▪ Mantle Convection: The Engine of Plate Tectonics

Everything discussed in this chapter so far has been a description of *how* plates move. The theory of mantle convection provides an explanation of *why* plates move.

As Arthur Holmes and other early advocates of continental drift realized, mantle convection is the "engine" that drives the large-scale tectonic processes operating on Earth's surface. In Chapter 1, we described hot mantle as a ductile solid capable of flowing like a sticky fluid. Heat escaping from Earth's deep interior causes this material to undergo convection (circulation upward and downward) at speeds of a few tens of millimeters per year.

Almost all scientists now accept that the lithospheric plates somehow participate in the flow of this mantle convection system. As is often the case, however, "the devil is in the details." Many different hypotheses have been advanced on the basis of one piece of evidence or another, but no one has yet come up with a satisfactory, comprehensive theory that ties everything together. In what follows, we will pose three questions that get at the heart of the matter and give you our current understanding of their answers. But the study of the mantle convection system remains a work in progress, and we may have to alter our views as new evidence becomes available.

Where Do the Plate-Driving Forces Originate?

Here's an experiment you can do in your kitchen: heat a pan of water until it is about to boil, then sprinkle some dry tea leaves in the center of the pan. You will notice that the leaves move across the surface of the water, dragged along by the convection currents in the hot water. Is this the way plates move about, passively dragged to and fro on the backs of convection currents rising up from the mantle?

The answer appears to be no. The main evidence comes from the rates of plate movement we discussed earlier in this chapter. In Figure 2.7, we can see that the faster-moving plates (the Pacific, Nazca, Cocos, Indian, and Australian plates) are being subducted along a large fraction of their boundaries. In contrast, the slower-moving plates (the North American, South American, African, Eurasian, and Antarctic plates) do not have significant attachments of descending lithospheric slabs. These observations suggest that the gravitational pull exerted by the cold (and thus dense) slabs of subducting lithosphere pulls the plates downward into the mantle. In other words, the plates are not dragged along by convection currents rising from the mantle, but rather "fall back" into the mantle under their own weight. According to this hypothesis, seafloor spreading is the passive upwelling of mantle material where the plates have been pulled apart by subduction forces.

But if the only important force in plate tectonics is the gravitational pull of subducting slabs, why did Pangaea break apart and the Atlantic Ocean open up? The only subducting slabs of lithosphere currently attached to the North and South American plates are found in the small island arcs that bound the Caribbean and Scotia seas, which are thought to be too small to drag the Atlantic apart. One possibility is that the overriding plates, as well as the subducting plates, are pulled toward their convergent boundaries. For example, as the Nazca Plate subducts beneath South America, it may cause the plate boundary at the Peru-Chile Trench to retreat toward the Pacific, "sucking" the South American Plate to the west.

ASSEMBLY OF PANGAEA

RODINIA (a) Late Proterozoic, 750 Ma

1 The supercontinent of Rodinia formed about 1.1 billion years ago and began to break up about 750 million years ago.

(b) Late Proterozoic, 650 Ma

(c) Middle Ordovician, 458 Ma

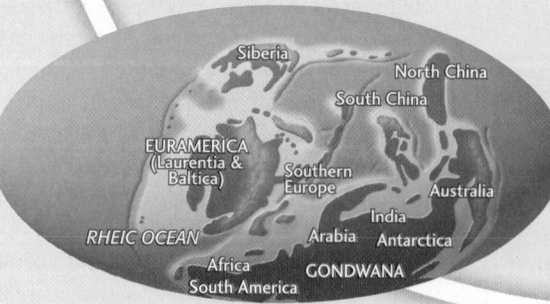

2 The supercontinent Pangaea was mostly assembled by 237 Ma, surrounded by a superocean called Panthalassa (Greek for "all seas"), the ancestral Pacific Ocean. The Tethys Ocean, between Africa and Eurasia, was the ancestor of the Mediterranean Sea.

(d) Early Devonian, 390 Ma

PANGAEA (e) Early Triassic, 237 Ma

FIGURE 2.16 ■ Continental rifting, drifting, and collisions assembled and then disassembled the supercontinent Pangaea.
[Paleogeographic maps by Christopher R. Scotese, 2003 PALEOMAP Project (www.scotese.com).]

BREAKUP OF PANGAEA

(f) Early Jurassic, 195 Ma

3 The breakup of Pangaea was signaled by the opening of rifts from which lava poured. Rock assemblages that are relics of this great event can be found today in 200-million-year-old volcanic rocks from Nova Scotia to North Carolina.

(g) Late Jurassic, 152 Ma

4 By about 150 million years ago, Pangaea was in the early stages of breakup. The Atlantic Ocean had partially opened, the Tethys Ocean had contracted, and the northern continents (Laurasia) had all but split away from the southern continents (Gondwana). India, Antarctica, and Australia began to split away from Africa.

5 By 66 million years ago, the South Atlantic had opened and widened. India was well on its way northward toward Asia, and the Tethys Ocean was closing to form the Mediterranean.

(h) Late Cretaceous, Early Tertiary, 66 Ma

THE PRESENT-DAY AND FUTURE WORLD

6 The modern world has been produced over the past 65 million years. India collided with Asia, ending its trip across the ocean, and is still pushing northward into Asia. Australia has separated from Antarctica.

(i) PRESENT-DAY WORLD

(j) 50 million years in the future

Other forces are evident from the history of plate movements. When the continents came together to form Pangaea, they acted as an insulating blanket, preventing heat from getting out of Earth's mantle (as it otherwise would through the process of seafloor spreading). That heat built up over time, forming hot bulges in the mantle beneath the supercontinent. These bulges raised Pangaea slightly and caused it to rift apart in a kind of "landslide" off the tops of the bulges. Gravitational forces continue to drive subsequent seafloor spreading as the plates "slide downhill" off the crest of the Mid-Atlantic Ridge. Earthquakes that sometimes occur in plate interiors provide direct evidence of the compression of plates by these ridge-related gravitational forces.

Convection in the mantle—the rising of hot matter in one place and the sinking of cold matter in another—is the driving force of plate tectonics. Although many questions remain, we can be reasonably sure of two things: first, that the plates themselves play an active role in this system, and second, that the forces associated with the sinking slabs and elevated ridges are probably the most important in governing the rates of plate movement (**Figure 2.17**). Scientists are attempting to resolve the other issues raised in this discussion by comparing their observations with detailed computer models of the mantle convection system. Some of their results will be discussed in Chapter 14.

How Deep Does Plate Recycling Occur?

For plate tectonics to work, the lithospheric material that descends into the mantle at subduction zones must be recycled through the mantle and eventually return to the crust as new lithosphere created at spreading centers. How deep into the mantle does this recycling process extend? That is, where is the lower boundary of the mantle convection system?

The lower boundary could be as deep as 2890 km below Earth's surface, where a sharp compositional boundary separates the mantle from the core (**Figure 2.18**). As we saw in Chapter 1, the iron-rich liquid below this core-mantle boundary is much denser than the solid rock of the mantle, preventing any significant exchange of material between the two layers. We can thus imagine a system of whole-mantle convection in which the material from the plates circulates all the way through the mantle, down as far as the core-mantle boundary (see Figure 2.18a).

However, some scientists think that the mantle might be divided into two layers: an upper mantle system above about 700 km, where the recycling of lithosphere takes place, and a lower mantle system from a depth of about 700 km to the core-mantle boundary, where convection is much more sluggish. According to this hypothesis, called stratified convection, the separation of the two systems is maintained because the upper system consists of lighter rock than the lower system and thus floats on top, in the same way the mantle floats on the core (see Figure 2.18b).

To test these two competing hypotheses, scientists have looked for "lithospheric graveyards" below convergent boundaries where plates have been subducted. Old subducted lithosphere is colder than the surrounding mantle and can therefore be "seen" using seismic waves. Moreover, there should be lots of it down there. From our knowledge of past plate movements, we can estimate that, just since the breakup of Pangaea, lithosphere equivalent to the surface area of Earth has been recycled into the mantle. Sure enough, scientists have found regions of colder material in the deep mantle under North and South America, eastern Asia, and other sites adjacent to convergent boundaries. These zones occur as extensions of descending lithospheric slabs, and some appear to go down as far as the core-mantle boundary. From this evidence, most scientists have

FIGURE 2.17 ■ A schematic cross section through the outer part of Earth, illustrating two of the forces thought to be important in driving plate tectonics: the pulling force of a sinking lithospheric slab and the pushing force of plates sliding off a mid-ocean ridge. [After D. Forsyth and S. Uyeda, *Geophysical Journal of the Royal Astronomical Society* 43 (1975): 163–200.]

FIGURE 2.18 ■ Two competing hypotheses on the depth extent of the mantle convection system that recycles the lithosphere.

concluded that plate recycling takes place through whole-mantle convection rather than stratified convection.

What Is the Nature of Rising Convection Currents?

What about the rising currents of hot mantle material needed to balance subduction? Are there concentrated, sheetlike upwellings directly beneath the mid-ocean ridges? Most scientists who study the problem think not. Instead, they believe that the rising currents are slower and spread out over broader regions. This view is consistent with the idea that seafloor spreading is a rather passive process: pull the plates apart almost anywhere, and you will generate a spreading center.

There is one exception, however: a type of narrow jet-like upwelling called a **mantle plume** (**Figure 2.19**). The best evidence for mantle plumes comes from regions

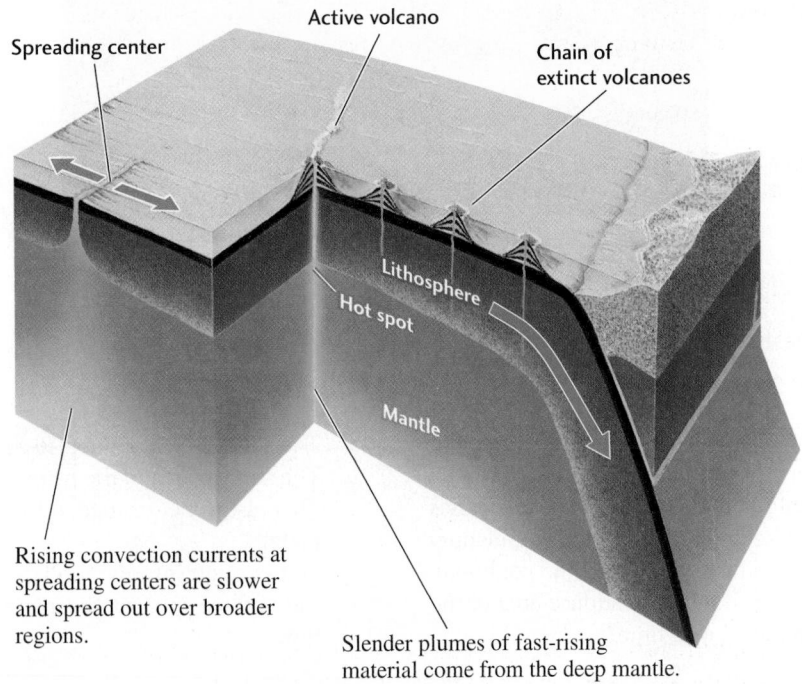

FIGURE 2.19 ■ An illustration of the mantle plume hypothesis.

of intense, localized volcanism (called *hot spots*), such as Hawaii, where huge volcanoes form in the middle of a plate, far from any spreading center. Mantle plumes are thought to be slender cylinders of fast-rising material, less than 100 km across, that come from the deep mantle (below the asthenosphere). They are intense enough to literally burn holes in the plates and produce tremendous volumes of lava. Mantle plumes may be responsible for outpourings of lava so massive that they may have changed Earth's climate and caused mass extinctions

Google Earth Project

This chapter focuses on the fundamental theory of plate tectonics and how that theory integrated previously independent geologic observations into a unified whole. To appreciate plate tectonics on a global scale, we'll begin by viewing Earth at an eye altitude of 11,000 km. Rotate the globe by using the virtual joystick in the upper right corner of the screen. Notice that this global view eliminates the distortion of a Mercator map at high latitudes and allows you to see the polar regions not shown on that type of map.

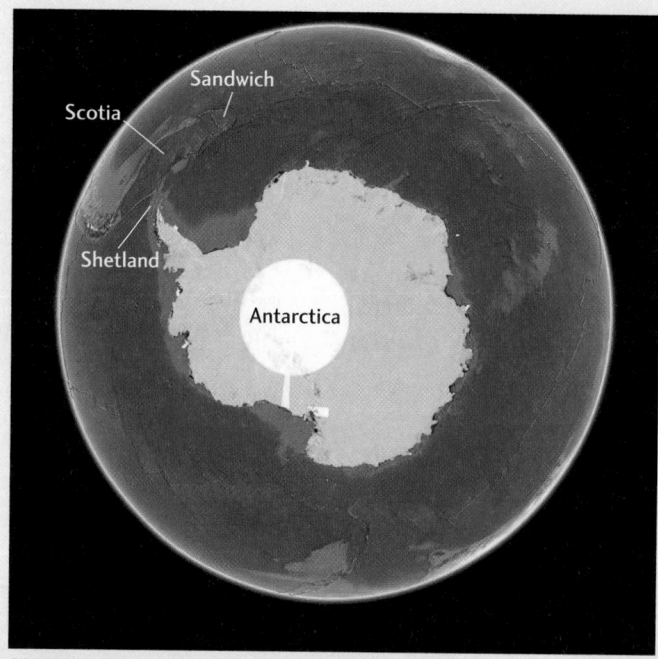

©2009 Google. Image U.S. Geological Survey Image ©2009 TerraMetrics Data SIO, NOAA, U.S. Navy, GEBCO Image NASA

LOCATION   Antarctica, Mid-Atlantic Ridge, and South Pacific Ocean

GOAL   Investigate divergent plate boundaries

LINKED   Figures 2.1, 2.7, and 2.15

1. Navigate to the South Pole and view the continent of Antarctica. From an eye altitude of about 7000 km, examine the continent and the plate boundaries that surround it. Use Figure 2.7 to help you identify the types of plate boundaries around the continent. Based on those types, how is the surface area of the Antarctic Plate changing over time?

   a. The surface area of the plate is decreasing.
   b. The surface area of the plate is increasing.
   c. There is no net change in the surface area of the plate.
   d. Not enough information is available to say how the surface area of the plate is changing over time.

(see Chapter 1). We will describe mantle plumes in more detail in Chapter 12.

The mantle plume hypothesis was first put forward in 1970, soon after the theory of plate tectonics had been established, by one of its founders, W. Jason Morgan of Princeton University. Like other aspects of the mantle convection system, however, the observations that bear on rising convection currents are indirect, and the mantle plume hypothesis remains very controversial.

2. Navigate northward into the Atlantic Ocean basin. Find the conspicuous undersea mountain range that runs from south to north through the middle of the ocean basin—the Mid-Atlantic Ridge—and consider its relationship to the continents on both sides. Notice that the submerged edges of eastern South America and western Africa would fit nicely together—evidence of continental drift. As you move north, focus on the section of the ridge between 15°N and 30°N from an eye altitude of about 2200 km. You may want to activate the "grid" option from the "View" tab along the top of the GE browser to find this location easily. Based on your observations, what is the best description of the plate boundary along this portion of the mid-Atlantic Ridge?

   a. A continuous divergent boundary
   b. A continuous convergent boundary
   c. A steplike pattern of spreading centers separated by perpendicular transform faults
   d. A steplike pattern of subduction zones separated by perpendicular transform faults

3. Now that you have familiarized yourself with the Mid-Atlantic Ridge system, let's use Google Earth to solve the Practicing Geology bonus problem. In that problem, we are asked to compare the average speed of continental drift between North America and Africa with the present-day rate of 23 mm/year, determined from GPS measurements. From the reconstruction of the supercontinent Pangaea in Figure 2.1, you can see that the North American margin just east of Charleston, South Carolina, once fit against the African margin just west of Dakar, Senegal. From the isochron map in Figure 2.15, you can estimate that the two continents began to rift apart about 200 million to 180 million years ago (see also Figure 2.16). Using the GE ruler tool to measure the ocean width at these locations, estimate the average rate at which the Atlantic has opened. What is that rate, and how does it compare with the present-day rate of continental drift?

   a. 5–10 mm/year, much slower than the present-day rate
   b. 15–20 mm/year, slower than the present-day rate
   c. 20–25 mm/year, comparable to the present-day rate
   d. 30–35 mm/year, faster than the present-day rate

4. Use the GE search window to locate Easter Island off the west coast of South America (it belongs to the country of Chile). By zooming out to an eye altitude of 5250 km, you can appreciate how small and remote this island is. By comparing topographic features on the seafloor with those in Figure 2.7, you should be able to locate Easter Island on the latter (it's not labeled). Which plate boundary is Easter Island nearest, and what is the present-day rate of seafloor spreading at that boundary?

   a. North American Plate–Pacific Plate boundary; 63 mm/year
   b. Pacific Plate–Nazca Plate boundary; 150 mm/year
   c. North American Plate–African Plate boundary; 24 mm/year
   d. Nazca Plate–South American Plate boundary, 79 mm/year

Optional Challenge Question

5. Locate Isla San Ambrosio, another tiny island off the west coast of Chile, at 26°20′34″ S, 79°53′19″ W, and measure its distance from Easter Island using the ruler tool. From the isochron map in Figure 2.15, you can see that the seafloor near Isla San Ambrosio is approximately 35 million years old. What is the average rate of seafloor spreading over those 35 million years, and how does it compare with the present-day rate near Easter Island? (*Hint:* Assume that seafloor spreading has been symmetrical across the Pacific Plate–Nazca Plate boundary over the last 35 million years.)

   a. 70–90 mm/year, much slower than the present-day rate
   b. 140–160 mm/year, comparable to the present-day rate
   c. 160–180 mm/year, slightly faster than the present-day rate
   d. 200–220 mm/year, much faster than the present-day rate

■ The Theory of Plate Tectonics and the Scientific Method

In Chapter 1, we considered the scientific method and how it guides the work of geologists. In the context of the scientific method, plate tectonics is a confirmed theory whose strength lies in its simplicity, its generality, and its consistency with many types of observations. Theories can always be overturned or modified. As we have seen, competing hypotheses about how mantle convection drives plate tectonics have been advanced. But the theory of plate tectonics itself—like the theories of Earth's age, the evolution of life, and genetics—explains so much so well, and has survived so many efforts to prove it false, that geologists treat it as fact.

The question remains, why wasn't plate tectonics discovered earlier? Why did it take the scientific establishment so long to move from skepticism about continental drift to acceptance of plate tectonics? Scientists approach their subjects differently. Scientists with particularly inquiring, uninhibited, and synthesizing minds are often the first to perceive great truths. Although their perceptions frequently turn out to be false (think of the mistakes Wegener made in proposing continental drift), these visionary people are often the first to see the great generalizations of science. Deservedly, they are the ones history remembers.

Most scientists, however, proceed more cautiously and wait out the slow process of gathering supporting evidence. Continental drift and seafloor spreading were slow to be accepted largely because these audacious ideas came far ahead of any firm evidence. Scientists had to explore the oceans, develop new instruments, and drill the seafloor before the majority could be convinced. Today, many scientists are still waiting for more evidence of how the mantle convection system really works.

SUMMARY

What is the theory of plate tectonics? According to the theory of plate tectonics, the lithosphere is broken into about a dozen rigid plates that move over Earth's surface. Three types of plate boundaries are defined by the direction of the movements of plates in relation to each other: divergent, convergent, and transform-fault boundaries. Earth's surface area does not change over time; therefore, the area of new lithosphere created at divergent boundaries equals the area of lithosphere recycled at convergent boundaries by subduction into the mantle.

What are some of the geologic characteristics of plate boundaries? Many geologic features develop at plate boundaries. Divergent boundaries are typically marked by volcanism and earthquakes at the crest of a mid-ocean ridge. Convergent boundaries are marked by deep-sea trenches, earthquakes, mountain building, and volcanism. Transform faults, along which plates slide horizontally past each other, can be recognized by earthquake activity and offsets in geologic features.

How can the age of the seafloor be determined? We can measure the age of the seafloor by using thermoremanent magnetization. Magnetic anomaly patterns mapped on the seafloor can be compared with a magnetic time scale that was established using the magnetic anomalies of lavas of known ages on land. Seafloor ages have been verified through dating of rock samples obtained by deep-sea drilling. Geologists can now draw isochron maps for most of the world's oceans, which allow them to reconstruct the history of seafloor spreading over the past 200 million years. Using this method and other geologic data, geologists have developed a detailed model of how Pangaea broke apart and the continents drifted into their present configuration.

What is the engine that drives plate tectonics? The plate tectonic system is driven by mantle convection, the energy for which comes from Earth's internal heat. Gravitational forces act on the cooling lithosphere as it slides downhill from spreading centers and sinks into the mantle at subduction zones. Subducted lithosphere extends as deep as the core-mantle boundary, indicating that the whole mantle is involved in the convection system that recycles the plates. Rising convection currents may include mantle plumes, intense jets of material from the deep mantle that cause localized volcanism at hot spots in the middle of plates.

KEY TERMS AND CONCEPTS

continental drift (p. 28)

convergent boundary (p. 34)

divergent boundary (p. 34)

geodesy (p. 42)

island arc (p. 37)

isochron (p. 43)

magnetic anomaly (p. 39)

magnetic time scale (p. 39)

mantle plume (p. 49)

mid-ocean ridge (p. 34)

Pangaea (p. 28)

plate tectonics (p. 31)

relative plate velocity (p. 41)

Rodinia (p. 44)

seafloor spreading (p. 29)

spreading center (p. 34)

subduction (p. 34)

transform fault (p. 34)

PRACTICING GEOLOGY EXERCISE

What Happened in Baja? How Geologists Reconstruct Plate Movements

Geographers and geologists have long puzzled over the unusual geography of Baja California. Why is the Gulf of California so long and thin? Why is the Baja California peninsula parallel to the Mexican coastline?

When the Spanish conquistador Hernando Cortés landed on the shores of California in 1535, he thought he had discovered an island. Decades passed before the Spanish realized that the northern half of *Isla California* was actually the west coast of North America, and that its lower half, Baja California, was a long, thin peninsula, separated from the continent by the narrow Gulf of California.

Four centuries later, plate tectonic theory provided a neat geologic answer to the Baja puzzle. To the north, in Alta California (a.k.a. the Golden State), the Pacific Plate is moving past the North American Plate along the San Andreas transform fault. To the south, the divergent boundary between the Pacific Plate and the small Rivera Plate forms part of the East Pacific Rise, a mid-ocean ridge that produces new oceanic crust as the two plates spread apart.

By mapping earthquake locations and undersea volcanoes, marine geologists were able to show that the San Andreas fault is connected to the East Pacific Rise by a

The Pacific Plate, on the left, is moving northwestward relative to the North American Plate, on the right, at a speed of about 50 mm/year, rifting the Baja California peninsula away from the Mexican mainland and opening the Gulf of California.

dozen small spreading centers offset by transform faults—a plate boundary that steps like stairs up the entire length of the Gulf of California. The relative movement of the Pacific and North American plates is thus shifting Baja California away from the mainland in a northwesterly direction, parallel to the transform faults, and the Gulf of California is being progressively widened by seafloor spreading.

How fast is this happening? An estimate can be made by using the equation

$$speed = distance \div time$$

We need two types of data to apply this equation:

▪ We can measure the *distance* by which Baja California has separated from Mexico directly from a seafloor map: about 250 km.

▪ We can estimate the *time* since the separation began from the pattern of magnetic anomalies across the East Pacific Rise. On both sides of that spreading center, the magnetic anomaly closest to the continental margin (and therefore the oldest) is the Gilbert reversed chron. Using the magnetic time scale in Figure 2.12c, we obtain a separation age of about 5 million years (My).

With this information, we can calculate the approximate speed of seafloor spreading in the Gulf of California:

$$speed = distance/time$$
$$= 250 \text{ km}/5 \text{ My}$$
$$= 50 \text{ km}/\text{My}$$

or 50 mm/year.

Of course, this is only an average speed. How steady has it been? The plate separation could have started slowly and gradually picked up speed, or started fast and then slowed down. If the former is true, then the present-day separation rate should be greater than the average rate; if the latter, it should be less.

Using GPS, geologists were able to test these hypotheses using a totally different type of measurement. In the decade from 1990 to 2000, they repeatedly surveyed the locations of points on both sides of the Gulf of California oriented parallel to the plate movements. They found that the distances between these points increased by half a meter; that is, by 500 millimeters in 10 years, or 50 mm/year. Thus, the present-day speed of movement is approximately the same as the average speed; no speedup or slowdown of plate movements is necessary to account for it.

Based on the agreement between these two measurements as well as other data, geologists came up with a simple story. Before 5 million years ago, when Baja California was part of the mainland, the boundary between the Pacific and North American plates lay somewhere west of the North American continent. About 5 million years ago, this boundary jumped inland, initiating seafloor spreading in the Gulf of California. The plate movement has been nearly steady at 50 mm/year ever since.

This theory has survived various tests. For example, it predicts that the current slipping along the San Andreas fault should also have begun about 5 million years ago, and that prediction agrees with the ages of rocks that have been displaced by the modern San Andreas fault.

The puzzle of Baja is no mere curiosity. As we will see in later chapters, the plate tectonic stories we learn through calculations like these help geologists calibrate earthquake hazards and search for mineral resources.

BONUS PROBLEM: Use a globe and the isochron map in Figure 2.15 to estimate the average speed of continental drift between North America and Africa. How well does this speed compare with the present-day value of 23 mm/year determined using GPS? [See this chapter's Google Earth Project, exercise 3.]

EXERCISES

1. Using Figure 2.7, trace the boundaries of the South American Plate on a sheet of paper and identify segments that are divergent, convergent, and transform-fault boundaries. Approximately what fraction of the plate area is occupied by the South American continent? Is the fraction of the South American Plate occupied by oceanic crust increasing or decreasing over time? Explain your answer using the principles of plate tectonics.

2. In Figure 2.7, identify an example of a transform-fault boundary that (a) connects a divergent plate boundary with a convergent plate boundary and one that (b) connects a convergent plate boundary with another convergent plate boundary.

3. Using the isochron map in Figure 2.15, estimate how long ago the continents of Australia and Antarctica were separated by seafloor spreading. Did this happen before or after South America separated from Africa?

4. Name three mountain belts formed by continental collisions that are occurring now or have occurred in the past.

5. Most active volcanoes are located on or near plate boundaries. Give an example of a volcano that is not on a plate boundary and describe a hypothesis consistent with plate tectonics that can explain its presence there.

THOUGHT QUESTIONS

1. Why are there active volcanoes along the Pacific coast in Washington and Oregon but not along the east coast of the United States?

2. What mistakes did Wegener make in formulating his theory of continental drift? Do you think the geologists of his era were justified in rejecting his theory?

3. Would you characterize plate tectonics as a hypothesis, a theory, or a fact? Why?

4. How do the differences between continental and oceanic crust affect the way lithospheric plates interact?

5. In Figure 2.15, the isochrons are symmetrically distributed in the Atlantic Ocean, but not in the Pacific Ocean. For example, seafloor as much as 180 million years old (in darkest blue) is found in the western Pacific, but not in the eastern Pacific. Why?

6. The theory of plate tectonics was not widely accepted until the banded patterns of magnetism on the ocean floor were discovered. In light of earlier observations—the jigsaw-puzzle fit of the continents, the occurrence of fossils of the same life-forms on both sides of the Atlantic, and the reconstruction of ancient climate conditions—why are these banded patterns of magnetism such a key piece of evidence?

MEDIA SUPPORT

 2-1 Animation: Divergent Boundaries

 2-3 Animation: Transform Fault Boundaries

 2-2 Animation: Convergent Boundaries

 2-1 Video: The Alpine Fault: A Plate Boundary You Can Touch

■ What Are Minerals? 58

■ The Structure of Matter 59

■ The Formation of Minerals 61

■ Classes of Rock-Forming Minerals 64

■ Physical Properties of Minerals 69

■ What Are Rocks? 74

■ The Rock Cycle: Interactions Between the Plate Tectonic and Climate Systems 79

■ Concentrations of Valuable Mineral Resources 81

Crystals of amethyst and quartz growing on top of epidote crystals (green). The planar surfaces are crystal faces, whose geometries are determined by the underlying arrangement of the atoms that make up the crystals. [John Grotzinger/Ramón Rivera-Moret/Harvard Mineralogical Museum.]

EARTH MATERIALS: MINERALS AND ROCKS

3

IN CHAPTER 2, WE saw how the plate tectonic system gives rise to Earth's large-scale structure and dynamics, but we touched only briefly on the wide variety of materials that appear in different plate tectonic settings. In this chapter, we focus on those materials: minerals and rocks. Minerals are the building blocks of rocks, which are, in turn, the records of geologic history. Rocks and minerals help determine the structure of Earth, much as concrete, steel, and plastic determine the structure, design, and architecture of large buildings.

To tell Earth's story, geologists often adopt a "Sherlock Holmes" approach: they use current evidence to deduce the processes and events that occurred in the past at some particular place. The kinds of minerals found in volcanic rocks, for example, provide evidence of eruptions that brought molten rock to Earth's surface, while the minerals in granite reveal that it crystallized deep in the crust under the very high temperatures and pressures produced when two continents collide. Understanding the geology of a region also allows us to make informed guesses about where undiscovered deposits of economically important mineral resources might lie.

This chapter begins with a description of minerals—what they are, how they form, and how they can be identified. We then turn our attention to the major groups of rocks formed from these minerals and the geologic environments in which they form.

■ What are Minerals?

Minerals are the building blocks of rocks. **Mineralogy** is the branch of geology that studies the composition, structure, appearance, stability, occurrence, and associations of minerals. With the proper tools, most rocks can be separated into their constituent minerals. A few kinds of rocks, such as limestone, are made up primarily of a single mineral (in this case, calcite). Other rocks, such as granite, are made up of several different minerals. To identify and classify the many kinds of rocks that compose Earth and understand how they are formed, we must know how minerals are formed.

Geologists define a **mineral** as a naturally occurring, solid crystalline substance, usually inorganic, with a specific chemical composition. Minerals are homogeneous: they cannot be divided mechanically into smaller components.

Let's examine each part of our definition of a mineral in a little more detail.

Naturally occurring: To qualify as a mineral, a substance must be found in nature. The diamonds mined in South Africa, for example, are minerals. The synthetic versions produced in industrial laboratories are not minerals, nor are the thousands of laboratory products invented by chemists.

Solid crystalline substance: Minerals are solid substances—they are neither liquids nor gases. When we say that a mineral is crystalline, we mean that the tiny particles of matter, or atoms, that compose it are arranged in an orderly, repeating, three-dimensional array. Solid materials that have no such orderly arrangement are referred to as glassy or amorphous (without form) and are not conventionally called minerals. Windowpane glass is amorphous, as are

some natural glasses formed during volcanic eruptions. Later in this chapter, we will explore in detail the process by which crystalline materials form.

Usually inorganic: Minerals are defined as inorganic substances and so exclude the organic materials that make up plant and animal bodies. Organic matter is composed of organic carbon, the form of carbon found in all organisms, living or dead. Decaying vegetation in a wetland may be geologically transformed into coal, which is also made of organic carbon, but although it is found in naturally occurring deposits, coal is not considered a mineral. Many minerals, however, are secreted by organisms. One such mineral, calcite (**Figure 3.1**), which forms the shells of oysters and many other marine organisms, contains inorganic carbon. These shells accumulate on the seafloor, where they may be geologically transformed into limestone. The calcite of these shells fits the definition of a mineral because it is inorganic and crystalline.

With a specific chemical composition: The key to understanding the composition of Earth materials lies in knowing how the chemical elements are organized into minerals. What makes each mineral unique is its chemical composition and the arrangement of its atoms in an internal structure. A mineral's chemical composition either is fixed or varies within defined limits. The mineral quartz, for example, has a fixed ratio of two atoms of oxygen to one atom of silicon. This ratio never varies, even though quartz is found in many different kinds of rocks. Similarly, the chemical elements that make up the mineral olivine—iron, magnesium, oxygen, and silicon—always have a fixed ratio. Although the numbers of iron and magnesium atoms may vary, the sum of those two atoms in relation to the number of silicon atoms always forms a fixed ratio.

(a)

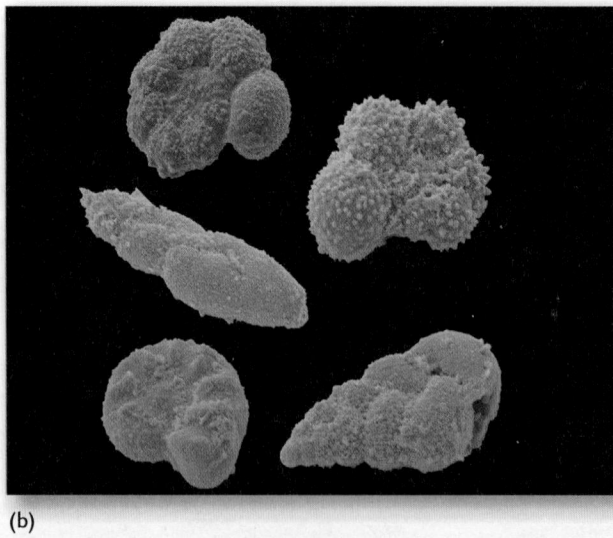

(b)

FIGURE 3.1 ■ Many minerals are secreted by organisms. (a) The mineral calcite contains inorganic carbon. (b) Calcite is found in the shells of many marine organisms, such as these foraminifera. [(a) John Grotzinger/Ramón Rivera-Moret/Harvard Mineralogical Museum; (b) Andrew Syred/Science Source.]

▪ The Structure of Matter

In 1805, the English chemist John Dalton hypothesized that each of the various chemical elements consists of a different kind of atom, that all atoms of any given element are identical, and that chemical compounds are formed by various combinations of atoms of different elements in definite proportions. By the early twentieth century, physicists, chemists, and mineralogists, building on Dalton's ideas, had come to understand the structure of matter much as we do today. We now know that an *atom* is the smallest unit of an element that retains the physical and chemical properties of that element. We also know that atoms are the small units of matter that combine in chemical reactions, and that atoms themselves are divisible into even smaller units.

The Structure of Atoms

Understanding the structure of atoms allows us to predict how chemical elements will react with one another and form new crystal structures. The structure of an atom is defined by a nucleus, which contains protons and neutrons and which is surrounded by electrons. (For a more detailed review of the structure of atoms, see Appendix 3.)

THE NUCLEUS: PROTONS AND NEUTRONS At the center of every atom is a dense *nucleus* containing virtually all the mass of the atom in two kinds of particles: protons and neutrons (**Figure 3.2**). A *proton* has a positive electrical charge of +1. A *neutron* is electrically neutral—that is, uncharged. Atoms of the same chemical element may have different numbers of neutrons, but the number of protons does not vary. For instance, all carbon atoms have six protons.

ELECTRONS Surrounding the nucleus is a cloud of moving particles called *electrons,* each with a mass so small that it is conventionally taken to be zero. Each electron carries a negative electrical charge of −1. The number of protons in the nucleus of any atom is balanced by the same number of electrons in the cloud surrounding the nucleus, so that the atom is electrically neutral. Thus, the nucleus of a carbon atom is surrounded by six electrons (see Figure 3.2).

Atomic Number and Atomic Mass

The number of protons in the nucleus of an atom is its **atomic number.** Because all atoms of the same element have the same number of protons, they also have the same atomic number. All atoms with six protons, for example, are carbon atoms (atomic number 6). In fact, the atomic number of an element can tell us so much about that element's behavior that the periodic table organizes elements according to their atomic number (see Appendix 3). Elements in

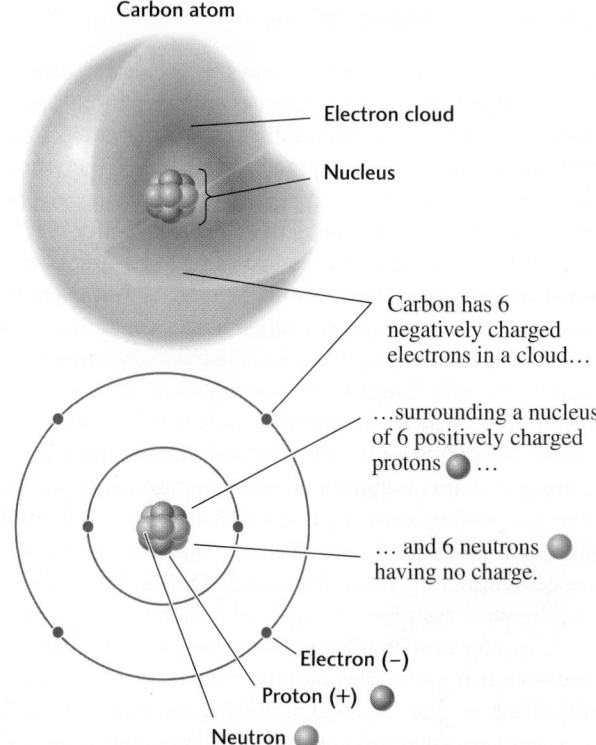

Carbon atom

Electron cloud

Nucleus

Carbon has 6 negatively charged electrons in a cloud…

…surrounding a nucleus of 6 positively charged protons ⬤ …

… and 6 neutrons ⬤ having no charge.

Electron (−)

Proton (+) ⬤

Neutron ⬤

FIGURE 3.2 ▪ Structure of the carbon atom (carbon-12). The six electrons, each with a charge of −1, are represented as a negatively charged cloud surrounding the nucleus, which contains six protons, each with a charge of +1, and six neutrons, each with no charge. The size of the nucleus is greatly exaggerated in these drawings; it is much too small to show at a true scale.

the same vertical column of the periodic table, such as carbon and silicon, tend to have similar chemical properties.

The **atomic mass** of an element is the sum of the masses of its protons and its neutrons. (Electrons, because they have so little mass, are not included in this sum.) Although atoms of the same element always have the same number of protons, they may have different numbers of neutrons, and therefore different atomic masses. Atoms of the same element with different numbers of neutrons are called **isotopes.** Isotopes of the element carbon, for example, all have six protons, but may have six, seven, or eight neutrons, giving atomic masses of 12, 13, and 14.

In nature, the chemical elements exist as mixtures of isotopes, so their average atomic masses are never whole numbers. The atomic mass of carbon, for example, is 12.011. It is close to 12 because the isotope carbon-12 is by far the most abundant. The relative abundances of the various isotopes of an element on Earth are determined by processes that enhance the abundances of some isotopes over others. Carbon-12, for example, is favored by some chemical reactions, such as photosynthesis, in which organic carbon compounds are produced from inorganic carbon compounds.

Chemical Reactions

The structure of an atom determines its chemical reactions with other atoms. *Chemical reactions* are interactions of the atoms of two or more chemical elements in certain fixed proportions that produce chemical compounds. For example, when two hydrogen atoms combine with one oxygen atom, they form a new chemical compound, water (H_2O). The properties of a chemical compound may be entirely different from those of its constituent elements. For example, when an atom of sodium, a metal, combines with an atom of chlorine, a noxious gas, they form the chemical compound sodium chloride, better known as table salt. We represent this compound by the chemical formula NaCl, in which the symbol Na stands for the element sodium and the symbol Cl for the element chlorine. (Every chemical element has been assigned its own symbol, which we use in a kind of shorthand for writing chemical formulas and equations; these symbols are given in the periodic table in Appendix 3.)

Chemical compounds, such as minerals, are formed either by **electron sharing** between the reacting atoms or by **electron transfer** between the reacting atoms. Carbon and silicon, two of the most abundant elements in Earth's crust, tend to form compounds by electron sharing. Diamond

The carbon atoms in diamond are arranged in regular tetrahedra…

…in which each atom shares an electron with each of its four neighbors.

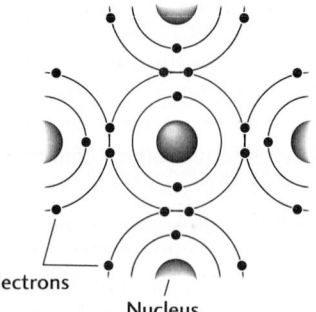

Carbon atoms   Electrons   Nucleus

FIGURE 3.3 ■ Some atoms share electrons to form covalent bonds.

is a compound composed entirely of carbon atoms sharing electrons (**Figure 3.3**).

In the reaction between sodium (Na) and chlorine (Cl) atoms to form sodium chloride (NaCl), electrons are transferred. The sodium atom loses one electron, which the chlorine atom gains (**Figure 3.4a**). An atom or group of atoms that has an electrical charge, either positive or negative,

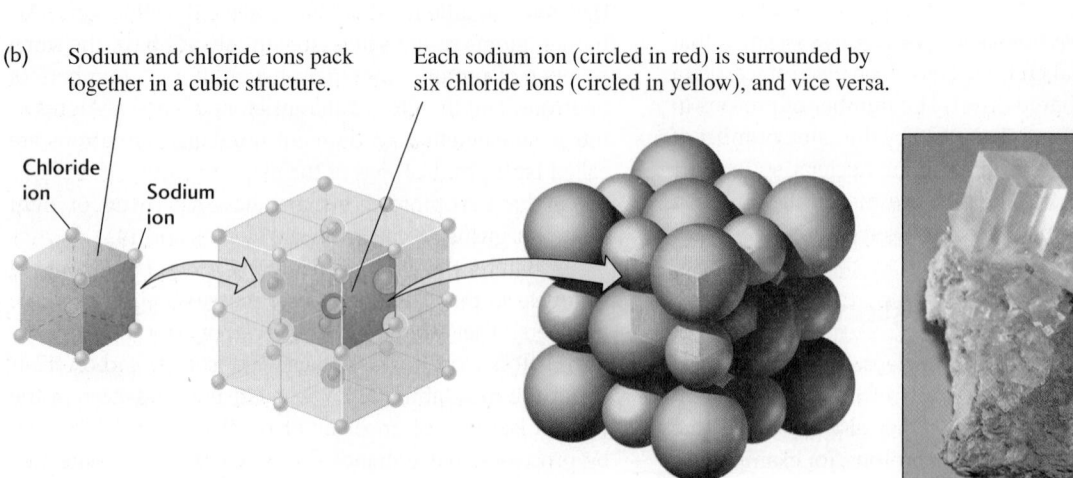

(a) When sodium (Na) and chlorine (Cl) react, the sodium atom loses one electron.

The chlorine atom gains that electron. Electrostatic attraction holds the two ions together.

Sodium atom
(1 electron in outer shell)

Chlorine atom
(7 electrons in outer shell)

Sodium ion (+)

Chloride ion (−)

(b) Sodium and chloride ions pack together in a cubic structure.

Each sodium ion (circled in red) is surrounded by six chloride ions (circled in yellow), and vice versa.

Chloride ion

Sodium ion

NaCl crystal

Halite (table salt)

FIGURE 3.4 ■ Some atoms transfer electrons to form ionic bonds. [Photo by John Grotzinger/Ramón Rivera-Moret/Harvard Mineralogical Museum.]

because of the loss or gain of one or more electrons is called an **ion.** Because the chlorine atom has gained a negatively charged electron, it is now a negatively charged ion, Cl^-. Likewise, the loss of an electron gives the sodium atom a positive charge, making it a sodium ion, Na^+. The compound NaCl itself remains electrically neutral because the positive charge on Na^+ is exactly balanced by the negative charge on Cl^-. A positively charged ion is called a **cation,** and a negatively charged ion is called an **anion.**

Chemical Bonds

When a chemical compound is formed either by electron sharing or by electron transfer, the ions or atoms that make up the compound are held together by electrostatic attraction between negatively charged electrons and positively charged protons. The attractions, or *chemical bonds,* between shared electrons or between gained and lost electrons may be strong or weak. Strong bonds keep a substance from decomposing into its elements or into other compounds. They also make minerals hard and keep them from cracking or splitting. Two major types of bonds are found in most rock-forming minerals: ionic bonds and covalent bonds.

IONIC BONDS The simplest form of chemical bond is the **ionic bond.** Bonds of this type form by electrostatic attraction between ions of opposite charge, such as Na^+ and Cl^- in sodium chloride (see Figure 3.4a), when electrons are transferred. This attraction is of exactly the same nature as the static electricity that can make nylon or silk clothing cling to the body. The strength of an ionic bond decreases greatly as the distance between ions increases, and it increases as the electrical charges of the ions increase. Ionic bonds are the dominant type of chemical bonds in mineral structures: about 90 percent of all minerals are essentially ionic compounds.

COVALENT BONDS Elements that do not readily gain or lose electrons to form ions, and instead form compounds by sharing electrons, are held together by **covalent bonds.** These bonds are generally stronger than ionic bonds. One mineral with a covalently bonded crystal structure is diamond, which consists of a single element, carbon. Each carbon atom can share four of its own electrons with other carbon atoms and can acquire another four electrons by sharing with other carbon atoms. In diamond, every carbon atom is surrounded by four others arranged in a four-sided pyramidal form (a *tetrahedron*), each side of which is a triangle (see Figure 3.3). In this configuration, each carbon atom shares an electron with each of its four neighbors, resulting in a very stable configuration. (Figure 3.10 shows a network of carbon tetrahedra linked together.)

METALLIC BONDS Atoms of metallic elements, which have strong tendencies to lose electrons, pack together as cations, and the freely mobile electrons are shared and dispersed among those cations. This free electron sharing results in a kind of covalent bond that we call a **metallic bond.** It is found in a small number of minerals, among them the metal copper and some sulfides.

The chemical bonds of some minerals are intermediate between pure ionic and pure covalent bonds because some electrons are exchanged and others are shared.

■ The Formation of Minerals

The orderly forms of minerals result from the chemical bonds we have just described. Minerals can be viewed in two complementary ways: as assemblages of submicroscopic atoms organized in an ordered three-dimensional array, and as crystals that we can see with the naked eye. In this section, we examine the crystal structures of minerals and the conditions under which minerals form. Later in this chapter, we will see how the crystal structures of minerals are manifested in their physical properties.

The Atomic Structure of Minerals

Minerals form by the process of **crystallization,** in which the atoms of a gas or liquid come together in the proper chemical proportions and in the proper arrangement to form a solid substance. (Remember that the atoms in a mineral are arranged in an orderly three-dimensional array.) The bonding of carbon atoms in diamond, a covalently bonded mineral, is one example of crystallization. Under the very high pressures and temperatures in Earth's mantle, carbon atoms bond together in tetrahedra, and each tetrahedron attaches to another, building up a regular three-dimensional structure from a great many atoms (see Figure 3.8). As a diamond crystal grows, it extends its tetrahedral structure in all directions, always adding new atoms in the proper geometric arrangement. Diamonds can be artificially synthesized from carbon under very high pressures and temperatures that mimic the conditions in Earth's mantle.

The sodium and chloride ions that make up sodium chloride, an ionically bonded mineral, also crystallize in an orderly three-dimensional array. In Figure 3.4b, we can see the geometry of their arrangement, with each ion of one kind surrounded by six ions of the other kind in a series of cubic structures extending in three directions. We can think of ions as solid spheres, packed together in close-fitting structural units. Figure 3.4b also shows the relative sizes of the ions in NaCl. The relative sizes of the sodium and chloride ions allow them to fit together in a closely packed arrangement.

Many of the cations of abundant minerals are relatively small, while most anions—including the most common anion on Earth, oxygen (O^{2-})—are large (**Figure 3.5**). Because anions tend to be larger than cations, most of the space of a crystal is occupied by the anions, and the cations fit into the

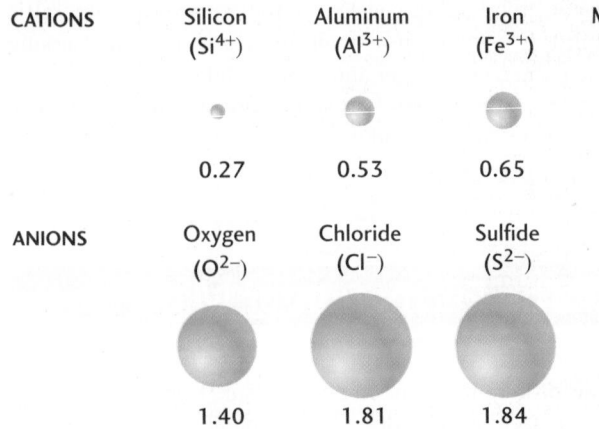

CATIONS	Silicon (Si^{4+})	Aluminum (Al^{3+})	Iron (Fe^{3+})	Magnesium (Mg^{2+})	Iron (Fe^{2+})	Sodium (Na^+)	Calcium (Ca^{2+})	Potassium (K^+)
	0.27	0.53	0.65	0.72	0.73	0.99	1.00	1.38

ANIONS	Oxygen (O^{2-})	Chloride (Cl^-)	Sulfide (S^{2-})
	1.40	1.81	1.84

FIGURE 3.5 ▪ Sizes of some ions commonly found in rock-forming minerals. Ionic radii are given in 10^{-8} cm. [After L. G. Berry, B. Mason, and R. V. Dietrich, *Mineralogy.* San Francisco: W. H. Freeman, 1983.]

spaces between them. As a result, crystal structures are determined largely by how the anions are arranged and how the cations fit between them.

Cations of similar sizes and charges tend to substitute for one another and to form compounds having the same crystal structure but differing in chemical composition. *Cation substitution* is common in minerals that contain the silicate ion (SiO_4^{4-}), such as olivine, which is abundant in many volcanic rocks. Iron (Fe^{2+}) and magnesium (Mg^{2+}) ions are similar to each other in size, and both have two positive charges, so they easily substitute for each other in the structure of olivine. The composition of pure magnesium olivine is Mg_2SiO_4; that of pure iron olivine is Fe_2SiO_4. The composition of olivine containing both iron and magnesium is given by the formula $(Mg,Fe)_2SiO_4$, which simply means that the number of iron and magnesium cations may vary, but their combined total (expressed as a subscript 2) in relation to the single SiO_4^{4-} ion does not vary. The proportion of iron to magnesium is determined by the relative abundances of the two elements in the molten material from which the olivine crystallizes. Similarly, aluminum (Al^{3+}) substitutes for silicon (Si^{4+}) in many silicate minerals. Aluminum and silicon ions are similar enough in size that aluminum can take the place of silicon in many crystal structures. In this case, the difference in charge between aluminum (3+) and silicon (4+) ions is balanced by an increase in the number of other cations, such as sodium (1+).

The Crystallization of Minerals

Crystallization starts with the formation of microscopic single **crystals,** orderly three-dimensional arrays of atoms in which the basic arrangement is repeated in all directions. The boundaries of crystals are natural flat (*planar*) surfaces called *crystal faces* (**Figure 3.6**). The crystal faces of a mineral are the external expression of the mineral's internal atomic structure. **Figure 3.7** pairs a drawing of a perfect quartz crystal with a photograph of the actual mineral. The six-sided (hexagonal) shape of the quartz crystal corresponds to its hexagonal internal atomic structure.

During crystallization, the initially microscopic crystals grow larger, maintaining their crystal faces as long as they are free to grow. Large crystals with well-defined faces form when growth is slow and steady and space is adequate to allow growth without interference from other crystals nearby (**Figure 3.8**). For this reason, most large mineral crystals form in open spaces in rocks, such as fractures or cavities.

FIGURE 3.6 ▪ Crystals of amethyst and quartz growing on top of epidote crystals (green). The planar surfaces are crystal faces and reflect the mineral's internal atomic structure. [John Grotzinger/Ramón Rivera-Moret/Harvard Mineralogical Museum.]

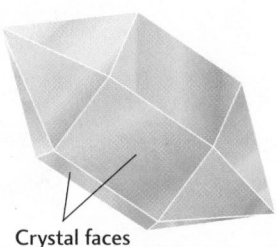

Crystal faces

A perfect quartz crystal

A natural quartz crystal

FIGURE 3.7 ■ A perfect crystal is rare in nature, but no matter how irregular the shapes of the crystal faces may be, the angles are always exactly the same. [Photo by Breck P. Kent.]

More often, however, the spaces between growing crystals fill in, or crystallization proceeds rapidly. Crystals then grow over one another and coalesce to become a solid mass of crystalline particles, or **grains.** In this case, few or no grains show crystal faces. Large crystals that can be seen with the naked eye are relatively unusual, but many minerals in rocks display crystal faces that can be seen under a microscope.

Unlike minerals, glassy materials—which solidify from liquids so quickly that they lack any internal atomic order— do not form crystals with planar faces. Instead, they are found as masses with curved, irregular surfaces. The most common natural glass is volcanic glass.

How Do Minerals Form?

Lowering the temperature of a liquid below its freezing point is one way to start the process of crystallization. In water,

for example, 0°C is the temperature below which crystals of ice—a mineral—start to form. Similarly, a **magma**—a mass of hot, molten liquid rock—crystallizes into solid minerals when it cools. As a magma falls below its melting point, which may be higher than 1000°C depending on the elements it contains, crystals of silicate minerals such as olivine or feldspar begin to form. (Geologists usually refer to melting points of magmas rather than freezing points, because freezing implies cold.)

Crystallization can also occur as liquids evaporate from a solution. A *solution* is a homogeneous mixture of one chemical substance with another, such as salt and water. As the water evaporates from a salt solution, the concentration of salt eventually gets so high that the solution can hold no more salt and is said to be *saturated.* If evaporation continues, the salt starts to **precipitate,** or drop out of solution as crystals. Deposits of table salt, or halite, form under just these conditions when seawater evaporates to the point of saturation in some hot, arid bays or arms of the ocean (**Figure 3.9**).

Diamond and graphite (the material used as the "lead" in pencils) exemplify the dramatic effects that temperature and pressure can have on mineral formation. These two minerals are **polymorphs,** minerals with alternative structures formed from the same chemical element or compound (**Figure 3.10**). They are both formed from carbon, but have different crystal structures and very different appearances. From experimentation and geologic observation, we know that diamond forms and remains stable at the very high pressures and temperatures found in Earth's mantle. High pressures force the atoms in diamond into a closely packed

FIGURE 3.8 ■ Giant crystals are sometimes found in caves, where they have room to grow. These selenite crystals are a gem-quality form of gypsum (calcium sulfate). [Javier Trueba/ MSF/Science Source.]

FIGURE 3.9 ■ Halite crystals precipitating within a modern hypersaline lagoon on San Salvador Island in the Bahamas. Note the cubic shape of the crystals. [John Grotzinger.]

structure. Diamond therefore has a higher **density** (mass per unit volume, usually expressed in grams per cubic centimeter, g/cm³) than graphite, which is less closely packed: diamond has a density of 3.5 g/cm³, while that of graphite is only 2.1 g/cm³. Graphite forms and is stable at moderate pressures and temperatures, such as those in Earth's crust.

Low temperatures can also produce close packing of atoms. Quartz and cristobalite, for example, are polymorphs of silica (SiO_2). Quartz forms at low temperatures and is relatively dense (2.7 g/cm³). Cristobalite, which forms at higher temperatures, has a more open structure and is therefore less dense (2.3 g/cm³).

■ Classes of Rock-Forming Minerals

All minerals on Earth have been grouped into seven classes according to their chemical composition (**Table 3.1**). Some minerals, such as copper, occur naturally as un-ionized pure elements; these minerals are classified as *native elements.* Most other minerals are classified by their anions. Olivine, for example, is classified as a silicate by its silicate anion, SiO_4^{4-}. Halite (sodium chloride, NaCl) is classified as a halide

Natural **diamond** is formed at very high pressures and temperatures in Earth's mantle.

Strong bonds connect closely packed carbon atoms in a tetrahedral structure.

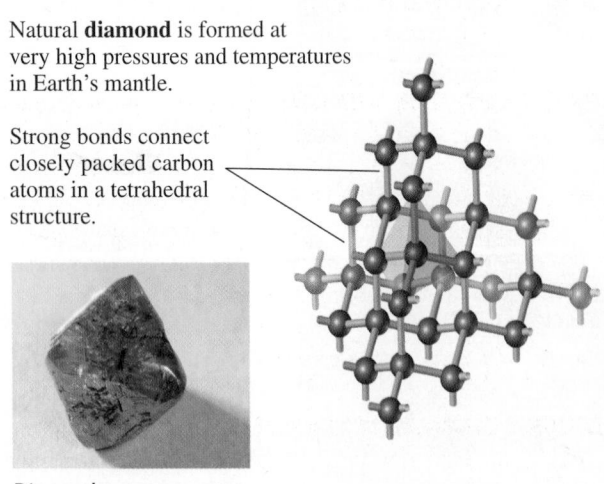

Diamond

Graphite is formed at lower pressures and temperatures than diamond. Strong bonds connect carbon atoms arranged in sheets.

Weak bonds connect carbon atoms between alternating sheets.

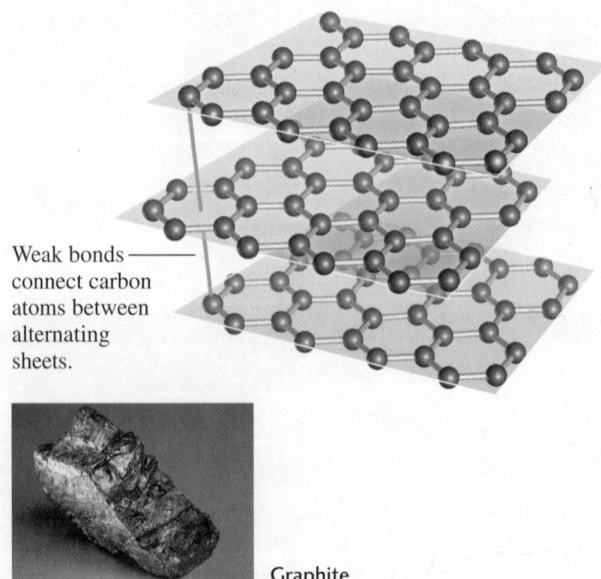

Graphite

FIGURE 3.10 ■ Graphite and diamond are polymorphs, alternative structures formed from the same chemical compound, carbon. [Photos by John Grotzinger/Ramón Rivera-Moret/Harvard Mineralogical Museum.]

TABLE 3-1   Some Chemical Classes of Minerals

Class	Defining Anions	Example
Native elements	None: no charged ions	Copper metal (Cu)
Oxides	Oxygen ion (O^{2-})	Hematite (Fe_2O_3)
Halides	Chloride (Cl^-), fluoride (F^-), bromide (Br^-), iodide (I^-)	Halite (NaCl)
Carbonates	Carbonate ion (CO_3^{2-})	Calcite ($CaCO_3$)
Sulfates	Sulfate ion (SO_4^{2-})	Anhydrite ($CaSO_4$)
Silicates	Silicate ion (SiO_4^{4-})	Olivine $(Mg,Fe)_2SiO_4$
Sulfides	Sulfide ion (S^{2-})	Pyrite (FeS^2)

by its chloride anion, Cl^-. So is its close relative, sylvite (potassium chloride, KCl).

Although many thousands of minerals are known, geologists commonly encounter only about 30 of them. These minerals are the building blocks of most crustal rocks and are called *rock-forming minerals*. Their relatively small number corresponds to the small number of elements that are abundant in Earth's crust.

In the following pages, we consider the five most common classes of rock-forming minerals:

■ **Silicates,** the most abundant class of minerals in Earth's crust, are composed of oxygen (O) and silicon (Si)—the two most abundant elements in the crust—mostly in combination with cations of other elements.

■ **Carbonates** are minerals composed of carbon and oxygen—in the form of the carbonate anion (CO_3^{2-})—in combination with calcium and magnesium. Calcite (calcium carbonate, $CaCO_3$) is one such mineral.

■ **Oxides** are compounds of the oxygen anion (O^{2-}) and metallic cations; an example is the mineral hematite (iron oxide, Fe_2O_3).

■ **Sulfides** are compounds of the sulfide anion (S^{2-}) and metallic cations; an example is the mineral pyrite (iron sulfide, FeS_2).

■ **Sulfates** are compounds of the sulfate anion (SO_4^{2-}) and metallic cations; an example is the mineral anhydrite (calcium sulfate, $CaSO_4$).

The other three chemical classes of minerals—native elements, hydroxides, and halides—are less common as rock-forming minerals.

Silicates

The basic building block of all silicate mineral structures is the silicate ion. It is a tetrahedron composed of a central silicon ion (Si^{4+}) surrounded by four oxygen ions (O^{2-}),

and thus has the formula SiO_4^{4-} (**Figure 3.11**). Because the silicate ion has a negative charge, it often bonds to cations to form minerals. The cations it typically bonds to include sodium (Na^+), potassium (K^+), calcium (Ca^{2+}), magnesium (Mg^{2+}), and iron (Fe^{2+}). Alternatively, the silicate ion can share oxygen ions with other silicate tetrahedra. Silicate tetrahedra can form a number of crystal structures: they may be isolated (linked only to cations), or they may be linked to other silicate tetrahedra in rings, single chains, double chains, sheets, or frameworks. Some of these structures are shown in Figure 3.11.

ISOLATED TETRAHEDRA Isolated tetrahedra are linked by the bonding of each oxygen ion of the tetrahedron to a cation (Figure 3.11a). The cations, in turn, bond to the oxygen ions of other tetrahedra. The tetrahedra are thus isolated from one another by cations on all sides. Olivine is a rock-forming mineral with this structure.

SINGLE-CHAIN STRUCTURES Single chains are formed by the sharing of oxygen ions. Two oxygen ions of each silicate tetrahedron bond to adjacent tetrahedra in an open-ended chain (Figure 3.11b). These single chains are linked to other chains by cations. Minerals of the pyroxene group are single-chain silicate minerals. Enstatite, a pyroxene, contains iron or magnesium ions, or both; the two cations may substitute for each other, as in olivine. The formula $(Mg,Fe)SiO_3$ represents this structure.

DOUBLE-CHAIN STRUCTURES Two single chains may combine to form double chains linked to each other by shared oxygen ions (Figure 3.11c). Adjacent double chains linked by cations form the structure of minerals in the amphibole group. Hornblende, a member of this group, is an extremely common mineral in both igneous and metamorphic rocks. It has a complex composition that includes calcium (Ca^{2+}), sodium (Na^+), magnesium (Mg^{2+}), iron (Fe^{2+}), and aluminum (Al^{3+}).

Silicate tetrahedra can be arranged into a number of different structures.

Mineral	Chemical formula	Cleavage planes and number of cleavage directions	Structure	Specimen
Olivine (a)	$(Mg,Fe)_2SiO_4$	1 plane	Isolated tetrahedra	
Pyroxene (b)	$(Mg,Fe)SiO_3$	2 planes at 90°	Single chains	
Amphibole (c)	$Ca_2(Mg,Fe)_5Si_8O_{22}(OH)_2$	2 planes at 60° and 120°	Double chains	
Mica (d)	Muscovite: $KAl_2(AlSi_3O_{10})(OH)_2$	1 plane	Sheets	
Feldspar (e)	Orthoclase feldspar: $KAlSi_3O_8$ Plagioclase feldspar: $(Ca,Na)\,AlSi_3O_8$	2 planes at 90°	Three-dimensional frameworks	

FIGURE 3.11 ■ The silicate ion is the basic building block of silicate minerals. [Photos by John Grotzinger/Ramón Rivera-Moret/Harvard Mineralogical Museum.]

SHEET STRUCTURES In sheet structures, each tetrahedron shares three of its oxygen ions with adjacent tetrahedra to build stacked sheets of tetrahedra (Figure 3.11d). Cations may be interlayered with the tetrahedral sheets. The micas and clay minerals are the most abundant sheet silicates. Muscovite, $KAl_2(AlSi_3O_{10})(OH)_2$, is one of the most common sheet silicates and is found in many types of rocks. It can be separated into extremely thin, transparent sheets. Kaolinite, $Al_2Si_2O_5(OH)_4$, which also has this structure, is a common clay mineral found in sediments and is the basic raw material for pottery.

FRAMEWORKS Three-dimensional frameworks may form as each tetrahedron shares all its oxygen ions with other tetrahedra. Feldspars, the most abundant minerals in Earth's crust, are framework silicates (Figure 3.11e), as is another of the most common minerals, quartz (SiO_2).

SILICATE COMPOSITIONS Chemically, the simplest silicate is silicon dioxide, also called silica (SiO_2), which is found most often as the mineral quartz. The silicate tetrahedra of quartz are linked, sharing two oxygen ions for each silicon ion, so the total formula adds up to SiO_2.

In other silicate minerals, as we have seen, the basic structural units—rings, chains, sheets, and frameworks—are bonded to cations such as sodium (Na^+), potassium (K^+), calcium (Ca^{2+}), magnesium (Mg^{2+}), and iron (Fe^{2+}). As noted in the discussion of cation substitution, aluminum (Al^{3+}) substitutes for silicon in many silicate minerals.

Carbonates

The basic building block of carbonate minerals is the carbonate ion (CO_3^{2-}), which consists of a carbon ion surrounded by, and covalently bonded to, three oxygen ions in a triangle (**Figure 3.12a**). Groups of carbonate ions are arranged in sheets somewhat like those of the sheet silicates, which are linked together by layers of cations. In calcite (calcium carbonate, $CaCO_3$), the sheets of carbonate ions are separated by layers of calcium ions (Figure 3.12b). Calcite is one of the most abundant minerals in Earth's crust and is the chief constituent of a group of rocks called limestones (Figure 3.12c). Dolomite ($CaMg(CO_3)_2$) is another major mineral of crustal rocks that is made up of the same carbonate sheets separated by alternating layers of calcium ions and magnesium ions.

Oxides

Oxide minerals are compounds in which oxygen is bonded to atoms or cations of other elements, usually metallic cations such as iron (Fe^{2+} or Fe^{3+}). Most oxide minerals are ionically bonded, and their structures vary with the size of the metallic cations. This class of minerals has great economic importance because it includes the ores containing many of the metals, such as chromium and titanium, used in the manufacture of metallic materials and devices. Hematite (Fe_2O_3) (**Figure 3.13a**) is a chief ore of iron.

Another group of abundant minerals in this class, the spinels (Figure 3.13b), are oxides of two metals, magnesium and aluminum ($MgAl_2O_4$). Spinels have a closely packed cubic structure and a high density (3.6 g/cm³), reflecting the conditions of high pressure and temperature under which they form. Transparent gem-quality spinels may resemble ruby or sapphire and are found in the crown jewels of England and Russia.

Sulfides

The chief ores of other valuable minerals—such as copper, zinc, and nickel—are members of the sulfide class. The basic building block of this class is the sulfide ion (S^{2-}), a sulfur atom that has gained two electrons. In the sulfide minerals, the sulfide ion is bonded to metallic cations. Most sulfide minerals look like metals, and almost all are opaque. The most common sulfide mineral is pyrite (FeS_2), often

(a) Carbonate ion (CO_3^{2-}) **(b) Calcium carbonate structure** **(c) Calcite**

FIGURE 3.12 ■ Carbonate minerals, such as calcite (calcium carbonate, $CaCO_3$), have a layered structure. (a) Top view of the carbonate ion, composed of a carbon ion surrounded by three oxygen ions in a triangle. (b) View of the alternating layers of calcium and carbonate ions in calcite. (c) Calcite. [Photo by John Grotzinger/Ramón Rivera-Moret/Harvard Mineralogical Museum.]

(a)

(b)

FIGURE 3.13 ■ Oxides include many economically valuable minerals. (a) Hematite. (b) Spinel. [John Grotzinger/Ramón Rivera-Moret/ Harvard Mineralogical Museum.]

called "fool's gold" because of its yellowish metallic appearance (**Figure 3.14**).

Sulfates

The basic building block of sulfates is the sulfate ion (SO_4^{2-}). It is a tetrahedron made up of a central sulfur atom surrounded by four oxygen ions (O^{2-}). One of the most abundant minerals of this class is gypsum (**Figure 3.15**), the primary component of plaster. Gypsum, a calcium sulfate, forms when seawater evaporates. During evaporation, Ca^{2+} and SO_4^{2-}, two ions that are abundant in seawater, combine and precipitate as layers of sediment, forming calcium sulfate ($CaSO_4 \cdot 2H_2O$). (The dot in this formula signifies that two water molecules are bonded to the calcium and sulfate ions.)

Another calcium sulfate, anhydrite ($CaSO_4$), differs from gypsum in that it contains no water. (Its name is derived from the word *anhydrous,* meaning "free from water.") Gypsum is stable at the low temperatures and pressures found at Earth's surface, whereas anhydrite is stable at the higher temperatures and pressures where sedimentary rocks are buried.

As scientists discovered in 2004, sulfate minerals precipitated from water and formed sedimentary layers early in the history of Mars. These minerals were precipitated by processes similar to those observed on Earth when lakes and shallow seas dried up. Many of these sulfate minerals, however, are quite different from the sulfate minerals commonly found on Earth and include strange iron-bearing sulfates that precipitated from very harsh, acidic waters (see Earth Issues 11.1).

FIGURE 3.14 ■ Pyrite, a sulfide mineral, is also known as "fool's gold." [John Grotzinger/Ramón Rivera-Moret/Harvard Mineralogical Museum.]

FIGURE 3.15 ■ Gypsum is a sulfate formed when seawater evaporates. [John Grotzinger/Ramón Rivera-Moret/Harvard Mineralogical Museum.]

TABLE 3-2	Mohs Scale of Hardness	
Mineral	Scale Number	Common Objects
Talc	1	
Gypsum	2	Fingernail
Calcite	3	Copper coin
Fluorite	4	
Apatite	5	Knife blade
Orthoclase	6	Window glass
Quartz	7	Steel file
Topaz	8	
Corundum	9	
Diamond	10	

■ Physical Properties of Minerals

Geologists use their knowledge of mineral composition and structure to understand the origins of rocks. First they must identify the minerals that make up a rock. To do so, they rely greatly on chemical and physical properties that can be observed relatively easily. In the nineteenth and early twentieth centuries, geologists carried field kits for rough chemical analyses of minerals that would help in their identification. One such test is the origin of the phrase "the acid test." It consists of dropping diluted hydrochloric acid (HCl) on a mineral to see if it fizzes (**Figure 3.16**). Fizzing indicates that carbon dioxide (CO_2) is escaping, which means that the mineral is likely to be calcite, a carbonate.

In this section, we review the physical properties of minerals, many of which contribute to their practical and decorative value.

Hardness

Hardness is a measure of the ease with which the surface of a mineral can be scratched. Just as diamond, the hardest mineral known, scratches glass, a quartz crystal, which is harder than feldspar, scratches a feldspar crystal. In 1822, Friedrich Mohs, an Austrian mineralogist, devised a scale (now known as the **Mohs scale of hardness**) based on the ability of one mineral to scratch another. At one extreme is the softest mineral (talc); at the other, the hardest (diamond) (**Table 3.2**). The Mohs scale is still one of the best practical tools for identifying an unknown mineral. With a knife blade and a few of the minerals on the hardness scale, a field geologist can gauge an unknown mineral's position

FIGURE 3.16 ■ The acid test. One easy but effective way to identify certain minerals is to drop diluted hydrochloric acid (HCl) on the substance of interest. If it fizzes, indicating the escape of carbon dioxide, the mineral is likely to be calcite. [Chip Clark/Fundamental Photographs.]

on the scale. If the unknown mineral is scratched by a piece of quartz but not by the knife, for example, it lies between 5 and 7 on the scale.

Recall that covalent bonds are generally stronger than ionic bonds. The hardness of any mineral depends on the strength of its chemical bonds: the stronger the bonds, the harder the mineral. Within the silicate class of minerals, hardness varies with crystal structure, from 1 in talc, a sheet silicate, to 8 in topaz, a silicate with isolated tetrahedra. Most silicates fall in the 5 to 7 range on the Mohs scale. Only sheet silicates are relatively soft, with hardnesses between 1 and 3.

Within groups of minerals that have similar crystal structures, hardness is related to other factors that also affect bond strength:

■ *Size:* The smaller the atoms or ions, the smaller the distance between them and the greater the electrostatic attraction—and thus the stronger the bond.

■ *Charge:* The larger the charge of ions, the greater the attraction between them, and thus the stronger the bond.

■ *Packing:* The closer the packing of atoms or ions, the smaller the distance between them, and thus the stronger the bond.

Size is an especially important factor for most metallic oxides and for most sulfides of metals with high atomic numbers, such as gold, silver, copper, and lead. Minerals of these groups are soft, with hardnesses of less than 3, because their metallic cations are so large. Carbonates and sulfates, whose structures are not closely packed, are also soft, with hardnesses of less than 5.

Cleavage

Cleavage is the tendency of a crystal to split along planar surfaces. The term *cleavage* is also used to describe the geometric pattern produced by such breakage. Cleavage varies inversely with bond strength: strong bonds produce poor cleavage, while weak bonds produce good cleavage. Because of their strength, covalent bonds generally produce poor or no cleavage. Ionic bonds are relatively weak, so they produce good cleavage. Even within a mineral that is entirely covalently bonded or entirely ionically bonded, however, bond strength varies along the different planes. For example, all of the bonds in diamond are covalent bonds, which are very strong, but some planes are more weakly bonded than others. Thus, diamond, the hardest mineral of all, can be cleaved along these weaker planes to produce perfect planar surfaces. Muscovite, a mica sheet silicate, splits along smooth, lustrous, flat, parallel surfaces, forming transparent sheets less than a millimeter thick. The excellent cleavage of micas results from the relative weakness of the bonds between its layers of cations sandwiched within sheets of silicate tetrahedra (**Figure 3.17**).

Cleavage is classified according to two primary sets of characteristics: the number of planes and pattern of cleavage, and the quality of surfaces and ease of cleaving.

NUMBER OF PLANES AND PATTERN OF CLEAVAGE

The number of planes and pattern of cleavage are identifying hallmarks of many rock-forming minerals. Muscovite, for example, has only one plane of cleavage, whereas calcite and dolomite crystals have three cleavage planes that give them a rhomboidal shape (**Figure 3.18**).

A crystal's structure determines its cleavage planes and its crystal faces. Crystals have fewer cleavage planes than

Silicate sheet

Aluminum hydroxide sheet

"Sandwich"

Cleavage occurs between aluminum hydroxide "sandwiches."

Potassium ions

"Sandwich"

FIGURE 3.17 ■ Cleavage of mica. The diagram shows the cleavage plane in the crystal structure, oriented perpendicular to the plane of the page. Horizontal lines mark the interfaces of silicate tetrahedral sheets and the sheets of aluminum hydroxide bonding the two tetrahedral sheets into a sandwich. Cleavage takes place between tetrahedral–aluminum hydroxide sandwiches. The photograph shows thin sheets of mica separating along the cleavage planes. [Chip Clark/ Fundamental Photographs.]

FIGURE 3.18 ■ Example of rhomboidal cleavage in calcite. [Charles D. Winters/Science Source.]

FIGURE 3.19 ■ Pyroxene and amphibole often look very much alike, but their differing angles of cleavage can be used to identify and classify them.

possible crystal faces. Faces may be formed along any of numerous planes defined by rows of atoms or ions. Cleavage occurs along any of those planes across which the bonding is weak. All crystals of a mineral exhibit its characteristic cleavage planes, whereas only some crystals display particular faces.

Distinctive angles of cleavage help identify two important groups of silicates, the pyroxenes and the amphiboles, that otherwise often look alike (**Figure 3.19**). Pyroxenes have a single-chain structure and are bonded so that their cleavage planes are almost at right angles (about 90°) to each other. In cross section, the cleavage pattern of pyroxenes is nearly a square. In contrast, amphiboles, which have a double-chain structure, are bonded so as to give two cleavage planes at about 60° and 120° to each other. They produce a diamond-shaped cross section.

QUALITY OF SURFACES AND EASE OF CLEAVING
A mineral's cleavage is assessed as perfect, excellent, good, fair, poor, or none according to the quality of surfaces produced and the ease of cleaving. A few examples are described below.

Muscovite can be cleaved easily, and it produces extremely smooth surfaces; its cleavage is *perfect*. Single-chain and double-chain silicates (the pyroxenes and amphiboles, respectively) show *good* cleavage. Although these minerals split easily along the cleavage plane, they also break across it, producing cleavage surfaces that are not as smooth as those of micas. *Fair* cleavage is shown by the ring silicate beryl. Beryl's cleavage is irregular, and the mineral breaks relatively easily along directions other than cleavage planes.

Many minerals are so strongly bonded that they lack even fair cleavage. Quartz, a framework silicate, is so strongly bonded in all directions that it breaks only along irregular surfaces. Garnet, a silicate with isolated tetrahedra, is also bonded strongly in all directions and so shows

no cleavage. This absence of a tendency to cleave is found in most framework and isolated tetrahedral silicates.

Fracture

Fracture is the tendency of a crystal to break along irregular surfaces other than cleavage planes. All minerals show fracture, either across cleavage planes or—in such minerals as quartz—with no cleavage in any direction. Fracture is related to how bond strengths are distributed in directions that cut across cleavage planes. Fractures may be *conchoidal,* showing smooth, curved surfaces like those of a thick piece of broken glass. Another common fracture surface with an appearance like split wood is described as *fibrous* or *splintery.* The shapes and appearances of fracture surfaces depend on the particular structure and composition of the mineral.

Luster

The way the surface of a mineral reflects light gives it a characteristic **luster.** Mineral lusters are described by the terms listed in **Table 3.3**. Luster is controlled by the kinds

TABLE 3-3	Mineral Lusters
Luster	Characteristics
Metallic	Strong reflections produced by opaque substances
Vitreous	Bright, as in glass
Resinous	Characteristic of resins, such as amber
Greasy	The appearance of being coated with an oily substance
Pearly	The whitish iridescence of such materials as pearl
Silky	The sheen of fibrous materials such as silk
Adamantine	The brilliant luster of diamond and similar minerals

FIGURE 3.20 ▪ Hematite may be black, red, or brown, but it always leaves a reddish brown streak when scraped along a ceramic streak plate. [Breck P. Kent.]

of atoms present and their bonding, both of which affect the way light passes through or is reflected by the mineral. Ionically bonded crystals tend to have a glassy, or vitreous, luster, but covalently bonded materials are more variable. Many have an adamantine luster, like that of diamond. Pure metals, such as gold, and many sulfides, such as galena (lead sulfide, PbS) have a metallic luster. A pearly luster results from multiple reflections of light from planes beneath the surfaces of translucent minerals, such as the mother-of-pearl inner surfaces of many clamshells, which are made of the mineral aragonite. Luster, although an important criterion for field classification, depends heavily on the visual perception of reflected light. Textbook descriptions fall short of the actual experience of holding the mineral in your hand.

Color

The **color** of a mineral is imparted by light, either transmitted through or reflected by crystals or irregular masses of the mineral. The color of a mineral may be distinctive, but it is not the most reliable clue to its identity. Some minerals always show the same color; others may have a range of colors. Many minerals show a characteristic color only on freshly broken surfaces or only on weathered surfaces. Some—precious opals, for example—show a stunning display of colors on reflecting surfaces. Others change color slightly with a change in the angle of the light shining on their surfaces. Many ionically bonded crystals are colorless.

Streak refers to the color of the fine deposit of mineral powder left on an abrasive surface, such as a tile of unglazed porcelain, when a mineral is scraped across it. Such a tile, called a *streak plate* (**Figure 3.20**), is a good identification

tool because the uniformly small grains of the mineral that are present in the powder are revealed on the plate. Hematite, for example, may look black, red, or brown, but this mineral will always leave a trail of reddish brown powder on a streak plate.

Color is a complex and not yet fully understood property of minerals. It is determined both by the kinds of ions found in the pure mineral and by trace elements.

IONS AND MINERAL COLOR The color of pure minerals depends on the presence of certain ions, such as iron or chromium, that strongly absorb portions of the light spectrum. Olivine that contains iron, for example, absorbs all colors except green, which it reflects, so we see this type of olivine as green. We see pure magnesium olivine as white (transparent and colorless).

TRACE ELEMENTS AND MINERAL COLOR All minerals contain impurities. Instruments can now measure even very small quantities of some elements—as little as a billionth of a gram in some cases. Elements that make up much less than 0.1 percent of a mineral are referred to as **trace elements.**

Some trace elements can be used to deduce the origins of the minerals in which they are found. Others, such as the traces of uranium in some granites, contribute to local natural radioactivity. Still others, such as small dispersed flakes of hematite that color a feldspar crystal brownish or reddish, are notable because they give a general color to an otherwise colorless mineral. Many of the gem varieties of minerals, such as emerald (green beryl) and sapphire (blue corundum), get their color from trace elements dissolved

FIGURE 3.21 ■ Trace elements give gems their colors. Sapphire (*left*) and ruby (*center*) are formed of the same common mineral, corundum (aluminum oxide). Small amounts of impurities produce the intense colors that we value. Ruby, for example, is red because of small amounts of chromium, the same element that gives emerald (*right*) its green color. [John Grotzinger/Ramón Rivera-Moret/Harvard Mineralogical Museum.]

in the solid crystal (**Figure 3.21**). Emerald derives its color from chromium; the sources of sapphire's blue color are iron and titanium.

Density

You can easily feel the difference in weight between a piece of hematite and a piece of sulfur of the same size by lifting the two pieces. A great many common rock-forming minerals, however, are too similar in density for such a simple test. Scientists therefore need some easy method to measure this property of minerals. A standard measure of density is **specific gravity,** which is the weight of a mineral divided by the weight of an equal volume of pure water at 4°C.

Density depends on the atomic mass of a mineral's atoms or ions and how closely they are packed in its crystal structure. Consider magnetite, an iron oxide with a density of 5.2 g/cm^3. Its high density results partly from the high atomic mass of iron and partly from the closely packed structure that magnetite shares with the other members of the spinel group of oxides. The density of iron olivine, at 4.4 g/cm^3, is lower than that of magnetite for two reasons. First, the atomic mass of silicon, one of the elements that make up olivines, is lower than that of iron. Second, iron olivine has a more openly packed structure than minerals of the spinel group. The density of magnesium olivine is even lower, at 3.32 g/cm^3, because magnesium's atomic mass is much lower than that of iron.

Increases in density caused by increases in pressure affect the way minerals transmit light, heat, and seismic waves. Experiments at extremely high pressures have shown that the structure of olivine converts into the denser structure of spinel olivine at pressures corresponding to a depth in Earth's mantle of 410 km. At a greater depth, 660 km, mantle materials are further transformed into silicate minerals with the even more densely packed structure of perovskite olivine. Because of the huge volume of the lower mantle, perovskite olivine is probably the most abundant mineral in Earth as a whole. Temperature also affects density: the higher the temperature, the more open and expanded the structure of the mineral, and thus the lower its density.

Crystal Habit

A mineral's **crystal habit** is the shape in which individual crystals or aggregates of crystals grow. Some minerals have such a distinctive crystal habit that they are easily recognizable. An example is quartz, with its six-sided column topped by a pyramid-like set of faces (see Figure 3.7). Crystal habits are often named after common geometric shapes, such as blades, plates, and needles. These shapes indicate not only the planes of the mineral's crystal structure, but also the typical speed and direction of crystal growth. Thus, a needlelike crystal is one that grows very quickly in one direction and very slowly in all other directions. In contrast, a plate-shaped crystal (often referred to as *platy*) grows fast in all directions that are perpendicular to its single direction of slow growth. *Fibrous* crystals take shape as multiple long, narrow fibers, essentially aggregates of long needles. *Asbestos* is a generic name for a group of silicate minerals with a more or less fibrous habit that allows the crystals to become embedded in the lungs if they are inhaled (**Figure 3.22**).

Table 3.4 summarizes the physical properties of minerals that we have discussed in this section.

FIGURE 3.22 ■ Chrysotile, a type of asbestos. Fibers are readily combed from the solid mineral. [Courtesy of Eurico Zimbres.]

TABLE 3-4	Physical Properties of Minerals
Property	Relation to Composition and Crystal Structure
Hardness	Strong chemical bonds result in hard minerals. Covalently bonded minerals are generally harder than ionically bonded minerals.
Cleavage	Cleavage is poor if bonds in crystal structure are strong, good if bonds are weak. Covalent bonds generally give poor or no cleavage; ionic bonds are weaker and so give good cleavage.
Fracture	Related to distribution of bond strengths across irregular surfaces other than cleavage planes.
Luster	Tends to be glassy for ionically bonded crystals, more variable for covalently bonded crystals.
Color	Determined by ions and trace elements. Many ionically bonded crystals are colorless. Iron tends to color strongly.
Streak	Color of fine mineral powder is more characteristic than that of massive mineral because of uniformly small size of grains.
Density	Depends on atomic weight of atoms or ions and their closeness of packing in crystal structure.
Crystal habit	Depends on the planes of a mineral's crystal structure and the typical speed and direction of crystal growth.

■ What Are Rocks?

A geologist's primary aim is to understand the properties of rocks and to deduce their geologic origins from those properties. Such deductions further our understanding of our planet, and they also provide important information about economically important resources. For example, knowing that oil forms in certain kinds of sedimentary rocks that are rich in organic matter allows us to explore for oil reserves more intelligently. Understanding how rocks form also guides us in solving environmental problems. For example, the underground storage of radioactive and other wastes depends on analysis of the rock to be used as a repository: Will this rock be prone to earthquake-triggered landslides? How might it transmit polluted waters in the ground?

Properties of Rocks

A **rock** is a naturally occurring solid aggregate of minerals or, in some cases, nonmineral solid matter. In an *aggregate,* minerals are joined in such a way that they retain their individual identity (**Figure 3.23**). A few rocks are composed of nonmineral matter. These rocks include the noncrystalline, glassy volcanic rocks obsidian and pumice as well as coal, which is made up of compacted plant remains.

What determines the physical appearance of a rock? Rocks vary in color, in the sizes of their crystals or grains, and in the kinds of minerals that compose them. Along a road cut, for example, we might find a rough white and pink speckled rock composed of interlocking crystals large enough to be seen with the naked eye. Nearby, we might see a grayish rock containing many large, glittering crystals of mica and some grains of quartz and feldspar. Overlying both

the white and pink rock and the gray one, we might see horizontal layers of a striped white and mauve rock that appear to be made up of sand grains cemented together. And these rocks might all be overlain by a dark, fine-grained rock with tiny white dots in it.

The identity of a rock is determined partly by its mineralogy and partly by its texture. Here, the term *mineralogy* refers to the relative proportions of a rock's constituent minerals. **Texture** describes the sizes and shapes of a rock's mineral crystals or grains and the way they are put together. If the crystals or grains, which are only a few millimeters in diameter in most rocks, are large enough to be seen with the naked eye, the rock is categorized as *coarse-grained.* If they are not large enough to be seen, the rock is categorized as *fine-grained.* The mineralogy and texture that determine a rock's appearance are themselves determined by the rock's geologic origin—where and how it formed (**Figure 3.24**).

The dark rock that caps the sequence of rocks in our road cut, called basalt, was formed by a volcanic eruption. Its mineralogy and texture were determined by the chemical composition of rocks that were melted deep within Earth. All rocks formed by the solidification of molten rock, such as basalt and granite, are called **igneous rocks.**

The striped white and mauve layers in the road cut are sandstone, formed as sand particles accumulated, perhaps on an ancient beach, and eventually were covered over, buried, and cemented together. All rocks formed as the burial products of layers of sediments (such as sand, mud, or the calcium carbonate shells of marine organisms), whether they were laid down on land or under the sea, are called **sedimentary rocks.**

The grayish rock of our road cut, a gneiss, contains crystals of mica, quartz, and feldspar. It formed deep in Earth's

Constituent minerals

| Orthoclase feldspar | Quartz | Biotite | Plagioclase feldspar |

Plagioclase feldspar
Orthoclase feldspar
Biotite
Quartz

FIGURE 3.23 ■ Rocks are naturally occurring aggregates of minerals. [John Grotzinger/Ramón Rivera-Moret/Harvard Mineralogical Museum.]

Rock (granite)

Type of rock and source material	Rock-forming process	Example
IGNEOUS Melting of rocks in hot, deep crust and upper mantle	Crystallization (solidification of magma or lava)	Granite
SEDIMENTARY Weathering and erosion of rocks exposed at surface	Deposition, burial, and lithification	Sandstone
METAMORPHIC Rocks under high temperatures and pressures in deep crust and upper mantle	Recrystallization of new minerals in solid state	Gneiss

FIGURE 3.24 ■ The three families of rocks are formed in different geologic environments by different geologic processes. [*granite and gneiss:* John Grotzinger/Ramón Rivera-Moret/ Harvard Mineralogical Museum; *sandstone:* John Grotzinger/Ramón Rivera-Moret/MIT.]

crust as high temperatures and pressures transformed the mineralogy and texture of buried sedimentary rock. All rocks formed by the transformation of preexisting solid rock under the influence of high temperatures and pressures are called **metamorphic rocks.**

The three types of rocks seen in our road cut represent the three great families of rock: igneous, sedimentary, and metamorphic. Let's take a closer look at each of these families and at the geologic processes that form them.

Igneous Rocks

Igneous rocks (from the Latin *ignis,* meaning "fire") form by crystallization from magma. When a body of magma cools slowly in Earth's interior, the minerals it contains begin to form microscopic crystals. As the magma cools below its melting point, some of these crystals have time to grow to several millimeters in diameter or larger before the whole mass crystallizes as a coarse-grained igneous rock. But when magma erupts from a volcano onto Earth's surface as lava, it cools and solidifies so rapidly that individual crystals have no time to grow gradually. In that case, many tiny crystals form simultaneously, and the result is a fine-grained igneous rock. Geologists distinguish two major types of igneous rocks—intrusive and extrusive—on the basis of the sizes of their crystals.

INTRUSIVE AND EXTRUSIVE IGNEOUS ROCKS *Intrusive igneous rocks* crystallize when magma intrudes into unmelted rock masses deep in Earth's crust. Large crystals grow as the magma slowly cools, producing coarse-grained rocks. Intrusive igneous rocks can be recognized by their large, interlocking crystals (**Figure 3.25**). Granite is an intrusive igneous rock.

TABLE 3-5	Some Common Minerals of Igneous, Sedimentary, and Metamorphic Rocks	
Igneous Rocks	Sedimentary Rocks	Metamorphic Rocks
Quartz	Quartz	Quartz
Feldspar	Clay minerals	Feldspar
Mica	Feldspar	Mica
Pyroxene	*Calcite	Garnet
Amphibole	*Dolomite	Pyroxene
Olivine	*Gypsum	Staurolite
	*Halite	Kyanite

*Nonsilicate minerals.

Extrusive igneous rocks form from magmas that erupt at Earth's surface as lava and cool rapidly. Extrusive igneous rocks, such as basalt, are easily recognized by their glassy or fine-grained texture.

COMMON MINERALS OF IGNEOUS ROCKS Most of the minerals of igneous rocks are silicates, partly because silicon is so abundant in Earth's crust and partly because many silicate minerals melt at the high temperatures and pressures reached in deeper parts of the crust and in the mantle. The silicate minerals most commonly found in igneous rocks include quartz, feldspars, micas, pyroxenes, amphiboles, and olivines (**Table 3.5**).

Extrusive igneous rocks form when lava erupts at the surface and cools rapidly.

The resulting rocks are fine-grained, like this basalt, or have a glassy texture.

Intrusive igneous rocks form when magma intrudes into unmelted rock and cools slowly.

The slow cooling allows large crystals to grow. The resulting rocks are coarse-grained, like this granite.

FIGURE 3.25 ■ Igneous rocks are formed by the crystallization of magma. [Photos by John Grotzinger/Ramón Rivera-Moret/Harvard Mineralogical Museum.]

Sedimentary Rocks

Sediments, the precursors of sedimentary rocks, are found at Earth's surface as layers of loose particles, such as sand, silt, and the shells of organisms. These particles originate in the processes of weathering and erosion. **Weathering** refers to all of the chemical and physical processes that break up and decay rocks into fragments and dissolved substances of various sizes. These particles are then transported by **erosion,** the set of processes that loosen soil and rock and move them downhill or downstream to the spot where they are deposited as layers of sediment (**Figure 3.26**).

Sediments are deposited in two ways:

- **Siliciclastic sediments** are made up of physically deposited particles, such as grains of quartz and feldspar derived from weathered granite. (*Clastic* is derived from the Greek word *klastos,* meaning "broken.") These sediments are laid down by running water, wind, and ice.

- **Chemical sediments** and **biological sediments** are new chemical substances that form by precipitation. Weathering dissolves some of a rock's components, which are carried in stream waters to the ocean. Halite

is a chemical sediment that precipitates directly from evaporating seawater. Calcite is precipitated by marine organisms to form shells or skeletons, which form biological sediments when the organisms die.

FROM SEDIMENT TO SOLID ROCK **Lithification** is the process that converts sediments into solid rock. It occurs in two ways:

- In *compaction,* particles are squeezed together by the weight of overlying sediments into a mass denser than the original.

- In *cementation,* minerals precipitate around deposited particles and bind them together.

Sediments are compacted and cemented after they are buried under additional layers of sediments. Sandstone forms by the lithification of sand particles, and limestone forms by the lithification of shells and other particles of calcite.

LAYERS OF SEDIMENT Sediments and sedimentary rocks are characterized by **bedding,** the formation of parallel layers of sediment as particles are deposited. Because

1 Particles and dissolved substances created by weathering…

2 …are transported downhill or downstream by erosion…

3 …and deposited as layers of sediment on land or in water,…

Lake

Beach

Igneous rock

Delta

Coral reefs

4 …where they form parallel layers, or bedding.

5 Buried sediments are lithified by compaction and cementation.

Metamorphic rock

6 Siliciclastic sediments, which are made up of rock fragments, form rocks like this sandstone.

7 Chemical and biological sediments may be precipitated directly from seawater or by organisms such as the corals that formed these fossilized skeletons.

FIGURE 3.26 ■ Sedimentary rocks are formed from particles of other rocks. [Photos by John Grotzinger/Ramón Rivera-Moret/MIT.]

sedimentary rocks are formed by surface processes, they cover much of Earth's land surface and seafloor. In terms of surface area, most rocks found at Earth's surface are sedimentary, but these rocks weather easily, so their volume is small compared with that of the igneous and metamorphic rocks that make up the main volume of the crust.

COMMON MINERALS OF SEDIMENTARY ROCKS

The most common minerals in siliciclastic sediments are silicates because silicate minerals predominate in the rocks that weather to form sedimentary particles (see Table 3.5). The most abundant silicate minerals in siliciclastic sedimentary rocks are quartz, feldspar, and clay minerals. Clay minerals are formed by the weathering and alteration of preexisting silicate minerals, such as feldspar.

The most abundant minerals of chemical and biological sediments are carbonates, such as calcite, the main constituent of limestone. Dolomite is a calcium-magnesium carbonate formed by precipitation during lithification. Two other chemical sediments—gypsum and halite—form by precipitation as seawater evaporates.

Metamorphic Rocks

Metamorphic rocks take their name from the Greek words for "change" (*meta*) and "form" (*morphe*). These rocks are produced when high temperatures and pressures deep within Earth cause changes in the mineralogy, texture, or chemical composition of any kind of preexisting rock—igneous, sedimentary, or other metamorphic rock—while maintaining its solid form. The temperatures of metamorphism are below the melting point of the rocks (about 700°C), but high enough (above 250°C) for the rocks to be changed by recrystallization and chemical reactions.

REGIONAL AND CONTACT METAMORPHISM Metamorphism may take place over a widespread area or a limited one (**Figure 3.27**). **Regional metamorphism** occurs where

Continental crust

Continental lithosphere

Asthenosphere

Trench

Oceanic crust

Oceanic lithosphere

Sediments

Hornfels formation   Magma

Hornfels

Eclogite

Micaschist

Blueschist

Contact metamorphism occurs in limited areas where heat from a magmatic intrusion metamorphoses neighboring rock.

Ultra-high-pressure metamorphism occurs deep in Earth's crust.

Regional metamorphism occurs where high pressures and temperatures extend over large regions.

High-pressure, low-temperature metamorphism occurs where oceanic crust is subducted beneath the leading edge of a continental plate.

FIGURE 3.27 ■ Metamorphic rocks form under high temperatures and pressures. [*hornfels:* Biophoto Associates/Science Source; *eclogite:* Julie Baldwin; *micaschist:* John Grotzinger; *blueschist:* Mark Cloos.]

high pressures and temperatures extend over large regions, as happens where plates collide. Regional metamorphism accompanies plate collisions that result in mountain building and the folding and breaking of sedimentary layers that were once horizontal. Where high temperatures are restricted to smaller areas, as in the rocks near and in contact with a magmatic intrusion, rocks are transformed by **contact metamorphism.** Other types of metamorphism, which we will describe in Chapter 6, include high-pressure metamorphism and ultra-high-pressure metamorphism.

Many regionally metamorphosed rocks, such as schists, have characteristic *foliation,* wavy or flat planes produced when the rock was folded. Granular textures are more typical of most contact metamorphic rocks and of some regional metamorphic rocks formed by very high pressures and temperatures.

COMMON MINERALS OF METAMORPHIC ROCKS

Silicates are the most abundant minerals in metamorphic rocks because most of the parent rocks from which they are formed are rich in silicates (see Table 3.5). Typical minerals of metamorphic rocks are quartz, feldspars, micas, pyroxenes, and amphiboles—the same kinds of silicates characteristic of igneous rocks. Several other silicates—kyanite, staurolite, and some varieties of garnet—are characteristic of metamorphic rocks alone. These minerals form under conditions of high pressure and temperature in the crust and are not characteristic of igneous rocks. They are therefore good indicators of metamorphism. Calcite is the main mineral of marbles, which are metamorphosed limestones.

▪ The Rock Cycle: Interactions Between the Plate Tectonic and Climate Systems

Earth scientists have known for over 200 years that the three families of rocks—igneous, metamorphic, and sedimentary—all can evolve from one to another. Their observations gave rise to the concept of a **rock cycle,** which explains how each type of rock is transformed into one of the other two types. The rock cycle is now known to be the result of interactions between two of the three global geosystems: the plate tectonic system and the climate system. Interactions between these two geosystems drive transfers of materials and energy among Earth's interior, the land surface, the ocean, and the atmosphere. For example, the formation of magmas at subduction zones results from processes operating within the plate tectonic system. When these magmas erupt, materials and energy are transferred to the land surface, where the materials (newly formed rocks) are subject to weathering by the climate system. The eruption process also injects volcanic ash and carbon dioxide

gas high into the atmosphere, where they may affect global climates. As global climates change, perhaps becoming warmer or cooler, the rate of weathering changes, which in turn influences the rate at which materials (sediments) are returned to Earth's interior.

Let's trace one turn of the rock cycle, beginning with the creation of new oceanic lithosphere at a mid-ocean ridge spreading center as two continents drift apart (**Figure 3.28**). The ocean gets wider and wider, until at some point the process reverses itself and the ocean closes. As the ocean basin closes, igneous rocks created at the mid-ocean ridge are eventually subducted beneath a continent. Sediments that were formed on the continent and deposited at its edge may also be dragged down into the subduction zone. Ultimately, the two continents, which were once drifting apart, may collide. As the igneous rocks and sediments that descend into the subduction zone go deeper and deeper into Earth's interior, they begin to melt to form a new generation of igneous rocks. The great heat associated with the intrusion of these igneous rocks, coupled with the heat and pressure that come with being pushed to levels deep within Earth, transforms these igneous rocks—and other surrounding rocks—into metamorphic rocks. When the continents collide, these igneous and metamorphic rocks are uplifted into a high mountain chain as a section of Earth's crust crumples, deforms, and undergoes further metamorphism.

The rocks of the uplifted mountains are exposed to the influences of the climate system, but they affect the climate system in turn, forcing moving air to rise, cool, and release precipitation. The rocks are slowly weathered, forming loose materials that erosion then strips away. Water and wind transport some of these materials across the continent and eventually to the edges of the continent, where they are deposited as sediments. The sediments laid down where the land meets the ocean are buried under successive layers of sediments, where they slowly lithify into sedimentary rock. These oceans, like those mentioned at the beginning of the cycle, were probably formed by seafloor spreading along mid-ocean ridges, thus completing the rock cycle.

The particular pathway illustrated here—that of a continent breaking apart, forming a new ocean basin, then closing back up again—is only one variation among many that may take place in the rock cycle. Any type of rock—igneous, sedimentary, or metamorphic—can be uplifted during a mountain-building event and then weathered and eroded to form new sediments. Some stages may be omitted: as a sedimentary rock is uplifted and eroded, for example, metamorphism and melting are skipped. In some cases, the rock cycle proceeds very slowly. For example, we know that some igneous and metamorphic rocks many kilometers deep in the crust may be uplifted or exposed to weathering and erosion only after billions of years have passed.

The rock cycle never ends. It is always operating at different stages in various parts of the world, forming and eroding mountains in one place and laying down and

1 The cycle begins with rifting within a continent. Sediments erode from the continental interior and are deposited in rift basins, where they are buried to form sedimentary rocks.

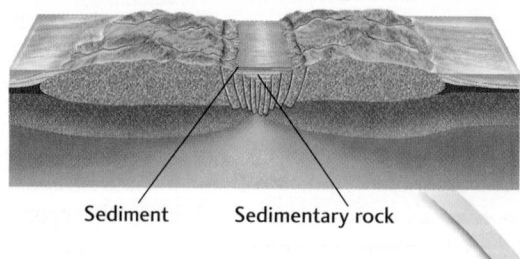

Sediment   Sedimentary rock

2 Rifting and spreading continue, and a new ocean basin develops. Magma rises from the asthenosphere at mid-ocean ridges and chills to form basalt, an igneous rock.

Oceanic crust   Magma   Oceanic
Continental crust   igneous rock
Continental lithosphere

6 Streams transport sediment away from collision zones to oceans, where it is deposited as layers of sand and silt. Layers of sediment are buried and lithify to form sedimentary rock.

Sediment   Sedimentary rock

3 Subsidence of the continental margin—sinking of Earth's lithosphere—leads to accumulation of sediment and formation of sedimentary rock during burial.

Sediment   Sedimentary rock   Subsidence

5 Further closing of the ocean basin leads to continental collision, forming high mountain ranges. Where continents collide, rocks are buried deeper or modified by heat and pressure, forming metamorphic rocks. Uplifted mountains force moisture-laden air to rise, cool, and release its moisture as precipitation. Weathering creates loose material—soils and sediment—that erosion strips away.

Metamorphic rock

4 Oceanic crust subducts beneath a continent, building a volcanic mountain chain. The subducting plate melts as it descends. Magma rises from the melting plate and mantle and cools to make granitic igneous rocks.

Continental
igneous rock

Magma

FIGURE 3.28 ■ The rock cycle results from the interaction of the plate tectonic and climate systems.

burying sediments in another. The rocks that make up the solid Earth are recycled continuously, but we can see only the surface parts of the cycle. We must deduce the recycling of the deep crust and the mantle from indirect evidence.

■ Concentrations of Valuable Mineral Resources

The rock cycle turns out to be crucial in creating economically important concentrations of the many valuable minerals found in Earth's crust. Minerals are not only sources of metals, which will be our focus here, but also provide us with stone for buildings and roads, phosphates for fertilizers, cement for construction, clays for ceramics, sand for silicon chips and fiber-optic cables, and many other items we use in our daily lives. Finding these minerals and extracting them is a vital job for Earth scientists, so we turn our attention next to how and where some of these geologic prizes are formed.

The chemical elements of Earth's crust are widely distributed in many kinds of minerals, and those minerals are found in a great variety of rocks. In most places, any given element will be found homogenized with other elements in amounts close to its average concentration in the crust.

An ordinary granitic rock, for example, may contain a small percentage of iron, close to the average concentration of iron in Earth's crust.

When an element is present in concentrations higher than the average, it means that the rock underwent some geologic process that concentrated larger quantities of that element than normal. The *concentration factor* of an element in a mineral deposit is the ratio of the element's abundance in the deposit to its average abundance in the crust. High concentrations of elements are found in a limited number of specific geologic settings. These settings are of economic interest because the higher the concentration of a resource in a given deposit, the lower the cost to recover it.

Ores are rich deposits of minerals from which valuable metals can be recovered profitably (see Practicing Geology). The minerals containing these metals are referred to as *ore minerals*. Ore minerals include sulfides (the largest group), oxides, and silicates. The ore minerals in each of these groups are compounds of metallic elements with sulfur, with oxygen, and with silicon and oxygen, respectively. The copper ore mineral covelite, for example, is a copper sulfide (CuS). The iron ore mineral hematite (Fe_2O_3) is an iron oxide. The nickel ore mineral garnierite is a nickel silicate ($Ni_3Si_2O_5(OH)_4$). In addition, some metals, such as gold, are found in their native state—that is, uncombined with other elements (**Figure 3.29**).

FIGURE 3.29 ■ Some metals are found in their native state. (a) A geologist examines rock samples in an underground gold mine in Zimbabwe, in southern Africa. (b) Native gold on a quartz crystal. [(a) Peter Bowater/Science Source; (b) 97-35023 by Chip Clark, Smithsonian.]

(a)

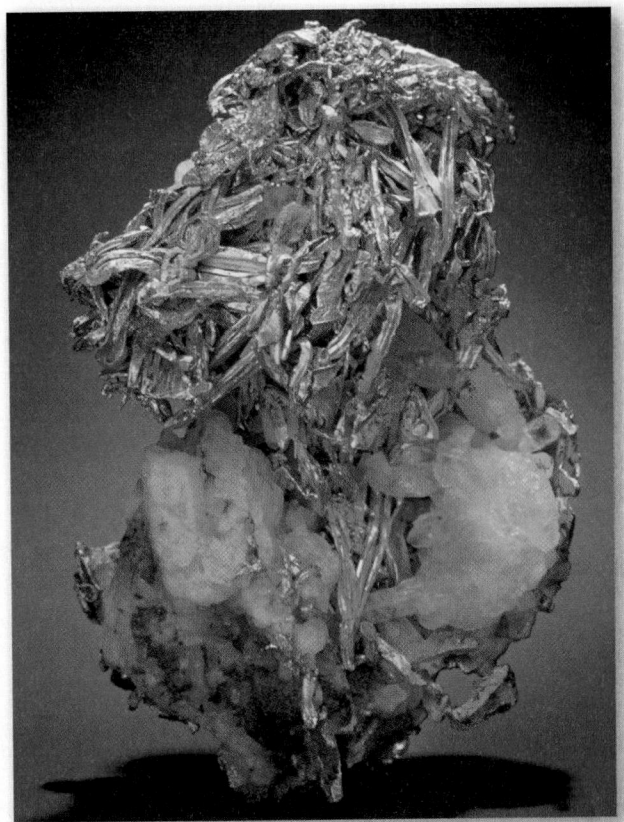

(b)

Hydrothermal Deposits

Many of the most valuable ores are formed in regions of volcanism by the interaction of igneous processes with the hydrosphere. Recall from our discussion of the rock cycle that subduction zones may be associated with the melting of oceanic lithosphere to form igneous rocks. Very large ore deposits can be formed in such plate tectonic settings when hot water solutions—also known as **hydrothermal solutions**—are formed around bodies of molten rock. This happens when circulating groundwater or seawater comes into contact with a magmatic intrusion, reacts with it, and carries off significant quantities of elements and ions released by the reaction. These elements and ions then interact with one another to form ore minerals, usually as the solution cools.

VEINS Hydrothermal solutions moving through rocks often deposit ore minerals (**Figure 3.30**). These fluids flow easily through fractures in the rocks, cooling rapidly in the process. Quick cooling causes rapid precipitation of the ore minerals. The resulting *tabular* (sheetlike) deposits of precipitated minerals in the fractures are called **veins.** Some ore minerals are found in veins; others are found in the rocks surrounding the veins, which are altered when the hydrothermal solutions heat and infiltrate those rocks. As the solutions react with the surrounding rocks, they may precipitate ore minerals together with quartz, calcite, or other common vein-filling minerals. Vein deposits are a major source of gold.

Hydrothermal vein deposits are among the most important sources of metallic ores. Typically, metallic ores exist as sulfides, such as iron sulfide (pyrite), lead sulfide (galena), zinc sulfide (sphalerite), and mercury sulfide (cinnabar) (**Figure 3.31**). Hydrothermal solutions reach the surface as hot springs and geysers, many of which precipitate metallic ores—including ores of lead, zinc, and mercury—as they cool.

DISSEMINATED DEPOSITS Deposits of ore minerals that are scattered through volumes of rock much larger than veins are called **disseminated deposits.** In both igneous and sedimentary rocks, minerals are disseminated along abundant cracks and fractures. Among the economically important disseminated deposits are the copper deposits of Chile and the southwestern United States. These deposits develop in geologic regions with abundant igneous rocks, usually emplaced as large intrusive bodies. In Chile, these intrusive igneous rocks are related to the subduction of oceanic lithosphere beneath the Andes (an event very similar to what was described in our example of the rock cycle). The most common copper mineral in these deposits is chalcopyrite, a copper sulfide (**Figure 3.32**). The copper was deposited when ore minerals were introduced into a great number of tiny fractures in granitic intrusive rocks and in the rocks surrounding the upper parts of the igneous intrusions. Some unknown process associated with the magmatic intrusion or its aftermath broke these rocks into millions of pieces. Hydrothermal solutions penetrated and re-cemented the rocks by precipitating ore minerals throughout the extensive network of tiny fractures. This

(a)   (b)

1 cm

FIGURE 3.30 ▪ Many deposits of ore minerals are found in veins formed by hydrothermal solutions. (a) Groundwater percolating through fractured rock dissolves metal oxides and sulfides. When heated by a magmatic intrusion, it rises, precipitating metallic ores in the rock fractures. (b) This quartz vein deposit (about 1 cm thick) in Oatman, Arizona, which contains gold and silver ores, was formed by such a process. [Photo by Peter Kresan.]

Galena
(lead sulfide)

Cinnabar
(mercury sulfide)

Pyrite
(iron sulfide)

Sphalerite
(zinc sulfide)

FIGURE 3.31 ■ Some metallic sulfide ores. Sulfides are the most common types of metallic ores. [Chip Clark/Fundamental Photographs]

widespread dispersal produced a low-grade but very large resource of many millions of tons of ore, which can be mined economically by large-scale methods (**Figure 3.33**).

The lead-zinc deposits of the Upper Mississippi Valley, which extend from southwestern Wisconsin to Kansas and Oklahoma, are found in sedimentary rocks. The ores in this disseminated hydrothermal deposit are not associated with a known magmatic intrusion that could have been a source of hydrothermal solutions, so their origin must be very different. Some geologists speculate that the ores were deposited by groundwater that was driven out of the ancestral Appalachian Mountains when they were much higher. A continent-continent collision between North America and Africa may have created a continental-scale squeegee that pushed fluids from deep within the collision zone all the way into the continental interior of North America. Groundwater may have penetrated hot crustal rocks at great depths and dissolved soluble ore minerals, then moved upward into the overlying sedimentary rocks, where it precipitated the minerals as fillings in cavities. In some cases, it appears that these solutions infiltrated limestone formations and dissolved some carbonates, then replaced the carbonates with equal volumes of sulfide crystals. The major minerals of these deposits are lead sulfide (galena) and zinc sulfide (sphalerite).

Igneous Deposits

The most important deposits of ore minerals in igneous rocks are found as segregations of ore minerals near the bottoms of magmatic intrusions (see Chapter 5, Practicing Geology). These deposits form when minerals with relatively high melting temperatures crystallize from a body of cooling magma, settle, and accumulate at the base of the magma. Most of the chromium and platinum ores of the world, such as the deposits in South Africa and Montana, are found as layered accumulations of minerals that formed in this way (**Figure 3.34**). One of the richest ore deposits ever found, at Sudbury, Ontario, is a large igneous intrusion containing great quantities of layered nickel, copper, and iron sulfides near its base. Geologists believe that these sulfide deposits formed from the crystallization of a dense, sulfide-rich liquid that separated from the rest of a cooling magmatic intrusion and sank to the bottom before it congealed.

Chalcopyrite
(a copper sulfide)

Malachite
(a copper carbonate)

Chalcocite
(a copper sulfide)

FIGURE 3.32 ■ Copper ores. Chalcopyrite and chalcocite are copper sulfide ores. Malachite is a carbonate of copper found in association with sulfides of copper. [Chip Clark/Fundamental Photographs.]

FIGURE 3.33 ▪ Kennecott Copper Mine, Utah, an open-pit mine. Open-pit mining is typical of the large-scale methods used to exploit disseminated ore deposits. [David R. Frazier/The Image Works.]

As the magma in a large granite-forming intrusion cools, the last material to crystallize forms *pegmatites*, extremely coarse-grained rocks in which minerals present in only trace amounts in the magma are concentrated. Pegmatites may contain rare ore minerals rich in elements such as beryllium, boron, fluorine, lithium, niobium, and uranium, as well as gem minerals such as tourmaline.

Sedimentary Deposits

Sedimentary deposits include some of the world's most valuable mineral sources. Many economically important minerals, such as copper, iron, and other metals, segregate as an ordinary result of sedimentary processes. These deposits are chemically precipitated in sedimentary environments

FIGURE 3.34 ▪ Chromite (chromium ore, visible as dark layers) in a layered igneous intrusion in the Bushveld Complex, South Africa. [Spencer Titley.]

(a)

(b)

FIGURE 3.35 ■ (a) Panning for gold was popularized by "forty-niners" during the California gold rush and is still popular in the San Gabriel River today. (b) Gold is denser than the other materials from the streambed, so it sinks to the bottom of the pan. [(a) Bo Zaunders/CORBIS; (b) David Butow/CORBIS SABA.]

to which large quantities of metals are transported in solution. Some of the important sedimentary copper ores, such as those of the Permian Kupferschiefer (German for "copper slate") beds of Germany, may have precipitated from hydrothermal solutions rich in metal sulfides that interacted with sediments on the seafloor. The plate tectonic setting of these deposits may have been something like the mid-ocean ridge described in our example of the rock cycle, except that it developed within a continent. Here, rifting of the continental crust led to development of a deep trough, where sediments and ore minerals were deposited in a very still, narrow sea.

Many rich deposits of gold, diamonds, and other heavy minerals such as magnetite and chromite are found in *placers*, sedimentary ore deposits that have been concentrated by the mechanical sorting action of river currents. These ore deposits originate where uplifted rocks weather to form grains of sediment, which are then sorted by weight when currents of water flow over them. Because heavy minerals settle out of a current more quickly than lighter minerals such as quartz and feldspar, they tend to accumulate on streambeds and sandbars. Similarly, ocean waves preferentially deposit heavy minerals on beaches or shallow offshore bars. A gold panner accomplishes the same thing: the shaking of a water-filled pan allows the lighter minerals to be washed away, leaving the heavier gold in the bottom of the pan (**Figure 3.35**).

Some placers can be traced upstream to the location of the original mineral deposit, usually of igneous origin, from which the minerals were eroded. Erosion of the Mother Lode, an extensive gold-bearing vein system lying along the western flanks of the Sierra Nevada, produced the placers that were discovered in 1848 and led to the California gold rush. The placers were found before their source was discovered. Placers also led to the discovery of the Kimberley diamond mines of South Africa two decades later.

SUMMARY

What is a mineral? Minerals, the building blocks of rocks, are naturally occurring, usually inorganic solids with specific crystal structures and chemical compositions. A mineral is constructed of atoms, the small units of matter that combine in chemical reactions. An atom is composed of a nucleus made up of protons and neutrons and surrounded by electrons. The atomic number of an element is the number of protons in its nucleus, and its atomic mass is the sum of the masses of its protons and neutrons.

How do atoms combine to form the crystal structures of minerals? Chemical elements react with one another to form compounds either by gaining or losing electrons to become ions or by sharing electrons. Ionic bonds, which form by electrostatic attraction between positive ions (cations) and negative ions (anions), are the dominant type of chemical bond in mineral structures. Atoms that form compounds by sharing electrons are held together by covalent bonds. When a mineral crystallizes, atoms or ions come together in the proper proportions to form a crystal structure—an orderly three-dimensional array in which the basic arrangement of the atoms is repeated in all directions.

What are the major classes of rock-forming minerals? Silicate minerals, the most abundant minerals in Earth's crust, are built of silicate ions that are linked in various ways. Silicate tetrahedra may be isolated (linked together only by cations) or bonded together in structures such as single chains, double chains, sheets, or frameworks. Carbonate minerals are made up of carbonate ions bonded to calcium, magnesium, or both. Oxide minerals are compounds of oxygen and metallic elements. Sulfide and sulfate minerals are composed of sulfide and sulfate ions, respectively, in combination with metallic elements.

What are the physical properties of minerals? Geologists use the physical properties of minerals to identify them. These physical properties include hardness—the ease with which a mineral's surface is scratched; cleavage—its tendency to split along planar surfaces; fracture—the way it breaks along irregular surfaces; luster—the way it reflects light; color—imparted by transmitted or reflected light to crystals or irregular masses or visible as streak (the color of a fine powder); density—mass per unit volume; and crystal habit—the shape in which individual crystals or aggregates of crystals grow.

What determines the properties of rocks? Mineralogy (the kinds and proportions of minerals that make up a rock) and texture (the sizes, shapes, and spatial arrangement of its crystals or grains) define a rock. The mineralogy and texture of a rock are determined by the geologic processes by which it formed.

What are the three families of rocks and how do they form? Igneous rocks form by the crystallization of magmas as they cool. Intrusive igneous rocks cool slowly in Earth's interior and have large crystals. Extrusive igneous rocks, which cool rapidly at Earth's surface, have a glassy or fine-grained texture. Sedimentary rocks form by the lithification of sediments after burial. Sediments are derived from the weathering and erosion of rocks at Earth's surface. Metamorphic rocks form when igneous, sedimentary, or other metamorphic rocks are subjected to high temperatures and pressures in Earth's interior that change their mineralogy, texture, or chemical composition.

How does the rock cycle explain the transformation of rocks from one type into another? The rock cycle relates geologic processes driven by the plate tectonic system and the climate system to the formation of the three families of rocks. We can view these processes by starting at any point in the cycle, such as the creation of new oceanic lithosphere at a spreading center as two continents drift apart. The ocean basin gets wider until at some point the process reverses itself. As the basin closes and igneous rocks and sediments are subducted beneath a continent, they begin to melt to form a new generation of igneous rocks. The heat and pressure associated with subduction and with the intrusion of these igneous rocks transforms surrounding rocks into metamorphic rocks. Ultimately, the two continents collide, and these igneous and metamorphic rocks are uplifted into a high mountain chain. The uplifted rocks slowly weather, and their fragments are deposited as sediments.

How do deposits of economically valuable minerals form? Ores are deposits of minerals from which valuable metals can be recovered profitably. Hydrothermal deposits of ore minerals are formed when groundwater or seawater reacts with a magmatic intrusion to form a hydrothermal solution. The heated water transports soluble minerals to cooler rocks, where they are precipitated in fractures. The resulting ores may be found in veins or in disseminated deposits. Igneous ore deposits typically form when minerals crystallize from cooling magma, settle, and accumulate at the base of the magma body. They are often found as layered accumulations of minerals. Other ore minerals are chemically precipitated in sedimentary environments to which metals are transported in solution.

KEY TERMS AND CONCEPTS

anion (p. 61)

atomic mass (p. 59)

atomic number (p. 59)

bedding (p. 77)

biological sediment
   (p. 77)

carbonate (p. 65)

cation (p. 61)

chemical sediment (p. 77)

cleavage (p. 70)

color (p. 72)

contact metamorphism
   (p. 79)

covalent bond (p. 61)

crystal (p. 62)

crystal habit (p. 73)

crystallization (p. 61)

density (p. 64)

disseminated deposit
   (p. 82)

electron sharing (p. 60)

electron transfer (p. 60)

erosion (p. 77)

fracture (p. 71)

grain (p. 63)

hardness (p. 69)

hydrothermal solution
   (p. 82)

igneous rock (p. 74)

ion (p. 61)

ionic bond (p. 61)

isotope (p. 59)

lithification (p. 77)

luster (p. 71)

magma (p. 63)

metallic bond (p. 61)

metamorphic rock (p. 76)

mineral (p. 58)

mineralogy (p. 58)

Mohs scale of hardness
   (p. 69)

ore (p. 81)

oxides (p. 65)

polymorph (p. 63)

precipitate (p. 63)

regional metamorphism (p. 78)

rock (p. 74)

rock cycle (p. 79)

sediment (p. 77)

sedimentary rock (p. 74)

silicate (p. 65)

siliciclastic sediment (p. 77)

specific gravity (p. 73)

streak (p. 72)

sulfate (p. 65)

sulfide (p. 65)

texture (p. 74)

trace element (p. 72)

vein (p. 82)

weathering (p. 77)

PRACTICING GEOLOGY EXERCISE

Is It Worth Mining?

Geologists employed by the Rocks-r-Us Corporation have discovered basaltic volcanic rocks laced with gold. The corporate executives ponder their figures and measurements and study their three-dimensional model of the ore mineral deposit, but in the end they have just one question: Should we open a mine?

Exploration for ore minerals is an important and challenging activity that employs many geologists. Finding a promising deposit is only the first step toward extracting useful materials, however. The shape of the deposit, and the distribution and concentration of the ore, must be estimated before mining begins. This is done by drilling closely spaced holes and obtaining continuous cores through the ore deposit and the surrounding rock. Information from the cores is used to create a three-dimensional model of the ore deposit. That model is then used to evaluate whether or not the deposit is large enough and has a high enough concentration of minerals to justify opening a mine. Geologists contribute key information of direct economic significance to this very practical decision-making process.

The planning of mining operations is typically based on chemical and mineralogical analyses of the extracted cores, from which two quantities are calculated:

- *Grade* refers to the concentration of ore minerals within economically valueless parent rock (referred to as *waste rock*).

- *Mass* refers to the amount of ore that could potentially be extracted from the deposit.

Both quantities are important because neither grade nor mass alone is sufficient to identify an economically valuable

Drill rig

The ore-containing core in a tube

The core

An ore deposit is drilled to provide core samples for geochemical and mineralogical analysis. A rotating metal tube, studded with diamond teeth, cuts into the deposit. The hollow space in the tube becomes filled with solid rock, which is extracted when the tube is pulled out of the rock. The core has the shape of a cylinder. [Photos by Ben Whiting, P. Geo.]

Approximate volume of gold deposit

deposit. For example, the grade could be locally very high in veins, but the overall mass could be low because the veins are rare. In another case, the mass might be high, but the ore minerals might be so dispersed within the waste rock that the costs of processing it to extract the ore would become too high. Thus, the ideal ore deposit is one that has both high grade and high mass.

Grade is calculated by determining the percentage of ore minerals within a volume of rock. Laboratory analysis of core samples provides this measurement. Mass is calculated by assigning the grade value determined for individual cores to the unknown volume of rock between drill holes. Mass is the amount of ore that could be extracted if all of it could be extracted from the rock, but it is rare that all of it can be. In the parlance of the mining industry, mass is often calculated in tons and referred to as *tonnage* because of the enormous volumes of rock that are involved.

Drilling and analysis of core samples has shown that the gold in the Rocks-r-Us deposit has an average grade of 0.02 percent across all cores. The deposit has been determined to have a rectangular geometry extending laterally for 50 meters in one direction and 1500 meters in the other, with a thickness of 2 meters.

What is the volume of the ore deposit?

$$V_{deposit} = \text{length} \times \text{width} \times \text{thickness}$$
$$= 50 \text{ m} \times 1500 \text{ m} \times 2 \text{ m}$$
$$= 150,000 \text{ m}^3$$

What is the volume of gold in the ore deposit?

$$V_{gold} = V_{deposit} \times \text{grade}$$
$$= 150,000 \text{ m}^3 \times 0.02\%$$
$$= 30 \text{ m}^3$$

Given that gold has a density of 19 g/cm^3 (about 6800 ounces/m^3), what is the mass of the gold in ounces?

$$\text{mass} = V_{gold} \times \text{density}$$
$$= 30 \text{ m}^3 \times 6800 \text{ ounces/m}^3$$
$$= 204,000 \text{ ounces}$$

The price of gold is typically reported in dollars per ounce. At the time of this writing, the price of gold is roughly $800/ounce. What is the potential value of this ore deposit?

$$\text{value} = \text{mass} \times \text{price}$$
$$= 204,000 \text{ ounces} \times \$800/\text{ounce}$$
$$= \$163,200,000$$

BONUS PROBLEM: You bring this information to a meeting with the Rocks-r-Us corporate executives. They calculate that, over its lifetime, the mine will cost about $120,000,000 to operate, including restoration of the land after mining is completed. Is the value of the gold worth it? What simple calculation will give you the answer?

EXERCISES

1. Define a mineral.

2. What is the difference between an atom and an ion?

3. Draw the atomic structure of sodium chloride.

4. Name two types of chemical bonds.

5. List the basic crystal structures of silicate minerals.

6. Name three classes of minerals other than silicates.

7. How would a field geologist measure hardness?

8. What is the difference between the carbonate minerals calcite and dolomite?

9. What are the differences between extrusive and intrusive igneous rocks?

10. What are the differences between regional and contact metamorphism?

11. What are the differences between siliciclastic and chemical or biological sediments?

12. Name a common silicate mineral found in each of the three families of rocks: igneous, sedimentary, and metamorphic.

13. Of the three families of rocks, which form at Earth's surface and which in Earth's interior?

14. What are the characteristics of an economically valuable ore deposit?

THOUGHT QUESTIONS

1. Describe the creation of an ore mineral deposit by hydrothermal activity.

2. Draw a simple diagram to show how silicon and oxygen in silicate minerals share electrons.

3. Diopside, a pyroxene, has the formula $(Ca,Mg)_2Si_2O_6$. What does this formula tell you about its crystal structure and cation substitution?

4. In some bodies of granite, we can find very large crystals, some as much as a meter across, yet these crystals tend to have few crystal faces. What can you deduce about the conditions under which these large crystals grew?

5. What physical properties of sheet silicates are related to their crystal structure and bond strength?

6. Choose two minerals from Appendix 4 that you think might make good abrasive or grinding stones for sharpening steel, and describe the physical property that causes you to believe they would be suitable for that purpose.

7. Aragonite, with a density of 2.9 g/cm3, has exactly the same chemical composition as calcite, which has a density of 2.7 g/cm^3. Other things being equal, which of these two minerals is more likely to have formed under high pressure?

8. There are at least seven physical properties one can use to identify an unknown mineral. Which ones are most useful in discriminating between minerals that look similar? Describe a strategy that would allow you to prove that an unknown clear calcite crystal is not the same mineral as a known clear crystal of quartz.

9. Coal, a natural organic substance that forms from decaying vegetation, is not considered to be a mineral. However, when coal is heated to high temperatures and buried under high pressures, it is transformed into the mineral graphite. Why is it, then, that coal is not considered a mineral, but graphite is? Explain your reasoning.

10. What geologic processes transform a sedimentary rock into an igneous rock?

11. Which igneous intrusion would you expect to have a wider contact metamorphic zone: one of a very hot magma or one consisting of a cooler magma?

12. Describe the geologic processes by which an igneous rock is transformed into a metamorphic rock and then exposed to erosion.

13. Using the rock cycle, trace the path from a magma to a granitic intrusion to a metamorphic gneiss to a sandstone. Be sure to include the roles of the plate tectonic and climate systems and the specific processes that create the rocks.

14. Where are igneous rocks most likely to be found? How could you be certain that the rocks were igneous and not sedimentary or metamorphic?

15. Back in the late 1800s, gold miners used to pan for gold by placing sediment from rivers in a pan and filtering water through the pan while swirling the pan's contents. The miners wanted to be certain that they had found real gold and not pyrite ("fool's gold"). Why did this method work? What mineral property does the process of panning for gold use? What is another possible method for distinguishing between gold and pyrite?

MEDIA SUPPORT

 3-1 Animation: Ionic Bonds

 3-3 Animation: Sedimentary Rocks

 3-2 Animation: Igneous Rocks

 3-4 Animation: Metamorphic Rocks

- How Do Igneous Rocks Differ from One Another?   92

- How Do Magmas Form?   97

- Magmatic Differentiation   99

- Forms of Igneous Intrusions   102

- Igneous Processes and Plate Tectonics   105

Granite, such as that shown in this image of Mt. Whitney, the tallest peak in the continental United States, makes up nearly all of the Sierra Nevada mountain range. [Courtesy Jennifer Griffes]

IGNEOUS ROCKS: SOLIDS FROM MELTS

4

MORE THAN 2000 YEARS AGO, the Greek scientist and geographer Strabo traveled to Sicily to view the eruptions of Mount Etna. He observed that the hot liquid lava spilling down from the volcano onto Earth's surface cooled and hardened into solid rock within a few hours. By the eighteenth century, geologists began to understand that some sheets of rock that cut across other rock formations had also been formed by the cooling and solidification of molten rock. In these cases, the magma had cooled much more slowly because it had remained buried in Earth's crust.

Today we know that deep in Earth's crust and mantle, rock melts and rises toward Earth's surface. Some magmas solidify before they reach the surface, and some break through and solidify on the surface. Both processes produce igneous rocks.

Understanding the processes that melt and resolidify rock is a key to understanding how Earth's crust forms. Although we still have much to learn about the exact *mechanisms* of melting and solidification, we do have good answers to some fundamental questions: How do types of igneous rock differ from one another? Where and how do magmas form? How do rocks solidify from those magmas?

In answering these questions, we will focus on the central role of igneous processes in the Earth system. Observations of igneous rocks by geologists from Strabo to today make sense only in light of plate tectonic theory. Specifically, igneous rocks form at spreading centers where plates move apart, along convergent boundaries where one plate descends beneath another, and at "hot spots" where hot mantle material ascends to the crust.

In this chapter, we will examine the wide range of igneous rock types, both intrusive and extrusive, and the processes by which they form. We will explore the forces that cause rock to melt and form magmas and the ways in which those magmas reach the locations at and below Earth's surface where they solidify. We will then take a more detailed look at the igneous processes associated with specific plate tectonic settings.

■ How Do Igneous Rocks Differ from One Another?

Geologists today classify igneous rock samples in the same way some geologists did in the late nineteenth century: by their texture and by their mineral and chemical composition.

Texture

Two hundred years ago, the first division of igneous rocks was made on the basis of texture, which largely reflects differences in mineral grain size: geologists classified rocks as either coarse-grained or fine-grained (see Chapter 3). Grain size is a simple characteristic that geologists can easily see in the field. A coarse-grained rock, such as granite, has distinct crystals that are easily visible to the naked eye. In contrast, the crystals of a fine-grained rock, such as basalt, are too small to be seen, even with a magnifying glass. Figure 4.1 shows samples of granite and basalt, accompanied by photomicrographs of very thin, transparent slices of each rock. *Photomicrographs*, which are simply photographs taken through a microscope, give us an enlarged view of minerals and their textures. Textural differences were clear to early geologists, but several more clues were needed to unravel the meaning of those differences.

FIRST CLUE: VOLCANIC ROCKS Early geologists observed volcanic rocks forming from lava during volcanic eruptions. (**Lava** is the term that we apply to magma flowing out onto Earth's surface.) They noted that where lava cooled rapidly, it formed either a fine-grained rock or a glassy one in which no crystals could be distinguished. Where lava cooled more slowly, as in the middle of a thick flow many meters high, somewhat larger crystals were formed.

SECOND CLUE: LABORATORY STUDIES OF CRYSTALLIZATION Just over a hundred years ago, experimental scientists began to understand the nature of crystallization. Anyone who has frozen a tray of ice cubes knows that water solidifies to ice in a few hours as its temperature drops below the freezing point. If you have ever attempted to retrieve your ice cubes before they were completely solid, you may have seen thin ice crystals forming at the surface and along the sides of the tray. During crystallization, the water molecules take up fixed positions in the solidifying crystal structure, and they are no longer able to move freely, as they did when the water was liquid. All other liquids, including magmas, crystallize in this way.

The first tiny crystals form a pattern. Other atoms or ions in the crystallizing liquid then attach themselves in such a way that the tiny crystals grow larger. It takes some time for the atoms or ions to "find" their correct places on a growing crystal, so crystals grow large only if they have time to grow slowly. If a liquid solidifies very quickly, as a magma does when it erupts onto the cool surface of Earth, the crystals have no time to grow. Instead, a large number of tiny crystals form simultaneously as the liquid cools and solidifies.

THIRD CLUE: GRANITE AS EVIDENCE OF SLOW COOLING By studying volcanoes, early geologists determined that fine-grained textures indicate quick cooling at Earth's surface and that fine-grained igneous rocks are evidence of former volcanism. But in the absence of direct observation, how could geologists deduce that coarse-grained rocks form by *slow* cooling deep in Earth's interior? Granite—

Granite

Basalt

FIGURE 4.1 ■ Igneous rocks were first classified by texture. Early geologists assessed rock texture with a small hand-held magnifying glass. Modern geologists have access to high-powered polarizing microscopes, which can produce photomicrographs of thin, transparent rock slices like those shown here. [Photos by John Grotzinger/Ramón Rivera-Moret/Harvard Mineralogical Museum; photomicrographs by Steven Chemtob.]

Seen with a magnifying glass
⊢ 1 cm ⊣

Seen through a polarizing microscope
⊢ 1 mm ⊣

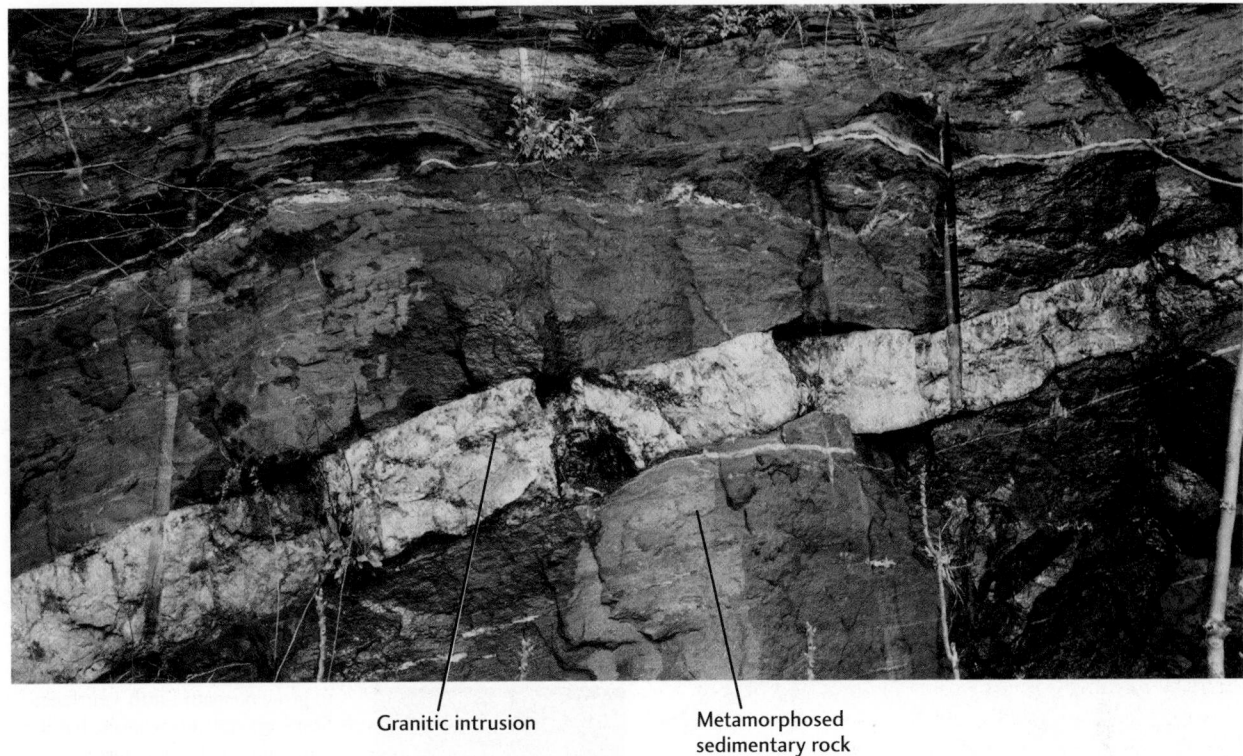

Granitic intrusion   Metamorphosed
                           sedimentary rock

FIGURE 4.2 ▪ Granite pegmatite sill or dike (the lighter-colored rock) in an outcrop of schist (darker colored rock) along the Harlem River, New York, suggests to geologists that the intruding rock had been forced into the fractures as a liquid. [Catherine Ursillo/Science Source.]

one of the commonest rocks of the continents—turned out to be the crucial clue (**Figure 4.2**). James Hutton, one of geology's founding fathers, saw granite cutting across and disrupting layers of sedimentary rock as he worked in the field in Scotland. He noticed that the granite had somehow fractured and invaded the sedimentary rock, as though the granite had been forced into the fractures as a liquid.

As Hutton looked at more and more granites, he began to focus on the sedimentary rocks bordering them. He observed that the minerals of the sedimentary rocks in contact with the granite were different from those found in sedimentary rocks at some distance from the granite. He concluded that the changes in the sedimentary rocks must have resulted from great heat, and that the heat must have come from the granite. Hutton also noted that the granite was composed of interlocked crystals (see Figure 4.1). By this time, chemists had established that a slow crystallization process produces this pattern.

With these three lines of evidence, Hutton proposed that granite forms from hot molten material that solidifies deep within Earth. The evidence was conclusive because no other explanation could accommodate all the facts. Other geologists, who saw the same characteristics of granites in widely separated places throughout the world, came to recognize that granite and many similar coarse-grained rocks were the products of magma that had crystallized slowly in Earth's interior.

INTRUSIVE AND EXTRUSIVE TEXTURES The full significance of an igneous rock's texture is now clear: it is linked to the rate, and therefore the place, of cooling. An **intrusive igneous rock** is one that has forced its way into the surrounding rock, called **country rock,** and solidified without reaching Earth's surface. Slow cooling of magma in Earth's interior allows adequate time for the growth of the large, interlocking crystals that characterize intrusive igneous rocks (**Figure 4.3**).

Rapid cooling at Earth's surface produces the fine-grained texture or glassy appearance of **extrusive igneous rocks** (see Figure 4.3). These rocks, formed partly or largely of volcanic glass, are formed from material that erupts from volcanoes. For this reason, they are also known as *volcanic rocks.* They fall into two major categories based on the type of erupted material from which they are formed:

▪ *Lavas:* Volcanic rocks formed from flowing lavas range in appearance from smooth and ropy to sharp, spiky, and jagged, depending on the conditions under which they are formed.

▪ *Pyroclasts:* In more violent eruptions, **pyroclasts** form when fragments of lava are thrown high into the air. **Volcanic ash** is made up of extremely small fragments, usually of glass, that form when escaping gases force a fine spray of magma from a volcano. **Bombs** are larger particles hurled from the volcano

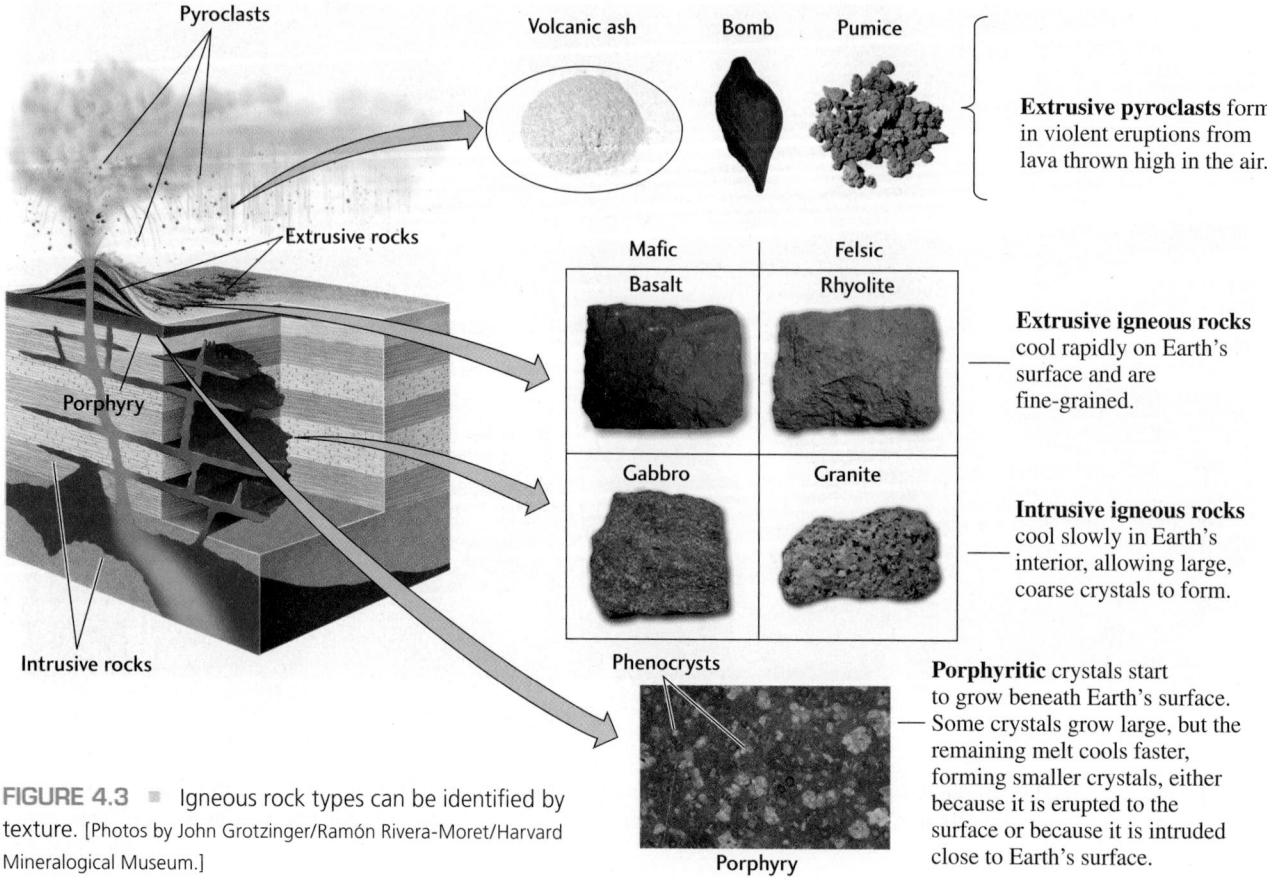

FIGURE 4.3 ■ Igneous rock types can be identified by texture. [Photos by John Grotzinger/Ramón Rivera-Moret/Harvard Mineralogical Museum.]

Extrusive pyroclasts form in violent eruptions from lava thrown high in the air.

Extrusive igneous rocks cool rapidly on Earth's surface and are fine-grained.

Intrusive igneous rocks cool slowly in Earth's interior, allowing large, coarse crystals to form.

Porphyritic crystals start to grow beneath Earth's surface. Some crystals grow large, but the remaining melt cools faster, forming smaller crystals, either because it is erupted to the surface or because it is intruded close to Earth's surface.

and streamlined by the air as they hurtle through it. As they fall to the ground and cool, these fragments of volcanic debris may stick together to form rocks.

One volcanic rock type is **pumice,** a frothy mass of volcanic glass in which a great number of spaces remain after trapped gas has escaped from the solidifying melt. Another wholly glassy volcanic rock type is **obsidian;** unlike pumice, it contains only tiny vesicles and so is solid and dense. Chipped or fragmented obsidian produces very sharp edges, and Native Americans and many other hunting groups used it for arrowheads and a variety of cutting tools.

A **porphyry** is an igneous rock that has a mixed texture in which large crystals "float" in a predominantly fine-grained matrix (see Figure 4.3). The large crystals, called *phenocrysts,* form in magma while it is still below Earth's surface. Then, before other crystals can grow, a volcanic eruption brings the magma to the surface, where it cools quickly to a finely crystalline mass. In some cases, porphyries form as intrusive igneous rocks; for example, they may form where magmas cool quickly at very shallow levels in the crust. Porphyry textures are important to geologists because they show that different minerals crystallize at different rates, a point that will be emphasized later in this chapter.

In Chapter 12, we will look more closely at how volcanic processes form extrusive igneous rocks. Now, however, we turn to the second way in which the family of igneous rocks is subdivided.

Chemical and Mineral Composition

We have just seen how igneous rocks can be subdivided according to their texture. They can also be classified on the basis of their chemical and mineral composition. Volcanic glass, which is formless even under a microscope, is often classified by chemical analysis alone. One of the earliest classifications of igneous rocks was based on a simple chemical analysis of their silica content. Silica (SiO_2) is abundant in most igneous rocks, accounting for 40 to 70 percent of their total weight.

Modern classifications group igneous rocks according to their relative proportions of silicate minerals (**Table 4.1;** see also Appendix 4).

The silicate minerals—quartz, feldspars, muscovite and biotite micas, amphiboles and pyroxenes, and olivine—form a systematic series. *Felsic* minerals are the highest in silica; *mafic* minerals are the lowest in silica. The adjectives *felsic* (from *fel*dspar and *silica*) and *mafic* (from *ma*gnesium and *ferric,* from the Latin *ferrum,* "iron") are applied both to minerals and to rocks containing large proportions of those minerals. Mafic minerals crystallize at higher

TABLE 4-1   Common Minerals of Igneous Rocks

Compositional Group	Mineral	Chemical Composition	Silicate Structure
FELSIC	Quartz	SiO_2	Frameworks
	Orthoclase feldspar	$KAlSi_3O_8$	
	Plagioclase feldspar	$NaAlSi_3O_8$; $CaAl_2Si_2O_8$	
	Muscovite (mica)	$KAl_3Si_3O_{10}(OH)_2$	Sheets
MAFIC	Biotite (mica)	K Mg Fe Al $Si_3O_{10}(OH)_2$	
	Amphibole group	Mg Fe Ca Na $Si_8O_{22}(OH)_2$	Double chains
	Pyroxene group	Mg Fe Ca Al SiO_3	Single chains
	Olivine	$(Mg,Fe)_2SiO_4$	Isolated tetrahedral

temperatures—that is, earlier in the cooling of a magma—than felsic minerals.

As the mineral and chemical compositions of igneous rocks became known, geologists soon noticed that some extrusive and intrusive rocks were identical in composition and differed only in texture. Basalt, for example, is an extrusive rock formed from lava. Gabbro has exactly the same mineral and chemical composition as basalt, but forms deep in Earth's crust (see Figure 4.3). Similarly, rhyolite and granite are identical in composition, but differ in texture. Thus, extrusive and intrusive rocks form two chemically and mineralogically parallel sets of igneous rocks. Conversely, most of the chemical and mineral compositions in the felsic-to-mafic series we have just described can appear in either extrusive or intrusive rocks. The only exceptions are very highly mafic rocks, which rarely appear as extrusive igneous rocks.

Figure 4.4 is a model that portrays these relationships. The horizontal axis plots silica content as a percentage of a given rock's weight. The percentages given—from high silica content at 70 percent to low silica content at 40 percent—cover the range found in igneous rocks. The vertical axis plots mineral content as a percentage of a given

rock's volume. This model can be used to classify an unknown rock sample with a known silica content: by finding its silica content on the horizontal axis, you can determine its mineral composition and, from that, the type of rock it is.

We can use Figure 4.4 to guide our discussion of intrusive and extrusive igneous rocks. We begin with the felsic rocks at the far left of the model.

FELSIC ROCKS **Felsic rocks** are poor in iron and magnesium and rich in felsic minerals that are high in silica. Such minerals include quartz, orthoclase feldspar, and plagioclase feldspar. Orthoclase feldspars, which contain potassium, are more abundant than plagioclase feldspars. Plagioclase feldspars contain varying amounts of calcium and sodium; as Figure 4.4 indicates, they are richer in sodium near the felsic end and richer in calcium near the mafic end of the scale. Thus, just as mafic minerals crystallize at higher temperatures than felsic minerals, calcium-rich plagioclases crystallize at higher temperatures than sodium-rich plagioclases.

Felsic rocks tend to be light in color. **Granite,** one of the most abundant intrusive igneous rocks, contains about 70 percent silica. Its mineral composition includes abundant

FIGURE 4.4 ■ Classification model for igneous rocks. The vertical axis shows the minerals contained in a given rock as a percentage of its volume. The horizontal axis shows the silica content of a given rock as a percentage of its weight. Thus, if you knew by chemical analysis that a coarsely textured rock sample was about 70 percent silica, you could deduce that its composition was about 6 percent amphibole, 3 percent biotite, 5 percent muscovite, 14 percent plagioclase feldspar, 22 percent quartz, and 50 percent orthoclase feldspar. Your rock would be granite. Although rhyolite has the same mineral composition, its fine texture would eliminate it from consideration.

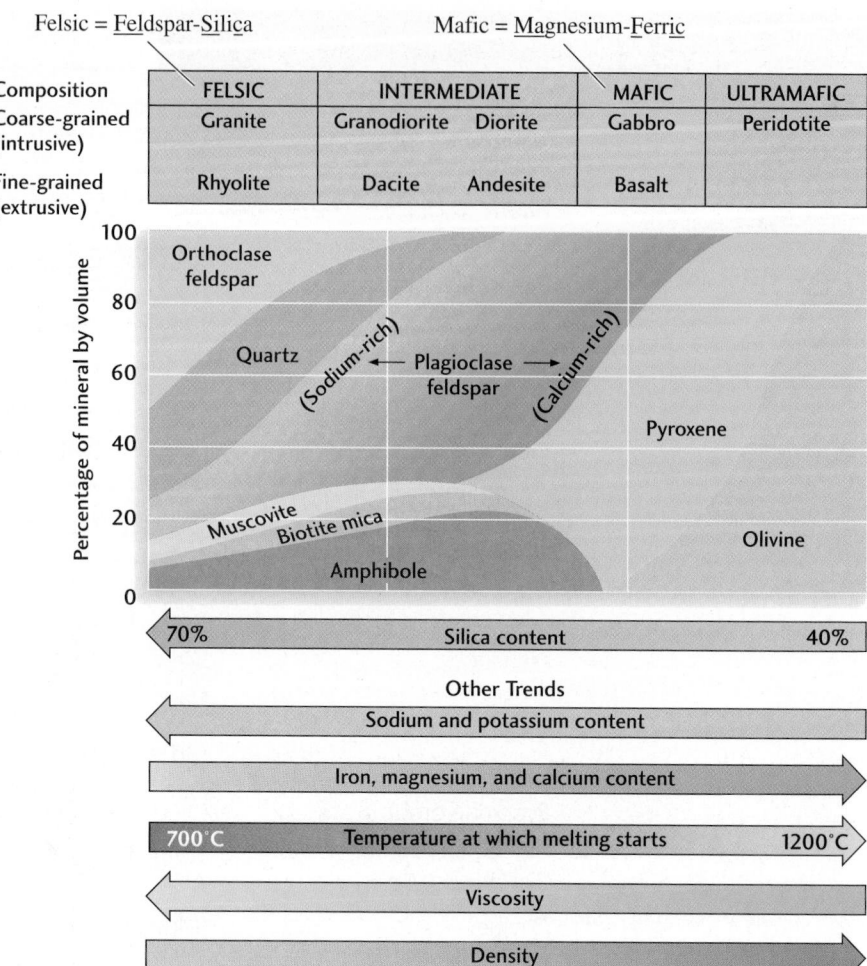

quartz and orthoclase feldspar and a smaller amount of plagioclase feldspar (see the far left of Figure 4.4). These light-colored felsic minerals give granite its pink or gray color. Granite also contains small amounts of muscovite and biotite micas and amphibole. **Rhyolite** is the extrusive equivalent of granite. This light brown to gray rock has the same felsic composition and light coloration as granite, but it is much more fine-grained. Many rhyolites are formed largely or entirely of volcanic glass.

INTERMEDIATE IGNEOUS ROCKS Midway between the felsic and mafic ends of the scale are the **intermediate igneous rocks.** As their name indicates, these rocks are neither as rich in silica as the felsic rocks nor as poor in it as the mafic rocks. We find the intermediate intrusive igneous rocks to the right of granite in Figure 4.4. The first is **granodiorite,** a light-colored rock that looks something like granite. It is also similar to granite in having abundant quartz, but its predominant feldspar is plagioclase, not orthoclase. To its right is **diorite,** which contains still less silica and is dominated by plagioclase feldspar, with little or no quartz. Diorites contain a moderate amount of the mafic

minerals biotite, amphibole, and pyroxene. They tend to be darker than granite or granodiorite.

The volcanic equivalent of granodiorite is **dacite.** To its right in the extrusive series is **andesite,** the volcanic equivalent of diorite. Andesite derives its name from the Andes, the volcanic mountain belt in South America.

MAFIC ROCKS **Mafic rocks** contain large proportions of pyroxenes and olivines. These minerals are relatively poor in silica but are rich in magnesium and iron, from which they get their characteristic dark colors. **Gabbro** is a coarse-grained, dark gray intrusive igneous rock. Gabbro has an abundance of mafic minerals, especially pyroxenes. It contains no quartz and only moderate amounts of calcium-rich plagioclase feldspar.

Basalt is the most abundant igneous rock of the crust, and it underlies virtually the entire seafloor. This dark gray to black rock is the fine-grained extrusive equivalent of gabbro. In some places, extensive thick sheets of basalt, called *flood basalts,* form large plateaus. The Columbia River basalts of Washington State and the remarkable formation known as the Giant's Causeway in Northern Ireland are two examples. The Deccan flood basalts of India and the Siberian flood basalts of northern Russia were formed by enormous outpourings of basalt that appear to coincide closely with two of the greatest periods of mass extinction in the fossil record.

ULTRAMAFIC ROCKS Ultramafic rocks consist primarily of mafic minerals and contain less than 10 percent feldspar. Here, at the far right of Figure 4.4, with a silica content of only about 45 percent, we find **peridotite,** a coarse-grained, dark greenish gray rock made up primarily of olivine with smaller amounts of pyroxene. Peridotites are the dominant rocks in Earth's mantle, and as we will see, they are the source of the basaltic magmas that form rocks at mid-ocean ridges. Ultramafic rocks are rarely found as extrusives. Because they solidify at such high temperatures, they are rarely liquid and hence do not form typical lavas.

TRENDS IN THE FELSIC-TO-MAFIC SERIES The names and exact compositions of the various rocks in the felsic-to-mafic series are less important to remember than the trends shown in Figure 4.4. There is a strong correlation between a rock's mineralogy and its temperature of crystallization or melting. As **Table 4.2** indicates, mafic minerals melt at higher temperatures than felsic minerals. At temperatures below their melting point, minerals crystallize; therefore, mafic minerals also crystallize at higher temperatures than felsic minerals. We can also see that silica content increases as we move from the mafic end to the felsic end of the series. Increasing silica content results in increasingly complex silicate structures (see Table 4.1), which interfere with a melted rock's ability to flow. Thus, **viscosity**—the measure of a liquid's *resistance* to flow—increases as silica content increases. Viscosity is an important factor in the behavior of lavas, as we will see in Chapter 12. Increasing silica content also results in decreasing density, as we saw in Chapter 1.

It is clear that an igneous rock's mineralogy provides a great deal of information about the conditions under which the rock's parent magma formed and crystallized. To interpret this information accurately, however, we must understand more about igneous processes. We turn to that topic next.

TABLE 4-2	Factors Affecting Melting Temperatures
Higher Melting Temperatures	Lower Melting Temperatures
◄——— Pressure increasing ———	
——— Water content increasing ———►	
Rock Composition	
◄——— More mafic	More felsic ———►

▪ How Do Magmas Form?

We know from the way Earth transmits seismic waves that the bulk of the planet is solid for thousands of kilometers down to the core-mantle boundary (see Chapter 1). The evidence of volcanic eruptions, however, tells us that there must also be liquid regions where magmas originate. How do we resolve this apparent contradiction? The answer lies in the processes that melt rocks and create magmas.

How Do Rocks Melt?

Although we do not yet understand the exact mechanisms of rock melting and solidification within Earth, we have learned a great deal from laboratory experiments using high-temperature furnaces (**Figure 4.5**). From these

FIGURE 4.5 ▪ Experimental device used to melt rocks in laboratory. [Sally Newman.]

experiments, we know that a rock's melting point depends on its chemical and mineral composition and on conditions of temperature and pressure (see Table 4.2).

TEMPERATURE AND MELTING A hundred years ago, geologists discovered that rock does not melt completely at a given temperature. Instead, rocks undergo **partial melting** because the minerals that compose them melt at different temperatures. As temperatures rise, some minerals melt and others remain solid. If the same conditions are maintained at any given temperature, the same mixture of solid and melted rock is maintained. The fraction of rock that has melted at a given temperature is called a *partial melt.* To visualize a partial melt, think of how a chocolate chip cookie would look if you heated it to the point at which the chocolate chips melted while the main part of the cookie stayed solid. The chips represent the partial melt, or magma.

The ratio of solid to partial melt depends on the proportions and melting temperatures of the minerals that make up the original rock. It also depends on the temperature at the depth in the crust or mantle where melting takes place. At the lower end of a rock's melting range, a partial melt might be less than 1 percent of the volume of the original rock. Much of the hot rock would still be solid, but significant amounts of liquid would be present as small droplets in the tiny spaces between crystals throughout the mass. In the upper mantle, for example, some basaltic magmas are produced by only 1 to 2 percent melting of peridotite. However, 15 to 20 percent melting of mantle peridotite to form basaltic magmas is common beneath mid-ocean ridges. At the high end of a rock's melting range, much of the rock would be liquid, containing lesser amounts of unmelted crystals. An example would be the reservoir of basaltic magma and crystals just beneath a volcano such as the island of Hawaii. Geologists have used this knowledge of partial melts to determine how different kinds of magmas form at different temperatures and in different regions of Earth's interior. As you can imagine, the composition of a magma formed from completely melted rock may be very different from that of a magma formed from rock in which only the minerals with the lowest melting points have melted. Thus, basaltic magmas that form in different regions of the mantle may have somewhat different compositions.

PRESSURE AND MELTING To get the whole story on melting, we must consider pressure as well as temperature. Pressure increases with depth within Earth as a result of the increasing weight of overlying rock. Geologists found that as they melted rocks under various pressures in the laboratory, higher pressures led to higher melting temperatures. Thus, rocks that would melt at a given temperature at Earth's surface would remain solid at the same temperature in Earth's interior. For example, a rock that melts at 1000°C at Earth's surface might have a much higher melting temperature, perhaps 1300°C, deep in the interior, where

pressures are many thousands of times greater than those at the surface. It is the effect of pressure that explains why the rocks in most of the crust and mantle do not melt. Rock can melt only when both temperature and pressure conditions are right.

Just as an increase in pressure can keep rock solid, a decrease in pressure can make rock melt, given a sufficiently high temperature. Because of convection currents in the mantle, mantle material rises to Earth's surface at mid-ocean ridges at a more or less constant temperature. As the material rises and the pressure on it decreases below a critical point, the solid rock melts spontaneously, without the introduction of any additional heat. This process, known as **decompression melting,** produces the greatest volume of magma anywhere on Earth. It is the process by which most basalts form on the seafloor.

WATER AND MELTING The many experiments on melting temperatures and partial melting of rocks paid other dividends as well. One of them was a better understanding of the role of water in melting. Geologists studying natural lavas in the field determined that water was present in some magmas. This finding gave them the idea of adding water to their experimental melts back in the laboratory. By adding small but varying amounts of water, they discovered that the compositions of partial melts varied not only with temperature and pressure, but also with the amount of water present.

Consider, for example, the effect of dissolved water on pure albite, a sodium-rich plagioclase feldspar, at the low pressures at Earth's surface. If only a small amount of water is present in the rock, the rock will remain solid at temperatures just over 1000°C, hundreds of degrees above the boiling point of water. At these temperatures, the water in the albite is present as a vapor (gas). If large amounts of water are dissolved in the albite, however, its melting temperature will decrease, dropping to as low as 800°C. This behavior follows the general rule that dissolving one substance (in this case, water vapor) in another (in this case, albite) lowers the melting temperature of the solution. If you live in a cold climate, you are probably familiar with this principle because you know that salt sprinkled on icy roads lowers the melting temperature of the ice. By the same principle, the melting temperature of albite—and of all silicate minerals—drops considerably in the presence of large amounts of water. The melting points of these minerals decrease in proportion to the amount of water dissolved in the molten silicate.

Melting of rock induced by the presence of water that lowers its melting point is referred to as **fluid-induced melting.** Water content is a significant factor in the melting of sedimentary rocks, which contain an especially large volume of water in their pore spaces, more than is found in igneous or metamorphic rocks. As we will see later in this chapter, the water in sedimentary rocks plays an important role in the melting that gives rise to much of the volcanic activity at subduction zones.

The Formation of Magma Chambers

Most substances are less dense in their liquid form than in their solid form. The density of melted rock is lower than the density of solid rock of the same composition. With this knowledge, geologists reasoned that large bodies of magma could form in the following way: If the less dense melted rock were given a chance to move, it would move upward—just as oil, which is less dense than water, rises to the surface of a mixture of oil and water. Being liquid, a partial melt could move slowly upward through pores and along the boundaries between crystals of the surrounding solid rock. As the hot drops of melted rock moved upward, they would mix with other drops, gradually forming larger pools of magma within Earth's solid interior.

The rise of magmas through the mantle and crust may be slow or rapid. Magmas rise at rates from 0.3 m/year to almost 50 m/year, over periods of tens of thousands or even hundreds of thousands of years. As they ascend, magmas may mix with other melts and may also melt portions of the crust. We now know that large pools of molten rock, called **magma chambers,** form in the lithosphere as rising magmas melt and push aside surrounding solid rock. We know that they exist because seismic waves have shown us the depth, size, and general outlines of the magma chambers underlying some active volcanoes. A magma chamber may encompass a volume as large as several cubic kilometers. We cannot yet say exactly how magma chambers form, nor exactly what they look like in three dimensions. We can think of them as large, liquid-filled cavities in solid rock, which expand as more of the surrounding rock melts or as magma migrates through cracks and other small openings. Magma chambers contract as they expel magma to the surface in volcanic eruptions.

Where Do Magmas Form?

Our understanding of igneous processes stems from geologic inferences as well as laboratory experimentation. One important source of information is volcanoes, which give us information about where magmas are located. Another is the record of temperatures measured in deep drill holes and mine shafts. This record shows that the temperature of Earth's interior increases with depth. Using these measurements, scientists have been able to estimate the rate at which temperature rises as depth increases.

The temperatures recorded at a given depth in some locations are much higher than the temperatures recorded at the same depth in other locations. These results indicate that some parts of Earth's mantle and crust are hotter than others. For example, the Great Basin of the western United States is an area where the North American continent is being stretched and thinned, with the result that the temperature increases with depth at an exceptionally rapid rate, reaching 1000°C at 40 km, not far below the base of the crust. This temperature is almost high enough to melt basalt. By contrast, in tectonically stable regions, such as the central parts of continents, the temperature increases much more slowly, reaching only 500°C at the same depth.

◾ Magmatic Differentiation

The processes we've discussed so far account for the melting of rocks to form magmas. But what accounts for the variety of igneous rocks? Are magmas of different chemical compositions made by the melting of different kinds of rock? Or do igneous processes produce a variety of rocks from an originally uniform parent material?

Again, the answers to these questions came from laboratory experiments. Geologists mixed chemical elements in proportions that simulated the compositions of natural igneous rocks, then melted those mixtures. As the melts cooled and solidified, the geologists observed and recorded the temperatures at which crystals formed, as well as the chemical compositions of those crystals. This research gave rise to the theory of **magmatic differentiation**, a process by which rocks of varying composition can arise from a uniform parent magma. Magmatic differentiation occurs because different minerals crystallize at different temperatures.

In a kind of mirror image of partial melting, the last minerals to melt are the first minerals to crystallize from a cooling magma. This initial crystallization withdraws chemical elements from the melt, changing the magma's composition. Continued cooling crystallizes the minerals that melted at the next lower temperature range. Again, the magma's chemical composition changes as various elements are withdrawn. Finally, as the magma solidifies completely, the last minerals to crystallize are the ones that melted first. Thus, the same parent magma, because of its changing chemical composition throughout the crystallization process, can give rise to different types of igneous rocks.

Fractional Crystallization: Laboratory and Field Observations

Fractional crystallization is the process by which the crystals formed in a cooling magma are segregated from the remaining liquid rock. This segregation happens in several ways, following a sequence commonly described as *Bowen's reaction series* (**Figure 4.6**). In the simplest scenario, crystals formed in a magma chamber settle to the chamber's floor and are thus removed from further reaction with the remaining liquid. Thus, crystals that form early are segregated from the remaining magma, which continues to crystallize as it cools.

The effects of fractional crystallization can be seen in the Palisades, a line of imposing cliffs that faces the city of New York on the west bank of the Hudson River (**Figure 4.7**). This igneous formation is about 80 km long and, in places, more than 300 m high. It formed as a magma of basaltic composition intruded into nearly horizontal layers of

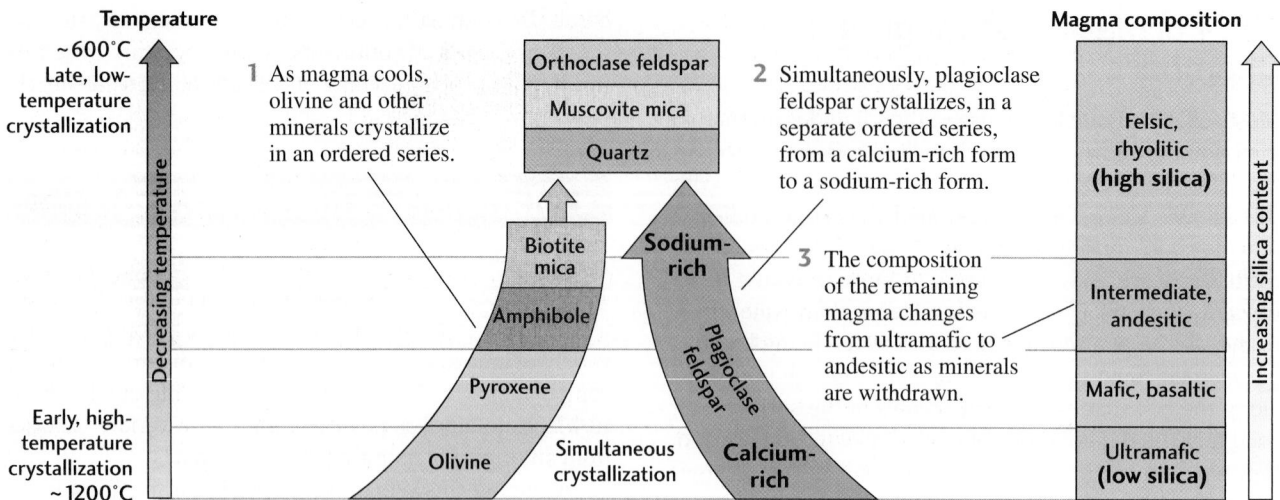

FIGURE 4.6 ▪ Bowen's reaction series provides a model of fractional crystallization.

sedimentary rock. It contains abundant olivine near the bottom, pyroxene and calcium-rich plagioclase feldspar in the middle, and mostly sodium-rich plagioclase feldspar near the top. This variation in mineral composition from bottom to top made the Palisades a perfect site for testing the theory of fractional crystallization.

Geologists melted rocks with about the same mineral compositions as those found in the Palisades intrusion and determined that the initial temperature of the magma from which it formed had to have been about 1200°C. The parts of the magma within a few meters of the relatively cold country rock above and below it cooled quickly. This quick cooling formed a fine-grained basalt and preserved the chemical composition of the original magma. The hot interior cooled more slowly, as evidenced by the slightly larger crystals found in the intrusion's interior.

The theory of fractional crystallization leads us to expect that the first mineral to crystallize from the slowly cooling interior of the Palisades intrusion would have been olivine, as this heavy mineral would sink through the melt to the bottom of the intrusion. It can be found today as a coarse-grained, olivine-rich layer just above the chilled, fine-grained basaltic layer along the bottom zone of contact with the underlying sedimentary rock. Plagioclase feldspar would have started to crystallize at about the same time; it has a lower density than olivine, however, and so

As magma cools, minerals crystallize at different temperatures and settle out of the magma in a particular order.

FIGURE 4.7 ▪ Fractional crystallization explains the composition of the basaltic intrusion that forms the Palisades. Minerals in the Palisades intrusion are ordered with olivine at the bottom, a gradient of pyroxene and calcium-rich plagioclase feldspar in the center, and sodium-rich plagioclase feldspar at the top. Layers of fine-grained basalt, which cooled quickly at the edges of the intrusion, surround the more slowly cooled interior. [© Breck Kent.]

would have settled to the bottom more slowly (see Practicing Geology). Continued cooling would have produced pyroxene crystals, which would have reached the bottom next, followed almost immediately by calcium-rich plagioclase feldspar. The abundance of plagioclase feldspar in the upper parts of the intrusion is evidence that the magma continued to change in composition until successive layers of settled crystals were topped off by a layer of mostly sodium-rich plagioclase feldspar. In addition to crystallizing at a lower temperature, sodium-rich plagioclase feldspar is less dense than either olivine or pyroxene, so it would have settled out last.

Being able to explain the layering of the Palisades intrusion as the result of fractional crystallization was an early success in understanding magmatic differentiation. It firmly tied field observations to laboratory results and was solidly based on chemical knowledge. We now know that this intrusion actually has a more complex history that includes several injections of magma and a more complicated process of olivine settling. Nevertheless, the Palisades intrusion remains a valid example of fractional crystallization.

Granite from Basalt: Complexities of Magmatic Differentiation

Studies of volcanic lavas have shown that basaltic magmas are common—far more common than the rhyolitic magmas that correspond in composition to granites. How, then, could granites have become so abundant in Earth's crust? The answer is that the process of magmatic differentiation is much more complex than geologists first thought.

The original theory of magmatic differentiation suggested that a basaltic magma would gradually cool and differentiate into a more felsic magma by fractional crystallization. The early stages of this differentiation would produce an andesitic magma, which might erupt to form andesitic lavas or solidify by slow crystallization to form dioritic intrusions. Intermediate stages would result in magmas of granodioritic composition. If the process were carried far enough, its late stages would form rhyolitic lavas and granitic intrusions. One line of research has shown, however, that so much time would be needed for small crystals of olivine to settle through a dense, viscous magma that they might never reach the bottom of a magma chamber. Other researchers have demonstrated that many layered intrusions—similar to but much larger than the Palisades intrusion—do not show the simple progression of layers predicted by the original theory.

The greatest sticking point in the original theory, however, was the source of granite. The great volume of granite found on Earth could not have formed from basaltic magmas by magmatic differentiation, because large quantities of liquid volume are lost by crystallization during successive stages of differentiation. To produce a given amount of granite, an initial volume of basaltic magma 10 times that of the granitic intrusion would be required. Based on that observation, there should be huge quantities of basalt underlying granitic intrusions. But geologists could not find anything like that amount of basalt. Even where great volumes of basalt are found—at mid-ocean ridges—there is no wholesale conversion into granite through magmatic differentiation.

Most in question is the original idea that all granitic rocks are formed from the differentiation of a single type of magma, a basaltic melt. Instead, geologists now believe that the melting of varied rock types in the upper mantle and crust is responsible for much of the observed variation in the composition of magmas:

1. Rocks in the upper mantle undergo partial melting to produce basaltic magmas.

2. Mixtures of sedimentary rock and basaltic oceanic crust, such as those found in subduction zones, melt to form andesitic magmas.

3. Mixtures of sedimentary, igneous, and metamorphic continental crustal rocks melt to produce granitic magmas.

Thus, the mechanisms of magmatic differentiation must be much more complex than first recognized in a number of ways:

■ Magmatic differentiation can begin with the partial melting of mantle and crustal rocks with a range of water contents over a range of temperatures.

■ Magmas do not cool uniformly; they may exist transiently at a range of temperatures within a magma chamber. Differences in temperature within and among magma chambers may cause the chemical composition of magma to vary from one region to another.

■ A few magma types are *immiscible*—they do not mix with one another, just as oil and water do not mix. When such magmas coexist in one magma chamber, each forms its own fractional crystallization series. Magmas that are *miscible*—that *do* mix—may follow a crystallization path different from that followed by any one magma type alone.

We now know more about the physical processes that interact with fractional crystallization within magma chambers (**Figure 4.8**). Magma at various temperatures in different parts of a magma chamber may flow turbulently, crystallizing as it circulates. Crystals may settle, then be caught up in currents again, and eventually be deposited on the chamber's walls. The margins of such a magma chamber may be a "mushy" zone of crystals and melt lying between the solid rock border of the chamber and the completely liquid magma within the heart of the chamber. And, at some mid-ocean ridges, such as the East Pacific Rise, a mushroom-shaped magma chamber may be surrounded by hot basaltic rock containing only small amounts (1 to 3 percent) of partial melt.

1 Partial melting of country rock creates a magma of a particular composition.

2 Cooling causes minerals to crystallize and settle.

3 A basaltic magma chamber breaks through, causing turbulent flow.

4 Mixing of two magmas results in andesitic magma.

Magma chamber A

Magma chamber B

Crystallizing minerals

Partial melting of country rock

Basaltic magma

5 Crystals formed in the mixed magma have a different composition, and may accumulate on the sides and roof of the magma chamber due to turbulence.

FIGURE 4.8 ■ Magmatic differentiation is a more complex process than first recognized. Some magmas derived from rocks of varying compositions may mix, whereas others are immiscible. Crystals may be transported to various parts of a magma chamber by turbulent flow in the liquid magma.

■ Forms of Igneous Intrusions

As noted earlier, we cannot directly observe the shapes of igneous intrusions. We can deduce their shapes and distributions only by observing parts of them where they have been uplifted and exposed by erosion, millions of years after the magma that formed them intruded and cooled.

We do have indirect evidence of current magmatic activity. Seismic waves, for example, have shown us the general outlines of the magma chambers that underlie some active volcanoes. In some nonvolcanic but tectonically active regions, such as an area near the Salton Sea in Southern California, measurements in deep drill holes reveal crustal temperatures much hotter than normal, which may be evidence of a magmatic intrusion nearby. But these methods cannot reveal the detailed shapes or sizes of intrusions.

Most of what we know about igneous intrusions is based on the work of field geologists who have examined and compared a wide variety of outcrops and have reconstructed their histories. In the following pages, we consider some of these forms. Figure 4.9 illustrates a variety of intrusive and extrusive structures.

Country rock   Volcano   Lava flow   Ash falls and pyroclastics   Eroded volcano with radiating dikes

Stock

Dike

Sill

Dike

Sill

Dike

Pluton

Batholith

FIGURE 4.9 ■ Basic forms of extrusive and intrusive igneous structures.

Dikes cut across layers of country rock…

…but sills run parallel to them.

Batholiths are the largest forms of plutons, covering at least 100 km².

Plutons

Plutons are large igneous bodies formed deep in Earth's crust. They range in size from a cubic kilometer to hundreds of cubic kilometers. We can study these large bodies when uplift and erosion uncover them or when mines or drill holes cut into them. Plutons are highly variable, not only in size but also in shape and in their relationship to the country rock.

This wide variation is due in part to the different ways in which magma makes space for itself as it rises through the crust. Most plutons intrude at depths greater than 8 to 10 km. At these depths, there are few holes or openings in the country rock because the pressure of the overlying rock would close them. But the upwelling magma overcomes even that great pressure.

Magma rising through the crust makes space for itself in three ways (**Figure 4.10**) that may be referred to collectively as *magmatic stoping:*

1. *Wedging open the overlying rock.* As the rising magma lifts the great weight of the overlying rock, it fractures that rock, penetrates the cracks, wedges them open, and so flows into the rock. Overlying rocks may bow upward during this process.

2. *Melting surrounding rock.* Magma also makes its way by melting country rock.

3. *Breaking off large blocks of rock.* Magma can push its way upward by breaking off blocks of the invaded crust. These blocks, known as *xenoliths* (from the Greek for "foreign rocks"), sink into the magma, where they may melt and blend into the liquid,

in some places changing the composition of the magma.

Most plutons show sharp zones of contact with country rock and other evidence of the intrusion of liquid magma into solid rock. Some plutons grade into country rock and contain structures vaguely resembling those of sedimentary rocks. The features of these plutons suggest that they formed by partial or complete melting of preexisting sedimentary rock.

Batholiths, the largest plutons, are great irregular masses of coarse-grained igneous rock that, by definition, cover at least 100 km^2 (see Figure 4.10). They are thick, horizontal, sheetlike or lobe-shaped bodies extending from a funnel-shaped central region. Their bottoms may extend 10 to 15 km deep, and a few are estimated to go even deeper. The coarse grain of batholiths results from slow cooling at great depths. Other, smaller plutons are called **stocks.** Both batholiths and stocks are **discordant intrusions;** that is, they cut across the layers of the country rock that they intrude.

Sills and Dikes

Sills and dikes are similar to plutons in many ways, but they are smaller and have a different relationship to the layering of the country rock (**Figure 4.11**). A **sill** is a sheetlike body formed by the injection of magma between parallel layers of bedded country rock. Sills are **concordant intrusions;** that is, their boundaries lie parallel to the country rock layers, whether or not those layers are horizontal. Sills range in thickness from a single centimeter to hundreds of meters, and they can extend over considerable areas. Figure 4.11a shows

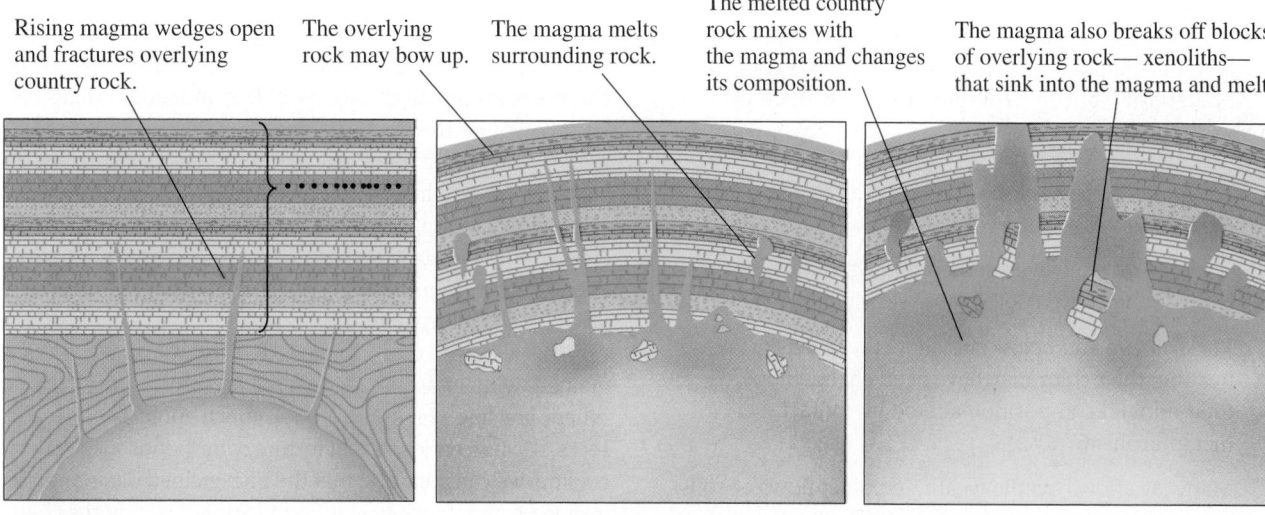

Rising magma wedges open and fractures overlying country rock.

The overlying rock may bow up.

The magma melts surrounding rock.

The melted country rock mixes with the magma and changes its composition.

The magma also breaks off blocks of overlying rock— xenoliths— that sink into the magma and melt.

FIGURE 4.10 ■ Magmas make their way into country rock in three basic ways: by invading cracks and wedging open overlying rock, by melting country rock, and by breaking off pieces of rock. Pieces of broken-off country rock, called xenoliths, can become completely dissolved in the magma. If the country rock differs in composition from the magma, the composition of the magma may change.

1 A sill runs parallel to country rock layers.

Sill {

(a)

(b)

2 A dike cuts across layers.

Dike {

FIGURE 4.11 ▪ Sills and dikes. (a) Sills are concordant intrusions. In Glacier National Park, Montana, the diorite sill intrudes into a sequence of sedimentary rocks. (b) Dikes are discordant intrusions. This dike of igneous rock (dark) intrudes into sedimentary rock in Grand Canyon National Park, Arizona. [(a) Marli Bryant Miller; (b) Asa Thorsen/Photo Researchers/Getty Images, Inc.]

a large sill at Finger Mountain, Antarctica. The 300-m-thick Palisades intrusion (see Figure 4.7) is another large sill.

Sills may superficially resemble lava flows and pyroclastic deposits, but they differ from those layers in four ways:

1. They lack the ropy, blocky, and vesicle-filled structures that characterize many volcanic rocks (see Chapter 12).

2. They are more coarse-grained than volcanic rocks because they have cooled more slowly.

3. Rocks above and below sills show the effects of heating: their color may have been changed or their mineral composition altered by contact metamorphism.

4. Many lava flows overlie weathered older flows or soils formed between successive flows; sills do not.

Dikes are the major route of magma transport in the crust. Dikes, like sills, are sheetlike igneous bodies, but dikes cut across the layers in bedded country rock (Figure 4.11b)

and so are discordant intrusions. Dikes sometimes form by forcing open existing fractures in the country rock, but more often they create channels through new cracks opened by the pressure of rising magma. Some individual dikes can be followed in the field for tens of kilometers. Their thicknesses vary from many meters to a few centimeters. In some dikes, xenoliths provide evidence of disruption of the country rock during the intrusion process. Dikes rarely exist alone; more typically, swarms of hundreds or thousands of dikes are found in a region that has been deformed by a large igneous intrusion.

The textures of dikes and sills vary. Many are coarse-grained, with an appearance typical of intrusive rocks. Many others are fine-grained and look much more like volcanic rocks. Because we know that texture corresponds to the rate of cooling, we can conclude that the fine-grained dikes and sills invaded country rock nearer Earth's surface, where the country rock was cold compared with the intrusions. Their fine texture is the result of rapid cooling. The coarse-grained ones formed at depths of many kilometers and invaded warmer rocks whose temperatures were much closer to their own.

FIGURE 4.12 ■ A granitic pegmatite vein. The center of the intrusion (*upper right*) cooled more slowly and developed coarser crystals. The margin of the intrusion (*lower left*) has finer crystals due to more rapid cooling. [John Grotzinger/Ramón Rivera-Moret/Harvard Mineralogical Museum.]

Veins

Veins are deposits of minerals found within a rock fracture that are foreign to the country rock. Irregular pencil-shaped or sheet-shaped veins branch off from the tops and sides of many igneous intrusions. They may be a few millimeters to several meters across, and they tend to be tens of meters to kilometers long or wide. The formation of veins is described in more detail in Chapter 3.

Veins of extremely coarse grained granite cutting across much finer grained country rock are called **pegmatites** (Figure 4.12). Pegmatites crystallize from a water-rich magma in the late stages of solidification.

Other veins are filled with hydrous minerals that are known to crystallize from hydrothermal solutions. From laboratory experiments, we know that these minerals typically crystallize at temperatures of 250°C to 350°C—high temperatures, but not nearly as high as the temperatures of magmas. The solubility and composition of the minerals in these hydrothermal veins indicate that abundant water was present as the veins formed. Some of the water may have come from the magma itself, but some may have been underground water in the cracks and pore spaces of the intruded rocks. On land, groundwaters originate as rainwater seeps into the soil and surface rocks. Hydrothermal veins are also abundant along mid-ocean ridges, where seawater infiltrates cracks in the newly formed seafloor, circulates down into hotter regions of the ridge, and reemerges at hydrothermal vents in the rift valley between the spreading plates. Hydrothermal processes at mid-ocean ridges are examined in more detail in Chapter 12.

■ Igneous Processes and Plate Tectonics

Geologists have observed that the facts and theories of igneous rock formation fit nicely into a framework based on plate tectonic theory. The geometry of plate movements is the link we need to tie tectonic activity and rock composition to igneous processes (Figure 4.13). Batholiths, for example, are found in the cores of many mountain ranges formed by the convergence of two plates. This observation implies a connection between pluton formation and the mountain-building process, and between both of those processes and plate movements. Similarly, our knowledge of the temperatures and pressures at which different kinds of rock melt gives us some idea of where melting takes place. For example, we know that mixtures of sedimentary rocks, because of their composition and water content, should melt at temperatures several hundred degrees below the melting point of basalt. This information leads us to predict that basalt will start to melt near the base of the crust in tectonically active regions of the upper mantle and that sedimentary rocks will melt at shallower depths.

Magma forms most abundantly in two plate tectonic settings: mid-ocean ridges, where two plates diverge and the seafloor spreads, and subduction zones, where one plate dives beneath another. Mantle plumes, though not associated with plate boundaries, also produce large amounts of magma.

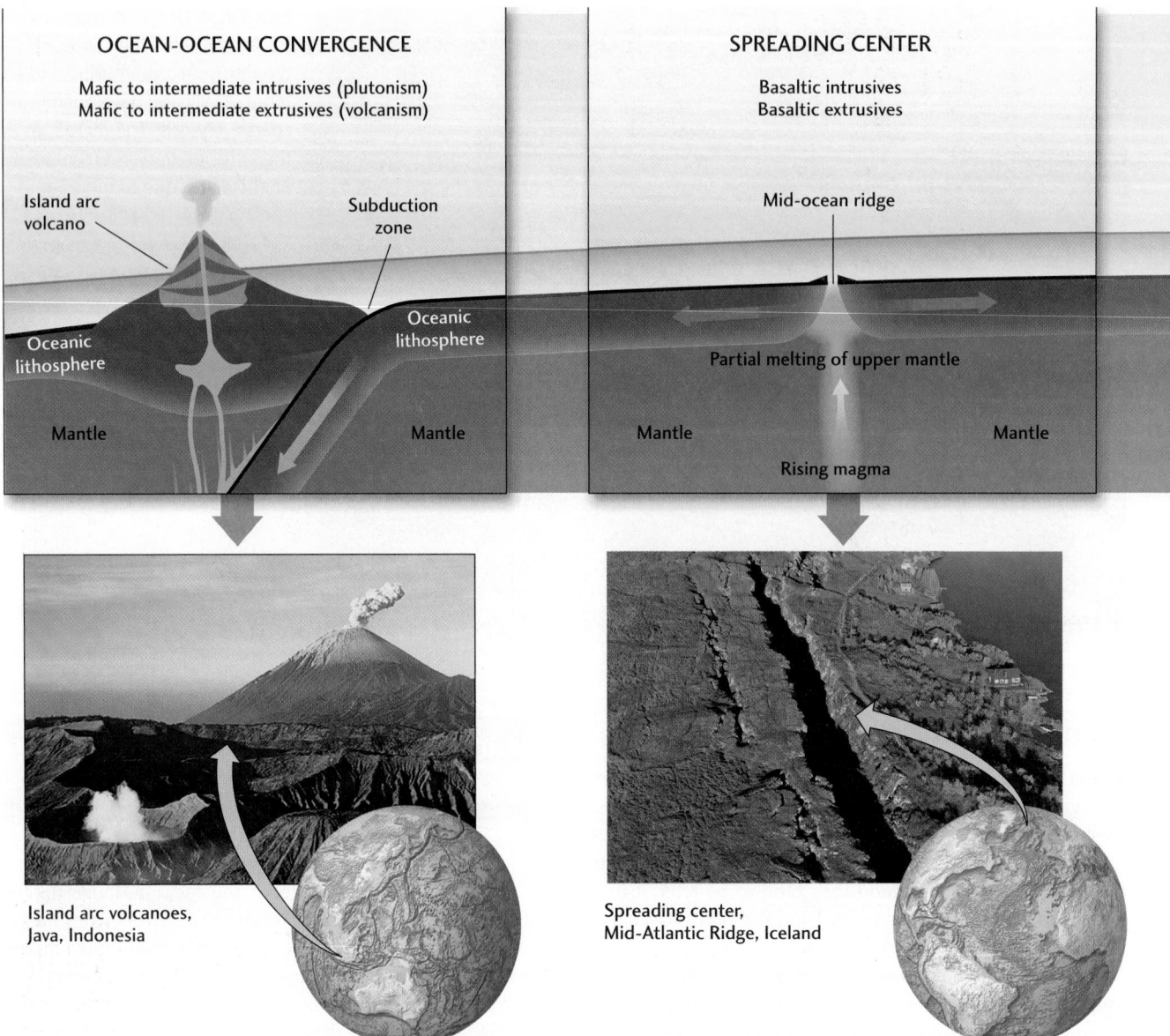

FIGURE 4.13 ■ Magmatic activity is related to plate tectonic settings. [Photos (left to right): Mark Lewis/Stone/Getty Images; Ragnar Th Sigurdsson/© ARCTIC Images/Alamy; G. Brad Lewis/Stone/Getty Images; © Michael Sedam/Age Fotostock.]

Spreading Centers as Magma Factories

Most igneous rocks are formed at mid-ocean ridges by the process of seafloor spreading. Each year, approximately 19 km³ of basaltic magma is produced along the mid-ocean ridges in the process of seafloor spreading—a truly enormous volume. In comparison, all the active volcanoes along convergent plate boundaries (about 400) generate volcanic rock at a rate of less than 1 km³/year. Enough magma has erupted during seafloor spreading over the past 200 million years to create all of the present-day seafloor, which covers nearly two-thirds of Earth's surface. Throughout the mid-ocean ridge network, decompression melting of mantle material that wells up along rising convection currents

creates magma chambers below the ridge axis. These magmas are extruded as lavas and form new seafloor. At the same time, intrusions of gabbro are emplaced at depth.

Before the advent of plate tectonic theory, geologists were puzzled by unusual assemblages of rocks that were characteristic of the seafloor but were found on land. Known as **ophiolite suites**, these assemblages consist of deep-sea sediments, submarine basaltic lavas, and mafic igneous intrusions (**Figure 4.14**). Using data gathered from deep-diving submarines, dredging, deep-sea drilling, and seismic exploration, geologists now explain these rocks as fragments of oceanic lithosphere that were transported by seafloor spreading and then raised above sea level and thrust onto a continent in a later episode of plate collision. On some of the more complete ophiolite suites preserved on

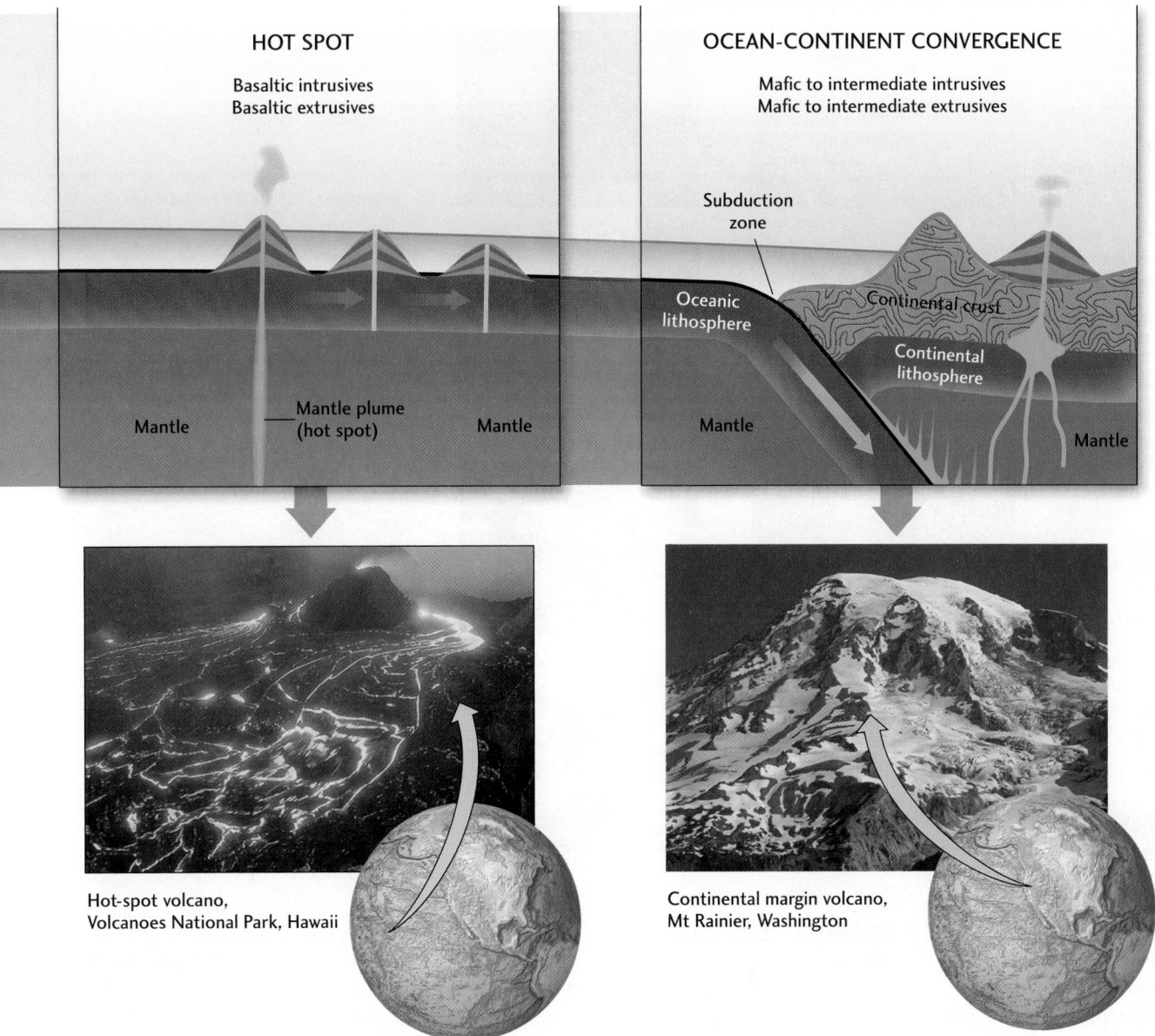

HOT SPOT

Basaltic intrusives
Basaltic extrusives

Mantle plume
(hot spot)

Mantle   Mantle

Hot-spot volcano,
Volcanoes National Park, Hawaii

OCEAN-CONTINENT CONVERGENCE

Mafic to intermediate intrusives
Mafic to intermediate extrusives

Subduction
zone

Oceanic
lithosphere

Continental crust

Continental
lithosphere

Mantle   Mantle

Continental margin volcano,
Mt Rainier, Washington

land, we can literally walk across rocks that used to lie along the boundary between Earth's oceanic crust and mantle.

How does seafloor spreading work? We can think of a spreading center as a huge factory that processes mantle material to produce oceanic crust. **Figure 4.15** is a highly schematic and simplified representation of what may be happening, based in part on studies of ophiolite suites found on land and on information gleaned from deep-sea drilling and seismic profiling. Deep-sea drilling has penetrated to the gabbro layer of the seafloor, but not to the crust-mantle boundary below. Seismic profiling has found several small magma chambers similar to the one shown in Figure 4.15.

INPUT MATERIAL: PERIDOTITE IN THE MANTLE The raw material fed into this magma factory comes from the convecting asthenosphere, in which the dominant rock type is peridotite. The mineral composition of the average peridotite in the mantle is chiefly olivine, with smaller amounts of pyroxene and garnet. Temperatures in the asthenosphere are hot enough to melt a small fraction of this peridotite (less than 1 percent), but not hot enough to generate substantial volumes of magma.

PROCESS: DECOMPRESSION MELTING Decompression melting is the process that generates great volumes of magma from peridotite at spreading centers. Recall that a decrease in pressure generally lowers a mineral's melting temperature. As the plates pull apart, the partially molten peridotite is sucked inward and upward toward the spreading center. The decrease in pressure as the peridotite rises causes a large fraction of the rock (up to 15 percent) to melt. The buoyancy of the melt causes it to rise faster than the denser surrounding rock. This process separates the liquid rock from the remaining crystal mush to produce large volumes of magma.

Sediment layers

OPHIOLITE SUITE

Deep-sea sediments: shales, limestone, chert, turbidites, fossils of pelagic marine organisms

Pillow lavas

Pillow lavas

Sheeted dike complex

Dikes

Thin section of gabbro

Gabbro (metamorphosed)

Peridotites and other ultramafic rocks (often metamorphosed)

Thin section of peridotite

FIGURE 4.14 ■ Idealized section of an ophiolite suite. The combination of deep-sea sediments, pillow lavas, sheeted dikes of gabbro, and mafic igneous intrusions indicates a deep-sea origin. [Photos courtesy of John Grotzinger. Thin sections courtesy of T. L. Grove.]

OUTPUT MATERIAL: OCEANIC CRUST PLUS MANTLE LITHOSPHERE The peridotites subjected to this process do not melt evenly: the garnet and pyroxenes they contain melt at lower temperatures than the olivine. For this reason, the magma generated by decompression melting is not peridotitic in composition; rather, it is enriched in silica and iron. This basaltic melt forms a magma chamber below the mid-ocean ridge crest, where it separates into three layers (see Figure 4.15):

1. Some magma rises through the narrow cracks that open where the plates are separating and erupts into the ocean, forming the basaltic *pillow lavas* that cover the seafloor.

2. Some magma solidifies in the cracks as vertical, sheeted dikes of gabbro.

3. The remaining magma solidifies as massive intrusions of gabbro as the underlying magma chamber is pulled apart by seafloor spreading.

These igneous units—pillow lavas, sheeted dikes, and massive gabbros—are the basic layers of the crust that geologists have found throughout the world's oceans.

Seafloor spreading results in another layer beneath this oceanic crust: the residual peridotite from which the basaltic magma was originally derived. Geologists consider this layer to be part of the mantle, but its composition is different from that of the convecting asthenosphere. In particular, the extraction of the basaltic melt makes the residual peridotite richer in olivine and stronger than ordinary mantle material. Geologists now believe it is this olivine-rich layer at the top of the mantle that gives the oceanic lithosphere its great rigidity.

Above the pillow lavas, a blanket of deep-sea sediment begins to cover the newly formed oceanic crust. As the seafloor spreads, these layers of sediment, pillow lavas, dikes, and gabbro are transported away from the mid-ocean ridge where this characteristic sequence of rocks is assembled, almost as if they were moving along a production line.

Subduction Zones as Magma Factories

Other types of magmas underlie regions where volcanoes are highly concentrated, such as the Andes of South America and the Aleutian Islands of Alaska. Both of these

2 A thin dike erupts, spilling lava on the ocean floor in characteristic "pillows."

1 Hot mantle rock rises, decompresses, and melts to a mush of crystals and basaltic magma.

Pillow lava

Dikes   Dikes intruding dikes

3 As the basalt mush cools, dikes intrude dikes to form sheeted dikes. Remnants of the spreading center move away laterally.

4 Sediments are deposited on the spreading seafloor.

0 km

Sheeted dikes in basalt

Oceanic crust

Moho

Mantle

Magma chamber

Gabbro

Peridotite layer

2

4

6

8

10

Spreading center   2   4   6   8   10 km

5 A gabbro layer is formed adjacent to the magma chamber.

Magma chamber

Peridotite layer

Mantle

6 In the magma chamber, crystals settle out of the magma, forming the peridotite layer.

FIGURE 4.15 ▪ Decompression melting creates magma at seafloor spreading centers.

regions lie over subduction zones, which are major magma factories (**Figure 4.16**). They generate magmas of varying composition, depending on how much and what kinds of materials are subducted.

Where oceanic lithosphere is subducted beneath a continent, the resulting volcanoes and volcanic rocks form a volcanic mountain belt on the continent. The Andes, which mark the subduction of the Nazca Plate beneath South America, are one such mountain belt. Similarly, subduction of the small Juan de Fuca Plate beneath western North America has generated the Cascade Range, with its active volcanoes, in northern California, Oregon, and Washington. Where oceanic lithosphere is subducted beneath oceanic lithosphere, a deep-sea trench and a volcanic island arc are formed.

INPUT MATERIAL: A MIXED BAG The variation in the chemical and mineral composition of magmas at subduction zones is a clue that the magma factories at convergent boundaries operate differently from those at spreading centers. The raw materials for these magma factories include mixtures of seafloor sediments, mixtures of basaltic

oceanic crust and felsic continental crust, mantle peridotite, and water.

PROCESS: FLUID-INDUCED MELTING The basic mechanism of magma formation at subduction zones is fluid-induced melting. The fluid involved is primarily water, which, as we have seen, lowers the melting temperature of rock. By the time oceanic lithosphere is subducted at a convergent plate boundary, a lot of water has been incorporated into its outer layers. We have already mentioned one of the processes responsible: hydrothermal activity during the formation of oceanic lithosphere. Some of the seawater that circulates through the crust near a spreading center reacts with basalt to form new hydrous minerals. In addition, as the lithosphere ages and is transported across the ocean basin, sediments containing water are deposited on its surface. The rocks formed from these sediments include shales, which contain large proportions of hydrous clay minerals. Some of these sediments get scraped off the subducting plate at the deep-sea trench, but much of this water-laden material is carried downward into the subduction zone.

6 …which can then erupt to form volcanoes.

5 The resulting magmas accumulate in magma chambers,…

Deep-sea trench

Oceanic sediments

Oceanic crust basalt

Oceanic mantle lithosphere

Magma chamber

Asthenosphere

1 Subducting oceanic crust carries sediments with it. Water remains trapped between sediment grains.

H_2O H_2O H_2O

4 The water and molten sediments move upward and melt parts of the overlying plate.

3 …causing the sedimentary rocks to melt at lower temperatures than surrounding dry mantle rocks.

Sediment grains

Water

2 The trapped water, as well as chemically bound water, is released as the temperature increases,…

FIGURE 4.16 ■ Fluid-induced melting creates magma in subduction zones.

As the lithospheric slab moves downward, it is subjected to increasing pressure. Water is squeezed out of the minerals in the outer layers of the descending crust and rises buoyantly into the mantle above the descending crust. At moderate depths of about 5 km, the temperature increases to about 150°C. Here, more water is released by metamorphic chemical reactions as basalt is converted to *amphibolite*, which is composed of amphibole and plagioclase feldspar (see Chapter 6). As other chemical reactions take place, additional water is released at depths ranging from 10 to 20 km. Finally, at depths greater than 100 km, the temperature increases to 1200°C–1500°C, and the subducted slab undergoes an additional metamorphic transition induced by the increased pressure. Amphibolite is converted to *eclogite*, which is composed of pyroxene and garnet (see Chapter 6). Here, the increase in both pressure and temperature in

the subducting slab releases all of its remaining water in addition to other materials.

During subduction, the released water induces melting in the descending basalt-rich oceanic crust and in the overlying peridotite-rich mantle material. Most of the resulting mafic magma accumulates at the base of the crust of the overriding plate, and some of it intrudes into the crust to form magma chambers, resulting in the formation of volcanoes.

OUTPUT: MAGMAS OF VARYING COMPOSITION

The magmas produced by fluid-induced melting at subduction zones are essentially basaltic, although their composition is more variable than that of mid-ocean ridge basalts. The composition of these magmas is further altered during their residence in the crust. Within magma chambers, the process of fractional crystallization increases the magma's silica content, producing eruptions of andesitic lavas. Where the overlying plate is continental, the heat from the magmas can melt the felsic rocks in the crust, forming magmas with an even higher silica content, such as dacitic and rhyolitic magmas. The contribution of lithospheric fluids to these magmas is suggested by the presence in the magmas of trace elements known to be present in oceanic crust and sediments.

Mantle Plumes as Magma Factories

Mantle plumes, like spreading centers, are sites of decompression melting, but they differ from spreading centers by forming within lithospheric plates rather than along the margins of plates. These plumes of hot mantle material rise from deep within Earth, possibly as deep as the core-mantle boundary. Mantle plumes that reach the surface, most of them far from plate boundaries, form the "hot spots" of Earth. At these locations, basaltic magmas produced by decompression melting of mantle material may erupt in huge outpourings to form islands, such as the Hawaiian Islands, or basalt plateaus, such as the Columbia Plateau in the Pacific Northwest of North America. Mantle plumes and hot spots are discussed in more detail in Chapter 12.

SUMMARY

How are igneous rocks classified? Igneous rocks can be divided into two broad textural classes: coarse-grained rocks, which are intrusive and therefore cooled slowly; and fine-grained rocks, which are extrusive and cooled rapidly. Igneous rocks can also be classified on the basis of their silica content using a scale that runs from felsic (rich in silica) to ultramafic (poor in silica).

How and where do magmas form? Magmas form at places in the lower crust and mantle where temperatures are high enough for partial melting of rock. Because the minerals within a rock melt at different temperatures, the composition of magmas varies with temperature. Pressure raises the melting temperature of rock, and the presence of water lowers it. Because melted rock is less dense than solid rock, magma rises through the surrounding rock, and drops of magma come together to form magma chambers.

How does magmatic differentiation account for the variety of igneous rocks? Because different minerals crystallize at different temperatures, the composition of magma changes as it cools and various minerals are withdrawn by crystallization.

What are the forms of igneous intrusions? Large intrusive igneous bodies are called plutons. The largest plutons are batholiths, which are thick horizontal masses extending from a funnel-shaped central region. Stocks are smaller plutons. Less massive than plutons are sills, which lie parallel to the layers of bedded country rock, and dikes, which cut across those layers. Veins form where water is abundant, either in the magma or in the surrounding country rock.

How do plate tectonic processes affect magma production? Magmas are produced at two types of plate boundaries. At spreading centers, peridotite rises from the mantle and undergoes decompression melting to form basaltic magma. At subduction zones, subducting oceanic lithosphere undergoes fluid-induced melting to generate magmas of varying composition. Mantle plumes within lithospheric plates are also sites of decompression melting that produce basaltic magmas.

KEY TERMS AND CONCEPTS

andesite (p. 96)

basalt (p. 96)

batholith (p. 103)

bomb (p. 93)

concordant intrusion
  (p. 103)

country rock (p. 93)

dacite (p. 96)

decompression melting
  (p. 98)

dike (p. 104)

diorite (p. 96)

discordant intrusion (p. 103)

extrusive igneous rock
  (p. 93)

felsic rock (p. 95)

fluid-induced melting (p. 98)

fractional crystallization
  (p. 99)

gabbro (p. 96)

granite (p. 95)

granodiorite (p. 96)

intermediate igneous rock
  (p. 96)

intrusive igneous rock
  (p. 93)

lava (p. 92)

mafic rock (p. 96)

magma chamber (p. 99)

magmatic differentiation
  (p. 99)

obsidian (p. 94)

ophiolite suite (p. 106)

partial melting (p. 98)

pegmatite (p. 105)

peridotite (p. 97)

pluton (p. 103)

porphyry (p. 94)

pumice (p. 94)

pyroclast (p. 93)

rhyolite (p. 96)

sill (p. 103)

stock (p. 103)

ultramafic rock (p. 97)

vein (p. 105)

viscosity (p. 97)

volcanic ash (p. 93)

PRACTICING GEOLOGY EXERCISE

How Do Valuable Metallic Ores Form? Magmatic Differentiation Through Crystal Setting

Fractional crystallization in the Palisades sill.

Some of the most economically important mineral deposits in the world are formed by differential settling of crystals in magma chambers. The Bushveld deposit in South Africa and the Stillwater deposit in Montana, just north of Yellowstone National Park, are two of the most famous examples. These deposits contain some of the world's largest reserves of metals in the platinum group—such as platinum and palladium—but vast quantities of iron, tin, chromium, and titanium are also found there. These deposits represent ancient magma chambers in which fractional crystallization led to the formation of different minerals over time, which settled to the bottom of the magma chamber in economically important concentrations. Geologists realized that the process of crystal settling was the key to understanding how the deposits formed.

Many geologic processes involve the movement of particles within fluids. We see the same basic principles in the movement of sand grains in rivers, in the hurtling of debris through the atmosphere when a volcano erupts, and in the settling of crystals through magmas. In each case, the movement of the particles is regulated by a number of factors.

In the example of the Palisades sill, we saw that as a basaltic magma cools, olivine crystallizes first, followed by pyroxene and plagioclase feldspar. Once crystallized, each mineral sinks through the remaining liquid magma to settle out at the bottom of the magma chamber. Thus, the Palisades sill is layered with olivine at its base, followed by pyroxene and plagioclase feldspar overlying it (see Figure 4.7).

Olivine not only crystallizes before feldspar, but also settles more quickly than feldspar due to its higher density.

Both fractional crystallization and crystal settling rates contribute to the segregation of minerals within magma chambers.

The rate at which crystals settle depends both on their density and size and on the viscosity of the remaining magma. That rate can be calculated using a mathematical relationship called *Stokes' law*:

$$V = \frac{gr^2(d_c - d_m)}{u}$$

where V is the velocity at which the crystals settle through the magma; g is the acceleration due to Earth's gravity (980 cm/s^2); r is the radius of the crystal; d_c is the density of the crystal; d_m is the density of the magma; and u is the viscosity of the magma.

Stokes' law states that three factors are important in regulating the velocity at which crystals will move through magma:

1. As crystals grow larger, their radius (r) increases. Because r is in the numerator of the equation, Stokes' law tells us that larger crystals will settle faster than smaller crystals. Furthermore, r is squared, which tells us that small increases in crystal size will result in much larger increases in their settling velocity.

2. Magma viscosity (u) is a measure of the resistance of the magma to flow or, in this case, to moving out of the way of a sinking crystal. Because u is in the denominator of the equation, it tells us that an increase in magma viscosity will result in a decrease in settling velocity.

3. Settling velocity (*V*) also depends on the difference between the density of the crystal (d_c) and the density of the magma (d_m). *V* will increase as the density of the crystal increases and as the density of the magma decreases. Thus, in a magma of constant density, crystals with higher density will settle faster than crystals with lower density.

If we now consider fractional crystallization in the Palisades sill, Stokes' law will help us determine the actual settling rates for particular minerals. Consider an olivine crystal with a radius of 0.1 cm and a density of 3.7 g/cm³. The magma through which the crystal settles has a density of 2.6 g/cm³ and a viscosity of 3000 poise (1 poise = 1 g/cm × s). How fast will this olivine crystal fall through the magma?

$$V = \frac{(980 \text{ cm/s}^2) \times (0.1)^2 \times (3.7 - 2.6 \text{ g/cm}^3)}{3000 \text{ g/cm}^3}$$

= 0.0036 cm/s

= 12.9 cm/hr

BONUS PROBLEM: Try the same calculation for a plagioclase feldspar crystal of the same size, which has a density of 2.7 g/cm³. Which mineral settles through the magma at a faster rate? How much faster does it settle?

EXERCISES

1. Why are intrusive igneous rocks coarse-grained and extrusive rocks fine-grained?

2. What kinds of minerals would you find in a mafic igneous rock?

3. What kinds of igneous rocks contain quartz?

4. Name two intrusive igneous rocks with a higher silica content than that of gabbro.

5. What is the difference between a magma formed by fractional crystallization and one formed by ordinary cooling?

6. How does fractional crystallization lead to magmatic differentiation?

7. Where in the crust, mantle, or core would you find a partial melt of basaltic composition?

8. In which plate tectonic settings would you expect magmas to form?

9. Why do melts migrate upward?

10. Where on the ocean floor would you find basaltic magmas being extruded?

THOUGHT QUESTIONS

1. How would you classify a coarse-grained igneous rock that contains about 50 percent pyroxene and 50 percent olivine?

2. What kind of rock would contain some plagioclase feldspar crystals about 5 mm long "floating" in a dark gray matrix of crystals of less than 1 mm?

3. What observations might you make to show that a pluton solidified during fractional crystallization?

4. Why are plutons more likely than dikes to show the effects of fractional crystallization?

5. What might be the origin of a rock composed almost entirely of olivine?

6. What processes create the unequal sizes of crystals in porphyries?

7. Water is abundant in the sedimentary rock and oceanic crust at subduction zones. How would that water affect melting in these zones?

MEDIA SUPPORT

 4-1 Animation: Intrusive Igneous Structures

 4-1 Video: Lava Flows and Features, Arizona

 4-2 Animation: Magma Chambers

 4-2 Video: Olinive: Igneous Rocks, Mantle Xenoliths, and Green Sand Beaches

■ Surface Processes
  of the Rock Cycle   116

■ Sedimentary Basins:
  The Sinks for
  Sediments   122

■ Sedimentary
  Environments   124

■ Sedimentary
  Structures   127

■ Burial and Diagenesis:
  From Sediment to
  Rock   130

■ Classification of
  Siliciclastic Sediments
  and Sedimentary
  Rocks   132

■ Classification of Chemical
  and Biological
  Sediments and
  Sedimentary Rocks 135

The large-scale cross-bedding visible in this sandstone records the history of its formation in an ancient desert. [John Grotzinger.]

SEDIMENTATION: ROCKS FORMED BY SURFACE PROCESSES

5

MUCH OF EARTH'S SURFACE, including its seafloor, is covered with sediments. These layers of loose particles have diverse origins. Most sediments are created by weathering of the continental crust. Some are the remains of mineral shells secreted by organisms. Still others consist of inorganic crystals that precipitated when dissolved chemicals in oceans and lakes combined to form new minerals.

Sedimentary rocks were once sediments, so they are records of the conditions at Earth's surface when and where those sediments were deposited. Geologists can work backward to infer the sources of the sediments from which the rocks were formed and the kinds of places in which the sediments were originally deposited. The top of Mount Everest, for example, is composed of fossiliferous (fossil-containing) limestones. Because we know that such limestones are formed from carbonate minerals in seawater, we can conclude that Mount Everest must once have been part of an ocean floor! This kind of analysis can be applied just as well to ancient shorelines, mountains, plains, deserts, and wetlands. By reconstructing such environments, we can map the continents and oceans of long ago.

Sedimentary rocks may also reveal former plate tectonic events and processes by their presence within or adjacent to volcanic island arcs, continental rift valleys, or collisional or volcanic mountain belts. In cases in which sediments and sedimentary rocks are derived from the weathering of preexisting rocks, we can form hypotheses about ancient climates and environments. We can also use sedimentary rocks formed by precipitation from seawater to read the history of changes in Earth's climate and seawater chemistry.

The study of sediments and sedimentary rocks has great practical value as well. Oil, natural gas, and coal, our most valuable fossil-fuel resources, are found in these rocks. A number of other important mineral resources are also sedimentary, such as the phosphate rock used for fertilizer and much of the world's iron ore. Knowing how these kinds of sediments form helps us to find and use these limited resources.

Finally, because virtually all sedimentary processes take place at or near Earth's surface where we humans live, they provide a background for our understanding of environmental problems. We once studied sedimentary rocks primarily to understand how to exploit the natural resources just mentioned. Increasingly, however, we study these rocks to improve our understanding of Earth's environment.

In this chapter, we will see how the surface processes of the rock cycle produce sediments and sedimentary rocks. We will describe the compositions, textures, and structures of sediments and sedimentary rocks and examine how they correlate with the kinds of environments in which the sediments and rocks are laid down. Throughout the chapter, we will apply our understanding of sediment origins to the study of human environmental problems and to the exploration for energy and mineral resources.

■ Surface Processes of the Rock Cycle

Sediments, and the sedimentary rocks formed from them, are produced by the surface processes of the rock cycle. These processes act on rocks after they have been moved from Earth's interior to its surface and uplifted by mountain building and before they are returned to Earth's interior by subduction. They move materials from a *source area*, where sediment particles are created, to a *sink area*, where they are deposited in layers. The path that sediment particles follow from source to sink may be a very long one—one that involves several important processes resulting from interactions between the plate tectonic and climate systems.

Let's look at the role of the Mississippi River in a typical sedimentary process. Plate movement lifts up rocks in the Rocky Mountains. Rainfall in those mountains—a source area—causes weathering of the rocks there. If precipitation increases in the mountains, weathering also increases. Faster weathering produces more sediment to be released into the river and transported downhill and downstream. At the same time, if the flow in the river also increases because of the higher rainfall, transportation of sediment down the length of the river increases, and the volume of sediment delivered to sink areas—sites of deposition, also known as *sedimentary basins*—in the Mississippi delta and the Gulf of Mexico increases as well. In these sedimentary basins, the sediments pile up on top of one another—layer after layer—and are eventually buried deep in Earth's crust, to depths where they may become filled with valuable oil and natural gas.

The surface processes of the rock cycle that are important in the formation of sedimentary rocks are reviewed in **Figure 5.1** and summarized here:

- *Weathering* is the general process by which rocks are broken down at Earth's surface to produce sediment particles. There are two types of weathering. **Physical weathering** takes place when solid rock is fragmented by mechanical processes, such as freezing and thawing or wedging by tree roots (**Figure 5.2**), that do not change its chemical composition. The rubble of broken stone often seen at the tops of mountains and hills is primarily the result of physical weathering. **Chemical weathering** refers to processes by which the minerals in a rock are chemically altered or dissolved. The blurring or disappearance of lettering on old gravestones and monuments is caused mainly by chemical weathering.

- *Erosion* refers to processes that dislodge particles of rock produced by weathering and move them away from the source area. Erosion occurs most commonly when rainwater runs downhill.

- *Transportation* refers to processes by which sediment particles are moved to sink areas. Transportation occurs when water, wind, or the moving ice of glaciers transport particles to new locations downhill or downstream.

- *Deposition* (also called *sedimentation*) refers to processes by which sediment particles settle out as water currents slow, winds die down, or glacier edges melt to form layers of sediment in sink areas. In aquatic environments, particles settle out, chemical precipitates form and are deposited, and the bodies and shells of dead organisms are broken up and deposited.

Weathering breaks down rocks physically and chemically.

Erosion carries away particles produced by weathering.

Transportation via streams, glaciers, and wind moves particles downhill.

Deposition (or **sedimentation**) occurs when particles settle out or dissolved minerals precipitate.

Burial occurs as layers of sediment accumulate and compact previous layers.

Diagenesis lithifies the sediment to make sedimentary rocks.

FIGURE 5.1 ■ Several surface processes of the rock cycle contribute to the formation of sedimentary rocks.

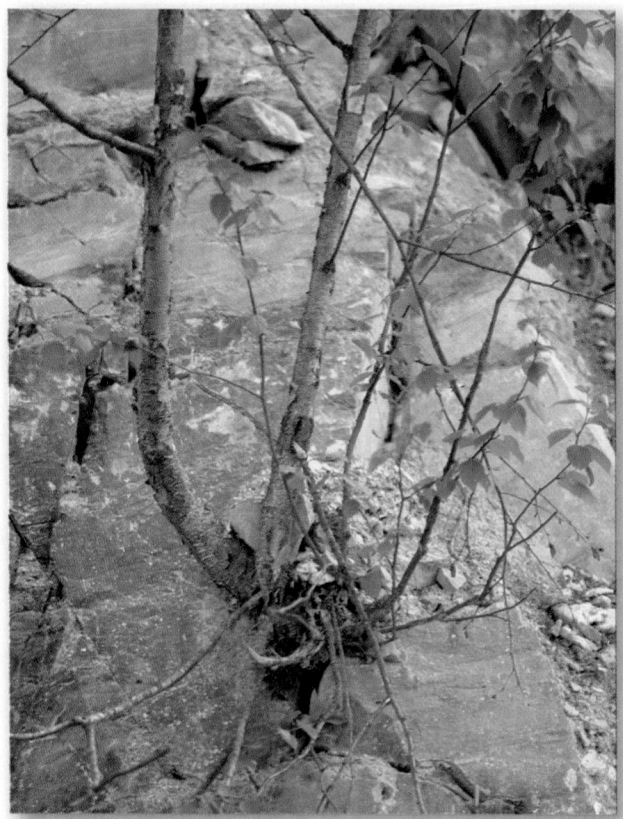

FIGURE 5.2 ■ Plant roots contribute to physical weathering by penetrating fractures and wedging rocks apart. [David R. Frazier/Science Source.]

■ *Burial* occurs as layers of sediment accumulate in sink areas on top of older, previously deposited sediments, which are compacted and progressively buried deep within a sedimentary basin. These sediments will remain at depth, as part of Earth's crust, until they are either uplifted again or subducted by plate tectonic processes.

■ *Diagenesis* refers to the physical and chemical changes—caused by pressure, heat, and chemical reactions—by which sediments buried within sedimentary basins are *lithified,* or converted into sedimentary rock.

Weathering and Erosion: The Source of Sediments

Chemical and physical weathering reinforce each other. Chemical weathering weakens rocks and makes them more susceptible to fragmentation. The smaller the fragments produced by physical weathering, the greater the surface area exposed to chemical weathering. Together, physical and chemical weathering of rock create both solid particles and dissolved products, and erosion carries them away. These end products can be classified as either siliciclastic sediments or chemical and biological sediments.

SILICICLASTIC SEDIMENTS Physical and chemical weathering of preexisting rocks forms *clastic particles* that are transported and deposited as sediments. Clastic particles range in size from boulders to particles of sand, silt, and clay. They also vary widely in shape. Natural breakage along bedding planes and fractures in the parent rock determine the shapes of boulders, cobbles, and pebbles. Sand grains are the remnants of individual crystals formerly interlocked in parent rock, and their shapes tend to reflect the shapes of those crystals.

Most clastic particles are produced by the weathering of common rocks composed largely of silicate minerals, so sediments formed from these particles are called **siliciclastic sediments.** The mixture of minerals in siliciclastic sediments varies. Minerals such as quartz resist weathering and thus are found chemically unaltered in siliciclastic sediments. These sediments may also contain partly altered fragments of minerals, such as feldspar, that are less resistant to weathering and therefore less stable. Still other minerals in siliciclastic sediments, such as clay minerals, are newly formed by chemical weathering. Varying intensities of weathering can produce different sets of minerals in sediments derived from the same parent rock. Where weathering is intense, the sediment will contain only clastic particles made of chemically stable minerals, mixed with clay minerals. Where weathering is slight, many minerals that are unstable under land surface conditions will survive as clastic particles in the sediment. Table 5.1 shows three possible sets of minerals in sediments derived from a typical granite outcrop.

CHEMICAL AND BIOLOGICAL SEDIMENTS Chemical weathering produces dissolved ions and molecules that accumulate in the waters of soils, rivers, lakes, and oceans. Chemical and biological reactions then precipitate these substances to form chemical and biological sediments. We distinguish between chemical and biological sediments mainly for convenience; in practice, many chemical and biological sediments overlap. **Chemical sediments** form at or near their place of deposition. The evaporation of

TABLE 5-1	Minerals Present in Sediments Derived from a Granite Outcrop Under Varying Intensities of Weathering		
Intensity of Weathering			
Low	Medium	High	
Quartz	Quartz	Quartz
Feldspar	Feldspar	Clay minerals
Mica	Mica	
Pyroxene	Clay minerals	
Amphibole		

seawater, for example, often leads to the precipitation of gypsum or halite (**Figure 5.3**).

Biological sediments also form near their place of deposition, but they are the result of mineral precipitation by organisms. Some organisms, such as mollusks and corals, precipitate minerals as they grow. After the organisms die, their shells or skeletons accumulate on the seafloor as sediments. In these cases, the organism *directly* controls mineral precipitation. However, in a second but equally important process, organisms control mineral precipitation only *indirectly*. Instead of taking up minerals from the water to form a shell, these organisms change the surrounding environment so that mineral precipitation occurs on the outside of the organism, or even away from the organism. Certain microorganisms are thought to enable

FIGURE 5.3 ■ Salts precipitate when water containing dissolved minerals evaporates, which has occurred here in Death Valley, California. [John G. Wilbanks/Age fotostock.]

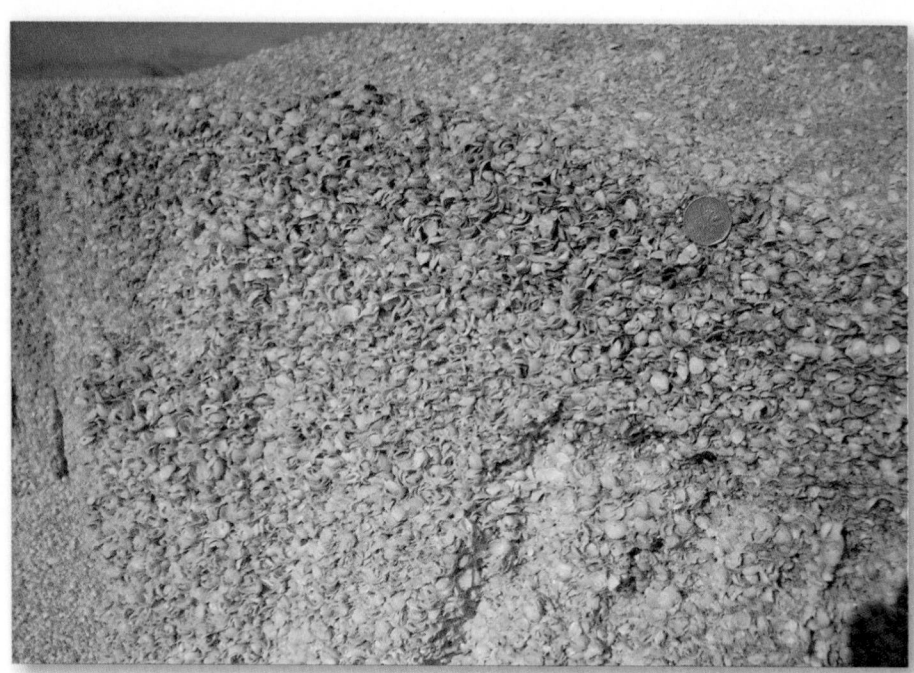

FIGURE 5.4 ■ One kind of sedimentary rock of biological origin is formed entirely of shell fragments. [John Grotzinger.]

the precipitation of pyrite (an iron sulfide mineral) in this way (see Chapter 11).

In shallow marine environments, directly precipitated biological sediments consist of layers of particles, such as whole or fragmented shells of marine organisms (Figure 5.4). Many different types of organisms, ranging from corals to clams to algae, may contribute their shells. Sometimes the shells are transported, further broken up, and deposited as **bioclastic sediments.** These shallow-water sediments consist predominantly of two calcium carbonate minerals, calcite and aragonite, in variable proportions. Other minerals, such as phosphates and sulfates, are only locally abundant in bioclastic sediments.

In the deep sea, biological sediments are made up of the shells of only a few kinds of planktonic organisms. Most of these organisms secrete shells composed primarily of calcite and aragonite, but a few species form silica shells, which are precipitated broadly over some parts of the deep seafloor. Because these biological particles accumulate in very deep water, where agitation by sediment-transporting currents is uncommon, they rarely form bioclastic sediments.

Transportation and Deposition: The Downhill Journey to Sedimentary Basins

After clastic particles and dissolved ions have been formed by weathering and dislodged by erosion, they start their journey to a sedimentary basin. This journey may be very long; for example, as we have seen, it might span the thousands of kilometers from the tributaries of the Mississippi River in the Rocky Mountains to the wetlands of the Mississippi delta.

Most agents of sediment transport carry sediments on a one-way trip downhill. Rocks falling from a cliff, a river flowing to the ocean carrying a load of sand, and glacial ice slowly dragging boulders downhill are all responses to gravity. Although wind may blow material from a low elevation to a higher one, in the long run the effects of gravity prevail. When a windblown particle drops into the ocean and settles through the water, it is trapped. It can be picked up again only by an ocean current, which can transport it to and deposit it in another site on the seafloor. Ocean currents transport sediments over shorter distances than do big rivers on land, and the short transportation distances of chemical and biological sediments contrast with the much greater distances over which siliciclastic sediments are transported. But eventually, all sediment transportation paths, as simple or complicated as they may be, lead downhill into a sedimentary basin.

CURRENTS AS TRANSPORT AGENTS Most sediments are transported by currents of air or water. The enormous quantities of all kinds of sediments found in the oceans result primarily from the transport capacities of rivers, which annually carry a solid and dissolved sediment load of about 25 billion tons (25×10^{15} g) (Figure 5.5). Air currents—winds—move sediments, too, but in far smaller quantities than rivers or ocean currents. As particles are lifted into the air or water, the current carries them downwind or downstream. The stronger the current—that is, the faster it flows—the larger the particles it can transport.

CURRENT STRENGTH, PARTICLE SIZE, AND SORTING Deposition starts where transportation stops. For clastic particles, gravity is the driving force of deposition. The tendency of particles to settle under the pull of gravity works against a current's ability to carry them. A particle's settling velocity is proportional to its density and its size (see Chapter 4, Practicing Geology, page 112).

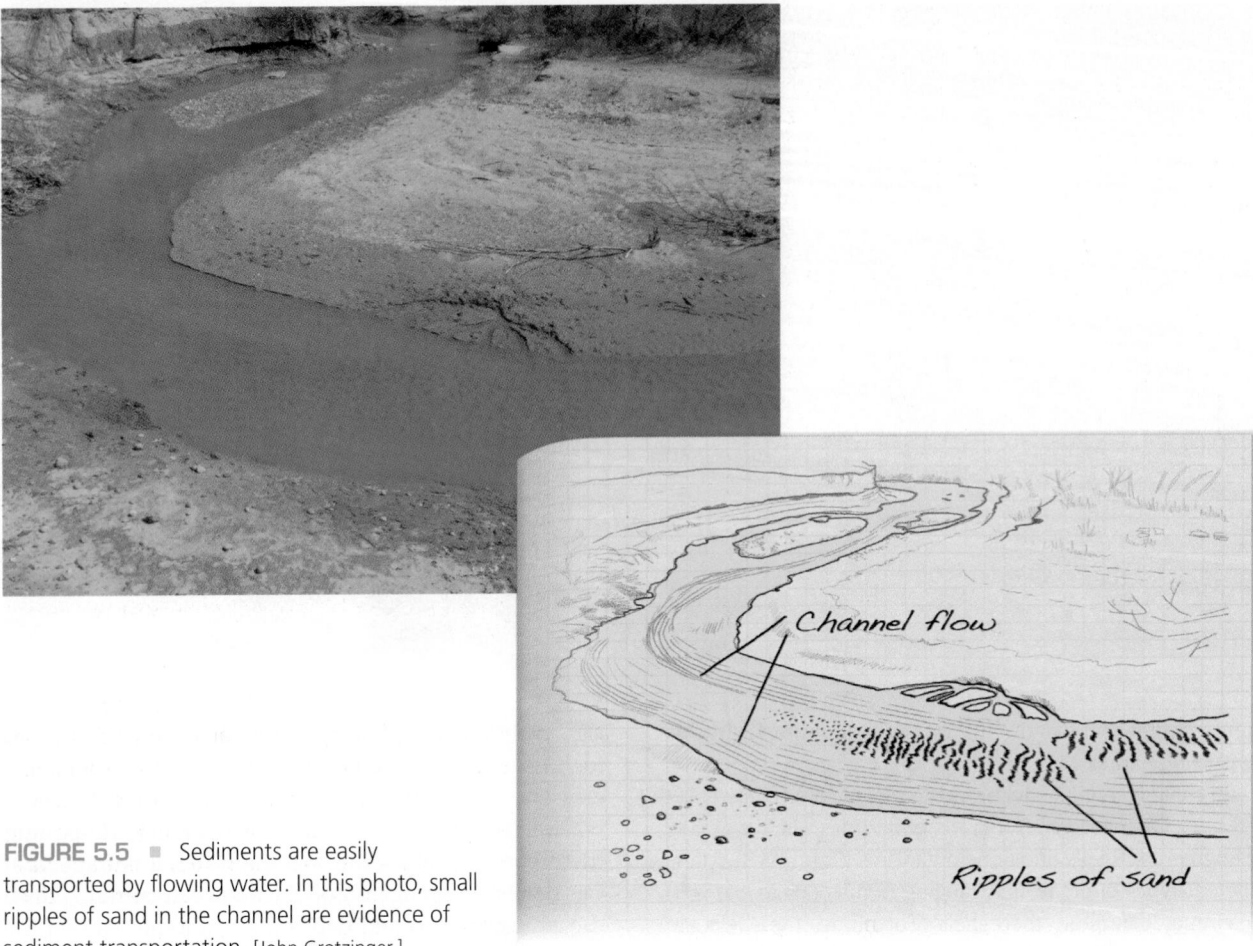

FIGURE 5.5 ■ Sediments are easily transported by flowing water. In this photo, small ripples of sand in the channel are evidence of sediment transportation. [John Grotzinger.]

Because all clastic particles have roughly the same density, we use particle size as the best indicator of how quickly a particle will settle. (We will take a more specific look at particle size categories later in this chapter.) In water, large particles settle faster than small ones. This is also true in air, but the difference is much smaller.

Current strength, which is directly related to current velocity, determines the size of the particles deposited in a particular place. As a wind or water current begins to slow, it can no longer keep the largest particles suspended, and those particles settle. As the current slows even more, smaller particles settle. When the current stops completely, even the smallest particles settle. Currents segregate particles in the following ways:

■ *Strong currents* (faster than 50 cm/s) carry gravel (which includes boulders, cobbles, and pebbles), along with an abundant supply of smaller particles. Such currents are common in swiftly flowing streams in mountainous terrain, where erosion is rapid. Beach gravels are deposited where ocean waves erode rocky shores.

■ *Moderately strong currents* (20–50 cm/s) lay down sand beds. Currents of moderate strength are common in most rivers, which carry and deposit sand in their channels. Rapidly flowing floodwaters

may spread sand over the width of a river valley. Waves and currents deposit sand on beaches and in the ocean. Winds also blow and deposit sand, especially in deserts. However, because air is much less dense than water, much higher current velocities are required for it to move sediments of the same size and density.

■ *Weak currents* (slower than 20 cm/s) carry muds composed of the finest clastic particles (silt and clay). Weak currents are found on the floor of a river valley when floodwaters recede slowly or stop flowing entirely. In the ocean, muds are generally deposited some distance from shore, where currents are too slow to keep even fine particles in suspension. Much of the floor of the open ocean is covered with mud particles originally transported by surface waves and currents or by wind. These particles slowly settle to depths where currents and waves are stilled and, ultimately, all the way to the bottom of the ocean.

As you can see, currents may begin by carrying particles of widely varying sizes, which then become separated as the strength of the current changes. A strong, fast current may lay down a bed of gravel while keeping sand and mud in suspension. If the current weakens and slows, it will lay

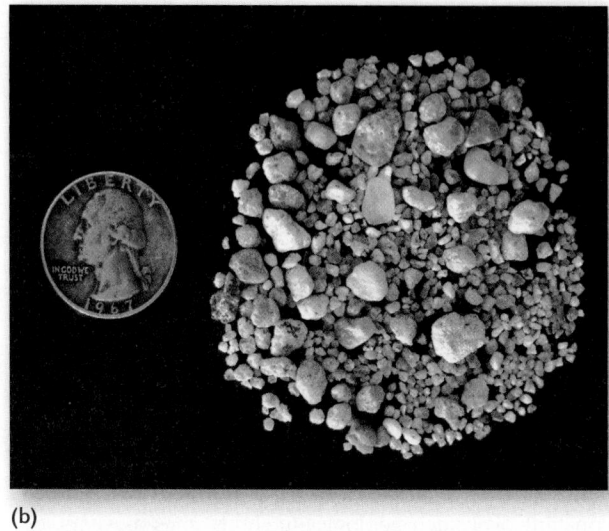

(a)

(b)

FIGURE 5.6 ■ As the strength of a current changes, sediments are sorted according to particle size. The relatively homogeneous group of sand grains on the left is well sorted; the group on the right is poorly sorted. [John Grotzinger.]

down a bed of sand on top of the gravel. If the current then stops altogether, it will deposit a layer of mud on top of the sand bed. This tendency for variations in current velocity to segregate sediments according to size is called **sorting.** A well-sorted sediment consists mostly of particles of a uniform size. A poorly sorted sediment contains particles of many sizes (**Figure 5.6**).

As cobbles, pebbles, and large sand grains are being transported by water or air currents, they tumble and strike one another or rub against the underlying rock. The resulting *abrasion* affects the particles in two ways: it reduces their size, and it rounds off their rough edges (**Figure 5.7**). These effects apply mostly to the larger particles; smaller sand grains and silt undergo little abrasion.

Particles are generally transported intermittently rather than steadily. A river may transport large quantities of

sand and gravel when it floods, then drop them as the flood recedes, only to pick them up again and carry them even farther in the next flood. Similarly, strong winds may carry large amounts of dust for a few days, then die down and deposit the dust as a layer of sediment. The strong tidal currents along some shorelines may transport broken shell fragments to places farther offshore and drop them there.

The total time it takes for clastic particles to be transported may be many hundreds or thousands of years, depending on the distance to the final sedimentary basin and the number of stops along the way. Clastic particles eroded by the headwaters of the Missouri River in the mountains of western Montana, for example, take hundreds of years to travel the 3200 km down the Missouri and Mississippi rivers to the Gulf of Mexico.

Distance of transport

Short   Moderate   Long

Larger, more angular

Smaller, more rounded

FIGURE 5.7 ■ Abrasion during transportation reduces the size and angularity of clastic particles. Particles become more rounded and slightly smaller as they are transported, although the general shape of the particle may not change significantly.

Oceans as Chemical Mixing Vats

The driving force of chemical and biological sedimentation is precipitation rather than gravity. Substances dissolved in water by chemical weathering are carried along with the water. These materials are part of the aqueous solution itself, so gravity cannot cause them to settle out. As the dissolved materials are carried down rivers, they ultimately enter the ocean.

Oceans may be thought of as huge chemical mixing vats. Rivers, rain, wind, and glaciers constantly bring in dissolved materials. Smaller quantities of dissolved materials enter the oceans through hydrothermal chemical reactions between seawater and hot basalt at mid-ocean ridges. The oceans lose water continuously by evaporation at the surface. The inflow and outflow of water are so exactly balanced that the amount of water in the oceans remains constant over such geologically short times as years, decades, or even centuries. Over a time scale of thousands to millions of years, however, the balance may shift. During the most recent ice age, for example, significant quantities of seawater were converted into glacial ice, and sea level was drawn down by more than 100 m.

The entry and exit of dissolved materials, too, are balanced. Each of the many dissolved components of seawater participates in some chemical or biological reaction that eventually precipitates it out of the water and onto the seafloor. As a result, the ocean's **salinity**—the total amount of dissolved material in a given volume of water—remains constant. Totaled over the world ocean, mineral precipitation balances the total inflow of dissolved material—yet another way in which the Earth system maintains balance.

We can better understand this chemical balance by considering the element calcium. Calcium is a component of the most abundant biological precipitate formed in the oceans: calcium carbonate ($CaCO_3$). On land, calcium dissolves when limestone and calcium-containing silicate minerals, such as some feldspars and pyroxenes, are weathered, and that calcium is transported to the ocean as dissolved calcium ions (Ca^{2+}). There, a wide variety of marine organisms take up calcium ions and combine them with carbonate ions (CO_3^{2-}), also present in seawater, to form their calcium carbonate shells. Thus, the calcium that entered the ocean as dissolved ions leaves it as solid sediment particles when the organisms die and their shells settle to the seafloor and accumulate there as calcium carbonate sediments. Ultimately, the calcium carbonate sediments will be buried and transformed into limestone. The chemical balance that keeps the concentrations of calcium dissolved in the ocean constant is thus controlled in part by the activities of organisms.

Nonbiological mechanisms also maintain chemical balance in the oceans. For example, sodium ions (Na^+) transported to the oceans react chemically with chloride ions (Cl^-) to form the precipitate sodium chloride (NaCl). This happens when evaporation raises the concentrations of sodium and chloride ions past the point of saturation. As we saw in Chapter 3, minerals precipitate when solutions become so saturated with dissolved materials that they can hold no more. The intense evaporation required to crystallize sodium chloride may take place in warm, shallow arms of the sea or in saline lakes.

■ Sedimentary Basins: The Sinks for Sediments

As we have seen, the currents that move sediments across Earth's surface generally flow downhill. Therefore, sediments tend to accumulate in depressions in Earth's crust. Such depressions are formed by **subsidence**, in which a broad area of the crust sinks (subsides) relative to the surrounding crust. Subsidence is induced partly by the weight of sediments on the crust, but is caused mainly by plate tectonic processes.

Sedimentary basins are regions of variable size where the combination of sedimentation and subsidence has formed thick accumulations of sediments and sedimentary rocks. Sedimentary basins are Earth's primary sources of oil and natural gas. Commercial exploration for these resources has helped us better understand the deep structure of sedimentary basins and of the continental lithosphere.

Rift Basins and Thermal Subsidence Basins

When plate separation begins within a continent, subsidence results from the stretching, thinning, and heating of the underlying lithosphere by the plate tectonic processes that are causing the separation (**Figure 5.8**). A long, narrow rift develops, bounded by great downdropped blocks of crustal rock. Hot, ductile mantle material rises, melts, and fills the space created by the thinned lithosphere and crust, initiating the eruption of basaltic lavas in the rift zone. Such **rift basins** are deep, narrow, and long, with thick successions of sedimentary rocks and extrusive and intrusive igneous rocks. The rift valleys of East Africa, the Rio Grande Valley, and the Jordan Valley in the Middle East are examples of rift basins.

At later stages of plate separation, when rifting has led to seafloor spreading and the newly separated continents are drifting away from each other, subsidence continues through the cooling of the lithosphere that was thinned and heated during the earlier rifting stage (see Figure 5.8). Cooling leads to an increase in the density of the lithosphere, which in turn leads to its subsidence below sea level, where sediments can accumulate. Because cooling of the lithosphere is the main process creating the sedimentary basins at this stage, they are called **thermal subsidence basins.** Sediments from erosion of the adjacent land fill the basin nearly to sea level along the edge of the continent, creating a **continental shelf.**

The continental shelf continues to receive sediments for a long time because the trailing edge of the drifting

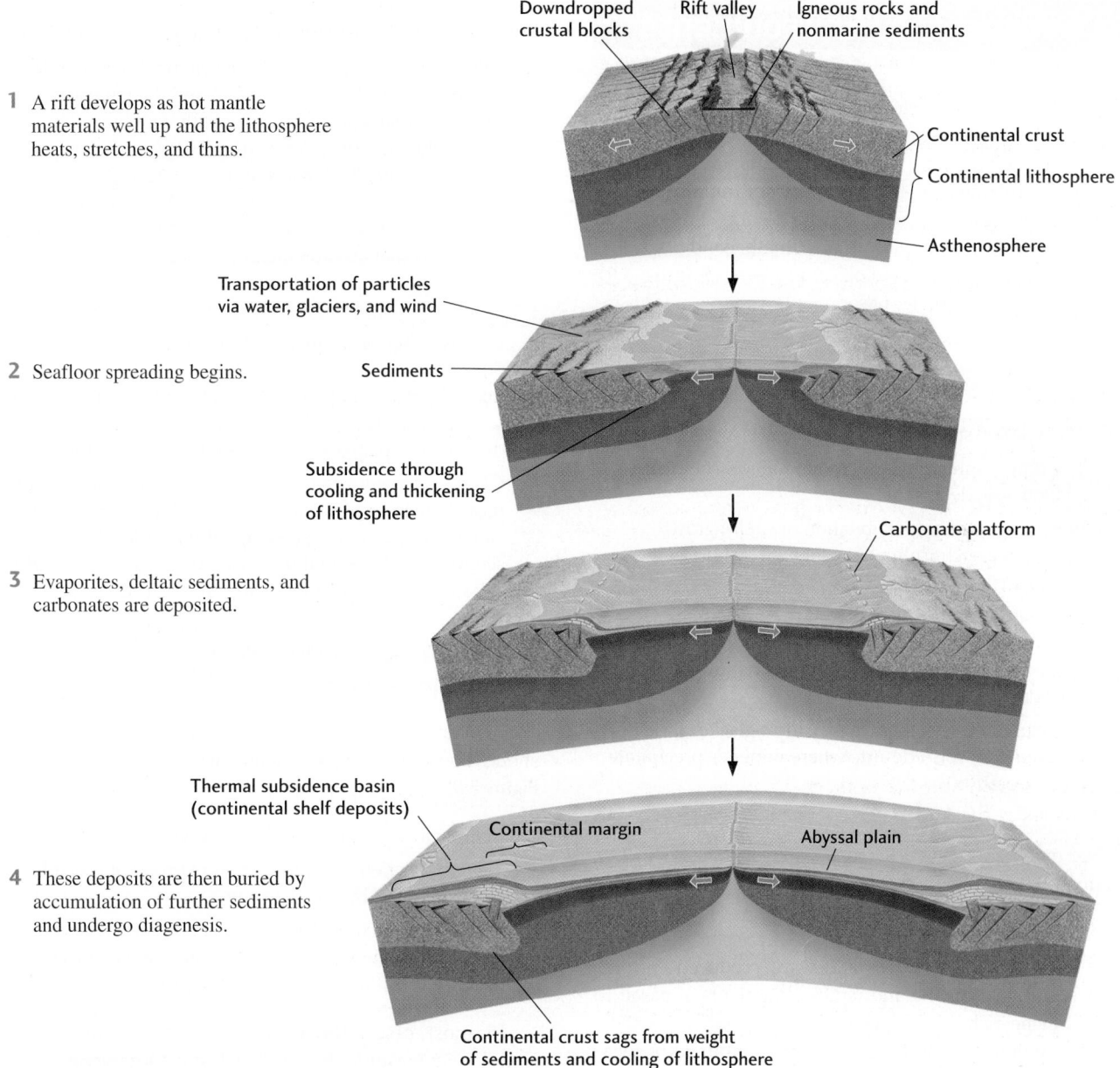

1 A rift develops as hot mantle materials well up and the lithosphere heats, stretches, and thins.

Downdropped crustal blocks

Rift valley

Igneous rocks and nonmarine sediments

Continental crust

Continental lithosphere

Asthenosphere

2 Seafloor spreading begins.

Transportation of particles via water, glaciers, and wind

Sediments

Subsidence through cooling and thickening of lithosphere

3 Evaporites, deltaic sediments, and carbonates are deposited.

Carbonate platform

4 These deposits are then buried by accumulation of further sediments and undergo diagenesis.

Thermal subsidence basin (continental shelf deposits)

Continental margin

Abyssal plain

Continental crust sags from weight of sediments and cooling of lithosphere

FIGURE 5.8 ▪ Sedimentary basins are formed by plate separation.

continent subsides slowly and because the continent provides a tremendous land area from which sediments can be derived. The load of the growing mass of sediment further depresses the crust, so that the basin can receive still more material from the land. As a result of this continuous subsidence and sediment transportation, continental shelf deposits can accumulate in an orderly fashion to thicknesses of 10 km or more. The continental shelves off the Atlantic coasts of North and South America, Europe, and Africa are good examples of thermal subsidence basins. These basins began to form when the supercontinent Pangaea split apart about 200 million years ago and the North American and South American plates separated from the Eurasian and African plates.

Flexural Basins

A third type of sedimentary basin develops at convergent plate boundaries where one plate pushes up over the other. The weight of the overriding plate causes the underlying plate to bend or flex down, producing a **flexural basin.** The Mesopotamian Basin in Iraq is a flexural basin formed when the Arabian Plate collided with and was subducted beneath the Eurasian Plate. The enormous oil reserves in Iraq (second only to Saudi Arabia's) owe their size to having the right ingredients in this important flexural basin. In effect, oil that had formed in the rocks now beneath the Zagros Mountains in Iran was squeezed out, forming several great pools of oil with volumes larger than 10 billion barrels.

▪ Sedimentary Environments

Between the source area where sediments are formed and the sedimentary basin where they are buried and converted to sedimentary rocks, sediments travel through many sedimentary environments. A **sedimentary environment** is an area of sediment deposition characterized by a particular combination of climate conditions and physical, chemical, and biological processes (**Figure 5.9**). Important characteristics of sedimentary environments include the following:

- The type and amount of water (ocean, lake, river, arid land)

- The type and strength of transport agents (water, wind, ice)

- The topography (lowland, mountain, coastal plain, shallow sea, deep sea)

- Biological activity (precipitation of shells, growth of coral reefs, churning of sediments by burrowing organisms)

- The plate tectonic settings of sediment source areas (volcanic mountain belt, continent-continent collision zone) and sedimentary basins (rift, thermal subsidence, flexural)

- The climate (cold climates may form glaciers; arid climates may form deserts where minerals precipitate by evaporation)

Consider the beaches of Hawaii, famous for their unusual green sands, which are a result of their distinctive sedimentary environment. The volcanic island of Hawaii is composed of olivine-bearing basalt, from which the olivine is released during weathering. Rivers transport the olivine to the beach, where waves and wave-produced currents concentrate the olivine and remove fragments of basalt to form olivine-rich sand deposits.

Sedimentary environments are often grouped by location: on continents, near shorelines, or in the ocean. This very general subdivision highlights the processes that give sedimentary environments their distinct identities.

Continental Sedimentary Environments

Sedimentary environments on continents are diverse due to the wide variation in temperature and rainfall over the land surface. These environments are built around lakes, rivers, deserts, and glaciers (see Figure 5.9).

- *Lake environments* include inland bodies of fresh or saline water in which the transport agents are relatively small waves and moderate currents. Chemical sedimentation of organic matter and carbonate minerals may occur in freshwater lakes. Saline lakes such as those found in deserts evaporate and precipitate a variety of *evaporite* minerals, such as halite. The Great Salt Lake in Utah is an example.

- *Alluvial environments* include the channel of a river, its borders and associated wetlands, and the flat valley floor on either side of the channel that is covered by water when the river floods (the *floodplain*). Rivers are present on all the continents except Antarctica, so alluvial deposits are widespread. Organisms are abundant in muddy flood deposits and produce organic sediments that accumulate in wetlands adjacent to river channels. Climates vary from arid to humid. An example is the Mississippi River and its floodplains.

- *Desert environments* are arid. Wind and the rivers that flow intermittently through deserts transport sand and dust. The dry climate inhibits abundant organic growth, so organisms have little effect on the sediments. Desert dune fields are an example of such an environment.

- *Glacial environments* are dominated by the dynamics of moving masses of ice and are characterized by a cold climate. Vegetation is present, but has little effect on sediments. At the melting border of a glacier, meltwater streams form a transitional alluvial environment.

Shoreline Sedimentary Environments

The dynamics of waves, tides, and river currents on sandy shores dominate shoreline sedimentary environments (see Figure 5.9):

- *Deltas,* where rivers enter lakes or oceans

- *Tidal flats,* where extensive areas exposed at low tide are dominated by tidal currents

- *Beaches,* where the strong waves approaching and breaking on the shore distribute sediments on the beach, depositing strips of sand or gravel

In most cases, the sediments that accumulate in shoreline environments are siliciclastic. Organisms affect these sediments mainly by burrowing into and mixing them. However, in some tropical and subtropical settings, sediment particles, particularly carbonate sediments, may be of biological origin. These biological sediments are also subject to transportation by waves and tidal currents.

Marine Sedimentary Environments

Marine sedimentary environments are usually classified by water depth, which determines the kinds of currents that are present (see Figure 5.9). Alternatively, they can be classified by distance from land.

- *Continental shelf environments* are located in the shallow waters off continental shores, where sedimentation is controlled by relatively gentle currents. Sediments may be composed of either siliciclastic or biological carbonate particles, depending

FIGURE 5.9 ■ Multiple factors interact to create sedimentary environments.

Continental Environments		1 Lake	2 Alluvial	3 Desert	4 Glacial
	Transport agent	Lake currents, waves	River currents	Wind	Ice, meltwater
	Sediments	Sand and mud, saline precipitates in arid climates	Sand, mud, and gravel	Sand and dust	Sand, mud, and gravel
	Climate	Arid to humid	Arid to humid	Arid	Cold
	Biological processes	Freshwater organisms and precipitates	Organic matter in muddy flood deposits and wetlands	Little biological activity	Little biological activity

Shoreline Environments		5 Delta	6 Beach	7 Tidal flats
	Transport agent	River currents, waves	Waves, tidal currents	Tidal currents
	Sediments	Sand and mud	Sand and gravel	Sand and mud
	Climate	Arid to humid	Arid to humid	Arid to humid
	Biological processes	Burial of plant debris	Little biological activity	Organisms mix sediments

Marine Environments		8 Deep sea	9 Continental shelf	10 Organic reefs	11 Continental margin/slope
	Transport agent	Ocean currents Turbidity currents	Waves and tides	Waves and tides	Ocean currents and waves
	Sediments	Mud and sand	Sand and mud	Calcified organisms	Mud and sand
	Biological processes	Deposition of remains of organisms	Deposition of remains of organisms	Secretion of carbonates by corals and other organisms	Deposition of remains of organisms

on how much siliciclastic sediment is supplied by rivers and on the abundance of carbonate-producing organisms. Sedimentation may also be chemical if the climate is arid and an arm of the sea becomes isolated and evaporates.

- *Organic reefs* are carbonate structures, formed by carbonate-secreting organisms, built up on continental shelves or on oceanic volcanic islands.

- *Continental margin and slope environments* are found in the deeper waters at and off the edges of continents, where sediments are deposited by turbidity currents. A *turbidity current* is a turbulent submarine avalanche of sediment and water that moves downslope. Most sediments deposited by turbidity currents are siliciclastic, but at sites where organisms produce abundant carbonate sediments, continental slope sediments may be rich in carbonates.

- *Deep-sea environments* are found far from continents, where the waters are much deeper than the reach of waves and tidal currents. These environments include the lower portion of the continental slope, which is built up by turbidity currents traveling far from continental margins; the abyssal plain of the deep sea, which accumulates carbonate sediments provided mostly by the shells of plankton; and the mid-ocean ridges.

Siliciclastic versus Chemical and Biological Sedimentary Environments

Sedimentary environments can be grouped not only by their location, but also according to the kinds of sediments found in them or according to the dominant sediment formation process. Grouping of sedimentary environments in this manner produces two broad classes: siliciclastic sedimentary environments and chemical and biological sedimentary environments.

Siliciclastic sedimentary environments are those dominated by siliciclastic sediments. They include all of the continental sedimentary environments as well as those shoreline environments that serve as transitional zones between continental and marine environments. They also include those marine environments of the continental shelf, continental margin and slope, and deep seafloor where siliciclastic sands and muds are deposited (**Figure 5.10**). The sediments of these siliciclastic environments are often called **terrigenous sediments** to indicate their origin on land.

Chemical and biological sedimentary environments are characterized principally by chemical and biological precipitation (**Table 5.2**).

FIGURE 5.10 ■ These sedimentary rocks exposed at El Capitan, in the Guadalupe Mountains of West Texas, were formed in an ancient ocean about 260 million years ago. The lower slopes of the mountains contain siliciclastic sedimentary rocks formed in deep-sea environments. The overlying cliffs of El Capitan are limestone and dolostone, formed from sediments deposited in a shallow sea when carbonate-secreting organisms died, leaving their shells in the form of a reef. [John Grotzinger.]

TABLE 5-2	Major Chemical and Biological Sedimentary Environments	
Environment	Agent of Precipitation	Sediments
Shoreline and Marine		
Carbonate (reefs, platforms, deep sea, etc.)	Shelled organisms, some algae; inorganic precipitation from seawater	Carbonate sands and muds, reefs
Evaporite	Evaporation of seawater	Gypsum, halite, other salts
Siliceous (deep sea)	Shelled organisms	Silica
Continental		
Evaporite	Evaporation of lake water	Halite, borates, nitrates, carbonates, other salts
Wetland	Vegetation	Peat

Carbonate environments are marine settings where calcium carbonate, mostly secreted by organisms, is the main sediment. They are by far the most abundant chemical and biological sedimentary environments. Hundreds of species of mollusks and other invertebrate animals, as well as calcareous (calcium-containing) algae and microorganisms, secrete carbonate shells or skeletons. Various populations of these organisms live at different depths, both in quiet areas and in places where waves and currents are strong. As they die, their shells and skeletons accumulate to form carbonate sediments.

Except for those of the deep sea, carbonate environments are found mostly in the warmer tropical and subtropical regions of the oceans, where carbonate-secreting organisms flourish. These environments include organic reefs, carbonate sand beaches, tidal flats, and shallow carbonate platforms. In a few places, carbonate sediments form in cooler waters that are supersaturated with carbonate ions—waters that are generally below 20°C, such as in some regions of the Southern Ocean south of Australia. These carbonate sediments are formed by a very limited group of organisms, most of which secrete calcite shells.

Siliceous environments are unique deep-sea sedimentary environments named for the silica shells deposited in them. The planktonic organisms that secrete these silica shells grow in surface waters where nutrients are abundant. When they die, their shells settle to the deep seafloor and accumulate as layers of siliceous sediments.

An *evaporite environment* is created when the warm seawater of an arid inlet or arm of the sea evaporates more rapidly than it can mix with seawater from the open ocean. The degree of evaporation and the length of time it has proceeded control the salinity of the evaporating seawater and thus the kinds of chemical sediments formed. Evaporite environments also form in lakes lacking river outlets. Such lakes may produce sediments of halite, borate, nitrates, and other salts.

■ Sedimentary Structures

Sedimentary structures include all kinds of features formed at the time of deposition. Sediments and sedimentary rocks are characterized by *bedding,* or *stratification,* which occurs when layers of sediment, or *beds,* with different particle sizes or compositions are deposited on top of one another. These beds range from only millimeters or centimeters thick to meters or even many meters thick. Most bedding is horizontal, or nearly so, at the time of deposition. Some types of bedding, however, form at a high angle relative to the horizontal.

Cross-Bedding

Cross-bedding consists of beds deposited by wind or water and inclined at angles as much as 35° from the horizontal (**Figure 5.11**). Cross-beds form when sediment particles are deposited on the steeper, downcurrent (leeward) slopes of sand dunes on land or sandbars in rivers and on the seafloor. Cross-bedding patterns in wind-deposited sand dunes may be complex as a result of rapidly changing wind directions (as in the photograph at the opening of this chapter). Cross-bedding is common in sandstones and is also found in gravels and some carbonate sediments. It is easier to see in sandstones than in sands, which must be excavated to see a cross section.

Graded Bedding

Graded bedding is most abundant in continental slope and deep-sea sediments deposited by dense, muddy turbidity currents, which hug the bottom of the ocean as they move downhill. Each bed progresses from large particles at the bottom to small particles at the top. As the current progressively slows, it drops progressively smaller

FIGURE 5.11 ■ Sediment particles transported down the steeper, downcurrent slope of a sand dune, sandbar, or ripple form cross-bedding.

particles. The grading indicates a weakening of the current that deposited the particles. A graded bed comprises one set of sediment particles, normally ranging from a few centimeters to several meters thick, that formed a horizontal or nearly horizontal layer at the time of deposition. Accumulations of many individual graded beds can reach a total thickness of hundreds of meters. A graded bed formed as a result of deposition by a turbidity current is called a *turbidite.*

Ripples

Ripples are very small ridges of sand or silt whose long dimension is at right angles to the current. They form low, narrow ridges, usually only a centimeter or two high, separated by wider troughs. These sedimentary structures are common in both modern sands and ancient sandstones (**Figure 5.12**). Ripples can be seen on the surfaces of windswept dunes, on underwater sandbars in shallow streams, and under the waves at beaches. Geologists can distinguish the symmetrical ripples made by waves moving back and forth on a beach from the asymmetrical ripples formed by currents moving in a single direction over river sandbars or windswept dunes (**Figure 5.13**).

Bioturbation Structures

In many sedimentary rocks, the bedding is broken or disrupted by roughly cylindrical tubes a few centimeters in diameter that extend vertically through several beds. These sedimentary structures are remnants of burrows and tunnels excavated by clams, worms, and other marine organisms that live on the ocean bottom. These organisms churn and burrow through muds and sands—a process called **bioturbation.** They ingest the sediment, digest the bits of organic matter it contains, and leave behind the reworked sediment, which fills the burrow (**Figure 5.14**). From bioturbation structures, geologists can determine the behavior of the organisms that burrowed in the sediment. Since the behavior of burrowing organisms is controlled partly by environmental factors, such as the strength of currents or the availability of nutrients, bioturbation structures can help us reconstruct past sedimentary environments.

(a)

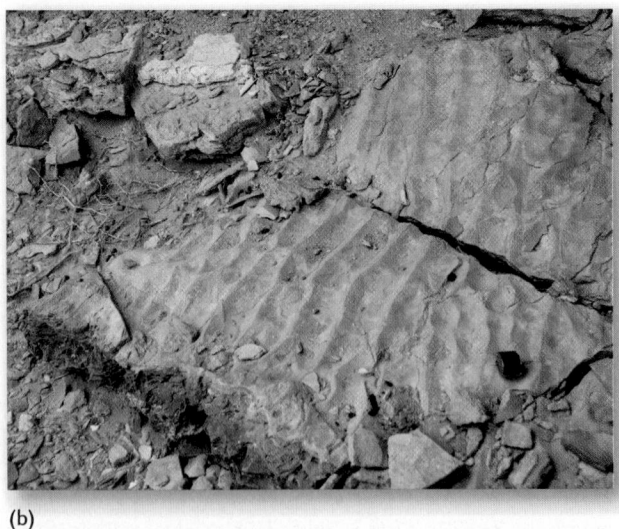

(b)

FIGURE 5.12 ■ Ripples. (a) Ripples in modern sand on a beach [Courtesy of John Grotzinger.] (b) Ancient ripple-marked sandstone. [John Grotzinger/Ramón Rivera-Moret/MIT.]

Symmetrical ripples (beach)

Waves   Equal slopes

(a)

Asymmetrical ripples (dune)

Gentler slope   Steeper slope   Wind or water

(b)

FIGURE 5.13 ■ Geologists can distinguish ripples formed by waves from ripples formed by currents. (a) The shapes of ripples on beach sand, produced by the back-and-forth movements of waves, are symmetrical. [John Grotzinger.] (b) Ripples on dunes and river sandbars, produced by the movement of a current in one direction, are asymmetrical. [John Grotzinger.]

Bedding Sequences

Bedding sequences are built of interbedded and vertically stacked layers of different sedimentary rock types. A bedding sequence might consist of cross-bedded sandstone, overlain by bioturbated siltstone, overlain in turn by rippled sandstone—in any combination of thicknesses for each rock type in the sequence.

Bedding sequences help geologists reconstruct the ways in which sediments were deposited and so provide insight into the history of geologic processes and events that occurred at Earth's surface long ago. **Figure 5.15** shows a

FIGURE 5.14 ■ Bioturbation structures. This rock is crisscrossed with fossilized tunnels originally made by organisms burrowing through the mud. [John Grotzinger/Ramón Rivera-Moret/Harvard Mineralogical Museum.]

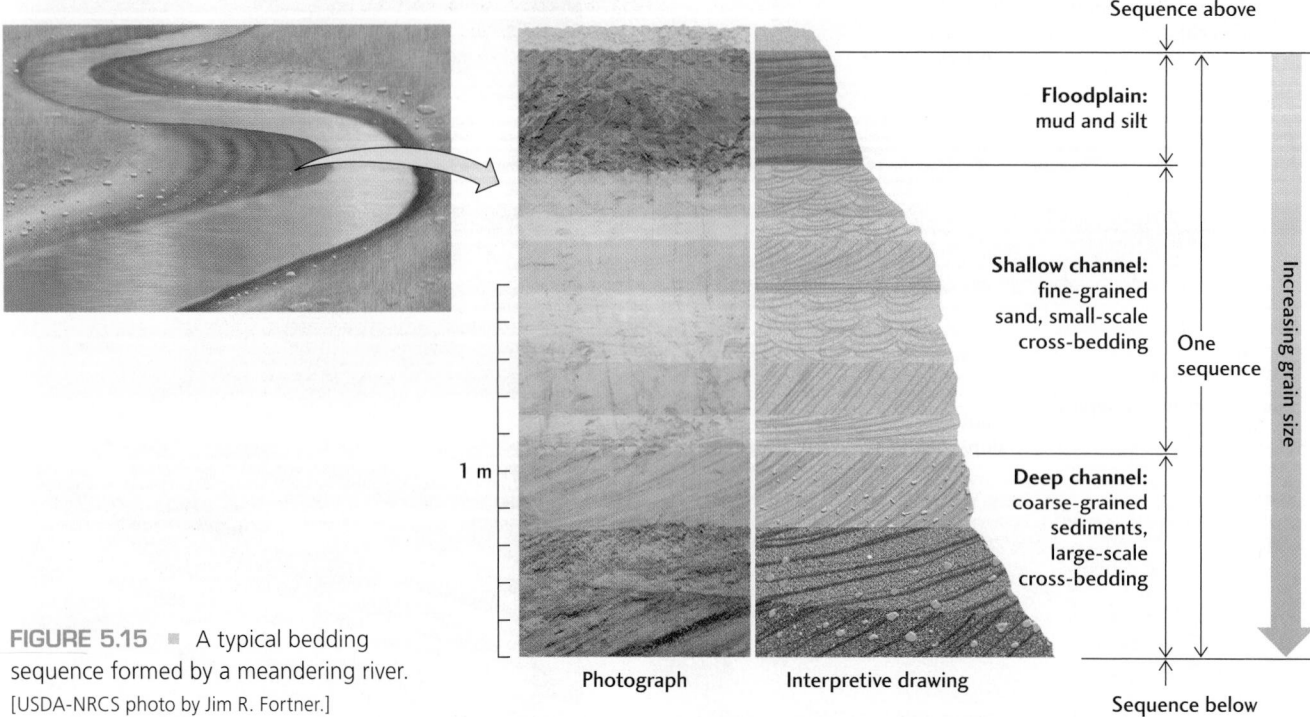

FIGURE 5.15 ■ A typical bedding sequence formed by a meandering river.
[USDA-NRCS photo by Jim R. Fortner.]

Photograph   Interpretive drawing

Sequence above

Floodplain: mud and silt

Shallow channel: fine-grained sand, small-scale cross-bedding

One sequence

Deep channel: coarse-grained sediments, large-scale cross-bedding

Sequence below

Increasing grain size

1 m

bedding sequence typically formed in alluvial sedimentary environments. A river lays down sediments as its channel meanders back and forth across the valley floor. Thus, the lower part of the sequence contains the beds deposited in the deepest part of the river channel, where the current was strongest. The middle part contains the beds deposited in the shallower parts of the channel, where the current was weaker, and the upper part contains the beds deposited on the floodplain. Typically, a bedding sequence formed in this manner consists of sediment particles that grade upward from large to small. This sequence may be repeated a number of times if the river meanders back and forth.

Most bedding sequences consist of a number of small-scale subdivisions. In the example shown in Figure 5.15, the basal layers contain cross-bedding. These layers are overlain by more cross-bedded layers, but the cross-beds are smaller in scale. Horizontal bedding occurs at the top of the bedding sequence. Today, computer models are used to analyze how bedding sequences of sands were deposited in alluvial environments.

■ Burial and Diagenesis: From Sediment to Rock

Most of the clastic particles produced by weathering on land end up deposited in various sedimentary basins in the oceans. A smaller amount of siliciclastic sediment is deposited in sedimentary environments on land. Most chemical and biological sediments are also deposited in ocean basins, although some are deposited in lakes and wetlands.

Burial

Once sediments reach the ocean floor, they are trapped there. The deep seafloor is the ultimate sedimentary basin and, for most sediments, their final resting place. As the sediments are buried under new layers of sediments, they are subjected to increasingly high temperatures and pressures as well as chemical changes.

Diagenesis

After sediments are deposited and buried, they are subject to **diagenesis**—the many physical and chemical changes that result from the increasing temperatures and pressures as they are buried ever deeper in Earth's crust. These changes continue until the sediment or sedimentary rock is either exposed to weathering or metamorphosed by more extreme heat and pressure (**Figure 5.16**).

Temperature increases with depth in Earth's crust at an average rate of 30°C for each kilometer of depth, although that rate varies somewhat among sedimentary basins. Thus, at a depth of 4 km, buried sediments may reach 120°C or more, the temperature at which certain types of organic matter may be converted to oil and natural gas (see Practicing Geology). Pressure also increases with depth—on average, about 1 atmosphere for each 4.4 m of depth. This increased pressure is responsible for the compaction of buried sediments.

1 Sediments are buried, compacted, and lithified at shallow depths in Earth's crust.

2 Diagenesis includes the processes —physical and chemical—that change sediments to sedimentary rocks.

Compaction
Compaction by burial squeezes out water.

50–60% water

↓

10–20% water

Cementation
Precipitation or addition of new minerals cements sediment particles.

↓

Lithification

3 Different sediments result in different sedimentary rocks.

Fine → Coarse

Mud → Shale

Sand → Sandstone

Gravel → Conglomerate

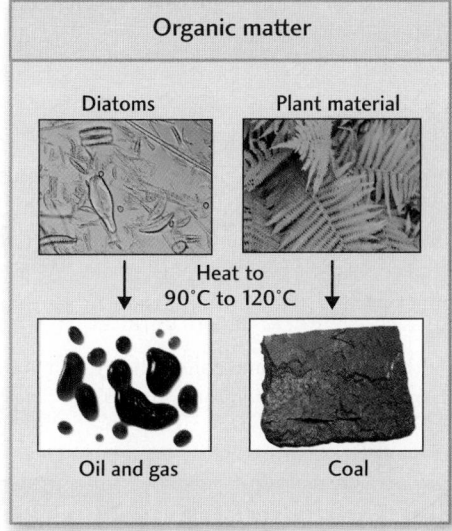

Organic matter

Diatoms

Plant material

Heat to 90°C to 120°C

Oil and gas

Coal

FIGURE 5.16 ■ Diagenesis is the set of physical and chemical changes that convert sediments into sedimentary rocks. [*mud, sand, gravel:* John Grotzinger; *shale:* John Grotzinger/Ramón Rivera-Moret/Harvard Mineralogical Museum; *sandstone, conglomerate, coal:* John Grotzinger/Ramón Rivera-Moret/MIT; *diatoms:* Mark B. Edlund, Ph.D./Science Museum of Minnesota; *plant material:* Roman Gorielov/Shutterstock; *oil and gas:* Wasabi/Alamy.]

Buried sediments are also continuously bathed in groundwater full of dissolved minerals. These minerals can precipitate in the pores between the sediment particles and bind them together—a chemical change called **cementation.** Cementation decreases **porosity,** the percentage of a rock's volume consisting of open pores between particles. In some sands, for example, calcium carbonate is precipitated as calcite, which acts as a cement that binds the grains and hardens the resulting mass into sandstone (**Figure 5.17**). Other minerals, such as quartz, may cement sands, muds, and gravels into sandstone, mudstone, or conglomerate.

The major physical diagenetic change is **compaction,** a decrease in the volume and porosity of a sediment. Compaction occurs as sediment particles are squeezed closer together by the weight of overlying sediments.

Quartz and grains   Calcite cement

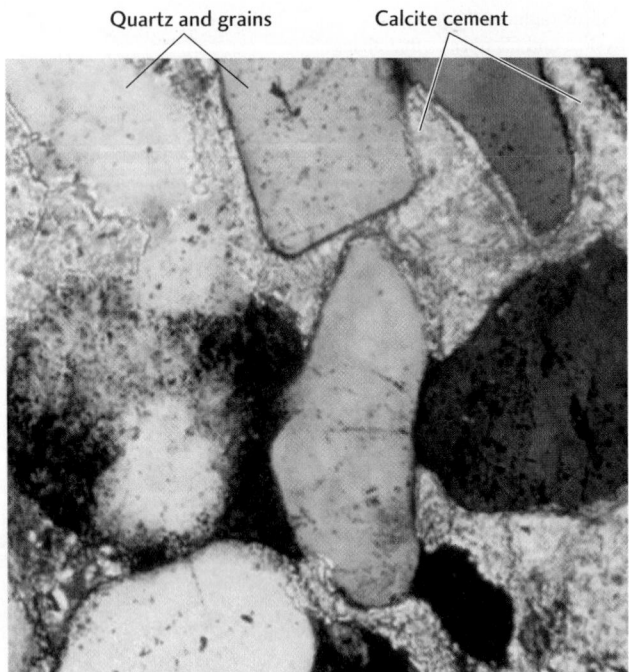

FIGURE 5.17 ■ This photomicrograph of sandstone shows quartz grains (white and gray) cemented by calcite (brightly colored and variegated) precipitated after deposition. [Peter Kresan.]

Sands are fairly well packed during deposition, so they do not compact much. However, newly deposited muds, including carbonate-containing muds, are highly porous. In many of these sediments, more than 60 percent of the volume consists of water in pore spaces. As a result, muds compact greatly after burial, losing more than half their water.

Both cementation and compaction result in **lithification,** the hardening of soft sediment into rock.

■ Classification of Siliciclastic Sediments and Sedimentary Rocks

We can now use our knowledge of sedimentary processes to classify sediments and their lithified counterparts, sedimentary rocks. As we have seen, the major divisions are the siliciclastic sediments and sedimentary rocks and the chemical and biological sediments and sedimentary rocks. Siliciclastic sediments and rocks constitute more than three-fourths of the total mass of all types of sediments and sedimentary rocks in Earth's crust (**Figure 5.18**). We therefore begin with them.

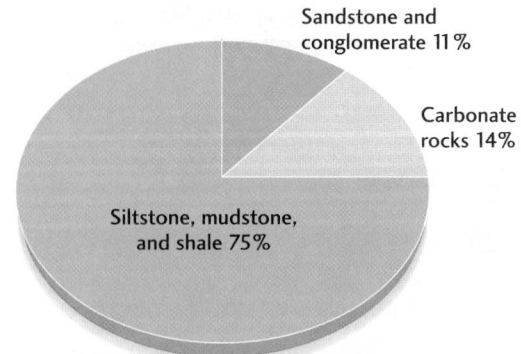

Sandstone and conglomerate 11%

Carbonate rocks 14%

Siltstone, mudstone, and shale 75%

FIGURE 5.18 ■ The relative abundances of the major sedimentary rock types. In comparison with these three types, all other sedimentary rock types—including evaporites, cherts, and other chemical sedimentary rocks— exist in only minor amounts.

Siliciclastic sediments and rocks are categorized primarily by particle size (**Table 5.3**):

■ *Coarse-grained:* gravel and conglomerate
■ *Medium-grained:* sand and sandstone
■ *Fine-grained:* silt and siltstone; mud, mudstone, and shale; clay and claystone

We classify siliciclastic sediments and rocks on the basis of their particle size because it distinguishes them by one of the most important conditions of sedimentation:

TABLE 5-3	Major Classes of Siliciclastic Sediments and Sedimentary Rocks	
Particle Size	Sediment	Rock
Coarse-Grained	**Gravel**	
Larger than 256 mm	Boulder	
256–64 mm	Cobble	Conglomerate
64–2 mm	Pebble	
Medium-Grained		
2–0.062 mm	Sand	Sandstone
Fine-Grained	**Mud**	
0.062–0.0039 mm	Silt	Siltstone
		Mudstone (blocky fracture)
Finer than 0.0039 mm	Clay	Shale (breaks along bedding)
		Claystone

current strength. As we have seen, the larger the particle, the stronger the current needed to transport and deposit it. This relationship between current strength and particle size is the reason like-sized particles tend to accumulate in sorted beds. In other words, most sand beds do not contain pebbles or mud, and most muds consist only of particles finer than sand.

Of the various types of siliciclastic sediments and sedimentary rocks, the fine-grained siliciclastics are by far the most abundant—about three times more common than the coarser-grained siliciclastics (see Figure 5.18). The abundance of the fine-grained siliciclastics, which contain large amounts of clay minerals, is due to the chemical weathering of the large quantities of feldspar and other silicate minerals in Earth's crust into clay minerals. We turn now to a consideration of each of the three major classes of siliciclastic sediments and sedimentary rocks in more detail.

Coarse-Grained Siliciclastics: Gravel and Conglomerate

Gravel is the coarsest siliciclastic sediment, consisting of particles larger than 2 mm in diameter and including pebbles, cobbles, and boulders. **Conglomerate** is the lithified equivalent of gravel (**Figure 5.19a**). Pebbles, cobbles, and boulders are easy to study and identify because of their large size, which tells us the strength of the currents that transported them. In addition, their composition can tell us about the nature of the distant terrain where they were produced.

There are relatively few sedimentary environments—mountain streams, rocky beaches with high waves, and glacier meltwaters—in which currents are strong enough to transport gravel. Strong currents also carry sand, and we almost always find sand between gravel particles. Some of it is deposited with the gravel, and some infiltrates the spaces between particles after the gravel is deposited.

Medium-Grained Siliciclastics: Sand and Sandstone

Sand consists of medium-sized particles, ranging from 0.062 to 2 mm in diameter. These particles can be moved by moderate currents, such as those of rivers, waves at shorelines, and the winds that blow sand into dunes. Sand grains are large enough to be seen with the naked eye, and many of their features are easily discerned with a low-power magnifying glass. The lithified equivalent of sand is **sandstone** (Figure 5.19b).

Both groundwater geologists and petroleum geologists have a special interest in sandstones. Groundwater geologists study the origin of sandstones to predict possible supplies of water in areas of porous sandstone, such as those found in the western plains of North America. Petroleum geologists must understand the porosity and cementation of sandstones because much of the oil and natural gas discovered in the past 150 years has been found in buried sandstones. In addition, much of the uranium used for nuclear power plants and weapons has come from uranium deposits precipitated in sandstones.

SIZES AND SHAPES OF SAND GRAINS Medium-sized siliciclastic particles—sand grains—are subdivided into fine, medium, and coarse grains. The average size of the grains in any one sandstone can be an important clue to both the strength of the current that carried them and the sizes of the crystals eroded from the parent rock. The range of grain sizes and their relative abundances are also significant. If all the grains are close to the average size, the sand is well sorted. If many grains are much larger or smaller than the average, the sand is poorly sorted (see Figure 5.6). The degree of sorting can help us distinguish, for example, between sands deposited on beaches (which tend to be well sorted) and sands deposited by glaciers (which tend to be muddy and poorly sorted). The shapes of sand grains can also be important clues to their origin. Sand grains,

(a) Conglomerate

(b) Sandstone

(c) Shale

FIGURE 5.19 ■ Examples of the three major classes of siliciclastic sedimentary rocks. [*conglomerate and sandstone:* John Grotzinger/Ramón Rivera-Moret/MIT; *shale:* John Grotzinger/Ramón Rivera-Moret/Harvard Mineralogical Museum.]

like pebbles and cobbles, are abraded and rounded during transportation. Angular grains imply short transport distances; rounded ones indicate long journeys down a large river system (see Figure 5.7).

MINERALOGY OF SANDS AND SANDSTONES

Siliciclastics can be further subdivided by their mineralogy, which can help identify the parent rocks. Thus, there are quartz-rich sandstones and feldspar-rich sandstones. Some sands are bioclastics, rather than siliciclastics; they are formed from materials such as carbonate minerals that were originally precipitated as shells, but then broken up and transported by currents. Thus, the mineralogy of sands and sandstones indicates the source areas and materials that were eroded to produce the sand grains. Sodium- and potassium-rich feldspars with abundant quartz, for example, might indicate that the sediments were eroded from a granitic terrain. Other minerals, as we will see in Chapter 6, might indicate metamorphic parent rocks.

The mineral content of sands and sandstones also indicates the plate tectonic setting of the parent rock. Sandstones containing abundant fragments of mafic volcanic rock, for example, might indicate that the sand grains were derived from a volcanic mountain belt at a subduction zone.

MAJOR KINDS OF SANDSTONES Sandstones can be divided into four major groups on the basis of their mineralogy and texture (**Figure 5.20**):

- **Quartz arenites** are made up almost entirely of quartz grains, usually well sorted and rounded. Pure quartz sands result from extensive weathering before and during transportation that removed everything but quartz, the most stable silicate mineral.

- **Arkoses** are more than 25 percent feldspar. Their grains tend to be more angular and less well sorted than those of quartz arenites. These feldspar-rich sandstones come from rapidly eroding granitic and metamorphic terrains where chemical weathering is subordinate to physical weathering.

- **Lithic sandstones** contain many particles derived from fine-grained rocks, mostly shales, volcanic rocks, and fine-grained metamorphic rocks.

- **Graywacke** is a heterogeneous mixture of rock fragments and angular grains of quartz and feldspar in which the sand grains are surrounded by a fine-grained clay matrix. Much of this matrix is formed from fragments of relatively soft rock, such as shale and some volcanic rocks, that are chemically altered and physically compacted after deep burial of the sandstone formation.

Fine-Grained Siliciclastics

The finest-grained siliciclastic sediments and sedimentary rocks are the silts and siltstones; the muds, mudstones, and shales; and the clays and claystones. All of them consist of

FIGURE 5.20 ▪ The mineralogy of four major groups of sandstones and the sedimentary environments where they are most likely to be found.

particles that are less than 0.062 mm in diameter, but they vary widely in their ranges of grain sizes and in their mineral compositions. Fine-grained sediments are deposited by the gentlest currents, which allow the finest sediment particles to settle slowly to the bottom in quiet waves.

SILT AND SILTSTONE **Siltstone** is the lithified equivalent of **silt,** a siliciclastic sediment in which most of the grains are between 0.0039 and 0.062 mm in diameter. Siltstone looks similar to mudstone or very fine grained sandstone.

MUD, MUDSTONE, AND SHALE **Mud** is a siliciclastic sediment containing water in which most of the particles are less than 0.062 mm in diameter. Thus, mud can be made of silt- or clay-sized sediment particles or varying quantities of both. The general term "mud" is very useful in fieldwork because it is often difficult to distinguish between silt- and clay-sized particles without a microscope.

Muds are deposited by rivers and tides. As a river recedes after flooding, the current slows, and mud, some of it containing abundant organic matter, settles on the floodplain. This mud contributes to the fertility of river floodplains. Muds are also left behind by ebbing tides along many tidal flats where wave action is mild. Much of the deep seafloor, where currents are weak or absent, is blanketed by muds.

The fine-grained rock equivalents of muds are mudstones and shales. **Mudstones** are blocky and show poor or no bedding. Distinct beds may have been present when the sediments were first deposited but then lost through bioturbation. **Shales** (Figure 5.19c) are composed of silt plus a significant component of clay, which causes them to break readily along bedding planes. Many muds contain more than 10 percent calcium carbonate sediments, forming calcareous mudstones and shales. Black, or organic,

shales contain abundant organic matter. Some, called oil shales, contain large quantities of oily organic material, which makes them a potentially important source of oil.

Hydraulic fracturing, also known as "fracking," is caused by the injection of highly pressurized fluids into shale. This creates new channels (fractures) in the rock, which link together tiny pores filled with oil and natural gas to create a flow of larger scale that is economically viable. The Marcellus Formation, found in the northeastern United States (see Figure 5.21), was named for Marcellus, New York. It is a unit of shale that had previously untapped natural gas reserves. In 2007, the Marcellus Shale was first drilled into, and using fracking methods, the extraction of natural gas became economically viable. The environmental impacts of fracking are debated, though, due to the effects of the chemicals used, the water supply, and the safety of drilling.

CLAY AND CLAYSTONE **Clay** is the most abundant component of fine-grained sediments and sedimentary rocks and consists largely of clay minerals. Clay-sized particles are less than 0.0039 mm in diameter. Rocks made up exclusively of clay-sized particles are called **claystones.**

■ Classification of Chemical and Biological Sediments and Sedimentary Rocks

Chemical and biological sediments and sedimentary rocks can be classified by their chemical composition (Table 5.4). Geologists distinguish between chemical sediments and biological sediments not only for convenience, but also to

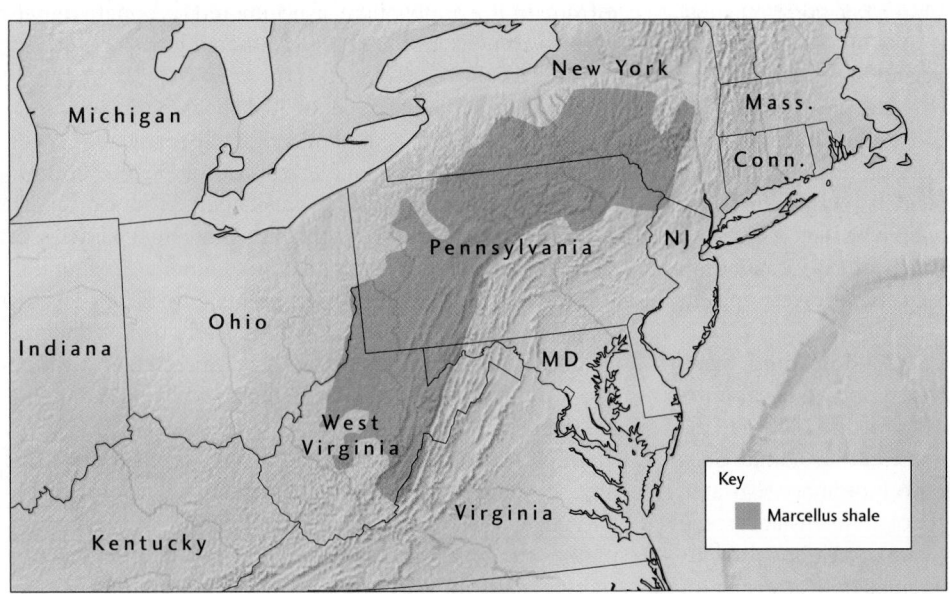

FIGURE 5.21 ■ The Marcellus Formation, found in the northeastern United States, has previously untapped natural gas reserves. The shaded area of the map indicates the most economically promising parts of the Marcellus shale.

TABLE 5-4	Classification of Biological and Chemical Sediments and Sedimentary Rocks		
Sediment	Rock	Chemical Composition	Minerals
Biological			
Sand and mud (primarily bioclastic)	Limestone	Calcium carbonate ($CaCO_3$)	Calcite, aragonite
Siliceous sediment	Chert	Silica (SiO_2)	Opal, chaldeony, quartz
Peat, organic matter	Organics	Carbon compounds; carbon compounded with oxygen and hydrogen	(Coal, oil, natural gas)
No primary sediment (formed by diagenesis)	Phosphorite	Calcium phosphate ($Ca_3(PO_4)_2$)	Apatite
Chemical			
No primary sediment (formed by diagenesis)	Dolostone	Calcium-magnesium carbonate ($CaMg(CO_3)_2$)	Dolomite
Iron oxide sediment	Iron formation	Iron silicate; oxide (Fe_2O_3); limonite, carbonate	Hematite, siderite
Evaporite sediment	Evaporite	Calcium sulfate ($CaSO_4$); sodium chloride (NaCl)	Gypsum, anhydrite, halite, other salts

emphasize the importance of organisms as the chief mediators of biological sedimentation. Both kinds of sediments can tell us about chemical conditions in the ocean, their predominant environment of deposition.

Carbonate Sediments and Rocks

Most **carbonate sediments** and **carbonate rocks** are formed by the accumulation and lithification of carbonate minerals that are directly or indirectly precipitated by organisms. The most abundant of these carbonate minerals is calcite (calcium carbonate, $CaCO_3$); in addition, most carbonate sediments contain aragonite, a less stable form of calcium carbonate. Some organisms precipitate calcite, some precipitate aragonite, and some precipitate both. During burial and diagenesis, carbonate sediments react with water to form a new suite of carbonate minerals.

The dominant biological sedimentary rock lithified from carbonate sediments is **limestone,** which is composed mainly of calcite (**Figure 5.22a**). Limestone is formed from carbonate sands and muds and, in some cases, ancient reefs (see Figure 5.10).

Another abundant carbonate rock is **dolostone,** made up of the mineral dolomite, which is composed of calcium–magnesium carbonate. Dolostones are diagenetically altered carbonate sediments and limestones. Dolomite does not form as a primary precipitate from ordinary seawater, and no organisms secrete shells of dolomite. Instead, some calcium ions in the calcite or aragonite of a carbonate sediment are exchanged for magnesium ions from seawater (or

magnesium-rich groundwater) slowly passing through the pores of the sediment. This exchange converts calcium carbonate ($CaCO_3$) into dolomite ($CaMg(CO_3)_2$).

DIRECT BIOLOGICAL PRECIPITATION OF CARBONATE SEDIMENTS Carbonate rocks are abundant because of the large amounts of calcium and carbonate minerals dissolved in seawater, which organisms can convert directly into shells. Calcium is supplied by the weathering of feldspars and other minerals in igneous and metamorphic rocks. Carbonate minerals are derived from the carbon dioxide in the atmosphere. Calcium and carbonate minerals also come from the easily weathered limestone on the continents.

Most carbonate sediments of shallow marine environments are bioclastic sediments originally secreted as shells by organisms living near the surface or on the bottom. After the organisms die, they break apart, producing shells or fragments of shells that constitute individual particles, or *clasts,* of carbonate sediment. These sediments are found in tropical and subtropical environments from Pacific islands to the Caribbean and the Bahamas. Carbonate sediments are most accessible for study in these spectacular vacation spots, but the deep sea is where most carbonate sediments are deposited today.

Most of the carbonate sediments deposited on the abyssal plain of the deep sea are derived from the calcite shells of **foraminifera** (see Figure 3.1b) and other planktonic organisms that live in the surface waters and secrete calcium carbonate. When the organisms die,

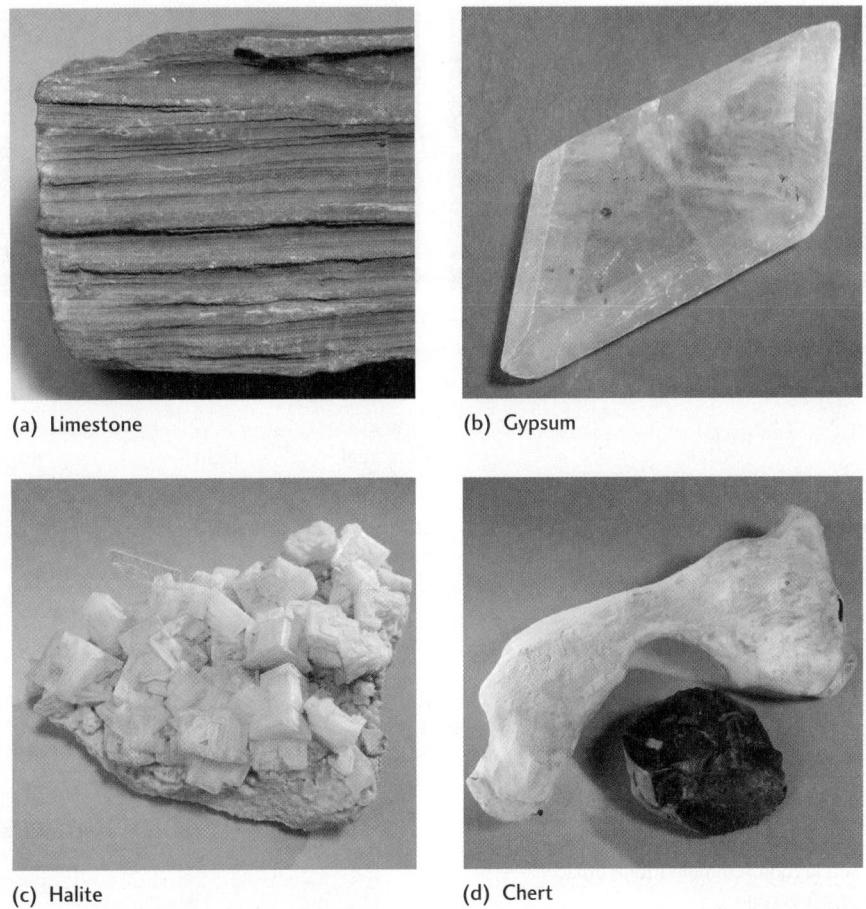

(a) Limestone

(b) Gypsum

(c) Halite

(d) Chert

FIGURE 5.22 ■ Chemical and biological sedimentary rocks: (a) limestone, lithified from carbonate sediments; (b) gypsum and (c) halite, marine evaporites that precipitate in shallow seawater basins; (d) chert, made up of siliceous sediments. [John Grotzinger/Ramón Rivera-Moret/Harvard Mineralogical Museum.]

their shells settle to the seafloor and accumulate there as sediments.

Reefs are moundlike or ridgelike organic structures composed of the carbonate skeletons and shells of millions of organisms. In the warm seas of the present, reefs are built mainly by corals, but hundreds of other organisms, such as algae, clams, and snails, also contribute. In contrast to the soft, loose sediments produced in other carbonate environments, the reef forms a rigid, wave-resistant structure of solid calcite and aragonite that is built up to and slightly above sea level. The solid calcite and aragonite of the reef is produced directly by the carbonate-cementing action of the organisms; there is no loose sediment stage.

Coral reefs may give rise to *carbonate platforms:* extensive flat, shallow areas, such as the Bahamas, where both biological and nonbiological carbonate sediments are deposited (**Figure 5.23**). Carbonate platforms are among the most important carbonate environments, both in past geologic ages and at present. The building of a carbonate platform results from interactions between the biosphere, hydrosphere, and lithosphere (see Earth Issues 5.1). The

process begins with a reef that encloses and shelters an area of shallow ocean water known as a *lagoon.* Carbonate-secreting organisms proliferate in and around the lagoon, and carbonate sediments accumulate rapidly, while in the open ocean outside the reef, sedimentation is much slower. At this point, the carbonate platform has a *ramp* morphology, with gentle slopes leading to deeper water. As sedimentation in the lagoon continues to outpace that outside the reef, the platform grows taller, developing a *rimmed shelf* morphology. Below the rims are steep slopes covered with loose carbonate sediments derived from the rim materials.

REEFS AND EVOLUTIONARY PROCESSES Today, reefs are constructed mainly by corals, but at earlier times in Earth's history, they were constructed by other organisms, such as a now-extinct variety of mollusk (**Figure 5.24**). Carbonate sediments and rocks formed from reefs record the diversification and extinction of reef-building organisms over geologic time. That record shows us how ecology and environmental change help to regulate the process of evolution.

1 The Bahamas are part of a carbonate platform system in the Atlantic Ocean east of Florida.

2 Reefs are built in warm, shallow seas by organisms that precipitate calcium carbonate.

Coral reef   Light   Lagoon   Open ocean

3 Within the shallow lagoon, growth of organisms is rapid, and carbonate sediment forms quickly...

4 ...whereas in the open ocean outside the reef, sedimentation is much slower.

Light   Ramp morphology

5 If the sea level rises, the reef continues to grow toward the light at sea level ... and lagoon sedimentation outpaces sedimentation in the open ocean.

Light

6 Eventually, the carbonate platform develops a rimmed shelf morphology, with steep sides falling away to the open ocean.

Loose carbonate fragments

FIGURE 5.23 ■ Marine organisms create carbonate platforms. [NASA; © Manfred Capale/Age Fotostock; © Stephen Frink/Corbis.]

Today, natural and anthropogenic changes threaten the growth of coral reefs, which are very sensitive to environmental change. In 1998, an El Niño event (described in Chapter 15) raised sea surface temperatures so much that many reefs in the western Indian Ocean were killed. The reefs of the Florida Keys are dying off for a completely different reason: they are getting too much of a good thing. Groundwaters originating in the farmlands of the Florida Peninsula are seeping out to the reefs and exposing them to lethal concentrations of nutrients.

INDIRECT BIOLOGICAL PRECIPITATION OF CARBONATE SEDIMENTS A significant fraction of the carbonate mud deposited in lagoons and on shallow carbonate platforms is precipitated indirectly from seawater. Microorganisms may be involved in this process, but their role is still uncertain. They may help to shift the balance of calcium (Ca^{2+}) and carbonate (CO_3^{2-}) ions in the seawater surrounding them so that calcium carbonate ($CaCO_3$) is formed. Microorganisms can precipitate carbonate minerals only if their external environment already contains

Earth Issues

5.1 Darwin's Coral Reefs and Atolls

For more than 200 years, coral reefs have attracted explorers and travel writers. Ever since Charles Darwin sailed the oceans on the *Beagle* from 1831 to 1836, these reefs have been a matter of scientific discussion as well. Darwin was one of the first scientists to analyze the geology of coral reefs, and his theory of the origin of one type of coral reef is still accepted today.

The coral reefs that Darwin studied were atolls, coral islands in the open ocean surrounding circular lagoons. The outermost part of an atoll is a slightly submerged, wave-resistant reef front: a steep slope facing the ocean. The reef front is composed of the interlaced skeletons of corals and calcareous algae, which form a tough, hard limestone. Behind the reef front is a flat platform extending into a shallow lagoon. An island may lie at the center of the lagoon. Parts of the reef, as well as the central island, are above sea level and may become forested. A great many plant and animal species inhabit the reef and the lagoon.

Coral reefs are generally limited to waters less than 20 m deep because, below that depth, seawater does not transmit enough light to enable reef-building organisms to grow. How, then, could an atoll be built up from the bottom of the deep, dark ocean? Darwin proposed that the process starts with a volcano building up to the sea surface from the seafloor and forming an island. As the volcano becomes dormant, temporarily or permanently, corals and algae colonize the shore of the island and build fringing reefs. Erosion may then lower the volcanic island almost to sea level.

Darwin reasoned that if such a volcanic island were to subside slowly beneath the waves, actively growing corals and algae might keep pace with its subsidence, continuously building up the reef over geologic time. In this way, the volcanic island would disappear, leaving an atoll in its place. More than 100 years after Darwin proposed his theory, deep drilling on several atolls found volcanic rock below their coralline limestone. And, some decades later, the theory of plate tectonics explained both volcanism and the subsidence that resulted from plate cooling and contraction.

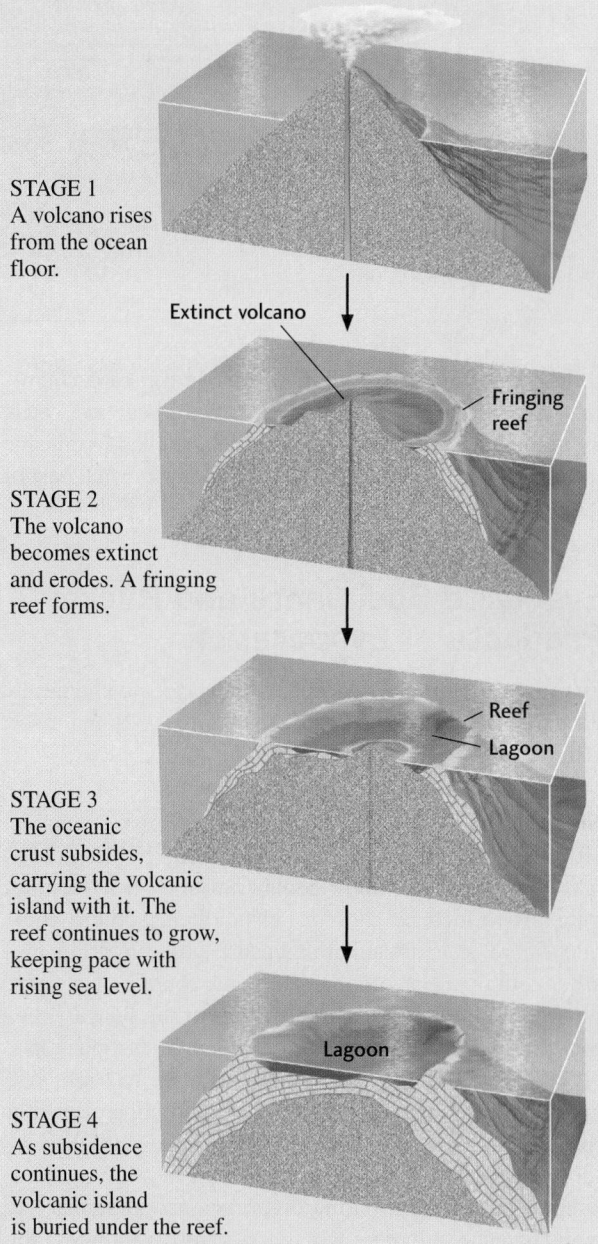

STAGE 1
A volcano rises from the ocean floor.

Extinct volcano

Fringing reef

STAGE 2
The volcano becomes extinct and erodes. A fringing reef forms.

Reef
Lagoon

STAGE 3
The oceanic crust subsides, carrying the volcanic island with it. The reef continues to grow, keeping pace with rising sea level.

Lagoon

STAGE 4
As subsidence continues, the volcanic island is buried under the reef.

Bora Bora atoll, South Pacific Ocean. Reef-building organisms have constructed a fringing reef around a volcanic island, forming a protected lagoon. [Jean-Marc Truchet/Stone/Getty Images.]

FIGURE 5.24 ■ Limestone formed from a reef constructed by now-extinct mollusks (rudists) in the Cretaceous Shuiba formation, Sultanate of Oman. [Courtesy of John Grotzinger.]

abundant calcium and carbonate ions. In this case, chemicals that the microorganisms emit into the seawater cause the minerals to precipitate. In contrast, shelled organisms secrete carbonate minerals continually as a normal part of their life cycle.

Evaporite Sediments and Rocks: Products of Evaporation

Evaporite sediments and **evaporite rocks** are chemically precipitated from evaporating seawater or, in some cases, lake water.

MARINE EVAPORITES Marine evaporites are chemical sediments and sedimentary rocks formed by the evaporation of seawater. These sediments and rocks contain minerals formed by the crystallization of sodium chloride (halite), calcium sulfate (gypsum and anhydrite), and other combinations of ions commonly found in seawater. As evaporation proceeds and the ions in the seawater become more concentrated, those minerals crystallize in a set sequence. As dissolved ions precipitate to form each mineral, the composition of the evaporating seawater changes.

Seawater has the same composition in all the oceans, which explains why marine evaporites are so similar the world over. No matter where seawater evaporates, the same sequence of minerals always forms. The study of evaporite sediments also shows us that the composition of the oceans has stayed more or less constant over the past 1.8 billion years. Before that time, however, the precipitation sequence may have been different, indicating that the composition of seawater may also have been different.

The great volume of many marine evaporites, some of which are hundreds of meters thick, shows that they could not have formed from the small amount of water that could be held in a small, shallow bay or pond. A huge amount of seawater must have evaporated to form them. The way in which such large quantities of seawater evaporate is very clear in bays or arms of the sea that meet the following conditions (**Figure 5.25**):

■ The freshwater supply from rivers is small.

■ Connections to the open ocean are constricted.

■ The climate is arid.

In such locations, water evaporates steadily, but the connections allow seawater to flow in to replenish the evaporating waters of the bay. As a result, those waters stay at a constant volume, but become more saline than the open ocean. The evaporating bay waters remain more or less constantly supersaturated and steadily deposit evaporite minerals on the floor of the bay.

As seawater evaporates, the first precipitates to form are the carbonates. Continued evaporation leads to the precipitation of gypsum, or calcium sulfate ($CaSO_4 \bullet 2H_2O$) (Figure 5.22b). By the time gypsum precipitates, almost no carbonate ions are left in the water. Gypsum is the principal component of plaster of Paris and is used in the manufacture of wallboard, which lines the walls of most new houses.

After still further evaporation, the mineral halite, or sodium chloride ($NaCl$)—one of the most common chemical sediments precipitated from evaporating seawater—starts to form (Figure 5.22c). Halite, as you may remember from Chapter 3, is table salt. Deep under the city of Detroit, Michigan, beds of salt laid down by an evaporating arm of an ancient ocean are commercially mined.

1 Salt water entered the Mediterranean through a narrow channel.

2 Evaporation removed more water than was replaced by freshwater inflow.

Evaporation

Freshwater inflow (small)

France

Italy

ATLANTIC OCEAN

Spain

MEDITERRANEAN SEA

Gypsum and halite crystals

Evaporite sediments

3 As the basin became more saline, gypsum and halite precipitated, forming evaporite sediments.

FIGURE 5.25 ▪ A marine evaporite environment of the past. The drier climate of the Miocene epoch made the Mediterranean Sea shallower than it is today, and its restricted connection to the open ocean created conditions suitable for evaporite formation. As seawater evaporated, gypsum precipitated to form evaporite sediments. A further increase in salinity led to the crystallization of halite. (The basin depth is greatly exaggerated in this diagram.)

In the final stages of evaporation, after the sodium chloride is gone, magnesium and potassium chlorides and sulfates precipitate from the water. The salt mines near Carlsbad, New Mexico, contain commercial quantities of potassium chloride. Potassium chloride is often used as a substitute for table salt by people with certain dietary restrictions.

This sequence of mineral precipitation from seawater has been studied in the laboratory and is matched by the bedding sequences found in certain natural evaporite formations. Most of the world's evaporites consist of thick sequences of dolomite, gypsum, and halite and do not contain the final-stage precipitates. Many do not go even as far as halite. The absence of the final stages indicates that the water did not evaporate completely, but was replenished by normal seawater as evaporation continued.

NONMARINE EVAPORITES Evaporite sediments also form in arid-region lakes that typically have few or no river outlets. In such lakes, evaporation controls the lake level, and incoming minerals derived from chemical weathering accumulate as sediments. The Great Salt Lake is one of the best known of these lakes (**Figure 5.26**). In the dry

FIGURE 5.26 ▪ The high concentrations of dissolved ions in the Great Salt Lake make it one of the saltiest bodies of water in the world—eight times more saline than seawater. Evaporite sediments form when these ions precipitate. [© Jon Mclean/Alamy.]

climate of Utah, evaporation has more than balanced the inflow of fresh water from rivers and rain. As a result, the concentrations of dissolved ions in the lake make it one of the saltiest bodies of water in the world—eight times more saline than seawater. Sediments form when these ions precipitate.

Small lakes in arid regions may precipitate unusual salts, such as borates (compounds of the element boron), and some become alkaline. The water in this kind of lake is poisonous. Economically valuable deposits of borates and nitrates (minerals containing the element nitrogen) are found in the sediments beneath some of these lakes.

Other Biological and Chemical Sediments

Carbonate minerals secreted by organisms are the principal source of biological sediments, and minerals precipitated from evaporating seawater are the principal

Southern limit of Great Barrier Reef

Data SIO, NOAA, U.S. Navy, NGA, GEBCO ©2009 Cnes/Spot Image Image ©2009 TerraMetrics Image ©2009 DigitalGlobe

Sediments are deposited in specific geologic environments where the formation of sedimentary rock takes place. Specifically, the processes that transform sediment into sedimentary rock happen near Earth's surface and always involve the presence of liquid water in some form. Thus, Google Earth is an ideal tool for interpreting and appreciating the spectrum of environments in which sedimentary rocks form.

Among these unique sedimentary environments are carbonate platforms. These platforms form when ions dissolved in seawater precipitate to form carbonate sediments. This process is often controlled by organisms. Consider the Great Barrier Reef off the northeastern coast of Australia. Here you will see a sedimentary environment driven by the life cycles of small marine organisms and the calcite-enriched water that they live in. What causes the distinctive blue-green color of the water here, and how are soluble minerals such as $CaCO_3$ precipitated as sediment on the ocean floor? Appreciate the geometry of the reef feature and its geographic limits to the north and south. Now compare the geometry of the Great Barrier Reef with a slightly different carbonate environment. Travel down to the equatorial waters of Bora Bora atoll in the South Pacific Ocean and see how the carbonate deposits there differ. Why does the reef take its circular shape, and what inspired it to begin growing here? These questions and many more can be explored though the GE interface.

source of chemical sediments. However, there are several less common biological and chemical sediments that are locally abundant. They include chert, coal, phosphorite, iron ore, and the organic carbon–rich sediments that produce oil and natural gas. The role of biological versus chemical processes in forming these sediments is variable.

SILICEOUS SEDIMENTS: SOURCES OF CHERT

One of the first sedimentary rocks to be used for practical purposes by our prehistoric ancestors was **chert,** which is composed of silica (SiO_2) (Figure 5.22d). Early hunters used it for arrowheads and other tools because it could be chipped and shaped to form hard, sharp implements. A common name for chert is *flint;* the two terms are virtually interchangeable. The silica in most cherts is in the form of extremely fine grained quartz. Some geologically young cherts consist of opal, a less fine grained form of silica.

Like calcium carbonate sediments, many siliceous sediments are precipitated biologically as silica shells

LOCATION	Great Barrier Reef, northeastern Australian coast, and Bora Bora Atoll, South Pacific Ocean
GOAL	Explore an area of significant sediment deposition in a modern sedimentary environment
LINKED	Figure 5.18 and Earth Issues 5.1

1. Navigate to the Great Barrier Reef on the northeastern coast of Australia and zoom in to an eye altitude of 20 km on Pipon Island, Queensland, Australia. Notice the white material around the island itself and along the coast of the peninsula just to the south (on mainland Australia). Based on your investigation of the images here (feel free to zoom in and out), how would you best characterize this white material? Be sure to consider any patterns you see in the distribution of this coastal material when choosing an answer.

   a. Unconsolidated carbonate sediment
   b. Large boulder deposits
   c. Cemented olivine sand
   d. Deltaic siliciclastic mudstone

2. From Pipon Island, zoom out to an eye altitude of 2300 km and appreciate the length of the offshore reef features paralleling the coast. Use the path measurement tool in GE to determine the approximate length of this reef system.

   a. 2000 km
   b. 1400 km
   c. 2800 km
   d. 750 km

3. Notice that the Great Barrier Reef provides some protection to the coastal environment where it is present. As you follow the reef to the south, it becomes less distinct and provides less coastal protection. At the southern end of the reef, waves from the open South Pacific are free to break on the Australian coast, and some of the best surfing in the world results. At approximately what southerly latitude does the reef system end?

   a. 10°30′23″ S; 143°30′06″ E
   b. 17°56′25″ S; 146°42′57″ E
   c. 24°39′49″ S; 153°15′18″ E
   d. 21°06′31″ S; 151°38′53″ E

4. Following up on questions 2 and 3, the Great Barrier Reef provides protection to the Australian coastline and allows for sedimentary processes to occur there. As one moves farther from the equator to latitudes of 25° S, it is clear that reef formation stops. Consider the conditions in which reef-building organisms precipitate calcium carbonate. What might be the primary climate-related factor controlling the southern limit of the Great Barrier Reef?

   a. Sea surface temperatures of less than 18°C
   b. The depth of ocean water along the coast to the south
   c. The amount of sediment on the beaches near Brisbane
   d. The color of the seawater along the coast south of 25°

Optional Challenge Question

5. Now let's travel to warmer climes by typing "Bora Bora atoll" into the GE search window and zooming in to an eye altitude of 20 km once you arrive there. In contrast to the Great Barrier Reef, this island in the South Pacific has a very limited reef system, yet that reef system has a unique geometry. The formation of an atoll like this one involves a unique relationship between biotic and geologic factors. From your observation and exploration of the atoll, which pair of biotic and abiotic factors properly reflects the relationship present here?

   a. Birds and quartz sand beaches
   b. Coral reefs and volcanic islands
   c. Foraminifera and outcrops of marine shale
   d. Whales and carbonate platforms

secreted by planktonic organisms that settle to the deep seafloor and accumulate as layers of sediment. After these sediments are buried by later sediments, they are cemented into chert. Chert may also form as nodules and irregular masses replacing carbonate in limestones and dolostones.

PHOSPHORITE SEDIMENTS Among the many other kinds of chemical and biological sediments deposited in the ocean is **phosphorite.** Sometimes called *phosphate rock,* phosphorite is composed of calcium phosphate precipitated from phosphate-rich seawater in places where currents of deep, cold water containing phosphate and other nutrients rise along continental margins. Organisms play an important role in creating phosphate-rich water, and bacteria that live on sulfur may play a key role in precipitating phosphate minerals. The phosphorite forms diagenetically by the interaction of calcium phosphate with muddy or carbonate sediments.

IRON OXIDE SEDIMENTS: SOURCE OF IRON FORMATIONS Iron formations are sedimentary rocks that usually contain more than 15 percent iron in the form of iron oxides and some iron silicates and iron carbonates. Iron oxides were once thought to be of chemical origin, but there is now some evidence that they may have been precipitated indirectly by microorganisms. Most of these rocks formed early in Earth's history, when there was less oxygen in the atmosphere and, as a result, iron dissolved more easily. Iron was transported to the ocean in soluble form, and where microorganisms were producing oxygen, it reacted with that oxygen and precipitated from solution as iron oxides (see Chapter 11).

ORGANIC SEDIMENTS: SOURCES OF COAL, OIL, AND NATURAL GAS Coal is a biological sedimentary rock composed almost entirely of organic carbon and formed by the diagenesis of wetland vegetation. In wetland environments, vegetation may be preserved from decay and accumulate as a rich organic material called **peat,** which contains more than 50 percent carbon. If peat is ultimately buried, it may be transformed into coal. Coal is classified as an **organic sedimentary rock,** a class that consists entirely or partly of organic carbon–rich deposits formed by the diagenesis of once-living material that has been buried.

In both lake and ocean waters, the remains of algae, bacteria, and other microscopic organisms may accumulate in fine-grained sediments as organic matter that can be transformed by diagenesis into oil and natural gas. **Crude oil** (petroleum) and **natural gas** are fluids that are not normally classed with sedimentary rocks. They can be considered organic sediments, however, because they form by the diagenesis of organic material in the pores of sedimentary rocks. Deep burial changes the organic matter

originally deposited along with inorganic sediments into a fluid that then escapes to porous rock formations and becomes trapped there. Oil and natural gas are found mainly in sandstones and limestones.

As supplies of oil and natural gas begin to diminish, the challenges for geologists increase. These challenges include finding new oil fields as well as squeezing out what is left behind in existing fields. Ultimately, it is the availability of organic sediments that limits how much oil and gas can be found. These sediments were more abundant in some periods of Earth's history, and they were formed more easily in certain parts of the world. So there are geologic constraints that we must learn to accept. But we can learn to be smarter about how we explore for what little oil is left, and the need for well-trained geologists has never been greater.

SUMMARY

What are the major processes that form sedimentary rock? Weathering breaks down rock into the particles that compose siliciclastic sediments and the dissolved ions and molecules that are precipitated to form chemical and biological sediments. Erosion mobilizes the particles produced by weathering. Currents of water and air and the movement of glaciers transport the sediments to their ultimate resting place in a sedimentary basin. Deposition (also called sedimentation) is the settling out of particles or precipitation of minerals to form layers of sediments. Burial and diagenesis compress and harden the sediments into sedimentary rock.

What are the two major types of sediments and sedimentary rocks? Sediments and the sedimentary rocks that form from them can be classified as one of two types: siliciclastic sediments or chemical and biological sediments. Siliciclastic sediments form from fragmentation of parent rock by physical and chemical weathering and are transported to sedimentary basins by water, wind, or ice. Chemical and biological sediments originate from minerals dissolved in and transported by water. Through chemical and biological reactions, these minerals are precipitated from solution to form sediments.

How are the major kinds of siliciclastic sediments and chemical and biological sediments classified? Siliciclastic sediments and sedimentary rocks are classified by particle size. The three major classes, in order of descending particle size, are coarse-grained siliciclastics (gravels and conglomerates); medium-grained siliciclastics (sands and sandstones); and fine-grained siliciclastics (silts and siltstones; muds, mudstones, and shales; and clays and claystones). This classification method emphasizes the importance of the strength of the current that transported the sediments. Chemical and biological sediments and sedimentary rocks are classified on the basis of their chemical composition. The most abundant of these rocks are the carbonate rocks: limestone and dolostone. Limestone is made up largely of biologically precipitated calcite. Dolostone is formed by the diage-netic alteration of limestone. Other chemical and biological sediments include evaporites; siliceous sediments such as chert; phosphorite; iron formations; and peat and other organic matter that is transformed into coal, oil, and natural gas.

KEY TERMS AND CONCEPTS

arkose (p. 134)

bedding sequence (p. 129)

bioclastic sediment (p. 119)

biological sediment (p. 118)

bioturbation (p. 128)

carbonate rock (p. 136)

carbonate sediment (p. 136)

cementation (p. 131)

chemical sediment (p. 118)

chemical weathering (p. 116)

chert (p. 143)

clay (p. 135)

claystone (p. 135)

coal (p. 144)

compaction (p. 131)

conglomerate (p. 133)

continental shelf (p. 122)

cross-bedding (p. 127)

crude oil (p. 144)

diagenesis (p. 130)

dolostone (p. 136)

evaporite rock (p. 140)

evaporite sediment (p. 140)

flexural basin (p. 123)

foraminifera (p. 136)

graded bedding (p. 127)

gravel (p. 133)

graywacke (p. 134)

iron formation (p. 144)

limestone (p. 136)

lithic sandstone (p. 134)

lithification (p. 132)

mud (p. 135)

mudstone (p. 135)

natural gas (p. 144)

organic sedimentary rock (p. 144)

peat (p. 144)

phosphorite (p. 144)

physical weathering (p. 116)

porosity (p. 131)

quartz arenite (p. 134)

reef (p. 137)

rift basin (p. 122)

ripple (p. 128)

salinity (p. 122)

sand (p. 133)

sandstone (p. 133)

sedimentary basin (p. 122)

sedimentary environment (p. 124)

sedimentary structure (p. 127)

shale (p. 135)

siliciclastic sediments (p. 118)

silt (p. 135)

siltstone (p. 135)

sorting (p. 121)

subsidence (p. 122)

terrigenous sediment (p. 126)

thermal subsidence basin (p. 122)

PRACTICING GEOLOGY EXERCISE

Where Do We Look for Oil and Gas?

The search for new deposits of oil and natural gas is taking on ever greater urgency as fuel supplies dwindle and geopolitical issues make nations eager to produce their own energy supplies. The search for these deposits must be guided by an understanding of how and where oil and gas form.

The first step in exploring for oil and gas is a search for sedimentary rocks formed from sediments that are likely to have been rich in organic matter. Once such rocks have been located, the next step is to determine how deeply they have been buried and the maximum temperature they might have achieved. These factors determine the prospectivity of the rocks—their likelihood of containing oil or gas.

Many fine-grained sediments and sedimentary rocks, such as shale, contain organic matter. Subsidence of sedimentary basins, coupled with deposition of overlying sedimentary layers, may result in deep burial of these organic-rich sediments. As they are buried progressively deeper, the sediments become increasingly hotter. The rate at which temperatures increase with depth is called the geothermal gradient (see Chapter 6).

Depending on the geothermal gradient in the sedimentary basin, organic-rich sedimentary rocks may eventually become hot enough that the organic matter they contain is transformed into oil or gas. That process of transformation (described in more detail in Chapter 23) is known as maturation. Maturation begins shortly after the sediments are deposited, but increases dramatically above 50°C. Oil is generated as the sediments are heated to temperatures between 60°C and 150°C. At higher temperatures, the oil becomes unstable and breaks down, or "cracks," to form natural gas.

Geologists have discovered organic-rich shales in the Rocknest basin, which has a geothermal gradient of 35°C/km. The accompanying diagram shows the relationship between depth of burial, temperature, and the relative amounts of oil and gas formed in shales in this sedimentary basin. Assuming that peak oil generation occurs at about 100°C, calculate the depth at which peak oil generation would occur in the Rocknest basin.

Depth of peak oil generation = Temperature of peak oil generation ÷ Geothermal gradient

= 100°C ÷ 35°C/km

= 2.85 km (2850 m)

If the organic-rich shales in the Rocknest basin were buried to depths of 2850 m or greater, then one might expect to find oil in the basin. However, if the depth of burial were shallower than 2850 m, then the prospectivity of the basin would be downgraded.

BONUS PROBLEM: The depth of peak gas generation in the Rocknest basin is 3575 m. Rearrange the equation above and solve for the temperature at which gas generation would peak.

EXERCISES

1. What processes change sediments into sedimentary rock?

2. How do siliciclastic sedimentary rocks differ from chemical and biological sedimentary rocks?

3. How and on what basis are the siliciclastic sedimentary rocks classified?

4. What kinds of sedimentary rocks are formed by the evaporation of seawater?

5. Define a sedimentary environment, and name three siliciclastic sedimentary environments.

6. Explain how plate tectonic processes control the development of sedimentary basins.

7. Name two kinds of carbonate rocks and explain how they differ.

8. How do organisms produce or modify sediments?

9. Name two ions that take part in the precipitation of calcium carbonate.

10. In what kinds of sedimentary rocks are oil and natural gas found?

THOUGHT QUESTIONS

1. Weathering of the continents has been much more widespread and intense in the past 10 million years than it was in earlier times. How might this observation be borne out in the sediments that now cover Earth's surface?

2. If you drilled one oil well into the bottom of a sedimentary basin that is 1 km deep and another that is 5 km deep, which would have the higher pressures and temperatures? Oil turns into natural gas at high basin temperatures. In which well would you expect to find more natural gas?

3. A geologist is heard to say that a particular sandstone was derived from a granite. What information could she have gleaned from the sandstone to lead her to that conclusion?

4. You are looking at a cross section of a rippled sandstone. What sedimentary structure would tell you the direction of the current that deposited the sand?

5. You discover a bedding sequence that has a conglomerate at the base; grades upward to a sandstone and then to a shale; and finally, at the top, grades to a limestone of cemented carbonate sand. What changes in the sediment's source area or in the sedimentary environment would have been responsible for this sequence?

6. From the base upward, a bedding sequence begins with a bioclastic limestone, passes upward into a dense carbonate rock made of carbonate-cementing organisms, and ends with beds of dolostone. Deduce the possible sedimentary environments represented by this sequence.

7. In what sedimentary environments would you expect to find carbonate muds?

8. How can you use the size and sorting of sediment particles to distinguish between sediments deposited in a glacial environment and those deposited in a desert?

9. Describe the beach sands that you would expect to be produced by the beating of waves on a coastal mountain range consisting largely of basalt.

10. What role do organisms play in the origin of some kinds of limestone? Compare the sediments formed in shallow environments with those formed in deep-sea environments.

11. Where are reefs likely to be found?

12. A bay is separated from the open ocean by a narrow, shallow inlet. What kind of sediment would you expect to find on the floor of the bay if the climate were warm and arid? What kind of sediment would you find if the climate were cool and humid?

13. How are chert and limestone similar in origin? Discuss the roles of biological versus chemical processes.

MEDIA SUPPORT

 5-1 Animation: Lithification

 5-2 Video: Limestone

 5-2 Animation: Cross Bedding

 5-3 Video: Sedimentary Bedding

 5-3 Animation: Sedimentary Basin Formation

 5-4 Video: The Rio Grande Rift

 5-1 Video: Original Horizontality, Superposition, and Sedimentary Structures

 5-5 Video: Natural Arches and Bridges

- Causes of Metamorphism 150

- Types of Metamorphism 153

- Metamorphic Textures 155

- Regional Metamorphism and Metamorphic Grade 159

- Plate Tectonics and Metamorphism 162

Connemara marble, found in western Ireland, shows strong deformation by folding during metamorphism. [Jennifer Griffes.]

METAMORPHISM: ALTERATION OF ROCKS BY TEMPERATURE AND PRESSURE

6

DURING THE ROCK CYCLE, rocks may be subjected to temperatures and pressures great enough to cause changes in their mineralogy, texture, or chemical composition. We are all familiar with some of the ways in which heat and pressure can transform materials. Cooking batter in a waffle iron not only heats up the batter but also puts pressure on it, transforming it into a rigid solid. In similar ways, rocks are transformed as they encounter high temperatures and pressures deep in Earth's crust.

Tens of kilometers below the surface, temperatures and pressures are high enough to cause chemical reactions and recrystallization that transform rock without being high enough to melt it. Increases in temperature and pressure and changes in the chemical environment can alter the mineral composition and crystalline texture of igneous and sedimentary rock, *even though it remains solid all the while.* The result is the third large class of rocks: metamorphic rocks, which have undergone changes in mineralogy, texture, chemical composition, or all three.

It is important to understand that most metamorphism is a dynamic process, not a static event. Earth's internal heat engine drives the plate tectonic processes that push rocks formed at Earth's surface down to great depths, thereby subjecting them to high pressures as well as high temperatures. But the transformed rocks return to Earth's surface eventually, and that process is largely powered by weathering and erosion—in other words, by the climate system.

This chapter examines the causes of metamorphism, the types of metamorphism that take place in certain geologic settings, and the origins of the various textures that characterize metamorphic rocks. It shows how geologists use characteristics of metamorphic rocks to understand how and where they were transformed, and it looks at what their journey through the rock cycle tells us about the processes that shape Earth's crust.

■ Causes of Metamorphism

Sediments and sedimentary rocks are products of Earth's surface environments, whereas igneous rocks are products of the magmas that originate in the lower crust and mantle. Metamorphic rocks are the products of processes acting on rocks at depths ranging from the upper to the lower crust.

When a rock is subjected to significant changes in temperature or pressure, it will, given enough time—short by geologic standards, but usually a million years or more—undergo changes in its chemical composition, mineralogy, and texture, or all three, until it is in equilibrium with the new temperature and pressure. A limestone filled with fossils, for example, may be transformed into a white marble in which no trace of fossils remains. The mineral and chemical composition of the rock may be unaltered, but its texture may have changed drastically, from small calcite crystals to large, interlocked calcite crystals that erase such former features as fossils. Shale, a well-bedded sedimentary rock so fine-grained that no individual crystal can be seen with the naked eye, may become schist, in which the original bedding is obscured and the texture is dominated by large crystals of mica. In this case, both mineralogy and texture have changed, but the overall chemical composition of the rock has remained the same.

Most metamorphic rocks are formed at depths of 10 to 30 km, in the middle to lower half of the crust. Only later are those rocks *exhumed,* or transported back to Earth's surface, where they may be exposed as outcrops. But metamorphism can also occur at Earth's surface. We can see metamorphic changes, for example, in the baked surfaces of soils and sediments just beneath volcanic lava flows.

The heat and pressure in Earth's interior and its fluid composition are the three principal factors that drive metamorphism. In much of Earth's crust, the temperature increases at a rate of 30°C per kilometer of depth, although that rate varies considerably among different regions, as we will see shortly. Thus, at a depth of 15 km, the temperature will be about 450°C—much higher than the average temperature at Earth's surface, which ranges from 10°C to

20°C in most regions. The contribution of pressure is the result of vertically oriented forces exerted by the weight of overlying rocks as well as horizontally oriented forces developed as the rocks are deformed by plate tectonic processes. The average pressure at a depth of 15 km amounts to about 4000 times the pressure at the surface.

As high as these temperatures and pressures may seem, they are only in the middle range of conditions for metamorphism, as **Figure 6.1** shows. A rock's *metamorphic grade* reflects the temperatures and pressures it was subjected to during metamorphism. We refer to metamorphic rocks formed under the lower temperatures and pressures of shallower crustal regions as *low-grade* metamorphic rocks and those formed under the higher temperatures and pressures at greater depths as *high-grade* metamorphic rocks.

As the grade of metamorphism changes, the assemblages of minerals within metamorphic rocks also change. Some silicate minerals are found mostly in metamorphic rocks: these minerals include kyanite, andalusite, sillimanite, staurolite, garnet, and epidote. Geologists use distinctive textures as well as mineral composition to help guide their studies of metamorphic rocks.

FIGURE 6.1 ■ Temperatures, pressures, and depths at which low-grade and high-grade metamorphic rocks form. The dark band shows the rates at which temperature and pressure increase with depth over much of the continental lithosphere.

The Role of Temperature

Heat can transform a rock's chemical composition, mineralogy, and texture by breaking chemical bonds and altering the existing crystal structures of the rock. When rock is moved from Earth's surface to its interior, where temperatures are higher, the rock adjusts to the new temperature. Its atoms and ions recrystallize, linking up in new arrangements and creating new mineral assemblages. Many new crystals grow larger than the crystals in the original rock.

The increase in temperature with increasing depth in Earth's interior is called the *geothermal gradient.* The geothermal gradient varies among plate tectonic settings, but on average it is about 30°C per kilometer of depth. In areas where the continental lithosphere has been stretched and thinned, such as Nevada's Great Basin, the geothermal gradient is *steep* (for example, 50°C per kilometer of depth). In areas where the continental lithosphere is old and thick, such as central North America, the geothermal gradient is *shallow* (for example, 20°C per kilometer of depth) (Figure 6.2).

Because different minerals crystallize and remain stable at different temperatures, we can use a rock's mineral composition as a kind of *geothermometer* to gauge the temperature at which it formed. For example, as sedimentary rocks containing clay minerals are buried deeper and deeper, the clay minerals begin to recrystallize and form new minerals, such as micas. With additional burial at greater depths and temperatures, the micas become unstable and begin to recrystallize into new minerals, such as garnet.

Plate tectonic processes such as subduction and continent-continent collision, which transport rocks and

FIGURE 6.2 ■ The geothermal gradient varies among plate tectonic settings, but pressure increases with depth at about the same rate everywhere. An *isotherm* is a line that connects zones of equal temperature.

In volcanic mountain belts, the 1300°C isotherm is about 50 km deep; …

…in plate extension areas, it is about 30 km deep; …

…and in stable continental lithosphere, it is about 150 km deep.

sediments into the hot depths of the crust, are the mechanisms that form most metamorphic rocks. In addition, limited metamorphism may occur where rocks are subjected to elevated temperatures near igneous intrusions. The heat is locally intense, but does not penetrate deeply; thus, the intrusions can metamorphose the surrounding country rock, but the effect is local in extent.

The Role of Pressure

Pressure, like temperature, changes a rock's chemical composition, mineralogy, and texture. Solid rock is subjected to two basic kinds of pressure, also called **stress:**

1. *Confining pressure* is a general force applied equally in all directions, like the pressure a swimmer feels under water. Just as a swimmer feels greater confining pressure when diving to greater depths, a rock descending to greater depths in Earth's interior is subjected to progressively increasing confining pressure in proportion to the weight of the overlying mass.

2. *Directed pressure,* or *differential stress,* is force exerted in a particular direction, as when you squeeze a ball of clay between your thumb and forefinger. Directed pressure is usually concentrated within particular zones or along discrete planes.

The compressive force exerted where lithospheric plates converge is a form of directed pressure, and it results in deformation of the rocks near the plate boundary. Heat reduces the strength of a rock, so directed pressure is likely to cause severe folding and other forms of ductile deformation, as well as metamorphism, where temperatures are high. Rocks subjected to differential stress may be severely distorted, becoming flattened in the direction the force is applied and elongated in the direction perpendicular to the force (**Figure 6.3**).

The minerals in a rock under pressure may be compressed, elongated, or rotated to line up in a particular direction, depending on the kind of stress applied to the rock. Thus, directed pressure guides the shape and orientation of the new crystals formed as minerals recrystallize under the influence of both heat and pressure. During the recrystallization of micas, for example, the crystals grow with the planes of their sheet silicate structures aligned perpendicular to the directed stress. The rock may develop a banded pattern as minerals of different compositions are segregated into separate planes.

Marble owes its remarkable strength to this recrystallization process. When limestone, a sedimentary rock, is heated to the very high temperatures that cause it to recrystallize, the original minerals and crystals become reoriented and tightly interlocked to form a very strong structure with no planes of weakness.

The pressure to which rock is subjected deep in Earth's crust is related to both the thickness and the density of the overlying rocks. Pressure, which is usually recorded in *kilobars* (1000 bars, abbreviated kbar), increases at a rate of 0.3 to 0.4 kbar per kilometer of depth (see Figure 6.1). One bar is approximately equivalent to the pressure of air at Earth's surface. A diver touring the deeper part of a coral reef at a depth of 10 m would experience an additional bar of pressure.

Minerals that are stable at the lower pressures near Earth's surface become unstable and recrystallize into new minerals under the increased pressures deep in Earth's crust. As we will see in Chapter 7, geologists have subjected rocks to extremely high pressures in the laboratory and recorded the pressures required to cause these changes. With these laboratory data

FIGURE 6.3 ■ These rocks in Sequoia National Forest, California, show both the banding and the folding characteristic of sedimentary rocks metamorphosed into marble, schist, and gneiss. [Gregory G. Dimijian/Science Source.]

in hand, we can examine the mineralogy and texture of metamorphic rock samples and infer what the pressures were in the area where they formed. Thus, metamorphic mineral assemblages can be used as pressure gauges, or *geobarometers*. Given a specific assemblage of minerals in a metamorphic rock, we can determine the range of pressures, and therefore the depth, at which the rock must have formed.

The Role of Fluids

Metamorphic processes can alter a rock's mineralogy by introducing or removing chemical components that are soluble in heated water. Hydrothermal fluids accelerate metamorphic chemical reactions because they carry dissolved carbon dioxide as well as other chemical substances—such as sodium, potassium, silica, copper, and zinc—that are soluble in hot water under pressure. As hydrothermal solutions percolate up to the shallower parts of the crust, they react with the rocks they penetrate, changing their chemical and mineral compositions and sometimes completely replacing one mineral with another without changing the rock's texture. This kind of change in a rock's composition by fluid transport of chemical substances into or out of it is called **metasomatism.** Many valuable deposits of copper, zinc, lead, and other metallic ores are

formed by this kind of chemical substitution, as we saw in Chapter 3.

Where do these chemically reactive fluids originate? Although most rocks appear to be completely dry and to have extremely low porosity, they characteristically contain water in minute pores (the spaces between grains). This water comes not from the pores of sedimentary rocks—from which most of the water is expelled during diagenesis—but rather from chemically bound water in clays. Water forms part of the crystal structure of metamorphic minerals such as micas and amphiboles. The carbon dioxide dissolved in hydrothermal fluids is derived largely from sedimentary carbonates: limestones and dolostones.

■ Types of Metamorphism

Geologists can duplicate metamorphic conditions in the laboratory and determine the precise combinations of pressure, temperature, and parent rock composition under which particular transformations might take place. But to understand when, where, and how such conditions came about in Earth's interior, we must categorize metamorphic rocks on the basis of their geologic settings (**Figure 6.4**).

1 Regional metamorphism
at convergent plate boundaries occurs at moderate to deep levels under moderate to ultra-high pressures and high temperatures.

2 High-pressure metamorphism
along linear belts of volcanic arcs, produced by continent-continent collision, occurs at high pressures.

6 Shock metamorphism,
which results from the heat and shock waves of a meteorite impact, transforms rock at impact site.

3 Contact metamorphism
affects a thin zone of country rock around an igneous intrusion.

Oceanic crust

Depth, km

0

35

75

Continental crust

Continental mantle lithosphere

Asthenosphere

Oceanic lithosphere

4 Seafloor metamorphism
occurs at mid-ocean ridges, where intruding magma drives seawater circulation through extruded basalts.

Water

5 Burial metamorphism
transforms sedimentary rocks at progressively increasing temperature and pressure.

FIGURE 6.4 ■ Different types of metamorphism occur in different geologic settings.

Regional Metamorphism

Regional metamorphism, the most widespread type of metamorphism, takes place where both high temperatures and high pressures are imposed over large parts of the crust. We use this term to distinguish this type of metamorphism from more localized transformations near igneous intrusions or faults. Regional metamorphism is a characteristic feature of convergent plate boundaries. It occurs in volcanic mountain belts, such as the Andes of South America, and in the cores of mountain chains produced by continent-continent collisions, such as the Himalaya of central Asia. These mountain chains are often linear features, so zones of regional metamorphism are often linear in their distribution. In fact, geologists usually interpret regionally extensive belts of metamorphic rocks as representing sites of former mountain chains that were eroded over millions of years, exposing the rocks at their core.

Some regional metamorphic belts are created by the high temperatures and moderate to high pressures near volcanic mountain belts formed where subducted plates sink deep into the mantle. Others are formed under the very high pressures and temperatures found deeper in the crust along boundaries where colliding continents deform rock and raise high mountain chains. In both cases, the metamorphosed rocks are typically transported to great depths in Earth's crust, then eventually uplifted, exposed, and eroded at Earth's surface. A full understanding of the patterns of regional metamorphism, including how rocks respond to systematic changes in temperature and pressure over time, depends on an understanding of the specific plate tectonic settings in which metamorphic rocks form. We will discuss that topic later in this chapter.

Contact Metamorphism

In **contact metamorphism,** the heat from an igneous intrusion metamorphoses the rock immediately surrounding it. This type of localized transformation normally affects only a thin zone of country rock along the zone of contact. In many contact metamorphic rocks, especially at the margins of shallow intrusions, the mineral and chemical transformations are largely related to the high temperature of the intruding magma. Pressure effects are important only where the magma is intruded at great depths. Here, the pressure results not from the intrusion forcing its way into the country rock, but from the presence of regional confining pressure. Contact metamorphism by volcanic deposits is limited to very thin zones because lavas cool quickly at Earth's surface and their heat has little time to penetrate the surrounding rocks deeply and cause metamorphic changes. Contact metamorphism may also affect xenoliths that are not completely melted. Blocks of rock up to several meters wide may be torn off the sides of magma chambers and completely surrounded by hot magma. Heat projects into these xenoliths from all directions, and they may become completely metamorphosed.

Seafloor Metamorphism

Another type of metamorphism, a form of metasomatism called **seafloor metamorphism,** is often associated with mid-ocean ridges (see Chapter 4). Hot basaltic lava at a seafloor spreading center heats infiltrating seawater, which starts to circulate through the newly forming oceanic crust by convection. The increase in temperature promotes chemical reactions between the seawater and the rock, forming altered basalts whose chemical compositions differ from that of the original basalt. Metasomatism resulting from percolation of high-temperature fluids also takes place on continents when hydrothermal solutions circulating near igneous intrusions metamorphose the rocks they intrude.

Other Types of Metamorphism

There are several other types of metamorphism that produce smaller amounts of metamorphic rock. Some of these types are extremely important in helping geologists understand conditions deep within Earth's crust.

BURIAL METAMORPHISM Recall from Chapter 5 that sedimentary rocks are transformed by diagenesis as they are gradually buried. Diagenesis grades into **burial metamorphism,** low-grade metamorphism that is caused by the progressive increase in pressure exerted by the growing layers of overlying sediments and sedimentary rocks and by the increase in heat associated with increased depth of burial.

Depending on the local geothermal gradient, burial metamorphism typically begins at depths of 6 to 10 km, where temperatures range between 100°C and 200°C and pressures are less than 3 kbar. This fact is of great importance to the oil and gas industry, which defines its "economic basement" as the depth where low-grade metamorphism begins. Oil and gas wells are rarely drilled below this depth because temperatures above 150°C convert organic matter trapped in sedimentary rocks into carbon dioxide rather than crude oil and natural gas.

HIGH-PRESSURE AND ULTRA-HIGH-PRESSURE METAMORPHISM Metamorphic rocks formed by **high-pressure metamorphism** (at 8 to 12 kbar) and **ultra-high-pressure metamorphism** (at pressures greater than 28 kbar) are rarely exposed at Earth's surface for geologists to study. These rocks are rare because they form at such great depths that it takes a very long time for them to be recycled to the surface. Most high-pressure metamorphic rocks form in subduction zones as sediments scraped from subducting oceanic crust are plunged to depths of over 30 km, where they experience pressures of up to 12 kbar.

Unusual metamorphic rocks once located at the base of Earth's crust can sometimes be found at Earth's surface. These rocks, called **eclogites** (see Figure 3.27), may contain minerals such as *coesite* (a very dense, high-pressure form of quartz) that indicate pressures of greater than 28 kbar, suggesting depths of over 80 km. Such rocks form at moderate to high temperatures, ranging from 800°C to 1000°C. In a few cases, these rocks contain *microscopic diamonds,* indicative of pressures greater than 40 kbar and depths greater than 120 km! Surprisingly, outcrop exposures of these ultra-high-pressure metamorphic rocks may cover areas greater than 400 km by 200 km. The only other two rocks known to come from these depths are diatremes and kimberlites (see Chapter 12), igneous rocks that form narrow pipes just a few hundred meters wide. Geologists agree that these latter rock types form by volcanic eruption, albeit from very unusual depths. In contrast, the mechanisms required to bring eclogites to the surface are hotly debated. It appears that these rocks represent pieces of the leading edges of continents that were subducted during continent-continent collisions and subsequently rebounded (via some unknown mechanism) to the surface before they had time to recrystallize at lower pressures.

SHOCK METAMORPHISM Shock metamorphism occurs when a meteorite collides with Earth. Upon impact, the energy represented by the meteorite's mass and velocity is transformed into heat and shock waves that pass through the impacted country rock. The country rock can be shattered and even partially melted to produce *tektites.* The smallest tektites look like droplets of glass. In some cases, quartz is transformed into coesite and *stishovite,* two of its high-pressure forms.

Most large impacts on Earth have left no trace of a meteorite because these bodies are usually destroyed in the collision with Earth. The occurrence of coesite and craters with distinctive fringing fractures, however, provides evidence of these collisions. Earth's dense atmosphere causes most meteorites to burn up before they strike its surface, so shock metamorphism is rare on Earth. On the surface of the Moon, however, shock metamorphism is pervasive. It is characterized by extremely high pressures of many tens to hundreds of kilobars.

▪ Metamorphic Textures

Metamorphism imprints new textures on the rocks it alters. The texture of a metamorphic rock is determined by the sizes, shapes, and arrangement of its constituent crystals. Some metamorphic rock textures depend on the particular kinds of minerals formed under metamorphic conditions. Variation in grain size is also important. In general, grain size increases as metamorphic grade increases. Each textural variety of metamorphic rock tells us something about the metamorphic process that created it. In this section, we examine those processes, then describe two major textural classes of metamorphic rocks: foliated rocks and granoblastic rocks.

Foliation and Cleavage

The most prominent textural feature of regionally metamorphosed rocks is **foliation,** a set of flat or wavy parallel cleavage planes produced by deformation of igneous and sedimentary rocks under directed pressure (**Figure 6.5**). These foliation planes may cut through the bedding of the original sedimentary rock at any angle or be parallel to the bedding (Figure 6.5). In general, as the grade of regional metamorphism increases, foliation becomes more pronounced.

A major cause of foliation is the formation of minerals with a platy crystal habit, chiefly the micas and chlorite. The planes of all the platy crystals are aligned parallel to the foliation, an alignment called the *preferred orientation* of the minerals (Figure 6.5). As platy minerals crystallize, their preferred orientation is usually perpendicular to the main direction of the forces squeezing the rock during metamorphism. Crystals of preexisting minerals may contribute to the foliation by rotating until they also lie parallel to the developing foliation plane.

The most familiar form of foliation is seen in slate, a common metamorphic rock, which is easily split into thin sheets along smooth, parallel surfaces. This *slaty cleavage* (not to be confused with the perfect cleavage of sheet silicates such as micas) develops at small, regular intervals in the rock.

Minerals with an elongate, needlelike crystal habit also tend to assume a preferred orientation during metamorphism: these crystals, too, normally line up parallel to the foliation plane. Rocks that contain abundant amphiboles (typically, metamorphosed mafic volcanic rocks) have this kind of texture.

Foliated Rocks

The **foliated rocks** are classified according to four main criteria:

1. Metamorphic grade
2. Grain (crystal) size
3. Type of foliation
4. Banding

Figure 6.6 shows examples of the major types of foliated rocks. In general, foliation progresses from one texture to another with increasing metamorphic grade. In this progression, as temperature and pressure increase, a shale may metamorphose first to a slate, then to a phyllite, then to a schist, then to a gneiss, and finally to a migmatite.

1 Directed pressure causes sedimentary rocks, such as shale, to form cleavage planes that may differ from their bedding planes.

2 The original bedding in a sample of shale is marked by the thin sandstone layers.

3 Regionally metamorphosed rocks show foliation caused by compressive forces.

4 Rocks develop foliation when they contain platy minerals that align along a preferred orientation. Depending on their preferred orientation, compressive forces cause the mineral crystals in the rock to grow or to align perpendicular to the compressive forces.

FIGURE 6.5 ■ Directed pressure on rocks containing platy minerals causes foliation. [(*top*) Marli Miller; (*bottom*) John Grotzinger.]

SLATE Slates are the lowest grade of foliated rocks. These rocks are so fine-grained that their individual crystals cannot be seen easily without a microscope. They are commonly produced by the metamorphism of shales or, less frequently, of volcanic ash deposits. Slates usually range from dark gray to black, colored by small amounts of organic material originally present in the parent shale. Slate splitters learned long ago to recognize foliation planes and use them to make thick or thin slabs for roofing tiles and blackboards. Flat slabs of slate are still used for flagstone walks in places where slate is abundant.

PHYLLITE Phyllites are rocks of a slightly higher grade than the slates, but are similar to them in character and

origin. They tend to have a more or less glossy sheen resulting from crystals of mica and chlorite that have grown a little larger than those of slates. Phyllites, like slates, tend to split into thin sheets, but less perfectly than slates.

SCHIST At low grades of metamorphism, the crystals of platy minerals are generally too small to be seen, and foliation planes are closely spaced. As rocks are subjected to higher temperatures and pressures, however, the platy crystals grow large enough to be visible to the naked eye, and the minerals tend to segregate into lighter and darker bands. This parallel arrangement of platy minerals produces the coarse, wavy foliation known as *schistosity*, which characterizes **schists.** Schists, which are intermediate-grade

As intensity of metamorphism increases, so does crystal size and coarseness of foliation.

Increasing intensity of metamorphism

Low grade	Intermediate grade	High grade

| Slate | Phyllite | Schist (abundant platy minerals) | Gneiss (fewer platy minerals) | Migmatite |

Increasing crystal size

Increasing coarseness of foliation

| Slaty cleavage | Slaty cleavage | Schistosity | Banding | Banding |

FIGURE 6.6 ■ Foliated rocks are classified by metamorphic grade, grain size, type of foliation, and banding. [*slate, phyllite, schist, gneiss:* John Grotzinger/Ramón Rivera-Moret/Harvard Mineralogical Museum; *migmatite:* Kip Hodges.]

rocks, are among the most abundant metamorphic rock types. They contain more than 50 percent platy minerals, mainly the micas muscovite and biotite. Schists may contain thin layers of quartz, feldspar, or both, depending on the quartz content of the parent shale.

GNEISS Even coarser foliation is shown by **gneisses,** light-colored rocks with coarse bands of light and dark minerals throughout the rock. This *gneissic foliation* results from the segregation of lighter-colored quartz and feldspar from darker-colored amphiboles and other mafic minerals. Gneisses are high-grade, coarse-grained metamorphic rocks in which the ratio of granular to platy minerals is higher than that in slate or schist. The result is poor foliation and thus little tendency to split. Under high pressures and temperatures, the mineral assemblages of lower-grade rocks containing micas and chlorite are transformed into new assemblages dominated by quartz and feldspars, with lesser amounts of micas and amphiboles.

MIGMATITE Temperatures higher than those necessary to produce gneiss may begin to melt the country rock. In this case, as with igneous rocks (see Chapter 4), the first minerals to melt will be those with the lowest melting temperatures. Therefore, only part of the country rock melts, and the melt migrates only a short distance before solidify-

ing again. Rocks produced in this way are badly deformed and contorted, and they are penetrated by many veins, small pods, and lenses of melted rock. The result is a mixture of igneous and metamorphic rock called **migmatite.** Some migmatites are mainly metamorphic, with only a small proportion of igneous material. Others have been so affected by melting that they are considered almost entirely igneous.

Granoblastic Rocks

Granoblastic rocks are nonfoliated metamorphic rocks composed mainly of crystals that grow in equant (equidimensional) shapes, such as cubes and spheres, rather than in platy or elongate shapes. These rocks result from metamorphic processes, such as contact metamorphism, in which directed pressure is absent, so foliation does not occur. Granoblastic rocks include hornfels, quartzite, marble, greenstone, amphibolite, and granulite (**Figure 6.7**). All granoblastic rocks except hornfels are defined by their mineralogy rather than their texture because all of them have a homogeneous granular texture.

Hornfels is a high-temperature contact metamorphic rock of uniform grain size that has undergone little or no deformation (see Figure 3.27). It is formed from fine-grained sedimentary rock and other types of rock containing an

Quartzite

Marble

FIGURE 6.7 ■ Granoblastic (nonfoliated) metamorphic rocks: quartzite [Breck P. Kent]; marble [Diego Lezama Orezzoli/ Corbis].

abundance of silicate minerals. Hornfels has a granular texture overall, even though it commonly contains pyroxene, which makes elongate crystals, and some micas. It is not foliated, and its platy or elongate crystals are oriented randomly.

Quartzites are very hard, white rocks derived from quartz-rich sandstones. Some quartzites are homogeneous, unbroken by preserved bedding or foliation (Figure 6.7a). Others contain thin bands of slate or schist, relics of former interbedded layers of clay or shale.

Marbles are the metamorphic products of heat and pressure acting on limestones and dolomites. Some white, pure marbles, such as the famous Italian Carrara marbles prized by sculptors, show a smooth, even texture of interlocked calcite crystals of uniform size. Other marbles show irregular banding or mottling from silicate and other mineral impurities in the original limestone (Figure 6.7b).

Greenstones are metamorphosed mafic volcanic rocks. Many of these low-grade metamorphic rocks form by seafloor metamorphism. Large areas of the seafloor are covered with basalts that have been slightly or extensively altered in this way at mid-ocean ridges. An abundance of chlorite gives these rocks their greenish cast.

Amphibolites are made up of amphibole and plagioclase feldspar. They are typically the product of medium- to high-grade metamorphism of mafic volcanic rocks. Foliated amphibolites can be produced by directed pressure.

Granulite, a high-grade metamorphic rock that is also referred to as *granofels,* has a homogeneous granular texture. It is a medium- to coarse-grained rock in which the crystals are equant and show only faint foliation at most. It is formed by the metamorphism of shale, impure sandstone, and many kinds of igneous rock.

Porphyroblasts

Newly formed metamorphic minerals may grow into large crystals surrounded by a much finer grained matrix of other minerals (**Figure 6.8**). These large crystals, called

FIGURE 6.8 ■ Garnet porphyroblasts in a schist matrix. The minerals in the matrix are continuously recrystallized as pressures and temperatures change and therefore grow to only a small size. In contrast, porphyroblasts grow to a large size because they are stable over a broad range of pressures and temperatures. [MSA 260 by Chip Clark, Smithsonian.]

TABLE 6-1   Classification of Metamorphic Rocks by Texture

Classification	Characteristics	Rock Name	Typical Parent Rock
Foliated	Distinguished by slaty cleavage, schistosity, or gneissic foliation; mineral grains show preferred orientation	Slate Phyllite Schist Gneiss	Shale, sandstone
Granoblastic (nonfoliated)	Granular, characterized by coarse or fine interlocking grains; little or no preferred orientation	Hornfels Quartzite Marble Argillite Greenstone Amphibolite Granulite	Shale, volcanics Quartz-rich sandstone Limestone, dolomite Shale Basalt Shale, basalt Shale, basalt
Porphyroblastic	Large crystals set in fine-grained matrix	Slate to gneiss	Shale

porphyroblasts, are found in rocks formed both by contact and by regional metamorphism. Porphyroblasts form from minerals that are stable over a broad range of pressures and temperatures. Crystals of these minerals grow large while the minerals of the matrix are being continuously recrystallized as pressures and temperatures change, so they replace parts of the matrix. Porphyroblasts vary in size, ranging from a few millimeters to several centimeters in diameter. Garnet and staurolite are two common minerals that form porphyroblasts, although many others are also found. The precise composition and distribution of porphyroblasts of these two minerals can be used to infer the pressures and temperatures that occurred during metamorphism, as we will see later in this chapter.

Table 6.1 summarizes the textural classes of metamorphic rocks and their main characteristics.

■ Regional Metamorphism and Metamorphic Grade

As we have seen, metamorphic rocks form under a wide range of conditions, and their mineralogies and textures are clues to the pressures and temperatures in the crust where and when they formed. Geologists who study the formation of metamorphic rocks constantly seek to determine the intensity and character of metamorphism more precisely than is indicated by a designation of "low grade" or "high grade." To make these finer distinctions, geologists "read" minerals as though they were pressure gauges and thermometers. These techniques are best illustrated by their application to regional metamorphism.

Mineral Isograds: Mapping Zones of Change

When we study a broad belt of regional metamorphism, we can see many outcrops, some showing one set of minerals, some showing others. Different zones within the belt may be distinguished by *index minerals:* abundant minerals that each form under a limited range of temperatures and pressures (**Figure 6.9**). For example, a zone of unmetamorphosed shales may lie next to a zone of weakly metamorphosed slates (Figure 6.9a). As we move from the shale zone into the slate zone, a new mineral—chlorite—appears. Chlorite is an index mineral marking the point at which we move into a new zone with a higher metamorphic grade. If laboratory studies have determined the temperature and pressure at which the index mineral forms, we can draw conclusions about the conditions that existed when the rocks in the zone were formed.

We can use the occurrences of index minerals to make a map of the boundaries between metamorphic zones. Geologists define these boundaries by drawing lines called *isograds* that plot the transitions from one zone to the next. Isograds are used in Figure 6.9a to show a series of mineral assemblages produced by the regional metamorphism of shale in New England. A pattern of isograds tends to follow the deformation features (folds and faults) of a region. An isograd based on a single index mineral, such as the chlorite isograd in Figure 6.9a, provides a good approximate measure of metamorphic pressure and temperature.

To determine metamorphic pressure and temperature more precisely, geologists can examine a group of two or three minerals that have crystallized together. For example, based on laboratory data, a geologist knows that a sillimanite zone that contains orthoclase feldspar and sillimanite must have formed by the reaction of muscovite and quartz

FIGURE 6.9 ◾ Index minerals define the different metamorphic zones within a belt of regional metamorphism. (a) Map of New England, showing metamorphic zones based on index minerals found in rocks metamorphosed from shale. (b) Rocks produced by the metamorphism of shale at different temperatures and pressures. [*slate, phyllite, schist, gneiss:* John Grotzinger/Ramón Rivera-Moret/Harvard Mineralogical Museum; *blueschist:* courtesy of Mark Cloos; *migmatite:* courtesy of Kip Hodges.]

at temperatures of about 600°C and pressures of about 5 kbar, liberating water (as water vapor) in the process. The sillimanite isograd records the following reaction:

muscovite $\quad + \quad$ quartz $\quad \rightarrow$
$KAl_3Si_3O_{10}(OH) \qquad SiO_2$

orthoclase feldspar $\quad +$ sillimanite $\quad +$ water
$\qquad KAlSi_3O_8 \qquad\qquad Al_2SiO_5 \qquad H_2O$

Because isograds reflect the pressures and temperatures at which the minerals in a regional metamorphic belt formed, the isograd sequence in one belt may differ from that in another. The reason for this variation is that, as we have seen, pressures and temperatures do not increase at the same rate in all plate tectonic settings.

Metamorphic Grade and Parent Rock Composition

The kind of metamorphic rock that results from a given grade of metamorphism depends partly on the mineralogy of the parent rock (**Figure 6.10**). The metamorphism of shale, as shown in Figure 6.9a, reveals the effects of

pressure and temperature on parent rock that is rich in clay minerals, quartz, and perhaps some carbonate minerals. The metamorphism of mafic volcanic rock, composed predominantly of feldspars and pyroxene, follows a different course (Figure 6.10b).

In the regional metamorphism of a basalt, for example, the lowest-grade rocks characteristically contain various **zeolite** minerals. The silicate minerals in the zeolite class contain water within their crystal structure. Zeolite minerals form at very low temperatures and pressures. Rocks that include this group of minerals are thus placed in the zeolite grade.

Overlapping with the zeolite grade is a higher grade of metamorphosed mafic volcanic rocks, the **greenschists,** whose abundant minerals include chlorite. Next are the amphibolites, which contain large amounts of amphiboles. The granulites, coarse-grained rocks containing pyroxene and calcium-rich plagioclase, constitute the highest grade of metamorphosed mafic volcanic rocks.

Rocks of the greenschist, amphibolite, and granulite grades are also formed during metamorphism of sedimentary rocks such as shale, as shown in Figure 6.10b. The pyroxene-bearing granulites are the products of high-grade metamorphism in which the temperature is high

(a) Changes in the mineral composition of shale during metamorphism

(b) Changes in the mineral composition of mafic rock during metamorphism

(c)

Metamorphic Facies	Minerals Produced from Shale Parent	Minerals Produced from Basalt Parent
Greenschist	Muscovite, chlorite, quartz, albite	Albite, epidote, chlorite
Amphibolite	Muscovite, biotite, garnet, quartz, albite, staurolite, kyanite, sillimanite	Amphibole, plagioclase feldspar
Granulite	Garnet, sillimanite, albite, orthoclase, quartz, biotite	Calcium-rich pyroxene, calcium-rich plagioclase feldspar
Eclogite	Garnet, sodium-rich pyroxene, quartz/coesite, kyanite	Sodium-rich pyroxene, garnet

FIGURE 6.10 ■ The kind of metamorphic rock that results from a given grade of metamorphism depends partly on the mineralogy of the parent rock. (a) Changes in the mineral composition of shale (a mafic volcanic rock) with increasing metamorphic grade. (b) Changes in the mineral composition of basalt (a sedimentary rock) with increasing metamorphic grade. (c) Major minerals of metamorphic facies produced from shale and basalt.

and the pressure is moderate. The opposite conditions, in which the pressure is high and the temperature moderate, produce rocks of the **blueschist** grade from parent rock of various starting compositions, from mafic volcanic rocks to shaley sedimentary rocks (see Figure 6.9b). The name comes from the abundance of glaucophane, a blue amphibole, in these rocks. Still another metamorphic rock, formed at extremely high pressures and moderate to high temperatures, is eclogite, which is rich in garnet and pyroxene.

Metamorphic Facies

We can plot this information about the grades of the metamorphic rocks in a regional metamorphic belt—derived from parent rocks of many different chemical compositions—on a graph of temperature and pressure (**Figure 6.11**). **Metamorphic facies** are groupings of rocks of various

mineral compositions formed under particular conditions of temperature and pressure from different parent rocks. By delineating metamorphic facies, we can be more specific about the grades of metamorphism observed in rocks. Two essential points characterize the concept of metamorphic facies:

1. Different kinds of metamorphic rocks of the same metamorphic grade form from parent rocks of different composition.

2. Different kinds of metamorphic rocks of different metamorphic grades form from parent rocks of the same composition.

Figure 6.10c shows the major minerals of the metamorphic facies produced from shale and basalt. Because parent rocks vary so greatly in composition, there are no sharp boundaries between metamorphic facies (see Figure 6.11).

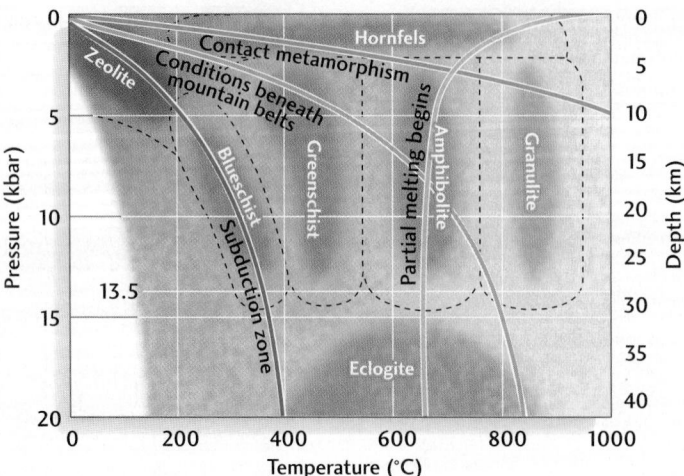

FIGURE 6.11 ■ Metamorphic facies correspond to particular combinations of temperature and pressure, which correspond to particular plate tectonic settings. The dashed lines indicate the overlapping nature of the boundaries between metamorphic facies.

Perhaps the most important reason for analyzing metamorphic facies is that they give us clues to the plate tectonic processes responsible for metamorphism, as we shall see next.

■ Plate Tectonics and Metamorphism

Soon after the theory of plate tectonics was proposed, geologists started to see how patterns of metamorphism fit into the larger framework of plate movements. Different types of metamorphism are likely to occur in different plate tectonic settings (see Figure 6.4):

- *Continental interiors.* Contact metamorphism, burial metamorphism, and perhaps regional metamorphism occur at different levels in the crust. Shock metamorphism is likely to be best preserved in continental interiors because their large areal extent provides a large target area to record rare meteorite impact events.
- *Divergent plate boundaries.* Seafloor metamorphism and contact metamorphism around plutons intruding into the oceanic crust occur at divergent plate boundaries.
- *Convergent plate boundaries.* Regional metamorphism, high-pressure and ultra-high-pressure metamorphism, and contact metamorphism.
- *Transform faults.* In oceanic settings, seafloor metamorphism may occur. In both oceanic and continental settings, we find extensive metamorphism caused by shearing forces along transform faults.

Metamorphic Pressure-Temperature Paths

As we have seen, the concept of metamorphic grade can inform us of the maximum pressure or temperature to which a metamorphic rock has been subjected, but it tells us nothing about where the rock encountered those conditions. Nor does it tell us anything about how the rock was **exhumed,** or transported back to Earth's surface.

Each metamorphic rock has a distinctive history of changing temperature and pressure that is reflected in its texture and mineralogy. This history is called a metamorphic pressure-temperature path, or **P-T path.** The P-T path can be a sensitive recorder of many important factors that influence metamorphism—such as sources of heat, which changes temperatures, and rates of tectonic transport (burial and exhumation), which changes pressures. Thus, P-T paths are characteristic of particular plate tectonic settings.

To obtain a P-T path, geologists must analyze specific minerals from metamorphic rock samples in the laboratory. One of the minerals most widely used for this purpose is garnet, a common porphyroblast that serves as a sort of P-T path recording device (**Figure 6.12**). Garnet crystals grow steadily during metamorphism, and as the pressure and temperature of the environment change, the chemical composition of the garnet changes. The oldest part of a garnet crystal is its core, and the youngest is its outer edge, so the variation in its composition from core to edge will yield the history of the metamorphic conditions under which it formed. Geologists can measure the chemical composition of a garnet porphyroblast in the laboratory and plot the corresponding pressure and temperature values as a P-T path (see Practicing Geology).

1 During metamorphism, a garnet crystal grows, and the composition of the growing crystal changes as the temperature and pressure around it change.

2 The composition of the crystal can be plotted on the P-T path as it grows from **❶** in its center to **❷** at its edge.

Thin section of garnet gneiss

Growth zoning in garnet

3 As rock is carried deeper in Earth's crust and is subjected to higher temperatures and pressures (the **prograde** path), the garnet crystal initially grows in a schist but ends up growing in a gneiss as metamorphism progresses.

4 The **retrograde** path indicates decreasing temperatures and pressures as rock is carried toward Earth's surface.

FIGURE 6.12 ■ Porphyroblasts such as garnet can be used to plot the P-T paths of metamorphic rocks. The P-T path that a metamorphic rock typically follows begins with an increase in pressure and temperature (the prograde path), followed by a decrease in pressure and temperature (the retrograde path). [Photos courtesy of Kip Hodges.]

P-T paths have two segments: a *prograde* segment that indicates increasing pressure and temperature, and a *retrograde* segment that indicates decreasing pressure and temperature. The P-T paths of some rock assemblages that form at convergent boundaries are shown in **Figure 6.13**.

Ocean-Continent Convergence

A distinct metamorphic assemblage forms when oceanic lithosphere is subducted beneath a plate carrying a continent on its leading edge (Figure 6.13a). Thick sediments eroded from the continent rapidly fill the deep-sea trench that forms a flexural basin at the subduction zone. As it

descends, the oceanic lithosphere stuffs the region below the inner wall of the trench (the wall closer to the continent) with these sediments, as well as with shreds of ophiolite suites scraped off the descending plate. The result is a chaotic mix known as a **mélange** (French for "mixture"). Assemblages of this type, located in the *forearc* region of a subduction zone—the area between the deep-sea trench and the volcanic mountain belt—are enormously complex and variable. The rocks formed there are all highly folded, intricately faulted, and metamorphosed (**Figure 6.14**). They are difficult to map in detail, but are recognizable by their distinctive combination of minerals and structural features.

FIGURE 6.13 ■ P-T paths indicate the trajectory of rocks during metamorphism. (a) Metamorphism of mélange at an ocean-continent convergence zone. (b) Metamorphism at a continent-continent convergence zone. The different P-T paths of rocks formed in these different plate tectonic settings indicate differences in geothermal gradients. Rocks transported to similar depths—and pressures—beneath mountain belts become much hotter than rocks transported to an equivalent depth by subduction.

SUBDUCTION-RELATED METAMORPHISM Blueschist—the metamorphic rock type whose minerals indicate that they were produced under high pressures but at relatively low temperatures (see Figure 6.9b)—forms from mélange in the forearc region of a subduction zone. Here, materials may be carried rapidly into the subduction zone to depths as great as 30 km. The cool subducting slab moves downward so quickly that there is little time for it to heat up, but pressure on the slab increases rapidly.

Eventually, as part of the subduction process, the material rises back to the surface. This exhumation results from two forces: buoyancy and circulation. Imagine trying to push a basketball below the surface of a swimming pool. The air-filled basketball has a lower density than the surrounding water, so it tends to rise back to the surface. In a

similar way, the subducted metamorphic rocks are driven upward by their inherent buoyancy relative to the denser crust that surrounds them. But what "pushes" the material down to begin with? A natural circulation pattern is set up in the subduction zone. You can think of a subduction zone as an eggbeater. As the eggbeater rotates, it moves the froth in a circular direction. What moves in one direction eventually moves in the opposite direction because of the circular motion. In an analogous way, the sinking slab in a subduction zone sets up a circular motion of material above the slab, first pulling it down to great depths, then returning it to the surface.

Figure 6.13a shows the typical P-T path of rocks subjected to blueschist-grade metamorphism during subduction and exhumation. Note that the P-T path forms a loop

FIGURE 6.14 ■ Mélange is a kind of breccia composed of rock fragments formed by churning in subduction zones. [John Platt.]

in this diagram. If we compare the graph in Figure 6.13a with the metamorphic facies diagram in Figure 6.11, we can see that the prograde segment of the path represents subduction, as shown by a rapid increase in pressure and only a relatively small increase in temperature. During exhumation, the path loops back around because temperature is still slowly increasing, but now pressure is rapidly decreasing. The retrograde segment of the P-T path represents the exhumation process described above.

EVIDENCE OF ANCIENT OCEAN-CONTINENT CONVERGENCE The essential elements of these subduction-related rock assemblages have been found at many places in the geologic record, particularly around the Pacific Ocean basin. One can see mélange in the Franciscan formation of the California Coast Ranges and in the parallel volcanic mountain belt in the Sierra Nevada to the east. These rocks mark the Mesozoic collision between the North American Plate and the Farallon Plate, which has nearly disappeared by subduction (see Figure 10.6). The location of mélange to the west and a volcanic mountain belt to the east shows that the Farallon Plate to the west was the subducted one. Analysis of the P-T paths of metamorphic minerals in the blueschist-grade Franciscan mélange reveals a loop similar to that shown in Figure 6.13a, indicating a rapid increase in pressure, which is characteristic of subduction.

Continent-Continent Collision

Because continental crust is buoyant, when a continent collides with another continent, both continents resist subduction and stay afloat on the mantle. As a result, a wide zone of intense deformation develops at the convergent boundary where the continents grind together. The remnant of such a boundary left behind in the geologic record is called a **suture.** The intense deformation results in a much-thickened continental crust in the collision zone, often producing high mountains. Ophiolite suites are often found near the suture.

As the lithosphere thickens, the deep parts of the continental crust heat up and undergo varying grades of metamorphism. In still deeper zones, melting may begin at the same time, forming magma chambers deep within the core of the mountain range. In this way, a complex mixture of metamorphic and igneous rock forms the core of the mountain belt. Millions of years afterward, when erosion has stripped off the surface layers of the mountains, their cores, containing schists, gneisses, and other metamorphic rocks, are exposed, providing a rock record of the metamorphic processes that formed them.

P-T paths of metamorphic rocks produced by continent-continent collision have a different shape from those of rocks produced by subduction alone. Continent-continent collision generates higher temperatures than subduction; therefore, as rock is pushed to greater depths, the temperature that corresponds to a given pressure will be higher (Figure 6.13b). The P-T path begins at the same place as the path for subduction, but shows a more rapid increase in temperature as greater pressures and depths are reached. Geologists generally interpret the prograde segment of a P-T path with this shape as indicating the burial of rocks beneath high mountains. The retrograde segment represents uplift and exhumation of the buried rocks during the collapse of mountains, either by erosion or by postcollision stretching and thinning of the continental crust.

The prime example of a continent-continent collision zone is the Himalaya, which began to form some 50 million years ago when the Indian continent collided with the Asian continent. That collision continues today: India is moving into Asia at a rate of a few centimeters per year, and the mountain building is still going on, together with faulting and very rapid erosion by rivers and glaciers.

Exhumation: A Link Between the Plate Tectonic and Climate Systems

Forty years ago, plate tectonic theory provided a ready explanation for how metamorphic rocks could be produced by seafloor spreading, subduction, and continent-continent collision. By the mid-1980s, the study of PT paths provided a clearer picture of the specific tectonic mechanisms involved in the deep burial and metamorphism of rocks. At the same time, it surprised geologists by providing an equally clear picture of the subsequent, and often very rapid, uplift and exhumation of these deeply buried rocks. Since the time of this discovery, geologists have been searching for exclusively tectonic mechanisms that could bring these rocks back to Earth's surface so quickly.

One popular idea is that mountains, having been built to great elevations during collisional crustal thickening, suddenly fail by gravitational collapse. The old saying "what goes up must go down" applies here, but with surprisingly fast results—so fast, in fact, that some geologists don't believe gravity is the only important mechanism involved. Other forces must also be at work.

As we will learn in Chapter 22, geologists who study landscapes have discovered that extremely high erosion rates can be produced by glaciers and streams in tectonically active mountainous regions. Over the past decade, these geologists have presented a new hypothesis that links rapid rates of uplift and exhumation to rapid erosion rates. The idea is that the climate system, not tectonic processes alone, drives the movement of rocks from the deep crust to the shallow crust through the process of rapid erosion. Thus, plate tectonic processes—which act through mountain building—and climate processes—which act through weathering and erosion—*interact* to control the flow of metamorphic rocks to Earth's surface. After decades of emphasis solely on plate tectonic explanations for regional and global geologic processes, it now seems that two apparently unrelated processes—metamorphism and erosion—are linked in an elegant way. As one geologist exclaimed: "Savor the irony should the metamorphic muscles that push mountains to the sky be driven by the pitter-patter of tiny raindrops."

SUMMARY

What are the causes of metamorphism? Metamorphism is alteration in the mineralogy, texture, or chemical composition of solid rock. It is caused by increases in pressure and temperature and by reactions with chemical components introduced by hydrothermal solutions. As rocks are pushed deep within the crust by plate tectonic processes and exposed to increasing temperatures and pressures, the chemical components of the parent rock rearrange themselves into a new set of minerals that are stable under the new conditions. Metamorphic rocks that form at relatively low temperatures and pressures are referred to as low-grade metamorphic rocks; those that form at high temperatures and pressures are high-grade metamorphic rocks. Chemical components may be added to or removed from a rock during metamorphism, usually by hydrothermal solutions.

What are the various types of metamorphism? The three most common types of metamorphism are regional metamorphism, during which rocks over large areas are metamorphosed by high pressures and temperatures generated during mountain building; contact metamorphism, during which country rock close to an igneous intrusion is transformed by the heat of the intruding magma; and seafloor metamorphism, during which hot fluids percolate through and metamorphose oceanic crust. Less common types are burial metamorphism, during which deeply buried sedimentary rocks are altered by pressures and temperatures higher than those that result in diagenesis; high-pressure and ultra-high-pressure metamorphism, which occur at great depths, as when sediments are subducted; and shock metamorphism, which results from meteorite impacts.

What are the chief kinds of metamorphic rocks? Metamorphic rocks fall into two major textural classes: foliated rocks (displaying foliation, a pattern of parallel cleavage planes resulting from a preferred orientation of crystals) and granoblastic, or nonfoliated, rocks. The kinds of rocks produced depend on the composition of the parent rock and the grade of metamorphism. Regional metamorphism of shale leads to zones of foliated rock of progressively higher grade, from slate to phyllite, schist, gneiss, and finally migmatite. Among granoblastic rocks, marble is derived from the metamorphism of limestone, quartzite from quartz-rich sandstone, and greenstone from basalt. Hornfels is the product of contact metamorphism of fine-grained sedimentary rocks and other types of rock containing an abundance of silicate minerals. Regional metamorphism of mafic volcanic rocks progresses from zeolite facies to greenschist facies and then to amphibolite and granulite facies.

What do metamorphic rocks reveal about the conditions under which they were formed? Zones of metamorphism can be mapped with isograds defined by the first appearance of an index mineral. The presence of an index mineral can indicate the temperature and pressure under which the rocks in the zone were formed. According to the concept of metamorphic facies, rocks of the same metamorphic grade may differ because of variations in the chemical composition of the parent rock, whereas rocks metamorphosed from the same parent rock may vary because they were subjected to different grades of metamorphism.

How are metamorphic rocks related to plate tectonic processes? During subduction and continent-continent collision at convergent plate boundaries, rocks and sediments are pushed to great depths in Earth's crust, where they are subjected to increasing pressures and temperatures that result in metamorphism. The shapes of metamorphic P-T paths provide insight into the plate tectonic settings where these rocks were metamorphosed. In the case of ocean-continent convergence, P-T paths indicate rapid subduction of rocks and sediments to environments with high pressures and relatively low temperatures. In continent-continent collision zones, rocks are pushed down to depths where pressures and temperatures are both high. In both settings, the P-T paths show that after the rocks experience the maximum pressures and temperatures, they are returned to shallow depths. This process of exhumation may be driven by weathering and erosion at Earth's surface as well as by plate tectonic processes.

KEY TERMS AND CONCEPTS

amphibolite (p. 158)

blueschist (p. 161)

burial metamorphism (p. 154)

contact metamorphism (p. 154)

eclogite (p. 155)

exhumation (p. 162)

foliated rock (p. 155)

foliation (p. 155)

gneiss (p. 157)

granoblastic rock (p. 157)

granulite (p. 158)

greenschist (p. 160)

greenstone (p. 158)

high-pressure metamorphism (p. 154)

hornfels (p. 157)

marble (p. 158)

mélange (p. 163)

metamorphic facies (p. 161)

metasomatism (p. 153)

migmatite (p. 157)

phyllite (p. 156)

porphyroblast (p. 159)

P-T path (p. 162)

quartzite (p. 158)

regional metamorphism (p. 154)

schist (p. 156)

seafloor metamorphism (p. 154)

shock metamorphism (p. 155)

slate (p. 156)

stress (p. 152)

suture (p. 165)

ultra-high-pressure metamorphism (p. 154)

zeolite (p. 160)

PRACTICING GEOLOGY EXERCISE

How Do We Read Geologic History in Crystals?

What can a tiny crystal of garnet tell us about the history of the place where it was found? Knowing the plate tectonic setting in which a rock sample was formed tells us what other kinds of minerals might be found there. Geologists use variations in the chemical composition of garnet porphyroblasts to deduce the relative rates at which the rocks containing them were buried and then exhumed. These rates, in turn, reflect particular plate tectonic settings, as shown in Figure 6.13.

The chemical composition of a garnet porphyroblast generally varies progressively from its center to its edges (see Figure 6.12). This progression gives us a sense of change in pressure or temperature as a function of time: the center of the crystal records earlier conditions, and the edge of the crystal records later conditions. Changes in the calcium content of garnet track changes in pressure, whereas its iron content is more sensitive to temperature changes. We have noted that the increase in pressure over a given range of temperature values is much higher during subduction at ocean-continent convergence zones than it is during mountain building at continent-continent convergence zones. Conversely, a rock heated by an igneous intrusion experiences an increase in temperature, but little change in pressure.

By analyzing the chemical composition of garnet crystals, we can distinguish between these different metamorphic processes. To do so, we can compare changes in the abundance of the element of interest (calcium or iron) with the sum of changes in the abundances of all the elements that can vary in garnet: calcium (Ca), iron (Fe), magnesium (Mg), and manganese (Mn). Calculating the actual changes in pressure and temperature that a rock experienced during

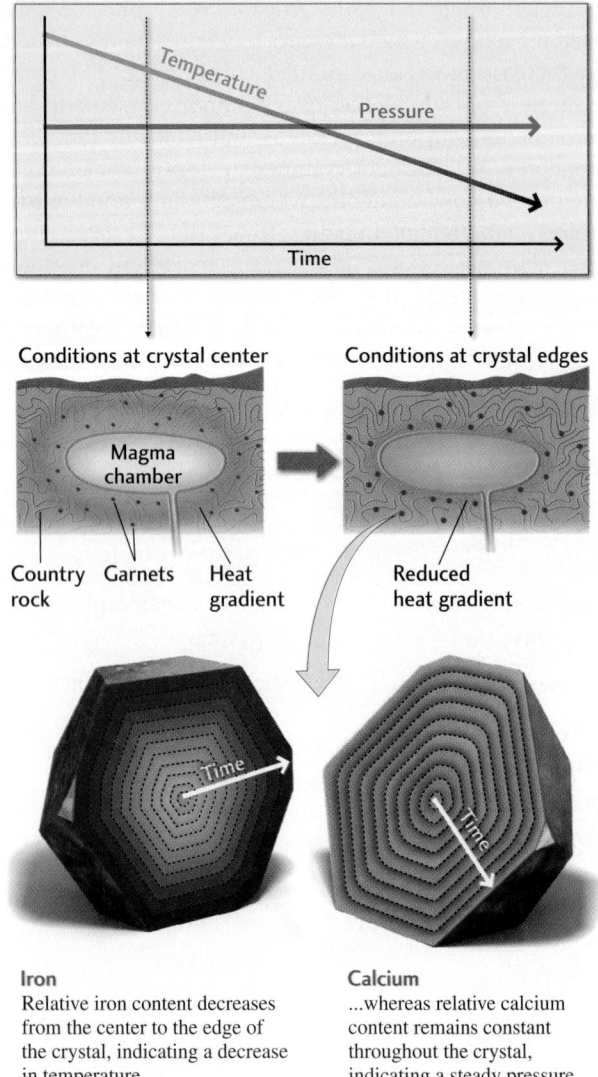

Element	Abundance at center	Abundance at edge
Ca	0.30	0.30
Fe	2.25	1.98
Mg	0.20	0.52
Mn	0.25	0.20

We first calculate the relative abundances of calcium and iron at the center of the crystal using the following ratios:

$$\left(\frac{Ca}{Ca + Fe + Mg + Mn}\right)_{center} = \frac{0.30}{0.30 + 1.98 + 0.52 + 0.20} = 0.10$$

$$\left(\frac{Fe}{Ca + Fe + Mg + Mn}\right)_{center} = \frac{2.25}{0.30 + 1.98 + 0.52 + 0.20} = 0.75$$

Then we do the same for the relative abundances of calcium and iron at the edge:

$$\left(\frac{Ca}{Ca + Fe + Mg + Mn}\right)_{edge} = \frac{0.30}{0.30 + 1.98 + 0.52 + 0.20} = 0.10$$

$$\left(\frac{Fe}{Ca + Fe + Mg + Mn}\right)_{edge} = \frac{1.98}{0.30 + 1.98 + 0.52 + 0.20} = 0.66$$

Based on these data, what can you say about the metamorphic event that caused this garnet crystal to grow? Was the rock carried down into a subduction zone, or was it sitting next to an igneous intrusion?

The decrease in iron content from 0.75 to 0.66 from center to edge is not associated with any change in calcium content. This observation indicates that metamorphism resulted mainly from a change in temperature with no change in pressure. These conditions are more consistent with metamorphism near an igneous intrusion than they are with subduction.

BONUS PROBLEM: Assume that the same calculations showed that iron content was constant, but calcium content changed significantly from center to edge. Would this pattern be consistent with transport of the rocks into a subduction zone?

Iron
Relative iron content decreases from the center to the edge of the crystal, indicating a decrease in temperature,...

Calcium
...whereas relative calcium content remains constant throughout the crystal, indicating a steady pressure.

metamorphism requires additional data, including the composition of the complete mineral assemblage. Even without those details, however, it is possible to make some rough estimates.

The following data were obtained by measuring the number of atoms of four elements in the center and at the edge of a garnet porphyroblast:

EXERCISES

1. What types of metamorphism are related to igneous intrusions?

2. What does preferred orientation refer to in a metamorphic rock? Think about how the alignment of minerals relates to metamorphic processes.

3. What is a porphyroblast?

4. Contrast the properties of a schist and a gneiss.

5. How are isograds related to metamorphic facies?

6. What is the difference between a granite and a slate?

7. How are metamorphic facies related to temperatures and pressures?

8. In which plate tectonic settings would you expect to find regional metamorphism?

9. What controls exhumation of metamorphic rocks?

10. What is the significance of eclogites at Earth's surface?

THOUGHT QUESTIONS

1. At what depths in Earth do metamorphic rocks form? What happens if temperatures get too high?

2. Why are there no metamorphic rocks formed under natural conditions of very low pressure and temperature, as shown in Figure 6.1?

3. How is slaty cleavage related to tectonic forces? What forces cause minerals to align with one another?

4. Would you choose to rely on chemical composition or type of foliation to determine metamorphic grade? Why?

5. You have mapped an area of regional metamorphism, such as the region in Figure 6.9a, and have observed a series of metamorphic zones, marked by north-south isograds, running from sillimanite in the east to chlorite in the west. Were metamorphic temperatures higher in the east or in the west?

6. Draw a P-T path for shock metamorphism of country rock during a meteorite impact.

7. Which kind of pluton would produce the highest grade of metamorphism, a granitic intrusion 20 km deep or a gabbro intrusion at a depth of 5 km?

8. Draw a sketch showing how seafloor metamorphism might take place.

9. Subduction zones are generally characterized by high pressure–low temperature metamorphism. In contrast, continent-continent collision zones are marked by moderate pressure–high temperature metamorphism. Which type of plate boundary has a higher geothermal gradient? Explain.

MEDIA SUPPORT

 6-1 Video: Gneiss: The Lewisian Complex of Scotland

 6-2 Video: Jade

- Plate Tectonic
  Forces   172

- Mapping Geologic
  Structure   172

- How Rocks
  Deform   175

- Basic Deformation
  Structures   177

- Styles of Continental
  Deformation   184

- Unraveling Geologic
  History   188

Aerial view of the Tree River Folds in northwestern
Canada. These large-scale folds have a wavelength
of about 1 km. [John Grotzinger]

DEFORMATION: MODIFICATION OF ROCKS BY FOLDING AND FRACTURING

7

WHEN ROCKS ARE CAUGHT UP IN PLATE BOUNDARIES, their textures and mineralogy can be transformed by metamorphism, as we saw in Chapter 6. Among the processes that cause regional metamorphism in continents, the most important is *deformation:* the modification of rocks by squeezing, stretching, folding, and faulting. On the scale of individual rocks, deformation can change granites into gneisses and sediments into schists. On a large scale, deformation can distort layers of sediments, which were deposited almost horizontally, into crazy-looking patterns.

Early geologists understood that most sedimentary rocks had originally been deposited as soft horizontal layers at the bottom of the ocean and had hardened over time. But what forces could have acted on these rocks, which seemed so strong and rigid, to produce the patterns they observed? Why were particular patterns of deformation repeated time and time again throughout geologic history? The discovery of plate tectonics in the 1960s provided the answers.

This chapter describes how rocks can be tilted, bent, and fractured into the patterns we see at the land surface. It concentrates on the processes of folding and faulting that deform continental rocks near plate boundaries. It shows how geologists collect and interpret field observations to make geologic maps, and then explores what those maps can tell us about the history of deformation, as well as the tectonic forces that caused it.

■ Plate Tectonic Forces

Deformation is a general term that encompasses the folding, faulting, shearing, compression, and extension of rock by plate tectonic forces. The types of deformation we see exposed at Earth's surface are caused mainly by the horizontal movements of the lithospheric plates relative to one another. For this reason, the tectonic forces that deform rocks at plate boundaries are predominantly *horizontally directed* and depend on the direction of relative plate movement:

- **Tensional forces,** which stretch and pull rock formations apart, dominate at divergent boundaries, where plates move away from each other.

- **Compressive forces,** which squeeze and shorten rock formations, dominate at convergent boundaries, where plates move toward each other.

- **Shearing forces,** which shear two parts of a rock formation in opposite directions, dominate at transform-fault boundaries, where plates slide past each other.

If plates were perfectly rigid, the plate boundaries would be sharp lineations, and points on either side of those boundaries would move at the relative plate velocity. This idealization is often a good approximation in the oceans, where rift valleys at mid-ocean ridges, deep-sea trenches, and nearly vertical transform faults form narrow plate boundary zones, often just a few kilometers wide.

Within continents, however, the deformation caused by plate movements can be "smeared out" across a plate boundary zone hundreds or even thousands of kilometers wide. The continental crust does not behave rigidly within these broad zones, so rocks at the surface are deformed by folding and faulting. Folds in rocks are like folds in clothing. Just as cloth pushed together from opposite sides bunches up in folds, layers of rock slowly compressed by tectonic forces in the crust can be pushed into folds (**Figure 7.1a**). Tectonic forces can also cause a rock formation to break and slip on both sides of a fracture, forming a fault (Figure 7.1b). When such a break occurs suddenly, the result is an earthquake. Active zones of continental deformation are marked by frequent earthquakes.

Geologic folds and faults can range in size from centimeters to meters (as in Figure 7.1) to tens of kilometers or more. Many mountain ranges are actually a series of large folds and faults that have been weathered and eroded. From the geologic record of deformation laid out on Earth's surface, geologists can deduce the directions of movement at ancient plate boundaries and reconstruct the tectonic history of the continental crust.

■ Mapping Geologic Structure

Faults and folds are examples of the basic features geologists observe and map to reconstruct crustal deformation. To better understand this process, we need information

(a)

(b)

FIGURE 7.1 ■ Rocks subjected to tectonic forces are deformed by folding and faulting. (a) An outcrop of originally horizontal rock layers has been bent into folds by compressive tectonic forces. (b) An outcrop of once-continuous rock layers has been displaced along small faults by tensional tectonic forces. [(a) Tony Waltham; (b) Marli Bryant Miller.]

FIGURE 7.2 ■ Dipping limestone and shale beds on the coast of Somerset County, England. The children are walking along the strike of the beds. The beds dip to the left at an angle of about 15°.
[Chris Pellant.]

about the geometry of faults and folds. The best place to find this information is at an *outcrop*, where the solid rock that underlies the ground surface—the *bedrock*—is exposed (not obscured by vegetation, soil, or loose boulders). At an outcrop, geologists can identify distinct **formations:** groups of rock layers that can be identified throughout a region by their physical properties. Some formations consist of a single rock type, such as limestone. Others are made up of thin, interlayered beds of different kinds of rock, such as sandstone and shale. However they vary, each formation comprises a distinct set of rock layers that can be recognized and mapped as a unit.

Figure 7.1a shows an outcrop in which the folding of sedimentary beds is clearly visible. Often, however, folded rocks are only partly exposed in an outcrop and can be seen only as inclined layers (**Figure 7.2**). The orientation of those layers is an important clue we can use to piece together a picture of the overall geologic structure. Two measurements describe the orientation of a rock layer exposed at an outcrop: the strike and the dip of the layer's surface.

Measuring Strike and Dip

The **strike** is the compass direction of a rock layer where it intersects with a horizontal surface. The **dip,** which is measured at right angles to the strike, is simply the amount of tilting—the angle at which the rock layer inclines from the horizontal. **Figure 7.3** shows how strike and dip are measured in the field. A geologist might describe the outcrop in this figure as "a bed of coarse-grained sandstone striking east-west and dipping 45° south."

Geologic Maps

Geologic maps are two-dimensional representations of the rock formations exposed at Earth's surface (**Figure 7.4**).

To create a geologic map, a geologist must choose an appropriate *scale*—the ratio of distance on the map to true surface distance. A common scale for geologic field mapping is 1:24,000 (pronounced "one to twenty-four thousand"), which means that 1 inch on the map corresponds to 24,000 inches (2000 feet) on Earth's surface. To depict the geology of an entire state, a geologist would choose a smaller scale: say, 1:1,000,000, where 1 cm represents 10 km and 1 inch almost 16 miles. The smaller the scale, the less detail can be depicted on the map.

FIGURE 7.3 ■ The strike and dip of a rock layer define its orientation at a particular place. The strike is the compass direction of a rock layer along the line of its intersection with a horizontal surface. The dip is the angle and direction of the steepest descent of a rock layer from the horizontal, measured at right angles to the strike. Here, the strike is east-west and the dip is 45° to the south. [After A. Maltman, *Geological Maps: An Introduction*, 2nd ed. New York: Van Nostrand Reinhold, 1998.]

Geologic map

8	Mixed sandstone and conglomerate
7	Shale
6	Sandstone
5	Siltstone
4	Shale
3	Shale and siltstone
2	Limestone and dolomite
1	Limestone

Susquehanna River

5 km

Geologic cross sections

PENNSYLVANIA

Map area

FIGURE 7.4 ■ A geologic map and cross sections are two-dimensional representations of a three-dimensional geologic structure. This figure shows a region of folded sedimentary rocks in central Pennsylvania east of the Susquehanna River. The rock formations exposed at the surface are labeled from the oldest (formation 1) to the youngest (formation 8).

Geologists keep track of different rock formations by assigning each formation a particular color on the map, usually keyed to the rock's type and age (see Figure 7.4). Many different rock formations may be exposed in highly deformed regions, so geologic maps can be very colorful!

Softer rocks, such as mudstones and other poorly consolidated sediments, are more easily eroded than harder rocks, such as limestones, sandstones, or metamorphic rocks. Consequently, rock types can exert a strong influence on the topography of the land surface and the exposure of rock formations (see Figure 7.4). The important relationships between geology and topography can be made clear by plotting the contours of the land surface on a geologic map.

Because geologic maps can represent such a huge amount of information, they have been called "textbooks on a piece of paper." To convey this information more concisely, geologic maps are annotated with special symbols that indicate the local strike and dip of rock formations and with special types of lines that mark faults and other significant features. For instance, the strike and dip of

rock formations are indicated on a geologic map by T-like symbols:

The top of the T indicates the strike direction, the shank of the T indicates the dip direction, and the number gives the dip angle in degrees. On a map where north is up, the symbol on the left thus describes the sandstone bed in Figure 7.3, which has an east-west strike and a dip of 45° to the south. The one on the right describes a formation that strikes northeast-southwest and dips to the southeast at an angle of 15°, such as the beds in Figure 7.2.

Of course, not every detail of surface geology can be represented on a map, so geologists must simplify the

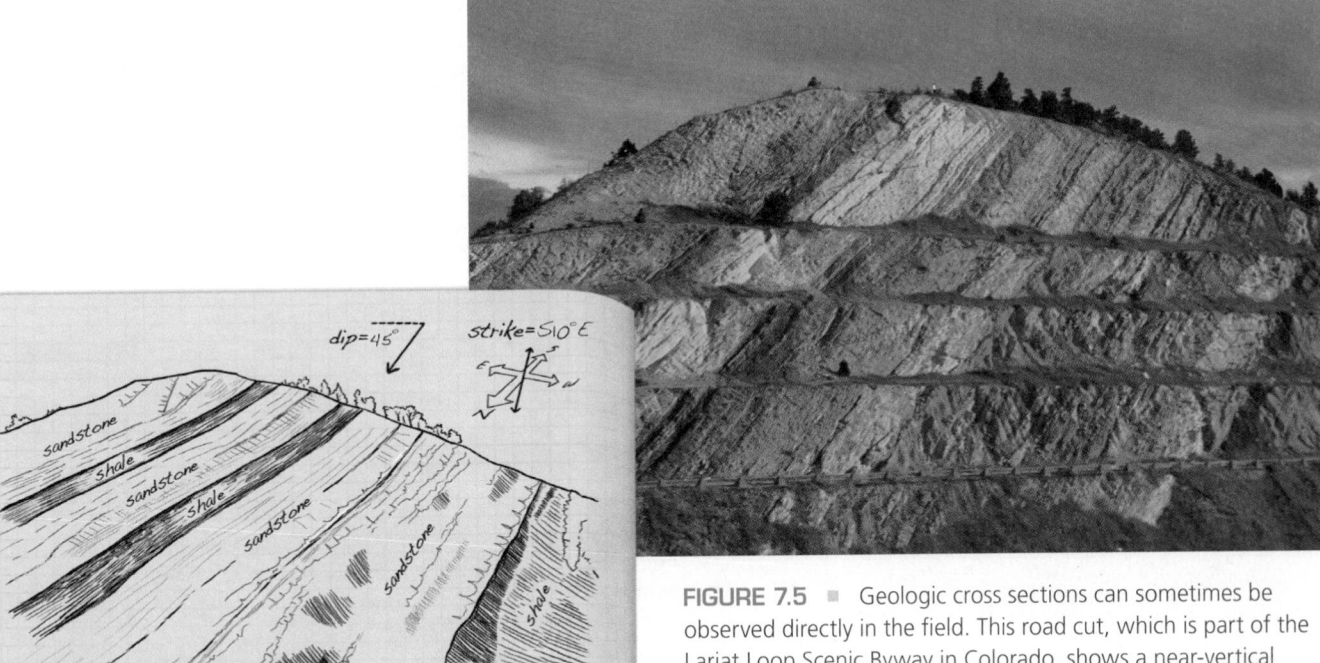

FIGURE 7.5 ∎ Geologic cross sections can sometimes be observed directly in the field. This road cut, which is part of the Lariat Loop Scenic Byway in Colorado, shows a near-vertical cross section through a sequence of sedimentary beds tilted by the uplift of the Rocky Mountains. [James Steinberg/Science Source.]

structures they see, perhaps by representing a complex zone of faulting as a single fault or by ignoring folds too small to show at the scale they have chosen. They may also "dust off" their maps by ignoring thin layers of soil and loose rock that cover up the geologic structure, portraying the structure as if outcrops existed everywhere. You should therefore think of a geologic map as a simplified *scientific model* of the surface geology.

Geologic Cross Sections

Once a region has been mapped, the two-dimensional geologic map must be interpreted in terms of the underlying three-dimensional geologic structure. How can the shapes of the rock layers be reconstructed, even when erosion has removed parts of a formation? The process is like putting together a three-dimensional jigsaw puzzle with some of the pieces missing. Common sense and intuition play important roles, as do basic geologic principles.

To piece together the puzzle, geologists construct **geologic cross sections**—diagrams showing the features that would be visible if vertical slices were made through part of the crust. Some small cross sections can actually be seen in the vertical faces of cliffs, quarries, and road cuts (**Figure 7.5**). Cross sections spanning much larger areas can be constructed from the information on a geologic map, including the strikes and dips observed at outcrops. The accuracy of cross sections based on surface mapping can be improved by drilling boreholes to collect rock samples as well as by seismic imaging. But drilling and seismic imaging are expensive, so data collected by these methods are usually available only for areas that have been explored for oil, water, or other valuable natural resources.

Figure 7.4 shows a geologic map of an area where originally horizontal beds of sedimentary rock were bent into a series of folds and eroded into a set of zigzagging ridges and valleys. We will explore some of the geologic relationships seen in this map later in this chapter. But first we will investigate the basic processes by which rocks deform.

∎ How Rocks Deform

Rocks deform in response to the tectonic forces acting on them. Whether they will respond to tectonic forces by folding, faulting, or some combination of the two depends on the orientation of the forces, the rock type, and the physical conditions (such as temperature and pressure) during deformation.

Brittle and Ductile Behavior of Rocks in the Laboratory

In the mid-1900s, geologists began to explore the forces of deformation by using powerful hydraulic rams to bend

This sample was compressed under conditions representative of the shallow crust. The fractures indicate that the marble is brittle at the laboratory equivalent of shallow depth.

This sample was compressed under conditions representative of the deeper crust. It deformed by bulging smoothly, indicating that marble is ductile at greater depth.

FIGURE 7.6 ▪ Results of laboratory experiments conducted to discover how rock—in this case, marble—is deformed by compressive forces. The marble samples are encased in translucent plastic jackets, which explains their shiny appearance. [Fred and Judith Chester/John Handin Rock Deformation Laboratory of the Center for Tectonophysics.]

Undeformed sample

and break small samples of rock. Engineers had invented such machines to measure the strength of concrete and other building materials, but geologists modified them to discover how rocks deform at pressures and temperatures high enough to simulate physical conditions deep in Earth's crust.

In one such experiment, the researchers applied compressive force by pushing down with a hydraulic ram on one end of a small cylinder of marble while at the same time maintaining confining pressure on the cylinder (**Figure 7.6**). Under low confining pressures, equivalent to those found at shallow depths in Earth's crust, the marble sample deformed only a small amount until the compressive force on its end was increased to the point that the entire sample suddenly broke (see Figure 7.6, left side). This experiment showed that marble behaves as a **brittle** material at the low confining pressures found in the shallow crust. Repeating the experiment under high confining pressures, equivalent to those that often accompany metamorphism, produced a different result: the marble sample slowly and steadily deformed into a shortened, bulging shape without fracturing (see Figure 7.6, right side). Marble thus behaves as a pliable, or **ductile,** material at the high confining pressures found deep in the crust.

Other experiments showed that when marble was heated to temperatures as high as those that accompany metamorphism, it acted as a ductile material at a lower confining pressure—just as heating wax changes it from a hard material that can break into a soft material that flows. The researchers concluded that the particular marble they were working with would deform by faulting at depths shallower than a few kilometers, but would deform by folding at the greater crustal depths where metamorphism generally occurs.

Brittle and Ductile Behavior of Rocks in Earth's Crust

Natural conditions in Earth's crust cannot be reproduced exactly in the laboratory. Tectonic forces are applied over millions of years, whereas laboratory experiments are rarely conducted over more than a few hours, or at most a few weeks. Nevertheless, the results of laboratory experiments can help us interpret what we see in the field. Geologists keep the following points in mind when mapping crustal folds and faults:

▪ The same rock can be brittle at shallow depths (where temperatures and pressures are relatively low) and ductile deep in the crust (where temperatures and pressures are higher). Metamorphism is usually accompanied by ductile deformation.

▪ Rock type affects deformation. In particular, the hard igneous and metamorphic rocks that form the crystalline *basement* of a continent (the crust beneath layers of sediments) often behave as brittle materials, fracturing along faults during deformation, while softer sedimentary rocks that overlie them often behave as ductile materials, folding gradually.

▪ A rock formation that would behave as a ductile material if deformed slowly may behave as a brittle material if deformed more rapidly. (Think of Silly Putty, which flows as a ductile clay when you squeeze it slowly, but breaks into pieces when you pull it apart very quickly.)

▪ Rocks break more easily when subjected to tensional (pulling and stretching) forces than when subjected to compressive forces. Sedimentary rock formations that will deform by folding during compression will often break along faults when subjected to tension.

■ Basic Deformation Structures

Geologists use the simple geometric concepts and measurements described earlier in this chapter to classify features such as faults and folds into different types of deformation structures.

Faults

A **fault** is a fracture that displaces the rock on either side of it. We can measure the orientation of the fracture surface, or *fault surface,* by its strike and dip, just as we do for other geologic surfaces (see Figure 7.3). The movement of the block of rock on one side of the fault with respect to that on the other side can be described by a *slip direction* and by the total displacement, or offset. For small faults, such as the ones pictured in Figure 7.1b, the offset might be only a couple of meters, whereas the offset along a major transform fault, such as the San Andreas fault, can amount to hundreds of kilometers (**Figure 7.7**).

Rocks on either side of a fault cannot interpenetrate one another, and at high pressures below the surface, they cannot open up, so the slip direction during faulting must be parallel to the fault surface. Faults can therefore be classified by their slip direction along this surface (**Figure 7.8**). A **dip-slip fault** is one on which there has been relative movement of blocks of rock up or down the dip of the fault plane. A **strike-slip fault** is a fault on which the movement of blocks has been horizontal, parallel to the strike of the fault plane. When blocks of rock move along the strike and simultaneously up or down the dip, the result is an *oblique-slip fault.* Dip-slip faults are caused by compressive or tensional forces, whereas strike-slip faults are the work of shearing forces. An oblique-slip fault results from shearing in combination with either compression or tension.

These fault types require further classification because the movement of rock can be up or down, or right or left. To describe these movements, geologists borrow some terminology used by miners, calling the block of rock above a dipping fault plane the **hanging wall** and the block of rock below it the **foot wall.** A dip-slip fault is called a **normal fault** if the hanging wall moves downward relative to the foot wall, extending the structure horizontally (Figure 7.8a). A dip-slip fault is called a *reverse fault* if the hanging wall moves upward relative to the foot wall, causing a shortening of the structure (Figure 7.8b)—the reverse of what geologists have (somewhat arbitrarily) chosen as "normal." A **thrust fault** is a low-angled reverse fault—that is, one with a dip of less than 45°, so that the movement is more horizontal than vertical (Figure 7.8c). When subjected to horizontal compression, brittle rocks of the continental crust usually break along thrust faults with dips of 30° or less, rather than along more steeply dipping reverse faults.

A strike-slip fault is a *left-lateral fault* if an observer on one side of the fault sees that the block on the opposite side has moved to the left (Figure 7.8d). It is a *right-lateral fault* if the block on the opposite side appears to have moved to the right (Figure 7.8e). As you can tell from the stream offset in Figure 7.7, the San Andreas fault is a right-lateral transform fault. Other faults show both strike-slip and dip-slip motions. These faults are known as **oblique slip faults** (Figure 7.8f).

Geologists can recognize faults in the field in several ways. A fault may form a *scarp* (a cliff) that marks where

FIGURE 7.7 ■ View of the San Andreas fault, showing the northwestward movement of the Pacific Plate with respect to the North American Plate. The map shows a formation of volcanic rocks 23 million years old that has been displaced by 315 km. The fault runs from top to bottom (dashed line) near the middle of the photograph. Note the offset of the stream (Wallace Creek) by 130 m as it crosses the fault.
[University of Washington Libraries, Special Collections, John Shelton Collection, KCN7-23.]

DIP-SLIP FAULTING

Normal faulting is caused by tensional forces that stretch a rock and tend to pull it apart.

Reverse faulting is caused by compressive forces that squeeze and shorten a rock.

A thrust fault is a reverse fault with a shallow-dipping fault plane.

STRIKE-SLIP FAULTING

OBLIQUE-SLIP FAULTING

Left-lateral strike-slip fault

Right-lateral strike-slip fault

Oblique-slip faulting is caused by a combination of forces, in this case left-lateral shearing with tension.

FIGURE 7.8 ■ The orientation of tectonic forces determines the style of faulting. Dip-slip faulting (a–c) is caused by compressive or tensional forces. Strike-slip faulting (d, e) is caused by shearing forces. Oblique-slip faulting (f) is caused by a combination of shearing and compressive or tensional forces.

the fault intersects the ground surface (**Figure 7.9**). If the offset has been large, as it is for transform faults such as the San Andreas, the rock formations facing each other across the fault may differ in type and age. When movements are smaller, offset features can be observed and measured. (As an exercise, try to match up the beds offset by the small faults in Figure 7.1b.) In establishing the time of faulting, geologists apply a simple rule: a fault must be younger than the youngest rocks it cuts (the rocks had to be there before they could break!) and older than the oldest undeformed formation that covers it.

On geologic maps, faults are represented by *fault traces:* lines that indicate the point where a fault intersects the ground surface. Normal faults are distinguished from thrust faults by the different types of "teeth" that annotate the fault trace:

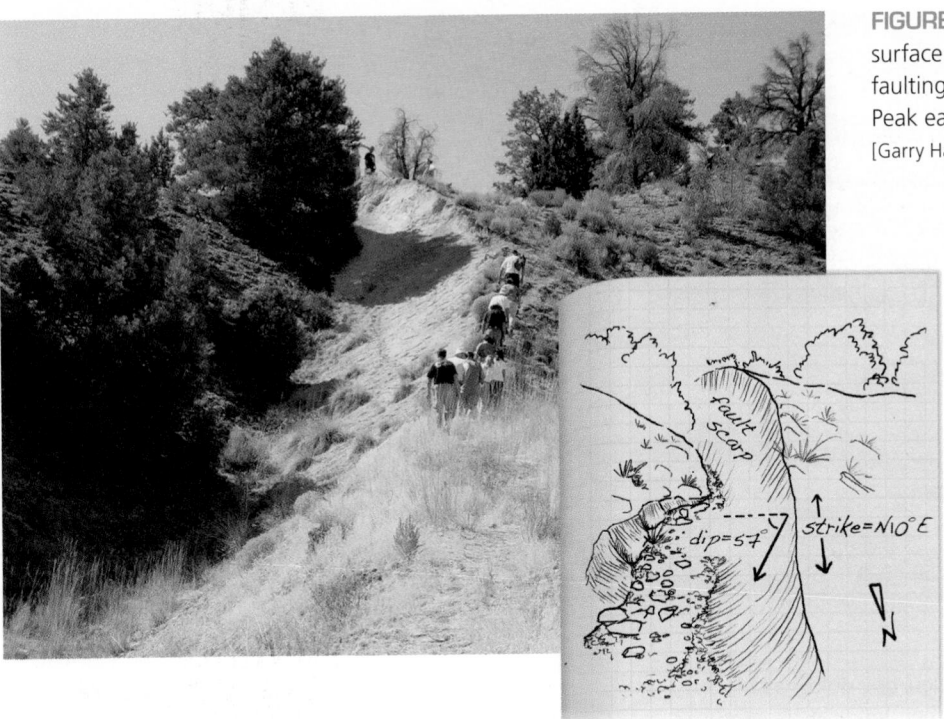

FIGURE 7.9 ■ This fault scarp is a fresh surface feature that formed by normal faulting during the 1954 Fairview Peak earthquake in Nevada.
[Garry Hayes/Geotripper Images.]

For both types of dip-slip faults, the teeth point toward the hanging wall. Examples of normal faults represented this way are shown in Figure 7.20; examples of thrust faults are shown in Figure 7.22. For strike-slip faults, the direction of movement, right-lateral or left-lateral, is indicated by a pair of arrows (yellow) that bracket the fault trace (see Figure 7.7):

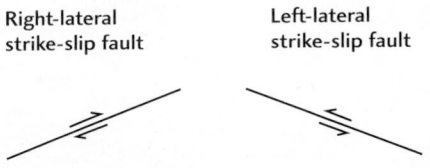

Right-lateral strike-slip fault   Left-lateral strike-slip fault

Folds

Folding is a common form of deformation observed in layered rocks (as in Figure 7.1a). **Folds** occur when an originally planar structure, such as a sedimentary bed, is bent into a curved structure. The bending can be produced by either horizontally or vertically directed forces in the crust, just as either pushing together the opposite edges of a piece of paper or pushing up or down on one side or the other can fold it.

Like faults, folds come in all sizes. In many mountain belts, majestic, sweeping folds can be traced over many kilometers (**Figure 7.10**). On a much smaller scale, very thin

FIGURE 7.10 ■ Large-scale folds in the sedimentary rocks that form the Kananaskis mountain range in Alberta, Canada. [Photoshot.]

sedimentary beds can be crumpled into folds a few centimeters long (**Figure 7.11**). The bending can be gentle or severe, depending on the magnitude of the applied forces, the length of time over which they were applied, and the resistance of the rocks to deformation.

Folds in which layered rocks are bent upward into arches are called **anticlines;** those in which rocks are bent downward into troughs are called **synclines** (**Figure 7.12**).

The two sides of a fold are its *limbs.* The *axial plane* of a fold is an imaginary surface that divides the fold as symmetrically as possible, with one limb on either side of the plane. The line made by the lengthwise intersection of the axial plane with the rock layers is the *fold axis.* A symmetrical horizontal fold has a horizontal fold axis and a vertical axial plane with limbs dipping symmetrically away from the axis.

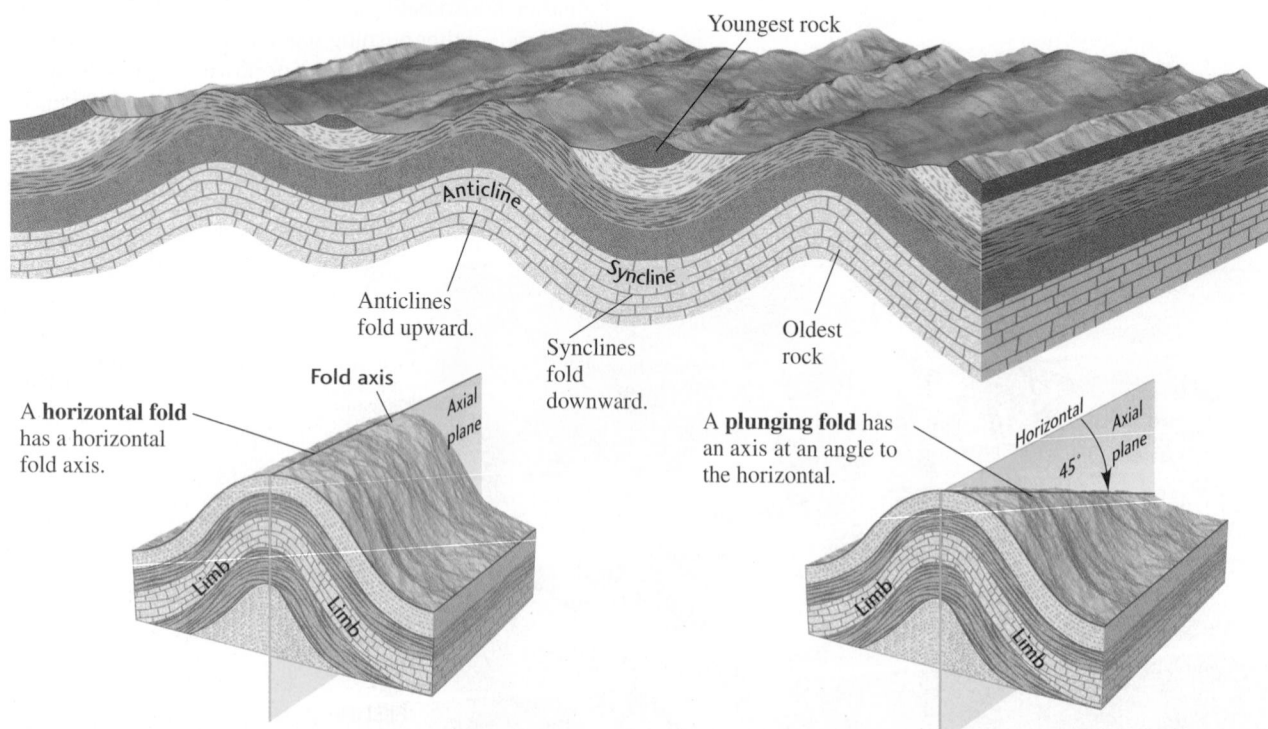

FIGURE 7.12 ■ The folding of rock layers is described by the direction of folding (upward or downward) and by the orientation of the fold axis and the axial plane.

Plunging anticline

Axial plane

Plunging syncline

FIGURE 7.13 ■ The geometry of plunging folds. Note the converging pattern of the layers of rock where they intersect the land surface.

Folds rarely stay horizontal, however. Follow the axis of any fold in the field, and sooner or later the fold dies out or appears to plunge into the ground. If a fold's axis is not horizontal, it is called a *plunging fold*. **Figure 7.13** diagrams the geometry of plunging anticlines and plunging synclines. In eroded mountain belts, a zigzag pattern of outcrops may appear in the field after erosion has removed much of the surface rock from the folds. The geologic map in Figure 7.4 shows this characteristic pattern.

Nor do folds usually remain symmetrical. With increasing amounts of deformation, folds can be pushed into asymmetrical shapes, with one limb dipping more steeply than the other (**Figure 7.14**). Such *asymmetrical folds* are common. When the deformation is so intense that one limb has been tilted beyond the vertical, the fold is called an *overturned fold*. Both limbs of an overturned fold dip in the same direction, but the order of the layers in the bottom limb is precisely the reverse of their original sequence—that is, older rocks are on top of younger rocks.

Observations in the field seldom provide complete information about folds. Bedrock may be obscured by

Symmetrical folds have limbs that dip symmetrically from the axial plane.

Asymmetrical folds have one limb that dips more steeply than the other.

Overturned folds have limbs that dip in the same direction, but one limb has been tilted beyond the vertical.

FIGURE 7.14 ■ With increasing deformation, folds are pushed into asymmetrical shapes. [(*left*) Tony Waltham; (*center*) © Photoshot; (*right*) Courtesy of John Grotzinger.]

overlying soils, or erosion may have removed much of the evidence of former structures. So geologists search for clues they can use to work out the relationship of one bed to another. For example, in the field or on a geologic map, an eroded anticline might be recognized as a strip of older rocks forming a core bordered on both sides by younger rocks dipping away from the core. An eroded syncline might appear as a core of younger rocks bordered on both sides by older rocks dipping toward the core. These relationships are illustrated in Figures 7.4 and 7.13. Determining the subsurface structure of folds by surface mapping has been an important method for finding oil, as described in the Practicing Geology exercise at the end of the chapter.

Circular Structures

Deformation along plate boundaries by horizontally directed forces usually results in linear faults and folds oriented nearly parallel to the plate boundary. Some types of deformation, however, are more symmetrical, forming nearly circular structures called basins and domes.

A **basin** is a synclinal structure, a bowl-shaped depression of rock layers in which the beds dip toward a central

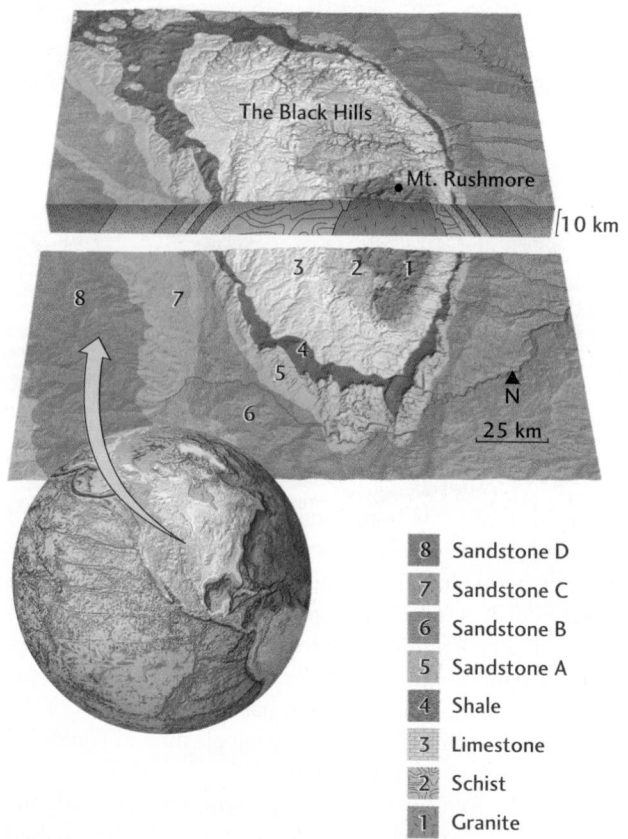

8	Sandstone D
7	Sandstone C
6	Sandstone B
5	Sandstone A
4	Shale
3	Limestone
2	Schist
1	Granite

FIGURE 7.16 ■ A geologic map and cross section of the Black Hills dome, which shows sedimentary rocks (formations 3–8) and metamorphic rocks (formation 2) that were uplifted and eroded by the intrusion of a granitic batholith (formation 1). At the Mount Rushmore National Memorial, the faces of four presidents—George Washington, Thomas Jefferson, Theodore Roosevelt, and Abraham Lincoln—are carved into these granitic rocks.

7	Red shale
6	Sandstone
5	Shale C
4	Shale B
3	Limestone
2	Dolomite
1	Shale A

FIGURE 7.15 ■ A geologic map and cross section of the Michigan Basin, which shows sedimentary layers deposited in a thick sequence from the oldest (formation 1) to the youngest (formation 7) during subsidence of the basin. The cross section has been vertically exaggerated by a factor of 5:1.

point (**Figure 7.15**). Sediments are often deposited in basins (see Chapter 5). In some cases, such as the Michigan Basin, shown in Figure 7.15, this deposition can produce sedimentary sequences many kilometers in thickness. A **dome** is an anticlinal structure, a broad circular or oval upward bulge of rock layers. The flanking beds of a dome encircle a central point and dip radially away from it (**Figure 7.16**). Domes, like other anticlines, are important in petroleum geology because oil is buoyant and tends to migrate upward through permeable rocks (see Practicing Geology exercise). If the rocks at the high point of a dome are impermeable to oil, the oil becomes trapped beneath them.

Domes and basins are typically many kilometers in diameter, and some extend for hundreds of kilometers. They are recognized in the field by outcrops that outline their characteristic circular or oval shapes. At these outcrops, the rock layers dip downward toward the center of the basin or upward toward the top of the dome (see Figures 7.15 and 7.16).

Some circular structures are formed by multiple episodes of deformation—for instance, when rocks are compressed in one direction and then again in a direction nearly perpendicular to the original direction. In many other cases, however, these structures result from the upward force of rising material or the downward force of sinking material, rather than the horizontally directed forces of plate tectonics. Not surprisingly, such circular structures tend to be more common in the interiors of plates, far away from active plate boundaries. There are many domes and basins, for example, in the central portion of the United States. Almost all of the Lower Peninsula of Michigan is a large sedimentary basin (see Figure 7.15); the Black Hills of South Dakota are an eroded dome (see Figure 7.16).

Several types of deformation can produce domes and basins. Some domes are formed by rising bodies of buoyant material—magma, hot igneous rock, or salt—that push the overlying sediments upward. As we saw in Chapter 5, some sedimentary basins form when a heated portion of the crust cools and contracts, causing the overlying sediments to subside (thermal subsidence basins). Others result when tectonic forces stretch and thin the crust (rift basins) or compress it downward (flexural basins). The weight of sediments deposited by a river delta can depress the crust into a sedimentary basin, such as the very large basin now forming at the mouth of the Mississippi River in the Gulf of Mexico.

Joints

As we have seen, a fracture that has displaced the rock on either side is called a fault. A second type of fracture is a **joint**—a crack in a rock formation along which there has been no appreciable movement (**Figure 7.17a**).

Joints are found in almost every outcrop. Some joints are caused by tectonic forces. Like any other brittle material, brittle rocks subjected to force fracture most easily at flaws or weak spots. These flaws can be tiny cracks, fragments of other materials, or even fossils. Regional tectonic forces—compressive, tensional, or shearing—may leave a set of joints as their imprint long after they have vanished.

The nontectonic expansion and contraction of rock can also form joints. Regular patterns of joints are often found in plutons and lavas that have cooled, contracted, and cracked. Erosion can strip away surface layers, releasing the confining pressure on underlying formations and allowing the rocks to expand and split at flaws.

Joints are usually only the beginning of a series of changes that greatly alter rock formations as they age. For example, joints provide channels through which water and air can reach deep into a formation and speed the weathering and internal weakening of its structure. If two or more sets of joints intersect, weathering may cause the formation to break into large columns or blocks (Figure 7.17b). The circulation of hydrothermal solutions through joints can deposit minerals such as quartz and calcite, forming veins, as we saw in Chapter 3.

Deformation Textures

Joints are examples of small features in rock formations that are best observed up close at an outcrop. Another type of small-scale deformation structure is the texture of a rock mass in areas of localized shearing, such as fault zones.

(a)

(b)

FIGURE 7.17 ■ Joint patterns. (a) Intersecting joints in a massive granite outcrop, Joshua Tree National Park, California. (b) Columnar joints in basalt, Giants Causeway, Northern Ireland. [(a) Sean Russell/Photolibrary; (b) Michael Brooke/Photolibrary/Getty Images, Inc.]

(a)

(b)

FIGURE 7.18 ■ (a) Fault breccia developed on a fault in eastern Nevada. The rust-colored breccia shows a cataclastic texture. The gray rocks on either side are limestones. (b) Mylonite developed in the Great Slave Lake shear zone, Northwest Territories, Canada. The rock was originally a granite. As a result of intense shearing forces, the large, originally angular potassium feldspar crystals have been rolled and transformed into smooth balls. [(a) Marli Bryant Miller; (b) Courtesy of John Grotzinger.]

As we have seen, tectonic forces cause the brittle parts of Earth's crust to crack and slip. As the rocks along a fault plane shear past each other, they grind and mechanically fragment solid rock. Where rocks behave as brittle materials (usually in the upper crust), shearing produces rocks with *cataclastic textures,* in which the grains are broken, angular fragments. One such rock type, called *fault breccia,* is shown in **Figure 7.18a**.

Deeper in the crust, where temperatures and pressures are high enough to allow ductile deformation, shearing forces can produce metamorphic rocks called *mylonites* (Figure 7.18b). The movement of rock surfaces against one another recrystallizes minerals and strings them out in bands or streaks. Development of mylonites typically occurs at the greenschist to amphibolite grades of metamorphism (see Chapter 6). The textural effects of deformation are most obvious in mylonites, but they are also prominent in cataclastic rocks.

The San Andreas fault of Southern California makes a good case study of how deformation textures might relate to changes in temperature and pressure with depth. This fault, which marks the boundary between the Pacific Plate and the North American Plate (see Figure 7.7), extends through the crust and probably down into the mantle. At depths up to about 20 km, the fault zone is thought to be very narrow and characterized by cataclastic textures, indicating brittle deformation. Earthquakes are generated in this zone. Below 20 km, however, earthquakes do not occur, and the fault is thought to be characterized by a broad zone of ductile deformation that produces mylonites.

■ Styles of Continental Deformation

If we look closely enough, we can find all the basic deformation structures—faults, folds, domes, basins, joints—in any zone of continental deformation. But when we view continental deformation at a regional scale, we see distinctive patterns of faulting and folding that relate directly to the tectonic forces causing the deformation. **Figure 7.19** depicts the deformation styles typical of the three main types of tectonic forces.

Tensional Tectonics

In brittle crust, the tensional forces that produce normal faulting may split a plate apart, resulting in a *rift valley*—a long, narrow trough formed when a block of rock drops downward relative to its two flanking blocks along nearly parallel, steeply dipping normal faults (Figure 7.19a). Well-known examples include the Rhine Valley, the rift valleys of East Africa (**Figure 7.20**), and the Red Sea, as well as the rift valleys of mid-ocean ridges. As we saw in Chapter 5, these structures form basins that fill with sediments eroded from the rift walls as well as with volcanic rocks extruded from tensional cracks in the crust.

Tension on the shallow continental crust usually produces normal faults with high dip angles, typically 60° or

(a) Tensional tectonics: Extension of continental crust produces normal faults with high dip angles in the upper crust that flatten with depth, forming curved fault surfaces.

Brittle upper crust

Normal faulting

Ductile lower crust

(b) Compressive tectonics: Compression of continental crust occurs on low-angled thrust faults.

Fold and thrust belt

Thrust faulting

(c) Shearing tectonics: Shearing of continental crust occurs on a nearly vertical strike-slip fault. The case shown here is for a right-lateral fault.

Strike-slip faulting

A left bend in the fault results in local compression.

A right bend in the fault results in local extension.

FIGURE 7.19 ▪ The orientation of tectonic forces—(a) tensional, (b) compressive, and (c) shearing—determines the style of continental deformation. On a regional scale, the basic types of faulting shown in the inset figures can lead to distinctive, complex patterns of deformation. [After John Suppe, *Principles of Structural Geology.* Upper Saddle River, NJ: Prentice Hall, 1985.]

more. Below a depth of about 20 km, however, crustal rocks are hot enough to behave as ductile materials, and deformation occurs by stretching rather than by fracturing. This change in rock behavior causes the dip of the faults to flatten with increasing depth, which results in normal faults with curved fault surfaces (called *listric* faults), as shown in Figure 7.19a. The crustal blocks moving along these curved faults are tilted backward as the stretching continues.

The Basin and Range province, which is centered on the Great Basin of Nevada and Utah, is a good example of a region defined by many adjacent rift valleys. This region, which is now more than 800 km wide, has been stretched and extended in a northwest-southeast direction by a factor of 2 during the last 15 million years. Here, normal faulting has created an immense landscape of eroded, rugged fault-block mountains and smooth, sediment-filled valleys, some covered with recent volcanic rocks (see Figure 10.5). This tensional deformation, which appears to be caused by upwelling convection currents beneath the Basin and Range province, continues today.

Compressive Tectonics

In subduction zones, oceanic lithosphere slips beneath an overriding plate along a huge thrust fault, or *megathrust.* The world's largest earthquakes, such as the great Tohoku, Japan, earthquake of March 11, 2011, which generated a disastrous tsunami that killed more than 19,000 people, are

FIGURE 7.20 ■ In East Africa, tensional forces are pulling the Somali subplate away from the African Plate, creating rift valleys bounded by normal faults (see Figure 2.8b). The rift valley shown here is filled by sediments and Lake Tanganyika, on the boundary between Tanzania and the Democratic Republic of Congo. The cross section has been vertically exaggerated by a factor of 2.5:1, which exaggerates the fault dips; the actual dips of the normal faults are about 60°.

caused by sudden slips on megathrusts. Thrust faulting is also the most common type of faulting within continents undergoing tectonic compression. Sheets of crust may glide over one another for tens of kilometers along nearly horizontal thrust faults, forming *overthrust* structures (**Figure 7.21**).

When two continents collide, the crust can be compressed across a wide zone, resulting in spectacular episodes of mountain building. During such collisions, the brittle basement rocks ride over one another by thrust faulting while the more ductile overlying sedimentary rocks are compressed into a series of great folds, forming a *fold and thrust belt* (Figure 7.19b). Large earthquakes are common in fold and thrust belts; a recent example is the great Wenchuan earthquake that hit Sichuan, China, on May 12, 2008, killing more than 80,000 people.

The ongoing collisions of Africa, Arabia, and India with the southern margin of the Eurasian continent have

Youngest rock
Oldest rock

1 Compressive forces fractured the rock layers…

2 …and thrust them horizontally over a section of the same rock.

3 Erosion of the topmost layers reveals the view we see today: Cambrian limestone over Jurassic sandstone that is 350 million years younger.

50 km

Keystone thrust fault, southern Nevada

FIGURE 7.21 ■ The Keystone thrust fault of southern Nevada is a large-scale overthrust structure of a kind formed during episodes of continental compression. Compressive forces have detached a sheet of rock layers (D, C, B) and thrust it a great distance horizontally over a section of the same rock layers (D, C, B, A).
[Marli Bryant Miller.]

FIGURE 7.22 ■ Space Shuttle photograph showing an oblique view of the San Andreas fault system. The annotations illustrate how the deviations of a transform fault's strike from the direction of plate motion can cause local extension and compression. Between the Gulf of California and the Salton Sea (near bottom of figure), the fault system jogs to the right in two major steps; the right-lateral fault segments (black lines), which are parallel to the Pacific-North America plate motion, are separated by rift valleys (in red) that are volcanically active, subsiding, and filling with sediments. As we look northward, the fault trace first bends to the left, away from the direction of plate motion, and then to the right, realigning with the plate motion in central California (near top of figure). This "Big Bend" in the San Andreas fault causes compression, which is taken up by reverse faulting in the Los Angeles region (middle of figure). [Image Science & Analysis Laboratory, NASA Johnson Space Center.]

created fold and thrust belts from the Alps to the Himalaya, many of which are still active. The great oil reservoirs of the Middle East are trapped in anticlines formed by this deformation. Compression across western North America, caused by that continent's westward movement during the opening of the Atlantic Ocean, created the fold and thrust belt of the Canadian Rockies. The Valley and Ridge province of the Appalachian Mountains is an ancient fold and thrust belt that dates back to the collisions that created the super-continent Pangaea.

Shearing Tectonics

A transform fault is a strike-slip fault that forms a plate boundary. Transform faults such as the San Andreas can off-set geologic formations by long distances (see Figure 7.7), but as long as they stay aligned with the direction of relative plate movement, the blocks on either side can slide past each other without much internal deformation. Long transform faults are rarely straight, however, so deformation patterns along these faults can be much more complicated. The faults may have bends and jogs that change the tectonic forces acting across portions of the plate boundary from shearing

forces to compressive or tensional forces. Those forces, in turn, cause secondary faulting and folding (Figure 7.19c).

A good example of these complications can be found in Southern California, where the right-lateral San Andreas fault bends first to the left and then to the right as one moves along its trace from south to north (**Figure 7.22**). The segments of the fault on both sides of this "Big Bend" are aligned with the direction of relative plate movement, so the blocks slip past each other there by simple strike-slip faulting. Within the Big Bend, however, the change in the fault orientation causes the plates to push against each other, which produces thrust faulting to the south of the fault. This thrusting has raised the San Gabriel and San Bernardino mountains to elevations exceeding 3000 m and, during the last half century, has produced a series of destructive earthquakes, including the 1994 Northridge quake, which caused more than $40 billion in damage to Los Angeles (see Chapter 13).

At the southern end of the San Andreas, between the Gulf of California and the Salton Sea, the boundary between the Pacific and North American plates jogs to the right in a series of steps. Within these jogs, the plate boundary is subjected to tensional forces, and normal faulting has

TIME 1
Horizontal layers of sediments are deposited on the seafloor.

TIME 2
Compressive forces cause folding, faulting, and uplift.

Compressive forces

Faults

TIME 3
Uplift is followed by erosion, which creates a new horizontal surface.

TIME 4
Volcanic eruptions flood the new surface with lava.

Lava flows

TIME 5
Tensional forces cause normal faults, creating downdropped blocks and breaking up earlier features.

Tensional forces

Normal faults

formed small rift valleys that are volcanically active, rapidly subsiding, and filling with sediments. This extension occurs within 200 km of the Big Bend compression, demonstrating how variable the tectonics along continental transform faults can be!

■ Unraveling Geologic History

The geologic history of a region is a succession of episodes of deformation and other geologic processes. Let's see how some of the concepts and methods introduced in this chapter can be used to reconstruct that history.

The cross sections in **Figure 7.23** represent a few tens of kilometers of a geologic region that underwent a succession of tectonic events. First, horizontal layers of sediment were deposited on the seafloor. Those layers were tilted and folded, and eventually uplifted above sea level, by horizontal compressive forces. There, erosion gave them a new horizontal surface. That surface was covered by lava when forces deep in Earth's interior caused a volcanic eruption. In the most recent stage, tensional forces resulted in normal faulting, which broke the crust into blocks.

Geologists see only the last stage, but visualize the entire sequence. They begin by identifying and determining the ages of the rock layers and recording the orientation of layers, folds, and faults on geologic maps. Then they use those maps to construct cross sections of the subsurface features. Once the geologists have identified sedimentary beds, they can start with the knowledge that the beds must originally have been horizontal and undeformed at the bottom of an ancient ocean. The succeeding events can then be reconstructed.

Present-day surface topography in young mountain ranges—such as the Alps, the Rocky Mountains, the Pacific Coast Ranges, and the Himalaya—can be traced in large part to deformation that has occurred over the past few tens of millions of years. These young systems still contain much of the information that geologists need to piece together the history of that deformation. Deformation that occurred hundreds of millions of years ago is no longer evident in the form of rugged mountains, however. Erosion has left behind only the remnants of folds and faults, expressed as low ridges and shallow valleys. As we will see in Chapter 10, even older episodes of mountain building are evident only in the twisted, highly metamorphosed formations that constitute the basement rocks of the interiors of continents.

FIGURE 7.23 ■ Stages in the development of a geologic region. A geologist sees only the last stage and attempts to reconstruct all the earlier stages from the observable structural features.

Google Earth Project

Image © 2009 DigitalGlobe Image USDA Farm Service Agency
Image © 2009 TerraMetrics Image © 2009 Province of British Columbia

Google Earth satellite image showing the northwestern United States and
the location of Ribbon Canyon, Wyoming (marked with a red dot).

Image © 2009 DigitalGlobe Image USDA Farm Service Agency

Google Earth three-dimensional satellite image of Ribbon Canyon,
Wyoming, showing beds dipping steeply in opposite directions.

When rocks deform, they may form folds. An anticline is a fold that is shaped like an arch, while a syncline is shaped like a trough. When examining a rock outcrop, we can recognize an anticline by a sequence of beds dipping, or tilting, away from the center of the fold. For a syncline, the beds would dip toward the center of the fold. Let's look at an example from Wyoming.

LOCATION   Sheep Mountain/Ribbon Canyon near Big Horn, Wyoming

GOAL   Learn to recognize simple fold structures

LINKED   Figures 7.3, 7.4, and 7.10

1. Type "Ribbon Canyon, Big Horn, Wyoming" into the GE search window. At the bottom left, under the "Layers" tab, click on "Terrain" to turn on the 3D viewing capability. Zoom in to an eye altitude of 8.75 km. What is the latitude and longitude of Ribbon Canyon, Wyoming?
   a. 44° 39' 12.26" N, 108° 11' 59.32" W
   b. 44° 31' 30.16" N, 107° 57' 24.39" W
   c. 45° 15' 17.18" N, 108° 59' 03.12" W
   d. 45° 29' 07.45" N, 107° 24' 18.94" W

2. In this chapter, we've traced various features that a geologist in the field might map. The map shown in the accompanying figure is centered at 44° 38' 21.52" N, 108° 10' 50.64" W, at an eye altitude of approximately 2.75 km. We have used the "Add path" tool (an icon at the top of the Google Earth screen) to trace some of the beds on both sides of the mound. Try mapping (tracing) these beds yourself. You can change the colors of your paths by clicking on "Style/Color" in the "Add path" window. Which way do the beds on each side of the mound appear to be dipping?
   a. The blue beds are dipping to the southwest, while the red beds are dipping to the northeast.
   b. The blue beds and the red beds are dipping to the southwest.
   c. The blue beds are dipping to the northeast, while the red beds are dipping to the southwest.
   d. The blue beds and the red beds are dipping to the northwest.

3. In which direction do the beds on each side of the mound appear to be tilted in relation to the fold? (You may need to rotate the perspective view to see the beds from different sides. To do this, at the top right of the Google Earth screen, drag the "N" around the circle to change direction. You can also scroll farther to the southeast to see a larger extent of this feature.) Given this information, do you think this area represents an anticline or a syncline? The beds are tilted
   a. away from the center of the fold, making it a syncline.
   b. toward the center of the fold, making it a syncline.
   c. away from the center of the fold, making it an anticline.
   d. toward the center of the fold, making it an anticline.

4. In this same area, what's the approximate highest elevation at the top of the mound?
   a. 1275 m
   b. 1320 m
   c. 1530 m
   d. 1610 m

Optional Challenge Question

5. Think about how anticlines and synclines are formed. Do you think the oldest beds are on the inside or the outside of the fold? (*Hint:* Imagine a stack of papers in which the papers on the bottom of the stack represent the oldest beds and the papers on top represent the youngest. If you fold the stack into a U shape [a syncline], are the "oldest" papers located on the inside or the outside of the fold? If you invert that stack into an upside-down U shape [an anticline], where are the oldest layers now?) At Ribbon Canyon,
   a. the oldest beds are on the outside of the fold.
   b. the oldest beds are on the inside of the fold.
   c. the oldest beds are all the same age.
   d. the beds are faulted, not folded.

SUMMARY

How do geologists represent geologic structure on maps and diagrams? Two important measures in geologic maps and diagrams are strike and dip. Strike is the compass direction of a rock layer along its intersection with a horizontal surface. Dip is the angle at which the rock layer inclines from the horizontal, measured at right angles to the strike. A geologic map is a two-dimensional model of the geologic features exposed at Earth's surface, showing various rock formations as well as other features such as faults. A geologic cross section is a diagram representing the geologic features that would be visible if a vertical slice were made through part of the crust. Geologic cross sections can be constructed from the information on a geologic map, although they can often be improved with subsurface data collected by drilling or seismic imaging.

What do laboratory experiments tell us about the way rocks deform? Laboratory studies show that rocks subjected to tectonic forces may behave as brittle materials or as ductile materials. These behaviors depend on temperature and pressure, the type of rock, the speed of deformation, and the orientation of tectonic forces. Rocks that are brittle at shallow depths can act as ductile material deep in the crust.

What are the basic deformation structures that geologists observe in the field? Among the geologic structures that result from deformation are folds, faults, circular structures, joints, and deformation textures caused by shearing. Fractures are known as faults if rocks are displaced across the fracture surface, and as joints if no displacement is observed.

What kinds of forces produce these deformation structures? Faults and folds are produced primarily by horizontally directed tectonic forces at plate boundaries. Horizontal tensional forces at divergent boundaries produce normal faults, horizontal compressive forces at convergent boundaries produce thrust faults, and horizontal shearing forces at transform-fault boundaries produce strike-slip faults. Folds are usually formed in layered rock by compressive forces, especially in regions where continents collide. Circular structures, such as domes and basins, can be produced by vertically directed forces far from plate boundaries. Some domes are caused by the rise of buoyant materials. Basins can form when tensional forces stretch the

crust or when a heated portion of the crust cools, contracts, and subsides. Joints can be caused by tectonic stresses or by the cooling and contraction of rock formations.

What are the main styles of continental deformation? There are three main styles of continental deformation. Tensional tectonics produces rift valleys with normal faulting; in continental regions undergoing extension, the dip angles of the normal faults flatten with depth, causing the fault blocks to tilt away from the rift as the faulting continues. Compressive tectonics produces thrust faulting; in the case of continent-continent collisions, compression may produce fold and thrust belts. Shearing tectonics produces strike-slip faulting, but bends and jogs in the fault may cause local thrust faulting and normal faulting.

How do we reconstruct the geologic history of a region? Geologists can observe only the end result of a succession of events: deposition, deformation, erosion, volcanism, and so forth. They deduce the deformational history of a region by identifying and determining the ages of rock layers, recording the geometric orientation of rock layers on geologic maps, mapping folds and faults, and constructing cross sections of subsurface structure consistent with their surface observations.

KEY TERMS AND CONCEPTS

anticline (p. 180)
basin (p. 182)
brittle (p. 176)
compressive force (p. 172)
deformation (p. 172)
dip (p. 173)

dip-slip fault (p. 177)
dome (p. 182)
ductile (p. 176)
fault (p. 177)
fold (p. 179)
foot wall (p. 177)
formation (p. 173)

geologic cross section (p. 175)
geologic map (p. 173)
hanging wall (p. 177)
joint (p. 183)
normal fault (p. 177)
oblique slip fault (p. 177)

shearing force (p. 172)
strike (p. 173)
strike-slip fault (p. 177)
syncline (p. 180)
tensional force (p. 172)
thrust fault (p. 177)

PRACTICING GEOLOGY EXERCISE
How Do We Use Geologic Maps to Find Oil?

Crude oil, or *petroleum* (from the Latin words for "rock oil"), has been collected from natural seeps at Earth's surface since ancient times. The foul-smelling, tarry substance was used as boat caulking, wheel grease, and medicine, but not commonly as a fuel until the process of oil refining was developed in the 1850s. Demand skyrocketed at that time, primarily because oil from whale blubber, the best fuel then available for lamps, had become terribly expensive ($60 per gallon in today's dollars!) as overfishing decimated whale populations.

The ability to refine clean lamp oil from petroleum set off North America's first oil boom. The mining of "black gold" was centered in areas around Lake Erie, where major petroleum seeps had been discovered—in northwestern Pennsylvania, northeastern Ohio, and southern Ontario. Early petroleum explorers, such as self-proclaimed "Colonel" Edwin Drake of Pennsylvania, simply drilled into the seeps, but this straightforward approach soon proved inadequate as a strategy for satisfying the new thirst for oil.

Could geologic knowledge be used to locate large petroleum reservoirs hidden underground—that is, in regions where no oil seeped to the surface? An affirmative answer was provided in 1861 by T. Sterry Hunt, a Connecticut-born geochemist. As a member of the Geological Survey of Canada, Hunt had been active in the new science of mapping natural resources. He documented the petroleum seeps of southern Ontario in 1850. As the oil production of the region increased, he noticed that seeps and successful wells tended to be aligned along the crests of geologic folds.

Hunt had also studied the physical and chemical properties of petroleum in the laboratory, and he knew it was formed when sedimentary rocks rich in organic material were subjected to heat and pressure (see Chapter 5). Petroleum is lighter than water; because of this buoyancy, it tends to rise toward the surface. Hunt hypothesized that the rising petroleum could accumulate in porous "reservoir rocks," such as sandstones, if such rocks were overlain by impermeable "cap rocks," such as shales, that prevented the petroleum from rising farther. Moreover, the most likely place to find large reservoirs would be along the fold axes of anticlines, where substantial amounts of petroleum could be trapped without escaping to the surface.

The accompanying figure illustrates a typical anticlinal trap, for which we can imagine the following narrative of geologic discovery. Erosion of the fold has exposed a sequence of sandstones, limestones, and shales. Mapping by an enterprising geologist shows that the axis of the anticline strikes to the northeast. Drilling at point A on the axis of the anticline first penetrates a thick sandstone layer exposed at the surface and then a thinner shale layer. Immediately below the shale the drilling crew encounters another sandstone layer containing gas and, below the gas, significant quantities of oil. The geologist infers that the shale is capping a major petroleum reservoir in the deeper sandstone layer, so he instructs his crew to move along the strike of the anticline and drill at point B. Bingo—another successful oil well!

Hunt's "anticlinal theory" allowed geologists to discover oil (and some to get rich) by mapping fold structures at the surface and, later, by the three-dimensional imaging of such structures using seismic techniques. The results have been impressive: most of the one trillion barrels of crude oil produced since 1861 have come from anticlinal oil traps of the type that Hunt first described.

BONUS PROBLEM: The company that manages the petroleum claim in the figure would like to expand its operations, and they propose to drill a new well further along the axis of the anticline at point C. As a consulting geologist, how would you rate their chances of bringing in another successful well? Illustrate your answer by sketching a geologic cross section.

■ Reconstructing Geologic History from the Stratigraphic Record　196

■ The Geologic Time Scale: Relative Ages　202

■ Measuring Absolute Time with Isotopic Clocks　207

■ The Geologic Time Scale: Absolute Ages　210

■ Recent Advances in Timing the Earth System　212

These trilobites are preserved as fossils in rocks about 365 million years old found in Ontario, Canada. [© Kevin Schafer/Alamy.]

EXERCISES

1. What type of fold is shown in Figure 7.1a? Is the small fault on the right side of Figure 7.1b a normal fault or a thrust fault? Estimate the fault's offset, expressing your answer in meters.

2. On a geologic map of 1:250,000 scale, how many centimeters would represent an actual distance of 2.5 km? What is the actual distance in miles of 1 inch on the same map?

3. The movement of the North American and Pacific plates along the San Andreas fault has offset the stream channel in Figure 7.7 by 130 m. Geologists have determined that this channel is 3800 years old. What is the slip rate along the San Andreas fault at this site, expressed in millimeters per year?

4. From Figure 2.7, estimate the direction of the plate tectonic forces that are causing the extension of the Red Sea.

5. Show that a left jog in a right-lateral strike-slip fault will produce compression, whereas a right jog in a right-lateral strike-slip fault will produce extension. Write a similar rule for left-lateral strike-slip faults.

6. Draw a geologic cross section that tells the following story: A series of marine sediments was deposited and subsequently deformed by compressive forces into a fold and thrust belt. The mountains of the fold and thrust belt eroded to sea level, and new sediments were deposited. The region then began to be extended, and lava intruded the new sediments to create a sill. In the latest stage, tensional forces broke the crust to form a rift valley bounded by steeply dipping normal faults.

THOUGHT QUESTIONS

1. In what sense is a geologic map a scientific model of the surface geology? Is it fair to say that geologic cross sections in combination with a geologic map constitute a scientific model of the three-dimensional geologic structure? (In formulating your answers, you may want to refer to the discussion of scientific models in Chapter 1.)

2. Why is it correct to say that "large-scale geologic structures should be represented on small-scale geologic maps"? How big a piece of paper would be required to make a map of the entire U.S. Rocky Mountains at 1:24,000 scale?

3. The submerged margin of a continent has a thick layer of sediments overlying metamorphic basement rocks. That continental margin collides with another continental mass, and the compressive forces deform it into a fold and thrust belt. During the deformation, which of the following geologic formations would be likely to behave as brittle materials and which as ductile materials: (a) the sedimentary formations in the upper few kilometers; (b) the metamorphic basement rocks at depths of 5 to 15 km; (c) lower crustal rocks at depths below 20 km? In which of these layers would you expect earthquakes?

4. It was the writer John McPhee who called geologic maps "textbooks on a piece of paper" in his epic narrative about a geologic traverse across North America, *Annals of the Former World* (p. 378). Can you locate a passage in this textbook that describes a geologic structure and sketch a geologic map consistent with McPhee's description?

5. Can you explain the geologic story in Exercise 6 in terms of plate tectonic events? Where in the United States do geologists think this sequence of events has taken place?

MEDIA SUPPORT

 7-1 Animation: Strike and Dip

 7-3 Animation: Reverse and Thrust Faults

 7-5 Animation: Rock Folding

 7-2 Animation: Normal Faults

 7-4 Animation: Strike-Slip Faults

 7-1 Video: Dome in the Desert: Upheaval Dome

CLOCKS IN ROCKS: TIMING THE GEOLOGIC RECORD

PHILOSOPHERS HAVE STRUGGLED WITH the notion of time throughout human history, but until fairly recently, they had very little data to constrain their speculations. The immensity of time—"deep time" measured in billions of years—was a great geologic discovery that changed our thinking about how Earth operates as a system.

Pioneering geologists such as James Hutton and Charles Lyell led us to understand that the planet was not shaped by a series of catastrophic events over a mere few thousand years, as many people had believed. Rather, what we see today is the product of ordinary geologic processes operating over much longer time intervals. Hutton stated this understanding as the principle of uniformitarianism, described in Chapter 1. Knowledge of geologic time helped Charles Darwin formulate his theory of evolution, and it has led to many other insights about the workings of the Earth system, the solar system, and the universe as a whole.

Geologic processes occur on time scales that range from seconds (meteorite impacts, volcanic explosions, earthquakes) to tens of millions of years (the recycling of oceanic lithosphere) and even billions of years (the tectonic evolution of continents). If we are careful enough, we can measure the rates of short-term processes, such as beach erosion or seasonal variations in the transport of sediments by rivers, in a few years. Precise surveying can monitor the slow movements of glaciers (meters per year), and with the Global Positioning System, we can track the even slower movements of the lithospheric plates (centimeters per year). Historical documents can provide certain types of geologic data, such as the dates of major earthquakes or volcanic eruptions, from hundreds or, in some cases, thousands of years ago.

However, the record of human observation is far too short for the study of many slow geologic processes (Figure 8.1). In fact, it's not even long enough to capture some types of rapid but infrequent events; for example, we have never witnessed a meteorite impact as big as the one that left the crater shown in Figure 1.7. We must rely instead on the geologic record: the information preserved in rocks that have survived erosion and subduction. Almost all oceanic crust older than 200 million years has been subducted back into the mantle, so most of Earth's history is documented only in the older rocks of the continents. Geologists can reconstruct subsidence from the record of sedimentation; uplift from the erosion of rock layers; and deformation from faults, folds, and metamorphic rocks. But to measure the pace of these processes and understand their common causes, we must be able to assign ages to events observed in the geologic record.

In this chapter, we will learn how geologists first plumbed the abyss of time by finding order in the geologic record. Then we will see how they used the discovery of radioactive "clocks in rocks" to develop a precise and detailed geologic time scale and to date the events that have occurred throughout Earth's 4.56-billion-year history.

■ Reconstructing Geologic History from the Stratigraphic Record

Geologists speak carefully about time. To them, *dating* refers not to a popular social activity, but to measuring the **absolute age** of an event in the geologic record: the number of years elapsed from that event until now. Before the twentieth century, no one knew much about absolute ages; geologists could determine only whether one event was earlier or later than another—their **relative ages.** They could say, for instance, that fish bones were first deposited in marine sediments before mammal bones first appeared in sediments on land, but they couldn't tell how many millions of years ago the first fish or mammals appeared.

FIGURE 8.1 ■ Two photographs of Bowknot Bend on the Green River in Utah, taken nearly 100 years apart, show that the configuration of rocks and geologic structures has changed very little in that time interval. [*left:* E.O. Beaman/USGS; *right:* H.G. Stevens/USGS.]

(a)

(b)

FIGURE 8.2 ■ Fossils are traces of living organisms preserved in the geologic record. (a) Ammonite fossils, ancient examples of a large group of invertebrate organisms that are now largely extinct. Their sole representative in the modern world is the chambered nautilus. (b) Petrified Forest, Arizona. These ancient logs are millions of years old. Their substance has been completely replaced by silica, which preserved the original details of their form. [(a) 34282_IrridAmmonites4x5 by Chip Clark, Smithsonian; (b) Thinkstock.]

The first geologic observations pertaining to the question of deep time came in the mid-seventeenth century from the study of fossils. A *fossil* is an artifact of life preserved in the geologic record (**Figure 8.2**). However, few people living in seventeenth-century Europe would have understood this definition. Most thought that the seashells and other lifelike forms they found in rocks dated from Earth's beginnings about 6000 years earlier, or grew there spontaneously.

In 1667, the Danish scientist Nicolaus Steno, who was working for the royal court in Florence, Italy, demonstrated that the peculiar "tongue stones" found in certain Mediterranean sedimentary rocks were essentially identical to the teeth of modern sharks (**Figure 8.3**). He concluded that tongue stones *really were* ancient shark teeth preserved in the rocks, and more generally, that fossils were the remains of ancient life deposited with sediments. To convince people of his ideas, Steno wrote a short but brilliant book about the geology of Tuscany, in which he laid the foundation for the modern science of **stratigraphy**—the study of *strata* (layers) in rocks.

Principles of Stratigraphy

Geologists still use the principles set forth by Steno to interpret sedimentary strata. Two of his basic rules are so simple they seem obvious to us today:

1. The **principle of original horizontality** states that sediments are deposited under the influence of gravity as nearly horizontal beds. Observations in a wide variety of sedimentary environments support this principle. If we find folded or faulted strata, we know that the beds were deformed by tectonic forces after the sediments were deposited.

2. The **principle of superposition** states that each layer of an undeformed sedimentary sequence is younger than the one beneath it and older than the one above it. A new layer cannot be deposited beneath an existing layer. Thus, strata can be vertically ordered in time from the lowest (oldest) bed to the uppermost (youngest) bed (**Figure 8.4**). A chronologically ordered set of strata is known as a **stratigraphic succession.**

FIGURE 8.3 ■ Nicolaus Steno was the first to demonstrate that fossils are the remains of ancient life. (a) A portrait of Nicolaus Steno (1638–1686). [SPL/Science Source.] (b) "Tongue stones" of the type found in sedimentary rocks in the Mediterranean region, where Steno worked. [Corbis RF/Alamy.] (c) This diagram is from Steno's 1667 book, which demonstrated that tongue stones are the fossilized teeth of ancient sharks. [Paul D. Stewart/SPL/Science Source.]

We can apply Steno's principles in the field to determine whether one sedimentary formation is older than another. Then, by piecing together the formations exposed in different outcrops, we can sort them into chronological order and thus construct the stratigraphic succession of a region—at least in principle.

In practice, there were two problems with this strategy. First, geologists almost always found gaps in a region's stratigraphic succession, indicating time intervals that had gone entirely unrecorded. Some of these intervals were short, such as periods of drought between floods; others lasted for millions of years—for example, periods of regional tectonic uplift when thick sequences of sedimentary rocks were removed by erosion. Second, it was difficult to determine the relative ages of two formations that were widely separated in space; stratigraphy alone couldn't determine whether a sequence of mudstones in, say, Tuscany was older, younger, or the same age as a similar sequence in England. It was necessary to expand Steno's ideas about the biological origin of fossils to solve these problems.

Fossils as Recorders of Geologic Time

In 1793, William Smith, a surveyor working on the construction of canals in southern England, recognized that fossils could help geologists determine the relative ages of sedimentary rocks. Smith was fascinated by the variety of

FIGURE 8.4 ■ Steno's principles guide the study of sedimentary strata. (a) Sediments are deposited in horizontal layers before being slowly transformed into sedimentary rock. If left undisturbed by tectonic processes, the youngest layers remain on the top and the oldest on the bottom. (b) Marble Canyon, part of the Grand Canyon, was cut by the Colorado River through what is now northern Arizona, revealing these undisturbed strata, which record millions of years of geologic history. [Fletcher & Baylis/Science Source.]

fossils he collected from the strata exposed along canal cuts. He observed that different layers contained different sets of fossils, and he was able to tell one layer from another by the characteristic fossils in each. He established a general order for the sequence of fossil assemblages and strata, from lowest (oldest) to uppermost (youngest). Regardless of its location, Smith could predict the stratigraphic position of any particular layer or formation in any outcrop in southern England based on its fossil assemblages. This stratigraphic ordering of the fossils of animal species (fauna) produces a sequence known as a *faunal succession.*

Smith's **principle of faunal succession** states that the sedimentary strata in an outcrop contain fossils in a definite sequence. The same sequence can be found in outcrops at other locations, so that strata in one location can be matched to strata in another location.

Using faunal successions, Smith was able to identify formations of the same age in different outcrops. By noting the vertical order in which the formations were found in each place, he compiled a composite stratigraphic succession for the entire region. His compilation showed how the complete succession would have looked if the formations at different levels in all the various outcrops could have been brought together at a single spot. **Figure 8.5** shows such a composite stratigraphic succession for two outcrops.

Smith kept track of his work by mapping outcrops using colors assigned to specific formations, thus inventing the geologic map (see Figure 7.4). In 1815, he summarized his lifelong research by publishing his *General Map of Strata in England and Wales,* a hand-colored masterpiece 8 feet tall and 6 feet wide—the first geologic map of an entire country. The original still hangs in the offices of the Geological Society of London.

The geologists who followed in Steno's and Smith's footsteps described and catalogued hundreds of fossils and their relationships to modern organisms, establishing the new science of *paleontology:* the historical study of ancient life-forms. The most common fossils they found were the shells of invertebrate animals. Some were similar to clams, oysters, and other living shellfish; others represented strange species with no living examples, such as the trilobites shown in the chapter opening photo. Less common were the bones of vertebrates, such as mammals, birds, and the huge extinct reptiles they called dinosaurs. Plant fossils were found to be abundant in some rocks, particularly in coal beds, where leaves, twigs, branches, and even whole tree trunks could be recognized. Fossils were not found in intrusive igneous rocks—no surprise, since any biological material would have been destroyed when the rocks melted—nor in high-grade metamorphic rocks—where any remains of organisms would have been distorted beyond recognition.

By the beginning of the nineteenth century, paleontology had become the single most important source of information about geologic history. The systematic study

1 Fossils found in some formations in outcrop A are the same as those found in some formations in outcrop B, some distance away.

Outcrop A

Outcrop B

Stratigraphic succession

Younger rocks

2 Formations with the same fossils are the same age.

Older rocks

3 A composite of the two outcrops would show formations I and II both overlying formation III.

FIGURE 8.5 ■ The principle of faunal succession can be used to correlate rock formations in different outcrops.

of fossils affected science far beyond geology, however. Charles Darwin studied paleontology as a young scientist, and he collected many unusual fossils on his famous voyage aboard the *Beagle* (1831–1836). During this world-circling tour, he also studied many unfamiliar animal and plant species in their native habitats. Darwin pondered what he had seen until 1859, when he proposed his theory of evolution by natural selection. His theory revolutionized the science of biology and provided a sound theoretical framework for paleontology: if organisms evolve progressively over time, then the fossils in each sedimentary bed must represent the organisms living when that bed was deposited.

Unconformities: Gaps in the Geologic Record

In compiling the stratigraphic succession of a region, geologists often find places in the geologic record where a formation is missing. Either no rock was ever deposited, or it was eroded away before the next strata were laid down. The surface between two beds that were laid down with a time gap between them—the boundary representing the missing time—is called an **unconformity** (Figure 8.6). A series of beds bounded above and below by unconformities is referred to as a *sedimentary sequence.* An unconformity, like a sedimentary sequence, represents the passage of time.

An unconformity may imply that tectonic forces raised the rock above sea level, where erosion removed some rock layers. Alternatively, the unconformity may have been produced by the erosion of newly exposed rock as sea level fell. As we will see in Chapter 21, global sea level can be lowered by hundreds of meters during ice ages, when water is withdrawn from the oceans to form continental ice sheets.

Unconformities are classified according to the relationships between the layers above and below them:

- A *disconformity* is an unconformity in which an upper sedimentary sequence overlies an erosional surface developed on an undeformed, still-horizontal lower sedimentary sequence (see Figure 8.6). Disconformities are often created when sea level drops or during broad tectonic uplifts.

- A *nonconformity* is an unconformity in which the upper sedimentary beds overlie metamorphic or igneous rocks (see Earth Issues 8.1, pages 204–205, for an example).

- An *angular unconformity* is an unconformity in which the upper beds overlie lower beds that have been folded by tectonic processes and then eroded to a more or less even plane. In an angular unconformity, the two sequences have bedding planes that are not parallel. Figure 8.7 depicts a dramatic angular unconformity found near the bottom of the Grand Canyon. The formation of an angular unconformity by tectonic processes is illustrated in Figure 8.8.

TIME 1
Beneath the ocean, sedimentary beds accumulate in layers A–D.

TIME 2
Later, tectonic forces cause uplift of the beds above sea level, exposing them to erosion.

Uplift

TIME 3
Erosion strips away layer D and part of C, leaving an irregular surface of hills and valleys.

TIME 4
Subsidence below sea level allows a new layer, E, to be deposited over C. The irregular surface of C is preserved as an unconformity.

Subsidence

Unconformity

FIGURE 8.6 ■ An unconformity is a surface between two rock layers representing a layer that never formed or was eroded away. The type of unconformity shown here, created through uplift and erosion followed by subsidence and another round of sedimentation on top of an undeformed surface, is called a disconformity.

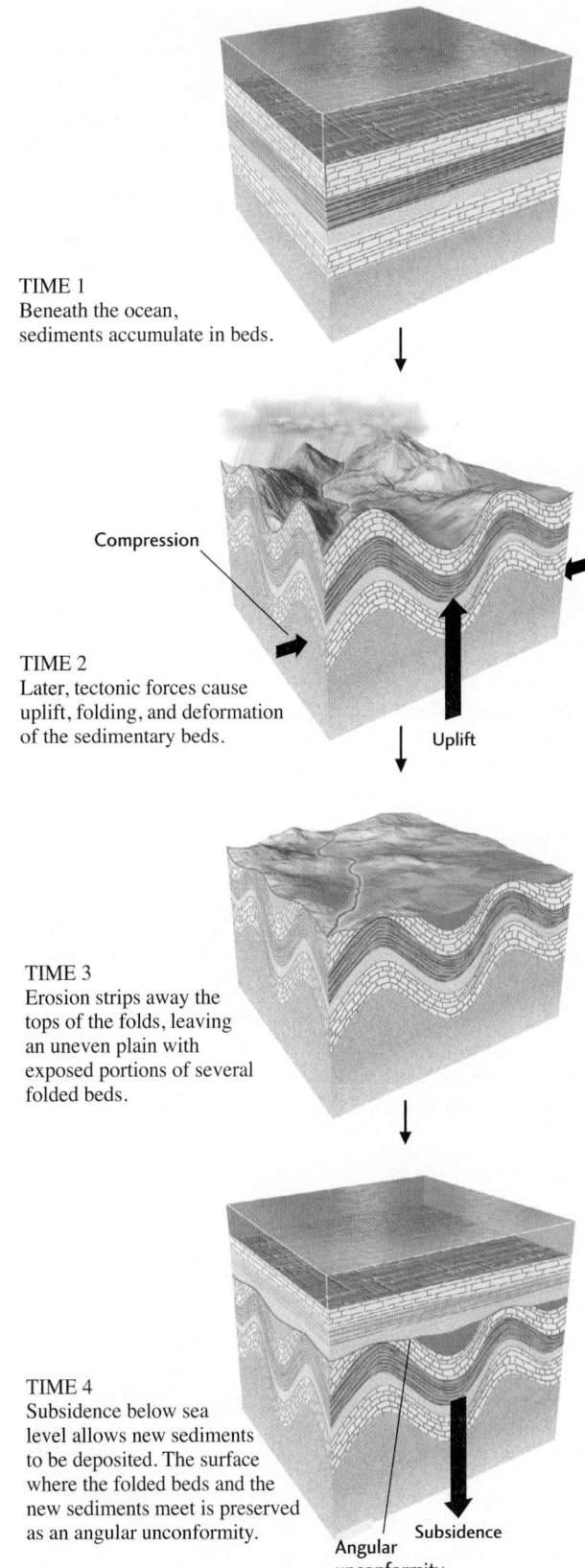

TIME 1
Beneath the ocean, sediments accumulate in beds.

TIME 2
Later, tectonic forces cause uplift, folding, and deformation of the sedimentary beds.

Compression

Uplift

TIME 3
Erosion strips away the tops of the folds, leaving an uneven plain with exposed portions of several folded beds.

TIME 4
Subsidence below sea level allows new sediments to be deposited. The surface where the folded beds and the new sediments meet is preserved as an angular unconformity.

Subsidence

Angular unconformity

FIGURE 8.7 ■ The Great Unconformity in the Grand Canyon, Arizona, is an angular unconformity between the horizontal Tapeats sandstone above and the steeply dipping Wapatai shale below. The Wapatai shale is part of the Grand Canyon Beds; the Tapeats sandstone formed during the Cambrian Period. [Ron Wolf.]

Section through the Grand Canyon strata

Angular unconformity

Cross-Cutting Relationships

Other disturbances of the layering of sedimentary strata also provide clues for determining the relative ages of rocks. Recall that dikes can cut through sedimentary beds; sills can be intruded parallel to bedding planes (see Chapter 4); and faults can displace bedding planes, dikes, and sills as they shift blocks of rock (see Chapter 7). These *cross-cutting relationships* can be used to establish the relative ages of igneous intrusions or faults within the stratigraphic succession. Because the deformation or intrusion events must have taken place after the affected sedimentary beds were deposited, those structures must be younger than the rocks they cut (**Figure 8.9**). If the intrusions or fault displacements are eroded and planed off at an unconformity and then overlaid by younger sedimentary beds, we know that those structures are older than the younger strata.

FIGURE 8.8 ■ An angular unconformity is a surface that separates two sedimentary sequences whose bedding planes are not parallel. This series of drawings shows how an angular unconformity can form.

TIME 1
Beneath the ocean,
sediments accumulate in beds.

TIME 2
Later, tectonic forces
cause uplift, folding,
and deformation of the
sedimentary beds.

TIME 3
A dike intrudes
the folded beds,
cutting across them.
Because the dike can be
seen to cut across the
folded beds, it is clear that
sedimentation and folding
preceded the intrusion.

Dike

Pluton

TIME 4
Faulting displaces
the beds and the
intruding dike.
Because both the
sedimentary beds and
the dike are displaced,
faulting must have taken
place after their formation.

Fault

FIGURE 8.9 ■ Cross-cutting relationships allow geologists
to establish the relative ages of igneous intrusions or faults
within a stratigraphic succession.

Geologists can combine field observations of cross-cutting relationships, unconformities, and stratigraphic successions to decipher the history of geologically complicated regions (Figure 8.10). Earth Issues 8.1 gives a more detailed example of how geologists work backward in time to determine the relative ages of the rocks in a region.

■ The Geologic Time Scale: Relative Ages

Early in the nineteenth century, geologists began to apply Steno's and Smith's stratigraphic principles to outcrops all over the world. The same distinctive fossils were discovered in similar formations on many continents. Moreover, faunal successions from different continents often displayed the same changes in fossil assemblages. By matching up faunal successions and using cross-cutting relationships, geologists could determine the relative ages of rock formations on a global basis. By the end of the century, they had pieced together a worldwide history of geologic events—a **geologic time scale.**

Intervals of Geologic Time

The geologic time scale divides Earth's history into intervals marked by distinct sets of fossils, and it places the boundaries of those intervals at times when those sets of fossils changed abruptly (Figure 8.11). The basic divisions of this time scale are the **eras:** the Paleozoic (from the Greek *paleo,* meaning "old," and *zoi,* meaning "life"), the Mesozoic ("middle life"), and the Cenozoic ("new life").

The eras are subdivided into **periods,** usually named for the locality in which the formations representing them were first or best described, or for some distinguishing characteristic of the formations. The Jurassic period, for example, is named for the Jura mountain range of France and Switzerland, and the Carboniferous period is named for the coal-bearing sedimentary rocks of Europe and North America. The Paleogene and Neogene periods of the Cenozoic are two exceptions: these Greek names mean "old origin" and "new origin," respectively.

Some periods are further subdivided into **epochs,** such as the Miocene, Pliocene, and Pleistocene epochs of the Neogene period (see Figure 8.11). Today we are living in the Holocene ("completely new") epoch of the Neogene period in the Cenozoic era.

Interval Boundaries Mark Mass Extinctions

Many of the major boundaries in the geologic time scale represent **mass extinctions:** short intervals during which

Geologists use cross sections based on field maps to understand the characteristics of the strata and the relationships between them.

F — Sandstones containing terrestrial fossils

E — Angular unconformity E

D

C — Sandstones, limestones, and shales containing marine fossils

A

B — Granitic intrusion

Unconformity C

Deformed metamorphosed sedimentary rocks

1 Sedimentary layers are deposited in flat, horizontal beds.

A

Sedimentary layers

2 The sedimentary beds are deformed and metamorphosed during tectonic uplift and compression.

A

3 An intrusion of granitic magma cuts across the previously deformed sedimentary beds.

B

4 An erosional surface develops on the deformed beds.

C

A

B

5 New layers of marine sediments form on the erosional surface during subsidence beneath sea level, resulting in an unconformity.

D

C

A

B

Unconformity

6 During a period of tectonic compression, the new marine beds become tilted and uplifted and start to erode.

D

C

A

B

Tilting

Sedimentary beds with marine fossils

7 Erosion wears the tilted beds down to a flat surface.

E

D

C

A

B

8 Finally, a river system deposits sandstones on the erosional surface, resulting in an angular unconformity.

Angular unconformity

Sedimentary beds with terrestrial fossils

F E D

C

A

B

FIGURE 8.10 ■ Geologists use stratigraphic principles and cross-cutting relationships to establish a relative chronology of geologic events.

Earth Issues

8.1 Stratigraphy of the Colorado Plateau: An Exercise in Relative Dating

We can use the strata exposed in the Grand Canyon and other parts of the Colorado Plateau to illustrate how relative dating works. These beds record a long history of sedimentation in a variety of environments, sometimes on land and sometimes under water. By matching the rock formations exposed at different localities, geologists have constructed a stratigraphic succession over a billion years long that spans both the Paleozoic and Mesozoic eras.

The lowest—and therefore oldest—rocks exposed in the Grand Canyon are dark igneous and metamorphic rocks that make up the Vishnu schist, a group of formations that have been shown to be about 1.8 billion years old.

Above the Vishnu schist are the younger Grand Canyon Beds. Although these sedimentary rocks contain fossils of single-celled microorganisms that provide evidence of early life, they do not contain the shelly fossils distinctive of the Cambrian and later periods, so they are categorized as Precambrian rocks.

A nonconformity separates the Vishnu schist and the Grand Canyon Beds. This structure indicates a period of deformation accompanying metamorphism of the Vishnu schist, and then a period of erosion, before the deposition of the Grand Canyon Beds. The tilting of the Grand Canyon Beds from their originally horizontal position shows that they, too, were folded after deposition and burial.

An angular unconformity divides the Grand Canyon Beds from the overlying horizontal Tapeats sandstone (see Figure 8.7). This unconformity indicates a long period of erosion after the lower rocks had been tilted. The Tapeats sandstone and Bright Angel shale can be dated as Cambrian by their fossils, many of which are trilobites.

Above the Bright Angel shale is a group of horizontal limestone and shale formations (Muav limestone, Temple Butte limestone, Redwall limestone) that represent a roughly 200-million-year history of marine sedimentation from the late Cambrian to the Carboniferous. There are so many time gaps represented by disconformities in these rocks that less than 40 percent of the Paleozoic is actually represented by rock strata (see Exercise 4).

The next set of strata, high up on the canyon wall, is the Supai formation (Carboniferous and Permian), which contains fossils of land plants like those found in the coal beds of North America and other continents. Overlying the Supai formation is the Hermit shale, a sandy red shale.

Continuing up the canyon wall, we find another continental deposit, the Coconino sandstone. This formation contains vertebrate animal tracks, which suggests that the Coconino was formed in a terrestrial environment during the Permian period. At the top of the cliffs at the canyon rim are two more formations of Permian age: the Toroweap, made mostly of limestone, is overlain by the Kaibab, a massive layer of sandy and cherty limestone. These two formations record subsidence below sea level and the deposition of marine sediments.

Above the Kaibab limestone and the canyon rim itself, but exposed within Grand Canyon National Park, is the Moenkopi formation, a red sandstone of Triassic age—the first appearance of rocks from the Mesozoic era in this stratigraphic succession.

The stratigraphic succession at the Grand Canyon, though picturesque and informative, represents an incomplete picture of Earth's history. Younger intervals of geologic time are not preserved there, and we must travel to nearby locations in Utah, such as Zion Canyon and Bryce Canyon, to fill in this more recent history. At Zion Canyon, we find equivalents of the Kaibab limestone and Moenkopi formation, which allow us to correlate this stratigraphic succession with the one at the Grand Canyon and establish a link. Unlike the Grand Canyon strata, however, the Zion strata extend upward to Jurassic time, including ancient sand dunes represented by the Navajo sandstone. In Bryce Canyon, to the east of Zion, we again find the Navajo sandstone, as well as strata that extend still farther upward, to the Wasatch formation of the Paleogene period.

The correlation of strata among these three areas of the Colorado Plateau shows how widely separated localities—each with an incomplete record of geologic time—can be pieced together to build a composite record of Earth's history.

Stratigraphic succession of the Colorado Plateau, reconstructed from strata exposed in Grand Canyon, Zion Canyon, and Bryce Canyon National Parks. [*Grand Canyon:* John Wang/Photo Disc/Getty Images; *Zion Canyon:* © Universal Images Group Limited/Alamy; *Bryce Canyon:* Tim Davis/Science Source.]

Paleogene	
Cretaceous	
Jurassic	
Triassic	
Permian	
Carboniferous	
Devonian	
Cambrian	
Precambrian	

Grand Canyon

Zion Canyon

Carmel fm
Navajo ss
Kayenta fm
Wingate ss
Chinle fm
Moenkopi fm
Kaibab ls
Older rocks not exposed

Moenkopi fm
Kaibab ls
Toroweap fm
Coconino ss
Hermit shale
Supai fm
Redwall ls
Temple Butte ls
Muav ls
Bright Angel shale
Tapeats ss
Colorado River
Grand Canyon Beds
Vishnu schist

Wasatch fm
Kaiparowits fm
Wahweap ss
Straight Cliffs ss
Tropic shale
Dakota ss
Winsor fm
Curtis ss
Entrada ss
Carmel fm
Navajo ss
Older rocks not exposed

Bryce Canyon

fm = Formation
ss = Sandstone
ls = Limestone

Location of section (below)

Zion National Park
NV
UT
AZ
B
C
Bryce Canyon National Park
CO
NM
N
Grand Canyon National Park
A

Grand Canyon

Zion Canyon

Bryce Canyon

Fault

Fault

A

B

C

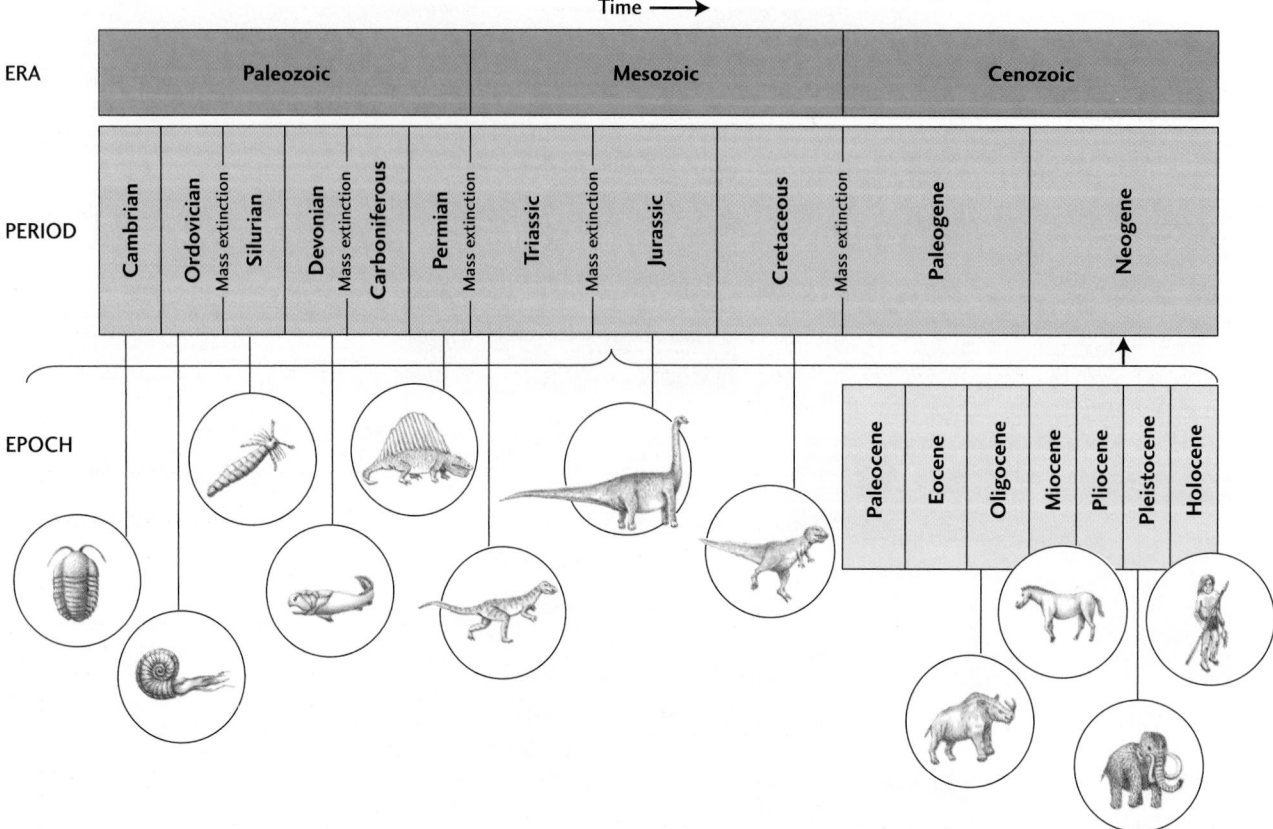

FIGURE 8.11 ■ The geologic time scale, showing the eras, periods, and epochs distinguished by assemblages of fossils. The boundaries of these intervals are marked by the abrupt disappearance of some life-forms and the appearance of new ones. The five most dramatic mass extinctions are indicated. Note that this diagram shows only the relative ages of the intervals.

a large proportion of the species living at the time simply disappeared from the fossil record, followed by the blossoming of many new species. These abrupt changes in faunal successions were a great mystery to the geologists who discovered them. Darwin's theory of evolution explained how new species could evolve, but what had caused the mass extinctions?

In some cases, we think we know. The mass extinction at the end of the Cretaceous period, which killed off 75 percent of the living species, including all the dinosaurs, was almost certainly the result of a large meteorite impact that darkened and poisoned the atmosphere and plunged Earth's climate into many years of bitter cold. This disaster marks the end of the Mesozoic era and the beginning of the Cenozoic. In other cases, we are still not sure. The largest mass extinction, at the end of the Permian period, which defines the Paleozoic-Mesozoic boundary, eliminated nearly 95 percent of all living species, but the cause of this event is still debated. The extreme events that separate intervals of geologic time are the subject of very active research, as we will see in Chapter 11.

Ages of Petroleum Source Rocks

Oil and natural gas come from organic matter that was buried in sedimentary rock formations at some time in the geologic past. The relative ages of these "petroleum source rocks" provide important clues about where to look for new oil and gas resources. Global surveys have shown that very little petroleum has come from Precambrian rocks, which makes sense, because the primitive organisms that existed before the Cambrian period generated little organic matter.

Petroleum source rocks were deposited during all three of the geologic eras following the Cambrian, although certain periods of geologic time have produced much more of this resource than others (**Figure 8.12**). The clear winners are the Jurassic and Cretaceous periods of the Mesozoic era, which together have accounted for almost 60% of the world's petroleum production. Sedimentary formations of Jurassic and Cretaceous age were the source rocks for the great oil fields of the Middle East, the Gulf of Mexico, Venezuela, and the North Slope of Alaska.

FIGURE 8.12 ■ Relative ages and amounts of sedimentary rocks that contained the organic matter now found as oil and natural gas. Bars in the lower graph show the percentage of these petroleum source rocks found worldwide (height of the bar) within a given age range (width of the bar). Almost 60% of the total inventory was deposited during the Jurassic and Cretaceous periods of the Mesozoic era.

If you examine Figure 2.16, you can see that, during these periods of geologic time, the supercontinent of Pangaea was breaking up into the modern continents. This tectonic activity formed many marine sedimentary basins and increased the rate at which sediments were deposited into these basins. During the Jurassic and Cretaceous periods, which comprise the Age of Dinosaurs, marine life was abundant, providing much of the organic matter that was buried in the sediments. This carbon-rich material has since been "cooked" and transported into the oil reservoirs, where we find it today.

■ Measuring Absolute Time with Isotopic Clocks

The geologic time scale based on stratigraphy and faunal successions is a relative time scale. It tells us whether one formation or fossil assemblage is older than another, but not how long the eras, periods, and epochs were in actual years.

Estimates of how long it takes mountains to erode and sediments to accumulate suggested that most geologic periods had lasted for millions of years, but nineteenth-century geologists did not know whether the duration of any specific period was 10 million years, 100 million years, or even longer.

They did know that the geologic time scale was incomplete. The earliest period of geologic history recorded by faunal successions was the Cambrian, when animal life, in the form of shelly fossils, suddenly appeared in the geologic record. Many rock formations were clearly older, however, because they occurred below Cambrian rocks in stratigraphic successions. But these formations contained no recognizable fossils, so there was no way to determine their relative ages. All such rocks were lumped into the general category *Precambrian*. What fraction of Earth's history was locked up in these cryptic rocks? How old was the oldest Precambrian rock? How old was Earth itself?

These questions sparked a huge debate in the latter half of the nineteenth century. Physicists and astronomers argued for a maximum age of less than 100 million years, but most geologists regarded this age as much too young, even though they had no precise data to back them up.

Discovery of Radioactivity

In 1896, a major advance in physics paved the way for reliable and accurate measurements of absolute ages. Henri Becquerel, a French physicist, discovered radioactivity in uranium. Within three years, the French chemist Marie Sklodowska-Curie discovered and isolated a new and highly radioactive element, radium.

In 1905, the physicist Ernest Rutherford suggested that the absolute age of a rock could be determined by measuring the decay of radioactive elements found in it. He calculated the age of one rock from measurements of its uranium content. This was the start of **isotopic dating,** the use of naturally occurring radioactive elements to determine the ages of rocks. Isotopic dating methods were refined over the next few years as more radioactive elements were discovered and the processes of radioactive decay became better understood. Within a decade of Rutherford's first attempt, geologists were able to show that some Precambrian rocks were billions of years old.

In 1956, the geochemist Clair Patterson measured the decay of uranium in meteorites and terrestrial rocks to determine that the solar system—and, by implication, Earth—was formed 4.56 billion years ago. That age has been modified by less than 10 million years since Patterson's original measurement, so we might say that he completed the discovery of geologic time.

Radioactive Isotopes: The Clocks in Rocks

How do geologists use radioactivity to determine the age of a rock? Recall that the nucleus of an atom consists of protons and neutrons. For a given element, the number of protons is constant, but the number of neutrons can vary among different *isotopes* of the same element (see Chapter 3). Most isotopes are stable, but the nucleus of a *radioactive* isotope can spontaneously disintegrate, or *decay*, emitting particles and transforming the atom into an atom of a different element. We call the original atom the *parent* and the product of decay its *daughter.*

One useful element for isotopic dating is rubidium, which has 37 protons and two naturally occurring isotopes: rubidium-85, which has 48 neutrons and is stable, and rubidium-87, which has 50 neutrons and is radioactive. A neutron in the nucleus of a rubidium-87 atom can spontaneously emit an electron, thus changing into a proton, which remains in the nucleus. The parent rubidium atom thus forms a daughter strontium-87 atom, with 38 protons and 49 neutrons (**Figure 8.13**).

A parent isotope decays into a daughter isotope at a constant rate. The rate of radioactive decay is measured by the isotope's **half-life:** the time required for one-half of the original number of parent atoms to be transformed into daughter atoms. At the end of the first

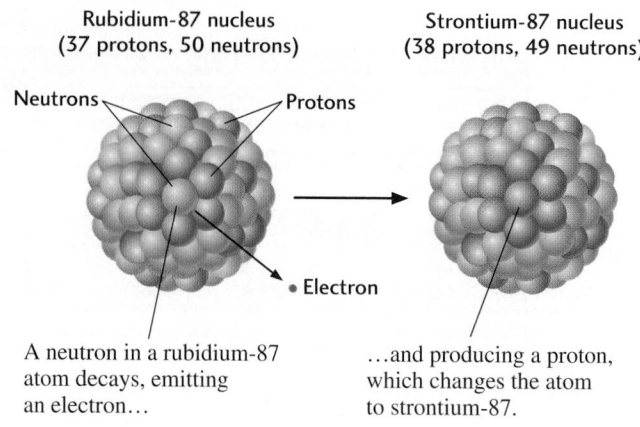

Rubidium-87 nucleus (37 protons, 50 neutrons) → Strontium-87 nucleus (38 protons, 49 neutrons)

Neutrons | Protons | Electron

A neutron in a rubidium-87 atom decays, emitting an electron…

…and producing a proton, which changes the atom to strontium-87.

FIGURE 8.13 ■ The radioactive decay of rubidium to strontium.

half-life, the number of parent atoms has decreased by a factor of two; at the end of the second half-life, by a factor of four; at the end of the third half-life, by a factor of eight, and so forth. As the parent decays, the amount of the daughter isotope increases, preserving the total number of atoms (**Figure 8.14**). The half-lives of radioactive elements commonly used for isotopic dating are given in **Table 8.1**.

Radioactive isotopes make good clocks because their half-lives do not vary with the changes in temperature, pressure, chemical environment, or other factors that can accompany geologic processes on Earth or other planets. So when atoms of a radioactive isotope are created anywhere in the universe, they start to act like a ticking clock, steadily transforming from one type of atom to another at a fixed rate.

We can measure the ratio of parent to daughter atoms in a rock sample with a mass spectrometer—a precise and sensitive instrument that can detect even minute quantities

FIGURE 8.14 ■ The fraction of atoms of the parent isotope declines at a constant rate over time. This rate of decay is measured by the half-life of the isotope. As the parent decays, the amount of the daughter isotope grows, preserving the total number of atoms.

TABLE 8-1	Major Radioactive Elements Used in Isotopic Dating			
Isotopes		Half-Life of Parent (years)	Effective Dating Range (years)	Examples of Minerals and Materials That Can Be Dated
Parent	Daughter			
Rubidium-87	Strontium-87	49 billion	10 million–4.6 billion	Muscovite, biotite, orthoclase feldspar
Uranium-238	Lead-206	4.5 billion	10 million–4.6 billion	Zircon, apatite
Potassium-40	Argon-40	1.3 billion	50,000–4.6 billion	Muscovite, biotite, hornblende
Uranium-235	Lead-207	0.7 billion	10 million–4.6 billion	Zircon, apatite
Carbon-14	Nitrogen-14	5730	100–70,000	Wood, charcoal, peat; bone and tissue; shells and other calcium carbonates

of isotopes—and determine how much of the daughter has been produced from the parent. Knowing the half-life, we can then calculate the time elapsed since the isotopic clock began to tick.

The isotopic age of a rock corresponds to the time since the isotopic clock was "reset" when the isotopes were locked into the minerals of the rock. This "locking" usually occurs when a mineral crystallizes from a magma or recrystallizes during metamorphism. During crystallization, however, the number of daughter atoms in a mineral is not necessarily reset to zero, so the initial number of daughter atoms must be taken into account when calculating isotopic age (see the Practicing Geology exercise at the end of the chapter).

Many other complications make isotopic dating a tricky business. A mineral can lose daughter isotopes by weathering or be contaminated by fluids circulating in the rock. Metamorphism of igneous rocks can reset the isotopic age of minerals in those rocks to a date much later than their crystallization age.

Isotopic Dating Methods

Isotopic dating is possible only if a measurable number of parent and daughter atoms remain in the sample being dated. For example, if a rock is very old and the decay rate of an isotope is fast, almost all the parent atoms will already have been transformed. In that case, we could determine that the isotopic clock had run down, but we would not be able to say when. Thus, isotopes that decay slowly over billions of years, such as rubidium-87, are most useful in measuring the ages of older rocks, whereas those that decay rapidly, such as carbon-14, can only be used to date younger rocks (see Table 8.1).

Carbon-14, which has a half-life of about 5700 years, is especially useful for dating bone, shell, wood, and other organic materials in sediments less than a few tens of thousands of years old. Carbon is an essential element in the living cells of all organisms. As green plants grow, they continuously incorporate carbon into their tissues from carbon dioxide in the atmosphere. When a plant dies, however, it stops absorbing carbon dioxide. At the moment of death, the ratio of carbon-14 to the stable isotope carbon-12 in the plant is identical to that in the atmosphere. Thereafter, the ratio decreases as the carbon-14 in the dead tissue decays. Nitrogen-14, the daughter isotope of carbon-14, is a gas and thus leaks from the material, so it cannot be measured to determine the time that has elapsed since the plant died. We can, however, estimate the absolute age of the plant material by comparing the ratio of carbon-14 left in the plant material with the ratio in the atmosphere at the time the plant died. The latter ratio can be estimated from carbon-14 ages calibrated using other measures of absolute time, such as dendrochronology (counting tree rings).

One of the most precise dating methods for old rocks is based on the decay of two related isotopes: the decay of uranium-238 to lead-206 and the decay of uranium-235 to lead-207. Isotopes of the same element behave similarly in the chemical reactions that alter rocks because the chemistry of an element depends mainly on its atomic number, not its atomic mass. The two uranium isotopes have different half-lives, however, so together they provide a consistency check that helps geologists compensate for the problems of weathering, contamination, and metamorphism discussed above. The lead isotopes from single crystals of zircon—zirconium silicate, a crustal mineral with a relatively high concentration of uranium—can be used to date the oldest rocks on Earth with an uncertainty of less than 1 percent. These formations turn out to be more than 4 billion years old.

■ The Geologic Time Scale: Absolute Ages

Armed with isotopic dating techniques, geologists of the twentieth century were able to nail down the absolute ages of the key events on which their predecessors had based the geologic time scale. More important, they were able to explore the early history of the planet recorded in Precambrian rocks. **Figure 8.15** presents the results of this century-long effort.

The assignment of absolute ages to the geologic time scale revealed great differences in the lengths of the geologic periods. The Cretaceous period (spanning 80 million years) turned out to be more than three times longer than the Neogene period (only 23 million years), and the Paleozoic era (291 million years) was found to be longer than the Mesozoic and Cenozoic eras combined. The biggest surprise was the Precambrian, which had a duration of over 4000 million years—almost nine-tenths of Earth's history!

Eons: The Longest Intervals of Geologic Time

To represent the rich history of the Precambrian, a division of the geologic time scale longer than the era, called the **eon,** was introduced. Four eons, based on the isotopic ages of terrestrial rocks and meteorites, are now recognized.

HADEAN EON The earliest eon, whose name comes from *Hades* (the Greek word for "hell"), began with the formation of Earth 4.56 billion years ago and ended about 3.9 billion years ago. During its first 650 million years, Earth was bombarded by chunks of material from the early solar system. Although very few rock formations survived this violent period, individual zircon grains with ages as great as 4.4 billion years have been found, indicating that Earth had a felsic crust within 200 million years of its formation. There is also evidence that some liquid water existed on Earth's surface at about this time, suggesting that the planet cooled rapidly. In Chapter 9, we will explore this early phase of Earth's history in more detail.

ARCHEAN EON The name of the next eon comes from *archaios* (the Greek word for "ancient"). Rocks of Archean age range from 3.9 billion to 2.5 billion years old. The geodynamo and the climate systems were established during the Archean eon, and felsic crust accumulated to form the first stable continental masses, as we will see in Chapter 10. The processes of plate tectonics were probably operating by the end of the Archean, although perhaps substantially differently from the way they did later in Earth's history. Life, in the form of primitive single-celled microorganisms, became established, as indicated by the fossils found in sedimentary rocks of this age.

PROTEROZOIC EON The last part of the Precambrian is the Proterozoic eon (from the Greek words *proteros* and *zoi,*

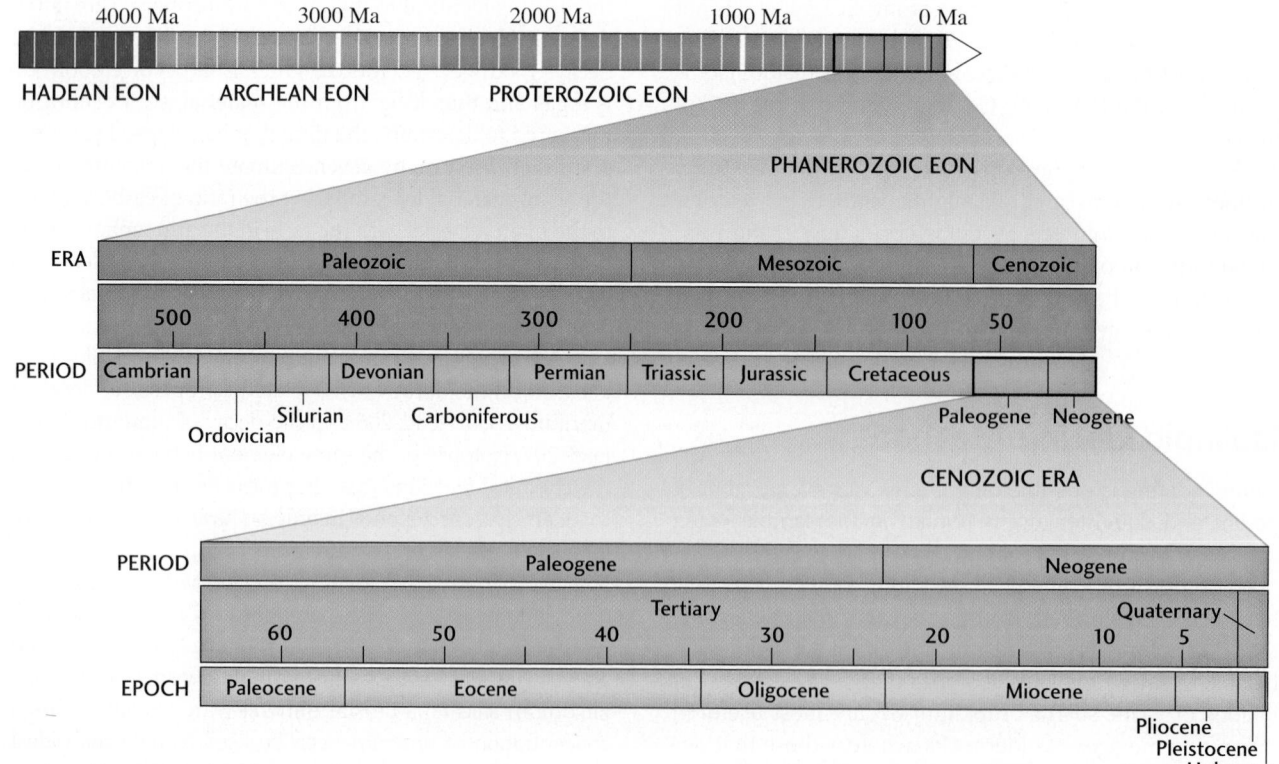

FIGURE 8.15 ■ The complete geologic time scale. (Ma, million years ago.) The intervals labeled "Tertiary" and "Quaternary" are older divisions that have been largely replaced by the Paleogene and Neogene periods, but are still sometimes used by geologists.

meaning "earlier life"), which spans the time interval from 2.5 billion to 542 million years ago. By the beginning of this eon, the plate tectonic and climate systems were working much as they do today. Throughout the Proterozoic, organisms that produced oxygen as a waste product (as plants do today) increased the amount of oxygen in Earth's atmosphere. We will explore the early evolution of life and its effects on the Earth system in Chapter 11.

PHANEROZOIC EON The start of the Phanerozoic eon is marked by the first appearance of shelly fossils at the beginning of the Cambrian period, now dated at 542 million years ago. The name of this eon—from the Greek *phaneros* and *zoi* ("visible life")—certainly fits, because it comprises all three eras originally recognized in the fossil record: the Paleozoic (542 million to 251 million years ago), the Mesozoic (251 million to 65 million years ago), and the Cenozoic (65 million years ago to the present).

Perspectives on Geologic Time

In the dusty sheep country of far western Australia stands a small promontory of ancient red rocks called the Jack Hills (**Figure 8.16**). Geologists have pulverized truckloads of these rocks to isolate a few sand-sized crystals of zircon. By measuring the lead-206 and lead-207 isotopes generated by the radioactive decay of uranium-238 and uranium-235, as described above, they have identified one small crystal fragment with an age of 4.4 billion years—the oldest mineral grain yet discovered in Earth's crust. How can we relate to such a mind-boggling span of time?

Imagine compressing the 4.56 billion years of Earth history into a single year, starting with the formation of Earth on January 1 and ending at midnight on December 31. Within the first week, Earth was organized into core, mantle, and crust. The oldest zircon grain from the Jack Hills crystallized on January 13. The first primitive organisms appeared in mid-March. By mid-June, stable continents had developed, and throughout the summer and early fall, the biological activity of evolving life increased the concentration of oxygen in the atmosphere. On November 18, at the beginning of the Cambrian period, complex organisms, including those with shells, appeared. On December 11, reptiles evolved, and late on Christmas Day, the dinosaurs became extinct. Modern humans, *Homo sapiens,* did not appear on the scene until 11:42 P.M. on New Year's Eve, and the most recent ice age did not end until 11:58 P.M. Three and a half seconds before midnight, Columbus landed on a West Indian island, and a couple of tenths of a second ago, you were born!

FIGURE 8.16 ■ The outcrop in the Jack Hills, in Western Australia, in which geologists have found zircon grains as old as 4.4 billion years. [Bruce Watson, Rensselaer Polytechnic Institute.] *Inset:* A zircon crystal ($ZrSiO_4$) from the Hadean eon extracted from the Jack Hills. [Dr. Martina Menneken.]

■ Recent Advances in Timing the Earth System

We have seen that the time scales of geologic processes are not uniform, but vary from seconds to billions of years. We must therefore use a variety of methods for timing the Earth system: some to determine the ages of very old rocks, others to measure rapid changes. New methods for determining the relative and absolute ages of Earth materials have steadily improved our understanding of how the Earth system works. To conclude our story of the geologic time scale, we will describe a few of these recent advances.

Sequence Stratigraphy

Until a few decades ago, geologists had to rely on rocks exposed at outcrops, in mines, and by drilling to map stratigraphic successions. As mentioned in Chapter 1 (and further described in Chapter 14), technological innovations in the field of seismic imaging now allow us to see below Earth's surface without actually digging holes. From recordings of seismic waves generated by controlled explosions or by natural earthquakes, we can construct three-dimensional images of deeply buried structures (**Figure 8.17**). Seismic imaging of sedimentary rocks allows geologists to identify sedimentary sequences and map their distribution in three dimensions, a type of geologic mapping called *sequence stratigraphy*.

Sedimentary sequences commonly form on the edges of continents; here, sediment deposition by rivers is modified by fluctuations in sea level. In the example shown in Figure 8.17, sediments were laid down in a delta where a river entered the ocean. As the sediments accumulated, the delta advanced seaward. When sea level fell because of continental glaciation, the deltaic deposits were exposed and eroded. Once the glaciers melted and sea level rose, the shoreline shifted inland, and a new deltaic sequence began to cover the old one, creating an unconformity.

Over millions of years, cycles such as this one can be repeated many times, producing a complex set of sedimentary sequences. Because sea level fluctuations are global, geologists can match sedimentary sequences of the same

(a) Seismic profile

(b) Sedimentary sequences

1 A sequence of delta sediments, B, accumulates over a previous sequence, A.

2 The sea level rises, and the shoreline recedes inland.

3 Another sequence of delta sediments, C, accumulates over sequence B.

FIGURE 8.17 ■ Sequence stratigraphy can be used to understand how sedimentary bedding patterns were created. (a) A seismic profile reveals individual sedimentary beds. (b) Geologists can group these beds into sedimentary sequences. (c) and (d) In this case, seismic imaging reveals a stratigraphic succession that is characteristic of a series of deltaic sequences.

age over large areas. The relative ages of these sequences can then be used to reconstruct the geologic history of a region, including any regional tectonic uplift or subsidence that contributed to sea level changes. Sequence stratigraphy has been especially effective in finding deeply buried oil and gas on continental margins, such as the Gulf of Mexico and the Atlantic margin of North America.

Chemical Stratigraphy

Many sedimentary beds contain minerals and chemicals that identify them as distinct units. For example, the amount of iron or manganese in carbonate sediments may vary from bed to bed if the composition of seawater changed during precipitation of the carbonate minerals. When the sediments are buried and transformed into sedimentary rocks, these chemical variations may be preserved, "fingerprinting" the formations. These chemical fingerprints may extend regionally or even globally, allowing us to match sedimentary rocks by *chemical stratigraphy* where no other features, such as fossils, are available.

Paleomagnetic Stratigraphy

Another technique for fingerprinting rock formations is *paleomagnetic stratigraphy*. As we saw in Chapter 1, Earth's magnetic field reverses itself at irregular intervals. These magnetic reversals are recorded by thermoremanent magnetization in volcanic rocks, which can be dated by isotopic methods. The resulting chronology of magnetic reversals—the magnetic time scale—allows us to "replay the magnetic tape" of seafloor spreading and determine the rates of plate movements, as we saw in Chapter 2. Even more detailed patterns of magnetic reversals can be observed in sediment cores, and these magnetic fingerprints can be dated using faunal successions. Paleomagnetic stratigraphy has recently become

one of the main methods for measuring sedimentation rates along the continental margins and in the deep sea. We will discuss paleomagnetic stratigraphy in more detail in Chapter 14.

Clocking the Climate System

The Pliocene and Pleistocene epochs were times of rapid and dramatic global climate change. We can chart these climate changes from the isotopes contained in shelly fossils buried in deep-sea sediments. Deep-sea drilling vessels such as the *JOIDES Resolution* (see Figure 2.13) have taken cores from sedimentary beds around the world's oceans. Geologists can use the carbon-14 dating method to estimate when the shells recovered from these sediment cores were formed, and they can measure the stable isotopes of oxygen to estimate temperature of the seawater in which the shell-producing organisms lived.

The carefully tabulation of both temperature and age estimates for many sedimentary layers has provided us with a precise record of global climate during the last 5 million years (**Figure 8.18**). The record shows a general cooling trend beginning about 3.5 million years ago and the subsequent development of rapid climate cycles that became especially large during the Pleistocene epoch. The low temperatures during these cycles, which were as much as 8° C below the average present-day temperature of Earth's surface, correspond to the Pleistocene "ice ages," when glaciers covered large areas of North America, Europe and Asia.

Repeated cycles of glaciation have occurred with dominant periods ranging from 40,000 to 100,000 years. Shorter-term cycles lasting a few thousand years or less are also evident. The effects of these climate cycles, such as rises and drops in sea level, can have profound effects on Earth's surface. We will explore glacial cycles and their causes in more detail in Chapters 15 and 21.

FIGURE 8.18 ∎ Changes in Earth's average surface temperature (jagged blue line) during the Pliocene and Pleistocene epochs, measured from temperature indicators in well-dated oceanic sediments. Zero change (dashed black line) corresponds to the average temperature during the Holocene epoch of the last 11,000 years. Note the rapid climate cycles since about 2.7 million years ago. The low temperatures during these cycles correspond to "ice ages." [Courtesy of L. E. Lisiecki and M. E. Raymo.]

Google Earth Project

In the Grand Canyon, in northern Arizona, the Colorado River cuts through sediments deposited over hundreds of millions of years of Earth history (see Earth Issues 8.1). This unimaginably long interval of time is marked by changes in the life-forms preserved in sedimentary rocks. Because plant and animal remains are deposited in and preserved in sediments, the age of a fossil is basically the same as that of the sedimentary bed in which it lies. This relationship, along with basic stratigraphic principles, allows us to determine the relative ages of sedimentary beds. Let's use this spectacular setting as a natural laboratory for understanding geologic time.

Image USDA Farm Service Agency Image © 2009 DigitalGlobe

LOCATION   Bright Angel Trail, Grand Canyon Visitor Center, Arizona, United States

GOAL   Visualize the Grand Canyon sedimentary sequence

LINKED   Earth Issues 8.1

SUMMARY

How do we know whether one rock is older than another? We can determine the relative ages of rocks by studying the stratigraphy, fossils, and cross-cutting relationships of rock formations observed at outcrops. According to Steno's principles, an undeformed sequence of sedimentary beds will be horizontal, with each bed younger than the beds beneath it and older than the beds above it. In addition, the fossils found in each bed reflect the organisms that were living when that bed was deposited. Knowing the faunal succession makes it easier to spot unconformities, which indicate time gaps in the stratigraphic record where no rock was deposited or where existing rock was eroded away before the next strata were laid down.

1. Type "Bright Angel Trail, Grand Canyon, Arizona" into the GE search window. Once you arrive there, zoom out to an eye altitude of 10 km. Use the cursor to measure the difference in elevation between the Bright Angel trailhead near the Grand Canyon Visitor Center and the Colorado River directly to the north. Which value best approximates this elevation difference?

   a. 30 m   c. 1300 m
   b. 100 m   d. 2500 m

2. Navigate down the canyon along the Bright Angel Trail from an eye altitude of about 2 km. Tilt your frame of view to the north so that you can gain an oblique view of the north wall of the canyon along the Colorado River. Trace the elevations of rock layers with distinctive colors across the landscape. What is the general orientation of the sedimentary rock layers near the surface?

   a. very nearly horizontal
   b. very nearly vertical
   c. tilted at about 45 degrees from the horizontal to the east
   d. tilted at about 45 degrees from the horizontal to the south

3. Viewing the north canyon wall from an eye altitude of about 2 km, locate the thin, white rock layer below the rim of the canyon and the thicker, tan rock layer just above the lowest exposures of red rock visible in the canyon. These formations are the Permian-aged Coconino Sandstone and the Cambrian-aged Tapeats Sandstone, respectively. Measure the vertical distance between these two formations and refer to Earth Issues 8.1. What is your estimate of the sediment thickness between the two formations, and which geologic periods are missing?

   a. 800 m of sediment with both the Ordovician and Silurian periods missing
   b. 400 m of sediment with no geologic periods missing
   c. 800 m of sediment with the Permian, Cambrian, and Devonian periods missing
   d. 200 m of sediment with the Carboniferous period missing

4. Based on the relationship of the layered rock exposed within the canyon walls and the canyon itself, which of the following must have formed first?

   a. The layer of rock nearest the bottom of the canyon
   b. The layer of rock at the rim of the canyon
   c. The Grand Canyon itself
   d. The smaller side canyon that the Bright Angel Trail follows

Optional Challenge Question

5. Navigate to the following latitude and longitude along the canyon: 36°10′56″ N; 113°06′52″ W. View it from an eye altitude of 30 km and zoom in as necessary. Below is a volcanic feature that has produced basaltic lava flows that interact with the Colorado River at the base of the canyon. Based on the principle of superposition and the visible cross-cutting relationships, which of the following sequences of events seems most likely? (It may be helpful to tilt the frame of view to the north along the river to gain a better perspective of the sequence of events.)

   a. A volcanic eruption produced lava flows, then the Colorado River cut through the lava flows, and finally layers of sedimentary rocks were deposited on either side of the river channel.
   b. Sedimentary rocks were deposited, then a volcanic eruption produced lava flows that covered the sediments, and finally, the Colorado River cut through the entire sequence to create a large canyon.
   c. Sedimentary rocks were deposited, then the Colorado River cut through them to create a large canyon, and finally a volcanic eruption produced lava flows that flowed into the river.
   d. A volcanic eruption produced lava flows on which sedimentary rocks were deposited, then the Colorado River cut through the entire sequence to create a large canyon.

How was a global geologic time scale created? By using faunal successions to match rocks in outcrops around the world, geologists compiled composite stratigraphic successions, from which they developed a relative time scale. The use of isotopic dating allowed them to assign absolute ages to the eons, eras, periods, and epochs that constitute the geologic time scale. Isotopic dating is based on the decay of radioactive isotopes, in which unstable parent atoms are transformed into stable daughter atoms at a constant rate. By measuring the amounts of parent and daughter atoms in a sample, geologists can calculate the absolute ages of rocks. The isotopic clock starts ticking when radioactive isotopes are locked into minerals as igneous rocks crystallize or metamorphic rocks recrystallize.

What are the principal divisions of the geologic time scale? The geologic time scale is divided into four eons: the Hadean (4.56 billion to 3.9 billion years ago), Archean (3.9 billion to 2.5 billion years ago), Proterozoic (2.5 billion to 542 million years ago), and Phanerozoic (542 million years ago to the present). The Phanerozoic eon is divided into three eras, the Paleozoic, Mesozoic, and Cenozoic, each of which is divided into shorter periods. The boundaries of the eras and periods are marked by abrupt changes in the fossil record; many correspond to mass extinctions.

What other methods are now being used to date the geologic record? The cyclical rise and fall of sea level produces complex sedimentary sequences on continental margins around the world that can be mapped using seismic imaging techniques and dated using fossils. Chemical fingerprints and magnetic reversals provide additional information about the ages of sedimentary sequences. Glacial cycles recorded in sediments can be dated using ice cores taken from the Antarctic and Greenland ice caps.

KEY TERMS AND CONCEPTS

absolute age (p. 196)

eon (p. 210)

epoch (p. 202)

era (p. 202)

geologic time scale
  (p. 202)

half-life (p. 208)

isotopic dating (p. 208)

mass extinction (p. 202)

period (p. 202)

principle of faunal
  succession (p. 199)

principle of original
horizontality (p. 197)

principle of superposition
  (p. 197)

relative age (p. 196)

stratigraphic succession
  (p. 197)

stratigraphy (p. 197)

unconformity (p. 200)

PRACTICING GEOLOGY EXERCISE:

How Do Isotopes Tell Us the Ages of Earth Materials?

Isotopic dating methods allow us to date many types of Earth materials for many practical purposes: rock formations in the search for minerals and petroleum; water samples to understand oceanic circulation; ice cores to chart climate variations; and even bubbles of air trapped in rocks and ice to measure changes in the composition of the atmosphere. So it's worthwhile to understand in more detail how geologists actually determine the ages of materials using isotopes.

Consider a mineral grain that was formed at time $T = 0$ and contains a certain amount of a parent isotope—say, 1000 atoms. If we measure the age of the mineral grain in half-lives of the parent isotope, the amount left at any age T will be $1000 \times 1/2^T$. In other words, in one half-life—that is, when $T = 1$—the initial amount of the parent isotope will be reduced to $1/2^1 = 1/2$ (500 atoms); in two half-lives, to $1/2^2 = 1/4$ (250 atoms); in three half-lives, to $1/2^3 = 1/8$ (125 atoms); and so on (see Figure 8.13).

The radioactive decay of each atom of the parent isotope generates one new atom of the daughter isotope. If the mineral grain remains a closed system (that is, if no isotopes are transferred into or out of the grain), the number of new daughter atoms produced from the parent atoms by age T must equal $1000 \times (1 - 1/2^T)$, because the new daughters and remaining parents must add up to the initial amount of the parent isotope (1000 atoms). The ratio of new daughters to remaining parents thus depends only on the age of the mineral grain:

$$\left(\frac{\text{number of new daughters}}{\text{number of remaining parents}} \right) = \frac{1 - 1/2^T}{1/2^T} = 2^T - 1$$

As the age of the mineral grain increases from 0 to 3 half-lives, for instance, this ratio increases from 0 to 7, independently of the initial number of parent atoms.

With a mass spectrometer, we can measure the parent and daughter isotopes precisely: today, such instruments can literally count the atoms in a small sample. But to determine the age of a mineral grain, we must account for any daughter isotope incorporated into the mineral grain at the time it crystallized. In our example, if there were 100 daughter atoms in the grain at $T = 0$, then the number of daughter atoms would increase to $500 + 100 = 600$ after one half-life; to $750 + 100 = 850$ after two half-lives; and to $875 + 100 = 975$ after three half-lives. The general expression for the total number of daughter atoms is therefore

number of daughters $= (2^T - 1) \times$ number of remaining parents
$+$ number of initial daughters

You may notice that this is an equation for a straight line with a slope of $(2^T - 1)$ and an intercept equal to the initial number of daughter atoms, as illustrated in part (a) of the accompanying illustration.

Although we can measure only the total amount of the daughter isotope, we can often infer the initial amount from another isotope of the same element. For example, strontium-87 is created by the decay of rubidium-87 (see Figure 8.13), but another isotope, strontium-86, is not produced by radioactive decay and is not itself radioactive. Thus, if a mineral grain remains a closed system after crystallization, the number of strontium-86 atoms will not change with age. The trick is to divide the daughter-parent relationship by the amount of strontium-86:

$$\left(\frac{\text{number of strontium-87}}{\text{number of strontium-86}}\right) =$$

$$(2^T - 1) \times \left(\frac{\text{number of rubidium-87}}{\text{number of strontium-86}}\right)$$

$$+ \left(\frac{\text{number of initial strontium-87}}{\text{number of strontium-86}}\right)$$

Different mineral grains in a rock will crystallize with differing initial amounts of strontium and rubidium. However, because the two strontium isotopes behave similarly in the chemical reactions that take place before crystallization, the strontium-87/strontium-86 ratio at crystallization will be the same for all the grains in the same rock. Therefore, by fitting a line to the data from several mineral grains, we can determine the initial strontium-87/strontium-86 ratio as well as the age T.

In part (b) of the accompanying figure, we apply this method to strontium and rubidium measurements from a famous stony meteorite, called Juvinas, that fell in southern France in 1821. The Juvinas meteorite, which is similar to the one shown in Figure 1.10a, is thought to have come from a planetary body that formed at the same time as Earth but was subsequently destroyed by planetary collisions (see Chapter 9). Using mass spectrometer measurements of four samples from this meteorite, we can plot the strontium-87/strontium-86 and rubidium-87/strontium-86

(a)

How parent and daughter isotopes evolve with time

(b)

Dating the Juvinas meteorite

(a) The number of parent atoms in a mineral grain decreases, and the number of daughter atoms increases, during radioactive decay. As a mineral grain ages, its representation on this plot moves continuously upward and to the left along the red line. The labeled points represent 0, 1, 2, and 3 half-lives. (b) A graph of the strontium-87/strontium-86 versus rubidium-87/strontium-86 ratios for the Juvinas meteorite. The data are obtained from mass spectrometer measurements of different minerals from the meteorite. [Martin Prinz/American Museum of Natural History.]

ratios along a line whose intercept gives an initial strontium-87/strontium-86 ratio of 0.699. That line is an isochron (a locus of equal time) with a slope of 0.067.

To solve for T, we begin with

$$(2^T - 1) = 0.067$$

Adding 1 to both sides of this equation and taking base-10 logarithms of both sides yields

$$T \log(2) = \log(1.067)$$

or

$$T = \frac{\log(1.067)}{\log(2)}$$

Using a scientific calculator (there's probably one on your smart phone), we find that $\log(1.067) = 0.0282$ and $\log(2) = 0.301$, which gives

$$T = \frac{0.0282}{0.301} = 0.094 \text{ half-lives}.$$

Multiplying the number of half-lives by the half-life of rubidium-87, 49 billion years (see Table 8.1), yields a meteorite age of

$$0.094 \times 49 \text{ billion years} = 4.59 \text{ billion years}$$

The uncertainty of this estimate is about 0.07 billion years, so it's consistent with the age of 4.56 billion years first obtained for Earth by Patterson in 1956.

BONUS PROBLEM: When plotted on a diagram like part (b) of the accompanying figure, the rubidium and strontium isotope ratios from several mineral grains collected from the same rock lie on a line with a slope of 0.0143. Assuming these mineral grains have been closed systems since they crystallized, calculate the age of the rock. *Hint:* log(1.0143) = 0.00617.

EXERCISES

1. Many fine-grained muds are deposited at a rate of about 1 cm/1000 years. At this rate, how long would it take to accumulate a sedimentary sequence half a kilometer thick?

2. Construct a cross section similar to the one at the top of Figure 8.10 to show the following sequence of geologic events: (a) deposition of a limestone formation; (b) uplift and folding of the limestone; (c) erosion of the folded rock; (d) subsidence and deposition of a sandstone formation.

3. How many formations can you count in the geologic cross section of the Grand Canyon in Earth Issues 8.1? How many are the same formations observed in Zion Canyon? Are any of the formations observed in both the Grand Canyon and the Bryce Canyon cross sections?

4. By comparing the sequence of formations illustrated in Earth Issues 8.1 with the relative time scale in Figure 8.11, identify a major disconformity in the Grand Canyon stratigraphic succession. Which periods of geologic time are missing? What is the minimum amount of geologic time, measured in millions of years, that is missing? (*Hint:* Consult Figure 8.15.)

5. What type of unconformity would probably be produced on a continental margin that was broadly uplifted above sea level and then subsided below sea level? What type of unconformity might separate young flat-lying sediments from older metamorphosed sediments?

6. Mass extinctions have been dated at 444 million years ago, 416 million years ago, and 359 million years ago. How are these events expressed in the geologic time scale of Figure 8.15?

7. A geologist discovers a distinctive set of fish fossils that dates from the Devonian period within a low-grade metamorphic rock. The rubidium-strontium isotopic age of the rock is determined to be only 70 million years. Give a possible explanation for the discrepancy.

8. Explain why the last eon of geologic time is named the Phanerozoic.

9. At the present rate of seafloor spreading, the entire seafloor is recycled every 200 million years. Assuming that the past rate of seafloor generation has been this fast or faster, calculate the minimum number of times the seafloor has been recycled since the end of the Archean eon.

THOUGHT QUESTIONS

1. As you pass by an excavation in the street, you see a cross section showing paving at the top, soil below the paving, and bedrock at the base. You also notice that a vertical water pipe extends through a hole in the street into a sewer in the soil. What can you say about the relative ages of the various layers and the water pipe?

■ Origin of the
Solar System   222

■ Early Earth:
Formation of a
Layered Planet   225

■ Diversity of the
Planets   228

■ What's in a Face?
The Age and
Complexion of
Planetary
Surfaces   230

■ Mars Rocks!   237

■ Exploring the
Solar System
and Beyond   245

Geologist-astronaut Harrison "Jack" Schmitt, *Apollo 17* lunar module pilot, uses an adjustable scoop to retrieve samples of lunar rock on the rim of Camelot Crater in the Taurus-Littrow Valley. [NASA.]

2. Why did the nineteenth-century geologists constructing the geologic time scale find sedimentary strata deposited in oceans and shallow seas more useful than strata deposited on land?

3. The theory of evolution suggests a "principle of floral (plant) succession" to complement Smith's principle of faunal succession. Why do you think Smith relied primarily on faunal fossils rather than floral fossils in his stratigraphic mapping?

4. In studying an area of tectonic compression, a geologist discovers a sequence of older, more deformed sedimentary rocks on top of a younger, less deformed sequence, separated by an angular unconformity. What plate tectonic processes might have created the angular unconformity?

5. A geologist documents a distinctive chemical signature caused by organisms of the Proterozoic eon that has been preserved in sedimentary rock. Would you consider this chemical signature to be a fossil?

6. Is carbon-14 a suitable isotope for dating geologic events in the Pliocene epoch?

7. How does determining the ages of igneous rocks help to date fossils?

MEDIA SUPPORT

 8-1 Animation: Geologic Time Scale

EARLY HISTORY OF THE TERRESTRIAL PLANETS

9

IN A SERIES OF SIX LANDINGS FROM 1969 THROUGH 1972, the astronauts of the Apollo missions explored the surface of the Moon. These astronauts, trained in geology, took photographs, mapped outcrops, conducted experiments, and collected dust and rock samples for analysis back on Earth. This unprecedented achievement was possible only through the close collaboration of scientists, engineers, and the funding agencies that recognized the importance of basic research in developing new technologies. Perhaps the most important ingredient of all was the irrepressible drive, inherent in all human beings, to explore the unknown. The desire to explore our universe has existed for as long as humans have been able to think. Edwin Powell Hubble best captured the spirit of space exploration when he modestly noted, "Equipped with his five senses, man explores the universe around him and calls the adventure science."

The modern era of space exploration began in the early 1900s when a handful of scientists with a yearning to escape the confines of Earth's gravity began to develop the first generation of rockets. By the late 1920s, these backyard rockets, powered by liquid propellants, were ready for use. Developments occurred rapidly over the next few decades, culminating in the fevered cold-war race between the United States and the Soviet Union to put the first rocket into space, the first satellite into Earth orbit, the first human on the Moon, and the first robot on Mars. By the mid-1970s—50 years after the first liquid-fueled rockets were invented—all of these goals had been achieved.

The scientific dividends of space exploration have been tremendous. The age of the solar system, evidence for liquid water on early Mars, and the thick atmosphere of Venus were all revealed by the mid-1970s. Since that time, we have carried on our exploration of the solar system and beyond. Using instruments in Earth orbit and on spacecraft sent to the far limits of our solar system, we have obtained a much better view of what, literally, is *way* out there! Of all these instruments, none has produced such visually spectacular images of deep space as the Hubble Space Telescope, named for Edwin Powell Hubble. Not since Galileo turned his telescope toward the heavens in 1609 has any instrument so changed our understanding of the universe.

The crater-marked surfaces of the Moon and our neighboring planets, and the occasional meteorite that crashes through Earth's atmosphere, remind us of a disorganized, chaotic time when the solar system was young and Earth's environment was much less hospitable. How did the solar system become the well-ordered place it is today, with planets moving in stately orbits around the Sun? How did Earth's rocky mass come together and differentiate into a core, mantle, and crust? Why does Earth's surface, with its blue oceans and wandering continents, look so different from those of its planetary neighbors? Geologists can draw from many lines of evidence to answer these questions. The rocks of continents preserve a record of geologic processes more than 4 billion years old, and even more ancient materials have been collected from meteorites. And now we can reach beyond Earth for the answers.

In this chapter, we will explore the solar system not only outward through the vast reaches of interplanetary space, but also backward in time to its earliest history. We will see how Earth and the other planets formed around the Sun and differentiated into layered bodies. We will compare the geologic processes that have shaped Earth with those that formed Mercury, Mars, Venus, and the Moon, and we will see how exploration of the solar system by spacecraft might answer fundamental questions about the development of our planet and the life it contains.

Origin of the Solar System

Our search for the origins of the universe—and of our own small part of it—goes back to the earliest recorded mythologies. Today, the generally accepted scientific explanation is the Big Bang theory, which holds that our universe began about 13.7 billion years ago with a cosmic "explosion." Before that moment, all matter and energy were compacted into a single, inconceivably dense point. Although we know little of what happened in the first fraction of a second after time began, astronomers have a general understanding of the billions of years that followed: in a process that still continues, the universe has expanded and thinned out to form galaxies and stars. Geology explores the latter third of that time: the past 4.5 billion years, during which our *solar system*—the star we call the Sun and the planets that orbit it—formed and evolved. In particular, geologists look to the solar system to understand the formation of Earth and the Earthlike planets.

The Nebular Hypothesis

In 1755, the German philosopher Immanuel Kant suggested that the origin of the solar system could be traced to a rotating cloud of gases and fine dust, an idea called the **nebular hypothesis.** We now know that outer space beyond the solar system is not as empty as we once thought. Astronomers have recorded many clouds of the type that Kant surmised, which they have named *nebulae* (plural of the Latin word for "fog" or "cloud") (**Figure 9.1**). They have also identified the materials that form these clouds. The gases are mostly hydrogen and helium, the two elements

(a)

(b)

FIGURE 9.1 ■ Space exploration has progressed from its modest beginnings to address fundamental questions such as the origin of the solar system. (a) Robert H. Goddard, one of the fathers of rocketry, fired this liquid oxygen–gasoline rocket on March 16, 1926, at Auburn, Massachusetts. (b) Seventy years later, on November 2, 1995, the Hubble Space Telescope (in orbit around Earth) took this stunning photograph of the Eagle Nebula. The dark, pillar-like structures are columns of cool hydrogen gas and dust that give birth to new stars. [(a) NASA; (b) NASA/ESA/STSci.]

that make up all but a small fraction of our Sun. The dust-sized particles are chemically similar to materials found on Earth.

How could the solar system take shape from such a cloud? The diffuse, slowly rotating nebula contracted under the force of gravity (**Figure 9.2**). Its contraction, in turn, accelerated the rotation of the particles (just as ice skaters spin more rapidly when they pull in their arms), and the faster rotation flattened the cloud into a disk.

The Sun Forms

Under the pull of gravity, matter began to drift toward the center of the nebula, accumulating into a protostar, the precursor of our present Sun. Compressed under its own weight, the material in the proto-Sun became dense and hot. Its internal temperature rose to millions of degrees, at which point nuclear fusion began. The Sun's nuclear fusion, which continues today, is the same nuclear reaction that occurs in a hydrogen bomb. In both cases, hydrogen atoms, under intense pressures and at high temperatures, combine (fuse) to form helium. Some mass is converted

into energy in the process. The Sun releases some of that energy as sunshine; an H-bomb releases it as an explosion.

The Planets Form

Although most of the matter in the original nebula was concentrated in the proto-Sun, a disk of gases and dust, called the **solar nebula,** remained to envelop it. The temperature of the solar nebula rose as it flattened into a disk. It became hotter in the inner region, where more of the matter accumulated, than in the less dense outer regions. Once formed, the disk began to cool, and many of the gases condensed—that is, they were transformed to their liquid or solid state, just as water vapor condenses into droplets on the outside of a cold glass and water solidifies into ice when it cools below the freezing point.

Gravitational attraction caused the dust and condensing material to clump together (accrete) into small, kilometer-sized chunks, or **planetesimals.** These planetesimals, in turn, collided and stuck together, forming larger, Moon-sized bodies (see Figure 9.2). In a final stage of cataclysmic impacts, a few of these larger bodies—with their stronger

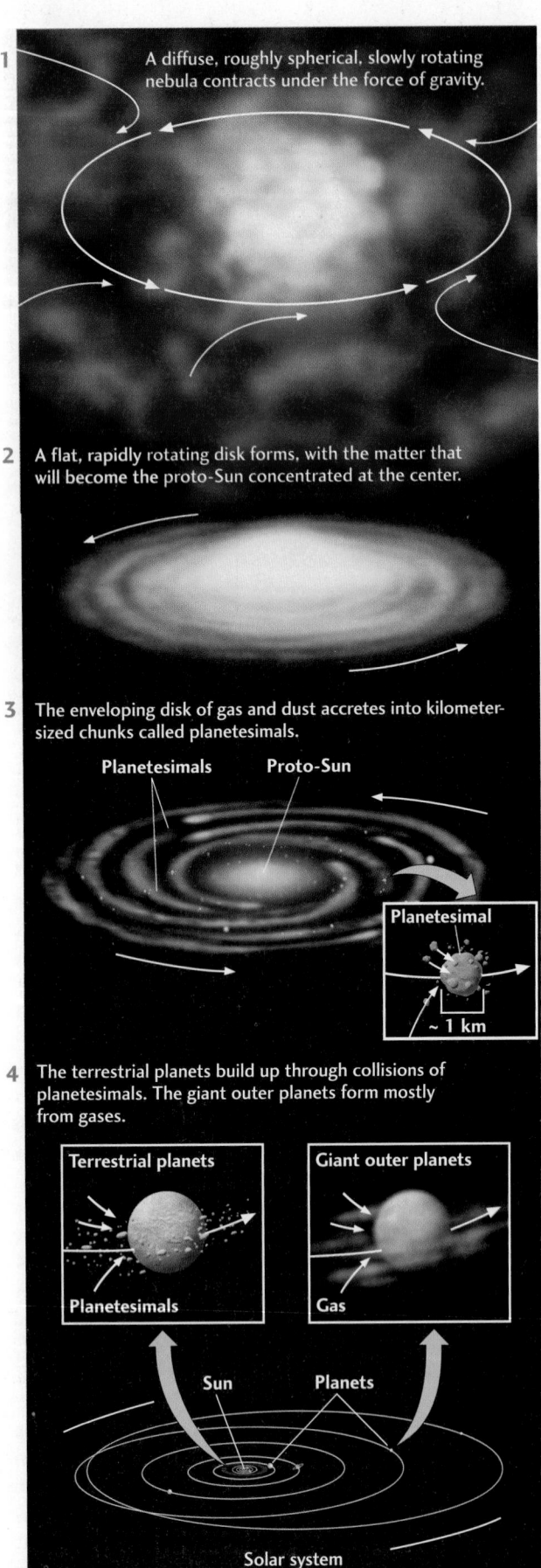

1 A diffuse, roughly spherical, slowly rotating nebula contracts under the force of gravity.

2 A flat, rapidly rotating disk forms, with the matter that will become the proto-Sun concentrated at the center.

3 The enveloping disk of gas and dust accretes into kilometer-sized chunks called planetesimals.

Planetesimals Proto-Sun

Planetesimal

~ 1 km

4 The terrestrial planets build up through collisions of planetesimals. The giant outer planets form mostly from gases.

Terrestrial planets

Giant outer planets

Planetesimals

Gas

Sun Planets

Solar system

FIGURE 9.2 ■ The nebular hypothesis explains the formation of the solar system.

gravitational attraction—swept up the others to form the planets in their present orbits. Planetary formation happened rapidly, probably within 10 million years after the condensation of the nebula.

As the planets formed, those in orbits close to the Sun and those in orbits farther from the Sun developed in markedly different ways. Thus, the composition of the inner planets is quite different from that of the outer planets.

TERRESTRIAL PLANETS The four inner planets, in order of closeness to the Sun, are Mercury, Venus, Earth, and Mars (**Figure 9.3**). They are also known as the Earth-like planets, or **terrestrial planets.** They formed close to the Sun, where conditions were so hot that most of their volatile materials (those that most easily become gases) boiled away. Radiation and matter streaming from the Sun—the solar wind—blew away most of the hydrogen, helium, water, and other light gases and liquids on these planets. Thus, the inner planets were formed mainly from the dense matter that was left behind, which included the rock-forming silicates as well as metals such as iron and nickel. From isotopic dating of the meteorites that occasionally strike Earth, which are believed to be remnants of this pre-planetary process, we know that the inner planets began to accrete about 4.56 billion years ago (see Chapter 8). Computer simulations indicate that they would have grown to planetary size in a remarkably short time—perhaps as quickly as 10 million years or less.

GIANT OUTER PLANETS Most of the volatile materials swept from the region of the terrestrial planets were carried to the cold outer reaches of the solar system to form the giant outer planets—Jupiter, Saturn, Uranus, and Neptune—and their satellites. These giant planets were big enough, and their gravitational attraction strong enough, to enable them to hold onto the lighter nebular materials. Thus, although they have rocky and metal-rich cores, they, like the Sun, are composed mostly of hydrogen and helium and the other light materials of the original nebula.

Small Bodies of the Solar System

Not all the material from the solar nebula ended up in planets. Some planetesimals collected between the orbits of Mars and Jupiter to form the *asteroid belt* (see Figure 9.3). This region now contains more than 10,000 **asteroids** with diameters larger than 10 km and about 300 larger than 100 km. The largest is Ceres, which has a diameter of 930 km. Most **meteorites**—chunks of material from outer space that strike Earth—are tiny pieces of asteroids ejected from the asteroid belt during collisions with one another. Astronomers originally thought the asteroids were the

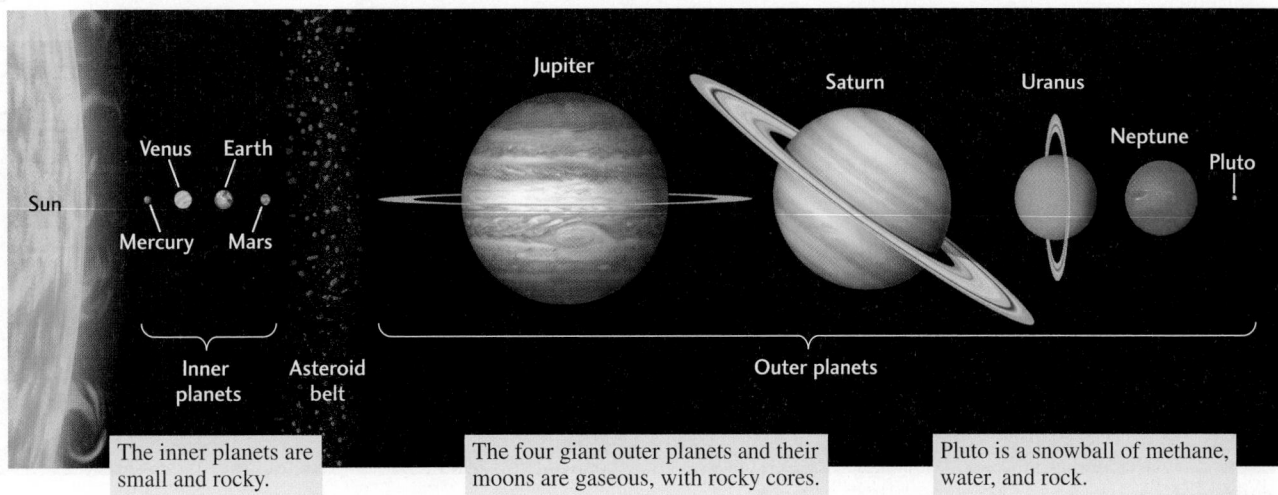

Venus Earth

Sun

Mercury Mars

Jupiter

Saturn

Uranus

Neptune
Pluto

Inner
planets

Asteroid
belt

Outer planets

The inner planets are
small and rocky.

The four giant outer planets and their
moons are gaseous, with rocky cores.

Pluto is a snowball of methane,
water, and rock.

FIGURE 9.3 ■ The solar system. This diagram shows the relative sizes of the planets as well as the asteroid belt separating the inner and outer planets. Although considered one of the nine planets since its discovery in 1930, Pluto was demoted from that status by the International Astronomical Union in 2006. With this revision, there are only eight true planets, not nine.

remains of a large planet that had broken apart early in the history of the solar system, but it now appears they are pieces that never coalesced into a planet, probably due to the gravitational influence of Jupiter.

Another important group of small, solid bodies is the *comets,* aggregations of dust and ice that condensed in the cooler outer reaches of the solar nebula. There are probably many millions of comets with diameters larger than 10 km. Most comets orbit the Sun far beyond the outer planets, forming concentric "halos" around the solar system. Occasionally, collisions or near misses throw a comet into an orbit that penetrates the inner solar system. We can then observe it as a bright object with a tail of gases blown away from the Sun by the solar wind. Perhaps the most famous of these is Halley's Comet, which has an orbital period of 76 years and was last seen in 1986. Comets are intriguing to geologists because they provide clues about the more volatile components of the solar nebula, including water and carbon-rich compounds, which they contain in abundance.

■ Early Earth: Formation of a Layered Planet

We know that Earth is a layered planet with a core, mantle, and crust surrounded by a fluid ocean and a gaseous atmosphere (see Chapter 1). How was Earth transformed from a hot, rocky mass into a living planet with continents, oceans, and a pleasant climate? The answer lies in **gravitational differentiation:** the transformation of random chunks of primordial matter into a body whose interior is divided into concentric layers that differ from one another both physically and chemically. Gravitational differentiation

occurred early in Earth's history, as soon as the planet got hot enough to melt.

Earth Heats Up and Melts

Although Earth probably started out as an accretion of planetesimals and other remnants of the solar nebula, it did not retain this form for long. To understand Earth's present layered structure, we must return to the time when Earth was still subject to violent impacts by planetesimals and larger bodies. As these objects crashed into the primitive planet, most of their energy of motion (kinetic energy) was converted into heat—another form of energy—and that heat caused melting. A planetesimal colliding with Earth at a typical velocity of 15 to 20 km/s would deliver as much kinetic energy as 100 times its weight in TNT. The impact energy of a body the size of Mars colliding with Earth would be equivalent to exploding several trillion 1-megaton hydrogen bombs (a single one of which would destroy a large city)—enough to eject a vast amount of debris into space and to melt most of what remained of Earth.

Many scientists now think that such a cataclysm did occur during the middle to late stages of Earth's accretion. A giant impact by a Mars-sized body created a shower of debris from both Earth and the impacting body and propelled it into space. The Moon aggregated from that debris (**Figure 9.4**). According to this theory, Earth re-formed as a body with an outer molten layer hundreds of kilometers thick—a *magma ocean.* The huge impact sped up Earth's rotation and changed the angle of its axis, knocking it from vertical with respect to Earth's orbital plane to its present 23° inclination. All this occurred about 4.51 billion years ago, between the beginning of Earth's accretion (4.56 billion years ago) and the formation of the oldest Moon rocks brought back by the Apollo astronauts (4.47 billion years ago).

1 During the middle to late stages of Earth's accretion, a Mars-sized body collided with Earth.

2 The giant impact quickly propelled a shower of debris from both the impacting body and Earth into space.

Impacter

Earth

4.2 min. after impact

8.4 min.

125 min.

3 The impact sped up Earth's rotation and tilted Earth's axis of rotation by 23°.

4 Earth re-formed as a largely molten body,...

5 ...and the Moon aggregated from the debris.

6 Moon rocks 4.47 billion years old brought back by the Apollo astronauts support this impact hypothesis.

FIGURE 9.4 ■ Computer simulation of the impact of a Mars-sized body on Earth. [*Solid-Earth Sciences and Society.* Washington, D.C.: National Research Council, 1993.]

Another source of heat that contributed to melting early in Earth's history was radioactivity. When radioactive elements decay, they emit heat. Although present in only small amounts, radioactive isotopes of uranium, thorium, and potassium have continued to keep Earth's interior hot.

Differentiation of Earth's Core, Mantle, and Crust

As a result of the tremendous impact energy absorbed during Earth's formation, its entire interior was heated to a "soft" state in which its components could move around. Heavy material sank to become the core, releasing gravitational energy and causing more melting, and lighter material floated to the surface and formed the crust. The rising lighter matter brought heat from the interior to the surface, where it could radiate into space. In this way, Earth differentiated into a layered planet with a central core, a mantle, and an outer crust (**Figure 9.5**).

EARTH'S CORE Iron, which is denser than most of the other elements, accounted for about a third of the primitive

planet's material (see Figure 1.12). This iron and other heavy elements, such as nickel, sank to form a central *core*, which begins at a depth of about 2890 km. By probing the core with seismic waves, scientists have found that it is molten on the outside but solid in a region called the *inner core,* which extends from a depth of about 5150 km to Earth's center at about 6370 km. Today the inner core is solid because the pressures deep in Earth's interior are too high for iron to melt.

EARTH'S CRUST Other molten materials that were less dense than iron and nickel floated toward the surface of the magma ocean. There they cooled to form Earth's solid *crust,* which today ranges in thickness from about 7 km on the seafloor to about 40 km on the continents. We know that oceanic crust is constantly generated by seafloor spreading and recycled into the mantle by subduction. In contrast, continental crust began to accumulate early in Earth's history from silicates of relatively low density with a felsic composition and low melting temperatures. This contrast between dense oceanic crust and less dense continental crust is what helps drive oceanic crust into subduction zones, while continental crust resists subduction.

FIGURE 9.5 ▪ Gravitational differentiation of early Earth resulted in a planet with three main layers.

Iron | Lighter matter | Crust (0–40 km) | Mantle (40–2890 km) | Liquid iron outer core (2890–5150 km) | Solid iron inner core (5150–6370 km)

During gravitational differentiation, iron sank to the center and lighter material floated upward...

...to give us Earth as a layered planet.

The 4.4-billion-year-old zircon grains recently found in Western Australia (see Chapter 8) are the oldest terrestrial material yet discovered. Chemical analysis indicates that they formed near Earth's surface under relatively cool conditions and in the presence of water. This finding suggests that Earth had cooled enough for a crust to exist only 100 million years after the planet re-formed following the giant impact that produced the Moon.

EARTH'S MANTLE Between the core and the crust lies the *mantle,* the layer that forms the bulk of the solid Earth. The mantle is made up of the material left in the middle zone after most of the denser material sank and the less dense material rose toward the surface. It is about 2850 km thick and consists of ultramafic silicate rocks containing more magnesium and iron than crustal silicates do. Convection in the mantle removes heat from Earth's interior (see Chapter 2).

Because the mantle was hotter early in Earth's history, it was probably convecting more vigorously than it does today. Some form of plate tectonics may have been operating even then, although the "plates" were probably much smaller and thinner, and the tectonic features were probably very different from the linear mountain belts and long mid-ocean ridges we now see on Earth's surface. Some scientists think that Venus today provides an analog for these long-vanished processes on Earth. We will compare tectonic processes on Earth and Venus shortly.

Earth's Oceans and Atmosphere Form

The oceans and atmosphere can be traced back to the "wet birth" of Earth itself. The planetesimals that aggregated into our planet contained ice, water, and other volatiles, such as nitrogen and carbon, locked up in minerals. As Earth differentiated, water vapor and other gases were freed from these minerals, carried to the surface by magmas, and released through volcanic activity.

The enormous volumes of gases spewed from volcanoes 4 billion years ago probably consisted of the same substances that are expelled from present-day volcanoes (though not necessarily in the same relative abundances): primarily hydrogen, carbon dioxide, nitrogen, water vapor, and a few others (**Figure 9.6**). Almost all of the hydrogen

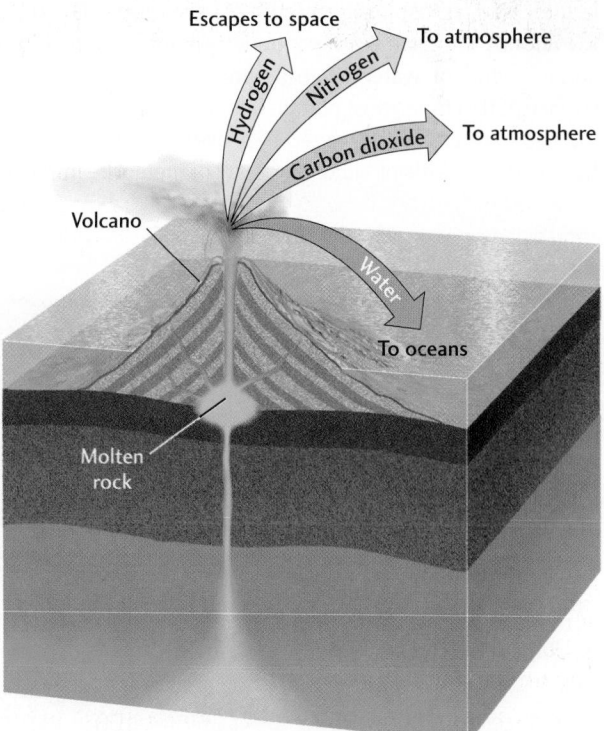

Escapes to space | To atmosphere | Hydrogen | Nitrogen | Carbon dioxide | To atmosphere | Volcano | Water | To oceans | Molten rock

FIGURE 9.6 ▪ Early volcanic activity contributed enormous amounts of water vapor, carbon dioxide, and nitrogen to the atmosphere and oceans. Hydrogen, because it is lighter, escaped into space.

escaped into space, while the heavier gases enveloped the planet. Some of the air and water may also have come from volatile-rich bodies from the outer solar system, such as comets, that struck the planet after it had formed. Countless comets may have bombarded Earth early in its history, contributing water, carbon dioxide, and gases to the early oceans and atmosphere. The early atmosphere lacked the oxygen that makes up 21 percent of the atmosphere today. Oxygen did not enter the atmosphere until oxygen-producing organisms evolved, as we will see in Chapter 11.

■ Diversity of the Planets

By about 4.4 billion years ago, in less than 200 million years since its origin, Earth had become a fully differentiated planet. The core was still hot and mostly molten, but the mantle was fairly well solidified, and a primitive crust and continents had begun to develop. Oceans and atmosphere had formed, and the geologic processes that we observe today had been set in motion. But what about the other terrestrial planets? Did they experience a similar early history? Information transmitted from space probes indicates that the four terrestrial planets have all undergone gravitational differentiation into layered structures with an iron-nickel core, a silicate mantle, and an outer crust (Table 9.1).

Mercury has a thin atmosphere consisting mostly of helium. The atmospheric pressure at its surface is less than a trillionth of Earth's atmospheric pressure. There is no surface wind or water to erode and smooth the ancient surface of this innermost planet. It looks like the Moon: it is intensely cratered and covered by a layer of rock debris, the fractured remnants of billions of years of meteorite impacts.

Because it is located close to the Sun and has essentially no atmosphere to protect it, the planet warms to a surface temperature of 470°C during the day and cools to −170°C at night—the largest temperature range for any planet.

Mercury's average density is nearly as great as Earth's, even though it is a much smaller planet. Accounting for differences in interior pressure (remember, higher pressures increase density), scientists have surmised that Mercury's iron-nickel core must make up about 70 percent of its mass, a record proportion for solar system planets (Earth's core is only one-third of its mass). Perhaps Mercury lost part of its silicate mantle in a giant impact. Alternatively, the Sun could have vaporized part of its mantle during an early phase of intense radiation. Scientists are still debating these hypotheses.

Venus developed into a planet with surface conditions surpassing most descriptions of hell. It is wrapped in a heavy, poisonous, incredibly hot (475°C) atmosphere composed mostly of carbon dioxide and clouds of corrosive sulfuric acid droplets. A human standing on its surface would be crushed by the atmospheric pressure, boiled by the heat, and eaten away by the sulfuric acid. At least 85 percent of Venus is covered by lava flows. The remaining surface is mostly mountainous—evidence that the planet has been tectonically active (Figure 9.7). Venus is close to Earth in mass and size, and its core seems to be about the same size as Earth's, with both liquid and solid portions. How it could develop into a planet so different from Earth is a question that intrigues planetary geologists.

Mars has undergone many of the same geologic processes that have shaped Earth (see Figure 9.7). The Red Planet is considerably smaller than Earth, with only about one-tenth of Earth's mass. However, the Martian core, like the cores of Earth and Venus, appears to have a radius of about half the planet's radius, and, like Earth's, it may have a liquid outer portion and a solid inner portion.

TABLE 9-1 · Characteristics of the Terrestrial Planets and Earth's Moon

	Mercury	Venus	Earth	Mars	Earth's Moon
Radius (km)	2440	6052	6370	3388	1737
Mass (Earth = 1)	0.06 (3.3×10^{23} kg)	0.81 (4.9×10^{24} kg)	1.00 (6.0×10^{24} kg)	0.11 (6.4×10^{23} kg)	0.01 (7.2×10^{22} kg)
Average density (g/cm^3)	5.43	5.24	5.52	3.94	3.34
Orbit period (Earth days)	88	224	365	687	27
Distance from Sun ($\times 10^6$ km)	57	108	148	228	
Moons	0	0	1	2	0

FIGURE 9.7 ■ A comparison of the surfaces of Earth, Mars, and Venus, all at the same scale. The topography of Mars, which shows the greatest range, was measured in 1998 and 1999 by a laser altimeter aboard the orbiting *Mars Global Surveyor* spacecraft. That of Venus, which shows the smallest range, was measured from 1990 to 1993 by a radar altimeter aboard the orbiting *Magellan* spacecraft. Earth's topography, which is intermediate in range and dominated by continents and oceans, has been synthesized from altimeter measurements of the land surface, ship-based measurements of ocean depth, and gravity-field measurements of the seafloor surface from Earth-orbiting spacecraft. [Courtesy of Greg Neumann/MIT/GSFC/NASA.]

Mars has a thin atmosphere composed almost entirely of carbon dioxide. No liquid water is present on its surface today; the planet is too cold, and its atmosphere is too thin, so any water on its surface would either freeze or evaporate. Several lines of evidence, however, indicate that liquid water was abundant on the surface of Mars before 3.5 billion years ago, and that large amounts of water ice may be stored below the surface and in polar ice caps today. Life might have formed on the wet Mars of billions of years ago and could exist today as microorganisms below the surface.

Most of the surface of Mars is older than 3 billion years. On Earth, in contrast, most surfaces older than about 500 million years have been obliterated through the combined activities of the plate tectonic and climate systems. Later in this chapter, we will compare surface processes on Earth and Mars in more detail.

Other than Earth itself, the *Moon* is the best-known body in the solar system because of its proximity to Earth and the manned and unmanned programs that have been designed to explore it. In bulk, its materials are lighter than Earth's, probably because the heavier matter of the giant impacting body remained embedded in Earth after the collision that formed it. The lunar core is therefore small, constituting only about 20 percent of the lunar mass.

The Moon has no atmosphere, and is mostly bone dry, having lost most of its water in the heat generated by the giant impact. There is some new evidence from spacecraft observations that water ice may be present in small amounts deep within sunless craters at the Moon's north and south poles. The heavily cratered lunar surface we see today is that of a very old, geologically dead body, dating back to a period early in the history of the solar system

when crater-forming impacts were very frequent. Once topography is created on any planetary body, plate tectonic and climate processes will work to "resurface" it, as they have on Venus and Mars. However, in the absence of these processes, the planet will remain pretty much the way it was just after its formation. Thus, the heavily cratered terrains of little-studied planetary bodies, such as Mercury, indicate that they lack both a convecting mantle and an atmosphere.

The giant gaseous outer planets—*Jupiter, Saturn, Uranus,* and *Neptune*—are likely to remain a puzzle for a long time. These huge gas balls are so chemically distinct and so large that their formation must have followed a course entirely different from that of the much smaller terrestrial planets. All four of the giant planets are thought to have rocky, silica-rich and iron-rich cores surrounded by thick shells of liquid hydrogen and helium. Inside Jupiter and Saturn, the pressures become so high that scientists believe the hydrogen turns into a metal.

Exactly what lies beyond the orbit of Neptune, the most distant giant planet, remains a mystery. Tiny *Pluto,* once regarded as the ninth planet, is a strange frozen mixture of gases, ice, and rock with an unusual orbit that sometimes brings it closer to the Sun than Neptune. Pluto, along with "2003 UB313" and two other bodies that share its attributes—tiny size, unusual orbit, rock-ice-gas composition—is now known as a **dwarf planet.** The dwarf planets lie within a belt of icy bodies that is the source region for the comets that periodically pass through the inner solar system. Other dwarf planet–sized objects are likely to be found as we explore the outer regions of the solar system. A spacecraft called *New Horizons* will visit Pluto beginning in 2015.

■ What's in a Face? The Age and Complexion of Planetary Surfaces

Like members of a family, the four terrestrial planets all bear a certain resemblance to one another. They are all differentiated planets, with an iron-nickel core, silicate mantle, and outer crust. But, as we have just seen, there are no twins in this family. Their different sizes and masses, and their variable distances from the Sun, make all four planets, and especially their surfaces, distinct.

Like human faces, the faces of planets reveal their ages. Instead of forming wrinkles as they get older, the surfaces of terrestrial planets are marked by craters. The surfaces of Mercury, Mars, and the Moon are heavily cratered and therefore obviously old. In contrast, Venus and Earth have very few craters because their surfaces are much younger. In this section, we will study planetary surfaces to learn about the tectonic and climate processes that have shaped them. Earth is excluded here because it is the subject of this textbook, and Mars will be mentioned only briefly because its surface is more thoroughly described in the following section.

The Man in the Moon: A Planetary Time Scale

If you look at the face of the Moon through binoculars on a clear night, you will see two distinct types of terrain: rough areas that appear light-colored with lots of big craters, and smooth, dark areas, usually circular in shape, where craters are small or nearly absent (**Figure 9.8**). The light-colored regions are the mountainous *lunar highlands*, which cover about 80 percent of the surface. The dark regions are low-lying plains called *lunar maria*, from the Latin for "seas," because they looked like seas to early Earth-bound observers. It is the contrast between highlands and maria that forms the pattern we can see from Earth as the "Man in the Moon."

In preparation for the Apollo missions to the Moon, geologists such as Gene Shoemaker (**Figure 9.9**) developed a relative time scale for the formation of lunar surfaces based on the following simple principles:

- Craters are absent on a new geologic surface; older surfaces have more craters than younger surfaces.
- Impacts by small bodies are more frequent than impacts by large bodies; thus, older surfaces have larger craters.
- More recent impact craters cross-cut or cover older craters.

FIGURE 9.8 ■ The Moon has two types of terrain: the lunar highlands, with many craters, and the lunar lowlands, or maria, with few craters. The maria appear darker due to the presence of widespread basalts that flowed across their surfaces over 3 billion years ago. [NASA/JPL.]

By applying these principles, and by mapping the numbers and sizes of craters—a procedure known as *crater counting*—geologists showed that the lunar highlands are older than the maria. They interpreted the maria to be basins formed by the impacts of asteroids or comets that were subsequently flooded with basalts, which "repaved" the basins. They were able to assign different parts of the Moon's surface to geologic intervals analogous to those in the relative time scale worked out by nineteenth-century geologists for Earth.

In the pre-Apollo days, no one knew the absolute ages of either the maria or the highlands, but the smart money held that both were very old. The intense cratering evident in the highlands and the big impacts that formed the maria were consistent with the results of theoretical models of the early solar system. These models predicted a period called the **Heavy Bombardment,** during which the planets collided frequently with the residual materials that still cluttered the solar system after they had been assembled (**Figure 9.10**). According to the models, the numbers and sizes of impacting objects would have been greatest just after the planets formed and would have quickly decreased as the materials were swept up by the planets.

By applying the isotopic dating methods described in Chapter 8 to rock samples brought back by the Apollo astronauts, geologists were able to calibrate the absolute time scale for the Moon that they had developed by crater counting. Sure enough, the highlands turned out to be very ancient (4.4 billion to 4.0 billion years old) and the maria

FIGURE 9.9 ■ Astrogeologist Eugene Shoemaker leads an astronaut training trip on the rim of Meteor Crater, Arizona, in May 1967. (An aerial view of this crater is shown in Figure 1.7b.) Shoemaker and other geologists used their observations of craters to develop a relative time scale for dating lunar surfaces. [USGS.]

younger (4.0 billion to 3.2 billion years old). **Figure 9.11** plots these ages on the ribbon of geologic time.

The relatively young ages of the maria turned out to be a puzzle, however. The best computer simulations of the Heavy Bombardment indicated that it should have been over rather quickly, perhaps in a few hundred million years or even less. Why, then, did some of the biggest impacts observed on the Moon—those that formed the maria—occur so late in lunar history?

The simulations missed an important event. The rate at which large objects struck the Moon did decrease quickly, as the simulations predicted, but then spiked up again in a period known as the *Late* Heavy Bombardment, which occurred between about 4.0 billion and 3.8 billion years ago (see Figure 9.10). The explanation of this event is still controversial, but it is likely that small changes in the orbits of Jupiter and Saturn about 4 billion years ago (caused by their gravitational interactions as they settled into their present orbits) perturbed the orbits of the asteroids. Some of the asteroids were thrown into the inner solar system, where they collided with the Moon and the terrestrial planets, including Earth. The Late Heavy Bombardment explains why so few rocks on Earth have ages greater than 3.9 billion years. It is the Late Heavy Bombardment that marks the end of the Hadean eon and the beginning of the Archean eon (see Figure 9.11).

The time scale first developed for the Moon by crater counting has been extended to other planets by taking into account the differences in impact rates resulting from each planet's mass and position in the solar system.

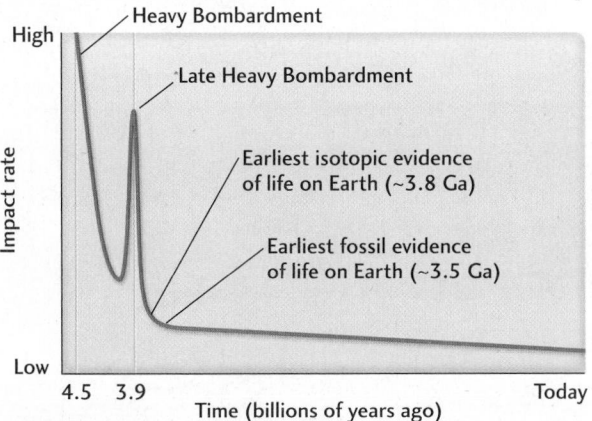

FIGURE 9.10 ■ The number of planetary impacts has varied over the history of the solar system. After the planets formed, they continued to collide with the residual matter that still cluttered the solar system. These collisions tapered off over the first 500 million years of planetary development. However, there was another period of frequent impacts, known as the Late Heavy Bombardment, which peaked about 3.9 billion years ago. (Ga, billion years ago.)

Mercury: The Ancient Planet

The topography of Mercury is poorly understood. *Mariner 10* was the first and only spacecraft to visit Mercury when it flew by the planet in March 1974. It mapped less than half

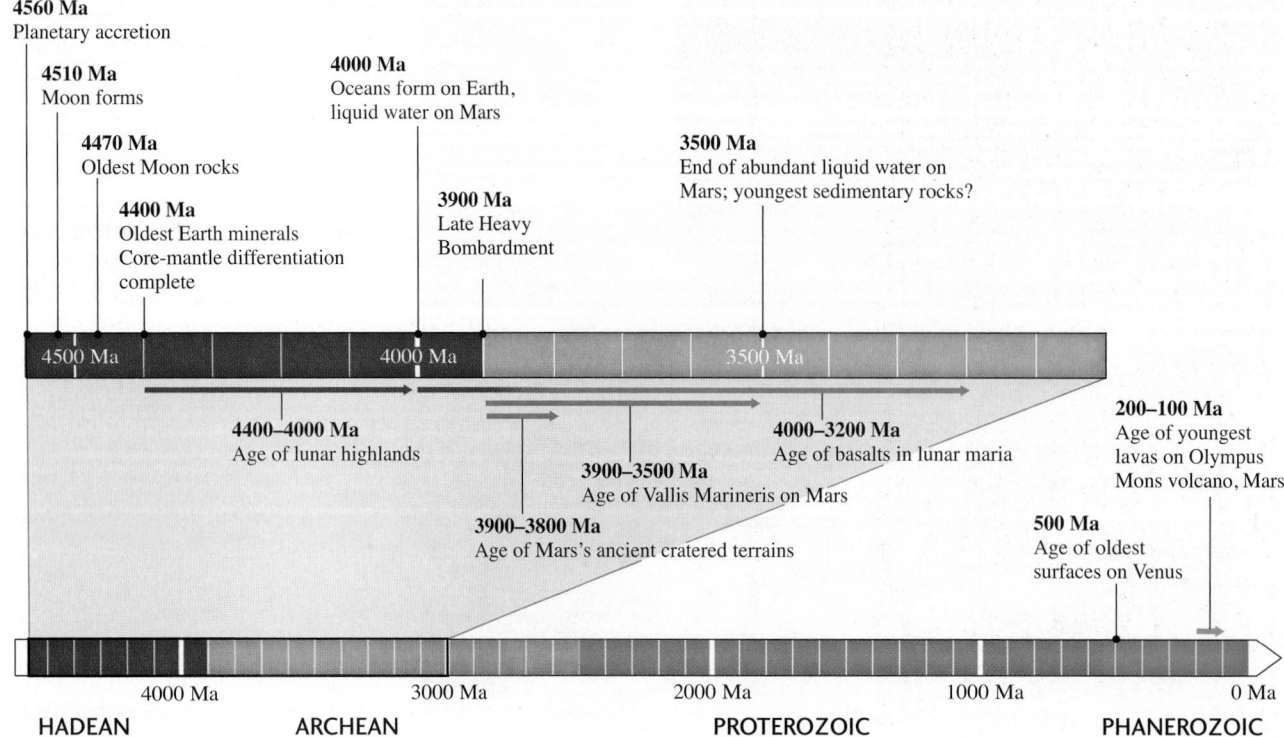

4560 Ma
Planetary accretion

4510 Ma
Moon forms

4470 Ma
Oldest Moon rocks

4400 Ma
Oldest Earth minerals
Core-mantle differentiation
complete

4000 Ma
Oceans form on Earth,
liquid water on Mars

3900 Ma
Late Heavy
Bombardment

3500 Ma
End of abundant liquid water on
Mars; youngest sedimentary rocks?

4500 Ma   4000 Ma   3500 Ma

4400–4000 Ma
Age of lunar highlands

3900–3500 Ma
Age of Vallis Marineris on Mars

3900–3800 Ma
Age of Mars's ancient cratered terrains

4000–3200 Ma
Age of basalts in lunar maria

200–100 Ma
Age of youngest
lavas on Olympus
Mons volcano, Mars

500 Ma
Age of oldest
surfaces on Venus

4000 Ma   3000 Ma   2000 Ma   1000 Ma   0 Ma

HADEAN   **ARCHEAN**   **PROTEROZOIC**   **PHANEROZOIC**

FIGURE 9.11 ■ By calibrating the relative time scale developed by crater counting with the absolute ages of lunar rocks, geologists have constructed a geologic time scale for the terrestrial planets. (Ma, million years ago.)

the planet, and we have little idea of what is on the other side.

Mariner 10 confirmed that Mercury has a geologically dormant, heavily cratered surface. It has the oldest surface of all the terrestrial planets (**Figure 9.12**). Between its large old craters lie younger plains, which are probably volcanic, like the lunar maria. The *Mariner 10* images show a difference in color between the craters and the plains, which supports this hypothesis. Unlike Earth and Venus, Mercury shows very few features that are clearly due to tectonic forces having reshaped its surface.

In many respects, the face of Mercury seems very similar to that of Earth's Moon. The two bodies are similar in size and mass, and most of their tectonic activity took place within the first billion years of their histories. There is one interesting difference, however. Mercury's face has several scars marked by scarps nearly 2 km high and up to 500 km long (**Figure 9.13**). Such features are common on Mercury, but rare on Mars and absent on the Moon. These cliffs appear to have resulted from horizontal compression of Mercury's brittle crust, which formed enormous thrust faults (see Chapter 7). Some geologists think they formed during the cooling of the planet's crust immediately after its formation.

On August 3, 2004, the first new mission to Mercury in 30 years was launched successfully. The *MESSENGER* spacecraft successfully entered an orbit around Mercury in

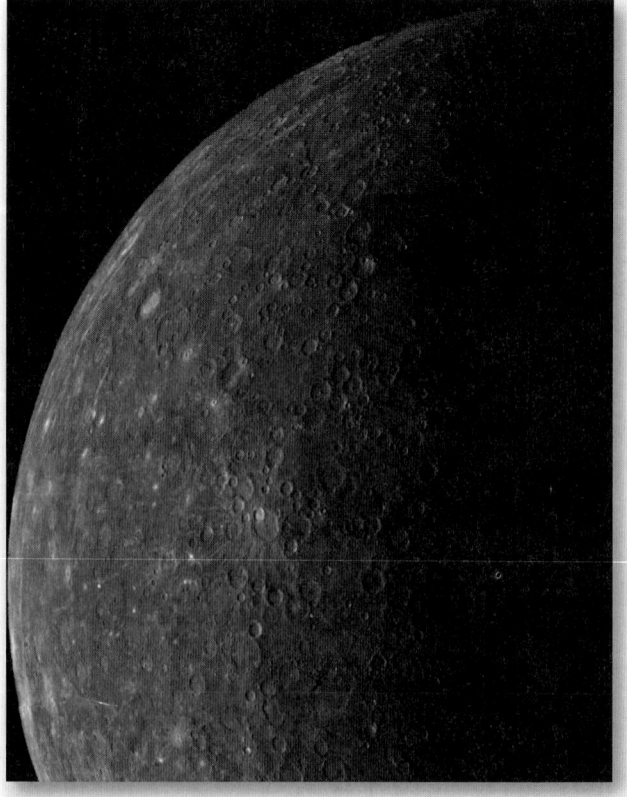

FIGURE 9.12 ■ Mercury has a heavily cratered surface similar to that of Earth's Moon. [NASA/JPL/Northwestern University.]

FIGURE 9.13 ■ The prominent scarp that snakes across this image of Mercury is thought to have formed as the planet's crust was compressed, possibly as it cooled following its formation. Note that the scarp must be younger than the craters it offsets. [NASA/JPL/Northwestern University.]

March 2011. *MESSENGER* has been providing information about Mercury's surface composition, its geologic history, and its core and mantle, and it will search for evidence of water ice and other frozen gases, such as carbon dioxide, at the planet's poles.

Venus: The Volcanic Planet

Venus is our closest planetary neighbor, often brightly visible in the sky just before sunset. Yet in the early decades of space exploration, Venus frustrated scientists. The entire planet is shrouded in a dense fog of carbon dioxide, water vapor, and sulfuric acid, which prevents scientists from studying its surface with ordinary telescopes and cameras. Although many spacecraft were sent to Venus, only a few were able to penetrate this acid fog, and the first ones that tried to land on its surface were crushed under the tremendous weight of its atmosphere.

It was not until August 10, 1990, after traveling 1.3 billion kilometers, that the *Magellan* spacecraft arrived at Venus and took the first high-resolution pictures of its surface (**Figure 9.14**). *Magellan* did this using *radar* (shorthand for *radio detection and ranging*) devices similar to the cameras that police officers use to enforce speed limits (they "see" at night, and through the fog, to clock your speed). Radar cameras form images by bouncing radio waves off stationary surfaces (like those of planets) or moving surfaces (like those of cars).

The images that *Magellan* returned to Earth show clearly that beneath its fog, Venus is a surprisingly diverse and tectonically active planet with mountains, plains, volcanoes, and rift valleys. The lowland plains of Venus—the blue regions in Figure 9.14—have far fewer craters than the Moon's youngest maria, indicating that they must be younger still. Estimates of their age range between 1600 million and 300 million years. Because there is no rain on Venus, there is very little erosion, and so the features we see today have been "locked in" for all that time. The relatively

Elevation (km)

0 2 4 6 8 10 12 14

FIGURE 9.14 ■ This topographic map of Venus is based on more than a decade of mapping, culminating in the 1990–1994 Magellan mission. Regional variations in elevation are illustrated by the highlands (tan colors), the uplands (green colors), and the lowlands (blue colors). Vast lava plains are found in the lowlands. [NASA/USGS.]

(a)

(b)

FIGURE 9.15 ■ Venus is a tectonically active planet with many surface features. (a) Maat Mons, a volcanic mountain that may be up to 3 km high and 500 km across. (b) Volcanic features called coronae are not observed on any other planet except Venus. The visible lines that define the coronae are fractures, faults, and folds produced when a large blob of hot lava collapsed like a fallen soufflé. Each corona is a few hundred kilometers across. [NASA/USGS.]

low number of craters suggests that many craters must have been covered by lava, and therefore that Venus must have been tectonically active relatively recently.

The young plains are dotted with hundreds of thousands of small volcanic domes 2 to 3 km across and

perhaps 100 m or so high, which formed over places where Venus's crust got very hot. There are larger, isolated volcanoes as well, up to 3 km high and 500 km across, similar to the shield volcanoes of the Hawaiian Islands (**Figure 9.15a**). *Magellan* also observed unusual

(a) Plate tectonics on Earth

1 Hot mantle material rises,…

2 …causing plates to form and diverge.

Plate

Plate

3 Where plates converge, a cooled plate sinks under the neighboring plate,…

4 …and its material warms and rises again.

(b) Flake tectonics on Venus

5 On Venus, more vigorous convection currents prevent thick crust from forming, and push and stretch the thin crust that does form.

6 The crust breaks up into flakes or crumples like a rug.

7 Blobs of hot magma bubble up to form large landmasses, mountains, and volcanic deposits.

FIGURE 9.16 ■ Flake tectonics on Venus is very different from plate tectonics on Earth, but could be similar to tectonic processes on early Earth.

circular features called *coronae* that appear to result from blobs of hot lava that rose, created a large bulge or dome in the surface, and then sank, collapsing the dome and leaving a wide ring that looks like a fallen soufflé (Figure 9.15b).

Because Venus has so much evidence of widespread volcanism, it has been called the Volcanic Planet. Venus has a convecting mantle like Earth's, in which hot material rises and cooler material sinks (**Figure 9.16a**), but unlike Earth, it does not appear to have thick plates of rigid lithosphere. Instead, only a thin crust of frozen lava overlies the convecting mantle. As the vigorous convection currents push and stretch the surface, the crust breaks up into flakes or crumples like a rug, and blobs of hot magma bubble up to form large landmasses and volcanic deposits (Figure 9.16b). Scientists have called this process **flake tectonics.** When Earth was younger and hotter, it is possible that flakes, rather than plates, were the main expression of its tectonic activity.

Mars: The Red Planet

Of all the planets, Mars has a surface most similar to Earth's. Mars has features suggesting that liquid water once flowed across its surface, and liquid water may still exist in its deep subsurface. And where there is water, there may be living organisms. No other planet in the solar system has as much chance of harboring extraterrestrial life as Mars.

The abundance of iron oxide minerals on the surface of Mars gives the Red Planet its name. Iron oxide minerals are common on Earth and tend to form where weathering of iron-bearing silicates occurs. We now know that many other minerals common on Earth, such as olivine and pyroxene, which form in basalt, are also present on Mars. But there are other relatively unusual minerals on Mars, such as sulfates, that record an earlier, wetter phase when liquid water may have been stable.

The topography of Mars shows a greater range of elevation than that of Earth or Venus (see Figure 9.7). Olympus Mons, at 25 km high, is a giant, recently active volcano—the tallest mountain in the solar system (**Figure 9.17a**). The Vallis Marineris canyon, 4000 km long and averaging 8 km deep, stretches the distance from New York to Los Angeles and is five times deeper than the Grand Canyon (Figure 9.17b). Recently, geologists have discovered evidence of past glacial processes, when ice sheets similar to the ones that covered North America during the most recent ice age flowed across the surface of Mars. Finally, like the Moon, Mercury, and Venus, Mars has both heavily cratered ancient highlands and younger lowlands.

(a) (b)

FIGURE 9.17 ▪ The topography of Mars shows a large range of elevation. (a) Olympus Mons is the tallest volcano in the solar system, with a summit almost 25 km above the surrounding plains. Encircling the volcano is an outward-facing scarp 550 km in diameter and several kilometers high. Beyond the scarp is a moat filled with lava, most likely derived from Olympus Mons. (b) Vallis Marineris is the longest (4000 km) and deepest (up to 10 km) canyon in the solar system. It is five times deeper than the Grand Canyon. In this image, the canyon is exposed as a series of fault-bounded basins whose sides have partly collapsed (as at upper left), leaving piles of rock debris. The walls of the canyon are 6 km high here. The layering of the canyon walls suggests deposition of sedimentary or volcanic rocks prior to faulting. [(a) NASA/USGS; (b) ESA/DLR/FU Berlin.]

However, unlike those of Mercury, Venus, and the Moon, the lowlands of Mars are created not only by lava flows, but also by sediments, sedimentary rocks, and accumulations of windblown dust.

The face of Mars may be sophisticated, but it has not always been easy to read, despite being visited and viewed more than any other planet except Earth. But, as we will see shortly, Mars's secrets are finally being revealed.

Earth: No Place Like Home

Every view of Earth underscores the unique beauty created by the overwhelming influences of plate tectonics, liquid water, and life. From its blue skies and oceans and its green vegetation to its rugged ice-covered mountains and moving continents, there truly is no place like home. Earth's remarkable appearance is maintained by the delicate balance of conditions necessary to support and sustain life.

The features that define the face of our planet are discussed throughout this textbook, but one process that is appropriate to review here is cratering. The impacts of meteorites and asteroids are preserved in the geologic record of all the terrestrial planets, but in contrast to the other planets, whose surfaces are essentially frozen in time, Earth preserves very few vestiges of its beginning. Recycling by plate tectonics, which is even more efficient than flake tectonics on Venus, has almost completely resurfaced our planet. Those craters that do remain are much younger than the end of the Late Heavy Bombardment, and they are preserved entirely on continents, which resist subduction (**Figure 9.18**).

Nevertheless, Earth accumulates a lot of junk from space. At present, some 40,000 tons of extraterrestrial material fall on Earth each year, mostly as dust and unnoticed small objects. Although the rate of impacts is now orders of magnitude lower than it was during the Heavy Bombardment, a large chunk of matter 1 to 2 km in size still collides with Earth every few million years or so. Although such collisions have become rare, telescopes are being assigned to search space and warn us in advance of sizable bodies that might slam into Earth. NASA astronomers recently predicted "with non-negligible probability" (1 chance in 300) that an asteroid 1 km in diameter will collide with Earth in March 2880. Such an event would threaten human civilization.

We already know that impacts with large objects can greatly upset the conditions that support life on Earth. As we will see in Chapter 11, a collision with a 10-km asteroid 65 million years ago caused the extinction of 75 percent of Earth's species, including all the dinosaurs. This event may have made it possible for mammals to become the dominant species and paved the way for humankind's emergence. **Table 9.2** describes the potential effects of impacts by objects of various sizes on our planet and its life-forms.

FIGURE 9.18 ■ Impact craters formed by meteorites and asteroids are rare on Earth compared with the other terrestrial planets. Recycling of Earth's crust by plate tectonics has erased almost all evidence of impacts. Those craters that remain (red dots) are preserved only on the continents. [NASA/JPL/ASU.]

TABLE 9-2	Impacts by Asteroids and Meteorites and Their Effects on Life on Earth			
	Example or Size Equivalent	Most Recent	Planetary Effects	Effects on Life
Supercolossal: radius (R) > 2000 km	Moon-forming event	4.51×10^9 years ago	Melts planet	Drives off volatiles; wipes out life on Earth
Colossal: R > 700 km	Pluto	More than 4.3×10^9 years ago	Melts crust	Wipes out life on Earth
Huge: R > 200 km	4 Vesta (large asteroid)	About 4.0×10^9 years ago	Vaporizes oceans	Life may survive below surface
Extra large: R > 70 km	Chiron (largest active comet)	3.8×10^9 years ago	Vaporizes upper 100 m of oceans	May wipe out photosynthesis
Large: R > 30 km	Comet Hale-Bopp	About 2×10^9 years ago	Heats atmosphere and surface to about 1000 K	Burns continents
Medium: R > 10 km	Cretaceous-Tertiary impactor; 433 Eros (largest near-Earth asteroid)	65×10^6 years ago	Causes fires, dust, darkness; chemical changes in ocean and atmosphere; large temperature swings	Cretaceous-Tertiary impact caused extinction of 75 percent of species and all dinosaurs
Small: R > 1 km	About size of near-Earth asteroids	About 300,000 years ago	Causes global dusty atmosphere for months	Interrupts photosynthesis; individuals die but few species extinct; threatens civilization
Very small: R > 100 m	Tunguska event (Siberia)	1908	Knocked over trees of kilometers away; caused minor hemispheric effects; dusty atmosphere	Newspaper headlines; romantic sunsets; increased birth rate

Source: J. D. Lissauer, *Nature* 402 (1998): C11–C14.

■ Mars Rocks!

We live in the golden age of Mars exploration. At the time of this writing, there are two robotic rovers operating on the surface of the planet, and three orbiters are circling it. These five spacecraft are returning a seemingly endless stream of new data that are leading to significant new discoveries. Our understanding of the history of past surface environments on Mars is changing dramatically. And the fun won't end soon: both NASA and the European Space Agency have promised to deliver an additional rover in the next few years, and plans are being made to return rock samples from Mars to Earth. All the scientists involved in these missions are grateful to live in a time of such adventure.

Our understanding of surface processes on Mars was dramatically improved when two golf-cart-sized robots landed on Mars in January 2004. The Mars Exploration Rovers, named *Spirit* and *Opportunity*, began their 300-million-kilometer journey from Cape Canaveral, Florida, to the Red Planet in June 2003, accompanied by the *Mars Express*, an orbiter equipped with geologic remote sensing tools. These missions succeeded beyond anyone's expectations, making 2004 and 2005 two of the greatest years in the history of space exploration. Another new orbiter, *Mars Reconnaissance Orbiter*, which started its mission in 2006, has collected a vast set of observations that show evidence of aqueous processes over broad regions of the planet. Its camera is taking stunning pictures of the surface of Mars at unprecedented resolution (25 cm/pixel). The *Phoenix* lander conducted operations in the polar region of Mars from June to November 2008 and confirmed the presence of water ice just a few centimeters below the dusty surface. In August 2012, the Mars Science Laboratory *Curiosity* rover made its spectacular landing in Gale crater on Mars (a video of the "Seven Minutes of Terror" can be found at: http://www.jpl.nasa.gov/video/index.php?id=1090). At the time of writing, it is currently in its two-year primary mission, driving toward the ~5-kilometer-high Mt. Sharp in the center of the crater that contains a rich record of the early environmental history of Mars when the planet may have been habitable. (One of the authors of this textbook, John Grotzinger, is the Chief Scientist for the *Curiosity* rover team.)

Missions to Mars: Flybys, Orbiters, Landers, and Rovers

Earlier missions to Mars helped lay the groundwork for the success of the current missions. All spacecraft sent to Mars since the early 1960s have worked in one of four ways. First, the pioneering Mars exploration spacecraft, such as *Mariner 4*, flew by Mars while quickly acquiring all the data they could before disappearing into deep space.

The second, and most common, mode of operation is to orbit Mars in the same way satellites orbit Earth. *Mariner 9*, launched in May 1971, was the first spacecraft to orbit another planet. Since that time, eight other orbiters have helped map the surface of Mars. *Mars Odyssey, Mars Express,* and *Mars Reconnaissance Orbiter* are still active today. After a very successful mission spanning over 10 years, *Mars Global Surveyor* ceased operations in 2006.

The third method of observing Mars involves landing a spacecraft on the Martian surface. The Viking mission deployed two spacecraft, each of which consisted of an orbiter and a lander. The *Viking 1* lander touched down on the surface of Mars on July 20, 1976, and became the first spacecraft to land on another planet and transmit useful data back to Earth. The Viking mission gave us our first look at the surface of another planet from the ground. It also provided our first chemical analyses of Martian rocks and performed the first life-detection experiments.

The fourth method of exploring Mars involves the use of a rover: a robotic vehicle that can move about on the surface of the planet. As exciting as the Viking mission was, it was two decades until another spacecraft landed safely on the surface of Mars. This time it was *Pathfinder*, which arrived on July 4, 1997. However, the *Pathfinder* lander also included a shoebox-sized rover—called *Sojourner*—that was able to ramble around on the surface, analyzing rocks and soils, within a few meters of the *Pathfinder* lander. *Sojourner* was the first mobile vehicle to operate successfully on another planet and became the prototype for the much larger and more capable Mars Exploration Rovers that landed in 2004.

EARLY MISSIONS: MARINER (1965–1971) AND VIKING (1976–1980) The Mariner and Viking missions returned the first detailed images of Mars to Earth. We saw a cratered Moon-like terrain over part of its surface. In other areas, we saw spectacular features, including enormous volcanoes and canyons, vast dune fields, ice caps at both poles, and the Martian moons Phobos and Deimos. Early images also confirmed global dust storms that had previously been observed from Earth. Orbiting spacecraft continue to monitor these dust storms (**Figure 9.19**).

In addition, extensive networks of stream channels were discovered, providing the first evidence that liquids—possibly water—may once have flowed across the surface

FIGURE 9.19 ■ Global dust storms occur on Mars. The storms begin locally and gradually expand to envelop the entire planet, as seen in these images. [NASA/JPL/ASU.]

FIGURE 9.20 ▪ Channel networks carved into the surface of Mars were revealed by the *Viking* orbiter. The complexity of these channels suggests that liquid water was probably the main force of erosion. [NASA/Washington University.]

of Mars (**Figure 9.20**). Collectively, these data also revealed something that had not been appreciated before: the planet is divisible into two main regions, northern low plains and southern cratered highlands.

The two *Viking* landers provided high-resolution views of the Martian terrain. Both landing sites were strewn with rocks, somewhat rounded by the effects of windblown sand. Chemical sensors showed that the rocks and soils were predominantly basaltic in composition. But all the rocks were loose, and there was no evidence of any exposed bedrock. An onboard biology experiment found no evidence of life at either site. These missions revealed that the Red Planet is red because of the presence of iron oxides in the soils, and that the color of the Martian sky is not blue, but pink, because of the high concentration of suspended iron oxide dust particles.

PATHFINDER (1997) *Pathfinder*'s camera returned images very similar to those from the *Viking* landers: the landing sites were rocky, with windblown sand forming tails behind some rocks, and there was no evidence of any exposed bedrock. However, in addition to evidence of basalts, the *Sojourner* rover detected evidence of andesites. The presence of andesites on Mars indicates that at least some parts of the Martian crust were formed by partial melting of previously formed basalts, suggesting a more complex history of crustal development than previously believed.

The *Pathfinder* instrument suite also included a magnet that collected dust from the atmosphere for analysis. The dust was found to contain magnetic minerals that form only in environments low in oxygen.

MARS GLOBAL SURVEYOR (1996–2006) AND MARS ODYSSEY (2001–) The vastly improved global mapping capabilities of *Mars Global Surveyor* and *Mars*

Odyssey resulted in a number of significant discoveries. *Mars Global Surveyor* carried a laser-based altimeter that surveyed Martian topography with unprecedented resolution. The new images provided the strongest evidence yet for liquid water, this time expressed as meandering stream deposits of loose sediment (**Figure 9.21**). The channels carved into bedrock observed by the *Viking* orbiters suggested flowing water; however, the presence of meandering stream deposits (see Chapter 18) is even stronger evidence for flowing water on the surface of Mars. But it was not until 2004 that the Mars Exploration Rovers first confirmed the presence of minerals that *require* liquid water to have been present.

Mars Global Surveyor and *Mars Odyssey* also showed that permafrost (soil rich in water ice; see Chapter 21) underlies the Martian soil from the mid-latitudes all the way to the poles. Widespread glaciers were also shown to have been present in the relatively recent past, suggesting that Mars—like Earth—may have experienced ice ages driven by changes in global climate. Finally, *Mars Global Surveyor*

FIGURE 9.21 ▪ This image acquired by *Mars Global Surveyor* shows clear evidence of meandering patterns within sediments deposited inside Eberswalde Crater. Liquid water appears to have flowed across the Martian surface and entered the crater, where it deposited sediments in meandering channels similar to those seen in the Mississippi River on Earth today (see Chapter 18). [NASA/JPL/MSSS.]

discovered rare patches of hematite (Fe_2O_3)—a mineral that often forms in water on Earth—scattered over the surface of Mars. As we will see, this discovery contributed to the success of the Mars Exploration Rovers mission.

Mars Exploration Rovers: *Spirit* and *Opportunity*

The Mars Exploration Rovers—*Spirit* and *Opportunity* (**Figure 9.22**)—were the first spacecraft sent to Mars that function almost as well as a human geologist would. Unlike orbiters, which look from afar, and landers, which cannot move from their landing site, *Spirit* and *Opportunity* can move around from rock to rock, picking and choosing which rocks to study in more detail. And when the right rock is found, the rover can look at it with a hand lens—just as geologists do here on Earth in the classroom and in the field. But unlike geologists on Earth, these rovers carry a mobile laboratory, so that the rocks can be analyzed on the spot without having to pay the enormous costs of flying them back to Earth. Because of this remarkable capability, *Spirit* and *Opportunity* have been dubbed the first *robotic geologists* on Mars.

The Mars Exploration Rovers were designed to survive 3 months under the hostile Martian surface conditions and to drive no farther than 300 m. They have since traveled more than 40 km in total across the Martian surface and *Opportunity* is still operating at the time of this writing in 2013 (*Spirit* stopped working in 2010). They have had to survive nighttime temperatures below −90°C, dust devils that could have tipped them over, global dust storms that diminished their solar power, and drives along rocky slopes of almost 30° and through piles of treacherous windblown dust. Despite all these obstacles, the rovers have discovered a treasure trove of geologic wonders.

FIGURE 9.22 ■ *Spirit* (*left*), one of the Mars Exploration Rovers, is about the size of a golf cart. *Spirit* is standing next to a twin of *Sojourner*, a rover that was sent to Mars in 1997. Mars Science Laboratory (*right*) is about the size of a small car and was sent to Mars in 2011. [NASA/JPL-Caltech.]

Mars Science Laboratory (MSL): *Curiosity*

Curiosity launched in 2011 and landed in Gale crater on Mars in August 2012. The overall scientific goals of the mission are to assess whether the field site where the rover landed ever had environmental conditions favorable to microbial life. *Curiosity* is similar to *Spirit* and *Opportunity*, but is about twice as long (3 meters or 10 feet) and five times as heavy—it literally weighs a ton (**Figure 9.23**). It also has the most sophisticated suite of instruments ever sent to another planet. While the MER rovers are solar powered, *Curiosity* is powered by a radioisotope power system, which gets its energy from the natural decay of plutonium.

Gale crater has a ~5-kilometer-high mountain in its center—Mt. Sharp—which is made up of sedimentary strata with a diverse suite of hydrated minerals, which indicate that these strata at least in part formed the presence of water. *Curiosity* has spent much of the first year of its

FIGURE 9.23 ■ The Mars Science Laboratory *Curiosity* rover took this self-portrait (composed of dozens of exposures) with its Mars Hand Lens Imager (MAHLI) during the 177th Martian day, or sol, of *Curiosity*'s work on Mars. The lower left quadrant of the image shows gray powder and two holes where *Curiosity* used its drill on the rock target "John Klein." [NASA/JPL-Caltech/MSSS.]

primary mission in an area called Yellowknife Bay, where it has already found evidence of a potentially habitable environment, preserved in rocks over 3 billion years old. Streams once flowed from the crater rim toward the base of Mt. Sharp where water pooled to form a lake that had low salinity and neutral pH—both favorable for life. *Curiosity* is currently on its way to investigate materials at the base of Mt. Sharp where it will continue the search for additional habitable environments that may have characterized this part of Mars over 3 billion years ago.

WHAT'S UNDER THE HOOD? *Spirit* and *Opportunity* both come equipped with six-wheel drive, a color stereo camera with human vision, front-and-back hazard avoidance cameras, a magnifying "hand lens" for close-up inspection of rocks and soils, and instruments to detect the chemical and mineral composition of rocks and soils. A sample of *Curiosity*'s instruments include HD-resolution color cameras with video capability, detectors that measure radiation that might be harmful to humans, and a weather station (wind direction and velocity, atmospheric pressure, humidity, etc.). At its core, *Curiosity* carries two laboratory instruments that provide information on the mineral composition of drilled samples (**Figure 9.24**), as well as their elemental and isotopic composition, and the presence of any organic compounds. The MER rovers are powered by solar energy, while MSL is powered by a nuclear power source, and all are controlled by scientists on Earth, who send daily command sequences to each rover via radio signals. Because it takes 10 minutes for these signals to travel between Earth and Mars, the rovers

have some self-controlled navigation and hazard avoidance capabilities. However, almost every other decision is made by a team of humans back on Earth. This arrangement ensures that the rovers "think" as geologists do. Onboard computers receive the command sequences from Earth that control each rover activity, including driving; taking pictures of the terrain, rocks, and soils; analyzing rocks and minerals; and studying the atmosphere and moons of Mars.

ROVER LANDING SITES The Mars Exploration Rovers mission was motivated by the search for evidence of liquid water on Mars. The rovers were built with this goal in mind and sent to two locations where data from *Mars Global Surveyor* and *Mars Odyssey* suggested that the chances of finding geologic evidence for water would be high. (Some of the best places, however, were eliminated from consideration because of the extreme risk of landing in a rocky terrain; see the Practicing Geology Exercise at the end of the chapter.) Two of several hundred possible sites were chosen, both near the Martian equator but on opposite sides of the planet. The equatorial positions provide the rovers' solar panels with maximum energy throughout the year. By the time that *Curiosity* landed 10 years later, engineers at the Jet Propulsion Laboratory in Pasadena, CA, had learned how to develop a landing system that could deliver the rover to a very specific site with very high science value. Four final sites were selected by the science team, and all four were accessible by the landing system. The finalist—Gale crater—was chosen because it had the greatest diversity of science targets. This turned out to be critically important and the first target explored—Yellowknife Bay—was successful in achieving the mission goal of a habitable ancient environment. Though *Curiosity* is extremely capable, the landing system was also critically important in helping to enable this discovery.

Spirit was sent to Gusev Crater, a large crater about 160 km in diameter that is thought to have once filled up with water to form a large lake (**Figure 9.25a**). *Opportunity* was sent to Meridiani Planum ("Plains of Meridiani"), where hematite had been detected by *Mars Global Surveyor* (Figure 9.25b). Since landing, *Spirit* trekked across a volcanic plain, ascended the Columbia Hills, and crawled down the other side to arrive at an outcrop whose distinctive shape won it the name of Home Plate. After this long and difficult trek, one of *Spirit*'s left front wheels locked up. But, by turning around and driving backward so it could drag rather than push the broken wheel, *Spirit* finally made it to a part of Home Plate where it made an important discovery: mineral deposits made up of more than 90 percent silica. These deposits indicate that heated waters that once flowed at or near the surface of Mars carried high concentrations of dissolved silica, which precipitated to form hard crusts, similar to what occurs in the hot springs at Yellowstone National Park on Earth today—a place where microorganisms are known to thrive (see the chapter opening photo in Chapter 11). Thus, *Spirit*'s discovery of high-silica rocks suggests the potential for a once-habitable environment

FIGURE 9.24 ■ *Curiosity* drilled into this rock target, "Cumberland," during the 279th Martian day, or sol, of the rover's work on Mars and collected a powdered sample of material from the rock's interior. *Curiosity* used the Mars Hand Lens Imager (MAHLI) camera on the rover's arm to capture this view of the hole in Cumberland on the same sol that the hole was drilled. The diameter of the hole is about 0.6 inch (1.6 cm). The depth of the hole is about 2.6 inches (6.6 cm). [NASA/JPL-Caltech/MSSS.]

(a)

100 km

0 % Hematite Abundance 20 %

(b)

FIGURE 9.25 ■ Mars Exploration Rover landing sites. (a) *Spirit* explored Gusev Crater, about 160 km in diameter, which is thought to have been filled with water, forming an ancient lake. A channel that might have supplied water to the crater is visible at the lower right. (b) *Opportunity* was sent to an area of Meridiani Planum where hematite—a mineral that often forms in water on Earth—is abundant. The image shows concentrations of hematite; the ellipse outlines the permissible landing area. [(a) NASA/JPL/ASU/MSSS; (b) NASA/ASU.]

that could be confirmed by a future mission with a set of instruments similar to *Curiosity*'s.

Opportunity landed in Eagle Crater (a small crater about 20 m in diameter), where it spent 60 days studying the first sedimentary rocks ever found on another planet and gathering evidence that they must have formed in water. *Opportunity* then moved on to another, larger crater (Endurance Crater, about 180 m in diameter), where it spent the next 6 months putting those sedimentary rocks into a broader context of environmental evolution. *Opportunity* then traveled 5 km to a much larger crater (Victoria Crater, about 1 km in diameter), where it has explored even more expansive outcrops of sedimentary rock. *Opportunity* discovered an ancient sandy desert where shallow pools of water once filled depressions between the sand dunes. These pools of water are thought to have been very acidic and also extremely saline. Microorganisms can survive in extremely acidic waters, as we will see in Chapter 11; however, if salinity becomes too high, the availability of water to the microorganisms becomes limited, and they cannot survive. (In a similar way, but substituting sugar for salt, that is why honey does not spoil, even without refrigeration

or addition of preservatives.) Thus, *Opportunity* has also discovered evidence of a potentially habitable environment, albeit one that would have required microorganisms to live in extreme conditions. *Opportunity* then drove over 10 kilometers to arrive at Endeavor crater where it discovered very ancient basaltic rocks that were altered to form clay-bearing deposits that indicate more neutral pH. This more ancient setting would have been more favorable for microorganisms had they ever originated on Mars.

Curiosity landed at the foot of Mt. Sharp—Gale's central mountain—near the end of an ancient alluvial fan that formed by sediments transported by streams from the crater rim. After landing it then drove ~600 meters to the east where it discovered fine-grained sedimentary rocks that preserve evidence of a past aqueous environment characterized by low salinity and neutral pH. At the time of this writing *Curiosity* is in the midst of a 10-km-long drive to the lower reaches of the mountain, where sedimentary rocks can be found that contain clay, sulfate, and iron-bearing minerals that likely formed in the presence of water. Gale is special because it contains a wide diversity of ancient aqueous environments. By exploring this diversity of geologic materials scientists hope to discover which kinds of rocks are favorable for preserving evidence of ancient habitable environments, as well as organic compounds that could be returned to Earth by future missions and analyzed for evidence of life.

Recent Missions: *Mars Reconnaissance Orbiter* (2006–) and *Phoenix* (May–November 2008)

Mars Reconnaissance Orbiter has been mapping the rocks and minerals of Mars at an unprecedented level of detail. Whereas the rovers are limited to a few kilometers of the Martian surface, the orbiter can map anywhere on the planet. It is equipped with several instruments, including a high-resolution stereo color camera capable of resolving objects on the surface of Mars as small as 1 m across. Another important device looks at the sunlight reflected from the Martian surface to reveal the presence of minerals that formed in water. One of the orbiter's most remarkable observations is the discovery of sedimentary layers that are so evenly bedded that they may preserve a record of periodic changes in the Martian climate that occurred billions of years ago (**Figure 9.26**).

FIGURE 9.26   ■   These sedimentary strata exposed in Becquerel Crater have a regular, almost periodic, appearance. Each bed is a few meters thick, and the beds are grouped in sequences a few tens of meters thick. These beds are thought to be composed of wind-deposited dust. The supply of sediment may have been regulated by periodic changes in climate. [NASA/JPL/University of Arizona.]

In May 2008, a new lander touched down on the surface of Mars. It was named *Phoenix* because it was the twin of another lander (*Mars Polar Lander*) that crashed on the surface of Mars in 1999. NASA scientists studied the causes of the malfunction and became confident that they could get it right with the remaining twin. The name *Phoenix* seemed appropriate for a project that was resurrected from the ashes of a former ruin. In Egyptian and Greek mythology, the phoenix is a bird that can periodically burn and regenerate itself.

Phoenix was sent to search for ice in the polar region of Mars. Equipped with solar panels to generate energy from the Sun, it was never designed to survive the dark Martian winter; it had a planned life span of only a few months. Its mission focused on analyzing the composition of several soil samples at the landing site. Within just a month of landing, it had accomplished its primary goal of demonstrating the presence of water ice within the soil. The presence of ice had been predicted by the *Mars Odyssey* orbiter, but it was important to confirm it on the ground.

In addition, *Phoenix* made its own surprising discovery concerning the surface environment of Mars. Based on data from the recent rovers and orbiters, a consensus had been developing that the global surface environment of Mars was likely to be very acidic. When *Phoenix* analyzed its first sample of polar soil, however, it found a neutral pH. This finding is another indicator of habitability, since most microorganisms prefer a neutral pH.

Recent Discoveries: The Environmental Evolution of Mars

The recent rover and orbiter missions to Mars have transformed our understanding of its early evolution. Like the Moon and the other terrestrial planets, Mars has ancient cratered terrains that preserve the record of the Late Heavy Bombardment. Therefore, these ancient terrains must be made of rocks older than 3.8 billion to 3.9 billion years (see Figure 9.11). Younger surfaces, which formed after the time of the Late Heavy Bombardment, are also widespread on Mars. Until recently, these younger surfaces were thought to be largely volcanic, as on Venus. However, data from the Mars Exploration Rovers, Mars Science Laboratory, and *Mars Express* show us that at least some—and perhaps many—of these younger surfaces are underlain by sedimentary rocks.

Some of these sedimentary rocks are composed of silicate minerals derived from erosion of old basaltic lavas and the pulverized basaltic rocks of the ancient cratered terrains. For example, the meandering stream deposits visible in Figure 9.21 may have formed largely by the accumulation of basaltic sediments. However, in most, if not all, of the sedimentary rocks beneath Meridiani Planum, where *Opportunity* has been exploring, sulfate minerals—which are chemical sediments—are mixed with silicate minerals. The sulfate minerals must have been precipitated

FIGURE 9.27 ▪ The first outcrop studied on another planet (Mars). This outcrop is made of sedimentary rocks formed partly from sulfate minerals, including jarosite. Jarosite can form only in water—and only in acid-rich water. The area shown in the photograph is about 50 cm in width. [NASA/JPL/Cornell.]

when water evaporated, probably in shallow lakes or seas. The water must have been very salty to precipitate these minerals, and it must have contained common sulfate minerals such as gypsum ($CaSO_4$). In addition, the presence of unusual sulfate minerals such as *jarosite* (**Figure 9.27**)—an

iron-rich sulfate mineral—tells us that the water must have been very acidic. On Mars, sulfuric acid probably formed when the abundant basaltic rocks interacted with water and were weathered, releasing their sulfur. The acid-rich water then flowed through rocks, heavily fractured from impacts, and over the surface to accumulate in lakes or shallow seas, where jarosite precipitated as a chemical sediment.

As we have seen in Chapters 5 and 8, sedimentary rocks are valuable records of Earth's history. The vertical succession of sedimentary rocks—their *stratigraphy*—tells us how environments change over time. One of the most exciting findings of the Mars Exploration Rovers mission so far has been the discovery of a stratigraphic record at Endurance Crater. Because the crater is so large, there is a lot of outcrop to observe, and it is mostly unaffected by the crater-forming impact. **Figure 9.28** shows the outcrop that contains all the stratigraphic clues. By using *Opportunity* to measure each layer, geologists were able to create a high-resolution stratigraphy (Figure 9.28b), the first of its kind generated for another planet. Remarkably, this interpretive drawing—from a planet 300 million miles away—provides the same level of understanding that is typically obtained here on Earth (as, for example, in Figure 5.15). *Curiosity* is now doing similar stratigraphic work at Gale crater and also hoping to learn about the time-ordered sequence of events that characterized the early environmental evolution of Mars.

Perhaps one day we will have enough understanding of the stratigraphy of Mars to be able to correlate sedimentary and volcanic rocks from one part of the planet with those from another. To do this, we will need to link observations

(a)   (b)

FIGURE 9.28 ▪ A sedimentary sequence exposed along the flank of Endurance Crater, photographed by the rover *Opportunity*. (a) An interpretive drawing showing each stage in the history of the outcrop. (b) The vertical succession of layers in the outcrop preserves an excellent record of early Martian environments. [NASA/JPL/Cornell.]

by the rovers on the ground with observations provided by orbiters overhead. The recent orbiters have shown that sulfates as well as clay minerals are abundant in several places on Mars—particularly in the Vallis Marineris, where they may form deposits up to several kilometers thick. This observation leads us to believe that their formation was related to a process that occurred globally, possibly over a long time. There is some evidence suggesting that the clay minerals may have formed before the sulfate minerals. However, we do not yet know whether these deposits were formed all at the same time, signaling a global environmental event that may have been unique in Mars's history, or whether they were formed at multiple times in different places. *Curiosity*'s initial discoveries at Yellowknife Bay, Gale crater, hints at this latter possibility. The conclusion would point to a more common process that operated throughout Mars's history wherever and whenever local conditions allowed.

The evidence is now compelling that at some point in Mars's history, there was liquid water on its surface and underground. The planet must have been warmer than it is today, unless the water was very short-lived, gushing to the surface briefly, then evaporating quickly or sinking back underground before being frozen, as would happen today. There are many questions left to answer. How much water was there? How long did it last? Did it ever rain, or was it all groundwater leaking to the surface? Did the water last long enough, and have the right composition, to allow life to get started? Only one thing is certain at this point: more missions are required to answer these questions.

▪ Exploring the Solar System and Beyond

An astronomer staring through a telescope is the first image that comes to mind when most people think of exploring the solar system. But most modern telescopes have no eyepiece at all, and instead record their images with digital cameras. Many telescopes, such as the Hubble Space Telescope, are not even located on Earth, but are positioned in space.

Regardless of how a telescope takes its photographs or where it is stationed, its purpose is the same: to gather more light than we can with the naked human eye. Its photographs can be processed to increase their brightness further or enhance their contrast; such techniques can reveal important planetary surface features such as craters and canyons. All the geologic surface features of the planets discussed so far in this chapter have been studied in this way.

However, the light gathered by telescopes and digital cameras such as those on *Spirit* and *Opportunity* can also be studied using a second technique. Once we have a record of the light coming from an object of interest—say, a planet,

a star, or an outcrop—we can study its spectrum. We are all familiar with how sunlight, when passed through a prism, splits into a rainbow of colors called its *spectrum* (plural *spectra*). The light generated by a star, or reflected off the surface of a planet or outcrop, also produces a spectrum. The colors of that spectrum can reveal the chemical composition of the light-producing or light-reflecting materials. Thus, geologists can look at the spectrum of light reflected from a planet and know which gases are in its atmosphere and which chemicals and minerals are in its rocks and soils.

Astronomers use this same principle to look at the light coming from faraway stars and galaxies. The spectra they see tell them the ages of those stars and galaxies, reveal how they evolved, and even provide mind-boggling insights into the origin and evolution of the universe.

Space Missions

Most observations of our solar system and beyond are still made from Earth. Over the past 50 years, however, we have sent all kinds of machines, robots, and even humans into space in our quest to explore the unknown. Space missions are a costly business, requiring a tremendous effort by hundreds and sometimes thousands of people at a cost of hundreds of millions to billions of dollars. The Mars Exploration Rovers mission cost on the order of $800 million for both rovers, and *Mars Science Laboratory*—about the size of a car—cost over $2 billion. And space missions are a risky business: fewer than half the missions sent to Mars have succeeded. As the space shuttle program reminds us, space exploration is risky for humans, too.

Are these efforts, costs, and risks worth it? For thousands of years, people have looked up at the skies and pondered the universe. What are the stars and planets made of? How did the universe form? Is there any life out there? To answer these questions, we have to look for clues, and most of those clues will be provided only by missions to space. The issue is not so much whether or not to explore space, but how. Most debates focus on whether it is essential to send humans into space or whether the Mars rovers missions have demonstrated the adequacy of robots.

We are actively exploring space in many different ways. Spacecraft have been sent to orbit planets, moons, and asteroids and to fly by planets and comets in the outer solar system and beyond. On other occasions, we have instructed landers and other probes to descend to planetary surfaces and make direct measurements of rocks, minerals, gases, and fluids. On July 3, 2005, a probe was released from the *Deep Impact* spacecraft and instructed to collide deliberately with the comet Tempel 1. The depth of the resulting crater and the light emitted at the time of the collision (**Figure 9.29**) revealed what the interior of the comet is made of. The comet was found to consist of a mixture of dust and ice; the dust component included clay, carbonate, and silicate minerals and was enriched in sodium, which is rare in space.

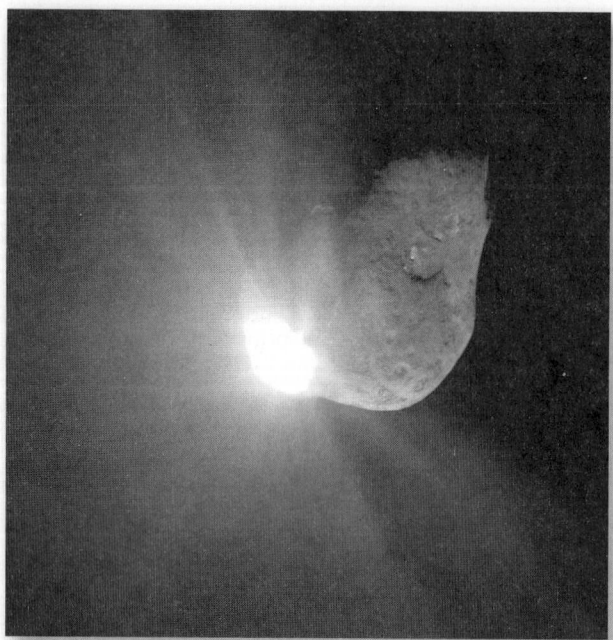

FIGURE 9.29 ■ The first moments after *Deep Impact*'s probe collided with the comet Tempel 1. Debris from the interior of the comet is expanding from the impact site. [NASA/JPL-Caltech/UMD.]

The Cassini-Huygens Mission to Saturn

An even more remarkable story of deep space exploration involves the Cassini-Huygens mission. In 2005, the *Huygens* lander became the farthest-traveled spacecraft to reach another planet and "live" to tell about it.

Cassini-Huygens is one of the most ambitious missions ever launched into space. The *Cassini-Huygens* spacecraft includes two components: the *Cassini* orbiter and the *Huygens* lander. The spacecraft was launched from Earth on October 15, 1997. After traveling over a billion kilometers across deep space in almost 7 years, *Cassini-Huygens* sailed through the rings of Saturn on July 1, 2004. Saturn's beautiful rings are what set it apart from the other planets (**Figure 9.30a**). It is the most extensive and complex ring system in the solar system, extending hundreds of thousands of kilometers from the planet. Made up of billions of particles of ice and rock—ranging in size from grains of sand to houses—the rings orbit Saturn at varying speeds. Understanding the nature and origin of these rings is a major goal of the Cassini-Huygens scientists.

On December 24, 2004, the *Huygens* lander was released from the orbiter and traveled over 5 million kilometers to

(a)

(b)

FIGURE 9.30 ■ (a) Saturn and its rings completely fill the field of view in this natural color image taken by the *Cassini-Huygens* spacecraft on March 27, 2004. Color variations in the rings reflect differences in the composition of the materials that make them up, such as ice and rock. The Cassini-Huygens scientists will investigate the nature and origin of the rings as the mission progresses. (b) The surface of Titan is strewn with "rocks" of ice composed of frozen methane and other carbon-containing compounds. [(a) NASA/ JPL/SSI/ESA/University of Arizona; (b) ESA/NASA/University of Arizona.]

Google Earth Project

ESA / DLR / FU Berlin (G.Neukum) Image NASA / JPL / University of Arizona © 2009 Google

With current technology, we are able to send "robotic geologists" to the surface of Mars to study its current and ancient environments until humans can go there one day. These rovers have many of the tools that human geologists would use, including cameras, microscopes, spectrometers, and tools to sample the rocks. The two Mars Exploration Rovers, *Spirit* and *Opportunity*, were launched in 2003 and landed on Mars in early 2004. In this exercise, we will use Google Mars to follow in *Opportunity*'s tracks as we explore Meridiani Planum, the region where it landed.

Open Google Earth and click on the planet icon at the top (that looks like Saturn). Select "Mars." In the "Fly To" search window on the left, type in "*Opportunity*." Google Mars will navigate to the area where *Opportunity* landed, and which it has been traversing since 2004.

As your cursor hovers over the Google Mars screen, a hand will appear. You can use the hand to click and drag the region of interest around. Click and drag so that the screen is centered on *Opportunity*'s landing site (marked by an American flag). Zoom in to see the traverse route the rover has taken (marked by a red line).

LOCATION   Meridiani Planum, the region of Mars where the rover *Opportunity* landed

GOAL   Use Google Mars tools to measure traverse distances and decipher crater morphology

LINKED   Figure 9.23

1. Where did *Opportunity* land? (You may have to zoom in on the landing site to see place names.)

    a. Eagle Crater   *c.* Victoria Crater
    b. Endurance Crater   *d.* Gusev Crater

2. What is the latitude and longitude of *Opportunity*'s landing site?

    a. 2°03′05″ S, 5°29′44″ W
    b. 1°56′51″ S, 5°30′30″ W
    c. 1°56′42″ S, 5°31′16″ W
    d. 2°25′01″ S, 5°30′10″ W

3. Which is the largest crater *Opportunity* has explored?

    a. Erebus Crater   *c.* Endeavour Crater
    b. Victoria Crater   *d.* Endurance Crater

4. In which direction overall has the rover driven?

    a. East   *c.* North
    b. West   *d.* South

5. What might you expect to see on the floor and in the walls of Victoria Crater?

    a. Dunes on the floor, rock outcrops in the walls
    b. Dunes on the floor, piles of dust in the walls
    c. Alien footprints on the floor, ancient riverbeds in the walls
    d. The spacecraft heat shield on the floor, plant roots in the walls

Optional Challenge Question

6. Using the GE ruler tool, select "path," and then select "meters" as the unit of measure. Draw a pathway following the traverse *Opportunity* has taken from Eagle Crater to the western rim of Victoria Crater (red line). (Ignore the rover paths within and around craters.) What is the approximate distance the rover traversed?

    a. 10,000 m   *c.* 4500 m
    b. 7300 m   *d.* 2700 m

reach Titan, one of Saturn's 18 moons. On January 14, 2005, it reached Titan's upper atmosphere, where it deployed a parachute and then plunged to the surface, where it landed successfully. Its cameras showed surface features that appear to be drainage networks, similar to those seen on Earth and Mars. The landing site was strewn with rocks up to 10–15 cm in diameter (Figure 9.30b). However, these "rocks" are probably ices made of methane (CH_4) and other organic compounds.

Bigger than the planet Mercury, Titan is of particular interest to scientists because it is one of the few moons in the solar system with its own atmosphere. It is cloaked in a thick, smog-like haze that scientists believe may be similar to Earth's atmosphere before life began more than 3.8 billion years ago (see Chapter 11). Organic compounds, including gases composed of methane, are plentiful on Titan. Further study of this moon promises to reveal much about planetary formation and, perhaps, about the early days of Earth.

Other Solar Systems

For ages, scientists and philosophers have speculated that there may be planets around stars other than our Sun. In the 1990s, astronomers discovered planets orbiting nearby Sun-like stars. In 1999, the first family of **exoplanets**—planets that lie outside the solar system—was found. These planets are too dim to be seen directly by telescopes, but their existence can be inferred from their slight gravitational pull on the stars they orbit, which causes to-and-fro movements of the star that can be measured. We see these movements recorded in the spectra of the starlight. Most planets found in this way are Jupiter-sized or larger and are close to their parent stars—many within scorching distance. Earth-sized planets have been discovered in recent years using other methods. By early 2009, astronomers had discovered over 300 new planets, organized in 249 solar systems. Spacecraft above Earth's atmosphere are able to search for the dimming of the parent star's light as an orbiting planet passes in front of it along the line of sight to Earth.

We are fascinated by planetary systems around other stars because of what they might teach us about our own origins. Our overriding interest, however, is in the profound scientific and philosophical implications posed by the question, "Is anyone else out there?" In about 15 years, a spacecraft named *Life Finder* could be carrying instruments to analyze the atmospheres of exoplanets in our galaxy for signs of the presence of some kind of life. Based on what we know about biological processes, life on an exoplanet would probably be carbon-based and require liquid water. The benign temperatures we enjoy on Earth—not too far outside the range between the freezing and boiling points of water—appear to be essential for life (see Chapter 11). An atmosphere is needed to filter harmful radiation from the parent star, so the planet must be large enough for its gravitational field to keep the atmosphere from escaping into space. For a habitable planet with complex life *as we know*

it to exist would require conditions even more limiting. For example, if the planet were too massive, delicate organisms such as humans would be too weak to withstand its gravitational force. Are these requirements too restrictive for life to exist elsewhere? Many scientists think not, considering the billions of Sun-like stars in our own galaxy.

SUMMARY

How did the solar system originate? According to the nebular hypothesis, the Sun and its planets formed when a cloud of gases and dust, known as the solar nebula, condensed about 4.5 billion years ago. The terrestrial inner planets, including Earth, differ from the giant outer planets in their composition.

How did Earth form and develop over time? Earth probably grew by the accretion of colliding planetesimals. Soon after it formed, it was struck by a large body about the size of Mars. Matter ejected into space from both Earth and the impacting body reassembled to form the Moon. The impact generated enough heat to melt most of what remained of Earth. Radioactivity and gravitational energy also contributed to this early heating and melting. Heavy matter, rich in iron, sank toward Earth's center to form the core, and lighter matter floated upward to form the crust. Still lighter gases formed Earth's atmosphere and oceans. In this way, Earth was transformed into a differentiated planet with distinct layers.

What are some major events in the early history of the solar system? The age of the solar system, as determined from isotopic dating of meteorites, is about 4.56 billion years. Earth and the other terrestrial planets had formed within about 10 million years. The impact that formed the Moon occurred about 4.51 billion years ago. Minerals as old as 4.4 billion years have survived in Earth's crust. The Late Heavy Bombardment, which peaked about 3.9 billion years ago, marks the end of the Hadean eon on Earth.

How can planetary surfaces be dated? Rocks returned from the surface of the Moon by the Apollo missions have been dated using isotopic methods. The lunar highlands show ages from 4.4 billion to about 4.0 billion years. The lunar maria show ages from 4.0 billion to 3.2 billion years. These isotopic ages allowed geologists to calibrate the relative time scale they had developed by crater counting.

Do other planets have plate tectonic systems? Venus is the only planet, other than Earth, that has features indicating tectonic activity resulting from mantle convection. But Venus does not appear to have thick lithospheric plates. Instead, it has a thin crust of frozen lava that breaks up into flakes or crumples like a rug as it is pushed and stretched by vigorous convection currents. This process, which geologists refer to as flake tectonics, may have occurred on Earth when it was younger and hotter.

How have Mars and the other planets been explored? Four types of spacecraft have been used to explore Mars and other planets. During a flyby, a spacecraft comes close to a planet only once. An orbiter circles a planet, making remote observations of its surface and interior. A lander can actually touch down on the surface of a planet to make local observations. A rover can leave the landing site and travel up to several kilometers to investigate new terrains.

Does water exist on Mars? Today, water is present on Mars only as ice caps at the Martian poles and as permafrost. In the past, it may have been present as a liquid, as geologic evidence shows that it ran across the surface to carve stream channels and deposit sediments in meandering streams. It also accumulated in shallow lakes or seas, where it evaporated and precipitated a variety of chemical sediments, including sulfate minerals.

How do we use light in exploring stars and the solar system? In some cases, we can use enhanced photographs from telescopes, which may reveal surface features of distant objects. In other cases, we can use information from the spectrum of light, which varies depending on the composition of the object that produces or reflects that light.

Is our solar system unique? We have evidence of more than 300 planets that circle other stars. In several cases, there is more than one planet in these solar systems. Because these new planets lie outside our solar system, they are called exoplanets.

KEY TERMS AND CONCEPTS

asteroid (p. 224)

dwarf planet (p. 229)

exoplanet (p. 248)

flake tectonics (p. 235)

gravitational differentiation (p. 225)

Heavy Bombardment (p. 230)

meteorite (p. 224)

nebular hypothesis (p. 222)

planetesimal (p. 223)

solar nebula (p. 223)

terrestrial planet (p. 224)

PRACTICING GEOLOGY EXERCISE

How Do We Land a Spacecraft on Mars? Seven Minutes of Terror

When we send a lander to Mars, how do we decide where to land it? The riskiest part of such a mission comes when the spacecraft enters the Martian atmosphere, descends through it, and lands on the planet's surface. This step, called *Entry, Descent, and Landing*, or "EDL," takes about 7 minutes. During that time, the lander decelerates from 12,000 to 0 miles per hour, and its heat shield becomes as hot as the surface of the Sun (about 1500°C) due to friction caused by the atmosphere. A lot can go wrong here, so EDL has been called "seven minutes of terror."

The shape and elevation of the Martian surface play key roles in lander design. The lander holds a limited amount of fuel to power its engines, so if the land surface varies too much in its elevation, the lander has to spend time (and fuel) maneuvering. Your task as the EDL team geologist is to choose a safe landing site—one that does not vary too much in elevation. At the same time, you will want to choose a site that provides interesting outcrops for the lander or rover to study. So your problem is to determine how much variation in elevation is "too much."

To solve this problem, we need the following information: The lander's engines start up when the radar determines that the lander is 1000 m above the Martian surface.

The engines slow the lander's descent, allowing it to descend at a rate of 50 meters per second (m/s) until it is 10 m above the ground. At that point, the lander descends at 2 m/s until touchdown. The engine's fuel consumption rate is 5 liters per second (L/s). The fuel tank holds 150 L.

First, how long does it take for the lander to descend to the surface? Note that two rates must be used here: one for the first 990 m of descent, and the other for the last 10 m of descent.

$$\text{time} = \text{distance} \div \text{lander descent rate}$$
$$= 900 \text{ m} \div 50 \text{ m/s}$$
$$= 20 \text{ s}$$

$$\text{time} = \text{distance} \div \text{lander descent rate}$$
$$= 10 \text{ m} \div 2 \text{ m/s}$$
$$= 5 \text{ s}$$

$$\text{total time} = 20 \text{ s} + 5 \text{ s} = 25 \text{ s}$$

Next, how much fuel is consumed during landing?

$$\text{fuel consumption} = \text{time} \times \text{fuel consumption rate}$$
$$= 25 \text{ s} \times 5 \text{ L/s}$$
$$= 125 \text{ L}$$

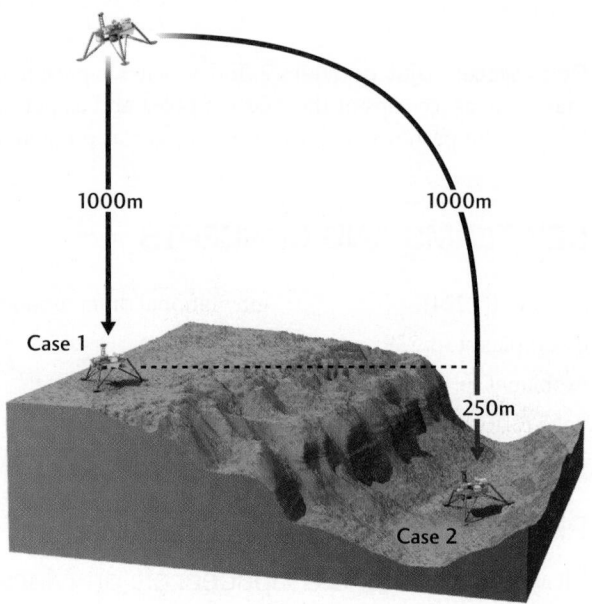

(*above*) Engineers building the Phoenix lander, which arrived at the surface of Mars in 2008. (*below*) Successful landing on the surface of Mars requires careful planning and consideration of the geologic environment of the surface, including variations in topography. [John Grotzinger.]

Given that 150 L of fuel are available, but only 125 L are used, there would be a reserve of 25 L remaining after landing. This calculation is for the "perfect" landing condition in which the total descent distance is 1000 m (see "Case 1" in the accompanying figure).

Now let's consider what would happen if the lander drifted sideways while descending because the wind was blowing and moved over a low spot on the Martian surface (see "Case 2" in the accompanying figure). In this case, the total descent distance would be greater than 1000 m. If the low spot were too low, the lander would be at risk of depleting all its fuel reserves before it ever landed and crashing to the surface. Therefore, we need to determine how much elevation change would use up this fuel reserve.

First, we need to determine how much reserve time is provided by the 25 L of reserve fuel:

time = reserve fuel volume ÷ fuel consumption rate
= 25 L ÷ 5 L/s = 5 s

Now we can determine the additional descent distance that could be safely traveled before the fuel reserve was used up:

descent distance = reserve time × lander descent rate
= 5 s × 50 m/s
= 250 m

The solution tells us how much elevation change is "too much" to tolerate for a safe landing site: anything more than 250 m is too much. The team geologist must find a landing site where elevation varies less than 250 m, yet which also provides interesting geologic features. In practice, there is a real trade-off between geologic interest and landing site safety.

BONUS PROBLEM: Determine the maximum variation in elevation at a landing site that could be tolerated by a lander with a fuel tank volume of 200 L. How much variation could be accepted if the final descent rate were 1 m/s rather than 2 m/s?

EXERCISES

1. How and why do the inner planets of the solar system differ from its outer planets?

2. What caused Earth to differentiate into a layered planet, and what was the result?

3. Mercury's average density is less than Earth's, but the relative size of its core is larger. How can you explain this?

4. What aspects of the geology of Earth and the Moon are consistent with high impact rates during the Late Heavy Bombardment?

5. What surface features would you look for on Mars if you were searching for evidence of liquid water in its geologic past?

6. What are the advantages and disadvantages of spacecraft that fly by or orbit a planetary body versus those that land on or move around on its surface?

THOUGHT QUESTIONS

1. If a giant impact such as the one that formed the Moon had occurred after life had arisen on Earth, what would have been the consequences?

2. If you were an astronaut landing on an unexplored planet, how would you decide whether the planet was differentiated and whether it was tectonically active?

3. Knowing how the Moon formed, what might you expect as a result if you were told that a large meteorite had collided with a planet twice its size? What could be the effect of this collision on the interior composition of this planet? How would the result of the impact differ if the meteorite were significantly smaller than the planet?

4. During a dust storm on Mars, sediments fill the atmosphere with dust. But Mars has an atmosphere much thinner than Earth's. To move sand, would the wind have to blow faster on Mars to compensate for this difference?

5. Many scientists think that water is present on Mars. Today it is frozen, but 4 billion years ago, it may have been liquid. What happened? Describe all the possible mechanisms for this change. What evidence would you search for to help decide among these possibilities?

6. How does the discovery of planets orbiting other stars contribute to the debate about the possibility of life elsewhere in the cosmos? What are the scientific and philosophical implications of the existence of life on the planets of other stars?

7. What might be the advantages and disadvantages of living on a differentiated planet? On a tectonically active planet?

MEDIA SUPPORT

 9-1 Animation: Solar System Formation and the Origin of the Moon

 9-1 Video: Barringer Meteorite Crater

- The Structure of North America   254

- Tectonic Provinces Around the World   259

- How Continents Grow   262

- How Continents Are Modified   266

- The Origins of Cratons   273

- The Deep Structure of Continents   276

Shaded relief map of the North American continent. [U.S. Geological Survey.]

HISTORY OF THE CONTINENTS

NEARLY TWO-THIRDS OF EARTH'S SURFACE—all of its oceanic crust—was created by seafloor spreading over the past 200 million years, an interval that spans a mere 4 percent of Earth's history. The stories of earlier events are entirely contained in the continental crust. Thus, to understand how Earth has evolved since its fiery beginnings, we must look to the continents, which contain rocks over 4 billion years old.

The geologic record of the continental crust is very complex, but our ability to read that record has improved immensely in just the last few years. Earth scientists use the theory of plate tectonics to interpret eroded mountain belts and ancient rock assemblages in terms of closing ocean basins and colliding continents. New geochemical tools, such as isotopic dating, help us decipher the history of continental rocks. We can now image the structure of the continents far below Earth's surface using networks of seismographs and other sensors.

In this chapter, we will describe the structure of Earth's continents, and we will look back through their 4-billion-year history to see what it tells us about the processes that formed them—and are still modifying them today.

We will see how plate tectonic processes have added new material to the continental crust, how plate convergence has thickened that crust into mountain belts, and how those mountains have been eroded to expose the metamorphic basement rocks found in many older regions of the continents. Then we will reach back into the earliest period of continental evolution, the Archean eon (3.9 billion to 2.5 billion years ago), to ponder two of the great puzzles of Earth's history: how did continents form, and how have they survived through billions of years of plate tectonics and continental drift?

Continents, like people, show a great variety of surface features that reflect their origins and experience over time. Yet, also like people, continents share many similarities in their basic structure and growth patterns. Before considering continents in general, let's begin by outlining the major features of one particular continent: North America.

■ The Structure of North America

The long-term tectonic history of North America is reflected in its **tectonic provinces**—large-scale regions formed by particular tectonic processes (**Figure 10.1**).

The oldest parts of North America's crust, built during the most ancient episodes of deformation, tend to be found in the northern interior of the continent. These regions, which include most of Canada and the closely connected landmass of Greenland, are *tectonically stable*. In other words, they have remained largely undisturbed by recent episodes of continental rifting, drift, and collision, and they have been eroded nearly flat. On the edges of these older tectonic provinces are younger metamorphic belts where most of the present-day mountain chains are found. These mountain chains form elongated topographic features near the margins of the continent. The two main examples are the *North American Cordillera,* which runs down the western edge of North America and includes the Rocky Mountains, and the *Appalachian fold belt,* which runs southwest to northeast on the continent's eastern margin. In our description of tectonic provinces, we will often refer to the geologic time scale shown in Figure 8.15, so you might want to bookmark this figure for reference.

The Stable Interior

Much of central and eastern Canada is a landscape of very old crystalline basement rock—a huge tectonic province (8 million km^2) called the *Canadian Shield* (**Figure 10.2**). It consists primarily of Precambrian granitic and metamorphic rocks, such as gneisses, together with highly deformed and metamorphosed sedimentary and volcanic rocks, and

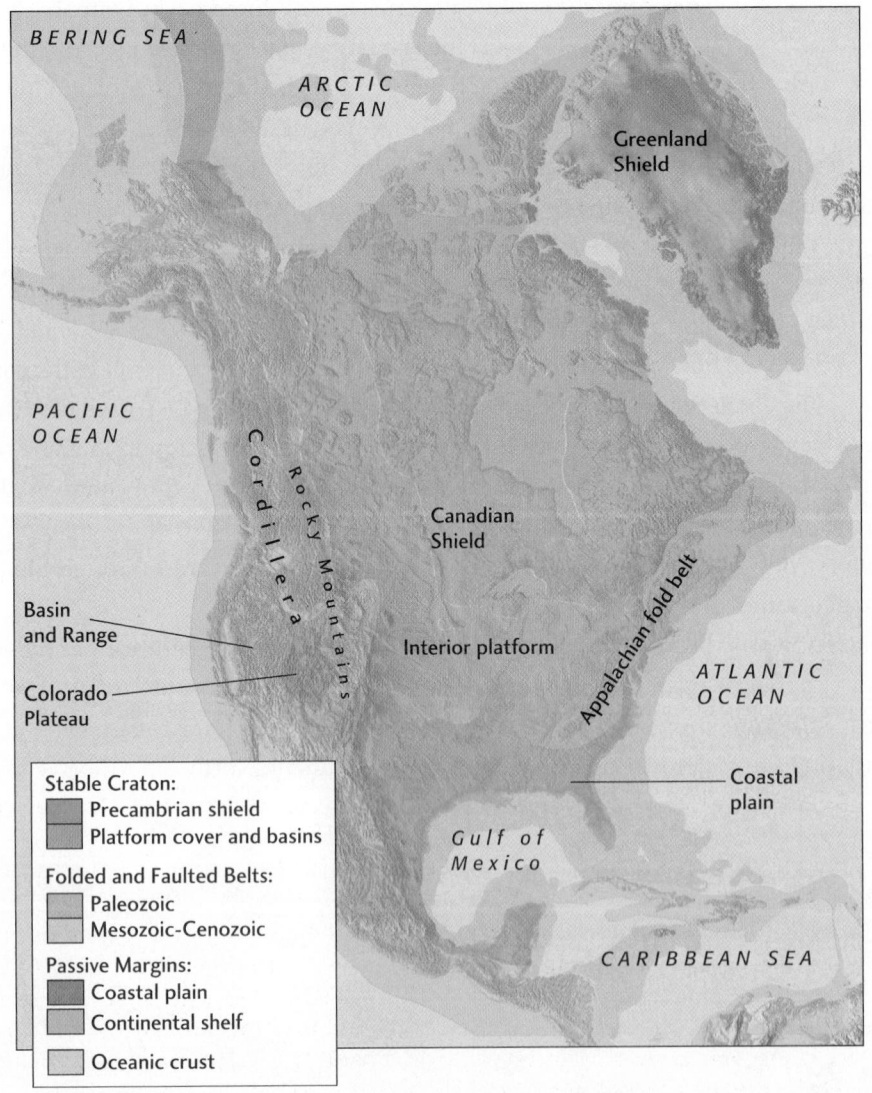

FIGURE 10.1 ■ The major tectonic provinces of North America reflect the processes that formed the continent.

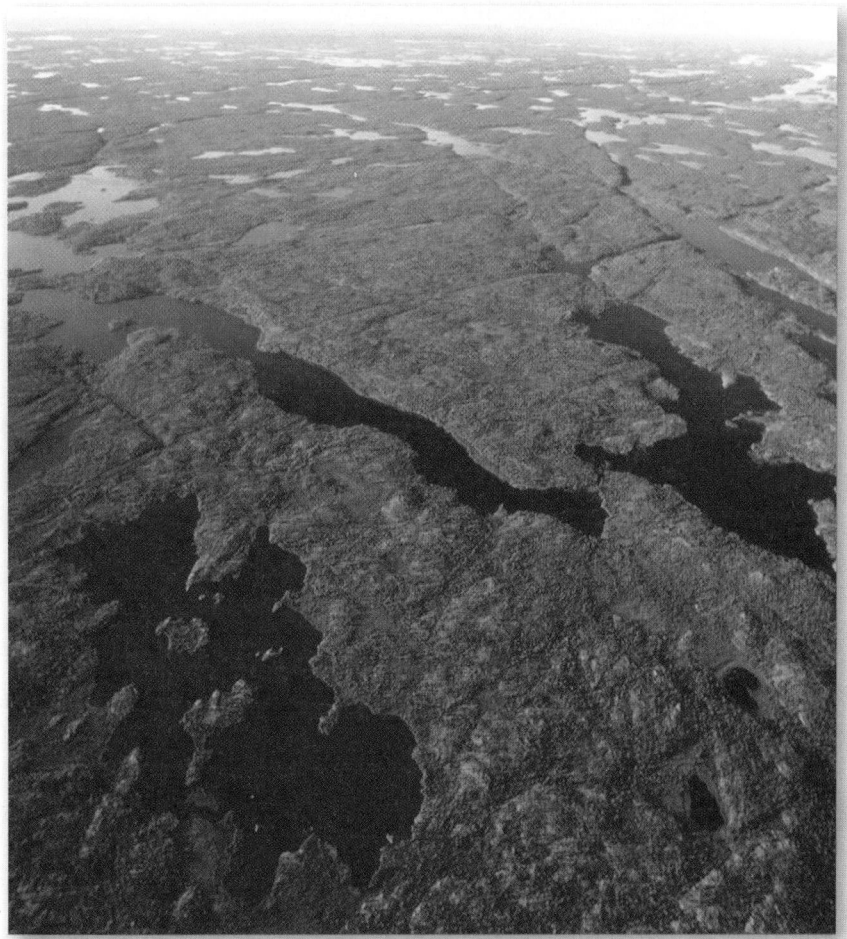

FIGURE 10.2 ■ An aerial view of ancient, eroded metamorphic basement rocks in Nunavut, Canada, exposed on the surface of the Canadian Shield in the Northwest Territories. [Paolo Koch/Science Source.]

it contains major deposits of iron, gold, copper, diamond, and nickel. Large portions of the shield were formed during the Archean eon, so it represents one of the oldest records of Earth's history. The nineteenth-century Austrian geologist Eduard Suess named these areas continental **shields** because they emerge from the surrounding sediments like a shield partly buried in the dirt of a battlefield.

In North America, extensive flat-lying (*platform*) sediments have been deposited on stable continental crust around the periphery of the Canadian Shield and near its center, beneath Hudson Bay (see Figure 10.1). The vast low-lying, sediment-covered region south and west of the Canadian Shield, which includes the Great Plains of Canada and the United States, is called the *interior platform*. The Precambrian basement rocks of the interior platform are a continuation of the Canadian Shield, although here they lie under nearly flat layers of Paleozoic sedimentary rocks, typically less than 2 km thick.

The North American platform sediments were laid down on the deformed and eroded Precambrian basement under a variety of conditions. Some rock formations

(marine sandstones, limestones, shales, deltaic deposits, evaporites) indicate sedimentation in extensive shallow inland seas. Others (nonmarine sediments, coal deposits) indicate deposition on floodplains or in lakes or wetlands.

Within the interior platform are a number of circular structures: broad *sedimentary basins,* roughly circular or oval depressions where the sediments are thicker than in the surrounding areas, and *domes,* areas where the platform sediments have been uplifted and eroded to expose the basement rock (**Figure 10.3**). Most of the basins are thermal subsidence basins; that is, they are regions that subsided when heated portions of the lithosphere cooled and contracted (see Chapter 5). An example is the Michigan Basin, a circular area of about 200,000 km² that covers most of the Lower Peninsula of Michigan (see Figure 7.15). This basin subsided throughout much of the Paleozoic era and received sediments more than 5 km thick in its central, deepest part. The sandstones and other sedimentary rocks of these basins, laid down under tectonically quiet conditions, have remained unmetamorphosed and only slightly deformed to this day. The interior platform basins contain

FIGURE 10.3 ■ A map of the interior platform of North America, showing its basin and dome structure. The basins are nearly circular regions of thick sediments. The domes are regions where the sediments are anomalously thin. Basement rocks are exposed on the tops of some domes, such as the Black Hills uplift and the Ozark Dome.

important deposits of uranium, coal, oil, and natural gas. Rich mineral deposits in the basement rocks lie close to the surface in the domes, and they may also become traps for oil and gas.

The Appalachian Fold Belt

Along the eastern side of North America's stable interior are the old, eroded Appalachian Mountains. This classic fold and thrust belt, which we first examined in Chapter 7, extends along eastern North America from Newfoundland to Alabama. The rock assemblages and structures of the Appalachians resulted from the continent-continent collisions that formed the supercontinent Pangaea 470 million to 270 million years ago. The western side of the Appalachian belt is bounded by the *Allegheny Plateau,* a region of slightly uplifted, mildly deformed sediments that is rich in coal and oil. Moving eastward, we encounter regions of increasing deformation (**Figure 10.4**):

■ *Valley and Ridge province.* Thick Paleozoic sedimentary rocks laid down on an ancient continental shelf were folded and thrust to the northwest by compressive forces from the southeast. The rocks show that the deformation occurred in three mountain-building episodes, one beginning in the middle Ordovician period (about 470 million years ago), one in the middle to late Devonian period (380 million to 360 million years ago), and one in the late Carboniferous and early Permian periods (320 million to 270 million years ago).

■ *Blue Ridge province.* These eroded mountains are composed largely of highly metamorphosed Precambrian and Cambrian crystalline rocks. The Blue Ridge rocks were not intruded and metamorphosed in place, but rather thrust as sheets over the sedimentary rocks of the Valley and Ridge province near the end of the Paleozoic era, about 300 million years ago.

■ *Piedmont.* This hilly region contains metamorphosed Precambrian and Paleozoic sedimentary and volcanic rocks intruded by granite, all now eroded to low relief. Volcanism began in late Precambrian time and continued into the Cambrian. The Piedmont rocks were thrust over Blue Ridge rocks along a major thrust fault, overriding them to the northwest. At least two episodes of deformation are evident, coinciding with the last two mountain-building episodes in the Valley and Ridge province.

The Coastal Plain and Continental Shelf

On the Atlantic coastal plain, east of the Appalachian fold belt, relatively undisturbed sediments of Jurassic age and younger are underlain by rocks similar to those of the Piedmont. The coastal plain and its offshore extension, the continental shelf (see Figure 10.1), began to develop in the Triassic period, about 180 million years ago, with the rifting that preceded the breakup of Pangaea and the opening of the modern Atlantic Ocean. Rift valleys

FIGURE 10.4 ■ The Appalachian fold belt province, shown in an aerial view to the northeast and an idealized cross section. The intensity of deformation increases from west to east. [After S. M. Stanley, *Earth System History.* New York: W. H. Freeman, 2005. Aerial view from NASA.]

formed basins that trapped a thick series of nonmarine sediments. As these deposits were accumulating, they were intruded by basaltic sills and dikes. The Connecticut River valley and the Bay of Fundy are such sediment-filled rift valleys.

In the early Cretaceous period, as seafloor spreading widened the Atlantic Ocean, the deeply eroded, sloping surface of the Atlantic coastal plain and continental shelf began to cool, subside, and receive sediments from the continent. Cretaceous and Tertiary sediments as much as 5 km thick filled this slowly developing thermal subsidence basin, and even more material was dumped into the deeper water at the continental margin. This still active basin continues to receive sediments. If the present stage of opening of the Atlantic is reversed some millions of years from now,

the sediments in this basin will be folded and faulted in the same kind of process that produced the Appalachians.

The coastal plain and continental shelf of the Gulf of Mexico are continuous extensions of the Atlantic coastal plain and shelf, interrupted only briefly by the Florida Peninsula, a large carbonate platform. The Mississippi, Rio Grande, and other rivers that drain the interior of the North American continent have delivered enough sediments to fill a basin some 10 to 15 km deep running parallel to the coast. The Gulf coastal plain and shelf are rich reservoirs of petroleum and natural gas.

The North American Cordillera

The stable interior platform of North America is bounded on the west by a younger complex of mountain ranges and deformation belts (**Figure 10.5**). This region is part of the North American Cordillera, a mountain belt extending from Alaska to Guatemala, and it contains some of the highest peaks on the continent. Across its middle section, between San Francisco and Denver, the Cordilleran system is about 1600 km wide and includes several different tectonic provinces: the Coast Ranges along the Pacific Ocean; the lofty Sierra Nevada; the Basin and Range province; the high tableland of the Colorado Plateau; and the rugged Rocky Mountains, which end abruptly at the edge of the Great Plains on the stable interior platform.

The history of the Cordillera is a complicated one that involves interactions among the Pacific, Farallon, and North American plates over the past 200 million years. Before the breakup of Pangaea, the Farallon Plate occupied most of the eastern Pacific Ocean. As North America moved westward, most of this plate's oceanic lithosphere was subducted eastward under the continent. The westward margin of the continent swept up island arcs and continental fragments, and the subduction zone eventually swallowed portions of the Pacific-Farallon spreading center, which converted the convergent boundary into the modern San Andreas transform-fault system (**Figure 10.6**). Today, all that is left of the Farallon Plate are small remnants, including the Juan de Fuca and Cocos plates, which are still subducting beneath North America.

The main phase of Cordilleran mountain building occurred in the last half of the Mesozoic era and in the early Paleogene period (150 million to 50 million years ago). The Cordilleran system is topographically higher than the Appalachians, which is not surprising, as there has been less time for erosion to wear it down. The form and height of the Cordillera that we see today resulted from even more recent events in the Neogene period, over the past 15 million or 20 million years, when the Pacific plate first encountered North America (see Figure 10.6). During these periods, the mountains underwent **rejuvenation;** that is, they were raised again and brought back to a more youthful stage. At that time, the central and southern Rockies attained much of their present height as a result of broad regional uplift. The Rockies were raised 1500 to 2000 m as Precambrian basement rocks and their veneer of later-deformed sediments were pushed above the level of their surroundings. Stream erosion accelerated, the mountain

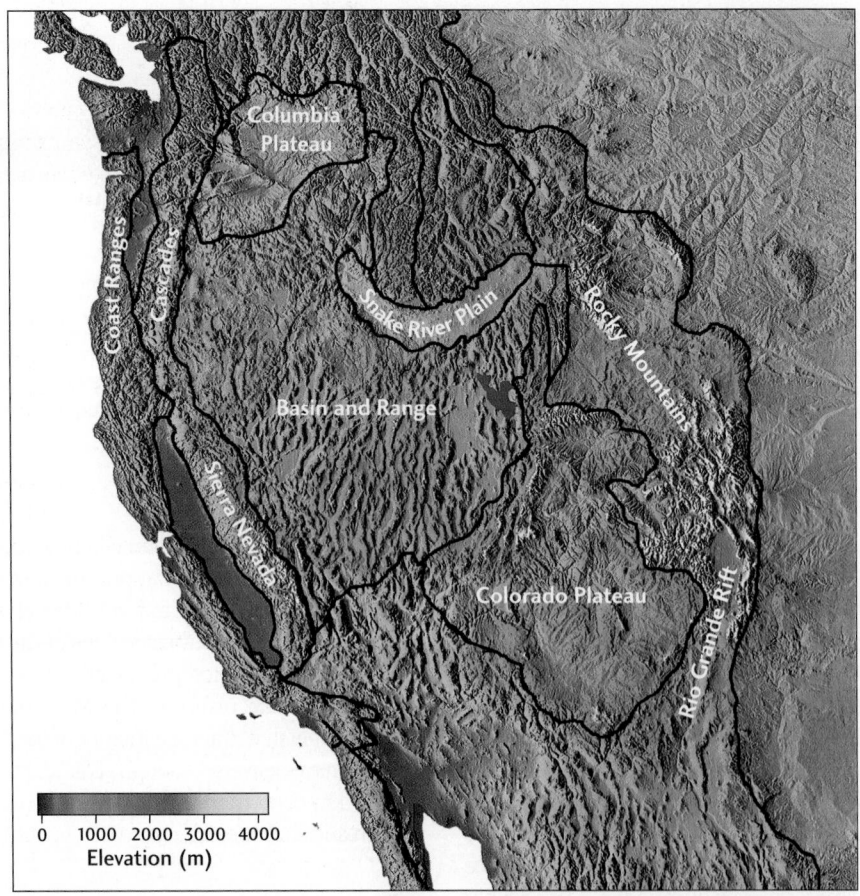

FIGURE 10.5 ■ Topography of the North American Cordillera in the western United States. Computer manipulation of digitized elevation data produced this color shaded relief map. The major tectonic provinces of the area are clearly visible, as if illuminated by a light source low in the west.

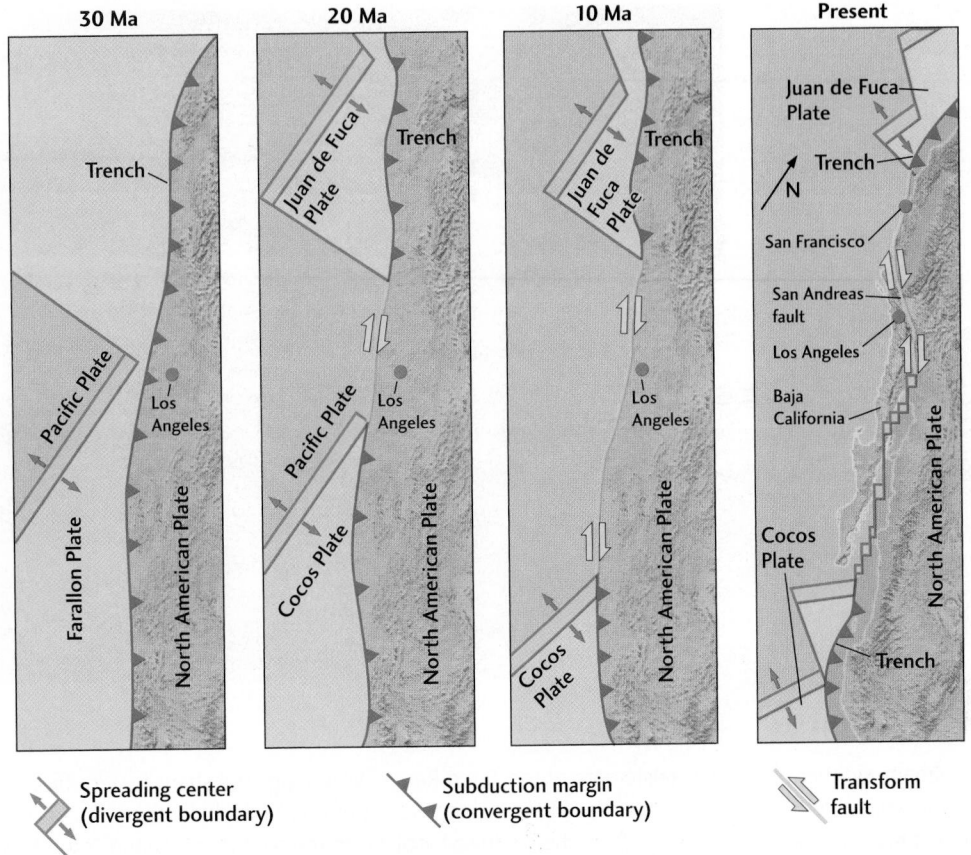

FIGURE 10.6 ■ The interaction of the west coast of North America with the shrinking Farallon Plate as it was progressively subducted beneath the North American Plate, leaving the present-day Juan de Fuca and Cocos plates as small remnants. Large solid arrows show the present-day direction of relative movement between the Pacific and North American plates. (Ma, million years ago.) [After W. J. Kious and R. I. Trilling, *This Dynamic Earth: The Story of Plate Tectonics.* Washington, D.C.: U.S. Geological Survey, 1996.]

topography sharpened, and the canyons deepened. As we will see in Chapter 22, rejuvenation is driven not only by plate tectonic processes, but also by interactions between the plate tectonic and climate systems. For example, some of the increase in the relief of the Cordilleran mountain chains may have occurred as a result of the onset of glacial cycles in the Pleistocene.

The *Basin and Range province* developed through the uplift and stretching of the crust in a northwest-southeast direction. This extension began with the heating of the lithosphere by upwelling convection currents in the mantle about 15 million years ago and continues to the present day (see Chapter 7). It has resulted in a wide zone of normal faulting extending from southern Oregon to Mexico and from eastern California to western Texas. The Basin and Range province is volcanically active and contains extensive hydrothermal deposits of gold, silver, copper, and other valuable metals. Thousands of steeply dipping normal faults have sliced the crust into a pattern of upheaved and down-dropped blocks, forming scores of rugged and nearly parallel mountain ranges separated by sediment-filled rift valleys. The Wasatch Range of Utah and the Teton

Range of Wyoming (**Figure 10.7**) are being uplifted on the eastern edge of the Basin and Range province, while the Sierra Nevada of California is being uplifted and tilted on the province's western edge.

The *Colorado Plateau* seems to be an island of stability that has experienced no major tension or compression since Precambrian time. The broad uplift of the plateau has allowed the Colorado River to cut through flat-lying sedimentary rock formations, creating the Grand Canyon. Geologists believe that this uplift was caused by the same type of lithospheric heating that is stretching the crust in the Basin and Range province.

■ Tectonic Provinces Around the World

We will now expand our view from North America to Earth's other continents. Each continent has its own distinctive features, but a general pattern becomes evident

FIGURE 10.7 ■ Image synthesized from satellite data of the Teton Range, Wyoming. The sharp eastern face of the mountain range, which has a vertical relief of more than 2000 m, is the result of normal faulting along the northeastern edge of the Basin and Range province. The view is from the northeast looking to the southwest. Grand Teton Mountain, near the center of the image, rises to an altitude of 4200 meters. [NASA/Goddard Space Flight Center, Landsat 7 Team.]

when continental geology is viewed on a global scale (Figure 10.8a). Continental shields and platforms make up the most stable parts of the continental lithosphere, called **cratons**, and contain the eroded remnants of ancient deformed rocks. The North American craton comprises the Canadian Shield and the interior platform (see Figure 10.1).

Around these cratons are elongated mountain belts, or **orogens** (from the Greek *oros*, meaning "mountain," and *gen*, "be produced"), that were formed by later episodes of compressive deformation. The youngest orogenic (mountain-building) systems, such as the North American Cordillera, are found along the **active margins** of continents, where tectonic processes caused by plate movements continue to deform the continental crust.

The **passive margins** of continents—those that are attached to oceanic crust as part of the same plate and thus are not near plate boundaries—are zones of extended crust, stretched during the rifting that broke older continents apart and initiated seafloor spreading. This rifting often occurred parallel to older mountain belts, such as the Appalachian fold belt.

Types of Tectonic Provinces

The general pattern of cratons bounded by orogens can be seen in Figure 10.8a, which summarizes the major tectonic provinces of the continents worldwide. The classifications portrayed on this map are closely related to those we used to describe the tectonic provinces of North America:

- *Shield.* A region of uplifted and exposed crystalline basement rocks of Precambrian age, which have remained undeformed throughout the Phanerozoic eon (542 million years ago to the present). Example: Canadian Shield.

- *Platform.* A region where Precambrian basement rocks are overlain by less than a few kilometers of relatively flat-lying sediments. Examples: interior platform of central North America, Hudson Bay.

- *Continental basin.* A region of prolonged subsidence where thick sediments have accumulated during the Phanerozoic, with beds dipping into the center of the basin. Example: Michigan Basin.

- *Phanerozoic orogen.* A region where mountain building has occurred during the Phanerozoic. Examples: Appalachian fold belt, North American Cordillera.

- *Extended crust.* A region where the most recent deformation has involved large-scale crustal extension. Examples: Basin and Range province, Atlantic coastal plain.

FIGURE 10.8 ▪ A global view of the continents, showing (a) their major tectonic provinces and (b) their tectonic ages. [W. Mooney/USGS.]

Tectonic Ages

The **tectonic age** of a rock is the time of the last major episode of crustal deformation of that rock (Figure 10.8b). Most continental basement rocks have survived a long and complex history of repeated deformation, melting, and metamorphism. We can often use isotopic dating tech-

niques and other age indicators (see Chapter 8) to extract more than one age to any particular rock. The tectonic age indicates the *last* time the isotopic clocks within a rock were reset by tectonic deformation and accompanying metamorphism of the upper crust. For example, many of the igneous rocks in the southwestern United States were originally derived from the melting of crust and mantle in

FIGURE 10.9 ■ The Vishnu schist, part of the middle Proterozoic basement (1.8 billion years old) found at the bottom of the Grand Canyon. [NPS photo by Erin Whittaker.]

the middle Proterozoic (1.9 billion to 1.6 billion years ago) (**Figure 10.9**). However, those rocks were substantially metamorphosed by subsequent tectonic activity, including several episodes of compressive deformation in the Mesozoic and rifting in the Cenozoic. Geologists thus assign this region to the youngest age category, Mesozoic-Cenozoic.

A Global Puzzle

The current distribution of continental tectonic provinces and their ages is like a giant puzzle in which the original pieces have been rearranged and reshaped by continental rifting, continental drift, and continent-continent collisions over billions of years. Only the past 200 million years of plate movements can be reliably determined from existing oceanic crust. Earlier plate movements must be inferred from the indirect evidence found in continental rocks. In Chapter 2, we saw that geologists have made amazing progress in reconstructing earlier configurations of the continents from paleomagnetic and paleoclimate data and from the signatures of deformation exposed in ancient mountain belts.

In the next section, we trace the history of the continents even further back into geologic time. Once again, we use the history of North America as our prime example, starting with its youngest provinces along its west coast and working backward in time to the Canadian Shield. We focus on three key questions about continental evolution: What geologic processes built the continents we see today? How do these processes fit into the theory of plate tectonics? Can plate tectonics explain the original formation of the cratons? As we will see, these questions have been only partially answered by geologic research.

■ How Continents Grow

Over the continents' 4-billion-year history, new crust has been added at an average rate of about 2 km³/year. Earth scientists continue to debate whether the growth of continental crust has occurred gradually over geologic time or was concentrated early in Earth's history. In the modern plate tectonic system, two basic processes work together to form new continental crust: magmatic addition and accretion.

Magmatic Addition

The process of magmatic differentiation of low-density, silica-rich rock in Earth's mantle and *vertical* transport of this buoyant, felsic material from the mantle to the crust is called **magmatic addition.**

Most new continental crust is born in subduction zones from magmas formed by fluid-induced melting of the subducting lithospheric slab and the mantle wedge above the slab (see Chapter 4). These magmas, which are of basaltic to andesitic composition, migrate toward the surface, pooling in magma chambers near the base of the crust. Here they incorporate crustal materials and differentiate further to form the felsic magmas that migrate into the upper crust, forming dioritic and granodioritic plutons capped by andesitic volcanoes.

Magmatic addition can emplace new crustal material directly at active continental margins. Subduction of the Farallon Plate beneath North America during the Cretaceous period, for example, created the batholiths along the western edge of the continent, including the rocks now exposed in Baja California and the Sierra Nevada. Subduction of the remnant Juan de Fuca Plate continues to add new material to the crust in the volcanically active Cascade Range of the Pacific Northwest, just as subduction of the Nazca Plate is building up the crust in the Andes of South America.

Buoyant felsic crust is also produced far away from continents, in volcanic island arcs at ocean-ocean convergence zones. Over time, these island arcs can merge into thick sections of silica-rich crust, such as those found today in the Philippines and other island groups of the southwestern Pacific (**Figure 10.10**). Plate movements transport these fragments of crust horizontally across the globe and eventually attach them to active continental margins by accretion.

Accretion

The integration of crustal material previously differentiated from mantle material into existing continental masses by *horizontal* transport during plate movements is called **accretion.**

Geologic evidence for accretion can be found on the active margins of North America. In the Pacific Northwest and Alaska, the crust consists of a mix of odd pieces—island arcs, seamounts (extinct underwater volcanoes), and remnants of basalt plateaus, old mountain ranges, and other slivers of continental crust—that were plastered onto the leading edge of the continent as it moved across Earth's surface. These pieces are sometimes referred to as **accreted terrains.** Geologists use this term to define a large piece of crust, tens to hundreds of kilometers in geographic extent, with common characteristics and a distinct origin, usually transported great distances by plate movements.

The geologic arrangement of accreted terrains can be chaotic (**Figure 10.11**). Adjacent blocks of crust can

FIGURE 10.10 ■ The Philippines and other island groups in the southwestern Pacific illustrate how island arcs can merge into thick sections of protocontinental crust at ocean-ocean convergence zones.

FIGURE 10.11 ■ Much of the North American Cordillera has been formed by terrain accretion over the past 200 million years. Wrangellia, for example, is a former basalt plateau that was transported to its present location from 5000 km away. Other accreted terrains are made up of island arcs, ancient seafloor, and continental fragments. [After D. R. Hutchison, "Continental Margins," *Oceanus* 35 (Winter 1992–1993): 34–44; modified from work of D. G. Howell, G. W. Moore, and T. J. Wiley.]

contrast sharply in their rock types, the nature of their folding and faulting, and their history of magmatic activity and metamorphism. Geologists often find fossils indicating that these blocks originated in different environments, and at different times, than the rocks of the surrounding area. For example, an accreted terrain comprising ophiolite suites (pieces of seafloor) that contain deep-water fossils might be surrounded by remnants of island arcs and continental fragments containing shallow-water fossils of a completely different age. The boundaries between accreted terrains are almost always faults that have undergone substantial slippage, although the nature of the faulting is often difficult to discern. Blocks of crust that seem completely out of place are called *exotic terrains.*

Before the discovery of plate tectonics, exotic terrains were a subject of fierce debate among geologists, who had difficulty coming up with reasonable explanations for their origins. Now accreted terrain analysis is a specialized field within plate tectonic research. Well over a hundred areas of the North American Cordillera have been identified as exotic terrains accreted during

ACCRETION OF A BUOYANT FRAGMENT TO A CONTINENT

Continental crust

A buoyant crustal fragment is carried into a subduction zone.

Fragment

Lithosphere

Asthenosphere

The fragment is more buoyant than the subducting lithosphere, so it is not subducted.

The fragment becomes welded to a continent on the overriding plate.

Accreted terrain

ACCRETION OF AN ISLAND ARC TO A CONTINENT

Continental crust

A plate carrying a continent is subducted beneath an island arc.

Island arc

The continental crust is more buoyant than the subducting lithosphere, so it is not subducted.

The sea closes, and the island arc becomes welded to the continent.

Accreted terrain

FIGURE 10.12 ■ Four distinct processes explain the accretion of exotic terrains.

the last 200 million years (many more than depicted in Figure 10.11). One such terrain, called Wrangellia, originally formed as a large basalt plateau (a region of oceanic crust thickened by a large outpouring of basaltic lava) and was then transported over 5000 km from the Southern Hemisphere to its current location in Alaska and western Canada. Extensive accreted terrains have also been mapped in Japan, Southeast Asia, China, and Siberia.

In only a few cases do we know precisely where these accreted terrains originated. We can begin to deci-

pher how the others came together by considering four distinct tectonic processes that can result in accretion (**Figure 10.12**):

1. A crustal fragment that is too buoyant to be subducted may be transferred from a subducting plate to a continent on the overriding plate. Such fragments can be small pieces of continental crust ("microcontinents") or thickened sections of oceanic crust (large seamounts, basalt plateaus).

ACCRETION ALONG A TRANSFORM FAULT

Two plates slide past each other along a transform fault.

Transform fault

Fragment

Plate A

Plate B

A crustal fragment on plate B is carried along the margin of plate A.

Fragment

When the fault becomes inactive, the fragment becomes welded to plate A in a position distant from its original position.

Accreted terrain

ACCRETION BY CONTINENTAL COLLISION AND RIFTING

One plate carrying a continent is subducted beneath another plate.

Continental plate A

Continental plate B

The continent is not subducted, so the two continents are sutured together along a set of thrust faults.

Thrust faults

Later, rifting and seafloor spreading carry the plates apart, leaving a fragment of one continent welded to the other.

Accreted terrain

2. A sea that separates an island arc from a continent may be closed as the thickened island arc crust collides with and becomes attached to the advancing edge of the continent.

3. Two plates that slide past each other along a transform fault may result in strike-slip faulting and the movement of a crustal fragment from one plate to the other. Today, the southwestern part of California, which is attached to the Pacific Plate, is moving northwestward relative to the North American Plate along the San Andreas transform fault. Strike-slip faulting landward of the deep-sea trench in oblique subduction zones can also transport terrains hundreds of kilometers.

4. Two continents may collide and be sutured together, then break apart later at a different location.

The fourth process explains some of the accreted terrains found on the passive eastern margin of North America. The Appalachian fold belt contains slices of ancient Europe and Africa as well as a variety of exotic terrains. Florida's oldest rocks and fossils are more like those in Africa than like those found in the rest of the United States, indicating that most of this peninsula was probably transported to North America when Pangaea was assembled and then left behind when North America and Africa split apart about 200 million years ago.

▪ How Continents Are Modified

The geology of the North American Cordillera, with its many exotic terrains, looks nothing like that of the ancient Canadian Shield, which lies directly east of the Cordillera. In particular, the accreted terrains of the youthful Cordilleran system do not show the high degree of melting or the high-grade metamorphism that characterize the Precambrian crust of the shield. Why such a difference? The answer lies in the tectonic processes that have repeatedly modified the older parts of the continent throughout its long history.

Orogeny: Modification by Plate Collision

Continental crust is profoundly altered by **orogeny**—the mountain-building processes of folding, faulting, magmatic addition, and metamorphism. Orogenic processes have repeatedly modified the edges of cratons. Most orogenies (episodes of mountain building) result from plate convergence. When one or both plates are made of oceanic lithosphere, their convergence usually results in subduction rather than orogeny. Orogenies can result when a continent rides forcibly over subducting oceanic crust, as in the Andean orogeny now under way in South America, but the most intense orogenies are caused by the collision of two or more continents. As we observed in Chapter 2, when continents collide, a basic tenet of plate tectonics—the rigidity of plates—must be modified.

Continental crust is much more buoyant than mantle material, so colliding continents resist being subducted with the plates that carry them. Instead, the continental crust deforms and breaks in a combination of intense folding and faulting that can extend hundreds of kilometers from the collision zone, as described in Chapter 7. Thrust faulting caused by the convergence can stack the upper part of the crust into overthrust sheets tens of kilometers thick, deforming and metamorphosing the rocks they contain (Figure 10.13). Continental shelf sediments can be scraped off the basement rock on which they were deposited and thrust inland. Horizontal compression throughout the crust can double its thickness, causing the rocks in the lower crust to melt. This melting can generate huge amounts of granitic magma, which rises to form extensive batholiths in the upper crust.

THE ALPINE-HIMALAYAN OROGENY To see orogeny in action today, we look to the great chains of high mountains that stretch from Europe through the Middle East and across

Overthrust sheets

FIGURE 10.13 ▪ When continents collide, the continental crust can break into overthrust sheets stacked one above the other.

Atlas Mountains   Alps   Caucasus   Zagros Mountains   Tien Shan   Tibetan Plateau   Himalaya

Earthquake depth
● ≤ 50 km deep (shallow focus)
● 50–300 km deep
● > 300 km deep (deep focus)

FIGURE 10.14 ■ The Alpine-Himalayan belt, showing the chains of high mountains built by the ongoing collision of the African, Arabian, and Indian plates with the Eurasian Plate. This orogeny is marked by intense earthquake activity.

Asia, known collectively as the *Alpine-Himalayan belt* (**Figure 10.14**). The breakup of Pangaea sent Africa, Arabia, and India northward, causing the Tethys Ocean to close as its lithosphere was subducted beneath Eurasia (see Figure 2.16). These former pieces of Gondwana collided with Eurasia in a complex sequence, beginning in the western part of Eurasia during the Cretaceous period and continuing eastward through the Tertiary, raising the Alps in central Europe, the Caucasus and Zagros mountains in the Middle East, and the Himalaya and other high mountain chains across central Asia.

The Himalaya, the world's highest mountains, are the most spectacular result of this modern episode of continent-continent collision (see the Practicing Geology exercise at the end of the chapter). About 50 million years ago, the Indian subcontinent, riding on the subducting Indian Plate, first encountered the island arcs and volcanic mountain belts that then bounded the Eurasian Plate (**Figure 10.15**). As the landmasses of India and Eurasia merged, the Tethys Ocean disappeared through subduction. Pieces of the oceanic crust were trapped along the suture zone between the converging continents and can be seen today as ophiolite suites along the Indus and Tsangpo river valleys that separate the high Himalaya from the Tibetan Plateau. The collision slowed India's advance, but the Indian Plate continued to drive northward. So far, India has penetrated over 2000 km into Eurasia, causing the largest and most intense orogeny of the Cenozoic era.

The Himalaya were formed from overthrust slices of the old northern portion of India, stacked one atop the other (see Figure 10.15). This process took up some of the compression. Horizontal compression and the formation of fold and thrust belts also thickened the crust north of India, caus-

ing the uplift of the huge Tibetan Plateau, which now has a crustal thickness of 60 to 70 km (almost twice the thickness of most continental crust) and stands nearly 5 km above sea level. These and other compression zones account for perhaps half of India's penetration into Eurasia. Further compression has pushed China and Mongolia eastward, out of India's way, like toothpaste squeezed from a tube. Most of this sideways movement has taken place along the Altyn Tagh fault and other major strike-slip faults shown on the map in **Figure 10.16**. The mountains, plateaus, faults, and great earthquakes of Asia, extending thousands of kilometers from the Indian-Eurasian suture, are all results of the Alpine-Himalayan orogeny, which continues today as India plows into Asia at a rate of 40 to 50 mm/year.

PALEOZOIC OROGENIES DURING THE ASSEMBLY OF PANGAEA If we go further back in geologic time, we find abundant evidence of older orogenies. We have already mentioned, for example, that at least three distinct orogenies were responsible for the Paleozoic deformation now exposed in the eroded Appalachian fold belt of eastern North America. These three episodes of mountain building were caused by plate convergence that led to the assembly of the supercontinent Pangaea near the end of the Paleozoic era.

The supercontinent Rodinia began to break up toward the end of the Proterozoic eon, forming several paleocontinents (see Figure 2.16). One was the large continent of *Gondwana*. Two of the others were *Laurentia*, which included the North American craton and Greenland, and *Baltica*, comprising what are now the lands around the Baltic Sea (Scandinavia, Finland, and the European part of Russia).

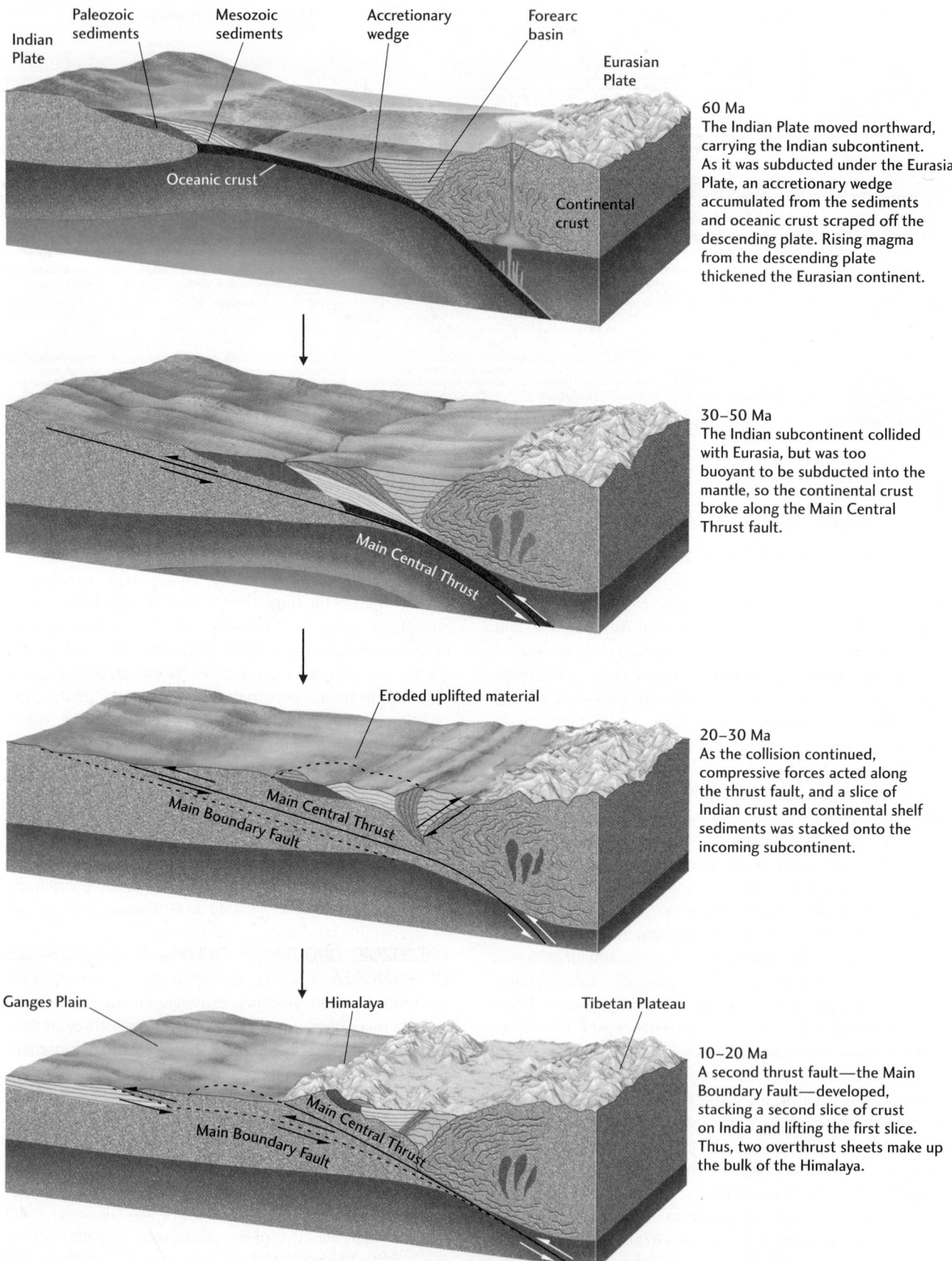

Indian Plate

Paleozoic sediments

Mesozoic sediments

Accretionary wedge

Forearc basin

Eurasian Plate

Oceanic crust

Continental crust

60 Ma
The Indian Plate moved northward, carrying the Indian subcontinent. As it was subducted under the Eurasian Plate, an accretionary wedge accumulated from the sediments and oceanic crust scraped off the descending plate. Rising magma from the descending plate thickened the Eurasian continent.

Main Central Thrust

30–50 Ma
The Indian subcontinent collided with Eurasia, but was too buoyant to be subducted into the mantle, so the continental crust broke along the Main Central Thrust fault.

Eroded uplifted material

Main Central Thrust

Main Boundary Fault

20–30 Ma
As the collision continued, compressive forces acted along the thrust fault, and a slice of Indian crust and continental shelf sediments was stacked onto the incoming subcontinent.

Ganges Plain

Himalaya

Tibetan Plateau

Main Central Thrust

Main Boundary Fault

10–20 Ma
A second thrust fault—the Main Boundary Fault—developed, stacking a second slice of crust on India and lifting the first slice. Thus, two overthrust sheets make up the bulk of the Himalaya.

FIGURE 10.15 ■ Cross sections showing the sequence of events that have caused the Himalayan orogeny, simplified and vertically exaggerated. (Ma, million years ago.) [After P. Molnar, "The Structure of Mountain Ranges," *Scientific American* (July 1986): 70.]

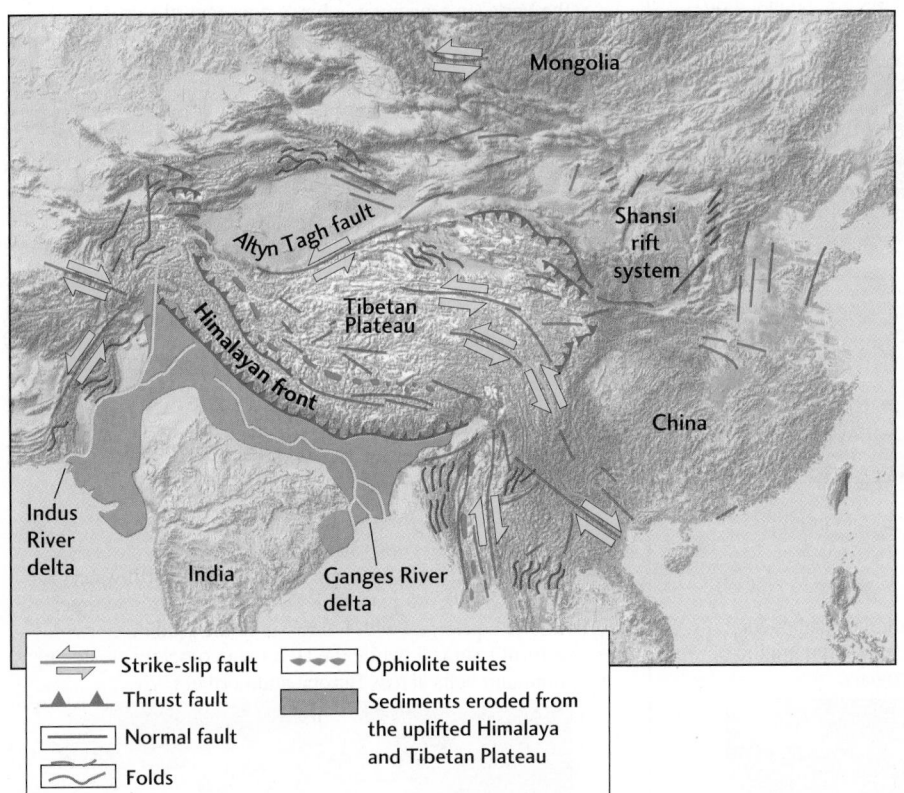

FIGURE 10.16 ■ The collision between India and Eurasia has produced many spectacular tectonic features, including large-scale faulting and uplift. [After P. Molnar and P. Tapponier, "The Collision Between India and Eurasia," *Scientific American* (April 1977): 30.]

In the Cambrian period, Laurentia was rotated almost 90° from its present orientation and straddled the equator; its southern (today, eastern) side was a passive continental margin. To its immediate south was the proto-Atlantic, or *Iapetus,* Ocean (in Greek mythology, Iapetus was the father of Atlantis), which was being subducted beneath a distant island arc. Baltica lay off to the southeast, and Gondwana was thousands of kilometers to the south. **Figure 10.17** shows the sequence of events as the three continents converged.

The island arc built up by the southward-directed subduction of Iapetus lithosphere collided with Laurentia in the middle to late Ordovician (470 million to 440 million years ago), causing the first episode of mountain building: the *Taconic orogeny.* (You can see some of the rocks accreted and deformed during this period if you drive the Taconic State Parkway, which runs east of the Hudson River for about 160 km north of New York City.) The second orogeny began when Baltica and a connected set of island arcs began to collide with Laurentia in the early Devonian (about 400 million years ago). The collision deformed southeastern Greenland, northwestern Norway, and Scotland in what European geologists refer to as the *Caledonian orogeny.* The deformation continued into present-day North America as the *Acadian orogeny,* as island arcs that would become the terrains of maritime Canada and New England accreted to Laurentia in the middle to late Devonian (380 million to 360 million years ago).

The grand finale in the assembly of Pangaea was the collision of the behemoth landmass of Gondwana with Laurasia and Baltica, by then joined into a continent named *Laurussia.* The collision began about 340 million years ago

with the *Variscan orogeny* in what is now central Europe and continued along the margin of the North American craton with the *Appalachian orogeny* (320 million to 270 million years ago). This latter phase of assembly pushed Gondwanan crust over Laurentia, lifting the Blue Ridge into a mountain chain that may have been as high as the modern Himalaya and causing much of the deformation now seen in the Appalachian fold belt. Also during this phase, Siberia and other Asian terrains converged with Laurussia in the *Ural orogeny,* forming the continent of *Laurasia* and pushing up the Ural Mountains. At the same time, extensive deformation created new mountain belts across Europe and northern Africa (the *Hercynian orogeny*).

The crunching together of all these continental masses profoundly altered the structure of the crust. The rigid cratons were little affected, but the younger accreted terrains caught in between were consolidated, thickened, and metamorphosed. The lower parts of this younger crust were partially melted, producing granitic magmas that rose to form batholiths in the upper crust and volcanoes at the surface. Uplifted mountains and plateaus were eroded, exposing high-grade metamorphic rocks that were once many kilometers deep and depositing thick sedimentary sequences. Sediments laid down following the first orogeny were deformed and metamorphosed by later mountain-building episodes.

EARLIER OROGENIES So far, we have investigated two major periods of mountain building: the Paleozoic orogenies associated with the assembly of Pangaea, and the Cenozoic Alpine-Himalayan orogeny. In Chapter 2, we

Middle Cambrian (510 Ma)
After the breakup of Rodinia, Laurentia straddled the equator. Its southern side was a passive continental margin, bounded on the south by the Iapetus Ocean.

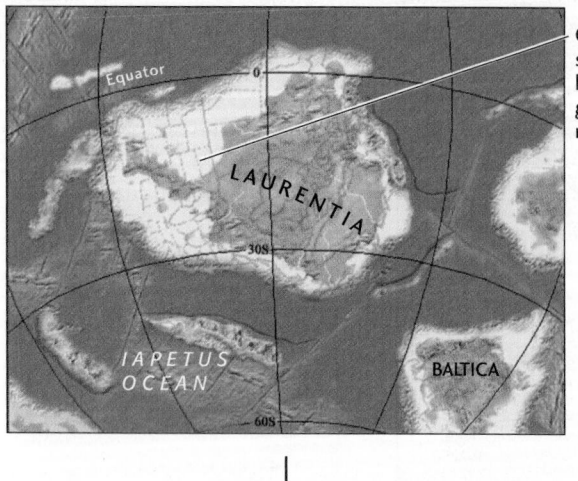

Outlines show U.S. state boundaries for geographic reference

Shelf and submerged continent

Early Carboniferous (340 Ma)
The collision of Gondwana with Laurussia began with the Variscan orogeny in what is now central Europe...

Late Ordovician (450 Ma)
The island arc built up by the southward-directed subduction of Iapetus lithosphere collided with Laurentia in the middle to late Ordovician, causing the Taconic orogeny.

Late Carboniferous (300 Ma)
... and continued along the margin of the North American craton with the Appalachian orogeny. At the same time, Siberia converged with Laurussia in the Ural orogeny to form Laurasia, while the Hercynian orogeny created new mountain belts across Europe and northern Africa.

Early Devonian (400 Ma)
The collision of Laurentia with Baltica caused the Caledonian orogeny and formed Laurussia. The southward continuation of the convergence caused the Acadian orogeny.

Early Permian (270 Ma)
The end product of these episodes of continental convergence was the supercontinent of Pangaea.

FIGURE 10.17 ■ Paleogeographic reconstructions of the present North Atlantic region, showing the sequence of orogenic episodes that resulted from the assembly of Pangaea. (Ma, million years ago.) [Ronald C. Blakey, Northern Arizona University, Flagstaff.]

discussed the assembly of the supercontinent Rodinia in the late Proterozoic eon. By now, it should not surprise you to learn that major orogenies accompanied the formation of that earlier supercontinent.

Some of the best evidence of these orogenies is found at the eastern and southern margins of the Canadian Shield in a broad belt known as the Grenville province, where new crustal material was added to the continent in the middle Proterozoic, about 1.1 billion to 1.0 billion years ago (see Figure 10.8b). Geologists believe that these rocks, which are now highly metamorphosed, originally consisted of volcanic mountain belt and island arc terrains that were accreted and compressed by the collision of Laurentia with the western part of Gondwana. They have drawn analogies between what happened during this *Grenville orogeny* and what is happening today in the Himalayan orogeny. A Tibet-like plateau was formed by compressive thickening of the crust through folding and thrust faulting, which metamorphosed the upper crust and partially melted large

parts of the lower crust. Once the orogeny ceased, erosion of the plateau thinned the crust and exposed crystalline rocks of high metamorphic grade. Geologists have found orogenic belts of similar age on continents worldwide. Although many of the details remain uncertain, they have reconstructed from this geologic record (which includes paleomagnetic data) a general picture of how Rodinia came together between 1.3 billion and 0.9 billion years ago.

The Wilson Cycle

From our brief look at the history of eastern North America, we can infer that the edges of many cratons have experienced multiple episodes of deformation in a general plate tectonic cycle that comprises four main phases (Figure 10.18):

1. Rifting during the breakup of a supercontinent
2. Passive margin cooling and sediment accumulation during seafloor spreading and ocean opening

1 Rifting within a continent splits the continent,...

7 The continent erodes, thinning the crust. Eventually the process may begin again.

2 ...leading to the opening of a new ocean basin and creation of new oceanic crust, starting the cycle.

6 As continents collide, orogeny thickens the crust and builds mountains, forming a new supercontinent.

3 As seafloor spreading continues and an ocean opens, passive margin cooling occurs and sediment accumulates.

5 Terrain accretion—from the sedimentary accretionary wedge or fragments carried by the subducting plate—welds material to the continent.

4 Convergence begins; oceanic crust is subducted beneath a continent, creating a volcanic mountain belt at the active margin.

FIGURE 10.18 ■ The Wilson cycle comprises the plate tectonic processes responsible for the formation and breakup of supercontinents and the opening and closing of ocean basins.

FIGURE 10.19 ■ The geologic time scale, showing some important events in the history of the continents. (Ma, million years ago.)

3. Active margin volcanism and terrain accretion during subduction and ocean closure

4. Orogeny during the continent-continent collision that forms the next supercontinent

This idealized sequence of events was named the **Wilson cycle** after the Canadian pioneer of plate tectonics, J. Tuzo Wilson, who first recognized its importance in the evolution of continents.

The geologic record suggests that the Wilson cycle has operated throughout the Proterozoic and Phanerozoic eons (Figure 10.19), resulting in the formation of at least two supercontinents prior to Rodinia. One of these supercontinents (named *Columbia)* formed about 1.9 billion to 1.7 billion years ago. An even earlier one, whose assembly marks the transition from the Archean eon to the Proterozoic eon, formed about 2.7 billion to 2.5 billion years ago. Did the Wilson cycle also operate in the Archean eon? We will return to that question shortly.

Epeirogeny: Modification by Vertical Movements

So far, our consideration of continental evolution has emphasized accretion and orogeny, processes that involve horizontal plate movements and are usually accompanied by deformation in the form of folding and faulting. Throughout the world, however, sedimentary rock sequences record another kind of movement that has modified continents: gradual downward and upward movements of broad regions of crust without significant folding or faulting. These vertical movements are referred to as **epeirogeny,** a term coined in 1890 by the American geologist Clarence Dutton (from the Greek *epeiros,* meaning "mainland").

Epeirogenic downward movements usually result in a sequence of relatively flat-lying sediments, such as those found in the stable interior platform of North America. Upward movements cause erosion and gaps in the sedimentary record seen as unconformities. Erosion can lead to

the exposure of crystalline basement rocks, such as those found on the Canadian Shield.

Geologists have identified several mechanisms of epeirogeny. One example is **glacial rebound** (Figure 10.20a) (see also Earth Issues 14.1). When large glaciers form, their weight depresses the continental crust. When they melt, the crust rebounds upward for tens of millennia. Glacial rebound explains the uplift of Finland and Scandinavia following the most recent glaciation, which ended about 17,000 years ago, as well as the raised beaches of northern Canada (Figure 10.21). Although glacial rebound seems slow by human standards, it is a rapid process, geologically speaking.

Heating and cooling of the continental lithosphere are important epeirogenic processes on longer time scales. Heating causes rocks to expand, decreasing their density and thus raising the continental surface (Figure 10.20b). A good example is the Colorado Plateau, which has been uplifted to about 2 km above sea level during the last 10 million years or so. Geologists think this heating results from active mantle upwelling, which is also stretching the crust in the Basin and Range province on the western and southern sides of the plateau.

Conversely, the cooling of lithosphere increases its density, making it sink under its own weight and creating a thermal subsidence basin (Figure 10.20c). Cooling of once-hot areas in the continental interior may explain the Michigan Basin and other deep basins in central North America (see Figure 10.3). When a new episode of seafloor spreading splits a continent apart, the uplifted edges are eroded and eventually subside as they cool, forming basins in which sediments are deposited and carbonate platforms accumulate (Figure 10.20d). This process has led to the formation of a thick continental shelf along the east coast of the United States.

One intriguing puzzle is the South African Plateau, where a craton has been uplifted during the Cenozoic to almost 2 km above sea level—more than twice the elevation of most cratons. However, the lithosphere in this part of the continent does not appear to be unusually

(a) GLACIAL REBOUND

The weight of glacial ice downwarps the continental lithosphere,...

...which rebounds once the ice is removed.

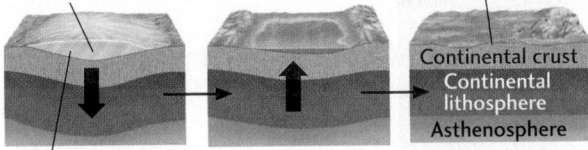

Continental glacier

(b) HEATING OF LITHOSPHERE

Upwelling of mantle material causes uplift and thinning of the continental lithosphere.

(c) COOLING OF LITHOSPHERE IN CONTINENTAL INTERIOR

Thermal subsidence basin

As the lithosphere cools and contracts, it subsides to form a basin within the continent.

(d) COOLING OF LITHOSPHERE ON CONTINENTAL MARGIN

Continental shelf sediments

When seafloor spreading splits a continent apart, the edges subside as they cool, accumulating thick sediments.

(e) HEATING OF DEEP MANTLE

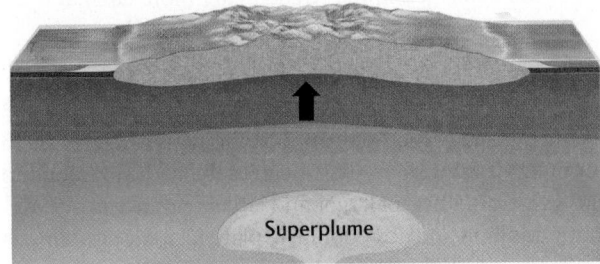

Superplume

A superplume rising from the deep mantle heats the lithosphere and raises the base of the continent, upwarping the surface over a broad area.

FIGURE 10.20 ■ Geologists have identified five major mechanisms of epeirogeny.

hot. One possible explanation is that the southern African craton may be uplifted by a hot, buoyant region of the lower mantle (see Chapter 14). This "superplume" could apply upward forces at the base of the lithosphere sufficient to raise the surface by about a kilometer (Figure 10.20e).

None of these proposed epeirogenic mechanisms, however, explains a central feature of continental cratons: the existence of raised continental shields and subsided platforms. These regions are too vast, and have persisted too long, to be explained by the plate tectonic processes we have discussed so far.

■ The Origins of Cratons

Every continental craton contains regions of ancient lithosphere that have been stable (i.e., undeformed) since the Archean eon (3.9 to 2.5 billion years ago). As we have seen, deformation has occurred at the edges of these stable landmasses, and new crust has accreted around them, during subsequent Wilson cycles. But how were these central parts of the cratons created in the first place?

We know that Earth was a hotter planet 4 billion years ago due to the heat generated by the decay of radioactive elements, which were more abundant then, as well as the energy released by differentiation and by impacts during the Heavy Bombardment (see Chapter 9). Evidence for a hotter mantle comes from a peculiar type of ultramafic volcanic rock found only in Archean crust, called *komatiite* (named after the Komati River in southeastern Africa, where it was first discovered). Komatiites contain a very high percentage (up to 33 percent) of magnesium oxide, so their formation would have required a much higher melting temperature than is found anywhere in the mantle today.

If the mantle was hotter during the Archean, then mantle convection may have been more vigorous. The plates may have been smaller and might have moved more rapidly. Volcanism was widespread, and the crust formed at spreading centers was probably thicker. Although lithosphere must surely have been recycled into the mantle, some geologists believe that the plates formed at this time were too thin and light to be subducted in the same way that oceanic lithosphere is consumed in modern subduction zones.

We do know that a silica-rich continental crust existed at this early stage in Earth's history. Formations as much as 3.8 billion years old have been found on many continents; most are metamorphic rocks evidently derived from even older continental crust. In a few places, small pieces of this early crust survive. The Acasta gneiss, in the northwestern part of the Canadian Shield, looks very similar to modern gneisses, although it has been dated at 4.0 billion years

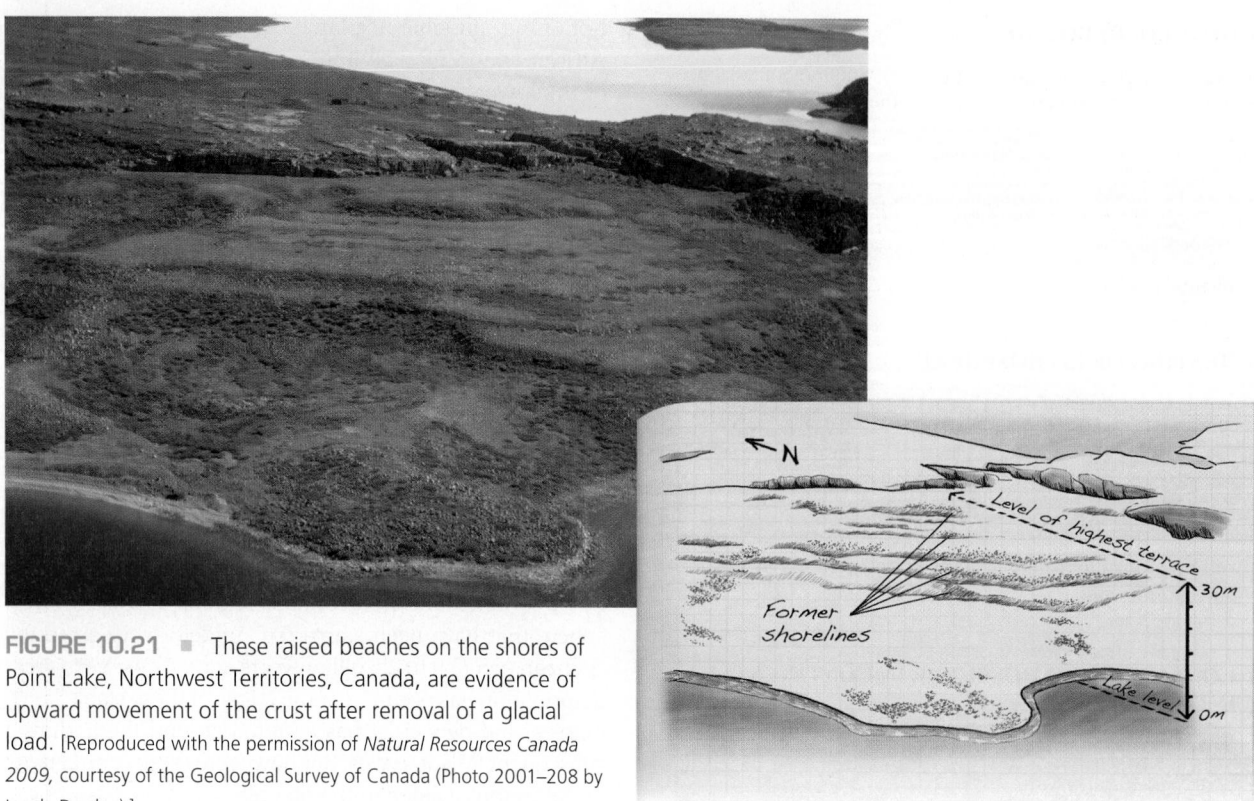

FIGURE 10.21 ■ These raised beaches on the shores of Point Lake, Northwest Territories, Canada, are evidence of upward movement of the crust after removal of a glacial load. [Reproduced with the permission of *Natural Resources Canada 2009*, courtesy of the Geological Survey of Canada (Photo 2001–208 by Lynda Dredge).]

ago (**Figure 10.22a**). Geologists recently discovered an even older rock formation, nearly 4.3 billion years old, in northern Quebec (Figure 10.22b). In Australia, single grains of zircon (a very hard mineral that survives erosion) have been dated as old as 4.4 billion years (see Chapter 8).

In the early part of the Archean, the continental crust that had differentiated from the mantle was very mobile. It may have been organized in small rafts that were rapidly pushed together and torn apart by intense tectonic activity—a version of the flake tectonic process that appears to be happening on Venus today. The first continental crust with long-term stability began to form about 3.3 billion to 3.0 billion years ago. In North America, the oldest surviving example is the central Slave province in northwestern Canada (where the Acasta gneiss is found), which stabilized about 3 billion years ago. Geologists have been able to show that this stabilization process involved not only the continental crust, but also chemical changes in the mantle portion of the continental lithosphere, as we will see shortly.

The rock formations in this Archean crust fall into two major groups (**Figure 10.23**):

1. **Granite-greenstone terrains** are areas of massive granitic intrusions that surround smaller pockets of greenstones, which in turn are capped with sediments. Greenstones, as we saw in Chapter 6, are low-grade metamorphic rocks derived from volcanic rocks, primarily of mafic composition. The origin of these greenstones is controversial, but many geologists think they were once pieces of oceanic crust formed at small spreading centers landward of island arcs accreted to the continents and later engulfed by the granitic intrusions.

2. **High-grade metamorphic terrains** are areas of high-grade (granulite facies) metamorphic rocks derived primarily from the compression, burial, and subsequent erosion of granitic crust. These areas look similar to the deeply eroded parts of modern orogenic belts, but the geometry of the deformation is different. Modern orogenies typically produce linear mountain belts where the edges of large cratons converge. Areas deformed in the Archean are more circular or S-shaped, reflecting the fact that the cratons were much smaller, with boundaries that were more curved.

By the end of the Archean, 2.5 billion years ago, enough continental lithosphere had been stabilized in cratons to allow the formation of larger and larger continents by magmatic addition and accretion. The plate tectonic system was probably operating much as it does today. It is at about this time that we see the first evidence of major continent-continent collisions and the assembly of supercontinents. From this point onward in Earth's history, the history of the continents was governed by the plate tectonic processes of the Wilson cycle.

(a)

(b)

FIGURE 10.22 ■ Newly discovered rocks show that continental crust existed on Earth's surface during the Hadean eon. (a) The Acasta gneiss from the Slave craton has been dated at 4.0 billion years old. (b) Amphibole-bearing rocks from the Nuvvuagittuq greenstone belt, northern Quebec, Canada, have been dated at 4.28 billion years ago, making them the oldest rock formation yet discovered. [(a) Courtesy of Sam Bowring, Massachusetts Institute of Technology; (b) Jonathan O'Neil.]

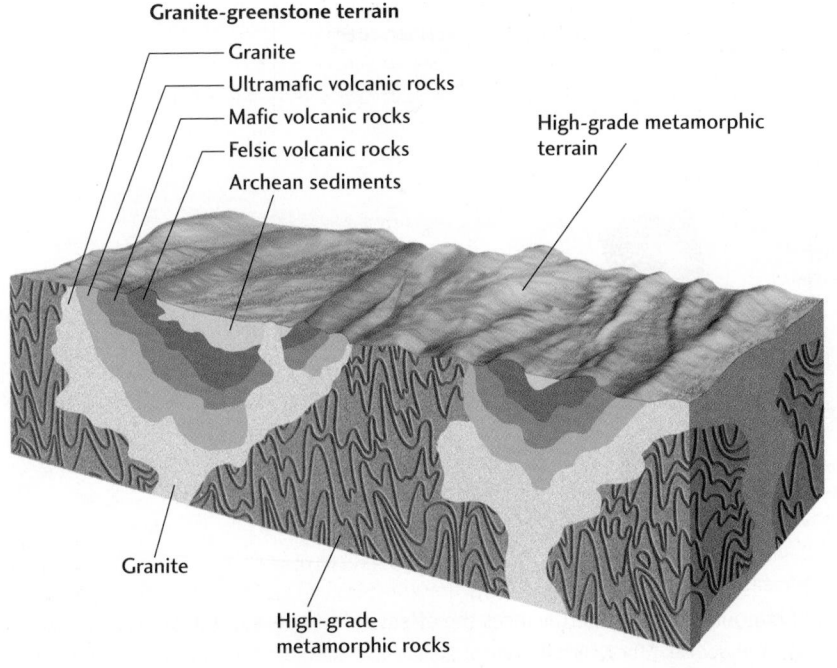

Granite-greenstone terrain

- Granite
- Ultramafic volcanic rocks
- Mafic volcanic rocks
- Felsic volcanic rocks
- Archean sediments

High-grade metamorphic terrain

Granite

High-grade metamorphic rocks

FIGURE 10.23 ■ Two major types of rock formations are found in Archean regions of continental cratons: granite-greenstone terrains and high-grade metamorphic terrains.

▪ The Deep Structure of Continents

In this chapter, we have surveyed the most important processes in the development of Earth's continental crust. However, we have not yet explained one very basic aspect of continental behavior: the long-term stability of the cratons. How have the cratons survived being knocked around by plate tectonic processes for billions of years? The answer to this question lies not in the crust, but in the lithospheric mantle below it.

Cratonic Keels

By using seismic waves to "see" into Earth's interior, geologists have discovered a remarkable fact: the continental cratons are underlain by a thick layer of mechanically strong mantle material that moves with the cratons as the continents drift. These thickened sections of lithosphere extend to depths of more than 200 km—more than twice the thickness of the oldest oceanic lithosphere.

At 100 to 200 km beneath oceanic crust (as well as beneath most younger regions of the continents), the mantle rocks are hot and weak. They are part of the ductile asthenosphere, which flows relatively easily, allowing the plates to slide across Earth's surface. The lithosphere beneath the cratons extends into this region like the hull of a boat into water, so we refer to these mantle structures as **cratonic keels,** as shown in **Figure 10.24.** All cratons on every continent appear to have such keels.

Cratonic keels present many puzzles that scientists are still trying to solve. Less heat is emitted from the mantle beneath the cratons than from the mantle beneath oceanic crust. This observation indicates that the keels are several

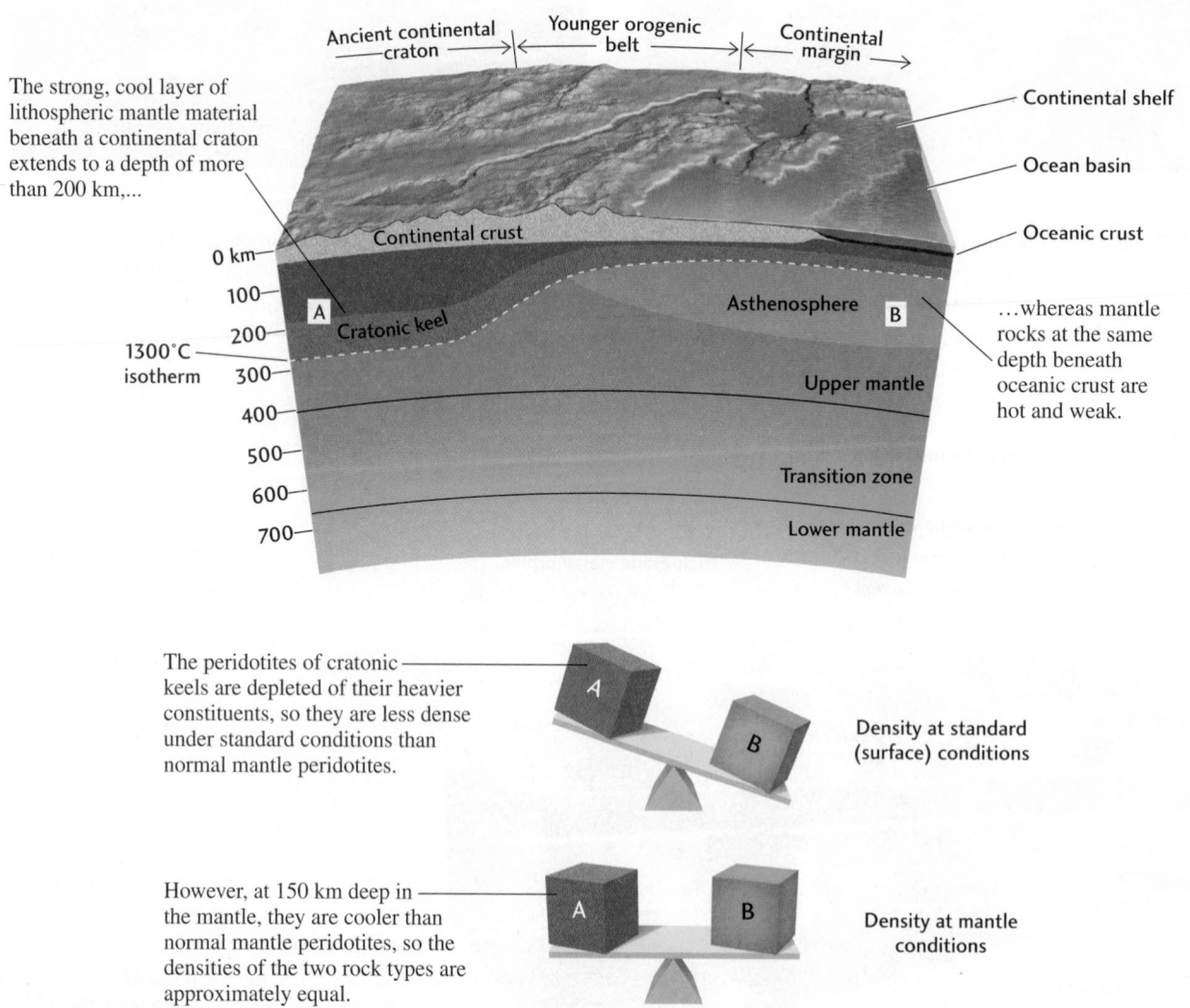

FIGURE 10.24 ▪ The chemical composition of cratonic keels counterbalances the effects of temperature to stabilize them against disruption by plate tectonic processes. [After T. H. Jordan, "The Deep Structure of Continents," *Scientific American* (January 1979): 92.]

hundred degrees cooler than the surrounding asthenosphere, which explains their strength. If the rocks of the mantle beneath the cratons are so cool, however, why don't they sink into the mantle under their own weight, as cold, heavy slabs of oceanic lithosphere do in subduction zones?

Composition of the Keels

The cratonic keels would indeed sink into the mantle if their chemical composition were the same as that of ordinary mantle peridotites. To get around this problem, geologists have hypothesized that the cratonic keels are made of rock with a different, less dense chemical composition (see Figure 10.24). Their lower density counteracts the increase in density resulting from their cooler temperature.

Strong evidence in support of this hypothesis has come from mantle samples found in kimberlite pipes—the same types of volcanic deposits that produce diamonds, as we will see in Chapter 12. Kimberlite pipes are the eroded necks of volcanoes that have erupted explosively from tremendous depths (**Figure 10.25**). Almost all kimberlites that contain diamonds are located within the Archean regions of cratons. A diamond will revert to graphite at depths shallower than 150 km unless its temperature drops quickly. Therefore, the presence of diamonds in these pipes shows that the kimberlite magmas came from deeper than 150 km, and that they erupted through the keels when magma fractured the lithosphere very rapidly.

During a violent kimberlite eruption, fragments of the cratonic keel, some containing diamonds, are ripped off and brought to the surface in the magma as mantle xenoliths. The majority of these xenoliths turn out to be peridotites containing less iron (a dense element) and less garnet (a dense mineral) than ordinary mantle peridotites. Such rocks can be produced by extraction of a basaltic

FIGURE 10.25 ■ Excavation of a kimberlite pipe at the Jwaneng diamond mine in Botswana. Diamonds are found in the dark-colored kimberlite rock in the center of the pit, which outlines the neck of an ancient, eroded volcano. The diamonds and other fragments of the African continental keel found at Jwaneng were erupted from depths of more than 150 km, and the analysis of these fragments supports the chemical stabilization hypothesis illustrated in Figure 10.24. Jwaneng is the world's richest diamond mine, producing 15.6 million carats (3120 kg) of diamonds worth over $2 billion in 2006. A major extension begun in 2010 is expected to yield 100 million carats worth approximately $15 billion over the life of the mine.
[Peter Essick/Aurora Photos.]

(or komatiitic) magma from the asthenosphere by partial melting. In other words, the mantle rock beneath the cratons is the depleted residue left over after melting sometime earlier in Earth's history. A cratonic keel made of such depleted rocks can still float atop the mantle, despite its cooler temperature (see Figure 10.24).

Age of the Keels

By analyzing xenoliths from kimberlites and the diamonds they contain, we have learned that the cratonic keels are about the same age as the Archean crust above them. (The diamond in your ring or necklace is likely to be several billion years old!) Therefore, the rocks now present in the cratonic keels must have been depleted by the extraction of a basaltic melt very early in Earth's history, and they must have been positioned beneath the Archean crust about the time that crust was stabilized.

In fact, keel formation was probably responsible for the tectonic stabilization of the cratons. The existence of a cool, mechanically strong keel explains why the cratons have managed to survive through many continental collisions, including at least four episodes of supercontinent formation, without much internal deformation.

Many aspects of this process are still not understood. How did the keels cool down? How did they achieve the density balance illustrated in Figure 10.24? Why are the regions of the cratons with the thickest keels of Archean age? Some scientists believe that the continents play a major role in the mantle convection that drives the plate tectonic system, but how the keels affect convection in the mantle is not completely understood. Indeed, many of the ideas presented in this chapter are hypotheses that have not yet been integrated into a fully accepted theory of continental evolution and deep structure. The search for such a theory remains a central focus of geologic research.

SUMMARY

What are the major tectonic provinces of North America? The continent's most ancient crust is exposed in the Canadian Shield. South of the Canadian Shield is the interior platform, where Precambrian basement rocks are covered by layers of Paleozoic sedimentary rocks. Around the edges of these provinces are elongated mountain chains. The Appalachian fold belt trends southwest to northeast on the eastern margin of the continent. The coastal plain and continental shelf of the Atlantic Ocean and Gulf of Mexico are parts of a passive continental margin that subsided after rifting during the breakup of Pangaea. The North American Cordillera is a mountainous region running down the western side of North America that contains several distinct tectonic provinces.

What types of tectonic provinces are found worldwide? The types of tectonic provinces seen in North America are found on other continents as well. Continental shields and platforms make up continental cratons, the oldest and most stable parts of continents. Around these cratons are orogens, the youngest of which are found at the active margins of continents, where tectonic deformation continues. The passive margins of continents are zones of crustal extension and sedimentation.

How do continents grow? Two plate tectonic processes, magmatic addition and accretion, add crust to continents. Buoyant silica-rich rocks are produced by magmatic differentiation, primarily in subduction zones, and added to the continental crust by vertical transport. Accretion occurs when preexisting crustal material is attached to existing continental masses by horizontal plate movement in one of four ways: the transfer of buoyant crustal fragments from a subducting plate to a continent on an overriding plate; the closure of a sea separating an island arc from a continent; the transport of crust laterally along continental margins by strike-slip faulting; or the collision and suturing of two continents and their subsequent rifting apart.

How do orogenies modify continents? Horizontal tectonic forces, arising mainly from plate convergence, can produce mountains by folding and faulting. Thrust faulting can stack the upper part of the crust into overthrust sheets tens of kilometers thick, pushing up high mountains. Compression can double the thickness of continental crust, causing the rocks in the lower crust to melt. This melting generates granitic magma, which rises to form extensive batholiths in the upper crust.

What is the Wilson cycle? The Wilson cycle is a sequence of tectonic events that occur during the assembly and breakup of supercontinents and the opening and closing of ocean basins. It has four main phases: rifting during the breakup of a supercontinent; passive margin cooling and sediment accumulation during seafloor spreading and ocean opening; active magmatic addition and accretion during subduction and ocean closure; and orogeny during continent-continent collision. Orogeny is followed by erosion, which thins the crust.

What are the mechanisms of epeirogeny? Epeirogeny is a downward or upward movement of a broad region of crust without folding or faulting. Epeirogenic upward movements can result from glacial rebound, heating of the lithosphere by upwelling mantle material, and possibly uplifting of the lithosphere by a "superplume" in the deep mantle. The cooling of previously heated lithosphere can cause epeirogenic downward movements in the interior of a continent or at the margins of two continents separated by rifting. These movements form thermal subsidence basins that become filled with sediments.

How have continental cratons survived billions of years of plate tectonic processes? The oldest regions of the cratons, formed in the Archean eon, are underlain by a layer of cool, strong mantle material more than 200 km thick that moves with the continents as they drift. These cratonic keels are probably made up of mantle peridotites that have been depleted of their denser chemical constituents by the extraction of magmas through partial melting. This process lowers the density of the keels and stabilizes them against disruption by plate tectonic processes.

KEY TERMS

accreted terrain (p. 263)

accretion (p. 263)

active margin (p. 260)

craton (p. 260)

cratonic keel (p. 276)

epeirogeny (p. 272)

glacial rebound (p. 272)

magmatic addition (p. 262)

orogen (p. 260)

orogeny (p. 266)

passive margin (p. 260)

rejuvenation (p. 258)

shield (p. 255)

tectonic age (p. 261)

tectonic province (p. 254)

Wilson cycle (p. 272)

PRACTICING GEOLOGY EXERCISE

How Fast Are the Himalaya Rising, and How Quickly Are They Eroding?

The Himalaya, the world's highest and most rugged mountains, are being raised by thrust faulting caused by the collision of India with Asia (see Figure 10.15). How rapidly are they rising, and how quickly are they being eroded away? The answers to these questions depend on accurate topographic mapping.

On February 6, 1800, Colonel William Lambton, of the 33rd Regiment of Foot of the British Army, received orders to begin the Great Trigonometrical Survey of India, the most ambitious scientific project of the nineteenth century. Over the next several decades, intrepid British explorers led by Lambton and his successor, George Everest, hauled bulky telescopes and heavy surveying equipment through the jungles of the Indian subcontinent, triangulating the positions of reference monuments established at high points in the terrain, from which they could accurately establish Earth's size and shape. Along the way, in 1852, the surveyors discovered that an obscure Himalayan peak, known on their maps only as "Peak XV," was the highest mountain on Earth. They promptly named it Mount Everest, in honor of their former boss. Its official Tibetan name, Chomolungma, means "Mother of the Universe."

On February 11, 2000, almost exactly 200 years after Lambton commenced his exploration, NASA launched another great survey, the Shuttle Radar Topography Mission (SRTM). The space shuttle *Endeavour* carried two large radar antennas into low Earth orbit, one in the cargo bay and the second mounted on a mast that could extend up to 60 m outward. Working together like a pair of eyes, these antennas mapped the height of the land

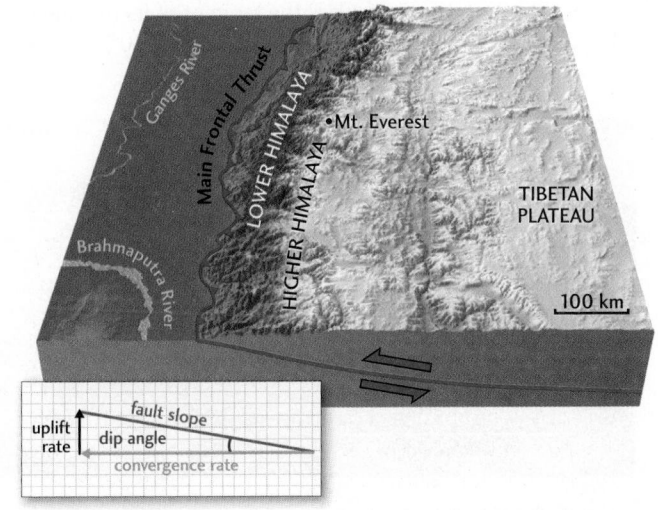

Cross section of the Himalaya, showing the approximate location of the thrust fault that is uplifting the mountains. The dip angle is about 10°.

surface below the shuttle on a grid of very dense geographic points, rendering the terrain in unprecedented three-dimensional detail. Remarkably, the height of Mount Everest as confirmed by the SRTM (8850 m, or 29,035 feet) turned out to be only 10 m more than the original 1852 estimate.

Although the accuracy of the Great Trigonometrical Survey was impressive, data collection was a slow process. It took the British over 70 years to measure the positions of 2700 stations across the Indian subcontinent, an average of about one position every 3 months. In comparison, the SRTM collected about 3000 position measurements *each second*. In just 11 days, the SRTM mapped 2.6 billion points covering 80 percent of Earth's land surface, including many remote areas of the continents that had not been previously surveyed. And, unlike the British surveyors, the shuttle crew did not have to contend with malaria or tigers!

The SRTM position measurements have been used to create a *digital elevation model*, or DEM, of the Himalaya, shown here as a topographic map. An analysis of the features on this map, which includes Earth's highest peaks and deepest gorges, indicates that the average height of the mountain range is staying approximately constant in time. In other words, the rate at which the Himalaya are rising is almost exactly balanced by the rate at which they are eroding:

$$\text{uplift rate} = \text{erosion rate}$$

As shown in the cross section, the geometry of the main thrust fault implies that

$$\text{thrust fault slope} = \text{uplift rate} \div \text{convergence rate}$$

Using GPS data, geologists have measured the convergence rate across the Himalaya to be about 20 mm/year. From earthquake locations, we know that the main thrust fault dips at an angle of about 10° below the mountain range. The slope of the fault is the tangent of its dip angle. Using a scientific calculator, we find tangent(10°) = 0.18.

The digital elevation model for the Mount Everest region of the Himalaya is derived from SRTM positions with a horizontal spacing of 90 m. [NASA images by Robert Simmon, based on SRTM data.].

Therefore, the erosion rate is

$$\text{erosion rate} = \text{thrust fault slope} \times \text{convergence rate}$$
$$= 0.18 \times 20 \text{ mm/year}$$
$$= 3.6 \text{ mm/year}$$

This estimate is consistent with the erosion rate of 3–4 mm/year obtained from the pressure-temperature paths of metamorphic rocks in the Himalaya exhumed by erosion, using the techniques described in Chapter 6.

BONUS PROBLEM: Given that the convergence rate between the Indian and Eurasian plates is about 54 mm/year (see Figure 2.7), what fraction of the relative plate movement is taken up by thrust faulting in the Himalaya? How is the remaining plate movement accommodated by deformation in Eurasia?

EXERCISES

1. Draw a rough topographic profile of the United States from San Francisco to Washington, D.C., and label the major tectonic provinces.

2. Why is the topography of the North American Cordillera higher than that of the Appalachian Mountains? How long ago were the Appalachians at their highest elevation?

3. Describe the tectonic province in which you live.

4. Are the interiors of continents usually younger or older than their margins? Explain your answer using the concept of the Wilson cycle.

5. Four processes of continental accretion are described in Figure 10.12. Illustrate two of them with examples of accreted terrains in North America.

6. Two continents collide, thickening the crust from 35 km to 70 km and forming a high plateau. After hundreds of millions of years, the plateau is eroded down to sea level. (a) What kinds of rocks might be exposed at the surface by this erosion? (b) Estimate the crustal thickness after the erosion has occurred. (c) Where in North America has this sequence of events been recorded in surface geology?

7. How many times have the continents been joined in a supercontinent since the end of the Archean eon? Use this number to estimate the typical duration of a Wilson cycle and the speed at which plate tectonic processes move continents.

8. How was orogeny in the Archean eon different from orogeny during the Proterozoic and Phanerozoic eons? What factors might explain these differences?

THOUGHT QUESTIONS

1. How would you recognize an accreted terrain? How could you tell whether it originated far away or nearby?

2. How would you identify a region where orogeny is taking place today? Give an example of such a region.

3. Would you prefer to live on a planet with orogenies or without them? Why?

4. Figure 10.8b shows more continental crust of Mesozoic-Cenozoic age than of any other tectonic age. Does this observation contradict the hypothesis that most of the continental crust was differentiated from the mantle in the first half of Earth's history?

5. Why are the ocean basins just about the right size to contain all the water on Earth's surface?

6. What would happen at Earth's surface if the cold keel beneath a craton were suddenly heated up? How might this effect be related to the formation of the Colorado Plateau?

MEDIA SUPPORT

 10-1 Animation: Major Tectonic Features of North America

 10-1 Video: The Wasatch Fault: Active Fault in the Rockies

- The Biosphere as a System   284

- Microorganisms: Nature's Tiny Chemists   288

- Geobiological Events in Earth's History   296

- Evolutionary Radiations and Mass Extinctions   300

- Astrobiology: The Search for Extraterrestrial Life   306

Grand Prismatic Hot Spring, Yellowstone National Park, Wyoming. The striking array of colors reflects different communities of microorganisms that are very sensitive to water temperature. Water flowing away from the center of the spring (blue) cools down, causing a given community of microorganisms to be replaced by a different community that grows best at the new, lower temperature. The boardwalk visible in the lower part of the photo allows tourists to peer into its depths and provides a sense of scale. [Luis Castañeda/age fotostock.]

GEOBIOLOGY: LIFE INTERACTS WITH EARTH

11

GEOLOGY IS THE STUDY of the physical and chemical processes that control the Earth system, today and in the past. Biology is the study of life and living organisms, including their structure, function, origin, and evolution. As separate as geology and biology may seem, organisms and their physical environment interact in many ways. We have long recognized that biology and geology are intimately related, but until recently, we have not known exactly how. Fortunately, technological advances in both Earth and life sciences now allow us to ask and answer questions that were previously beyond our scope. Over the past decade, scientists working at the frontiers of both fields have begun to understand how several important geobiological processes work.

We know that organisms can change Earth. For example, Earth's atmosphere is distinct from that of every other planet in having a significant concentration of oxygen—the result of the evolution of oxygen-producing microorganisms billions of years ago. Organisms also contribute to the weathering of rocks by releasing chemicals that help break down minerals; through this process, they obtain nutrients essential to their growth. Similarly, geologic processes can change life, as when an asteroid struck Earth 65 million years ago, causing a mass extinction that killed off the dinosaurs.

This chapter explores the links between organisms and Earth's physical environment. It describes how the biosphere works as a system and what gives Earth its ability to support life. Next, it explores the remarkable roles microorganisms play in geologic processes, and it discusses some of the major geobiological events that have changed our planet. Finally, it considers the key ingredients for sustaining life and ponders the eternal question posed by astrobiologists: Is there life out there?

■ The Biosphere as a System

Life is everywhere on Earth. The **biosphere** is that part of our planet that contains all of its living organisms. It includes the plants and animals with which we are most familiar as well as the nearly invisible microorganisms that live in some of the most extreme environments on Earth. These organisms live on Earth's surface, in its atmosphere and ocean, and within its upper crust, and they interact continuously with all of these environments. Because the biosphere intersects with the lithosphere, hydrosphere, and atmosphere, it can influence or even control basic geologic and climate processes. **Geobiology** is the study of these interactions between the biosphere and Earth's physical environment.

The biosphere is a system of interacting components that exchanges energy and matter with its surroundings. Inputs into the biosphere include energy (usually in the form of sunlight) and matter (such as carbon, nutrients, and water). Organisms use these inputs to function and grow. In the process, they create an amazing variety of outputs, some of which have important influences on geologic processes. At a local scale—such as that of a water-filled pore within loose sediment particles—a small group of organisms may have a geologic effect that is limited to a particular sedimentary environment. At larger scales, the activities of organisms may influence the concentrations of gases in the atmosphere or the cycling of certain elements through Earth's crust.

Ecosystems

Think of a class project in which each member of a team has special skills that allow the team as a whole to exceed the capabilities of individuals working alone. Groups of organisms act in similar ways: individual organisms play roles that contribute to the survival of other organisms as well as their own. In the case of human groups, we accomplish this teamwork as a result of conscious decisions. For the organisms living together in a particular environment—referred to as a *community*—it happens through trial and error and involves feedbacks between the community and individuals. These feedbacks determine the structure and functioning of the community.

Whether at local, regional, or global scales, the interactions of biological communities with their environments define organizational units known as **ecosystems.** Ecosystems are composed of biological and physical components that function in a balanced, interrelated fashion. Ecosystems occur at many different scales (**Figure 11.1**). They may be separated by geologic barriers such as mountains, deserts, or oceans at the largest scale, or by barriers such as different water temperatures within a single hot spring at a much smaller scale (see the chapter opening photo). But no matter how large or small they are, all ecosystems are characterized by a flow, or *flux*, of energy and matter between organisms and their environment.

A typical ecosystem might involve, say, a river and its surroundings, where different groups of organisms are adapted to live in the water (fish), in the sediment

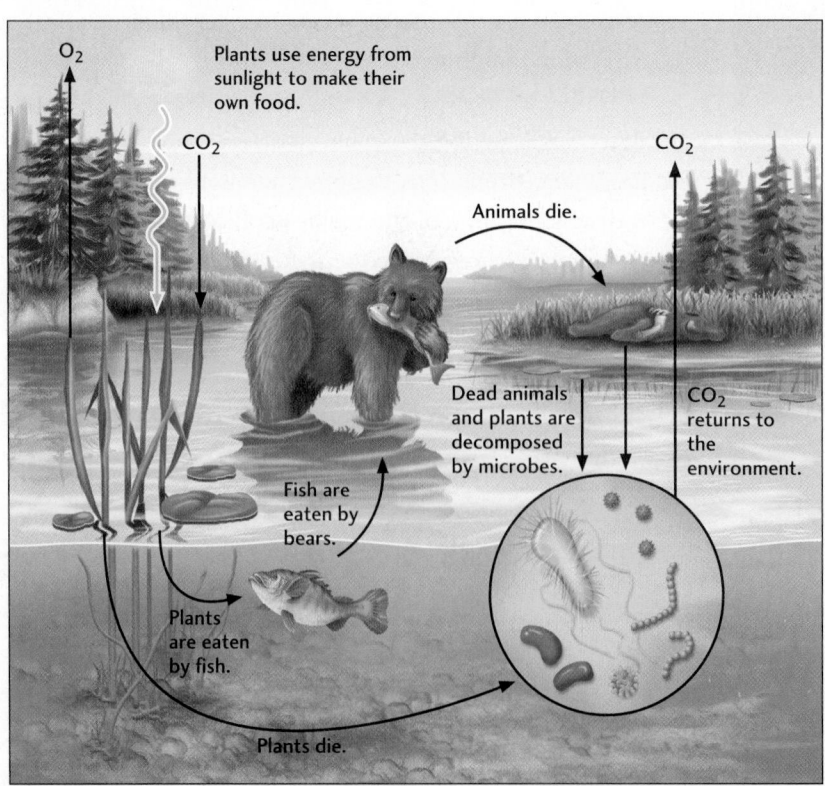

FIGURE 11.1 ■ Ecosystems are characterized by a flux of energy and matter between organisms and their environment. In this example, sunlight is used as an energy source by plants, which are eaten by fish, which are eaten by bears. The plants, fish, and bears eventually die and are decomposed by microorganisms. In this way, the matter that made up those organisms is returned to the physical environment, where it can be used again.

(worms, snails), on the banks (grass, trees, muskrats), and in the sky above (birds, insects). In one sense, the river controls where the organisms live by supplying the ecosystem with water, sediment, and dissolved mineral nutrients. Conversely, the organisms influence how the river behaves; for example, grass and trees stabilize the riverbanks against the destructive effects of floods. The balance between such biologically controlled and geologically controlled processes ensures the long-term stability of the ecosystem.

Ecosystems respond sensitively to biological changes, such as the introduction of new groups of organisms. When severe imbalances in ecosystems occur, responses are often dramatic. Consider the effects of introducing a new organism into your neighborhood environment, such as a pretty new plant for your garden. In all too many cases, the new organism is better suited for its new environment than the current inhabitants and becomes *invasive,* multiplying rapidly and squeezing out the previous inhabitants (**Figure 11.2**). A successful invader often comes from a place where the physical environment is similar but where biological competition is less intense, so it is likely to win the battle for nutrients and space in its new home. Organisms that are squeezed out may become *extinct* if they are outcompeted by the invader in all the regions they formerly occupied.

Earth's history shows us that ecosystems respond sensitively to geologic processes as well. Impacts by meteorites, huge volcanic eruptions, and rapid global warming are just a few of the processes that have contributed to the extinction of major groups of organisms. We will explore some of their effects later in this chapter.

Inputs: The Stuff Life Is Made Of

The organisms of any ecosystem can be subdivided into producers and consumers according to the way they obtain their *food,* which is their source of energy and nutrients (**Table 11.1**). Producers, or **autotrophs,** are organisms that make their own food. They use energy from sunlight or, in some cases, energy derived from chemicals in their environment to manufacture organic compounds such as carbohydrates. Consumers, or **heterotrophs,** get their food by feeding directly or indirectly on producers.

It is often said that you are what you eat, and that statement is true not only for humans, but for all organisms. Our foods are all made up of more or less the same materials: molecules composed of carbon, hydrogen, oxygen, nitrogen, phosphorus, and sulfur. So it doesn't matter whether the organism is an autotroph or a heterotroph, it still utilizes the same six elements as food. What differs is the form (that is, the molecular structure) their food comes in. When heterotrophs such as humans eat bread, we are feeding ourselves on *carbohydrates:* large molecules formed of carbon, hydrogen, and oxygen. Even the lowliest microorganisms dine on carbon-bearing molecules such as

(a)

(b)

(c)

FIGURE 11.2 ■ Invasive organisms create problems by dominating their local ecosystems. (a) Kudzu, introduced into North America to stop highway erosion, rapidly overgrows other plants. (b) Purple loosestrife, introduced from Europe as a garden flower, has invaded many North American wetlands. (c) Zebra mussels aggressively colonize and overwhelm ordinary mussels. In 2002, the U.S. Fish and Wildlife Service estimated that $5 billion was spent by electric utility companies just to unclog water intake pipes blocked by zebra mussels. [(a) Kerry Britton/Forest Service, USDA; (b) © RalphWilliam/Alamy; (c) courtesy of U.S. Fish and Wildlife Service/ Washington, DC, Library.]

TABLE 11-1	Organisms as Producers and Consumers		
Type	Energy Source	Carbon Source	Example
Photoautotroph	Sun	CO_2	Cyanobacteria
Photoheterotroph	Sun	Organic compounds	Purple bacteria
Chemoautotroph	Chemicals	CO_2	H, S, Fe bacteria
Chemoheterotroph	Chemicals	Organic compounds	Most bacteria, fungi, and animals, including humans

carbon dioxide (CO_2) or methane (CH_4). The difference is that what is food for a microorganism may not seem like food to us!

CARBON The fundamental building block of all life on Earth is carbon. On our planet, wherever life is present, carbon must be present. If water is removed, the composition of all organisms, including humans, is dominated by carbon. No other chemical element can match carbon in the variety and complexity of the compounds it can form. Part of the reason for its versatility is that it can form four covalent bonds with itself and other elements (see Figure 3.3), which allows for a wide variety of structures. Carbon acts as the template around which all organic molecules, such as carbohydrates and proteins, are built. Thus, carbon is critically important to organisms because it goes into manufacturing every part of them, from genes to body structures.

The biosphere largely controls the flux of carbon through the Earth system. Marine organisms extract carbon—which is present in seawater as dissolved CO_2—to form carbonate shells and skeletons. When the organisms die, their skeletons settle to the seafloor, where they accumulate as sediments, effectively moving carbon from the biosphere to the lithosphere. The accumulation of the organic remains of organisms in freshwater wetlands and on the seafloor also moves carbon from the biosphere to the lithosphere. Over geologic time, these organic remains are transformed into oil, natural gas, and coal deposits. Today, when we extract and burn these deposits, we are moving carbon from the lithosphere to the atmosphere in the form of CO_2 emissions.

NUTRIENTS Nutrients are chemical elements or compounds that organisms require to live and grow. Common plant nutrients include the elements phosphorus, nitrogen, and potassium—the ones most commonly found in garden fertilizer. Other organisms also depend on iron and calcium. Some organisms can manufacture their own nutrients, but others must obtain them in their diets from materials in their environment. Some specialized microorganisms can obtain nutrients by dissolving minerals.

WATER Life as we know it requires water (H_2O). All organisms on Earth, including humans, are composed primarily of water, typically 50 to 95 percent. It is well known that humans can live for weeks without food, but most will perish in a few days without water. Even microorganisms that live in the atmosphere must obtain water from tiny droplets that condense around dust particles, and viruses must obtain water from their hosts.

Water is the habitat in which life first emerged and in which much of it still thrives. Water's chemical properties and the way it responds to changes in temperature make it an ideal medium for biological activity. The cells of all organisms are made up primarily of an aqueous solution that promotes the chemical reactions of life. Water also helps moderate Earth's climate, which has supported life for at least 3.5 billion years (see Chapter 15). Water is such an important ingredient for life that the search for extraterrestrial life must begin with the search for water, as we will see at the end of this chapter.

ENERGY All organisms need energy to live and grow. Some of the simplest organisms, such as single-celled algae, obtain energy from sunlight. Others acquire energy by breaking down chemicals in their environment. Heterotrophs get energy by feeding on other organisms. Energy is important because it fuels the conversion of simple molecules such as carbon dioxide and water into larger molecules, such as carbohydrates and proteins, which are essential for life.

Processes and Outputs: How Organisms Live and Grow

Metabolism encompasses all the processes organisms use to convert inputs into outputs. In one type of metabolic process, organisms use small molecules, such as CO_2, H_2O, and CH_4, and energy to create larger molecules, such as proteins and certain types of carbohydrates, that enable them to function and grow. Other carbohydrates—for example, a sugar called glucose—are stored for later use as an energy source—that is, food. In another type of metabolic process, organisms break down food to release the energy it contains.

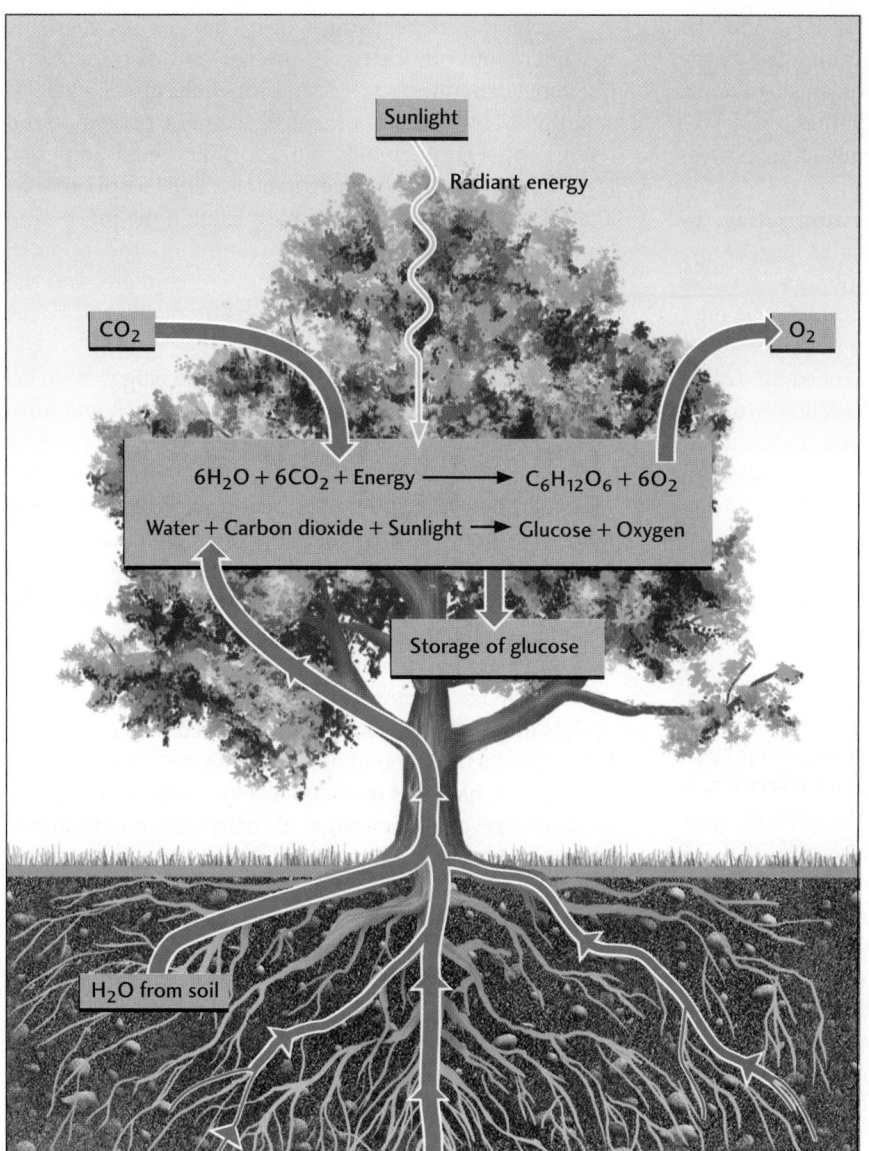

One particularly well-known metabolic process is **photosynthesis** (Figure 11.3). Through this process, organisms such as plants and algae use energy from sunlight to convert water and carbon dioxide into carbohydrates (such as glucose) and oxygen (Table 11.2). This reaction proceeds as follows:

$$\text{water} + \text{carbon dioxide} + \text{sunlight} \rightarrow$$
$$6\,H_2O + 6\,CO_2 + \text{energy} \rightarrow$$
$$\text{glucose} + \text{oxygen}$$
$$C_6H_{12}O_6 + 6\,O_2$$

TABLE 11-2	Comparison of Photosynthesis and Respiration
Photosynthesis	**Respiration**
Stores energy as carbohydrates	Releases energy from carbohydrates
Uses CO_2 and H_2O	Releases CO_2 and H_2O
Increases mass	Decreases mass
Produces oxygen	Consumes oxygen

The oxygen is released into the atmosphere, and the glucose is stored as an energy source for future use by the organism. An important group of microorganisms known as the **cyanobacteria** also use photosynthesis to make carbohydrates; in fact, they probably originated the process early in life's history.

The other key metabolic process is **respiration,** by which organisms release the energy stored in carbohydrates such as glucose (see Table 11.2). All organisms use oxygen to burn, or *respire,* carbohydrates to release energy, but different organisms respire in different ways. For example, humans and many other organisms consume oxygen gas (O_2) from the atmosphere to metabolize carbohydrates, and they release carbon dioxide and water as by-products. In this case, the reaction is the reverse of photosynthesis:

glucose + oxygen →
$C_6H_{12}O_6$ + 6 O_2 →

water + carbon dioxide + energy
6 H_2O + 6 CO_2 + energy

But other organisms, such as microorganisms that live in environments where oxygen is absent, have a more difficult task. They must break down oxygen-containing compounds dissolved in water, such as sulfate (SO_4^{-2}), to obtain oxygen. During the course of these reactions, various gases—such as hydrogen (H_2), hydrogen sulfide (H_2S), and methane (CH_4)—may be produced as by-products.

The metabolism of organisms affects the geologic components of their environment. For example, the oxygen released by photosynthesis reacts with iron-bearing silicate minerals such as pyroxene and amphibole to form iron-bearing oxide minerals such as hematite (see Chapter 16). When organisms produce CO_2 and CH_4, they escape to the atmosphere and contribute to global warming. Conversely, when organisms consume these gases, they contribute to global cooling.

Biogeochemical Cycles

In the course of living and dying, organisms continuously exchange energy and matter with their environment. This exchange occurs at the scale of the individual organism, the ecosystem of which it is a part, and the global biosphere. The metabolic consumption and production of gases such as CO_2 and CH_4 is a good example of how organisms may exert global controls on Earth's climate. Carbon dioxide and methane are *greenhouse gases:* gases that absorb heat emitted by Earth and trap it in the atmosphere. When organisms produce more CO_2 and CH_4 than they consume, the climate will warm; when they consume more CO_2 and CH_4 than they produce, the climate will cool. The concentration of greenhouse gases in the atmosphere is not the only control on global climates, as we will learn in Chapter 15, but it is an important one that directly involves the biosphere.

Geobiologists keep track of exchanges between the biosphere and other parts of the Earth system by studying biogeochemical cycles. A **biogeochemical cycle** is a pathway by which a chemical element or compound moves between the biological ("bio") and environmental ("geo") components of an ecosystem. The biosphere participates in biogeochemical cycles through the inflow and outflow of atmospheric gases by respiration, the inflow of nutrients from the lithosphere and hydrosphere, and the outflow of those nutrients through the death and decay of organisms.

Because ecosystems vary in scale, so do biogeochemical cycles. Phosphorus, for example, may cycle back and forth between the water and the microorganisms in the pores of sediments, or it may cycle back and forth between uplifted rocks in mountains and the sediments deposited along the margins of ocean basins (**Figure 11.4**). In either case, when phosphorus-containing organisms die, the phosphorus may accumulate in a temporary repository before being recycled. Sediments and sedimentary rocks are an important repository for this element.

Knowledge of biogeochemical cycles is important for understanding the mechanisms associated with major geobiological events throughout Earth's history, as we will see later in this chapter. It is also critical for understanding how elements and compounds that humans emit into the atmosphere and ocean are interacting with the biosphere, as we will see in Chapters 15 and 23.

■ Microorganisms: Nature's Tiny Chemists

Single-celled organisms, which include bacteria, archaea, some fungi, some algae, and most protists, are known as **microorganisms,** or *microbes.* Wherever there is water, there are microorganisms. Microorganisms, like other organisms, need water to live and reproduce. Microorganisms can be as small as a few microns in size (1 micron = 10^{-6} m) and can inhabit almost any nook or cranny you can think of, from at least 5 km beneath Earth's surface to more than 10 km high in the atmosphere. They live in air, in soil, on and in rocks, inside roots, in piles of toxic waste, on frozen snowfields, and in water bodies of every type, including boiling hot springs. They live at temperatures that range from lower than −20°C to higher than the boiling point of water (100°C).

People have exploited the useful effects of microbial metabolism for thousands of years to produce bread, wine, and cheese. Today, people also use microorganisms to produce antibiotics and other valuable drugs. Geobiologists study microorganisms to understand their roles in biogeochemical cycles and to understand the early evolution of the biosphere before the advent of more complex organisms.

Tectonic processes uplift phosphate-containing rocks.

Phosphate-containing rocks are weathered and eroded.

Plants take up phosphorus from soil.

Animals eat plants.

Runoff carries sediment to rivers, lakes, and oceans.

Phosphate-bearing compounds in fertilizers dissolve in water.

Phosphate-containing minerals accumulate to form phosphate-containing rock.

Decomposers break down animal and plant matter and return phosphorus to soil.

Phosphorus leaches from soil into water.

FIGURE 11.4 ■ The biosphere plays key roles in the biogeochemical cycle of phosphorus.

Abundance and Diversity of Microorganisms

Microorganisms dominate Earth in terms of numbers of individuals. Concentrations ranging from 10^3 to 10^9 microorganisms/cm^3 have been reported from soils, sediments, and natural waters. Every time you walk on the ground, you step on billions of microorganisms! In some cases, surfaces become coated with dense concentrations of microorganisms called *biofilms*, which may contain as many as 10^8 individuals/cm^2 of surface area.

More important, microorganisms are the most genetically diverse group of organisms on Earth. **Genes** are large molecules within the cells of every organism that encode all of the information that determines what that organism will look like, how it will live and reproduce, and how it differs from all other organisms. Genes are also the basic

hereditary units passed on from generation to generation. The genetic diversity of microorganisms is important because it has allowed them to colonize, adapt to, and thrive in environments that would be lethal to most other organisms. These abilities, in turn, are important because they allow microorganisms to recycle important materials in a broad—even extreme—range of geologic environments.

THE UNIVERSAL TREE OF LIFE Biologists have learned how to use the genetic information contained in living organisms to understand which forms of life are most closely related to one another. This knowledge has allowed them to organize the hierarchy of ancestors and descendants into a universal tree of life (**Figure 11.5**). About 30 years ago, a startling discovery was made when the first family trees for microorganisms were constructed. When the genes for *all types of microorganisms* were compared, it was shown that,

FIGURE 11.5 ■ The universal tree of life shows how all organisms are related to one another. Organisms are subdivided into three great domains: the Bacteria, Archaea, and Eukaryota. These domains are all descended from a universal common ancestor. All three domains are dominated by microorganisms. Note that animals appear at the tip of the eukaryote branch. (Ma, million years ago.)

despite their similar sizes (tiny) and shapes (simple rods and ellipses), there were enormous differences in their genetic content. Furthermore, when the genes of *all types of organisms,* including plants and animals, were compared, it was revealed that the differences among groups of microorganisms were much greater than the differences between plants and animals, including humans.

THE THREE DOMAINS OF LIFE The single root of the universal tree of life shown in Figure 11.5 is called the *universal ancestor.* This universal ancestor gave rise to three major groups, or domains, of descendants: the Bacteria, the Archaea, and the Eukaryota. The Bacteria and Archaea appear to have evolved first; all of their descendants have remained single-celled microorganisms. The Eukaryota, thought to be the youngest branch of the tree, have a more complex cellular structure, which includes a cell nucleus that contains the genes. This structure made it possible for eukaryotes to evolve from small, single-celled organisms into larger, multicellular organisms—an essential step in the evolution of animals and plants.

Precambrian microorganisms, like those living today, were tiny. The traces of individual microorganisms preserved in rocks are therefore called **microfossils.** Needless to say, such features are much harder to find than the macroscopic fossils of shells, bones, and twigs used by geologists to study the evolution of animals and plants during the Phanerozoic eon (recall that *Phanerozoic* means "visible life").

For geobiologists, the universal tree of life is a map that reveals how microorganisms relate to one another and interact with Earth. The names of microorganisms, such as *Halobacterium, Thermococcus,* and *Methanopyrus,*

suggest that these organisms can live in extreme environments that are very salty (*halo,* "halite"), or hot (*thermo*), or high in methane (*methano*). Microorganisms that live in extreme environments are almost exclusively archaea and bacteria.

EXTREMOPHILES: MICROORGANISMS THAT LIVE ON THE EDGE **Extremophiles** are microorganisms that live in environments that would kill other organisms (Table 11.3). The suffix *phile* is derived from the Latin word *philus,* which means "to have a strong affinity or preference for." Extremophiles live on all kinds of foods, including oil and toxic wastes. Some use substances other than oxygen, such as nitric acid, sulfuric acid, iron, arsenic, or uranium, for respiration.

Acidophiles are microorganisms that thrive in acidic environments. Acidophiles can tolerate pH levels low enough to kill other organisms. These microorganisms live by eating sulfide! They are able to survive in such acidic habitats because they have developed a way to pump out the acid that accumulates inside their cells. Such extremely acidic habitats occur naturally (see Earth Issues 11.1, page 292), but are more commonly associated with mining.

Thermophiles are microorganisms that live and grow in extremely hot environments. They grow best in temperatures that are between 50°C and 70°C and can tolerate temperatures up to 120°C. They will not grow if the temperature drops to 20°C. Thermophiles live in geothermal habitats, such as hot springs and hydrothermal vents at mid-ocean ridges, and in environments that create their own heat, such as compost piles and garbage landfills. The microorganisms

TABLE 11-3   Characteristics of Extremophiles

Type	Tolerance	Environment	Example
Halophile	High salinity	Playa lakes Marine evaporites	Great Salt Lake, Utah
Acidophile	High acidity	Mine drainage Water near volcanoes	Rio Tinto, Spain
Thermophile	High temperature	Hot springs Mid-ocean ridge vents	Yellowstone National Park
Anaerobe	No oxygen	Pores of wet sediments Groundwater Microbial mats Mid-ocean ridge vents	Cape Cod Bay sediments

that cover the bottom of Grand Prismatic Hot Spring (see the chapter opening photo) are dominated by thermophiles. Of the three domains of life, the Eukaryota (which include humans) are generally the least tolerant of high temperatures (60°C seems to be their upper limit). The Bacteria are more tolerant (with an upper limit close to 90°C), and the Archaea are the most tolerant, able to withstand temperatures of up to 120°C. Microorganisms that can stand temperatures above 80°C are called *hyperthermophiles*.

Halophiles are microorganisms that live and grow in highly saline environments. They can tolerate salt concentrations up to 10 times that of normal ocean water. Halophiles live in naturally hypersaline playa lakes such as the Great Salt Lake and the Dead Sea (see Chapter 19) and in some parts of the ocean, such as the southern end of San Francisco Bay, where seawater is commercially evaporated

to extract salt (**Figure 11.6**). These microorganisms can control the salt concentration inside their cells by expelling extra salt from their cells into the environment.

Anaerobes are microorganisms that live in environments completely devoid of oxygen. At the bottoms of most lakes, streams, and oceans, the fluids in the pores of sediments just a few millimeters or centimeters below the sediment-water interface are starved of oxygen. Microorganisms that live at the sediment-water interface use up all the oxygen during respiration, creating an *anaerobic* (oxygen-free) *zone* beneath them in the sediment, where only anaerobes thrive. The oxygen-rich upper sediment layer is known as the *aerobic zone*. Many microorganisms that live in the aerobic zone could not survive in the anaerobic zone, and vice versa. The boundary between these zones is often very sharp, as shown in **Figure 11.7**.

FIGURE 11.6 ▪ Humans have dammed off parts of the ocean to create ponds where seawater can evaporate to precipitate halite for table salt and other uses. The halophilic bacteria that thrive in these hypersaline environments produce a distinctive pigment that turns the ponds pink. [Yann Arthus-Bertrand/CORBIS.]

Earth Issues

11.1 Sulfide Minerals React to Form Acidic Waters on Earth and Mars

Many economically significant mineral deposits are associated with high concentrations of sulfide minerals. When water comes into contact with sulfide minerals, the sulfide they contain reacts with oxygen to form sulfuric acid. Thus, during the course of mining and afterward, rainwater and groundwater may interact with these minerals to produce highly acidic surface water and groundwater. Unfortunately, these acidic waters are lethal to most organisms. As they spread throughout the environment, extensive devastation may result. In some cases, the only organisms that survive are acidophilic extremophiles.

In a few places on Earth, where sulfide minerals occur in high enough concentrations, acidic waters are produced naturally. One of these places is the Rio Tinto in Spain. Here, geologists have been able to study a system in which a naturally occurring ore deposit, almost 400 million years old, interacts with groundwater that flows through it by hydrothermal circulation. With the help of mineral-dissolving acidophilic microorganisms, sulfide minerals such as pyrite (FeS_2) in the ore deposit react with oxygen in the groundwater to produce sulfuric acid, sulfate ions (SO_4^{-2}), and iron ions (Fe^{3+}). The warm spring water that flows out of the deposit as a river (*rio* in Spanish) is extremely acidic. Your skin would dissolve if you went swimming in that water.

The river is red (*tinto* in Spanish) because of the dissolved Fe^{3+} ions. The Fe^{3+} ions combine with oxygen to produce the iron oxide minerals goethite and hematite, which may be reddish or brownish in color. In addition, unusual iron sulfate minerals such as jarosite (yellow-brown in color) form abundantly in the Rio Tinto. When geologists encounter this mineral on Earth, they know that the water from which it precipitated must have been extremely acidic.

What is a rare—and environmentally damaging—geologic setting on Earth may once have been widespread on Mars. As we saw in Chapter 9, past exploration of Mars has revealed abundant sulfate minerals similar to those found in the Rio Tinto, including jarosite. Understanding how this unusual mineral forms on Earth allows geologists to make inferences about past environments on Mars. In this case, the presence of jarosite indicates that some ancient waters on Mars were very acidic, perhaps because of the interaction of groundwater with igneous rocks composed of basalt with trace amounts of sulfide.

This scenario has implications for how we think about the possibility of life—past or present—on other planets. Environments such as the Rio Tinto on Earth show that microorganisms have learned to adapt to highly acidic conditions, and they help motivate the search for ancient life on Mars. Some scientists, however, think that although life may have learned to adapt to such harsh conditions, it may not have been able to originate under those conditions. In any event, the search for life on other planets will be strongly guided by our understanding of rocks, minerals, and extreme environments on Earth.

Microorganisms thrive in the acidic water of the Rio Tinto, Spain. [Courtesy of Andrew H. Knoll.]

Microorganism-Mineral Interactions

Microorganisms play a critical role in many geologic processes, including mineral precipitation, mineral dissolution, and the flux of elements through Earth's crust in biogeochemical cycles. As we will learn later in this chapter, they have also been crucial factors in the evolutionary history of larger, more complex organisms.

MINERAL PRECIPITATION Microorganisms precipitate minerals in two distinct ways: *indirectly* by influencing the composition of the water surrounding them and *directly* in their cells as a result of their metabolism. Indirect precipitation occurs when dissolved minerals in an oversaturated solution precipitate on the surfaces of individual microorganisms. This happens because the surface of a microorganism has sites that bind dissolved mineral-forming elements. Mineral precipitation often leads to the complete encrustation of the microorganisms, which are effectively buried alive. Microbial precipitation of carbonate minerals and silica in hot springs are good examples of this type of microbial biomineralization (**Figure 11.8a**). Thermophiles may become completely overgrown by the mineral deposits they help to precipitate.

Minerals are directly precipitated by the metabolic activities of some microorganisms. For example, microbial respiration causes precipitation of pyrite (**Figure 11.9**) in

FIGURE 11.7 ■
Microorganisms can form layered deposits called microbial mats. The top part of the mat, which is exposed to the Sun, contains photosynthetic autotrophic microorganisms, as revealed by the green color. Farther down in the mat, but still within the aerobic zone, are nonphotosynthetic autotrophs, as revealed by the purple color. Deeper in the mat, the color turns gray, revealing the anaerobic zone where heterotrophs live.
[John Grotzinger.]

the anaerobic zone of sediments that contain iron-bearing minerals and water in which sulfate is dissolved. As we have learned, all organisms—including microorganisms—need oxygen for respiration. In the anaerobic zone, however, O_2 is not available. Some microbial decomposers have adapted to this harsh, but very common, environment by evolving ways to obtain oxygen from other sources. These microorganisms can remove the oxygen contained in sulfate (SO_4), which is abundant in most sediment pore fluids. In the process, they make hydrogen sulfide gas (H_2S), which produces the unpleasant odor of rotten eggs that is released when you dig into sandy or muddy sediments at low tide. In the final step of the process, hydrogen sulfide reacts with iron, which replaces the hydrogen to form pyrite (FeS_2). Pyrite is remarkably abundant in sedimentary rocks that contain organic matter, such as shales. Another example of direct precipitation is the formation of tiny particles of magnetite inside some bacteria (Figure 11.8b), which use these crystals to navigate by sensing Earth's magnetic field.

(a) Indirect precipitation of calcium carbonate

(b) Direct precipitation of magnetite

FIGURE 11.8 ■ Microorganisms can precipitate minerals indirectly or directly. (a) The precipitation of calcium carbonate on the surfaces of bacteria is an example of indirect precipitation. (b) Intracellular production of magnetite (Fe_3O_4) crystals by some bacteria is an example of direct precipitation. Some organisms use these crystals to find their way by sensing Earth's magnetic field. [(a) Grant Ferris, University of Toronto; (b) Richard B. Frankel, Ph.D., California Polytechnic State University.]

FIGURE 11.9 ▪ Pyrite commonly forms small globules in the pore fluids of anaerobic sediments. [Courtesy of Dr. Jüergen Schieber.]

MINERAL DISSOLUTION Some elements that are essential for microbial metabolism, such as sulfur and nitrogen, are readily available from natural waters in dissolved form, but others, such as iron and phosphorus, must be actively scavenged from minerals by microorganisms. All microorganisms need iron, but iron concentrations in near-surface waters are generally so low that the microorganisms must obtain it by dissolving nearby minerals. In a similar way, some microorganisms obtain phosphorus—required for construction of biologically important molecules—by dissolving minerals such as apatite (calcium phosphate). Some autotrophs derive their energy not from sunlight, but from the chemicals produced when minerals are dissolved. These organisms are known as **chemoautotrophs** (see Table 11.1). For example, manganese (Mn^{2+}), iron (Fe^{2+}), sulfur (S), ammonium (NH_4^+), and hydrogen (H_2) supply microorganisms with energy when they are released from minerals.

Microorganisms dissolve minerals by producing organic molecules that react with those minerals to liberate ions from mineral surfaces. Rates of mineral dissolution are normally slow, but may be enhanced where minerals containing nutrient elements are coated by microbial biofilms. Mineral-dissolving acidophiles thrive in waters where mineral dissolution results in prolific acid formation.

MICROORGANISMS AND BIOGEOCHEMICAL CYCLES Pyrite precipitation by microorganisms plays an important role in the global biogeochemical cycling of sulfur (**Figure 11.10**). As we have seen, iron and sulfur are precipitated as pyrite, which accumulates abundantly within sediments. As layers of sediment are deposited, the pyrite becomes buried and encapsulated in sedimentary rocks. The pyrite remains buried until the rocks are returned to Earth's surface by tectonic uplift. When the rocks are weathered, the iron and sulfur in the pyrite are dissolved as ions in water or become incorporated into new minerals that accumulate in sediments, starting the biogeochemical cycle over again.

On a global scale, microorganisms play roles in several other biogeochemical cycles. Microbial precipitation of phosphate minerals contributes to the flow of phosphorus into sediments, particularly along the west coasts of South America and Africa, where phosphorus-rich deep ocean water that rises to the surface is available to microorganisms that live in shallower water, as we saw in Chapter 5. The chemical weathering of continental rocks is influenced by microorganisms that can increase the acidity of soils, leading to faster weathering rates. And finally, as we also saw in Chapter 5, the precipitation of carbonate minerals in marine environments is stimulated by microbial processes. This last example is especially important because carbonate minerals serve as a sink for atmospheric CO_2 and for cations such as Ca^{2+} and Mg^{2+} released during weathering of silicate minerals.

MICROBIAL MATS Microbial mats are layered microbial communities. The microbial mats you are most likely to see are those that are exposed to the Sun (see Figure 11.7). They commonly occur in tidal flats, hypersaline lagoons, and hot springs. On the top, you will usually find a layer of oxygen-producing cyanobacteria that use energy from sunlight for photosynthesis. This uppermost layer is green because cyanobacteria contain the same light-absorbing pigment that plants and algae have. This layer may be as thin as 1 mm, yet it can be as effective in producing energy from the Sun as a hardwood forest or grassland. This uppermost green layer defines the aerobic zone of the mat. The anaerobic zone occurs below the cyanobacterial layer and is often a dark gray color. Although this anaerobic part of the mat contains no oxygen, it still can be very active. The anaerobic heterotrophs in this layer derive their food from the organic matter produced by the cyanobacteria. Their respiration often results in the precipitation of pyrite, as described earlier in this chapter.

Microbial mats are miniature models of the same biogeochemical cycles that occur at regional or even global scales. In a microbial mat, photosynthetic autotrophs use energy from sunlight to convert carbon in atmospheric CO_2 into carbon in larger molecules such as carbohydrates. After the photoautotrophs die, the heterotrophs use the carbon in their bodies as an energy source. In the process, the heterotrophs convert some of this carbon into CO_2, which is returned to the atmosphere, where it can be used by the next generation of photoautotrophs, and so on. In the case of microorganisms, this cycle is confined to the very small scale of a layer of sediment, but it is directly analogous to the process by which rain forests—at a global scale—extract CO_2 from the atmosphere during photosynthesis. Although individual trees do the actual work, one can think of a rain forest as a giant photosynthesis machine that removes enormous quantities of CO_2 from the atmosphere and produces enormous quantities of carbohydrates. When the trees die, their organic matter is used by heterotrophs on the forest floor to produce energy. This

Volcanoes release hydrogen sulfide gas.

Humans burn fossil fuels, emitting sulfur compounds to the atmosphere.

Tectonic processes uplift rocks, and weathering breaks down sulfur-bearing minerals to release sulfur compounds to the atmosphere.

Rain combines with hydrogen sulfide to form sulfuric acid.

Sulfuric acid increases weathering of rocks.

Rivers transport sediment and dissolved sulfur to water bodies.

Plants take up sulfur-bearing compounds from soil.

Animals eat plants.

Dissolved sulfur precipitates in lakes and oceans as sulfate (e.g., gypsum) and sulfide (e.g., pyrite) minerals.

Sulfur leaches from soil into water.

Decomposers in anaerobic sediments break down animal and plant matter and produce hydrogen sulfide, which reacts with iron to produce pyrite.

FIGURE 11.10 ■ Pyrite precipitation by microorganisms is a key component of the sulfur cycle.

process returns enormous amounts of carbon—in the familiar form of CO_2—to the atmosphere.

STROMATOLITES Today, microbial mats are restricted to places on Earth where plants and animals cannot interfere with their growth. Before the evolution of plants and animals, however, microbial mats were widespread, and they are one of the most common features preserved in Precambrian sedimentary rocks formed in aquatic environments. **Stromatolites**—rocks with distinctive thin layers—are believed to have been formed from ancient microbial mats. Stromatolites range in shape from flat sheets to dome-shaped structures with complex branching patterns (**Figure 11.11**). They are one of the most ancient types of fossils on Earth and give us a glimpse of a world once ruled by microorganisms.

Most stromatolites probably formed when sediment raining down on microbial mats was trapped and bound by microorganisms living on the surfaces of the mats (Figure 11.11d). Once covered with sediment, the micro-

organisms grew upward between the sediment particles and spread laterally to bind the particles in place. Each stromatolite layer corresponds to the deposition of a sediment layer followed by the trapping and binding of that layer. Microbial communities can be observed building such structures today in intertidal environments such as Shark Bay, Western Australia (Figure 11.11a).

In other cases, however, stromatolites form by mineral precipitation, rather than by trapping and binding of sediment by microorganisms. That mineral precipitation may be indirectly controlled by microorganisms, or it may simply be the result of oversaturation of the surrounding water. As we saw in Chapter 5, the ocean contains abundant calcium and carbonate, which react to form the minerals calcite and aragonite. These minerals are important for the growth of stromatolites formed by mineral precipitation.

The potential role of microorganisms in stromatolite formation is important to understand because these layered, dome-shaped structures have been used as evidence

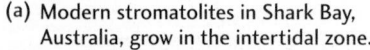
(a) Modern stromatolites in Shark Bay, Australia, grow in the intertidal zone.

(b) In northern Siberia, ancient stromatolites (over 1 billion years old) in cross section form columns.

(c) A cross section of a living stromatolite reveals layering similar to that seen in ancient stromatolites.

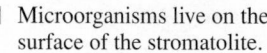
(d) The layering reveals how both modern and ancient stromatolites grow.

1 Microorganisms live on the surface of the stromatolite.

2 Sediment is deposited on the microorganisms,...

3 ...which react by growing upward through the sediment, forming a new layer.

FIGURE 11.11 ■ Stromatolites are sedimentary features that result from the interaction of microorganisms with their environment. [Images from John Grotzinger.]

for life on early Earth. But if stromatolites can be built by nonmicrobial mineral precipitation, their use as evidence for early life is uncertain. Only by carefully studying the processes by which microorganisms interact with minerals and sediments, and the chemical and textural fingerprints of these interactions, will we be able to determine whether the formation of stromatolites on early Earth required the presence of microorganisms.

■ Geobiologic Events in Earth's History

The geologic time scale divides time based on the comings and goings of fossil assemblages (see Chapter 8). These biological patterns provide a convenient ruler for subdivid-

ing Earth's history, but they were almost always associated with global environmental changes. At many of the major boundaries in the geologic time scale, Earth experienced a one-time event that caused dramatic changes in conditions for life. Some of these changes were triggered by organisms themselves, others by geologic events, and still others by forces from outside the Earth system.

We will now study a few of these dramatic events in Earth's history—events in which the link between life and the physical environment is clearly visible. **Figure 11.12** shows the great antiquity of life on Earth and the timing of several of these major events.

Origin of Life and the Oldest Fossils

When Earth first formed some 4.5 billion years ago, it was lifeless and inhospitable. A billion years later, it was teeming with microorganisms. How did life begin? Along with

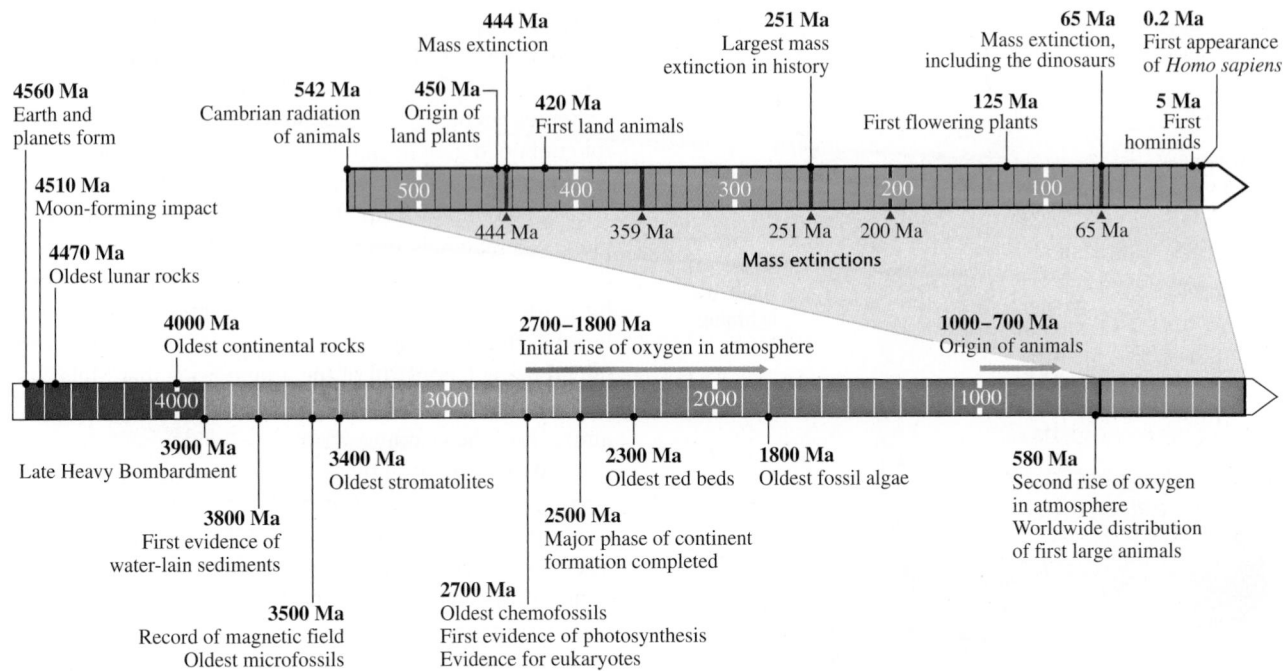

FIGURE 11.12 ▪ The geologic time scale, showing major events in the history of life. (Ma, million years ago.)

other grand puzzles such as the origin of the universe, this question remains one of science's greatest mysteries.

The question of *how* life may have originated is very different from the question of *why* life originated. Science offers an approach only to understanding the "how" part of this mystery because, as you may recall from Chapter 1, it uses observations and experiments to create testable hypotheses. These hypotheses may explain the series of steps involved in the origin and evolution of life, and they can be tested by searching for evidence in the fossil and geologic records. However, observations and experiments do not provide a testable approach to the question of why life evolved.

The fossil record tells us that single-celled microorganisms were the earliest forms of life, and that they evolved into all the multicellular organisms that are found in the younger parts of the geologic record. The fossil record also shows us that most of life's history involved the evolution of microorganisms. We can find microfossils in rocks 3.5 billion years old, yet we can conclusively identify fossils of multicellular organisms only in rocks younger than 1 billion years. It therefore appears that microorganisms were the only organisms on Earth for at least 2.5 billion years!

The theory of evolution predicts that these first microorganisms—and all life that came after them—evolved from a universal ancestor (see Figure 11.5). What did this universal ancestor look like? We really don't know, but most geobiologists agree that it must have had several important characteristics. The most crucial of these would be genetic information: instructions for growth and reproduction. Otherwise, it would have had no descendants. The universal ancestor must also have been composed of carbon-rich

compounds. As we have seen, all organic substances, including organisms, are made principally of carbon.

How did the universal ancestor arise? One approach to answering this question would be to search for clues in rocks. However, well-preserved fossils are found only in sedimentary rocks that have not been significantly affected by metamorphism or deformation. There are no well-preserved sedimentary rocks from the time when life first evolved, so scientists must use other approaches. Laboratory chemists have played an important role here.

Prebiotic Soup: The Original Experiment on the Origin of Life

In laboratory experiments that probe the origin of life, scientists have tried to recreate some of the environmental conditions thought to have existed on Earth before life arose. In the early 1950s, Stanley Miller, a graduate student at the University of Chicago, did the first experiment designed to explore life-building chemical reactions on early Earth. His experiment was amazingly simple (**Figure 11.13**). At the bottom of a flask, he created an "ocean" of water that he then heated to create water vapor. The water vapor emitted from the ocean was mixed with other gases to create an "atmosphere" containing some of the compounds thought to be most abundant in Earth's early atmosphere: methane (CH_4), ammonia (NH_3), hydrogen (H_2), and the water vapor. Oxygen—an important gas in Earth's atmosphere today—was probably absent at that time. In the next step, Miller exposed this atmosphere to electrical sparks ("lightning"), which caused the gases to react with one another and with the water in the ocean.

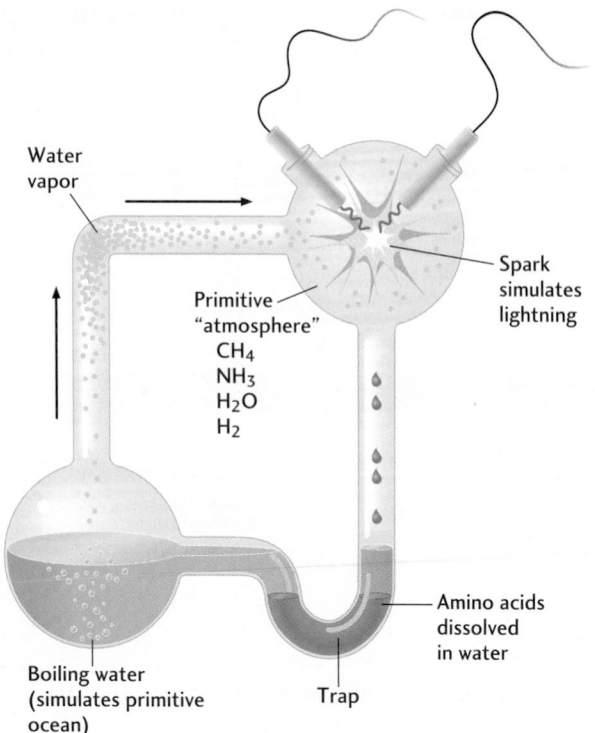

FIGURE 11.13 ■ Stanley Miller used this simple experimental design to explore the origin of life. In this apparatus, ammonia (NH_3), hydrogen (H_2), water vapor (H_2O), and small carbon-bearing molecules such as methane (CH_4) were converted into amino acids—a key component of living organisms.

The results were impressive. The experiment yielded compounds called *amino acids* in addition to other carbon-bearing compounds. Amino acids are the fundamental building blocks of protein molecules, which are essential for life. Thus, if you want to build an organism, creating amino acids is a good place to start. Miller's discovery was exciting because it showed that amino acids could

have been abundant on early Earth. It led to the hypothesis that Earth's ocean and atmosphere formed a sort of "prebiotic soup" of amino acids in which life originated. Other researchers have suggested that our universal ancestor contained genetic material that enabled amino acids to form proteins, which it then relied on for self-perpetuation.

The "prebiotic soup" hypothesis predicted that early planetary materials might contain amino acids. That prediction was borne out years later when, in 1969, a meteorite hit Earth near Murchison, Australia. When geologists analyzed it, they discovered that the *Murchison meteorite* contained many (about 20) of the amino acids that Miller had created in the laboratory! In fact, it even had similar relative amounts of those amino acids.

The message of all these discoveries is the same: amino acids could have formed on a planet without oxygen. But the opposite is also true: where oxygen is present, amino acids do not form, or are present only in tiny amounts. This is one of several reasons why Earth scientists think early Earth was a planet without oxygen.

The Oldest Fossils and Early Life

Whatever the processes by which life originated, the oldest potential fossils on Earth suggest that it had originated by 3.5 billion years ago. Stromatolites shaped like small cones provide some of the best available evidence for life at this time (**Figure 11.14a**). Stromatolites are common in continental cratons and have been identified in sedimentary rocks of early Archean age. In addition, the ratios of carbon isotopes found in some early Archean rocks show values that could have been produced only by biological processes (see the Practicing Geology Exercise at the end of the chapter, p. 310). The oldest fossils that preserve possible morphological evidence for life are tiny threads that are similar in size and appearance to modern microorganisms, encased in chert. These features were found in

(a)

(b)

FIGURE 11.14 ■ (a) Early Archean (3.4-billion-year-old) stromatolites in the Warrawoona formation, Western Australia. The conical shapes suggest that the microbial mats that formed these rocks grew toward the sunlight. (b) Abundant microfossils are well preserved in the 2.1-billion-year-old Gunflint formation of southern Ontario, Canada. [Images courtesy of H. J. Hofmann.]

formations in Western Australia that may be as old as 3.5 billion years, although their interpretation as microfossils remains controversial. Younger, better-preserved microfossils occur in the 3.2-billion-year-old Fig Tree formation of South Africa and in the 2.1-billion-year-old Gunflint formation of southern Canada (Figure 11.14b). The Gunflint fossils, discovered in 1954, were the first ever discovered in Precambrian rock, and they set off a tidal wave of research that continues to this day. In the past 50 years, we have seen, in many new localities, just how ancient life on Earth is and how well it can be preserved under the right geologic circumstances.

Most geobiologists agree that there was life on Earth 3.5 billion years ago, but are uncertain about how those early organisms functioned or obtained energy and nutrients. Some scientists argue that the oldest organisms on the universal tree of life were chemoautotrophic, obtaining their energy directly from chemicals in the environment. Furthermore, those oldest organisms may have been hyperthermophilic. This possibility suggests that life may have originated in very hot water, such as that in hot springs or hydrothermal vents on mid-ocean ridges, where sunlight was unavailable as an energy source, but chemicals were abundant (**Figure 11.15**).

CHEMOFOSSILS AND EUKARYOTES Form and size alone are not enough to allow us to deduce the function of microorganisms, so microfossils are ultimately limited in the information they can provide. Additional information can be gleaned from **chemofossils,** the chemical remains of organic compounds made by ancient microorganisms while they were alive. When an organism dies, most of the organic compounds in its body are quickly broken down into much smaller molecules, usually by heterotrophs.

Some of these molecules, however, are very stable and resist recycling. *Cholestane,* for example, is a remarkably durable substance, made only by eukaryotes, that is very similar to the well-known compound cholesterol. Cholestane chemofossils have been identified in 2.7-billion-year-old rocks from Western Australia. The presence of these chemofossils tells us that single-celled eukaryotic microorganisms must have emerged by that time. It is the eukaryotes that would eventually evolve into multicellular organisms, including animals, but not until much later.

Origin of Earth's Oxygenated Atmosphere

The rise of oxygen—the stuff we breathe—is another important milepost in the history of interactions between life and its environment. As we learned in Chapter 9, Earth's early atmosphere contained little oxygen. Our current oxygen-rich atmosphere was produced by early life through photosynthesis. Remarkably, the same Australian rocks that preserve chemofossil evidence of eukaryotes also preserve chemofossil evidence of cyanobacteria. Because of this evidence, geologists believe that photosynthesis had become an important metabolic process by 2.7 billion years ago. Thus, one group of organisms (cyanobacteria) permanently altered Earth's environment by changing the composition of its atmosphere, while another group of organisms (eukaryotes) was influenced by that change to evolve in new directions.

The oxygenation of Earth's atmosphere probably occurred in two main steps, separated by more than a billion years. The first major increase began with the evolution of the cyanobacteria. The oxygen they produced reacted with iron dissolved in seawater, causing iron oxide minerals, such as magnetite and hematite, and silica-rich minerals, such as

FIGURE 11.15 ■ Hot water released from hydrothermal vents along mid-ocean ridges (visible here as a plume of what looks like black smoke) is full of mineral nutrients from which chemoautotrophic microorganisms obtain their energy. It is possible that life originated in such environments. [Dr. Ken MacDonald/SPL/Science Source.]

(a)

(b)

(c)

FIGURE 11.16 ■ Unusual sedimentary rocks and new, larger eukaryotes mark the rise of oxygen concentrations in the atmosphere between 2.7 billion and 2.1 billion years ago. (a) A banded iron formation. (b) These fossils of *Grypania*, a type of eukaryotic algae, are visible with the naked eye. (c) Red beds are made up of sandstones and shales cemented together by iron oxide minerals. [(a) Francois Gohier/ Science Source; (b) courtesy of H. J. Hofmann; (c) John Grotzinger.]

chert and iron silicates, to precipitate and sink to the seafloor. These minerals accumulated in thin, alternating layers of sediments called **banded iron formations** (Figure 11.16a). Iron is soluble in water when oxygen concentrations are low, as would have been the case on Earth before cyanobacteria evolved. When oxygen concentrations are high, however, iron reacts with oxygen to form highly insoluble compounds. Therefore, the oxygen produced by cyanobacteria would have immediately caused iron to precipitate from seawater and sink to the seafloor. This process would have continued until most of the dissolved iron was used up, allowing oxygen to accumulate in the ocean and atmosphere.

Atmospheric oxygen concentrations began to build about 2.4 billion years ago and reached an initial plateau about 2.1 billion to 1.8 billion years ago, when the first eukaryotic fossils, of a type of algae, entered the geologic record (Figure 11.16b). The large size of these organisms— at least 10 times larger than anything that came before them—is thought to be a consequence of the oxygen increase. This time also marks the first appearance of **red beds,** unusual stream deposits of sandstones and shales bound together by iron oxide cement, which gives them their red color (Figure 11.16c). The presence of iron oxides in these deposits indicates that oxygen must have been present in the atmosphere to precipitate them.

After eukaryotic algae came on the scene, not much happened for over a billion years. Then, about 580 million years ago, atmospheric oxygen concentrations rose dramatically, almost to their modern level. The reason for this second increase is still not understood, though it may be related to an increase in the burial of organic carbon by sedimentation. In a process somewhat similar to the one that produces banded iron formations, oxygen reacts easily with organic matter, usually with the help of microorganisms. Thus, as long as there is organic matter around, oxygen will be used up. If organic matter is removed from the system by burial in sediments, however, it cannot react with the available oxygen. Thus, the second step in the rise of atmospheric oxygen might have been related to an increase in sediment production. Such an increase may have occurred when mountains were built—and then eroded—during global tectonic events, such as the assembly of supercontinents. In any case, the consequences were dramatic: the first large multicellular animals suddenly appeared, and all the modern groups of animals evolved shortly thereafter, ushering in the Phanerozoic eon with its wonderfully complex and diverse organisms.

■ Evolutionary Radiations and Mass Extinctions

In most cases, the boundaries of the eras and periods within the Phanerozoic eon are marked by the demise, or *extinction*, of a particular group of organisms, followed by the rise,

FIGURE 11.17 ■ The diversity of animal fossils reveals both mass extinctions and radiations. This graph shows the number of "shelly" animal families found in the fossil record during the last 600 million years; each family comprises many species. During a radiation, such as the Cambrian explosion, the number of new families increases. During a mass extinction, such as the one at the end of the Cretaceous period, the number of families decreases. (Ma, million years ago.)

or *radiation,* of a new group of organisms. When groups of organisms are no longer able to adapt to changing environmental conditions or compete with more successful groups of organisms, they become extinct. An interval when many groups of organisms become extinct at the same time is called a *mass extinction* (**Figure 11.17**) (see Chapter 8). In a few cases, the boundaries of the geologic time scale are marked by environmental catastrophes of truly global magnitude. Radiations are stimulated by the availability of new habitats when a mass extinction eliminates highly competitive and established groups of organisms.

Radiation of Life: The Cambrian Explosion

Perhaps the most remarkable geobiological event in Earth's history, aside from the origin of life itself, was the sudden appearance of large animals with shells and skeletons at the end of Precambrian time (**Figure 11.18**). This rapid development of new types of organisms from a common ancestor—what biologists call an **evolutionary radiation**—had such an extraordinary effect on the fossil record that its culmination 542 million years ago is used to mark the most profound boundary of the geologic time scale: the beginning of the Phanerozoic eon. This boundary also coincides with the start of the Paleozoic era and the Cambrian period (see Chapter 8 and Figure 11.12).

Evolutionary radiations are rapid by nature; if they were not, they would not be noticed in the fossil record. However, the radiation of animals during the early Cambrian, after almost 3 billion years of very slow evolution, was so fast that it is often called the **Cambrian explosion,** or biology's Big Bang. Every major animal group that exists on Earth today, as well as a few more that have since become extinct, appeared within less than 10 million years. All the major

branches (*phyla*) on the animal tree of life (**Figure 11.19**) originated during the Cambrian explosion. Note, however, that as impressive as this tree of animals seems, it is a single, short branch of the universal tree of life (see Figure 11.5).

Geobiologists have raised two major questions about the Cambrian explosion. First, what allowed these early animals to develop such complex body forms so rapidly, and therefore to become so diverse? Systematic change in organisms over many generations is referred to as **evolution.** Evolution is driven by **natural selection,** the process by which populations of organisms adapt to changes in their environment. The theory of *evolution by natural selection* states that, over many generations, individuals with the most favorable traits are most likely to survive and reproduce, passing those traits on to their offspring. If environmental conditions change over time, the traits that are favored change as well. This process can lead eventually to the emergence of new species.

One hypothesis for the cause of the Cambrian explosion is that the genes of these early animals changed in some way that made it possible for them to exceed some sort of evolutionary barrier. The stage was set by the development of multicellularity in late Precambrian time (**Figure 11.20**), which opened up new evolutionary possibilities. It is also possible that the ancestral animals had to reach a certain size before they could diversify. Some Precambrian animals, such as the fossil animal embryo shown in Figure 11.20, are so small they can be seen only with a microscope. The development of shells and skeletons might have been an important trigger of further diversification: once one group of animals had evolved hard parts, the others had to as well, or they would have been eliminated through competition.

The second riddle of the Cambrian explosion is why these animals differentiated *when* they did. Geobiologists have puzzled over the timing of the Cambrian explosion for

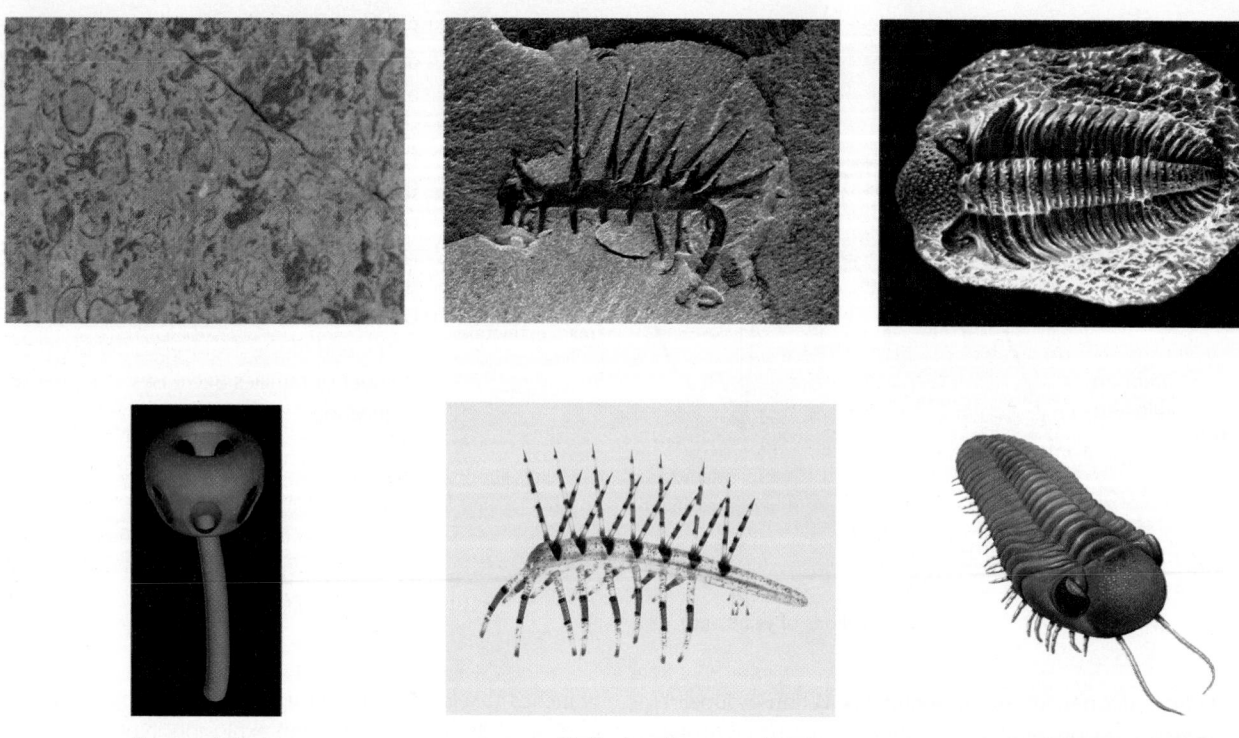

Namacalathus   ***Hallucigenia***   **Trilobites**

FIGURE 11.18 ■ Fossils that record the Cambrian explosion. Precambrian organisms such as *Namacalathus* (*left*) were the first organisms to use calcite in making shells. These organisms became extinct at the Precambrian-Cambrian boundary. Their extinction paved the way for a strange new group of organisms, including *Hallucigenia* (*center*) and the more familiar trilobites (*right*), that formed weak shells made of organic material similar to fingernails. In each example, the fossils are shown on top and the reconstructed organism is shown on the bottom. [*left, top:* John Grotzinger; *left, bottom:* W. A. Watters; *center, top:* Burgess Shale Hallucigenia 18-5 by Chip Clark, Smithsonian; *center, bottom:* Chase Studio/Science Source; *right, top:* courtesy of Musée cantonal de géologie, Lausanne. Photo by Stéphane Ansermat; *right, bottom:* Chase Studio/Science Source.]

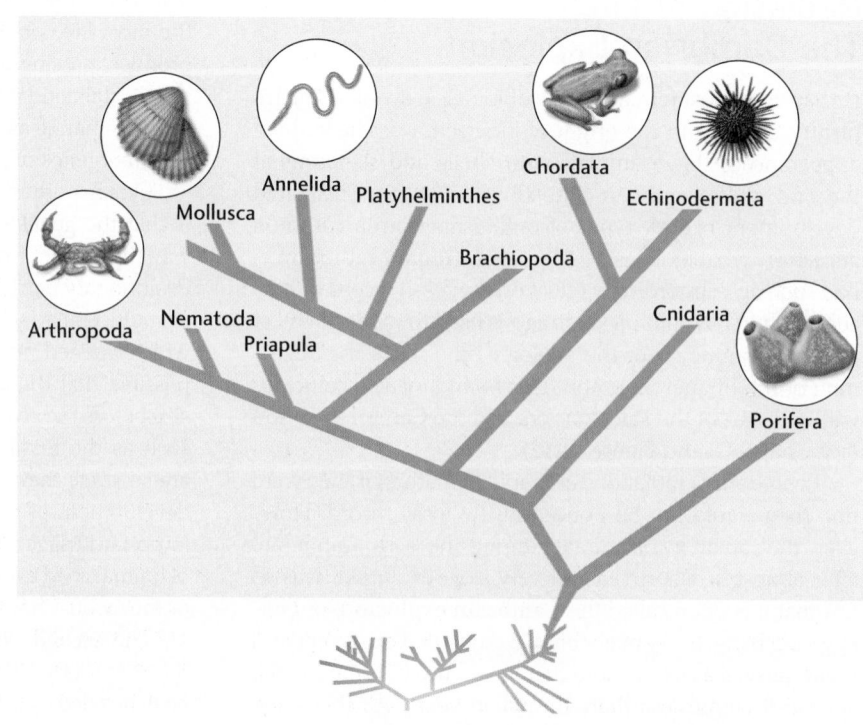

FIGURE 11.19 ■ Every major group of animals alive today originated during a great evolutionary radiation in the early Cambrian known as the Cambrian explosion.

FIGURE 11.20 ■ A fossilized animal embryo from the latest part of Precambrian time. Such fossils show that multicellular animals had evolved before the Cambrian period and are the ancestors of the animals that evolved during the Cambrian explosion. [Courtesy Shuhai Xiao, Virginia Tech.]

more than 150 years. Back in the days of Charles Darwin, it wasn't clear whether the Cambrian explosion represented the origin of life itself. But the sudden appearance of complex and diverse animal fossils in the geologic record presented a challenge to Darwin's theory of natural selection. His theory predicted slow changes in the form and function of organisms; hence, it predicted that less complex life-forms should have occurred before the first animals, and it could not easily accommodate these complex creatures that apparently had no simpler ancestors. Therefore, Darwin hypothesized that the expected ancestors must be absent from the record because the rocks containing the Cambrian fossils must lie above an unconformity. He predicted that rocks from the time of the proposed unconformity would eventually be discovered, and that those rocks would contain the "missing" ancestors. Darwin turned out to be right, but it was only in the past several decades that geobiologists discovered the fossils described earlier in this chapter, proving that animals did indeed originate before the Cambrian explosion.

So it seems clear that the Cambrian animals did have ancestors, perhaps lurking between tiny grains of sand at the bottom of shallow seas. However, isotopic dating techniques show that these tiny animals were probably less than 100 million years older than their Cambrian descendants. Other dating techniques, based on studies of the genes of modern organisms, suggest that the origin of animals may have predated the Cambrian explosion by several hundred million years. But even these estimates hardly matter compared with the billions of years that passed before the Cambrian explosion occurred.

Most geobiologists agree that once animals had evolved, they could have radiated at any time. Why, then, did they radiate about 542 million years ago and not at some other time? Perhaps the timing of the Cambrian explosion was driven by the dramatic environmental changes that occurred near the end of Precambrian time. To human eyes, Earth at that time would have seemed a very strange place: long chains of great mountains were forming as the pieces of the giant continent Gondwana were being fused together, and the climate was in turmoil, flipping between frigid periods when the entire Earth may have been covered in ice and extremely warm, ice-free periods (see Chapter 21). Oxygen concentrations in the oceans and atmosphere were increasing as erosion of the rising mountains produced sediments, which buried the organic matter whose decomposition would otherwise have consumed that oxygen. This last change may have been the most important. Without sufficient oxygen, animals simply cannot grow large.

Whatever the ultimate cause of the Cambrian explosion, one point stands clear: evolutionary radiations are the result of genetic possibility combined with environmental opportunity. The radiation of organisms is not just the result of having the right genes, and it is not just the result of living in the right environment. Organisms must take advantage of both to evolve.

Tail of the Devil: The Demise of Dinosaurs

The mass extinction that marks the Cretaceous-Tertiary boundary and the end of the Mesozoic era (about 65 million years ago; see Figures 8.11 and 8.15) represents one of the greatest such events in Earth's history. Entire global ecosystems were obliterated, and about 75 percent of all species on Earth, both on land and in the ocean, were extinguished forever. The dinosaurs are only one of several groups that became extinct at the end of the Cretaceous period, but they are certainly the most prominent. Other groups, such as ammonites, marine reptiles, certain types of clams, and many types of plants and plankton, also perished.

In contrast to the Cambrian explosion, almost all scientists agree on the cause of the Cretaceous-Tertiary mass extinction. We are now virtually certain that the cause was a gigantic asteroid impact. In 1980, geologists discovered a thin layer of dust containing *iridium*—an element that is typical of extraterrestrial materials—in sediments deposited at the end of the Cretaceous in Italy (**Figure 11.21**).

This extraterrestrial dust was subsequently found at many other locations around the world, on every continent and in every ocean, but always exactly at the Cretaceous-Tertiary boundary. The geologists argued that the accumulation of this much iridium-bearing dust would require an asteroid about 10 km in diameter to hit Earth, explode, and send its cosmic detritus across the globe. Publication of this hypothesis spurred a search for the impact crater. That search was bound to be difficult for two reasons. First, most of Earth's surface is covered by oceans, so the crater could easily have been under water. Second, since the crater would be 65 million years old, it could have been eroded or filled in with sediments and sedimentary rock. In the early 1990s,

FIGURE 11.21 ■ The pocket knife marks a light-colored layer of clay, containing both extraterrestrial materials and materials from local rocks at the Chicxulub impact site, which accumulated in the Raton Basin of the southwestern United States. Such deposits have been found worldwide. [Dr. David A. Kring.]

however, geologists found a huge crater, almost 200 km in diameter and 1.5 km deep, buried under sediments near a town on Mexico's Yucatán Peninsula, called Chicxulub.

Geologic evidence from Chicxulub, as well as from the surrounding region and around the world, has allowed geologists to paint a picture of what happened there. The name *Chicxulub* means "tail of the devil" in the local Mayan language, and the immediate aftermath of the impact would have been hellish indeed. The asteroid struck Chicxulub at Mach 40, coming in from the south at an angle of about 20° to 30° from the horizontal. Its explosion would have produced a blast 6 million times more powerful than the 1980 eruption of Mount St. Helens. It would have created winds of unimaginable fury and a tsunami as high as 1 km (100 times higher than the great Indian Ocean tsunami of 2004). The sky would have turned black with massive amounts of dust and vapor. A global firestorm may have resulted as the flaming fragments from the blast fell back to Earth (**Figure 11.22**).

Materials from the impact crater spread out in a radial kill zone focused toward western and central North America. Creatures around at that time, assuming they weren't in the kill zone, might have witnessed the following events: a brilliant flash as the asteroid rammed into Chicxulub, vaporizing Earth's upper crust at temperatures up to 10,000°C; an arc of flaming hot rocks that bolted across the sky at speeds of up to 40,000 km/hour, then crashed into North America; and a plume of debris, gas, and molten material that heated part of the atmosphere to several hundred degrees, punched into space, and then collapsed back to Earth. Over the next several days or weeks, the finer materials in this plume would have settled across Earth's entire surface.

The direct effects of the impact would have been devastating for many organisms. But worse yet would have been the aftermath for months and years to come, which scientists think led to the actual mass extinction. The high concentration of debris in the atmosphere would have blocked out the Sun, vastly reducing the light available for photosynthesis. In addition to solid particles of debris, poisonous sulfur- and nitrogen-bearing gases would have been injected into the atmosphere, where they would have reacted with water vapor to form toxic sulfuric and nitric acids that would have rained down on Earth. The combination of these two effects, and others, would have been devastating to plants and other photosynthetic autotrophs, and thus to both marine and terrestrial ecosystems that depended on them as the base of the food chain. Heterotrophs, including the dinosaurs, would have been next; once their food sources died off, they would have died off as well. A cascading series of such effects leading to the collapse of ecosystems was probably the ultimate cause of the mass extinction.

Global Warming Disaster: The Paleocene-Eocene Mass Extinction

The mass extinction at the Paleocene-Eocene boundary (about 55 million years ago; see Figure 8.11) was not one of the largest such events. It was an important event in the evolution of life, however, because it paved the way for the mammals, including the primates, to radiate into an important group. Unlike the mass extinction that wiped out the dinosaurs, it had no extraterrestrial cause. Instead, it was caused by abrupt global warming. Earth scientists are very interested in the details of what happened because global warming—this time produced by human activities—may threaten ecosystems in the coming decades (as we will see in Chapter 23).

We now believe that the global warming at the end of the Paleocene epoch occurred when the oceans suddenly belched an enormous amount of methane—a potent

FIGURE 11.22 ■ An artist's rendition of the Cretaceous-Tertiary scene after the asteroid impact. [Richard Bizley/Science Source.]

greenhouse gas—into the atmosphere. The resulting global warming was the primary cause of the mass extinction. But where did all that methane come from? To unravel this mystery, we must weave together many of the processes we have learned about in this chapter, including microbial metabolism, biogeochemical cycles, and the global behavior of the biosphere.

MICROORGANISMS SOW THE SEEDS OF DISASTER The story begins with the biogeochemical cycle of carbon, which will be described in more detail in Chapter 15. Normally, carbon is removed from the atmosphere by photoautotrophs, including algae and cyanobacteria in the oceans. After these marine organisms die, they slowly settle to the seafloor, where they accumulate as organic debris. Some of this carbon-rich debris is buried in sediments, but some is consumed by heterotrophic microorganisms as food. As you may recall, some heterotrophic microorganisms that live in anaerobic environments produce methane as a by-product of respiration. The methane produced by these anaerobes accumulates in the pores of seafloor sediments. If the seafloor is as cold as it is in our present climate (about 3°C), the methane combines with water to form a frozen solid (a methane-water ice), which remains within the sediments. Geologists searching for oil and natural gas have found layers with abundant methane-water ices in the upper 1500 m of sediments along many continental margins. If temperatures rise by even a few degrees, however, the methane-water ices melt, and the methane is quickly transformed into a gas.

THE OCEANS BUBBLE METHANE At the end of the Paleocene epoch, average temperatures in the deep sea may have risen by as much as 6°C. Once the first methane-water ices thawed and were transformed back into gases, they bubbled up through the oceans and entered the atmosphere, where they reinforced the greenhouse effect. This effect raised temperatures on the seafloor even further, which accelerated the rate of thawing. These positive feedbacks eventually resulted in a sudden—and catastrophic—release of methane that caused average global temperatures to rise dramatically. As much as 2 *trillion* tons of carbon, in the form of methane, may have escaped to the atmosphere over a period as short as 10,000 years or less!

Because methane easily reacts with oxygen to produce carbon dioxide, the release of methane also caused oxygen concentrations in the oceans to plummet. Marine organisms were essentially suffocated when oxygen concentrations dropped below a critical level. The oxygen decrease and temperature rise were devastating to seafloor ecosystems, and up to 80 percent of bottom feeders, such as clams, became extinct.

RECOVERY AND THE EVOLUTION OF MODERN MAMMALS Following the catastrophe, it took about 100,000 years for Earth to return to its previous state. During this time, temperatures remained unusually high until Earth was able to absorb all the extra carbon that had been released into the atmosphere. The warmer temperatures allowed rapid expansion of forests into higher latitudes. Redwoods—related to the giant sequoias of California—grew as far north as 80°, rain forests were widespread in Montana and the Dakotas, and tropical palms flourished near London, England. Primitive mammals rapidly evolved into the ancestors of today's modern mammals, which adapted to cope with the high temperatures of that time. One particular group of mammals—the *primates*—eventually gave rise to humans.

METHANE DEPOSITS TODAY: A TICKING TIME BOMB? Could we see a repeat of the Paleocene-Eocene global warming disaster today? In the frozen tundra of northern Canada and other Arctic regions of the world, there may be as much as half a trillion tons of frozen methane, and deep-sea sediments around the world contain much more. The global inventory of methane deposits is estimated to be 10 trillion to 20 trillion tons of carbon present as methane, far more than what was released to cause the Paleocene-Eocene mass extinction. Human activities are adding greenhouse gases to the atmosphere at an unprecedented rate, causing the climate to warm significantly. If this trend continues and the oceans warm up, it is possible that those methane deposits could thaw. We would be wise to pay attention to the lessons of our geologic history.

THE MOTHER OF ALL MASS EXTINCTIONS: WHODUNIT? The Cretaceous-Tertiary and Paleocene-Eocene extinctions are clear-cut examples of dramatic changes in Earth's environment that caused the catastrophic collapse of ecosystems and led to mass extinction. Those events were big, but not the biggest. In the mass extinction that marked the end of the Permian period and the Paleozoic era (see Figure 11.17), 95 percent of all species on Earth became extinct.

In this case, it seems unlikely that something as straightforward as an asteroid impact could explain how almost every species on Earth was killed. Not surprisingly, the absence of clear-cut evidence for any single cause has resulted in a long list of hypotheses, as we saw in Chapter 1. Some scientists point to extraterrestrial events, such as a comet impact or an increase in the solar wind. Others argue for events generated by Earth itself, such as an increase in volcanism, depletion of oxygen in the oceans, or a sudden release of carbon dioxide from the oceans. As in the Paleocene-Eocene extinction, a sudden release of methane from the oceans has also been proposed.

Recently, it has been shown that the mass extinction at the end of the Permian occurred exactly 251 million years ago. Perhaps it is no coincidence that the age of an enormous deposit of flood basalts in Siberia is also 251 million years. *Flood basalts,* as we will see in Chapter 12, are extrusive igneous rocks formed from huge volumes of lava that pour out across the surface of Earth in a relatively short time. In Siberia, volcanic fissures spewed out some 3 million cubic kilometers of basaltic lava, covering an area of 4 million square kilometers, almost twice the size of Alaska. Isotopic dating of the basalt shows that all of it was formed within 1 million years or less. It is hard to escape the conclusion that the Permian mass extinction was somehow related to this catastrophic eruption, which would have injected enormous amounts of carbon dioxide and sulfur dioxide gases into the atmosphere. Carbon dioxide contributes to global warming, and sulfur dioxide is the principal source of acid rain. Both are harmful to life if atmospheric concentrations get too high.

More work is required to test all these hypotheses. For example, the Deccan basalts of India are about 65 million years old, and it is possible that the massive outpouring of lava that formed them enhanced the Cretaceous-Tertiary mass extinction. However, equally large outpourings have occurred at other times in Earth's history without such apparently devastating effects.

Whatever the cause of the Permian mass extinction, one point is clear: just as in the Cretaceous-Tertiary and Paleocene-Eocene mass extinctions, the ultimate cause was the collapse of ecosystems. We know that this collapse occurred, although we don't know exactly how. The message that we should take away from this history lesson is that history may repeat itself. The environmental changes that humans are making today will inevitably influence ecosystems—we just don't know exactly how, at least not yet.

∎ Astrobiology: The Search for Extraterrestrial Life

Looking up at the stars on a clear night, it's hard not to wonder whether we are alone in the universe. As we have learned, the activities of life on our planet create distinctive biogeochemical signatures. Some of these signatures of life could be detected remotely, such as the presence of oxygen in the atmosphere of a planet in another solar system. In other cases, we might be able to land a spacecraft equipped with instruments to detect chemofossils or morphological fossils preserved in rocks.

In the past few decades, **astrobiologists** have begun to search systematically for evidence of life on other worlds. Although no organisms have yet been discovered beyond Earth, we should be encouraged to pursue this quest. Life may have gotten started somewhere, even if it failed to flourish. In our own solar system, Mars and Europa (a moon of Jupiter) are tantalizing targets because they are similar to Earth in several important ways. In addition, new discoveries of planets orbiting other stars have allowed us to extend this search to other solar systems.

The search for life on other worlds requires a patient, systematic, scientific approach. The most widely accepted approach has been to recognize that life, as we know it here on Earth, is based on liquid water and carbon-bearing organic compounds. Therefore, a sensible strategy might begin with a search for these two principal components of life. Compounds made of carbon are common throughout the universe; astronomers find evidence for them everywhere, from interstellar gases and dust particles to meteorites that land on Earth (**Figure 11.23**). Therefore, astrobiologists have focused on searching for liquid water. The Mars Exploration Rover mission, described in Chapter 9, was designed to search for evidence of water on

FIGURE 11.23 ▪ The Allende meteorite, which fell to Earth near Allende, Mexico, in 1969, is full of carbon compounds. Such findings provide evidence that these compounds, one of the two key components of life, are common throughout the universe. [John Grotzinger/Ramón Rivera-Moret/Harvard Mineralogical Museum.]

the surface of Mars, while the Mars Science Laboratory mission was designed to look for habitable environments. If the two rovers, *Spirit* and *Opportunity,* had failed to detect any evidence of water on Mars, any future plans to search for life or habitable environments on Mars (such as the Mars Science Laboratory mission) might well have been abandoned.

Of course, there is some risk in this "life as we know it" approach to searching for extraterrestrial life. We might miss forms of life we know nothing about. One could imagine a whole host of other elements and compounds that life could be based on. In general, however, these alternative schemes mainly provide fuel for science fiction writers. At least for the time being, carbon and water are regarded as the key components of all life in the universe.

Habitable Zones Around Stars

At the broadest scale, we assume that life is restricted to surfaces of planets and moons that orbit stars (**Figure 11.24**). The trick is to identify planets where water could remain stable as a liquid for a long enough time that life could originate. That could take hundreds of millions of years, based on our experience on Earth. If the surface of a planet is too close to its star, water will boil off and become a gas. That is what happened on Venus, which is 30 percent closer to the Sun than Earth and whose surface temperature is 475°C. If the surface of a planet is too far from its star, the water will freeze and become a solid. That is the case on Mars today, which is 50 percent farther from the Sun than Earth and whose surface temperature may fall below −150°C. Earth is in the middle zone, where water is stable as a liquid and surface temperatures are just right for life. For every star, there is a **habitable zone,** marked by the distances from the star at which water is stable as a liquid. If a planet is within the habitable zone, there a chance that life might have originated there.

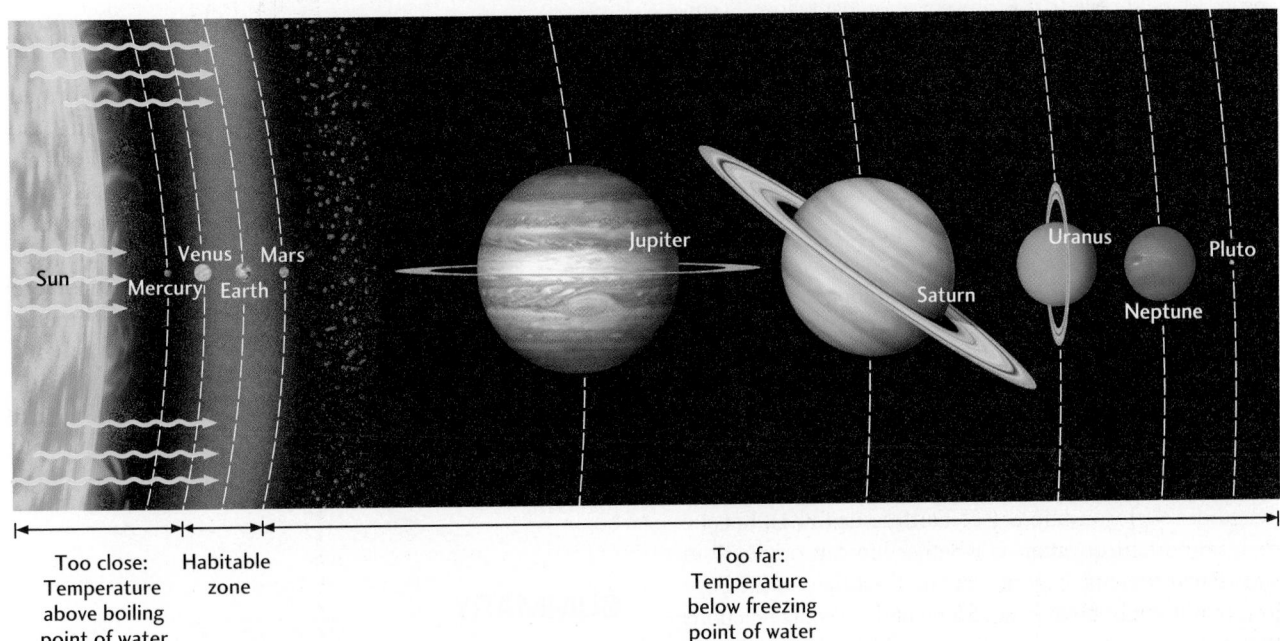

Too close:
Temperature
above boiling
point of water

Habitable
zone

Too far:
Temperature
below freezing
point of water

FIGURE 11.24 ▪ Stars have habitable zones where life on an orbiting planet could exist. The habitable zone is determined by distance from the star; it extends from the point at which water would boil away (too close to the star) to the point at which water would freeze solid (too far from the star).

Greenhouse gases such as carbon dioxide and methane also play an important role in determining the habitable zone. The Martian atmosphere may have had high concentrations of greenhouse gases early in its history. Thus, even though Mars is farther from the Sun than Earth, it might have been warmed through the greenhouse effect, as Earth is today. Indeed, new discoveries suggest that liquid water was once present on the surface of Mars, although we don't know how long it might have been stable. Thus, it is possible that Mars was habitable at some time in the past. But once the greenhouse gases were lost, Mars was transformed into the icy desert it is today.

Habitable Environments on Mars

People have long wondered about life on Mars. Mars is the planet most closely resembling Earth and is therefore the most likely planet in our solar system to host, or to have hosted, life. As we saw in Chapter 9, the Mars Exploration Rovers and Mars Science Laboratory found clear evidence of liquid water on the Martian surface at some point in the past. Based on their estimates of the ages of surface features, geologists estimate that water on Mars was stable 3 billion years ago, when it carved deep canyons across the planet's surface, dissolved rocks and minerals, and then precipitated them in a variety of basins where the water evaporated.

Water is present on Mars today, but only as ice. Any life that evolved early on would have had to seek refuge deep beneath the surface from the frigid modern climate. Any organisms that had remained on the surface would now be thoroughly frozen. However, the interior of Mars, like that of Earth, is warmed by radioactive decay, so at some depth within Mars, the ice that is present at or just below its surface must turn into liquid water. It is therefore possible that organisms—perhaps microbial extremophiles—live within a watery zone located a few hundred meters to a few kilometers below the surface of Mars.

Unfortunately, the lack of liquid water is not the only challenge that modern or ancient life would have to face on Mars. As we saw in Chapter 9, the sedimentary rocks discovered by the Mars Exploration Rover *Opportunity* are full of jarosite, an unusual iron sulfate mineral that precipitates from highly acidic water (**Figure 11.25**). On Earth, jarosite accumulates in some of the most acidic waters ever observed in natural environments.

Thus, it seems that life on Mars would have to cope not only with limited water, but possibly with very acidic water. The encouraging news is that extremophiles on Earth can live under such conditions (see Earth Issues 11.1). But the more important question is whether life can originate in such environments. Experiments on the origin of life suggest that it might be difficult. Some of the simple reactions that Miller observed in the 1950s would not be possible in an ocean of highly acidic water.

Not all environments on Mars may be highly acidic, however. The *Curiosity* rover has recently discovered a

FIGURE 11.25 ▪ Sedimentary rocks recently discovered on Mars contain a variety of sulfate minerals that form by precipitation from water. The presence of jarosite shows that the waters from which they precipitated were extremely acidic. Extremophiles can live in these conditions, but it is not yet clear whether they could originate in such acidic waters. The holes in the rocks were drilled in 2004 by *Opportunity,* one of the Mars Exploration Rovers, to analyze their composition. [NASA/JPL/Cornell.]

habitable lake environment represented by rocks formed over 3 billion years ago whose chemistry indicates the presence of more neutral to alkaline conditions. Furthermore, this ancient environment was not very salty, in strong contrast to the extremely salty environment discovered by *Opportunity*—the same environment that was also very acidic. The findings of the *Curiosity* mission are encouraging, as they suggest that Mars has environments that might be favorable for life as well as environments that might challenge life. The stunning discoveries made by *Opportunity, Spirit, Phoenix,* and *Curiosity* confirm that Mars may well have been habitable at some time. But only continued exploration will show whether life ever originated there.

SUMMARY

What is geobiology? Geobiology is the study of how organisms have influenced and been influenced by Earth's physical environment.

What is the biosphere? The biosphere is the part of our planet that contains all of its living organisms. Because the biosphere intersects with the lithosphere, hydrosphere, and atmosphere, it can influence or even control basic geologic and climate processes. The biosphere is a system of interacting components that exchanges energy and matter with its surroundings. Organisms use inputs of energy and matter to function and grow. In the process, they generate outputs such as oxygen and certain sedimentary minerals.

How do organisms interact with their physical environment? The activities of organisms influence the concentrations of gases in the atmosphere and the cycling of elements through Earth's crust. Organisms contribute to the weathering of rocks by releasing chemicals that help break down minerals, precipitate minerals in sedimentary environments, and modify the composition of the oceans. The oxygen in Earth's atmosphere is the result of the metabolism of photosynthetic microorganisms that evolved billions of years ago. In a similar way, the physical environment influences life. Geologic barriers such as mountains, deserts, and oceans help determine how ecosystems are divided. Some geologic processes can cause mass extinction events that permanently change life.

What is metabolism? Metabolism is a process that organisms use to convert inputs to outputs. Photosynthesis is a metabolic process in which organisms use energy from sunlight to convert water and carbon dioxide into carbohydrates, releasing oxygen as a by-product. Respiration is a metabolic process in which organisms use oxygen to release the stored energy of carbohydrates. Many organisms take up oxygen from the atmosphere and release carbon dioxide and water as by-products of respiration. Others, such as microorganisms that live in environments where oxygen is absent, must obtain oxygen by breaking down oxygen-containing compounds, producing substances such as hydrogen, hydrogen sulfide, or methane as by-products of respiration.

What are some ways in which metabolism affects the physical environment? When organisms produce oxygen, it is released into the atmosphere, where it can react with other elements and compounds. When organisms release carbon dioxide or methane, which are both greenhouse gases, they contribute to global warming. Conversely, when organisms consume these gases, they contribute to global cooling.

How do microorganisms interact with the physical environment? Microorganisms are the most abundant and the most diverse organisms on Earth. Some microorganisms, called extremophiles, can live in extremely hot, acidic, salty, oxygen-depleted, or otherwise inhospitable environments. Microorganisms are involved in many geologic processes, such as weathering, mineral precipitation, mineral dissolution, and the release of gases into the atmosphere. In these ways, they play critical roles in the flux of elements through the Earth system in biogeochemical cycles.

How did life originate? Experiments show that compounds thought to be abundant on early Earth, such as methane, ammonia, and water, could have combined to form amino acids, which could then have combined to form proteins and genetic materials. These results have been supported by the finding of meteorites that are rich in amino acids and other carbon-bearing compounds. The oldest potential fossils on Earth are 3.5 billion years old and appear to be the remnants of microorganisms, based on their shape and size. Chemofossils from about 2.7 billion years ago suggest that photosynthetic bacteria and eukaryotes were both present at that time. Banded iron formations, red beds, and the appearance of eukaryotic algae testify to an initial rise in atmospheric oxygen by about 2.1 billion years ago. A second, more dramatic rise in oxygen occurred near the end of Precambrian time and may have triggered the evolution of animals.

What is the difference between radiation and extinction? When groups of organisms are no longer able to adapt to changing environmental conditions or to compete with more successful groups, they become extinct. In a mass extinction, many groups of organisms become extinct at the same time. An evolutionary radiation is the relatively rapid evolution of new types of organisms from a common ancestor. Radiations may be stimulated by the availability of new habitats when a mass extinction eliminates highly competitive and established groups. The greatest radiation of animals in Earth's history occurred during the early Cambrian period, when all the animal phyla living today evolved. Several mass extinctions have occurred throughout the Phanerozoic eon. A major mass extinction occurred at the end of the Cretaceous period, when an asteroid hit Earth and 75 percent of all species were wiped out. Global warming resulting from a release of methane caused a mass extinction at the Paleocene-Eocene boundary. The cause of the greatest mass extinction of all time, which wiped out 95 percent of all species at the end of the Permian period, is unknown.

How can we search for life on other worlds? Astrobiologists searching for extraterrestrial life recognize that life as we know it on Earth is based on carbon-containing compounds and liquid water. There is ample evidence that carbon compounds are common throughout the universe, so astrobiologists search for evidence of the presence of liquid water, today or in the past. There is a habitable zone at a certain distance from every star where liquid water could be stable. If a planet is within the habitable zone, there is a chance that life might have originated there. There is unambiguous evidence that Mars had liquid water on its surface, and thus may have been habitable, at some time in the past.

KEY TERMS AND CONCEPTS

astrobiologist (p. 306)

autotroph (p. 285)

banded iron formation
   (p. 300)

biogeochemical cycle
   (p. 288)

biosphere (p. 284)

Cambrian explosion
   (p. 301)

chemoautotroph
   (p. 294)

chemofossil (p. 299)

cyanobacteria (p. 288)

ecosystem (p. 284)

evolution (p. 301)

evolutionary radiation
   (p. 301)

extremophile (p. 290)

gene (p. 289)

geobiology (p. 284)

habitable zone
   (p. 307)

heterotroph (p. 285)

metabolism (p. 286)

microbial mat (p. 294)

microfossil (p. 290)

microorganism (p. 288)

natural selection
   (p. 301)

photosynthesis
   (p. 287)

red bed (p. 300)

respiration (p. 288)

stromatolite (p. 295)

PRACTICING GEOLOGY EXERCISE

How Do Geobiologists Find Evidence of Early Life in Rocks?

Perhaps the most important question a geobiologist can ask is, "What evidence of life is preserved in rocks?" If fossils of animal shells and skeletons are present in rocks, then this question is easy to answer. In many cases, however, the geologic processes that turn sediments into sedimentary rocks destroy the materials that could have become fossils. Furthermore, most organisms do not have hard shells or skeletons that are easily preserved, so we would not expect them to become fossils. And on early Earth, before the advent of animals with hard shells and skeletons, most organisms were microscopic in size. Simply put, how can the former presence of life on Earth be detected in situations in which fossils are not preserved?

One approach that geobiologists often depend on is the use of chemical signatures of former life. Carbon provides the most obvious example of an element that might have been concentrated by biological processes. Not all concentrations of carbon are of biological origin, however, so additional tests must be applied.

One of these tests asks whether the carbon present has a distinctive *isotopic* composition. Recall from Chapter 3 that isotopes are atoms of the same element with different numbers of neutrons. Many elements of low atomic weight have two or more stable (nonradioactive) isotopes. A carbon atom has six protons, but may have six, seven, or eight neutrons, which give it atomic masses of 12, 13, or 14, respectively. Carbon-12 is by far the most common isotope, so samples of carbon from ancient rocks or modern sediments will yield mostly carbon-12.

Fortunately, it turns out that metabolic processes such as photosynthesis use carbon-12 and carbon-13 differently. The difference in atomic mass between carbon-12 (often denoted ^{12}C) and carbon-13 (^{13}C) results in differences in their uptake by organisms. Photosynthetic organisms, for example, use carbon dioxide and water to form carbohydrates. They use carbon dioxide molecules that contain ^{12}C preferentially over those that contain ^{13}C. As a result, photosynthetic organisms become enriched in ^{12}C relative to the environment from which they draw carbon dioxide.

We can therefore use carbon isotopes as a tool to detect ancient life by measuring the amounts of ^{12}C and ^{13}C present in sedimentary rocks. If sediments are formed in the presence of organic matter that is enriched in ^{12}C (or in any particular isotope), this enrichment may be passed along to the sediments, and then on to the resulting rock. Thus, a shale that might be billions of years old can preserve a "signature" of life recorded by its carbon isotope composition.

We begin by measuring the amounts of ^{12}C and ^{13}C in a rock sample and calculating the ratio between them ($^{12}C/^{13}C$). We then compare that ratio with the $^{12}C/^{13}C$ ratio in a *standard*. The standard is a material (often a pure mineral, such as calcite) whose $^{12}C/^{13}C$ ratio is precisely known and varies very little. The standard can be compared over and over again with samples of other carbon-bearing rocks and sediments, as well as with living organisms and other natural substances. By comparing both rock samples and organisms with the standard, we can search for similarities that link rock samples to particular biological processes.

The following table gives the $^{12}C/^{13}C$ ratios for a standard, three rock samples, and two natural substances—plant material and methane gas:

Standard	Rock A	Rock B	Rock C	Plant Material	Methane Gas
1000	995	1020	1050	1025	1060

The following equation* allows us to compare these data:

$$R_{(sample)} = [^{12}C/^{13}C \text{ of the standard}] - [^{12}C/^{13}C \text{ of the sample}]$$

where R stands for the value of the difference between the two ratios.

For most rocks, the value of R will be close to zero, but could be slightly positive or negative. In contrast, if photosynthesis was involved in the formation of organic matter incorporated into a sedimentary rock—for example, a shale—the value of R for a sample of that shale could be very negative—close to

---

*This equation is simplified from what is normally used in practice. It neglects the standard "delta" notation, which normalizes the actual abundances of the isotopes in the sample to those in the standard.

Plants take up carbon dioxide during photosynthesis. Because they take up CO_2 molecules containing ^{12}C more easily than those containing ^{13}C, plants become enriched in ^{12}C relative to their environment.

−20. Some chemoautotrophic microorganisms that consume methane gas produce carbonate rocks with an extremely negative R value, on the order of −50.

Using the data and equation above, let's try to identify which of our rock samples formed in the presence of biological processes. Let's begin with rock B:

$$R_{(rock\,B)} = [^{12}C/^{13}C \text{ of the standard}] - [^{12}C/^{13}C \text{ of the rock B}]$$
$$= 1000 - 1020$$
$$= -20$$

This result shows that rock B has an R value that is substantially different from zero, suggesting that ancient biological processes may have been at work when the rock formed. We can confirm that its value of −20 is a close match for the R value predicted for photosynthesis by calculating the R value for plant material:

$$R_{(plant)} = [^{12}C/^{13}C \text{ of the standard}] - [^{12}C/^{13}C \text{ of the plant material}]$$
$$= 1000 - 1025$$
$$= -25$$

The R value for the plant material, at −25, is close enough to the R value for rock B, at −20, that our hypothesis of ancient photosynthesis is strengthened.

BONUS PROBLEM: Try calculating R values for rocks A and C. Which rock does not record a distinctive signature of biological processes? Is there a rock among our samples that might have formed in the presence of methane-consuming microorganisms? If so, how can you check this result?

EXERCISES

1. Can the biosphere be considered an Earth system? How would that system be described?

2. How do autotrophs differ from heterotrophs?

3. What is metabolism?

4. In what environments might you find extremophiles? Can humans live under extreme conditions?

5. What is the difference between photosynthesis and respiration?

6. Explain how life is related to water. What would happen if all the water on Earth turned to ice?

7. Draw a diagram of a biogeochemical cycle. What are the inputs and outputs? What are the processes that power the cycle?

8. Carbon is regarded as the starting point for all of life. What else is important?

9. In the diagram shown in Figure 11.12, how many period boundaries of the geologic time scale are marked by mass extinctions? How many era boundaries are marked by mass extinctions?

10. What controls the habitable zone around stars? Is Neptune in the habitable zone of our solar system?

THOUGHT QUESTIONS

1. How does the biogeochemical cycle of carbon affect global climates?

2. During an evolutionary radiation, organisms evolve rapidly. What would the geologic record look like if an evolutionary radiation occurred during an interval represented by an unconformity? How would you distinguish between an evolutionary radiation and the effects of the unconformity?

3. Carbon and water are the basis for all life as we know it. If a giraffe made of silicon walked past one of the Mars Exploration Rovers, how would we know it was alive?

MEDIA SUPPORT

  11-1 Animation: Ecosystems

- Volcanoes as Geosystems   314

- Lavas and Other Volcanic Deposits   315

- Eruptive Styles and Landforms   320

- Interactions of Volcanoes with Other Geosystems   326

- The Global Pattern of Volcanism   329

- Volcanism and Human Affairs   334

Grand Canyon of the Yellowstone, where the Yellowstone River cuts down 250 m through brightly colored rhyolitic lavas. The lavas were deposited by a huge volcanic eruption less than a million years ago. [Richard Nowitz/Photodisc/Getty Images.]

VOLCANOES

THE NORTHWESTERN CORNER OF WYOMING is a geologic wonderland of geysers, hot springs, and steam vents—the visible signs of a vast active volcano that stretches across the wilderness of Yellowstone National Park. Every day, this volcano expels more energy in the form of heat than is consumed as electric power in the three surrounding states of Wyoming, Idaho, and Montana combined. This energy is not released steadily; some of it builds up in hot magma chambers until the volcano blows its top. A cataclysmic eruption of the Yellowstone volcano 630,000 years ago ejected 1000 km^3 of rock into the air, covering regions as far away as Texas and California with a layer of volcanic ash.

The geologic record shows that volcanic explosions nearly this big, or even bigger, have occurred in the western United States at least six times during the last 2 million years, so we can be fairly certain that such an eruption will happen again. We can only imagine what it might do to human civilization. Hot ash would snuff out all life within 100 km or more, and cooler but choking ash would blanket the ground more than 1000 km away. Ash thrown high into the stratosphere would dim the Sun for several years, dropping temperatures and plunging the Northern Hemisphere into an extended volcanic winter.

The hazards volcanoes pose to human society, as well as the mineral resources and energy they provide, are certainly good enough reasons to study them. In addition, volcanoes are fascinating because they are windows through which we can look into Earth's deep interior to understand the igneous and plate tectonic processes that have generated its oceanic and continental crust.

In this chapter, we will examine how magma rises through Earth's crust, emerges onto its surface as lava, and cools into solid volcanic rock. We will see how plate tectonic processes and mantle convection explain volcanism at plate boundaries and at "hot spots" within plates. We will see how volcanoes interact with other components of the Earth system, particularly the hydrosphere and the atmosphere. Finally, we will consider their destructive power as well as the potential benefits they can provide for human society.

■ Volcanoes as Geosystems

The geologic processes that give rise to volcanoes and volcanic rocks are known collectively as *volcanism.* We had a glimpse of some of these processes when we examined the formation of igneous rocks in Chapter 4, but we will take a more detailed look at them here.

Ancient philosophers were awed by volcanoes and their fearsome eruptions of molten rock. In their efforts to explain volcanoes, they spun myths about a hot, hellish underworld below Earth's surface. Basically, they had the right idea. Modern researchers also see evidence of Earth's internal heat in volcanoes. Temperature readings of rocks as far down as humans have drilled (about 10 km) show that Earth does indeed get hotter with depth. We now believe that temperatures at depths of 100 km and more—within the asthenosphere—reach at least 1300°C, high enough for the rocks there to begin to melt. For this reason, we identify the asthenosphere as a main source of *magma,* the molten rock that we call *lava* after it rises to the surface and erupts.

Portions of the solid lithosphere that ride above the asthenosphere may also melt to form magma.

Because magma is liquid, it is less dense than the rocks that produce it. Therefore, as magma accumulates, it begins to float upward through the lithosphere. In some places, the magma may find a path to the surface by fracturing the lithosphere along zones of weakness. In other places, the rising magma melts its way toward the surface. Most of the magma freezes at depth, but some fraction, probably only 10–30%, eventually reaches the surface and erupts as lava. A **volcano** is a hill or mountain constructed from the accumulation of lava and other erupted materials.

Taken together, the rocks, magmas, and processes needed to describe the entire sequence of events from melting to eruption constitute a **volcanic geosystem.** This type of geosystem can be viewed as a chemical factory that processes the input material (magmas from the asthenosphere) and transports the end product (lava) to the surface through an internal plumbing system.

Figure 12.1 is a simplified diagram of a volcano, showing the plumbing system through which magma travels to

Central vent

Side vent

Lava flows

Dike

Sill

Magma chamber

Lithospheric mantle

5 …and gases are injected into the atmosphere.

4 Lava accumulates on the surface to form a volcano,…

3 Lava erupts from the magma chamber through a central vent and side vents.

2 …rises through the lithosphere to form a magma chamber.

1 Magma, which originates in the asthenosphere,…

FIGURE 12.1 ■ Volcanoes transport magma from Earth's interior to its surface, where rocks are formed and gases are injected into the atmosphere (or hydrosphere, in the case of an underwater eruption).

the surface. Magmas rising buoyantly through the lithosphere pool together in a magma chamber, usually at shallow depths in the crust. This chamber periodically empties through a pipelike feeder channel to a central vent on the surface in repeated cycles of *central eruptions.* Lava can also erupt from vertical cracks and other vents on the flanks of a volcano.

As we saw in Chapter 4, only a small fraction of the asthenosphere melts in the first place. The resulting magma gains chemical components as it melts the surrounding rocks while rising through the lithosphere. It loses other components as crystals settle out during transport or in shallow magma chambers. And its gaseous constituents escape to the atmosphere or ocean as it erupts at the surface. By accounting for these changes, we can extract clues to the chemical composition and physical state of the upper mantle where the lavas originated. We can also learn about eruptions that occurred millions or even billions of years ago by using isotopic dating (see Chapter 8) to determine the ages of lavas.

■ Lavas and Other Volcanic Deposits

Lavas of different types produce different landforms. The differences depend on the chemical composition, gas content, and temperature of the lavas. The higher the silica content and the lower the temperature, for example, the more viscous the lava is, and the more slowly it moves. The more gas a lava contains, the more violent its eruption is likely to be.

Types of Lava

Erupted lavas, the end products of volcanic geosystems, usually solidify into one of three major types of igneous rock (see Chapter 4): basalt, andesite, or rhyolite.

BASALTIC LAVAS Basalt is an extrusive igneous rock of mafic composition (high in magnesium, iron, and calcium) and has the lowest silica content of the three igneous rock types; its intrusive equivalent is gabbro. Basaltic magma, the product of mantle melting, is the most common magma type. It is produced along mid-ocean ridges and at hot spots within plates, as well as in continental rift valleys and other zones of extension. The volcanic island of Hawaii, which is made up primarily of basaltic lava, lies above a hot spot.

Basaltic lavas erupt when hot, fluid magmas fill up a volcano's plumbing system and overflow (**Figure 12.2**). Basaltic eruptions are rarely explosive. On land, a basaltic eruption sends lava down the flanks of the volcano in great streams that can engulf everything in their path (**Figure 12.3**). When cool, these lavas are black or dark gray, but at their high eruption temperatures (1000°C to 1200°C), they glow in reds and yellows. Because their temperatures are high and their silica content low, they are extremely fluid and can flow downhill fast and far. Lava streams flowing as fast as 100 km/hour have been observed, although velocities of a few kilometers per hour are more common. In 1938, two daring Russian volcanologists measured temperatures and collected gas samples while floating down a river of molten basalt on a raft of colder solidified lava.

FIGURE 12.2 ■ A central vent eruption from Kilauea, a shield volcano on the island of Hawaii, produces a river of hot, fast-flowing basaltic lava. [J. D. Griggs/USGS.]

FIGURE 12.3 ■ A partly buried school bus in Kalapana, Hawaii. The village was buried by a basaltic lava flow from Kilauea. [© Roger Ressmeyer/CORBIS.]

The surface temperature of the raft was 300°C, and the river temperature was 870°C. Lava streams have been observed to travel more than 50 km from their sources.

Basaltic lava flows take on different forms depending on how they cool. On land, they solidify as pahoehoe (pronounced pa-hoh-ee-hoh-ee) or aa (ah-ah) (**Figure 12.4**).

Pahoehoe (the word is Hawaiian for "ropy") forms when a highly fluid lava spreads in sheets and a thin, glassy, elastic skin congeals on its surface as it cools. As the molten liquid continues to flow below the surface, the skin is dragged and twisted into coiled folds that resemble rope.

"Aa" is what the unwary exclaim after venturing barefoot onto lava that looks like clumps of moist, freshly plowed earth. Aa forms when lava loses its gases and consequently flows more slowly than pahoehoe, allowing a thick skin to form. As the flow continues to move, the thick skin breaks into rough, jagged blocks. The blocks pile up in a steep front of angular boulders that advances like a tractor tread. Aa is truly treacherous to cross. A good pair of boots may last about a week on it, and the traveler can count on cut knees and elbows.

A single downhill basaltic flow commonly has the features of pahoehoe near its source, where the lava is still fluid and hot, and of aa farther downstream, where the flow's surface—having been exposed to cool air longer—has developed a thicker outer skin.

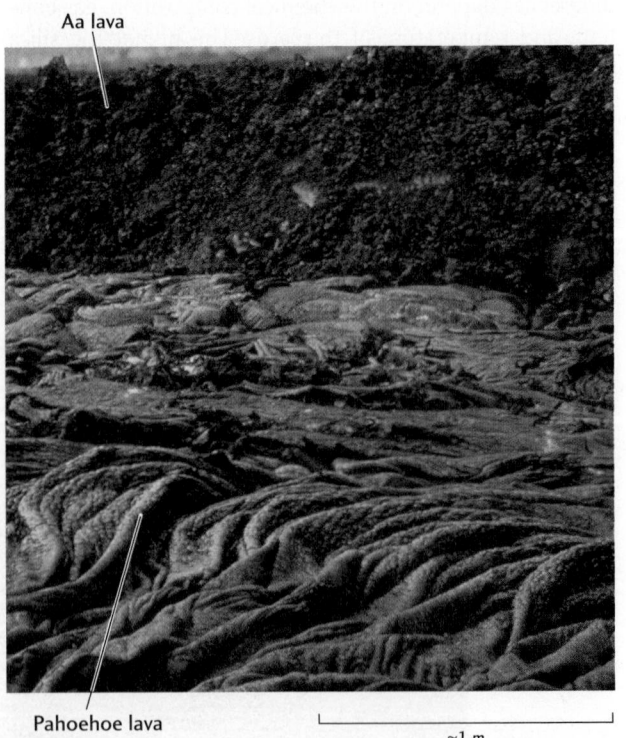

Aa lava

Pahoehoe lava

~1 m

FIGURE 12.4 ■ The two forms of basaltic lava are shown here: the jagged aa lava flow is moving over a pahoehoe lava flow on the island of Hawaii. [© Corbis.]

(a)

FIGURE 12.5 ■ These bulbous pillow lavas, which were recently extruded on the Mid-Atlantic Ridge, were photographed from the deep-sea submersible *Alvin*. [OAR/National Undersea Research Program/NOAA.]

Basaltic lava that cools under water forms *pillow lavas:* piles of ellipsoidal, pillowlike blocks of basalt about a meter wide (**Figure 12.5**). Pillow lavas are an important indicator that a region on dry land was once under water. Scuba-diving geologists have actually observed pillow lavas forming on the ocean floor off Hawaii. Tongues of molten basaltic lava develop a tough, plastic skin on contact with the cold ocean water. Because the lava inside the skin cools more slowly, the pillow's interior develops a crystalline texture, whereas the quickly chilled skin solidifies to a crystal-less glass.

ANDESITIC LAVAS Andesite is an extrusive igneous rock with an intermediate silica content; its intrusive equivalent is diorite. Andesitic magmas are produced mainly in the volcanic mountain belts above subduction zones. The name comes from a prime example: the Andes of South America.

The temperatures of **andesitic lavas** are lower than those of basalts, and because their silica content is higher, they flow more slowly and lump up in sticky masses. If one of these sticky masses plugs the central vent of a volcano, gases can build up beneath the plug and eventually blow off the top of the volcano. The explosive eruption of Mount St. Helens in 1980 (**Figure 12.6**) is a famous example.

(b)

(c)

FIGURE 12.6 ■ Mount St. Helens, an andesitic volcano in southwestern Washington State, before, during, and after its cataclysmic eruption in May 1980, which ejected about 1 km³ of pyroclastic material. The collapsed northern flank can be seen in the bottom photo [*Before*: U.S. Forest Service/USGS. *Erupting*: U.S. Geological Survey. *After*: Lyn Topinka/USGS.]

FIGURE 12.7 ■ A phreatic eruption of an island-arc volcano spews out plumes of steam into the atmosphere. The volcano, about 6 miles off the Tongan island of Tongatau, is one of about 36 in the area. [Dana Stephenson/Getty Images.]

Some of the most destructive volcanic eruptions in history have been *phreatic,* or steam, explosions, which occur when hot, gas-charged magma encounters groundwater or sea-water, generating vast quantities of superheated steam (**Figure 12.7**). The island of Krakatau, an andesitic volcano in Indonesia, was destroyed by a phreatic explosion in 1883. This legendary eruption was heard thousands of kilometers away, and it generated a tsunami that killed more than 40,000 people.

RHYOLITIC LAVAS Rhyolite is an extrusive igneous rock of felsic composition (high in sodium and potassium) with a silica content greater than 68 percent; its intrusive equivalent is granite. It is light in color, often a pretty pink. Rhyolitic magmas are produced in zones where heat from the mantle has melted large volumes of continental crust. Today, the Yellowstone volcano is producing huge amounts of rhyolitic magma that are building up in shallow chambers.

Rhyolite has a lower melting point than andesite, becoming liquid at temperatures of only 600°C to 800°C. Because **rhyolitic lavas** are richer in silica than any other lava type, they are the most viscous. A rhyolitic flow typically moves more than 10 times more slowly than a basaltic flow, and it tends to pile up in thick, bulbous deposits (**Figure 12.8**). Gases are easily trapped beneath rhyolitic lavas, and large rhyolitic volcanoes such as Yellowstone produce the most explosive of all volcanic eruptions.

FIGURE 12.8 ■ Aerial view of a rhyolite dome erupted about 1300 years ago in Newberry Caldera, Oregon. The light colored rhyolite flow stands out against the trees with Paulina Peak in the background. Its dome shape indicates that the lava was very viscous. [William Scott/USGS.]

Textures of Volcanic Rocks

The textures of volcanic rocks, like the surfaces of solidified lava flows, reflect the conditions under which they solidified. Coarse-grained textures with visible crystals can result if lavas cool slowly. Lavas that cool quickly tend to have fine-grained textures. If they are silica-rich, rapidly cooled lavas can form *obsidian,* a volcanic glass.

Volcanic rock often contains little bubbles, created as gases are released during an eruption. As we have seen, magma is typically charged with gas, like soda in an unopened bottle. When magma rises toward Earth's surface, the pressure on it decreases, just as the pressure on the soda drops when the bottle cap is removed. And just as the carbon dioxide in the soda forms bubbles when the pressure is released, the water vapor and other dissolved gases escaping from lava as it erupts create gas cavities, or *vesicles* (**Figure 12.9**). *Pumice* is an extremely vesicular volcanic rock, usually rhyolitic in composition. Some pumice has so many vesicles that it is light enough to float on water.

Pyroclastic Deposits

Water and gases in magma can have even more dramatic effects than bubble formation. Before magma erupts, the confining pressure of the overlying rock keeps these volatiles from escaping. When the magma rises close to the surface and the pressure drops, the volatiles may be released with explosive force, shattering the lava and any overlying solidified rock and sending fragments of various sizes, shapes, and textures into the air (**Figure 12.10**). These fragments, known as *pyroclasts,* are classified according to their size.

VOLCANIC EJECTA The finest pyroclasts are fragments less than 2 mm in diameter, which are classified as *volcanic ash.* Volcanic eruptions can spray ash high into the atmosphere,

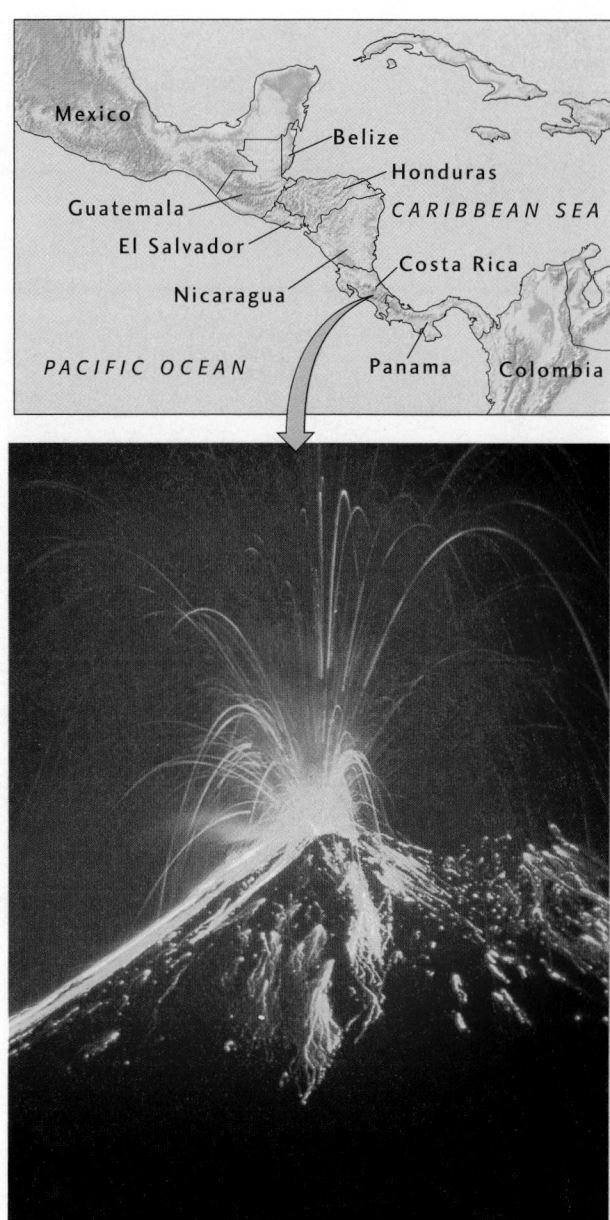

FIGURE 12.10 ■ An explosive eruption at Arenal volcano, Costa Rica, hurls pyroclasts into the air. [Gregory G. Dimijian/Science Source.]

FIGURE 12.9 ■ A sample of vesicular basalt.
[John Grotzinger.]

where ash that is fine enough to stay aloft can be carried great distances. Within 2 weeks of the 1991 eruption of Mount Pinatubo in the Philippines, for example, its ash was traced all the way around the world by Earth-orbiting satellites.

Fragments ejected as blobs of lava that cool in flight and become rounded, or as chunks torn loose from previously solidified volcanic rock, can be much larger. These fragments are called *volcanic bombs* (**Figure 12.11**). Volcanic bombs as large as houses have been thrown more than 10 km by explosive eruptions.

FIGURE 12.11 ■ Volcanologist Katia Krafft examines a volcanic bomb ejected from Asama volcano, Japan. Krafft was later killed by a pyroclastic flow on Mount Unzen (see Figure 12.13). [Science Source.]

(a)

~0.3 m

(b)

FIGURE 12.12 ■ (a) Welded tuff from an ash-flow deposit in the Great Basin of northern Nevada. (b) Volcanic breccia.
[(a) John Grotzinger; (b) Fletcher & Baylis/Science Source.]

Sooner or later, these pyroclasts fall to Earth, building the largest deposits near their source. As they cool, the hot, sticky fragments become welded together (lithified). Rocks created from smaller fragments are called **tuffs;** those formed from larger fragments are called **breccias** (Figure 12.12).

PYROCLASTIC FLOWS **Pyroclastic flows,** which are particularly spectacular and often deadly, occur when a volcano ejects hot ash and gases in a glowing cloud that rolls downhill at high speeds. The solid particles are buoyed up by the hot gases, so there is little frictional resistance to their movement.

In 1902, with very little warning, a pyroclastic flow with an internal temperature of 800°C exploded from the side of Mont Pelée, on the Caribbean island of Martinique. The avalanche of choking hot gas and glowing volcanic ash plunged down the slopes at a hurricane speed of 160 km/hour. In 1 minute and with hardly a sound, the searing emulsion of gas and ash enveloped the town of St. Pierre and killed 29,000 people. It is sobering to recall the statement of one Professor Landes, issued the day

before the cataclysm: "The Montagne Pelée presents no more danger to the inhabitants of St. Pierre than does Vesuvius to those of Naples." Professor Landes perished with the others. In 1991, French volcanologists Maurice and Katia Krafft were killed by a pyroclastic flow on Mount Unzen, Japan (Figure 12.13).

■ Eruptive Styles and Landforms

The surface features produced by a volcano as it ejects material vary with the properties of the magma, especially its chemical composition and gas content, the type of

FIGURE 12.13 ■ This pyroclastic flow plunged down the slopes of Mount Unzen, Japan, in June 1991. Note the firefighter and fire engine in the foreground, trying to outrun the hot ash cloud descending on them. Three scientists who were studying this volcano died when they were engulfed by a similar flow. [AP/Wide World Photos.]

material (lava versus pyroclasts) erupted, and the environmental conditions under which it erupts—on land or under the sea. Volcanic landforms also depend on the rate at which lava is produced and the plumbing system that gets it to the surface (**Figure 12.14**).

Central Eruptions

Central eruptions discharge lava or pyroclasts from a *central vent,* an opening atop a pipelike feeder channel rising from the magma chamber. The magma ascends through this channel to erupt at Earth's surface. Central eruptions create the most familiar of all volcanic features: the volcanic mountain, shaped like a cone.

SHIELD VOLCANOES A *lava cone* is built by successive flows of lava from a central vent. If the lava is basaltic, it flows easily and spreads widely. If flows are copious and frequent, they create a broad, shield-shaped volcano 2 or more kilometers high and many tens of kilometers in circumference, with relatively gentle slopes. Mauna Loa, on the island of Hawaii, is the classic example of such a **shield volcano** (Figure 12.14a). Although it rises only 4 km above sea level, it is actually the world's tallest mountain: measured from its base on the seafloor, Mauna Loa is 10 km high, taller than Mount Everest! It grew to this enormous size by the accumulation of thousands of lava flows, each only a few meters thick, over a period of about a million years. The island of Hawaii actually consists of the overlapping tops of several active shield volcanoes emerging from the ocean.

VOLCANIC DOMES In contrast to basaltic lavas, andesitic and rhyolitic lavas are so viscous they can barely flow. They often produce a *volcanic dome,* a bulbous, steep-sided mass of rock (see Figure 12.8). Domes look as though the lava has been squeezed out of a vent like toothpaste, with very little lateral spreading. Domes often plug vents, trapping gases beneath them (Figure 12.14b). Pressure can increase until an explosion occurs, blasting the dome into fragments.

CINDER CONES When volcanic vents discharge pyroclasts, these solid fragments can build up to create *cinder cones.* The profile of a cinder cone is determined by the *angle of repose* of the fragments: the maximum angle at which the fragments will remain stable rather than sliding downhill. The larger fragments, which fall near the vent, form very steep but stable slopes. Finer particles are carried farther from the vent and form gentler slopes at the base of the cone. The classic concave-shaped volcanic cone with its central vent at the summit develops in this way (Figure 12.14c).

STRATOVOLCANOES When a volcano emits lava as well as pyroclasts, alternating lava flows and beds of pyroclasts build a concave-shaped composite volcano, or **stratovolcano** (Figure 12.14d). Lava that solidifies in the central feeder channel and in radiating dikes strengthens the cone structure. Stratovolcanoes are common above subduction zones. Famous examples are Mount Fuji in Japan, Mount Vesuvius and Mount Etna in Italy, and Mount Rainier in Washington State. Mount St. Helens had a near-perfect stratovolcano shape until its eruption in 1980 destroyed its northern flank (see Figure 12.6).

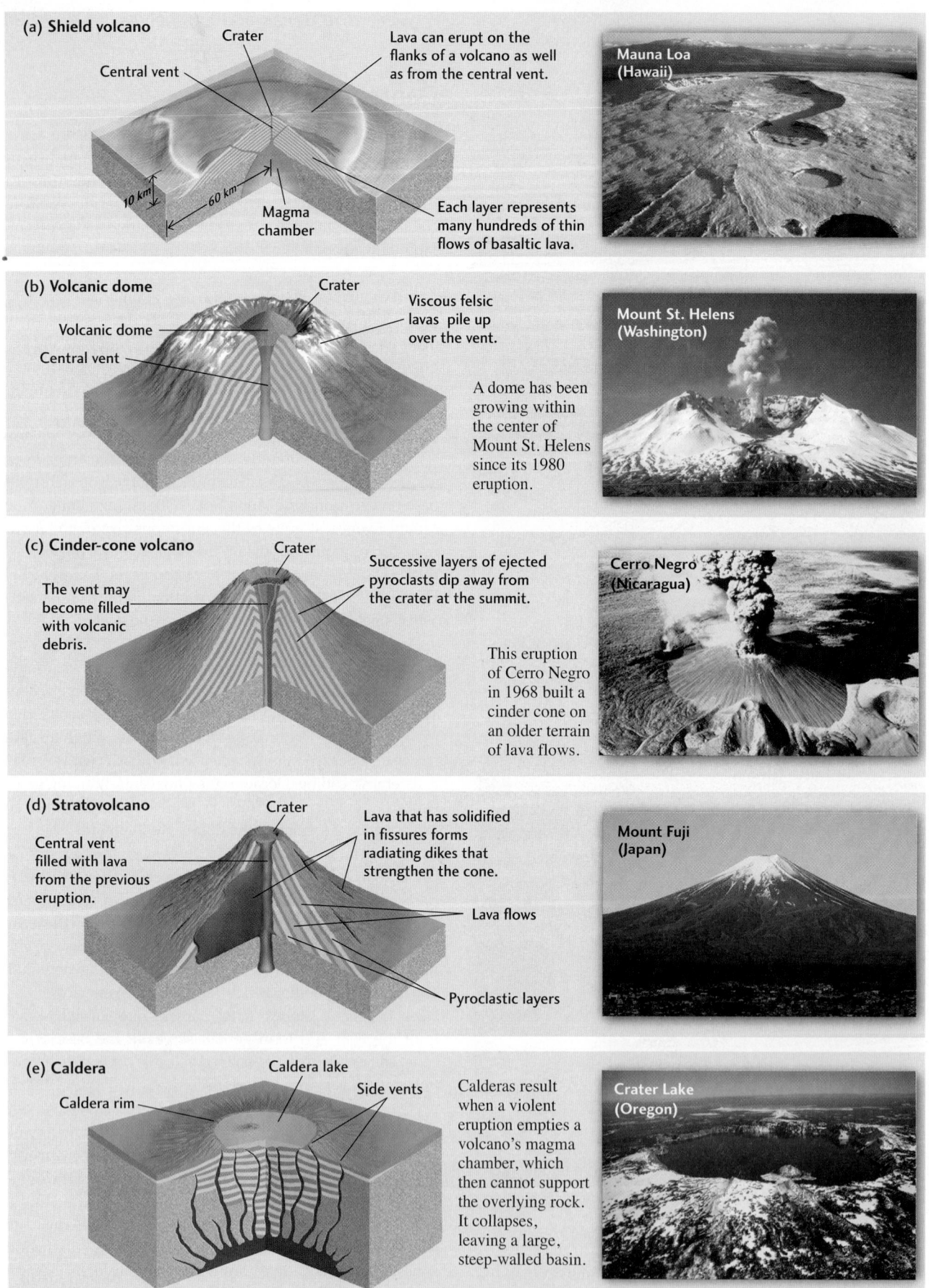

(a) Shield volcano

Crater

Central vent

Lava can erupt on the flanks of a volcano as well as from the central vent.

10 km

60 km

Magma chamber

Each layer represents many hundreds of thin flows of basaltic lava.

Mauna Loa (Hawaii)

(b) Volcanic dome

Crater

Volcanic dome

Central vent

Viscous felsic lavas pile up over the vent.

A dome has been growing within the center of Mount St. Helens since its 1980 eruption.

Mount St. Helens (Washington)

(c) Cinder-cone volcano

Crater

The vent may become filled with volcanic debris.

Successive layers of ejected pyroclasts dip away from the crater at the summit.

This eruption of Cerro Negro in 1968 built a cinder cone on an older terrain of lava flows.

Cerro Negro (Nicaragua)

(d) Stratovolcano

Crater

Central vent filled with lava from the previous eruption.

Lava that has solidified in fissures forms radiating dikes that strengthen the cone.

Lava flows

Pyroclastic layers

Mount Fuji (Japan)

(e) Caldera

Caldera lake

Caldera rim

Side vents

Calderas result when a violent eruption empties a volcano's magma chamber, which then cannot support the overlying rock. It collapses, leaving a large, steep-walled basin.

Crater Lake (Oregon)

FIGURE 12.14 ■ The eruptive styles of volcanoes and the landforms they create are determined principally by the composition of magma. [(a) U.S. Geological Survey; (b) Lyn Topinka/USGS Cascades Volcano Observatory; (c) Smithsonian; (d) CORBIS; (e) Bates Littlehales/National Geographic/Getty Images.]

CRATERS A bowl-shaped pit, or **crater,** is found at the summit of most volcanic mountains, surrounding the central vent. During an eruption, the upwelling lava overflows the crater walls. When the eruption ceases, the lava that remains in the crater often sinks back into the vent and solidifies, and the crater may become partly filled by rock fragments that fall back into it. When the next eruption occurs, that material may be blasted out of the crater. Because a crater's walls are steep, they may cave in or become eroded over time. In this way, a crater can grow to several times the diameter of the vent and hundreds of meters deep. The crater of Mount Etna in Sicily, for example, is currently 300 m in diameter.

CALDERAS When great volumes of magma are discharged rapidly from a large magma chamber, the chamber can no longer support its roof. In such cases, the overlying volcanic structure can collapse catastrophically, leaving a large, steep-walled, basin-shaped depression much larger than a crater, called a **caldera** (Figure 12.14e). The development of the caldera that forms Crater Lake in Oregon is shown in Figure 12.15. Calderas can be impressive features, ranging in diameter from a few kilometers to 50 km or more. Owing to their size and high-volume eruptions, large caldera systems are sometimes called "supervolcanoes". The Yellowstone supervolcano, which is the largest active volcano in the United States, has a caldera with an area greater than Rhode Island.

After some hundreds of thousands of years, enough fresh magma may reenter the collapsed magma chamber to reinflate it, forcing the caldera floor to dome upward again and creating a *resurgent caldera.* The cycle of eruption, collapse, and resurgence may occur repeatedly over geologic time. Three times over the last 2 million years, the Yellowstone supervolcano has erupted catastrophically, in each instance ejecting hundreds or thousands of times more material than the 1980 Mount St. Helens eruption and depositing ash over much of what is now the western United States. Other resurgent calderas are Valles Caldera in New Mexico and Long Valley Caldera in California, which last erupted about 1.2 million and 760,000 years ago, respectively.

DIATREMES When magma from Earth's deep interior escapes explosively, the vent and the feeder channel below it are often left filled with volcanic breccia as the eruption wanes. The resulting structure is called a **diatreme.** Shiprock, a formation that towers over the surrounding plain in New Mexico, is a diatreme exposed by erosion of the sedimentary rocks through which it erupted. To transcontinental air travelers, Shiprock looks like a gigantic black skyscraper in the red desert (Figure 12.16).

FIGURE 12.15 ■ Stages in the formation of the Crater Lake caldera. The stage-3 collapse occurred about 7,700 years ago.

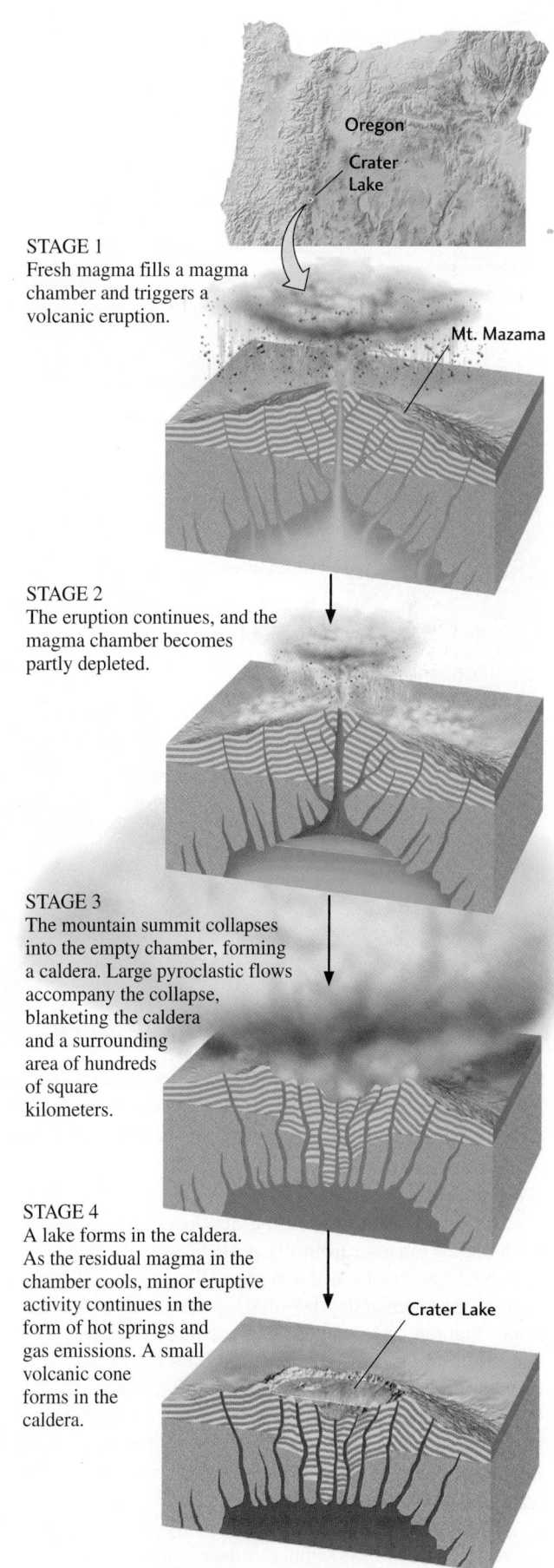

STAGE 1
Fresh magma fills a magma chamber and triggers a volcanic eruption.

Oregon
Crater Lake

Mt. Mazama

STAGE 2
The eruption continues, and the magma chamber becomes partly depleted.

STAGE 3
The mountain summit collapses into the empty chamber, forming a caldera. Large pyroclastic flows accompany the collapse, blanketing the caldera and a surrounding area of hundreds of square kilometers.

STAGE 4
A lake forms in the caldera. As the residual magma in the chamber cools, minor eruptive activity continues in the form of hot springs and gas emissions. A small volcanic cone forms in the caldera.

Crater Lake

(a)

1 Gas-charged magma from deep in the mantle forces its way upward, fracturing the lithosphere.

2 Rapidly ascending magma breaks off and carries crust and mantle fragments as it explodes at supersonic speed.

3 After the eruption, the feeder channel forms a diatreme made up of solidified magma and these rock fragments, or breccia.

4 The softer sediments of the cone and surface of the crust erode, leaving the diatreme core and radiating dikes we see today.

Time

Former volcanic cone   Diatreme   Dike

Mantle and crust fragments

0 km

Crust

Lithosphere

100 km

Gas-charged magmas

Asthenosphere

(b)

FIGURE 12.16 ■ (a) The formation of a diatreme. (b) Shiprock, towering 515 m above the surrounding flat-lying sediments of New Mexico, is a diatreme that has been exposed by erosion of the softer sedimentary rocks that once enclosed it. Note the vertical dike, one of six radiating from the central vent. [Airphoto—Jim Wark.]

The eruptive mechanism that produces diatremes has been pieced together from the geologic record. The kinds of minerals and rocks found in some diatremes could have formed only at great depths—100 km or so, well within the upper mantle. Gas-charged magmas force their way upward from these depths by fracturing the lithosphere and exploding into the atmosphere, ejecting gases and solid fragments torn from the deep crust and mantle, sometimes at supersonic speed. Such an eruption would probably look like the exhaust jet of a giant rocket upside down in the ground blowing rocks and gases into the air.

Perhaps the most exotic diatremes are *kimberlite pipes,* named after the fabled Kimberley diamond mines of South Africa. Kimberlite is a volcanic type of peridotite—an ultramafic rock composed primarily of olivine. Kimberlite pipes also contain a variety of mantle fragments, including diamonds that were pulled into the magma as it exploded toward the surface (see Figure 10.25). The extremely high pressures needed to squeeze carbon into the mineral diamond are reached only at depths greater than 150 km. From careful studies of diamonds and other mantle fragments found in kimberlite pipes, geologists have been able to reconstruct sections of the mantle as if they had had been able to drill down to 200 km or more. These studies provide strong support for the theory that the upper mantle is made primarily of peridotite.

FIGURE 12.17 ■ A fissure eruption generates a "curtain of fire" on Kilauea, Hawaii, in 1992. [U.S. Geological Survey.]

Fissure Eruptions

The largest volcanic eruptions do not come from a central vent, but through large, nearly vertical cracks in Earth's surface, sometimes tens of kilometers long (Figure 12.17). Such **fissure eruptions** are the main style of volcanism along mid-ocean ridges, where new oceanic crust is formed.

A moderate-sized fissure eruption occurred in 1783 on a segment of the Mid-Atlantic Ridge that comes ashore in Iceland (Figure 12.18). A fissure 32 km long opened and, in six months, spewed out 12 km^3 of basalt, enough to cover Manhattan to a height about halfway up the Empire State Building. The eruption also released more than 100 megatons of sulfur dioxide, creating a poisonous blue haze that

1 Highly fluid basalt erupting from fissures…

2 …forms widespread layers rather than mountains.

Earlier flows

Cinder cones

Lava

Fissures

(a)

(b)

FIGURE 12.18 ■ (a) In a fissure eruption, highly fluid basaltic lava flows rapidly away from the fissures. (b) These volcanic cones lie along the Laki fissure in Iceland, which opened in 1783 and erupted the largest flow of lava on land in recorded history. [(a) After R. S. Fiske/USGS; (b) Tony Waltham.]

hung over Iceland for more than a year. The resulting crop failures caused three-quarters of the island's livestock and one-fifth of its human population to die of starvation. Sulphuric aerosols from the Laki eruption were transported by the prevailing winds across Europe, causing crop damage and respiratory illnesses in many countries.

FLOOD BASALTS Highly fluid basaltic lavas erupting from fissures on continents can spread out in sheets over flat terrain. Successive flows often pile up into immense basalt plateaus, called **flood basalts,** rather than forming a shield volcano as they do when the eruption is confined to a central vent. In North America, a huge eruption of flood basalts about 16 million years ago buried 160,000 km^2 of preexisting topography in what is now Washington, Oregon, and Idaho to form the Columbia Plateau (**Figure 12.19**). Individual flows were more than 100 m thick, and some were so fluid that they traveled more than 500 km from their source. An entirely new landscape with new river valleys has since developed atop the lava that buried the old surface. Plateaus formed by flood basalts are found on every continent as well as on the seafloor.

ASH-FLOW DEPOSITS Eruptions of pyroclasts on continents have produced extensive sheets of hard volcanic tuffs called **ash-flow deposits.** A succession of forests in Yellowstone National Park have been buried under such ash flows. Some of the largest pyroclastic deposits on the planet are the ash flows erupted in the mid-Cenozoic era, 45 million to 30 million years ago, through fissures in what is now the Basin and Range province of the western United States. The amount of material released during this pyroclastic flare-up was a staggering 500,000 km^3—enough to cover the entire state of Nevada with a layer of rock nearly 2 km thick! Humans have never witnessed one of these spectacular events.

■ Interactions of Volcanoes with Other Geosystems

Volcanoes are chemical factories that produce gases as well as solid materials. Courageous volcanologists have collected volcanic gases during eruptions and analyzed them to determine their composition. Water vapor is the main constituent of volcanic gases (70 to 95 percent), followed by carbon dioxide, sulfur dioxide, and traces of nitrogen, hydrogen, carbon monoxide, sulfur, and chlorine. Volcanic eruptions can release enormous amounts of these gases. Some volcanic gases may come from deep within Earth, making their way to the surface for the first time. Some may be recycled groundwater and ocean water, recycled atmospheric gases, or gases that were trapped in earlier generations of rocks.

(a)

(b)

FIGURE 12.19 ■ (a) The Columbia Plateau covers 160,000 km^2 in Washington, Oregon, and Idaho. (b) Successive flows of flood basalts piled up to build this immense plateau, here cut by the Palouse River. [© Charles Bolin/Alamy.]

As we have seen, volcanic gases released at Earth's surface have a number of effects on other geosystems. The emission of volcanic gases during Earth's early history is thought to have created the oceans and the atmosphere, and volcanic gas emissions continue to influence those components of the Earth system today. Periods of intense volcanic activity have affected Earth's climate repeatedly, and they may have been responsible for some of the mass extinctions documented in the geologic record.

Volcanism and the Hydrosphere

Volcanic activity does not stop when lava or pyroclastic materials cease to flow. For decades or even centuries after a major eruption, volcanoes continue to emit steam and other gases through small vents called *fumaroles* (**Figure 12.20**). These emanations contain dissolved materials that precipitate onto surrounding surfaces as the water evaporates or cools, forming various encrusting deposits. Some of these precipitates contain valuable minerals.

Fumaroles are a surface manifestation of **hydrothermal activity:** the circulation of water through hot volcanic rocks and magmas. Circulating groundwater that comes into contact with buried magma (which may remain hot for hundreds of thousands of years) is heated and returned to the surface as hot springs and geysers. A *geyser* is a hot-water fountain that spouts intermittently with great force, frequently accompanied by a thunderous roar. The best-known geyser in the United States is Old Faithful in Yellowstone National Park, which erupts about every 90 minutes, sending a jet of hot water as high as 60 m into the air (**Figure 12.21**). We'll take a closer look at the mechanisms that drive hot springs and geysers in Chapter 17.

Hydrothermal activity is especially intense in the spreading centers at mid-ocean ridges, where huge volumes of water and magma come into contact. Fissures created by tensional forces allow seawater to circulate throughout

FIGURE 12.21 ■ Old Faithful geyser, in Yellowstone National Park, erupts regularly about every 65 minutes.
[Simon Fraser/SPL/Science Source.]

the newly formed oceanic crust. Heat from the hot volcanic rocks and deeper magmas drives a vigorous convection current that pulls cold seawater into the crust, heats it, and expels the hot water back into the overlying ocean through vents on the rift valley floor (**Figure 12.22**).

Given the common occurrence of hot springs and geysers in volcanic geosystems on land, the evidence for pervasive hydrothermal activity at spreading centers immersed in deep water should come as no surprise. Nevertheless, geologists were amazed once they recognized the intensity of the convection and discovered some of its chemical and biological consequences. The most spectacular manifestations of this process were first found in the eastern Pacific Ocean in 1977. Plumes of hot, mineral-laden water with temperatures as high as 350°C were seen spouting through hydrothermal vents at the crest of the East Pacific Rise (see Figure 11.15). The rates of fluid flow turned out to be very high. Marine geologists have estimated that the entire volume of the ocean's water is circulated through the cracks and vents of Earth's spreading centers in only 10 million years.

FIGURE 12.20 ■ A fumarole encrusted with sulfur deposits on the Merapi volcano in Indonesia.
[R. L. Christiansen/USGS.]

FIGURE 12.22 ■ Near spreading centers, seawater circulates through the oceanic crust, is heated by magma, and is reinjected into the ocean, forming black smokers and depositing minerals on the seafloor.

Scientists have come to realize that the interactions between the lithosphere and the hydrosphere at spreading centers profoundly affect the geology, chemistry, and biology of the oceans in a number of ways:

■ The creation of new lithosphere accounts for almost 60 percent of the energy flowing out of Earth's interior. Circulating seawater cools the new lithosphere very efficiently and therefore plays a major role in the outward transport of Earth's internal heat.

■ Hydrothermal activity leaches metals and other elements from the new crust, injecting them into the oceans. These elements contribute as much to seawater chemistry as the mineral components dumped into the oceans by all the world's rivers.

■ Metal-rich minerals precipitate out of the circulating seawater and form ores of zinc, copper, and iron in shallow parts of the oceanic crust. These ores form when seawater sinks through porous volcanic rocks, is heated, and leaches these elements from the new crust. When the heated seawater, enriched with dissolved minerals, rises and reenters the cold ocean, the ore-forming minerals precipitate.

The energy and nutrients at hydrothermal vents feed unusual colonies of strange organisms whose energy comes from Earth's interior rather than from sunlight (see Figure 11.15). Chemoautotrophic hyperthermophiles, similar to those that populate hot springs on land, form the base of complex ecosystems, providing food for giant clams and tube worms up to several meters long. Some scientists have speculated that life on Earth may have begun in the energetic, chemically rich environments of hydrothermal vents (see Chapter 11).

Volcanism and the Atmosphere

Volcanism in the lithosphere affects weather and climate by changing the composition and properties of the atmosphere. Large eruptions can inject sulfurous gases into the atmosphere tens of kilometers above Earth (**Figure 12.23**).

FIGURE 12.23 ■ Satellite image of the huge ash cloud spewing from the erupting Cordón Caulle volcano in central Chile on June 13, 2011. The ash plume extends 800 km from the snow-covered Andes mountains (on left side of photo) to the Argentine city of Buenos Aires (center right of photo). This ash cloud encircled the planet, closing airports in Australia and New Zealand. [NASA image courtesy Jeff Schmaltz, MODIS Rapid Response Team at NASA GSFC.]

Through various chemical reactions, these gases form an aerosol (a fine airborne mist) containing tens of millions of metric tons of sulfuric acid. Such aerosols may block enough of the Sun's radiation from reaching Earth's surface to lower global temperatures for a year or two. The eruption of Mount Pinatubo, one of the largest explosive eruptions of the twentieth century, led to a global cooling of at least 0.5°C in 1992. (Chlorine emissions from Mount Pinatubo also hastened the loss of ozone in the atmosphere, nature's shield that protects the biosphere from the Sun's ultraviolet radiation.)

The debris lofted into the atmosphere during the 1815 eruption of Mount Tambora in Indonesia resulted in even greater cooling. The next year, the Northern Hemisphere suffered a very cold summer; according to a diarist in Vermont, "no month passed without a frost, nor one without a snow." The drop in temperature and the ash fall caused widespread crop failures. More than 90,000 people perished in that "year without a summer," which inspired Lord Byron's gloomy poem, "Darkness":

I had a dream, which was not all a dream.
The bright sun was extinguish'd, and the stars
Did wander darkling in the eternal space,
Rayless, and pathless, and the icy earth
Swung blind and blackening in the moonless air;
Morn came and went—and came, and brought no day.

And men forgot their passions in the dread
Of this their desolation; and all hearts
Were chill'd into a selfish prayer for light.

■ The Global Pattern of Volcanism

Before the advent of plate tectonic theory, geologists noted a concentration of volcanoes around the rim of the Pacific Ocean and nicknamed it the Ring of Fire (see Figure 2.6). The explanation of the Ring of Fire in terms of subduction zones was one of the great successes of the new theory. As we will see in this section, plate tectonics can explain essentially all major features in the global pattern of volcanism (**Figure 12.24**).

Figure 12.25 shows the locations of the world's active volcanoes that occur on land or above the ocean surface. About 80 percent are found at convergent plate boundaries, 15 percent at divergent plate boundaries, and the remaining few within plate interiors. There are many more active volcanoes than shown on this map, however. Most of the lava erupted on Earth's surface comes from vents beneath the oceans, located at spreading centers on mid-ocean ridges.

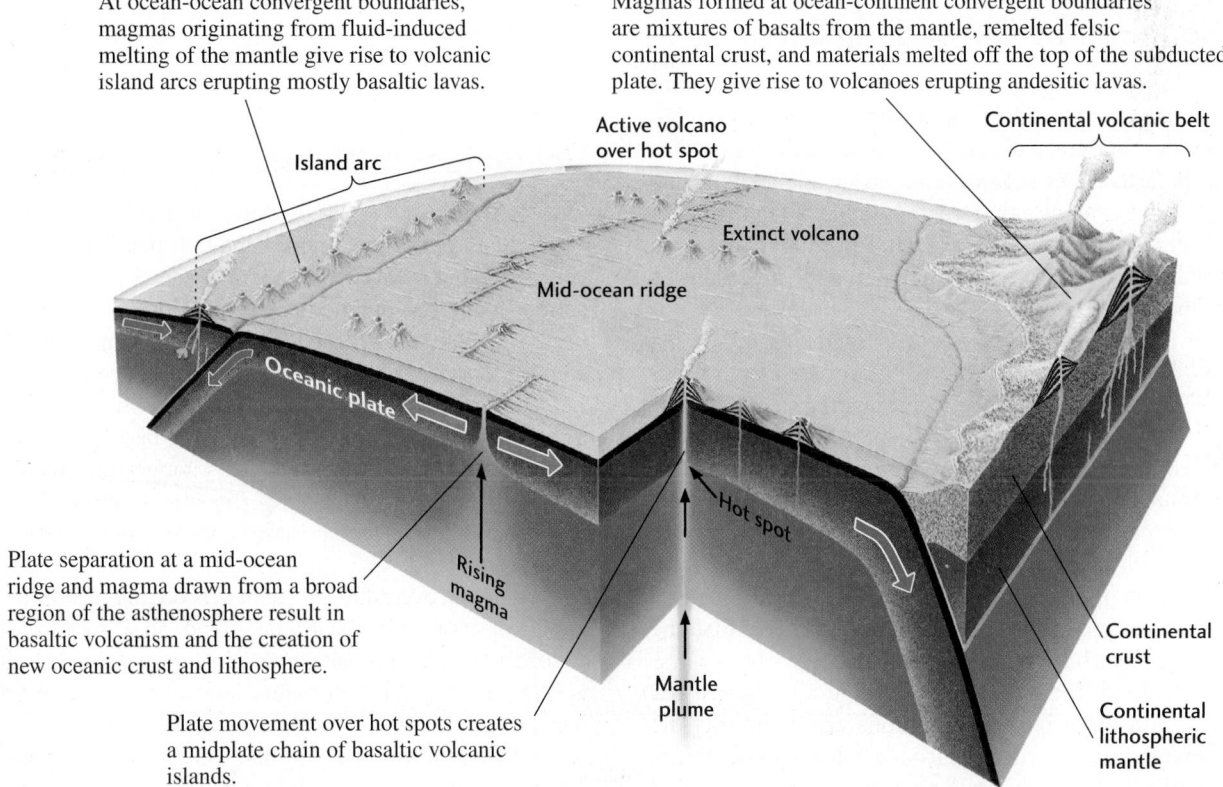

At ocean-ocean convergent boundaries, magmas originating from fluid-induced melting of the mantle give rise to volcanic island arcs erupting mostly basaltic lavas.

Magmas formed at ocean-continent convergent boundaries are mixtures of basalts from the mantle, remelted felsic continental crust, and materials melted off the top of the subducted plate. They give rise to volcanoes erupting andesitic lavas.

Continental volcanic belt

Active volcano over hot spot

Island arc

Extinct volcano

Mid-ocean ridge

Oceanic plate

Rising magma

Hot spot

Plate separation at a mid-ocean ridge and magma drawn from a broad region of the asthenosphere result in basaltic volcanism and the creation of new oceanic crust and lithosphere.

Mantle plume

Continental crust

Continental lithospheric mantle

Plate movement over hot spots creates a midplate chain of basaltic volcanic islands.

FIGURE 12.24 ■ Plate tectonic processes explain the global pattern of volcanism.

FIGURE 12.25 ■ The active volcanoes of the world with vents on land or above the ocean surface are represented on this map by red dots. Black lines represent plate boundaries. Not shown on this map are the numerous vents of the mid-ocean ridge system below the ocean surface.

Volcanism at Spreading Centers

As we have seen, enormous volumes of basaltic lava erupt continually along the global network of mid-ocean ridges—enough to have created all of the present-day seafloor. This "crustal factory" lies beneath a rift valley a few kilometers wide, and it extends along the thousands of kilometers of mid-ocean ridges (see Figure 12.24). The erupted magma is formed by decompression melting of mantle peridotite, as described in Chapter 4.

Divergent boundaries comprise segments of a mid-ocean ridge offset in a zigzag pattern by transform faults (see Figure 2.7). Detailed geologic mapping of the seafloor has revealed that the ridge segments can themselves be quite complex. They are often composed of shorter, parallel spreading centers that are offset by a few kilometers and may partly overlap. Each of these spreading centers is an "axial volcano" that erupts basaltic lava at variable rates along its length. Basalts from nearby axial volcanoes often show slight geochemical differences, indicating that the axial volcanoes have separate plumbing systems.

In Iceland, the Mid-Atlantic Ridge rises above the ocean, and large basaltic eruptions are common. The most recent major eruption, from a volcano beneath the Eyjafjallajökull ice cap on the southern coast of Iceland in 2010, ejected massive amounts of very fine-grained ash high into the atmosphere, which disrupted air traffic across western Europe for many weeks (see Earth Issues 12.1).

Volcanism in Subduction Zones

One of the most striking features of a subduction zone is the chain of volcanic mountains that parallels the convergent boundary above the sinking slab of oceanic lithosphere, regardless of whether the overriding lithosphere is oceanic or continental (see Figure 12.24). The magmas that feed subduction-zone volcanoes are produced by fluid-induced melting (see Chapter 4) and are more varied in their chemical composition than the basaltic magmas produced at mid-ocean ridges. They range from mafic to felsic—that is, from basaltic to rhyolitic—although intermediate (andesitic) compositions are the most common observed on land.

Where the overriding lithosphere is oceanic, subduction-zone volcanoes form volcanic island arcs, such as the Aleutian Islands of Alaska and the Mariana Islands of the western Pacific. Where oceanic lithosphere is subducted beneath a continent, the volcanoes and volcanic rocks coalesce to form a volcanic mountain belt on land, such as the Andes, which mark the subduction of the oceanic Nazca Plate beneath continental South America.

Earth Issues

12.1 Volcanic Ash Clouds over Europe

On April 14, 2010, Eyjafjallajökull volcano in Iceland began a series of eruptions that shut down air travel over western and northern Europe for a period of six days. [According to the joke, "Eyjafjallajökull" is Icelandic for "name that no one can pronounce" Actually, it's pronounced: aye-ya-fyah-dla-jow-kudl and means "island-mountain glacier."] These eruptions led to the closure of most of Europe's larger airports with caused the cancelation of many flights to and from Europe, resulting in the highest level of air traffic disruption since World War II. Many people were stranded for days with little comfort as flights were sequentially canceled, leaving people to struggle to find alternative means of transportation or accommodations. In the week following the eruption it is estimated that 250,000 British, French and Irish citizens were stranded abroad, the European economy may have lost almost 2 billion dollars, and the aviation industry lost up to 250 million dollars per day.

The eruptions were predicted well in advance. Seismic activity in and around Eyjafjallajökull began in late 2009 and increased in intensity and frequency until March 20, 2010, when a small eruption occurred. A second, much larger eruption occurred on April 14th ejecting 250 million cubic meters of volcanic ash. The ash cloud rose to elevations of 9,000 meters and, though not as large as the 1980 Mount St. Helens eruption in Oregon, it was high enough to enter the jet stream, which flowed directly over Iceland at the time. The westward flow of the jet stream transported the ash to Europe where it spread out over a large part of the continent.

Much of this volcanic ash was produced by the interaction of hot magma with glacial ice and water, which made it very

In this April 16, 2010, photo, the Eyjafjallajökull volcano in southern Iceland sends ash into the air just prior to sunset. [AP Photo/Brynjar Gauti.]

fined grained, less than 2 mm in size. When ash of this size gets caught up in jet engines, the high temperatures of the jet (up to 2000°C) can remelt it, recreating a sticky lava that can cause engine failure. In extreme cases, planes have had to literally glide their way out of the ash cloud before engines can be restarted.

The Eyjafjallajökull eruptions lasted for only one month; by June, 2010, very little ash was being ejected. But future eruptions in Iceland are inevitable, and the agricultural and environmental consequences for Europe are potentially dire. Right now, geologists are carefully monitoring the nearby Katla volcano, whose historical eruptions have often followed those of Eyjafjallajökull. In the long term.

The terrain of Japan is a prime example of the complex of intrusive and extrusive igneous rock that may evolve over many millions of years at a subduction zone. Everywhere in this small country are all kinds of extrusive igneous rocks of various ages, mixed in with mafic and intermediate intrusives, metamorphosed volcanic rocks, and sedimentary rocks derived from erosion of the igneous rocks. The erosion of these various rocks has contributed to the distinctive landscapes portrayed in so many classic and modern Japanese paintings.

Intraplate Volcanism: The Mantle Plume Hypothesis

Decompression melting explains volcanism at spreading centers, and fluid-induced melting can account for the volcanism above subduction zones, but how can plate tectonic

theory explain *intraplate volcanism*—that is, volcanoes far from plate boundaries? Geologists have found a clue in the ages of such volcanoes.

HOT SPOTS AND MANTLE PLUMES Consider the Hawaiian Islands, which stretch across the middle of the Pacific Plate. This island chain begins with the active volcanoes on the island of Hawaii and continues to the northwest as a string of progressively older, extinct, eroded, and submerged volcanic mountains and ridges. In contrast to the seismically active mid-ocean ridges, the Hawaiian island chain is not marked by frequent large earthquakes (except near the active volcanoes). It is essentially aseismic (without earthquakes), and is therefore called an *aseismic ridge.* Active volcanoes at the beginnings of progressively older aseismic ridges can be found elsewhere in the Pacific and in other large ocean basins. Two examples are the active volcanoes of Tahiti, at the southeastern end of the Society

Islands, and the Galápagos Islands, at the western end of the aseismic Nazca Ridge (see Figure 12.25).

Once the general pattern of plate movements had been worked out, geologists were able to show that these aseismic ridges approximated the paths that the plates would take over a set of volcanically active **hot spots** that were fixed relative to one another, as if they were blowtorches anchored in Earth's mantle (**Figure 12.26**). Based on this evidence, they hypothesized that hot spots were caused by hot, solid material rising in narrow, cylindrical jets from deep within the mantle (perhaps as deep as the core-mantle boundary), called **mantle plumes.** According to the mantle plume hypothesis, when peridotites transported upward in a mantle plume reach lower pressures at shallower depths, they begin to melt, producing basaltic magma. The magma

penetrates the lithosphere and erupts at the surface. The current position of a plate over the hot spot is marked by an active volcano, which becomes inactive as plate movement carries it away from the hot spot. The movement of the plate thus generates a trail of extinct, progressively older volcanoes. As shown in Figure 12.26a, the Hawaiian Islands fit this pattern well. Dating of the volcanoes yields a rate of movement of the Pacific Plate over the Hawaiian hot spot of about 100 mm/year.

Some aspects of intraplate volcanism within continents can also been explained by the mantle plume hypothesis. The modern Yellowstone caldera, only 630,000 years old, is still volcanically active, as evidenced by the geysers, hot springs, uplift, and earthquakes observed in the area. It is the youngest member of a chain of sequentially older and

(a)

The Pacific Plate has moved northwest over the Hawaiian hot spot,…   …resulting in a chain of volcanic islands and seamounts.   The ages of the volcanoes are consistent with plate movement of about 100 mm/year.

(b)

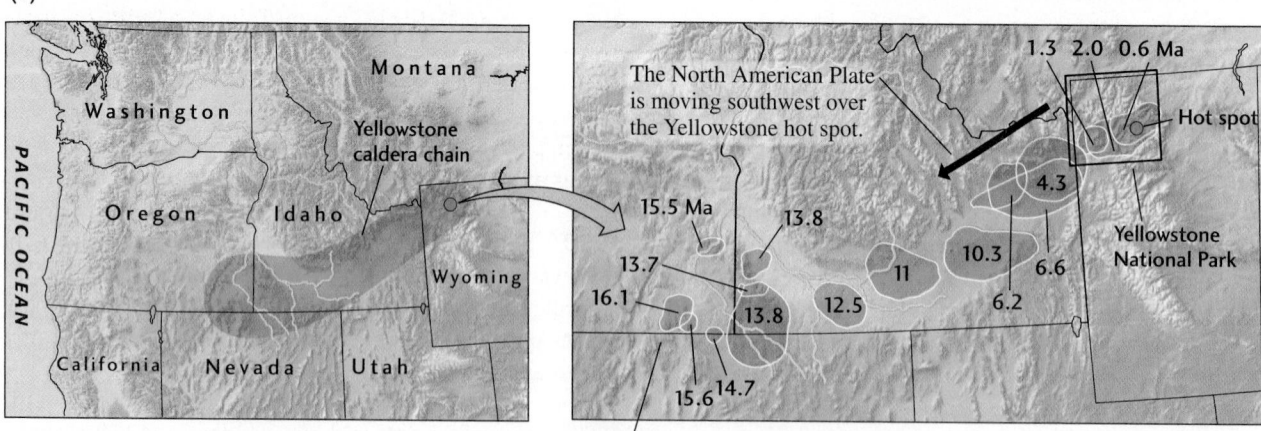

The ages of the calderas are consistent with plate movement of about 25 mm/year.

FIGURE 12.26 ■ The movement of a plate over a hot spot generates a trail of progressively older volcanoes. (a) The volcanoes of the Hawaiian island chain and its extension into the northwestern Pacific (the Emperor seamounts) show a northwestward trend toward progressively older ages. (b) A chain of progressively older calderas marks the movement of the North American Plate over a continental hot spot during the past 16 million years. (Ma, million years ago.)
[Wheeling Jesuit University/NASA Classroom of the Future.]

now-extinct calderas that supposedly mark the movement of the North American Plate over the Yellowstone hot spot (Figure 12.26b). The oldest member of the chain, a volcanic area in Oregon, erupted about 16 million years ago, producing some of the flood basalts of the Columbia Plateau. The North American Plate has moved over the Yellowstone hot spot to the southwest at a rate of about 25 mm/year during the past 16 million years. Accounting for the relative movement of the Pacific and North American plates, this rate and direction are consistent with the plate movements inferred from the Hawaiian Islands.

MEASURING PLATE MOVEMENTS USING HOT-SPOT TRACKS

Assuming that hot spots are anchored by plumes rising from the deep mantle, geologists can use the worldwide distribution of their volcanic tracks to compute how the global system of plates is moving with respect to the deep mantle. The results are sometimes called "absolute plate movements" to distinguish them from the movements of plates relative to each other. The absolute plate movements calculated from hot-spot tracks have helped geologists understand the forces driving the plates. Plates that are being subducted along large fractions of their boundaries—such as the Pacific, Nazca, Cocos, Indian, and Australian plates—are moving rapidly with respect to the hot spots, whereas plates without much subducting slab—such as the Eurasian and African plates—are moving slowly. This observation supports the hypothesis that the gravitational pull of the dense sinking slabs is an important force driving plate movements (see Chapter 2).

The use of hot-spot tracks to reconstruct absolute plate movements works fairly well for recent plate movements. Over longer periods, however, a number of problems arise. For instance, according to the fixed-hot-spot hypothesis, the sharp bend in the Hawaiian aseismic ridge (where it becomes the north-trending Emperor seamount chain; see Figure 12.26a), dated at about 43 million years ago, should coincide with an abrupt shift in the direction of the Pacific Plate. However, no sign of such a shift is evident in magnetic isochron maps, leading some geologists to question the fixed-hot-spot hypothesis. Others have pointed out that, in a convecting mantle, plumes would not necessarily remain fixed relative to one another, but might be moved about by shifting convection currents.

LARGE IGNEOUS PROVINCES

The origin of fissure eruptions on continents—such as those that formed the Columbia Plateau and even larger basalt plateaus in Brazil and Paraguay, India, and Siberia—is a major puzzle. The geologic record shows that these eruptions can release immense amounts of lava—up to several million cubic kilometers—in a period as short as a million years.

Flood basalts are not limited to continents; they also create large oceanic plateaus, such as the Ontong Java Plateau on the northern side of the island of New Guinea and major parts of the Kerguelen Plateau in the southern Indian Ocean. These features are all examples of what geologists call **large igneous provinces** (LIPs) (Figure 12.27). LIPs are large volumes of predominantly mafic extrusive and intrusive igneous rock whose origins lie in processes other than normal seafloor spreading. LIPs include continental flood basalts and associated intrusive rocks, oceanic basalt plateaus, and the aseismic ridges produced by hot spots.

The fissure eruption that covered much of Siberia with basaltic lava is of special interest to geobiologists because it happened at the same time as the greatest mass extinction in the geologic record, which occurred at the end of the Permian period, about 251 million years ago (see Chapter 11). Some geologists think that the eruption caused the mass extinction, perhaps by polluting the atmosphere with volcanic gases that triggered major climate changes (see the Practicing Geology Exercise at the end of the chapter).

Many geologists believe that almost all LIPs were created at hot spots by mantle plumes. However, the amount of lava erupting from the most active hot spot on Earth

FIGURE 12.27 ▪ The global distribution of large igneous provinces on continents and in ocean basins. These provinces are marked by unusually large deposits of basaltic magma. [After M. Coffin and O. Eldholm, *Reviews of Geophysics* 32 (1994): 1–36, Figure 1.]

1 Instability at the core-mantle boundary causes a mantle plume to arise, led by a hot, turbulent plume head.

2 When the plume head reaches the top of the mantle, basaltic magma produced by decompression melting penetrates the lithosphere and erupts as flood basalts.

3 As the plate moves over the remains of the plume, the plume tail—now a hot spot— may form a hot-spot volcano.

4 Continued plate movement over the hot spot creates a chain of hot-spot volcanoes.

FIGURE 12.28 ■ A speculative model for the formation of flood basalts and other large igneous provinces. A new mantle plume rises from the core-mantle boundary, led by a hot, turbulent plume head. When the plume head reaches the top of the mantle, it flattens, generating a huge volume of basaltic magma, which erupts as flood basalts.

today, Hawaii, is paltry compared with the enormous out-pourings of fissure eruptions. What explains these unusual bursts of basaltic magma from the mantle? Some geologists speculate that they result when a new plume rises from the core-mantle boundary. According to this hypothesis, a large, turbulent blob of hot material—a "plume head"—leads the way. When this plume head reaches the top of the mantle, it generates a huge quantity of magma by decompression melting, which erupts in massive flood basalts (**Figure 12.28**). Others dispute this hypothesis, pointing out that continental flood basalts often seem to be associated with preexisting zones of weakness in the continental crust and suggesting that the magmas are generated by convective processes localized in the upper mantle. Sorting out the origins of LIPs is one of the most exciting areas of current geologic research.

■ Volcanism and Human Affairs

Large volcanic eruptions are not just of academic interest to geologists. More than 600 million people live close enough to active volcanoes to be directly affected by eruptions. A repeat of the largest eruptions observed in the geologic record could disrupt or even destroy civilization itself. We must understand volcanic hazards to reduce the risks they pose. But in a world of growing human consumption, we also need to understand and appreciate the benefits volcanism provides to society in the form of mineral resources, fertile soils, and thermal energy.

Volcanic Hazards

Volcanic eruptions have a prominent place in human history and mythology. The myth of the lost continent of Atlantis may have its source in the explosion of Thera, a volcanic island in the Aegean Sea (also known as Santorini). The eruption, which has been dated at 1623 B.C., formed a caldera 7 km by 10 km in diameter, visible today as a lagoon up to 500 m deep with two small active volcanoes in the center. The eruption and the tsunami that followed it destroyed dozens of coastal settlements over a large part of the eastern Mediterranean. Some scientists have attributed the mysterious demise of the Minoan civilization to this ancient catastrophe.

Of the 500 to 600 active volcanoes that rise above sea level (see Figure 12.25), at least one in six is known to have claimed human lives. So far in this century, only about 600 people have died in volcanic eruptions, more than half of them in the 2010 eruptions of Mount Merapi in Indonesia. But history teaches us that this luck will not hold over the longer term. In the past 500 years alone, more than 250,000 people have been killed by volcanic eruptions (**Figure 12.29a**). Volcanoes can kill people and damage property in many ways, some of which are listed in Figure 12.28b and depicted in **Figure 12.30**. We have already mentioned some of these hazards, including pyroclastic flows and tsunamis. Several additional volcanic hazards are of special concern.

(a)

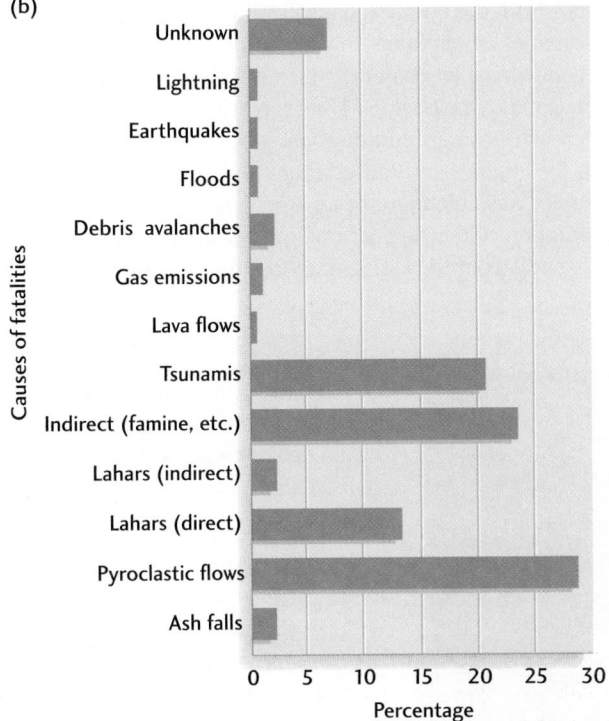

(b)

FIGURE 12.29 ■ (a) Cumulative statistics on fatalities caused by volcanoes since A.D. 1500. The seven eruptions that dominate the record, each of which claimed 10,000 or more victims, are named. These eruptions account for two-thirds of the total deaths. (b) Specific causes of volcano fatalities since A.D. 1500. [After T. Simkin, L. Siebert, and R. Blong, *Science* 291 (2001): 255.]

LAHARS Among the most dangerous volcanic events are the torrential flows of wet volcanic debris called **lahars.** They can occur when a pyroclastic flow meets a river or a snowbank; when the wall of a water-filled crater breaks; when a lava flow melts glacial ice; or when heavy rainfall transforms new ash deposits into mud. One extensive layer of volcanic debris in the Sierra Nevada of California contains 8000 km³ of material of lahar origin, enough to cover all of Delaware with a deposit more than a kilometer thick. Lahars have been known to carry huge boulders for tens of kilometers. When Nevado del Ruiz in the Colombian Andes erupted in 1985, lahars triggered by the melting of glacial ice near the summit plunged down the slopes and buried the town of Armero 50 km away, killing more than 25,000 people (**Figure 12.31**). In volcanic terrains beneath icecaps, a common danger is the torrential release of floodwater when magma melts large volumes of glacial ice; this very fluidized type of lahar is called a *jökulhlaup* (YOU-KYL-LOOP) in Icelandic.

FLANK COLLAPSE A volcanic mountain is constructed from thousands of deposits of lava or pyroclasts or both—not the best way to build a stable structure. The volcano's sides may become too steep and break or slip off. In recent years, volcanologists have discovered many prehistoric examples of catastrophic structural failures in which a big piece of a volcano broke off, perhaps because of an earthquake, and slid downhill in a massive, destructive landslide. On a worldwide basis, such *flank collapses* occur at an average rate of about four times per century. The collapse of one side of Mount St. Helens was the most damaging part of its 1980 eruption (see Figure 12.6).

Surveys of the seafloor off the Hawaiian Islands have revealed many giant landslides on the underwater flanks of the Hawaiian Ridge. When they occurred, these massive earth movements probably triggered huge tsunamis. In fact, coral-bearing marine sediments have been found some 300 m above sea level on one of the Hawaiian islands. These sediments were probably deposited by a giant tsunami caused by a prehistoric flank collapse.

The southern flank of Kilauea, on the island of Hawaii, is advancing toward the ocean at a rate of 100 mm/year, which is relatively fast, geologically speaking. This advance became even more worrisome when it suddenly accelerated by a factor of several hundred on November 8, 2000. A network of motion sensors detected an ominous surge in velocity of about 50 mm/day. The surge lasted for 36 hours, after which the normal motion was reestablished. Since then, similar surge events, though variable in size, have been observed every 2–3 years. Someday—maybe thousands of years from now, but perhaps sooner—the southern flank of the volcano is likely to break off and slide into the ocean. This catastrophic event would trigger a tsunami that could prove disastrous for Hawaii, California, and other Pacific coastal areas.

CALDERA COLLAPSE Although infrequent, collapses of large calderas are some of the most destructive natural

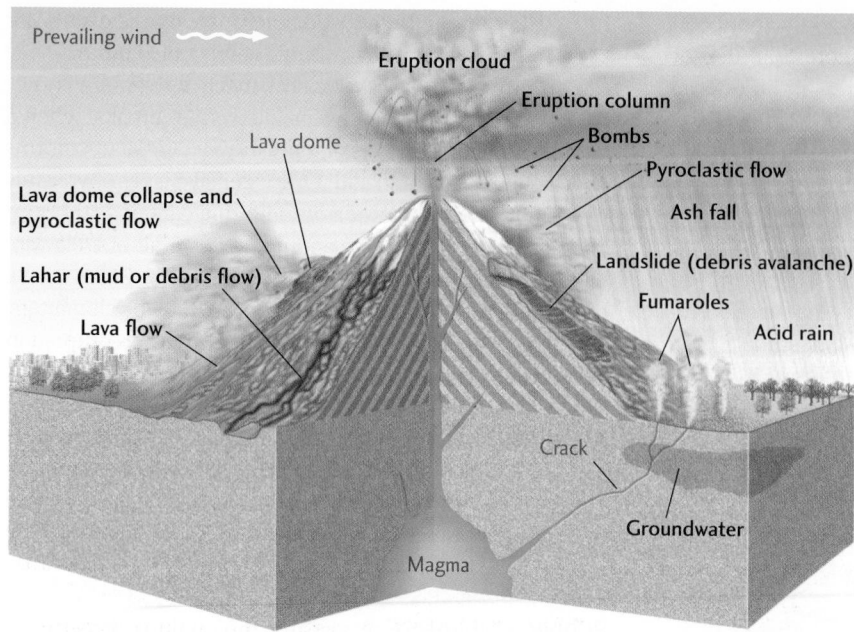

FIGURE 12.30 ■ Some of the volcanic hazards that can kill people and destroy property.
[B. Meyers et al./USGS.]

phenomena on Earth. Monitoring the activity of calderas is very important because of their long-term potential for widespread destruction. Fortunately, no catastrophic collapses have occurred in North America during recorded history, but geologists are concerned about an increase in small earthquakes in the Yellowstone and Long Valley Calderas as well as other indications of activity in their underlying magma chambers. For example, carbon dioxide leaking into the soil from magma in the crust has been killing trees since 1992 on Mammoth Mountain, a volcano on the boundary of Long Valley Caldera. Regions of the Yellowstone caldera have been rising at rates as high as 7 cm/year since 2004, and a swarm of more than a thousand small earthquakes occurred near the center of the caldera in a 2-week period

from December 2008 to January 2009. As in the case of the Long Valley Caldera, these observations are consistent with the injection of magma at mid-crustal depths.

ERUPTION CLOUDS A less deadly but still costly hazard comes from the eruption of ash clouds that can damage the jet engines of airplanes that fly through them. More than 60 commercial jet passenger planes have been damaged by such clouds. One Boeing 747 temporarily lost all four engines when ash from an erupting volcano in Alaska was sucked into the engines and caused them to flame out. Fortunately, the pilot was able to make an emergency landing. Warnings of eruption clouds near air traffic lanes are now being issued by several countries. The eruptions of the Eyjafjallajökull,

FIGURE 12.31 ■ Armero, Colombia, was submerged by lahars after an eruption of the long-dormant Nevado del Ruiz volcano in 1985.
[STF/ASP/Getty Images.]

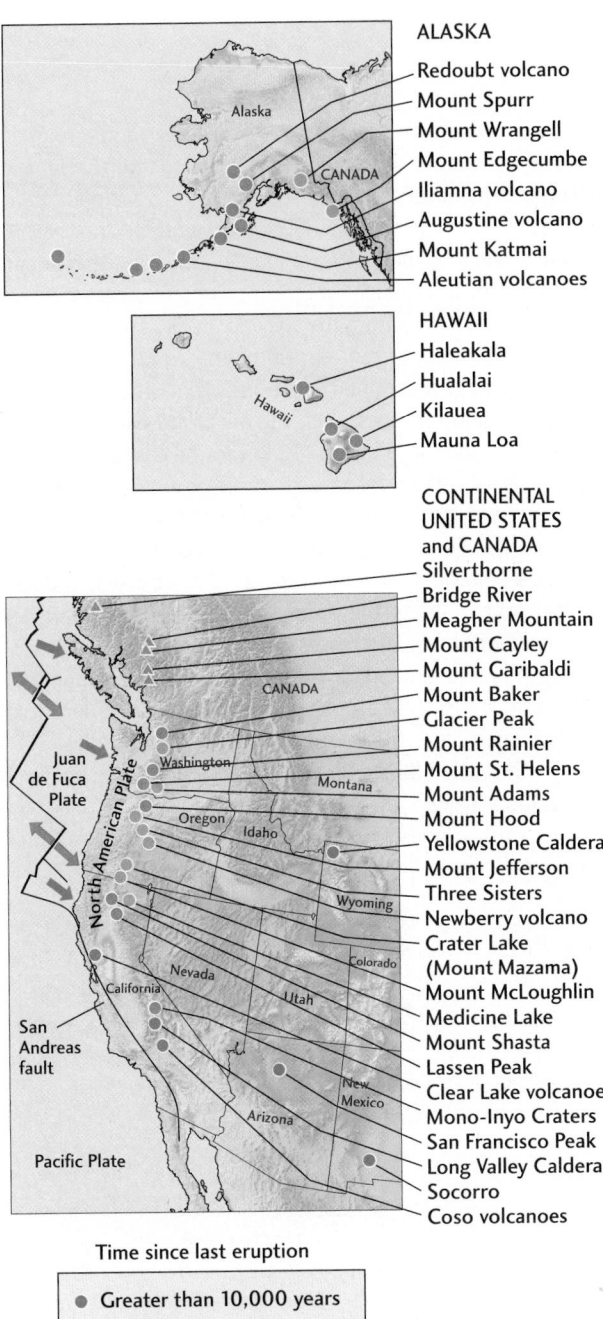

FIGURE 12.32 ■ Locations of potentially hazardous volcanoes in the United States and Canada. Volcanoes within each U.S. group are color-coded by time since their last eruption; those that have erupted most recently are thought to present the greatest cause for concern. (These classifications are subject to revision as studies progress, and are not available for Canadian volcanoes.) Note the relationship between the volcanoes extending from northern California to British Columbia and the convergent boundary between the North American Plate and the Juan de Fuca Plate. [After R. A. Bailey, P. R. Beauchemin, F. P. Kapinos, and D. W. Klick/USGS.]

Iceland, volcano in April and May of 2010, disrupted air traffic across the North Atlantic, resulting in over a billion dollars of losses to commercial airlines (see Earth Issues 12.1).

Reducing the Risks of Volcanic Hazards

There are about 100 high-risk volcanoes in the world today, and some 50 volcanic eruptions occur each year. These volcanic eruptions cannot be prevented, but their catastrophic effects can be greatly reduced by a combination of science and enlightened public policy. Volcanology has progressed to the point that we can identify the world's dangerous volcanoes and characterize their potential hazards by studying deposits laid down in earlier eruptions. Some potentially dangerous volcanoes in the United States and Canada are identified in **Figure 12.32**. Assessments of their hazards can be used to guide zoning regulations to restrict land use—the most effective measure to reduce property losses and casualties.

Such studies indicate that Mount Rainier, because of its proximity to the heavily populated cities of Seattle and Tacoma, probably poses the greatest volcanic risk in the United States (**Figure 12.33**). At least 80,000 people and their homes are at risk in Mount Rainier's lahar-hazard zones. An eruption could kill thousands of people and cripple the economy of the Pacific Northwest.

FIGURE 12.33 ■ Mount Rainier, seen from Tacoma, Washington. [Patrick Lynch/Alamy.]

Earth Issues

12.2 Mount St. Helens: Dangerous but Predictable

Mount St. Helens, in the Cascade Range of the Pacific Northwest, is the most active and explosive volcano in the contiguous United States (see Figure 12.6). It has a documented 4500-year history of destructive lava flows, pyroclastic flows, lahars, and distant ash falls. Beginning on March 20, 1980, a series of small to moderate earthquakes under the volcano signaled the start of a new eruptive phase after 123 years of dormancy, motivating the U.S. Geological Survey to issue a formal hazard alert. The first outburst of ash and steam erupted from a newly opened crater on the summit one week later.

Throughout April, the seismic tremors increased, indicating that magma was moving beneath the summit, and instruments detected an ominous swelling of the northeastern flank of the mountain. The USGS issued a more serious warning, and people were ordered out of the vicinity. On May 18, the main eruption began abruptly. A large earthquake apparently triggered the collapse of the north side of the mountain, loosening a massive landslide, the largest ever recorded anywhere. As this huge debris avalanche plummeted down the mountain, gas and steam under high pressure were released in a tremendous lateral blast that blew out the northern flank of the mountain.

USGS geologist David A. Johnston was monitoring the volcano from his observation post 8 km to the north. He must have seen the advancing blast wave before he radioed his last message: "Vancouver, Vancouver, this is it!" A northward-directed jet of superheated (500°C) ash, gas, and steam roared out of the breach with hurricane force, devastating a zone 20 km outward from the volcano and 30 km wide. A vertical eruption column sent an ash plume 25 km into the sky, twice as high as a commercial jet flies. The ash plume drifted to the east and northeast with the prevailing winds, bringing darkness at noon to an area 250 km to the east and depositing a layer of ash as deep as 10 cm over much of Washington, northern Idaho, and western Montana. The energy of the blast was equivalent to about 25 million tons of TNT. The volcano's summit was destroyed, its elevation was reduced by 400 m, and its northern flank disappeared. In effect, the mountain was hollowed out.

Earthquakes and magmatic activity have continued off and on since the 1980 eruption. After more than a decade of relative quiescence, the volcano reawoke in September 2004 with a series of minor steam and ash eruptions that continued into 2005. Growth of the central volcanic dome (see Figure 12.14b) suggests that the current phase of eruptive activity may persist for some time into the future.

PREDICTING ERUPTIONS Can volcanic eruptions be predicted? In many cases, the answer is yes. Instrumented monitoring can detect signals such as earthquakes, swelling of the volcano, and gas emissions that warn of impending eruptions. People at risk can be evacuated if the authorities are organized and prepared. Scientists monitoring Mount St. Helens were able to warn people before its eruption in 1980 (see Earth Issues 12.2). Government infrastructure was in place to evaluate the warnings and to enforce evacuation orders, so very few people were killed.

Another successful warning was issued a few days before the cataclysmic eruption of Mount Pinatubo in the Philippines on June 15, 1991. A quarter of a million people were evacuated, including some 16,000 residents of the nearby U.S. Clark Air Force Base (which was heavily damaged by the eruption and has since been permanently abandoned). Tens of thousands of lives were saved from the lahars that destroyed everything in their paths. Casualties were limited to the few who disregarded the evacuation order. And in 1994, 30,000 residents of Rabaul, Papua New Guinea, were successfully evacuated by land and sea hours before the two volcanoes on either side of the town erupted, destroying or damaging most of it. Many owe their lives to the government, which conducted evacuation drills, and to scientists at the local volcano observatory, who issued a warning when their seismographs

The eruption of Mount St. Helens on May 18, 1980, sent an ash plume into the stratosphere and an avalanche and blast wave toward the north.

(May 17, 3 P.M.) View of Mount St. Helens the day before its eruption. The north side of the volcano has bulged outward from magma intruded at shallow levels during the previous two months. [Keith Ronnholm.]

(May 18, 8:33 A.M.) An earthquake and massive landslide "uncork" the volcano, releasing an ash plume and a powerful lateral blast wave. [Keith Ronnholm.]

recorded the ground tremors that signaled magma moving toward the surface.

CONTROLLING ERUPTIONS Can we go further by actually controlling volcanic eruptions? Not likely, because large volcanoes release energy on a scale that dwarfs our capabilities for control. Under special circumstances and on a small scale, however, the damage can be reduced. Perhaps the most successful attempt to manage volcanic activity was made on the Icelandic island of Heimaey in January 1973. By spraying advancing lava with seawater, residents cooled and slowed the flow, preventing the lava from blocking the entrance to their

harbor and saving some homes from destruction. The best focus for our efforts, however, will be the establishment of more warning and evacuation systems and more rigorous restriction of settlements in potentially dangerous locations.

Natural Resources from Volcanoes

In this chapter, we have seen something of the beauty of volcanoes and something of their destructiveness. But it should be kept in mind that volcanoes contribute to our well-being in many, though often indirect, ways. Soils

Google Earth Project

Some of the most spectacular and dangerous volcanoes occur in the island arcs and volcanic mountain belts above subduction zones. Google Earth is a good tool for observing the sizes and shapes of these volcanoes. We will use it to investigate a famous example, Mt. Fuji, on the Japanese island of Honshu.

LOCATION   Mt. Fuji, Japan, and Sarychev Peak, Kurile Islands

GOAL   Observe the sizes and shapes of active stratovolcanoes

LINKED   Figure 12.14

Google Earth view of Mt. Fuji, Japan.

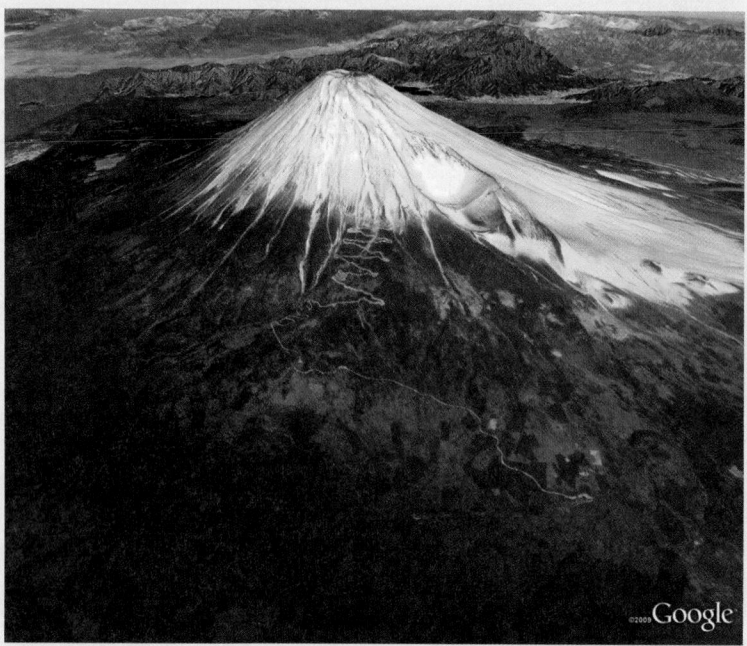

Image © 2009 DigitalGlobe, Image © 2009 TerraMetrics, Image © 2009 Digital Earth Technology, Image © 2009 GeoEye

1. Type "Mt. Fuji, Japan" into the GE search window; once you arrive there, tilt your frame of view to the north and observe the topography of the mountain from an eye altitude of several kilometers. Use the cursor to measure the peak height above sea level. Which of the answers below best describes the general shape of Mt. Fuji?

derived from volcanic materials are exceptionally fertile because of the mineral nutrients they contain. Volcanic rock, gases, and steam are also sources of important industrial materials and chemicals, such as pumice, boric acid, ammonia, sulfur, carbon dioxide, and some metals. Hydrothermal activity is responsible for the deposition of unusual minerals that concentrate relatively rare elements, particularly metals, into ore deposits of great economic value. Seawater circulating through mid-ocean ridges is a major factor in the formation of such ores and in the maintenance of the chemical balance of the oceans.

In some regions where geothermal gradients are steep, Earth's internal heat can be tapped to heat homes and drive electric generators. **Geothermal energy** depends on the heating of water as it passes through a region of hot rock (a *heat reservoir*) that may be hundreds or thousands of meters beneath Earth's surface. Hot water or steam can be brought to the surface through boreholes drilled for the purpose. Usually, the water is naturally occurring groundwater that seeps downward along fractures in rock. Less typically, the water is artificially introduced by pumping from the surface.

a. A large linear fissure in Earth's surface

b. A low-relief, very broad shield volcano

c. A steep-sided, low-elevation cinder cone

d. A high-elevation, steep-sided stratovolcano

2. Based on your observations of Mt. Fuji and the surrounding area, what single feature convinces you that you are looking at a volcano?

a. The number of trees and the amount of snow present on the mountainside

b. The presence of a crater at the top of the mountain

c. The steepness of the mountain slopes and the large landslide on the south slope

d. The proximity of the mountain to the coastline of Japan and its distance from China

3. After considering the visible characteristics of Mt. Fuji from various angles, how would you classify its level of volcanic activity at the time the satellite photo was taken?

a. The eroded shapes of the landscape around the volcano indicate that it is now extinct, a conclusion further supported by the presence of snow.

b. The steep slopes, circular shape, and well-defined crater indicate recent volcanic activity, but the presence of abundant snow near the crater rim suggests that the volcano is not currently erupting.

c. The steep slopes, circular shape, and well-defined crater, combined with fresh lava on the snowfields of the main summit, indicate that the volcano is active and currently erupting.

d. The circular shape and well-defined crater suggest a once-active volcano, but the lack of fresh lava and the presence of snow indicate that the volcano is extinct.

4. Tokyo, Japan, one of the largest cities on Earth, is home to more than 12 million people. To assess the hazard to Tokyo from Mt. Fuji, consider that the prevailing winds are expected to blow the cloud from a major eruption to the east, dumping up to a meter of ash more than 100 km from the volcano. Measure the distance and direction from the volcano to the urban center of Tokyo. Which of the following statements is most consistent with this information?

a. Mt. Fuji is too far away from Tokyo to pose a significant hazard.

b. The volcano poses a significant hazard to Tokyo because it is close to the city and because the prevailing winds are likely to blow an eruption cloud in its direction.

c. The volcano poses only a moderate hazard to Tokyo; it is close enough, but the prevailing winds are likely blow any eruption cloud away from the city.

d. The volcano is not a hazard to Tokyo because it is extinct and not expected to erupt.

Optional Challenge Question

5. Zoom out to an eye altitude of 3000 km. Look for the deep-sea trench that marks a subduction zone east of Mt. Fuji. Move along the subduction zone to the northeast until you encounter Matua Island in the Kurile Islands chain, which belongs to Russia. This island is dominated by Sarychev Peak, one of the most active volcanoes of the Kurile Islands. Measure the height of the volcano and observe its features. Which of the following statements best describes your observations?

a. Sarychev Peak is an island arc volcano, smaller but currently more active than Mt. Fuji.

b. Sarychev Peak is located in a continental volcanic mountain belt; it is smaller and currently less active than Mt. Fuji.

c. Sarychev Peak is a mid-ocean ridge volcano, larger and currently more active than Mt. Fuji.

d. Sarychev Peak is a hot spot shield volcano, smaller but currently more active than Mt. Fuji.

By far the most abundant source of geothermal energy is naturally occurring groundwater that has been heated to temperatures of 80°C to 180°C. Water at these relatively low temperatures is used for residential, commercial, and industrial heating. Warm groundwater drawn from a heat reservoir in the Paris sedimentary basin now heats more than 20,000 apartments in France. Reykjavik, the capital of Iceland, which sits atop the Mid-Atlantic Ridge, is almost entirely heated by geothermal energy.

Heat reservoirs with temperatures above 180°C are useful for generating electricity. They are present primarily in regions of recent volcanism as hot, dry rock, natural hot water, or natural steam. Naturally occurring water heated above the boiling point and naturally occurring steam are highly prized resources. The world's largest facility for producing electricity from natural steam, located at The Geysers, 120 km north of San Francisco, generates more than 600 megawatts of electricity (**Figure 12.34**). Some 70 geothermal electricity-generating plants operate in California, Utah, Nevada, and Hawaii, producing 2800 megawatts of power—enough to supply about a million people.

FIGURE 12.34 ■ The Geysers, one of the world's largest supplies of natural steam. The geothermal energy is converted into electricity for San Francisco, 120 km to the south. [© Charles E. Rotkin/Corbis.]

SUMMARY

What are the major types of volcanic deposits? Lavas are classified as basaltic (mafic), andesitic (intermediate), or rhyolitic (felsic) on the basis of their content of silica and other minerals. Basaltic lavas are relatively fluid and flow freely; andesitic and rhyolitic lavas are more viscous. Lavas differ from pyroclasts, which are formed by explosive eruptions and vary in size from fine ash particles to house-sized bombs.

How are volcanic landforms shaped? The chemical composition and gas content of magma are important factors in a volcano's eruptive style and in the shape of the landforms it creates. A shield volcano grows from repeated eruptions of basaltic lava from a central vent. Andesitic and rhyolitic lavas tend to erupt explosively. The erupted pyroclasts may pile up into a cinder cone. A stratovolcano is built of alternating layers of lava flows and pyroclastic deposits. The rapid ejection of magma from a large magma chamber, followed by collapse of the chamber's roof, results in a large depression, or caldera. Basaltic lavas can erupt from fissures along mid-ocean ridges as well as on continents, where they flow over the landscape in sheets to form flood basalts. Pyroclastic eruptions from fissures can cover an extensive area with ash-flow deposits.

How is the global pattern of volcanism related to plate tectonics? The huge volumes of basaltic magma that form oceanic crust are produced by decompression melting and erupted at spreading centers on mid-ocean ridges. Andesitic lavas are the most common lava type in the volcanic mountain belts of ocean-continent subduction zones. Rhyolitic lavas are produced by the melting of felsic continental crust. Within plates, basaltic volcanism occurs above hot spots, which are manifestations of rising plumes of hot mantle material.

What are some hazards and beneficial effects of volcanism? Volcanic hazards that can kill people and damage property include pyroclastic flows, tsunamis, lahars, flank collapses, caldera collapses, eruption clouds, and ash falls. Volcanic eruptions have killed about 250,000 people in the past 500 years. On the positive side, volcanic materials produce nutrient-rich soils, and hydrothermal processes are important in the formation of many economically valuable mineral ores. Geothermal heat drawn from areas of hydrothermal activity is a useful source of energy in some regions.

KEY TERMS AND CONCEPTS

andesitic lava (p. 317)
ash-flow deposit (p. 326)
basaltic lava (p. 315)
breccia (p. 320)
caldera (p. 323)
crater (p. 323)

diatreme (p. 323)
fissure eruption (p. 325)
flood basalt (p. 326)
geothermal energy (p. 340)
hot spot (p. 332)

hydrothermal activity (p. 327)
lahar (p. 335)
large igneous province (p. 333)
mantle plume (p. 332)
pyroclastic flow (p. 320)

rhyolitic lava (p. 318)
shield volcano (p. 321)
stratovolcano (p. 321)
tuff (p. 320)
volcanic geosystem (p. 314)
volcano (p. 314)

PRACTICING GEOLOGY EXERCISE

Are the Siberian Traps a Smoking Gun of Mass Extinction?

The mass extinction at the end of the Permian period, dated at 251 million years ago, marks the transition from the Paleozoic era to the Mesozoic era, as described in Chapter 8. The flood basalts of Siberia—the product of the largest continental volcanic eruption in the Phanerozoic eon—have also been dated at 251 million years ago. Is this just a coincidence, or was the eruption of the flood basalts responsible for the end-Permian mass extinction?

Let's first consider the size and rate of the Siberian eruption. Geologic mapping of these flood basalts, called the Siberian Traps, shows that they once extended across much of the Siberian platform and craton, covering an area exceeding 4 million square kilometers. Although much has been eroded away or buried beneath younger sediments, the total volume of the basalts must have originally exceeded 2 million cubic kilometers and may have been as much as 4 million cubic kilometers. Isotopic dating indicates that the basalts were extruded over a period of about 1 million years, implying an average eruption rate of 2 to 4 km³/year.

To appreciate how large this rate really is, we can compare it with the volcanism at rapidly diverging plate boundaries. Enough basalt is extruded along mid-ocean ridges to form the entire oceanic crust, so the production rate of seafloor spreading is given by the formula

$$\text{production rate} = \text{spreading rate} \times \text{crustal thickness} \times \text{ridge length}$$

The fastest spreading we see today is along the East Pacific Rise near the equator, where the Pacific Plate is separating from the Nazca Plate at an average rate of about 140 mm/year, or 1.4×10^{-4} km/year (see Figure 2.7), creating a basaltic crust with an average thickness of

The Siberian Traps are flood basalts that cover an area almost twice the size of Alaska. The basalts exposed on the Siberian craton reach thicknesses of more than 6 km and have been heavily eroded since their eruption 251 million years ago. A vast area of these flood basalts is now buried beneath the sediments of the Siberian platform. [Sergey Anatolievich Pristyazhnyuk/123RF.com.]

7 km. The length of the Pacific-Nazca plate boundary is about 3600 km, so the production rate along this spreading center is

$$1.4 \times 10^{-4} \text{ km/year} \times 7 \text{ km} \times 3600 \text{ km} = 3.5 \text{ km}^3/\text{year}$$

From this calculation, we see that the Siberian eruption produced basalt at a rate comparable to that of the entire Pacific-Nazca plate boundary, the largest magma factory on Earth today!

You can sail on the tropical sea surface over the Pacific-Nazca plate boundary and be completely unaware of the magmatic activity deep beneath you. Most of the magma generated by seafloor spreading solidifies as igneous intrusions to form the basaltic dikes and massive gabbros of the oceanic crust (see Figure 4.15). The basalts that are extruded onto the seafloor are quickly quenched by sea-water to produce pillow lavas, and the gases that are emitted dissolve into the ocean.

But if you were visiting Siberia about 251 million years ago, you would probably not be so comfortable. The Siberian basalts were erupted directly onto the land surface through fissures in the continental crust, flooding millions of square kilometers. This exceptionally rapid extrusion of lavas would have generated huge pyroclastic deposits—much more than typical flood basalt eruptions, such as those of the Columbia Plateau—and it would also have discharged massive amounts of ash and gases,

including carbon dioxide and methane, into the atmosphere. Such an eruption could have triggered changes in Earth's climate of a magnitude that might have led to the end-Permian mass extinction, in which 95 percent of the species living at the time were completely wiped out (see Chapter 11).

Some geologists have argued for years that the end-Permian mass extinction was the result of this intense Siberian volcanism, possibly caused by the sudden arrival of a "plume head" at Earth's surface (see Figure 12.28). Others have preferred alternative hypotheses, such as a meteorite impact or a sudden release of gases from the ocean. However, recent isotopic dating with improved techniques has shown that the Siberian volcanism occurred immediately before or during the end-Permian mass extinction. The finding that these extreme events so precisely coincide has convinced many more geologists that the Siberian Traps are the "smoking gun" behind the largest killing of species in Earth history.

PROBLEM: The Big Island of Hawaii, which has a total rock volume of about 100,000 km³, has been formed by a series of basaltic eruptions over the last 1 million years. Calculate the production rate of the Hawaiian basalts and compare it with that of the Siberian Traps. What length of the Nazca-Pacific Plate boundary produces basalt at a rate equivalent to the Hawaiian hot spot?

EXERCISES

1. On Earth's surface as a whole, what process generates the greater volume of volcanic rock, decompression melting or fluid-induced melting? Which of these processes creates the more dangerous volcanoes?

2. What is the difference between magma and lava? Describe a geologic situation in which a magma does not form a lava.

3. What are the three major types of volcanic rocks and their intrusive counterparts? Is kimberlite one of these three types?

4. What type of volcano is the Arenal volcano, shown in Figure 12.10?

5. Most volcanism occurs near plate boundaries. What type of plate boundary can produce large amounts of rhyolitic lavas?

6. What evidence suggests that the Yellowstone volcano was produced by a hot spot?

7. How do scientists predict volcanic eruptions?

THOUGHT QUESTIONS

1. What might be the effects on civilization of a Yellowstone-type caldera eruption, such as the one described at the opening of this chapter?

2. Give a few examples of what geologists have learned about Earth's interior by studying volcanoes and volcanic rocks.

3. Why are eruptions of stratovolcanoes generally more explosive than those of shield volcanoes?

4. While on a field trip, you come across a volcanic formation that resembles a field of sandbags. The individual

ellipsoidal forms have a smooth, glassy surface texture. What type of lava is this, and what information does this give you about its history?

5. Why are the volcanoes on the northwestern side of the island of Hawaii dormant whereas those on the southeastern side are more active?

6. How do interactions between volcanic geosystems and the climate system increase volcanic hazards?

MEDIA SUPPORT

 12-1 Animation: Types of Volcanoes

 12-4 Video: White Island, New Zealand: Hydrothermal Features

 12-2 Animation: Evolution of Crater Lake

 12-5 Video: White Island, New Zealand: Stratovolcano in the Pacific

 12-3 Animation: The Formation of Shiprock

 12-6 Video: The Naples Metropolis

 12-4 Animation: Volcanism and Plate Tectonics

 12-7 Video: Mount Vesuvius and the Plinian Eruption of 79 A.D.

 12-1 Video: Cinder Cones of Northern Arizona: Sunset and SP Craters

 12-8 Video: Aeolian Islands

 12-2 Video: Crater Lake: Caldera in the Cascades

 12-9 Video: Hawaii: The Hot Spot and Volcanoes

 12-3 Video: Mount Etna: Europe's Largest Active Volcano

 12-10 Video: Hawaiian Lava Flows

- What Is an Earthquake? 348

- How Do We Study Earthquakes? 353

- Earthquakes and Patterns of Faulting 362

- Earthquake Hazards and Risks 365

- Can Earthquakes Be Predicted? 375

The tsunami from the 2011 Tohoku earthquake crashes over a seawall designed to protect the city of Miyako from destructive sea waves.
[AP Photo/Mainichi Shimbun, Tomohiko Kano.]

EARTHQUAKES

EARTHQUAKES RIVAL ALL OTHER natural disasters in the threat they pose to human life and property. Our fragile "built environment" is necessarily anchored in Earth's active crust, which makes it extremely vulnerable to earthquakes and their secondary effects, such as ground ruptures, landslides, and tsunamis. Some events of the past century are sobering illustrations of this fact.

On a fine April morning in 1906, the citizens of northern California were awakened by the roar and violent shaking caused by the breaking of the San Andreas transform fault, which created the most destructive earthquake the United States has yet experienced. The ensuing fires destroyed the city of San Francisco; by the time the flames died away, nearly 3000 of its inhabitants were dead (Figure 13.1).

Almost a century later, on December 26, 2004, a much larger fault—a subduction-zone megathrust—broke west of the Indonesian island of Sumatra, lifting up the seafloor and sending a huge tsunami across the Indian Ocean. This monster wave drowned more than 220,000 people living on coastlines from Thailand to Africa. Another megathrust ruptured off Japan on March 11, 2011, creating an even larger tsunami, pictured here, that drowned almost 20,000. Between the 1906 earthquake and now, earthquakes have killed more than 2 million people around the world.

To cope with the all-too-frequent death and destruction caused by earthquakes, we have long sought to improve our ability to predict where and when these events might occur and our understanding of what happens when they do. Science has shown that seismic activity can be understood in terms of the basic machinery of plate tectonics. As a result, attempts to reduce earthquake risk have become increasingly fused with the quest for a more fundamental understanding of the geologically active Earth.

This chapter will examine what happens during earthquakes, how scientists use seismic waves to locate and measure earthquakes, and what can be done to reduce the casualties and economic damage caused by earthquakes. We cannot yet reliably predict when large earthquakes will occur, but we can take measures to reduce their destructiveness. We can use our geologic knowledge to identify places where large earthquakes are likely; work with engineers to design buildings, dams, bridges, and other structures that can withstand seismic shaking; and help endangered communities prepare for and respond to seismic events.

■ What Is an Earthquake?

The movement of lithospheric plates generates enormous forces at the boundaries between those plates, as we saw in Chapter 7. These forces deform brittle crustal rocks in ways that can be described by the concepts of stress, strain, and strength. *Stress* is the local force per unit area that causes rocks to deform. *Strain* is the relative amount of deformation, expressed as the percentage of distortion (for example, compression of a rock by 1 percent of its length). Rocks *fail*—that is, they lose cohesion and break into two or more parts—when they are stressed beyond a critical value, called their *strength*.

An **earthquake** occurs when rocks under stress suddenly fail along a geologic fault. Most large earthquakes are caused by ruptures of preexisting faults, where past earthquakes have already weakened the rocks on the

FIGURE 13.1 ■ This photograph, taken from a balloon by George Lawrence 5 weeks after the San Francisco earthquake of April 18, 1906, shows the devastation of the city caused by the quake and subsequent fires. The view is looking over Nob Hill toward the business district. [Corbis.]

fault surface. The two blocks of rock on either side of the fault slip suddenly, releasing energy in the form of seismic waves, which we feel as ground shaking. When the fault slips, the stress is reduced, dropping to a value below the rock strength. After the earthquake, the stress begins to increase again, eventually leading to another large earthquake. The faults involved in this repeated earthquake cycle are called *active faults,* and they are concentrated in the zones that form plate boundaries, where most of the stress and strain caused by plate movements is concentrated.

The Elastic Rebound Theory

The earthquake on the San Andreas fault that devastated San Francisco in 1906 was studied more than any earthquake had been up to that time. By mapping the displacement of the ground across the fault and analyzing seismic recordings of the earthquake, geologists showed that this rupture began just west of the Golden Gate and propagated along the fault strike for over 400 km, southeastward to the mission town of San Juan Bautista and northwestward all the way to Cape Mendocino (**Figure 13.2**). In 1910, one of the scientists who studied this rupture, Henry Fielding Reid of Johns Hopkins University, advanced the **elastic rebound theory** to explain how earthquakes recur on active faults in Earth's crust.

Picture a strike-slip fault between two crustal blocks, and imagine that surveyors have painted straight lines on the ground, running perpendicular to the fault and extending from one block to the other, as in **Figure 13.3a**. The two blocks are being pushed in opposite directions by plate movements. The weight of the overlying rock

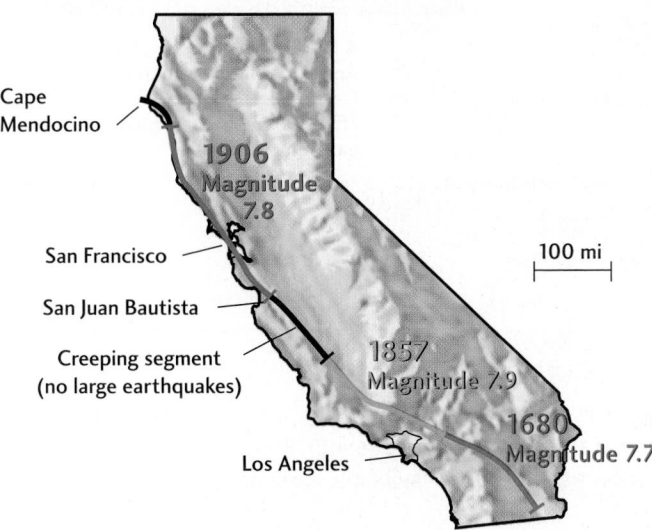

FIGURE 13.2 ■ Map of California, showing the segments of the San Andreas fault that ruptured in 1680, 1857, and 1906. [Southern California Earthquake Center.]

presses them together, however, so friction locks them in place along the fault. They do not move, just as a pushed car does not move when the emergency brake is engaged. Instead of slipping along the fault as stress builds up, the blocks are strained elastically near the fault, as shown by the bent lines in Figure 13.3b. By *elastically,* we mean that the blocks would spring back and return to their undeformed, stress-free shape if the fault were suddenly to unlock.

As the slow plate movements continue to push the blocks in opposite directions, the elastic strain in the rocks—measured by the bending of the survey lines—continues to build up over decades, centuries, or even millennia (Figure 13.3c). At some point, the strength of the rocks is exceeded. Somewhere along the fault surface, the frictional bond that locks the fault can no longer hold, and it breaks. The blocks slip suddenly in a rupture that starts at a point but quickly extends over a section of the fault.

Figure 13.3d shows how the two blocks have rebounded—sprung back to their undeformed state—after the earthquake. The bent survey lines have straightened, and the two blocks have been displaced. (Note that a fence built just before the rupture has been bent during the rebound.) The distance of the displacement is called the **fault slip.** In the inset photograph of Figure 13.3, you can see that the fault slip during the 1906 earthquake was about 4 m. The peak velocity of the slipping at any point on the fault is about 1 m/s, so the entire episode of fault slipping at a given point takes only a few seconds. After the fault has slipped, it locks up again. The steady movement of the blocks on either side of the fault then causes stress to rise again, and the earthquake cycle repeats.

The energy that is slowly built up by elastic strain as the two blocks are pushed in opposite directions is like the elastic energy stored in a rubber band when it is slowly stretched. The sudden release of energy when a fault slips is like the violent backlash, or *rebound,* that occurs when the rubber band breaks. Some of this elastic energy is radiated in seismic waves, which can cause violent shaking many kilometers away from the fault.

The elastic rebound theory implies that there should be a periodic buildup and release of elastic energy at faults, and that the time between ruptures, called the **recurrence interval,** should be a constant (as shown in the bottom panel of Figure 13.3). The recurrence interval can be calculated by dividing the fault slip in each rupture by the long-term slip rate. The long-term slip rate on the San Andreas fault is about 30 mm/year, for example, so earthquakes with 4 m of slip should occur with a recurrence interval of about once every 130 years.

However, most active faults, including the San Andreas, do not conform to this simple theory. For instance, all of the strain accumulated since the last earthquake may not be released in the next—that is, the rebound may be incomplete—or the stress on one fault may change because

ROCKS DEFORM ELASTICALLY, THEN REBOUND DURING AN EARTHQUAKE RUPTURE

A A farmer builds a stone wall across a right-lateral strike-slip fault a few years after its last rupture.

B Over the next 150 years, the relative motion of the blocks on either side of the fault causes the ground and the stone wall to deform.

C Just before the next rupture, a new fence is built across the already deformed land.

D The fault slips, lowering the stress, and the elastic rebound restores the blocks to their prestressed state. Both the rock wall and the fence are shifted equal amounts along the fault.

40 km

Strike-slip fault

20 km

Rocks deform as strain develops

STRESS BUILDS UNTIL IT EXCEEDS ROCK STRENGTH

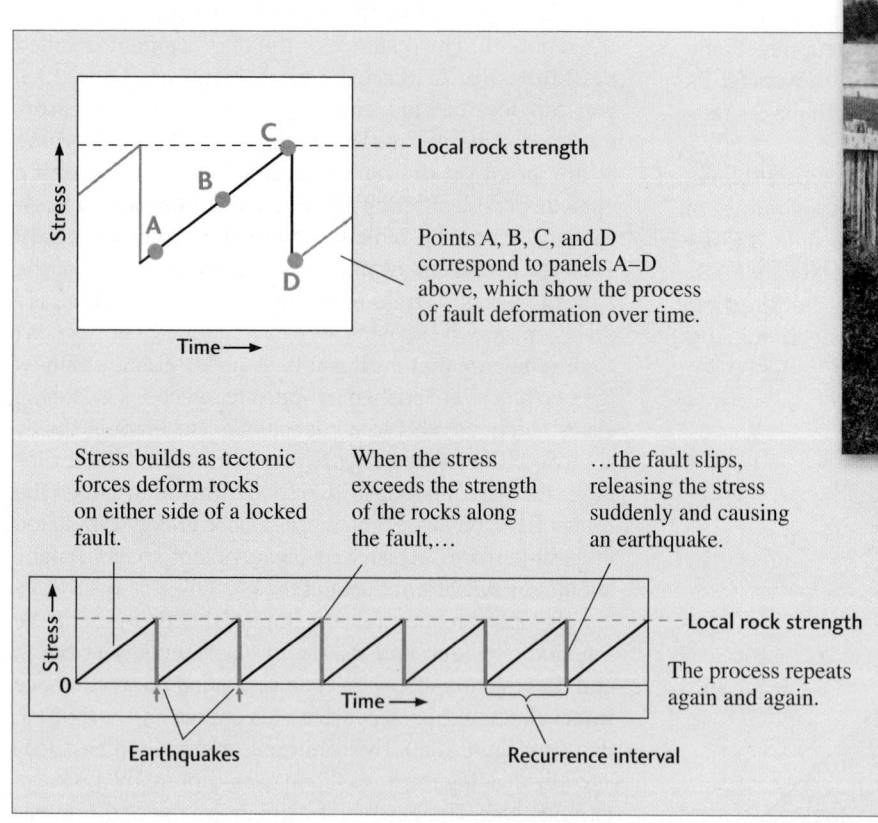

Local rock strength

Stress

Time

Points A, B, C, and D correspond to panels A–D above, which show the process of fault deformation over time.

Stress builds as tectonic forces deform rocks on either side of a locked fault.

When the stress exceeds the strength of the rocks along the fault,…

…the fault slips, releasing the stress suddenly and causing an earthquake.

Stress

0

Time

Earthquakes

Local rock strength

The process repeats again and again.

Recurrence interval

A fence built across the San Andreas fault near Bolinas, California, is offset by nearly 4 m after the great San Francisco earthquake of 1906.

FIGURE 13.3 ■ The elastic rebound theory explains the earthquake cycle. According to the theory, stress on rocks builds up over time as a result of plate movements. Earthquakes occur when that stress exceeds rock strength. Rocks under stress deform elastically, then rebound when an earthquake occurs. Panels A–D show deformation at the points labeled A–D in the bottom panel. [G. K. Gilbert/USGS.]

FIGURE 13.4 ▪ Irregularities in the earthquake cycle can be caused by incomplete stress release, changes in stress caused by earthquakes on nearby faults, and local variations in rock strength.

of earthquakes on nearby faults (**Figure 13.4**). Over the long term, the strength of the fault rocks themselves may change. These irregularities are some of the reasons why earthquakes are so difficult to predict.

Fault Rupture During Earthquakes

The point at which fault slipping begins is the **focus** of an earthquake (**Figure 13.5**). The **epicenter** is the geographic point on Earth's surface directly above the focus. For example, you might hear in a news report, "The U.S. Geological Survey reports that the epicenter of last night's destructive earthquake in California was located 6 kilometers east of Los Angeles City Hall. The depth of the focus was 10 kilometers."

The focal depths of most earthquakes in continental crust range from 2 to 20 km. Continental earthquakes below 20 km are rare because under the high temperatures and pressures found at those greater depths, continental crust behaves as a ductile rather than a brittle material (just as hot wax flows when stressed, whereas cold wax breaks; see Chapter 7). In subduction zones, however, where cold oceanic lithosphere plunges into the mantle, earthquakes can originate at depths of almost 700 km.

The fault rupture does not happen all at once. It begins at the focus and expands outward along the fault surface, typically at 2 to 3 km/s (see Figure 13.5). The rupture stops where the stresses become insufficient to continue breaking the fault (where the rocks are stronger) or where the rupture enters ductile material in which it can no longer propagate. As we will see later in this chapter, the magnitude of an earthquake is related to the total area of fault rupture. Most earthquakes are very small, with rupture dimensions much less than the depth of the focus, so that the rupture never breaks the ground surface.

In large, destructive earthquakes, however, surface breaks are common. The 1906 San Francisco earthquake caused surface displacements averaging about 4 m along the 470-km section of the San Andreas fault that ruptured in that event (see the inset of Figure 13.3). Fault ruptures in the largest earthquakes can extend for more than 1000 km, and the fault slip can be tens of meters. Generally, the longer the fault rupture, the greater the fault slip.

As we have seen, the sudden slipping of the blocks at the time of the earthquake reduces the stress on the fault and releases much of the stored elastic energy. Most of this stored energy is converted to frictional heat in the fault zone or dissipated by rock fracturing, but part of it is released as seismic waves that travel outward from the rupture, much as waves ripple outward from the spot where a stone is dropped into a still pond. The focus of the earthquake generates the first seismic waves, but slipping parts of the fault continue to generate waves until the rupture stops. In a large event, the propagating rupture continues to produce waves for many tens of seconds. These waves can cause destruction in regions all along the fault rupture, even far from the epicenter. Towns along the San Andreas fault far north of San Francisco were badly damaged in the 1906 earthquake.

Foreshocks and Aftershocks

Almost all large earthquakes trigger smaller earthquakes called **aftershocks.** Aftershocks follow the triggering event, or *mainshock,* in sequences, and their foci are distributed in and around the rupture surface of the mainshock (**Figure 13.6**). Aftershock sequences exemplify the complexities of earthquakes that cannot be described by simple elastic rebound theory. Although fault slipping during the mainshock decreases the stress along most of the rupture surface, it can increase the stress on parts of the fault surface that did not slip, or where the slip was incomplete, as well as in surrounding regions. Aftershocks happen where that stress exceeds the rock strength.

The number and sizes of aftershocks depend on the magnitude of the mainshock, and their frequencies decrease with time after the mainshock. The Aftershocks of a magnitude 5 earthquake might last for only a few weeks, whereas those of a magnitude 7 earthquake occur over a larger region and can continue for several years. The size of the largest aftershock is generally about one magnitude unit smaller than the mainshock. According to this rough rule of thumb, a magnitude 7 earthquake might have an aftershock as large as magnitude 6.

In populated regions, the shaking from large aftershocks can be very dangerous, compounding the damage caused by the mainshock. On September 4, 2010, a magnitude 7.1 earthquake west of Christchurch, the second largest city in New Zealand, caused extensive damage, but nobody was killed and only a few people were injured. However, a magnitude 6.3 aftershock struck right beneath the center of Christchurch on February 22, 2011, collapsing

C

BEFORE EARTHQUAKE

1

Epicenter

Focus

15 km

120 km

0 Seconds
The rupture begins at the focus and expands outward along the fault surface, sending out seismic waves in all directions.

2

Fault cracks at surface

5 Seconds
The rocks on either side of the fault begin to rebound from their deformed state.

3

Fault crack extends

10 Seconds
The rupture progresses along the fault surface, reducing the stress and allowing the rocks on either side to rebound. Seismic waves continue to be emitted in all directions as the fault propagates.

4

D

Stress →
Time →

A
B
C
1
2
3
4
D

Numbers 1–4 on the graph correspond to panels 1–4, which show snapshots of the fault rupture.

20 Seconds
The rupture has progressed along the entire length of this fault segment, and the displacement across it ceases. However, the rupture may continue to propagate further along the fault beyond this particular segment.

AFTER EARTHQUAKE

FIGURE 13.5 ■ During an earthquake, fault slipping begins at the focus and spreads out along the fault surface. Panels 1–4 are snapshots of the fault rupture corresponding to the numbered points on the graph.

JUST BEFORE EARTHQUAKE

DURING EARTHQUAKE

JUST AFTER EARTHQUAKE

FIGURE 13.6 ■ Aftershocks are smaller earthquakes that follow a large earthquake (the mainshock). Foreshocks occur shortly before the mainshock, near its focus.

buildings and killing 185 people (**Figure 13.7**). The economic losses from this aftershock, estimated at $15 billion, were several times greater than the losses caused by the mainshock five months before. Other strong aftershocks hit the city on June 13 and December, 2011, injuring dozens and causing $4 billion in additional damages. More aftershocks can be expected in the years to come.

FIGURE 13.7 ■ Ruins of the Christchurch Catholic Cathedral, one of many buildings in downtown Christchurch, New Zealand, destroyed by an earthquake on February 22, 2011. This event was an aftershock of a larger but less damaging earthquake that occurred west of Christchurch on September 4, 2010. [EPA/DAVID WETHEY/Landov.]

A **foreshock** is a small earthquake that occurs shortly before a mainshock, near its focus (see Figure 13.6). One or more foreshocks have preceded many large earthquakes, so scientists have tried to use foreshocks to predict when and where large earthquakes might happen. Unfortunately, it is usually very hard to distinguish foreshocks from other small earthquakes that occur randomly and frequently on active faults, so this method has only rarely proved successful. The magnitude-9 Tohoku earthquake that caused a great tsunami to hit Honshu, Japan on March 11, 2011 (see Earth Issues 13.1), was preceded by a foreshock of magnitude 7.2 about 50 hours before the mainshock. In some sense, the mainshock was an anomalously big "aftershock" associated with the first event. But what turned out to be a foreshock was considered at the time to be only an ordinary magnitude-7.2 earthquake in the subduction zone.

As this example illustrates, foreshocks, mainshocks, and aftershocks can be classified in a definitive way only after the earthquake sequence has ended; during the sequence we cannot be sure whether the mainshock—the biggest event in the sequence—is yet to come.

■ How Do We Study Earthquakes?

As in any experimental science, instruments and field observations provide the basic data used to study earthquakes. These data enable investigators to locate earthquakes, determine their sizes and numbers, and understand their relationships to faults.

Seismographs

The **seismograph,** an instrument that records the seismic waves, is to the Earth scientist what the telescope is to the astronomer: a tool for peering into inaccessible regions (**Figure 13.8**). The ideal seismograph would be a device affixed to a stationary frame that is not attached to Earth. When the ground shook, the seismograph would measure the changing distance between the frame, which did not move, and the vibrating ground, which did. As yet, we have no way to position a seismograph that is not attached to Earth—although Global Positioning System (GPS) technology is beginning to remove this limitation. So we compromise. We attach a dense mass, such as a piece of steel, to Earth so loosely that the ground can vibrate up and down or side to side without causing much movement of the mass.

This loose attachment can be achieved by suspending the mass from a spring (Figure 13.8a). When seismic waves move the ground up and down, the mass tends to remain stationary because of its inertia (an object at rest tends to stay at rest), but the mass and the ground move relative to each other because the spring can compress or stretch. In this way, the vertical displacement of the ground caused by seismic waves can be recorded by a pen on chart paper or, almost always these days, digitally on a computer. Such a record is called a *seismogram.*

Loose attachment of the mass can also be achieved with a hinge. A seismograph that has its mass suspended on hinges, like a swinging gate (Figure 13.8b), can record the horizontal displacement of the ground.

A typical seismographic station has seismographs set up to measure three components of ground movement: vertical, horizontal east-west, and horizontal north-south. Modern seismographs can detect ground oscillations of less than a billionth of a meter—an astounding feat, considering that such small displacements are of atomic size!

(a) **Seismograph designed to detect vertical movement**

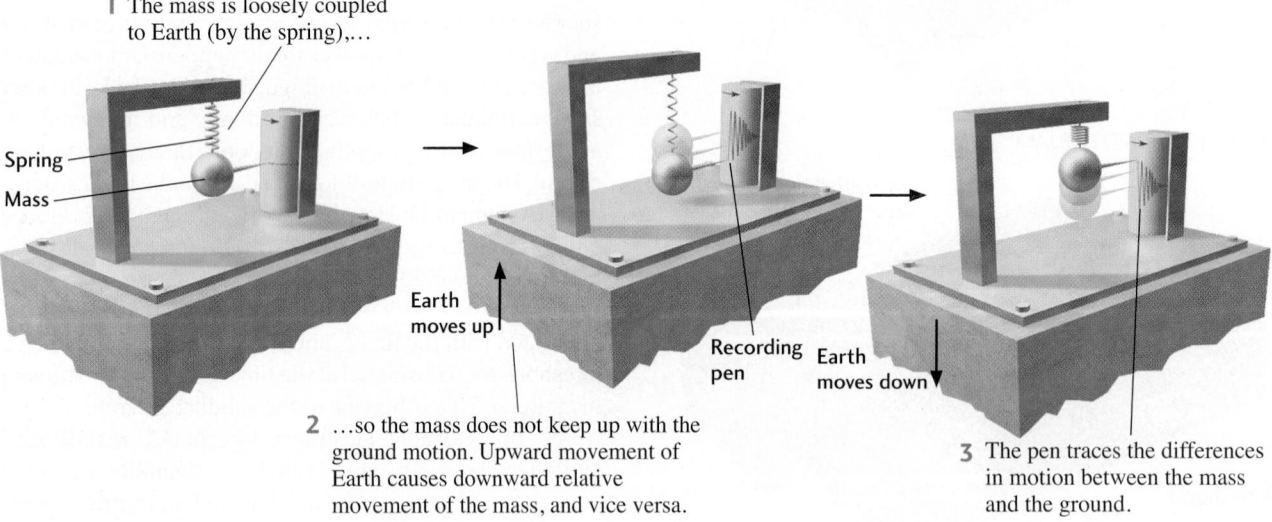

1 The mass is loosely coupled to Earth (by the spring),...

Spring
Mass

Earth moves up

2 ...so the mass does not keep up with the ground motion. Upward movement of Earth causes downward relative movement of the mass, and vice versa.

Recording pen   Earth moves down

3 The pen traces the differences in motion between the mass and the ground.

(b) **Seismograph designed to detect horizontal movement**

Earth moves side to side

Mass

Hinge

FIGURE 13.8 ■ A seismograph consists of a dense mass (such as a steel ball) attached to a recording device. Because of its inertia and its loose coupling to Earth through (a) a spring or (b) a hinge, the mass does not keep up with the motion of the ground.

Earth Issues

13.1 The Tsunami Stone of Aneyoshi

On a hill along the Tohoku coastline of northeastern Honshu, in the fishing hamlet of Aneyoshi, sits a stone monument of uncertain age, inscribed with Japanese characters that read, "High dwellings are the peace and harmony of our descendants. Remember the calamity of the great tsunamis. Do not build any homes below this point." Aneyoshi, now part of the city of Miyako, was once more conveniently located down by the sea where the fishermen tied their boats, but only four of its residents survived the tsunami of 1896 and only two survived the tsunami of 1933. The stone reminds people why they now live on higher ground.

History became prophecy at 2:46 P.M. on March 11, 2011, when the offshore thrust fault that separates Japan from the Pacific plate began to slip. The rupture started on a small patch of the fault surface 30 kilometers beneath the ocean, about 100 km southeast of Aneyoshi, and accelerated outward like a crack through glass, reaching speeds of nearly 3 km/s (more than 6000 miles/hour). By the time it stopped several minutes later, the Pacific Plate had moved under Japan by as much as 40 m along a fault surface the size of South Carolina. Seismic waves from this Tohoku megaquake, which measured magnitude 9, propagated over Earth's surface and through its deep interior, causing the planet to ring like a bell for many days.

The thrusting of Honshu eastward and upward over the Pacific plate raised the seafloor as much as 10 m almost instantly, displacing several hundred billion tons of water, which flowed away from the fault in a huge tsunami. In less than an hour, the water waves, slower than the seismic waves but much more deadly, passed into the bays and inlets of the Japanese coastline like an undulating monster, gaining height as they approached the shore. Funneled into harbors, the waves created immense walls of water—tsunami is Japanese for "harbor wave"—which inundated the near-shore communities, sweeping up boats, cars, and buildings, in some places travelling several kilometers inland.

The fast-moving swath of devastation was captured on horrific videos from helicopters overhead and by survivors who made it to high ground and the tops of buildings. The tsunami overran the seawalls designed to protect the city center of Miyako, destroying all but 30 of the 1,000 boats in its famous fishing fleet and killing many hundreds who could not, or did not, escape in time. Though the exact number remains uncertain, the death toll along the Tohoku coastline was almost 20,000. One of the highest levels reached by the enormous wave—the greatest in recent Japanese history—was 39 m (128 ft) above the shoreline, just below the Aneyoshi tsunami stone. The residents in their houses above the stone were safe.

The tsunami stone at Aneyoshi. [Ko Sasaki/The New York Times/Redux.]

Seismic Waves

Install a seismograph almost anywhere, and within a few hours it will record the passage of seismic waves generated by an earthquake somewhere on Earth. The waves travel from the earthquake focus through Earth and arrive at the seismograph in three distinct groups (**Figure 13.9a**). The first waves to arrive are called primary waves, or **P waves.** The secondary waves, or **S waves,** follow. Both P waves and S waves travel through Earth's interior. Afterwards come the slower **surface waves,** which travel around Earth's surface.

P waves in rock are similar to sound waves in air, except that P waves travel through the solid rock of Earth's crust at about 6 km/s, which is about 20 times faster than sound waves travel through air. Like sound waves, P waves are *compressional waves,* so called because they travel through solid, liquid, or gaseous materials as a succession

(a) Seismic waves generated at an earthquake focus arrive at a seismograph far from the earthquake.

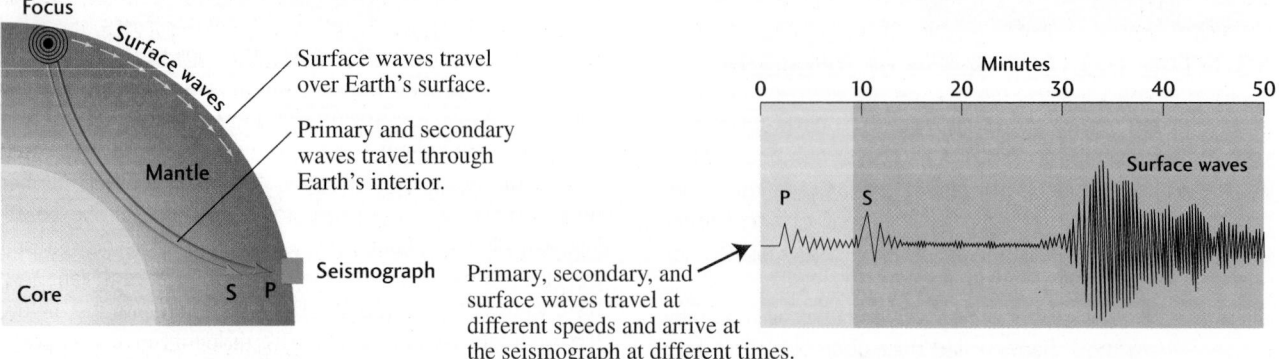

Focus

Surface waves

Mantle

Core   S P

Seismograph

Surface waves travel over Earth's surface.

Primary and secondary waves travel through Earth's interior.

Primary, secondary, and surface waves travel at different speeds and arrive at the seismograph at different times.

Minutes
0   10   20   30   40   50

P   S   Surface waves

(b) Seismic waves are characterized by distinct types of ground deformation.

P-wave motion

P waves (primary waves) are compressional waves that travel quickly through rock.

Compressional-wave crest

P waves travel as a series of contractions and expansions, pushing and pulling particles in the direction of their path of travel.

The red square charts the contraction and expansion of a section of rock.

Wave direction

S-wave motion

S waves (secondary waves) travel at about half the speed of P waves.

Shear-wave crest

S waves are shear waves that push material at right angles to their path of travel.

The red square shows how a section of rock shears from a square to a parallelogram as the S wave passes.

Wave direction

Surface-wave motion

Surface waves ripple across Earth's surface, where air above the surface allows free movement. There are two types of surface waves.

In one type, the ground surface moves in a rolling, elliptical motion that decreases with depth beneath the surface.

Wave direction

In the second type, the ground shakes sideways, with no vertical motion.

Wave direction

FIGURE 13.9 ■ (a) The three types of waves travel by different routes and at different speeds to a seismograph that records them. (b) The three types of seismic waves are characterized by distinct types of ground deformation. The red squares show the distortion of a section of rock as a wave passes through it.

of compressions and expansions (Figure 13.9b). P waves can be thought of as push-pull waves: they push or pull particles of matter in the direction of their path of travel.

S waves travel through solid rock at a little more than half the velocity of P waves. They are *shear waves* that displace material at right angles to their path of travel (see Figure 13.9b). Shear waves cannot travel through liquids or gases.

The velocities at which P and S waves travel are higher when the resistance to their movement is greater. It takes more force to compress solids than to shear them, so P waves always travel faster than S waves through a solid, which is why the P waves from an earthquake arrive at a seismograph before the S waves. This physical principle also explains why S waves cannot travel through air, water, or Earth's liquid outer core: gases and liquids put up no resistance to shear.

Surface waves are confined to Earth's surface and outer layers, like waves on the ocean. Their velocity is slightly less than that of S waves. One type of surface wave sets up a rolling motion in the ground; another type shakes the ground sideways (see Figure 13. 9b). Surface waves are usually the most destructive waves in a large, shallow-focus earthquake, especially in sedimentary basins, where reverberations in the soft near-surface sediments can increase their amplitudes, causing much stronger shaking than in the basement rocks.

People have felt seismic waves and witnessed their destructiveness throughout history, but not until the close of the nineteenth century were scientists able to devise seismographs to record them accurately. Seismic waves enable us to study earthquakes but also provide our most important means of probing Earth's deep interior, as we will see in Chapter 14.

Locating the Focus

Locating an earthquake's focus is like deducing the distance of a lightning strike from the time interval between the flash of light and the sound of thunder: the greater the distance to the lightning bolt, the longer the time interval. Light travels faster than sound, so the lightning flash may be likened to the P waves of an earthquake and the thunder to the slower S waves.

The time interval between the arrival of P waves and S waves depends on the distance the waves have traveled from the focus: the longer the interval, the longer the distance the waves have traveled. Seismologists have used networks of sensitive seismographs around the world and highly accurate clocks to time the arrival of seismic waves from earthquakes as well as from underground nuclear test sites at known locations. From the results, they have constructed *travel-time curves*, which show how long it takes seismic waves of each type to travel a certain distance.

To estimate the distance to a new quake's focus, seismologists read from a seismogram the amount of time between the arrival of the first P waves and the arrival of the first S waves. Then they use travel-time curves, like the ones shown in Figure 13.10, to determine the distance from the seismograph to the focus. If they can determine the distances from three or more seismographic stations, they can locate the focus (see Figure 13.10). They can also deduce the time of the quake at the focus—the earthquake's *origin time*—because the arrival time of the first P waves at each station is known, and it is possible to determine from the travel-time curves how long those waves took to reach the station. Today, this entire process is done automatically by computers, which use data from a large network of seismographs to determine each earthquake's epicenter, depth of focus, and origin time.

Measuring the Size of an Earthquake

Locating earthquakes is only one step on the way to understanding them. We must also determine their sizes, or *magnitudes*. Other things being equal (such as distance from the focus and regional geology), an earthquake's magnitude is the main factor that determines the intensity of the seismic waves it produces, and thus the earthquake's potential destructiveness.

RICHTER MAGNITUDE In 1935, Charles Richter, a California seismologist, devised a simple procedure that assigned a numerical size to each earthquake, now called the *Richter magnitude* (Figure 13.11). Richter studied astronomy as a young man and learned how astronomers use a logarithmic scale to measure the brightnesses of stars, which vary over a huge range of values. Adapting this idea to earthquakes, Richter took as his measure of earthquake size the logarithm of the largest ground movement registered by a standard type of seismograph at a standard distance, thus defining a **magnitude scale.**

On Richter's magnitude scale, two earthquakes at the same distance from a seismograph differ by one magnitude unit if the peak amplitude of the ground movement they produce differs by a factor of 10. The ground movement of an earthquake of magnitude 3, therefore, is 10 times that of an earthquake of magnitude 2. Similarly, a magnitude 6 earthquake produces ground movements that are 100 times greater than those of a magnitude 4 earthquake. The energy released as seismic waves increases even more with earthquake magnitude, by a factor of about 32 for each magnitude unit. A magnitude 7 earthquake releases 32 × 32, or about 1000, times the energy of a magnitude 5 earthquake. According to this energy scale, the Tohoku megaquake was a million times more powerful than a magnitude 5 event!

Seismic waves gradually weaken as they move away from the focus, so to make his procedure work for any seismograph, Richter had to find a way to correct the

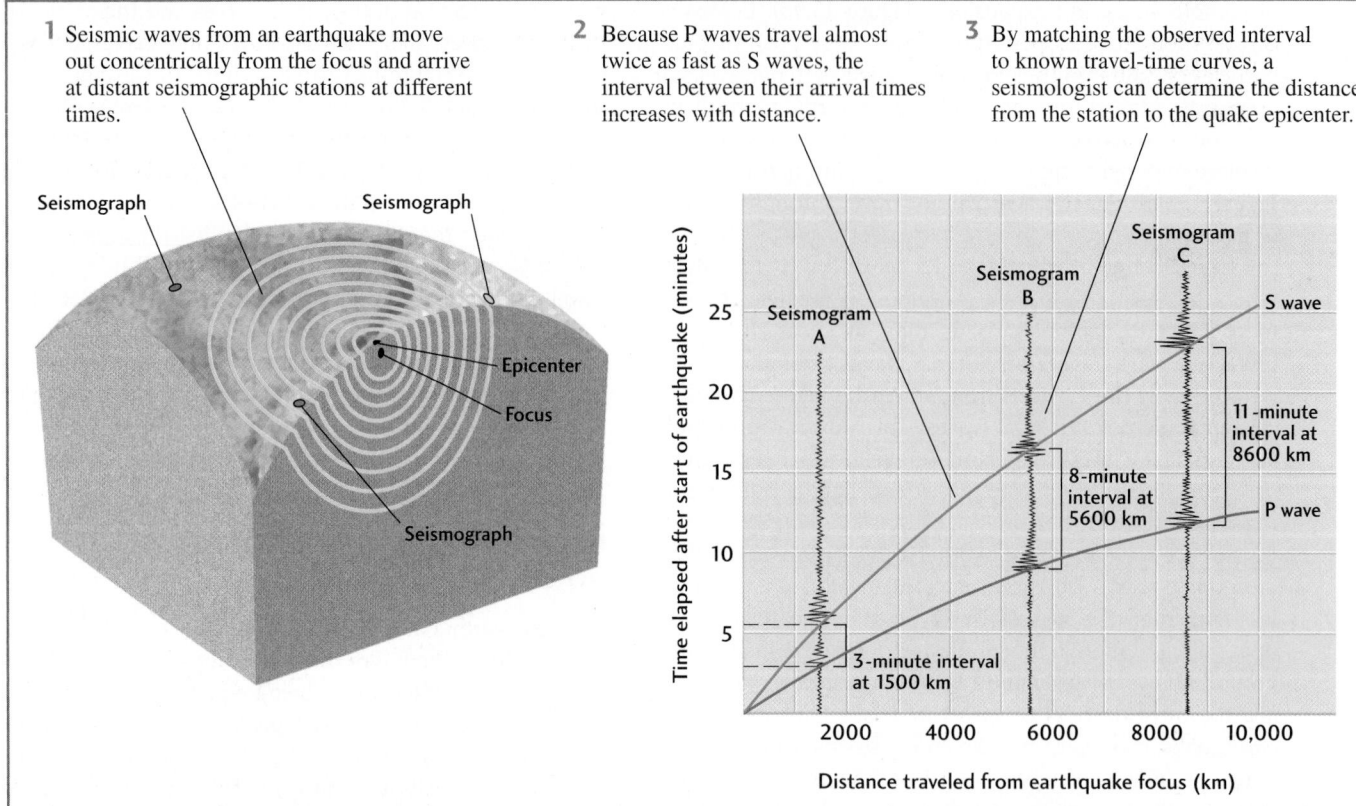

1 Seismic waves from an earthquake move out concentrically from the focus and arrive at distant seismographic stations at different times.

2 Because P waves travel almost twice as fast as S waves, the interval between their arrival times increases with distance.

3 By matching the observed interval to known travel-time curves, a seismologist can determine the distance from the station to the quake epicenter.

FIGURE 13.10 ▪ Readings from three or more seismographic stations can be used to determine the location of an earthquake's focus.

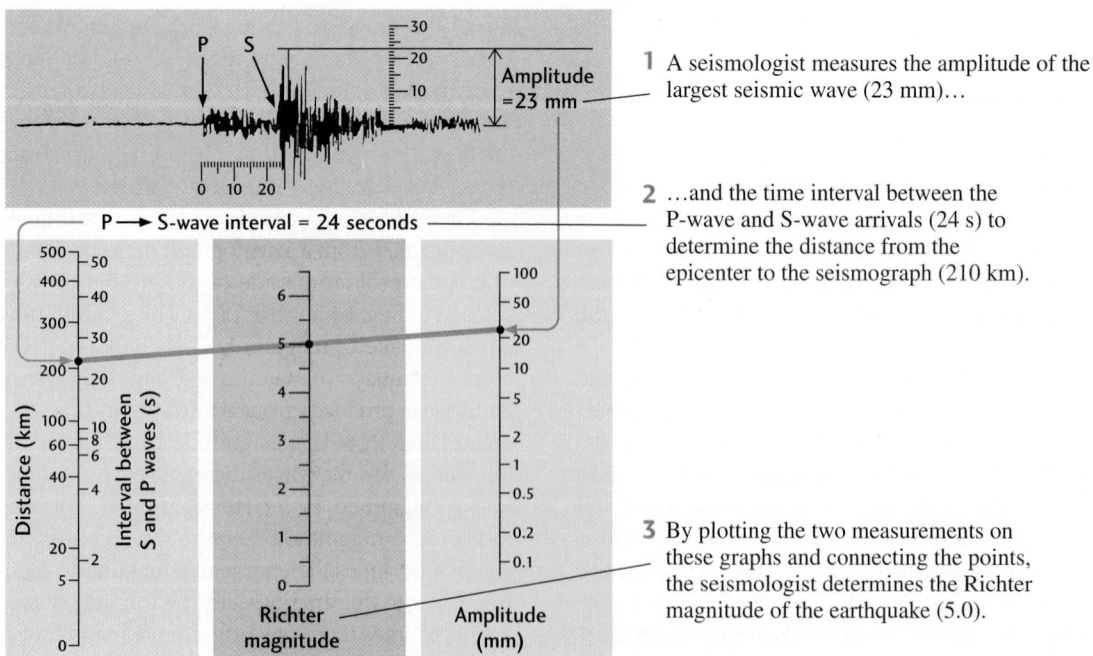

1 A seismologist measures the amplitude of the largest seismic wave (23 mm)…

2 …and the time interval between the P-wave and S-wave arrivals (24 s) to determine the distance from the epicenter to the seismograph (210 km).

3 By plotting the two measurements on these graphs and connecting the points, the seismologist determines the Richter magnitude of the earthquake (5.0).

FIGURE 13.11 ▪ The maximum amplitude of ground movement, corrected by the P-S wave interval, is used to assign a Richter magnitude to an earthquake. [California Institute of Technology.]

4 If the seismologist then draws a circle with a radius calculated from the travel-time curves around each seismographic station,…

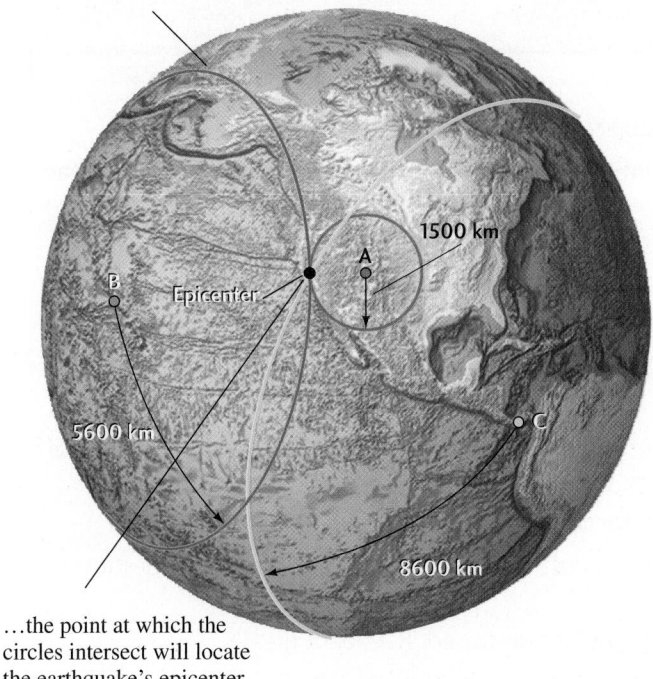

1500 km

A

B

Epicenter

5600 km

C

8600 km

5 …the point at which the circles intersect will locate the earthquake's epicenter.

measurement of ground movement for the distance between the seismograph and the focus. He devised a simple graph that allowed seismologists at different locations to quickly come up with nearly the same value for the magnitude of an earthquake no matter how far their instruments were from the focus (see Figure 13.11). His procedure came to be used throughout the world.

MOMENT MAGNITUDE Although "Richter scale" has become a household term, seismologists prefer a measure of earthquake size more directly related to the physical properties of the faulting that causes an earthquake. The *seismic moment* of an earthquake is defined as a number proportional to the product of the area of faulting and the average fault slip. The corresponding *moment magnitude* increases by about one unit for every tenfold increase in the area of faulting (see the Practicing Geology exercise at the end of the chapter).

Although Richter's method and the moment method produce roughly the same numerical values, the moment magnitude can be measured more accurately from seismographs, and it can sometimes be determined directly from field measurements of the fault.

MAGNITUDE AND FREQUENCY Large earthquakes occur much less often than small ones. This observation can be expressed by a simple relationship between earthquake frequency and magnitude (**Figure 13.12**). Worldwide,

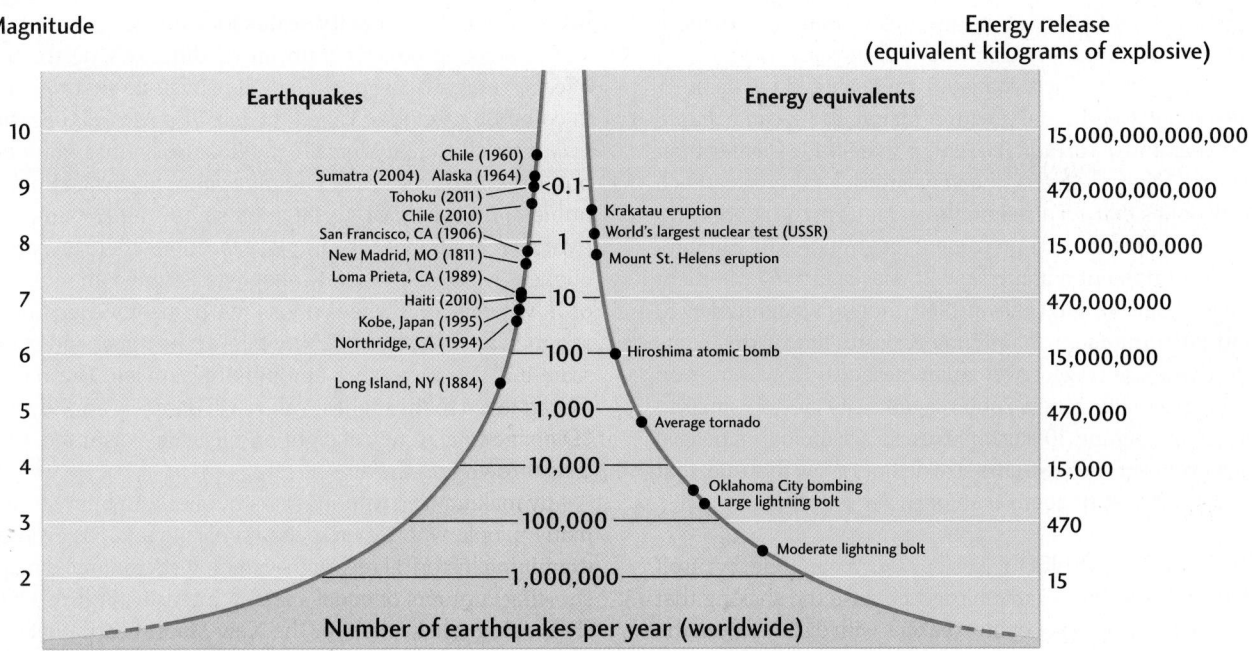

Magnitude

Energy release
(equivalent kilograms of explosive)

Earthquakes

Energy equivalents

Magnitude	Earthquake	Number per year	Energy equivalent	Energy release
10				15,000,000,000,000
9	Chile (1960)	<0.1		470,000,000,000
	Sumatra (2004) Alaska (1964)		Krakatau eruption	
	Tohoku (2011)		World's largest nuclear test (USSR)	
8	Chile (2010) San Francisco, CA (1906)	1	Mount St. Helens eruption	15,000,000,000
	New Madrid, MO (1811)			
	Loma Prieta, CA (1989)			
7	Haiti (2010)	10		470,000,000
	Kobe, Japan (1995)			
	Northridge, CA (1994)			
6		100	Hiroshima atomic bomb	15,000,000
5	Long Island, NY (1884)	1,000	Average tornado	470,000
4		10,000		15,000
			Oklahoma City bombing	
3		100,000	Large lightning bolt	470
2		1,000,000	Moderate lightning bolt	15

Number of earthquakes per year (worldwide)

FIGURE 13.12 ■ Relationship between moment magnitude, seismic energy release, and number of earthquakes per year worldwide. Examples of earthquakes of various magnitudes and of other large sources of sudden energy release are included for comparison. [Adapted from IRIS Consortium, http://www.iris.edu.]

TABLE 13-1 Modified Mercalli Intensity Scale

Intensity Level	Description
I	Not felt.
II	Felt only by a few people at rest. Suspended objects may swing.
III	Felt noticeably indoors. Many people do not recognize it as an earthquake. Parked cars may rock slightly.
IV	Felt indoors by many, outdoors by few. Dishes, windows, doors rattle. Parked cars rock noticeably.
V	Felt by most; many awakened. Some dishes, windows broken. Unstable objects overturned.
VI	Felt by all. Some heavy furniture moves. Damage slight.
VII	Slight to moderate damage in well-built structures; considerable damage in poorly built structures; some chimneys broken.
VIII	Considerable damage in well-built structures. Damage great in poorly built structures. Fall of chimneys, factory stacks, columns, monuments, walls.
IX	Damage great in well-built structures, with partial collapse. Buildings shifted off foundations.
X	Some well-built wooden structures destroyed; most masonry and frame structures destroyed. Rails bent.
XI	Few if any masonry structures remain standing. Bridges destroyed. Rails bent greatly.
XII	Damage total. Lines of sight and level are distorted. Objects thrown into the air.

approximately 1,000,000 earthquakes with magnitudes greater than 2 take place each year. This number decreases by a factor of 10 for each magnitude unit. Hence, there are about 100,000 earthquakes with magnitudes greater than 3, about 1000 with magnitudes greater than 5, and about 10 with magnitudes greater than 7 each year.

According to these statistics, there should be, on average, about 1 earthquake with a magnitude greater than 8 per year and 1 earthquake with a magnitude greater than 9 every 10 years. In fact, the very largest earthquakes, such as the ones that occurred on thrust faults in the subduction zones off Japan in 2011 (moment magnitude 9.0), Sumatra in 2004 (moment magnitude 9.2), Alaska in 1964 (moment magnitude 9.1), and Chile in 1960 (moment magnitude 9.5) and 2010 (moment magnitude 8.8) are almost this common when averaged over many decades. However, even the largest subduction zone megathrusts are too small to create magnitude 10 earthquakes, so seismologists believe that events of such extreme size do not follow this rule; that is, they are less frequent than once per century.

SHAKING INTENSITY Earthquake magnitude by itself does not describe seismic hazard because the shaking that causes destruction generally weakens with distance from the fault rupture. A magnitude 8 earthquake in a remote area far from the nearest city might cause no human or economic losses, whereas a magnitude 6 quake immediately beneath a city would likely cause serious damage. The destruction in

Christchurch by the earthquakes of February 22 and June 13, 2011, illustrates this important point (see Figure 13.7).

In the late nineteenth century, before Richter invented his magnitude scale, seismologists and earthquake engineers developed **intensity scales** for estimating the intensity of shaking directly from an earthquake's destructive effects. Table 13.1 shows one intensity scale that remains in common use today, called the *modified Mercalli intensity scale* after Giuseppe Mercalli, the Italian scientist who proposed it in 1902. This scale assigns a value, given as a Roman numeral from I to XII, to the intensity of the shaking at a particular location. For example, a location where an earthquake is barely felt by a few people is assigned an intensity of II, whereas one where it was felt by nearly everyone is assigned an intensity of V. Numbers at the upper end of the scale indicate increasing amounts of damage. The narrative attached to the highest value, XII, is tersely apocalyptic: "Damage total. Lines of sight and level are distorted. Objects thrown into the air."

By making observations at many sites and interviewing many people who experienced an earthquake, or even by examining historical records, seismologists can make maps showing contours of equal shaking intensity. Figure 13.13 shows an intensity map for the New Madrid earthquake of December 16, 1811, a magnitude 7.7 event near the southern tip of Missouri, which was felt as far away as Boston. Although earthquake intensities are generally highest near the fault rupture, they also depend on the local geology.

FIGURE 13.13 ■ Measurements of modified Mercalli intensities for the New Madrid earthquake of December 16, 1811, a magnitude 7.7 event near the juncture of Missouri, Arkansas, and Tennessee. Regions near the epicenter experienced intensities greater than IX, and intensities as high as VI were observed 200 km from the epicenter (see Table 13.1). [Carl W. Stover and Jerry L. Cossman, USGS Professional Paper 1527, 1993.]

For example, when sites at equal distances from the rupture are compared, the shaking tends to be more intense on soft sediments (especially water-saturated sediments near shorelines) than on hard basement rock. Intensity maps thus provide engineers with crucial data for designing structures to withstand seismic shaking.

Determining Fault Mechanisms

The pattern of ground shaking also depends on the orientation of the fault rupture and the direction of slipping, which together specify the **fault mechanism** of an earthquake. The fault mechanism tells us whether the rupture was on a *normal, reverse,* or *strike-slip* fault. If the rupture was on a strike-slip fault, the fault mechanism also tells us whether the movement was *right-lateral* or *left-lateral* (see Figure 7.8 for the definitions of these terms). We can then use this information to infer the regional pattern of tectonic forces (Figure 13.14).

For shallow ruptures that break the surface, we can sometimes determine the fault mechanism from field observations of the fault scarp. As we have seen, however, most ruptures are too deep to break the surface, so we must deduce the fault mechanism from seismograms.

For large earthquakes at any depth, this task turns out to be easy, because there are enough seismographic stations around the world to surround any earthquake's focus. In some directions from the focus, the very first movement of the ground recorded by a seismograph—a P wave—is a *push away* from the focus, causing upward movement on a vertical seismograph. In other directions, the initial ground movement is a *pull toward* the focus, causing downward movement on a vertical seismograph. For strike-slip ruptures, the directions of the largest pushes lie on an axis rotated 45° from the fault plane and are perpendicular to the directions of the largest pulls (Figure 13.15). The locations of pushes and pulls can therefore be plotted and divided into four sections based on the positions of the seismographic stations. One of the two boundaries between those sections will be the fault orientation; the other will be a plane perpendicular to the fault. The slip direction on the fault plane can also be determined from the arrangement of pushes and pulls. In this manner, without surface evidence, seismologists can deduce whether the horizontal crustal forces that triggered an earthquake were tensional, compressive, or shearing forces.

FIGURE 13.14 ■ The three main types of fault mechanisms that initiate earthquakes and the stresses that cause them. (a) A fault before movement takes place. (b) Normal faulting due to tensional stress. (c) Reverse faulting due to compressive stress. (d) Strike-slip faulting due to shearing stress (in this case, left-lateral).

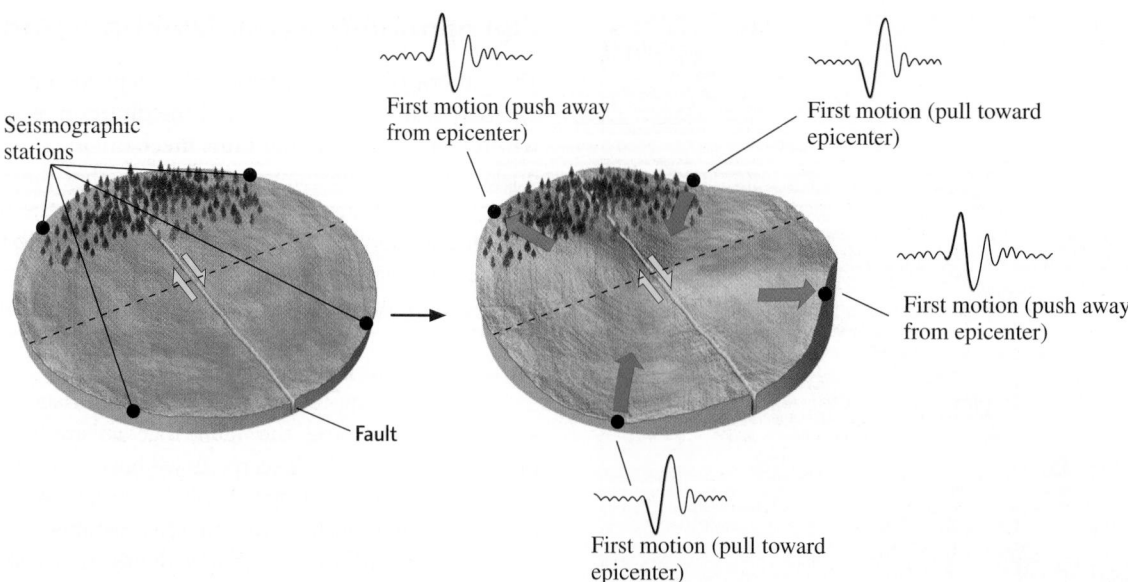

FIGURE 13.15 ■ The movement marking the first P wave arriving from a fault rupture at each of several seismographic stations is used to determine the orientation of the fault and the direction of fault slipping. The case shown here is the rupture of a right-lateral strike-slip fault. Note that the alternating pattern of pushes and pulls would remain the same if a plane perpendicular to the fault ruptured with left-lateral displacement. Seismologists can usually choose between the two possibilities using additional information, such as field mapping of the fault scarp or the alignment of aftershocks along the fault.

GPS Measurements and "Silent" Earthquakes

As we saw in Chapter 2, GPS receivers can record the slow movements of lithospheric plates. These instruments can also measure the strain that builds up from such movements, as well as the sudden slipping on a fault when it ruptures in an earthquake.

Seismologists use GPS observations to study another kind of movement along active faults. It has been known for many years that a section of the San Andreas fault in central California creeps continuously, rather than rupturing suddenly. This creep slowly deforms structures and cracks pavements that cross the fault trace. More recently, GPS receivers have found surface movements at convergent plate boundaries that reflect short-lived creep events, which commonly last days to weeks at a time. They have been named *silent earthquakes* because these gradual movements do not trigger destructive seismic waves. Nevertheless, they can release large amounts of stored strain energy.

These observations raise many questions that geologists are now trying to answer: What causes faults to stick and slip catastrophically in some places and creep in others? Will the release of strain energy by silent earthquakes make destructive earthquakes in those regions less frequent or less severe? Can silent earthquakes be used to predict potentially destructive earthquakes?

■ Earthquakes and Patterns of Faulting

As we have seen, seismologists are using networks of sensitive seismographs to locate earthquakes around the world, measure their magnitudes, and deduce their fault mechanisms. These methods are revealing new information about tectonic processes on scales much smaller than the plates themselves. In this section, we summarize the global pattern of earthquake occurrence from the perspective of plate tectonics and show how regional studies of active fault systems are improving our understanding of earthquakes along plate boundaries and within plate interiors.

The Big Picture: Earthquakes and Plate Tectonics

The *seismicity map* in **Figure 13.16** shows the epicenters of earthquakes recorded around the world since 1976. The most obvious features of this map, known to seismologists for many decades, are the belts of seismic activity that mark the major plate boundaries. The fault mechanisms observed for earthquakes in these belts (**Figure 13.17**) are consistent with the types of faulting along different types of plate boundaries that we described in Chapter 7.

FIGURE 13.16 ■ Global map of seismic activity from January, 1976, through October, 2013. Each dot represents the epicenter of an earthquake larger than magnitude 5. Colors indicate the focal depth. Note the concentration of earthquakes along the boundaries between major lithospheric plates. [Map based on data from Global CMT catalog; plot by M. Boettcher.]

FIGURE 13.17 ■ The fault mechanisms observed at different types of plate boundaries conform to the predictions of plate tectonic theory.

Mid-ocean ridge (divergence)

Normal faulting

Transform fault (strike-slip faulting)

Rift valley (divergence)

Deep-sea trench (convergence)

Lithosphere

Asthenosphere

Lithosphere

Asthenosphere

Shallow earthquakes coincide with normal faulting at divergent boundaries and with strike-slip faulting at transform-fault boundaries.

Large shallow earthquakes occur mainly on thrust faults at the plate boundary.

Intermediate-focus earthquakes occur in the descending slab.

Deep-focus earthquakes also occur in the descending slab.

DIVERGENT BOUNDARIES The narrow belts of shallow earthquakes that run through ocean basins coincide with mid-ocean ridge crests and their offsets on transform faults. The P waves recorded from the ridge-crest quakes indicate that they are caused by normal faulting. The faults strike parallel to the ridge and dip toward the rift valley at the ridge crest. Normal faulting implies that tensional forces are at work as the plates are pulled apart during seafloor spreading. Earthquakes also have normal fault mechanisms in zones where continental crust is being pulled apart, such as in the East African rift valleys and in the Basin and Range province of western North America.

TRANSFORM-FAULT BOUNDARIES Earthquake activity is even greater along the transform-fault boundaries that offset mid-ocean ridge segments. These earthquakes have strike-slip fault mechanisms, just as one would expect where plates create horizontal shearing forces as they slide past each other in opposite directions. Moreover, for earthquakes along these transform faults, the slip direction indicated by the fault mechanisms is left-lateral where the ridge crest steps right and right-lateral where it steps left. These directions are the opposite of what would be needed to create the offsets of the ridge crest, but are consistent with the direction of slip predicted by seafloor spreading. In the mid-1960s, seismologists used this property of the transform faults to support the hypothesis of seafloor spreading. Slip directions on transform faults that run through continental crust, such as California's San Andreas fault and New Zealand's Alpine fault (both right-lateral), also agree with the predictions of plate tectonics.

CONVERGENT BOUNDARIES The world's largest earthquakes occur at convergent plate boundaries. The four greatest earthquakes of the last hundred years were of this type: the 2011 Tohoku earthquake (magnitude 9.0); the Sumatra earthquake of 2004; the Alaska earthquake of 1964; and the largest of all, the 1960 earthquake in the subduction zone west of Chile. During the Chile earthquake, the crust of the Nazca Plate slipped an average of 20 m beneath the crust of the South American Plate on a fault rupture with an area larger than Arizona! The fault mechanisms of these great earthquakes show that they were caused by horizontal compression along *megathrusts*, the huge thrust faults that form the boundaries where one plate is subducted beneath another. All three of these earthquakes displaced the seafloor, generating tsunamis that devastated coastlines.

Earth's deepest earthquakes also occur at convergent boundaries. Almost all earthquakes originating below 100 km rupture the descending plate in a subduction zone. The fault mechanisms of these deep earthquakes show a variety of orientations, but they are consistent with the deformation expected within the descending plate as gravity pulls it back into the convecting mantle. The deepest earthquakes take place in the oldest—and therefore

coldest—descending plates, such as those beneath South America, Japan, and the island arcs of the western Pacific Ocean.

INTRAPLATE EARTHQUAKES Although most earthquakes occur at plate boundaries, a small percentage of global seismic activity originates within plate interiors. The foci of these *intraplate earthquakes* are relatively shallow, and most occur on continents. Among these earthquakes are some of the most famous in American history: a sequence of three large events near New Madrid, Missouri, in 1811–1812; the Charleston, South Carolina, earthquake of 1886; and the Cape Ann earthquake, near Boston, Massachusetts, in 1755. Many intraplate earthquakes occur on old faults that were once parts of ancient plate boundaries. These faults no longer form plate boundaries, but they remain zones of crustal weakness that concentrate and release intraplate stresses.

One of the deadliest intraplate earthquakes (magnitude 7.6) occurred near Bhuj, in the state of Gujarat, in western India, in 2001. It is estimated that some 20,000 lives were lost. The Bhuj earthquake occurred on a previously unknown thrust fault 1000 km south of the boundary between the Indian and Eurasian plates, but the compressive stresses responsible for this rupture resulted from the ongoing collision of these two plates. Intraplate earthquakes show that strong crustal forces can develop and cause faulting within a lithospheric plate far from modern plate boundaries.

Regional Fault Systems

Although the fault mechanisms of most major earthquakes conform to the predictions of plate tectonic theory, a plate boundary can rarely be described as a single fault, particularly when the boundary involves continental crust. Rather, the zone of deformation between two moving plates usually comprises a network of interacting faults—a *fault system*. The fault system in Southern California provides an interesting example (**Figure 13.18**).

The "master fault" of this system is our old nemesis, the San Andreas—a right-lateral strike-slip fault that runs northwestward through California from the Salton Sea near the Mexican border until it goes offshore in the northern part of the state (see Figure 7.7). There are a number of subsidiary faults on either side of the San Andreas, however, that generate large earthquakes. In fact, most of the damaging earthquakes in Southern California during the past century have occurred on those subsidiary faults.

Why is the San Andreas fault system so complex? Part of the explanation has to do with the geometry of the San Andreas fault itself. A bend in the fault creates compressive forces that cause thrust faulting in the area north of Los Angeles (see Figure 7.22). Thrust faulting in this "Big Bend" was responsible for two recent deadly earthquakes, the San Fernando earthquake of 1971 (magnitude 6.6, 65 people killed) and the Northridge earthquake of 1994 (magnitude 6.7,

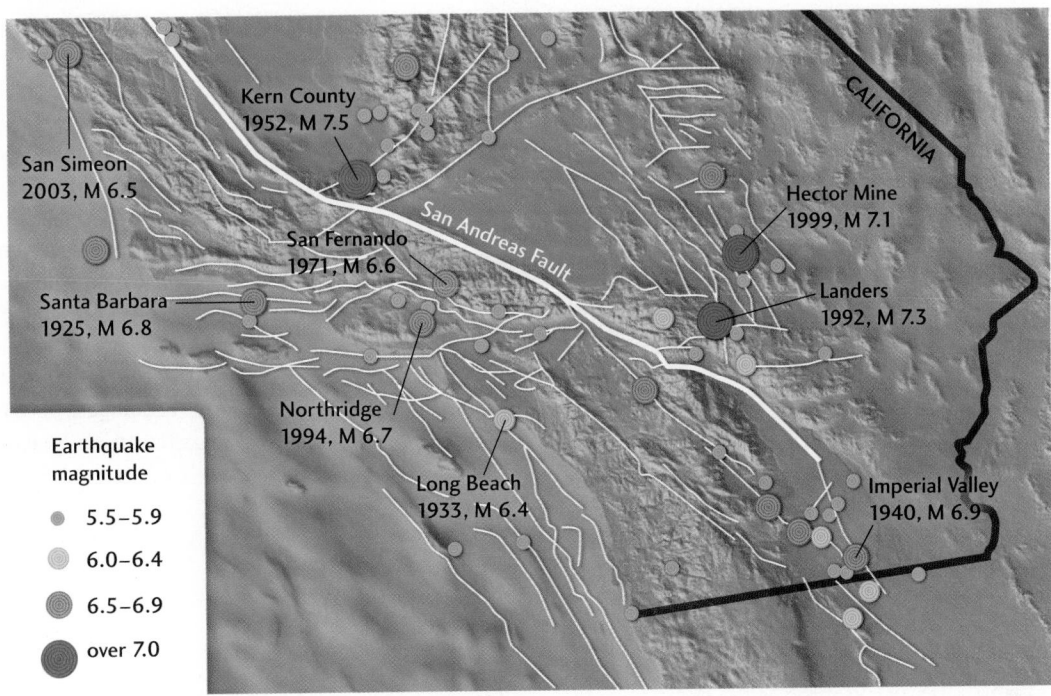

FIGURE 13.18 ■ A map of the fault system of Southern California, showing the surface traces of the San Andreas fault (thick white line) and its subsidiary faults (thin white lines). Colored circles show the epicenters of earthquakes with magnitudes greater than 5.5 during the twentieth century. Significant earthquakes are labeled with their names, years, and magnitudes. [Courtesy of Southern California Earthquake Center.]

58 people killed) (see Figure 13.18). Over the past several million years, this thrust faulting has raised the San Gabriel Mountains to elevations of 1800 to 3000 m.

Another complication is the plate extension taking place east of California in the Basin and Range province, which spans the state of Nevada and much of Utah and Arizona (see Chapters 7 and 10). This broad zone of extensional deformation is connected with the San Andreas fault system through a series of faults that run along the eastern side of the Sierra Nevada and through the Mojave Desert. Faults of this system were responsible for the 1992 Landers earthquake (magnitude 7.3) and the 1999 Hector Mine earthquake (magnitude 7.1), as well as the 1872 Owens Valley earthquake (magnitude 7.6).

■ Earthquake Hazards and Risks

In just the last decade, earthquakes worldwide have killed more than 700,000 people and disrupted the economies of entire regions. The United States has been relatively lucky, although two earthquakes on the San Andreas fault—the 1989 Loma Prieta earthquake (magnitude 7.1), which occurred 80 km south of San Francisco, and the 1994 Northridge earthquake in Los Angeles—were among the costliest

disasters in the nation's history. Damage amounted to more than $10 billion in the Loma Prieta quake and $40 billion in the Northridge quake because of their proximity to urban centers. About 60 people died in each event, but the death toll would have been many times higher if stringent building codes had not been in place (**Figure 13.19**).

Destructive earthquakes are even more frequent in Japan than in California. The recorded history of destructive earthquakes in Japan, going back 2000 years, has left an indelible impression on the Japanese people. Perhaps that is why Japan is the best prepared of any nation in the world to deal with earthquakes. It has impressive public education campaigns, building codes, and warning systems. Despite this preparedness, more than 5600 people were killed in a devastating (magnitude 6.9) earthquake that struck the city of Kobe on January 16, 1995 (**Figure 13.20**). The large numbers of casualties and structural failures (50,000 buildings destroyed) resulted partly from the less stringent building codes that were in effect before 1980, when much of the city was built, and partly from the location of the earthquake rupture, which was very close to the city. The tsunami of the 2011 Tohoku earthquake caused an even greater loss of life (almost 20,000). The disaster was compounded by meltdowns and explosions at the Fukushima-Daiichi power plant, one of the world's largest nuclear facilities (see Figure 13.30). Although the economic costs are still being counted, the Tohoku earthquake is already the most expensive natural disaster in recorded history.

FIGURE 13.19 ■ Sixteen people died in the Northridge Meadows apartment building in Los Angeles during the 1994 Northridge earthquake. The victims lived on the first floor and were crushed when the upper levels collapsed. Many more buildings would have collapsed if the newer buildings in the area had not been constructed according to stringent building codes. [Nick Ut, Files/AP Photo.]

FIGURE 13.20 ■ This elevated expressway in Kobe, Japan, was overturned during the earthquake of 1995. [Tom Wagner/SABA/Corbis.]

How Earthquakes Cause Damage

Earthquakes proceed as chain reactions in which the primary effects of earthquakes—faulting and ground shaking—trigger secondary effects, which include landslides and tsunamis as well as destructive processes within the built environment, such as collapsing structures and fires.

FAULTING AND SHAKING The *primary hazards* of earthquakes are the ruptures in the ground surface that occur when faults break the surface, the permanent subsidence and uplift of the ground surface caused by faulting, and the ground shaking caused by seismic waves radiated during the quake. Seismic waves can shake structures so hard that they collapse. The ground accelerations near the epicenter of a large earthquake can approach and even exceed the acceleration of gravity, so an object lying on the ground surface can literally be thrown into the air. Very few structures built by human hands can survive such severe shaking, and those that do are severely damaged.

The collapse of buildings and other structures is the leading cause of casualties and economic damage during earthquakes. In cities, most casualties are caused by falling buildings and their contents. Death tolls can be especially high in densely populated areas of developing countries, where buildings are often constructed from bricks and mortar without steel reinforcement. A magnitude 7 earthquake on January 12, 2010, destroyed 250,000 residences and 30,000 commercial buildings in Haiti's capital city, Port-au-Prince, killing more than 230,000 people, making it the fifth deadliest seismic disaster ever recorded (**Figure 13.21**). Improving construction

FIGURE 13.21 ▪ Homes in Port-au-Prince destroyed by the Haiti earthquake of January 12, 2010. [© Cameron Davidson/Corbis.]

practices so that buildings are able to withstand shaking is the key to avoiding such tragedies.

LANDSLIDES AND OTHER TYPES OF GROUND FAILURE The primary hazards of faulting and ground shaking generate a number of *secondary hazards*. Among secondary hazards are landslides and other forms of ground failure that give rise to mass movements of Earth materials (described in Chapter 16). When seismic waves shake water-saturated soils, those soils can behave like a liquid and become unstable. The ground simply flows away, taking buildings, bridges, and everything else with it. Such soil *liquefaction* destroyed the residential area of Turnagain Heights near Anchorage, Alaska, in the 1964 earthquake (see Figure 16.16); the Nimitz Freeway near San Francisco in the 1989 Loma Prieta earthquake; and areas of Kobe in the 1995 earthquake. Liquefaction was responsible for much of the damage during the 2010–2011 earthquake sequence in Christchurch, New Zealand, destroying many homes and causing massive damage to underground water and sewer systems throughout the city.

In some instances, ground failure can cause more damage than the ground shaking itself. A 1970 earthquake in Peru triggered an immense avalanche of rock and snow (up to 50 million cubic meters) that destroyed the mountain towns of Yungay and Ranrahirca (see Figure 16.25).

Of the more than 66,000 people killed in the earthquake, about 18,000 of them died in the avalanche.

TSUNAMIS A large earthquake that occurs beneath the ocean can generate a destructive sea wave, sometimes called a "tidal wave," but more accurately named a **tsunami** (Japanese for "harbor wave"), since it has nothing to do with tides. Tsunamis are by far the deadliest and most destructive hazards associated with the world's largest earthquakes: the megathrust events that occur in subduction zones.

When a megathrust ruptures, it can push the seafloor landward of the deep-sea trench upward by as much as 10 m, displacing a large mass of the overlying ocean water. This disturbance flows outward in waves that travel across the ocean at speeds of up to 800 km/hour (about as fast as a commercial jetliner). In the deep sea, a tsunami is hardly noticeable, but when it approaches shallow coastal waters, the waves slow down and pile up, inundating the shoreline in walls of water that can reach heights of tens of meters (**Figure 13.22**). This "run-up" can propagate inland for hundreds of meters or even kilometers, depending on the slope of the land surface.

Tsunamis caused by megathrust events are most common in the Pacific Ocean, which is ringed with very active subduction zones. The destructive power of a great tsunami was brought home by the terrifying video images captured on March 11, 2011 as the Tohoku tsunami swept over

TSUNAMI GENERATION

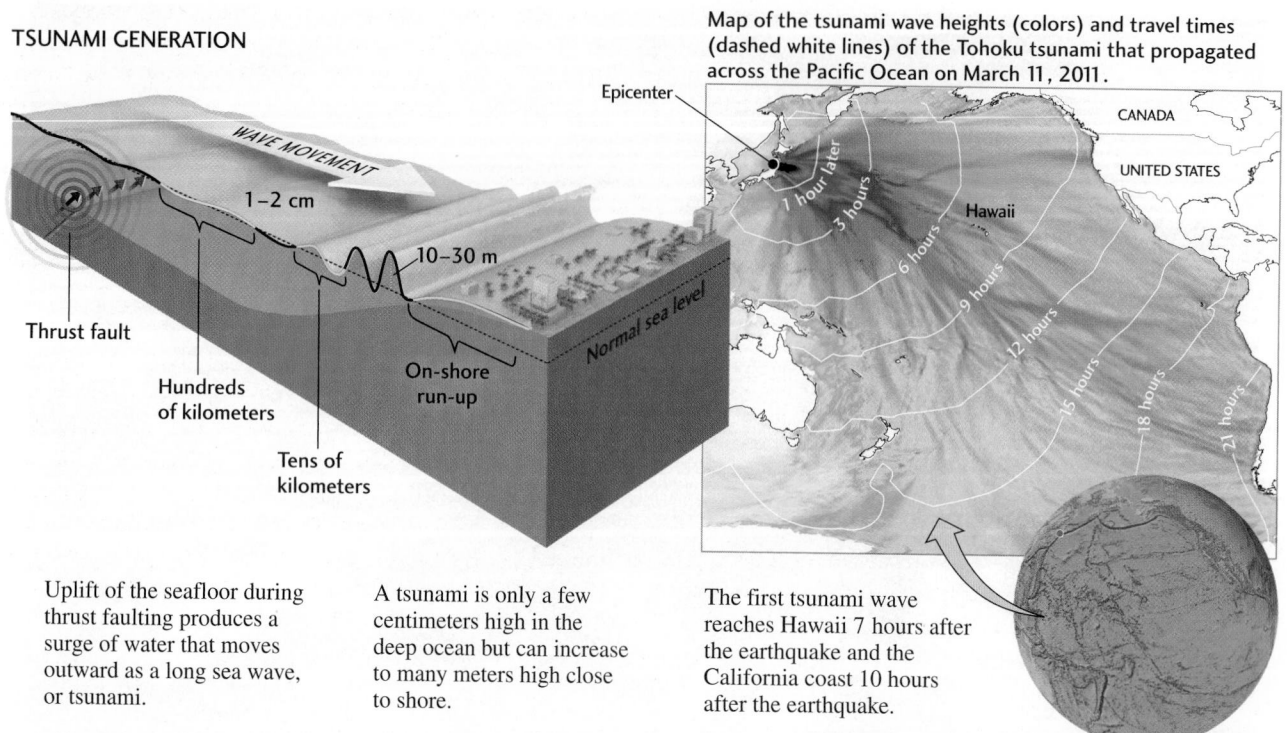

Map of the tsunami wave heights (colors) and travel times (dashed white lines) of the Tohoku tsunami that propagated across the Pacific Ocean on March 11, 2011.

Uplift of the seafloor during thrust faulting produces a surge of water that moves outward as a long sea wave, or tsunami.

A tsunami is only a few centimeters high in the deep ocean but can increase to many meters high close to shore.

The first tsunami wave reaches Hawaii 7 hours after the earthquake and the California coast 10 hours after the earthquake.

FIGURE 13.22 ▪ Earthquakes on megathrusts may generate tsunamis that can propagate across ocean basins. [Map by NOAA, Pacific Marine Environmental Laboratory.]

the shoreline of northeastern Japan. In the coastal city of Miyako, the height of the water mass reached an astounding 38 m (123 ft!) above normal sea level, destroying nearly everything in its path (see Earth Issues 13.1). In low-lying regions near the port city of Sendai, the tsunami traveled up to 10 km inland, transporting huge floating debris fields of buildings, boats, cars, and trucks (**Figure 13.23**). The waves propagated across the entire Pacific Ocean, attaining heights of more than 2 m along the coast of Chile, 16,000 km away.

Tsunami warning systems in Japan and the circum-Pacific region worked according to design. The warning times along the Japanese coast nearest the earthquake were too short for complete evacuation. Nevertheless, the system is credited with saving many thousands of lives.

No tsunami warning system was in place when the magnitude 9.2 Sumatra earthquake of September 26, 2004 unleashed an ocean-wide tsunami that swept over low-lying coastal areas from Indonesia and Thailand to

FIGURE 13.23 ▪ Video image taken from a helicopter, showing the tsunami surge carrying debris across farmland near Sendai, Japan, following the Tohoku earthquake of March 11, 2011. [AP Photo/NHK TV.]

FIGURE 13.24 ■ The tsunami caused by the 2004 Sumatra earthquake struck without warning on a beach in Phuket, Thailand. [Courtesy David Rydevik.]

Sri Lanka, India, and the east coast of Africa (**Figure 13.24**). Within 15 minutes, the first wave ran up the Sumatran coastline. Few eyewitnesses survived there, but geologic investigations after the tsunami indicated that the maximum wave height on the beaches of the west-facing coast was about 15m, and the run-up attained heights of 25 to 35 m,

reaching inland up to 2 km and wiping out most building structures, vegetation, and life in its path (**Figure 13.25**). It is believed that more than 150,000 people perished along the Sumatran coastline, though no one will ever be sure because many bodies were washed out to sea.

Disturbances of the seafloor caused by landslides or volcanic eruptions can also produce tsunamis. The 1883 explosion of Krakatau, a volcano in Indonesia, generated a tsunami that reached 40 m in height and drowned 36,000 people on nearby coasts.

FIRES The secondary hazards of earthquakes also include destructive processes that stem from the nature of the built environment itself, such as the fires ignited by ruptured gas lines or downed electric power lines. Damage to water mains in an earthquake can make firefighting all but impossible—a circumstance that contributed to the burning of San Francisco after the 1906 earthquake (see Figure 13.1). Most of the 140,000 fatalities in the 1923 Kanto earthquake, one of Japan's greatest disasters, resulted from fires in the cities of Tokyo and Yokohama.

Reducing Earthquake Risk

In assessing the possibility of damage from earthquakes, or from any type of natural disaster, it is important to distinguish between *hazard* and *risk*. In the case of earthquakes,

FIGURE 13.25 ■ This small headland near Banda Aceh, on the west coast of Sumatra, was previously covered by dense vegetation to the waterline, but was stripped clean to a height of about 15 m by the 2004 tsunami. [Courtesy Jose Borrero, University of Southern California Tsunami Research Group.]

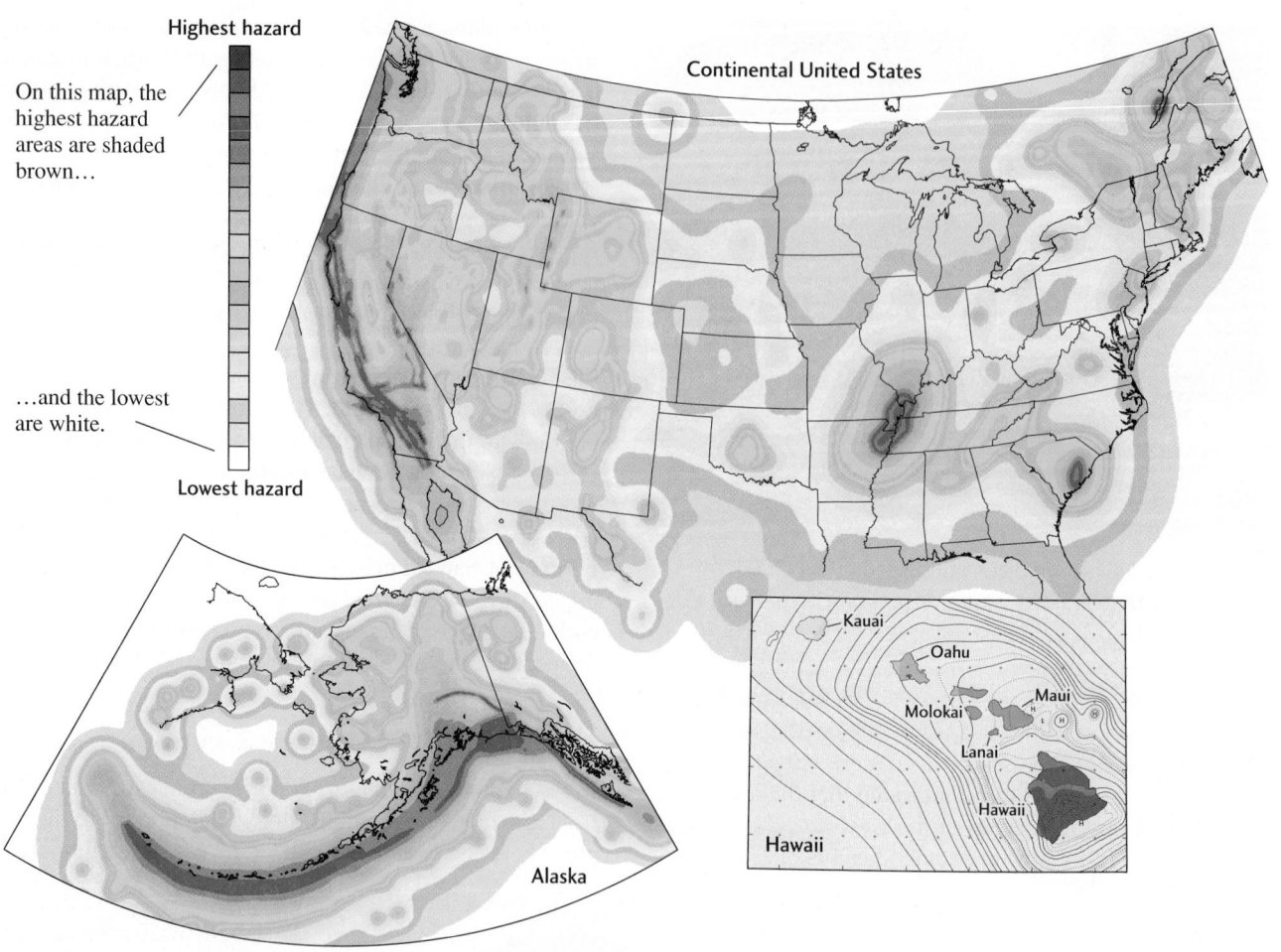

Highest hazard

On this map, the highest hazard areas are shaded brown…

…and the lowest are white.

Lowest hazard

Continental United States

Alaska

Kauai

Oahu

Molokai

Maui

Lanai

Hawaii

Hawaii

FIGURE 13.26 ■ Seismic hazard map for the United States. The regions of highest hazard lie along the plate boundaries of the West Coast and Alaska and on the southern side of the Big Island of Hawaii. In the central and eastern United States, the areas of highest hazard are near New Madrid, Missouri, and Charleston, South Carolina; in eastern Tennessee; and in portions of the Northeast. [U.S. Geological Survey, http://geohazards.cr.usgs.gov/eq/.]

seismic hazard describes the frequency and intensity of earthquake shaking and ground disruption that can be expected over the long term at some specified location. Seismic hazard, which depends on the proximity of the site to active faults that might generate earthquakes, can be expressed in the form of a seismic hazard map. **Figure 13.26** displays the national seismic hazard map produced by the U.S. Geological Survey.

In contrast, **seismic risk** describes the *damage* that can be expected over the long term in a specified region, such as a county or state, usually measured in terms of casualties and dollar losses per year. A region's risk depends not only on its seismic hazard, but also on its exposure to seismic damage (its population and density of buildings and other infrastructure) and its fragility (the vulnerability of its built environment to seismic shaking). Because so many geologic and economic variables must be considered, estimating seismic risk is a complex job. The results of the first comprehensive study of seismic risk in the United States, published by the Federal Emergency Management Agency in 2001, are presented in **Figure 13.27**.

The differences between seismic hazard and seismic risk can be appreciated by comparing the two types of national maps. For instance, although the seismic hazard levels in Alaska and California are both high, California's exposure to seismic damage is much greater, yielding a much larger total risk. California leads the nation in seismic risk, with about 75 percent of the national total; in fact, a single county, Los Angeles, accounts for 25 percent. Nonetheless, the problem is truly national: 46 million people in several metropolitan areas outside of California face substantial earthquake risks. Those areas include Hilo, Honolulu, Anchorage, Seattle, Tacoma, Portland, Salt Lake City, Reno, Las Vegas, Albuquerque, Charleston, Memphis, Atlanta, St. Louis, New York, Boston, and Philadelphia.

Not much can be done about seismic hazard because we have no way to prevent or control earthquakes. However, there are many important steps that society can take to reduce seismic risk if the hazard is properly characterized.

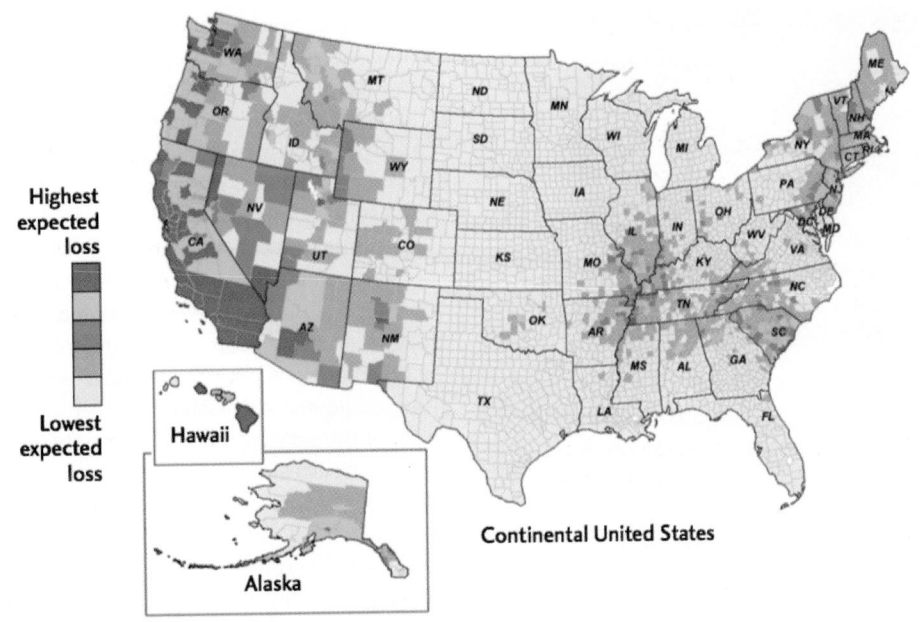

FIGURE 13.27 ■ Seismic risk map for the United States. The map shows current annualized earthquake losses on a county-by-county basis. [Federal Emergency Management Agency, Report 366, Washington, D.C., 2001.]

HAZARD CHARACTERIZATION The first step is to follow the advice of the old proverb, "Know thy enemy." We still have much to learn about the sizes and frequencies of ruptures on active faults. For example, it is only in the past couple of decades that we have come to appreciate that an earthquake in the Cascadia subduction zone, which stretches from northern California through Oregon and Washington to British Columbia, could produce a tsunami as large as the one that devastated the Indian Ocean region in 2004 and Japan in 2011. These dangers became apparent when geologists found evidence of a magnitude 9 earthquake that occurred in 1700, before any written historical accounts of the area existed. This monstrous rupture caused major ground subsidence along the Cascadia coastline and left a record of flooded, dead coastal forests. A tsunami at least 5 m high hit Japan, where historical records pin down its exact date (January 26, 1700). Geologists know that the Juan de Fuca Plate is being subducted under the North American Plate at a rate of about 40 mm/year. They have debated whether this movement occurs seismically, by continuous creep, or perhaps by silent earthquakes, but current opinion pegs the average time between magnitude 9 earthquakes in the Cascadia subduction zone at 500 to 600 years.

Although we have a good understanding of seismic hazard in some parts of the world—the United States and Japan in particular—we know much less about other regions. During the 1990s, the United Nations sponsored an effort to map seismic hazard worldwide as part of the International

FIGURE 13.28 ■ Global seismic hazard map. [K. M. Shedlock et al., *Seismological Research Letters* 71(2000): 679–686.]

FIGURE 13.29 ■ Housing tracts constructed within the San Andreas fault zone, on the San Francisco Peninsula, before the state of California passed legislation restricting this practice. The red line indicates the approximate fault trace, along which the ground ruptured and slipped about 2 m during the earthquake of 1906. [Michael Rymer.]

Decade of Natural Disaster Reduction. This effort resulted in the first global seismic hazard map, shown in **Figure 13.28**. The map is based primarily on historical earthquakes, so it may underestimate the hazard in some regions where the historical record is short. Much more needs to be done to characterize seismic hazard on a global scale.

LAND-USE POLICIES The exposure of buildings and other structures to seismic hazard can be reduced by policies that restrict land uses in high-hazard areas. This approach works well where the hazard is localized, as in the case of known faults. Erecting buildings on active faults, as was done in the case of the residential developments pictured in **Figure 13.29**, is clearly unwise, because few buildings can withstand the deformation to which they might be

subjected during a quake. In the 1971 San Fernando earthquake, a fault ruptured under a densely populated area of Los Angeles, destroying almost 100 structures. The state of California responded in 1972 with a law that restricts the construction of new buildings across an active fault. If an existing residence is on or very near a fault, the owners and real estate agents are required to disclose that information to potential buyers. A notable omission is that the act does not cover publicly owned or industrial facilities.

Siting of nuclear power plants and other critical industrial facilities to avoid seismic and tsunamic hazards would seem to be an obvious priority, but the experience in Japan shows how additional considerations, such as the need for water to cool the reactors, can lead to unwise compromises. Two nuclear facilities along the Japanese coast have been severely damaged by the earthquakes in the last few years, the Kashiwazaki-Kariwa facility in 2007 and the Fukushima-Daiichi facility in 2011 (**Figure 13.30**). The heightened public concern about the earthquake safety of nuclear power plants has led the Japanese government to shut down a number of nuclear plants, raising concerns about future power shortages.

EARTHQUAKE ENGINEERING Although land-use policies help to reduce the risk from localized hazards such as ground ruptures and soil liquefaction, they are less effective where seismic shaking is distributed across large regions. The risks from seismic shaking can best be reduced by good engineering and construction. Standards for the design and construction of new buildings are regulated by building codes enacted by state and local governments. A **building code** specifies the forces a structure must be able to withstand, based on the maximum intensity of shaking expected in the area. In the aftermath of an earthquake, engineers study buildings that were damaged and

FIGURE 13.30 ■ Aerial photo taken by a small, unmanned drone on March 24, 2011, showing the reactor containment buildings of the Fukushima-Daiichi nuclear power plant that were damaged by explosion after the Tohoku tsunami crippled the plant. [AP Photo/Air Photo Service.]

recommend modifications to building codes that could reduce future damage from similar earthquakes.

U.S. building codes have been largely successful in preventing loss of life during earthquakes. In the 20-year period from 1981 to 2012 for example, 146 people died in 11 severe earthquakes in the western United States, whereas more than 1 million people were killed by earthquakes worldwide. Nevertheless, more can be done. Damage from inevitable earthquakes could be reduce by retrofitting older, more vulnerable structures to be seismically safe, as well as by using specialized construction materials and advanced engineering methods in new construction, such as putting entire buildings on movable supports to isolate them from seismic shaking.

EMERGENCY PREPAREDNESS AND RESPONSE

Public authorities must plan ahead and be prepared with emergency supplies, rescue teams, evacuation procedures, firefighting plans, and other steps to minimize the consequences of a severe earthquake. For individuals, earthquake preparedness begins at home. Earth Issues 13.2 summarizes some of the steps you can take to protect yourself and your family from earthquakes.

Once an earthquake happens, networks of seismographs can transmit signals automatically to central processing facilities. In a fraction of a minute, computers can pinpoint the earthquake's focus, measure its magnitude, and determine its fault mechanism. If equipped with strong-motion sensors that accurately record the most violent shaking, these automated systems can also deliver accurate maps in nearly real time showing where the ground shaking was strong enough to cause significant damage. Such "ShakeMaps" can help emergency managers and other officials deploy equipment and personnel as quickly as possible to save people trapped in rubble and to reduce further economic losses from fires and other secondary hazards. Bulletins about the magnitude and area of the shaking can also be channeled through the mass media to reduce public confusion during disasters or to allay the fears aroused by minor tremors.

EARTHQUAKE EARLY WARNING SYSTEMS With

the technology just described, it is possible to detect earthquakes in the early stages of fault rupture, rapidly predict the intensity of the future ground motions, and warn people before they experience the intense shaking that that might be damaging. Earthquake early warning systems detect strong shaking near an earthquake's epicenter and transmit alerts ahead of the seismic waves. Potential warning times depend primarily on the distance between the user and the earthquake epicenter. There is a "blind zone" near an earthquake epicenter where early warning is not feasible, but at more distant sites, warnings can be issued from a few seconds up to about one minute prior to strong ground shaking.

Earthquake early warning systems have already been deployed in at least 5 countries—Japan, Romania, Taiwan,

and Turkey—and a prototype system is being developed in the United States. Japan is the only country with a nationwide system that provides public alerts. A national seismic network of nearly 1000 seismographs is used to detect earthquakes and issue warnings, which are transmitted via the Internet, satellite, cell phones, as well as automated control systems that do such things as stop trains and place sensitive equipment in a safe mode. Earthquake early warning systems are being developed in the states of California and Washington, but their full deployment has not yet been funded by the U.S. Congress or state legislatures.

TSUNAMI WARNING SYSTEMS The great tsunamis

generated by the Sumatra earthquake of 2004 and the Tohoku earthquake of 2011 illustrate the issues associated with tsunami warning. Because tsunami waves travel 10 times more slowly than seismic waves, there is enough time after a large suboceanic quake occurs, sometimes many hours, to warn people on distant shorelines of an impending disaster. Warnings broadcast by the Pacific Tsunami Warning Center, based in Hawaii, after the Tohoku earthquake allowed islands such as Hawaii and the western coastlines of the Americas to be evacuated prior to the tsunami arrival (see Figure 13.22). Unfortunately, no such system had been installed in the Indian Ocean, so the 2004 tsunami struck with essentially no warning, killing tens of thousands.

The most difficult situations arise in areas located close to active offshore faults, where tsunamis arrive so quickly that there is no time for a warning. One such place is Papua New Guinea, where in 1998, a tsunami killed as many as 3000 people in coastal villages near the epicenter of the quake that caused it. Such communities could be protected by building barrier walls to block inundation by ocean water, but this type of construction is expensive and has been tried only in Japan with mixed results (**Figure 13.31**).

FIGURE 13.31 ■ This tsunami barrier was designed to protect the town of Taro, Japan, but it was breached by the tsunami from the great Tohoku earthquake. [Carlos Barria/Reuters/Landov.]

Earth Issues

13.2 Seven Steps to Earthquake Safety

Individuals living in seismically active areas need to prepare for earthquakes and know how to respond when one strikes. Here are seven steps to earthquake safety, recommended by the Southern California Earthquake Center, that you can use to protect yourself and your family.

Before an earthquake occurs:

1. *Identify potential hazards in your home and begin to fix them.* As buildings are becoming better designed to withstand seismic shaking, more of the damage and injuries that occur are resulting from falling objects. You should secure items in your home that are heavy enough to cause damage or injury if they fall or valuable enough to be a significant loss if they break.

2. *Create your disaster plan.* With your family or housemates, plan now what you will do before, during, and after an earthquake. The plan should include safe spots you can go to during the shaking, such as under sturdy desks and tables; a safe spot outside your home where you can meet after the shaking stops; and contact phone numbers, including someone outside the area who can be called to relay information in case local communications are disrupted.

3. *Create your disaster kit.* Stock your disaster kit with essential items. Your personal kit should include medications, a first-aid kit, a whistle, sturdy shoes, high-energy snacks, a flashlight with extra batteries, and personal hygiene supplies. Your home kit should include a fire extinguisher, wrenches to turn off gas and water mains, a portable radio, drinking water, food supplies, and extra clothing.

4. *Identify your building's potential weaknesses and begin to fix them.* Consult a building inspector or contractor to identify potential safety problems. Common problems include inadequate foundations, unbraced cripple walls, weak first stories, unreinforced masonry, and vulnerable pipes.

During an earthquake:

5. *Drop, cover, and hold on.* During an earthquake or severe aftershock, drop to the floor, take cover under a sturdy desk or table, and hold on to it so that it doesn't move away from you. Wait there until the shaking stops. Stay away from danger zones, such as those near the exterior walls of buildings, near windows, and under architectural façades.

After an earthquake:

6. *After the shaking stops, check for damage and injuries needing immediate attention.* Take care of your own situation first; get to a safe location and remember your disaster plan. If you are trapped, protect your mouth, nose, and eyes from dust; signal for help using a cell phone or whistle or by knocking loudly on a solid part of the building three times every few minutes (rescuers will be listening for such knocks). Check for injuries and treat people needing assistance. Check for fires, gas leaks, damaged electrical systems, and spills. Stay away from damaged structures.

7. *When safe, follow your disaster plan.* Be in communication by turning on your radio and listening for advisories; check your phones, call your out-of-area contact to report your status, and then stay off the phone except for emergencies. Check your food and water supplies and check on your neighbors.

For further information, see *Putting Down Roots in Earthquake Country,* Southern California Earthquake Center, available online at http://www.earthquakecountry.info/roots/index.php.

Before
1 Identify potential hazards in your home and begin to fix them
2 Create your disaster plan
3 Create your disaster kit
4 Identify your building's potential weaknesses and begin to fix them

During
5 During the earthquake and aftershocks: Drop, cover, and hold on

After
6 After the shaking stops, check for damage and injuries needing immediate attention
7 When safe, follow your disaster plan

In these places, the best warning system is a very simple one: if you feel a strong earthquake, move quickly away from the coastal lowlands to higher ground!

▪ Can Earthquakes Be Predicted?

If we could predict earthquakes reliably, communities could be prepared, people could be evacuated from dangerous locations, and many aspects of the impending disaster might be averted. How well can we predict earthquakes?

Predicting an earthquake means specifying its time, location, and size. By combining plate tectonic theory with detailed geologic mapping of regional fault systems, geologists can reliably predict which faults are likely to produce earthquakes over the long term. However, specifying precisely *when* a particular fault will rupture in a large earthquake has turned out to be very difficult.

Long-Term Forecasting

Ask a seismologist to predict the time of the next large earthquake at a particular location and the response is likely to be, "The longer the time since the last big quake, the sooner the next one will be." As we have seen, the recurrence interval—the time required to accumulate the strain that will be released by fault slipping in a future earthquake—can be calculated from the rate of relative plate movement and the expected fault slip, as estimated from the displacements observed in past earthquakes. Geologists can also estimate the intervals between large earthquakes up to several thousand years in the past by finding and dating soil layers that were offset by fault displacements (**Figure 13.32**).

Although these two methods usually give similar results, the uncertainty of the predictions turns out to be large—as much as 100 percent of the recurrence interval. In Southern California, for example, the recurrence interval for the San Andreas fault is estimated to be 110 to 180 years, but the observed intervals between individual earthquakes can be appreciably shorter or longer than this average value. One part of this fault experienced a large earthquake in 1857, whereas another part (the southernmost) appears to have remained locked since a large earthquake that occurred around 1680 (see Figure 13.2). Therefore, an earthquake can be expected there at any time—tomorrow, or decades from now.

Because the prediction intervals are decades to centuries, these methods of earthquake prediction are referred to as *long-term forecasting* to distinguish them from what most people would really want: a *short-term prediction* of a large rupture on a specific fault accurate to within days or even hours of the actual event.

FIGURE 13.32 ▪ Geologist Gordon Seitz examines layers of rock and peat that have been disturbed by prehistoric earthquakes in a trench crossing the San Jacinto fault, a major strand of the San Andreas fault system in Southern California. By dating the peat layers using the carbon-14 method, geologists can reconstruct the history of large earthquakes on this fault. Such information helps scientists to forecast future events. [Courtesy of Tom Rockwell, San Diego State University.]

Short-Term Prediction

There have been a few successful short-term earthquake predictions. In 1975, an earthquake was predicted only hours before it occurred near Haicheng, in northeastern China. Chinese seismologists used what they considered to be *precursors* to make their predictions: swarms of tiny earthquakes and a rapid deformation of the ground several hours before the mainshock. Almost a million people, prepared in advance by a public education campaign, evacuated their homes and factories in the hours before the quake. Although many towns and villages were destroyed and several hundred people were killed, it appears that many were saved. The very next year, however, an unpredicted earthquake struck the Chinese city of Tangshan, killing more than 240,000 people. Obvious precursors such as those seen in Haicheng have not been repeated in subsequent large events.

Although many schemes have been proposed, we have not yet found a reliable method of predicting earthquakes minutes to weeks ahead of time. We cannot say that short-term earthquake prediction is impossible, but seismologists do not expect that it will be feasible in the near future.

We do have some useful guidelines about how the earthquake probabilities change over time. We know that earthquakes tend to cluster together in both space and

Earth Issues

13.3 Italian Scientists Convicted of Manslaughter for Miscommunication of Risk Before 2009 L'Aquila Earthquake

On April 6, 2009, a magnitude 6.3 earthquake devastated the mountain city of L'Aquila, Italy, killing 309 people, injuring more than 1500, and leaving tens of thousands homeless. In the wake of this disaster, a local prosecutor indicted the vice-director of the Italian Department of Civil Protection (DCP) and six scientific advisors from Italy's Major Risk Commission, a high-level advisory body, on charges of criminal manslaughter for statements made *before* the earthquake.

The case quickly became a cause célèbre among scientists. The indictments appeared to blame the scientists for not alerting the local population of an impending earthquake—for a "failure-to-predict." It is well known that large earthquakes cannot be accurately predicted in the short term. Why would an Italian court punish scientists for not doing something they didn't (and still don't) know how to do?

Scientific organizations from around the world sent letters of protest to the Italian president. Nonetheless, after a year-long trial, an Italian court found all seven guilty as charged; it sentenced them to six years in prison and levied fines totaling more than 10 million euros.

So what really happened in L'Aquila?

Seismic activity in this part of Italy increased in January 2009. A number of small shocks, part of a "seismic swarm," were widely felt and prompted school evacuations and other preparedness measures. In February and March, media coverage was inflamed by a series of earthquake predictions issued by a L'Aquila resident named Gioacchino Giuliani, who worked as technician in a national physics laboratory. These predictions had no official auspices and turned out to be false alarms, but they were widely

Rubble of the L'Aquila city hall after the devastating earthquake of April 6, 2009. [© Alessandro Bianchi/Reuters/Corbis.]

reported by the media and caused some people to panic and evacuate their homes.

Government scientists responded to this chaotic situation by stating that there were no accurate methods for earthquake prediction, that earthquake swarm activity was common in this part of Italy, and that the probability of substantially larger earthquakes remained small. But these assurances did not dispel public concern caused by Giuliani's continuing predictions, so the government hastily convened its Major Risk Commission in L'Aquila on March 31. The commission concluded that "there is no reason to say that a sequence of small-magnitude events can be considered a sure predictor of a strong event." This statement was scientifically correct—most seismic swarms in Italy do not lead up to a much larger earthquake—but it underplayed a fact accepted by most seismologists: the chances of larger earthquakes do increase during a swarm.

At a press conference following the meeting, the DCP vice-director, who was not a seismologist, said that "the scientific community tells us there is no danger, because there is an

time—for example, large earthquakes have nearby after-shocks—and seismologists have shown how the chances of a potentially damaging earthquake tend to go up during periods of increased seismic activity. Interpreting this type of information can be tricky, however, because, even when the seismic activity is high, accurate predictions of large earthquakes are still not possible. During seismic crises, it is easy for the public to become confused about how the hazard is changing. For example, a miscommunication of short-term earthquake probabilities before the damaging L'Aquila earthquake of April 6, 2009, led to the criminal prosecution of scientific advisors to the Italian

government on charges of manslaughter (see Earth Issues 13.3). Forecasts based on earthquakes clustering are now being deployed to help Italians assess how the seismic hazards are changing. These short-term forecasting methods are also being developed in other regions, including California.

Medium-Term Forecasting

Uncertainties in long-term forecasting can be reduced by studying the behavior of regional fault systems. One strategy is to generalize the elastic rebound theory. The ongoing discharge of energy. The situation looks favorable." This statement was *not* scientifically correct, because, even during an intense seismic swarm, small earthquakes cannot relieve the regional tectonic stress that leads to large earthquakes (see the Practicing Geology exercise at the end of this chapter).

The tremors continued into April, prompting more school evacuations. Shortly before 11 P.M. on April 5, just a few hours before the mainshock, a strong, magnitude-3.9 earthquake shook the city. In an interview in *Nature Magazine*, Vincenzo Vittorini describes how he debated with his wife and his terrified nine-year-old daughter whether to spend the rest of the night outside—a customary response to seismic activity in this part of Italy. Recalling official statements claiming that each shock diminished the potential for a major earthquake, he persuaded his family to remain in their apartment building. The building collapsed in the mainshock at 3:32 A.M., and his wife and daughter and five others were killed. Nearly everyone in L'Aquila, including the prosecutor, lost relatives or friends. Tragic testimony like Vittorini's constituted much of the prosecution's case, which charged that the Major Risk Commission had given "incomplete, imprecise, and contradictory information about the nature, causes, and future developments of the seismic hazards."

With hindsight it is clear that the Italian scientists got trapped by a simple yes-no question, "Will we be hit by a major earthquake?" From what the scientists could have known a week before the earthquake, a big shock was not very likely, probably less than a 1-in-100 chance. Even so, seismic activity had increased the probability of a large earthquake above the long-term average—large earthquakes are more likely during seismic swarms than in times of no seismic activity. Distracted by Giuliani's predictions, the authorities did not emphasize this increase in hazard, nor did they focus on advising the people of L'Aquila about preparatory measures warranted by the seismic crisis. Instead, they tried to calm the population by making reassuring statements that were widely interpreted as a firm prediction: "no major earthquake will occur."

Few scientists would argue the merit of prosecuting public servants who were trying in good faith to protect the public under chaotic circumstances. With hindsight the failure of the defendants to highlight the increased hazard may be regrettable, but the inactions of a stressed risk advisory system and misstatements by nonscientists representing that system can hardly be construed as criminal acts on the part of individual scientists. The L'Aquila verdicts are currently under appeal. One can only hope that judicial sanity will prevail.

A few weeks after the L'Aquila disaster, the government appointed an international panel of experts, which one of us (THJ) chaired, to suggest guidelines for improving earthquake forecasting procedures in Italy. Our report reaffirmed the infeasibility of high-probability earthquake prediction by any known method and addressed how short-term forecasts—in which the probabilities of large local shocks are invariably low—could be publicly utilized. Authoritative statements about what is, and is not, known about the current hazard are needed to educate the public and fill information vacuums that lead to informal predictions and misinformation. Alert protocols should be standardized to facilitate decisions at different levels of government, based in part on objective analysis of costs and benefits but also on the less tangible aspects of value-of-information, such as gains in psychological preparedness and resilience.

Our review found the Italian system wanting, but we could point to no country where operational earthquake forecasting was done much better. Regions that face high seismic risk can learn lessons from L'Aquila. Among them is the need to separate the role of science advisors, whose job is to provide objective information about natural hazards, with that of civil decision makers, who must weigh the social, economic, and political benefits of protective actions against the costs of mistakes. The L'Aquila prosecution has misconstrued these roles.

simple version of the theory depicted in Figure 13.3 describes how the tectonic stress that builds steadily on an isolated fault segment is released in a periodic sequence of fault ruptures. However, as we have seen in the case of Southern California (see Figure 13.18), faults are rarely isolated. Instead, they are connected to one another in complex networks. Thus, a rupture on one fault segment changes the stresses throughout the surrounding region (see Figure 13.4). Depending on the geometry of the fault system, these changes can either increase or decrease the likelihood of earthquakes on nearby fault segments. In other words, when and where earthquakes happen in one part of a fault system influences when and where they happen elsewhere in the system.

If Earth scientists can understand how variations in stress raise or lower the frequency of small seismic events, they might be able to predict earthquakes over time intervals as short as a few years, or maybe even a few months, although still with substantial uncertainties. Monitoring of such events on networks of seismographs could then provide a regional "stress gauge." Someday you might hear a news report that says, "The National Earthquake Prediction Evaluation Council estimates that, during the next year, there is a 50 percent probability of a magnitude 7 or larger earthquake on the southern segment of the San Andreas fault."

The ability to issue such *medium-term forecasts* would raise some difficult questions, however. How should society respond to a threat that is neither imminent nor long-term? A medium-term forecast would give the probability of an earthquake only on time scales of months to years—not precisely enough to evacuate regions that might be damaged. False alarms would be common. What effect would such predictions have on property values and other investments in the threatened region? These questions would have to be addressed by both policy makers and scientists.

SUMMARY

What is an earthquake? An earthquake is a shaking of the ground that occurs when brittle rocks being stressed by tectonic forces break suddenly along a fault. When they break, the elastic energy built up over years of slow deformation is released rapidly, and some of it is radiated as seismic waves. The focus of an earthquake is the point at which the fault first breaks; the epicenter is the point on Earth's surface directly above the focus. The foci of most continental earthquakes are shallow. In subduction zones, however, earthquakes can occur at depths as great as 690 km.

What are the three types of seismic waves? Earthquakes generate three types of seismic waves that can be recorded by seismographs. Two types of waves travel through Earth's interior: P (primary) waves, which are transmitted by all forms of matter and move fastest, and S (secondary) waves, which are transmitted only by solids and move at a little more than half the velocity of P waves. P waves are compressional waves that travel as a succession of compressions and expansions. S waves are shear waves that displace material at right angles to their path of travel. Surface waves are confined to Earth's surface and outer layers. They travel slightly more slowly than S waves.

What is earthquake magnitude and how is it measured? Earthquake magnitude is a measure of the size of an earthquake. Richter magnitude is proportional to the logarithm of the amplitude of the largest ground movement recorded by seismographs. Seismologists now prefer to use moment magnitude because it is derived from the physical properties of the faulting that causes the earthquake: the area of faulting and the average fault slip.

How frequently do earthquakes occur? About 1,000,000 earthquakes with magnitudes greater than 2 take place each year. This number decreases by a factor of 10 for each magnitude unit. Hence, there are about 100,000 earthquakes with magnitudes greater than 3, about 1000 with magnitudes greater than 5, and about 10 with magnitudes greater than 7. The largest earthquakes, with magnitudes of 9 to 9.5, are rare and are confined to thrust faults in subduction zones.

What governs the type of faulting that occurs in an earthquake? The fault mechanism of an earthquake is determined by the type of plate boundary at which it occurs. Normal faulting, caused by tensional forces, occurs at divergent boundaries. Strike-slip faulting, caused by shearing forces, occurs along transform-fault boundaries. The largest earthquakes, caused by compressive forces, occur on megathrusts at convergent boundaries. A small number of earthquakes occur far from plate boundaries, mostly on continents.

What are the hazards of earthquakes? Faulting and ground shaking during an earthquake can damage or destroy buildings and other infrastructure. They can also trigger secondary hazards, such as landslides and fires. Earthquakes on the seafloor can trigger tsunamis, which may cause widespread destruction when they reach shallow coastal waters.

What can be done to reduce the risk of earthquakes? Land-use regulations can restrict new building near active fault zones, and construction in high-hazard areas can be regulated by building codes so that buildings and other structures will be strong enough to withstand the expected intensity of seismic shaking. Systems using networks of seismographs and other sensors are being developed to provide early warnings of earthquakes and tsunamis. Public authorities can plan ahead, be prepared, and put early warning systems in place. People living in earthquake-prone areas can be informed about how to prepare and what to do when an earthquake occurs.

Can scientists predict earthquakes? Scientists can characterize the level of seismic hazard in a region, but they cannot consistently predict earthquakes with the accuracy that would be needed to alert a population hours to weeks in advance. The best hope of making such predictions in the future may lie in a better understanding of how variations in stress raise or lower the frequency of seismic events in a regional fault system.

KEY TERMS AND CONCEPTS

aftershock (p. 351)

building code (p. 372)

earthquake (p. 348)

elastic rebound theory
(p. 349)

epicenter (p. 351)

fault mechanism (p. 361)

fault slip (p. 349)

focus (p. 351)

foreshock (p. 353)

intensity scale (p. 360)

magnitude scale (p. 357)

P wave (p. 355)

recurrence interval (p. 349)

S wave (p. 357)

seismic hazard (p. 370)

seismic risk (p. 370)

seismograph (p. 353)

surface wave (p. 357)

tsunami (p. 367)

PRACTICING GEOLOGY EXERCISE

Can Earthquakes Be Controlled?

Earthquakes of magnitude 4 rarely result in much damage to nearby communities, whereas quakes of magnitude 8 can be incredibly destructive. Would it somehow be possible for humans to control the slip on a fault to keep earthquakes small?

Experiments in oil and gas fields have shown that small earthquakes can be caused by injecting water or other fluids into fault zones through deep drill holes. The fluid lubricates the fault, reducing the friction that keeps it from slipping. Pump and pop! You get an earthquake. Why not

TYPICAL AREA OF FAULT RUPTURE

TYPICAL SLIP DISTANCE OF FAULT RUPTURE

The top panel shows how the rupture area of the San Andreas fault increases with earthquake magnitude. The lower panel shows how the slip distance of the fault rupture increases with magnitude.

control the sizes of earthquakes by using this fluid injection technique to release energy on a fault only in ruptures smaller than magnitude 4?

The feasibility of this method depends on how many events of magnitude 4 would produce the same fault slip over the same area as one event of magnitude 8. From observations of many earthquakes, seismologists have discovered two simple rules about moment magnitude that can guide this calculation:

1. *Area rule:* The area of faulting increases by a factor of 10 for each unit of moment magnitude. Therefore, a magnitude 8 rupture has 10,000 times the area of a magnitude 4 rupture (because $10^{(8-4)} = 10^4$).

2. *Slip rule:* The average slip of a fault rupture increases by a factor of 10 for each two units of moment magnitude. Therefore, the slip of a magnitude 8 rupture is 100 times that of a magnitude 4 rupture (because $10^{(8-4)/2} = 10^2$).

The area of a magnitude 8 rupture is typically about 10,000 km², and the average slip is about 5 meters per event.

▪ The area rule implies that the area of a magnitude 4 rupture will be 10,000 times smaller than that of a magnitude 8 rupture, or 1 km².

▪ The slip rule implies that the slip of a magnitude 4 rupture will be 100 times smaller than the slip of a magnitude 8 rupture, or 0.05 m (5 cm).

Therefore, the number of magnitude 4 events needed to equal a single magnitude 8 event is

$$10,000 \times 100 = 1,000,000$$

This calculation shows that small earthquakes don't add up to much of the displacement that occurs across a fault; the big ones are what really count. On a fault like the San Andreas, which has earthquakes of nearly magnitude 8 every 100 years or so, we would have to generate magnitude 4 earthquakes at a rate of almost 10,000 per year to take up the same amount of fault movement.

Injecting faults with fluids to increase the rate of small earthquakes would be a bad idea for at least two reasons. It would be prohibitively expensive: drilling thousands of holes along the fault and pumping all that water down to earthquake focal depths would cost billions of dollars. It would also be dangerous: one of the ruptures induced by the fluid injection could become a much larger earthquake than intended. An effort to control earthquakes could end up causing a big one!

BONUS PROBLEM: How many magnitude 4 earthquakes would provide the same slip over the same area as one magnitude 6 earthquake?

EXERCISES

1. Seismographic stations report the following S-wave–P-wave arrival time intervals for an earthquake: Dallas, S-P = 3 minutes; Los Angeles, S-P = 2 minutes; San Francisco, S-P = 2 minutes. Use a map of the United States and the travel-time curves in Figure 13.9 to obtain a rough location for the epicenter of the earthquake.

2. Describe two scales for measuring the size of an earthquake. Which is the more appropriate scale for measuring the amount of faulting that caused the earthquake? Which is more appropriate for measuring the amount of shaking experienced by a particular observer?

3. How much more energy is released by a magnitude 7.5 earthquake than by a magnitude 6.5 earthquake?

4. In Southern California, a magnitude 5 earthquake occurs about once per year. Approximately how many magnitude 4 earthquakes would you expect each year? How many magnitude 2 earthquakes?

5. What are the fault mechanisms of earthquakes at the three types of plate boundaries?

6. Destructive earthquakes occasionally occur within lithospheric plates, far from plate boundaries. Why?

7. At a location along the boundary between the Nazca Plate and the South American Plate, the relative plate movement is 80 mm/year. The last large earthquake there, in 1880, showed a fault slip of 12 m. Should local residents begin to worry about another large earthquake?

THOUGHT QUESTIONS

1. The belts of shallow-focus earthquakes shown by the blue dots in Figure 13.16 are wider and more diffuse on the continents than in the oceans. Why? (*Hint:* You might want to review Chapter 7.)

2. Why are earthquakes with focal depths greater than 20 km infrequent in continental lithosphere?

3. Why do the largest earthquakes occur on megathrusts at subduction zones and not, say, on continental strike-slip faults?

4. How big would a fault have to be to produce a magnitude-10 earthquake? Do you think such a large earthquake could occur by the rupturing of subduction-zone megathrusts?

5. In Figure 13.3, the right-lateral fault offsets the fence line to the right. In Figure 13.17, the mid-ocean ridge crest is also offset to the right. Why, then, is the transform fault in Figure 13.17 left-lateral?

6. Would you support legislation to prevent landowners from building structures close to active faults?

7. Taking into account the possibility of false alarms, mass hysteria, economic depression, and other possible negative consequences of earthquake prediction, do you think the objective of predicting earthquakes should have a high priority?

MEDIA SUPPORT

 13-1 Animation: Earthquakes and Plate Boundaries

 13-1 Video: The San Andreas Fault

 13-2 Animation: Tsunamis

- Exploring Earth's Interior with Seismic Waves   384
- Layering and Composition of Earth's Interior   388
- Earth's Internal Temperature   390
- Visualizing Earth's Three-Dimensional Structure   394
- Earth's Magnetic Field and the Geodynamo   396

Seismic waves can be used to map features generated by dynamic processes in Earth's interior. This image shows variations in shear wave speed on cross sections through the mantle and on the surface of the inner core. Yellow lines on the surface of the globe are the plate boundaries. [Courtesy of J. H. Woodhouse, Oxford University.]

EXPLORING EARTH'S INTERIOR

14

HUMANS HAVE BURROWED in mines to depths as great as 4 km to extract gold and other minerals, and they have drilled down to more than 10 km in search of petroleum. But these efforts, heroic though they are, have barely scratched the surface of our massive planet. The crushing pressures and red-hot temperatures of Earth's deeper layers make the planet's interior inaccessible to us for the foreseeable future. Nevertheless, we can learn much about the structure and composition of Earth's interior from our position on its surface.

Some of the best information comes from seismology. Chapter 13 described the terrible shaking and destruction that can be wrought by seismic waves. Yet this same seismic energy can be harnessed to illuminate Earth's deepest regions, allowing us to construct three-dimensional images of geologic features in the lower crust, the rising and falling of convection currents in the mantle, and even the workings of the outer and inner core. Our understanding of Earth's interior has been further enriched by material erupted from volcanoes, by the behavior of Earth materials under high temperatures and pressures in the laboratory, and by the information contained in Earth's gravitational and magnetic fields.

In this chapter, we will explore Earth's interior down to its center, nearly 6400 km beneath our feet. We will see how seismic waves have been used to image the structure of Earth's crust, mantle, and core. We will investigate temperatures deep inside Earth and the machinery of the two great geosystems driven by its internal heat engine: the plate tectonic system, which is driven by convection in the mantle, and the geodynamo in the outer core, which generates Earth's magnetic field.

■ Exploring Earth's Interior with Seismic Waves

Different types of waves—light, sound, and seismic—have a common characteristic: the velocity at which they travel depends on the material through which they are passing. Light waves travel fastest through a vacuum, more slowly through air, and even more slowly through water. Sound waves, on the other hand, travel faster through water than through air and not at all through a vacuum. Why?

Sound waves are simply propagating variations in pressure. Without something to compress, such as air or water, they cannot exist. The more force it takes to compress a material, the faster sound will travel through it. The speed of sound in air—Mach 1, in the jargon of jet pilots—is typically 0.34 km/s, or about 760 miles per hour, at Earth's surface. Water resists compression much more than air, so the speed of sound waves in water is correspondingly higher, about 1.5 km/s. Solid materials are even more resistant to compression, so sound waves travel through them at even higher speeds. Sound travels through granite at about 6 km/s—nearly 13,500 miles per hour!

Basic Types of Waves

As we saw in Chapter 13, some of the seismic waves created by earthquakes are **compressional waves** (like sound waves), which travel with a push-pull motion, while others are **shear waves,** which travel with a side-to-side motion, displacing material at right angles to their path of travel (see Figure 13.10). Solids are more resistant to compression than to shearing, so compressional waves always travel through solids faster than shear waves do. This physical principle explains a relationship we discussed in Chapter 13: compressional waves are always the first arrivals at a seismographic station (and hence are called primary, or P waves), and shear waves are the secondary arrivals (S waves). It also explains why the speed of shear waves in gases and liquids is zero: those materials have no resistance to shearing. Shear waves cannot propagate through any fluid—air, water, or the liquid iron in Earth's outer core.

Geologists can calculate the speed of a P or S wave by dividing the distance traveled by the travel time. These wave speeds can then be used to infer which materials the waves encountered along their paths.

The concepts of travel times and wave paths sound simple enough, but complications arise when waves pass through more than one type of material. At the boundary between two different materials, some of the waves bounce off—that is, they are *reflected*—and others are transmitted into the second material, just as light is partly reflected and partly transmitted when it strikes a windowpane. The waves that cross the boundary between two materials are bent, or *refracted,* as their velocity changes from that in the

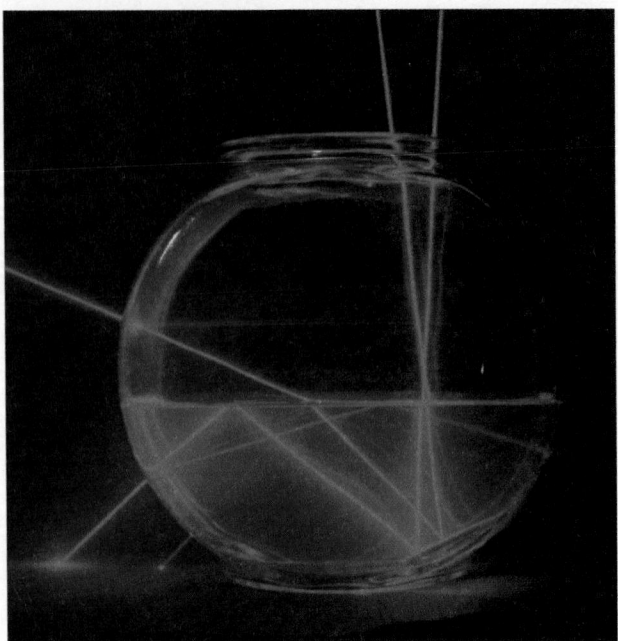

FIGURE 14.1 ■ In this experiment, two beams of laser light enter a bowl of water from the top at slightly different angles. Both beams are reflected from a mirror on the bottom of the bowl. One is then reflected at the water-air boundary and passes through the bowl to make a bright spot on the table. Most of the light in the other beam is bent (refracted) as it passes from the water to the air, although a small amount is reflected to form a second spot on the table. [Susan Schwartzenberg/The Exploratorium.]

first material to that in the second. **Figure 14.1** shows a laser light beam whose path bends as it goes from water into air, much as a P or an S wave bends as it travels from one material to another. By studying the speeds at which seismic waves travel and how they are refracted and reflected at Earth's internal boundaries, seismologists have been able to model the layering of Earth's crust, mantle, and core with great precision (see Figure 1.12).

Paths of Seismic Waves Through Earth

If Earth were made of a single material with constant properties from the surface to the center, P and S waves would travel from the focus of an earthquake to a distant seismographic station along straight lines through Earth's interior, just as the Sun's rays travel in straight lines through outer space. When the first global networks of seismographs were installed about a century ago, however, seismologists discovered that **seismic ray paths** are not straight lines; the waves are refracted and reflected by Earth's layered structure.

WAVES REFRACTED THROUGH EARTH'S INTERIOR

From their observations of travel times and the amount of upward refraction of the ray paths, seismologists were able to conclude that P waves travel much faster through rock deep within Earth than they do through rock at Earth's surface. This was hardly surprising, because rock subjected to the great pressures in Earth's interior would be squeezed into tighter crystal structures. The atoms in these tighter structures would be more resistant to further compression, which would cause P waves to travel through them more quickly.

Seismologists were very surprised, however, by what they found at progressively greater distances from an earthquake focus (**Figure 14.2**). After the P and S waves had traveled beyond about 11,600 km, they suddenly disappeared! Like airplane pilots and ship captains, seismologists prefer to measure distances traveled over Earth's surface in angular degrees—from 0° at the earthquake focus to 180° at a point on the opposite side of Earth. Each degree measures 111 km at the surface, so 11,600 km corresponds to an angular distance of 105°, as shown in Figure 14.2. When they looked at seismograms recorded beyond 105° from the

focus, they did not see the distinct P- and S-wave arrivals that were so clear on seismograms recorded at shorter distances. Beyond about 15,800 km from the focus (142°), the P waves suddenly reappeared, although they were much delayed compared with their expected travel times. The S waves never reappeared.

In 1906, the British seismologist R. D. Oldham put these observations together to provide the first evidence that Earth has a liquid outer core. S waves cannot travel through the outer core, he argued, because it is liquid, and liquids have no resistance to shearing. Thus, there is an S-wave **shadow zone** beyond 105° from the earthquake focus, where S-wave ray paths encounter the core-mantle boundary (see Figure 14.2b). The propagation of P waves is more complicated (see Figure 14.2a). At 105°, their ray paths also encounter the core-mantle boundary. At that boundary, P-wave velocity drops by almost a factor of two. Therefore, the waves are refracted downward into the core and emerge at greater angular distances after the delay caused by their detour through the core. This refraction effect forms a P-wave shadow zone at angular distances between 105° and 142°.

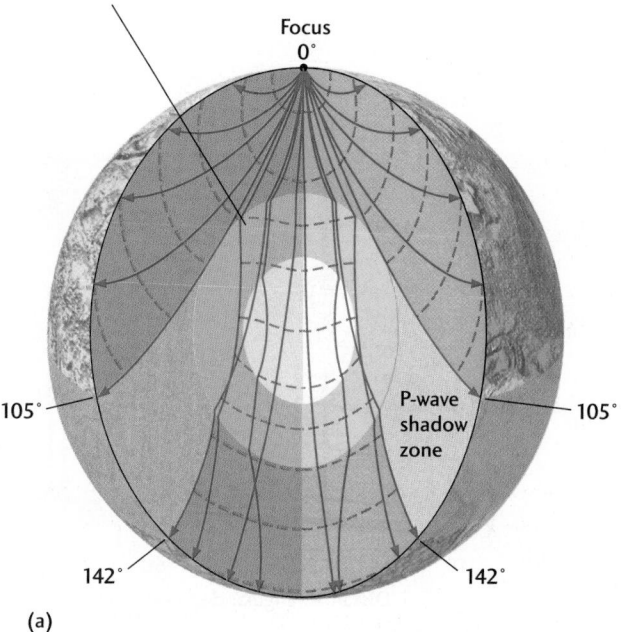

P waves cannot reach the surface within the shadow zone because of the way they are refracted when they enter and leave the core.

(a)

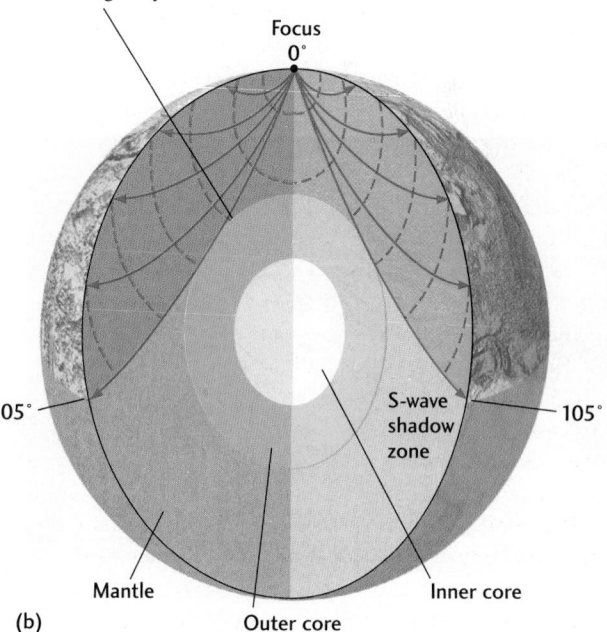

Although S waves reach the core, they cannot travel through its liquid outer region, and therefore never emerge beyond 105° from the focus.

(b)

FIGURE 14.2 ■ Earth's core creates P-wave and S-wave shadow zones. The ray paths of the seismic waves from an earthquake focus through Earth's interior are shown by solid lines (blue for P-waves, green for S-waves). The dashed lines show the progress of the waves at 2-minute intervals. Distances are measured in angular degrees from the earthquake focus. (a) The P wave shadow zone extends from 105° to 142°. (b) The larger S-wave shadow zone extends from 105° to 180°.

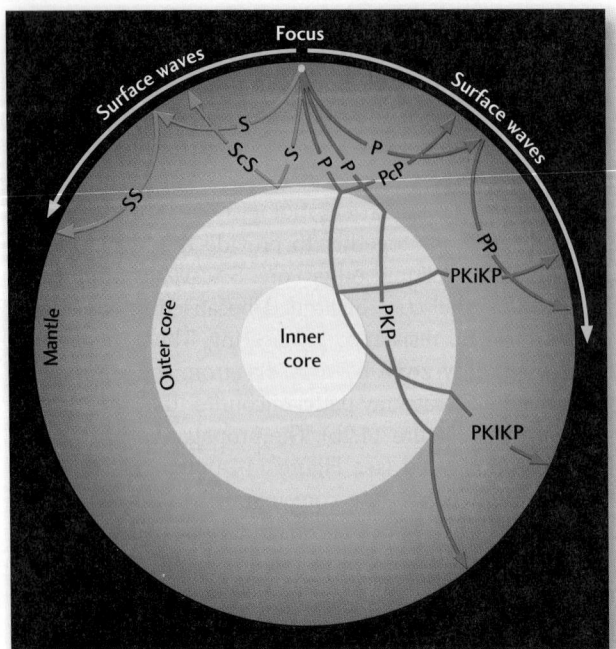

FIGURE 14.3 ■ Seismologists use a simple labeling scheme to describe the various ray paths taken by seismic waves. PcP and ScS are compressional and shear waves, respectively, that are reflected by the core. PP and SS waves are internally reflected from Earth's surface. A PKP wave travels through the liquid outer core, a PKIKP wave travels through the solid inner core, and a PKiKP wave is reflected by the inner core. Surface waves propagate along Earth's outer surface, like waves on the surface of a pond.

WAVES REFLECTED BY EARTH'S INTERNAL BOUNDARIES When seismologists looked at records of seismic waves made at angular distances of less than 105° from an earthquake focus, they found waves that must have been reflected from the core-mantle boundary. They labeled a compressional wave reflected from the top of the outer core PcP and a shear wave ScS. (The lower-case *c* indicates a core reflection.) In 1914, a German seismologist, Beno Gutenberg, used the travel times of these reflected waves to determine the depth of the core-mantle boundary, which modern estimates put at about 2890 km. **Figure 14.3** shows examples of the ray paths taken by these core-reflected waves.

Figure 14.3 also shows the ray paths of some other prominent wave types seen on seismograms, along with the labels seismologists have attached to them. For example, a compressional wave reflected once at Earth's surface is labeled PP, and a shear wave with a similar path is labeled SS. **Figure 14.4** shows the ray paths of these wave types and their internal reflections on seismograms recorded at different angular distances from an earthquake focus.

The ray path of a compressional wave through the outer core is labeled with a *K* (from the German word for "core"), so PKP designates a compressional wave that propagates through the crust and mantle, into the outer core, and back through the mantle and crust to a seismograph at Earth's surface. In 1936, Danish seismologist Inge Lehmann (**Figure 14.5**) discovered Earth's inner core by observing compressional waves refracted by its outer boundary, which she

FIGURE 14.4 ■ (a) P and S waves are refracted upward in the mantle and also can be reflected from Earth's surface. A seismic wave that has been reflected once from Earth's surface is labeled with a double letter (PP or SS). (b) Seismograms recorded at various distances from an earthquake focus in the Aleutian Islands, Alaska. The colored lines identify the arrival times of the P and S waves, the surface waves, and the PP and SS waves reflected from Earth's surface.

determined to be at a depth of about 5150 km. Ray paths through the inner core are labeled with an *I*, so the waves Lehmann observed are labeled PKIKP. Other seismologists have since observed compressional waves (labeled PKiKP) reflected from the top side of the inner core–outer core boundary (the lowercase *i* indicates a reflection rather than a refraction; see Figure 14.3).

Seismic Exploration of Near-Surface Layering

Seismic waves can also be used to probe the shallow parts of Earth's crust. This technique, called *seismic profiling,* has a number of practical applications. Seismic waves generated by artificial sources, such as dynamite explosions on land and compressed-air explosions at sea, are reflected by geologic structures at shallow depths in the crust (**Figure 14.6**). Recording of these reflections has proved to be the most successful method for finding deeply buried oil and gas reservoirs. This type of seismic exploration is now a multibillion-dollar industry. Reflected seismic waves are also used to measure the depth of water tables and the

FIGURE 14.5 ■ Danish seismologist Inge Lehmann discovered Earth's inner core in 1936. [SPL/Science Source.]

(a)

(b)

FIGURE 14.6 ■ (a) The *Geco Topaz,* a vessel operated by WesternGeco Inc., towing hydrophones, conducting a three-dimensional seismic survey. The bubbles behind the ship are compressed-air explosions that send out compressional waves; the reflections of those waves from the rocks below are recorded by seismographs pulled on cables behind the ship to produce an image of the subsurface structure. (b) A three-dimensional image produced by a seismic survey. The colors represent layers of sediment beneath the seafloor, some of which may trap oil and natural gas. [(a) © John Lawrence Photography/Alamy; (b) courtesy of BP.]

thickness of glaciers. At sea, compressional waves can be generated by mechanical devices similar to loudspeakers, and oceanographic ships routinely use the underwater sound they produce to measure the depth of the ocean and the thickness of sediments on the seafloor.

■ Layering and Composition of Earth's Interior

By measuring the travel times of compressional and shear waves from earthquakes around the world, geologists have developed a detailed model of how the wave speeds change with depth from Earth's surface to its center. We will explore this Earth model, which is shown in **Figure 14.7**, by taking an imaginary downward journey through Earth's interior, from its outer crust to its inner core.

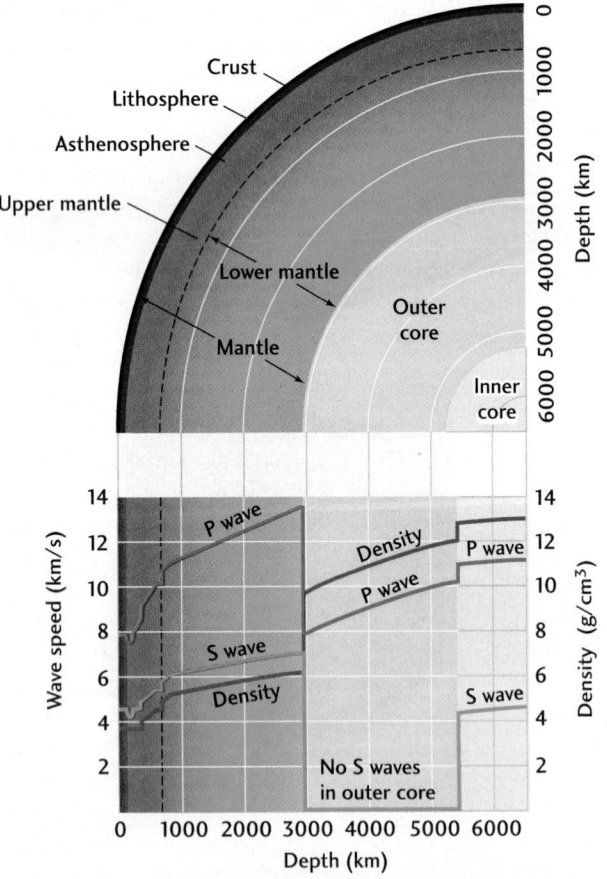

FIGURE 14.7 ■ Earth's layering as revealed by seismology. The lower diagram shows changes in P-wave and S-wave velocities and rock densities with depth. The upper diagram is a cross section through Earth on the same depth scale, showing how those changes are related to the major layers (see also Figure 1.12).

The Crust

By measuring the velocities of seismic waves passing through samples of various materials in the laboratory, seismologists have compiled a library of seismic wave velocities through different rock types. Rough values for P-wave velocities in igneous rocks, for example, are as follows:

■ Felsic rocks typical of the upper continental crust (granite): 6 km/s

■ Mafic rocks typical of oceanic crust or the lower continental crust (gabbro): 7 km/s

■ Ultramafic rocks typical of the upper mantle (peridotite): 8 km/s

These velocities vary because they depend on a rock's density and its resistance to compression and shear, which depend on chemical composition and crystal structure. In general, higher densities correspond to higher P-wave velocities; typical densities for granite, gabbro, and peridotite are 2.6 g/cm³, 2.9 g/cm³, and 3.3 g/cm³, respectively.

We know from measurements of P-wave velocities that the upper part of the continental crust is made mostly of low-density granitic rocks. The measurements also show that no granite exists on the deep seafloor; the crust there consists entirely of basalt and gabbro overlain by sediments. The velocity of P waves increases abruptly to 8 km/s at the **Mohorovičić discontinuity,** or *Moho,* which marks the base of the crust (see Chapter 1). That velocity indicates that the mantle below the Moho is made primarily of dense peridotite.

Seismic data show that Earth's crust is thin (about 7 km) under the oceans, thicker (about 33 km) under the stable, flat-lying continents, and thickest (as much as 70 km) under the high mountains of orogenic zones. The elevations of continents relative to the deep seafloor can be explained by the principle of **isostasy,** which states that the total weight of lithospheric columns should be same for continents and oceans (see the Practicing Geology exercise at the end of the chapter).

The Mantle

The **upper mantle,** which extends from the Moho to 410 km, is made primarily of peridotite, a dense ultramafic rock composed primarily of olivine and pyroxene. These minerals contain less silica and more magnesium and iron than those in typical crustal rocks (see Chapter 4). S-wave velocities have been used to explore the layering of the mantle (**Figure 14.8**). The layering of the upper mantle is caused primarily by the effects of increasing temperature and pressure on peridotite. Olivine and pyroxene undergo partial melting under the conditions found in the uppermost part of the mantle. At greater depths, increasing pressure forces the atoms of these minerals closer together into more compact crystal structures.

The mantle just below the Moho is relatively cold. Like the crust, it is part of the lithosphere, the rigid layer that

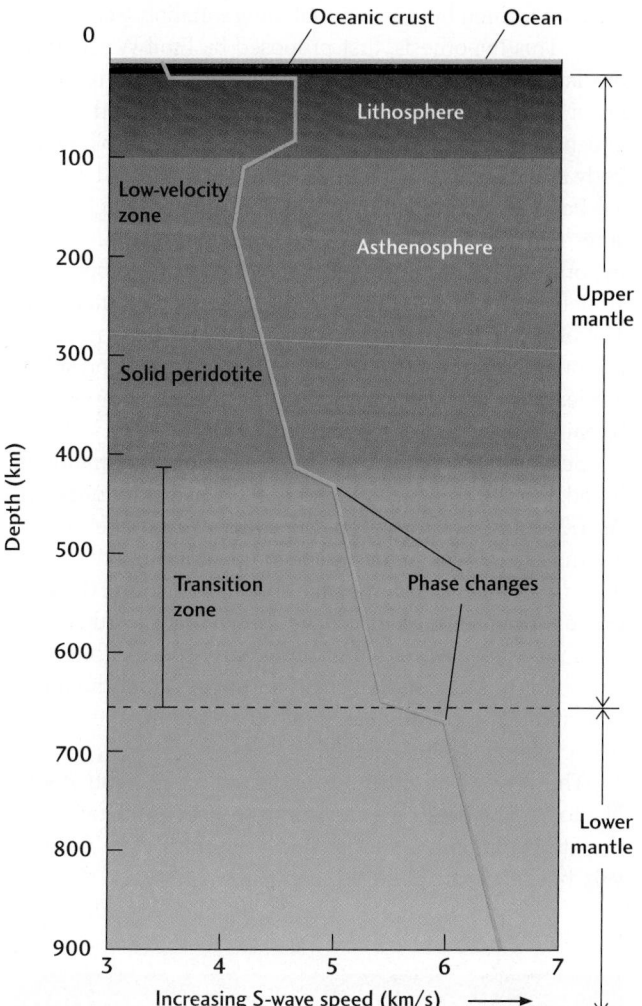

FIGURE 14.8 ■ The structure of the mantle beneath old oceanic lithosphere, showing S-wave velocities to a depth of 900 km. Changes in S-wave velocity mark the strong, brittle lithosphere, the weak, ductile asthenosphere, and a transition zone, in which increasing pressure forces rearrangements of atoms into denser and more compact crystal structures (phase changes).

forms the plates (see Chapter 1). On average, the thickness of the lithosphere is about 100 km, but it is highly variable geographically, ranging from almost no thickness near spreading centers, where new oceanic lithosphere is forming from hot, rising mantle material, to over 200 km beneath the cold, stable continental cratons.

Near the base of the lithosphere, S-wave velocity abruptly decreases, marking the start of a **low-velocity zone.** At about 100 km, the temperature approaches the melting temperature of peridotite, partially melting some of its minerals. Although the amount of melting is small (in most places, less than 1 percent), it is sufficient to decrease the rigidity of the rock, which slows the S waves passing through it. Because partial melting also allows the rock to flow more easily, geologists identify the low-velocity zone with the top part of the asthenosphere: the weak, ductile

layer on which the rigid lithospheric plates slide. This idea fits nicely with evidence that the asthenosphere is the source of most basaltic magmas (see Chapters 4 and 12).

The base of the low-velocity zone lies at about 200 to 250 km below oceanic crust, where S-wave velocity increases to a value consistent with solid peridotite. The low-velocity zone is not as well defined under stable continental cratons, where colder lithospheric mantle extends to these depths.

At depths of about 200 to 400 km, the S-wave velocity again increases with depth. Within this zone, pressure continues to increase with depth, but the temperature does not rise as rapidly as it does near the surface, due to the effects of convection within the asthenosphere. (We will discuss why this is so in the next section.) The combined effects of pressure and temperature decrease the amount of melting with depth and cause rock rigidity—and thus S-wave velocity—to rise.

About 400 km below the surface, S-wave velocity increases by about 10 percent within a narrow zone less than 20 km thick. This jump in S-wave velocity can be explained by a **phase change** in olivine, the major mineral constituent of the upper mantle, whose ordinary crystal structure is transformed into a denser, more closely packed structure at high pressures. When olivine is subjected to high pressures in the laboratory, the atoms that form its crystal structure shift into a more compact arrangement at the temperatures and pressures corresponding to depths of about 410 km. Moreover, the jumps in the P- and S-wave velocities measured in the laboratory match the increase observed for seismic waves at this depth.

Below 410 km, mantle properties change slowly as depth increases, but at a depth of about 660 km, the S-wave velocity abruptly increases again, indicating a second major phase change in olivine to an even more closely packed crystal structure. Laboratory experiments have confirmed the existence of another major mineralogical phase change at the pressures and temperatures found at this depth.

Because it contains at least two major phase changes, the layer between 410 km and 670 km in depth is called the **transition zone.** Phase changes are transitions in a rock's mineralogy, but not in its chemical composition. Some geologists have argued, however, that the increase in seismic wave velocities at 660 km comes in part from a change in the chemical composition of mantle rock. This debate was critical to understanding the plate tectonic system, because a chemical change would imply that the convection that drives plate tectonics does not penetrate much beyond this depth—in other words, that convection in the mantle is stratified (as shown in Figure 2.18b). Evidence from detailed studies of mantle structure now indicates very little, if any, chemical change in this region of the mantle.

Below the transition zone at 660 km, seismic wave velocities increase gradually and do not show any more unusual features until close to the core-mantle boundary. This relatively homogeneous region, more than 2000 km thick, is called the **lower mantle.** The lower mantle is convecting and exchanges mass with the upper mantle, driven in part by slabs of oceanic lithosphere subducting through the upper mantle into the lower mantle.

The Core-Mantle Boundary

At the **core-mantle boundary,** about 2890 km below the surface, we encounter the most extreme change in properties found anywhere in Earth's interior. From the way seismic waves are reflected by this boundary, seismologists can tell that it is a very sharp interface. The material changes abruptly from solid silicate rock to a liquid iron alloy. Because of the complete loss of rigidity, the S-wave velocity drops from about 7.5 km/s to zero, and the P-wave velocity drops from more than 13 km/s to about 8 km/s, resulting in the core shadow zones. Density, on the other hand, increases by about 4 g/cm^3 (see Figure 14.7). This large density difference, which is even greater than the increase in density from atmosphere to lithosphere at Earth's solid surface, keeps the core-mantle boundary very flat (you could probably skateboard on it!) and prevents any large-scale mixing of the mantle and core.

The core-mantle boundary appears to be a very active place. Heat conducted out of the core increases the temperatures at the base of the mantle by as much as 1000°C (see Figure 14.10). Indeed, the paths of seismic waves that pass near the base of the mantle show peculiar complications, suggesting a region of exceptional geologic activity. In a thin layer above the core-mantle boundary, there is a steep (10 percent or more) decrease in seismic wave velocities, which may be an indication that the mantle in contact with the core is partially molten, at least in some places. As we noted in Chapter 12, some geologists believe this hot region to be the source of mantle plumes that rise all the way to Earth's surface, creating volcanic hot spots such as Hawaii and Yellowstone.

The lowest boundary layer of the mantle, a region about 300 km thick, may be the ultimate graveyard of some subducted lithospheric material, such as the dense, iron-rich parts of the oceanic crust. It is possible that this zone experiences an upside-down version of the tectonics we see at Earth's surface. For example, accumulations of heavy, iron-rich material might form chemically distinct "anticontinents" that are constantly pushed to and fro across the core-mantle boundary by convection currents. We still have a lot to learn about the geologic processes that might be active in this strange place.

The Core

Many lines of evidence support the hypothesis that Earth's core is made of iron and nickel. These metals are abundant in the universe (see Chapter 1); in addition, they are dense enough to explain the mass of the core (about one-third of Earth's total mass) and to be consistent with the theory that the core formed by gravitational differentiation (see Chapter 9). This hypothesis, first proposed by Emil Wiechert in the late nineteenth century, was buttressed by discoveries of meteorites made almost entirely of iron and nickel, which presumably came from the breakup of a planetary body that also had an iron-nickel core (see Figure 1.10).

Laboratory measurements at appropriately high pressures and temperatures have led to a slight revision of this hypothesis. A pure iron-nickel alloy turns out to be about 10 percent too dense to match the data for the outer core. Therefore, it has been proposed that the core includes minor amounts of some lighter element. Oxygen and sulfur are leading candidates, although the precise composition remains the subject of research and debate.

Seismology tells us that the core below the mantle is liquid, but the core is not liquid to the very center of Earth. As Lehmann first discovered, P waves that penetrate to depths of 5150 km suddenly speed up, indicating the presence of an *inner core*, a metallic sphere two-thirds the size of the Moon. Seismologists have shown that the inner core transmits shear waves, confirming early speculations that it is solid. In fact, some calculations suggest that the inner core spins at a slightly faster rate than the mantle, acting like a "planet within a planet."

The very center of the planet is not a place you would like to be. The pressures are immense, over 4 million times the atmospheric pressure at Earth's surface. And it's also very hot, as we are about to see.

■ Earth's Internal Temperature

The evidence of Earth's internal heat is everywhere: volcanoes, hot springs, and the elevated temperatures measured in mines and boreholes. Earth's internal heat fuels convection in the mantle, which drives the plate tectonic system, as well as the geodynamo in the core, which produces Earth's magnetic field.

Earth's internal heat engine is powered by several sources. During the planet's violent origin, kinetic energy released by impacts with planetesimals heated its outer regions, while gravitational energy released by differentiation of the core heated its deep interior (see Chapter 9). The decay of radioactive isotopes in Earth's interior continues to generate heat.

After Earth formed, it began to cool, and it is cooling to this day as heat flows from the hot interior to the cool surface. The temperatures in the planet's interior result from a balance between heat gained and heat lost.

Heat Flow Through Earth's Interior

Earth cools in two main ways: through the slow transport of heat by conduction and through the more rapid transport of heat by convection. Conduction dominates in the lithosphere, whereas convection is more important throughout most of Earth's interior.

CONDUCTION THROUGH THE LITHOSPHERE

Heat energy exists in a material as vibrations of atoms; the higher the temperature, the more intense the vibrations. The **conduction** of heat occurs when thermally agitated atoms and molecules jostle one another, mechanically transferring kinetic energy from a hot region to a cool one. Heat is transferred from regions of high temperature to regions of low temperature by this process.

Materials vary in their ability to conduct heat. Metal is a better conductor than plastic (think of how rapidly a metal handle on a frying pan heats up compared with one made of plastic). Rock and soil are very poor heat conductors, which is why underground pipes are less susceptible to freezing than those above ground. Rock conducts heat so poorly that a lava flow 100 m thick takes about 300 years to cool from 1000°C to ground surface temperatures. Moreover, the cooling time of a layer increases with the square of its thickness, so a lava flow twice as thick (200 m) would take four times as long to cool (about 1200 years).

The conduction of heat through the outer surface of the lithosphere causes the lithosphere to cool slowly over time. As it cools, its thickness increases, just as the cold crust on a bowl of hot wax thickens over time. Rock, like wax, contracts and becomes denser with decreasing temperature, so the average density of the lithosphere must increase over time, and therefore, according to the principle of isostasy, its surface must sink to lower levels. Thus, the mid-ocean ridges stand high because the lithosphere there is young, hot, and thin, whereas the abyssal plains are deep because the lithosphere there is old, cold, thick, and dense.

From these principles, geologists have constructed a simple but precise theory of seafloor topography that uses conductive cooling to explain the large-scale features of ocean basins. The theory predicts that ocean depth should depend primarily on the age of the seafloor. Because the cooling depth goes as the square-root of cooling time, ocean depth should increase as the square-root of seafloor age. In other words, seafloor that is 40 million years old should have subsided twice as much as seafloor that is only 10 million years old (because $\sqrt{40/10} = \sqrt{4} = 2$). This simple mathematical relationship matches seafloor topography near the mid-ocean ridge crests amazingly well, as demonstrated in **Figure 14.9**.

Conductive cooling of the lithosphere accounts for a wide variety of other geologic phenomena, including the subsidence of passive continental margins and thermal subsidence basins (see Chapter 5). It explains why the amount of heat flowing out of oceanic lithosphere is high near spreading centers and decreases as the oceanic lithosphere gets older, and it tells us why the average thickness of the oceanic lithosphere is about 100 km. The

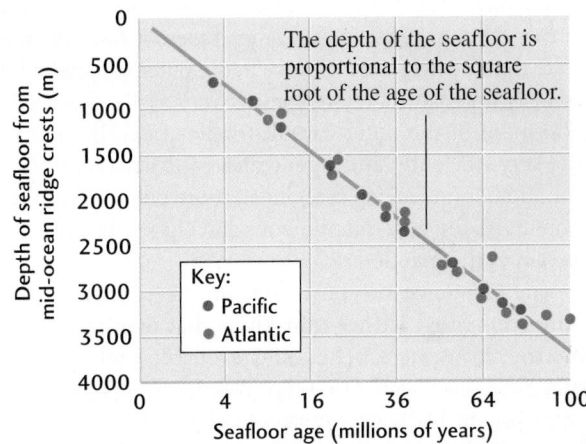

FIGURE 14.9 ■ Topography of mid-ocean ridges in the Atlantic and Pacific oceans, showing how ocean depth increases in proportion to the square root of lithosphere age as plates move away from spreading centers. The same theoretical curve, derived by assuming that the lithosphere cools by conduction, matches the data for both ocean basins, even though seafloor spreading is much faster in the Pacific than in the Atlantic.

establishment of this theory was one of the great successes of plate tectonics.

Conductive cooling does not explain all aspects of heat flow through Earth's outer surface, however. Marine geologists have found that seafloor older than about 100 million years does not continue to subside as the simple theory would predict. Moreover, simple conductive cooling is far too inefficient to account for the cooling of Earth over its entire history. It can be shown that if the 4.5-billion-year-old Earth cooled by conduction alone, very little of the heat from depths greater than about 500 km would have reached the surface. The mantle, which was molten in Earth's early history, would be far hotter than it is now. To understand these observations, we must consider the second mode of heat transport, convection, which is more efficient than conduction in getting heat out of Earth's interior.

CONVECTION IN THE MANTLE AND CORE When a fluid—either liquid or gas—is heated, it expands and rises because it has become less dense than the surrounding material. The upward movement of the heated fluid displaces cooler fluid downward, where it is heated and then rises to continue the cycle. This process, called **convection,** transfers heat more efficiently than conduction because the heated material itself moves, carrying its heat with it. Convection is the same process by which water is heated in a kettle on the stove (see Figure 1.16). Liquids conduct heat poorly, so a kettle of water would take a long time to boil if convection did not distribute the heat rapidly. Convection is what moves heat when a chimney draws, when warm tobacco smoke rises, or when thunderclouds form on a hot day.

We have already seen how seismic waves revealed that Earth's outer core is liquid. Other types of data demonstrate that the iron-rich material in the outer core has a low viscosity and can therefore convect very easily. Convective movement in the outer core distributes heat through the core very efficiently, and it generates Earth's magnetic field, a phenomenon we will examine in more detail later in this chapter. At the core-mantle boundary, heat from the core flows into the mantle.

The existence of convection in the solid mantle is more surprising, but we now know that mantle rock below the lithosphere behaves as a ductile material; over long periods, it can flow like a very viscous fluid (see Earth Issues 14.1). As we saw in Chapters 1 and 2, seafloor spreading and plate movements are direct evidence of this solid-state convection at work. The hot mantle material that rises under mid-ocean ridges builds new lithosphere, which cools as it spreads away. In time, it sinks back into the mantle at subduction zones, where it is eventually resorbed and reheated. Through this process, heat is carried from Earth's interior to its surface.

Temperatures Inside Earth

Geologists have many reasons for wanting to understand the *geothermal gradient*—the increase in temperature with depth—in Earth's interior. Temperature and pressure determine the state of matter (solid or molten), its viscosity (resistance to flow), and how its atoms are packed together in crystals. A curve that describes the geothermal gradient in Earth's interior is called a **geotherm.** In **Figure 14.10,** we compare one possible geotherm (in yellow) with the melting curves for mantle and core materials (in red). The melting curves show how the onset of melting depends on pressure, which increases with depth.

Geologists at Earth's surface can directly measure temperatures at depths of up to 4 km in mines and more than 10 km in boreholes. They have found that the geothermal gradient is 20°C to 30°C per kilometer in most continental crust. Conditions below the crust can be inferred from the properties of lavas and rocks erupted by volcanoes. These data indicate that temperatures near the base of the litho-

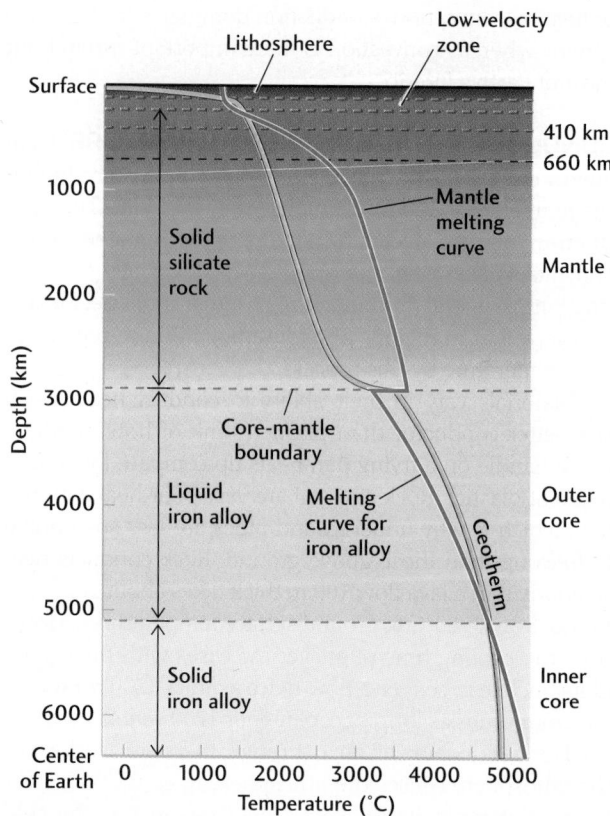

FIGURE 14.10 ■ An estimate of Earth's geotherm, which describes the increase in temperature with depth (yellow line). The geotherm first rises above the melting curve—the temperature at which peridotite begins to melt (red line)—in the upper mantle, forming the partially molten low-velocity zone. It does so again in the outer core, where the iron-nickel alloy is in a liquid state. The geotherm falls below the melting curve throughout most of the mantle and in the solid inner core.

sphere range from 1300°C to 1400°C. As Figure 14.10 shows, it is at these temperatures that the geotherm rises above the melting point of mantle rock. The geotherm intersects the melting curve at about 100 km beneath most oceanic crust, and somewhat deeper (150 to 200 km) beneath most continental crust. From this depth to where the geotherm drops below the melting curve, at depths of 200 to 250 km, mantle material is partially molten. These observations are consistent with the existence of a shear-wave low-velocity zone (see Figure 14.8), as well as with widespread evidence suggesting that basaltic magmas are produced by partial melting in the upper part of the asthenosphere.

The steep geothermal gradient near Earth's surface tells us that heat is transported through the lithosphere by conduction. Below the lithosphere, the temperature does not rise as rapidly. If it did, temperatures in the deeper parts of the mantle would be so high (tens of thousands of degrees) that the lower mantle would be entirely molten, which is

Earth Issues

14.1 Glacial Rebound: Nature's Experiment with Isostasy

If you depress a cork floating in water with your finger and then release it, the cork pops up almost instantly. A cork floating in molasses would rise more slowly because the drag of the viscous fluid would slow the process. How convenient it would be if we could push Earth's crust down somewhere, remove the depressive force, and then sit back and watch the depressed area rise. From its response, we could learn much more about how isostasy works—in particular, about the viscosity of the mantle and how it affects rates of epeirogenic movement (uplift and subsidence).

Nature has actually done this experiment for us. The depressive force is the weight of a continental glacier—an ice sheet 2 to 3 km thick. During the onset of an ice age, ice sheets can form in only a few thousand years. The immense ice load depresses the crust, and a downward bulge develops under the ice sheet, displacing enough mantle to provide buoyant support. Using the information in the Practicing Geology exercise and the densities of ice (0.92 g/cm³) and mantle material (3.3 g/cm³), we can calculate how much downwarping a 3-km ice sheet requires to achieve isostatic equilibrium:

$$(0.92 \text{ g/cm}^3 \div 3.3 \text{ g/cm}^3) \times 3.0 \text{ km} = 0.84 \text{ km}$$

At the onset of a warming trend, the ice sheet melts rapidly. With the removal of its weight, the depressed crust begins to rebound, eventually rising back to its original level—in this case, 840 m higher than when it was under the full glacial load. Such *glacial rebound* has occurred in Norway, Sweden, Finland, Canada, and elsewhere in formerly glaciated regions. The most recent ice sheet retreated from those areas some 12,000 years ago, and the land has been rising ever since.

We can measure the rate of uplift by dating ancient beaches that were once at sea level and have since been uplifted. Figure 10.21 shows a series of raised beaches in northern Canada that have allowed geologists to measure the speed of glacial rebound, and thus to infer the viscosities of mantle materials. Those viscosities are very high. Even the asthenosphere—the weak layer where most of the mantle flow during glacial rebound takes place—is 10 orders of magnitude more viscous than silica glass is at mantle temperatures.

The principle of isostasy explains glacial rebound. Rates of uplift measured from raised beaches allow geologists to infer mantle viscosities, including that of the asthenosphere, where most of the mantle flow takes place during uplift.

TIME 1
At the start of an ice age, a continental ice sheet forms and continues to thicken over a few thousand years.

Continental crust

Ice

Lithosphere

Asthenosphere

TIME 2
The continental crust bends downward under the ice load to the extent needed to provide buoyant support.

TIME 3
At the end of the ice age, rapid warming melts the glacier. The depressed crust begins to rebound.

TIME 4
Rebound continues long after the glacier has melted, slowly returning the crust to its pre-ice age elevation.

Raised beaches

inconsistent with seismological observations. Instead, the change in temperature with depth drops to about 0.5°C per kilometer, which is the geothermal gradient expected in a convecting mantle. This drop occurs because convection mixes cooler material near the top of the mantle with warmer material at greater depths, averaging out the temperature differences (just as temperatures are evened out when you stir your bathwater).

Phase changes—observed as steep increases in seismic wave velocities—occur in the transition zone at depths of 410 km and 660 km (see Figure 14.8). Seismology can accurately determine the depths (and thus the pressures) of these phase changes, so the temperatures at which the phase changes take place can be calibrated using high-pressure laboratory experiments. The values obtained from the laboratory data are consistent with the geotherm shown in Figure 14.10.

We have more limited information about the temperatures at greater depths. Most geologists agree that convection extends throughout the mantle, vertically mixing material and keeping the geothermal gradient low. Near the base of the mantle, however, we would expect temperatures to increase more rapidly, because the core-mantle boundary restricts vertical mixing. Convective movements near the core-mantle boundary, like those near the crust, are primarily horizontal rather than vertical. Close to the boundary, heat is transported from the core into the mantle mainly by conduction, and the geothermal gradient should therefore be high, as it is in the lithosphere.

Seismology tells us that the outer core is liquid, which means that its temperature is high enough to melt the iron alloy that constitutes it. Laboratory data indicate that this temperature is probably greater than 3000°C, and that estimate is consistent with the high geothermal gradient at the base of the mantle predicted by convection models. The inner core, on the other hand, is solid. Because its iron-nickel composition is nearly the same as that of the outer core, the boundary between the inner core and outer core should correspond to the depth where the geotherm crosses the melting curve for the core material. This hypothesis implies that the temperature at Earth's center is slightly less than 5000°C.

Many aspects of this story can be debated, however, especially in regard to the deeper parts of the geotherm. For example, some geologists believe that the temperature at Earth's center may be as high as 6000°C to 7000°C. More laboratory experiments and better calculations are required to reconcile these differences.

▪ Visualizing Earth's Three-Dimensional Structure

So far, we have investigated how the properties of Earth's materials vary with depth. Such a one-dimensional description would suffice if our planet were a perfectly symmetrical sphere, but of course it is not. At the surface, we can see *lateral variations* (geographic differences) in Earth's structure associated with oceans and continents and with the basic features of plate tectonics: mid-ocean ridges at spreading centers, deep-sea trenches at subduction zones, and mountain belts uplifted by continent-continent collisions.

Below the crust, we can expect that convection will cause variations in temperature from one part of the mantle to another. Downwelling currents, such as those associated with subducted lithospheric plates, will be relatively cold, whereas upwelling currents, such as those associated with mantle plumes, will be relatively hot. Computer models tell us that lateral variations in temperature due to mantle convection should be on the order of several hundred degrees. From laboratory experiments on rocks, we know that such temperature differences should cause small variations in seismic wave velocities from place to place. For example, a temperature increase of 100°C reduces the speed of an S wave traveling through mantle peridotite by about 1 percent (or even more if the rock is close to its melting temperature). If the mantle is indeed convecting, then seismic wave velocities should vary by several percentage points from place to place. Seismologists can make three-dimensional maps of these small lateral variations in wave velocities using the techniques of seismic tomography.

Seismic Tomography

Seismic tomography is an adaptation of a medical technique commonly used to map the human body, called computerized axial tomography (CAT). CAT scanners construct three-dimensional images of organs by measuring small differences in X rays that sweep the body in many directions. Similarly, we can use the travel times of seismic waves from earthquakes, as recorded on thousands of seismographs all over the world, to sweep Earth's interior in many different directions and construct a three-dimensional image of what's inside. A reasonable hypothesis, consistent with the results of laboratory experiments, is that regions where seismic waves speed up are composed of relatively cool, dense rock (for example, subducted plates), whereas regions where seismic waves slow down are composed of relatively hot, buoyant rock (for example, rising plumes).

Seismic tomography has revealed features in the mantle that are clearly associated with mantle convection. In the 1990s, researchers at Harvard University constructed a tomographic model of the mantle. Their model is displayed in **Figure 14.11** as a cross section of Earth and as a series of global maps at depths ranging from just below the crust down to the core-mantle boundary. Near Earth's surface (Figure 14.11b), you can clearly see the structures of plate tectonics. The upwelling of hot mantle material along the mid-ocean ridges is visible in warm colors; the cold lithosphere in old ocean basins and beneath the continental cratons is visible in cool colors.

At greater depths, the features become more variable and less coherent with surface tectonic features, reflecting

(a) Tomographic cross section

A tomographic cross section through Earth reveals hot regions, such as a superplume rising from Earth's core beneath South Africa,…

North America

Africa

…and cooler regions, such as the descending remnants of the Farallon Plate under the North American Plate.

(b–e) Global maps at four different depths

(b)

70 km

Near Earth's surface, hot rocks in the asthenosphere lie beneath oceanic spreading centers.

(c)

200 km

Moving deeper, we see the cold lithosphere of stable continental cratons and the warmer asthenosphere beneath ocean basins.

(d)

500 km

Deeper in the mantle, the features no longer match the continental positions.

(e)

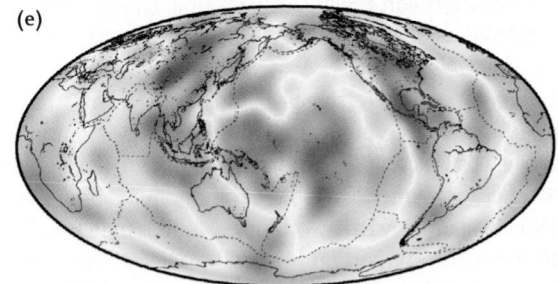

2800 km (near core-mantle boundary)

Near the core-mantle boundary, the colder regions around the Pacific may be the "graveyards" of sinking lithospheric slabs.

FIGURE 14.11 ■ A three-dimensional model of Earth's mantle created by seismic tomography. Regions with faster S-wave velocities (blue and purple) indicate colder, denser rock; regions with slower S-wave velocities (red and yellow) indicate hotter, less dense rock. (a) Cross section of Earth. (b–e) Global maps at four different depths. [S-wave velocities by G. Ekström and A. Dziewonski, Harvard University; cross section (a) from M. Gurnis, *Scientific American* (March 2001): 40; maps (b–e) by L. Chen and T. Jordan, University of Southern California.]

what is probably a complex pattern of mantle convection. Some large-scale features stand out particularly well. You will notice that just above the core-mantle boundary (Figure 14.11e), there is a red region of relatively low S-wave velocities beneath the central Pacific Ocean, surrounded by a broad blue ring of higher S-wave velocities. Seismologists have speculated that the high velocities represent a "graveyard" of cold oceanic lithosphere subducted beneath the Pacific's volcanic island arcs and mountain belts—the Ring of Fire—during the last 100 million years or so.

The cross section through the mantle (Figure 14.11a) clearly reveals material from the once-large Farallon Plate, which has been almost completely subducted under North America (see Chapter 10). The obliquely sinking slab material (in blue) appears to have penetrated the entire mantle. The image also indicates sinking colder rock beneath Indonesia, another subduction zone. In addition, a large yellow blob of hotter rock, thought to be a "superplume," can be seen rising at an angle from the core-mantle boundary to a position beneath southern Africa. This hot, buoyant mass pushing up the cooler material above it may explain the uplifted, mile-high plateaus of South Africa (see Figure 10.20e). The other visible blobs of hotter and cooler material may be evidence of material exchanges among the lithosphere, the mantle, and the layer of hotter material at the core-mantle boundary.

Earth's Gravitational Field

The same temperature variations that speed up and slow down seismic waves also influence the densities of mantle rocks. Laboratory experiments have shown that the expansion of rock caused by a 300°C increase in temperature reduces its density by about 1 percent. This might seem to be a small effect, but the mass of Earth's mantle is enormous (about 4 billion trillion tons!), so even small changes in the distribution of its mass can lead to observable variations in the pull of Earth's gravity.

Geologists can determine features of Earth's mass distribution by observing variations in the gravitational field above its surface, as well as from bulges and dimples in the shape of the planet. They have been able to show that the shape measured by Earth-orbiting satellites matches the pattern of mantle convection imaged by seismic tomography (see Earth Issues 14.2). This agreement has allowed us to refine our models of the mantle convection system.

■ Earth's Magnetic Field and The Geodynamo

Like the mantle, Earth's outer core transports most of its heat by convection. But the same techniques that have revealed so much about mantle convection—seismic tomography and the study of Earth's gravitational field—have provided almost no information about convection in the core. Why not?

The problem has to do with the fluidity of the outer core. The mantle is a viscous solid that flows very slowly. As a result, convection creates regions where temperatures are significantly higher or lower than the average mantle geotherm. We can see these regions in Figure 14.11 as places where seismic wave velocities are lower or higher than the average at that depth. The outer core, in contrast, has a very low viscosity: its molten material can flow as easily as water or liquid mercury. Even small density variations caused by convection are quickly smoothed out by the rapid flow under the force of gravity. Any lateral variations in seismic wave velocity caused by convection will be much too small for us to see using seismic tomography, and they will not cause measurable distortions in the shape of the planet.

We can, however, investigate convection in the outer core through observations of Earth's magnetic field. In Chapter 1, we briefly described the magnetic field and its generation by the geodynamo in the outer core. In Chapter 2, we discussed magnetic reversals and the use of magnetic anomalies in volcanic rocks to measure seafloor spreading. In this section, we will further explore the nature of Earth's magnetic field and its origin in the geodynamo.

The Dipole Field

The most basic instrument used for sensing Earth's magnetic field is the magnetic compass, invented by the Chinese more than 22 centuries ago. For hundreds of years, explorers and ship captains used magnetic compasses to navigate, but they had little understanding of how these ancient devices actually worked. In 1600, William Gilbert, physician to Queen Elizabeth I, provided a scientific explanation. He proposed that "the whole Earth is itself a great magnet" whose field acts on the small magnet of a compass needle to align it in the direction of the north magnetic pole.

Scientists of Gilbert's day had begun to visualize a magnetic field as lines of force, such as those revealed by the alignment of iron filings on a piece of paper above a bar magnet. Gilbert showed that Earth's magnetic lines of force point into the ground at the north magnetic pole and outward at the south magnetic pole, as if a powerful bar magnet were located at Earth's center and oriented along an axis inclined about 11° from Earth's axis of rotation (see Figure 1.17). In other words, the lines of force revealed a **dipole** (two-pole) magnetic field.

Complexity of the Magnetic Field

Gilbert solved an important problem for a seafaring nation dependent on the compass for navigation, but his explanation was only partly correct. We now know that the source of the magnetic field is a geodynamo powered by core convection, not a permanent magnet at Earth's center (which

Earth Issues

14.2 The Geoid: The Shape of Planet Earth

The surface of the ocean is warped upward in places where the pull of Earth's gravitational field is stronger and downward where the pull is weaker. The shape of the ocean's surface can be precisely measured by radar altimeters mounted on satellites. By averaging out wave motions and other fluctuations, oceanographers can map the small-scale variations in gravity caused by geologic features on the seafloor, such as faults and seamounts (see Chapter 20). Variations in gravity are also produced by the much larger features caused by mantle convection currents.

A perfectly still ocean would have a surface that conforms to what geologists call the *geoid*. The surface of a still body of water is perfectly "flat" in the sense that the pull of gravity is perpendicular to that surface—otherwise, the water would flow "downhill" to make the surface flatter. The geoid is defined as an imaginary surface at some reference height above Earth adjusted to be everywhere perpendicular to the local gravitational force. Because the ocean surface approximates the geoid, we usually take the reference height to be sea level. When we measure the height of a mountain relative to sea level, we are actually measuring its height above the geoid at that point. In this sense, the geoid is simply the shape of Earth. Geologists can use the geoid to calculate the strength and direction of the gravitational force at any point on the planet's surface and infer how rock density varies in Earth's interior.

Radar altimeters can easily map the geoid over the oceans, but how can we get this information on dry land? It turns out that the geoid can be measured for the entire Earth by tracking orbiting satellites. Three-dimensional mass variations in the mantle exert a small gravitational pull on the satellites, shifting their orbits slightly. By monitoring these shifts over long periods, scientists can create a two-dimensional map of the geoid over continents as well as oceans.

A smoothed version of the observed geoid reveals the large-scale features of Earth's gravitational field. Relative to what sea level would be on an Earth without any lateral variation in mass, the elevation of the geoid varies from a low of about −110 m at a point near the coast of Antarctica to a high of just over 100 m on the island of New Guinea in the western Pacific.

The geoid shows some similarities to the large-scale features of the deeper parts of the mantle, as you can see by comparing the geoid map with Figures 14.11d and e. This agreement suggests that the three-dimensional variations in both rock density and S-wave velocity are related to temperature differences arising from mantle convection.

Geophysicists Brad Hager and Mark Richards tested this hypothesis in the mid-1980s. Using laboratory data for calibration, they first calculated three-dimensional density variations from the seismic wave velocity variations mapped by seismic tomography. They then constructed a computer model of convective flow by assuming that the heavier parts of the mantle are sinking while the lighter parts are rising. Finally, they calculated what the geoid shape should be according to this convection model. You can see that their model results match the observed geoid quite well, especially for the largest features. This agreement has given geologists confidence that temperature variations within the mantle convection system can explain what we see both in seismic images and in the gravitational field.

A smoothed map of the geoid, or "shape of Earth," derived from satellite observations. The contours, given here in meters, show how the observed sea level deviates from that on an ideal Earth without any lateral variation in rock density. (b) A map of the geoid computed from a model of mantle convection that is consistent with the temperature structure of the mantle derived from seismic tomography. By matching the observed geoid with such theoretical models, geologists have improved their understanding of the mantle convection system. [(a) NASA; (b) model by B. Hager, Massachusetts Institute of Technology; maps by L. Chen and T. Jordan, University of Southern California.]

would be quickly destroyed by the high temperatures in the core). The geodynamo is formed by rapid convective movements in the liquid, iron-rich, electrically conducting outer core. The magnetic field produced by the geodynamo is considerably more complex than a simple dipole field, and it is constantly changing over time due to these fluid movements.

Within a few decades after Gilbert's famous pronouncement, careful observers had realized that the magnetic field varies over time. Not surprisingly, some of the best evidence for these changes came from the compass measurements systematically recorded by the British navy. Navigators had to correct their compass bearings to account for the displacement of the north magnetic pole (magnetic north) from the north rotational pole (true north), and these corrections showed that the north magnetic pole was moving at rates of 5° to 10° per century (**Figure 14.12**). Little did the British sailors know that these changes were caused by convective movements deep in Earth's core!

THE NONDIPOLE FIELD Measurements at Earth's surface have revealed that only about 90 percent of Earth's magnetic field can be described by the simple dipole field illustrated in Figure 1.17. The remaining 10 percent, which geologists refer to as the *nondipole field,* has a more complex structure. This structure can be seen by comparing the magnetic field strengths calculated for a simple dipole field (**Figure 14.13a**) with those of the observed field (Figure 14.13b). If we extrapolate the observed

lines of force down to the core-mantle boundary using a computer model, the size of the nondipole field actually increases relative to the size of the dipole field, as indicated by the bumpiness of the orange and blue colors on the map in Figure 14.13c. The poorly conducting mantle tends to smooth out complexities in the magnetic field, making the dipole field seem bigger than it really is.

SECULAR VARIATION Magnetic records for the last 300 years (many from the British navy) show that both the dipole and nondipole components of Earth's magnetic field are changing over time, but that this *secular* (time-related) *variation* is fastest for the nondipole component. Secular variation is evident when we compare a map of today's magnetic field at the core-mantle boundary (Figure 14.13c) with maps reconstructed for previous centuries (Figure 14.13d, e). Changes in field strength occur on time scales of decades and indicate that fluid movements within the geodynamo are on the order of millimeters per second.

Scientists can use this secular variation to help them understand convection in the outer core. With high-performance computers, they have been able to simulate the complex convective movements and electromagnetic interactions in the outer core that might be creating the geodynamo. The magnetic lines of force from one such simulation are shown in Figure 14.14. Away from the core, the lines of force can be approximated by a dipole field, but they become more complicated near the core-mantle boundary. Within the core itself, they are hopelessly entangled by the strong convective movements.

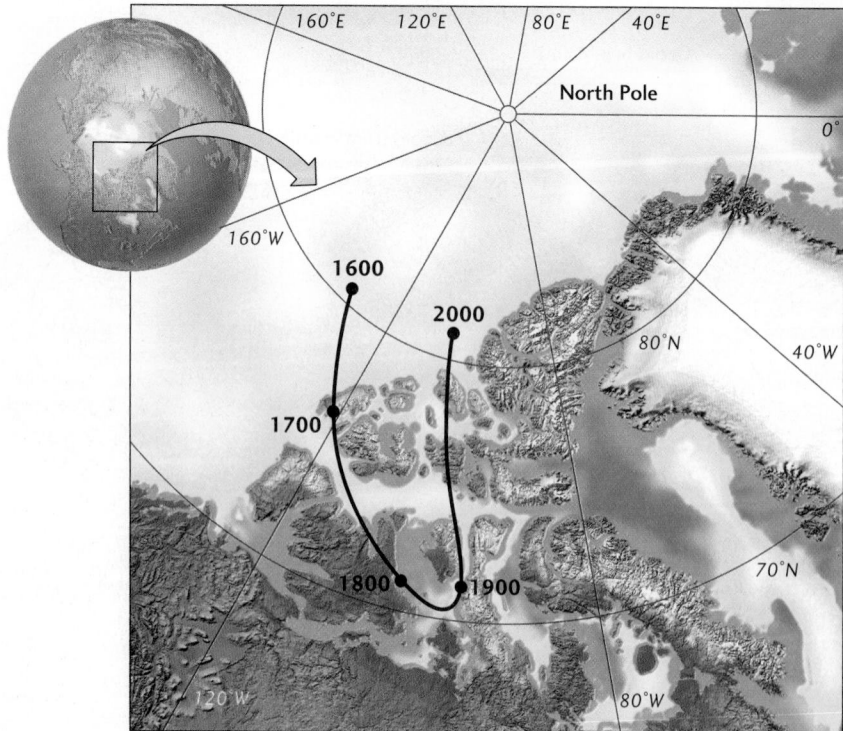

FIGURE 14.12 ■ Path of the north magnetic pole, as mapped by compass readings and other measurements of Earth's magnetic field since 1600. Changes in the pole location are caused by convective movements within Earth's fluid outer core.

(a) Map of ideal dipole field tilted by 11°

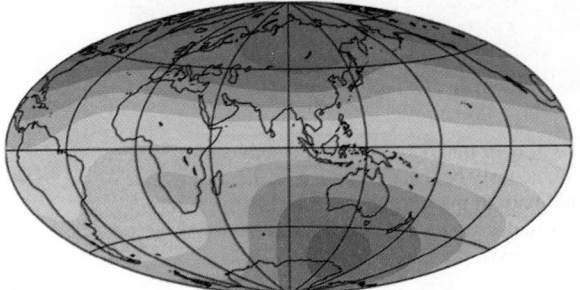

(b) Magnetic field mapped at surface in 2000

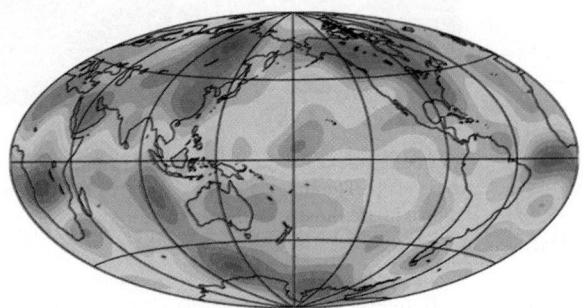

(c) Magnetic field mapped at core-mantle boundary in 2000

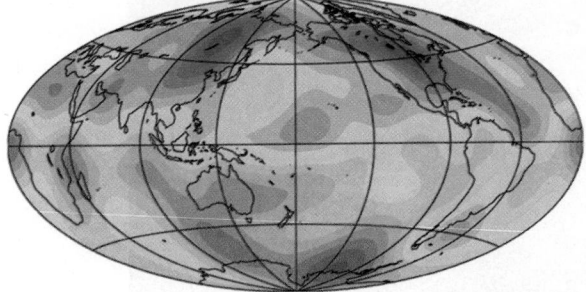

(d) Magnetic field mapped at core-mantle boundary in 1900

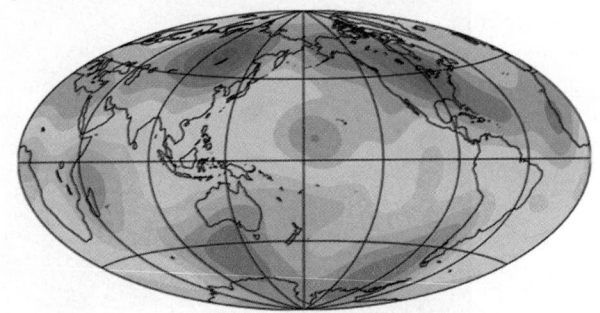

(e) Magnetic field mapped at core-mantle boundary in 1800

FIGURE 14.13 ▪ Earth's magnetic field changes over time. Blue colors show the strength of the inward-pointing field, and orange colors show the strength of the outward-pointing field. The magnetic field mapped at the surface (b) is more complex than a simple dipole (a), and it shows more complications when extrapolated to the core-mantle boundary (c). The features on the nondipole field change over time, as seen in (c) through (e), owing to convection in Earth's fluid outer core. [Maps courtesy of J. Bloxham, Harvard University.]

MAGNETIC REVERSALS These same computer simulations also allow us to understand a remarkable behavior of the geodynamo: spontaneous reversals of the magnetic field. As discussed in Chapter 2, the magnetic field reverses its direction at irregular intervals (ranging from tens of thousands to millions of years), exchanging the north and south magnetic poles as if the magnet depicted in Figure 1.16 were flipped 180°. Recent computer simulations of the geodynamo were able to reproduce these sporadic reversals in the absence of any external triggers (**Figure 14.14**). In other words, it is possible for Earth's magnetic field to reverse itself spontaneously, purely through internal interactions.

This behavior illustrates a fundamental difference between the geodynamo and the dynamos used in power plants. A steam-powered dynamo is an artificial system engineered by humans to do a particular job. The geodynamo, in contrast, exemplifies a *self-organized natural system*—one whose behavior is not predetermined by external constraints, but emerges from interactions within the system. The other two global geosystems, the plate tectonic and climate systems, also display a wide variety of self-organized behaviors. Understanding how these natural systems organize themselves is one of the greatest challenges to geoscience. We will return to this subject when we discuss the climate system in Chapter 15.

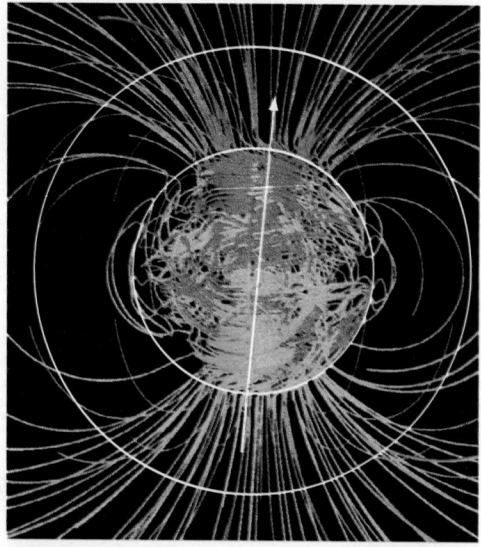

Time 1

Magnetic lines of force with normal orientation prior to reversal. The magnetic lines of force in the mantle approximate those of a dipole field.

Time 2

Beginning of magnetic reversal. Geodynamo spontaneously begins to reorganize, increasing the complexity of the lines of force within the outer core and decreasing the strength of the dipole component of the magnetic field.

Time 3

Reversal continues with rapid changes in the structure of the magnetic field, which continues to have a weak dipole component.

Time 4

Reversal nearly complete. Dipole field restrengthens with its north pole now pointing south.

FIGURE 14.14 ■ Computer models have shown that spontaneous changes in the geodynamo could cause magnetic reversals. [Courtesy of G. Glatzmaier, University of California, Santa Cruz.]

Paleomagnetism

We have seen repeatedly how the geologic record of ancient magnetism, or **paleomagnetism,** has provided crucial information for understanding Earth's history. Magnetic anomalies mapped on oceanic crust confirmed the existence of seafloor spreading, and they still provide the best data for tracking plate movements since the breakup of Pangaea 200 million years ago (see Chapter 2). Paleomagnetic data from old continental rocks have been essential for establishing the existence of earlier supercontinents, such as Rodinia (see Chapter 10).

Scientists have also used paleomagnetic data to reconstruct the history of Earth's magnetic field. The oldest magnetized rocks found so far, which formed about 3.5 billion years ago, indicate that Earth had a magnetic

30,000 years ago   Today

FIGURE 14.15 ■ The orientation of Earth's magnetic field 30,000 years ago was the reverse of today's, as evidenced by magnetized rocks found in the fireplace of an ancient campsite. The rocks, when cooling after the last fire, became magnetized in the direction of the ancient magnetic field, leaving a permanent record of its orientation.

field similar to the present one at that time. The presence of magnetization in the most ancient rocks is consistent with the ideas about Earth's differentiation discussed in Chapter 1, which imply that a convecting liquid core must have been established very early in Earth's 4.5-billion-year history.

Let's delve a little more deeply into the rock-forming processes that have allowed geologists to draw these remarkable conclusions. You may find it helpful to consult the material in Figure 2.12 and its accompanying text as you read this section.

THERMOREMANENT MAGNETIZATION In the early 1960s, an Australian graduate student found a fireplace in an ancient campsite where Aborigines had cooked their meals. He carefully removed several stones that had been baked by the fires, first noting their physical orientation. Then he measured the direction of the stones' magnetization and found that it was exactly the reverse of Earth's present magnetic field. He proposed to his disbelieving professor that, as recently as 30,000 years ago, when the campsite was occupied, the direction of the magnetic field was the reverse of the present one—that is, a compass needle would have pointed south rather than north.

Recall that high temperatures destroy magnetization. An important property of many magnetizable materials is that, as they cool below about 500°C, they become magnetized in the direction of the surrounding magnetic field. This happens because groups of atoms of the material align themselves in the direction of the magnetic field when the material is hot. When the material has cooled, these atoms are locked in place. This process is called **thermoremanent magnetization,** because the magnetization caused by heating and cooling is "remembered" by the rock long after the magnetizing

field has disappeared. Thus, the Australian student was able to determine the direction of Earth's magnetic field at the time the stones cooled after the last campfire (**Figure 14.15**).

Thermoremanent magnetization is the same process that magnetizes lava flows and newly formed oceanic crust, as described in Chapter 2. The discovery of magnetic reversals in these igneous rock types was a key ingredient in formulating the theory of plate tectonics.

DEPOSITIONAL REMANENT MAGNETIZATION Some sedimentary rocks can take on a different type of remanent magnetization. Marine sedimentary rocks form when particles of sediment that have settled to the seafloor become lithified. Magnetic grains among the particles—chips of the mineral magnetite (Fe_3O_4), for example—become aligned in the direction of Earth's magnetic field as they fall through the water, and this orientation may be incorporated into the rock when the sediments become lithified. The **depositional remanent magnetization** found in some sedimentary rocks results from the parallel alignment of all these tiny magnets, as if they were compasses pointing in the direction of the magnetic field prevailing at the time of deposition (**Figure 14.16**).

PALEOMAGNETIC STRATIGRAPHY Geologists have used paleomagnetism in combination with isotopic dating methods to work out the time sequence of magnetic reversals over the last 170 million years (**Figure 14.17**). This information can be used, in turn, to date new rock formations. Paleomagnetic stratigraphy is useful to archaeologists and anthropologists as well as to geologists. For example, the paleomagnetic stratigraphy of continental sediments has been used to date sediments containing the remains of predecessors of our own species.

1 Magnetic mineral grains transported to the ocean with other sediments become aligned with Earth's magnetic field while settling through the water.

Direction of magnetic field

Ocean

2 This orientation is preserved in the lithified sediments, which thus "remember" the field that existed at the time of deposition.

Magnetic particles in marine sediment

FIGURE 14.16 ■ Newly formed sediments can become magnetized in the direction of the magnetic field at the time of their deposition.

As we saw in Chapter 2, periods of "normal" (same as today) and reversed magnetic field orientation, which are called *magnetic chrons,* have irregular lengths, but on average they last about a half-million years. Superimposed on the chrons are transient, short-lived reversals known as *subchrons,* which may last anywhere from several thousand years to tens of millions of years. The reversal found in the rocks remagnetized by the Australian Aborigines' campfire (see Figure 14.15) can be interpreted as a reversed subchron within the present normal magnetic chron.

The Magnetic Field and the Biosphere

From the rock record, we know that the geodynamo began to operate early in Earth's history, and that life therefore evolved within a strong magnetic field. The consequences turn out to be rather surprising. For example, many types of organisms—pigeons, sea turtles, whales, and even bacteria—have evolved sensory systems that use the magnetic field for navigation (**Figure 14.18**). Their basic sensors are small crystals of magnetite that become magnetized by Earth's magnetic field as they are biologically precipitated within the organism (see Figure 11.8b). These crystals act as tiny compasses to orient the organism within the magnetic field. Geobiologists have discovered that some animals can even use arrays of magnetite crystals to sense the strength of the magnetic field, which provides them with additional information for navigation.

The magnetic field is not just a convenient frame of reference for flying and swimming species. It constitutes a part of the Earth system that is essential for sustaining a rich and delicate biosphere at the planet's surface. Although the machinery of the geodynamo operates deep within the core, its magnetic lines of force reach far into outer space, forming a barrier that shields Earth's surface from the damaging radiation of the solar wind (see Figure 1.18). Without the protection of a strong magnetic field, this intense stream of high-energy, electrically charged particles would be lethal to many organisms.

Moreover, if the geodynamo were to stop producing a magnetic field, bombardment by the solar wind would gradually strip away Earth's atmosphere, further degrading the terrestrial environment. This appears to have happened in the case of Mars. Paleomagnetism in the ancient Martian

This long period of normal polarity is called "the Cretaceous quiet zone."

Magnetic chrons have irregular lengths, but on average they span about a half-million years.

Time (millions of years ago)

Jurassic | Cretaceous | Eocene | Miocene

Paleocene | Oligocene | Pliocene | Pleistocene | Present

4000 Ma | 3000 Ma | 2000 Ma | 1000 Ma | 0 Ma

HADEAN EON | ARCHEAN EON | PROTEROZOIC EON | PHANEROZOIC EON

FIGURE 14.17 ■ The paleomagnetic time scale from 170 million years ago to the present, showing normal chrons (black bands) and reversed chrons (white bands).

FIGURE 14.18 ■ A flock of homing pigeons preparing to land at their coop in Cuba's Villaclara province after a 240-km flight across the island from Havana. These birds use Earth's magnetic field to navigate the long flights back to their homes. Recent evidence suggests that homing pigeons sense the magnetic field using receptors in their inner ears and beaks. [© Desmond Boylan/Reuters/Corbis.]

crust has been detected by orbiting spacecraft, so we know the Red Planet once had an active geodynamo that generated a strong magnetic field. Sometime early in the planet's history, however, its geodynamo ceased to operate, perhaps because the Martian core cooled enough to freeze. Exposure to the solar wind subsequently eroded its atmosphere to the tenuous state we observe today.

SUMMARY

What do seismic waves reveal about the layering of Earth's crust and mantle? Correlations of seismic wave velocities with rock types have made it possible to use seismic waves to explore the composition of Earth's interior. These explorations have revealed that the continental crust is made mostly of low-density granitic rock, and that the deep seafloor is composed of basalt and gabbro. The crust and outer part of the mantle make up the rigid lithosphere. Beneath the lithosphere lies the asthenosphere, the weak, ductile layer of the mantle on which the lithospheric plates slide. At the top of the asthenosphere, the temperature is high enough to partially melt peridotite, forming an S-wave low-velocity zone. Below 200 to 250 km, S-wave velocities again increase with depth. At two depths in the mantle, 410 km and 660 km below the surface, S-wave velocities show jumps caused by phase changes in mantle minerals. Below 660 km lies the lower mantle, a layer 2000 km thick, in which seismic wave velocities increase gradually.

What do seismic waves tell us about the layering of Earth's core? Seismic waves reflected from the core-mantle boundary locate this sharp boundary at a depth of 2890 km. The failure of S waves to penetrate below the core-mantle boundary indicates that the outer core is liquid. A jump in P-wave velocity marks the boundary between the liquid outer core and the solid inner core at a depth of 5150 km. Several lines of evidence suggest that the core is composed mostly of iron and nickel, with minor amounts of some lighter element, such as oxygen or sulfur.

How hot does it get in Earth's interior? Earth's interior is hot because it still retains much of the heat generated by its violent formation as well as heat currently being generated by the decay of radioactive isotopes. It has cooled over geologic time, primarily by convection in the mantle and core but also by conduction of heat through the lithosphere. A geotherm is a curve that describes how temperature increases with depth. Within most continental crust, it increases at a rate of 20°C to 30°C per kilometer. Temperatures near the base of the lithosphere reach 1300°C to 1400°C, which is hot enough to begin to melt mantle peridotite. The temperature in the liquid core is probably greater than 3000°C. The temperature at Earth's center is probably about 5000°C.

What has seismic tomography revealed about structures in the mantle? Seismologists can use seismic tomography to create three-dimensional images of Earth's interior. Regions where seismic wave velocities increase indicate relatively cool, dense rock; regions where they decrease indicate relatively hot, less dense rock. Tomographic images reveal the structures of plate tectonics close to Earth's surface, from the upwelling of hot mantle material under mid-ocean ridges to the cold lithosphere that extends deep beneath continental cratons. They also reveal many features of mantle convection, such as the sinking of lithospheric slabs into the lower mantle and the rising of plumes from deep within the mantle.

What does Earth's gravitational field tell us about its interior? Variations in the strength of gravity over Earth's surface and corresponding distortions in its shape can be measured by satellites. These variations arise primarily from the temperature variations caused by mantle convection, which affect the density of rock (higher temperatures reduce densities). The observed gravitational field is in agreement with the pattern of mantle convection inferred from seismic tomography.

What does Earth's magnetic field tell us about the liquid outer core? Convective movements in the outer core stir its electrically conducting iron-rich liquid, forming a geodynamo that produces the magnetic field. At Earth's surface, the magnetic field produced by the geodynamo is primarily a dipole field, but it has a small nondipole component. Maps of the magnetic field derived from compass readings show that the pattern of magnetic field strengths at Earth's surface has changed over the last several centuries. All of these observations tell us something about the nature of the rapid convective movements that drive the geodynamo.

What is paleomagnetism and what is its importance? Geologists have discovered that minerals in some rock types align themselves in the direction of Earth's magnetic field at the time the rocks form. This remanent magnetization can be preserved in rocks for millions of years. Paleomagnetic stratigraphy tells us that Earth's magnetic field has reversed (flipped back and forth) over geologic time. The chronology of reversals has been worked out, so that the direction of remanent magnetization of a rock formation can be used as an indicator of its age.

KEY TERMS AND CONCEPTS

compressional wave (p. 384)

conduction (p. 391)

convection (p. 392)

core-mantle boundary (p. 390)

depositional remanent magnetization (p. 401)

dipole (p. 396)

geotherm (p. 392)

isostasy (p. 388)

low-velocity zone (p. 389)

lower mantle (p. 390)

Mohorovičić discontinuity (p. 388)

paleomagnetism (p. 400)

phase change (p. 389)

seismic ray path (p. 384)

seismic tomography (p. 384)

shadow zone (p. 385)

shear wave (p. 384)

thermoremanent magnetization (p. 401)

transition zone (p. 389)

upper mantle (p. 388)

PRACTICING GEOLOGY EXERCISE

The Principle of Isostasy: Why Are Oceans Deep and Mountains High?

Earth's topography is dominated by continents, which typically have elevations of 0 to 1 km above sea level, and ocean basins, which typically have elevations of 4 to 5 km below sea level. Why the difference? The answer comes from the principle of isostasy, which relates the elevations of continents and oceans to the densities of crustal and mantle rocks. This amazingly useful principle not only explains much of Earth's topography, but also allows scientists to use changes in crustal elevation over time to investigate the properties of the mantle (see Earth Issues 14.1 on page 493).

Isostasy (from the Greek for "equal standing") is based on Archimedes' principle, which states that the weight of a floating solid is equal to the weight of the fluid it displaces. (According to legend, the Greek philosopher Archimedes discovered this principle over 2200 years ago while sitting in his bath; boggled by its implications, he rushed naked into the street, yelling "Eureka, I have found it!" Major discoveries rarely provoke such enthusiastic responses from modern scientists.)

Consider a block of wood floating in water. In each unit of area, the block's mass is its density times its thickness, whereas the mass of the water it displaces is the density of water times a reduced thickness, given by the block's thickness minus its elevation above the water. Archimedes' principle states that these two must be equal:

$$\text{wood density} \times \text{wood thickness} =$$
$$\text{water density} \times \text{water thickness} =$$
$$\text{water density} \times (\text{wood thickness} - \text{wood elevation})$$

We can solve the last algebraic equation to find the elevation:

$$\text{Wood elevation} = \left(1 - \frac{\text{wood density}}{\text{water density}}\right) \times \text{wood thickness}$$

The expression in brackets is called the "buoyancy factor" because it tells us what fraction of the wood will rise above the water surface. A light wood, such as young pine, has only half the density of water, so its buoyancy factor is

$$\frac{1\ \text{g/cm}^3 - 0.5\ \text{g/cm}^3}{1\ \text{g/cm}^3 = 0.5} = 0.5$$

The pine block will float high, with half of its volume out of the water. However, in the case of old oak, which has a density of 0.9 g/cm^3, the buoyancy factor is only 0.1, so the block will float low, with only one-tenth of its thickness above the water.

If continental crust (density = 2.8 g/cm^3) floated alone on top of mantle material (3.3 g/cm^3), the previous equation could be modified to give continental elevation by simply replacing "wood" with "continent" and "water" with "mantle." However, we must account for the oceanic crust (2.9 g/cm^3) and the ocean water (1.0 g/cm^3) that also float on the mantle. Since these two layers fill up the basins around the continents, we must subtract from the continental elevation the height that each of those layers alone would float above the mantle, given by its buoyancy factor times its thickness. The isostatic equation for continents therefore has three terms, one positive and two negative:

$$\begin{aligned}
\text{Continent elevation} = & \left(1 - \frac{\text{continental crust density}}{\text{mantle density}}\right) \times \text{continental thickness} \\
& - \left(1 - \frac{\text{oceanic crust density}}{\text{mantle density}}\right) \times \text{oceanic crust thickness} \\
& - \left(1 - \frac{\text{oceanic water density}}{\text{mantle density}}\right) \times \text{oceanic water thickness}
\end{aligned}$$

Using thicknesses of 33 km and 7 km for continental and oceanic crust, respectively, and a water depth of 4.5 km, we obtain

The principle of isostasy explains how high a wood block floats in water and how high a continent floats above sea level.

continent elevation
= (0.15 × 33 km) − (0.12 × 7.0 km)− (0.70 × 4.5 km)
= 0.96 km above sea level

This result is consistent with the overall distribution of Earth's topography (see Figure 1.8).

Because of isostasy, elevation is a sensitive indicator of crustal thickness, so regions of lower elevation must have thinner crust (or higher average density), whereas regions of higher elevation, such as the Tibetan Plateau (see Figure 10.16), must have thicker crust (or lower average density).

BONUS PROBLEM: The average elevation of the Tibetan Plateau is about 5 km above sea level. Use the isostatic equation to compute the average thickness of the crust in this region, assuming that its average density is 2.8 g/cm³.

EXERCISES

1. The velocity of compressional waves in the lower part of the oceanic crust averages about 7 km/s. What rock type is most consistent with this observation as well as with your other knowledge of the oceanic crust?

2. What evidence suggests that the asthenosphere is partially molten?

3. What evidence indicates that Earth's outer core is molten and composed mostly of iron and nickel?

4. What is the depth to the core, and how do we know it?

5. What is the difference between conduction and convection? Which process is more efficient in transporting heat through the mantle?

6. Would the temperature at the Moho beneath a continental craton be hotter or cooler than the temperature at the Moho beneath an ocean basin?

7. How can features of mantle convection, such as rising and descending convection currents, be seen by seismic tomography?

8. How can a mountain float on the mantle when both are composed of rock?

9. How do igneous rocks become magnetized when they form? How does the magnetization of sedimentary rocks differ from this process?

10. What evidence supports the hypothesis that Earth's magnetic field is generated by a geodynamo in its outer core?

11. Does the magnetic field change by an observable amount over the span of a human lifetime? What does the answer suggest about convective movements in the outer core?

THOUGHT QUESTIONS

1. The Moon shows no evidence of plate tectonic processes, nor has it been volcanically active for billions of years. What does this observation imply about the state and temperature of the interior of that planetary body?

2. How do the existence of Earth's magnetic field, iron meteorites, and the abundance of iron in the cosmos support the ideas that Earth's core is mostly iron and that the outer core is liquid?

3. How would you use seismic waves to find a chamber of molten magma in the crust?

4. How does seismic tomography answer the question, "How deep do subducted slabs go before they are recycled?"

5. Where in the mantle might you look to find regions of anomalously low S-wave velocities?

MEDIA SUPPORT

 14-1 Animation: P + S Waves

- Components of the Climate System   408

- The Greenhouse Effect   413

- Climate Variation   416

- The Carbon Cycle   423

- Twentieth-Century Warming: Fingerprints of Anthropogenic Global Change   428

A snapshot of the climate system taken by sensors on several spacecraft, showing cloud cover (in white), variations in sea surface temperature (from the warmest in red to the coolest in dark blue), and land surface properties, including density of vegetation (from the lowest density in brown to the highest in green). [R. B. Husar, Washington University/NASA Visible Earth.]

THE CLIMATE SYSTEM

15

IN THE LAST SEVERAL CHAPTERS, we descended into Earth's deep interior to explore the internal heat engine that drives the plate tectonic system and the geodynamo. In this chapter, we return to Earth's surface to examine a global geosystem powered not by Earth's internal heat engine, but by external heat from the Sun: the climate system.

No aspect of Earth science is more important to our continued well-being than the study of the climate system. Throughout geologic time, evolutionary radiations and extinctions of organisms have been closely connected to changes in climate. Even the short history of our own species is deeply imprinted by climate change: agricultural societies began to flourish only about 11,700 years ago, when the harsh climate of the most recent ice age was rapidly transformed into the mild and steady climate of the Holocene epoch. Now, a globalized human society based on a petroleum-fueled economy is injecting greenhouse gases into the atmosphere at an ever-increasing rate, with potentially dire consequences: global warming, sea level rise, and unfavorable changes in weather patterns. The climate system is a huge, incredibly complex machine, and, like it or not, our hands are on the controls. We're in the driver's seat, with pedal to the metal, so we had better understand how the machinery works!

In this chapter, we will examine the main components of the climate system and the ways in which those components interact to produce the climate we live in today. We will investigate the geologic record of climate change and discuss the important role of the carbon cycle in regulating climate. Finally, we will look at the evidence for recent global warming and its relationship to changes in the composition of the atmosphere caused by human activities.

An understanding of the climate system will equip us to study the wide range of geologic processes that shape the face of our planet—weathering, erosion, sediment transport, and the interaction of the plate tectonic and climate systems—which will be the topics of the next seven chapters. The material presented here will also prepare us for the final topic of this textbook: a geologic perspective on the resource needs and environmental impacts of human society.

Components of the Climate System

At any point on Earth's surface, the amount of energy received from the Sun changes on daily, yearly, and longer-term cycles associated with Earth's movement through the solar system. This cyclical variation in the input of solar energy, known as **solar forcing,** causes changes in the surface environment: temperatures rise during the day and fall at night, and they rise in summer and fall in winter. The term *climate* refers to the average conditions at a point on Earth's surface and their variation during these cycles of solar forcing.

Climate is described by daily and seasonal statistics on the atmospheric temperature near Earth's surface (the *surface temperature*) as well as surface humidity, cloud cover, rate of rainfall, wind speed, and other weather conditions. **Table 15.1** gives an example of seasonal temperature statistics for New York City, which include measures of temperature variation (record highs and lows) as well as average values. In addition to these common weather statistics, a full scientific description of climate includes the nonatmospheric components of the surface environment, such as soil moisture and streamflow on land as well as sea surface temperature and the velocity of currents in the oceans.

The *climate system* includes all the components of the Earth system and all the interactions among those components that determine how climate varies in space and time (**Figure 15.1**). The main components of the climate system are the atmosphere, hydrosphere, cryosphere, lithosphere, and biosphere. Each component plays a different role in the climate system, and that role depends on its ability to store and transport mass and energy.

The Atmosphere

Earth's atmosphere is the most mobile and rapidly changing part of the climate system. Like Earth's interior, the atmosphere is layered (**Figure 15.2**). About three-fourths of its mass is concentrated in the layer closest to Earth's surface, the **troposphere,** which has an average thickness of 11 km. Above the troposphere is the **stratosphere,** a dryer layer that extends to an altitude of about 50 km. The outer atmosphere, above the stratosphere, has no abrupt cutoff; it slowly becomes thinner and fades away into outer space.

The troposphere convects vigorously due to the uneven heating of Earth's surface by the Sun (*tropos* is the Greek word for "turn" or "mix"). When air is warmed, it expands, becomes less dense than cooler air, and tends to rise; conversely, cool air tends to sink. The resulting convection patterns in the troposphere (which we'll examine more closely in Chapter 19), combined with Earth's rotation, set up a

TABLE 15-1	Seasonal Temperatures (°F) in Central Park, New York City			
Data Type*	January 1	April 1	July 1	October 1
Record high	62	83	100	88
Average high	39	56	82	69
Average low	28	39	67	55
Record low	−4	12	52	36

*Temperatures are averages for the date shown for the 30-year period 1971–2000; record temperatures are those for the period 1869–2011.

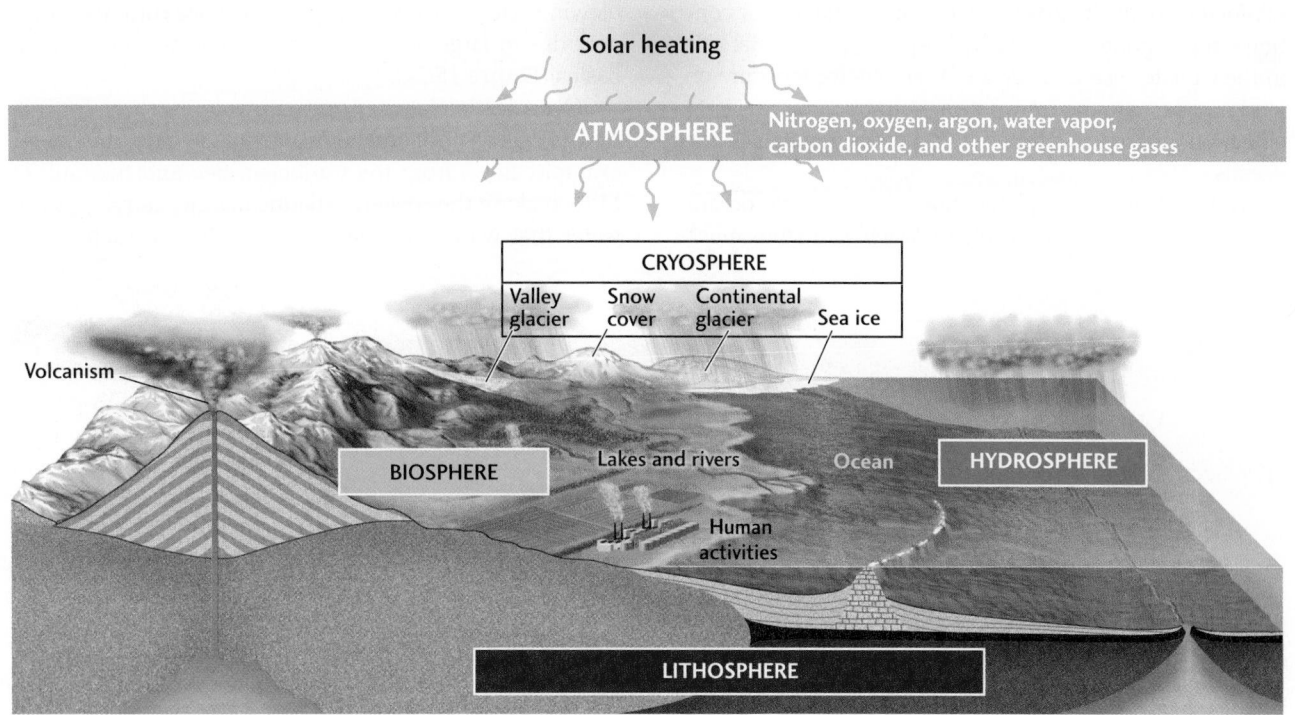

FIGURE 15.1 ■ Earth's climate system involves complex interactions among many components.

series of prevailing wind belts. In temperate regions, the prevailing winds have a generally eastward flow, such that they transport a typical parcel of air eastward around the globe in about a month (which is why it takes a few days for storms to blow across the continental United States). The spiral-like global circulation of air in these wind belts also transports heat energy from the warmer equatorial regions to the cooler polar regions.

FIGURE 15.2 ■ Layers of the atmosphere, showing variations in temperature (indicated by the blue line) and in pressure (which decreases rapidly with altitude).

The atmosphere is a mixture of gases, mainly nitrogen (78 percent by volume in dry air) and oxygen (21 percent by volume). The remaining 1 percent consists of argon (0.93 percent), carbon dioxide (0.035 percent), and other minor gases (0.035 percent), including methane and ozone. Water vapor is concentrated in the troposphere in highly variable amounts (up to 3 percent, but typically about 1 percent). Water vapor and carbon dioxide are the principal greenhouse gases in the atmosphere.

Ozone (O_3^+) is a highly reactive greenhouse gas produced primarily by the ionization of molecular oxygen by ultraviolet radiation from the Sun. In the lower part of the atmosphere, ozone exists in only tiny amounts, although it is a strong enough greenhouse gas to play a significant role in regulating Earth's surface temperature. Most atmospheric ozone is found in the stratosphere, where its concentration reaches a maximum at an altitude of 25 to 30 km (see Figure 15.2). This stratospheric ozone layer filters out incoming ultraviolet radiation, protecting the biosphere at Earth's surface from its potentially damaging effects.

The Hydrosphere

The *hydrosphere* comprises all the liquid water on, over, and under Earth's surface, including oceans, lakes, streams, and groundwater. Almost all of that liquid water is in the oceans (1350 million cubic kilometers); lakes, streams, and groundwater constitute a mere 1 percent (15 million cubic kilometers) of the hydrosphere. However small, these continental components of the hydrosphere play a vital role in the climate system. They are reservoirs for moisture on land and provide the transport system for returning precipitation and transporting salt and other minerals to the oceans.

Although water circulates more slowly in the oceans than air does in the atmosphere, water can store much more heat energy than air. For that reason, ocean currents transport heat energy very effectively. Prevailing winds blowing across the oceans generate surface currents, which give rise to large-scale circulation patterns within ocean basins (**Figure 15.3a**).

Oceanic circulation patterns involve vertical convection as well as horizontal movement. The Gulf Stream, for example, flows from the Caribbean Sea and the Gulf of Mexico along the western Atlantic margin, carrying warm water that warms the climate of the North Atlantic and

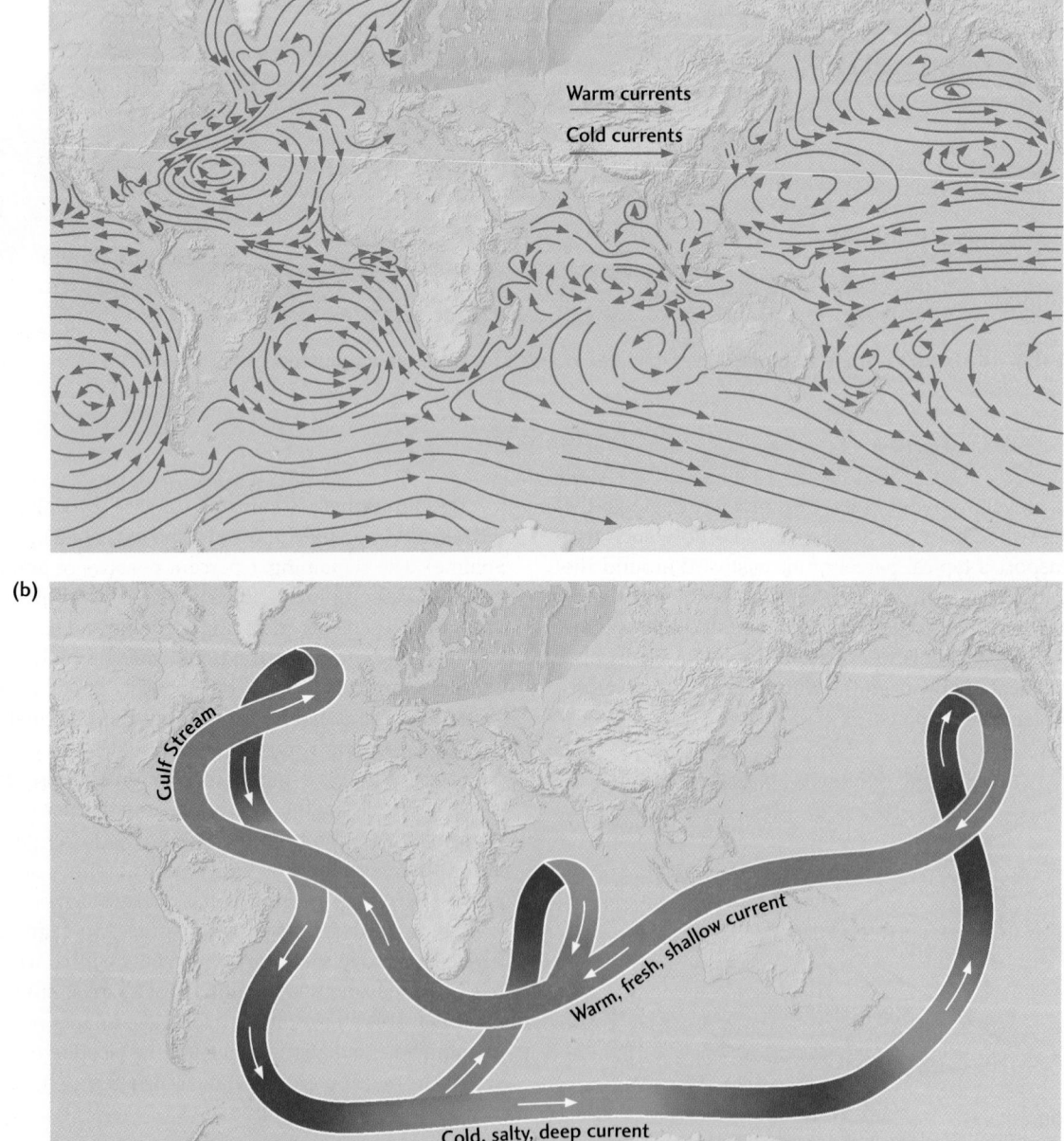

FIGURE 15.3 ■ Two major circulation systems in the oceans. (a) Currents at the surface of the oceans are generated by winds. [U.S. Naval Oceanographic Office.] (b) A schematic representation of thermohaline circulation, which acts like a conveyor belt to transport heat from warm equatorial regions to cool polar regions.

Europe. In the North Atlantic, that water cools and becomes more saline (because less fresh water enters the oceans from rivers at high latitudes than evaporates from the ocean surface). Cool water is denser than warm water, and salty water is denser than fresh water; therefore, this cooler, saltier water sinks. In this way, a subsurface cold current is created that flows southward as part of a global pattern of **thermohaline circulation**—so called because it is driven by differences in temperature and salinity. On a global scale, thermohaline circulation acts like an enormous conveyor belt running through the oceans that moves heat from the equatorial regions toward the poles (Figure 15.3b). Changes in this circulation pattern can strongly influence global climate.

The Cryosphere

The ice component of the climate system is called the *cryosphere.* It comprises 33 million cubic kilometers of ice, primarily in the ice caps of the polar regions. Today, continental glaciers cover about 10 percent of the land surface (15 million square kilometers), storing about 75 percent of the world's fresh water. Floating ice includes *sea ice* in the open ocean, as well as frozen lake and river water. The role of the cryosphere in the climate system differs from that of the liquid hydrosphere because ice is relatively immobile and because it reflects almost all of the solar energy that falls on it.

The seasonal exchange of water between the cryosphere and the hydrosphere is an important process of the climate system. During winter, sea ice typically covers 14 million to 16 million square kilometers of the Arctic Ocean (**Figure 15.4**) and 17 million to 20 million square kilometers of the Southern Ocean, shrinking to about one-third of that area in summer. About one-third of the land surface is covered by seasonal snows, almost entirely (all but 2 percent) in the Northern Hemisphere. Melting snow is the source of much of the fresh water in the hydrosphere. In the U.S. Sierra Nevada and Rocky Mountains, for example, 60 to 70 percent of annual precipitation is snowfall, which is later released as water during spring snowmelt and stream runoff. Much larger amounts of water are exchanged between the cryosphere and the hydrosphere during glacial cycles. At the peak of the most recent ice age, 20,000 years ago, sea level was about 130 m lower than it is today, and the volume of the cryosphere was three times larger.

The Lithosphere

The part of the lithosphere that is most important to the climate system is the land surface, which makes up about 30 percent of Earth's total surface area. The composition of the land surface affects the way it absorbs solar energy or releases it to the atmosphere. As the temperature of the land surface rises, more heat energy is radiated back into

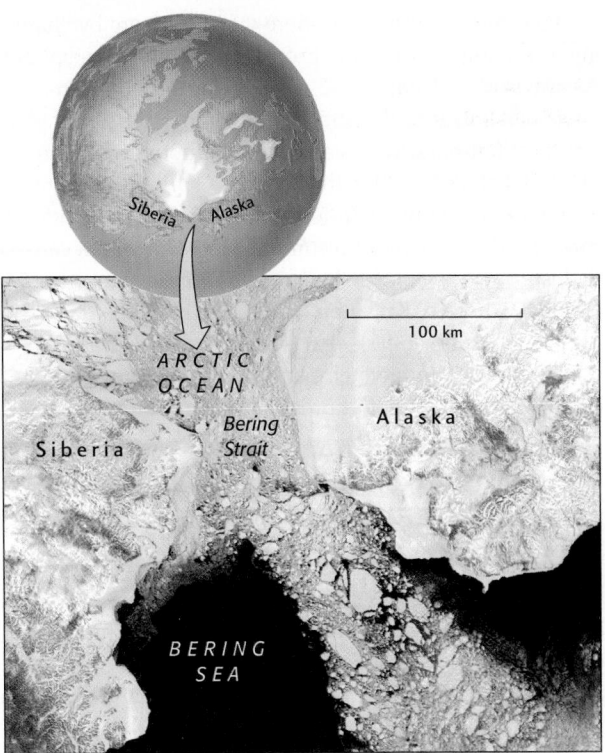

FIGURE 15.4 ■ The volume of sea ice varies seasonally. This satellite image shows Arctic sea ice breaking up and flowing through the Bering Strait in May 2002. [NASA MODIS Satellite.]

the atmosphere, and more water evaporates from the land surface and enters the atmosphere. Because evaporation uses considerable energy, it causes the land surface to cool. Consequently, soil moisture and other factors that influence rates of evaporation—such as vegetation cover and the subsurface flow of water—are very important in controlling atmospheric temperatures.

Topography has a direct effect on climate through its influence on atmospheric circulation. Air masses that flow over mountain ranges dump rain on the windward side, creating a rain shadow on the leeward side of the mountains (see Figure 17.3). At much longer time scales, geologists have documented many changes in the climate system that result from plate tectonic processes. The overall asymmetry of the continents—a direct consequence of plate movements—induces hemispheric asymmetries in the global climate system. Changes in the shape of the seafloor due to seafloor spreading cause changes in sea level, and the drift of continents over the poles leads to the growth of continental glaciers. The movements of continents can also block ocean currents or open gateways through which they can flow, inhibiting or facilitating the global transfer of heat. For example, if future tectonic activity were to close the narrow channel between the Bahamas and Florida through which the Gulf Stream flows, average temperatures in western Europe might drop drastically.

Volcanism in the lithosphere affects climate by changing the composition and properties of the atmosphere. As we saw in Chapter 12, large volcanic eruptions can inject aerosols into the stratosphere, blocking solar radiation and temporarily lowering atmospheric temperatures on a global scale. After the massive April 1815 eruption of Mount Tambora in Indonesia, New England suffered through a "year without a summer" in 1816, which caused widespread crop failures. Recent large volcanic eruptions—including those of Krakatau (1883), El Chichón (1982), and Mount Pinatubo (1991)—each produced an average dip of 0.3°C in global surface temperatures about 14 months after the eruption. Temperatures returned to normal in about 4 years.

The Biosphere

The *biosphere* comprises all the organisms living on and beneath Earth's surface, in its atmosphere, and in its waters. Life is found almost everywhere on Earth, but the amount of life at any location depends on local climate conditions, as we can see from the satellite image of plant and algal biomass in **Figure 15.5**.

The total energy contained and transported by living organisms is relatively small on a global scale: less than 0.1 percent of incoming solar energy is used by plants in photosynthesis and thus enters the biosphere. The biosphere, however, is strongly coupled to the other components of the climate system by the metabolic processes described in Chapter 11. For example, terrestrial vegetation can affect atmospheric temperature, because plants absorb solar radiation for photosynthesis and release it as heat during respiration, and atmospheric moisture, because they take up groundwater and release it as water vapor. Organisms also regulate the composition of the atmosphere by taking up or releasing greenhouse gases such as carbon dioxide (CO_2) and methane (CH_4). Through photosynthesis, plants and algae transfer CO_2 from the atmosphere to the biosphere. Some of the carbon from that CO_2 moves from the biosphere to the lithosphere when it is precipitated as calcium carbonate shells or buried as organic matter in marine sediments. The biosphere thus plays a central role in the carbon cycle.

Humans, of course, are part of the biosphere, though hardly an ordinary part. Our influence over the biosphere is growing rapidly, and we have become the most active agents of environmental change. As an organized society, we behave in fundamentally different ways from other species. For example, we can study climate change scientifically and modify our actions according to what we have learned.

One of the anthropogenic changes in the climate system that is of greatest concern is a recent increase in atmospheric concentrations of greenhouse gases. We'll look next at the factors that regulate Earth's surface temperature and at the role greenhouse gases play in that process.

FIGURE 15.5 ■ The biosphere, represented by the global distribution of algal and plant biomass in the oceans and on land, as mapped by NASA's SeaWiFS satellite. [NASA/Goddard Space Flight Center.]

■ The Greenhouse Effect

The Sun is a yellow star that emits about half its radiant energy as visible light. The other half is split between infrared waves, which have longer wavelengths and lower energy intensities than visible light (and which we perceive as heat), and ultraviolet waves, which have shorter wavelengths and higher energy intensities than visible light. The average amount of solar radiation Earth's surface receives throughout the year is 340 watts per square meter of surface area (340 W/m², 1 *watt* = 1 joule per second, and *joule* is a unit of energy or heat). In comparison, the average amount of heat flowing out of Earth's deep interior by mantle convection is minuscule, only 0.06 W/m². Essentially all the energy driving the climate system ultimately comes from the Sun (**Figure 15.6**).

We know that the global surface temperature, averaged over daily and seasonal cycles, remains constant. Therefore, Earth's surface must be radiating energy back into space at a rate of precisely 340 W/m². Any less would cause the surface to heat up; any more would cause it to cool down. In other words, Earth maintains a *radiation balance:* an equilibrium between incoming and outgoing radiant energy. How is this equilibrium achieved?

A Planet Without Greenhouse Gases

Suppose Earth were a rocky sphere like the Moon, with no atmosphere at all. Some of the sunlight falling on the surface would be reflected back into space, and some would be absorbed by the rocks, depending on the color of the surface. A perfectly white planet would reflect all the solar energy falling on it, whereas a perfectly black planet would absorb it all. The fraction of the solar energy reflected by a surface is called its **albedo** (from the Latin word *albus,* meaning "white"). Although the full Moon looks bright to us, the rocks on its surface are mainly dark basalts, so its albedo is only about 7 percent. In other words, the Moon is dark gray—very nearly black.

The energy radiated by a black body increases rapidly as its temperature increases. A cold bar of iron is black and gives off little heat. If you heat the bar to 100°C, it gives off warmth in the form of infrared radiation (like a steam radiator). If you heat the bar to 1000°C, it becomes bright orange, radiating heat at visible wavelengths (like the burner on an electric stove).

A black body exposed to the Sun heats up until its temperature is at just the right value for it to radiate the incoming solar energy back into space. The same principle applies to a "gray body" like the Moon, except that the reflected energy must be excluded from the radiation balance. And, in the case of rotating bodies like the Moon and Earth, day-and-night cycles must be taken into account. The Moon's daytime temperatures rise to 130°C, and its nighttime temperatures drop to −170°C. Not a pleasant environment!

Earth rotates much faster than the Moon (once per day rather than once per month), which evens out the day-and-night extremes of temperature. Earth's albedo, at about 29 percent, is much higher than the Moon's because Earth's blue oceans, white clouds, and ice caps are more reflective than dark lunar basalts. If our atmosphere did not contain greenhouse gases, the average surface temperature

1 Solar energy input to Earth's surface is, on average, 342 W/m².

2 Heat flowing out of Earth's deep interior is much smaller — only 0.06 W/m².

3 To maintain a constant temperature, heat radiating from Earth must balance solar input.

Sun

FIGURE 15.6 ■ Earth's energy balance is achieved by the radiation of incoming solar energy back into space. Heat gain from Earth's interior is negligible in comparison to that from solar energy.

required to balance the absorbed solar radiation would be about −19°C (−2°F), cold enough to freeze all the water on the planet. Instead, Earth's average surface temperature remains a balmy 14°C (57°F). The difference of 33°C is a result of the greenhouse effect.

Earth's Greenhouse Atmosphere

Greenhouse gases, such as water vapor, carbon dioxide, methane, and ozone, absorb energy—coming directly from the Sun as well as radiated by Earth's surface—and reradiate it as infrared energy in all directions, including downward to the surface. In this way, they act like the glass in a greenhouse, allowing light energy to pass through, but trapping heat in the atmosphere. This trapping of heat, which increases the temperature at the surface relative to the temperature higher in the atmosphere, is known as the **greenhouse effect.**

How Earth's atmosphere balances incoming and outgoing radiation is illustrated in **Figure 15.7**. Incoming solar radiation that is not directly reflected is absorbed by Earth's atmosphere and surface. To achieve radiation balance, Earth radiates this same amount of energy back into space as infrared energy. Because of the heat trapped by greenhouse gases, the amount of energy transported away from Earth's surface, both by radiation and by the flow of warm air and moisture from the surface, is significantly larger than the amount Earth receives as direct solar radiation. The excess is exactly the energy received as Earthward infrared radiation from the greenhouse gases. It is this "back radiation" that causes Earth's surface to be 33°C warmer than it would be if the atmosphere contained no greenhouse gases.

Balancing the Climate System Through Feedbacks

How does the climate system actually achieve the radiation balance illustrated in Figure 15.7? Why does the greenhouse effect yield an overall warming of 33°C and not some larger or smaller amount? The answers to these questions are not simple because they depend on interactions among the many components of the climate system. The most important of those interactions involve *feedbacks.*

FIGURE 15.7 ■ To maintain radiation balance, Earth radiates as much energy into outer space, on average, as it receives from the Sun (340 W/m²). Of the incoming radiation, 100 W/m² (29 percent) is reflected, 161 W/m² is absorbed by Earth's surface, and 79 W/m² is absorbed by Earth's atmosphere. Radiation and flows of warm air and moisture transport more energy away from Earth's surface (502 W/m²) than it receives. The greenhouse gases in the atmosphere reflect most of this energy (342 W/m²) back to Earth's surface as infrared radiation. [IPCC, *Climate Change 2013: The Physical Science Basis.*]

Feedbacks come in two basic types: **positive feedbacks,** in which a change in one component is *enhanced* by the changes it induces in other components, and **negative feedbacks,** in which a change in one component is *reduced* by the changes it induces in other components. Positive feedbacks tend to amplify changes in a system, whereas negative feedbacks tend to stabilize the system against change.

Here are some of the feedbacks within the climate system that affect the surface temperature achieved by radiation balance:

■ *Water vapor feedback.* A rise in temperature increases the amount of water vapor that moves from Earth's surface into the atmosphere through evaporation. Water vapor is a greenhouse gas, so this increase enhances the greenhouse effect, and the temperature rises further—a positive feedback.

■ *Albedo feedback.* A rise in temperature reduces the accumulation of ice and snow in the cryosphere, which decreases Earth's albedo and increases the energy its surface absorbs. This increased warming enhances the temperature rise—another positive feedback.

■ *Radiative damping.* A rise in atmospheric temperature increases the amount of infrared energy radiated back into space, which moderates the temperature rise—a negative feedback. This "radiative damping" stabilizes Earth's climate against major temperature changes, keeping the oceans from freezing up or boiling off and thus maintaining an equable habitat for water-loving life.

■ *Plant growth feedback.* Increasing atmospheric CO_2 concentrations stimulate plant growth. Growing plants remove CO_2 from the atmosphere by converting it into carbon-rich organic matter, thus reducing the greenhouse effect—another negative feedback.

Feedbacks can involve much more complex interactions among components of the climate system. For example, an increase in atmospheric water vapor produces more clouds. Because clouds reflect solar energy, they increase the planetary albedo, which sets up a negative feedback between atmospheric water vapor and temperature. On the other hand, clouds absorb infrared radiation efficiently, so increasing the cloud cover enhances the greenhouse effect, thus providing a positive feedback between atmospheric water vapor and temperature. Does the net effect of clouds produce a positive or a negative feedback?

Scientists have found it surprisingly difficult to answer such questions. The components of our climate system are joined through an amazingly complex web of interactions on a scale far beyond experimental control. Consequently, it is often impossible to gather data that isolate one type of feedback from all the others. Scientists must therefore turn to computer models to understand the inner workings of the climate system.

Climate Models and Their Limitations

Generally speaking, a **climate model** is any representation of the climate system that can reproduce one or more aspects of its behavior. Some models are designed to study local or regional climate processes, such as the relationships between water vapor and clouds, but the most interesting representations are global models that describe how climate has changed in the past or predict how it might change in the future.

At the heart of such global climate models are schemes for computing movements within the atmosphere and oceans based on the fundamental laws of physics. These *general circulation models* represent the currents of air and water driven by solar energy on scales ranging from small disturbances (storms in the atmosphere, eddies in the oceans) to global circulation patterns (wind belts in the atmosphere, thermohaline circulation in the oceans). Scientists represent the basic physical variables (temperature, pressure, density, velocity, and so forth) on three-dimensional grids comprising millions, or even billions, of geographic points. They use supercomputers to solve mathematical equations that describe how the variables change over time at each of these points (**Figure 15.8**). You see the results of these calculations whenever you

FIGURE 15.8 ■ Numerical climate models are used to predict future climate change. This global climate model, developed with support from the U.S. Department of Energy, portrays interactions among the atmosphere, hydrosphere, cryosphere, and land surface. The colors in the oceans represent sea surface temperatures, and the arrows represent surface wind velocities. [Warren Washington and Gary Strand/National Center for Atmospheric Research.]

tune in the weather report on your favorite TV station. These days, most weather predictions are made by entering the current conditions observed at thousands of weather stations into a general circulation model and running it forward in time. Weather predictions thus use the same basic computer programs that are used for climate modeling.

Climate modeling is more difficult than weather prediction, however. In predicting weather a few days from now, scientists can ignore such slow processes as changes in atmospheric greenhouse gas concentrations or in oceanic circulation. Climate predictions, on the other hand, require that we properly model these slow processes, including all important feedbacks, in addition to the rapid movements of air masses. Moreover, the simulation must be extended years or decades into the future. Such enormous calculations require weeks of time on the world's largest supercomputers.

Because current climate models are complex and subject to error, their predictions must be viewed with some skepticism. Many questions remain about how the climate system works—for instance, how clouds affect atmospheric temperatures. The predictions of climate models have been a topic of much debate among experts and government authorities who must understand and deal with the consequences of human-induced climate change. We will take a closer look at those predictions in Chapter 23.

▪ Climate Variation

Earth's climate varies considerably from place to place: its poles are frigid and arid, its tropics sweltering and humid. Comparable variations in climate can also occur over time. The geologic record shows us that periods of global warmth have alternated with periods of glacial cold many times in the past. This climate variation is erratic; dramatic changes can happen in just a few decades or evolve over time scales of many millions of years.

1 During normal years, warm surface waters pool in the western tropical Pacific.

2 Trade winds blow from east to west, pushing warm surface waters westward.

3 Cold water wells up from the depths in the eastern tropical Pacific.

Normal July 1996

Low pressure   High pressure

Equator   NE trades

Indonesia   SE trades   South America

Equator

Warm water pool   Upwelling

4 During an El Niño event, the warm waters shift eastward.

5 The trade winds slacken, or may even reverse.

El Niño July 1997

High pressure   Low pressure

Indonesia   South America

Warm water pool

La Niña July 1998

6 During a La Niña event, surface waters in the eastern Pacific are colder than normal, and the trade winds strengthen.

FIGURE 15.9 ▪ The El Niño–Southern Oscillation (ENSO) is a natural variation in the exchange of heat between the atmosphere and the surface waters of the tropical Pacific Ocean. [U.S.–French TOPEX/Poseidon mission.]

Some climate variation can be attributed to factors outside the climate system, such as solar forcing and changes in the distribution of land and sea surfaces caused by continental drift. Others result from changes within the climate system itself, such as the growth of continental glaciers that increase Earth's albedo. Both types of variations, external and internal, can be amplified or suppressed by feedbacks. In this section, we will examine several types of climate variation and discuss their causes, beginning with short-term variation on a regional scale.

Short-Term Regional Variations

Local and regional climates are much more variable than the average global climate: averaging over large surface areas, like averaging over time, tends to smooth out small-scale fluctuations. Over periods of years to decades, the predominant regional variations result from interactions between atmospheric circulation and sea and land surfaces. They generally occur in distinct geographic patterns, although their timing and amplitudes can be highly irregular.

One of the best-known examples is a warming of the eastern Pacific Ocean that occurs every 3 to 7 years and lasts for a year or so. Peruvian fishermen call such an event **El Niño** ("the boy child" in Spanish) because the warming typically reaches the surface waters off the coast of South America around Christmastime. El Niño events can decimate fish populations, which depend on the upwelling of cold water for their nutrient supply, and can thus be disastrous for coastal human populations that depend on fishing.

Scientists have shown that El Niño and a complementary cooling event, known as *La Niña* ("the girl child"), are part of a natural variation in the exchange of heat between the atmosphere and the tropical Pacific Ocean. This variation is known as the El Niño–Southern Oscillation, or **ENSO** (Figure 15.9).

Normally, atmospheric pressure gradients cause prevailing winds, the trade winds, to blow from east to west, pushing the warm tropical waters westward. This movement of water causes colder water to well up from the ocean depths off Peru. Sporadically, the trade winds weaken or occasionally even reverse direction, cutting off the upwelling and equalizing water temperatures across the tropical Pacific (an El Niño event). At other times, the trade winds strengthen, enhancing the temperature difference between the eastern and western Pacific (a La Niña event). This recurring swing in air pressure is called the Southern Oscillation.

In addition to disrupting the eastern Pacific fishery, El Nino has been implicated in triggering changes in wind and rain patterns over much of the globe. The 1997–1998 El Niño was the strongest on record, and it contributed to droughts in Australia and Indonesia; heavy rains and flash floods in Peru, Ecuador, and Kenya; and storms in California that caused landslides and floods (Figure 15.10).

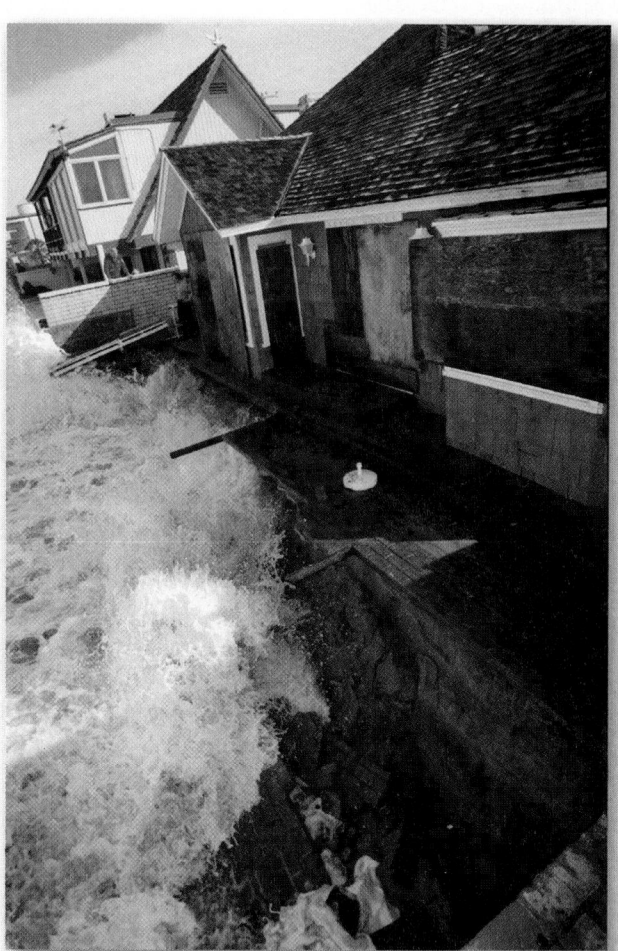

FIGURE 15.10 ■ Storm waves associated with the 1997–1998 El Niño attacking homes along the Pacific coastline in Del Mar, California. [Reuters/Landov.]

Crops failed and fisheries were decimated in many areas. According to one estimate, the global disruption in weather patterns and ecosystems may have cost 23,000 lives and caused $33 billion in damage.

Climate scientists have identified similar patterns of weather and climate variation in other regions. One example is the North Atlantic Oscillation, a highly irregular fluctuation in the balance of atmospheric pressures between Iceland and the Azores that has a strong influence on the movement of storms across the North Atlantic and thus affects weather conditions throughout Europe and parts of Asia. A better understanding of these patterns is improving long-range weather forecasting and may provide important information about the regional effects of human-induced climate change.

Long-Term Global Variations: The Pleistocene Ice Ages

Some of the most dramatic climate variations that can be seen in the geologic record are the glacial cycles of the

FIGURE 15.11 ■ At the last glacial maximum around 20,000 years ago, continental glaciers covered most of North America. The continental shelves were exposed by the lowering of the sea level, illustrated here by the expanded coastline of Florida. [Wm. Robert Johnston.]

Pleistocene epoch, which began 1.8 million years ago. A **glacial cycle** begins with a gradual decline in temperature of about 6°C to 8°C from a warm **interglacial period** to a cold *glacial period,* or **ice age.** As the climate cools, water is transferred from the hydrosphere to the cryosphere. The amount of sea ice increases, and more snow falls on the continents in winter than melts in summer, increasing the volume and area of polar ice caps and decreasing the volume of the oceans. As the ice caps expand into lower latitudes, they reflect more solar energy back into space, and Earth's surface temperatures fall further—an example of albedo feedback. Sea level falls, exposing areas of the continental shelves that are normally under water. At the peak of the ice age—the *glacial maximum*—great continental glaciers up to 2 or 3 m thick cover vast land areas (**Figure 15.11**). The ice age ends abruptly with a rapid rise in temperature. Water is transferred from the cryosphere to the hydrosphere as the ice caps melt and sea level rises.

TIMING THE PLEISTOCENE ICE AGES A precise record of Pleistocene temperature variations can be obtained by measuring oxygen isotopes preserved in marine

FIGURE 15.12 ■ Changes in global climate over the last 1.8 million years, as inferred from oxygen isotope ratios in marine sediments. The peaks indicate interglacial periods (high temperatures, low ice volumes, high sea level), and the valleys indicate ice ages (low temperatures, high ice volumes, low sea level). [L. E. Lisiecki and M. E. Raymo, *Paleoceanography* 20 (2005): 1003.]

sediments and in glacial ice. Pleistocene marine sediments contain numerous fossils of foraminifera: small, single-celled marine organisms that secrete shells of calcite ($CaCO_3$). The proportions of oxygen isotopes incorporated into these shells depend on the oxygen isotope ratio of the seawater in which the organisms lived. Water (H_2O) containing the lighter and more common isotope, oxygen-16 (^{16}O), has a greater tendency to evaporate than water containing the heavier oxygen-18 (^{18}O). Therefore, during ice ages, ^{18}O is preferentially left behind in the oceans as water containing ^{16}O evaporates from the ocean surface and is trapped in glacial ice, and the $^{18}O/^{16}O$ ratio in the oceans rises. Paleoclimatologists can use $^{18}O/^{16}O$ ratios in marine sediment beds to estimate sea surface temperature and ice volume at the time the beds were deposited. **Figure 15.12** shows changes in global climate over the last 1.8 million years as inferred from these estimates.

As $^{18}O/^{16}O$ ratios in the oceans increase during ice ages, those in the layers of ice that form the growing glaciers

Earth Issues

15.1 Ice-Core Drilling in Antarctica and Greenland

At the Vostok Station in the frozen Antarctic, Russian and French scientists have worked for decades to uncover the climatological history of Earth hidden in glacial ice. In the 1970s and 1980s, they drilled boreholes 2000 m deep into the East Antarctic ice sheet and brought up a set of ice cores for detailed laboratory study. The cores contained layers of ice produced by annual cycles of ice formation from snow. By carefully counting the layers, working from the top down, the researchers were able to infer the age of the ice with depth, much as tree rings are used to reveal the age of a tree. They measured oxygen isotope ratios in the ice as well as the gas composition of small bubbles trapped in the ice. From this stratigraphic record, they produced a detailed history of glacial cycles over the last 160,000 years.

By 1998, the ice borers at Vostok had drilled to a depth of 3600 m, penetrating ice accumulated over the past four glacial cycles and extending the climate record to more than 400,000 years ago. The data supported other evidence suggesting that variations in Earth's orbit—Milankovitch cycles—control the alternation of ice ages and interglacial periods, and they showed that surface temperatures were correlated with concentrations of greenhouse gases in the atmosphere (see Figure 15.12). The Vostok results have been confirmed by ice core drilling at a number of other locations on the Antarctic and Greenland ice sheets.

These triumphs have not been won easily. The Vostok Station, located at an elevation of 3500 m near the center of Antarctica (its location is marked on the map of Figure 21.6), is an especially grueling place to do research. Its average annual temperature is only −55°C, and the lowest reliably measured temperature on Earth's surface, −89.2°C, was recorded there in 1983. The scientists not only had to endure these extreme conditions, but also had to be careful not to melt and contaminate the ice cores while drilling them, transporting them to

Russian scientists at the Vostok Station carefully remove an ice core from a drill. The layers produced by annual cycles of ice formation are visible in the core. [Alexey Ekaikin/Reuters/Landov.]

laboratories, and storing them. And they had to guard against misleading results—caused, for example, by the reaction of CO_2 with impurities in the ice. It is a tribute to the patience and ingenuity of these hardy bands of researchers that glacial ice cores have contributed so much to our understanding of the history of global climate change.

decrease. The best records of climate variation during the last half-million years come from ice cores drilled in the East Antarctic ice sheet by Russian scientists at the Vostok Station and in the Greenland ice sheet by European and U.S. teams (see Earth Issues 15.1). The oxygen isotope ratios of the ice layers in the cores can be used to estimate atmospheric temperatures at the time the ice formed. The composition of the atmosphere, including the concentrations of carbon dioxide and methane, can also be measured in tiny bubbles of air trapped at the time the ice formed. **Figure 15.13** displays these three types of measurements from the Vostok ice core.

MILANKOVITCH CYCLES The major ups and downs in the marine sediment record (see Figure 15.12) during the Pleistocene epoch match the glacial cycles in the ice core record (see Figure 15.13). Why does the climate fluctuate in such a pattern? Solar forcing is an obvious possibility.

We know that it gets cold in winter because the amount of sunlight that falls at a particular latitude decreases due to the tilt of Earth's axis. Could periods of glacial cold be explained by decreases in solar energy input over much longer time scales?

The answer appears to be yes. There are indeed small periodic variations in the amount of radiation Earth receives from the Sun. These variations are caused by **Milankovitch cycles,** periodic variations in Earth's movement around the Sun, named after the Serbian geophysicist who first calculated them in the early twentieth century. Three kinds of Milankovitch cycles can be correlated with global climate variation (**Figure 15.14**).

First, the shape of Earth's orbit around the Sun changes periodically, becoming more circular at some times and more elliptical at others. The degree of ellipticity of Earth's orbit around the Sun is known as its *eccentricity*. A nearly circular orbit has low eccentricity, and a more elliptical orbit has high eccentricity (Figure 15.14a). The amount of solar radiation Earth receives, averaged over its surface, varies slightly with eccentricity. The length of one cycle of variation in eccentricity is about 100,000 years.

Second, the angle or *tilt* of Earth's axis of rotation changes periodically. Today this angle is 23.5°, but it cycles between 21.5° and 24.5° with a period of about 41,000 years. These variations also slightly change the amount of radiation Earth receives from the Sun (Figure 15.14b).

Third, Earth's axis of rotation wobbles like a top, giving rise to a pattern of variation called *precession* with a period of about 23,000 years (Figure 15.14c). Precession, too, modifies the amount of radiation Earth receives from the Sun, though by less than variations in eccentricity and tilt.

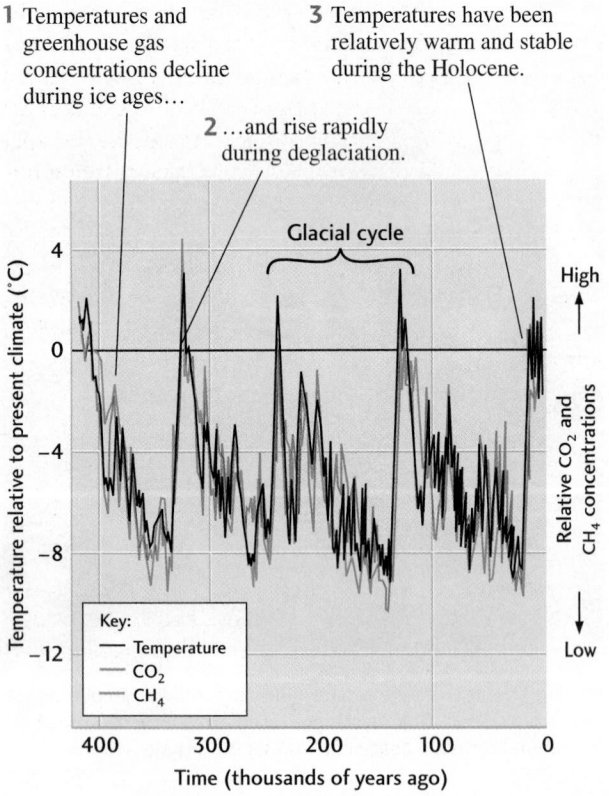

FIGURE 15.13 ■ Three types of data were recovered from the Vostok Station ice cores, which were drilled to a depth of 3600 m in the East Antarctic ice sheet. Temperatures were estimated from oxygen isotope ratios. Carbon dioxide and methane concentrations came from measurements of air samples trapped as tiny bubbles within the Antarctic ice. [IPCC, *Climate Change 2001: The Scientific Basis*.]

CORRELATIONS BETWEEN MILANKOVITCH CYCLES AND GLACIAL CYCLES You can see lots of small ups and downs in the record in Figure 15.12, but in the last half-million years, the record reveals a sawtooth pattern of major glacial cycles that looks roughly like this:

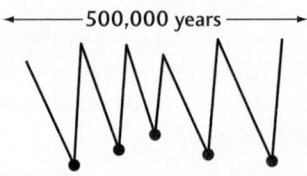

In particular, you can count five glacial maxima, in which ice volumes are high and temperatures are low (shown in the sketch above as black dots), revealing an average time interval between glacial maxima of about 100,000 years. This 100,000-year spacing of minimum temperatures closely matches the times of high orbital eccentricity, when Earth received slightly less radiation from the Sun—a Milankovitch cycle.

(a) Eccentricity (100,000 years)

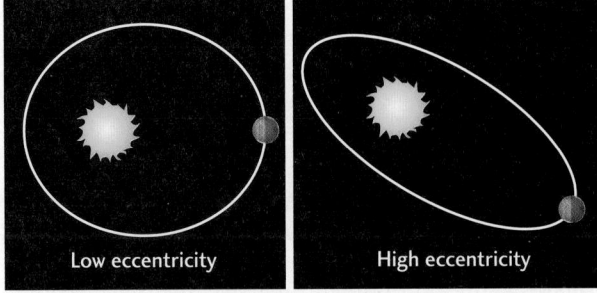

Low eccentricity   High eccentricity

(b) Tilt (41,000 years)

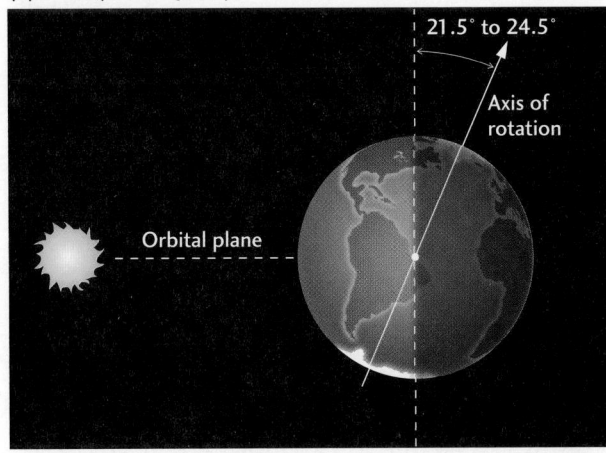

21.5° to 24.5°

Axis of rotation

Orbital plane

(c) Precession (23,000 years)

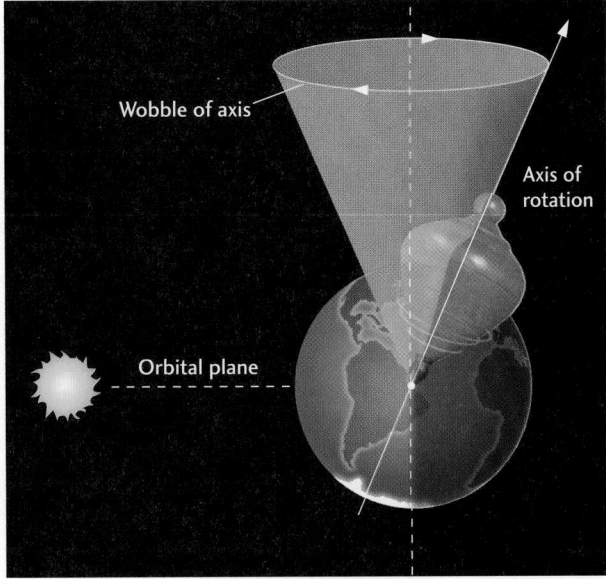

Wobble of axis

Axis of rotation

Orbital plane

FIGURE 15.14 ▪ Three kinds of Milankovitch cycles (much exaggerated in these diagrams) affect the amount of solar radiation Earth receives. (a) Eccentricity is the degree of ellipticity of Earth's orbit. (b) Tilt is the angle between Earth's axis of rotation and the angle perpendicular to the orbital plane. (c) Precession is the wobble of the axis of rotation. One can imagine this motion by thinking of the wobble of a spinning top.

Now let's move backward in time to examine the first half-million years of the record, in Figure 15.12, from 1.8 million to 1.3 million years ago. Again, we see many small fluctuations, but major maxima and minima occur more frequently than they do in the later record, as approximated in the following sketch:

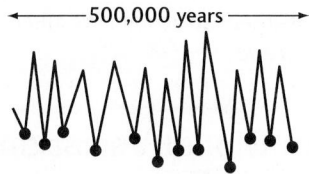

←————— 500,000 years —————→

During this period, we find 12 glacial maxima, with an average spacing between them of about 41,000 years (500,000 years/12 cycles = 41,667 years per cycle). This shorter interval is very close to the 41,000-year cycle of variation in the tilt of Earth's axis—another Milankovitch cycle! Like the variation in eccentricity, the variation in tilt is very small—only about 3° (see Figure 15.14b)—but it's evidently enough to trigger ice ages.

The small changes in solar radiation caused by Milankovitch cycles cannot by themselves explain the large drops in Earth's surface temperature from interglacial periods to ice ages. Some type of positive feedback must be operating within the climate system to amplify the solar forcing. The data in Figure 15.13 strongly suggest that this feedback involves greenhouse gases. Atmospheric concentrations of carbon dioxide and methane precisely track the temperature variations throughout the glacial cycles: warm interglacial periods are marked by high concentrations, cold glacial periods by low concentrations. Exactly how this feedback works has not yet been fully explained, but it demonstrates the importance of the greenhouse effect in long-term climate variations.

Many other aspects of this story are not yet understood. For example, you will notice in Figure 15.12 that the 41,000-year periodicity continued to dominate the climate record up to about 1 million years ago. Then, the highs and lows became more variable, eventually shifting to the 100,000-year periodicity after about 700,000 years ago. What caused this transition? Climate scientists are still scratching their heads.

In fact, we don't really know what triggered the Pleistocene ice ages. The climate record shows that the 41,000-year glacial cycles were not confined to the Pleistocene, but extended back at least into the Pliocene epoch (5.3 million to 1.8 million years ago), when Antarctica became covered in ice. The global cooling of Earth's climate that preceded these glaciations began during the Miocene epoch (23 million to 5.3 million years ago). Its cause continues to be debated, although most geologists believe it is somehow related to continental drift. According to one hypothesis, the collision of the Indian subcontinent with Eurasia and the resulting Himalayan orogeny led to an increase in the weathering of silicate rocks, and the chemical reactions of weathering decreased the amount of CO_2 in the atmosphere. Other hypotheses are based on changes in oceanic circulation associated with the opening of the Drake Passage between South America and Antarctica (25 million to 20 million years ago) or the closing of the Isthmus of Panama between North and South America (about 5 million years ago). Perhaps the cooling resulted from a combination of these events.

Long-Term Global Variations: Paleozoic and Proterozoic Ice Ages

In addition to the Pleistocene ice ages, there is good evidence in the geologic record for earlier episodes of continental glaciation during the Permian-Carboniferous and Ordovician periods and at least twice in the Proterozoic eon. In most cases, these events can be explained by plate tectonic processes, coupled with albedo feedback and other feedbacks in the climate system.

For most of Earth's history, there were no extensive land areas in the polar regions, and there were no ice caps. Oceanic circulation extended from equatorial regions into polar regions, transporting heat and helping the atmosphere to distribute temperatures fairly evenly over Earth's surface. When large land areas drifted to positions that obstructed this efficient transport of heat, the differences in temperature between the poles and the equator increased. As the poles cooled, ice caps formed. Some geologists believe that at one time in the late Proterozoic, Earth was completely covered by ice, and that only greenhouse gases emitted into the atmosphere by volcanoes allowed it to warm up again. We'll take a closer look at this "Snowball Earth hypothesis" in Chapter 21.

Variations During the Most Recent Glacial Cycle

Within glacial cycles, temperatures do not vary smoothly over time (see Figure 15.13). Superimposed on the 100,000-year glacial cycles are climate fluctuations of shorter duration, some nearly as large as the changes from glacial to interglacial periods. Geologists have combined information from cores in continental and valley glaciers, lake sediments, and deep-sea sediments to reconstruct a decade-by-decade—and in some cases, a year-by-year—history of short-term climate variations during the most recent glacial cycle.

The most recent ice age is known as the *Wisconsin glaciation.* Temperatures began to drop about 120,000 years ago, but reached their lowest values only about 21,000 to 18,000

years ago (the Wisconsin glacial maximum). Temperatures then rebounded to warm interglacial levels 11,700 years ago, marking the end of the Pleistocene and the beginning of the Holocene. Here we summarize some of the basic features of this remarkable chronicle.

- During the Wisconsin glaciation, Earth's climate was highly variable, with shorter (1000-year) temperature oscillations occurring within longer (10,000-year) cycles. The most extreme variations appear to have been in the North Atlantic region, where average local temperatures rose and fell by as much as 15°C. Each 10,000-year cycle comprised a set of progressively cooler 1000-year oscillations and ended with an abrupt warming. Massive discharges of icebergs and fresh water resulting from these sudden warmings altered thermohaline circulation in the oceans and dumped large amounts of glacial material into deep-sea sediments.

- The transition from the Wisconsin glaciation to the current interglacial period, the Holocene, also involved rapid climate fluctuations. The climate abruptly warmed around 14,500 years ago. It then cooled back to glacial conditions in an ice age called the "Younger Dryas," and finally warmed to nearly present-day conditions about 11,700 years ago. Both warming periods were very rapid; broad regions of Earth experienced almost simultaneous changes from ice age to interglacial temperatures during intervals as short as 30 to 50 years. Evidently, the entire climate system can flip from one state (glacial cold) to another (interglacial warmth) in less than a human lifetime! This observation raises the possibility that anthropogenic changes could trigger abrupt shifts to a new (and unknown) climate state, rather than just a gradual warming.

- The Holocene has been unusually long and stable when compared with the previous interglacial periods of the Pleistocene epoch. Regional temperatures have fluctuated by about 5°C on time scales of 1000 years or so, but the global changes during this period have been much smaller, with a total range of only 2°C. These equable Holocene conditions were no doubt favorable for the rapid rise of agriculture and civilization that followed the end of the Wisconsin glaciation.

Some scientists think that if human civilization had not come along, Earth's climate might by now be plunging into another ice age, driven by decreasing amounts of solar energy due to Milankovitch cycles and accompanied by decreasing atmospheric concentrations of greenhouse gases. According to one hypothesis, the expansion of civilization began to release significant amounts of greenhouse gases into the atmosphere as early as 8000 years ago, primarily through deforestation and the rise of agriculture, extending the warm interglacial period beyond its natural limit.

Whatever the reason, measurements from ice cores indicate that, from the end of the Pleistocene until the dawn of the industrial age, atmospheric concentrations of the major greenhouse gases stayed relatively constant. The average CO_2 concentration, for example, fluctuated only between 260 and 280 ppm—less than a 10 percent variation over that entire period. But that situation ended early in the nineteenth century with the beginning of the industrial revolution, when human emissions of greenhouse gases shot upward.

■ The Carbon Cycle

In the past 200 years, atmospheric CO_2 concentrations have risen by nearly 50 percent, from about 270 ppm to over 400 ppm (reached in mid-2013). Earth's atmosphere has not contained this much CO_2 for at least the last 400,000 years, and probably for the last 20 million years. Atmospheric CO_2 concentrations are now increasing at an unprecedented rate of 0.5 percent per year, faster than at any time in recent geologic history.

Yet the situation could be worse. Over the decade 2000–2009, human activities injected an average of 8.9 gigatons (Gt) of carbon into the atmosphere each year. (A gigaton, or 1 billion tons, is 10^{12} kg, the mass of 1 km^3 of water. Note that emissions are calculated in gigatons of carbon, not carbon dioxide. See Exercise 4 at the end of this chapter.) Fossil-fuel burning and other industrial activities emitted about 7.8 Gt/yr of carbon, and the burning of forests and other changes in land use emitted an additional 1.1 Gt/yr. If all of that carbon had stayed in the air, the atmospheric CO_2 increase would have been over 1 percent per year, more than twice the observed rate. Instead, 4.9 Gt of carbon was removed from the atmosphere each year by natural processes. Where did all that carbon go?

We will address this question by examining the **carbon cycle:** the continual movement of carbon between different components of the Earth system. We touched on the carbon cycle when we discussed biogeochemical cycles—geochemical cycles that involve the biosphere—in Chapter 11. Let's begin with a broader look at geochemical cycles.

Geochemical Cycles and How They Work

Geochemical cycles are patterns of flow, or *flux*, of chemicals from one component of the Earth system to another. In discussing geochemical cycles, we view components of the Earth system—atmosphere, hydrosphere, cryosphere, lithosphere, and biosphere—as **geochemical reservoirs** for the storage of carbon and other chemicals, linked by processes that transport chemicals among them. By quantifying the amounts of the chemicals that are stored in and moved among the various reservoirs, we can gain new insights into the workings of the Earth system.

RESIDENCE TIME Reservoirs gain chemicals from inflows and lose chemicals from outflows. If inflow equals outflow, the amount of the chemical in the reservoir stays the same, even though the chemical is constantly entering and leaving it. On average, a molecule of the chemical spends a certain amount of time, called the **residence time,** in a reservoir.

Think of a crowded bar where many more people want to get in than are allowed by the fire code. After the room fills up, or reaches its *capacity,* the bouncer begins stopping people at the door. During the busiest hours, when people are waiting to get in, the bar is filled to capacity, or *saturated,* and is at a steady state, with the number of people going in exactly balancing the number of people coming out. Even though some people come early and stay late and others leave after only a short time, we can calculate an average length of time between arrival and departure—the residence time—by dividing the capacity of the room by the rate of arrivals (*inflow*) or departures (*outflow*). If the room's capacity is 30 people and a new person is let in every 2 minutes on average, the residence time is 60 minutes.

Similarly, we can visualize a chemical's residence time in the ocean as the average time that elapses between the entry of a molecule of that chemical into the ocean and its removal through sedimentation or some other process. For example, the residence time of sodium in the ocean is extremely long—about 48 million years—because sodium is highly soluble in seawater (that is, the capacity of the reservoir to store sodium is high) and because rivers contain relatively small amounts of sodium (its inflow into the reservoir is low). In contrast, iron has a residence time in the ocean of only about 100 years because its solubility in seawater is very low and the inflow from rivers is relatively high.

Residence times of chemicals in the atmosphere are usually shorter than those in the ocean because the atmosphere is a smaller reservoir than the ocean and fluxes into and out of the atmosphere can be larger. Sulfur dioxide, for example, has a residence time in the atmosphere of hours to weeks, and oxygen, which makes up about 21 percent of the atmosphere, has a residence time of 6000 years. Atmospheric nitrogen gas is abundant (about 78 percent of the atmosphere) and stable and so its residence time is almost 400 million years. A molecule of nitrogen that entered the atmosphere in the late Paleozoic era, about 300 million years ago, is still likely to be there!

CHEMICAL REACTIONS In many cases, reactions with other chemicals govern a chemical's residence time in a reservoir. For example, as we learned in Chapter 5, a calcium ion (Ca^{2+}) can be removed from solution in seawater

Carbon dioxide from
fossil-fuel burning
dissolves in the ocean,...

...where it combines
with water to form
carbonic acid,...

...which dissociates into
hydrogen and bicarbonate
ions.

The hydrogen ions react
with carbonate ions to
form more bicarbonate.

Available
carbonate

The net effect is to reduce the
carbonate available to marine
organisms that precipitate
calcium carbonate.

FIGURE 15.15 ■ Increasing CO_2 concentrations in the atmosphere drive a series of chemical reactions in seawater, causing ocean acidification and reducing the ability of marine organisms to form shells and skeletons of calcium carbonate.

by reacting with a carbonate ion (CO_3^{2-}) to form calcium carbonate ($CaCO_3$), which can precipitate as carbonate sediment. The amount of calcium that remains dissolved in seawater thus depends on the availability of carbonate ions, which in turn depends on the influx of carbon dioxide (CO_2) into the ocean. When carbon dioxide dissolves in water, most of it reacts with the water to form carbonic acid (H_2CO_3), which can dissociate into hydrogen (H^+) and bicarbonate ions (HCO_3^-). Some of the hydrogen ions then react with carbonate ions to form more bicarbonate ions (**Figure 15.15**). The net effect is to increase the acid content of seawater and decrease the concentration of carbonate ions. The decrease in carbonate affects the ability of marine organisms such as corals, clams, and foraminifera to build their shells and skeletons by precipitating calcium carbonate. As we shall see, this process of **ocean acidification** is one of the most threatening aspects of anthropogenic global change.

TRANSPORT ACROSS INTERFACES Fluxes between reservoirs are governed by processes that transport chemicals into and out of them (**Figure 15.16**). For example, volcanoes transport gases, aerosols, and dust from the lithosphere into the atmosphere. Wind lifts dust from the lithosphere into the atmosphere, and gravity pulls it back to Earth's surface. Windborne dust is also an important mechanism for transporting minerals from the lithosphere

to the hydrosphere, although by far the largest flux between those two reservoirs comes from minerals dissolved or suspended in rivers.

Evaporation and precipitation transport huge amounts of water between the atmosphere and the surfaces of both land and ocean. At the sea surface, gas molecules and salts, in the form of tiny crystals, escape from their dissolved state in the water and enter the atmosphere. That flux is balanced by the dissolution of atmospheric constituents that return to the oceans as rain and by the dissolution of gases directly across the ocean surface.

Sedimentation is the great flux that keeps the ocean in a steady state, primarily by counterbalancing the influx of chemicals in river water. As seafloor sediments are buried, they become part of the oceanic crust. There they stay until they move into the mantle through subduction or become part of the continental crust through accretion. Over the long term, tectonic uplift exposes crustal rocks to weathering and erosion, maintaining the balance of fluxes among the reservoirs.

As we saw in Chapter 11, the biosphere is a unique reservoir because each individual organism interacts constantly with its environment. The most important fluxes into and out of the biosphere are the inflow and outflow of atmospheric gases by respiration, the inflow of nutrients from the lithosphere and hydrosphere, and the outflow of nutrients through the death and decay of organisms. The

FIGURE 15.16 ■ A number of processes result in fluxes of chemicals between components of the climate system.

carbon cycle, which depends critically on the pumping of carbon into and out of the atmosphere by living organisms, is clearly a biogeochemical cycle.

EXAMPLE: THE CALCIUM CYCLE Before we examine the carbon cycle in more detail, let's take a look at the calcium cycle, which provides a simpler illustration of the concepts involved in geochemical cycles (**Figure 15.17**).

The ocean contains about 560,000 Gt of calcium dissolved in a total ocean mass of about 1.4×10^9 Gt. Calcium steadily enters this reservoir in rivers, which transport large quantities of dissolved and suspended calcium. That calcium is derived from the weathering of carbonate rocks and other minerals such as gypsum and calcium-rich plagioclase feldspar. A much smaller amount of calcium enters the ocean via transport by windblown dust. If the ocean received this

FIGURE 15.17 ■ The calcium cycle, emphasizing fluxes into and out of the ocean. Fluxes are given in gigatons (Gt; 10^{12} kg) per year. The inflow of calcium into the ocean approximately balances the outflow.

continuous inflow of calcium without there being any way to remove it, the ocean would quickly become supersaturated with calcium. The flux that keeps the amount of calcium in the ocean relatively constant is the precipitation of calcium carbonate, as described above. A smaller amount of calcium is precipitated as gypsum in evaporites. Over much longer time scales, calcium-rich sediments are uplifted and weathered, and the calcium they contain is returned to the ocean.

The amount of calcium the ocean can hold is much larger than the inflow and outflow of calcium, so calcium has a fairly long residence time in the ocean. By dividing the total annual influx (0.9 Gt/year) by the ocean's calcium capacity (560,000 Gt), we obtain a residence time of about 600,000 years.

The Cycling of Carbon

Carbon cycles among four main reservoirs: the atmosphere; the oceans, including marine organisms; the land surface, including plants and soils; and the deeper lithosphere (**Figure 15.18**). We can describe the flux of carbon among these reservoirs in terms of several basic subcycles. During times when Earth's climate is stable, each subcycle can be characterized by a constant flux.

ATMOSPHERE-OCEAN GAS EXCHANGE The exchange of CO_2 directly across the interface between the oceans and the atmosphere amounts to an average carbon flux of about 80 Gt per year. The flux through this subcycle depends on many factors, including air and sea temperatures and the composition of the seawater, but it is particularly sensitive to wind velocity, which increases the transfer of CO_2 and other gases by stirring up the surface water and generating sea spray. Carbon dioxide dissolved in seawater escapes from solution and enters the atmosphere by evaporating from sea spray, while atmospheric CO_2 enters the ocean by dissolving in sea spray and rain or directly across the sea surface.

ATMOSPHERE-BIOSPHERE GAS EXCHANGE The subcycle with the greatest carbon flux, 120 Gt per year, is the exchange of CO_2 between the terrestrial biosphere and the atmosphere by photosynthesis, respiration, and decomposition. Plants take in this entire amount of CO_2 during photosynthesis and respire about half of it back into the atmosphere. The other half is incorporated into plant tissues—leaves, wood, and roots—as organic carbon. Animals eat the plants, and microorganisms decompose them; both processes result in the breakdown of plant tissues and the respiration of CO_2. Much of the organic carbon released by these processes—about three times the total plant mass—is stored in soils. A significant fraction (about 4 Gt/year) reenters the atmosphere through direct oxidation by forest fires and other combustion of plant material.

A small fraction of the CO_2 incorporated into plant tissues (0.4 Gt/year) is dissolved in surface waters and

FIGURE 15.18 ▪ The carbon cycle describes the fluxes of carbon between the atmosphere and its other principal reservoirs. Amounts of carbon stored in each reservoir are given in gigatons; fluxes are given in gigatons per year. [IPCC, *Climate Change 2001: The Scientific Basis*, updated according to IPCC, *Climate Change 2013: The Physical Science Basis*.]

transported by rivers to the ocean, where it is respired back into the atmosphere by marine organisms and eventually taken up again by plants through photosynthesis.

LITHOSPHERE-ATMOSPHERE GAS EXCHANGE The weathering of carbonate rock removes about 0.2 Gt of carbon per year from the lithosphere and an equal amount from the atmosphere. The CO_2 dissolved in rainwater forms carbonic acid, which reacts with carbonates in the rock, releasing carbonate and bicarbonate ions, which are transported by rivers to the ocean. Here, shell-forming marine organisms reverse the weathering reaction, precipitating calcium carbonate and releasing an equal amount of carbon back into the atmosphere as CO_2. This subcycle illustrates one way in which the carbon cycle is linked to the calcium cycle.

Another such linkage is through the weathering of silicate rocks, most of which contain significant amounts of calcium. Silicate weathering releases calcium into surface waters, which flow to the ocean, where the calcium ions combine with carbonate ions to form calcium carbonate, thus removing CO_2 from the atmosphere. The net flux of carbon from silicate weathering is relatively small (less than 0.1 Gt/year), so, like volcanism (which releases minor amounts of CO_2 into the atmosphere), it is usually neglected in short-term climate modeling. Over the long term, however, the effects of silicate weathering can be substantial, because, unlike carbonate weathering, it removes CO_2 from the atmosphere and stores it, semi-permanently, in the lithosphere. For example, the uplifting of the Himalaya and the Tibetan Plateau, which began about 40 million years ago, may have increased weathering rates enough to reduce the concentration of CO_2 in the atmosphere, contributing to the subsequent climate cooling that led to the Pleistocene glaciations (see Earth Issues 22.1).

Human Perturbations of the Carbon Cycle

With this background, let's return to the fate of anthropogenic carbon emissions. **Figure 15.19** shows what happened to the carbon that was added to the atmosphere by human activities in the decade 2000-2009. Out of a total of 8.9 Gt/yr injected into the atmosphere by human activities,

1 Human activities release a total of 8.9 Gt of carbon into the atmosphere each year.

2 New plant growth and air-sea gas exchange remove 4.9 Gt/yr,...

3 ...yielding a net atmospheric increase of 4.0 Gt/yr.

ATMOSPHERE

Fossil-fuel burning (7.7 Gt/yr)

Land-use change: deforestation, agriculture (1.1 Gt/yr)

Land uptake by new plant growth (2.6 Gt/yr)

Ocean uptake by air-sea gas exchange (2.3 Gt/yr)

Terrestrial biosphere
LAND

OCEAN

Cement production (0.1 Gt/yr)

Fossil organic carbon

Rock carbonates

GEOLOGIC RESERVOIRS

FIGURE 15.19 ■ Much of the CO_2 emitted into the atmosphere by human activities is absorbed by the oceans and by plant growth on land. The remainder stays in the atmosphere, increasing the concentration of CO_2. The fluxes shown in this figure (given in gigatons per year) are for the decade 2000–2009. [IPCC, *Climate Change 2013: The Physical Science Basis.*]

FIGURE 15.20 ■ Marine organisms that form their shells or skeletons by precipitating calcium carbonate, such as these corals in the Great Barrier Reef of Australia, are threatened by ocean acidification. [© Charles Stirling (Diving)/Alamy.]

only 45 percent (4.0 Gt/year) remained in the atmosphere as CO_2. The rest was absorbed in nearly equal amounts by the oceans (2.3 Gt/year) and the land surface (2.6 Gt/year). Through the carbon cycle, the hydrosphere and lithosphere have clearly been doing their fair share of absorbing our increasing carbon emissions!

Although this removal of carbon from the atmosphere acts to reduce the rate of global warming—a good thing, no doubt—its effects on marine life can be deadly. Anthropogenic carbon emissions are being absorbed by the oceans, making seawater more acidic, and this ocean acidification is increasing the solubility of calcium in seawater, making it more difficult for key marine organisms to form their calcium carbonate shells and skeletons (see Figure 15.15). Coral reefs are already in trouble (**Figure 15.20**), and if the present trends continue, ocean acidification could cause population declines in common marine organisms such as starfish and mollusks within the next few decades. In fact, some biologists believe that this type of global change has already contributed to massive die-offs of starfish recently reported on both the east and west coasts of North America.

What will happen on land is less clear. In fact, exactly what is happening to the huge amount of carbon dioxide being pulled out of the atmosphere by terrestrial plants has been a real puzzle (see the Practicing Geology exercise at the end of the chapter).

■ Twentieth-Century Warming: Fingerprints of Anthropogenic Global Change

How do we know that Earth's climate is changing, or that the changes are the result of our own activities? Humans have been tracking global temperatures for some time. The most basic device for measuring climate, the thermometer, was invented in the early seventeenth century, and Daniel Fahrenheit set up the first standard temperature scale in 1724. By 1880, temperatures around the world were being reported by enough meteorological stations on land and on ships at sea to allow accurate estimation of Earth's average annual surface temperature.

Although the average annual surface temperature fluctuates substantially from year to year and from decade to decade, the overall trend has been upward (**Figure 15.21**). Between the end of the nineteenth century and the beginning of the twenty-first, the average annual surface temperature rose by about 0.6°C (Figure 15.20a). This increase is referred to as the **twentieth-century warming.**

The twentieth-century warming was not uniform over the globe. **Figure 15.22** shows the geographic variation of the yearly average temperatures for 1912, 1962, and 2012, colored according to the temperature differences relative to the baseline period 1951-1980. Globally averaged, the difference between 1912 and 2012 is about 0.8°C, consistent with the twentieth-century warming (compare with Figure 15.21). But some of the regional differences are larger, and some are smaller. In the arctic region, for example, the temperature rise has been several times higher than the mean value, whereas in the central Pacific Ocean, there has been very little. In general, the land surfaces have warmed more than the oceans. Most of the warming has occurred during the last 50 years. In large regions of the northern continents, the temperature rise between 1962 and 2012 has exceeded 1°C.

We know that human activities are responsible for the increasing concentrations of CO_2 in the atmosphere because the carbon isotopes of fossil fuels have a distinctive ratio that precisely matches the changing isotopic composition of atmospheric carbon. But how certain can we be that the twentieth-century warming was a direct consequence of the anthropogenic CO_2 increase—that is, a result of an *enhanced greenhouse effect*—and not some other kind of change associated with natural climate variation?

To answer this and other questions about how Earth's climate is changing, the United Nations has set up a special scientific organization, the Intergovernmental Panel on Climate Change (IPCC), to review all research on climate and climate change. The IPCC is charged with developing a consensus, science-based view on how Earth's climate has changed in the past and what might happen in the future, including the potential environmental and socioeconomic impacts of anthropogenic climate change. Much of the information about the climate system described in this textbook has been gleaned from the IPCC *Assessment Reports* (see Earth Issues 15.2).

The twentieth-century warming lies within the range of temperature variations that have been inferred for the Holocene. In fact, average temperatures in many regions of the world were probably warmer 10,000 to 8000 years ago

(a)

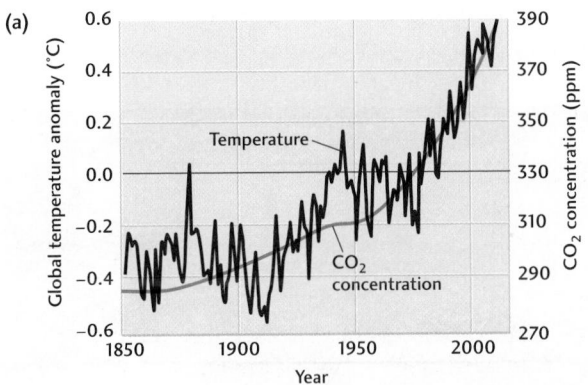

(b) 1000–2000

The twentieth-century warming is clearly anomalous when compared with climate variation over the last millennium.

FIGURE 15.21 ■ A comparison of average annual surface temperature anomalies (black lines) with atmospheric CO_2 concentrations (blue lines) shows a recent warming trend that is correlated with increases in atmospheric CO_2 concentrations. (a) Average global annual surface temperature anomalies, calculated from thermometer measurements, and CO_2 concentrations between 1850 and 2010. (b) Average annual surface temperature anomalies for the Northern Hemisphere, estimated from tree rings, ice cores, and other climate indicators, and atmospheric CO_2 concentrations for the last millennium. In both of these figures, the temperature anomaly is defined as the difference between the observed temperature and the temperature average for the period 1961–1990. [IPCC, *Climate Change 2001: The Scientific Basis*, and IPCC, *Climate Change 2013: The Physical Science Basis*.]

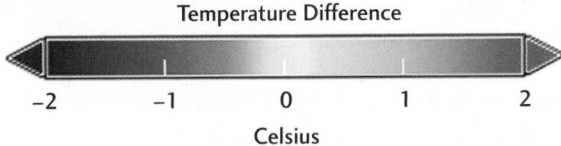

Temperature Difference

Celsius

FIGURE 15.22 ■ Surface temperature anomalies for the years 1912 (top), 1962 (middle), and 2012 (bottom) measured relative to the mean local temperatures for the baseline period 1951–1980. The globally averaged difference between 1912 and 2012 is about 0.8°C, consistent with the twentieth century warming (see Figure 15.20). In the arctic region, the warming has been several times higher than this mean value, whereas in the central Pacific Ocean, it has been very small. [NASA's Goddard Space Flight Center Scientific Visualization Studio.]

than they are today. The twentieth-century record is clearly anomalous, however, when compared with the pattern and rate of climate change documented during the last millennium. Although direct temperature measurements are not available from before the nineteenth century, climate indicators such as ice cores and tree rings have allowed climatologists to reconstruct a temperature record for the Northern Hemisphere during that period (Figure 15.21b). That record

shows an irregular but steady global cooling of about 0.2°C in the nine centuries between 1000 and 1900. It also shows that fluctuations in average surface temperature during each of these centuries were less than a few tenths of a degree.

Earth Issues

15.2 The Intergovernmental Panel on Climate Change

Earth's climate system is incredibly complex, so predicting its response to anthropogenic emissions of greenhouse gases is hardly a straightforward task. No one person can keep up with the vast amount of climate-change research that is being conducted worldwide by thousands of scientists, and even experts disagree on key points. In 1988, the United Nations (UN) and the World Meteorological Organization (WMO) established the Intergovernmental Panel on Climate Change (IPCC) to provide government leaders and the public at large with a clear scientific view of current knowledge about climate change and its potential environmental and socioeconomic impacts.

The IPCC is open to all UN and WMO members, and 195 countries are currently participating. The main product of the IPCC has been a series of *Assessment Reports* released every five to six years since 1990. Thousands of scientists from all over the world have contributed to the work of the IPCC on a voluntary basis as authors, contributors, and reviewers of these major reports. Each report in succession has laid out the most definitive scientific summaries of how climate has changed in the past and how it might change in the future.

IPCC's *First Assessment Report*, published in 1990, played a key role in the creation of the United Nations Framework Convention on Climate Change, the main international treaty to reduce global warming and deal with the consequences of climate change. The IPCC *Second Assessment Report* of 1995 provided important material drawn on by negotiators of the Kyoto Protocol in 1997. The *Third Assessment Report* came out in 2001, and the *Fourth* in 2007.

A meeting of lead authors of the IPCC's *Fifth Assessment Report* (AR5), held in Changwon, Korea, in June, 2011. [Benjamin Kriemann/IPCC.]

The *Fifth Assessment Report*, which is currently in production, will comprise subreports from the three IPCC working groups. The first, entitled *The Physical Science Basis of Climate Change*, was released in draft form in September 2013 and runs to more than 2000 pages. Many of the basic data on climate change described in this chapter and elsewhere in this textbook have been updated according to this 2013 IPCC assessment. The final version of the *Fifth Assessment*, which will include reports on *Climate Change Impacts, Adaptation and Vulnerability* and *Mitigation of Climate Change*, is scheduled for release in 2014.

In 2007, the Nobel Peace Prize was awarded jointly to the IPCC and Al Gore "for their efforts to build up and disseminate greater knowledge about man-made climate change, and to lay the foundations for the measures that are needed to counteract such change."

The second argument, and to many scientists a more compelling one, comes from the agreement between the observed pattern of warming and the pattern predicted by the best climate models. Models that include changes in atmospheric greenhouse gas concentrations not only reproduce the twentieth-century warming, but also reproduce the observed patterns of temperature change both geographically and with altitude in the atmosphere—what some scientists have called the "fingerprints" of the enhanced greenhouse effect. For example, these models predict that as enhanced greenhouse warming occurs, nighttime low temperatures at Earth's surface should increase more rapidly than daytime high temperatures, thus reducing daily temperature variation. Climate data for the last century confirm this prediction.

Another fingerprint of global warmer has been the changes seen in mountain glaciers at lower latitudes. Glaciers found above 5000 m in Africa, South America, and Tibet (**Figure 15.23**) have been shrinking during the last hundred years, an observation that is also consistent with the predictions of climate models.

As we emphasized earlier in this chapter, aspects of the climate system that are still poorly understood may introduce substantial errors into the predictions of climate models. Nevertheless, the consistency of the measured trends with the basic physics of the enhanced greenhouse effect lends powerful support to the hypothesis that we ourselves are the agents responsible for the recent global warming. We will discuss global warming further, and look at the societal problems it poses, in Chapter 23.

FIGURE 15.23 ■ Glaciologist Lonnie Thompson at an altitude of 5300 m (17,390 ft) on Tibet's Dasuopu Glacier. Ice coring on this glacier provides evidence of abnormal global warming during the twentieth century. [Lonnie Thompson/Byrd Polar Research Center, Ohio State University.]

SUMMARY

What is the climate system? The climate system includes all of the components of the Earth system, and all of the interactions among those components, that determine how climate varies in space and time. The main components of the climate system are the atmosphere, hydrosphere, cryosphere, lithosphere, and biosphere. Each component plays a role in the climate system that depends on its ability to store and transport mass and energy.

What is the greenhouse effect? When Earth's surface is warmed by the Sun, it radiates heat back into the atmosphere. Carbon dioxide and other greenhouse gases absorb some of this infrared radiation and reradiate it in all directions, including downward to Earth's surface. This radiation maintains the atmosphere at a warmer temperature than it would be if there were no greenhouse gases, similar to the warmer air temperature maintained in a greenhouse.

How has Earth's climate changed over time? Natural variations in climate occur on a wide range of scales in both time and space. Some variations result from factors outside the climate system, such as solar forcing and changes in the distribution of land and sea surfaces caused by continental drift. Others result from variations within the climate system itself. Short-term regional climate variations include the El Niño–Southern Oscillation. Long-term global climate variations are exemplified by the Pleistocene glacial cycles, during which average surface temperatures changed by as much as 6°C to 8°C.

What are ice ages, and what causes them? Studies of the geologic ages of glacial deposits on land and in marine sediments show that continental ice sheets advanced and retreated many times during the Pliocene and Pleistocene epochs. Each ice age involved a massive transfer of water from the hydrosphere to the cryosphere, resulting in expansion of glaciers and a lowering of sea level. The favored explanation is that these glacial cycles are being driven by Milankovitch cycles, small periodic variations in Earth's movement through the solar system that alter the amount of solar radiation received at Earth's surface. These variations have been amplified by positive feedbacks involving atmospheric concentrations of greenhouse gases. The global cooling that initiated the Pleistocene glacial cycles may have resulted from continental movements that changed oceanic circulation patterns.

What are geochemical cycles? Geochemical cycles are fluxes of chemicals from one component of the Earth system to another. The atmosphere, hydrosphere, cryosphere, lithosphere, and biosphere act as geochemical reservoirs and are linked by processes that transport chemicals among them. If a reservoir is at a steady state, inflow balances outflow, and the residence time of the chemical can be calculated as the total amount of the chemical in the reservoir divided by the inflow.

What is the carbon cycle? The carbon cycle is the flux of carbon among its four principal reservoirs: the atmosphere, lithosphere, oceans, and terrestrial biosphere. Major fluxes of carbon between these reservoirs include gas exchange between the atmosphere and the ocean surface; the movement of carbon dioxide between the biosphere and the atmosphere through photosynthesis, respiration, and direct oxidation; the transport of dissolved organic carbon in surface waters to the ocean; and the weathering and precipitation of calcium carbonate.

What are the effects of anthropogenic carbon emissions? Human emissions of carbon are enhancing the greenhouse effect by increasing the concentration of carbon dioxide in the atmosphere. Some of this carbon dioxide dissolves in the oceans, where it combines with water to form carbonic acid. The resulting ocean acidification acts to increase the concentration of bicarbonate ions at the expense of carbonate ions, making it harder for marine organisms to precipitate shells and skeletons of calcium carbonate.

Was the twentieth-century warming caused by human activities? The observed increase of about 0.6°C in Earth's average annual surface temperature during the twentieth century is correlated with a significant rise in atmospheric concentrations of CO_2 and other greenhouse gases. The changing isotope ratios of atmospheric carbon show that much of it is being produced by fossil-fuel burning. Most experts on Earth's climate are now convinced that the twentieth-century warming was human-induced and that the warming will continue into the twenty-first century as atmospheric concentrations of greenhouse gases continue to rise.

KEY TERMS AND CONCEPTS

albedo (p. 413)

carbon cycle (p. 423)

climate model (p. 415)

El Niño (p. 417)

ENSO (p. 417)

geochemical cycle
 (p. 423)

geochemical reservoir
 (p. 423)

glacial cycle (p. 418)

greenhouse effect
 (p. 414)

greenhouse gas (p. 414)

ice age (p. 418)

interglacial period
 (p. 418)

Milankovitch cycle (p. 420)

negative feedback (p. 415)

ocean acidification (p. 424)

positive feedback (p. 415)

residence time (p. 423)

solar forcing (p. 408)

stratosphere (p. 408)

thermohaline circulation
 (p. 411)

troposphere (p. 408)

twentieth-century warming
 (p. 428)

PRACTICING GEOLOGY EXERCISE

Balancing Carbon Emission with Carbon Accumulation: The Case of the Missing Sink

Understanding how humans are changing the carbon cycle is one of the most pressing issues in Earth science today because it holds the key to learning to manage anthropogenic global change. We can see from Figure 15.19 that, of the 8.9 Gt/year of carbon emitted by humans in 2000–2009, 2.6 Gt/year—almost a third—was absorbed by the land surface. Plant photosynthesis and respiration dominate the exchange of CO_2 between the atmosphere and the land surface, so an increase in the rate of photosynthesis by land plants must clearly be the cause. But where on Earth is this happening? This question was so hard to answer that scientists for years called it the "missing sink" problem. The answer turns out to be important, because future treaties in which nations agree to regulate their carbon emissions will need to take into account all carbon sources and sinks within each nation's boundaries.

As shown in the accompanying figure, the total anthropogenic carbon emissions rose from an average of 6.9 Gt/year in the 1980s, to 8.0 Gt/year in the 1990s, to 8.9 Gt/year in the 2000s. The rate at which these atmospheric emissions were absorbed into the ocean also increased, so that the percentage captured by the oceans remained nearly constant. However, the fraction of anthropogenic carbon accumulating in the atmosphere was not nearly so steady; in fact, the rate actually went down between the 1980s and 1990s from 3.4 Gt/year to 3.1 Gt/year, and then jumped to 4.0 Gt/year in the 2000s.

Atmospheric accumulation varies inversely with the carbon absorbed by the missing sink. The total amount of carbon is conserved; therefore, when summed over all geochemical reservoirs, carbon accumulation must balance carbon emission, as shown graphically by the bar charts in the figure. This balance allows us to calculate the amount of carbon absorbed by the missing sink:

$$\frac{missing}{sink} = \frac{total\ carbon}{emissions} - \frac{atmospheric}{accumulation} - \frac{ocean}{accumulation}$$

The values are given by the red bars in the figure. Carbon absorption by the missing sink increased by 80 percent from 1.5 Gt/year in the 1980s to 2.7 Gt/year in the 1990s, and then decreased slightly to 2.6 Gt/year in the 2000s.

Comparison of the global carbon balance for the last three decades. [IPCC, *Climate Change 2013: The Physical Science Basis.*]

An obvious place to look for the missing carbon is in the world's forests, which account for about half the annual terrestrial uptake of CO_2 by photosynthesis. Forests are classified according to the average annual temperature where they grow as boreal ($-5°C$ to $5°C$), temperate ($5°C$ to $20°C$), or tropical ($20°C$ to $30°C$). Early models suggested that the growth of temperate forests might account for most of the missing carbon. However, data in the IPCC's *Fifth Assessment Report* indicate that the current carbon accumulation is almost as large in the boreal regions and actually greater in the tropical regions.

$$
\begin{array}{c}
\text{missing} \\
\text{sink} \\
\text{(2.6 Gt/} \\
\text{year)}
\end{array}
=
\begin{array}{c}
\text{boreal} \\
\text{accumulation} \\
\text{(0.5 Gt/} \\
\text{year)}
\end{array}
+
\begin{array}{c}
\text{temperate} \\
\text{forest growth} \\
\text{(0.8 Gt/} \\
\text{year)}
\end{array}
+
\begin{array}{c}
\text{tropical} \\
\text{forest growth} \\
\text{(1.3 Gt/} \\
\text{year)}
\end{array}
$$

This newer model suggests that tropical forest growth (1.3 Gt/year) is more than offsetting tropical deforestation, which accounts for most of the carbon emissions from land-use changes (1.1 Gt/year).

Although progress is being made, the carbon-balance problem is far from solved. Because of the measurement difficulties, all estimates of where carbon is accumulating have large uncertainties, so more research is needed to pin down the numbers. Nevertheless, these estimates not only demonstrate the importance of our forests as carbon sinks, but also raise major policy issues about how they should be managed. For example, how much "carbon credit" should nations such as the United States and Brazil receive for the carbon taken up by their forests? Such issues will figure prominently in the negotiation of international treaties to deal with anthropogenic global change.

BONUS PROBLEM: From the data shown in the figure, calculate the net carbon flux from the land surface by balancing the carbon emitted by land-use changes with the carbon absorbed by the missing sink. In the three decades shown, was this net carbon flux from the land surface positive (net emission) or negative (net absorption)? What factors might explain why the magnitude of this net flux increased steadily from the 1980s to the 2000s?

EXERCISES

1. What is a greenhouse gas, and how does it affect Earth's climate?

2. Why is it incorrect to assert that greenhouse gases prevent heat energy from escaping to outer space?

3. From the information given in Figure 15.18, estimate the residence time of carbon dioxide (a) in the ocean and (b) in the atmosphere.

4. Between 2000 and 2010, emissions of carbon into the atmosphere from fossil-fuel burning and other industrial activities averaged about 7.8 Gt/year, almost all of it in the form of carbon dioxide. What was the mass of carbon dioxide emitted?

5. List three causes of climate change that result from plate tectonic processes.

6. What is the role of continental glaciers in climate variation?

7. What information about glacial cycles has been obtained by studying ice cores?

THOUGHT QUESTIONS

1. Give an example not discussed in this chapter of a positive feedback and a negative feedback in the climate system.

2. Do Milankovitch cycles fully explain the warming and cooling of the global climate during the Pleistocene glacial cycles?

3. How would the calcium cycle be affected by a global increase in chemical weathering?

4. Draw a simple diagram of the geochemical cycle for the element sodium, which is found in marine evaporites (halite) and clay minerals as well as dissolved in seawater.

5. What would the carbon cycle have looked like after life had originated, but before photosynthesis had evolved?

6. If human activities keep pumping CO_2 into the atmosphere at a steadily increasing rate and Earth's climate warms significantly in the next 100 years, how might the global carbon cycle be affected?

7. Why are scientists reasonably sure that most of the twentieth-century warming was due to changes in the climate system caused by human activities?

MEDIA SUPPORT

 15-1 Animation: Climate Systems

 15-2 Animation: Current Systems

 15-3 Animation: Carbon Cycle

- Weathering, Erosion, Mass Wasting, and the Rock Cycle   436

- Controls on Weathering   436

- Chemical Weathering   438

- Physical Weathering   443

- Soils: The Residue of Weathering   445

- Mass Wasting   447

- Classification of Mass Movements   453

- Understanding the Origins of Mass Movements   460

WEATHERING, EROSION, AND MASS WASTING:

16

Interactions Between the Climate and Plate Tectonic Systems

As solid as the hardest rocks may seem, all rocks—like rusting old automobiles and yellowed old newspapers—eventually weaken and crumble when exposed to water and the gases of the atmosphere. Unlike cars and newspapers, however, rocks may take thousands of years to disintegrate.

In this chapter, we will describe three geologic processes that break down rocks and transport the products over short distances: weathering, erosion, and mass wasting. These three processes result from interactions between the climate and plate tectonic systems.

Weathering is the first step in flattening the mountains that have been uplifted by plate tectonic processes. Even as mountains are being uplifted, chemical decay and physical fragmentation join with rainfall, wind, ice, and snow to wear them away. Erosion and mass wasting are the processes that loosen weathered soil and rock and transport it downhill or downwind. *Erosion* generally refers to processes that move Earth materials on a grain-by-grain basis. *Mass wasting* refers to processes that cause large masses of material to collapse and move downslope. Both processes carry weathered material away from its source, exposing fresh, unaltered rock surfaces to weathering.

■ Weathering, Erosion, Mass Wasting, and the Rock Cycle

Weathering, as we saw in Chapter 5, is the general process by which rocks are broken down at Earth's surface. Weathering produces all the clays and soils of the world, as well as the dissolved substances that streams carry to the ocean. *Chemical weathering* occurs when the minerals in a rock are chemically altered or dissolved. *Physical weathering* takes place when solid rock is fragmented by mechanical processes that do not change its chemical composition. Chemical and physical weathering reinforce each other. Chemical weathering weakens rocks and makes them more susceptible to physical weathering. The smaller the pieces produced by physical weathering, the greater the surface area available for chemical weathering.

Once weathering reduces rocks to particles, they may accumulate as soil, or they may be removed by erosion, transported, and deposited elsewhere as sediments. *Erosion* is the process by which particles produced by weathering are dislodged and removed from their source, usually by means of currents of water or air. Erosion moves particles from hillslopes to the starting points of stream channels. **Mass wasting** includes all the processes by which weathered and unweathered Earth materials move downslope in larger amounts and in large single events, usually by means of gravity. The products of mass wasting—particles released by weathering as well as large masses of unweathered rock—are also transported to the starting points of stream channels. Once these materials reach stream channels, streams and rivers can efficiently transport them farther downslope, perhaps across continents and all the way to the ocean. The transport of sediments by streams from their source areas in mountains to their sinks in the world ocean will be covered in more detail in Chapter 18.

Weathering is one of the major processes of the rock cycle. It shapes Earth's surface topography and alters rock materials, converting all kinds of rocks into sediments and soils. The early sections of this chapter emphasize chemical weathering because it is in some ways the fundamental driving force of the weathering process. For example, the effects of physical weathering, which we will examine next, depend largely on the chemical decay of minerals. Before we look at either type of weathering in detail, however, let's examine the factors that control weathering.

■ Controls on Weathering

All rocks weather, but the manner and rates of their weathering vary. The four key factors that control rates of weathering are the properties of the parent rock, the climate, the presence or absence of soil, and the length of time the rocks are exposed to the atmosphere. These four factors are summarized in **Table 16.1**.

The Properties of Parent Rock

The mineralogy and crystal structure of parent rock affect weathering because different minerals weather at

TABLE 16-1	Major Factors Controlling Rates of Weathering		
	Weathering Rate		
	Slow ———————————————————————→ Fast		
PROPERTIES OF PARENT ROCK			
Mineral solubility in water	Low (e.g., quartz)	Moderate (e.g., pyroxene, feldspar)	High (e.g., calcite)
Rock structure	Massive	Some zones of weakness	Very fractured or thinly bedded
CLIMATE			
Rainfall	Low	Moderate	High
Temperature	Cold	Temperate	Hot
PRESENCE OR ABSENCE OF SOIL AND VEGETATION			
Thickness of soil layer	None—bare rock	Thin to moderate	Thick
Organic content	Low	Moderate	High
LENGTH OF EXPOSURE	Short	Moderate	Long

different rates and because a rock's crystal structure affects its susceptibility to cracking and fragmentation. Old inscriptions on gravestones are evidence of the varying rates at which rocks weather. The carved letters on a recently erected gravestone stand out clearly from the stone's polished surface. After a hundred years in a moderately rainy climate, however, the surface of a limestone monument will be dull, and the letters inscribed on it will have almost melted away, much as the brand name on a bar of soap disappears after a few washes (**Figure 16.1**). Slate or granite, on the other hand, will show only minor changes. The differences in the weathering of limestone and slate result from their different mineral compositions. Given enough time, however, even a resistant rock will ultimately decay. After several hundred years, a granite monument will also have weathered appreciably, and its surface and letters will be somewhat dulled and blurred.

Climate: Rainfall and Temperature

The rates of both chemical and physical weathering vary not only with the properties of parent rock, but also with the climate—especially the temperature and amount of rainfall—where that parent rock is located. High temperatures and heavy rainfall promote faster chemical weathering; cold and dry climates slow the process. Water in cold climates cannot dissolve minerals when it is frozen. In arid regions, water is relatively unavailable.

On the other hand, climates that minimize chemical weathering may enhance physical weathering. For example, freezing water may act as a wedge, widening cracks and pushing a rock apart. In temperate climates, the alternating freezing and thawing that accompany changes in temperature cause rocks to contract and expand, helping to break them apart.

The Presence or Absence of Soil

Although soil is itself a product of weathering, its presence or absence affects the chemical and physical weathering of other materials. Soil production is a *positive feedback process*—that is, soil formation advances more soil formation. Once soil starts to form, it works as a geologic agent to weather rock more rapidly. The soil retains rainwater, and it hosts a variety of vegetation, bacteria, and other organisms. The metabolism of those organisms creates an acidic environment that, in combination with moisture, promotes chemical weathering. Plant roots and organisms tunneling through the soil promote physical weathering by helping to create fractures in rock. Chemical and physical weathering, in turn, lead to the production of more soil.

The Length of Exposure

The longer a rock weathers, the greater its chemical alteration, dissolution, and physical fragmentation. Rocks that have been exposed at Earth's surface for many thousands

FIGURE 16.1 ■ These early-nineteenth-century gravestones at Wellfleet, Massachusetts, show the results of chemical weathering. The stone on the right, which is limestone, has become so weathered that it is unreadable. The stone on the left, which is slate, has remained legible under the same conditions. [Courtesy of Raymond Siever.]

of years form a *rind*—an external layer of weathered material ranging from several millimeters to several centimeters thick—that surrounds fresh, unaltered rock. In dry climates, some rinds have grown as slowly as 0.006 mm per 1000 years.

Now that we have examined the factors that control rates of weathering, we can consider the two types of weathering—chemical and physical—in more detail.

◾ Chemical Weathering

Chemical weathering occurs when minerals react with air and water. In these chemical reactions, some minerals dissolve. Others combine with water and atmospheric components such as oxygen and carbon dioxide to form new minerals. We begin our investigation by examining the chemical weathering of feldspar, the most abundant mineral in Earth's crust.

The Role of Water: Feldspar and Other Silicates

Feldspar is one of many silicates that are altered by chemical reactions to form clay minerals. Feldspar's behavior during weathering helps us understand the weathering process in general, for two reasons:

1. Feldspar is a key mineral in a great many igneous, sedimentary, and metamorphic rocks and is one of the most abundant minerals in Earth's crust.

2. The chemical processes that characterize feldspar weathering also characterize weathering in many other kinds of minerals.

Feldspar is one component of granite, which, as you will recall, is made up of several different minerals, all of which decay at different rates. A sample of unweathered granite is hard and solid because an interlocking network of feldspar, quartz, and other crystals holds it tightly together. When the feldspar is transformed by weathering into a loosely adhering clay, the network is weakened and the mineral grains are separated (**Figure 16.2**). In this instance, chemical weathering, by producing the clay, also promotes physical weathering because the rock now fragments easily along widening cracks at the boundaries between grains.

The white to cream-colored clay produced by the weathering of feldspar is called **kaolinite,** named for Gaoling, a hill in southwestern China where it was first obtained. Chinese artisans had used pure kaolinite as the raw material of pottery and porcelain for centuries before Europeans borrowed the idea in the eighteenth century.

Only in the arid climates of some deserts and polar regions does feldspar remain relatively unweathered. This observation points to water as an essential component of the chemical reaction by which feldspar becomes kaolinite. Kaolinite is a hydrous aluminum silicate. In the reaction that produces it, the solid feldspar undergoes *hydrolysis* (a decomposition reaction involving water; from *hydro*, "water," and *lysis*, "to loosen"). The feldspar is broken down and also loses several chemical components, while kaolinite gains water.

The only part of a solid that reacts with a fluid is the solid's surface, so as we increase the surface area of the solid, we speed up the reaction. For example, as we grind coffee beans into finer and finer particles, we increase the ratio of their surface area to their volume. The finer the coffee beans are ground, the faster their reaction with water, and the stronger the brew becomes. Similarly, the smaller the fragments of minerals and rocks, the greater their surface area. The ratio of surface area to volume increases greatly as the average particle size decreases, as shown in **Figure 16.3**.

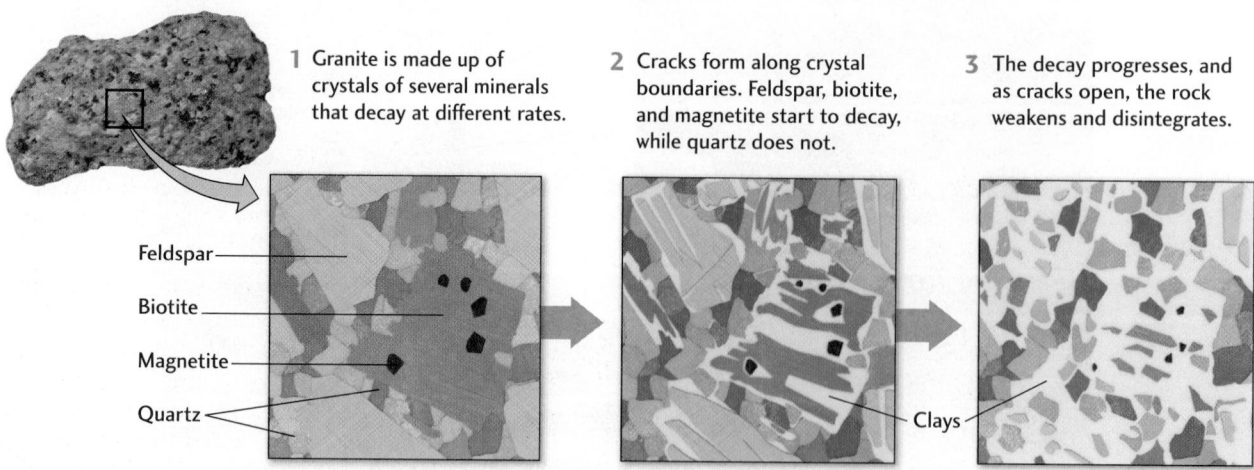

1 Granite is made up of crystals of several minerals that decay at different rates.

2 Cracks form along crystal boundaries. Feldspar, biotite, and magnetite start to decay, while quartz does not.

3 The decay progresses, and as cracks open, the rock weakens and disintegrates.

Feldspar

Biotite

Magnetite

Quartz

Clays

FIGURE 16.2 ◾ Diagrammatic microscopic views of stages in the disintegration of granite. [John Grotzinger/Ramón Rivera-Moret/Harvard Mineralogical Museum.]

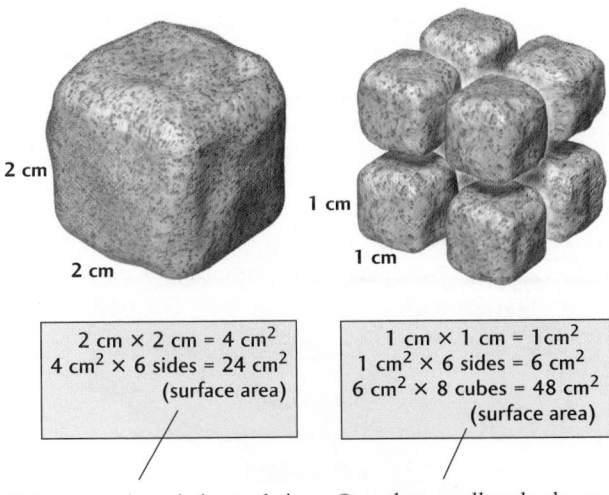

2 cm

2 cm

1 cm

1 cm

2 cm × 2 cm = 4 cm^2
4 cm^2 × 6 sides = 24 cm^2
(surface area)

1 cm × 1 cm = 1cm^2
1 cm^2 × 6 sides = 6 cm^2
6 cm^2 × 8 cubes = 48 cm^2
(surface area)

1 Large rocks, relative to their mass, have less surface area for chemical weathering…

2 …than small rocks do, so smaller rocks weather more quickly.

FIGURE 16.3 ■ As a rock mass breaks into smaller pieces, more surface area becomes exposed to the chemical reactions of weathering.

Carbon Dioxide, Weathering, and the Climate System

Carbon dioxide, like water, is involved in the chemical reactions of weathering. Thus, variation in the atmospheric concentration of CO_2 leads to corresponding variation in the rate of weathering (**Figure 16.4**). Higher concentrations of CO_2 in the atmosphere lead to higher concentrations in the soil, which accelerate the weathering of rocks. As we saw in Chapter 15, increases in atmospheric CO_2, a greenhouse gas, make Earth's climate warmer and thus promote weathering in that way as well. The weathering of calcium-rich rock, in turn, removes CO_2 from the atmosphere, making global climates cooler. In this way, chemical weathering links the plate tectonic system to the climate system. As more and more CO_2 is used up through weathering, the climate cools, and weathering decreases. As weathering decreases, the concentration of CO_2 in the atmosphere builds up again, and the climate warms, thus continuing the cycle.

THE ROLE OF CARBON DIOXIDE IN WEATHERING

The reaction of feldspar with pure water in a laboratory is an extremely slow process that would take thousands of years to break down even a small amount of feldspar completely. We can speed weathering by adding a strong acid (such as hydrochloric acid) to the water, in which case the feldspar will break down in a few days. An *acid* is a substance that releases hydrogen ions (H^+) to a solution. A strong acid produces abundant hydrogen ions; a weak one, relatively few. The strong tendency of hydrogen ions to combine chemically with other substances makes acids excellent solvents.

On Earth's surface, the most common acid—and the one most responsible for increasing weathering rates—is carbonic acid (H_2CO_3). This weak acid forms when carbon dioxide (CO_2) gas from the atmosphere dissolves in rainwater:

$$\text{carbon dioxide} + \text{water} \rightarrow \text{carbonic acid}$$
$$CO_2 \qquad H_2O \qquad H_2CO_3$$

The amount of carbon dioxide dissolved in rainwater is small because the amount of CO_2 gas in the atmosphere is small. About 0.03 percent of the molecules in Earth's atmosphere are carbon dioxide. Thus, the amount of carbonic acid formed in rainwater is also very small, only about 0.0006 g/L.

As human activities increase the concentration of carbon dioxide in the atmosphere, the amount of carbonic acid in rainwater is increasing slightly. Acid rain accelerates weathering, but most of the acidity of acid rain comes not from carbon dioxide, but from sulfur dioxide and nitrogen gases, which react with water to form strong sulfuric and nitric acids, respectively. These acids promote weathering to a greater degree than carbonic acid does. Volcanoes and coastal wetlands emit gaseous forms of carbon, sulfur, and nitrogen into the atmosphere, but by far the largest source of these gases is industrial pollution.

Although rainwater contains only a relatively small amount of carbonic acid, that amount is enough to dissolve great quantities of rock over long periods. The chemical reaction for the weathering of feldspar is

$$\text{feldspar} + \text{carbonic acid} + \text{water} \rightarrow$$
$$2KAlSi_3O_8 \qquad 2H_2CO_3 \qquad H_2O$$

		dissolved	dissolved
dissolved +	dissolved +	potassium +	bicarbonate
kaolinite	silica	ions	ions
$Al_2Si_2O_5(OH)_4$	$4SiO_2$	$2K^+$	$2HCO_3^-$

This simple weathering reaction illustrates the three main effects of chemical weathering on silicates:

1. It *leaches,* or dissolves away, cations and silica.

2. It *hydrates,* or adds water to, the minerals.

3. It makes solutions less acidic.

Specifically, the carbonic acid in rainwater helps to weather feldspar in the following way (see Figure 16.4):

■ A small proportion of the carbonic acid molecules in rainwater ionize, forming hydrogen ions (H^+) and bicarbonate ions (HCO_3^-), thus making the water slightly acidic.

■ The slightly acidic water dissolves potassium ions and silica from feldspar, leaving a residue of kaolinite, a solid clay. The hydrogen ions from the acidic water combine with the oxygen atoms of the feldspar to form the water in the kaolinite structure. The kaolinite becomes part of the soil or is carried away as the dissolved silica, potassium ions, and bicarbonate ions are carried away by rain and stream waters and are ultimately transported to the ocean.

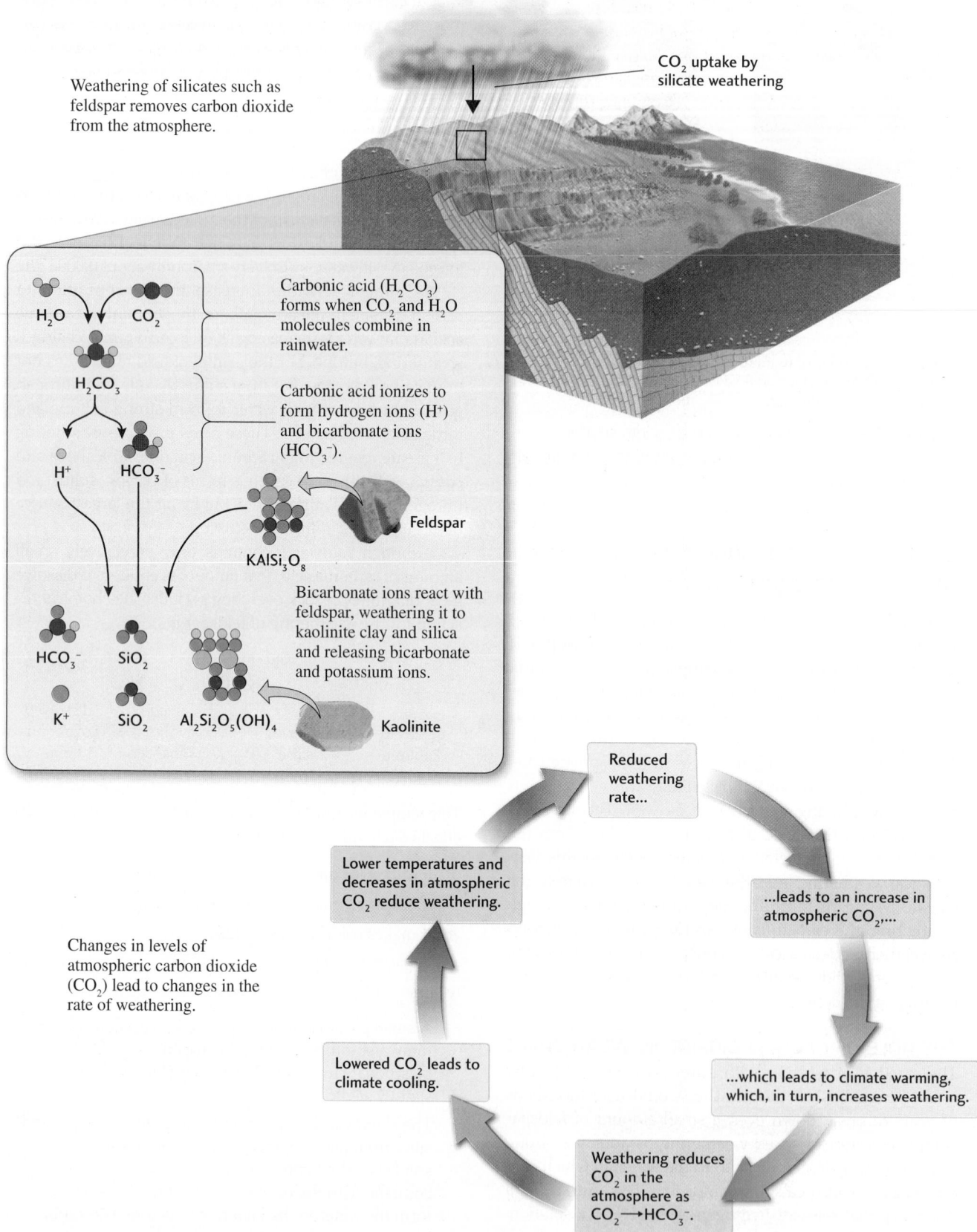

Weathering of silicates such as feldspar removes carbon dioxide from the atmosphere.

CO$_2$ uptake by silicate weathering

Carbonic acid (H$_2$CO$_3$) forms when CO$_2$ and H$_2$O molecules combine in rainwater.

Carbonic acid ionizes to form hydrogen ions (H$^+$) and bicarbonate ions (HCO$_3^-$).

H$_2$O   CO$_2$

H$_2$CO$_3$

H$^+$ HCO$_3^-$

KAlSi$_3$O$_8$ Feldspar

Bicarbonate ions react with feldspar, weathering it to kaolinite clay and silica and releasing bicarbonate and potassium ions.

HCO$_3^-$ SiO$_2$

K$^+$ SiO$_2$ Al$_2$Si$_2$O$_5$(OH)$_4$ Kaolinite

Changes in levels of atmospheric carbon dioxide (CO$_2$) lead to changes in the rate of weathering.

Reduced weathering rate...

...leads to an increase in atmospheric CO$_2$,...

...which leads to climate warming, which, in turn, increases weathering.

Weathering reduces CO$_2$ in the atmosphere as CO$_2$ → HCO$_3^-$.

Lowered CO$_2$ leads to climate cooling.

Lower temperatures and decreases in atmospheric CO$_2$ reduce weathering.

FIGURE 16.4 ■ Variation in atmospheric carbon dioxide concentrations leads to corresponding variation in rates of weathering as well as in global temperatures, which also influence weathering. In this way, the lithosphere and the climate system are linked.

THE ROLE OF SOIL IN WEATHERING Now that we understand how acidic water weathers feldspar, we can better understand why feldspars on bare rock outcrops are better preserved than those buried in damp soils. The chemical reactions of weathering give us two separate but related clues to the factors responsible: the amount of water and the amount of acid available for those reactions. The exposed feldspar weathers only while the rock is moist with rainwater. During dry periods, the only moisture that touches bare rock is dew. The feldspar in damp soil, however, is constantly in contact with the small amounts of water retained in the pores between soil particles. Thus, feldspar weathers continuously in moist soil.

There is more acid in soil moisture than there is in falling rain. Rainwater carries its original carbonic acid into the soil. As water filters through the soil, it picks up additional carbonic acid and other acids produced by the roots of plants, by the many insects and other animals that live in the soil, and by the bacteria that degrade plant and animal remains. Recently, it was discovered that some bacteria release organic acids, even in waters hundreds of meters deep in the ground. These organic acids weather feldspar and other minerals in rocks below the surface. Bacterial respiration in soil may increase the soil's carbon dioxide concentration to as much as 100 times the atmospheric concentration!

Rock weathers more rapidly in the tropics than in temperate and cold climates, mainly because plants and bacteria grow more quickly in warm, humid climates, contributing more of the carbonic acid and other acids that promote weathering. Additionally, most chemical reactions, weathering included, speed up with an increase in temperature.

The Role of Oxygen: From Iron Silicates to Iron Oxides

Iron is one of the eight most abundant chemical elements in Earth's crust, but metallic iron, the element in its pure form, is rarely found in nature. It is present only in certain kinds of meteorites that fall to Earth from other places in the solar system. Most of the iron ores used for the production of iron and steel are formed by weathering. These ores are composed of iron oxide minerals originally produced by the weathering of iron-rich silicate minerals, such as pyroxene and olivine. The iron released by dissolution of the silicate minerals combines with oxygen from the atmosphere and hydrosphere to form iron oxide minerals.

The iron in minerals may be present in one of three forms: metallic iron, ferrous iron, or ferric iron. In the metallic iron found in meteorites (and in manufactured items), the iron atoms are uncharged: they have neither gained nor lost electrons by reacting with another element. In the *ferrous iron* (Fe^{2+}) found in silicate minerals, the iron atoms have lost two of the electrons they have in the metallic form and have thus become ions. In the *ferric iron* (Fe^{3+}) found

in iron oxide minerals, the iron atoms have lost three electrons. The electrons lost by the iron are gained by oxygen atoms in a process called *oxidation*. Oxygen atoms in air and water oxidize ferrous iron to form ferric iron. Thus, all the iron oxides formed at Earth's surface, the most abundant of which is **hematite** (Fe_2O_3), are ferric. Oxidation, like hydrolysis, is one of the important reactions of chemical weathering.

When an iron-rich silicate mineral such as pyroxene is exposed to water, its silicate structure breaks down, releasing silica and ferrous iron into the water. In solution, the ferrous iron is oxidized to the ferric form (**Figure 16.5**). The strong chemical bonds that form between ferric iron and oxygen make ferric iron insoluble in most natural surface waters. It therefore precipitates from solution, forming a

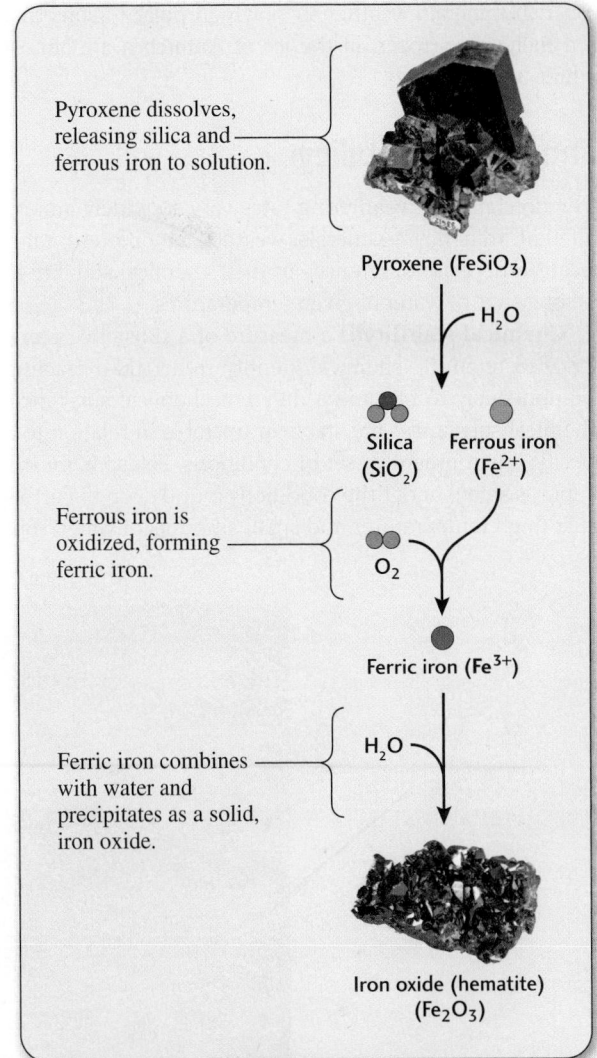

Pyroxene dissolves, releasing silica and ferrous iron to solution.

Pyroxene ($FeSiO_3$)

H_2O

Silica (SiO_2)　Ferrous iron (Fe^{2+})

Ferrous iron is oxidized, forming ferric iron.

O_2

Ferric iron (Fe^{3+})

Ferric iron combines with water and precipitates as a solid, iron oxide.

H_2O

Iron oxide (hematite) (Fe_2O_3)

FIGURE 16.5 ■ The general course of the chemical reactions by which an iron-rich silicate mineral, such as pyroxene, weathers in the presence of oxygen and water. [John Grotzinger/Ramón Rivera-Moret/Harvard Mineralogical Museum.]

solid ferric iron oxide. We are familiar with ferric iron oxide in another form: rust, which is produced when metallic iron in manufactured items is exposed to the atmosphere.

We can show this overall weathering reaction with the following example:

iron-rich + oxygen → hematite + dissolved
pyroxene   silica

$4FeSiO_3$ O_2 $2Fe_2O_3$ $4SiO_2$

Although the equation does not show it explicitly, water is required for this reaction to proceed.

Iron-containing minerals, which are widespread, weather to the characteristic red and brown colors of oxidized iron (**Figure 16.6**). Iron oxides are found as coatings and encrustations that color soils and weathered surfaces of iron-containing rocks. The red soils of Georgia and other warm, humid regions are colored by iron oxides. In contrast, iron-rich minerals weather so slowly in polar regions that iron meteorites frozen in the ice of Antarctica are almost entirely unweathered.

Chemical Stability

Why do chemical weathering rates vary so widely among different minerals? Minerals weather at different rates because there are differences in their chemical stability in the presence of water at given temperatures.

Chemical stability is a measure of a substance's tendency to retain its chemical identity rather than reacting spontaneously to become a different chemical substance. Chemical substances are stable or unstable in relation to a specific environment or set of conditions. Feldspar, for example, is stable under the conditions found deep in Earth's crust (high temperatures and small amounts of water), but

unstable under the conditions at Earth's surface (lower temperatures and abundant water). Two characteristics of a mineral—its solubility and its rate of dissolution—help determine its chemical stability.

SOLUBILITY The *solubility* of a mineral is measured by the amount of that mineral dissolved in water when the solution is saturated. *Saturation* is the point at which the water cannot hold any more of the dissolved substance. The higher a mineral's solubility, the lower its chemical stability under most weathering conditions. Halite-containing evaporites, for example, are unstable under most weathering conditions. Their solubility is high (about 350 g/L), and they are leached from soil by even small amounts of water. Quartz, in contrast, is stable under most weathering conditions. Its solubility in water is very low (only about 0.008 g/L), and it is not easily leached from soil.

RATE OF DISSOLUTION A mineral's rate of dissolution is measured by the amount of that mineral that dissolves in an unsaturated solution in a given time. The faster the mineral dissolves, the less stable it is. Feldspar dissolves at a much faster rate than quartz, and primarily for that reason, it is less stable than quartz under weathering conditions.

RELATIVE STABILITIES OF COMMON ROCK-FORMING MINERALS The relative chemical stabilities of various minerals can be used to determine the intensity of weathering in a given area. In a tropical rain forest, only the most stable minerals will be left on an outcrop or in the soil, so we know that the weathering in that environment is intense. In an arid region such as the desert of North Africa, where weathering is minimal, alabaster (gypsum) monuments remain intact, as do many other unstable minerals. **Table 16.2** shows the

FIGURE 16.6 ■ Red and brown iron oxides color weathering rocks in Monument Valley, Arizona.
[© Charles & Josette Lenars/ Corbis/Aurora Photos.]

TABLE 16-2	Relative Stabilities of Common Minerals	
Stability of Minerals	**Rate of Weathering**	
MOST STABLE	Slowest	
Iron oxides (hematite)		
Aluminum hydroxides (gibbsite)		
Quartz		
Clay minerals		
Muscovite mica		
Orthoclase feldspar		
Biotite mica		
Sodium-rich plagioclase feldspar (albite)		
Amphiboles		
Pyroxenes		
Calcium-rich plagioclase feldspar (anorthite)		
Olivine		
Calcite		
Halite		
LEAST STABLE	Fastest	

relative stabilities of all the common rock-forming minerals. Salt and carbonate minerals are the least stable, iron oxides the most stable.

■ Physical Weathering

Now that we have surveyed chemical weathering, we can turn to its partner process, physical weathering. We can see the workings of physical weathering most clearly by examining its role in arid regions, where chemical weathering is minimal.

How Do Rocks Break?

Rocks can break for a variety of reasons, including stress along natural zones of weakness and biological and chemical activity.

NATURAL ZONES OF WEAKNESS Rocks have natural zones of weakness along which they tend to crack. In sedimentary rocks such as sandstone and shale, these zones are the bedding planes formed by successive layers of solidified sediment. Foliated metamorphic rocks such as slate form

FIGURE 16.7 ■ Weathered, enlarged joint patterns have developed in two directions in these rocks at Point Lobos State Reserve, California. [Jeff Foott/Discovery Channel Images/Getty Images, Inc.]

parallel cleavage planes, which enable them to be split easily. Granites and other nonfoliated rocks are often referred to as *massive,* which in this case means that they contain no preexisting planes of weakness. Massive rocks tend to crack along regular fractures spaced one to several meters apart, called *joints* (**Figure 16.7**). As we saw in Chapter 7, joints and less regular fractures result from deformation and from cooling and contraction while rocks are still deeply buried in Earth's crust. Through uplift and erosion, the rocks eventually rise to Earth's surface. There, freed from the weight of overlying rock, the fractures open slightly. Once the fractures open a little, both chemical and physical weathering work to widen them.

ACTIVITIES OF ORGANISMS The activities of organisms contribute to physical as well as chemical weathering. Bacteria and algae may invade cracks in rock, producing microfractures. These organisms, both those in cracks and those that may encrust the rock, produce acids, which then promote chemical weathering. In some regions, acid-producing fungi are active in soils, contributing to chemical weathering. Many of us have seen a crack in a rock that has been widened by a tree root (see Figure 5.2). Animals burrowing or moving through cracks can also break rock.

FROST WEDGING One of the most efficient mechanisms for widening cracks in rock is **frost wedging:** breakage resulting from the expansion of freezing water. As water freezes, it expands, exerting an outward force strong enough to wedge open a crack and split a rock (**Figure 16.8**). This same process can crack the engine block of a car that is not protected by antifreeze. Frost wedging is most important where water episodically freezes and thaws, as in temperate climates and in mountainous regions.

FIGURE 16.8 ■ Granite boulder split by frost in the Sierra Mountains, California. [Susan Rayfield/Science Source.]

EXFOLIATION One form of rock breakage is not directly related to preexisting zones of weakness. **Exfoliation** is a physical weathering process in which large flat or curved sheets of rock are detached from an outcrop. These sheets may look like the layers peeled from a large onion (**Figure 16.9**). Even though exfoliation is common, no generally accepted explanation for it has yet emerged. Some geologists have suggested that exfoliation results from an uneven distribution of expansion and contraction caused by chemical weathering and temperature changes.

Interactions Between Weathering and Erosion

Weathering and erosion are closely related, interacting processes, as we saw in Chapter 5. Physical weathering and erosion are tied by the processes by which wind, water, and ice work to dislodge particles of rock and move them away from their source. Physical weathering breaks a large mass of rock into smaller pieces, which are more easily eroded and transported.

The steepness of slopes affects both physical and chemical weathering, which in turn affect erosion. Weathering and erosion are more intense on steep slopes, and their action makes slopes gentler. Flows of rainwater are the primary agent of erosion, but wind may blow away the finest particles, and glacial ice can carry away large blocks torn from bedrock.

Chemical weathering rates are low at high elevations, where temperatures are generally low, soil is thin or absent, and vegetation is sparse. Physical weathering is greater at high elevations and in glacial terrains, where ice tears apart rock. We can see that erosional processes are closely related to the sizes of the fragments formed by physical weathering. As weathered material is eroded and transported,

FIGURE 16.9 ■ Exfoliation on Half Dome, Yosemite National Park, California. [Tony Waltham.]

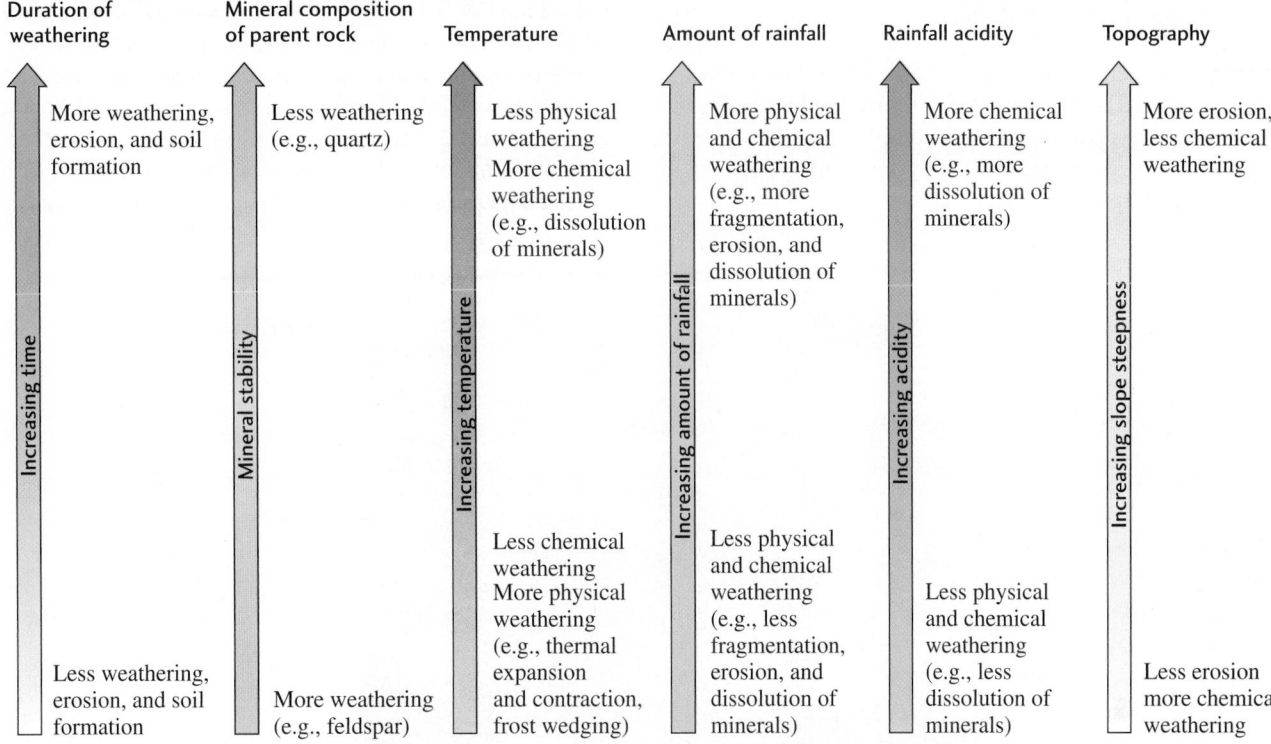

Duration of weathering	Mineral composition of parent rock	Temperature	Amount of rainfall	Rainfall acidity	Topography
More weathering, erosion, and soil formation	Less weathering (e.g., quartz)	Less physical weathering More chemical weathering (e.g., dissolution of minerals)	More physical and chemical weathering (e.g., more fragmentation, erosion, and dissolution of minerals)	More chemical weathering (e.g., more dissolution of minerals)	More erosion, less chemical weathering
Increasing time	*Mineral stability*	*Increasing temperature*	*Increasing amount of rainfall*	*Increasing acidity*	*Increasing slope steepness*
Less weathering, erosion, and soil formation	More weathering (e.g., feldspar)	Less chemical weathering More physical weathering (e.g., thermal expansion and contraction, frost wedging)	Less physical and chemical weathering (e.g., less fragmentation, erosion, and dissolution of minerals)	Less physical and chemical weathering (e.g., less dissolution of minerals)	Less erosion more chemical weathering

FIGURE 16.10 ■ Summary of the factors that influence weathering and erosion.

its size and shape may change further, and its composition may change as a result of chemical weathering. When transportation stops, deposition of the sediment formed by weathering begins.

Figure 16.10 is a chart summarizing the factors that influence weathering and erosion.

■ Soils: The Residue of Weathering

On moderate and gentle slopes, plains, and lowlands, where erosion is less intense, a layer of loose, heterogeneous weathered material remains overlying the bedrock. It may include particles of weathered and unweathered parent rock, clay minerals, iron and other metal oxides, and other products of weathering. Geologists use the term **soil** to describe layers of material, initially created by fragmentation of rock during weathering, that experience additions of new materials, losses of original materials, and modification through physical mixing and chemical reactions. Organic matter, called **humus,** is an important component of most of Earth's soils; it consists of the remains and waste products of the many organisms that live in soil. Leaf litter contributes significantly to the soil of forests. In addition, most soils have the ability to support rooted plants. Not all soils support life, however, and soils occur in places, such as Antarctica and Mars, where life is limited or possibly absent altogether.

Soils vary in color, from the brilliant reds and browns of iron-rich soils to the black of soils rich in organic matter. Soils also vary in texture. Some are full of pebbles and sand; others are composed entirely of clay. Soils are easily eroded, so they do not form on very steep slopes or where high altitude or frigid climate prevents the growth of plants that would hold them in place and contribute organic matter. Soil scientists, agronomists, and engineers, as well as geologists, study the composition and origin of soils, their suitability for agriculture and construction, and their value as a guide to climate conditions in the past.

Soils form at the interface between the climate and plate tectonic systems. They are crucial to life on Earth's continents, and they are one of human society's most valuable natural resources. Soils are the primary reservoir of nutrients for agriculture and the ecological systems that produce renewable natural resources. They filter our water and recycle our wastes, and they provide the necessary substratum for our buildings and infrastructure. In addition, they help regulate the global climate by storing and releasing carbon dioxide. Soils contain twice as much carbon as the atmosphere and three times more than all of the world's vegetation.

Soils as Geosystems

As we have seen, the concept of Earth as a set of interacting geosystems is of great value in understanding geologic processes. Soils, like many other components of the Earth system, can be described as a geosystem with inputs, processes, and outputs (**Figure 16.11**).

INPUTS: WEATHERED ROCK, ORGANISMS, AND DUST Soils develop from weathered rock, with additional inputs of organic matter from the biosphere and dust from the atmosphere. As discussed earlier, physical weathering breaks down rock into smaller pieces, and chemical weathering transforms minerals in that rock (such as feldspar) into other minerals (such as clays). Plants and other organisms

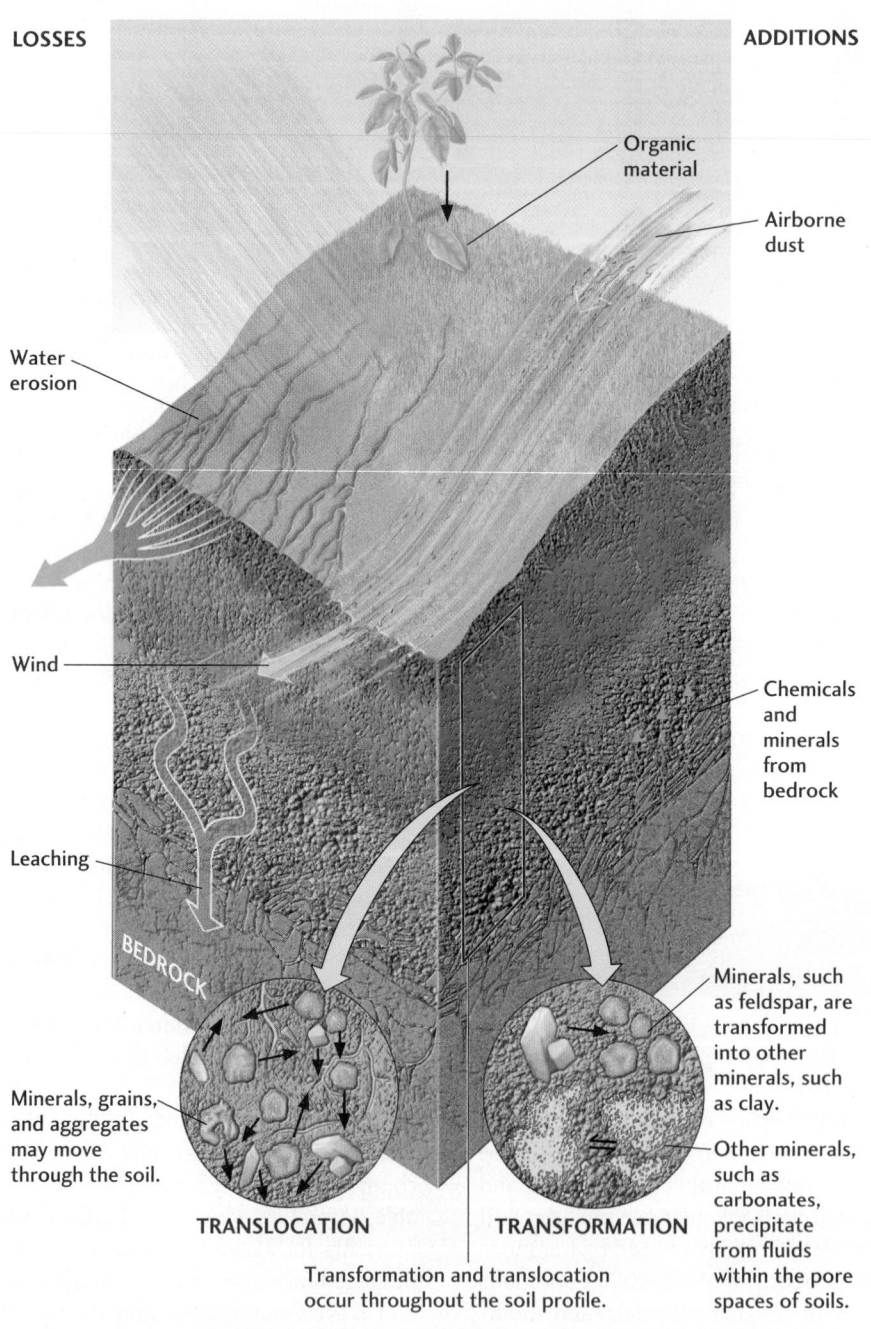

LOSSES   ADDITIONS

Organic material

Airborne dust

Water erosion

Wind

Chemicals and minerals from bedrock

Leaching

BEDROCK

Minerals, grains, and aggregates may move through the soil.

Minerals, such as feldspar, are transformed into other minerals, such as clay.

Other minerals, such as carbonates, precipitate from fluids within the pore spaces of soils.

TRANSLOCATION   **TRANSFORMATION**

Transformation and translocation occur throughout the soil profile.

FIGURE 16.11 ■ Soils are geosystems that develop through inputs of new materials, losses of original materials, and modification through physical mixing and chemical reactions. Soil modification processes can be divided into two basic types: translocations and transformations. The distinct soil horizons that make up the soil profile are also visible in this diagram.

may colonize the soil, and when they die, their tissues decompose to form humus. The atmosphere also contributes matter to the soil, but this material is predominantly inorganic dust.

PROCESSES: TRANSFORMATIONS AND TRANSLOCATIONS As soil ages and matures, the materials added to or removed from it cause it to undergo a set of *transformations*. The addition of humus, for instance, provides a source of nutrients that encourage further plant growth and add more humus—a positive feedback process within the soil geosystem. Many soil transformations involve chemical weathering of feldspar and other minerals to form clay minerals.

Translocations are lateral and vertical movements of materials within the developing soil. Water is the main agent of translocation, usually transporting dissolved salts. Water selectively removes some materials as it percolates down through the soil after rainfall in a process called *leaching*. However, it may also rise from below the soil surface when temperatures increase and evaporation draws more water to the surface. Organisms also play an important role in translocation by moving components of the soil as they burrow through it.

Soils are dynamic and respond to changes in climate, interactions with organisms, and perturbations by humans. Five factors are important in their formation and development:

1. *Parent material:* the solubility of minerals, the sizes of grains, and the patterns of fragmentation, such as joints and cleavage, in the bedrock

2. *Climate:* temperatures, precipitation levels, and their seasonal patterns of variation

3. *Topography:* slope steepness and the direction the slope faces: gentler slopes that face toward the Sun promote better soil development

4. *Organisms:* the diversity and abundance of organisms living in the soil

5. *Time:* the amount of time that a soil has to form

OUTPUTS: SOIL PROFILES Most soils form distinct layers as they develop. The composition and appearance of a soil is known as a **soil profile.** Soil profiles consist of up to six *horizons:* distinct layers of varying color and texture, usually parallel to the land surface, that are visible in vertical sections of exposed soils (see Figure 16.11).

The soil's topmost layer, called the *O-horizon*, is usually thin and consists of loose leaves and organic detritus. Beneath this topmost layer is the *A-horizon*, typically not much more than a meter or two thick and usually the darkest layer because it contains the highest concentration of humus. Next down is the *E-horizon*, which consists mostly of clay and insoluble minerals such as quartz, as soluble minerals will have been leached from this layer. Beneath the E-horizon is the *B-horizon*, in which organic matter is sparse. Soluble minerals and iron oxides accumulate in this layer. The climate influences the specific types of minerals that accumulate in the B-horizon; carbonate minerals and gypsum, for example, are found there in arid climates. The next layer, the *C-horizon*, is slightly altered bedrock, broken and decayed, mixed with clay produced by chemical weathering. Unaltered bedrock forms the lowest level, or *R-horizon*.

The five soil development factors listed above interact to create 12 different soil types, each with a distinct profile, that are recognized by scientists who study soils (Table 16.3).

Paleosols: Working Backward from Soil to Climate

Recently, there has been much interest in ancient soils that have been preserved as rock in the geologic record. These *paleosols*, as they are called, are being studied as guides to ancient climates and even to the amounts of carbon dioxide and oxygen in the atmosphere in former times. The mineralogy of paleosols billions of years old, for example, provides evidence that there was no oxidation of soils at that early stage of Earth's history, and therefore that oxygen had not yet become a major component of Earth's atmosphere.

Soil formation is just one step in the evolution of a landscape. Weathering and rock fragmentation often destabilize topographic features and lead to the more dramatic changes caused by mass wasting. This process is an important part of the general erosion of the land, especially in hilly and mountainous regions.

■ Mass Wasting

On the morning of June 1, 2005, as residents of Laguna Beach, California, were waking up and enjoying their morning coffee, the hillside broke loose beneath their feet and collapsed. Seven multi-million-dollar homes were destroyed as a large mass of soil and weathered bedrock gave way and slid downhill. Twelve other homes were badly damaged, and hundreds more were evacuated as residents waited anxiously for geologists to evaluate their home sites and determine whether it was safe to return. Some houses completely collapsed; others literally broke in half; and still others were left stranded at the top of the hill, where they jutted out over a large gash formed where the sliding mass of earth broke away (see the chapter opening photo).

This mass wasting event was triggered by very high seasonal rainfall—the second highest ever recorded for that part

TABLE 16-3   Twelve Recognized Soil Types

Soil Type	Description	Most Important Formation Factors[a]
Alfisols	Soils of humid and subhumid climates with a subsurface horizon of clay accumulation, not strongly leached, common in forested areas	Climate, organisms
Andisols	Soils that formed in volcanic ash and contain compounds rich in organic matter and aluminum	Parent material
Aridisols	Soils formed in dry climates, low in organic matter and often having subsurface horizons with salt accumulation	Climate
Entisols	Soils lacking subsurface horizons because the parent material accumulated recently or because of constant erosion; common on floodplains, mountains, and badlands (highly eroded, rocky areas)	Time, topography
Gelisols	Weakly weathered soils formed in areas that contain permafrost (frozen soil) within the soil profile	Climate
Histosols	Soils with a thick upper layer very rich in organic matter (>25%) and containing relatively little mineral material	Topography
Inceptisols	Soils with weakly developed subsurface horizons and little or no subsoil clay accumulation because the soil is young or the climate does not promote rapid weathering	Time, climate
Mollisols	Mineral soils of semiarid and subhumid midlatitude grasslands that have a dark, organic-rich A-horizon and are not strongly leached	Climate, organisms
Oxisols	Very old, highly leached soils with subsurface accumulations of iron and aluminum oxides, commonly found in humid tropical environments	Climate, time
Spodosols	Soils formed in cold, moist climates that have a well-developed B-horizon with accumulation of aluminum and iron oxides, formed under pine vegetation in sandy parent material	Parent material, organisms, climate
Ultisols	Soils with a subsurface horizon of clay accumulation, highly leached (but not as highly as oxisols), commonly found in humid tropical and subtropical climates	Climate, time, organisms
Vertisols	Soils that develop deep, wide cracks when dry (shrink and swell) due to high clay content (>35%) and are not highly leached	Parent material

[a]All five soil formation factors (climate, organisms, parent material, topography, time) combine to create these soils, but only the most important factors are listed for each soil type.

Source: Adapted from E. C. Brevik, *Journal of Geoscience Education* 50 (2002): 541.

of Southern California—which saturated the soil and bedrock and created the necessary conditions in an already unstable geologic environment to tip the scales in favor of disaster. Earlier in the same year, high rainfall had triggered similar events, including one that killed 10 people when homes were buried in La Conchita, California (**Figure 16.12**).

These events in Southern California represent just one of many kinds of downslope movements of masses of soil, rock, mud, or other materials under the force of gravity, known collectively as **mass movements.** These masses are not pulled downslope primarily by the action of an

erosional agent, such as wind, flowing water, or glacial ice. Instead, mass movements occur when the force of gravity exceeds the strength of the slope materials. The materials then move down the slope, sometimes very slowly, sometimes as a large, sudden, catastrophic movement. Mass movements can displace small, almost imperceptible amounts of soil down a gentle hillside, or they can be huge landslides that dump tons of earth and rock on valley floors below steep mountain slopes.

Every year, mass movements take a toll of lives and property throughout the world. In late October and early

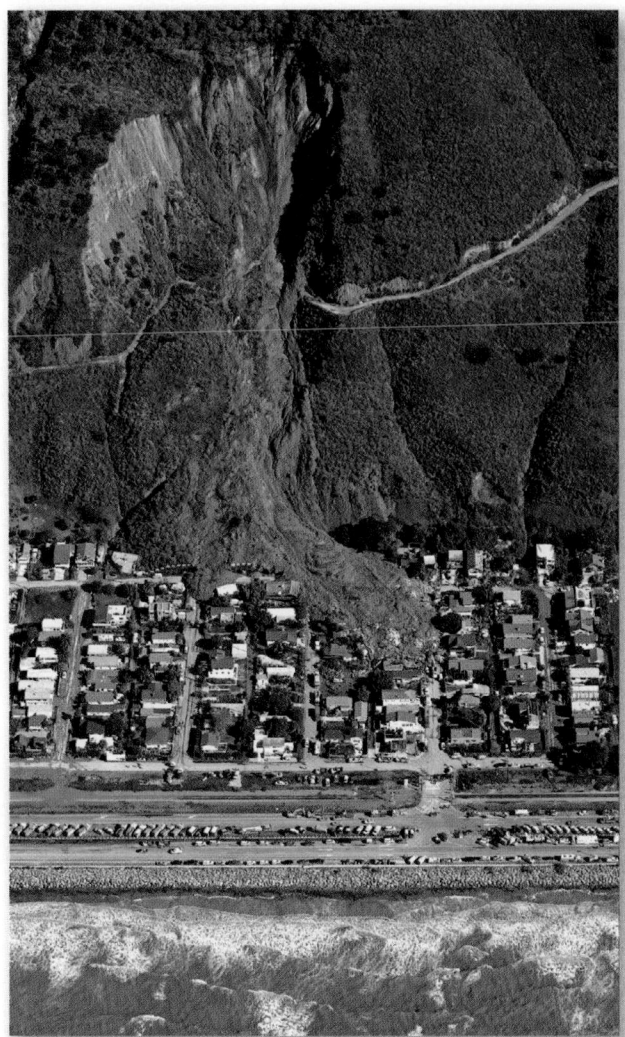

FIGURE 16.12 ■ A massive mudslide buried homes in La Conchita, California, in 2005. [AP Photo/Kevork Djansezian.]

interference with natural processes. We cannot prevent most natural mass movements, but we can plan construction and land development so as to minimize losses. Mass movements change the landscape by scarring mountainsides as great masses of material fall or slide away from the slopes. The material that moves ends up as tongues or wedges of debris on the valley floor, sometimes piling up and damming a stream running through the valley. Such scars and debris deposits, mapped in the field or from aerial photographs, are clues to past mass movements. By reading these clues, geologists may be able to predict and issue timely warnings about new movements likely to occur in the future.

Mass wasting is influenced by three primary factors (Table 16.4):

1. *The nature of the slope materials.* Slopes may be made up of **unconsolidated materials,** which are loose and uncemented, or **consolidated materials,** which are compacted or bound together by mineral cements.

2. *The amount of water in the materials.* This factor depends on how porous the materials are and on how much rain or other water they have been exposed to.

3. *The steepness of the slope.* This factor contributes to the tendency of materials to fall, slide, or flow under various conditions.

All three factors operate in nature, but slope steepness and water content are most strongly influenced by human activity, such as excavation for building and highway construction. All three factors produce the same result: they lower resistance to movement. The force of gravity takes over, and the slope materials begin to fall, slide, or flow.

Slope Materials

Slope materials vary greatly because they are so dependent on the physical properties of the local terrain. Thus, the foliated bedrock of one hillside may be prone to extensive fracturing whereas another slope only a few hundred meters away may be composed of massive granite. Slopes of unconsolidated material are the least stable of all.

UNCONSOLIDATED SAND AND SILT The behavior of loose, dry sand and silt illustrates how the steepness of slopes influences mass wasting. Children's sandboxes have made nearly everyone familiar with the characteristic slope of a pile of dry sand. The angle between the slope of any pile of sand or silt and the horizontal is the same, whether the pile is a few centimeters or several meters high. For most sands and silts, that angle is about 35°. If you scoop some sand from the base of the pile very slowly and carefully, you can increase the slope angle a little, and it will hold temporarily. If you then jump on the ground near the sand pile, however, the sand will cascade down the side of the pile, which will again assume its original slope of 35°.

November of 1998, for example, one of the most catastrophic hurricanes of the twentieth century, Hurricane Mitch, dropped torrential rains on Central America, saturating the ground and causing raging floods and landslides. At least 9000 people were killed, and billions of dollars in damage was done as the floods and slides laid waste to once-fertile land and crops of corn, beans, coffee, and peanuts. One of the hardest-hit areas was near the Nicaragua-Honduras border, where a series of landslides and mudflows buried at least 1500 people. Dozens of villages were simply obliterated, engulfed by a sea of mud. The walls of a crater on Casita volcano collapsed and started a series of slides and flows that were described as a moving wall of mud more than 7 m high. Those in the direct path of the avalanche could not escape, and many were buried alive as they tried to outrun the fast-moving mud.

Because mass movements are responsible for so much destruction, we want to be able to predict them, and we certainly want to refrain from provoking them by unwise

TABLE 16-4   Factors that Influence Mass Movements

Nature of Slope Material	Water Content	Steepness of Slope	Stability of Slope
UNCONSOLIDATED			
Loose sand or sandy silt	Dry Wet	Angle of repose	High Moderate
Unconsolidated mixture of sand, silt, soil, and rock fragments	Dry Wet	Moderate	High Low
	Dry Wet	Steep	High Low
CONSOLIDATED			
Rock, jointed and deformed	Dry or wet	Moderate to steep	Moderate
Rock, massive	Dry or wet Dry or wet	Moderate Steep	High Moderate

The slope angle of the sand pile is its **angle of repose,** the maximum angle at which a slope of loose material will lie without cascading down. A slope that is steeper than the angle of repose is unstable and will tend to collapse to the stable angle of repose. Sand or silt grains will form piles with slopes at and below the angle of repose because friction between the individual grains holds them in place. However, as more and more grains are placed on the pile and the slope steepens, the ability of the frictional forces to prevent sliding decreases, and the pile suddenly collapses.

The angle of repose varies with a number of factors, one of which is the size and shape of the particles (**Figure 16.13a**). Larger, flatter, and more angular pieces of loose material remain stable on steeper slopes. The angle of repose also varies with the amount of moisture between particles. The angle of repose of wet sand is higher than that of dry sand because the small amount of moisture between the grains tends to bind them together so that they resist movement. The source of this binding tendency is *surface tension:* the attractive force between molecules at a surface (Figure 16.13b). Surface tension is what makes water drops spherical and allows a razor blade or paper clip to float on a smooth water surface. Too much water between particles, on the other hand, separates the particles and allows them to move freely over one another. Saturated sand, in which all the pore space is occupied by water, runs like a fluid and collapses to a flat pancake shape (Figure 16.13c). The surface tension that binds damp sand allows beach sculptors to create elaborate sand castles (**Figure 16.14**), but when the tide comes in and saturates the sand, the structures collapse. Similarly, but at a much larger scale, landslides on hillslopes depend on water abundances in soils. Heavy rainfall may saturate pore spaces on a slope, causing catastrophic ground failure.

UNCONSOLIDATED MIXTURES Slope materials composed of mixtures of unconsolidated sand, silt, clay, soil, and fragments of rock (often called *debris*) will form slopes with moderate to steep angles (see Table 16.4). The platy shape of clay minerals, the organic content of soils, and the rigidity of rock fragments are all key factors that affect the ability of the material to form slopes with a specific angle.

CONSOLIDATED MATERIALS Slopes of consolidated dry materials—such as rock, compacted or cemented sediments, and vegetated soils bound together by plant roots—may be steeper and less regular than slopes consisting of unconsolidated materials. They can become unstable when they are oversteepened or denuded of vegetation. The particles of some consolidated sediments, such as dense clays, are bound together by cohesive forces associated with tightly packed particles. *Cohesion* is an attractive force between particles of a solid material that are close together. The greater the cohesive forces in a material, the greater the resistance to movement.

Water Content

The effect of water on consolidated materials is similar to its effect on unconsolidated materials. Mass movements of consolidated materials can usually be traced to the effects of moisture, often in combination with other factors, such as the loss of vegetation or the steepening of a slope. When the ground becomes saturated with water, the planes of weakness within the solid material are lubricated, the friction between particles is lowered, and the particles or larger aggregates can move past one another more easily, so that the material may start to flow like a fluid. This process is called **liquefaction.**

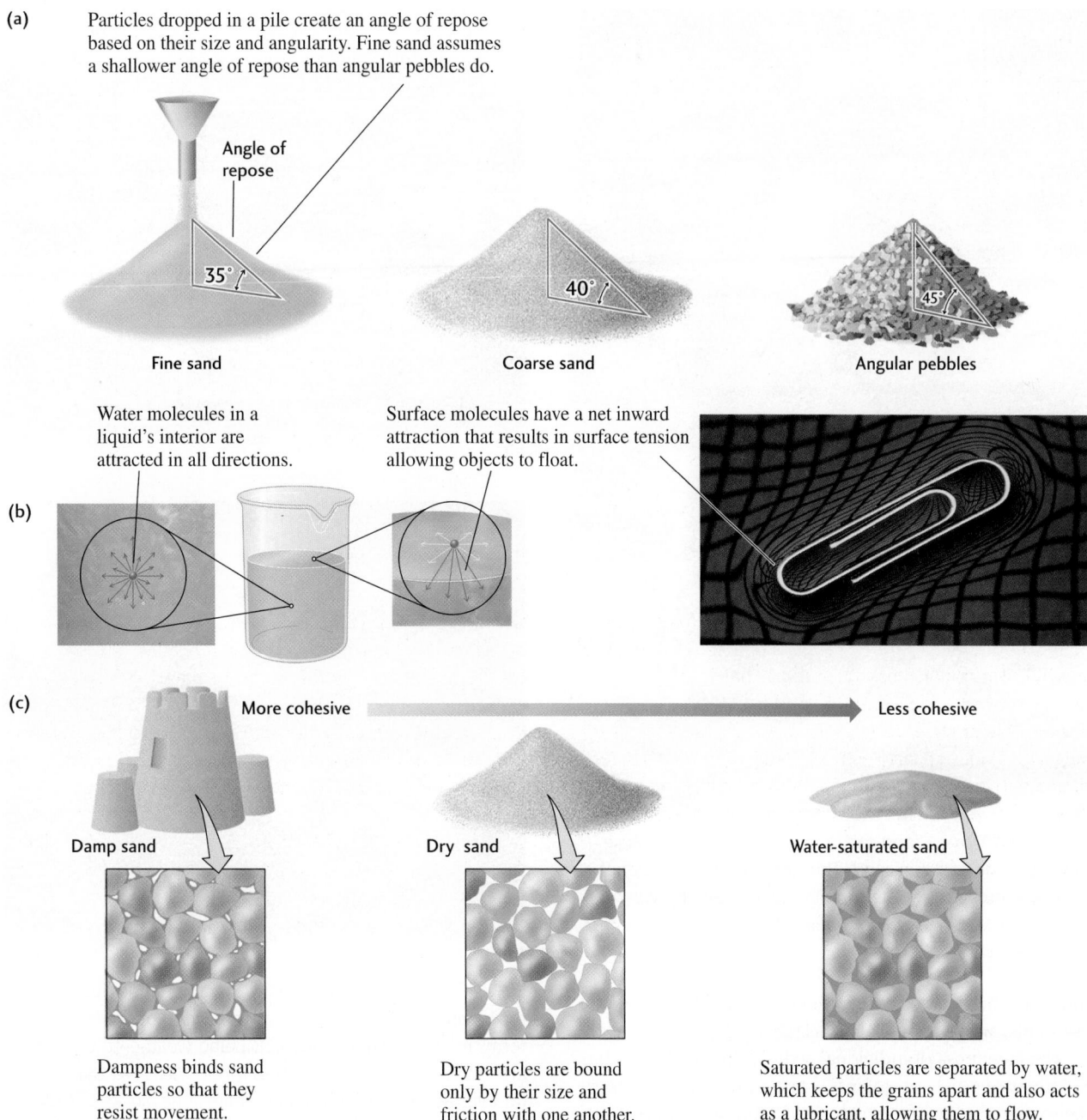

(a) Particles dropped in a pile create an angle of repose based on their size and angularity. Fine sand assumes a shallower angle of repose than angular pebbles do.

Angle of repose

35° — Fine sand

40° — Coarse sand

45° — Angular pebbles

(b) Water molecules in a liquid's interior are attracted in all directions.

Surface molecules have a net inward attraction that results in surface tension allowing objects to float.

(c) More cohesive ———→ Less cohesive

Damp sand

Dry sand

Water-saturated sand

Dampness binds sand particles so that they resist movement.

Dry particles are bound only by their size and friction with one another.

Saturated particles are separated by water, which keeps the grains apart and also acts as a lubricant, allowing them to flow.

FIGURE 16.13 ■ The angle of repose of a pile of unconsolidated material depends on the shape of the particles and the water content of the pile.

Slope Steepness

Rock slopes range from the relatively gentle inclines of easily weathered shales and volcanic ash beds to vertical cliffs of massive granite. The stability of rock slopes depends on the weathering and fragmentation of the rock. Shales, for example, tend to weather easily and fragment into small pieces that form a thin layer of loose, angular rock fragments (often called *rubble*) covering the bedrock (**Figure 16.15a**). The resulting angle of repose is similar to that of loose, coarse sand. The weathered rubble gradually

builds up beyond the angle of repose and becomes unstable, and eventually some of it slides downhill.

In contrast, limestones and hard, cemented sandstones in arid environments resist erosion and break into large blocks, forming steep, bare bedrock slopes above and gentler slopes covered with broken rock (often called **talus**) below (Figure 16.15b). The bedrock cliffs are fairly stable, except for the occasional mass of rock that falls and rolls down to the rock-covered slope below. Where such limestones or sandstones are interbedded with shale, slopes may be stepped (see Figure 8.12). As shale slides from

FIGURE 16.14 ■ Sand castles hold their shape because they are made of damp sand. The steepness of their slopes is maintained by surface tension in the moisture between grains. [Kelly Mooney Photography/Corbis.]

(a)

(b)

FIGURE 16.15 ■ The stability of a rock slope depends on the weathering and fragmentation pattern of the rock that forms it. (a) This small outcrop is being weathered to form broken blocks of rock known as rubble. [John Grotzinger.] (b) Talus accumulates on slopes where large blocks of rock fall or roll downhill to form a cone-shaped pile. [Phil Stoffer/US Geological Survey.]

under the harder beds, those beds are undercut, become less stable, and eventually fall as large blocks.

The steepness of individual sedimentary beds also has an influence on slope stability. Mass movements are most likely when the dip of the beds close to the surface is parallel to the slope.

Triggers of Mass Movements

When the right combination of materials, moisture, and steepness makes a slope unstable, a mass movement is inevitable. All that is needed is a trigger. Sometimes a landslide, like the one at Laguna Beach, is provoked by a heavy rainstorm. Many mass movements are set off by vibrations, such as those that occur in earthquakes. Others may be precipitated by gradual steepening due to erosion that eventually results in sudden collapse of the slope.

Geologic reports can help to minimize the human cost of mass movements (see Practicing Geology), but only if municipal planners and individual home buyers heed those reports and avoid building or buying in unstable areas. The devastating mass movements in Southern California during 2005 were clearly related to the unusually high seasonal rainfall during the winter of 2004–2005. That rainfall, however, was related to El Niño conditions (described in Chapter 15), which Earth scientists now understand to recur on a regular basis.

Similarly, most of the damage in the great Alaska earthquake of March 27, 1964, was caused by the slides it triggered. Mass movements of rock, earth, and snow wreaked havoc in residential areas of Anchorage, and there were major submarine slides along lakeshores and the seacoast. Huge landslides took place along the flat plains below the 30–35-m-high bluffs along the coast. The bluffs were composed of interbedded clays and silts. During the earthquake, the ground shook so hard that unstable, water-saturated sandy layers in the clay were transformed into fluid slurries. Enormous blocks of clay and silt were shaken down from the bluffs and slid along the flat ground with the liquefied sediments, leaving a completely disrupted terrain of jumbled blocks and broken buildings (**Figure 16.16**). Houses and roads were carried along by the slides and destroyed. The whole process took only 5 minutes, beginning about 2 minutes after the first shock of the earthquake. At one locality, three people were killed and 75 homes were destroyed.

Studies of slope stability in both California and Alaska, and of the likelihood of repeated high rainfall or earthquakes in those regions, had indicated that both areas were prime candidates for landslides. A geologic report issued more than a decade before the quake had warned of the hazards of development in the part of Alaska that suffered the most damage, but the scenic beauty of the area overwhelmed people's judgment. The same is true of Southern California. In Alaska, people paid the price with their lives. Fortunately, at Laguna Beach, the cost was only the value of people's homes, but even that was steep in an area where the average price of a house was well over a million dollars.

■ Classification of Mass Movements

Although the popular press often refers to any mass movement as a "landslide," there are many different kinds of mass movements, each with its own characteristics. In this textbook, we use the term *landslide* only in its popular sense, to refer to mass movements in general.

Geologists classify mass movements in accordance with three characteristics, as summarized in **Figure 16.17**:

1. The nature of the moving material (for example, whether it is rock or unconsolidated material)

2. The velocity of the movement (from a few centimeters per year to many kilometers per hour)

3. The nature of the movement: whether it is sliding (the bulk of the material moves more or less as a unit) or flowing (the material moves as if it were a fluid)

The nature and velocity of the movement are greatly influenced by the water or air content of the moving material.

Some movements have characteristics that are intermediate between sliding and flowing. Most of the mass may move by sliding, for example, but parts of it along the base may move as a fluid. A movement is called a *flow* if that is the main type of motion. It is not always easy to tell the

(a)

(b)

Before earthquake

Sand and gravel

Clay

Clay

Water-saturated sandy layer

After earthquake

Pre-earthquake profile

FIGURE 16.16 ■ (a) A landslide triggered by the great Alaska earthquake of 1964. [NOAA/Tehrkot/Landov.] (b) Cross sections of the bluffs at Anchorage, Alaska, before and after the earthquake.

FIGURE 16.17 ■ Mass movements are classified according to the nature of the moving material, the velocity of the movement, and the nature of the movement.

a cliff or steep mountainside (**Figure 16.18**). Weathering weakens bedrock along joints until the slightest pressure—often exerted by frost wedging—is enough to trigger a rockfall. The velocities of rockfalls are the fastest of all rock movements, but the travel distances are the shortest, generally only meters to hundreds of meters. Evidence for the origin of rockfalls is clear in the blocks seen in the accumulation of talus at the foot of a steep bedrock cliff, which can be matched with rock outcrops on the cliff. Talus accumulates slowly, building up into blocky slopes along the base of a cliff over long periods.

In *rockslides*, rocks do not fall freely, but rather slide down a slope. Although these movements are fast, they are slower than rockfalls because masses of bedrock slide more or less as a unit, often along downward-sloping bedding or joint planes (**Figure 16.19**).

(a)

Ice wedging often breaks rocks along joints, preparing them to loosen and fall away.

Jointed bedrock

Individual blocks free-fall down slope.

(b)

FIGURE 16.18 ■ (a) Rockfall in Grindewald, Switzerland. [Pascal Lauener/Reuters Schweiz/Landov.] (b) In a rockfall, individual blocks plummet in free fall from a cliff or steep mountainside.

exact nature of a mass movement, however; the only evidence may be the debris deposited after the movement is over.

Mass Movements of Rock

Rock movements include rockfalls, rockslides, and rock avalanches. These movements may involve small blocks or larger masses of bedrock. During a *rockfall*, newly detached individual blocks of rock plummet suddenly in free fall from

Frost wedging has loosened jointed bedrock layers,…

…which move downhill more or less as a unit.

(a)

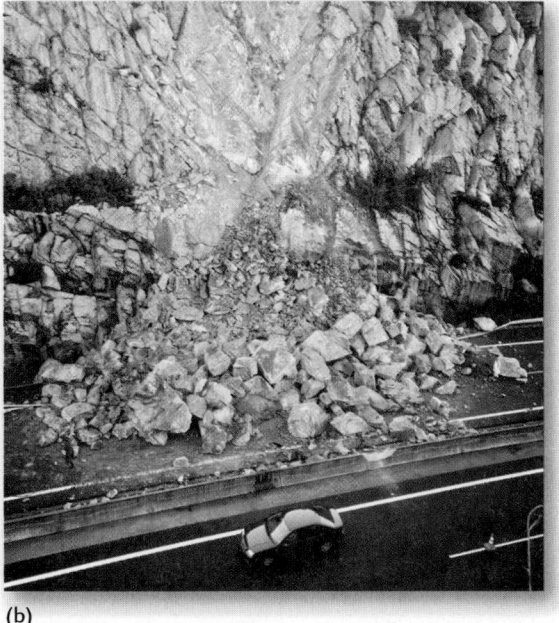

(b)

FIGURE 16.19 ■ (a) In a rockslide, large masses of bedrock move more or less as a unit in a fast, downward slide. (b) Rockslide in Northeastern Spain. [Cabalar/EPA/Newscom.]

(a)

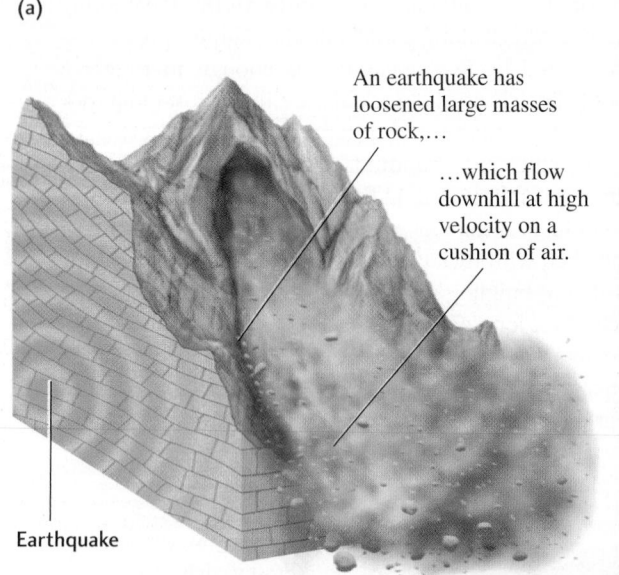

An earthquake has loosened large masses of rock,...

...which flow downhill at high velocity on a cushion of air.

Earthquake

(b)

FIGURE 16.20 ■ (a) In a rock avalanche, large masses of broken rocky material flow, rather than slide, downslope at high velocity. (b) The two rock avalanches shown in the photo were triggered by the November 3, 2002, earthquake along the Denali fault in Alaska. The rock avalanches traveled down south-facing mountains, moved across the 1.5-mile-wide Black Rapids Glacier, and flowed partway up the opposite slope. [Photo by Dennis Trabant, USGS, mosaic by Rod March, USGS.]

Rock avalanches differ from rockslides in their much greater velocities and travel distances (**Figure 16.20**). They are composed of large masses of rocky material that break up into smaller pieces when they fall or slide. The pieces then flow farther downhill at velocities of tens to hundreds of kilometers per hour, riding on a cushion of air. Rock avalanches are typically triggered by earthquakes. They are some of the most destructive mass movements because of their large volume (many contain more than a half million cubic meters of material) and their capacity to transport materials for thousands of meters at high velocities.

Most rock mass movements occur in high mountainous regions; they are rare in low hilly areas. Rock masses tend to move where weathering fragments rocks already predisposed to breakage by deformation features such as faults and joints, relatively weak bedding planes, or foliation. In many such regions, extensive talus accumulations have been built by infrequent but large-scale rockfalls and rockslides.

Mass Movements of Unconsolidated Material

Mass movements of unconsolidated material include various mixtures of sand, silt, clay, soil, and fragmented bedrock, as well as trees and shrubs and materials of human construction, from fences to cars and houses. Most mass movements of unconsolidated materials are slower than most rock movements, largely because of the lower slope angles at which these materials become unstable. Although some unconsolidated materials move as coherent units, many flow like very viscous fluids. (Viscosity, as you will recall from Chapter 4, is a measure of a fluid's resistance to flow.)

The slowest type of unconsolidated mass movement is **creep**: the gradual downhill movement of soil or other debris (**Figure 16.21**). Rates of creep range from about 1 to 10 mm/year, depending on the soil type, the climate, the steepness of the slope, and the density of the vegetation. The movement is a very slow deformation of the soil, with the upper layers of soil moving down the slope faster than the lower layers. Such slow movements may cause trees, telephone poles, and fences to lean or move slightly downslope. The weight of the masses of soil creeping downhill can break poorly supported retaining walls and crack the walls and foundations of buildings. In icy regions where the deeper layers of soil are permanently frozen, a type of creep called *solifluction* occurs when water in the surface layers of soil alternately freezes and thaws, causing the soil to ooze downhill, carrying broken rocks and other debris with it.

Earthflows and debris flows are fluid mass movements that occur when rainfall soaks and loosens permeable material overlying a layer of less permeable rock. They usually travel faster than creep, as much as a few kilometers per hour, primarily because the moving materials are saturated with water and thus have little resistance to flow. An *earthflow* is a fluid mass movement of relatively fine grained materials, such as soils, weathered

(a)

Building foundations shear and crack

Trees grow with curved trunks

Road cracks

Objects lean

(b)

1 Rock layers meet the surface at an angle.

2 As the rock weathers, it moves slowly downhill in the overlying soil.

3 The movement of the soil surface is faster than that of deeper soil and rock.

4 Thus, it pushes parts of objects at the surface faster than buried parts, causing them to lean downhill.

FIGURE 16.21 ■ (a) Creep is the downhill movement of soil or other debris at a rate of about 1 to 10 mm/year. (b) A fence offset by creep in Marin County, California. [Travis Amos.]

shales, and clay (**Figure 16.22**). A *debris flow* is a fluid mass movement of rock fragments supported by a muddy matrix (**Figure 16.23**). Debris flows are made up mostly of material coarser than sand and tend to move more rapidly than earthflows. The slide at Laguna Beach, California, described above, was classified as a debris flow. In some cases, debris flows may reach velocities of 100 km/hour.

Mudflows are flowing masses made up mostly of material finer than sand, along with some rock debris, and containing large amounts of water (**Figure 16.24**). The mud offers little resistance to flow because of its high water content and thus tends to move faster than earth or debris. Many mudflows move at several kilometers per hour. Most common in hilly and semiarid regions, mudflows occur

(a)

Rain has soaked fine-grained permeable soils,...

...which loosen quickly...

Water-permeable soil

Water-impermeable rocks

...and flow downhill over impermeable rocks at moderate speeds.

(b)

FIGURE 16.22 ■ (a) An earthflow is a fluid movement of relatively fine grained material that may travel as fast as a few kilometers per hour. (b) Earthflow in the Buller Valley on the South Island of New Zealand. [G. R. Roberts/Science Source.]

(a)

Rain has soaked shale-derived muds and rubble overlying less porous bedrock,...

Clear-cut slopes

Shale

Jointed bedrock

...which loosen quickly, resulting in a flow of mixed mud, rock, and surface debris.

(b)

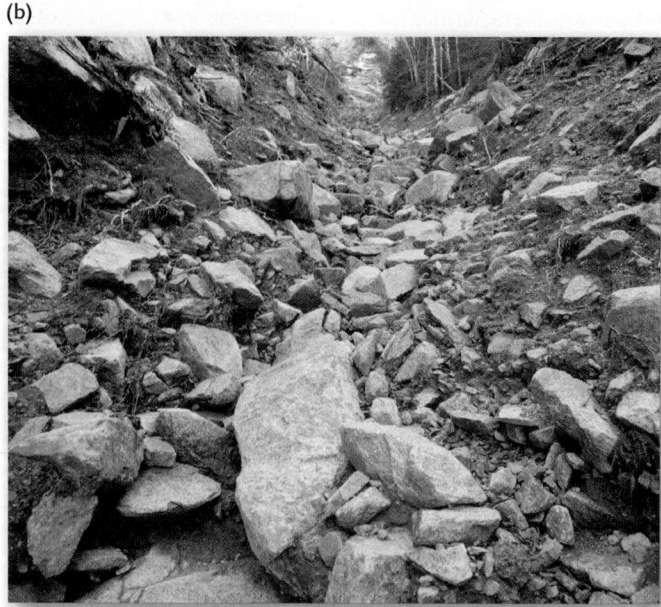

FIGURE 16.23 ▪ (a) A debris flow contains material that is coarser than sand and travels at rates from a few kilometers per hour to many tens of kilometers per hour. (b) Debris flow in Upper Bavaria, Germany. [© Erin Paul Donovan/Alamy.]

when fine-grained material becomes saturated. Mudflows of wet pyroclastic material, called *lahars*, may be triggered by volcanic eruptions, as when a lava flow melts snow and ice (see Chapter 12). Similarly, mudflows may start when dry, cracked mud on a slope is subjected to infrequent, sometimes prolonged, rains. If the mud keeps absorbing water as the rain continues, its physical properties change: its internal friction decreases, and the mass becomes much less resistant

to movement. The slope, which is stable when dry, becomes unstable, and any disturbance, such as an earthquake, triggers movement of waterlogged masses of mud. Mudflows may travel down tributary valleys on upper slopes and merge on the main valley floor. Where mudflows exit from confined upper valleys into broader, lower valley slopes and flats, they may splay out to cover large areas with wet debris. Mudflows can carry huge boulders, trees, and even houses.

(a)

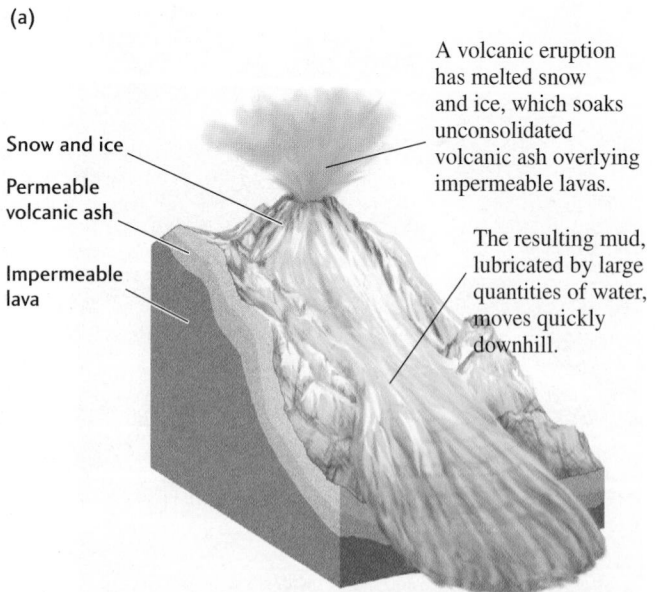

Snow and ice

Permeable volcanic ash

Impermeable lava

A volcanic eruption has melted snow and ice, which soaks unconsolidated volcanic ash overlying impermeable lavas.

The resulting mud, lubricated by large quantities of water, moves quickly downhill.

(b)

FIGURE 16.24 ▪ (a) Mudflows tend to move faster than earthflows or debris flows because they contain large quantities of water. (b) An earthquake in Tadzhikistan in January 1989 produced 15-m-high mudflows on slopes weakened by rain. [Washington State DOT/Seattle Times/MCT/Newscom.]

Debris avalanches (**Figure 16.25**) are fast downhill movements of soil and rock that usually occur in humid mountainous regions. Their speed results from the combination of high water content and steep slopes. Water-saturated debris may move as fast as 70 km/hour, a speed comparable to that of water flowing down a moderately steep slope. A debris avalanche carries with it everything in its path.

In 1962, a debris avalanche on Nevado de Huascarán, Peru, one of the highest mountains in the Andes, traveled almost 15 km in about 7 minutes, engulfing most of eight towns and killing 3500 people. Eight years later, on May 31, 1970, an earthquake toppled a large mass of glacial ice at the top of the same mountain. As the ice broke up, it mixed with the debris of the high slopes and became an ice-debris avalanche. The avalanche picked up additional debris as it raced downhill, increasing its speed to an almost unbelievable 280 km/hour. Up to 50 million cubic meters of muddy debris roared down into the valleys, killing 18,000 people and wiping out scores of villages (Figure 16.25b). On May 30, 1990, an earthquake shook another mountainous area in northern Peru, in the same active subduction zone, again setting off mudflows and debris avalanches. It was the day before a memorial ceremony scheduled to commemorate

(a) Unconsolidated ash and rock move down a steep slope at high speeds, lubricated by high water or air content.

(b) Towns of Yungay and Ranrahirca before an earthquake-induced debris avalanche on Mount Huascarán, Peru, buried these towns.

(c) Aftermath of the avalanche.

FIGURE 16.25 ■ (a) A debris avalanche is the fastest type of unconsolidated flow because of its high water content and movement down steep slopes. (b, c) In 1970, an earthquake-induced debris avalanche on Mount Huascarán, Peru, buried the towns of Yungay and Ranrahirca, killing some 18,000 people. The avalanche traveled 17 km at a speed of up to 280 km/hour and is estimated to have consisted of up to 50 million cubic meters of water, mud, and rocks. [Lloyd Cluff/Corbis.]

(a)

Scar

A mass of unconsolidated material slowly slides as a unit. Sliding is quick, but only for a short distance.

(b)

Scar

FIGURE 16.26 ■ (a) A slump is a slow slide of unconsolidated material that travels as a unit. (b) Soil slump, Northern California. [Marli Bryant Miller.]

the disaster that had occurred 20 years earlier. In regions close to convergent plate boundaries, where uplift and volcanism build up unstable slopes and earthquakes are frequent, there can be no doubt about the necessity of learning how to predict both earthquakes and the dangerous mass movements that follow.

In a **slump,** a mass of unconsolidated material slides slowly downslope as a unit, leaving a scar at its source (**Figure 16.26**). In most places, the slump slips along a basal surface that forms a concave-upward shape, like a spoon. Faster than slumps are *debris slides* (**Figure 16.27**), in which rock material and soil move largely as one or more units along planes of weakness, such as a waterlogged clay zone either within or at the base of the debris. During the slide, some of the debris may behave as a chaotic, jumbled flow. Such a slide may become predominantly a flow as it

moves rapidly downhill and most of the material mixes as if it were a fluid.

■ Understanding the Origins of Mass Movements

To understand how slope steepness, the nature of the slope materials, and their water content interact to create mass movements, geologists study both natural mass movements and those provoked by human activities. They investigate the causes of a recent mass movement by combining eyewitness reports with geologic studies of the distribution and nature of the moved material as well as its source. They

(a)

Rock, soil, and surface features (such as trees) move as units downhill along a plane of weakness.

(b)

FIGURE 16.27 ■ (a) A debris slide travels as one or more units and moves more quickly than a slump. (b) The Hope Princeton debris slide, which occurred in 1965 in British Columbia. [Joy Spurr/© Bruce Coleman/Photoshot.]

can infer the causes of a prehistoric mass movement from geologic evidence alone where the material is still present and can be analyzed for size, shape, and composition.

Natural Causes of Mass Movements

The 1925 landslide in the Gros Ventre River valley of western Wyoming illustrates how water, the nature of slope materials, and slope steepness interact to produce mass movements (Figure 16.28). In the spring of that year, melting snow and heavy rains swelled streams and saturated the ground in the valley. One local rancher, out on horseback, looked up to see the whole side of the valley racing toward his ranch. From the gate to his property, he watched the slide hurtle past him at about 80 km/hour and bury everything he owned.

About 37 million cubic meters of rock and soil slid down one side of the valley that day, then surged more than 30 m up the opposite side and fell back to the valley floor. Most of the rockslide was a confused mass of blocks of sandstone, shale, and soil, but one large section of the side of the valley, covered with soil and a forest of pine, slid down as a unit. The rockslide dammed the river, and a large lake grew over the next 2 years. Then the lake overflowed, breaking the dam and rapidly flooding the valley below.

The causes of the Gros Ventre slide were all natural. In fact, the stratigraphy and structure of the valley made a slide almost inevitable. On the side of the valley where the slide occurred, a permeable, erosion-resistant sandstone

STEP 1
A layer of soft, impermeable shale, overlain by permeable sandstone, dipped toward the Gros Ventre River at the same angle as the surface slope.

STEP 2
The sandstone layer has been eroded by the river, and was unsupported at its lower edge.

STEP 3
Heavy spring rain and snowmelt saturated the sandstone and made the shale slippery.

STEP 4
Loss of friction between sandstone and shale caused the sandstone to slide downslope into the river.

STEP 5
The slide formed a debris dam that created a large lake.

STEP 6
The lake broke through the unconsolidated debris, causing sudden, disastrous downstream flooding.

(a)   (b)

FIGURE 16.28 ■ The 1925 Gros Ventre slide. (a) The scar left by the Gros Ventre landslide can still be seen at Grand Teton National Park, Wyoming. [Garry Hayes/Geotripper Images.] (b) How the slide occurred. [After W. C. Alden, "Landslide and Flood at Gros Ventre, Wyoming," *Transactions of the American Institute of Mining, Metallurgical, and Petroleum Enginees* (1928): 345–361.]

formation dipped about 20° toward the river, paralleling the slope of the valley wall. Under the sandstone were beds of soft, impermeable shale that became slippery when wet. The conditions became ideal for a slide when the river channel cut through most of the sandstone at the bottom of the valley wall and left it with virtually no support. Only friction along the bedding plane between the shale and the sandstone kept the layer of sandstone from sliding. The river's undercutting of the sandstone's support was equivalent to scooping sand from the base of a sand pile: both cause oversteepening. The heavy rains and meltwater saturated the sandstone and the surface of the underlying shale, creating a slippery surface along the bedding planes at the top of the shale. No one knows what triggered the Gros Ventre slide, but at some point shear stress at the base of the slide mass exceeded the shear strength, and almost all of the sandstone slid downslope along the water-lubricated surface of the shale.

The formation of a dam in a stream and the growth of a lake are common consequences of a mass movement. Because most slide materials are permeable and weak, such a dam is soon breached when the lake water reaches a high level or overflows. Then the lake drains suddenly, releasing a catastrophic torrent of water (see Figure 16.28).

Human Activities That Promote or Trigger Mass Movements

Although the vast majority of mass wasting is natural, human activities can trigger landslides or make them more likely in vulnerable areas. Construction and excavation activities that result in the steepening or undercutting of slopes, and activities such as clear-cutting that remove vegetation cover, can increase the likelihood of landslides. Careful engineering of drainage systems in vulnerable areas can keep water from making slope materials more unstable. Some geologic settings are so susceptible to landslides, however, that we should forgo construction projects in these areas entirely.

One such place was Vaiont, a valley in the Italian Alps bordered by steep walls of interbedded limestone and shale. A large reservoir had been formed in the valley by a concrete dam (the second highest in the world at the time, at 265 m). On the night of October 9, 1963, a great debris slide of 240 million cubic meters (2 km long, 1.6 km wide, and more than 150 m thick) plunged into the deep water of the reservoir. The debris filled the reservoir for a distance of 2 km upstream of the dam and created a giant spillover. In the violent torrent that hurtled downstream as a 70-m-high flood wave, 3000 people died.

Engineers had ignored three warning signs at Vaiont (**Figure 16.29**):

1. The weakness of the cracked and deformed layers of limestone and shale that made up the steep walls of the reservoir

2. The scar of an ancient slide on the valley walls above the reservoir

3. A forewarning of danger signaled by a small rockslide in 1960, just 3 years earlier

Although the 1963 landslide was natural and could not have been prevented, its consequences could have been much less severe. If the reservoir had been located in a geologically safer place, where the water was less likely to spill over the dam, damage might have been limited to a lesser loss of property and far fewer deaths. We cannot prevent most natural mass movements, but we can minimize our losses through more careful planning of construction and land development.

FIGURE 16.29 ■ The tragic effects of a mass movement at the Vaiont Dam reservoir, in the Italian Alps, should have been predictable and preventable. A small rockslide in 1960 warned of the danger of mass movement above the reservoir. In October 1963, a massive debris slide caused the water in the reservoir to overflow the dam, flooding the downstream areas and killing 3000 people.

Google Earth Project

Weathering, erosion, and mass wasting control the form of Earth's surface as well as the supply of sedimentary materials that flow with streams and groundwater into oceans and lakes. Landforms, products of surface weathering, and mass movement features are all easily observed using Google Earth. The three examples we will study here illustrate these processes and products. Monument Valley, Arizona, shows us how the color of rocks and the shape of landforms can be related to weathering, whereas the results of downslope mass movements are visible at La Conchita, California, and Gros Ventre, Wyoming.

Data SIO, NOAA, U.S. Navy, NGA, GEBCO, Image © 2009 DigitalGlobe, Image NMRGIS, Image USDA Farm Service Agency

Google Earth satellite image showing the southwestern United States and location of sites discussed in the exercise.

LOCATION   Monument Valley, Arizona; La Conchita, California; Gros Ventre, Wyoming

GOAL   Observe weathering features, erosional processes, and mass wasting events at Earth's surface using Google Earth imagery

LINKED   Figures 16.6, 16.12, 16.17, and 16.28

1. Navigate to Monument Valley, Arizona, at 36°56′45″ N, 110°08′01″ W, and zoom to an eye altitude of 15 km. Given the color of the sedimentary material you see in the landscape, which of the following do you think represents the most prominent form of weathering taking place here?

    a. Frost wedging by ice

    b. Exfoliation of granite

    c. Oxidation of iron

    d. Dissolution of calcium carbonate

2. The unique geologic features in Monument Valley all seem to have flat upper surfaces. The development of flat surfaces may be due to the presence of a protective cap rock that weathers along bedding planes. To test the hypothesis that these surfaces are flat, scan over them, noting a number of elevations at different points, and determine whether the range of elevations you measure is consistent with a flat surface. Begin your study by examining the feature at 36°58′00″ N; 110°06′50″ W, at an eye altitude of 7 km. What range of elevations do you find?

    a. A narrow range of elevations, consistently around 2000 m

    b. A broad range of elevations, spanning 500–4000 m

    c. A moderate range of elevations, spanning 500–1500

    d. All the surfaces are below sea level

3. Navigate to 34°21′53″ N; 19°26′43″ W and take a look at the aftermath of the 2005 La Conchita mass movement event. This mass movement was triggered by excessive rainfall. Investigate the area by tilting the frame of view to the northeast. Based on

your investigation, how would you best describe the distinct landscape feature you see here?

a. A mudflow
b. A rockslide
c. A stable slope
d. A vertical cliff

4. Before leaving the La Conchita site, use the measuring tool to determine the approximate downslope length of the mass that has moved. Now navigate east to Gros Ventre, Wyoming (83001). Zoom to an eye altitude of 7 km and rotate your frame of view to the southeast. Here you will find another example of a mass movement, but in a different geologic setting. Locate the landslide and use the same measurement tool to measure the downslope length of the mass that has moved. Which of the comparisons below best matches your measurements?

a. The Gros Ventre mass is 2200 m longer than that at La Conchita.
b. The Gros Ventre mass is 1300 m longer than that at La Conchita.

c. The Gros Ventre mass is 1200 m shorter than that at La Conchita.
d. The Gros Ventre mass is 700 m shorter than that at La Conchita.

Optional Challenge Question

5. The Gros Ventre landslide occurred in a steep section of the Rocky Mountains adjacent to a river valley. Note the lake upstream from the river. How did the lake form?

a. Lava, erupting from a nearby volcano, poured across the river, creating a dam.
b. The landslide permanently removed all the vegetation from the mountain range, preventing soil development.
c. The landslide exposed springs that fed water into a depression behind it.
d. The landslide moved down the mountain and across the river, forming a dam.

SUMMARY

What is weathering and how is it controlled?
Rocks are broken down at Earth's surface by chemical weathering—the chemical alteration or dissolution of minerals—and by physical weathering—the fragmentation of rocks by mechanical processes. Erosion dislodges the products of weathering, which are the raw materials of sediments, and moves them away from their source. The properties of the parent rock affect weathering because different minerals weather at different rates and have differing susceptibilities to fracturing. Climate strongly affects weathering: warmth and heavy rainfall speed weathering; cold and dryness slow it down. The presence of soil accelerates weathering by providing moisture and acids secreted by organisms. The longer a rock weathers, the more completely it breaks down.

What are the processes of chemical weathering?
The weathering of feldspar, the most abundant silicate mineral, serves as an example of the processes that weather most silicate minerals. In the presence of water, feldspar undergoes hydrolysis to form kaolinite. Carbon dioxide (CO_2) dissolved in water promotes chemical weathering by reacting with the water to form carbonic acid (H_2CO_3). The slightly acidic water dissolves away potassium ions and silica, leaving kaolinite. Iron (Fe), which is found in ferrous form in many silicate minerals, weathers by oxidation, producing ferric iron oxides. These processes operate at varying rates, depending on the chemical stability of the minerals involved under various weathering conditions.

What are the processes of physical weathering?
Physical weathering breaks rocks into fragments along preexisting zones of weakness or along joints and other fractures in massive rock. Physical weathering is promoted by frost wedging and by burrowing and tunneling by animals and tree roots, all of which expand cracks. Microorganisms contribute to both physical and chemical weathering. Patterns of breakage such as exfoliation probably result from interactions between chemical weathering and temperature changes.

What factors are important in soil development?
Soil is a mixture of rock particles, clay minerals, and other products of weathering, as well as humus. It develops through inputs of new materials, losses of original materials, and modification through physical mixing and chemical reactions. The five key factors that affect soil development are parent material, climate, topography, organisms, and time.

What are mass movements, and what kinds of materials do they move? Mass movements are slides, flows, or falls of large masses of material downslope in response to the force of gravity. The movements may be imperceptibly slow or too fast for a human to outrun. The masses may consist of consolidated material, including rock and compacted or cemented sediments; or unconsolidated material. Mass movements of rock include rockfalls, rockslides, and rock avalanches. Mass movements of unconsolidated material include creep, slumps, debris slides, debris avalanches, earthflows, mudflows, and debris flows.

What factors are responsible for mass movements, and how are such movements triggered? The three factors that have the greatest bearing on the predisposition of material to move down a slope are the nature of the slope material, the water content of the material, and the steepness of the slope. Slopes made up of unconsolidated material become unstable when they are steeper than the angle of repose, the maximum slope angle that the material will assume without cascading downslope. Slopes made up of consolidated material may also become unstable when they are steepened or denuded of vegetation. Water absorbed by slope material contributes to instability by lowering internal friction and by lubricating planes of weakness in the material. Mass movements may be triggered by earthquakes, heavy rainfall, or gradual steepening of a slope due to erosion.

KEY TERMS AND CONCEPTS

angle of repose (p. 450)

chemical stability (p. 442)

consolidated material
   (p. 449)

creep (p. 456)

exfoliation (p. 444)

frost wedging (p. 443)

hematite (p. 441)

humus (p. 445)

kaolinite (p. 438)

liquefaction (p. 450)

mass movement (p. 448)

mass wasting (p. 436)

slump (p. 460)

soil (p. 445)

soil profile (p. 447)

talus (p. 451)

unconsolidated material
   (p. 449)

PRACTICING GEOLOGY EXERCISE

What Makes a Slope Too Unstable to Build On?

On hills with low slope, the ground is stable because the perpendicular component of gravity is large and tends to keep rocks and soils firmly in place.

On hills with moderate slope, the ground may become unstable because the component of gravity parallel to the hill slope is increased. This tends to push rocks and soils downslope and failure may occur.

On hills with steep slope, the ground is unstable because the parallel component of gravity is greatly increased and so the chance of failure is also increased.

Forces acting on a block of soil or rock at different slopes.

How can the destruction of homes and other buildings by landslides be avoided? Landslides are most likely in areas where steep topography coincides with other key factors, such as episodic heavy rainfall or earthquakes. Understanding the risk associated with buying or building a home in such areas begins with an assessment of the terrain and its likelihood to undergo mass wasting. Geologists play an important role in making such assessments and in advising potential homeowners and local planners about what kinds of real estate are more likely than others to experience a landslide.

Intuition tells us that if we build a structure on a slope that is too steep, it will slide downhill. Recall the pile of sand we described in the text: as the pile gets steeper, less of it stays in place, and at some point, no matter how much sand we put on the pile, it will just keep slip-sliding away. The same thing is true of masses of soil and rock. They, too, will slide downhill if the slope gets too steep. The important question is, how steep is too steep? We use the three primary factors involved in mass movements to determine which slopes are too steep to build on (see p. 451).

The most important factor is the steepness of the slope. Other things being equal, a structure on a steeper slope will slide sooner than a structure of the same size on a gentler slope. The second most important factor is the nature of the slope materials. The better these materials stick together, the more stable the slope will be. The third factor is the presence of water in the slope materials. During heavy rainfall, soil and rock absorb water, their cohesiveness is reduced, and a landslide may be triggered, as some homeowners in Southern California discovered in 2005 (see the chapter opening photo and Figure 16.12).

The accompanying diagram (see page 451) illustrates the forces acting on a mass of soil or rock—a potential slide mass. The main force responsible for mass movements is gravity. Gravity acts everywhere on Earth's surface, pulling everything toward the center of Earth. When the slide mass rests on a horizontal surface, gravity exerts a downward force on it, holding it firmly in place. On a slope, however, the force of gravity is directed across the base of the slide mass at an angle because gravity pulls toward the center of Earth, no matter what the position of the slide mass is. In this case, the force of gravity can be divided into two components: a force perpendicular to the base of the slide mass and a force parallel to its base.

What do these forces tell us about the likelihood of a landslide? The parallel component of gravity creates *shear stress* parallel to the base of the slide mass, which pulls it in the downslope direction. The perpendicular component of gravity, known as *shear strength,* acts to resist the mass's downward slide. Friction at the base of the slide mass and cohesion between the particles of the slide mass contribute to shear strength.

Sliding tends to occur on steeper slopes because shear stress increases and shear strength decreases. When shear stress becomes greater than shear strength, the mass will slide downslope. Thus, landslides are most likely where shear stress is high (on steeper slopes) and shear strength is low (as on a slope saturated by high rainfall).

A simple equation known as the *safety factor,* F_s, can be used to predict where and when mass movement will occur:

$$F_s = \frac{\text{shear strength}}{\text{shear stress}}$$

If the safety factor for a slope is less than 1, then mass movement is expected to occur there.

We can consult Tables A and B for values of shear strength and shear stress to determine which combinations of slope steepness and slope material would provide safe building sites.

Table A

Slope	Shear Stress
5°	1
20°	5
30°	25

Table B

Material	Shear Strength
Loose soil	3
Slate	10
Granite	50

Let's calculate the safety factor for a building site on slate on a moderate slope of 20°.

$$F_s = \frac{\text{shear strength for slate}}{\text{shear stress for 20° slope}} = 10/5 = 2$$

The slope is expected to be stable, but not by much. In many municipalities, this slope would be considered so marginally stable that it might not be permitted for building without meeting extremely expensive engineering standards.

BONUS PROBLEM: Fill in the blanks in Table C for the remaining combinations of slope and material (e.g., starting with a 5° slope on loose soil). Which slopes would be stable enough to build on?

Table C

	Safety Factor (F_s)		
Loose	Soil	Slate	Granite
5°	____	____	____
20°	____	____	____
30°	____	____	____

EXERCISES

1. What do the various kinds of rocks used for gravestones tell us about weathering?

2. What rock-forming minerals found in igneous rocks weather to clay minerals?

3. How does abundant rainfall affect weathering?

4. Which weathers faster, a granite or a limestone?

5. How does physical weathering affect chemical weathering?

6. How does climate affect chemical weathering?

7. What role do earthquakes play in the occurrence of landslides?

8. What kinds of mass movements advance so rapidly that a person could not outrun them?

9. How does the absorption of water weaken unconsolidated slope materials?

10. What is the angle of repose, and how does it vary with water content?

11. How does the steepness of a slope affect mass wasting?

12. What is a mudflow, and how is it produced?

THOUGHT QUESTIONS

1. In northern Illinois, you can find two soils developed on the same kind of bedrock: one is 10,000 years old and the other is 40,000 years old. What differences would you expect to find in their compositions or profiles?

2. Which igneous rock would you expect to weather faster, a granite or a basalt? What factors influenced your answer?

3. Assume that a granite with grains about 4 mm across and a rectangular system of joints spaced about 0.5 to 1 m apart is weathering at Earth's surface. What size would you ordinarily expect the largest weathered particle to be?

4. Why do you think a road built of concrete, an artificial rock, tends to crack and develop a rough, uneven surface in a cold, wet region even when it is not subjected to heavy traffic?

5. Pyrite is a mineral in which ferrous iron is combined with sulfide ions. What major chemical process weathers pyrite?

6. Rank the following rocks in the order of the rapidity with which they would weather in a warm, humid climate: a sandstone made of pure quartz, a limestone made of pure calcite, a granite, an evaporite deposit of halite.

7. What would a planet look like if there were no weathering at its surface?

8. Would a prolonged drought affect the potential for landslides? How?

9. What geologic conditions might you want to investigate before you bought a house at the base of a steep hill of bedrock covered by a thick layer of soil?

10. What evidence would you look for to indicate that a mountainous area had undergone a great many prehistoric landslides?

11. What factors would make the potential for mass movements in a mountainous terrain in the humid tropics greater or less than the potential in a similar terrain in a desert?

12. What kind(s) of mass movements would you expect on a steep hillside with a thick layer of soil overlying unconsolidated sands and muds after a prolonged period of heavy rain?

13. What factors weaken rock and enable gravity to start a mass movement?

MEDIA SUPPORT

 16-1 Animation: Mass Wasting

 16-1 Video: Just Passing Through: A Rockfall in Glenwood Springs, Colorado

 16-2 Video: Spheroidal Weathering

 16-3 Video: Mass Movement I

 16-4 Video: Mass Movement II

- The Geologic Cycling of Water   470

- Hydrology and Climate   472

- The Hydrology of Groundwater   477

- Erosion by Groundwater   488

- Water Quality   490

- Water Deep in the Crust   492

Angel Falls, Venezuela. Known in Venezuela as Salto Angel, the falls cascade 3185 feet from Auyun Tepui, a towering sandstone plateau. The falls were named for pilot Jimmy Angel, who crash-landed atop the tepui in the 1930s. [Miquel Gonzalez/laif/Redux.]

THE HYDROLOGIC CYCLE AND GROUNDWATER

17

MOST OF US HAVE HEARD the lines from Samuel Taylor Coleridge's "Rime of the Ancient Mariner": "Water, water everywhere, Nor any drop to drink." About 71 percent of Earth's surface is covered in water, but only a fraction of that water is available for human consumption. Humans cannot survive more than a few days without water. The amount of water consumed by modern society, however, far exceeds what we need for simple physical survival. We use water in immense quantities for industry, agriculture, and urban needs such as sewage systems. The study of the cycling of water is becoming ever more important as the demand on our limited water supplies increases.

In the previous two chapters, we have seen that water is essential to a wide variety of geologic processes. We saw in Chapter 15 that water exchanged between the oceans and the atmosphere forms a critical link in Earth's climate system; climate scientists now recognize that understanding the cycling of water is one of the most important steps in climate prediction. We saw in Chapter 16 that water is also important in weathering and erosion, both as a solvent of minerals in rock and soil and as a transport agent that carries away dissolved and weathered materials. The cycling of water links all of these processes. In Chapters 18 and 20, we will see how the streams and rivers formed by runoff and the glacial ice of the cryosphere help shape the landscapes of the continents. This chapter focuses on the water that sinks into Earth's crust to forms large reservoirs of groundwater.

In this chapter, we will study the distribution, movements, and characteristics of water on, over, and under Earth's surface. Then we will follow the path of water in more detail as it sinks into the ground and flows through underground reservoirs. As we do so, we will see what makes groundwater a limited resource that must be carefully managed.

■ The Geologic Cycling of Water

At what rates can we pump water from underground reservoirs without depleting them? What will be the effects of climate change on water supplies? Informed decision making about the conservation and management of water resources requires knowledge of how water moves on, over, and under Earth's surface and how this flow responds to natural changes and human modifications. This field of study is known as **hydrology.**

Flow and Reservoirs

We can see water in Earth's lakes, oceans, and polar ice caps, and we can see water moving over Earth's surface in streams and glaciers. It is harder to see the massive amounts of water stored in the atmosphere and underground, or the flows of water into and out of these storage places. As water evaporates, it moves into the atmosphere as water vapor. As it falls from the sky as rain and sinks into the ground, it becomes **groundwater**—the mass of water that flows beneath Earth's surface. Because organisms use water, small amounts of water are also stored in the biosphere.

Each place that stores water is referred to as a *reservoir.* Earth's largest natural reservoirs, in the order of their size, are the oceans; glaciers and polar ice; groundwater; lakes and rivers; the atmosphere; and the biosphere. **Figure 17.1** shows the distribution of water among these reservoirs. Although the total amount of water in rivers and lakes is very small compared with the amounts in the oceans and even in groundwater, these reservoirs of water are important to human populations because they do not contain salt or high concentrations of other dissolved materials.

Reservoirs gain water from inflows, such as rain and streams running in, and lose water from outflows, such as evaporation and streams running out. If inflow equals outflow, the size of the reservoir stays the same, even though water is constantly entering and leaving it. These flows mean that any given quantity of water spends a certain average time, called the *residence time,* in a reservoir.

How Much Water Is There?

Earth's total water supply is enormous—about 1.4 billion cubic kilometers distributed among the various reservoirs. If all of that water covered the land area of the United States, it would submerge the 50 states under a layer of water about 145 km deep. This total is constant, even though the flows from one reservoir to another may vary from day to day, year to year, and century to century. Over these geologically short time intervals, there is neither a net gain nor a net loss of water to or from Earth's interior, nor any significant loss from the atmosphere to outer space.

The Hydrologic Cycle

All water on Earth cycles among the various reservoirs in the oceans, atmosphere, and on and under the land surface. The cyclical movement of water—from the oceans to the atmosphere by evaporation, to Earth's surface by precipitation, to streams through runoff over and under the ground, and back to the oceans—is called the **hydrologic cycle** (**Figure 17.2**).

Within the range of temperatures found at Earth's surface, water shifts among the three states of matter: liquid (water), gas (water vapor), and solid (ice). These transformations power some of the main fluxes from one reservoir to another. Earth's external heat engine, powered by the Sun, drives the hydrologic cycle, mainly by evaporating water from the oceans and transporting it as water vapor in the atmosphere.

Under the right temperature and humidity conditions, water vapor condenses into the tiny droplets of water that form clouds and eventually fall as rain or snow—together known as **precipitation.** Some of the precipitation that falls on land soaks into the ground by **infiltration,** a process by which water enters rock or soil through cracks or small pores between particles. Part of this groundwater evaporates through the soil surface and returns to the atmosphere as water vapor. Another part moves through the biosphere as it is absorbed by plant roots, carried up to the leaves, and returned to the atmosphere by a process called *transpiration.* Most of this groundwater, however, flows slowly underground. The residence time of water in groundwater reservoirs is long, but it eventually returns to the surface in springs that empty into rivers and lakes, and thus returns to the oceans.

The precipitation that does not infiltrate the ground runs off the land surface, gradually collecting into streams. The sum of all precipitation that flows over the land surface, including not only streams but also the fraction that temporarily infiltrates near-surface soil and rock and then flows back to the surface, is called **runoff.** Some runoff may later seep into the ground or evaporate from rivers and lakes, but most of it eventually flows into the oceans.

Snowfall may be converted into ice in glaciers, which return water to the oceans by melting and runoff, and to the atmosphere by *sublimation,* the transformation of water from a solid (ice) directly into a gas (water vapor).

Most of the water that evaporates from the oceans returns to them as precipitation. The remainder falls over land and either evaporates or returns to the oceans as runoff. Figure 17.2 shows how the total flows among reservoirs balance one another in the hydrologic cycle. The land surface, for example, gains water from precipitation and loses the same amount of water by evaporation and runoff. The oceans gain water from runoff and precipitation and lose the same amount by evaporation. More water evaporates from the oceans than falls

SALT WATER 95.96%
Oceans and seas
$(1.40 \times 10^9 \text{ km}^3)$

FRESH WATER 4.04%

Glaciers and polar ice 2.97%
$(4.34 \times 10^7 \text{ km}^3)$

Atmosphere 0.001%
$(1.5 \times 10^4 \text{ km}^3)$

Lakes and rivers 0.009%
$(1.27 \times 10^5 \text{ km}^3)$

Underground waters 1.05%
$(1.54 \times 10^7 \text{ km}^3)$

Biosphere 0.0001%
$(2 \times 10^3 \text{ km}^3)$

(a)

(b)

(c)

(d)

(e)

FIGURE 17.1 ■ The distribution of water on Earth. [(b) John Grotzinger; (c) Ron Niebrugge/wildnatureimages; (d) Viktor Lyagushkin/National Geographic Creative; (e) Charlie Munsey/Corbis.]

on them as precipitation. This loss is balanced by the water returned as runoff from the continents. Thus, on a global scale, the size of each reservoir stays constant. Variations in climate, however, produce local variations in the balance among evaporation, precipitation, runoff, and infiltration.

How Much Water Can We Use?

Only a very small proportion of Earth's enormous supply of water is useful to human society. The global hydrologic cycle ultimately controls our water supplies. For example,

the 96 percent of Earth's water that resides in the oceans is essentially off limits to us. Almost all the water we use is *fresh water*—water that is not salty. Artificial desalination (the removal of salt) is now producing small but steadily growing amounts of fresh water from seawater in areas such as the arid Middle East. In the natural world, however, fresh water is supplied only by rain, rivers, lakes, groundwater, and water melted from snow or ice on land. All these waters are ultimately supplied by precipitation. Therefore, the practical limit to the amount of natural fresh water that we can ever envision using is the amount steadily supplied to the continents by precipitation.

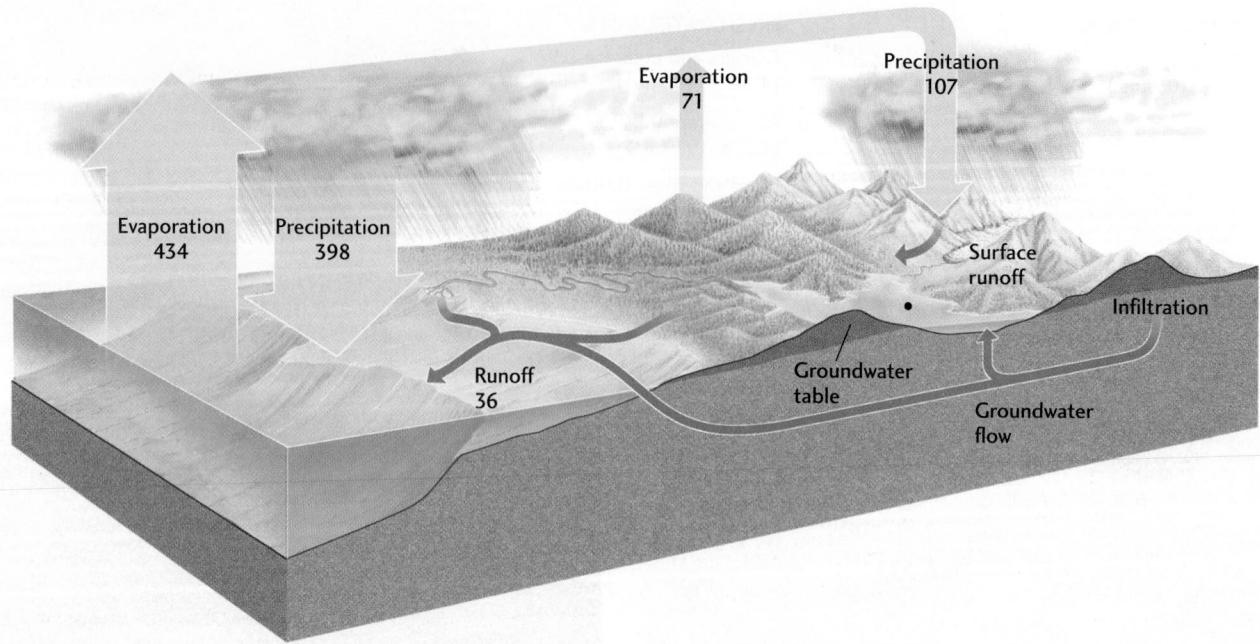

FIGURE 17.2 ■ The hydrologic cycle is the movement of water through Earth's crust, atmosphere, oceans, lakes, and streams. The numbers indicate the amounts of water (in thousands of cubic kilometers per year) that flow between these reservoirs annually.

■ Hydrology and Climate

For most practical purposes, geologists focus on local hydrology—the amount of water in the reservoirs of a region and how it flows from one reservoir to another—rather than global hydrology. The strongest influence on local hydrology is the local climate, especially temperature and precipitation levels. In warm areas where rain falls frequently throughout the year, water supplies—both at the surface and underground—are abundant. In warm arid or semiarid regions, it rarely rains, and water is a precious resource. People who live in icy climates rely on meltwaters from snow and ice. In some parts of the world, seasons of heavy rain, called *monsoons,* alternate with long dry seasons during which water supplies shrink, the ground dries out, and vegetation shrivels.

Humidity, Rainfall, and Landscape

Many geographic variations in climate are related to the average temperature of the air and the average amount of water vapor it contains, both of which affect levels of precipitation. The **relative humidity** is the amount of water vapor in the air, expressed as a percentage of the total amount of water the air could hold at the same temperature if it were saturated. When the relative humidity is 50 percent

and the temperature is 15°C, for example, the amount of moisture in the air is one-half the maximum amount the air could hold at 15°C.

Warm air can hold much more water vapor than cold air. When unsaturated warm air cools enough, it becomes supersaturated, and some of its water vapor condenses into water droplets. Those condensed water droplets form clouds. We can see clouds because they are made up of visible water droplets rather than invisible water vapor. When enough moisture has condensed and the droplets have grown too heavy to stay suspended by air currents, they fall as rain.

Most of the world's rain falls in warm, humid regions near the equator, where both the air and the surface waters of the ocean are warmed by the Sun. Under these conditions, a great deal of ocean water evaporates, resulting in high relative humidity. When air is warmed, it expands, becomes less dense, and rises. When the humid air over tropical oceans rises to high altitudes and blows over nearby continents, it cools, condenses, and becomes supersaturated. The result is heavy rainfall over the land, even at great distances from the coast.

At about 30°N and 30°S latitude, the air that has dropped its precipitation in the tropics begins to sink back toward Earth's surface. This cold, dry air warms and absorbs moisture as it sinks, producing clear skies and arid climates. Many of the world's deserts are located at these latitudes.

Polar climates also tend to be very dry. The polar oceans and the air above them are cold, so the air can hold little moisture, and little ocean water evaporates. Between the tropical and polar extremes are the temperate climates, where rainfall and temperatures are moderate.

The climate patterns we have just described are driven by patterns of air circulation in the atmosphere, as we'll see in Chapter 19. Plate tectonic processes also influence climate processes. The uplifting of mountain ranges, for example, forms **rain shadows,** areas of low precipitation on their leeward (downwind) slopes. Humid winds rising over high mountains cool and precipitate on the windward slopes, losing much of their moisture by the time they reach the leeward slopes (**Figure 17.3**). The air warms again as it drops to lower elevations on the other side of the mountain range. Because the warmer air can hold more moisture, relative humidity declines, decreasing the likelihood of precipitation even more. The Cascade Range of Oregon, uplifted by the subduction of the Pacific Plate under the North American Plate, creates a rain shadow. The prevailing winds that blow inland from the Pacific Ocean release heavy rainfall on the mountains' western slopes, supporting a lush forest ecosystem. The eastern slopes, on the other side of the range, are dry and barren.

Just as landscape features can alter precipitation patterns, the resulting variations in precipitation patterns control rates of weathering and erosion, which shape the landscape. In Chapter 22, we will explore further how the climate system and the plate tectonic system act together to control hydrologic patterns involved in landscape development.

Droughts

Droughts—periods of months or years when precipitation is much lower than normal—can occur in all climates, but arid regions are especially vulnerable to their effects. Lacking replenishment from precipitation, streams may shrink and dry up, ponds and lakes may evaporate, and the soil may dry and crack while vegetation dies. As human populations grow, demands on water supplies increase, so a drought can deplete already inadequate supplies.

The severest drought of the past few decades has affected a region of Africa known as the Sahel, along the southern border of the Sahara (**Figure 17.4**). This long drought has expanded the desert, as we'll see in Chapter 19, and has effectively destroyed farming and grazing. Hundreds of thousands of lives have been lost to famine in the area.

Another prolonged but less severe drought affected most of California from 1987 until February 1993, when torrential rains arrived. During the drought, groundwater and surface reservoirs dropped to their lowest levels in 15 years. Some restrictions on water use were instituted, but a move to reduce the extensive use of water supplies for irrigation encountered strong political resistance from farmers and the agricultural industry. As threats of water shortages loom, the use of water enters the arena of public policy debate (see Earth Policy 17.1).

An example of a shorter term, but high impact event is the 2013 drought in New Zealand. That country experienced a severe, widespread drought from late 2012 until April 2013. The extent of this drought made it unusual—it simultaneously occurred across the entire North Island

1 Prevailing winds carry warm air over oceans, where it gathers moisture as water vapor.

2 When moist air encounters mountains, it rises, cools, and condenses, precipitating rain or snow.

3 The result is a rainy windward slope.

4 As the air mass passes over the mountains, the cool air—now depleted of moisture—sinks and warms. Its relative humidity decreases,…

5 …and a dry leeward slope, or rain shadow, is formed.

Ocean

Wind

Desert

FIGURE 17.3 ▪ Rain shadows are areas of low rainfall on the leeward (downwind) slopes of a mountain range.

FIGURE 17.4 ■ This millet field in Mali, at the edge of the Sahara, shows the effects of a long drought on soil and crops. This photo was taken in 1984–1985, but the drought continues today. [Tomas van Houtryve/VII Network/Corbis.]

Earth Policy

17.1 Water Is a Precious Resource: Who Should Get It?

Until recently, most people in the United States have taken their water supply for granted. In the near future, however, because of climate change and population growth, particularly in arid regions, many areas of the country will experience water shortages more and more frequently. These shortages will create conflict among several sectors of society—residential, industrial, agricultural, and recreational—over who has the greatest right to the water supply.

In recent years, widely publicized droughts and restrictions on water use in California, Florida, Colorado, and many other places have made the public aware that the nation faces major water shortages. Public concern waxes and wanes, however, as periods of drought and abundant rainfall come and go and governments fail to pursue long-term solutions with the urgency they deserve. Here are some facts to ponder:

■ A human can survive with about 2 liters of water per day. In the United States, per capita water use by individuals is about 250 liters per day. If uses of water for industrial, agricultural, and energy production are considered, then per capita use rises to about 6000 liters per day.

■ Industry uses about 38 percent and agriculture about 43 percent of the water withdrawn from U.S. reservoirs.

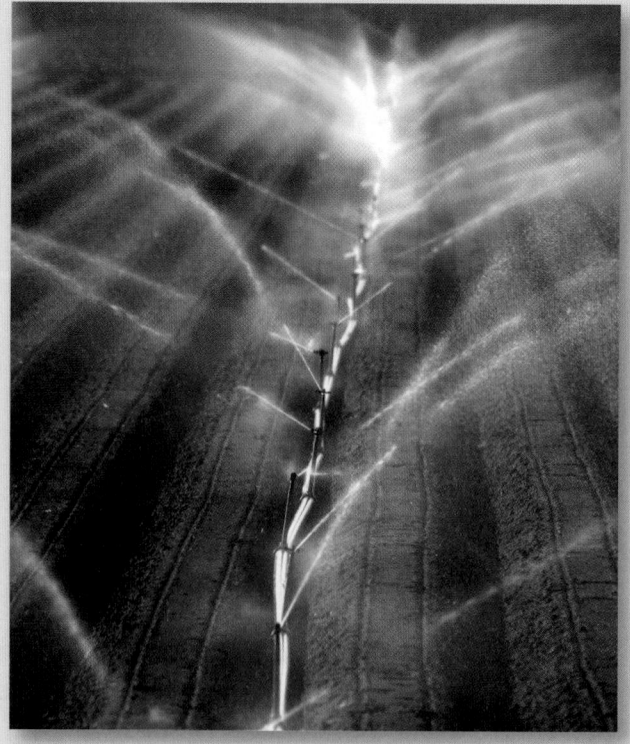

Irrigation in California's Imperial Valley, a natural desert. [David McNew/Getty Images.]

and parts of the South Island. Many of the pastures are not irrigated and depend on rainfall. The lack of rain caused crop failures and affected pastures in a country where agriculture is a major industry.

Our climate history can give us perspective on the severity of droughts. The southwestern United States, for example, has been experiencing a recent drought. During the 400-year period from 1500 to 1900, however, the Southwest was drier, on average, than it has been during the last century. Moreover, the geologic record shows droughts that were more severe and of longer duration than the present drought has been (at least so far). Are the recent droughts just short-term fluctuations in climate, or do they signal a return to an extended dry period? How will global climate change affect rainfall in the Southwest? By exploring the past, geologists and climate scientists may find information that will help them predict the future.

The Hydrology of Runoff

How much of the precipitation that falls on a land area ends up as runoff? A dramatic short-term example of how precipitation levels affect local stream and river runoff can be seen when flash flooding occurs after torrential rains. When levels of precipitation and runoff are measured over a larger area (such as all the states drained by a major river) and over a longer period (such as a year), the relationship between them is less direct, but still strong. The maps in **Figure 17.5** illustrate this relationship. When we compare them, we see that in areas of low precipitation—such as Southern California, Arizona, and New Mexico—only a small fraction of precipitation ends up as runoff. In these dry regions, much of the precipitation leaves the land surface by evaporation and infiltration. In more humid areas, such as the southeastern United States, a much higher proportion of the precipitation runs off in streams. A large river may carry large amounts of water from an area with high rainfall to an area with low rainfall. The Colorado River, for example, begins in an area of moderate rainfall in Colorado and then carries its water through arid western Arizona and Southern California.

Rivers and streams carry most of the world's runoff. The millions of small and medium-sized streams carry about half the world's runoff. About 70 major rivers carry the other half, and the Amazon River of South America carries almost half of that. The Amazon carries about 10 times more water than the Mississippi, the largest river

■ Per capita domestic water use in the United States is two to four times greater than in western Europe, where consumers pay as much as 350 percent more for their water.

■ Although the western United States receives one-fourth of the country's rainfall, per capita water use in the western states (mostly for irrigation) is 10 times greater than that in the eastern states, and water prices are much lower there. In California, for example, which imports most of its water, 85 percent of water use is for irrigation, 10 percent for municipalities and personal consumption, and 5 percent for industry. A 15 percent reduction in irrigation use would almost double the amounts of water available for use by cities and industries.

■ The fresh water used in the United States eventually returns to the hydrologic cycle, but it may return to a reservoir that is not well located for human use, and its quality may be degraded. Recycled irrigation water is often saltier than natural fresh water and is loaded with pesticides. Polluted urban waste water ends up in the oceans.

■ The traditional ways of increasing water supplies, such as building dams and artificial reservoirs and drilling wells, have become extremely costly because most of the best (and therefore cheapest) sites have already been used. Furthermore, the building of more dams to hold larger reservoirs carries environmental costs, such as the flooding of inhabited areas, detrimental changes in river flows above and below the dams, and the disturbance of fish and other wildlife habitats. Factoring in these costs has led to delays in dam projects and rejection of proposals for new dams.

U.S. water use by category in 2005. [Data from U.S. Geological Survey, USGS Circular 1344.]

(a) Average annual precipitation

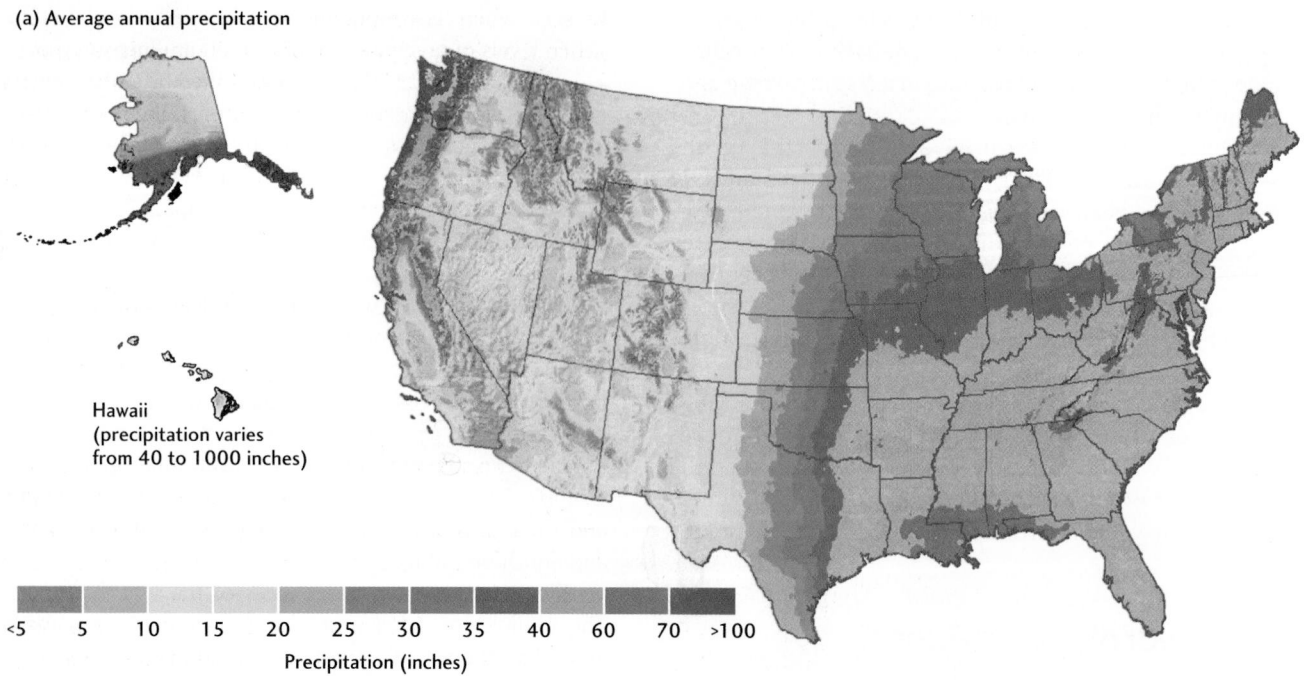

Hawaii
(precipitation varies
from 40 to 1000 inches)

Precipitation (inches)

<5 5 10 15 20 25 30 35 40 60 70 >100

(b) **Average annual runoff**

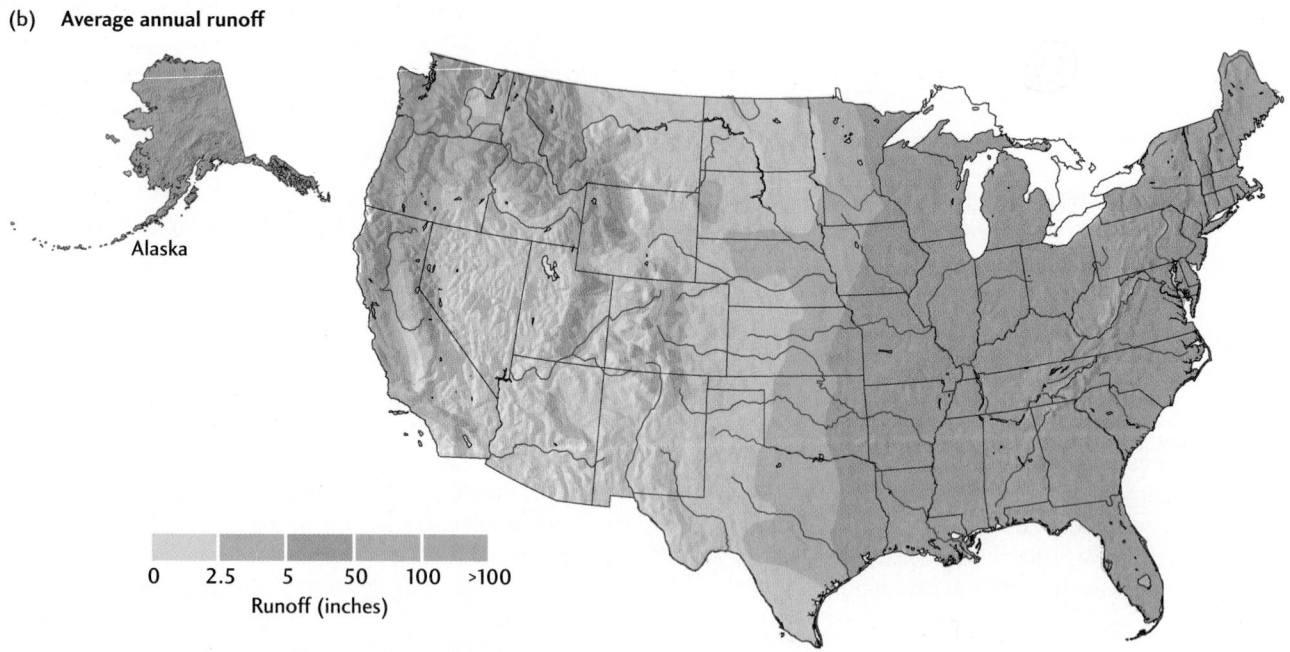

Alaska

Runoff (inches)

0 2.5 5 50 100 >100

FIGURE 17.5 ■ (a) Average annual precipitation in the United States. (b) Average annual runoff in the United States. [(a) Data from U.S. Department of Commerce, *Climatic Atlas of the United States,* 1968; (b) data from USGS Professional Paper 1240-A, 1979.]

of North America (Table 17.1). The major rivers transport great volumes of water because they collect it from large networks of streams and rivers that cover very large areas. The Mississippi, for example, collects its water from a network of streams that covers about two-thirds of the United States (Figure 17.6).

Runoff collects and is stored in natural lakes as well as in artificial reservoirs created by the damming of streams. Wetlands, such as swamps and marshes, also act as reservoirs for runoff (Figure 17.7). If these reservoirs are large enough, they can absorb short-term inflows from major rainfall events, holding some of the water that

TABLE 17-1	Water Flows of Some Major Rivers
River	Water Flow (m³/s)
Amazon, South America	175,000
La Plata, South America	79,300
Congo, Africa	39,600
Yangtze, Asia	21,800
Brahmaputra, Asia	19,800
Ganges, Asia	18,700
Mississippi, North America	17,500

would otherwise spill over riverbanks. During dry seasons or droughts, these reservoirs release water to streams or to water systems built for human use. Thus, they help to control flooding by smoothing out seasonal or yearly variations in runoff and releasing steady flows of water downstream.

In addition to these roles, wetlands are important to biological diversity because they are breeding grounds for a great many types of plants and animals. For all these reasons, many governments have laws that regulate the artificial draining of wetlands for real estate development. Nevertheless, wetlands are disappearing rapidly as land development continues. In the United States, more than half the wetlands that existed before European settlement are now gone. California and Ohio have kept only 10 percent of their original wetlands.

■ The Hydrology of Groundwater

Groundwater forms as raindrops and melting snow infiltrate soil and other unconsolidated surface materials and even sink into the cracks and crevices of bedrock. This groundwater, formed from recent atmospheric precipitation, is known as **meteoric water** (from the Greek *meteoron*, "phenomenon in the sky," which also gives us the word *meteorology*). The enormous reservoir of groundwater stored beneath Earth's surface equals about 29 percent of all the fresh water stored in lakes and rivers, glaciers and polar ice, and the atmosphere. For thousands of years, people have drawn on this resource, either by digging shallow wells or by storing water that flows out onto the surface at natural springs.

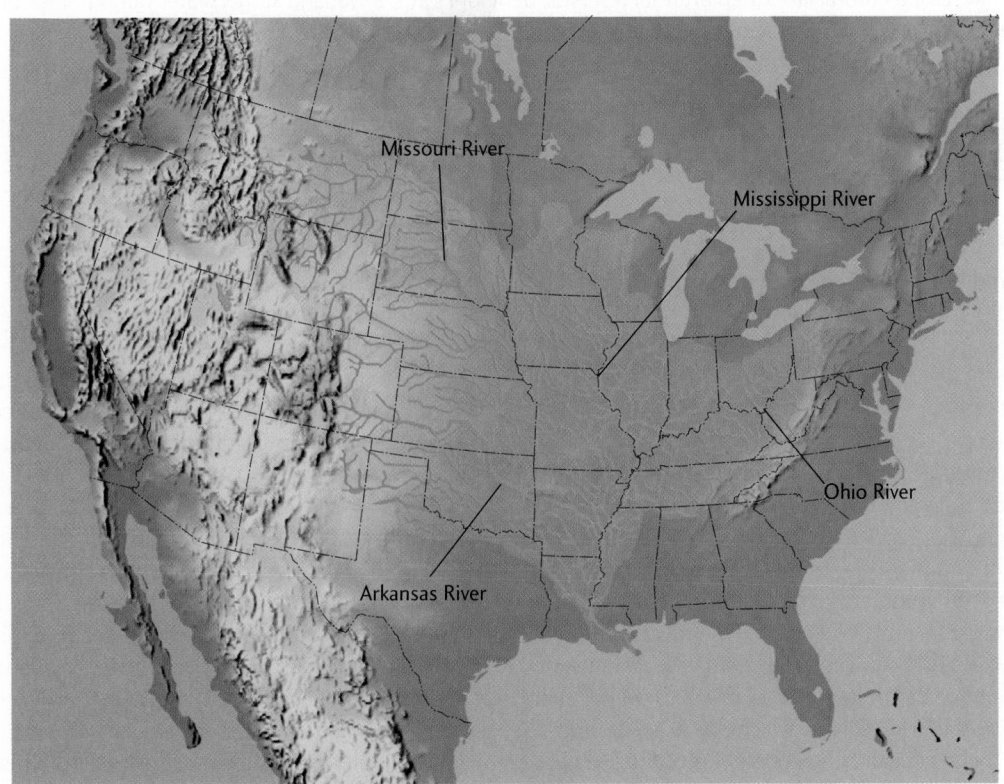

FIGURE 17.6 ■ The Mississippi River and its tributaries form the largest drainage network in the United States.

DRY PERIOD: LOW RUNOFF   **WET PERIOD: HIGH RUNOFF**

In dry periods, streams bring in small amounts of water…

…and carry away small amounts.

In wet periods, streams bring in large amounts of water,…

…which is stored…

…and slowly released during dry periods.

FIGURE 17.7 ■ Like a natural lake or an artificial reservoir behind a dam, a wetland stores water during times of rapid runoff and slowly releases it during periods of little runoff.

These springs are direct evidence of water moving below the surface (**Figure 17.8**).

Porosity and Permeability

When water moves into and through the ground, what determines where and how fast it flows? With the exception of caves, there are no large open spaces underground for pools or rivers of water. The only spaces available for water are pores and cracks in soil and bedrock. Some pores, however small and few, are found in every kind of rock and soil, but large amounts of pore space are most often found in sandstones and limestones.

Recall from Chapter 5 that the amount of pore space in rock, soil, or sediment determines its *porosity:* the percentage of its total volume that is taken up by pores. This pore space consists mainly of space between grains and in cracks (**Figure 17.9**). It can vary from a small percentage of the total volume of the material to as much as 50 percent where rock has been dissolved by chemical weathering. Sedimentary rocks typically have porosities of 5 to 15 percent. Most metamorphic and igneous rocks have little pore space, except where fracturing has occurred.

There are three types of pores: spaces between grains (*intergranular porosity*), spaces in fractures (*fracture porosity*), and spaces created by dissolution (*vuggy porosity*). Intergranular porosity, which characterizes soils, sediments, and sedimentary rocks, depends on the size and shape of the grains that make up those materials and on how they are packed together. The more loosely packed the grains, the greater the pore space between them. The smaller the particles, and the more they vary in shape and size, the more tightly they fit together. Minerals that cement grains together reduce intergranular porosity. Intergranular porosity varies from 10 to 40 percent.

Porosity is lower in igneous and metamorphic rocks, in which pore space is created mostly by fractures, including joints and cleavage at natural zones of weakness. Fracture porosity values are commonly as low as 1 to 2 percent, though some fractured rocks contain appreciable pore space—as much as 10 percent of the rock volume—in their many cracks.

Pore space in limestones and other highly soluble rocks such as evaporites may be created when groundwater interacts with the rock and partly dissolves it, leaving irregular voids known as *vugs*. Vuggy porosity can be very

FIGURE 17.8 ■ Groundwater flows from a cliff in Vasey's Paradise, Marble Canyon, Grand Canyon National Park, Arizona, where hilly topography allows it to flow out onto the surface in natural springs. [© Inge Johnsson/Alamy.]

MORE POROUS vs. **LESS POROUS**
Uncemented sandstone · Cemented sandstone
Sand grain · Cementing mineral · Pore space
Well-sorted sandstone · Poorly sorted sandstone
Fractured shale · Unfractured shale
Impermeable rock · Small amounts of pore space along cracks · Silt grains · Clay · Very small amounts of pore space between clays and silt grains

FIGURE 17.9 ■ Porosity in rocks depends on several factors. In sandstones, the extent of cementation and the degree of grain sorting are both important. In shales, porosity is limited due to the small spaces between the tiny grains, but can be enhanced by fracturing.

high (over 50 percent); caves are examples of extremely large vugs.

Although a rock's porosity tells us how much water it can hold if all its pores are filled, it gives us no information about how rapidly water can flow through those pores. Water travels through a porous material by winding between grains and through cracks. The smaller the pore spaces, and the more complex the path, the more slowly the water travels. The capacity of a solid to allow fluids to pass through it is its **permeability.** Generally, permeability increases as porosity increases, but permeability also depends on the sizes of the pores, how well they are connected, and how tortuous a path the fluid must travel to pass through the material. Vuggy pore networks in carbonate rocks may have extremely high permeabilities. Cave systems are so permeable that they allow people as well as water to move through them!

Both porosity and permeability are important considerations for geologists searching for groundwater supplies. In general, a good groundwater reservoir is a body of rock, sediment, or soil that has both high porosity (so that it can hold large amounts of water) and high permeability (so that the water can be pumped from it easily). Well drillers in temperate climates, for example, know that they are most likely to find a good supply of water if they drill into porous sand or sandstone beds not far below the surface. A rock with high porosity but low permeability may contain a great deal of water, but because the water flows so slowly, it is hard to pump it out of the rock. **Table 17.2** summarizes the porosities and permeabilities of various rock types.

TABLE 17-2 Porosity and Permeability of Aquifer Rock and Sediment Types		
Rock or Sediment Type	Porosity	Permeability
Gravel	Very high	Very high
Coarse- to medium-grained sand	High	High
Fine-grained sand and silt	Moderate	Moderate to low
Sandstone, moderately cemented	Moderate to low	Low
Fractured shale or metamorphic rock	Low	Very low
Unfractured shale	Very low	Very low

The Groundwater Table

As well drillers bore deeper into soil or rock, the samples they bring up become wetter. At shallow depths, the material is unsaturated: the pores contain some air and are not completely filled with water. This level is called the **unsaturated zone** (often termed the *vadose zone*). Below it is the **saturated zone** (often termed the *phreatic zone*), in which the pores are completely filled with water. The unsaturated and saturated zones may be in unconsolidated material or in bedrock. The boundary between the two zones is the **groundwater table,** usually shortened to *water table* (**Figure 17.10**). When a hole is dug below the water table, water from the saturated zone flows into the hole and fills it to the level of the water table.

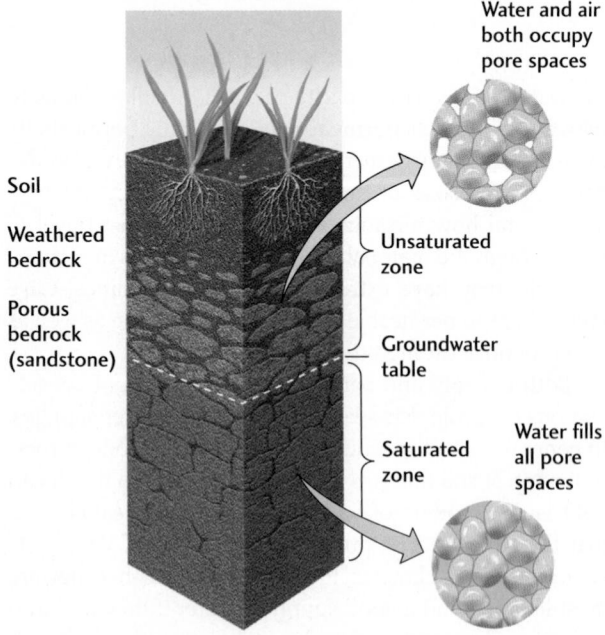

Soil

Weathered bedrock

Porous bedrock (sandstone)

Water and air both occupy pore spaces

Unsaturated zone

Groundwater table

Saturated zone

Water fills all pore spaces

FIGURE 17.10 ■ The groundwater table is the boundary between the unsaturated zone and the saturated zone. The saturated and unsaturated zones may be in unconsolidated material or in bedrock.

Groundwater moves under the force of gravity, so some of the water in the unsaturated zone may be on its way down to the water table. A fraction of that water, however, remains in the unsaturated zone, held in small pore spaces by surface tension. Surface tension, as you will recall from Chapter 16, is what keeps the sand on a beach moist. Evaporation of this water into pore spaces in the unsaturated zone is slowed both by the effect of surface tension and by the relative humidity of the air in the pore spaces, which can be close to 100 percent.

If we were to drill wells at several sites and measure the elevations of the water levels in those wells, we could construct a map of the water table. A cross section of the landscape might look like the one shown in **Figure 17.11a**. The water table follows the general shape of the surface topography, but its slopes are gentler. It is exposed at the land surface in river and lake beds and at springs. Under the influence of gravity, groundwater moves downhill from places where the water table elevation is high—under a hill, for example—to places where the water table elevation is low—such as a spring where groundwater flows out onto the surface.

Water enters and leaves the saturated zone through recharge and discharge (Figure 17.11b). **Recharge** is the infiltration of water into any subsurface formation. Rain and melting snow are the most common sources of recharge. **Discharge,** the movement of groundwater to the surface, is the opposite of recharge. Groundwater is discharged by evaporation, through springs, and by pumping from artificial wells.

Water may also enter and leave the saturated zone through streams. Recharge may take place through the bottom of a stream whose stream channel lies at an elevation above that of the water table. Streams that recharge groundwater in this way are called *influent streams,* and they are most characteristic of arid conditions, in which the water table is deep. Conversely, when a stream channel lies at an elevation below that of the water table, water is discharged from the groundwater into the stream. Such an *effluent stream* is typical of humid conditions. Effluent streams continue to flow long after runoff has stopped because they are fed by groundwater. Thus, the reservoir of groundwater may be increased by influent streams and depleted by effluent streams.

(a)

1 Rainwater infiltrates porous soil and rocks...

Water table

Unsaturated zone

2 ...and flows underground toward lakes and streams.

Movement of groundwater

Saturated zone

(b)

During wet periods, the water table is high, and both deep and shallow wells can be pumped.

1 Abundant rainfall recharges groundwater...

Effluent stream

Water table high

2 ...and groundwater is discharged as it moves into lakes and streams.

During dry periods, the water table is lower and only deep wells can be pumped.

1 During dry periods, evaporation discharges groundwater in soils,...

Influent stream

2 ...springs stop flowing, streams dry up,...

Water table low

3 ...the water table falls, and water from streams and lakes infiltrates and recharges the surface soil and rock.

FIGURE 17.11 ■ Dynamics of the groundwater table in a permeable shallow formation in a temperate climate. (a) The groundwater table follows the general shape of the surface topography, but its slopes are gentler. (b) The elevation of the water table fluctuates in response to the balance between water added by precipitation (recharge) and water lost by evaporation and from springs, streams, and wells (discharge).

Aquifers

Rock formations through which groundwater flows in sufficient quantity to supply wells are called **aquifers.** Groundwater may flow in unconfined or confined aquifers. In *unconfined aquifers,* the water travels through formations of more or less uniform permeability that extend to the surface. The level of the groundwater reservoir in an unconfined aquifer is the same as the height of the water table (as in Figure 17.11a).

Many permeable formations, however—typically sandstones—are bounded above and below by low-permeability beds, such as shales. These relatively impermeable formations are called **aquicludes.** Groundwater either cannot flow through them or flows through them very slowly. When aquicludes lie both over and under an aquifer, they form a *confined aquifer* (**Figure 17.12**).

The aquicludes above a confined aquifer prevent rainwater from infiltrating the aquifer directly. Instead, a confined aquifer is recharged by precipitation over a *recharge area,* often a topographically higher upland characterized by outcrops of permeable rock. Here there is no aquiclude preventing infiltration, so the rainwater travels down to and through the aquifer underground.

Water moving through a confined aquifer—known as **artesian flow**—is under pressure. At any point in the aquifer, that pressure is equivalent to the weight of all the water in the aquifer above that point. If we drill a well into

a confined aquifer at a point where the elevation of the ground surface is lower than that of the water table in the recharge area, the water will flow out of the well under its own pressure (**Figure 17.13**). Such wells are called *artesian wells,* and they are extremely desirable because no energy is required to pump the water to the surface.

In more complex geologic environments, the water table may be more complicated. For example, if a relatively impermeable mudstone layer forms an aquiclude within an otherwise permeable sandstone formation, the aquiclude may lie below the water table of a shallow aquifer and above the water table of a deeper aquifer (**Figure 17.14**). The water table in the shallow aquifer is called a *perched water table* because it is "perched" above the main water table in the deeper aquifer. Many perched water tables are small, only a few meters thick and restricted in area, but some extend for hundreds of square kilometers.

Balancing Recharge and Discharge

When recharge and discharge are balanced, the groundwater reservoir in an aquifer and the elevation of the water table remain constant, even though water is continually flowing through the aquifer. For recharge to balance discharge, rainfall must be frequent enough to compensate for runoff in streams and the outflow from springs and wells.

1 A confined aquifer is situated between two aquicludes (beds of low permeability).

Upland recharge area

Average height of water table in recharge area

Water table

Flowing artesian well

Difference in elevations

Height of entry of water into well

Aquiclude
Confined aquifer
Aquiclude

3 If the well were as high as the water table in the recharge area, there would be no pressure difference and thus no flow.

2 The flow of an artesian well is governed by the difference in pressure caused by the difference between the height of the water table in the recharge area and the height of the top of the well.

FIGURE 17.12 ▪ A permeable formation situated between two aquicludes forms a confined aquifer, through which water flows under pressure.

But recharge and discharge are rarely equal because recharge varies with rainfall from season to season. Typically, the water table drops in dry seasons and rises in wet seasons (see Figure 17.11b). A longer period of low recharge, such as a prolonged drought, will be followed by a longer-term imbalance and a greater lowering of the water table.

An increase in discharge, usually due to increased pumping from wells, can also produce a long-term imbalance and a lowering of the water table. Shallow wells may end up in the unsaturated zone and go dry. When a well pumps water from an aquifer faster than recharge can replenish it, the water table is lowered in a cone-shaped area around the well, called a *cone of depression* (**Figure 17.15**). The water level in the well is lowered to the depressed level of the water table. If the cone of depression extends below the bottom of the well, the well goes dry. If the bottom of the well is above the base of the aquifer, extending the well deeper into the aquifer may allow more water to be withdrawn, even at continued high pumping rates. If the rate of pumping is maintained and the well is deepened so much that the full thickness of the aquifer is tapped, however, the cone of depression can reach the bottom of the aquifer and deplete it. The aquifer will recover only if the pumping rate is reduced enough to give it time to recharge.

Excessive withdrawals of water may not only deplete the aquifer, but may also cause another undesirable environmental effect. As water pressure in the pore spaces falls, the ground surface overlying the aquifer may subside,

FIGURE 17.13 ■ Water flows from an artesian well under its own pressure. [John Dominis/Time Life Pictures/Getty Images.]

FIGURE 17.14 ■ A perched water table forms in some geologically complex situations—in this case, where a mudstone aquiclude is located above the main water table in a sandstone aquifer. The dynamics of the perched water table's recharge and discharge may be different from those of the main water table.

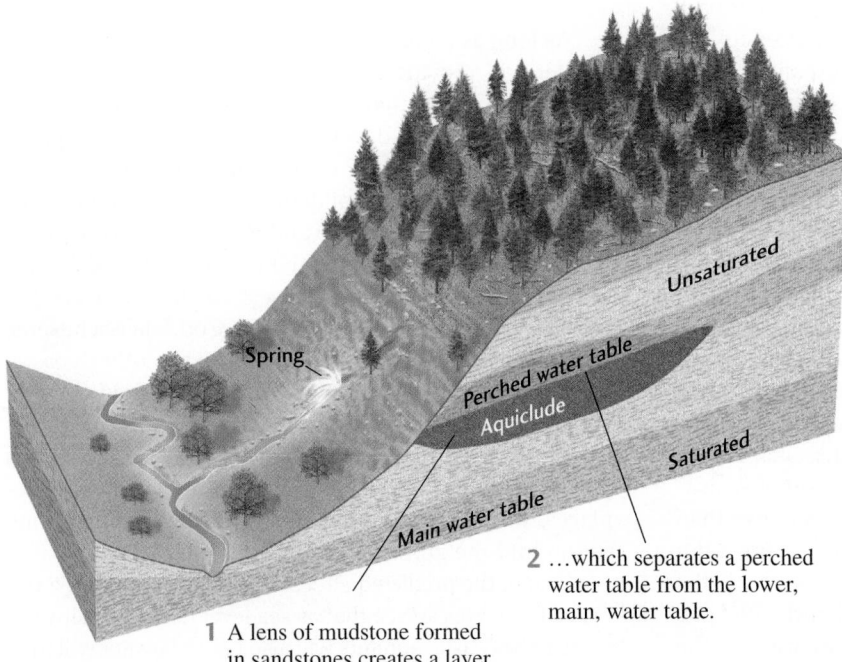

Spring

Unsaturated

Perched water table

Aquiclude

Saturated

Main water table

1 A lens of mudstone formed in sandstones creates a layer of very low permeability. This layer forms a shallow aquiclude,…

2 …which separates a perched water table from the lower, main, water table.

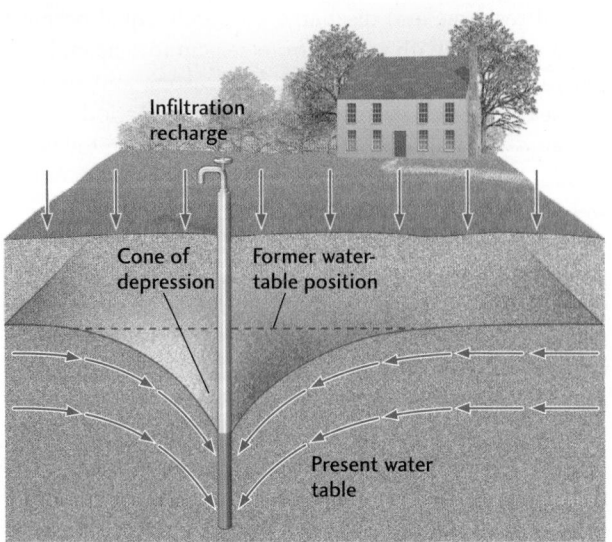

FIGURE 17.15 ■ When discharge from a well exceeds recharge, the water table is lowered in a cone of depression. The water level in the well is lowered to the depressed level of the water table.

FIGURE 17.16 ■ Excessive pumping of groundwater in Antelope Valley, California, has led to fissures and sinklike depressions on Rogers Lakebed at Edwards Air Force Base. This fissure, formed in January 1991, is about 625 m long. [James W. Borchers/USGS.]

creating sinklike depressions (**Figure 17.16**). As water in some types of sediments is removed, those sediments compact, and the loss of volume lowers the ground surface, a phenomenon known as *subsidence*. Subsidence caused by excessive pumping has occurred in Mexico City and in Venice, Italy, as well as in many other regions of heavy pumping, such as the San Joaquin Valley in California. In these places, the rate of subsidence has reached almost 1 m every 3 years. Although there have been a few attempts to reverse the subsidence by pumping water back into the ground, they have not been very successful because most compacted materials do not expand to their former state. The best way to halt further subsidence is to restrict pumping.

People who live near the ocean's edge may face a different problem when rates of discharge from an aquifer are high in relation to recharge: a flow of salt water into the aquifer. Near shorelines, or a little offshore, an underground boundary separates salty groundwater under the sea from fresh groundwater under the land. This *saltwater margin* slopes downward and inland from the shoreline in such a way that salt water underlies the fresh water of the aquifer (**Figure 17.17a**). Under many oceanic islands, a lens of fresh groundwater (shaped like a simple double-convex lens) floats on a base of seawater. The fresh water floats because it is less dense than the seawater (1.00 g/cm³ versus 1.02 g/cm³, a small but significant difference). Normally, the pressure of the fresh water keeps the saltwater margin slightly offshore. The balance between recharge and discharge in the fresh-water aquifer maintains this freshwater-seawater boundary.

As long as recharge is at least equal to discharge, the aquifer will provide fresh water. If water is withdrawn from a well faster than it can be recharged, however, a cone of depression develops at the top of the aquifer, mirrored by an inverted cone rising from the saltwater margin below. The cone of depression makes it more difficult to pump fresh water, and the inverted cone leads to an intake of salt water at the bottom of the well (Figure 17.17b). People living closest to the coast are the first affected. Some towns on Cape Cod, Massachusetts, on Long Island, New York, and in many other coastal areas have had to post notices that town drinking water contains more salt than is considered healthful by environmental agencies. There is no ready solution to this problem other than to slow the pumping or, in some places, to recharge the aquifer artificially by funneling runoff into the ground.

One of the predicted effects of global warming is a rise in sea level. We can see that as sea level rises, the saltwater margins of coastal aquifers will also rise. Seawater will then invade coastal aquifers and turn fresh groundwater into salt water.

(a)

1 The boundary between fresh and salty groundwater along shorelines is determined by the balance between recharge and discharge in the freshwater aquifers.

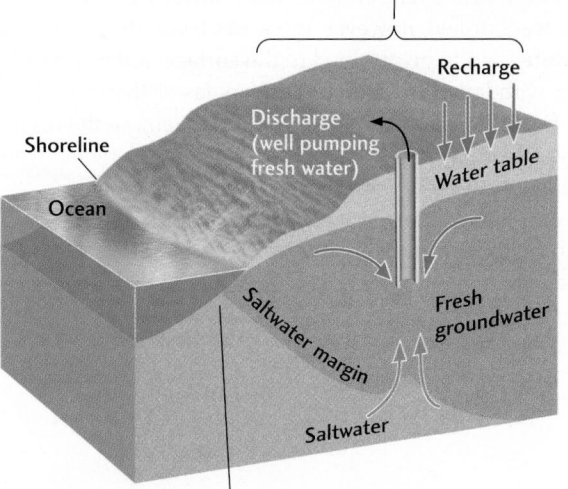

2 Normally, the pressure of fresh water keeps the saltwater margin slightly offshore.

(b)

1 Extensive pumping lowers the pressure of the fresh water, allowing the saltwater margin to move inland.

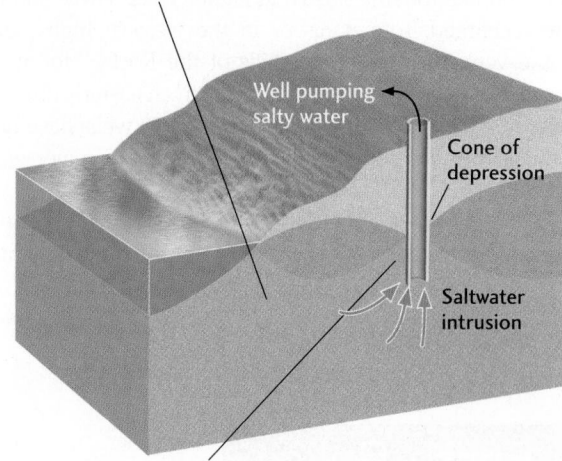

2 This movement creates both a cone of depression and an inverted cone of depression that brings salty water into the well. A well that formerly pumped fresh water now pumps salty water.

FIGURE 17.17 ■ The balance between recharge and discharge maintains the saltwater margin of a coastal aquifer.

The Speed of Groundwater Flows

The speed at which water moves underground strongly affects the balance between discharge and recharge. Most groundwaters flow slowly—a fact of nature that is responsible for our groundwater supplies. If groundwater flowed as rapidly as streams, aquifers would run dry after a period without rain, just as many small streams do. But the slow flow of groundwater also makes rapid recharge impossible if groundwater levels are lowered by excessive pumping.

Although all groundwaters flow through aquifers slowly, some flow more slowly than others. In the middle of the nineteenth century, Henri Darcy, town engineer of Dijon, France, proposed an explanation for the difference in flow rates. While studying the town's water supply, Darcy measured the elevations of water in various wells and mapped the water table in the district. He calculated the distances that the water traveled from well to well and measured the permeabilities of the aquifers. Here are his findings:

■ For a given aquifer and a given distance of travel, the rate at which water flows from one point to another is directly proportional to the vertical drop in elevation of the water table between the two points: as the vertical drop increases, the rate of flow increases.

■ For a given aquifer and a given vertical drop, the rate of flow is inversely proportional to the distance the water travels: as the distance increases, the rate of flow decreases. The ratio of the vertical drop to the flow distance is known as the **hydraulic gradient.**

Darcy reasoned that the relationship between the rate of flow and the hydraulic gradient should hold whether the water is moving through a well-sorted gravel aquifer or a less permeable silty sandstone aquifer. As you might guess, water moves faster through the large pore spaces of the well-sorted gravel than through the torturous twists and turns of the finer-grained and less permeable silty sandstone. Darcy recognized the importance of permeability and included a measure of permeability in his final explanation of how groundwater flows. So, other things being equal, the greater the permeability, and thus the greater the ease of flow, the faster the flow. The simple equation Darcy developed from these observations, now known as **Darcy's law,** can be used to predict the behavior of groundwater and thus has important applications in the management of water resources, as discussed in the Practicing Geology exercise.

Groundwater Resources and Their Management

Large parts of North America rely solely on groundwater for all their water needs. The demand for groundwater resources has grown as populations have increased and uses such as irrigation have expanded (**Figure 17.18**). Many areas of the Great Plains and other parts of the Midwest rest on sandstone formations, most of which are confined aquifers that function like the one shown in Figure 17.12. These aquifers are recharged from outcrops in the western high plains, some very close to the foothills of the Rocky Mountains. From there, the water flows downhill in an easterly direction over hundreds of kilometers. Thousands of wells have been drilled into these aquifers, which constitute a major water resource.

Darcy's law tells us that water flows at a rate proportional to the slope of an aquifer between its recharge area and a given well. In the Great Plains, the slopes are gentle, and water moves through the aquifers slowly, recharging them at low rates. At first, many of the wells drilled in these aquifers were artesian, and the water flowed freely. As more wells were drilled, however, the water levels dropped, and the water had to be pumped to the surface. Today, water is being withdrawn from these aquifers faster than the slow flow from distant recharge areas can fill them, so the reservoirs of groundwater they contain are being depleted (see Earth Policy 17.2).

A variety of innovative approaches are being used to enhance the sustainability of groundwater resources. In some areas, efforts to reduce excessive discharge have

Earth Policy

17.2 The Ogallala Aquifer: An Endangered Groundwater Resource

For more than 100 years, groundwater from the Ogallala aquifer, a formation of sand and gravel, has supplied fresh water to the cities, towns, ranches, and farms of much of the southern Great Plains. The population of the region has climbed from a few thousand people late in the nineteenth century to about a million today. Pumping of water from the aquifer, primarily for irrigation, has been so extensive—about 6 billion cubic meters of water per year from 170,000 wells—that recharge from rainfall cannot keep up. Water pressure in the wells has declined steadily, and the water table has dropped by 30 m or more.

Natural recharge of the Ogallala aquifer is very slow because rainfall on the southern Great Plains is sparse, the degree of evaporation is high, and the recharge area is small. The waters in the Ogallala aquifer today may have been supplied as much as 10,000 years ago, during the Wisconsin glaciation, when the climate of the Great Plains was wetter. At current rates of recharge, if all pumping were to stop now, it would take several thousand years for the water table to recover its original elevation and for well pressure to be restored. Some scientists have attempted to recharge the aquifer artificially by injecting water from shallow lakes that form in wet seasons on the high plains. These experiments have managed to increase recharge, but the aquifer is still in danger over the long term.

It is estimated that the remaining supplies of groundwater in the Ogallala aquifer will last only into the early decades of the twenty-first century. If the water it produces cannot be replaced, about 5.1 million acres of irrigated land in western Texas and eastern New Mexico will dry up—and so will 12 percent of the country's supply of cotton, corn, sorghum,

Much of the southern Great Plains region is underlain by the Ogallala aquifer. The blue area represents the aquifer. The general recharge area is located along the western margin of the aquifer. [U.S. Geological Survey.]

FIGURE 17.18 ▪ Groundwater withdrawals, United States, 1950–2005.
[U.S. Geological Survey.]

and wheat and a significant fraction of the feedlots for the nation's cattle.

Other aquifers in the northern Great Plains and elsewhere in North America are in a similar condition. In three major areas of the United States—Arizona, the high plains, and California—groundwater supplies have been significantly depleted.

The Kenya Aquifer

A recent study was done from July 2012 to July 2013 of potential groundwater resources in Kenya. The study scanned an area of more than 36,000 square kilometers in north and central Turkana, which is located in northwest Kenya, to assess the available groundwater resources in an area plagued by drought and famine. The approach used satellite and radar imagery, along with geographical surveys, climate maps, and seismic reflection data to create precise groundwater maps of the region. Five aquifers were discovered in the arid Turkana region: the Lotikipi Basin Aquifer (roughly the size of Rhode Island), the smaller Lodwar Basin Aquifer, and three smaller aquifers that need to be confirmed by drilling. The minimum reserve is estimated at 250 billion meters cubed of water (~66 trillion gallons). The geology of northwest Kenya is a mix of sandstone and volcanics, and porous rocks are ideal for storing groundwater. This potentially enormous groundwater supply can improve the lives of the people in Kenya, where drought, famine, and poverty are prevalent. Not only can the water be used for drinking, but for irrigation of crops and for livestock. Further study is needed to assess if the water quality is safe for drinking, how much water there actually is, how easy it is to access, and how expensive it is to tap. The replenishment rate must be studied so the aquifer isn't depleted faster than it can be recharged.

Ancient Water

In one of the deepest mines on Earth geologists recently discovered pockets of ancient water trapped in rock that is more than 1.5 billion years old. The discovery was based on a novel dating technique that exploits isotopes of xenon. Xenon and other noble gases accurately record when fluid masses last were in contact with the atmosphere.

The only waters older than this are minute, pinhead-size inclusions within minerals found in rocks that are over 3 billion years old. But water this abundant that actually flows from the rock has never been known before. The water occurs within open fractures, formed billions of years ago when tectonic forces related to continental formation created extensive fracture systems within metamorphic rocks. Some of these fractures became filled with economically valuable minerals but others were just filled with water that has never been in contact with the atmosphere.

This discovery has implications for the habitability of deep crustal environments. The water contains hydrogen and methane that could be used by microorganisms adapted to live in extreme environments (see Chapter 11, "Geobiology: Life Interacts with Earth"). If scientists were able to prove that microbes also live in this environment, then it would show that they too, along with the water, have been evolving in isolation for potentially billions of years. And, as we turn our attention increasingly to wonder about the potential habitability of Mars, it allows for the possibility that similar microbes could also occupy similar subterranean fractures systems that exist on planetary timescales.

been supplemented by attempts to increase the recharge of aquifers artificially. On Long Island, for example, the water authority drilled a large system of recharge wells to pump treated wastewater into the ground. The water authority also constructed large, shallow basins over natural recharge areas to increase infiltration by catching and diverting runoff, including stormwater and industrial waste drainage. The officials in charge of the program knew that urban development can decrease recharge by interfering with infiltration. As urbanization progresses, the impermeable materials used to pave large areas for streets, sidewalks, and parking lots increase runoff and prevent water from infiltrating the ground. Such decreases in natural infiltration may deprive aquifers of much of their recharge. One remedy is to catch and use stormwater runoff in a systematic program of artificial recharge, as the Long Island water authority did. The multiple efforts of the water authority have helped to rebuild the Long Island aquifer, though not to its original levels.

Orange County, near Los Angeles, California, receives only about 15 inches of rainfall per year, yet this water must supply a population of 2.5 million people. Groundwater pumped from beneath the western part of the county meets about 75 percent of its requirements. The water table is gradually dropping, however, threatening to diminish this supply. To help replenish the supply, the Orange County Water District operates 23 wells that inject treated wastewater, mixed with groundwater from a second aquifer that is located beneath the county's main aquifer. The recycled water meets drinking water standards with additional treatment, but most of the contaminants are filtered out by the aquifer's pore network.

■ Erosion by Groundwater

Every year, thousands of people visit caves, either on tours of popular attractions such as Mammoth Cave, Kentucky, or in adventurous explorations of little-known caves. These underground open spaces are actually enormous vugs produced by the dissolution of limestone—or, rarely, of other soluble rocks such as evaporites—by groundwater. Huge amounts of limestone have been dissolved to make some caves. Mammoth Cave, for example, has tens of kilometers of large and small interconnected chambers. The Big Room at Carlsbad Caverns, New Mexico, is more than 1200 m long, 200 m wide, and 100 m high.

Limestone is widespread in the upper parts of Earth's crust, but caves form only where limestone is located at or near the ground surface and where enough carbon dioxide–rich or sulfur dioxide–rich water infiltrates the surface to dissolve extensive areas of this relatively soluble rock. As we saw in Chapter 16, atmospheric carbon dioxide dissolved in rainwater forms carbonic acid, which enhances the dissolution of limestone. Water that infiltrates

soil may pick up even more carbon dioxide from plant roots, microorganisms, and other soil-dwelling organisms that give off this gas. As this carbon dioxide–rich water moves through the unsaturated zone to the saturated zone, it creates openings as it dissolves carbonate minerals. These openings are enlarged as the limestone dissolves along joints and fractures, forming a network of rooms and passages. Extensive cave networks form in the saturated zone, where—because the caves are filled with water—dissolution takes place over all surfaces, including floors, walls, and ceilings.

We can explore caves that were once below the water table but are now in the unsaturated zone because the water table has dropped. In these caves, now air-filled, water saturated with calcium carbonate may seep through the ceiling. As each drop of water drips from the cave's ceiling, some of its dissolved carbon dioxide evaporates, escaping to the cave's atmosphere. Its evaporation makes the calcium carbonate in the groundwater solution less soluble, so each water droplet precipitates a small amount of calcium carbonate on the ceiling. These deposits accumulate, just as an icicle grows, in a long, narrow spike of carbonate, called a *stalactite*, suspended from the ceiling. When the drop of water falls to the cave floor, more carbon dioxide escapes, and another small amount of calcium carbonate is precipitated on the floor below the stalactite. These deposits also accumulate, forming a *stalagmite*. Eventually, a stalactite and a stalagmite may grow together to form a column (**Figure 17.19**).

Microbial extremophiles (see Chapter 11) have been discovered living in caves, despite the lack of sunlight and highly acidic conditions that prevent most organisms from living in these environments. Some geologists think these microorganisms contributed to the formation of Carlsbad Caverns by using sulfates dissolved from gypsum ($CaSO_4$)

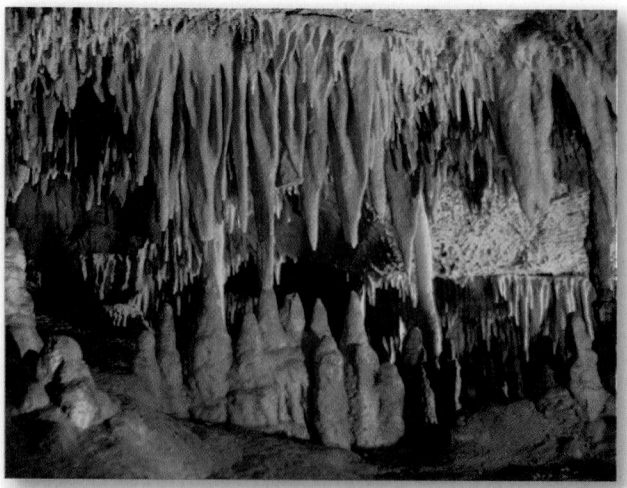

FIGURE 17.19 ■ Luray Caverns, Virginia. Stalactites from the ceiling and stalagmites from the floor have joined to form a column. [© Ivan Vdovin/Alamy.]

evaporites as an energy source and releasing sulfuric acid as a by-product. The sulfuric acid then helped to dissolve limestone to form the caves.

In some places, dissolution may thin the roof of a limestone cave so much that it collapses suddenly, producing a **sinkhole:** a small, steep depression in the land surface above the cave (**Figure 17.20**). Sinkholes are characteristic of a distinct type of topography known as *karst,* named for a region in the northern part of Slovenia. **Karst topography** is an irregular, hilly type of terrain characterized by sinkholes, caves, and a lack of surface streams (**Figure 17.21**). Underground drainage channels replace the normal surface drainage system of small and large streams. The short, scarce streams often end in sinkholes, detouring underground and sometimes reappearing miles away.

Karst topography is found in regions with three characteristics:

1. A humid climate with abundant vegetation (providing carbon dioxide–rich waters)

2. Extensively jointed limestone formations

3. Appreciable hydraulic gradients

In North and Central America, karst topography is found in limestone terrains of Indiana, Kentucky, and Florida and on the Yucatán Peninsula of Mexico. It is also well developed on uplifted coral limestone terrains formed from tropical island arcs in the late Cenozoic era.

FIGURE 17.20 ■ A large sinkhole formed by the collapse of a shallow underground cavern in Winter Park, Florida. Such collapses can occur so suddenly that moving cars are buried. [AP Photo.]

Karst terrains often have environmental problems, including the potential for catastrophic cave-ins and surface subsidence due to the collapse of underground spaces. The spectacular tower karst of southeastern China formed when cave networks collapsed to form sinkholes, which then expanded and merged, leaving "towers" behind (**Figure 17.22**).

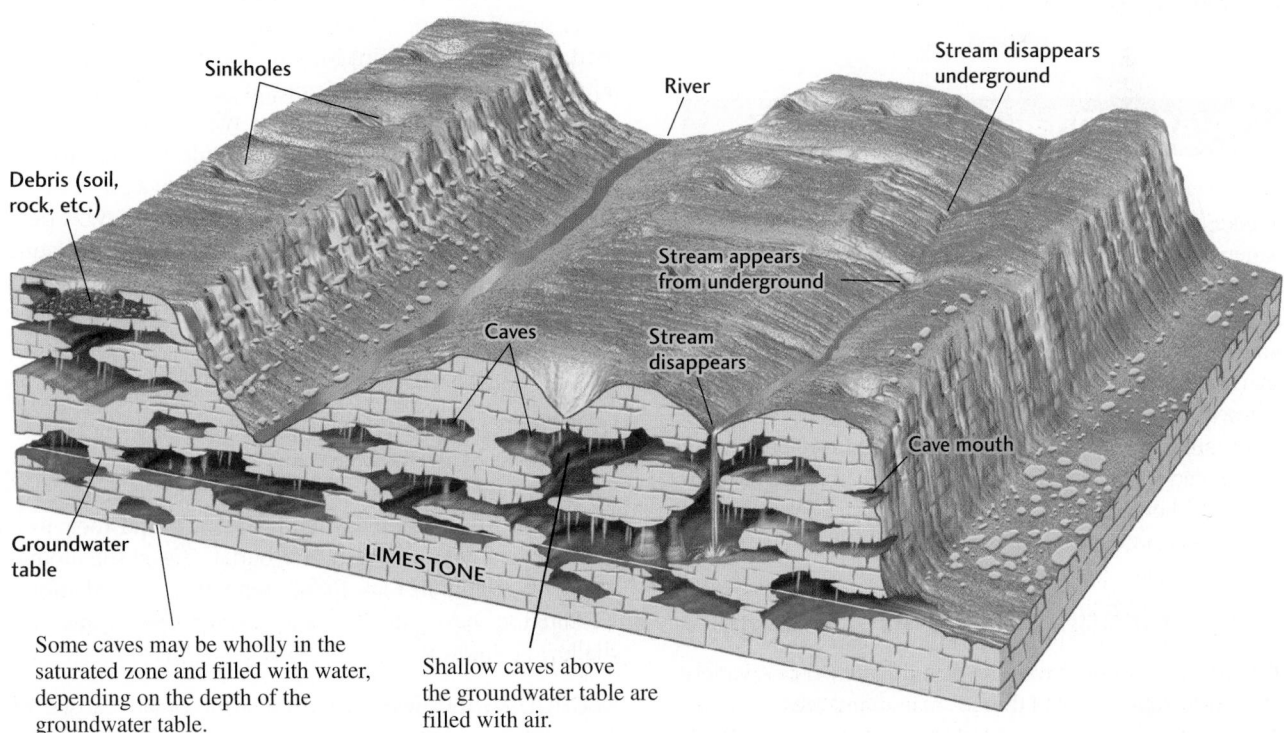

Sinkholes

River

Stream disappears underground

Debris (soil, rock, etc.)

Stream appears from underground

Caves

Stream disappears

Cave mouth

Groundwater table

LIMESTONE

Some caves may be wholly in the saturated zone and filled with water, depending on the depth of the groundwater table.

Shallow caves above the groundwater table are filled with air.

FIGURE 17.21 ■ Some major features of karst topography are caves, sinkholes, and disappearing streams.

FIGURE 17.22 ■ The tower karst of southeastern China is a spectacular terrain that features isolated hills with nearly vertical slopes. [Dennis Cox/Alamy.]

■ Water Quality

Unlike people in many other parts of the world, North Americans are fortunate in that almost all of their public water supplies are free of bacterial contamination, and the vast majority are free enough of chemical contaminants to drink safely. Yet as more rivers become polluted and more aquifers are contaminated by toxic wastes, North Americans are likely to see changes in water quality. Most residents of the United States are beginning to see their supply of fresh, pure water as a limited resource. Many people now travel with their own supply of bottled water, supplied by either home-installed purification systems or commercially available spring water.

Contamination of the Water Supply

The quality of groundwater is often threatened by a variety of contaminants. Most of these contaminants are chemicals, though microorganisms in water can also have negative effects on human health under certain conditions.

LEAD POLLUTION Lead is a well-known pollutant produced by industrial processes that inject contaminants into the atmosphere. When water vapor condenses in the atmosphere, lead is incorporated into precipitation, which then transports it to Earth's surface. Lead is routinely eliminated from public water supplies by chemical treatment before the water is distributed through the water mains. In older homes with lead pipes, however, lead can leach into the water. Even in newer construction, lead solder used to connect copper pipes and metals used in faucets may be sources of contamination. Replacing old lead pipes with durable plastic pipes can reduce lead contamination. Even letting the water run for a few minutes to clear the pipes can help.

OTHER CHEMICAL CONTAMINANTS A number of human activities produce chemicals that can contaminate groundwater (**Figure 17.23**). Some decades ago, when we knew much less about the health and environmental effects of toxic wastes, industrial, mining, and military wastes now known to be hazardous were dumped on the ground, disposed of in lakes and streams, or discharged underground. Even though many of these sources of pollution are being addressed, the contaminants are still making their way through aquifers by the slow flow of groundwater, and toxic chemicals are still entering groundwater from a number of other sources.

The disposal of chlorinated solvents—such as trichloroethylene (TCE), widely used as a cleaner in industrial processes—poses a formidable problem. These solvents persist in the environment because they are difficult to remove from contaminated waters. The burning of coal and the incineration of municipal and medical waste emit mercury into the atmosphere, which then contaminates water supplies. Buried gasoline storage tanks can leak, and road salt inevitably drains into the soil and ultimately into aquifers. Rain can wash agricultural pesticides, herbicides, and fertilizers into the soil, from which they percolate downward into aquifers. In some agricultural areas where nitrate fertilizers are heavily used, groundwaters contain high concentrations of nitrate. In one recent study, 21 percent of the shallow wells sampled exceeded the maximum amounts of nitrate (10 ppm) allowed in drinking water in the United States. Such high nitrate levels pose a danger of "blue baby" syndrome (an inability to maintain healthy oxygen levels) to infants 6 months old and younger.

RADIOACTIVE WASTES There is no easy solution to the problem of groundwater contamination by radioactive wastes. When radioactive wastes are buried underground, they may be leached by groundwater and find their way into aquifers. Storage tanks and burial sites at the atomic weaponry plants in Oak Ridge, Tennessee, and Hanford, Washington, have already leaked radioactive wastes into shallow groundwaters.

MICROORGANISMS The most widespread causes of groundwater contamination by microorganisms are leaky residential septic tanks and cesspools. These containers,

FIGURE 17.23 ■ Many human activities can contaminate groundwater. Contaminants from surface sources such as dumps and from subsurface sources such as septic tanks and cesspools enter aquifers through normal groundwater flow. Contaminants may be introduced into water supplies through pumping wells. [After U.S. Environmental Protection Agency.]

widely used in neighborhoods that lack full sewer networks, are buried settling tanks in which bacteria decompose the solid wastes from household sewage. To prevent contamination of drinking water, cesspools should be replaced by septic tanks, which must be installed at sufficient distance from water wells in shallow aquifers.

Reversing Contamination

Can we reverse the contamination of groundwater supplies? Yes, but the process is costly and very slow. The faster an aquifer recharges, the easier it is to decontaminate. If the recharge rate is rapid, fresh water moves into the aquifer as soon as we close off the sources of contamination, and in a relatively short time, the water quality is restored. Even a fast recovery, however, can take a few years.

The contamination of slowly recharging aquifers is more difficult to reverse. The rate of groundwater movement may be so slow that contamination from a distant source takes a long time to appear. By the time it does, it is too late for rapid remediation. Even after the recharge area has been cleaned up, some contaminated deep aquifers extending hundreds of kilometers from the recharge area may not be free of contaminants for many decades.

When public water supplies are polluted, we can pump the water and then treat it chemically to make it safe, but that is an expensive procedure. Alternatively, we can try to treat the water while it remains underground. In one moderately successful experimental procedure, contaminated water was funneled into a buried bunker full of iron filings, which detoxified the water by reacting with the contaminants. The reactions produced new, nontoxic compounds that attached themselves to the iron filings.

Is the Water Drinkable?

Much of the water in groundwater reserves is unusable not because it has been contaminated by human activities, but because it naturally contains large quantities of dissolved materials. Water that tastes agreeable and is not dangerous to human health is called **potable** water. The amounts of dissolved materials in potable waters are very small, usually measured by weight in parts per million (ppm). Potable groundwaters of good quality typically contain about 150 ppm total dissolved materials because even the purest natural waters contain some dissolved substances derived from weathering. Only distilled water contains less than 1 ppm dissolved materials.

Geologic studies of streams and aquifers allow us to improve the quality of our water resources as well as their quantity. The many cases of groundwater contamination caused by human activity have led to the establishment of water quality standards based on medical studies. These studies have concentrated on the effects of ingesting average amounts of water containing various quantities of contaminants, both natural and anthropogenic. For example, the U.S. Environmental Protection Agency has set the maximum allowable concentration of arsenic, a well-known poison, at 0.05 ppm (Figure 17.24). Natural contamination of groundwater by arsenic is particularly acute in Bangladesh, where groundwater provides 97 percent of the drinking water supply. Geologists are helping to guide the placement of new wells that draw water with acceptable concentrations of arsenic.

Groundwater is almost always free of solid particles when it seeps into a well from a sand or sandstone aquifer. The complex passageways of the pore networks in the

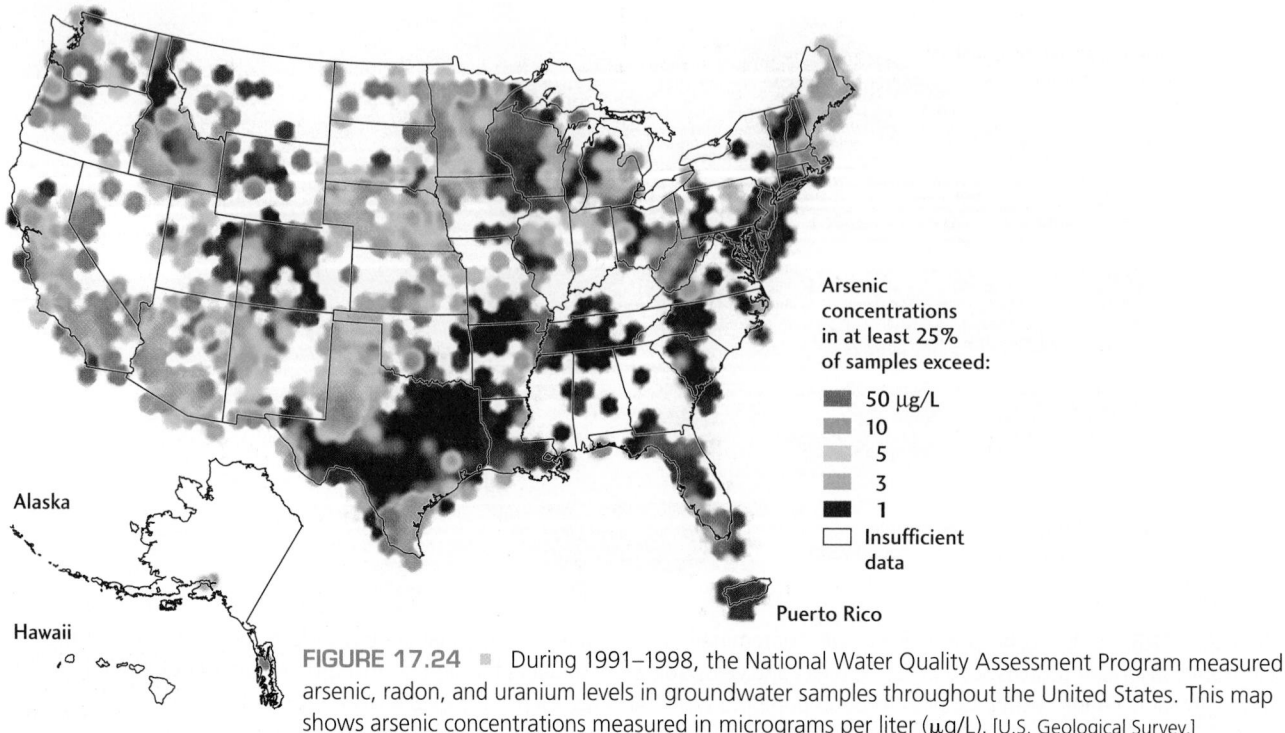

Arsenic concentrations in at least 25% of samples exceed:

- 50 μg/L
- 10
- 5
- 3
- 1
- ☐ Insufficient data

Alaska

Hawaii

Puerto Rico

FIGURE 17.24 ■ During 1991–1998, the National Water Quality Assessment Program measured arsenic, radon, and uranium levels in groundwater samples throughout the United States. This map shows arsenic concentrations measured in micrograms per liter (μg/L). [U.S. Geological Survey.]

rock or sand act as a fine filter, removing small particles of clay and other solids and even straining out microorganisms and some large viruses. Limestone aquifers may have larger pores and so may filter water less efficiently. Any microbial contamination found at the bottom of a well is usually introduced from nearby underground sewage disposal systems, often when septic tanks leak or are located too close to the well.

Some groundwaters, although perfectly safe to drink, simply taste bad. Some have a disagreeable taste of "iron" or are slightly sour. Groundwaters passing through limestone dissolve carbonate minerals and carry away calcium, magnesium, and bicarbonate ions, making the water "hard." Hard water may taste fine, but it does not lather readily when used with soap. Water passing through waterlogged forests or swampy soils may contain dissolved organic compounds and hydrogen sulfide, which give the water a disagreeable smell similar to rotten eggs.

How do these differences in taste and quality arise in safe drinking waters? Some of the highest-quality, best-tasting public water supplies come from lakes and artificial surface reservoirs, many of which are simply collecting places for rainwater. Some groundwaters taste just as good; these tend to be waters that pass through rocks that weather only slightly. Sandstones made up largely of quartz, for example, contribute little in dissolved materials, and thus waters passing through them have a pleasant taste.

As we have seen, the contamination of groundwater in relatively shallow aquifers is a serious problem, and remediation is difficult. But are there deeper groundwaters that we can use?

■ Water Deep in the Crust

Most crustal rocks below the groundwater table are saturated with water. Even in the deepest wells drilled for oil, some 8 or 9 km deep, geologists find water in permeable formations. At these depths, groundwaters move so slowly—probably less than a centimeter per year—that they have plenty of time to dissolve minerals from the rocks through which they pass. Thus, dissolved materials become more concentrated in these waters than in near-surface waters, making them unpotable. For example, deep groundwaters that pass through salt beds, which dissolve quickly, tend to contain large concentrations of sodium chloride.

At depths greater than 12 to 15 km, deep in the basement igneous and metamorphic rocks that underlie the sedimentary formations of the upper crust, porosities and permeabilities are very low due to the tremendous weight of the overlying rocks. Although these rocks contain very little water, they are saturated (Figure 17.25). Even some mantle rocks are presumed to contain water, although in minute quantities.

Hydrothermal Waters

In some regions of the crust, such as along subduction zones, hot waters containing dissolved carbon dioxide play an important role in the chemical reactions of metamorphism, as we saw in Chapter 6. These *hydrothermal waters* dissolve some minerals and precipitate others.

Most hydrothermal waters of the continents come from meteoric waters that percolate downward to deeper regions

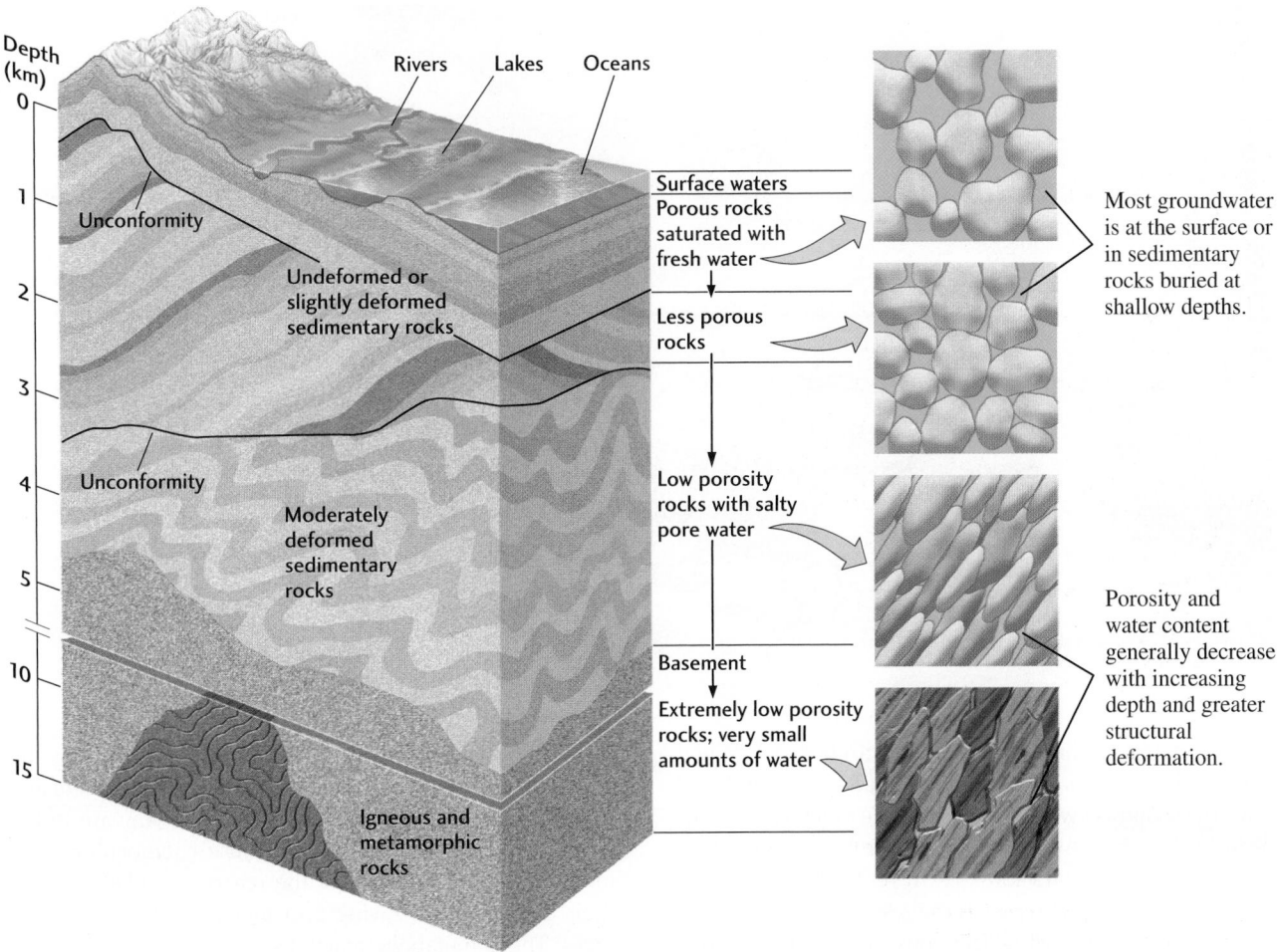

FIGURE 17.25 ■ Porosity and permeability, and therefore water content, generally decrease with increasing depth in Earth's crust.

of the crust. The percolation rates for meteoric waters deep in the crust are very low, and thus the water may be very old. It has been determined that the water at Hot Springs, Arkansas, derives from rain and snow that fell more than 4000 years ago and slowly infiltrated the ground. Water that escapes from magma can also contribute to hydrothermal waters. In areas of igneous activity, sinking meteoric waters are heated as they encounter hot masses of rock. The hot meteoric waters then mix with water released from the nearby magma.

Hydrothermal waters are loaded with chemical substances dissolved from rocks at high temperatures. As long as the water remains hot, the dissolved material stays in solution. However, as hydrothermal waters reach the surface, where they cool quickly, they may precipitate various minerals, such as opal (a form of silica) and calcite or aragonite (forms of calcium carbonate). Crusts of calcium carbonate produced at some hot springs build up to form the rock travertine, which can form impressive deposits such as those seen at Mammoth Hot Spring in Yellowstone National Park (**Figure 17.26**). Amazingly, microbial extremophiles that can withstand temperatures above the boiling

point of water have been discovered in these environments, where they may contribute to the formation of calcium carbonate crusts. Hydrothermal waters that cool slowly below the surface deposit some of the world's richest metallic ores, as we learned in Chapter 3.

Hot springs and geysers exist where hydrothermal waters migrate rapidly upward without losing much heat and emerge at the surface, sometimes at boiling temperatures. Hot springs flow steadily; geysers erupt hot water and steam intermittently (see Figure 12.21).

The theory explaining the intermittent eruption of geysers is an example of geologic deduction. We cannot observe the process directly because the dynamics of underground hydrothermal systems are hidden from sight hundreds of meters below the surface. Geologists hypothesized that geysers are connected to the surface by a system of very irregular and crooked fractures, recesses, and openings, in contrast to the more regular and direct plumbing of hot springs (**Figure 17.27**). The irregular fractures sequester some water in recesses, thus helping to prevent the deepest waters from mixing with shallower waters and cooling. The

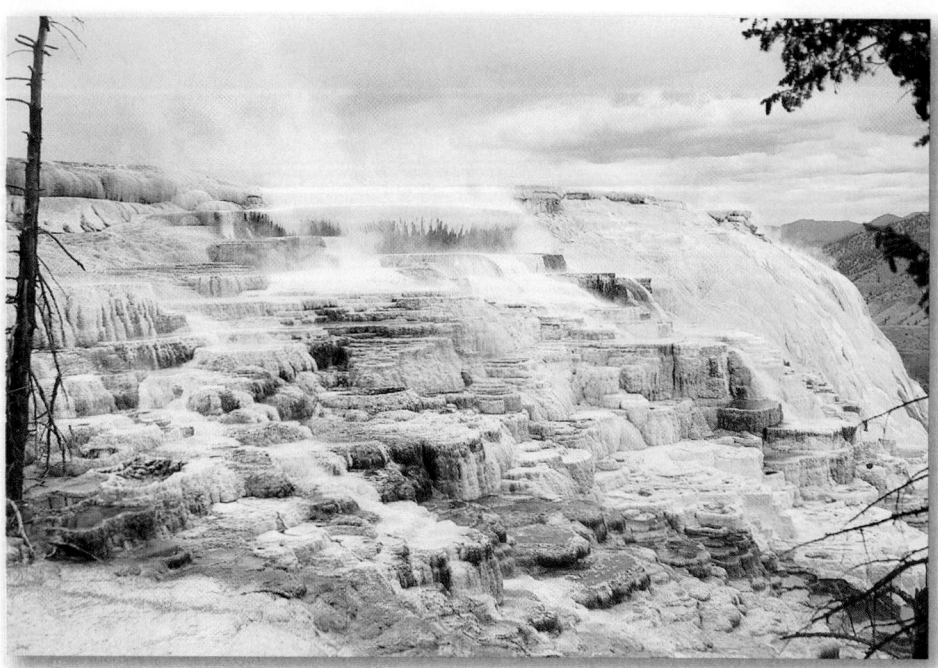

FIGURE 17.26 ■ Travertine deposits at Mammoth Hot Springs, Yellowstone National Park, form large lobelike masses made of aragonite and calcite. [John Grotzinger.]

deepest waters are heated by contact with hot rock. When they reach the boiling point, steam starts to ascend and heats the shallower waters, increasing the pressure and triggering an eruption. After the pressure is released, the geyser becomes quiet as the fractures slowly refill with water.

In 1997, geologists reported the results of a novel technique used to study geysers. They lowered a miniature video camera to about 7 m below the surface of a geyser. They found that the geyser shaft was constricted at that point. Farther down, the shaft widened to a large chamber containing a wildly boiling mixture of steam, water, and what appeared to be carbon dioxide bubbles. These direct observations dramatically confirmed the previous theory of how geysers work.

Although hydrothermal waters are useful to human society as sources of geothermal energy and metallic ores, these waters do not contribute to surface water supplies, primarily because they contain so much dissolved material.

Ancient Microorganisms in Deep Aquifers

In recent years, geologists have explored aquifers deep underground (as much as several thousand meters) in search of potable groundwater. They failed to find it, but they did unveil a remarkable interaction between the biosphere and the lithosphere. They found microorganisms living in the groundwater in huge numbers. These chemoautotrophic microorganisms, well out of the reach of sunlight, derive their energy by dissolving and metabolizing minerals in rocks. These metabolic reactions, aside from serving as a source of energy for the microorganisms, continue the weathering process underground. The chemicals released by these reactions make the water unpotable.

Geobiologists think that the ancestors of these microorganisms were enclosed within the pores of sediments, which were then buried at great depths, where they became sealed off from the surface. In some cases, these deep aquifers may not have been in contact with Earth's surface for hundreds of millions of years. Yet the microorganisms persisted, living solely on chemicals provided by the dissolution of minerals and evolving new generations of descendants without interference from any other organisms. These ecosystems, involving only microorganisms, are probably the most ancient on Earth and testify to the remarkable balance that can be achieved between life and environment.

SUMMARY

How does water move through the hydrologic cycle? The water movements of the hydrologic cycle maintain a balance among the major reservoirs of water on Earth. Evaporation from the oceans, evaporation and transpiration from the continents, and sublimation from glaciers transfer water to the atmosphere. Precipitation returns water from the atmosphere to the oceans and the land surface. Runoff returns part of the precipitation that falls on land to the ocean. The remainder infiltrates the ground and forms groundwater. Differences in climate produce local variations in the balance among evaporation, precipitation, runoff, and infiltration.

1 Meteoric water infiltrates the soil and filters down through permeable rocks.

Sinking cold water

Hot springs

Geyser

3 Hot springs occur where heated groundwater is discharged at the surface.

4 Geysers occur where an irregular network of pores and cracks slows the flow of water. Steam and boiling water are released to the surface under pressure, resulting in intermittent eruptions.

Permeable volcanic rock

Impermeable volcanic rock

Fault zone

Rising hot water

2 As the water approaches the magma, it heats up and becomes less dense, setting up a circulation system that returns it to the surface.

Magma

FIGURE 17.27 ■ Circulation of water over magma or hot rock deep in the crust produces geysers and hot springs.

How does water move below the ground? Groundwater forms as precipitation infiltrates the ground and travels through porous and permeable formations. The groundwater table is the boundary between the unsaturated and saturated zones. Groundwater moves downhill under the influence of gravity, eventually emerging at springs where the water table intersects the ground surface. Groundwater may flow through unconfined aquifers in formations of uniform permeability or in confined aquifers, which are bounded by aquicludes. Confined aquifers produce artesian flows and spontaneously flowing artesian wells. Darcy's law describes the rate of groundwater flow in relation to the hydraulic gradient and the permeability of the aquifer.

What factors govern human use of groundwater resources? As the human population grows, the demand for groundwater increases greatly, particularly where irrigation is widespread. As discharge exceeds recharge, many aquifers, such as those of the Great Plains of North America, are being depleted, and there is no prospect of their renewal for many years. Artificial recharge may help to renew some aquifers. The contamination of groundwater by industrial wastes, radioactive wastes, and sewage further reduces supplies of potable groundwater.

What geologic processes are affected by groundwater? Erosion by groundwater in limestone terrains produces karst topography, characterized by caves, sinkholes, and disappearing streams. At great depths in the crust, rocks contain extremely small quantities of water because their porosities are very low. The heating of these waters forms hydrothermal waters, which may return to the surface as geysers and hot springs.

KEY TERMS AND CONCEPTS

aquiclude (p. 482)

aquifer (p. 482)

artesian flow (p. 482)

Darcy's law (p. 485)

discharge (p. 480)

drought (p. 473)

groundwater (p. 470)

groundwater table (p. 480)

hydraulic gradient (p. 485)

hydrologic cycle (p. 470)

hydrology (p. 470)

infiltration (p. 470)

karst topography (p. 489)

meteoric water (p. 477)

permeability (p. 479)

potable (p. 491)

precipitation (p. 470)

rain shadow (p. 473)

recharge (p. 480)

relative humidity (p. 472)

runoff (p. 470)

saturated zone (p. 480)

sinkhole (p. 489)

unsaturated zone (p. 480)

PRACTICING GEOLOGY EXERCISE

How Much Water Can Our Well Produce?

The most important question anyone who is considering drilling a well can ask is whether that well will produce enough water to satisfy their needs. A well drilled in one kind of formation might produce plenty of water, while another not far away, but in a different formation, might not. How can we predict groundwater behavior well enough to know how much water a well in a certain location will produce?

Henri Darcy was able to turn his conceptual understanding of the principles of groundwater flow into a simple and very useful mathematical equation. This equation—*Darcy's law*—shows how geologic factors control the rate of water flow through an aquifer:

$$Q = A\left[\frac{K(h_a - h_b)}{l}\right]$$

What this equation says is that the volume of water flowing in a certain time (Q) is proportional to the cross-sectional area of the aquifer through which the volume of water flows (A); the *hydraulic conductivity* of the aquifer, a measure of the permeability of the rock or soil that composes it (K); and the hydraulic gradient. The hydraulic gradient can be determined by installing test wells at two points, a and b, measuring the difference in the elevation of the water table between them ($h_a - h_b$), and then dividing the result by the distance between them (l). The water flow will increase if the cross-sectional area of the aquifer increases, if the hydraulic gradient increases, or if the hydraulic conductivity increases.

In rural and many suburban parts of the United States, it is still common practice to drill wells for the family water supply. When choosing a home site, a family must be careful to take into account the geology of the site and whether it is suitable for sufficient water flow. Will the water flowing through the well at point B in the accompanying diagram produce enough water for a family's needs? It depends on a number of factors, including the type of formation in which the well is drilled. We can use Darcy's law to evaluate the effects of hydraulic conductivity on the amount of water that will flow through the well.

Using the measurements provided in the diagram, we find the following values:

Cross-sectional area of well pipe = $A = 0.25$ m²

$$\text{Hydraulic gradient} = \left[\frac{h_a - h_b}{l}\right]$$

$$= \left[\frac{440 \text{ m} - 415 \text{ m}}{1250 \text{ m}}\right]$$

$$= \left[\frac{25 \text{ m}}{1250 \text{ m}}\right]$$

Next, we find the value of Q for different values of K representing different Earth materials: clay, silty sand, well-sorted sand, and well-sorted gravel.

Material	Hydraulic conductivity (K)
Clay	0.001 m/day
Silty sand	0.3 m/day
Well-sorted sand	40 m/day
Well-sorted gravel	3750 m/day

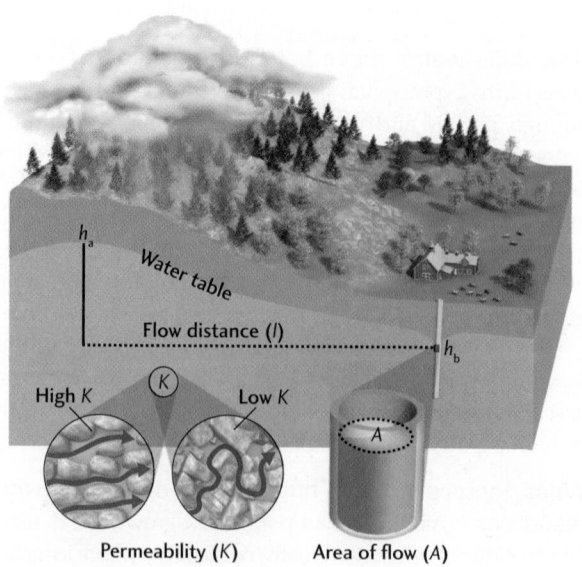

Permeability (K)   Area of flow (A)

- The Form of
  Streams   500

- Where Do
  Channels Begin?
  How Running
  Water Erodes
  Soil and Rock   507

- How Currents Flow
  and Transport
  Sediment   509

- Deltas: The Mouths
  of Rivers   513

- Streams as
  Geosystems   515

Aerial view of the meandering Adelaide River in Australia. Its appearance is typical of meandering streams in lowland environments. [Peter Bowater/Science Source.]

Now we can use Darcy's law to determine Q for well-sorted sand:

$$Q = A\left[\frac{K(h_a - h_b)}{l}\right]$$

$$= 0.25 \text{ m}^2 \times 40 \text{ m/day} \times 0.02$$

$$= 0.2 \text{ m}^3/\text{day (about 50 gallons/day)}$$

and for clay:

$$Q = A\left[\frac{K(h_a - h_b)}{l}\right]$$

$$= 0.25 \text{ m}^2 \times 0.001 \text{ m/day} \times 0.02$$

$$= 0.000005 \text{ m}^3/\text{day (about 1 teaspoon/day)}$$

It is clear based on these calculations that if the well has been drilled in well-sorted sand, it might provide just enough water for a family of four, assuming each person uses 10 gallons per day for drinking, showering, toilet use, cooking, cleaning, and yard maintenance. In contrast, if the well has been drilled in clay, it is likely to be a severe disappointment.

BONUS PROBLEM: Use Darcy's law to find the volume of water that could flow through the well in one day if it were drilled in silty sand or in well-sorted gravel.

Flow velocities calculated by Darcy's law have been confirmed experimentally by measuring how long it takes a harmless dye introduced into one well to reach another well. In most aquifers, groundwater moves at a rate of a few centimeters per day. In very permeable gravel beds near the surface, groundwater may travel as much as 15 cm/day. (This speed is still much slower than the speeds of 20 to 50 cm/s typical of river flows.)

EXERCISES

1. What are the main reservoirs of water at and near Earth's surface?

2. How do mountains form rain shadows?

3. What is an aquifer?

4. What is the difference between the saturated and unsaturated zones of an aquifer?

5. How do aquicludes form a confined aquifer?

6. Why does water from an artesian well flow to Earth's surface without pumping?

7. How are recharge and discharge balanced to keep the groundwater table at a constant level?

8. How does Darcy's law relate groundwater movement to permeability?

9. How does groundwater create karst topography?

10. What are the sources of water in hot springs?

11. What are some common contaminants in groundwater?

12. How do microorganisms survive deep in Earth's crust?

THOUGHT QUESTIONS

1. If global warming caused evaporation from the oceans to increase greatly, how would the hydrologic cycle of today be altered?

2. If you lived near the seashore and started to notice that your well water had a slightly salty taste, how would you explain the change in water quality?

3. Why would you recommend against extensive development and urbanization of the recharge area of an aquifer that serves your community?

4. If it were discovered that radioactive waste had seeped into groundwater from a nuclear processing plant, what kind of information would you need to predict how long it would take for the radioactivity to appear in well water 10 km from the plant?

5. What geologic processes would you infer are taking place below the surface at Yellowstone National Park, which has many hot springs and geysers?

6. Why should communities ensure that septic tanks are maintained in good condition?

7. Why are more and more communities in cold climates restricting the use of salt to melt snow and ice on highways?

8. Your new house is built on soil-covered granitic bedrock. Although you think that prospects for drilling a successful water well are poor, a well driller who is familiar with the area says he has drilled many good water wells in this granite. What arguments might each of you offer to convince the other?

9. How might the hydrologic cycle have been different 18,000 years ago, at the Wisconsin glacial maximum, when much of North America, Europe, and Asia were covered with ice?

10. You are exploring a cave and notice a small stream flowing on the cave floor. Where could the water be coming from?

MEDIA SUPPORT

 17-1 Animation: Water Cycle

STREAM TRANSPORT: FROM MOUNTAINS TO OCEANS

18

BEFORE CARS AND AIRPLANES EXISTED, people traveled on rivers. In 1803, the United States purchased the Louisiana Territory from France. It was a huge tract of over 2 million square kilometers, taking in portions of what today is Texas and Louisiana and extending up to Montana and North Dakota. In 1804, President Thomas Jefferson asked Meriwether Lewis and William Clark to lead an expedition across this new territory and into western North America. One of their most important goals was to map the western rivers, which provided the key to opening up this uncharted frontier. Lewis and Clark decided to follow the Missouri River and its headwaters to their source. They then crossed the Rocky Mountains and followed the Columbia River westward to the Pacific Ocean. The total trip was 6000 km—with the section along the Missouri River alone extending over 3200 km—and upstream all the way.

The writings and maps produced by Lewis and Clark created a body of knowledge that could have been obtained only by following one of the great rivers that drain the interior of North America. On other continents and in other countries, other big rivers evoke a similar sense of adventure: in South America, the Amazon; in Asia, the Yangtze and Indus; and in Africa, the fabled Nile. Yet streams and rivers are not only the access routes for legendary explorations, but also the places where people settle and make their homes. A body of water flows through almost every town and city in most parts of the world. These streams have served as commercial waterways for barges and steamers and as water resources for resident populations and industries. The sediments they have deposited during floods have built fertile lands for agriculture. Living near a river also entails risks, however. When rivers flood, they destroy lives and property, sometimes on a huge scale.

Streams are the lifelines of the continents. Their appearance is a record of the interaction of climate and plate tectonic processes. Tectonic processes lift up the land, producing the steep topography and slopes of mountainous regions. Climate determines where rain and snow will fall. Rainwater runs downhill, eroding the rocks and soils of the mountains, forming channels and carving out valleys as it gathers into streams. Streams carry back to the sea the bulk of the precipitation that falls on land and much of the sediment produced by erosion of the land surface. Streams are so important to understanding the role of climate and water on Earth that their discovery on Mars has fueled a generation of missions to search for evidence of water—and a different climate in the planet's ancient past.

In this chapter, we focus on how streams form and how they accomplish their geologic work: how, on a large scale, streams carve valleys and develop vast networks of channels; and how, at a smaller scale, streams break up and erode solid rock. We examine how water flows in currents and how currents carry sediment. Then we return to a larger scale to look at streams as geosystems shaped by interactions between the plate tectonic and climate systems.

■ The Form of Streams

We use the word **stream** for any body of water, large or small, that flows over the land surface, and **river** for the major branches of a large stream system. Most streams run through well-defined troughs called *channels*, which allow water to flow over long distances. As streams move across Earth's surface—in some places over bedrock, in others over unconsolidated sediments—they erode these materials and create *valleys*.

Identifying and mapping stream valleys were essential tasks for Lewis and Clark during their mission 200 years ago. As they traveled upstream and the river branched, they had to choose which branch was the larger of the two. They used two observations to help them make this choice: the width of the stream valley and the depth of the stream channel. Was the valley wide enough, and the channel deep enough, for their boats? Narrow valleys and shallow channels would mean that the branch led into a much shorter, and therefore less desirable, route; wider valleys and deeper channels, on the other hand, promised a longer passage up the main branch of the river.

Stream Valleys

A stream **valley** encompasses the entire area between the tops of the slopes on both sides of the stream. The cross-sectional profile of many stream valleys is V-shaped, but many other stream valleys have a broad, low profile like that shown in **Figure 18.1**. At the bottom of the valley is the **channel,** the trough through which the water runs. The channel carries all the water during normal, non-flood times. At low water levels, the stream may run only along the bottom of the channel. At high water levels, the stream occupies most of the channel. In broad valleys, a **floodplain**—a flat area about level with the top of the channel—lies on either side of the channel. It is this part of the valley that is flooded when the stream spills over its banks, carrying with it silt and sand from the channel.

In high mountains, stream valleys are narrow and have steep walls, and the channel may occupy most or all of the valley bottom (**Figure 18.2**). A small floodplain may be visible only at low water levels. In such valleys, the stream is actively cutting into the bedrock, a process that is characteristic of newly uplifted highlands in tectonically active areas. Its erosion of the valley walls is helped by chemical weathering and mass wasting. In lowlands, where tectonic uplift has long since ceased, the stream shapes its valley by eroding sediment particles and transporting them downstream. With a long time to operate, these processes produce gentle slopes and floodplains many kilometers wide.

Channel Patterns

As a stream channel makes its way along the bottom of a valley, it may run straight in some stretches and take a snaking, irregular path in others, sometimes splitting into

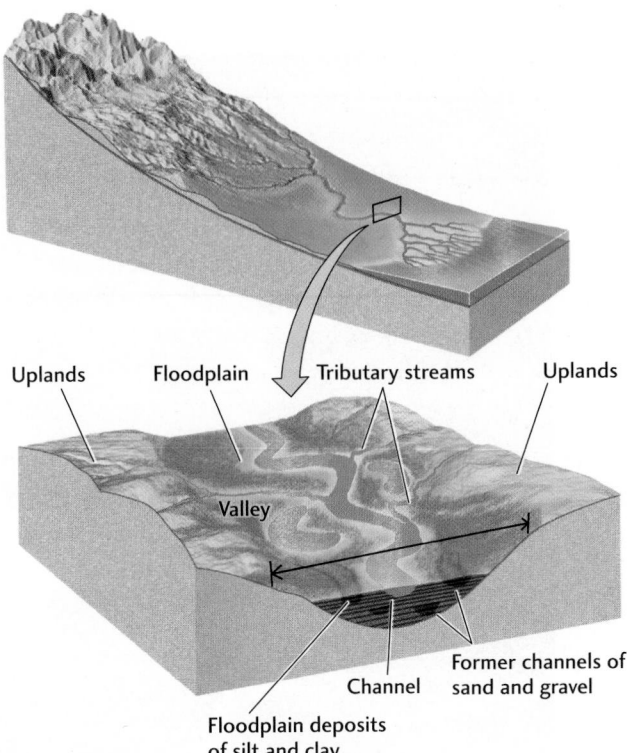

FIGURE 18.1 ■ A stream flows in a channel that moves over a broad, flat floodplain in a wide valley. Floodplains may be narrow or absent in steep-walled valleys.

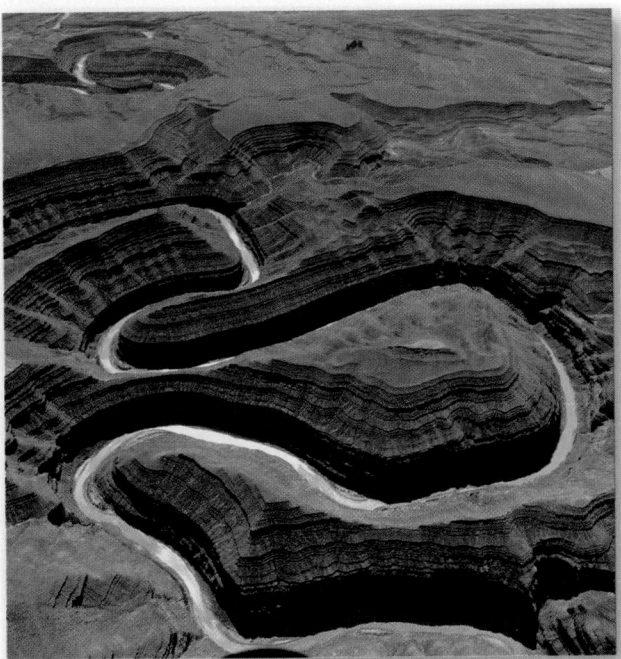

FIGURE 18.2 ■ This section of the San Juan River, Utah, is a good example of an incised meander belt, a deeply eroded, meandering, V-shaped valley with virtually no floodplain. [DEA/PUBBLIC AER FOTO/De Agostini/Getty Images.]

multiple channels. The channel may run along the center of the floodplain or hug one edge of the valley.

MEANDERS On a great many floodplains, stream channels follow curves and bends called **meanders,** named for the Maiandros (now Menderes) River in Turkey, known in ancient times for its winding, twisting course. Meanders are the normal pattern for low-velocity streams flowing through gently sloping or nearly flat plains or lowlands, where their channels typically cut through unconsolidated sediments—fine sand, silt, or mud—or easily eroded bedrock. Meanders are less pronounced, but still common, in streams flowing down slightly steeper slopes over harder bedrock. In such terrain, meandering stretches may alternate with long, relatively straight ones.

A stream that has cut deeply into the curves and bends of its channel may produce incised meanders (see Figure 18.2). Other streams meander on somewhat wider floodplains bounded by steep, rocky valley walls. We are not sure why these two different patterns appear. We do know that meandering is widespread not only in streams, but also in many other kinds of flows. For example, the Gulf Stream, a powerful current in the western North Atlantic Ocean, meanders. Lava flows on Earth meander, and planetary geologists have found meanders in former water channels on Mars (see Figures 9.20 and 9.21) as well as in lava flows on Mars and Venus.

Meanders on a floodplain migrate over periods of many years as the stream erodes the outside banks of bends, where the current is strongest (**Figure 18.3a**). As the outside banks are eroded, sediments are deposited to form curved sandbars, called **point bars,** along the inside banks, where the current is weaker (Figure 18.3b). In this way, meanders slowly shift position from side to side, as well as downstream, in a snaking motion something like that of a long rope being snapped. This migration may be quite rapid: some meanders on the Mississippi River shift as much as 20 m/year. As meanders move, so do the point bars, building up an accumulation of sand and silt over the part of the floodplain across which the channel migrated.

As meanders migrate, sometimes unevenly, the bends may grow closer and closer together, until finally the stream bypasses one of them, often during a major flood. The stream then takes a new, shorter course. In its abandoned path, it leaves behind an **oxbow lake:** a crescent-shaped, water-filled loop (Figure 18.3c).

Engineers sometimes artificially straighten and confine a meandering river, channelizing it along a straight path with the aid of artificial levees made of concrete. The Army Corps of Engineers has been channelizing the Mississippi River since 1878. In 13 years, it decreased the length of the lower Mississippi by 243 km. Part of the severity of the disastrous Mississippi flood of 1993 was ascribed to channelization. Without channelization, floods are more frequent, but less damaging. With it, damage may be catastrophic when a flood breaches the artificial levees, as it did in 1993. Channelization has also been criticized for destroying wetlands and much of

(a)

(b)

1 Meanders shift from side to side in a snaking motion.

2 The current is faster at outside banks, which are eroded,…

3 …and sediments get deposited at inside banks where the current is slower, forming point bars.

Point bar

(c)

Point bars

4 As the erosion and deposition process continues, the bends grow closer and the point bars bigger.

5 During a major flood, when velocity and water volume increase, the river takes a new, shorter course, cutting across the loop.

6 The abandoned loop remains as an oxbow lake.

Oxbow lake

FIGURE 18.3 ■ Meanders migrate over a period of many years. (a) How meanders move. (b) Meanders in an Alaskan river. (c) Oxbow lake in Blackfoot River valley, Montana. [(b) Peter Kresan; (c) James Steinberg/Science Source.]

the natural life of the floodplain by cutting off the supply of sediments deposited by small, frequent floods.

Such environmental concerns stimulated action to restore one channelized river, the Kissimmee in central Florida, to its original meandering course. Today, this restoration project is well under way. If left to its own natural processes, the Kissimmee might have taken many decades or hundreds of years to restore itself.

BRAIDED STREAMS Some streams have many channels instead of a single one. A **braided stream** is a stream whose channel divides into an interlacing network of channels, which then rejoin in a pattern resembling braids of hair (**Figure 18.4**). Braided streams are found in many settings, from broad lowland valleys to wide, sediment-filled rift valleys adjacent to mountain ranges. Braids tend to form in streams with large variations in volume of flow

FIGURE 18.4 ■ This section of the Joekulgilkvisl river in Iceland is a braided stream. [Dirk Bleyer/imagebrok/imagebroker.net/SuperStock.]

combined with a high sediment load and banks that are easily eroded. They are well developed, for example, in the sediment-choked streams formed at the edges of melting glaciers. Currents in braided streams are usually swiftly flowing, in contrast to those in meandering streams.

Stream Floodplains

A stream channel migrating over the floor of a valley creates a floodplain. Point bars formed during migration build up the surface of the floodplain, as does sediment deposited by floodwaters. Erosional floodplains, covered with a thin layer of sediment, can form when a stream erodes bedrock or unconsolidated sediment as it migrates.

When a stream overflows its banks and floodwaters spread out over the floodplain, the velocity of the water slows, and the current loses its ability to carry sediment. The floodwater velocity drops most quickly along the immediate borders of the channel. As a result, the current deposits large amounts of coarse sediment, typically sand and gravel, along a narrow strip at the edge of the channel. Successive floods build up **natural levees,** ridges of coarse material that confine the stream within its banks between floods, even when water levels are high (**Figure 18.5**). Where natural levees have reached a height of several meters and the stream almost fills the channel, the floodplain level is below the stream level. You can walk the streets of an old river town built on a floodplain, such as Vicksburg, Mississippi, and look up at the levee, knowing that the river waters are rushing by above your head.

During floods, fine sediments—silts and muds—are carried well beyond the stream's banks, often over the entire floodplain, and are deposited there as floodwaters continue to lose velocity. Receding floodwaters leave behind standing ponds and pools of water. The finest clays are deposited there as the standing water slowly disappears by evaporation and infiltration. These fine-grained floodplain deposits, which are rich in mineral and organic nutrients, have been a major resource for agriculture since ancient times. The fertility of the floodplains of the Nile and other rivers of the Middle East, for example, contributed to the evolution of the cultures that flourished there thousands of years ago. Today, the great, broad floodplain of the Ganges in northern India continues to play an important role in India's life and agriculture. Many ancient and modern cities are sited on floodplains (see Earth Issues 18.1).

Drainage Basins

Every topographic rise between two streams, whether it measures a few meters or a thousand, forms a **divide:** a ridge of high ground along which all rainfall runs off down one side or the other. A **drainage basin** is an area of land, bounded by divides, that funnels all its water into the network of streams draining that area (**Figure 18.6**). Drainage basins occur at many scales, from a ravine surrounding a small stream to a great region drained by a major river and its tributaries (**Figure 18.7**).

(a)

1 In between floods, sediments are deposited only within the channel.

Floodplain

Stream channel

2 During a flood, water spreads over the floodplain, rapidly loses velocity, and drops its sediments along the immediate borders of the channel, forming levees.

Sediment-laden floodwater

Natural levee

3 After successive floods, levees confine the stream to its channel between floods, even when the water is high.

(b)

Levee

FIGURE 18.5 ■ (a) Floods form natural levees along the banks of a stream. (b) These natural levees lie along the main channel of the Mississippi River near South Pass, Louisiana.
[(b) U.S. Geological Survey National Wetlands Research Center.]

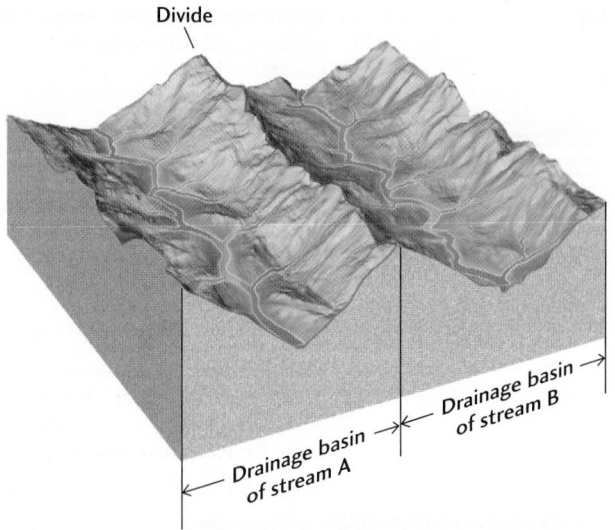

FIGURE 18.6 ■ Drainage basins are separated by divides.

A continent has several major drainage basins separated by major divides. In North America, the continental divide formed by the Rocky Mountains separates all waters flowing into the Pacific Ocean from all those entering the Atlantic. Lewis and Clark followed the Missouri River upstream to its headwaters at the continental divide in western Montana. After crossing over the divide, they found the headwaters of the Columbia River, which they followed to the Pacific Ocean.

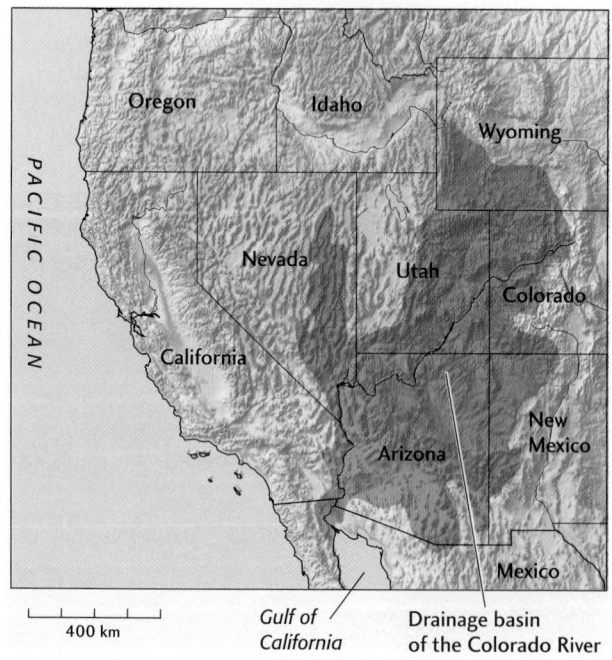

FIGURE 18.7 ■ The drainage basin of the Colorado River covers about 630,000 km², constituting a large part of the southwestern United States. The basin is surrounded by divides that separate it from the neighboring drainage basins. [After U.S. Geological Survey.]

Dendritic drainage is characterized by branches similar to the limbs of a tree.

Rectangular drainage, developed on a strongly jointed rocky terrain, tends to follow the joint pattern.

Trellis drainage develops in valley and ridge terrain, where rocks of varying resistance to erosion are folded into anticlines and synclines.

Radial drainage patterns develop on a single large peak, such as a large dormant volcano.

FIGURE 18.8 ■ Some typical drainage network patterns.

Drainage Networks

A map showing the courses of all the large and small streams in a drainage basin reveals a pattern of connections called a **drainage network.** If you followed a stream from its mouth (where it ends) to its headwaters (where it begins), you would see that it steadily divides into smaller and smaller **tributaries,** forming a drainage network that shows a characteristic branching pattern.

Branching is a general property of many kinds of networks in which material is collected and distributed. Perhaps the most familiar branching networks are those of tree branches and roots. Most rivers follow the same kind of irregular branching pattern, called **dendritic drainage** (from the Greek *dendron,* meaning "tree"). This fairly random drainage pattern is typical of terrains where the bedrock is of a uniform type, such as horizontally bedded sedimentary rock or massive igneous or metamorphic rock.

Other drainage patterns are described as *rectangular, trellis,* and *radial* (**Figure 18.8**).

Drainage Patterns and Geologic History

We can observe directly, or determine from historical or geologic records, how most stream drainage patterns developed. Some streams, for example, cut through erosion-resistant bedrock ridges to form steep-walled notches or gorges. What could cause a stream to cut a narrow valley directly through a ridge rather than running along the lowland on either side of it? The geologic history of the region provides the answers.

If a ridge is formed by deformation while a preexisting stream is flowing over it, the stream may erode the rising ridge to form a steep-walled gorge, as in **Figure 18.9**. Such a stream is called an **antecedent stream** because it existed

(a) **1** An antecedent stream flowed on horizontal sedimentary rocks. **2** Slow tectonic uplift folded the rocks in an anticline. **3** The stream maintained its course, cutting through the anticline as it developed,...

4 ...and it now flows through a steep-walled gorge of its own making.

(b)

FIGURE 18.9 ■ (a) How an antecedent stream cuts a steep-walled gorge. (b) The Delaware Water Gap, located between Pennsylvania and New Jersey. At this point, the Delaware River is an antecedent stream. [(b) Michael P. Gadomski/Science Source.]

Earth Issues

18.1 The Development of Cities on Floodplains

Floodplains have attracted human settlement since the beginning of civilization. Floodplains are natural sites for urban settlements because they combine easy transportation along a river with access to fertile agricultural lands. Such sites, however, remain subject to the floods that formed the floodplains. Small floods are common and usually cause little damage, but the larger episodes that happen every few decades or so can be quite destructive.

About 4000 years ago, cities began to dot floodplains in Egypt along the Nile, in the ancient land of Mesopotamia along the Tigris and Euphrates rivers, and in Asia along the Indus River of India and the Yangtze and Huang Ho of China. Later, many of the capital cities of Europe were built on floodplains: Rome on the Tiber, London on the Thames, Paris on the Seine. Floodplain cities in North America include St. Louis on the Mississippi, Cincinnati on the Ohio, and Montreal on the St. Lawrence. Floods periodically destroyed sections of these ancient and modern cities on the lower parts of the floodplains, but each time the inhabitants rebuilt them.

Today, most large cities are protected by artificial levees that strengthen and heighten the river's natural levees. In addition, extensive systems of dams help control the flooding that would affect these cities. But these structures cannot eliminate the risk entirely. In 1973, for example, the Mississippi River went on a rampage with a flood that continued for 77 consecutive days at St. Louis, Missouri. The river reached a record 4.03 m above flood stage. In 1993, the Mississippi and its tributaries broke loose again, shattering the old record in a disastrous flood that has been officially designated the second worst flood in U.S. history (behind the flooding of New Orleans by the storm surge from Hurricane Katrina in 2005). The flood resulted in 487 deaths and more than $15 billion in property damage. At St. Louis, the Mississippi stayed above flood stage for 144 of the 183 days between April and September. In an unexpected secondary effect, the floodwaters leached agricultural chemicals from farmlands and deposited them in the flooded areas, causing widespread pollution.

Figuring out how to protect society from floods presents some knotty problems. Some geologists believe that the construction of artificial levees to confine the Mississippi contributed to its record floods. A river hemmed in by artificial levees can no longer erode its banks and widen its channel to accommodate additional water during periods of high flow. In addition, the floodplain no longer receives deposits of sediment. In the case of New Orleans, the floodplain has sunk below the level of the Mississippi River, making future flooding more likely.

What are cities and towns in this position to do? Some have urged a halt to all construction and development on the lowest parts of floodplains. Some have called for the elimination of federally subsidized disaster funds for rebuilding in such areas. Harrisburg, Pennsylvania, hit hard by a flood in 1972, turned some of its devastated riverfront area into a park. In a dramatic step after the 1993 Mississippi flood, the citizens of Valmeyer, Illinois, voted to move the entire town to high ground several miles away. The new site was chosen with the help of a team of geologists from the Illinois Geological Survey. Yet the benefits of living on floodplains continue to attract people to those sites, and some people who have lived on floodplains all their lives want to stay there and are prepared to live with the risk. The costs of protecting some floodplains are prohibitive, and these places will continue to pose public policy problems.

Like many cities built on river floodplains, Liuzhou, China, is subject to flooding. This flood, in July 1996, was the largest recorded in the city's 500-year history. [Xie Jiahua/China Features/Corbis Sygma.]

1 A superposed dendritic stream developed on horizontal beds.

Folded beds
Horizontal beds
Unconformity
Anticline

2 Most horizontal beds were stripped away by erosion.

3 A stream cut a gorge—or water gap—through resistant beds of a buried anticline.

FIGURE 18.10 ▪ How a superposed stream maintains its course.

before the present topography was created. The stream maintains its original course despite changes in the underlying rocks and in the topography.

In another geologic situation, a stream may flow in a dendritic drainage pattern over horizontal beds of sedimentary rock that overlie folded and faulted rocks with varying resistance to erosion. Over time, as the softer beds are stripped away by erosion, the stream cuts into a harder bed of underlying rock and erodes a gorge in that resistant bed (**Figure 18.10**). Such a **superposed stream** flows through resistant formations because its course was established at a higher level, on uniform rock, before downcutting began. A superposed stream tends to continue the pattern that it developed earlier rather than adjusting to its new conditions.

▪ Where Do Channels Begin? How Running Water Erodes Soil and Rock

Stream channels begin where rainwater, running off the surface of the land, flows so fast that it abrades the soil and bedrock and carves into it to form a *gully* (essentially a small valley). Once a gully forms, it captures more of the runoff, and thus the tendency of the stream to cut downward increases. As the gully progressively deepens, the rate of downcutting increases as more water is captured (**Figure 18.11**).

The erosion of unconsolidated material is relatively easy to observe. We can easily see a stream picking up loose sand from its bed and carrying it away. At high water levels

FIGURE 18.11 ▪ Streams create gullies when the action of water flowing across Earth's surface causes erosion. The smallest gullies converge to form larger stream channels, and farther downslope these become river channels. These gullies were formed in the desert of Oman by occasional rainstorms that inundate the surface with rapidly flowing water, which erodes the bedrock. [Courtesy of Petroleum Development Oman.]

and during floods, we can even see a stream scouring and cutting into its banks, which then slump into the flow and are carried away. Streams progressively cut their channels upstream into higher land. This process, called *headward erosion,* commonly accompanies widening and deepening of valleys. Its progress may be extremely rapid—as much as several meters in a few years in easily erodible soils. Downstream erosion is much less common and is best expressed in rare catastrophic events, such as when an earthquake collapses a natural dam and sends scouring waters plunging downstream.

We cannot so easily see the erosion of solid rock. Running water erodes solid rock by abrasion, by chemical and physical weathering, and by the undercutting action of currents.

Abrasion

One of the major ways a stream breaks apart and erodes rock is by **abrasion.** The sand and pebbles the stream carries create a sandblasting action that wears away even the hardest rock. In some streambeds, pebbles and cobbles rotating inside swirling eddies grind deep **potholes** into the bedrock (**Figure 18.12**). At low water, pebbles and sand can be seen lying quietly at the bottom of exposed potholes.

FIGURE 18.12 ■ Bourke's Luck Potholes in river rock in Blyde River Canyon Nature Reserve, South Africa. Flowing water rotates the pebbles inside the potholes, grinding deep holes in the bedrock. [© Walter G. Allgöwer/Age Fotostock.]

Chemical and Physical Weathering

Chemical weathering breaks down rocks in streambeds just as it does on the land surface. Physical weathering in streams can be violent, as the crashes of boulders and the constant smaller impacts of pebbles and sand split rock along natural zones of weakness. Such impacts in a stream channel break up rock much faster than slow weathering on a gently sloping hillside does. When these processes have loosened large blocks of bedrock, strong upward eddies may pull them up and out in a sudden violent plucking action.

Physical weathering is particularly strong at rapids and waterfalls. *Rapids* are places in a stream where the flow velocity increases because the slope of the streambed suddenly steepens, typically at rocky ledges. The high speed and turbulence of the water quickly break rocks into smaller pieces that are carried away by the strong current.

The Undercutting Action of Waterfalls

Waterfalls develop where hard rocks resist erosion or where faulting offsets the streambed. The tremendous impact of huge volumes of plunging water and tumbling boulders

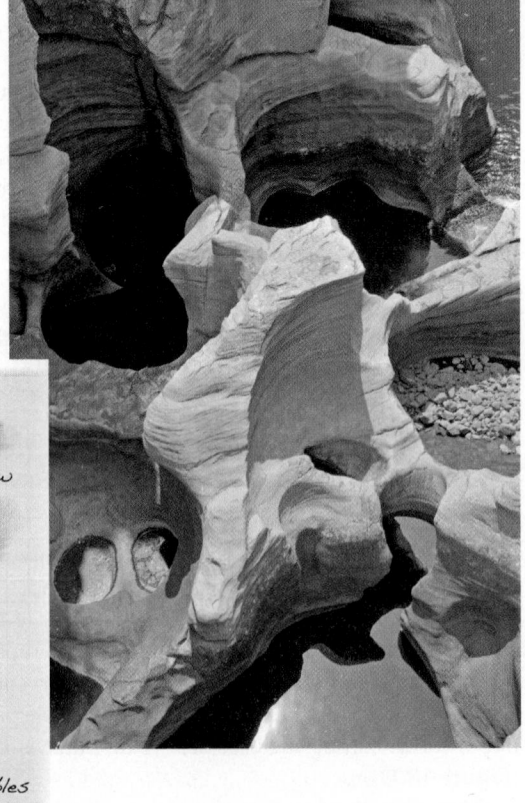

Water flow

Potholes

Swirling of pebbles causes abrasion to form the pothole

FIGURE 18.13 ■ This waterfall on the Iguaçú River, Brazil, is retreating upstream as falling water and sediment pound the cliff's base and undercut it. Looking downstream, from the center to the upper left of the photo, one can see steep walls, the remnants of the waterfall's retreat upstream. [Donald Nausbaum/Getty Images, Inc.]

quickly erodes streambeds below waterfalls. Waterfalls also erode the underlying rock of the cliffs that form the falls. As erosion undercuts these cliffs, the upper streambed collapses, and the falls recede upstream (**Figure 18.13**). Erosion by falls is fastest where the rock layers are horizontal, with erosion-resistant rocks at the top and softer rocks, such as shales, making up the lower layers. Historical records show that the main section of Niagara Falls, perhaps the best-known falls in North America, has been moving upstream in this way at a rate of a meter per year.

■ How Currents Flow and Transport Sediment

All currents, whether in water or in air, share the basic characteristics of fluid flow. We can illustrate two kinds of fluid flow by using lines of motion called *streamlines* (**Figure 18.14**). In **laminar flow,** the simplest kind of fluid movement, straight or gently curved streamlines run parallel to one another without mixing or crossing between layers. The slow movement of thick syrup over a pancake, with strands of unmixed melted butter flowing in parallel but separate paths, is a laminar flow. **Turbulent flow** has a more complex pattern of movement, in which streamlines mix, cross, and form swirls and eddies. Fast-moving stream waters typically show this kind of motion. Turbulence—the degree to which there are irregularities and eddies in the flow—may be low or high.

Whether a fluid flow is laminar or turbulent depends primarily on three factors:

1. Its velocity (rate of movement)
2. Its geometry (primarily its depth)
3. Its viscosity (resistance to flow)

Viscosity arises from the attractive forces between the molecules of a fluid. These forces tend to impede the slipping and sliding of molecules past one another. The greater the attractive forces, the greater the resistance to mixing with neighboring molecules, and the higher the viscosity. For example, when cold syrup or viscous cooking oil is poured, its flow is sluggish and laminar. The viscosity of most fluids, including water, decreases as their temperature increases. Given enough heat, a fluid's viscosity may decrease sufficiently to change a laminar flow into a turbulent one.

Water has low viscosity in the range of temperatures found at Earth's surface. For this reason alone, most streams in nature tend toward turbulent flow. In addition,

Laminar flow

Turbulent flow

In laminar flows, straight or gently curved streamlines run parallel without mixing or crossing.

In turbulent flows, streamlines mix, cross, and form swirls and eddies.

Flow

FIGURE 18.14 ■ The two basic patterns of fluid flow: laminar flow and turbulent flow. The photograph shows the transition from laminar to turbulent flow in water along a flat plate, revealed by the injection of a dye. Flow is from left to right. [Henri Werlé, Onera The French Aerospace Lab.]

the velocity and geometry of most natural streams make them turbulent. In nature, we are likely to see laminar flows of water only in thin sheets of rain runoff flowing slowly down nearly level slopes. In cities, we may see small laminar flows in street gutters.

Because most streams and rivers are broad and deep and flow quickly, their flows are almost always turbulent. A stream may show turbulent flow over much of its width and laminar flow along its edges, where the water is shallower and is moving more slowly. In general, the flow velocity is highest near the center of a stream; where a stream meanders, the flow velocity is highest at the outsides of the bends. We commonly refer to a rapid flow of water as a strong current.

Erosion and Sediment Transport

Currents vary in their ability to erode and carry sand grains and other sediments. Laminar flows of water can lift and carry only the smallest, lightest clay-sized particles. Turbulent flows, depending on their velocity, can move particles ranging in size from clay to pebbles and cobbles. As turbulence lifts particles into a flow, the flow carries them downstream. Turbulence also rolls and slides larger particles along the bottom of the channel. A current's **suspended load** includes all the material temporarily or permanently

suspended in the flow. Its **bed load** is the material the current carries along the bed by sliding and rolling (**Figure 18.15**). The *bed*, in this context, is the layer of unconsolidated material in the channel that interacts with the current.

The greater the velocity of a current, the larger the particles it can carry as suspended load and bed load. A flow's ability to carry material of a given particle size is its **competence.** As current strength increases and coarser particles are suspended, the suspended load grows. At the same time, more of the bed material is in motion, and the bed load also increases. As we would expect, the larger the volume of a flow, the greater the sediment load (suspended load plus bed load) it can carry. The total sediment load carried by a flow is its **capacity.**

The velocity and the volume of a flow affect both the competence and the capacity of a stream. The Mississippi River, for example, flows at moderate speeds along most of its length and carries only fine to medium-sized particles (clay to sand), but it carries huge quantities of them. A small, steep, fast-flowing mountain stream, in contrast, may carry boulders, but only a few of them.

A current's suspended load depends on a balance between turbulence, which lifts particles, and the competing downward pull of gravity, which makes them settle out of the current and become part of the bed. The speed with which suspended particles of various weights settle to the

1 Current flowing over a bed of gravel, sand, silt, and clay carries a **suspended load** of finer particles…

3 As current velocity increases, the suspended load grows,…

5 Particles move by saltation, jumping along a bed. At a given current velocity, smaller particles jump higher and travel farther than larger particles.

2 …and a **bed load** of material sliding and rolling along the bottom.

4 …and the increased force of the flow generates an increase in the bed load.

FIGURE 18.15 ■ A current flowing over a bed of unconsolidated material can transport particles in three ways.

bed is called their **settling velocity.** Small grains of silt and clay are easily lifted into the stream and settle slowly, so they tend to stay in suspension. The settling velocities of larger particles, such as medium- and coarse-grained sand, are much higher. Most larger grains therefore stay suspended in the current only a short time before they settle.

As current velocity increases, sediment particles in the bed load begin to move by a third process, known as **saltation:** an intermittent jumping motion along the bed. Sand grains are most likely to move by saltation because they are light enough to be picked up from the bed, yet heavy enough not to be transported in suspension. The grains are sucked up into the flow by turbulent eddies, move with the current for a short distance, and then fall back to the bed (see Figure 18.15). If you were to stand in a rapidly flowing sandy stream, you might see a cloud of saltating sand grains moving around your ankles. The bigger the grain, the longer it will tend to remain on the bed before it is picked up. Once a large grain is in the current, it will settle quickly. The smaller the grain, the more frequently it will be picked up, the higher it will "jump," and the longer it will take to settle.

Worldwide, streams carry about 25 billion tons of siliciclastic sediments and an additional 2 billion to 4 billion tons of dissolved matter each year. Humans are responsible for much of the present sediment load. According to some estimates, prehuman sediment transport was about 9 billion tons per year, less than half the present value. In some places, we increase the sediment load of streams through agriculture and accelerated erosion. In other places, we decrease the sediment load by constructing dams, which trap sediment behind them.

To study how a particular stream carries sediments, geologists and hydraulic engineers measure the relationship between particle size and the force the flow exerts on the particles in the suspended and bed loads. This relationship shows them how much sediment a particular flow can move and how rapidly it can move it. That information allows them to design dams and bridges or to estimate how quickly artificial reservoirs behind dams will fill with sediments. As we saw in Chapter 5, geologists can infer the velocities of ancient currents from the sizes of grains in sedimentary rocks.

Figure 18.16 graphs the relationship between the sizes of sediment particles on the streambed and the flow velocities required to erode them. You will notice that in this graph, contrary to our earlier discussion of competence, the current velocity required to erode some kinds of particles from the bed actually increases as particle size decreases. This relationship exists because it is easier for the current to lift noncohesive particles (particles that do not stick together) than cohesive particles (particles that stick together, as many clay minerals do). The finer the cohesive particles, the greater the velocity of the flow required to erode them. Settling velocities for these small particles, however, are so slow that even a gentle current, about 20 cm/s, can keep them in suspension and transport them over long distances.

Sediment Bed Forms: Dunes and Ripples

When a current transports sand grains by saltation, the sand tends to form dunes and ripples (see Chapter 5). **Dunes** are elongated ridges of sand up to many meters high that form in flows of wind or water over a sandy bed. **Ripples** are very small dunes—with heights ranging from less than a centimeter to several centimeters—whose long dimension is formed at right angles to the current. Although underwater ripples and dunes produced by water currents are harder to observe than those produced on land by air currents, they form in the same way and are just as common.

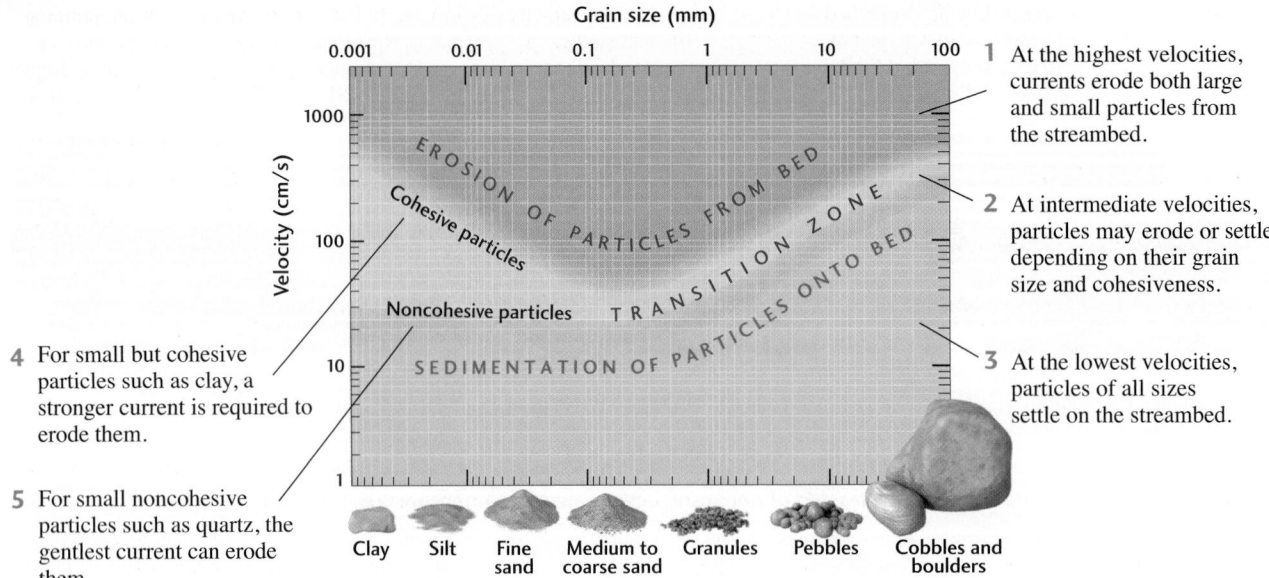

FIGURE 18.16 ■ Relationship of current velocity to the erosion and settling of particles of different sizes. The blue area represents velocities at which particles are eroded from the streambed; the gray area, velocities at which particles may either be eroded or settle, and the brown area, velocities at which particles settle onto the bed. [After F. Hjulstrom, as modified by A. Sundborg, "The River Klarälven," *Geografisk Annaler* 1956.]

As a current moves sand grains by saltation, they are eroded from the upstream side of ripples and dunes and deposited on the downstream side. The steady downstream transfer of grains across the ridges causes the ripple and dune forms to migrate downstream. The speed of this migration is much slower than the movement of individual grains and very much slower than the current. (We will look at ripple and dune migration in more detail in Chapter 19.)

The shapes of ripples and dunes and their migration speeds change as the velocity of the current increases. At the lowest current velocities, few grains are saltating, and the sediment bed is flat. At slightly higher velocities, the number of saltating grains increases. A rippled bed forms, and the ripples migrate downstream (**Figure 18.17**). As the velocity increases further, the ripples grow larger and migrate faster until, at a certain point, dunes replace the ripples. Both ripples and dunes have a cross-bedded structure (see Figure 5.11). As the current flows over their tops, it can actually reverse and flow backward along their downstream side. As the dunes grow larger, small ripples form on them. These ripples tend to climb over the backs of the dunes because they migrate more quickly than the dunes. Very high current velocities will erase the dunes and form a flat bed below a dense cloud of rapidly saltating sand grains. Most of these grains hardly settle to the bottom before they are picked up again. Some are in permanent suspension.

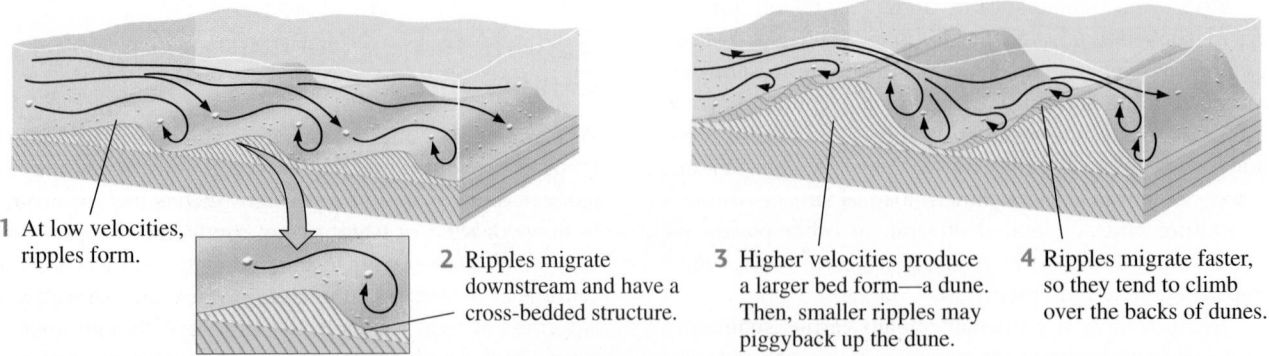

1 At low velocities, ripples form.

2 Ripples migrate downstream and have a cross-bedded structure.

3 Higher velocities produce a larger bed form—a dune. Then, smaller ripples may piggyback up the dune.

4 Ripples migrate faster, so they tend to climb over the backs of dunes.

FIGURE 18.17 ■ The form of a sediment bed changes with current velocity. [After D. A. Simmons and E. V. Richardson, "Forms of Bed Roughness in Alluvial Channels," *American Society of Civil Engineers Proceedings* 87 (1961): 87–105.]

■ Deltas: The Mouths of Rivers

Sooner or later, all rivers end as they flow into a lake or an ocean, mix with the surrounding water, and—no longer able to travel downslope—gradually lose their forward momentum. The largest rivers, such as the Amazon and the Mississippi, can maintain some current many kilometers out to sea. Where smaller rivers enter a turbulent, wave-swept sea, the current disappears almost immediately beyond the river's mouth.

Delta Sedimentation

As its current gradually dies out, a river progressively loses its power to transport sediments. The coarsest material, typically sand, is dropped first, right at the mouths of most rivers. Finer sands are dropped farther out, followed by silt and, still farther out, by clay. As the floor of the lake or ocean slopes to deeper water away from the shore, the deposited materials build up a large, flat-topped deposit called a **delta.** (We owe the name *delta* to the Greek historian Herodotus, who traveled through Egypt about 450 B.C. The roughly triangular shape of the sediment deposit at the mouth of the Nile prompted him to name it after the Greek letter Δ, delta.)

As a river approaches its delta, where its slope is almost level with the ocean surface, it reverses its upstream-branching drainage pattern. Instead of collecting more water from tributaries, it discharges water into **distributaries:** smaller streams that receive water and sediments from the main channel, branch off *downstream,* and thus distribute the water and sediment into many channels. Materials deposited on top of the delta, typically sand, make up

horizontal **topset beds.** Downstream, on the outer front of the delta, fine-grained sand and silt are deposited to form gently inclined **foreset beds,** which resemble large-scale cross-beds. Spread out on the seafloor seaward of the foreset beds are thin, horizontal **bottomset beds** of mud, which are eventually buried as the delta continues to grow. **Figure 18.18** shows how these structures form in a typical large marine delta.

The Growth of Deltas

As a delta builds outward into the ocean, the mouth of its river advances seaward, leaving new land in its wake. Much of this land is a *delta plain* just a few meters above sea level. At the seaward edge of the plain, broad depressions between distributary channels lie below sea level and form shallow bays that fill with fine-grained sediments. With time, they fill further and ultimately become salt marshes (see Figure 18.18).

As a delta grows, the river flow shifts from some distributaries to others that provide shorter routes to the sea. As a result of such shifts, the delta may grow in one direction for some hundreds or thousands of years, then the main stream may break out into a new distributary, sending sediments into the ocean in another direction. In this way, a major river may form a large delta thousands of square kilometers in area. The delta of the Mississippi has been growing for millions of years. About 150 million years ago, it started out around what is now the junction of the Ohio and the Mississippi rivers, at the southern tip of Illinois. It has advanced about 1600 km since then, creating almost the entire states of Louisiana and Mississippi as well as major parts of adjacent states. **Figure 18.19** shows the growth of the Mississippi delta over the past 6000 years, as well as the direction its growth is likely to take in the future.

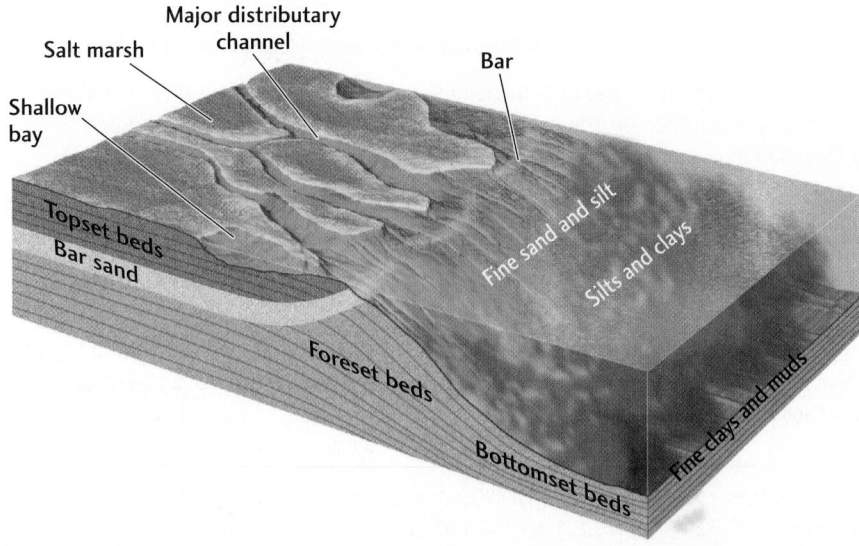

FIGURE 18.18 ■ A typical large marine delta, many kilometers in extent, in which the fine-grained foreset beds are deposited at a very low angle, typically only 4° to 5° or less. Sandbars form at the mouths of the distributaries, where the current velocity suddenly decreases. The delta builds forward by the advance of these bars and the topset, foreset, and bottomset beds. Between distributary channels, shallow bays fill with fine-grained sediments and become salt marshes. This general structure is found on the Mississippi delta.

(a)

(b)

(c)

Silt carried by Atchafalaya River discharge...

(d)

...will increase as the delta relocates here in the future.

FIGURE 18.19 ■ Over the past 6000 years, the Mississippi River has built its delta first in one direction and then in another as water flow has shifted from one major distributary to another. (a) The modern delta was preceded by deltas deposited to the east and west. (b) The infrared-sensitive film used to record this satellite image of the Mississippi delta causes vegetation to appear red, relatively clear water to appear dark blue, and water with suspended sediment to appear light blue. At the upper left are New Orleans and Lake Pontchartrain. Well-defined natural levees and point bars can be seen at the center. At the lower left are beaches and islands that formed as ocean waves and currents transported river-deposited sand from the delta. (c) Satellite photograph of the Mississippi delta. (d) This image shows the discharge of sediment into the Gulf of Mexico from the Mississippi River delta and the Atchafalaya River. A major flood could divert the main flow of the Mississippi into the Atchafalaya, causing a new delta to form. Construction of artificial levees by the Army Corps of Engineers has prevented this so far. [(b) From G. T. Moore, "Mississippi River Delta from Landsat2," *Bulletin of the American Association of Petroleum Geologists,* 1979; (c) NASA; (d) U.S. Geological Survey National Wetlands Research Center.]

Deltas grow by the addition of sediment, and they sink as the sediment becomes compacted and Earth's crust subsides under the weight of the sediment load. Venice, built on part of the Po River delta in northern Italy, has been sinking steadily for many years. Both crustal subsidence and depression of the ground attributable to the pumping of water from aquifers beneath the city are responsible for its sinking.

Human Effects on Deltas

The extensive wetlands found in delta plains are valuable natural resources because, like all wetlands, they store floodwaters and provide habitat for many diverse species of plants and animals, as noted in Chapter 17. The wetlands of the Mississippi River delta, like delta wetlands in many other areas, have suffered a two-pronged attack. First, the extensive flood-control dams built on the river since the 1930s have decreased the volume of sediment brought to the delta, thereby reducing the sediment supply to the wetlands. Second, massive artificial levees have prevented the small but frequent floods that nourish the delta wetlands with sediments. At New Orleans, the Mississippi River floodplain has sunk below the level of the river, making future catastrophic flooding more likely.

The Effects of Ocean Currents, Tides, and Plate Tectonic Processes

Strong waves, ocean currents, and tides affect the growth and shape of marine deltas. Waves and ocean currents may move sediment along the shore almost as rapidly as it is dropped there by a river. The delta front then becomes a long beach with only a slight seaward bulge at the river mouth. Where tidal currents move in and out, they redistribute delta sediments into elongate bars parallel to the direction of the current, which in most places is at approximately right angles to the shore (see Figure 18.19b).

In some places, waves and tides are strong enough to prevent a delta from forming. Instead, the sediments a river transports to the ocean are dispersed along the shoreline as beaches and bars and transported into deeper waters offshore. The east coast of North America lacks deltas for this reason. The Mississippi has been able to build its delta because neither waves nor tides are very strong in the Gulf of Mexico.

Plate tectonic processes also exert some control over where deltas form because delta formation has two other preconditions: uplift in the drainage basin, which provides abundant sediments, and crustal subsidence in the delta region to accommodate the great weight and volume of those sediments. Two of the world's large deltas—those of the Mississippi and the Rhône (in France)—derive their sediments primarily from distant mountain ranges: the Rockies for the Mississippi and the Alps for the Rhône. Both are in the same type of plate tectonic setting: a passive margin originally formed by continental rifting. The active continent-continent convergence that is elevating the Himalaya has also formed the great deltas of the Indus and Ganges rivers.

Few large deltas are associated with active subduction zones. The reason may be that it is unusual for a large river (such as the Columbia River of the Pacific Northwest) to carry abundant sediment through a volcanic mountain belt (such as the Cascade Range) to the sea. Furthermore, these rapidly uplifting areas are too unstable for large deltas to develop. Oceanic island arcs are too small in land area to provide much siliciclastic sediment to their streams.

■ Streams as Geosystems

Streams are dynamic geosystems that are continually changing in response to the influences of climate and plate tectonic processes, and those changes, in turn, influence the transport of water and sediments. The flow of a stream may appear steady when you view it from a bridge for a few minutes or canoe along it for a few hours, but its volume and velocity may change appreciably from month to month and season to season. At any one location, the stream is constantly changing, shifting from low water to flood stage over a few hours or days and reshaping its valley over longer periods (Figure 18.20). The flow and channel dimensions of a stream also change as it moves downslope, from narrow valleys in its upland headwaters to broader floodplains in its middle and lower courses. Most of these longer-term changes are adjustments in the normal (nonflood) volume and velocity of the flow as well as in the depth and width of the channel.

From their headwaters to their mouths, all streams react to changes in climate (such as changes in precipitation) and to plate tectonic processes (such as uplift or subsidence of Earth's crust). As we have seen, streams gather into ever larger streams, eventually forming a single large stream, as in the case of the Mississippi River. Precipitation in the headwaters may affect streamflow far downriver, where a river's volume may exceed the volume of the channel and then spill over the banks to create a flood. In this way, processes and events in one part of the stream network are propagated through the system to affect the behavior of a different part of the system.

Sediment transport changes in a similar way, though over a longer time scale. If precipitation in the headwaters increases over a long period—say, because the climate becomes rainier—or tectonic uplift rates increase, erosion rates and sediment yields increase. The stream network propagates a "wave" of sediment that eventually reaches the delta, where it may be preserved in the rock record as an interval of unusually high sediment accumulation. We will further explore these relationships and their effects on landscape development in Chapter 22.

FIGURE 18.20 ■ Stream networks transport water and sediments from their headwaters to the ocean.

Several factors are important in controlling how water and sediments move through stream geosystems. These factors include the stream's discharge, its longitudinal profile, and changes in its base level.

Discharge

We can measure the size of a stream's flow by its **discharge:** the volume of water that passes a given point in a given time as it flows through a channel of a certain width and depth. (In Chapter 17, we defined *discharge* as the volume of water leaving an aquifer in a given time. These definitions are consistent because they both describe volume of flow per unit of time.) Stream discharge is usually measured in cubic

meters per second (m³/s) or cubic feet per second. Discharge in small streams may vary from about 0.25 to 300 m³/s. The discharge of a well-studied medium-sized river in Sweden, the Klarälven, varies from 500 m³/s at low water levels to 1320 m³/s at high water levels. The discharge of the Mississippi River can vary from as little as 1400 m³/s at low water levels to more than 57,000 m³/s during floods.

Discharge is matched by recharge at any location where rainfall or discharging groundwater contributes to the stream. When recharge is less than discharge, such as during a drought, stream levels may fall dramatically. When recharge is greater than discharge, stream levels will rise, and flooding will occur if the imbalance between discharge and recharge becomes too great.

To calculate discharge, we multiply the cross-sectional area (the width multiplied by the depth of the part of the channel occupied by water) by the velocity of the flow (distance traveled per second):

$$\underset{\text{(width} \times \text{depth)}}{\text{discharge}} = \underset{\text{(distance traveled per second)}}{\text{cross section} \times \text{velocity}}$$

Figure 18.21 illustrates this relationship. If discharge is to increase, then either velocity or cross-sectional area, or both, must increase. Think of increasing the discharge of a garden hose by opening the valve more, which increases the velocity of the water coming out the end of the hose. The cross-sectional area of the hose cannot change, so the discharge must increase. In a stream, as discharge at a particular point increases, both the velocity and the cross-sectional area of the flow tend to increase. (The velocity is also affected by the slope of the channel and the roughness of the bottom and sides, which we can neglect for the

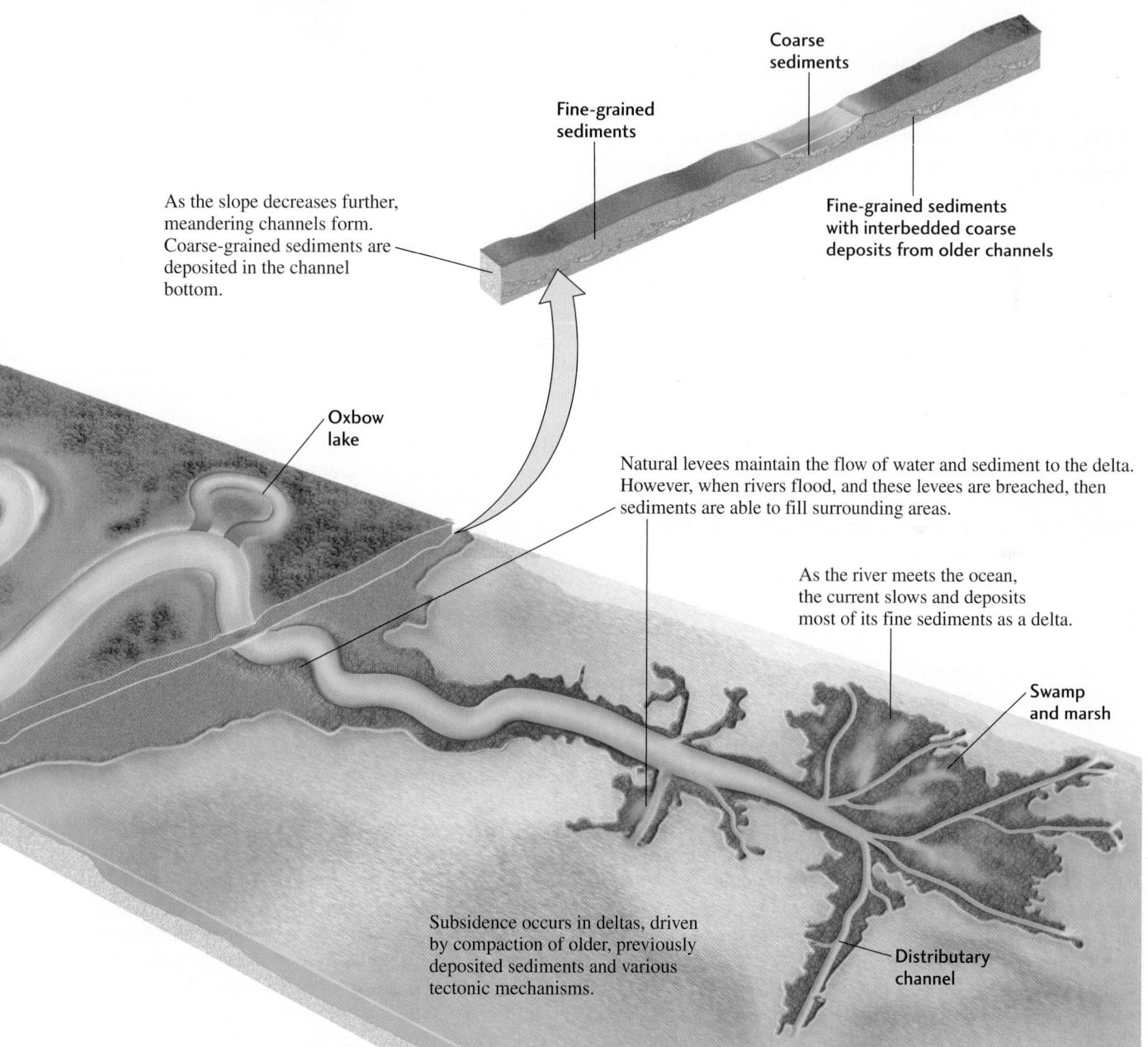

As the slope decreases further, meandering channels form. Coarse-grained sediments are deposited in the channel bottom.

Coarse sediments

Fine-grained sediments

Fine-grained sediments with interbedded coarse deposits from older channels

Oxbow lake

Natural levees maintain the flow of water and sediment to the delta. However, when rivers flood, and these levees are breached, then sediments are able to fill surrounding areas.

As the river meets the ocean, the current slows and deposits most of its fine sediments as a delta.

Swamp and marsh

Subsidence occurs in deltas, driven by compaction of older, previously deposited sediments and various tectonic mechanisms.

Distributary channel

A stream with a smaller cross-sectional area and a lower velocity has a lower discharge (3 m × 10 m = 30 m² × 1 m/s = 30 m³/s discharge)…

…than a stream with a larger cross-sectional area and a higher velocity (9 m × 10 m = 90 m² × 2 m/s = 180 m³/s).

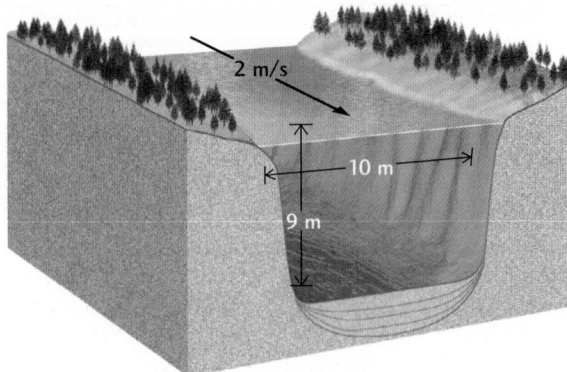

FIGURE 18.21 ■ A stream's discharge depends on the velocity and the cross-sectional area of its flow. [Data from T. Dunne and L. B. Leopold, *Water in Environmental Planning.* San Francisco: W. H. Freeman, 1978.]

moment.) The cross-sectional area increases because the flow occupies more of the channel's width and depth.

Discharge in most rivers increases downstream as more and more water flows in from tributaries. Increased discharge means that width, depth, or velocity must increase as well. Velocity does not increase downstream as much as the increase in discharge might lead us to expect, however, because the slope along the lower courses of a stream decreases (and decreasing slope reduces velocity). Where discharge does not increase significantly downstream and slope decreases greatly, a river will flow more slowly.

Floods

A **flood** is an extreme case of increased discharge that results from a short-term imbalance between inflow and outflow. As discharge increases, the flow velocity in the channel increases, and the water gradually fills the channel. As discharge continues to increase, the stream reaches *flood stage* (the point at which water first spills over the banks).

Some streams overflow their banks almost every year when the snows melt or rains arrive; others flood at irregular intervals. Some floods bring very high water levels that inundate the floodplain for days. At the other extreme are minor floods that barely break out from the channel before they recede. Small floods are more frequent, occurring every 2 or 3 years on average. Large floods are generally less frequent, usually occurring every 10, 20, or 30 years.

No one can know exactly how severe—either in water height or in discharge—a flood will be in any given year, so predictions are stated as *probabilities*, not certainties. For a particular stream, we might say that there is a 20 percent probability that a flood of a given discharge—say, 1500 m³/s—will occur in any one year. That probability

corresponds to an average recurrence interval—in this case, 5 years (1 in 5 = 20 percent)—that we expect between two floods with a discharge of 1500 m³/s. We speak of a flood of this discharge as a 5-year flood. A flood of greater magnitude—say, 2600 m³/s—on the same stream is likely to happen only once every 50 years, and it would therefore be called a 50-year flood. As with earthquakes, floods of greater magnitudes have longer recurrence intervals. A graph of the annual probabilities and recurrence intervals of floods with a range of discharges is known as a *flood frequency curve.*

The recurrence interval of floods of a certain discharge depends on three factors:

1. The climate of the region
2. The width of the floodplain
3. The size of the channel

For a stream in a dry climate, for example, the recurrence interval of 2600 m³/s floods may be much longer than the recurrence interval of 2600 m³/s floods for a similar stream in an area that gets intermittent rain. For this reason, flood frequency curves for individual rivers are necessary if towns along those rivers are to be prepared to cope with flooding. A flood frequency curve for one river—the Skykomish, in Washington State—is shown in **Figure 18.22**.

The prediction of river floods and their heights has become much more reliable as automated rainfall gauges and streamgauges, coupled with new computer models, have come into use. Geologists can now forecast the rise and fall of river levels as much as several months in advance, and they can issue reliable flood warnings days in advance. This information is useful for many other purposes as well, from water resource management to the planning of recreational river trips (see the Practicing Geology Exercise at the end of this chapter).

FIGURE 18.22 ▪ The flood frequency curve for the Skykomish River at Gold Bar, Washington. This curve predicts the probability that a flood of a certain discharge will occur in any given year. [After T. Dunne and L. B. Leopold, *Water in Environmental Planning.* San Francisco: W. H. Freeman, 1978.]

Longitudinal Profiles

We have seen that streamflow at any locality balances inflow and outflow, which become temporarily out of balance during floods. Studies of changes in discharge, flow velocity, channel dimensions, and topography along the entire length of a stream, from its headwaters to its mouth, reveal a larger-scale and longer-term balance. A stream is in dynamic equilibrium when erosion of the streambed balances sedimentation in the channel and floodplain over its entire length. This equilibrium is controlled by five factors:

1. Topography (especially slope)
2. Climate
3. Streamflow (including both discharge and velocity)
4. The resistance of rock in the streambed to weathering and erosion
5. Sediment load

A particular combination of these factors—such as steep slope, humid climate, high discharge and velocity, hard rock, and low sediment load—would mean that the stream is eroding a steep valley into bedrock and carrying downstream all sediment derived from that erosion. Downstream, where the slope is lower and the stream flows over easily erodible sediments, it would deposit sandbars and floodplain sediments, building up the elevation of the streambed by sedimentation.

We can describe the **longitudinal profile** of a stream from headwaters to mouth by plotting the elevation of its streambed against distance from its headwaters. **Figure 18.23** plots the slope of the Platte and South Platte rivers from the headwaters of the South Platte in central Colorado to the mouth of the Platte where it enters the Missouri River in Nebraska. The longitudinal profiles of all streams, from small rills to large rivers, form a similar smooth, concave-upward curve, from notably steep near the stream's headwaters to low, almost level, near its mouth.

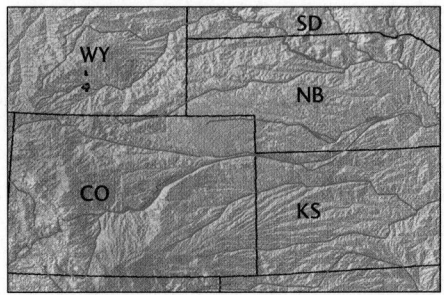

FIGURE 18.23 ▪ (a) A generalized longitudinal profile of a typical stream. (b) The longitudinal profile of the Platte and South Platte rivers from the headwaters of the South Platte in central Colorado to the mouth of the Platte at the Missouri River in Nebraska. [Data from H. Gannett, in *Profiles of Rivers in the United States.* USGS Water Supply Paper 44, 1901.]

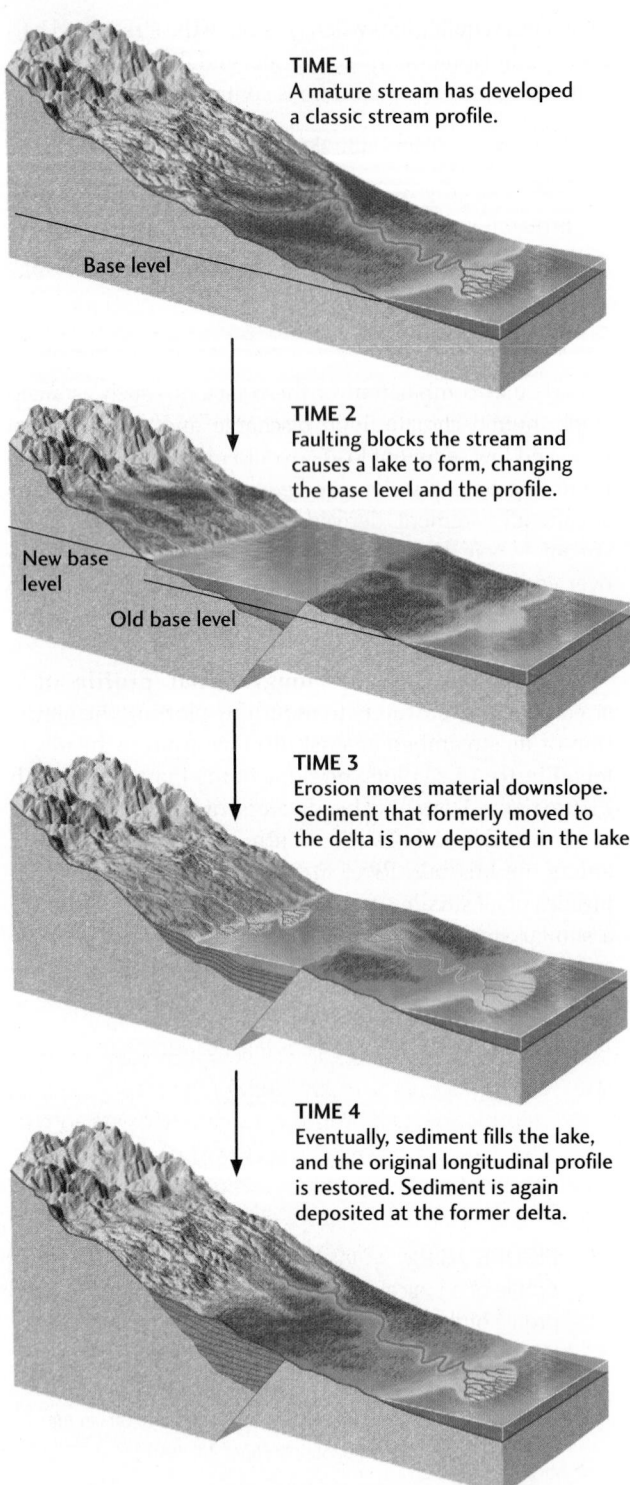

TIME 1
A mature stream has developed a classic stream profile.

Base level

TIME 2
Faulting blocks the stream and causes a lake to form, changing the base level and the profile.

New base level

Old base level

TIME 3
Erosion moves material downslope. Sediment that formerly moved to the delta is now deposited in the lake.

TIME 4
Eventually, sediment fills the lake, and the original longitudinal profile is restored. Sediment is again deposited at the former delta.

FIGURE 18.24 ■ The base level of a stream controls the lower end of its longitudinal profile. If the base level of a stream changes, the longitudinal profile adjusts to the new base level over time.

Why do all streams follow this profile? The answer lies in the combination of factors that control erosion and sedimentation. All streams run downslope from their headwaters to their mouths. Erosion is greater in the higher parts of a stream's course than in the lower parts because

slopes are steeper and flow velocities are higher, and both of these factors have an important influence on the erosion of bedrock (as we'll see in Chapter 22). In a stream's lower courses, where it carries sediments derived from erosion of the upper courses, erosion decreases and sedimentation increases. Differences in topography and the other factors listed above may make the longitudinal profile of an individual stream steeper or shallower, but the general shape remains a concave-upward curve.

BASE LEVELS The longitudinal profile of a stream is controlled at its lower end by the stream's **base level:** the elevation at which it ends by entering a large standing body of water, such as a lake or ocean, or another stream. Streams cannot cut below their base level because the base level is the "bottom of the hill"—the lower limit of the longitudinal profile.

When tectonic processes change the base level of a stream, that change affects its longitudinal profile in predictable ways. If the regional base level rises—perhaps because of faulting—the profile will show the effects of sedimentation as the stream builds up channel and floodplain deposits to reach the new, higher regional base-level elevation (**Figure 18.24**).

Damming a stream artificially can create a new local base level, with similar effects on the longitudinal profile (**Figure 18.25**). The slope of the stream upstream from the dam decreases because the new local base level artificially flattens the stream's profile at the location of the reservoir formed behind the dam. The decrease in slope lowers the stream's velocity, decreasing its ability to transport sediment. The stream deposits some of the sediment on the bed, which makes the concavity somewhat shallower than it was before the dam was built. Below the dam, the stream, now carrying much less sediment, adjusts its profile to the new conditions and typically erodes its channel in the section just below the dam.

This kind of erosion has severely affected sandbars and beaches in Grand Canyon National Park downstream from the Glen Canyon Dam. The erosion threatens animal habitats and archaeological sites as well as beaches used for recreation. River specialists calculated that if discharge during floods were increased by a certain amount, enough sand would be deposited to prevent depletion by erosion. This calculation was confirmed by an experiment in which a controlled flood was staged at the Glen Canyon Dam in 1996. As the gates of the dam were opened, about 38 billion liters of water spilled into the canyon at a rate fast enough to fill Chicago's 100-story Willis (formerly Sears) Tower in 17 minutes. This experiment showed that eroded areas could be restored by sedimentation during floods.

Falling sea level also alters regional base levels and longitudinal profiles. The regional base levels of all streams flowing into the ocean are lowered, and their valleys are cut into former stream deposits. When the drop in sea level is large, as it was during the last ice age, rivers erode steep valleys into coastal plains and continental shelves.

This stream has a typical longitudinal profile.

A dam is built, impounding a reservoir behind it.

The stream dumps sediments in the reservoir, creating a shallower profile.

The higher-velocity water below the dam, robbed of its sediments, erodes the channel, creating a new, steeper profile.

FIGURE 18.25 ■ A change in the base level of a stream caused by human activities, such as the construction of a dam, alters the stream's longitudinal profile.

GRADED STREAMS Over the years, a stream's longitudinal profile becomes stable as the stream gradually fills in low spots and erodes high spots, thereby producing the smooth, concave-upward curve that represents the balance between erosion and sedimentation. That balance is governed not only by the stream's base level, but also by the elevation of its headwaters and by all the other factors controlling the dynamic equilibrium of the stream. At equilibrium, the stream is a **graded stream:** one in which the slope, velocity, and discharge combine to transport its sediment load, with neither net sedimentation nor net erosion in the stream or its floodplain. If the conditions that produce a particular graded stream change, the stream's longitudinal profile will change to reach a new equilibrium. Such changes may include changes in depositional and erosional patterns and alterations in the shape of the channel.

In places where the regional base level is constant over geologic time, the longitudinal profile represents the balance between tectonic uplift and erosion on the one hand and sediment transport and deposition on the other—in other words, the stream is graded. If uplift is dominant, typically in the upper courses of a stream, the longitudinal profile is steep and expresses the dominance of erosion and transport. As uplift slows and the headwater region is eroded, the profile becomes shallower.

EFFECTS OF CLIMATE Climate affects the longitudinal profile of a stream, primarily through the influences of temperature and precipitation on weathering and erosion (see Chapter 16). Warm temperatures and high rainfall promote weathering and erosion of soils and hillslopes and thus enhance sediment transport by streams. High rainfall also leads to greater discharge, which results in more erosion of the streambed. An analysis of sediment transport over the entire United States provides evidence that global climate change over the past 50 years is responsible for a general increase in streamflow. Short-term buildup of sediment or erosion may be the result of climate change, primarily variations in temperature.

ALLUVIAL FANS Plate tectonic processes can force changes in the longitudinal profile of a stream in a number of ways. One place where a stream must adjust suddenly to changed conditions is at a *mountain front,* where a mountain range abruptly meets gently sloping plains. Here, streams leave narrow mountain valleys and enter broad, relatively flat valleys at lower elevations. Along sharply defined mountain fronts, typically at steep fault scarps, streams drop large amounts of sediment in cone- or fan-shaped accumulations called **alluvial fans** (Figure 18.26). This deposition results from the sudden decrease in current velocity that occurs as the channel widens abruptly. To a minor extent, the lowering of the slope below the front also slows the stream. The surface of the alluvial fan typically has a concave-upward shape connecting the steeper mountain profile with the gentler valley profile. Coarse materials, from boulders to sand, dominate on the steep upper slopes of the fan. Lower down, finer sands, silts, and muds are deposited. Fans from many adjacent streams along a mountain front may merge to form a long wedge of sediment whose appearance may mask the outlines of the individual fans that make it up.

TERRACES Tectonic uplift can result in flat, steplike surfaces in a stream valley that line the stream above the floodplain. These **terraces** mark former floodplains that existed at a higher level before regional uplift or an increase in discharge caused the stream to erode into its former floodplain. Terraces are made up of floodplain deposits and are often paired, one on each side of the stream, at the same elevation (Figure 18.27). Terrace formation starts when a stream creates a floodplain. Rapid uplift then changes the stream's equilibrium, causing it to cut down into the

FIGURE 18.26 ■ An alluvial fan (Tucki Wash) in Death Valley, California. Alluvial fans are large cone- or fan-shaped accumulations of sediment deposited when stream velocity slows, as at a mountain front. [Marli Miller.]

floodplain. In time, the stream reestablishes a new equilibrium at a lower level. It may then build another floodplain, which will also undergo uplift and be sculpted into another, lower pair of terraces.

Lakes

Lakes are accidents of a stream's longitudinal profile, as we can see easily where a lake has formed behind a dam (see Figure 18.25). Lakes form wherever the flow of a stream is obstructed. They range in size from ponds only 100 m across to the world's largest and deepest lake, Lake Baikal in southwestern Siberia, which holds approximately 20 percent of the total fresh water in the world's lakes and rivers. It is located in a continental rift zone, a typical plate tectonic

setting for lakes. The damming that takes place in a rift valley results from faulting that blocks a normal exit of water (see Figure 18.24). Streams can flow into a rift valley easily, but cannot flow out until water builds up to a high enough level to allow them to exit. Similarly, a great many lakes formed in the northern United States and Canada when glacial ice and glacial sediments disrupted normal stream drainage. Sooner or later, if plate tectonics and climate remain stable, such lakes will drain away as new outlets form and the longitudinal profile of the streams becomes smoother.

Because lakes are so much smaller than oceans, they are more likely than oceans to be affected by water pollution. Chemical and other industries have polluted Lake Baikal. In North America, Lake Erie has been polluted for many years, although there has been some improvement recently.

FIGURE 18.27 ■ Terraces form when the land surface is uplifted, causing a stream to erode into its floodplain and establish a new floodplain at a lower level. The terraces are remnants of the former floodplain.

Google Earth Project

Water, one of the most prolific weathering and transport agents on Earth, is constantly moving material from one location to another. Google Earth is an ideal tool for interpreting and appreciating this uniquely surficial process. Large rivers such as the Mississippi illustrate how efficiently river systems can gather sediment from mountainous regions of a continent (a source area) and transport it to the ocean, where deltas form (a sink area). What kinds of drainage and channel patterns do you find in the Mississippi drainage basin? How does the slope of the river channel change as one moves downstream? These questions and many more can be explored though the GE interface.

Image © USDA Farm Service Agency Image © 2009 TerraMetrics Data SIO, NOAA, U.S. Navy, GEBCO

This image shows the continental scale of the Mississippi River, from near its point of origin (Ft. Benton) to where it enters the Gulf of Mexico near New Orleans.

LOCATION   Missouri-Mississippi drainage basin, United States

GOAL   Understand source-to-sink transportation of sediment by river systems; observe meandering rivers with point bars, eroded outside banks, and oxbow lakes

LINKED   Figure 18.20

1. Type "Ft. Benton, Montana, United States" into the GE search window. Once you arrive there, zoom out to an eye altitude of 35 km. You will be looking down on the Missouri River, the longest tributary of the Mississippi River. Examine the stretch of river that flows from the southwest to the northeast through town and describe the channel pattern you see.

    a. distributary
    b. braided
    c. meandering
    d. artificially straightened

2. Using the cursor, determine the change in the elevation of the Missouri River channel over the 525 km between Ft. Benton, Montana, and Williston, North Dakota. Now compare that value with the change in the elevation of the Mississippi River channel over the same distance between Memphis, Tennessee, and Baton Rouge, Louisiana, to the south. Which relationship is most accurate?

    a. The slope of the Missouri River is steeper than that of the Mississippi River.
    b. The slope of the Mississippi River is steeper than that of the Missouri River.

c. The slopes of the two rivers are nearly equal.

d. The slopes of the rivers cannot be compared from the information given.

3. From an eye altitude of about 500 km, follow the Mississippi River south from its inception in Lake Itasca, Minnesota. The Mississippi River was used to denote the boundaries along portions of the states of Wisconsin, Iowa, Illinois, Missouri, Kentucky, Arkansas, Tennessee, and Mississippi. Notice that with the "Borders and Labels" layer activated, you can compare the locations of state lines (determined by the original surveyed location of the river channel) with the location of the modern river. How has the river changed over time? (*Hint:* Refer to Figure 18.3. One can also view changing channel patterns with the GE time function.)

a. The river channel has straightened its course in all locations.

b. The river has shortened its path by widening its channel.

c. The river channel has become more sinuous over its entire length.

d. The river has cut off meanders in some places and lengthened them in others.

4. Based on your inspection of the river's characteristics at each of the locations below, which of the following cities seems most vulnerable to seasonal flooding by the Mississippi River? (*Hint:* Look for evidence of levees and at the proximity of the channel to each city.)

a. Cairo, Illinois

b. Biloxi, Mississippi

c. St. Louis, Missouri

d. Memphis, Tennessee

Optional Challenge Question

5. Using the GE search window, navigate to New Orleans, Louisiana. Zoom out to an eye altitude of 310 km to appreciate the close relationship between the city and the Mississippi delta. Now zoom in on the delta itself to observe the deposition of sediment from the Mississippi River channel in the Gulf of Mexico. Why is sediment deposited in this location in particular, and why in such large quantities?

a. Hurricanes originating in the Gulf of Mexico drive sediment from Florida over to Louisiana.

b. The Mississippi River meets the ocean here, so the river current slows down, and this decrease in current velocity causes the sediments in the current to drop out and be deposited.

c. The Army Corps of Engineers dumped all the sediment here when it dredged the Mississippi River.

d. Sediments were shed from the Appalachian Mountains when they were uplifted in the Cretaceous period.

SUMMARY

How do stream valleys and their channels and floodplains develop? As a stream flows, it carves a valley and creates a floodplain on either side of its channel. The valley may have steep to gently sloping walls. The channel may be straight, meandering, or braided. During normal, nonflood periods, the channel carries the flow of water and sediments. During floods, the sediment-laden water overflows the banks of the channel and inundates the floodplain. The velocity of the floodwater decreases as it spreads over the floodplain. The water drops sediments, which build up natural levees and floodplain deposits.

How do drainage networks work as collection systems and deltas as distribution systems for water and sediment? A stream and its tributaries constitute an upstream-branching drainage network that collects water and sediments from a drainage basin. Each drainage basin is separated from other drainage basins by a divide. Drainage networks show various branching patterns—dendritic, rectangular, trellis, or radial. Where a river enters a lake or ocean, it may drop its sediments to form a delta. At the delta, the river tends to branch downstream to form distributaries, which drop the river's sediment load in topset, foreset, and bottomset beds. Deltas are modified or absent where waves, tides, and shoreline currents are strong. Plate tectonic processes influence delta formation by providing uplift in the drainage basin and subsidence in the delta region.

How does flowing water in streams erode solid rock and transport and deposit sediment? Any fluid can move in either laminar or turbulent flow, depending on its velocity, viscosity, and flow geometry. The flows of natural streams are almost always turbulent. These flows are responsible for transporting sediment in suspension, by rolling and sliding along the bed, and by saltation. The settling velocity measures the speed with which suspended particles settle to the streambed. Running water erodes solid rock by abrasion; by chemical weathering; by physical weathering as sand, pebbles, and boulders crash against rock; and by the plucking and undercutting actions of currents. When a current transports sand grains by saltation, cross-bedded dunes and ripples may form on the streambed.

How does a stream's longitudinal profile represent an equilibrium between erosion and sedimentation? A stream is in dynamic equilibrium when erosion balances sedimentation over its entire length. Topography, climate, discharge and velocity, resistance to erosion, and sediment load affect this equilibrium. A stream's longitudinal profile is a plot of the stream's elevation from its headwaters to its base level.

KEY TERMS AND CONCEPTS

abrasion (p. 508)	dendritic drainage (p. 505)	laminar flow (p. 509)	settling velocity (p. 511)
alluvial fan (p. 521)	discharge (p. 516)	longitudinal profile (p. 519)	stream (p. 500)
antecedent stream (p. 505)	distributary (p. 513)	meander (p. 501)	superposed stream (p. 507)
base level (p. 520)	divide (p. 503)	natural levee (p. 503)	suspended load (p. 510)
bed load (p. 510)	drainage basin (p. 503)	oxbow lake (p. 501)	terrace (p. 521)
bottomset bed (p. 513)	drainage network (p. 505)	point bar (p. 501)	topset bed (p. 513)
braided stream (p. 502)	dune (p. 511)	pothole (p. 508)	tributary (p. 505)
capacity (p. 510)	flood (p. 518)	ripple (p. 511)	turbulent flow (p. 509)
channel (p. 500)	floodplain (p. 500)	river (p. 500)	valley (p. 500)
competence (p. 510)	foreset bed (p. 513)	saltation (p. 511)	
delta (p. 513)	graded stream (p. 521)		

PRACTICING GEOLOGY EXERCISE

Can We Paddle Today? Using Streamgauge Data to Plan a Safe and Enjoyable River Trip

How do geologists measure and record the flow of water in streams and rivers? In the United States, the U.S. Geological Survey (USGS) has been tracking water flow information for over a hundred years. The USGS uses *streamgauges* to measure and record the height of the water surface at repeated intervals (hourly, daily, weekly, or longer). In 2007, it operated and maintained more than 7400 streamgauges on rivers and streams across the nation. Geologists use USGS streamgauge data to manage the nation's water resources in a number of ways: to forecast floods and droughts, to manage and operate dams and artificial reservoirs, and to protect water quality, among others.

The ready availability of USGS streamgauge data can also make for safer and more enjoyable outings for anglers, kayakers, canoeists, and rafters. Many USGS-maintained streamgauges transmit near-real-time data through a satellite or telephone network directly to a Web site. Streamgauge data are updated at intervals of 4 hours or less and are available to the public on the Internet at http://water.usgs.gov.

Checking streamgauge data before a river trip can prevent the disappointment and loss of time associated with a long drive to a favorite paddling spot, only to find the water too low or too fast to paddle. On the other hand, paddlers may be willing to travel farther for a river run when they know that flow conditions on their favorite river are optimal. The USGS data allow paddlers to match the conditions of the water to their own abilities.

Streamgauges record the height of the water surface, often known as *stage*. The use of stage alone, however, can be misleading. "Stage" refers to the water surface elevation above a fixed reference point near the streamgauge and may not correspond directly to water depth. Never assume a stage reading is the equivalent of the distance between the water's surface and the streambed. Discharge is a more reliable indicator of the conditions that river enthusiasts will encounter.

As discussed in the chapter text, the discharge of a stream is determined by measuring the cross-sectional area (width × depth) and velocity of the streamflow. The depth of the stream below the fixed reference point and its width at each streamgauge are known, so discharge can be estimated if velocity and stage are known. The discharge values from each measurement can be plotted against the stage recorded at the same time to develop a rating curve for each streamgauge. Paddlers can find the streamgauge reading for their favorite river run, then read the rating curve to get an estimate of the discharge.

A river's discharge typically increases as rainfall or snowmelt in the watershed increases. The way people keep track of this is by measuring the stage. Stage is the height for the river relative to an arbitrary refernce point. As the stage and discharge increase, the river will become increasingly turbulent, and potentially dangerous.

Suppose you consult the USGS Web site and see that the most recent streamgauge reading for the river you want to paddle is 3 feet.

▪ Find 3 feet on the vertical axis of the accompanying graph and read across to the rating curve.

▪ Then read down to find the discharge at a stage of 3 feet: 500 cubic feet per second (CFS).

As the discharge increases at a particular spot, a river will become increasingly challenging and eventually dangerous to recreationists. How high the discharge must become before extra caution is warranted can be known only through experience with the stretch of river that is to be used. Paddlers should consider keeping notes or a logbook on the conditions they encounter at various discharges on a particular stretch of the river to learn the river and to plan future trips.

Streamgauge and discharge data available on the USGS Web site also allow river recreationists to project likely conditions on the river over several days. For example, anglers may be interested in knowing when it will be safe to wade a river as flows decline following a heavy rainstorm or snowmelt. By monitoring near-real-time hydrographs (graphs of discharge over time) on the site, recreationists who are interested in river flows can monitor changing conditions to determine when the water is ideal for their sport and skill level.

BONUS PROBLEM: Try reading the rating curve yourself. What is the discharge that corresponds to a stage of 10 feet? A stage of 25 feet?

A rating curve records the relationship between stage and discharge at a particular streamgauge.

EXERCISES

1. How does velocity determine whether a given flow is laminar or turbulent?

2. How does the size of a sediment grain affect the speed with which it settles to the bottom of a flow?

3. What kind of bedding characterizes a ripple or a dune?

4. How do braided and meandering stream channels differ?

5. Why is a floodplain so named?

6. What is a natural levee, and how is it formed?

7. What is the discharge of a stream, and how does it vary with velocity?

8. How is a stream's longitudinal profile defined?

9. What is the most common kind of drainage network developed over horizontally bedded sedimentary rocks?

10. What is a distributary channel?

THOUGHT QUESTIONS

1. Why might the flow of a very small, shallow stream be laminar in winter and turbulent in summer?

2. Describe and compare the floodplain and valley above and below the waterfall shown in Figure 18.13.

3. You live in a town on a meander bend of a major river. An engineer proposes that your town invest in new, higher artificial levees to prevent the meander from being cut off. Give the arguments, pro and con, for and against this investment.

4. In some places, engineers have artificially straightened a meandering stream. If such a straightened stream were then left free to adjust its course naturally, what changes would you expect?

5. If global warming produces a significant rise in sea level as polar ice caps melt, how will the longitudinal profiles of the world's rivers be affected?

6. In the first few years after a dam was built on it, a stream severely eroded its channel downstream of the dam. Could this erosion have been predicted?

7. Your hometown, built on a river floodplain, experienced a 50-year flood last year. What are the chances that another flood of that magnitude will occur next year?

8. Define the drainage basin where you live in terms of the divides and drainage networks.

9. What kind of drainage network do you think is being established on Mount St. Helens since its violent eruption in 1980?

10. The Delaware Water Gap is a steep, narrow valley cut through a structurally deformed high ridge in the Appalachian Mountains. How could it have formed?

11. A major river, which carries a heavy sediment load, has no delta where it enters the ocean. What conditions might be responsible for the lack of a delta?

MEDIA SUPPORT

 18-1 Animation: Channel Patterns

 18-2 Video: Great Floods on the Mississippi River

 18-2 Animation: Drainage Basins

 18-3 Video: The Mississippi River

 18-1 Video: Geology and Warfare: The Battle of Monte Cassino

- Global Wind
  Patterns   530

- Wind as a
  Transport Agent   531

- Wind as an
  Agent of Erosion 534

- Wind as a
  Depositional
  Agent   535

- The Desert
  Environment   540

Sand dunes in the Namib Desert, in southwestern Africa, are among the tallest in the world. [John Grotzinger.]

WINDS AND DESERTS

19

AT ONE TIME OR ANOTHER, we've all been caught in a wind strong enough to have blown us over, had we not leaned into it or held onto something solid. London, England, which rarely gets strong winds, experienced a major windstorm on January 25, 1990. Winds blowing at more than 175 km/hour ripped roofs off buildings, blew trucks over, and made it virtually impossible to walk on the streets. In deserts, strong winds are much more common, often howling for days on end. Dust storms are frequent, and many winds are strong enough to blow sand grains into the air, creating sandstorms.

Recently, concern over the expansion of Earth's deserts has increased. Conditions in southern Spain, for example, have become so dry that people there are increasingly wondering whether the Sahara has jumped the Mediterranean Sea and is now encroaching on southern Europe. The process of *desertification,* in which nondesert land is transformed into desert, has become a major focus of scientists trying to understand Earth's climate system.

Wind is much like water in its ability to erode, transport, and deposit sediment, and it is capable of moving enormous quantities of sand and dust over large regions of continents and oceans. That is not surprising, because the general laws of fluid motion that govern liquids also govern gases. The much lower density of air makes wind currents less powerful than water currents, even though wind speeds are often much greater than those in currents of water. And there are other differences. Unlike a stream, whose discharge is increased by rainfall, wind transports sediments most effectively in the absence of rain.

In this chapter, we will look at the role of erosion, transportation, and deposition by wind in shaping the surface of the land. We will focus particularly on deserts because so many of the geologic processes that shape those arid environments are related to the work of wind. We will also look at the elements that make up desert landscapes, and how those landscapes are spreading across the globe.

■ Global Wind Patterns

Wind is a natural flow of air that is parallel to the surface of the rotating planet. The ancient Greeks called the god of winds Aeolus, and geologists today use the term **eolian** for geologic processes powered by wind. Although winds obey all the laws of fluid flow that apply to water in streams (see Chapter 18), there are some differences between wind and water flows. Unlike flows of water in stream channels, winds are generally unconfined by solid boundaries, except for the ground surface and the walls of narrow valleys. Air flows are free to spread out in all directions, including upward into the atmosphere.

The winds at any location vary in speed and direction from day to day. Over the long term, however, they tend to come mainly from one direction because Earth's atmosphere flows in global prevailing *wind belts* (**Figure 19.1**). In the temperate latitudes, which are located between 30° and 60° N and 30° and 60° S, the prevailing winds come from the west and so are referred to as the *westerlies*. In the tropics,

which are between 30° S and 30° N of the equator, the *trade winds* (named for an archaic use of the word *trade* to mean "track" or "course") blow from the east.

These prevailing wind belts arise because the Sun warms a given amount of land surface most intensely at the equator, where its rays are almost perpendicular to Earth's surface. The Sun heats the land less intensely at high latitudes and at the poles because there its rays strike Earth's surface at an angle. At the equator, hot air, which is less dense than cold air, rises and flows toward the poles, gradually sinking as it cools. The sinking air reaches ground level in the subtropics, at about 30° S and 30° N, then flows back along Earth's surface toward the equator to form the trade winds. These air movements set up a global pattern of air circulation between the North and South Poles.

This simple circulatory pattern of air flow is complicated by Earth's rotation, which deflects any current of air or water to the right in the Northern Hemisphere and to the left in the Southern Hemisphere. This effect is called the *Coriolis effect,* named after its discoverer. The Coriolis effect on atmospheric circulation deflects warm and cold air flows moving both northward and southward in both

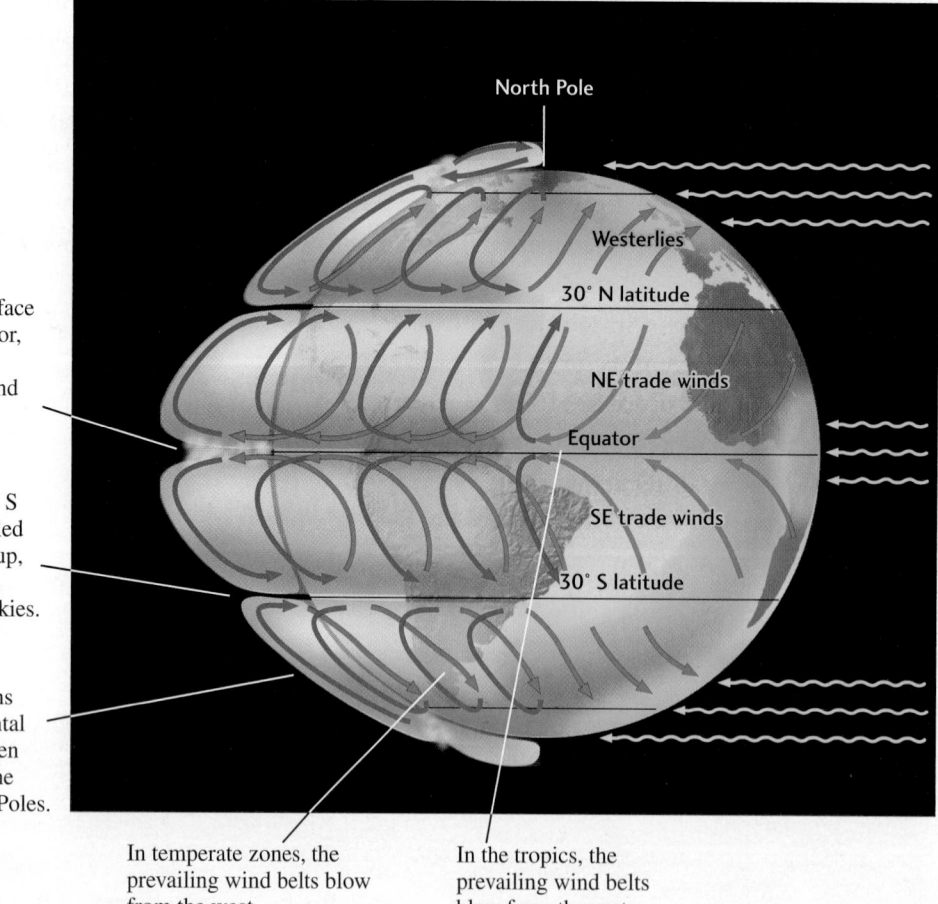

At the poles, the Sun's rays strike the surface at an angle and so are spread out over greater areas, yielding colder temperatures.

At the equator, the Sun's rays are almost perpendicular to the surface, concentrating heat in this region.

There is little surface wind at the equator, and the air rises, forming clouds and rain as it cools.

At 30° N and 30° S latitudes, the cooled air sinks, warms up, absorbs moisture, and yields clear skies.

These two motions set up the horizontal circulation between the equator and the North and South Poles.

North Pole

Westerlies

30° N latitude

NE trade winds

Equator

SE trade winds

30° S latitude

In temperate zones, the prevailing wind belts blow from the west.

In the tropics, the prevailing wind belts blow from the east.

FIGURE 19.1 ■ Earth's atmosphere circulates in prevailing wind belts created by variable solar heating and by Earth's rotation.

hemispheres. For example, as surface winds in the Northern Hemisphere blow southward into the hot equatorial belt, they are deflected to the right (westward) and hence blow from the northeast rather than from the north. These are the northeast trade winds. The Northern Hemisphere westerlies are flows of air that initially moved northward, but are deflected to the right (eastward) and thus blow from the southwest. Near the equator, air movement is mostly upward, so there is little wind at Earth's surface.

As hot air rises at the equator, it cools and releases its moisture, causing the cloudiness and abundant rain of the tropics. This air, now cool and dry, warms and absorbs moisture as it sinks at about 30° N and 30° S latitude. Many of the world's great deserts, such as the Sahara, lie at these latitudes. As the global climate changes, these belts of sinking dry air may also change, expanding and shifting their margins in some places and contracting them in others. In this way, a region bordering a desert—perhaps already suffering from a shortage of rain—may begin to emerge as a persistent desertlike environment. Eventually, the region may become part of the desert.

■ Wind as a Transport Agent

Most North Americans are familiar with rainstorms or snowstorms: high winds accompanied by heavy precipitation. We may have less experience with dry storms, during which high winds blowing for days on end carry enormous amounts of sand and dust. The amount of material the wind can carry depends on the strength of the wind, the sizes of the particles, and the surface materials of the area over which the wind blows.

Wind Strength

Figure 19.2 shows the relative amounts of sand that winds of various speeds can erode from a 1-m-wide strip across a sand dune's surface. A strong wind of 48 km/hour can move half a ton of sand (a volume roughly equivalent to two large suitcases) from this small surface area in a single day. At higher wind speeds, the amounts of sand that can be moved increase rapidly. No wonder entire houses can be buried by a sandstorm lasting several days!

Particle Size

The wind exerts the same kind of force on particles on the land surface as a stream exerts on particles on its bed. Like flows of water in streams, air flows are nearly always turbulent. As we saw in Chapter 18, turbulence depends on three characteristics of a fluid: velocity, flow depth, and viscosity. The extremely low density and viscosity of air make it turbulent even at the velocity of a light breeze. Thus, turbulence

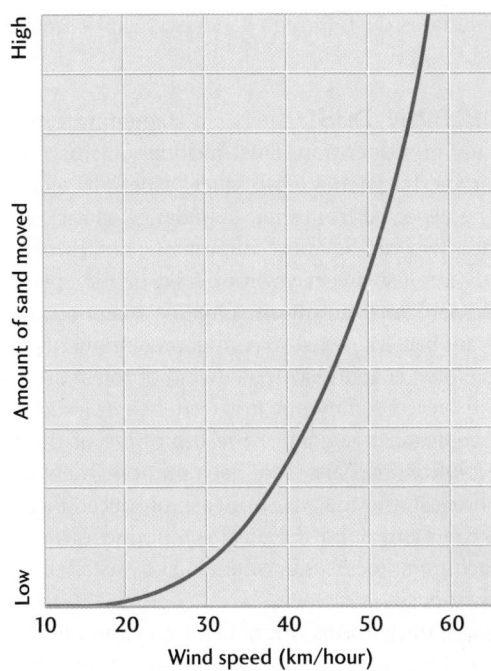

FIGURE 19.2 ■ The amount of sand moved daily across each meter of width of a dune's surface varies with wind speed. High-speed winds blowing for several days can move enormous quantities of sand. [After R. A. Bagnold, *The Physics of Blown Sand and Desert Dunes.* London: Methuen, 1941.]

and forward motion combine to lift particles into the wind and carry them along, at least temporarily.

Even the lightest breezes carry dust, the finest-grained material. **Dust** usually consists of particles less than 0.01 mm in diameter (including silt and clay), but often includes somewhat larger particles. Moderate winds can carry dust to heights of many kilometers, but only strong winds can carry particles larger than 0.06 mm in diameter, such as sand grains. Moderate breezes can roll and slide these grains along a sandy bed, but it takes a stronger wind to lift them into the air current. Wind usually cannot transport particles larger than sand, however, because air has such low viscosity and density. Even though winds can be very strong, only rarely can they move pebbles in the way that rapidly flowing streams do.

Surface Conditions

Wind can lift sand and dust only from dry surface materials, such as dry soil, sediment, or bedrock. It cannot erode and transport wet soils because they are too cohesive. Wind can carry sand grains weathered from a loosely cemented sandstone, but it cannot erode grains from granite or basalt.

Materials Carried by Wind

As air moves, it picks up loose particles and transports them over surprisingly long distances. As we have seen, most of this material is dust, though sand can also be transported by wind.

WINDBLOWN DUST Air has a staggering capacity to hold dust in suspension. Dust includes microscopic rock and mineral fragments of all kinds, especially silicates, as might be expected from their abundance as rock-forming minerals. Two of the most important sources of silicate minerals in dust are clays from soils on dry plains and volcanic ash from eruptions. Organic materials, such as pollen and bacteria, are also common components of dust. Charcoal dust is abundant downwind of forest fires; when found in buried sediments, it is evidence of forest fires in earlier geologic times. Since the beginning of the industrial revolution, humans have been pumping new kinds of synthetic dust into the air—from ash produced by burning coal to the many solid chemical compounds produced by manufacturing processes, incineration of wastes, and motor vehicle exhausts.

In large dust storms, 1 km³ of air may carry as much as a thousand tons of dust, equivalent to the volume of a small house. When such storms cover hundreds of square kilometers, they may carry more than 100 million tons of dust and deposit it in layers several meters thick. (See Earth Issues 19.1 for a discussion of similar dust storms on Mars.) Fine-grained particles from the Sahara have been found as far away as England and have been traced across the Atlantic Ocean to Florida. Wind annually transports about 260 million tons of material, mostly dust, from the Sahara to the Atlantic Ocean. Scientists on oceanographic research vessels have measured airborne dust far out to sea, and today it can be observed directly from space (**Figure 19.3**). Comparison of the composition of this dust with that of deep-sea sediments in the same region indicates that windblown dust is an important contributor to marine sediments, supplying up to a billion tons of material each year. A large part of this dust comes from volcanoes, and there are individual ash beds on the seafloor marking very large eruptions.

Volcanic ash is an abundant component of dust because much of it is very fine grained and is erupted high into the atmosphere, where it can travel farther than nonvolcanic dust blown by winds closer to Earth's surface. Volcanic explosions inject huge quantities of dust into the atmosphere. The volcanic dust from the 1991 eruption of Mount Pinatubo in the Philippines encircled Earth, and most of the finest-grained particles did not settle until 1994 or 1995.

Mineral dust in the atmosphere increases when agriculture, deforestation, erosion, or other land-use changes disrupt soils. A large amount of the mineral dust in the atmosphere today may be coming from the Sahel, a semi-arid region on the southern border of the Sahara where drought and overgrazing are responsible for a heavy load of dust.

Windblown dust has complex effects on the global climate. Mineral dust in the atmosphere scatters incoming visible light from the Sun and absorbs infrared energy radiated outward by Earth's surface. Thus, mineral dust has

FIGURE 19.3 ■ Satellite photograph of a dust storm originating in the Namib Desert in September 2002. Dust and sand are being transported from right (east) to left (west) by strong winds blowing out to sea. These sediments can be transported for hundreds to thousands of kilometers across the ocean. [NASA.]

a net cooling effect in the visible portion of the spectrum and a net warming effect in the infrared portion.

WINDBLOWN SAND The sand transported by wind may consist of almost any kind of mineral grain produced by weathering. Quartz grains are by far the most common because quartz is such an abundant constituent of many surface rocks, especially sandstones. Many windblown quartz grains have a frosted or matte (roughened and dull) surface (**Figure 19.4**) like the inside of a frosted light bulb. Some of the frosting is produced by wind-driven impacts, but most of it results from slow dissolution by dew. Even the tiny amounts of dew found in arid climates are enough to etch microscopic pits and hollows into sand grains, creating the frosted appearance. Frosting is found only in eolian environments, so it is good evidence that a sand grain has been blown by the wind.

Most windblown sands are locally derived. Sand grains are typically buried in dunes after traveling a relatively short distance (usually no more than a few hundred kilometers), mainly by saltation near the ground. The extensive sand

Earth Issues

19.1 Martian Dust Storms and Dust Devils

Of all the planets in the solar system, Mars is the most like Earth. Although Mars has a thinner atmosphere, it has weather that changes seasonally and an Earthlike day of 24 hours and 37 minutes. Mars also has a complex surface environment including ice, soil, and sediment. As we saw in Chapter 9, it is almost certain that water once flowed on the Martian surface.

Today, Mars is cold and dry, and its surface environment is dominated by various eolian processes. These processes have created a variety of wind-sculpted landforms and deposited a wide variety of sediments. Sediments range from very widespread dust layers to more localized dune fields made up of sand and silt. Coarser deposits, composed of basaltic and hematitic granules, have been observed by the Mars Exploration Rovers. Planetary geologists estimate that the winds that produced these eolian deposits blew at velocities of up to 30 m/s (108 km/hour) during great dust storms that covered the entire planet (see Figure 9.19). Although these winds are not as fast as Earth's strongest winds, and although Mars's atmosphere is less dense than Earth's, Martian winds are still strong enough to form an array of eolian depositional and erosional features identical to those observed on Earth. Even the pink color of the Martian atmosphere owes its origin to large quantities of windblown dust lifted into the atmosphere during dust storms.

As the seasons change on Mars, so does the weather. When the Mars Exploration Rover *Spirit* landed in January 2004, it explored the surface of Mars for months on end without seeing any evidence of recent eolian activity. However, in March 2005—more than a year after *Spirit*'s landing—its cameras began to observe dust devils in action, a sign that the seasons were changing. During the windy and dusty season, global dust storms are accompanied by local dust devils, which occur when the Sun heats the planet's surface. Warmed soil and rocks heat the layer of the atmosphere closest to the surface, and the warm air rises in a whirling motion, stirring up the dust from the surface like a miniature tornado.

Martian dust storms and dust devils directly affect our ability to study the Martian surface. The rovers we've sent to Mars depend on solar power. Ultimately, the life span of the rovers is limited by the time it takes for enough windblown dust to settle on their solar panels and terminate power generation. Global dust storms contribute to the deposition of dust on the solar panels. Dust devils, however, are thought to help clean the dust off the solar panels when they move over the rovers, prolonging their useful lives.

View of two dust devils on the floor of Gusev Crater, taken by Spirit from the summit of Husband Hill on August 21, 2005. These dust devils move across the Martian landscape at about 10 to 15 km/hour. [NASA/JPL.]

dunes of such major deserts as the Sahara and the wastes of Saudi Arabia are exceptions. In those great sandy regions, sand grains may have traveled more than 1000 km.

Windblown calcium carbonate grains accumulate where there are abundant fragments of shells and coral, such as in Bermuda and on many coral islands in the Pacific Ocean. The White Sands National Monument in New Mexico is a prominent example of sand dunes made of gypsum sand grains eroded from evaporite deposits formed in nearby playa lakes (whose formation we will describe later in this chapter).

FIGURE 19.4 ▪ Photomicrograph of frosted and rounded sand grains from Oman. [John Grotzinger.]

■ Wind as an Agent of Erosion

By itself, wind can do little to erode large masses of solid rock exposed at Earth's surface. Only when rock is fragmented by chemical and physical weathering can wind pick up the resulting particles. In addition, the particles must be dry, because wet soils and damp fragmented rock are held together by cohesion. Thus, wind erodes most effectively in arid climates, where winds are strong and dry and any moisture quickly evaporates.

Sandblasting

Windblown sand is an effective natural **sandblasting** agent. The common method of cleaning buildings and monuments with compressed air and sand works on exactly the same principle: the high-speed impact of sand particles wears away the solid surface. Natural sandblasting mainly works close to the ground, where most sand grains are carried. Sandblasting rounds and erodes rock outcrops, boulders, and pebbles and frosts the occasional glass bottle.

Ventifacts are wind-faceted pebbles that have several curved or almost flat surfaces that meet at sharp ridges (**Figure 19.5**). Each surface or facet is formed by sandblasting of the pebble's windward side. Occasional storms roll or rotate the pebbles, exposing a new windward side to be sandblasted. Many ventifacts are found in deserts and in glacial gravel deposits, where the necessary combination of gravel, sand, and strong winds is present.

Deflation

As particles of clay, silt, and sand become loose and dry, blowing winds can lift and carry them away, gradually

FIGURE 19.5 ■ These wind-blown ventifacts from the Taylor Valley, Antarctica have been shaped by windblown sand in a frigid environment. [Ronald Sletten.]

eroding the ground surface in a process called **deflation** (**Figure 19.6**). Deflation, which can scoop out shallow depressions or hollows, occurs on dry plains and deserts and on temporarily dry river floodplains and lake beds. Firmly established vegetation—even the sparse vegetation of arid and semiarid regions—retards it. Deflation occurs slowly in areas with plants because their roots bind the soil and their stems and leaves disrupt air flows and shelter the ground surface. Deflation is rapid where the vegetation cover is broken, either naturally by killing drought or artificially by cultivation, construction, or motor vehicle tracks.

When deflation removes the finer-grained particles from a mixture of pebbles, sand, and silt, it produces a remnant surface of pebbles too large for the wind to transport. Over thousands of years, as deflation removes the finer-grained particles, the pebbles accumulate as a layer of **desert pavement**: a coarse, gravelly ground surface that protects the soil or sediments below from further erosion.

FIGURE 19.6 ■ A shallow deflation hollow in the San Luis Valley, Colorado. Wind has scoured the surface and eroded it to a slightly lower elevation. Deflation occurs in dry areas where vegetation cover is absent or broken. [Breck P. Kent.]

(a)

1 Desert pavement formation begins when the wind blows fine-grained materials into heterogeneous soil or sediment.

2 During rainstorms, the fine, windblown sediments infiltrate beneath the coarse layer of pebbles.

3 Microbes living beneath the pebbles produce bubbles that help raise the pebbles and maintain their position at the surface.

Stays consistent thickness

Gets thicker as more windblown dust is supplied over the millennia

4 Over time, these processes lead to thickening of the dust accumulating beneath the pebble layer.

5 A continued supply of windblown dust makes the deposit thicker.

(b)

FIGURE 19.7 ∎ According to a recent hypothesis, desert pavement is formed by the interaction of climate and microorganisms with windblown sediment and soil. [John Grotzinger.]

This theory of desert pavement formation is not completely accepted because a number of pavements seem not to have formed in this way. A new hypothesis proposes that some of them are formed by the deposition of windblown dust. The coarse pebble pavement stays at the surface, while the windblown dust infiltrates below the surface layer of pebbles, is modified by soil-forming processes, and accumulates there (**Figure 19.7**).

from low knolls to huge hills more than 100 m high (see the chapter opening photo). The finer dust falls to the ground as a more or less uniform blanket of silt and clay. By observing these depositional processes working today, geologists have been able to link them with observed characteristics, such as bedding and texture, in sandstones and dust deposits to deduce past climates and wind patterns.

∎ Wind as a Depositional Agent

When the wind dies down, it can no longer transport the sand and dust it carries. The coarser material is deposited in sand dunes of various shapes, ranging in size

Where Sand Dunes Form

Sand dunes occur in relatively few environmental settings. Many North Americans have seen the dunes that form behind ocean beaches or along large lakes. Some dunes are found on the sandy floodplains of large rivers in semiarid and arid regions. Most spectacular are the fields of dunes that cover large expanses of some deserts (**Figure 19.8**).

FIGURE 19.8 ■ Linear dunes in the Qaidam Basin, China. [David Rubin.]

Such dunes may reach heights of 250 m, truly mountains of sand.

Dunes form only in settings that have a ready supply of loose sand: sandy beaches along coasts, sandy river bars or floodplain deposits in river valleys, and sandy bedrock formations in deserts. Another common factor in dune formation is wind power. On oceans and lakes, strong winds blow onshore from the water. Strong winds, sometimes of long duration, are common in deserts.

As we have seen, wind cannot pick up wet materials easily, so most dunes are found in dry climates. The exception is beaches along a coast, where sand is so abundant, and dries so quickly in the wind, that dunes can form even in humid climates. In such climates, soil and vegetation begin to cover the dunes only a little way inland from the beaches, and the winds no longer pick up the sand there.

Dunes may stabilize and become vegetated when the climate grows more humid, then start moving again when an arid climate returns. There is geologic evidence that during droughts two to three centuries ago and earlier, sand dunes in the western high plains of North America were reactivated and migrated over the plains.

How Sand Dunes Form and Move

The wind moves sand by sliding and rolling it along the surface and by *saltation*, the jumping motion that temporarily suspends grains in a current of water or air. Saltation in air flows works the same way as it does in streams (see Figure 18.15), except that the jumps in air flows are higher and longer. Sand grains suspended in an air current may rise to heights of 50 cm over a sand bed and 2 m over a pebbly surface—much higher than grains of the same size can jump in water. The difference arises partly from the fact that air is less viscous than water and therefore does not inhibit the bouncing of the grains as much as water does. In addition, the impact of grains falling in air induces higher jumps as they hit other grains on the surface. These collisions, which the air barely cushions, kick surface grains into the air in a sort of splashing effect. As saltating grains strike a sand bed, they can push forward grains too large to be thrown into the air, causing the bed to creep in the direction of the wind. A sand grain striking the surface at high speed can propel another grain as far as six times its own diameter.

When wind moves sand along a bed, it almost inevitably produces ripples and dunes much like those formed by water (**Figure 19.9**). Ripples in dry sand, like those under water, are *transverse;* that is, at right angles to the current. At low to moderate wind speeds, small ripples form. As the wind speed increases, the ripples become larger. Ripples migrate in the direction of the wind over the backs of larger dunes. Some wind is almost always blowing, so a sand bed is almost always rippled to some extent.

Given enough sand and wind, any obstacle—such as a large rock or a clump of vegetation—can start a dune. Streamlines of wind, like those of water, separate around

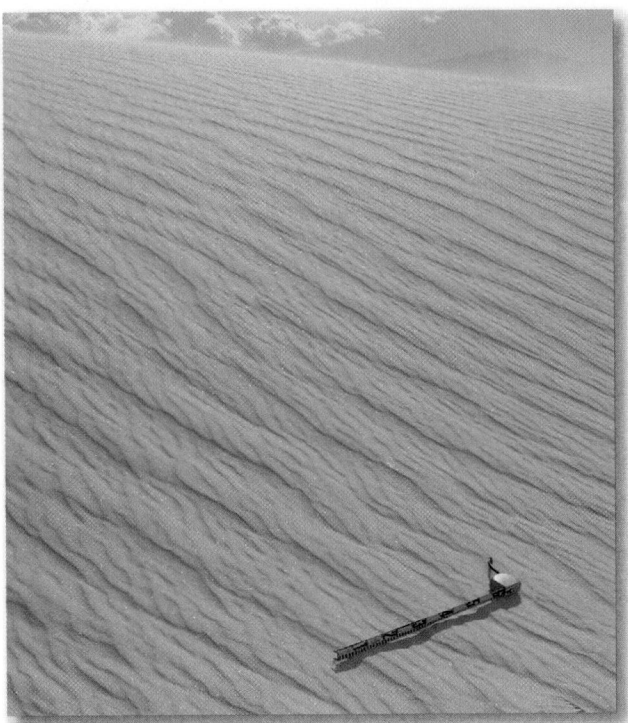

FIGURE 19.9 ■ Wind ripples in White Sands National Monument, New Mexico. Although complex in form, these ripples are always transverse (at right angles) to the wind direction. [John Grotzinger.]

the obstacle and rejoin downwind, creating a wind shadow downstream of the obstacle. Wind velocity is much lower in the wind shadow than in the main flow around the obstacle. In fact, it is low enough to allow sand grains blown into the wind shadow to settle there. The wind there is moving so slowly that it can no longer pick up these grains, and they accumulate as a *sand drift,* a small pile of sand in the lee of the obstacle (**Figure 19.10**). As the process continues, the sand drift itself becomes an obstacle. If there is enough sand, and the wind continues to blow in the same direction long enough, the sand drift grows into a dune. Dunes may also grow by the enlargement of ripples, just as underwater dunes do.

As a dune grows, it starts to migrate downwind through the combined movements of a host of individual grains. Sand grains constantly saltate to the top of the low-angled windward slope and then fall over into the wind shadow on the leeward slope, as shown in **Figure 19.11**. These grains gradually build up a steep, unstable accumulation on the upper part of the leeward slope. Periodically, the accumulation gives way and spontaneously slips or cascades down this **slip face,** as it is called, to a new slope at a lower angle. If we overlook these short-term, unstable steepenings of the slope, the slip face maintains a stable, constant slope angle—its angle of repose. As we saw in Chapter 16, the angle of repose increases with the size and angularity of the particles.

(a)

Early stage: small sand drifts form in wind shadow

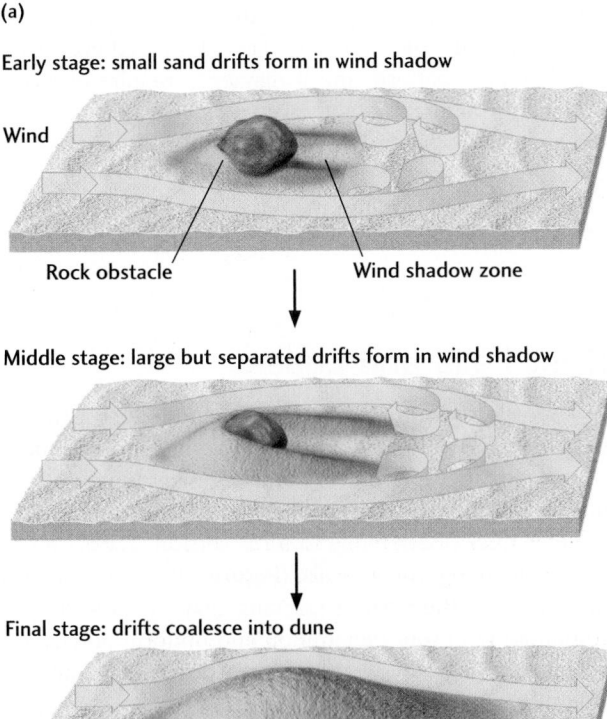

Wind

Rock obstacle   Wind shadow zone

Middle stage: large but separated drifts form in wind shadow

Final stage: drifts coalesce into dune

(b)

FIGURE 19.10 ■ Sand dunes may form in the lee of a rock or other obstacle. (a) By separating the wind streamlines, the obstacle creates a wind shadow in which the eddies are weaker than the main flow. Windborne sand grains are thus able to settle in the wind shadow, where they pile up in drifts that eventually coalesce into a dune. (b) Sand drifts, Owens Lake, California. [(a) After R. A. Bagnold, *The Physics of Blown Sand and Desert Dunes.* London: Methuen, 1941; (b) Marli Miller.]

1 A ripple or dune advances by the movements of individual grains of sand. The whole form moves forward slowly as sand erodes from the windward slope and is deposited on the leeward slope.

2 Particles of sand arriving on the windward slope of the dune move by saltation over the crest,...

3 ...where the wind velocity decreases and the sand deposited slips down the leeward slope.

4 This process acts like a conveyor belt that moves the dune forward.

5 The dune stops growing vertically when it reaches a height at which the wind is so fast that it blows the sand grains off the dune as quickly as they are brought up.

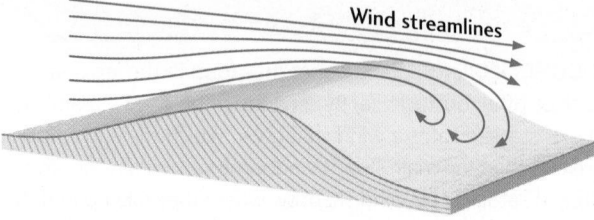

FIGURE 19.11 ◼ Sand dunes grow and move as wind transports sand particles by saltation.

Successive slip face deposits at the angle of repose create the cross-bedding that is the hallmark of windblown dunes (see Figure 5.11). As dunes accumulate, interfere with one another, and become buried in a sedimentary sequence, the cross-bedding is preserved even though the original shapes of the dunes are lost. Sets of sandstone cross-beds many meters thick are evidence of high windblown dunes. From the directions of these eolian cross-beds, geologists can reconstruct wind directions of the past. Cross-bedding preserved on Mars (see Figure 9.28b) provides evidence of ancient windblown dunes there.

As more sand accumulates on the windward slope of a dune than blows off onto the slip face, the dune grows in height. Most dunes are meters to tens of meters in height, but the huge dunes of Saudi Arabia may reach 250 m, which seems to be the limit. The limit on dune height results from the relationship between wind streamline behavior, wind velocity, and topography. Wind streamlines advancing over the back of a dune become more compressed as the dune grows higher (see Figure 19.11). As more air rushes through a smaller space, the wind velocity increases. Ultimately, the air speed at the top of the dune becomes so great that sand grains blow off the top of the dune as quickly as they are brought up the windward slope. When this equilibrium is reached, the height of the dune remains constant.

Dune Types

A person standing in the middle of a large expanse of dunes might be bewildered by the seemingly orderless array of undulating slopes. It takes a practiced eye to see the dominant pattern, and it may even require observation from the air. The general shapes and arrangements of sand dunes depend on the amount of sand available and the direction, duration, and strength of the wind. Geologists recognize four main types of dunes: barchans, blowout dunes, transverse dunes, and longitudinal dunes (**Figure 19.12**).

Dust Falls and Loess

As the velocity of the wind decreases, the dust it carries in suspension settles to form **loess,** a blanket of sediment composed of fine-grained particles. Beds of loess lack internal stratification. In compacted deposits more than a meter thick, loess tends to form vertical cracks and to break off along sheer walls (**Figure 19.13**). Geologists theorize that the vertical cracking may be caused by a combination of root penetration and uniform downward percolation of groundwater, but the exact mechanisms are still unknown.

Loess covers as much as 10 percent of Earth's land surface. The largest loess deposits are found in China and North America. China has more than a million square

Barchans are crescent-shaped dunes, usually but not always found in groups. The horns of the crescent point downwind. Barchans are the products of limited sand supply and unidirectional winds.

Wind

Blowout dunes are almost the reverse of barchans. The slip face of a blowout dune is convex downwind, whereas the barchan's is concave downwind.

Transverse dunes are long ridges oriented at right angles to the wind direction. These dunes form in arid regions where there is abundant sand and vegetation is absent. Typically, sand-dune belts behind beaches are transverse dunes formed by strong onshore winds.

Longitudinal dunes are long ridges of sand whose orientation is parallel to the wind direction. These dunes may reach heights of 100 m and extend many kilometers. Most areas covered by longitudinal dunes have a moderate sand supply, a rough pavement, and winds that are always in the same general direction.

FIGURE 19.12 ■ The general shapes and arrangements of sand dunes depend on the amount of sand available and the direction, duration, and velocity of the wind.

kilometers of loess deposits (**Figure 19.14**). Its greatest deposits extend over wide areas in the northwest; most are 30 to 100 m thick, although some exceed 300 m. The winds blowing over the Gobi Desert and the arid regions of central Asia provided the dust, which still blows over eastern Asia and the Chinese interior. Some of the loess deposits in China are over 2 million years old. They formed after an increase in the elevation of the Himalaya and related mountain belts in western China introduced rain shadows and dry climates to the continental interior. The uplift of these mountain belts was responsible for the cold, dry climates of the Pleistocene epoch in much of Asia. These climates

FIGURE 19.13 ■ Stacked layers of loess in Elba, Nebraska.
[Daniel R. Muhs, U.S. Geological Survey.]

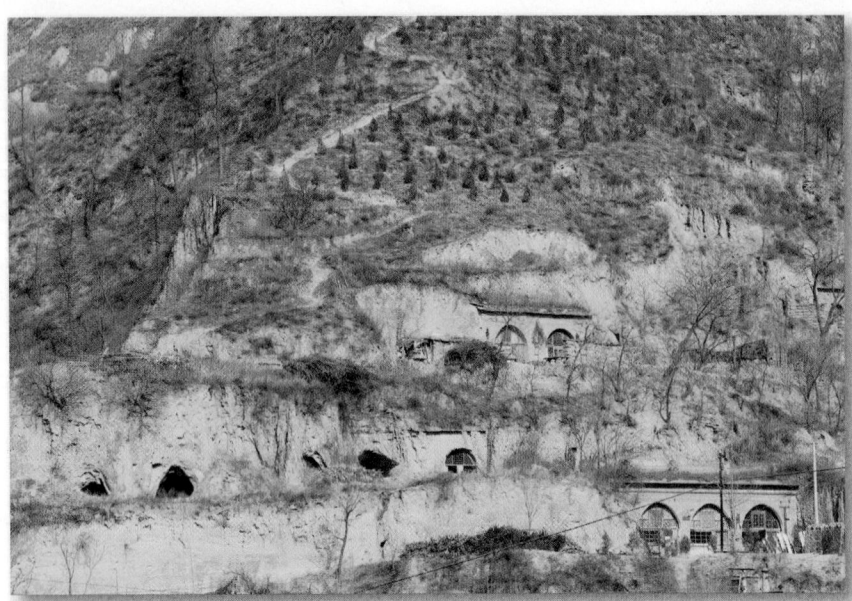

FIGURE 19.14 ■ Ancient cave houses dug into loess deposits in Shanxi province, northern China. [© Ashley Cooper/Age Fotostock.]

inhibited vegetation and dried out soils, causing extensive wind erosion and transportation.

The best-known loess deposit in North America is in the upper Mississippi River valley. It originated as silt and clay deposited on the extensive floodplains of streams draining the edges of melting glaciers in the Pleistocene epoch. Strong winds dried the floodplains, whose frigid climate and rapid rates of sedimentation inhibited vegetation, and blew up tremendous amounts of dust, which then settled to the east. Geologists recognize that this loess deposit is distributed as a blanket of more or less uniform thickness on both hills and valleys, all in or near formerly glaciated areas. Changes in the regional thickness of the loess in relation to the prevailing westerly winds confirm its eolian origin. Its thickness on the eastern sides of major river floodplains is 8 to 30 m, greater than on the western sides, and decreases rapidly downwind to 1 to 2 m farther east of the floodplains.

Soils formed on loess are fertile and highly productive. Their cultivation poses environmental problems, however, because they are easily eroded into gullies by small streams and deflated by wind when they are poorly managed.

■ The Desert Environment

Of all Earth's environments, the desert is where wind is best able to do its work of erosion, transportation, and deposition. The deserts of the world are among the most hostile environments for humans. Yet many of us are fascinated by these hot, dry, apparently lifeless zones, full of bare rocks and sand dunes. The dry climate of deserts creates harsh yet fragile conditions, where human impacts last for decades.

All told, arid regions amount to one-fifth of Earth's land area, about 27.5 million square kilometers. Semiarid plains account for an additional one-seventh. Given the reasons for the existence of large areas of desert in the modern world—the effects of Earth's wind belts on climates, mountain building, and continental drift—we can be confident that, according to the principle of uniformitarianism, extensive deserts have existed throughout geologic time. Conversely, today's deserts may have been wet regions in the past, but may have dried out as a result of long-term climate change.

Where Deserts Are Found

The locations of the world's great deserts are determined by rainfall, which in turn is determined by a number of factors (Figure 19.15). The Sahara and Kalahari deserts of Africa and the Great Australian Desert get extremely low amounts of rainfall, normally less than 25 mm/year and in some places less than 5 mm/year. These subtropical deserts are found at about 30° N and 30° S, where prevailing wind patterns cause dry air to sink to ground level (see Figure 19.1). Because the relative humidity is extremely low in these zones of sinking air, clouds are rare, and the chance of precipitation is very small. The Sun beats down week after week.

Deserts also exist at temperate latitudes—between 30° N and 50° N and between 30° S and 50° S—in regions where rainfall is low because moisture-laden winds either are blocked by mountain ranges or must travel great distances from the ocean, their source of moisture. The Great Basin and Mojave deserts of the western United States, for example, lie in rain shadows created by the western coastal mountains. The Gobi and other deserts of central Asia are so deep in the continental interior that the winds reaching them have precipitated all their ocean-derived moisture long before they arrive there.

FIGURE 19.15 ■ Major desert areas of the world (exclusive of polar deserts). Notice the relationships of their locations to prevailing wind belts and major mountain ranges. Notice, too, that sand dunes make up only a small proportion of the total desert area. [After K. W. Glennie, *Desert Sedimentary Environments*. New York: Elsevier, 1970.]

Another kind of desert is found in polar regions. There is little precipitation in these cold, dry areas because the frigid air can hold little moisture. The dry valley region of southern Victoria Land in Antarctica is so dry and cold that its environment resembles that of Mars.

THE ROLE OF PLATE TECTONICS In a sense, deserts are a result of plate tectonic processes. The mountains that create rain shadows are raised at convergent plate boundaries. The great distance separating central Asia from the oceans is a consequence of the size of the Asian continent, a huge landmass assembled from smaller landmasses by continental drift. Large deserts are found at low latitudes because continental drift moved continents there from higher latitudes. If, in some future plate tectonic scenario, the North American continent were to move south by 2000 km or so, the northern Great Plains of the United States and Canada would become a hot, dry desert. Something like that happened to Australia. About 20 million years ago, Australia was far to the south of its present position, and its interior had a warm, humid climate. Since then, Australia has moved northward into an arid subtropical zone, and its interior has become a desert.

THE ROLE OF CLIMATE CHANGE Changes in a region's climate may transform semiarid lands into deserts, a process called **desertification.** Climate changes that we do

not fully understand may decrease precipitation for decades or even centuries. After such a dry period, a region may return to a milder, wetter climate. Over the past 10,000 years, the climate of the Sahara appears to have oscillated between drier and wetter conditions. We have evidence from orbiting satellites that an extensive system of river channels existed there a few thousand years ago (**Figure 19.16**). Now dry and buried by more recent sand deposits, these ancient drainage systems carried abundant running water across the northern Sahara during wetter periods.

The Sahara may now be expanding northward (see the Practicing Geology exercise). The Desert Watch project, led by the European Space Agency, reports that over 300,000 km² of Europe's Mediterranean coast—an area almost as large as the state of New York, with a population of 16 million—has been enduring the longest drought in recorded history. During 2005 and 2012, fires raged along the southern Spanish coast, and temperatures set new record highs for weeks on end. Were these merely long, hot summers, or are these the initial symptoms of desertification, made worse by overpopulation and overdevelopment within the fragile ecosystems of dry landscapes?

Evidence to support the latter scenario is building. Soils have been loosened by the prolonged dryness, making them more susceptible to wind transportation and deflation. Groundwater levels have reached new lows. And there is little question that Europe is getting warmer:

(a)

(b)

FIGURE 19.16 ■ The climate of the Sahara was not always as arid as it is today. (a) Remote sensing techniques that look only at Earth's surface see nothing but sand in the Sahara. (b) Remote sensing techniques that penetrate a few meters below the surface, however, see a dense network of buried riverbeds. [NASA/JPL Imaging Radar Team.]

during the twentieth century, its average temperatures increased about 0.7°C. The 1990s marked the hottest decade since record keeping began in the mid-1800s, registering two of the five hottest years ever recorded.

THE ROLE OF HUMAN ACTIVITIES Climate oscillations occur naturally in the Sahara and in other deserts, but human activities are responsible for some of the desertification occurring today. The growth of human populations in semiarid regions, along with increased agriculture and animal grazing, may result in the expansion of deserts. When population growth and periods of drought coincide, the results can be disastrous. In Spain, the greatest urban and agricultural expansion is taking place on the Mediterranean coast—the nation's driest region. Former farmlands have been stripped of vegetation due to overfarming (up to four crops per year), which depletes water and strips soils. A tourism boom and its accompanying development are literally paving over the dry lands and desiccating the countryside that is left. In 2004, more than 350,000 new homes were built on the Mediterranean coast, many with backyard swimming pools and nearby golf courses requiring large amounts of water. In isolation, any one of these human activities might not have a negative effect. Together, however, they add up to desertification.

"Making the desert bloom," the opposite of desertification, has been a slogan of some countries with desert lands.

They irrigate on a massive scale to convert semiarid or arid areas into productive farmlands. The Central Valley of California, where many of North America's fruits and vegetables are grown, is one example. If the waters used for irrigation contain dissolved substances (as almost all natural waters do), then, with time, these waters evaporate and deposit the dissolved substances as salts. Thus, ironically, irrigation in an arid or semiarid climate can eventually cause desertification through the slow accumulation of salts.

Desert Weathering and Erosion

As unique as deserts are, the same geologic processes operate there as elsewhere. Physical and chemical weathering work the same way in deserts as they do everywhere, but the balance between the two processes is different: in deserts, physical weathering predominates over chemical weathering. Chemical weathering of feldspars and other silicates into clay minerals proceeds slowly because the water required for those reactions is scarce. The little clay that does form is usually blown away by strong winds before it can accumulate. Slow chemical weathering and rapid wind transportation combine to prevent the buildup of any significant thickness of soil, even where sparse vegetation binds some of the weathered particles. Thus, desert soils are thin and patchy. Sand, gravel, rock fragments of many sizes, and bare bedrock are characteristic of much of the desert surface.

FIGURE 19.17 ■ Petroglyphs scratched in desert varnish by Native Americans at Newspaper Rock, Canyonlands, Utah. The scratches are several hundred years old, but appear fresh on the varnish, which has accumulated over thousands of years. [Peter Kresan.]

THE COLORS OF THE DESERT The rusty, orange-brown colors of many weathered surfaces in the desert come from the ferric iron oxide minerals hematite and limonite. These minerals are produced by the slow chemical weathering of iron silicate minerals such as pyroxene. Even when present in only small amounts, they stain the surfaces of sands, gravels, and clays.

Desert varnish is a distinctive dark brown, sometimes shiny, coating found on many rock surfaces in the desert. It is a mixture of clay minerals with smaller amounts of manganese and iron oxides. Desert varnish probably forms when dew causes chemical weathering of primary minerals on an exposed rock surface to form clay minerals and iron and manganese oxides. In addition, tiny quantities of windblown dust may adhere to the rock surface. The process is so slow that Native American inscriptions scratched in desert varnish hundreds of years ago still appear fresh, with a stark contrast between the dark varnish and the light unweathered rock beneath (**Figure 19.17**). Desert varnish requires thousands of years to form, and some particularly ancient varnishes in North America are of Miocene age. However, recognizing desert varnish as such on ancient sandstones is difficult.

STREAMS: THE PRIMARY AGENTS OF EROSION
Wind plays a larger role in erosion in deserts than it does elsewhere, but it cannot compete with the erosive power of streams. Even though it rains so seldom that most streams flow only intermittently, streams do most of the erosional work in the desert when they do flow.

Even the driest desert gets occasional rain. In sandy and gravelly areas of deserts, rainfall infiltrates soil and permeable bedrock and temporarily replenishes groundwater in the unsaturated zone. There, some of it evaporates very slowly into pore spaces between particles. A smaller amount eventually reaches the groundwater table far below—in some places, as much as hundreds of meters below the surface. Desert oases form where the groundwater table comes close enough to the surface that the roots of palms and other plants can reach it.

When rain occurs in heavy cloudbursts, so much water falls in such a short time that infiltration cannot keep pace, and the bulk of the water runs off into streams. Unhindered by vegetation, the runoff is rapid and may cause flash floods along valley floors that have been dry for years. Thus, a large proportion of streamflows in deserts consist of floods (**Figure 19.18a**). When floods occur in deserts, they have great erosive power because little of the loose sediment is held in place by vegetation. Streams may become so choked with sediment that they look more like fast-moving mudflows. The abrasiveness of this sediment load moving at flood velocities makes desert streams efficient eroders of bedrock valleys.

Desert Sediments and Sedimentation

Deserts are composed of a diverse set of sedimentary environments. These environments may change dramatically when rain suddenly forms raging rivers and widespread lakes. Prolonged dry periods intervene, during which sediments are blown into sand dunes.

ALLUVIAL SEDIMENTS As sediment-laden flash floods dry up, they leave distinctive alluvial deposits on the floors of desert valleys. In many cases, a flat fill of coarse sediment covers the entire valley floor, and the ordinary differentiation of the stream into channel, natural levees, and floodplain is absent (Figure 19.18b). The sediments of many other desert valleys clearly show the intermixing of stream-deposited channel and floodplain sediments with

(a)

(b)

FIGURE 19.18 ■ A large proportion of the streamflows in deserts occur as floods. (a) A desert valley during a summer thunderstorm at Saguaro National Park, Arizona. (b) The same valley a day after the storm. The coarse sediment deposited by such sudden desert floods may cover the entire valley floor. [Peter Kresan.]

eolian sediments. This combination of alluvial and eolian processes in the past formed extensive layers of eolian sandstones separated by channel sediments and ancient floodplain sandstones.

Large alluvial fans are prominent features at mountain fronts in deserts because desert streams deposit much of their sediment load on the fans (see Figure 18.26). The rapid infiltration of stream water into the permeable sediments that make up the fan deprives the stream of the water required to carry the sediment load any farther downstream. Debris flows and mudflows make up large parts of the alluvial fans of arid, mountainous regions.

EOLIAN SEDIMENTS By far the most dramatic sedimentary accumulations in deserts are the sand dunes we have described above. Dune fields range in size from a few square kilometers to the "seas of sand" found on the

FIGURE 19.19 ■ A desert playa lake in Death Valley, California. [Robert Harding Picture Library/Superstock.]

Arabian Peninsula (see Figure 19.8). These sand seas—or *ergs*—may cover as much as 500,000 km², twice the area of the state of Nevada.

Although film and television portrayals might lead one to think that deserts are mostly sand, only one-fifth of the world's desert area is actually covered by sand (see Figure 19.15). The other four-fifths are rocky or covered with desert pavement. Sand covers only a little more than one-tenth of the Sahara, and sand dunes are even less common in the deserts of the southwestern United States.

EVAPORITE SEDIMENTS **Playa lakes** are permanent or temporary lakes that form in arid mountain valleys or basins where water is trapped after rainstorms (**Figure 19.19**). Desert streams carry large amounts of dissolved minerals, and those minerals accumulate in playa lakes. As the lake water evaporates, the minerals are concentrated and gradually precipitated. Playa lakes are sources of evaporite minerals such as sodium carbonate, borax (sodium borate), and other unusual salts. If evaporation is complete, the lakes become **playas,** flat beds of clay that are sometimes encrusted with precipitated salts.

Desert Landscapes

Desert landscapes are some of the most varied on Earth. Large low, flat areas are covered by playas, desert pavements, and dune fields. Uplands are rocky, cut in many places by steep stream valleys and gorges. The lack of vegetation and soil makes everything seem sharper and harsher than it would in a landscape in a more humid climate. In contrast to the rounded, soil-covered, vegetated slopes found in most humid regions, the coarse fragments of varying size produced by desert weathering form steep cliffs with masses of angular talus at their bases (**Figure 19.20**).

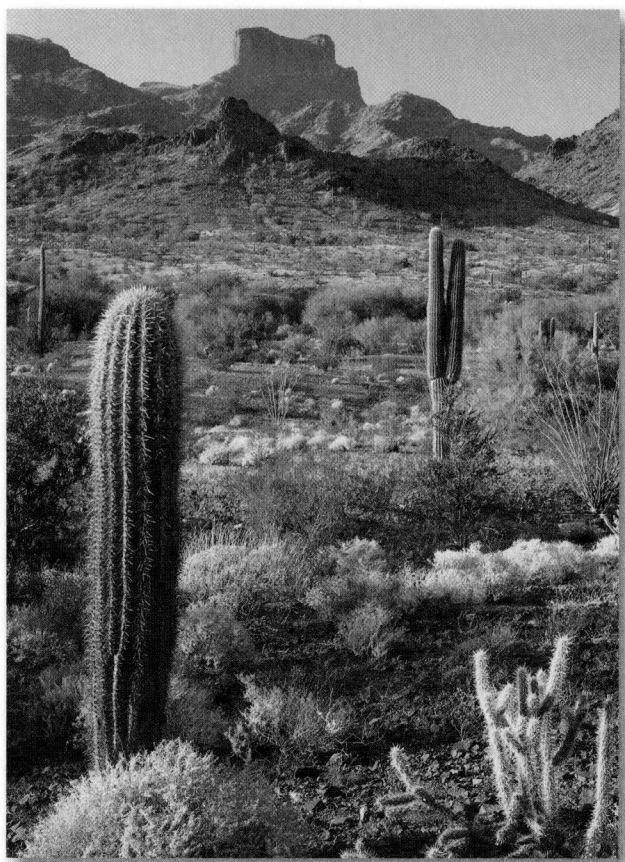

FIGURE 19.20 ■ This desert landscape at Kofa Butte, Kofa National Wildlife Refuge, Arizona, shows the steep cliffs and masses of talus produced by desert weathering. [Peter Kresan.]

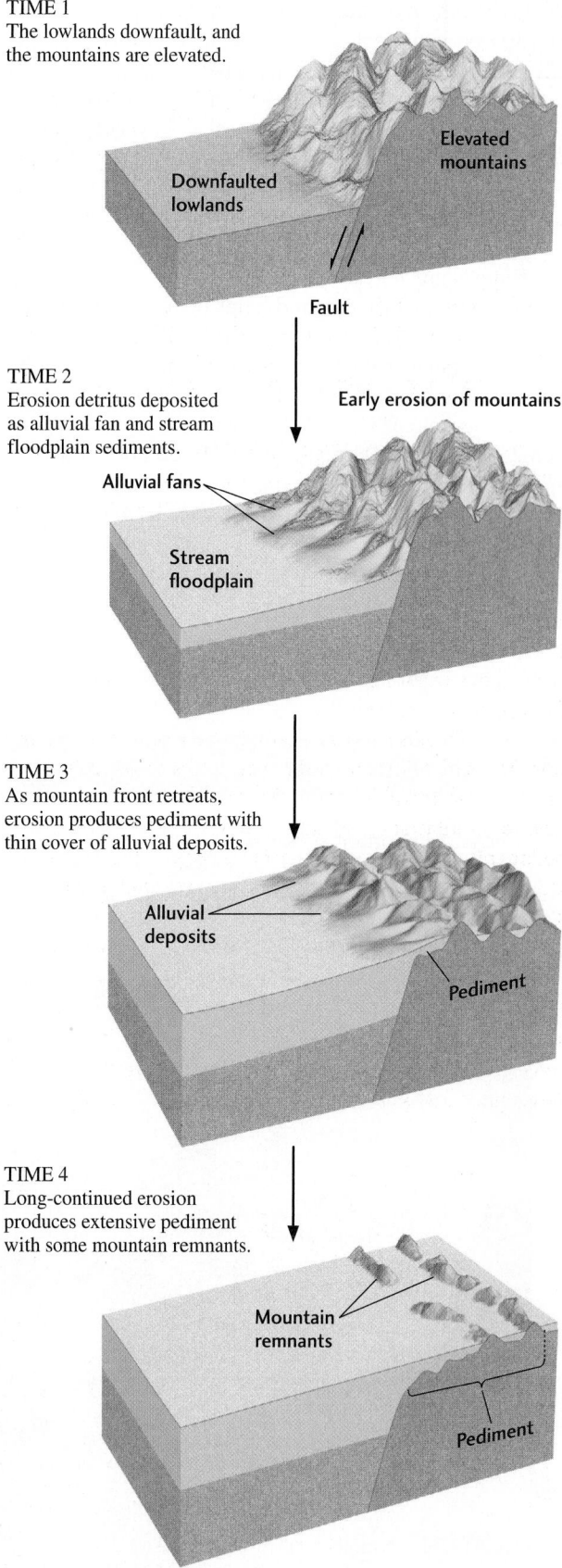

TIME 1
The lowlands downfault, and the mountains are elevated.

Downfaulted lowlands

Elevated mountains

Fault

TIME 2
Erosion detritus deposited as alluvial fan and stream floodplain sediments.

Early erosion of mountains

Alluvial fans

Stream floodplain

TIME 3
As mountain front retreats, erosion produces pediment with thin cover of alluvial deposits.

Alluvial deposits

Pediment

TIME 4
Long-continued erosion produces extensive pediment with some mountain remnants.

Mountain remnants

Pediment

FIGURE 19.21 ■ Pediments form as mountain fronts erode and retreat.

Much of the landscape of deserts is shaped by streams, but their valleys—called **dry washes** in the western United States and **wadis** in the Middle East—are dry most of the time. Stream valleys in deserts have the same range of profiles as valleys elsewhere. Far more of them, however, have steep walls because of the rapid erosion caused by stream flooding, combined with the lack of rainfall that might soften the slopes of the valley walls between flood events.

Desert streams are widely spaced because of the relatively infrequent rainfall. Drainage patterns in deserts are generally similar to those in other terrains, with one important difference: many desert streams die out before they can reach across the desert to join larger rivers flowing to the oceans. Most terminate at the base of alluvial fans. Damming by dunes or confinement within closed valleys with no outlet may lead to the development of playa lakes.

A special type of eroded bedrock surface, called a **pediment,** is a characteristic landform of the desert. Pediments are broad, gently sloping platforms of bedrock left behind as a mountain front erodes and retreats from its valley (Figure 19.21). The pediment forms like an apron around the base of the mountains as thin alluvial deposits of sand and gravel accumulate. Long-continued erosion eventually forms an extensive pediment below a few mountain remnants (**Figure 19.22**). A cross section of a typical pediment and its mountains would reveal a fairly steep mountain slope abruptly leveling into the gentle pediment slope. Alluvial fans deposited at the lower edge of the pediment merge with the sedimentary fill of the valley below the pediment.

There is much evidence that pediments are formed by running water, which cuts and forms the pediment surface as well as transporting and depositing sediments to create an apron of alluvial fans. At the same time, the mountain slopes at the head of the pediment maintain their steepness as they retreat, instead of becoming the rounded, gentler slopes found in humid regions. We do not know how specific rock types and erosional processes interact in an arid environment to keep the slopes steep as the pediment is enlarged.

FIGURE 19.22 ▪ Cima Dome is a pediment in the Mojave Desert. The surface of the dome is covered by a thin veneer of alluvial sediments. The two knobs, on the left and right sides of the dome, are regarded as the final remnants of the former mountain. [Marli Miller.]

Google Earth Project

Winds are important agents of landscape development in deserts. Winds can remove sediment from one portion of a landscape, such as a beach, and transport it to another location, where it may accumulate as dunes. One of the best places on Earth to see large-scale sand dunes on the move is in the Namib Desert in the country of Namibia, located in southwestern Africa.

Data SIO, NOAA, U.S. Navy, GEBCO © 2009 Cnes/Spot Image, © 2009 DigitalGlobe

The Namib Desert is located in southwestern Africa and contains well-defined longitudinal dunes oriented in a north-south direction. The desert is bounded by uplifted bedrock to the east and the Atlantic Ocean to the west. Note the location of places discussed in the exercise.

LOCATION   Namibia, southwestern Africa

GOAL   Observe sand seas and a variety of wind-derived landforms

LINKED   Chapter opening photo, Figure 19.3, and Figure 19.8

SUMMARY

How do prevailing winds form and where do they flow? Earth is encircled by belts of prevailing winds that develop because the Sun warms Earth most intensely at the equator, causing air to rise there and flow toward the poles. As the air moves toward the poles, it gradually cools and begins to sink. This cool, dense air then flows back along Earth's surface to the equator. The Coriolis effect, produced by Earth's rotation, deflects these prevailing winds to the right in the Northern Hemisphere and to the left in the Southern Hemisphere.

How do winds transport and erode sand and finer-grained sediments? Winds can pick up and transport dry particles in a manner similar to flowing water. Air flows are limited, however, in the size of particles they can carry (rarely larger than sand grains) and in their ability to keep particles in suspension. These limitations result from air's low viscosity and density. Windblown materials include volcanic ash, quartz grains, and other mineral fragments such as clay, as well as organic materials such as pollen and bacteria. Wind can carry great amounts of sand and dust. It moves sand grains primarily by saltation and carries finer-grained dust particles in suspension. Sandblasting and deflation are the primary ways in which winds erode Earth's surface.

How do winds deposit sand dunes and dust? When winds die down, they deposit sand in dunes of various shapes and sizes. Dunes form in sandy desert regions, behind beaches, and along sandy floodplains, all of which are places with a ready supply of loose sand and moderate to strong winds. Dunes start as sand drifts in the lee of obstacles and may grow to heights of up to 250 meters, though most are tens of meters in height. Dunes migrate downwind as sand grains saltate up their gentler windward slopes and fall over onto their steeper downwind slip faces. The shapes and arrangements of sand dunes are determined by the direction, duration, and strength of the wind and by the abundance of sand. As the velocity of dust-laden winds decreases, the dust settles to form loess, a thick blanket of fine particles. Loess layers have been deposited in many formerly glaciated areas by winds blowing

1. Type "Namib Desert, Africa" into the GE search window and zoom out to an eye altitude of 1000 km once you arrive there. At this altitude, the Namib Desert dune field is marked by a light brown patch along the Atlantic Ocean. Using the path measurement tool, measure the perimeter of this sand sea. What is your result?

    a. 300 km
    b. 600 km
    c. 1000 km
    d. 800 km

2. If you zoom in to an eye altitude of 55 km at 24°12′00″ S, 15°07′00″ E, what type of sand dunes do you see?

    a. Blowout dunes
    b. Longitudinal dunes
    c. Transverse dunes
    d. Barchans

3. Moving 40 km directly north of your previous location, you will notice that the sand sea ends abruptly at the Kuiseb River. Zoom in on the river to see its channel and the vegetation fringing it. No sand dunes occur north of this river. Why?

    a. Sand grains are produced only south of the river.
    b. The wind blows from north to south and transports sediment from the river into the sand sea.
    c. The wind blows from south to north and transports sediment into the river, which then transports it to the Atlantic Ocean.

    d. Sand moving to the north is dissolved by vegetation along the riverbank.

4. Now zoom back out to an eye altitude of 55 km and move to 24°44′00″ S, 15°20′10″ E. Note the patch of white material, which marks the position of a playa lake called Sossusvlei. How did these playa lake sediments accumulate?

    a. Winds blowing from west to east transported salts from the Atlantic Ocean to Sossusvlei.
    b. Flash floods originating in the mountainous regions to the east transported dissolved minerals to Sossusvlei, where the waters ponded against the dunes and evaporated.
    c. Salty groundwater moved uphill from the Atlantic Ocean to Sossusvlei, where it emerged and evaporated, leaving salts behind.
    d. The white material is dust that was transported by wind from the interior of the African continent.

Optional Challenge Question

5. Considering your answers to the previous questions and your personal investigation into the details of the area, where do you think all this sand is coming from?

    a. Beach sand carried northward by winds from the area near the border between Namibia and South Africa
    b. Flood deposits by inland lahars flowing off actively erupting volcanoes
    c. Sand washed ashore by a tsunami
    d. Sand shaken from nearby mountaintops as a result of intense earthquake-induced ground motion

over the floodplains of streams formed by glacial meltwater. Loess can accumulate to great thicknesses downwind of dusty desert regions.

How do wind and water combine to shape the desert environment and its landscape? Deserts occur in subtropical zones of sinking air, in the rain shadows of mountain ranges, and in the interiors of some continents. In all these places, the air is dry, and rainfall is rare. In deserts, physical weathering is predominant, whereas chemical weathering is minimal because of the lack of water. Most desert soils are thin, and bare rock surfaces are common. Wind plays a larger role in shaping the landscape in deserts than it does elsewhere, but streams are responsible for most erosion in deserts even though they flow only intermittently. Playa lakes, which form in arid mountain valleys or basins, deposit evaporite minerals as they dry up. Among the prominent features of desert landscapes are pediments, which are broad, gently sloping platforms eroded from bedrock as mountains retreat while maintaining the steepness of their slopes.

KEY TERMS AND CONCEPTS

deflation (p. 534)	dry wash (p. 545)	pediment (p. 545)	slip face (p. 537)
desert pavement (p. 534)	dust (p. 531)	playa (p. 544)	ventifact (p. 534)
desert varnish (p. 543)	eolian (p. 530)	playa lake (p. 544)	wadi (p. 545)
desertification (p. 541)	loess (p. 538)	sandblasting (p. 534)	

PRACTICING GEOLOGY EXERCISE

Can We Predict the Extent of Desertification?

In regions of the world with arid to semiarid climates, farmlands and grazing lands are being lost to desertification at an alarming rate. Two questions are important to land managers hoping to prevent further degradation of environmentally sensitive dry lands: First, what processes are leading to degradation and desertification? Second, how widespread is this desertification likely to be?

Map of northern Africa, showing deserts, areas considered to be susceptible to desertification, and areas farther from deserts that are regarded as environmentally stable.

Desertification occurs whenever a nondesert area starts to exhibit the characteristics of a true desert. The term was coined by the United Nations in 1977 to describe changes that were particularly evident in northern Africa at that time. Over the past 50 years, a semiarid area the size of Texas at the southern edge of the Sahara, called the Sahel, has started to become desert. The same fate now threatens more than one-third of the African continent. Desertification is most pronounced in northern Africa, but affects every continent except Antarctica. North Americans should be aware that the areas that fringe the deserts of the southwestern United States may also be susceptible to desertification if not managed properly.

The main cause of desertification is not drought, but mismanagement of land, including overgrazing, overly intensive cultivation, and felling of trees and brushwood for fuel. The processes that lead to desertification include erosion of soil by both water and wind, long-term reductions in the amount or diversity of natural vegetation, and, on irrigated farmlands, the accumulation of salts in soils by evaporation of groundwater used for irrigation.

The accompanying figure is a regional map of northern Africa showing the current extent of the Sahara. It also shows the regions adjacent to the desert that currently support agriculture, but which are highly susceptible to desertification. The grid superimposed on the map subdivides it into squares, each of which represents 100 km². We can use this grid to measure the minimum area that is susceptible to desertification.

1. Find the areas on the map that have been identified as susceptible to desertification.

2. Count the grid squares that correspond to these identified areas. Count only those squares that contain *just* the areas that are susceptible to desertification—not those squares that include boundaries between those

areas and the more environmentally stable adjacent land. This way you will obtain a minimum estimate of the potential increase in desert area.

3. Find the area by multiplying the total number of squares and the value for each square. Remember that each square represents 100 km^2.

$$\text{area of desertification} =$$
(total number of squares) $\times$ 100 km^2 (area per square)

This result is the minimum area of desertification because the squares that include boundaries between susceptible lands and environmentally stable lands were not included in our calculation.

BONUS PROBLEM: Now, try the calculation yourself, but this time including the squares that include the boundaries. The result will provide a sense of what the *maximum area* of desertification might be.

EXERCISES

1. What types of materials and sizes of particles can the wind move?

2. What is the difference between the way wind transports dust and the way it transports sand?

3. How is the wind's ability to transport sedimentary particles linked to climate?

4. What are the main features of wind erosion?

5. Where do sand dunes form?

6. Name three types of sand dunes and show the relationship of each to wind direction.

7. What typical desert landforms are composed of sediment?

8. What are the geologic processes that form playa lakes?

9. What is desertification?

10. Where are loess deposits found?

THOUGHT QUESTIONS

1. You have just driven a truck through a sandstorm and discover that the paint has been stripped from the lower parts of the truck, but the upper parts have been barely scratched. What process is responsible, and why is it restricted to the lower parts of the truck?

2. What evidence might you find in an ancient sandstone that would point to its eolian origin?

3. Compare the heights to which sand and dust are carried in the atmosphere and explain the differences or similarities.

4. Trucks continually have to haul away sand covering a coastal highway. What do you think might be the source of the sand? Could its encroachment be stopped?

5. What features of a desert landscape would lead you to believe it was formed mainly by streams, with secondary contributions from eolian processes?

6. Which of the following would be a more reliable indication of the direction of the wind that formed a barchan:

cross-bedding or the orientation of the dune's shape on a map? Why?

7. What factors determine whether sand dunes will form on a stream floodplain?

8. There are large areas of sand dunes on Mars. From this fact alone, what can you infer about conditions on the Martian surface?

9. What aspects of an ancient sandstone would you study to show that it was originally a desert sand dune?

10. What kinds of landscape features would you ascribe to the work of the wind, to the work of streams, or to both?

11. How does desert weathering differ from or resemble weathering in more humid climates?

12. What evidence would cause you to infer that dust storms and strong winds were common in glacial times?

MEDIA SUPPORT

 19-1 Animation: Desert Pavement Formation

 19-1 Video: Desert Processes I

 19-2 Video: Desert Processes II

- How Ocean Basins Differ from Continents   552

- Coastal Processes   552

- The Shaping of Shorelines   564

- Continental Margins   570

- Topography of the Deep Seafloor   572

- Ocean Sedimentation   578

Damage to an amusement park and beachfront property in Seaside Heights, N.J. after Hurricane Sandy. November 18, 2012. [Marcus Yam/The New York Times/Redux.]

COASTLINES AND OCEAN BASINS

FOR MOST OF HUMAN HISTORY, the 71 percent of Earth's surface covered by oceans was a mystery. The large populations who lived at the edge of the sea knew well the impact of the waves, the rise and fall of the tides, and the devastating effects of powerful storms. But they could only guess at the forces that caused these processes. We now know that they result from interactions within the climate system and the solar system. Tides are caused by gravitational interactions between Earth and the Sun and Moon, and coastal surf and storms result from interactions between the atmosphere and the hydrosphere.

And what about the deep sea, which is invisible to humans without the aid of remote observation tools? The nature of the seafloor beyond the shallowest coastal waters remained a mystery until the middle of the nineteenth century. In 1872, the *Challenger*, a small wooden British warship converted and fitted out for scientific study, became the first research vessel to explore the oceans scientifically. The *Challenger* expedition discovered great areas of submerged hills and flat plains, extraordinarily deep trenches, and submarine volcanoes.

Today, Earth scientists still search for answers to the questions first raised by these early discoveries. What forces raised the submarine mountain ranges and depressed the trenches? Why are some areas of the seafloor flat and others hilly? Although oceanographers made many important discoveries in the first half of the twentieth century, the answers to most of these questions had to await the plate tectonic revolution of the late 1960s. As we saw in Chapter 2, it was geological observations of the seafloor, not of the continents, that led to the theory of plate tectonics.

In this chapter, we examine the processes that affect shorelines and coastal areas and consider the effects of waves, tides, and damaging storms. Then we move farther offshore to examine the submerged margins of the continents that bound ocean basins, and finish with a discussion of the deep seafloor.

How Ocean Basins Differ from Continents

Plate tectonic theory has provided us with a basic understanding of the differences between the geology of continents and the geology of ocean basins. Away from continental margins, the deep seafloor has no folded and faulted mountains like those on the continents. Instead, deformation is largely restricted to the faulting and volcanism found at mid-ocean ridges and subduction zones. Moreover, the weathering and erosion processes described in previous chapters are much less important in the oceans than on land because the oceans lack efficient fragmentation processes, such as freezing and thawing, and major erosive agents, such as streams and glaciers. Deep-sea currents can erode and transport sediments, but cannot effectively attack the plateaus and hills of basaltic rock that form the oceanic crust.

Because deformation, weathering, and erosion are minimal over much of the seafloor, volcanism and sedimentation dominate the geology of ocean basins. Volcanism creates mid-ocean ridges, island groups (such as the Hawaiian Islands) in the middle of an ocean, and island arcs near deep-sea trenches. Sedimentation shapes much of the rest of the seafloor. Soft sediments of mud and calcium carbonate blanket the low hills and plains of the seafloor. Sediments begin to accumulate on oceanic crust as soon as it is formed at mid-ocean ridges. As the crust spreads farther and farther from the ridge, it accumulates more and more sediments. Deep-sea sedimentation is more continuous than the sedimentation in most continental environments, and it therefore preserves a better record of geologic events—for example, as we have seen, it provides a more detailed history of Earth's climate changes.

The marine sediment record is limited, however, because subduction is continually recycling oceanic crust, thereby destroying marine sediments by metamorphism and melting. On average, it takes only a few tens of millions of years for the crust created at a mid-ocean ridge to spread across an ocean and come to a subduction zone. As we saw in Chapter 2, the oldest parts of today's seafloor were formed in the Jurassic period, about 180 million years ago; they are currently found near the western edge of the Pacific Plate (see Figure 2.15). In the next 10 million years or so, the sedimentary record that lies atop this crust will disappear into the mantle.

The five major oceans (Atlantic, Pacific, Indian, Arctic, and Southern) form a single connected body of water sometimes referred to as the *world ocean.* The term *sea* is often used to refer to smaller bodies of water set off somewhat from the oceans. The Mediterranean Sea, for example, is narrowly connected with the Atlantic Ocean by the Strait of Gibraltar and with the Indian Ocean by the Suez Canal. Other seas are more broadly connected, as is the North Sea with the Atlantic Ocean. Seawater—the salty water of the oceans and seas—is remarkably constant in its general chemical composition from year to year and from place to place. The chemical equilibrium maintained by the oceans is determined by the composition of river waters entering the oceans, the composition of the sediments they transport to the oceans, and the formation of new sediments in the oceans.

Coastal Processes

Coastlines are the broad regions where land and streams meet the ocean. Environmental problems such as coastal erosion and pollution of shallow waters make the geology of coastlines an active area of research. The landscapes of coastlines, even within a single continent, present striking contrasts (**Figure 20.1**). On the coast of North Carolina, for example, long, straight, sandy beaches stretch for miles along low coastal plains (Figure 20.1a). Here, tectonic activity is limited, and it is the currents produced by breaking waves that mold the coastline. The Oregon coastline, on the other hand, is dominated by rocky cliffs. Even though the effect of waves is considerable, it is tectonic uplift that shapes this landscape. Many of the seaward edges of

FIGURE 20.1 ■ Coastlines exhibit a variety of geologic forms. (a) Long, straight, sandy beach, Pea Island, North Carolina. (b) Rocky coastline, Mount Desert Island, Maine. This formerly glaciated coastline has rebounded since the end of the last ice age, about 11,000 years ago. (c) The Twelve Apostles, Port Campbell, Australia, a group of stacks that developed from cliffs of sedimentary rock. These remnants of shoreline erosion are left as the shoreline retreats under the action of waves. (d) Coral reef along the Florida coastline. [(a) Courtesy of Bill Birkemeier/U.S. Army Corps. of Engineers; (b) Neil Rabinowitz/Corbis; (c) Christopher Groenhout/ Getty Images, Inc.; (d) Dr. Hays Cummins, Interdisciplinary Studies, Miami University.]

islands in the tropics are coral reefs, shaped by biological sedimentation (Figure 20.1d). As we will see, plate tectonic processes, erosion, and sedimentation work together to create this great variety of coastline shapes and materials.

The major geologic forces operating at the **shoreline**— the line where the water surface meets the land surface—are ocean currents created by waves and tides. These currents eventually erode even the most resistant rocky shores. They also transport the sediments produced by erosion and deposit them on beaches and in shallow waters along the shore.

As we have seen in earlier chapters, currents are the key to understanding geologic processes at Earth's surface, and coastal processes are no exception. Let's examine the various types of currents that shape our shorelines.

Wave Motion: The Key to Shoreline Dynamics

Centuries of observation have taught us that waves are constantly changing. In quiet weather, waves with calm troughs between them roll regularly into shore. In the high winds of a storm, however, waves move in a confusion of shapes and sizes. They may be low and gentle far from shore, yet become high and steep as they approach land. High waves can break on the shore with fearful violence, shattering concrete seawalls and tearing apart houses built along the beach. To understand the dynamics of shorelines, and to make sensible decisions about coastal development, we need to understand how waves work.

Wind blowing over the surface of the ocean creates waves by transferring its energy of motion from air to water. As a gentle breeze of 5 to 20 km/hour starts to blow over a calm sea surface, ripples—little waves less than a centimeter high—take shape. As the speed of the wind increases to about 30 km/hour, the ripples grow to full-sized waves. Stronger winds create larger waves and blow off their tops to make whitecaps. The height of waves depends on three factors:

▪ The wind speed

▪ The length of time over which the wind blows

▪ The distance the wind travels over water

Storms blow up large, irregular waves that radiate outward from the storm center, like the ripples moving outward from a pebble dropped into a pond. As the waves travel outward in ever-widening circles, they become more regular, changing into low, broad, rounded waves called *swell*, which can travel hundreds of kilometers. Several storms at different distances from a shoreline, each producing its own pattern of swell, may account for the often irregular intervals between waves approaching the shore.

If you have seen waves in an ocean or a large lake, you have probably noticed how a piece of wood floating on the water moves a little forward as the crest of a wave passes and then a little backward as the trough between waves passes. Although it moves back and forth, the wood stays in roughly the same place—and so does the water around it. The water molecules move in a circle, even though the waves are moving toward the shore.

We can describe a wave form in terms of the following three characteristics (**Figure 20.2**):

1. *Wavelength,* the distance between wave crests

2. *Wave height,* the vertical distance between the crest and the trough

3. *Period,* the time it takes for two successive wave crests to pass a fixed point

We can measure the velocity at which a wave moves forward by using a simple equation:

$$V = \frac{L}{T}$$

where V is the velocity, L is the wavelength, and T is the period. Thus, a typical wave with a length of 24 m and a period of 8 s would have a velocity of 3 m/s. The periods of waves range from just a few seconds to as long as 15 or 20 s, and their wavelengths vary from about 6 m to as much as 600 m. Consequently, wave velocities vary from 3 to 30 m/s. Wave motion becomes very small below a depth equal to about one-half the wavelength. That is why deep divers and submarines are unaffected by the waves at the surface.

The Surf Zone

Swell becomes higher as it approaches the shore, where it assumes the familiar sharp-crested wave shape. These waves are called *breakers* because, as they come closer to shore, they break and form surf—a foamy, bubbly surface. The *surf zone* is the belt along which breaking waves collapse as they approach the shore.

The transformation from swell to breakers starts where water depth decreases to less than one-half the wavelength of the swell. At that point, the wave motion just above the bottom becomes restricted because the water can only move back and forth horizontally. Above that, the water can move vertically just a little (see Figure 20.2). The restricted motion of the water molecules slows the whole wave. Its period remains the same, however, because the swell keeps coming in from deeper water at the same rate. From the wave equation, we know that if the period remains constant but the wavelength decreases, then the velocity must also decrease.

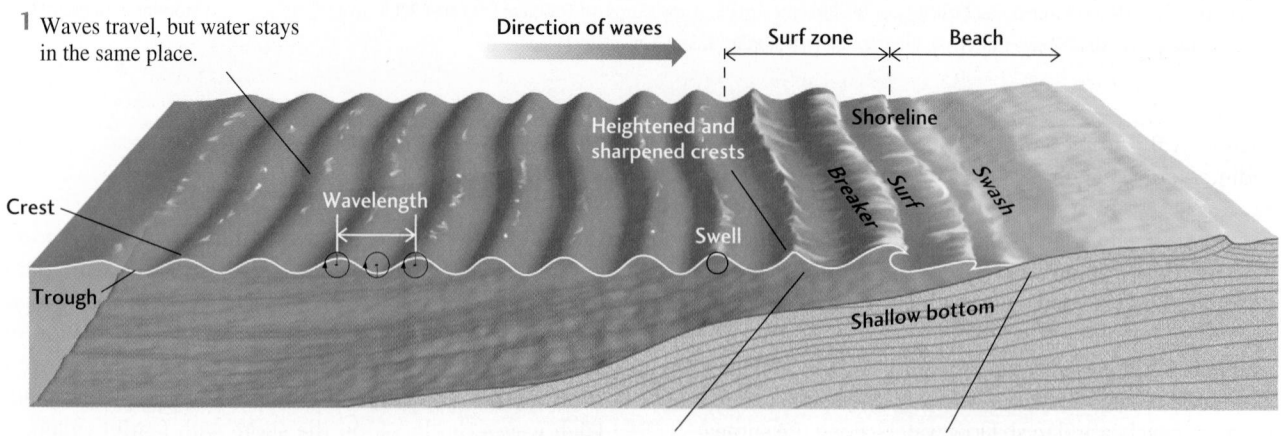

1 Waves travel, but water stays in the same place.

Direction of waves

Surf zone   Beach

Heightened and sharpened crests

Shoreline

Crest

Wavelength

Breaker   Surf   Swash

Trough

Swell

Shallow bottom

2 When the bottom shallows to about one-half the wavelength, the wave slows.

3 As waves approach the shore, they become too steep to support themselves and break in the surf zone, running up the beach in a swash.

FIGURE 20.2 ▪ Wave motion at the shoreline is influenced by water depth and the shape of the bottom.

The typical wave that we used as our example earlier might keep the same period of 8 s while its length decreased to 16 m, in which case it would have a velocity of 2 m/s. Thus, as waves approach the shore, they become more closely spaced, higher, and steeper, and their wave crests become sharper.

As a wave rolls toward the shore, it becomes so steep that the water can no longer support itself, and the wave breaks with a crash in the surf zone (see Figure 20.2). Gently sloping bottoms cause waves to break farther from shore; steeply sloping bottoms make waves break closer to shore. Where rocky shores are bordered by deep water, the waves break directly on the rocks with a force amounting to tons per square meter, throwing water high into the air (see the chapter opening photo). It is not surprising that concrete seawalls built to protect structures along the shore quickly start to crack and must be repaired constantly.

After breaking in the surf zone, the waves, now reduced in height, continue to move in, breaking again at the shoreline. They run up onto the sloping front of the beach, forming an uprush of water called *swash*. The water then runs back down again as *backwash*. Swash can carry sand and even large pebbles and cobbles onto the beach if the waves are high enough. The backwash carries the particles seaward again.

The back-and-forth motion of the water just offshore is strong enough to carry sand grains and even gravel. Wave action in water as deep as 20 m can move fine sand. Large waves caused by intense storms can scour the bottom at much greater depths, down to 50 m or more. At shallower depths, storms transport sediments in an offshore direction, often depleting beaches of their fine sand.

Wave Refraction

Far from shore, the lines of wave crests are parallel to one another, but are usually at some angle to the shoreline. As the waves approach the shore over a shallowing bottom, they gradually bend to a direction more parallel to the shore (**Figure 20.3a**). This bending is called *wave refraction*. It is similar to the bending of light rays in optical refraction, which makes a pencil half in and half out of water appear to bend at the water surface. Wave refraction begins as the part of a wave closest to the shore encounters the shallowing bottom first, and the front of the wave slows. Then the next part of the wave meets the bottom and it, too, slows. Meanwhile, the parts closest to shore have moved into even shallower water and have slowed even more. Thus, in a continuous transition along the wave crest, the line of waves bends toward the shore as it slows (Figure 20.3b).

FIGURE 20.3 ■ Wave refraction. (a) Waves approach the shore at an angle. (b) As waves move closer to the shore, the angle of the wave crests becomes more parallel to the shoreline. (c) Wave refraction increases the erosion of projecting headlands. (d) Wave refraction gives rise to longshore drift and longshore currents. [Carol Barrington–Destination Ph/Aurora Photos.]

(a)

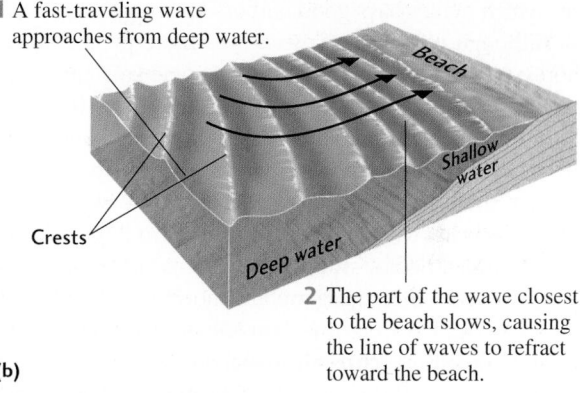

1 A fast-traveling wave approaches from deep water.

Beach

Shallow water

Crests

Deep water

2 The part of the wave closest to the beach slows, causing the line of waves to refract toward the beach.

(b)

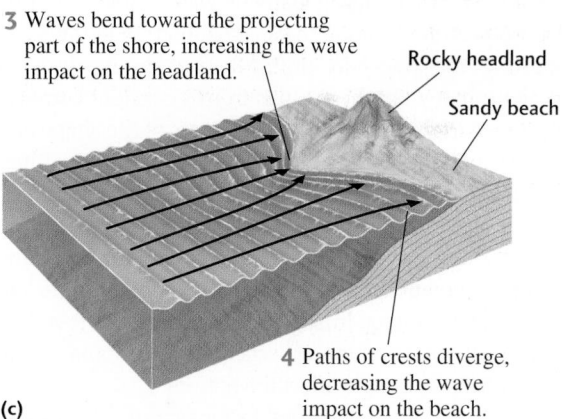

3 Waves bend toward the projecting part of the shore, increasing the wave impact on the headland.

Rocky headland

Sandy beach

4 Paths of crests diverge, decreasing the wave impact on the beach.

(c)

5 Waves approach the shore at an angle, causing a longshore current parallel to the shoreline.

Path of sand

Path of water

Rip current

Longshore current

6 Longshore drift results from movement of sand particles by swash and backwash.

(d)

Wave refraction results in more intense wave action on projecting headlands (Figure 20.3c) and less intense action in indented bays. The bottom becomes shallow more quickly around headlands than in the surrounding deeper water on either side. Thus, waves are refracted around headlands— that is, they are bent toward the projecting part of the shore on both sides. The waves converge around the headland and expend proportionally more of their energy breaking there than at other places along the shore. Because of this concentration of wave energy at headlands, they tend to be eroded more quickly than straight sections of shoreline.

The opposite happens as a result of wave refraction in a bay. The waters in the center of the bay are deeper, so the waves are refracted on either side into shallower water. The energy of wave motion is diminished at the center of the bay, which makes bays good harbors for ships.

Although refraction makes waves more parallel to the shore, most waves still approach it at some small angle. As the waves break on the shore, the swash moves up the beach slope in a direction perpendicular to that small angle. The backwash runs down the slope in the opposite direction at a similar small angle. The combination of these two motions moves the water a short way down the beach (Figure 20.3d). Sand grains carried by swash and backwash are thus moved along the beach in a zigzag motion known as *longshore drift*.

Waves approaching the shoreline at an angle can also cause a **longshore current,** a shallow-water current that flows parallel to the shore. The movement of swash and backwash creates a zigzag path of water molecules that transports sediments along the shallow bottom in the same direction as the longshore drift. Much of the transport of sand along many beaches results from longshore currents. Longshore currents are prime determiners of the shape and extent of sandbars and other depositional shoreline features. At the same time, because of their ability to erode loose sand, longshore currents may remove large amounts of sand from a beach. Longshore drift and longshore currents, working together, are potent agents of sand transport on beaches and in very shallow waters. In slightly deeper waters (less than 50 m), longshore currents—especially those running during intense storms—strongly affect the bottom.

Some types of flows related to longshore currents can pose a threat to unwary swimmers. A *rip current*, for example, is a strong flow of water moving seaward at right angles to the shore (see Figure 20.3d). It occurs when a longshore current builds up along the shore and the water piles up imperceptibly until a critical point is reached. At that point, the water breaks out to sea, flowing through the oncoming waves in a fast current. Swimmers can avoid being carried out to sea by swimming parallel to the shore to get out of the rip.

Tides

For thousands of years, mariners and coastal dwellers have observed the twice-daily rise and fall of the ocean that we call **tides.** Many observers noticed a relationship between the position and phases of the Moon, the heights of the tides, and the times of day at which the water reaches high tide. Not until the seventeenth century, however, when Isaac Newton formulated the laws of gravitation, did we begin to understand that tides result from the gravitational pull of the Moon and the Sun on the water of the oceans.

The gravitational attraction between any two bodies decreases as they get farther apart. Thus, the strength of this attraction varies across Earth's surface. On the side of Earth closest to the Moon, the ocean water experiences a greater gravitational attraction than the average for the whole of the solid Earth. This pull produces a bulge in the water. On the side of Earth farthest from the Moon, the solid Earth, being closer to the Moon than the water, is pulled toward the Moon more than the water is, and the water therefore appears to be pulled away from Earth as another bulge. Thus, two bulges of water are formed in Earth's oceans: one on the side nearest the Moon, and the other on the side farthest from the Moon (**Figure 20.4a**). As Earth rotates, these bulges of water stay approximately aligned: one always faces the Moon, and the other is always directly opposite the Moon. These bulges of water passing over the rotating Earth are the high tides.

The Sun, although much farther away, has so much mass (and thus so much gravitational force) that it, too, causes tides. Sun tides are a little less than half the height of Moon tides, and they are not synchronous with Moon tides. Sun tides occur as Earth rotates once every 24 hours, the length of a solar day. The rotation of Earth with respect to the Moon is a little longer because the Moon is moving around Earth, resulting in a lunar day of 24 hours and 50 minutes. In that lunar day, there are two high tides, with two low tides between them.

When the Moon, Earth, and Sun line up, the gravitational forces of the Sun and the Moon reinforce each other. This alignment produces the *spring tides*, which are the highest tides; their name is not related to the season, but to the German verb *springen*, meaning "to leap up." They appear every 2 weeks, at the full and new Moon. The lowest tides, the *neap tides*, come in between, at the first- and third-quarter Moon, when the Sun and Moon are at right angles to each other with respect to Earth (Figure 20.4b).

Although tides occur regularly everywhere, the difference between high and low tides varies in different parts of the ocean. As Earth rotates, the tidal bulges of water move along the surface of the ocean, encountering obstacles, such as continents and islands, which hinder the flow of the water. In the middle of the Pacific Ocean—in the Hawaiian Islands, for example, where there is little to obstruct the flow of the tides—the difference between low and high tides is only 0.5 m. Near Seattle, where the shape of the shoreline along Puget Sound is very irregular and the tidal flow must move through narrow passageways, the difference between low and high tides is about 3 m. Extraordinary tides occur in a few places, such as the Bay of Fundy in eastern Canada, where the tidal range can be more than 12 m. Many coastal

(a)

(b)

FIGURE 20.4 ∎ Tides are caused by the gravitational attraction of Earth, the Moon, and the Sun. (a) The Moon's gravitational pull forms two bulges of ocean water, one on the side of Earth nearest the Moon and the other on the side farthest from the Moon. As Earth rotates, these bulges remain aligned with the Moon and pass over Earth's surface, creating the high tides. (b) At the new and full Moon, Sun and Moon tides reinforce each other, causing the highest (spring) tides. At the first- and third-quarter Moon, Sun and Moon tides are in opposition, causing the lowest (neap) tides.

residents need to know when tides will occur, so governments publish tide tables showing predicted tide heights and times. These tables combine local knowledge of water flow patterns with knowledge of the astronomical movements of Earth and the Moon with respect to the Sun.

Tides moving near shorelines cause currents that can reach speeds of a few kilometers per hour. As the tide rises, water flows in toward the shore as a *flood tide,* moving through narrow passages into inlets and bays, into shallow coastal wetlands, and up small streams. As the tide passes the high stage and starts to fall, the water flows out as an *ebb tide,* and low-lying coastal areas are exposed. Tidal currents meander across **tidal flats,** muddy or sandy areas that are exposed at low tide but flooded at high tide (**Figure 20.5**). Where obstacles restrict tidal flow and increase tidal range, current velocities may become very high. Large sand ridges many meters high may be formed in these tidal channels.

Hurricanes and Coastal Storm Surges

Hurricanes are the greatest storms on Earth, swirling masses of dense clouds hundreds of kilometers across that suck their energy from the warm surface waters of tropical oceans. The term *hurricane* originates from the name *Huracan,* a god of storms to the Mayan people of Central America. In

FIGURE 20.5 ∎ Tidal flats, such as this one at Mont-Saint-Michel, France, may be extensive areas covering many square kilometers, but most often are narrow strips seaward of the beach. When a very high tide advances on a broad tidal flat, it may move so rapidly that some areas are flooded faster than a person can run. The beachcomber is well advised to learn the local tides before wandering. [Thierry Prat/Corbis-Sygma.]

the western Pacific and China Sea, hurricanes are known as *typhoons,* from the Cantonese word *tai-fung,* meaning "great wind." In Australia, Bangladesh, Pakistan, and India, they are known as *cyclones;* in the Philippines, they are called *baguios.*

Whatever you call them, these intense tropical storms can wreak havoc. For example, a catastrophic cyclone struck the coastal lowlands of Bangladesh in 1970, drowning as many as 500,000 people—perhaps the deadliest natural disaster of modern times. Another cyclone hit the same region in 1991, drowning at least 140,000 (**Figure 20.6**). The 1991 storm was more intense, but its death toll was lower due to better disaster preparations; 2 million people were evacuated.

The damaging effects of a hurricane's extremely high, sustained winds and torrential rains are intuitively easy to understand. However, the associated storm surge, which may flood major regions of the coastline, is potentially the most destructive effect of a hurricane. When Hurricane Katrina struck New Orleans, Louisiana, on August 29, 2005, the disaster that followed was not so much the result of the direct impact of the hurricane itself as of the storm surge, which ultimately caused several sections of the artificial levee system protecting New Orleans to collapse (see Earth Policy 20.1 on pages 562–563). Subsequent flooding of parts of the city claimed hundreds of lives and left the city submerged and abandoned for almost a month. Hurricane Sandy was the second costliest hurricane in United States history (second to Hurricane Katrina) and hit the east coast of the United States in late October 2012. While Sandy caused water levels to rise along the entire east coast, it caused a catastrophic storm surge along the New Jersey, New York, and Connecticut coastlines.

HURRICANE FORMATION Hurricanes form over tropical parts of Earth's oceans, between 8° and 20° latitude, in areas of high humidity, light winds, and warm sea surface temperatures (typically 26°C or greater). These conditions usually occur in the summer and early fall in the tropical North Atlantic and North Pacific. For this reason, hurricane "season" in the Northern Hemisphere runs from June through November (**Figure 20.7**).

The first sign of hurricane development is the appearance of a cluster of thunderstorms over the tropical ocean in a region where the trade winds converge. Occasionally, one of these clusters breaks out from this convergence zone and becomes better organized. Most hurricanes that affect the Atlantic Ocean and Gulf of Mexico originate in a convergence zone just off the coast of West Africa and intensify as they break out and move westward across the tropical Atlantic.

As the hurricane develops, water vapor condenses to form rain, which releases heat energy. In response to this atmospheric heating, the surrounding air becomes less dense and begins to rise, and the atmospheric pressure at sea level drops in the region of heating. As the warm air rises, it triggers more condensation and rainfall, which in turn releases more heat. At this point, a positive feedback process is set in motion, as the rising temperatures in the center of the storm cause surface pressures to fall to progressively lower levels. In the Northern Hemisphere,

FIGURE 20.6 ■ Devastation caused by a cyclone in Chittagong, Bangladesh, in 1991. [Peter Charlesworth/LightRocket via Getty Images.]

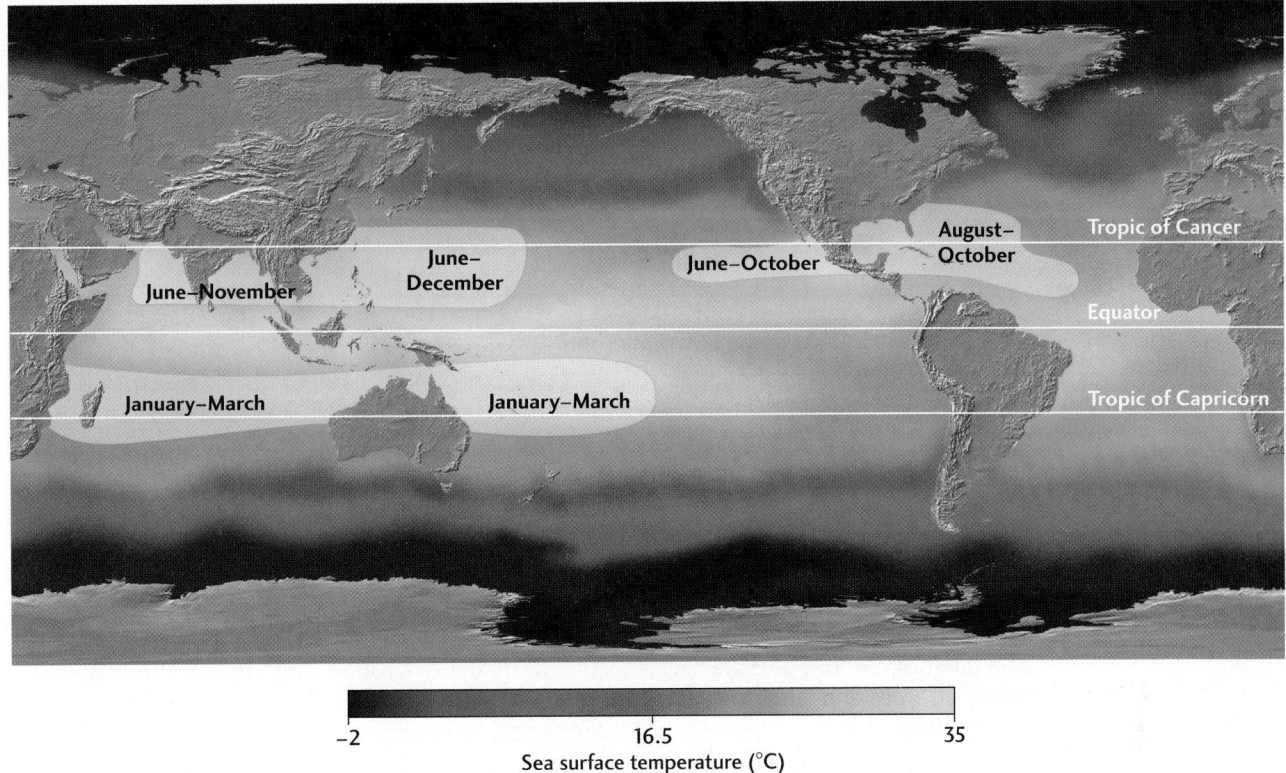

FIGURE 20.7 ▪ Hurricanes arise in summer and early fall when ocean temperatures are warmest. The light-shaded areas indicate places where hurricanes are most common. The times of year when they are most frequent are also shown. [NASA/GSFC.]

because of the Coriolis effect (see Chapter 19), the increasing winds begin to circulate in a counterclockwise pattern around the storm's area of lowest pressure, which ultimately becomes the "eye" of the hurricane (**Figure 20.8**).

Once sustained wind speeds reach 37 km/hour (23 miles/hour), the storm system is called a *tropical depression*. As winds increase to 63 km/hour (39 miles/hour), the system is called a *tropical storm* and receives a name. This naming tradition started with the use of World War II code names, such as Andrew, Bonnie, Charlie, and so forth. Finally, when wind speeds reach 119 km/hour (74 miles/hour), the storm is classified as a hurricane. Once it becomes a hurricane, the storm is assigned a 1–5 rating based on the *Saffir-Simpson hurricane intensity scale* (**Table 20.1**). This scale is used to estimate the potential property damage and flooding expected along the coast from hurricane landfall. It is analogous to the Mercalli intensity scale for earthquakes (see Table 13.1).

FIGURE 20.8 ▪ Hurricane Katrina on August 28, 2005, a few hours before it struck New Orleans. In the Northern Hemisphere, winds circulate in a counterclockwise direction around the "eye" of the hurricane, which is the location of lowest atmospheric pressure. [NASA/Jeff Schmaltz, MODIS Land Rapid Response Team.]

TABLE 20-1 The Saffir-Simpson Hurricane Intensity Scale

Storm Classification	Description
Category 1	Winds 119–153 km/hour (74–95 miles/hour). Storm surge generally 1–1.5 m (4–5 feet) above normal. No real damage to building structures. Damage primarily to unanchored mobile homes, shrubbery, and trees. Some damage to poorly constructed signs. Some coastal road flooding and minor pier damage.
Category 2	Winds 154–177 km/hour (96–110 miles/hour). Storm surge generally 2–2.5 m (6–8 feet) above normal. Some roofing material, door, and window damage to buildings. Considerable damage to shrubbery and trees, with some trees blown down. Considerable damage to mobile homes, poorly constructed signs, and piers. Coastal and low-lying escape routes flood 2–4 hours before arrival of the hurricane center. Small craft in unprotected anchorages break moorings. Hurricane Frances of 2004 made landfall over the southern end of Hutchinson Island, Florida, as a Category 2 hurricane.
Category 3	Winds 178–209 km/hour (111–130 miles/hour). Storm surge generally 2.5–3.5 m (9–12 feet) above normal. Some structural damage to small residences and utility buildings with a minor amount of curtain-wall failure. Damage to shrubbery and trees with foliage blown off trees and large trees blown down. Mobile homes and poorly constructed signs are destroyed. Low-lying escape routes are cut off by rising water 3–5 hours before arrival of the hurricane center. Flooding near the coast destroys smaller structures, with larger structures damaged by battering from floating debris. Terrain continuously lower than 1.5 m (5 feet) above sea level may be flooded 3 m (10 feet) inland or more. Evacuation of low-lying residences within several blocks of the shoreline may be required. Hurricanes Jeanne and Ivan of 2004 were Category 3 hurricanes when they made landfall in Florida and Alabama, respectively. Hurricane Katrina of 2005 made landfall near Buras-Triumph, Louisiana, with winds of 204 km/hour (127 miles/hour). Katrina will prove to be the costliest hurricane on record, with estimates of more than $200 billion in losses.
Category 4	Winds 210–250 km/hour (131–155 miles/hour). Storm surge generally 3.5–5 m (13–18 feet) above normal. More extensive curtain-wall failures with some complete roof structure failures on small residences. Shrubs, trees, and all signs are blown down. Complete destruction of mobile homes. Extensive damage to doors and windows. Low-lying escape routes may be cut off by rising water 3–5 hours before arrival of the hurricane center. Major damage to lower floors of structures near the shore. Terrain lower than 3 m (10 feet) above sea level may be flooded, requiring massive evacuation of residential areas as far inland as 15 km (9 miles).
Category 5	Winds greater than 250 km/hour (155 miles/hour). Storm surge generally greater than 5 m (18 feet) above normal. Complete roof failure on many residences and industrial buildings. Some complete building failures with small utility buildings blown over or away. All shrubs, trees, and signs blown down. Complete destruction of mobile homes. Severe and extensive window and door damage. Low-lying escape routes are cut off by rising water 3–5 hours before arrival of the hurricane center. Major damage to lower floors of all structures located less than 4.5 m (15 feet) above sea level and within 500 m (1650 feet) of the shoreline. Massive evacuation of residential areas on low ground within 15–20 km (9–12 miles) of the shoreline may be required. Only three Category 5 hurricanes have made landfall in the United States since records began. Hurricane Andrew of 1992 made landfall over southern Miami-Dade County, Florida, causing $26.5 billion in losses—the second costliest hurricane on record.

FIGURE 20.9 ■
Hurricane storm surges along coastlines may result in the complete destruction of residential buildings, which pile up as lines of debris well inland of the shoreline. The damage seen here was caused by Hurricane Katrina in 2005.
[U.S. Navy/Getty Images.]

STORM SURGES As a hurricane intensifies, a dome of seawater—known as a **storm surge**—rises above the level of the surrounding ocean surface. The height of the storm surge is directly related to the atmospheric pressure in the eye of the hurricane and the strength of the winds that encircle it. Large swells, high surf, and wind-driven waves ride atop the surge. As the hurricane nears land, the surge moves ashore and floods coastal land areas, causing extensive damage to structures and the shoreline environment (**Figure 20.9**). Any landmass in the path of a storm surge will be affected to a greater or lesser extent, depending on a number of factors. The stronger the storm and the shallower the offshore waters, the higher the storm surge. When the effects of a storm surge coincide with a normal high tide, the result is known as a *storm tide* (**Figure 20.10**).

FIGURE 20.10 ■ A storm tide is the combination of a storm surge and a normal high tide. If a storm surge arrives at the same time as a high tide, the water height will be increased. For example, if a normal high tide 1 m above sea level is combined with a storm surge of 5 m, the resulting storm tide will be 6 m in height.

Earth Policy

20.1 The Great New Orleans Flood

On August 25, 2005, Hurricane Katrina struck southern Florida as a Category 1 storm, killing 11 people. Three days later, in the Gulf of Mexico, the hurricane grew to a monster Category 5 storm, with maximum sustained winds of up to 280 km/hour (175 miles/hour) and gusts up to 360 km/hour (225 miles/hour). On August 28, the National Weather Service issued a bulletin predicting "devastating" damage to the Gulf Coast, and the mayor of New Orleans ordered an unprecedented mandatory evacuation of the city.

When Katrina made landfall just south of New Orleans on August 29, it was a nearly Category 4 storm, with sustained winds of 204 km/hour (127 miles/hour). It had a minimum atmospheric pressure of 918 millibars (27.108 inches), making it the third strongest hurricane on record to make landfall in the United States. More than 100 people lost their lives during the early morning hours of August 29 as a result of the direct impact of the storm.

A 5- to 9-m storm surge came ashore over virtually the entire coastline of Louisiana, Mississippi, Alabama, and the Florida panhandle. The 9-m storm surge at Biloxi, Mississippi, was the highest ever recorded in the United States. The effects of this storm surge on New Orleans were devastating and unprecedented. Lake Pontchartrain, which is really a coastal embayment that is easily influenced by ocean conditions, was inundated by the storm surge. By midday on August 29, several sections of the levee system that held back the waters of Lake Pontchartrain from New Orleans had collapsed. Subsequent flooding of the city to depths of up to 7 or 8 m left 80 percent of New Orleans under water. The effects of flooding claimed at least another 300 lives, and by September 21, the death toll exceeded 1500 as disease

Water spills over a levee along the Inner Harbor Navigational Canal and floods the inner city of New Orleans. [Vincent Laforet-Pool/Getty Images.]

and malnourishment indirectly caused by the flooding took effect.

Hurricane Katrina has surpassed Hurricane Andrew as the costliest natural disaster in U.S. history, with damages reaching almost $200 billion. In addition to the thousands of lives lost, over 150,000 homes were destroyed, and over a million people were displaced—a humanitarian crisis unrivaled in the United States since the Great Depression.

The storm surge is the deadliest of a hurricane's associated hazards, as underscored by Hurricane Katrina in 2005 and Hurricane Sandy in 2012. The magnitude of a hurricane is usually described in terms of its wind speed (see Table 20.1), but coastal flooding causes many more deaths than high wind. Boats ripped from their moorings, utility poles, and other debris floating atop a storm surge often demolish those buildings that are not destroyed by the winds. Even without the weight of floating debris, a storm surge can severely erode beaches and highways and undermine bridges. Because much of the United States' densely populated Atlantic and Gulf Coast shorelines lie less than 3 m above sea level, the danger from storm surges there is tremendous.

Hurricane Sandy was the second costliest hurricane in United States history, estimated at over $68 billion (second to Hurricane Katrina) and hit the east coast of the United States in late October 2012. As the storm formed, it affected much of the Caribbean, including Jamaica, Haiti, Dominican Republic, Puerto Rico, Cuba, and the Bahamas, then worked its way up the eastern coastline of the United States, affecting 24 states. The storm made landfall near Brigantine, New Jersey, and while Sandy caused water levels to rise along the entire east coast, it caused catastrophic storm surge along the New Jersey, New York, and Connecticut shorelines. The surge flooded streets, tunnels, and subway lines and cut power to much of the region. The path was correctly predicted nearly

Residents wade through a flooded street in New Orleans in the aftermath of Hurricane Katrina. [James Nielsen/AFP/Getty Images.]

What happened, and what could have been done to limit the damage? As with most natural disasters, the outcome was a result of rare but powerful geologic forces coupled with a lack of human preparation. No one had anticipated and planned for the worst-case scenario. Earth scientists had predicted for decades that a Category 4 or 5 hurricane would strike New Orleans eventually. The historical record of hurricanes made it clear that such an event was almost certain to happen. As Figure 20.11 shows, New Orleans is about in the middle of the "catcher's mitt" for hurricane landfalls in the United States. But the city was prepared to resist the damaging effects of only a Category 3 or smaller hurricane. Federal budget cuts had left only token support available to maintain and reinforce the east bank of hurricane levees that held back Lake Pontchartrain. This complex network of concrete walls, metal gates, and giant earthen berms was never completed, leaving the city vulnerable. Furthermore, it is not easy to protect a city from hurricane storm surges when its sidewalks and houses are, on average, 4 m below sea level. New Orleans is equally vulnerable to unusually large floods of the Mississippi River, which is also held back by an artificial levee system.

When large catastrophic events are rare, it is natural to question whether they are worth worrying about, and human memory may fail to provide the necessary guidance. Over the short term, we may escape these threats by random good luck. But over the long term, recorded history and the geologic record show that these rare and devastating forces will eventually take their toll if we are not adequately prepared.

8 days before it hit the eastern seaboard, so preparations were underway days before. Subway entrances and grates were covered in New York City, although flooding still occurred. There were mandatory evacuations, schools were closed, mass transit was shut down, airports were closed, and bus and rail services were suspended. Despite the preparation, communities were flooded with water and sand, houses were washed from their foundations, the New Jersey boardwalk and pier was destroyed, and cars and boats were tossed about. The storm also caused devastating fires in New York. Trees fell on power lines, transformers exploded, and wires fell into water, creating a dangerous fire that was hard to access and fight due to flooded communities.

HURRICANE LANDFALL Because hurricanes form over and move across tropical waters, most make landfall at low latitudes. Most North Atlantic hurricanes make landfall in Florida and the northern Gulf of Mexico (**Figure 20.11**). However, because there is a tendency for winds to be deflected northward (due to the Coriolis effect), hurricanes sometimes make landfall farther up the Atlantic coast. In rare cases, they may reach New England, but are always of lower intensity there because of the lower ocean surface temperatures. The most powerful Category 4 and 5 hurricanes are restricted to lower latitudes.

The tropical storms that grow into hurricanes can be tracked by satellites, and weather conditions inside the storms can be probed by airplanes. By feeding many types

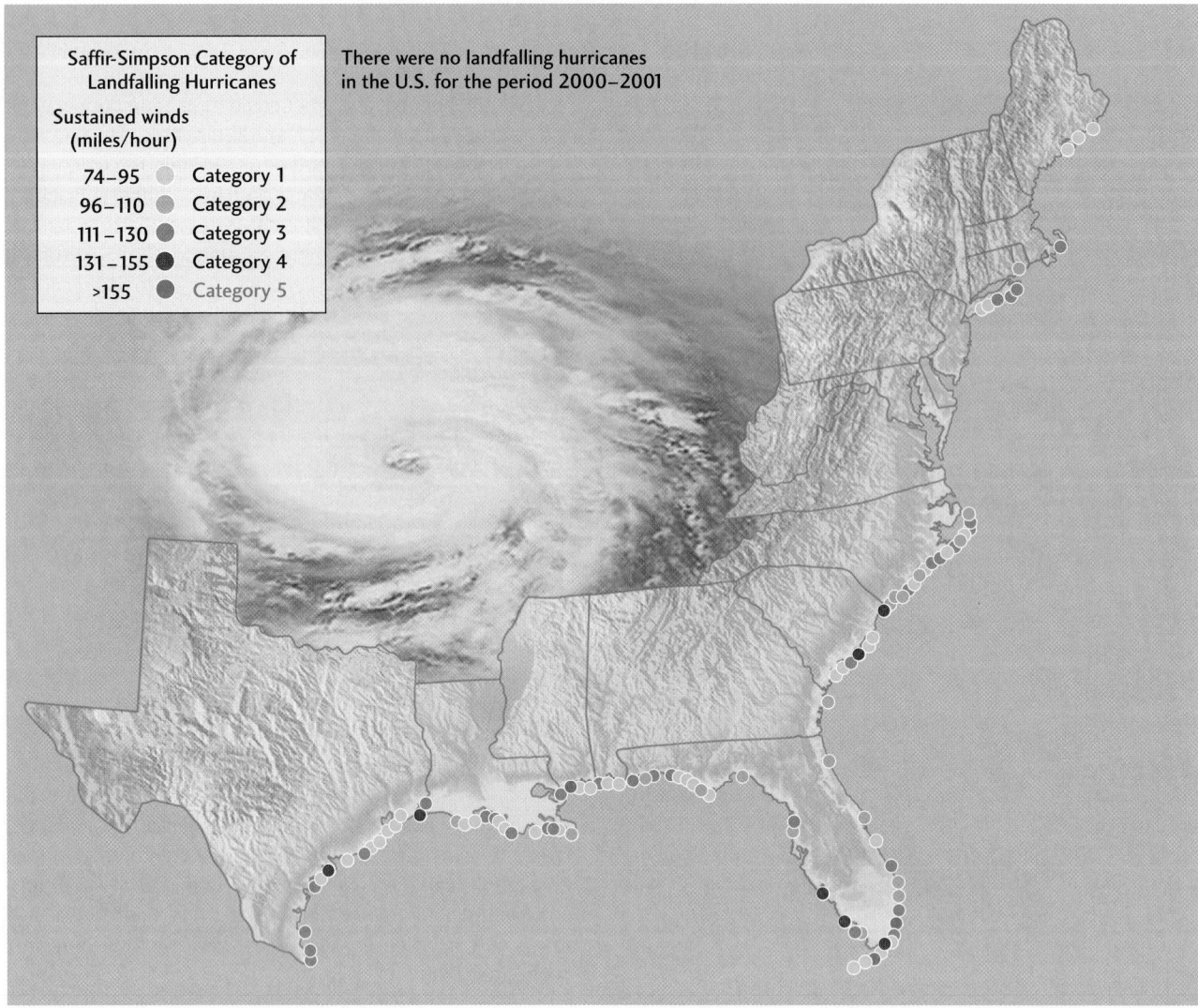

Saffir-Simpson Category of Landfalling Hurricanes

Sustained winds (miles/hour)

74–95	●	Category 1
96–110	●	Category 2
111–130	●	Category 3
131–155	●	Category 4
>155	●	Category 5

There were no landfalling hurricanes in the U.S. for the period 2000–2001

FIGURE 20.11 ■ Hurricanes originating in the North Atlantic Ocean usually make landfall in the coastal areas of the southeastern United States, including the Gulf Coast states. Hurricanes lose energy as they move across cold water, so the number of hurricanes that make landfall drops dramatically for the central and northeastern states. [NOAA.]

of data into computer models, meteorologists can predict a storm's track and changes in its intensity up to several days in advance of landfall with reasonable accuracy. The National Hurricane Center accurately predicted that Katrina would hit New Orleans as a severe hurricane 3 days before it actually did.

■ The Shaping of Shorelines

The effects of the coastal processes we have just described are best observed at shorelines. Waves, longshore currents, tidal currents, and storm surges interact with plate tectonic processes and with the geologic structures of the coast to shape shorelines into a multitude of forms. We can see these factors at work in the most popular of shoreline environments: beaches.

Beaches

A **beach** is a shoreline environment made up of sand and pebbles. The shape of a beach may change from day to day, week to week, season to season, and year to year. Waves and tides sometimes broaden and extend a beach by depositing sand and sometimes narrow it by carrying sand away.

Many beaches are straight stretches of sand ranging from 1 km to more than 100 km long; others are smaller crescents of sand between rocky headlands. Belts of dunes border the landward edge of many beaches; bluffs

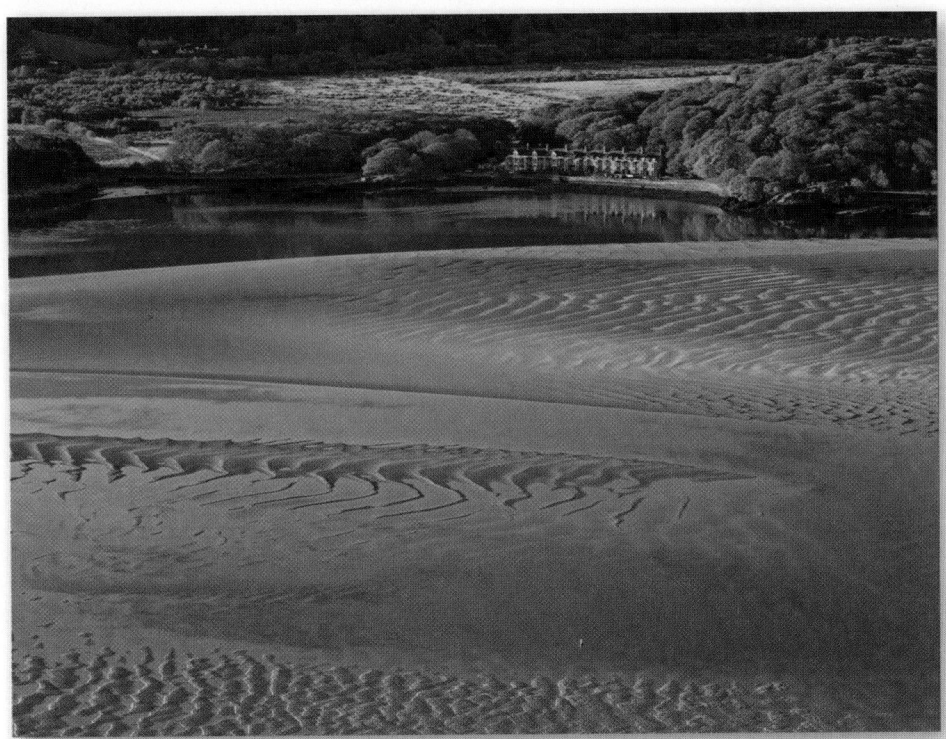

FIGURE 20.12 ■ A tide terrace exposed at low tide. This shallow depression between an outer ridge (a sandbar at high tide) and the upper beach is rippled by the tidal flow in many places.
[© David Hall/Alamy.]

or cliffs of sediment or rock border others. A beach may have a *tide terrace*—a flat, shallow area between the upper beach and an outer bar of sand—on its seaward side (**Figure 20.12**).

THE STRUCTURE OF A BEACH **Figure 20.13** shows the major parts of a beach. These parts may not all be present at all times on any particular beach. Farthest out is the *offshore*, which is bounded by the surf zone, where the bottom begins to become shallow enough for waves to break. The *foreshore* includes the surf zone; the tide terrace; and, right at the shoreline, the *swash zone*, a slope dominated by the swash and backwash of the waves. The *backshore* extends from the swash zone up to the highest level of the beach.

THE SAND BUDGET OF A BEACH A beach is a scene of constant movement. Each wave moves sand back and

forth with its swash and backwash. Both longshore drift and longshore currents move sand down the beach. At the end of the beach, and to some extent along it, sand is removed and deposited in deep water. In the backshore or along sea cliffs, sand and pebbles are freed by erosion and replenish the beach. Winds that blow over the beach transport sand, sometimes offshore into the water and sometimes onshore onto the land.

All these processes together maintain a balance between addition and removal of sand, resulting in a beach that may appear to be stable but is actually exchanging its material with the environments on all sides. **Figure 20.14** illustrates the sand budget of a beach: the inputs and outputs caused by erosion, sedimentation, and transport. At any point along a beach, the beach gains sand from a number of sources: material eroded from the backshore; sand brought to the beach by longshore drift and longshore

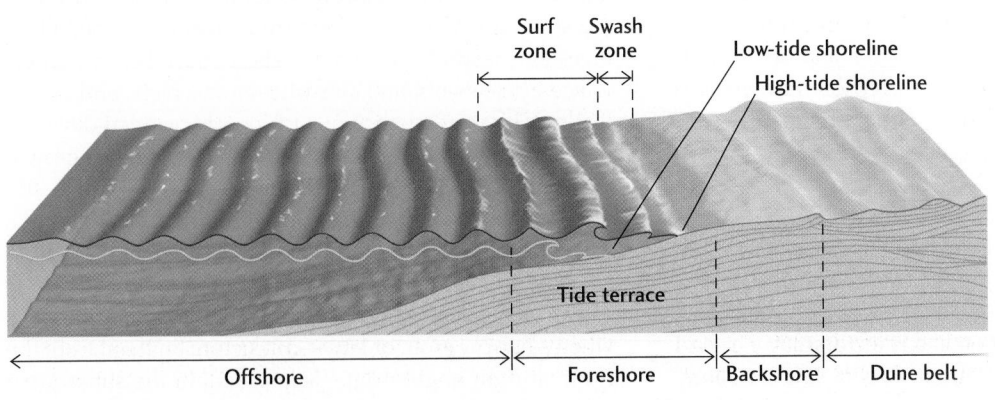

Surf zone Swash zone

Low-tide shoreline

High-tide shoreline

Tide terrace

Offshore Foreshore Backshore Dune belt

FIGURE 20.13 ■ A profile of a beach, showing its major parts.

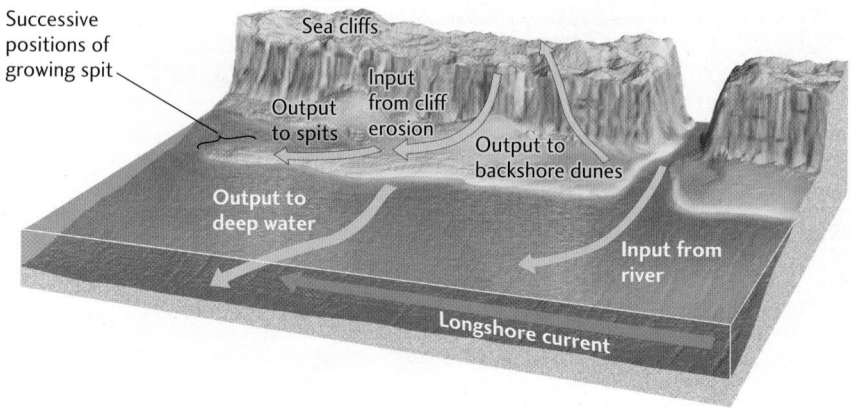

SAND BUDGET	
INPUTS	**OUTPUTS**
Sediments eroded from backshore cliffs by waves	Sediments transported to backshore dunes by offshore winds
Sediments eroded from upcurrent beach by longshore drift and current	Sediments transported downcurrent by longshore drift and current
Sediments brought in by rivers	Sediments transported to deep water by tidal currents and waves

FIGURE 20.14 ▪ The sand budget of a beach is a balance between inputs and outputs of sand due to erosion, sedimentation, and transport.

currents; and sediments carried to the shoreline by rivers. The beach also loses sand in a number of ways: winds carry sand to backshore dunes, longshore drift and longshore currents carry it downcurrent, and deep-water currents and waves transport it during storms.

If the total sand input balances the total sand output, the beach is in dynamic equilibrium, and it keeps the same general form. If input and output are not balanced, the beach grows or shrinks. Temporary imbalances are natural over weeks, months, or even years. A series of intense storms, for example, might move large amounts of sand from the beach to deeper waters offshore, narrowing the beach. Then, weeks of mild weather and low waves might move sand onto the shore and rebuild a wide beach. Without this constant shifting of sand, beaches might be unable to recover from the effects of trash, litter, and other kinds of pollution. Within a year or two, even oil from spills is transported or buried out of sight, although the tarry residue may later be uncovered in spots.

SOME COMMON FORMS OF BEACHES Long, wide, sandy beaches grow where sand inputs are abundant, often where soft sediments make up the coast. Where the backshore is low and winds blow onshore, wide dune belts border the beach. Where the shoreline is tectonically elevated and the coast is made up of hard rock, cliffs line the shore,

and any small beaches that form are composed of material eroded from those cliffs. Where the coast is low-lying, sand is abundant, and tidal currents are strong, extensive tidal flats are laid down and are exposed at low tide.

PRESERVATION OF BEACHES What happens if one of the inputs to a beach is cut off—for example, by a concrete seawall built along a beach to prevent erosion? Because erosion supplies sand to the beach, preventing erosion cuts the sand supply and so shrinks the beach. Such attempts to save a beach, undertaken without an understanding of its dynamic equilibrium, may actually destroy it.

Humans are altering the dynamic equilibrium of more and more beaches by placing buildings on them and erecting structures to protect them from erosion. We build cottages and resort hotels on the shore; pave beach parking lots; erect seawalls; and construct groins, piers, and breakwaters. The consequence of such poorly planned development is shrinkage of the beach in one place and its growth in another. As landowners and developers bring suit against one another and against state governments, trial lawyers take the issue of "sand rights"—the beach's right to the sand that it naturally contains—into the courts.

To use a classic example, let's examine what happens when a narrow groin or jetty—a structure built out from the shore at right angles to it—is installed. In the subsequent

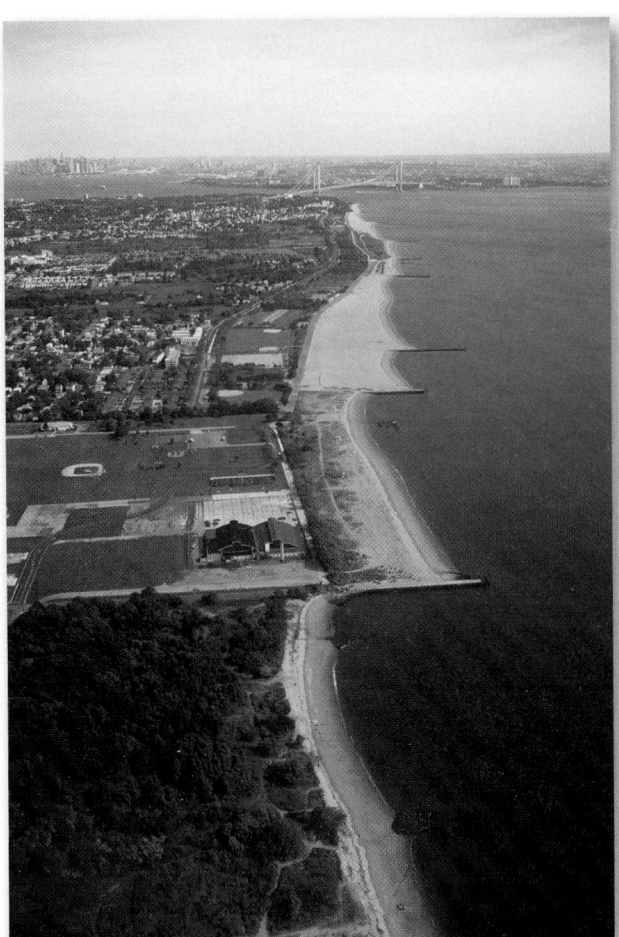

FIGURE 20.15 ■ Construction of a groin to control beach erosion may result in erosion downcurrent of the groin and loss of parts of the beach there, while sand piles up on the other side of the groin. [Airphoto—Jim Wark.]

months and years, the sand disappears from the beach on one side of the groin and greatly enlarges the beach on the other side (**Figure 20.15**). These changes are the predictable result of normal coastal processes. The waves, longshore current, and longshore drift bring sand toward the groin from the upcurrent direction (usually the prevailing wind direction). Stopped at the groin, they dump the sand there. On the downcurrent side of the groin, the current and drift pick up again and erode the beach. On this side, however, replenishment of sand is sparse because the groin blocks inputs of sand. As a result, the sand budget is out of balance, and the beach shrinks. If the groin is removed, the beach returns to its former state.

The only way to preserve a beach is to leave it alone. Groins and seawalls are only temporary solutions to the problem of beach erosion, and even if they are kept in repair with large expenditures of money—many times at public expense—the beach itself will suffer. Beach restoration projects, which involve pumping large volumes of sand from offshore, have had some success (see Practicing Geology exercise), but they, too, are extremely costly. Sooner or later, we must learn to let beaches remain in their natural state.

Erosion and Deposition at Shorelines

The topography of shorelines is a product of the same forces that shape the continental interior: plate tectonic processes that elevate or depress Earth's crust, erosional processes that wear it down, and sedimentation that fills in the low spots. Thus, several factors are directly at work:

■ Tectonic uplift of the coastal region, which leads to erosional coastal forms

■ Tectonic subsidence of the coastal region, which leads to depositional coastal forms

■ The nature of the rocks or sediments at the shoreline

■ Changes in sea level, which affect the submergence or emergence of a shoreline

■ The average and storm wave heights, which affect erosion

■ The heights of the tides, which affect both erosion and sedimentation

EROSIONAL COASTAL FORMS Erosion is an important process along uplifted rocky coasts. Along these coasts, prominent cliffs and headlands jut into the sea, alternating with narrow inlets and irregular bays with small beaches. Waves crash against the rocky shorelines, undercutting cliffs and causing huge blocks of rock to fall into the water, where they are gradually eroded away. As the sea cliffs retreat, isolated remnants called *stacks* may be left standing in the sea, far from the shore (see Figure 20.1c). Erosion by waves also planes the rocky surface beneath the surf zone and creates a **wave-cut terrace,** which is sometimes visible at low tide (**Figure 20.16**). Wave erosion over long periods may straighten shorelines as headlands are eroded faster than recesses and bays.

Where relatively soft sediments or sedimentary rocks make up the coastal region, slopes are gentler and the heights of shoreline bluffs are lower. Waves erode these softer materials efficiently, and erosion of bluffs on such shores may be extraordinarily rapid. The high sea cliffs of soft glacial sediments at the Cape Cod National Seashore in Massachusetts, for instance, are retreating about a meter each year. Since Henry David Thoreau walked the entire length of the beach below those cliffs in the mid-nineteenth century and wrote of his travels in *Cape Cod*, about 6 km^2 of coastal land have been eaten away by the ocean, equivalent to about 150 m of beach retreat.

Our discussion of beaches illustrates the importance of erosional processes in those soft-sediment environments. In recent decades, more than 70 percent of the total length of the world's sand beaches has retreated at a rate of at

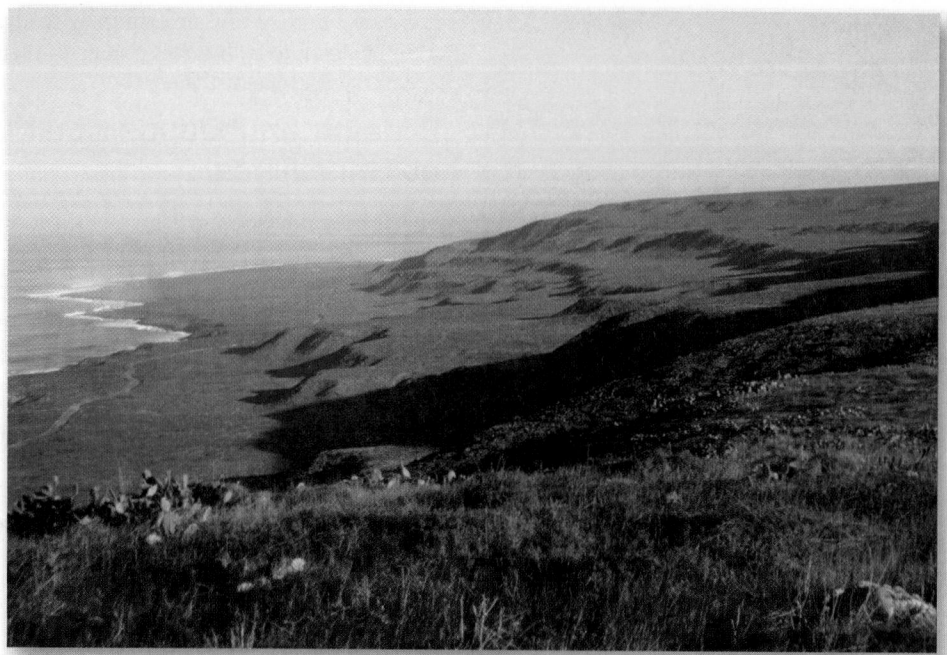

FIGURE 20.16 ▪ Multiple wave-cut terraces on the California coastline. Each terrace records a distinctly different sea level elevation. Sea level is controlled in turn by glacial ice volumes (see Chapter 15); when ice volumes are stable, sea level is fixed, and waves erode bedrock. [Photo by Dan Muhs/USGS, Muhs, Daniel R., Simmons, Kathleen, R., Kennedy, George L., and Rockwell, Thomas K. The last interglacial period on the Pacific Coast of North America: Timing and paleoclimate. Geological Society of America.]

least 10 cm/year, and 20 percent has retreated at a rate of more than 1 m/year. Much of this loss can be traced to the damming of rivers, which decreases sediment input to the shoreline.

DEPOSITIONAL COASTAL FORMS Sediments build up in areas where subsidence depresses Earth's crust along a coastline. Such coastlines are characterized by wide, low-lying coastal plains of sedimentary rock and by long, wide beaches. Shoreline forms along these coastlines include sandbars, low-lying sandy islands, and extensive tidal flats. Long beaches grow longer as longshore currents carry sand to the downcurrent end of the beach. There it builds up, first forming a submerged sandbar, then rising above the surface and extending the beach by a narrow addition called a **spit.**

Offshore, long sandbars may build up into **barrier islands** that form a barricade between open-ocean waves and the main shoreline. Barrier islands are common, especially along low-lying coasts composed of small sediment particles that are easily eroded and transported, or of poorly cemented sedimentary rocks where longshore currents are strong. As the sand builds up above the waves, vegetation takes hold, stabilizing the islands and helping them resist wave erosion during storms. Barrier islands are separated from the coast by tidal flats or shallow lagoons. Like beaches on the main shore, barrier islands are maintained at a dynamic equilibrium by the forces shaping them. That equilibrium can be disturbed by natural changes in climate or in wave and current patterns, as well as by human activities. Disruption or devegetation can lead to increased erosion, and barrier islands may even disappear beneath the sea surface. Barrier islands may also grow larger and more stable if sedimentation increases.

Over hundreds of years, sandy shorelines may undergo significant changes. Hurricanes and other intense storms may form new inlets, elongate spits, or breach existing spits and barrier islands. Such changes have been documented by aerial photographs taken at various time intervals. The shoreline of Chatham, Massachusetts, at the elbow of Cape Cod, has changed enough in the past 160 years or so that a lighthouse has had to be moved. **Figure 20.17** illustrates the many changes that have taken place in the configuration of the barrier islands to the north and to the long spit of Monomoy Island, including several breaches of the barrier islands. Many homes in Chatham are now at risk, but there is little that the residents or the state can do to prevent coastal processes from taking their natural course.

Effects of Sea Level Change

The shorelines of the world serve as barometers for impending changes caused by many types of human activities. The pollution of our inland waterways sooner or later arrives at our beaches, just as sewage from city dumping and oil from oceangoing tankers wash up on the shore.

(a) Beach near Chatham Light

The 1987 breach in the barrier spit, shown at the right below, closed again before this photo was taken.

Plymouth
Cape Cod Bay
Area of detail

(b)

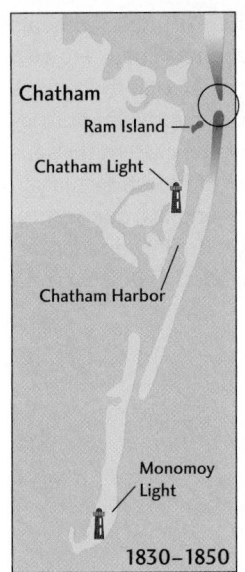

Chatham
Ram Island
Chatham Light
Chatham Harbor
Monomoy Light
1830–1850

1870–1890

1910–1930

1950–1970

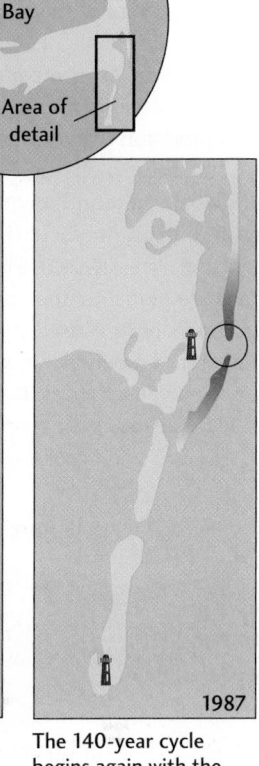

1987

The circle shows the approximate location of the 1846 breach in barrier island. Ram Island later disappears.

The beach south of the inlet breaks up and migrates southwest toward the mainland and Monomoy.

The southern beach has disappeared, and its remnants soon will connect Monomoy to the mainland.

The northern beach steadily grows with cliff sediment; Monomoy breaks from the mainland.

The 140-year cycle begins again with the Jan. 2 breach in the barrier spit across from the Chatham Light (circle).

FIGURE 20.17 ■ Migrating barrier islands at Chatham, Massachusetts, at the southern tip of Cape Cod. (a) Aerial view of Monomoy Point. This spit has advanced into deep water to the south (*foreground*) from barrier islands along the main body of the Cape to the north (*background*). (b) Changes in the shoreline at Chatham over the past 160 years. [(a) Steve Dunwell/The Image Bank/Getty Images; (b) after Cindy Daniels, *Boston Globe* (February 23, 1987).]

As real estate development and construction along shorelines expands, we will see the continuing contraction, and even the disappearance, of some of our finest beaches. As global warming and glacial melting cause sea level to rise, we will see the effects of that change on our shorelines as well.

Shorelines are particularly sensitive to changes in sea level, which can alter tidal heights, change the approach patterns of waves, and affect the paths of longshore currents. The rise and fall of sea level at a shoreline can be local—the result of tectonic subsidence or uplift—or global—the result of glacial melting or growth. One of the primary concerns about human-induced global warming is

its potential for causing sea level rises that will flood coastal cities, as we will see in Chapter 21.

In periods of globally lowered sea level, areas that were offshore are exposed to agents of erosion. Rivers extend their channels over formerly submerged regions and cut valleys into newly exposed coastal plains. When sea level rises, flooding the backshore, marine sediments build up along former land areas, erosion is replaced by sedimentation, and river valleys are submerged. Today, long fingers of the sea indent many of the shorelines of the northern and central Atlantic coast of North America. These long indentations are former river valleys that were flooded as the last ice age ended about 11,000 years ago and sea level rose.

Sea level variations on geologic time scales can be measured by studies of wave-cut terraces (see Figure 20.16), but detecting global sea level changes on shorter (human) time scales is more difficult. Local changes can be measured by using a tide gauge that records sea level relative to a land-based benchmark. The major problem with this approach is that the land itself moves vertically as a result of deformation, sedimentation, and other geologic processes, and this movement is incorporated into the tide-gauge observations. A newer method of tracking sea level changes makes use of an altimeter mounted on a satellite. The altimeter sends pulses of radar beams that are reflected off the ocean surface, providing measurements of the elevation of the ocean surface relative to the orbit of the satellite with a precision of a few centimeters.

Using these methods, oceanographers have determined that global sea level has risen by 17 cm over the last century, and is continuing to rise about 3 mm/year. This recent increase in sea level correlates with a worldwide increase in average global temperatures, which most scientists now believe has been caused, at least in part, by anthropogenic emissions of greenhouse gases (see Chapter 23). Some of the rise may result from short-term variations, but the magnitude of the rise is consistent with climate models that take greenhouse warming into account. These models predict that without significant worldwide efforts to reduce greenhouse gas emissions, sea level will rise by another 3100 cm during the twenty-first century.

■ Continental Margins

Offshore, beyond the shoreline, lies the continental shelf. At its edge is a **continental slope** that descends more or less steeply into the depths of the ocean. At the foot of the slope is a **continental rise,** a gently sloping apron of muddy and sandy sediments extending to the flat **abyssal plain** at the bottom of the ocean basin (**Figure 20.18**).

The shorelines, shelves, and slopes of the continents are together called **continental margins.** There are two basic types of continental margins, passive and active. A **passive margin** is formed when seafloor spreading carries a continent far from a plate boundary; the eastern margins of North America and Australia and the western margin of Europe are examples. The name implies quiescence: volcanoes are absent, and earthquakes are few and far between. In contrast, **active margins,** such as the western

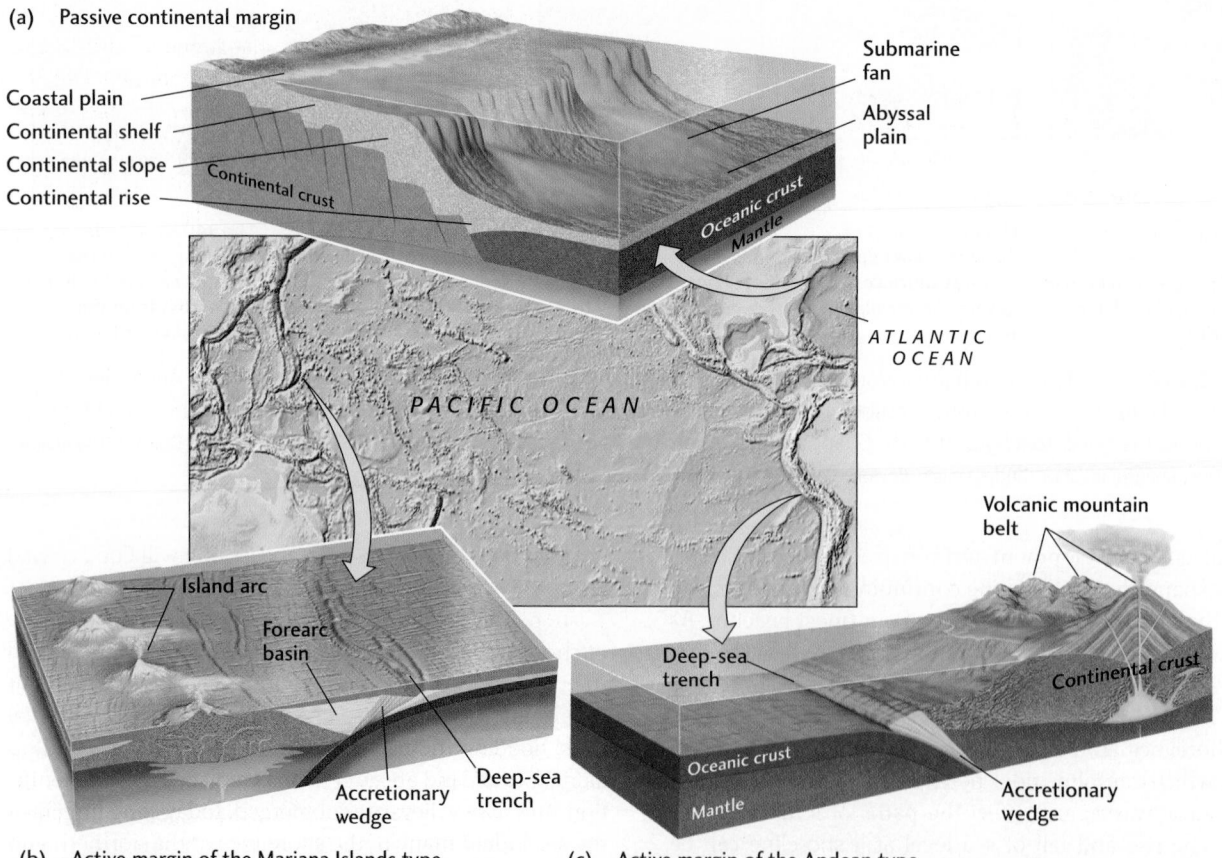

(a) Passive continental margin

Coastal plain
Continental shelf
Continental slope
Continental rise
Continental crust
Oceanic crust
Mantle

Submarine fan
Abyssal plain

ATLANTIC OCEAN

PACIFIC OCEAN

Island arc
Forearc basin
Accretionary wedge
Deep-sea trench

(b) Active margin of the Mariana Islands type

Volcanic mountain belt
Deep-sea trench
Continental crust
Oceanic crust
Mantle
Accretionary wedge

(c) Active margin of the Andean type

FIGURE 20.18 ■ Schematic profiles of three types of continental margins: (a) Passive margin. (b) Active margin of the Mariana Islands type. (c) Active margin of the Andean type.

margin of South America, result from subduction near a continent. Occasionally, active margins are associated with transform faulting. Volcanic activity and frequent earthquakes give these continental margins their name. Active margins at subduction zones include an offshore trench and an active volcanic island arc or mountain belt.

The continental shelves of passive margins consist of essentially flat-lying shallow-water sediments, both terrigenous and carbonate, several kilometers thick (Figure 20.18a). Although the same kinds of sediments can be found on the shelves of active margins, they are more likely to be structurally deformed and to include ash and other volcanic materials as well as deep-sea sediments. Most active margins on the eastern side of the Pacific Ocean (for example, west of the Andes in South America) feature a narrow continental shelf that falls off sharply into a deep-sea trench without much accumulation of sediments (Figure 20.18c). Those on the western side of the Pacific (for example, off the Mariana Islands) have a wider shelf lying between the continent and the subduction zone. The trench forms a substantial *forearc basin,* where thick sequences of sediment are deposited (Figure 20.18b). Most of these sediments come from erosion of the uplifted island arc, but some sediments are scraped off the subducting oceanic crust, forming an *accretionary wedge.*

The Continental Shelf

The continental shelf is one of the most economically valuable parts of the ocean. Georges Bank off New England and the Grand Banks of Newfoundland, both on the continental shelf off North America, are among the world's most productive fishing grounds. Recently, oil-drilling platforms have been used to extract huge quantities of oil and gas from beneath continental shelves, especially off the Gulf Coast of Louisiana and Texas.

Because continental shelves lie at shallow depths, they are subject to emergence and submergence as a result of changes in sea level. During the Pleistocene ice ages, all the shelves now at depths of less than 100 m were above sea level, and many of their features were formed then. Shelves at high latitudes were glaciated, and thus have an irregular topography of shallow valleys, basins, and ridges. Those at lower latitudes are more regular, cut by occasional stream valleys.

The Continental Slope and Rise

The waters of the continental slope and rise are too deep for the seafloor to be affected by waves and tidal currents. As a consequence, sediments that have been carried across the continental shelf by waves and tides come to rest as they are draped over the continental slope. Geologists have observed signs of sediment slumping and erosional scars in the form of gullies and submarine canyons on continental

slopes. In addition, deposits of sand, silt, and mud on both the continental slope and the continental rise indicate active transport of sediments into deeper waters. For some time, geologists puzzled over what kind of current might cause both erosion and sedimentation on the slope and rise at such great depths.

The answer proved to be **turbidity currents:** turbulent flows of muddy water down the slope (**Figure 20.19**). Turbidity currents can both erode and transport sediment. They begin when sediments draped over the edge of the continental shelf slump onto the continental slope. Such a sudden submarine landslide throws mud into suspension, creating a dense, turbid layer of water. Because of its suspended load of mud, the turbid water is denser than the overlying clear water and flows beneath it, accelerating down the slope. As the turbidity current reaches the foot of the slope, it slows. Some of the coarser sandy sediment starts to settle out, often forming a *submarine fan*—a deposit something like an alluvial fan on land. Stronger turbidity currents may continue across the continental rise, cutting channels into submarine fans. Where these currents reach the abyssal plain, they spread out and come to rest in graded beds of sand, silt, and mud called *turbidites.*

According to current research, submarine landslides that start turbidity currents are common. Some of them may be huge. One slide generated 8- to 10-m-thick turbidites—consisting of 500 km^3 of sediments—over a large area of the western Mediterranean. Submarine slides may occur spontaneously or be triggered by an earthquake. They may also be caused by thawing of methane-water ices: crystalline solids composed of methane and water. Methane-water ices are stable at the high pressures and low temperatures of many large areas of the oceans. In deeply buried sediments, they turn into methane gas (see Chapter 11). If sea level is lowered, as it is during ice ages, the pressure at the ocean bottom is reduced, and the ices may gasify, triggering a landslide. The quantities of gas produced are enormous, and geologists have speculated about the possibility of exploiting these subsea methane-water ices as fuels.

Submarine Canyons

Submarine canyons are deep valleys eroded into the continental shelf and slope. They were discovered near the beginning of the twentieth century and were first mapped in detail in 1937. At first, some geologists thought they might have been formed by rivers. There is no question that the shallower parts of some canyons were river channels during periods when sea level was low. But this hypothesis could not provide a complete explanation. Most of the canyon floors are thousands of meters deep. Even at the minimum sea levels of the Pleistocene ice ages, rivers could have eroded the seafloor only to a depth of about 100 m.

(a)

(b) Slump

FIGURE 20.19 ■ Turbidity currents transport sediments from the continental shelf into deeper waters. (a) Slumps on the continental slope generate turbidity currents, which flow down the continental slope and continental rise to the abyssal plain. (b) A slump at the head of a submarine canyon at the edge of the continental shelf. [U.S. Navy.]

Although other types of currents have been proposed, turbidity currents are now the favored explanation for the deeper parts of submarine canyons (see Figure 20.19). Evidence supporting this conclusion comes in part from a comparison of deposits within modern submarine canyons with well-preserved similar deposits of the past, particularly turbidites deposited in canyons and channels on submarine fans.

■ Topography of the Deep Seafloor

A topographic map of Earth's ocean basins (Figure 20.20) reveals the most important geologic features submerged beneath the oceans: mid-ocean ridges, volcanic tracks of hot spots, deep-sea trenches, island arcs, and continental margins.

Making a map of the deep seafloor is no easy task. Because sunlight can penetrate only 100 m or so below the ocean surface, the deep ocean is a very dark place. It is not possible to map the seafloor using visible light, nor can we use radar beamed from spacecraft, which we have used to map the surface of cloud-shrouded Venus. Ironically, remote sensing from spacecraft has allowed us to map the surfaces of our planetary neighbors with much higher

resolution than we have been able to do for Earth's deep seafloor, even to this day.

Probing the Seafloor from Surface Ships

It is possible for scientists to view the seafloor directly from a deep-diving submersible. These small submarines can observe and photograph the seafloor at great depths (Figure 20.21). With their mechanical arms, they can break off pieces of rock, sample soft sediments, and capture specimens of exotic deep-sea animals. Newer robotic submersibles can be guided by scientists on a mother ship at the surface. But submersibles of either type are expensive to build and operate, and they cover small areas at best.

For most work, today's oceanographers use instrumentation on ships at the surface to sense the seafloor topography indirectly. Echo sounders (also known as *sonar*), developed in the early part of the twentieth century, send out pulses of sound waves; when those waves are reflected back from the ocean bottom, they are picked up by sensitive microphones in the water. Oceanographers can measure depth by determining the interval between the time a pulse leaves the device and the time it returns as a reflection. The result is an automatically traced profile of the seafloor. Powerful echo sounders are also used to probe the stratigraphy of sedimentary layers beneath the seafloor (see Figure 14.6).

FIGURE 20.20 ■ A topographic map of Earth's ocean basins, showing the major features of the deep seafloor.

| *ALVIN*
(manned
submersible) | ROV
(remotely
operated vehicle) | *SeaBeam*
(hull-mounted
swath-mapping sonar) | *JOIDES Resolution*
drilling ship | Permanent seafloor
observatory |

FIGURE 20.21 ■ High-technology methods for exploring the deep seafloor. The manned deep submersible *Alvin* and a remotely operated vehicle (ROV) are directed from a surface ship. *SeaBeam,* a hull-mounted multibeam echo sounder, continuously maps seafloor topography in a wide swath as a ship steams across the ocean surface. The drilling ship *JOIDES Resolution* (see also Figure 2.13), part of the Ocean Drilling Program, uses bottom transponders to maneuver a drill pipe into a reentry hole on the seafloor. Permanent unmanned seafloor observatories can monitor processes in the subsurface and the overlying water column for extended periods.

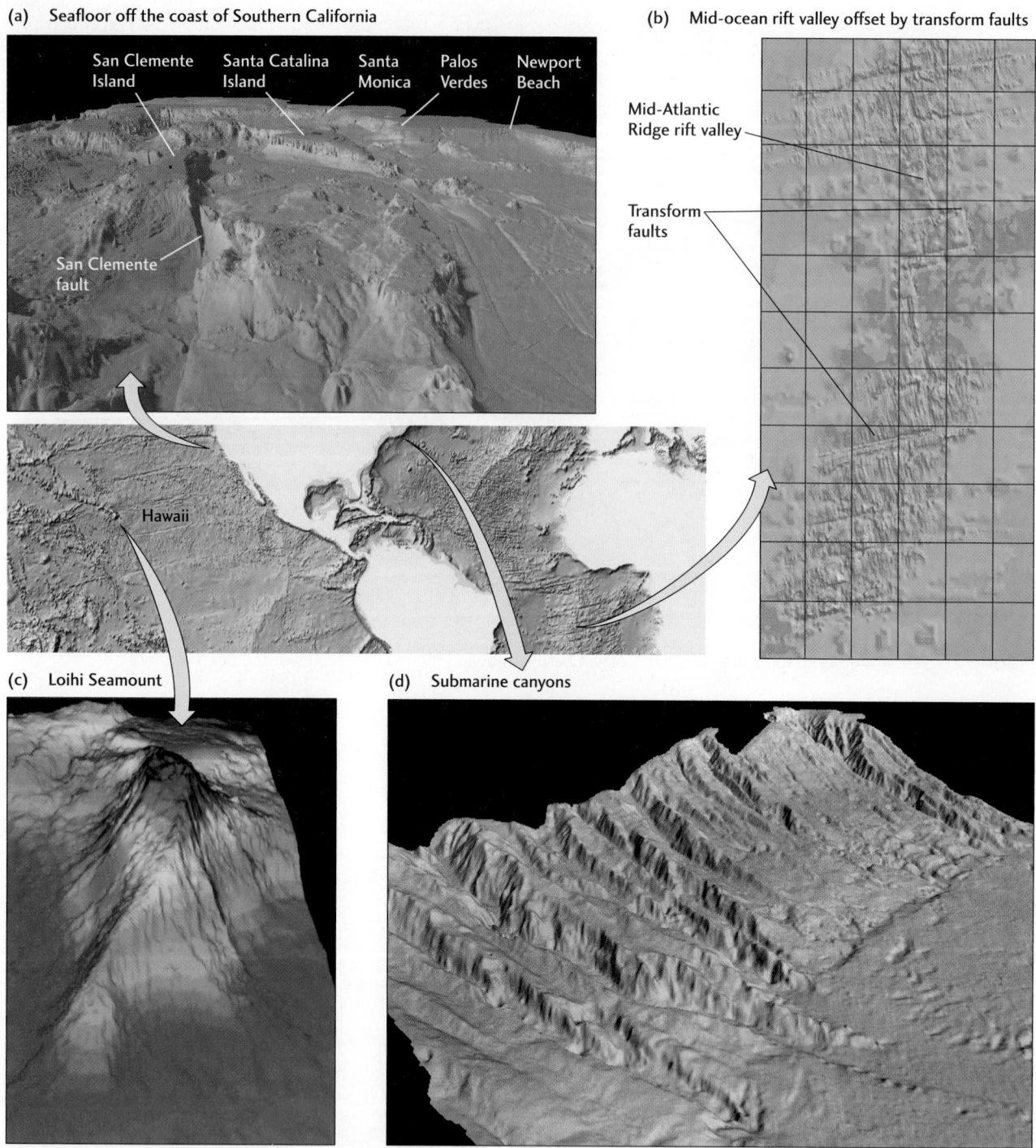

(a) Seafloor off the coast of Southern California

San Clemente Island Santa Catalina Island Santa Monica Palos Verdes Newport Beach

San Clemente fault

Hawaii

(b) Mid-ocean rift valley offset by transform faults

Mid-Atlantic Ridge rift valley

Transform faults

(c) Loihi Seamount

(d) Submarine canyons

FIGURE 20.22 ■ Four examples of seafloor topography mapped by high-resolution swath-mapping echo sounder arrays on surface ships and rendered by computer processing as three-dimensional images. (a) The seafloor off the coast of Southern California, showing fault-bounded structures of a geologic province known as the California Borderland. (b) The Mid-Atlantic Ridge between 25° and 36° S latitudes, showing the southeast-trending rift valley offset by northeast-trending transform faults. (c) Loihi Seamount, just south and east of the island of Hawaii, the newest in the string of hot-spot volcanoes that form the Hawaiian Islands. (d) The continental shelf (*top*), slope (*central and upper area*), and rise (*lower left*) off the coast of New England. Note the deep submarine canyons that incise this continental margin. [(a) Chris Goldfinger and Jason Chaytor, Oregon State University; (b) IEDA: Global Multi-Resolution Topography Synthesis; (c) Dr. Robert Tyce, University of Rhode Island; (d) Courtesy of Lincoln Pratson & William Haxby, Lamont Doherty Observatory of Columbia University, Palisades, NY, 10964. *Geology*, January 1996.]

Many of today's oceanographic vessels are outfitted with hull-mounted sonar arrays that can reconstruct a detailed image of seafloor topography in a swath extending as much as 10 km on either side of the ship as it steams along (see Figure 20.21). These systems can map the seafloor with unprecedented resolution of small-scale geologic features, such as undersea volcanoes, canyons, and faults. Figure 20.22 shows several impressive images of the seafloor derived by this type of mapping.

Other types of instruments can be towed behind a ship or lowered to the bottom to evaluate the magnetism of the seafloor, the shapes of undersea cliffs and mountains, and the heat coming from the crust. Underwater cameras on sleds towed near the ocean bottom can photograph the details of the seafloor and the organisms that inhabit the deep sea. Since 1968, the U.S. Deep Sea Drilling Project and its successor, the international Ocean Drilling Program, have sunk hundreds of drill holes to depths of many hundreds of meters below the seafloor. Cores obtained from these drill holes have provided geologists with sediment and rock samples for detailed physical and chemical studies.

Charting the Seafloor by Satellite

Despite all this fancy gear, there are still many regions of the oceans that have not been surveyed, and our knowledge of the seafloor remains fragmentary. Recently, however, satellite altimeters have been used to chart the global topography of the seafloor indirectly. The elevation of the ocean surface depends not only on waves and ocean currents, but also on variations in gravity caused by the topography and composition of the underlying seafloor. The gravitational attraction of a seamount, for example, can cause water to "pile up" above it, producing a bulge in the ocean surface of as much as 2 m above average sea level. Similarly, the diminished gravity over a deep-sea trench is evident as a depression in the ocean surface of as much as 60 m below average sea level.

This method has allowed us to infer features of the seafloor from satellite data and display them as if the seas were drained away. Marine geologists have used this technique to map new features of the seafloor not revealed by ship surveys, especially in the poorly surveyed southern oceans. The satellite data have also revealed deeper structures below the oceanic crust, including gravity anomalies associated with convection currents in the mantle, as described in Chapter 14.

Profiles Across Two Oceans

To gain an appreciation of the geologic features that lie beneath the oceans, let's take a brief tour across two of Earth's major ocean basins, the Atlantic and the Pacific. We'll travel across these ocean basins as if we were driving a deep-diving submersible along the seafloor.

AN ATLANTIC PROFILE The Atlantic profile shown in Figure 20.23 extends from North America to Gibraltar.

FIGURE 20.23 ■ A topographic profile of the Atlantic Ocean basin along a route from New England (*left*) to Gibraltar (*right*).

Starting from the coast of New England, we descend from the shoreline to depths of 50 to 200 m and travel eastward across the continental shelf. After traveling about 50 to 100 km down a very gently inclined surface, we find ourselves at the edge of the shelf. There, we start down a steeper incline, the continental slope. This mud-covered slope descends at an angle of about 4°, a drop of 70 m over a horizontal distance of 1 km, which would be a noticeable grade if we were driving on land.

The continental slope is irregular and is marked by gullies and submarine canyons eroded into the slope and the shelf behind it (see Figure 20.22d). On the lower parts of the slope, at depths of about 2000 to 3000 m, the downward incline becomes gentler as we reach the continental rise.

The continental rise is tens to hundreds of kilometers wide, and it grades imperceptibly into the wide, flat abyssal plain at depths of about 4000 to 6000 m. The plain is broken by occasional submerged volcanoes, mostly extinct, called **seamounts.** As we travel along the abyssal plain, we gradually climb into a province of low **abyssal hills** whose slopes are covered with fine-grained sediments. As we continue up the hills, the sediment layer becomes thinner, and outcrops of basalt appear beneath it. As we rise along this steep, irregular topography to depths of about 3000 m, we are climbing the flanks and then the mountains of the Mid-Atlantic Ridge.

Abruptly, we come to the edge of a deep, narrow valley a few kilometers wide at the top of the ridge (**Figure 20.24**). This narrow cleft, marked by active volcanism, is the rift valley where two plates separate. As we cross the valley and climb its eastern side, we move from the North American Plate to the Eurasian Plate. Continuing eastward, we encounter topography similar to that on

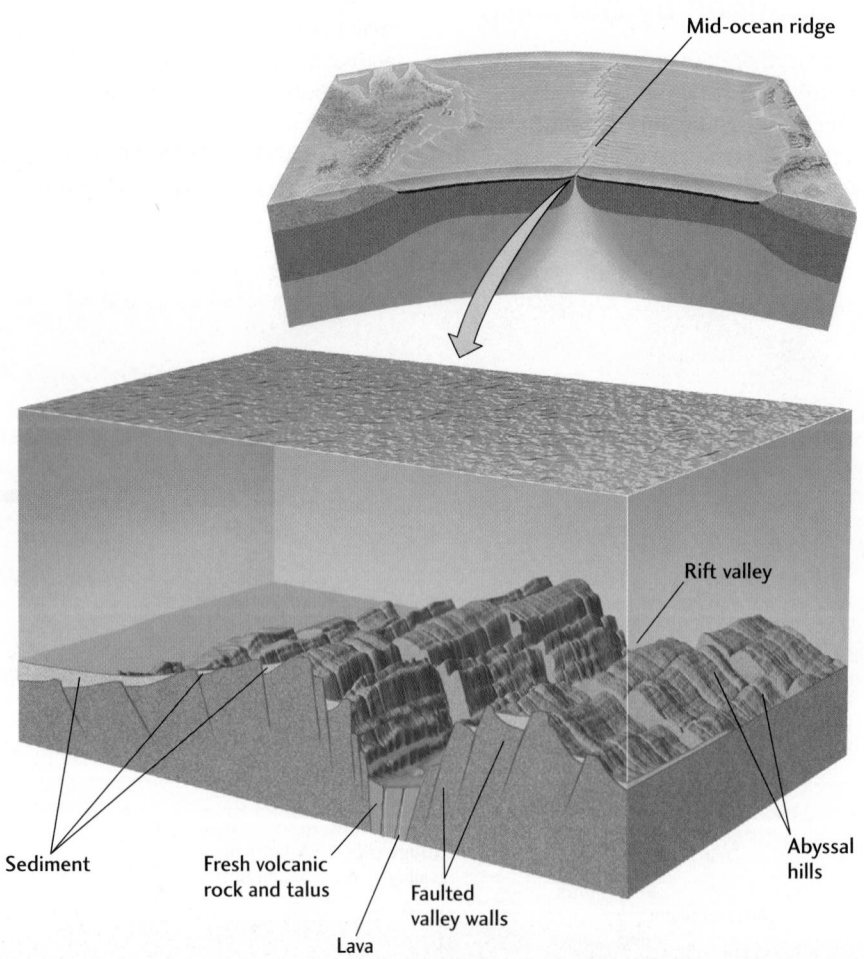

FIGURE 20.24 ■ A profile of the central rift valley of the Mid-Atlantic Ridge in the FAMOUS (French-American Mid-Ocean Undersea Study) area southwest of the Azores. This deep valley is where most of the basin's new oceanic crust is extruded. [After ARCYANA, "Transform Fault and Rift Valley from Bathyscaph and Diving Saucer," *Science* 190 (1975): 108.]

the western side of the ridge, only in reverse order, because the seafloor is more or less symmetrical on either side of the ridge. Passing over the rough topography of the abyssal hills on the flank of the Mid-Atlantic Ridge, we descend to the abyssal plain and then travel up the continental rise, slope, and shelf off the coast of Europe. On the path we have taken, this symmetry is disturbed by some large seamounts and the volcanic islands of the Azores, which mark an active hot spot, perhaps caused by the heat from an upwelling mantle plume.

A PACIFIC PROFILE Our second virtual tour moves westward across the Pacific, from South America to Australia (**Figure 20.25**). Beginning on the west coast of Chile, we cross a narrow continental shelf only a few tens of kilometers wide. At the edge of the shelf, we plunge down a continental slope that is much steeper than the one we traversed

in the Atlantic, extending down to 8000 m as we enter the Peru-Chile Trench. This long, deep, narrow depression of the seafloor is the surface expression of the subduction of the Nazca Plate under the South American Plate.

Continuing across the trench and up onto the abyssal hills of the Nazca Plate, we come to the East Pacific Rise, an active mid-ocean ridge. The East Pacific Rise is lower than the Mid-Atlantic Ridge, and its rate of seafloor spreading is the world's fastest—at about 150 mm/year, more than six times faster than that of the Mid-Atlantic Ridge—but it has the characteristic central rift valley and outcrops of fresh basalt. We cross over to the Pacific Plate, on the western side of the East Pacific Rise, and travel westward over its broad central regions, which are studded with seamounts and volcanic islands.

We eventually arrive at another subduction zone, marked by the Tonga Trench, where the Pacific Plate sinks

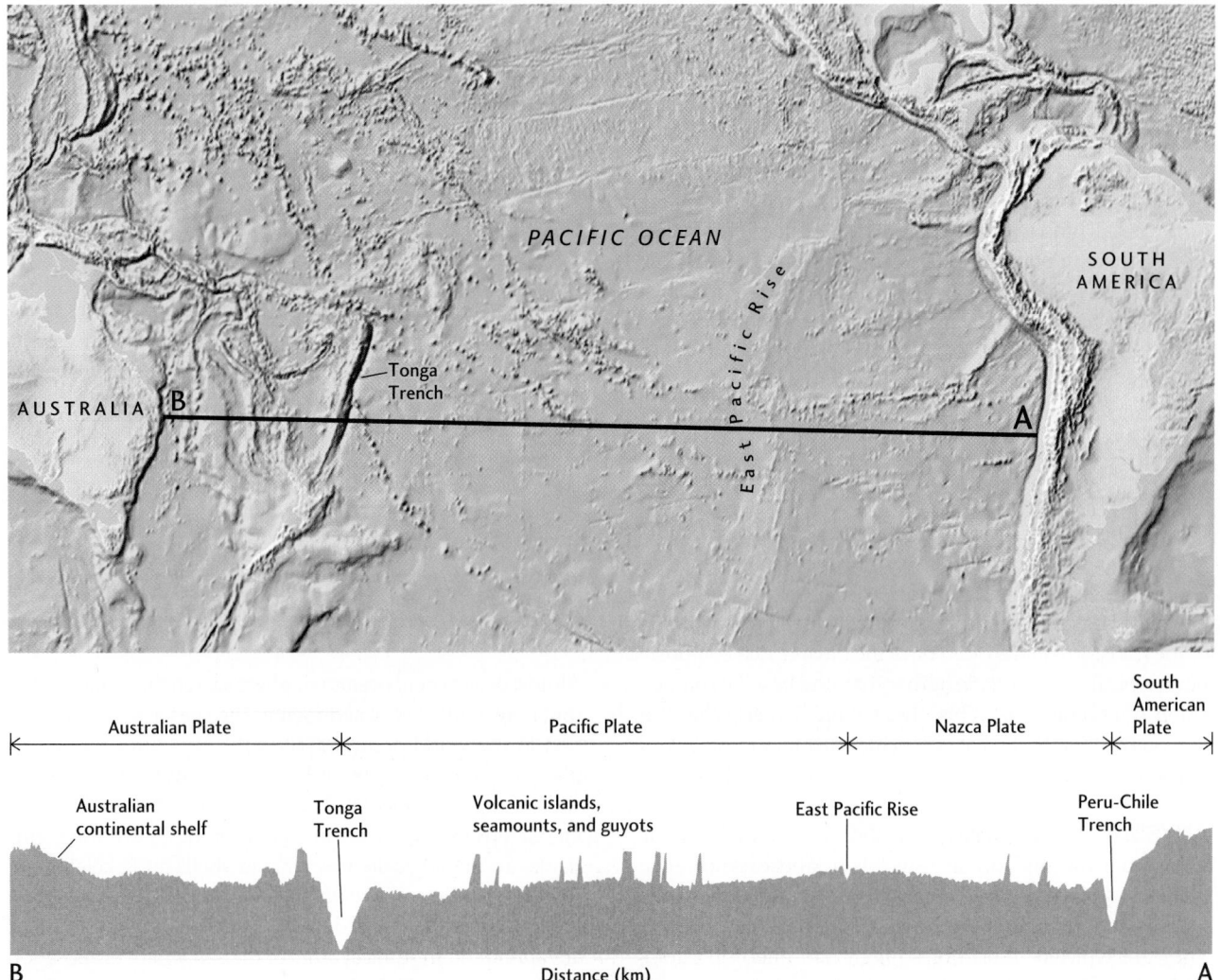

FIGURE 20.25 ■ A topographic profile of the Pacific Ocean basin along a route from South America (*right*) to Australia (*left*).

into the mantle beneath the Australian Plate. This is one of the deepest places in the world ocean, almost 11,000 m below the ocean surface. On the western side of the trench, the seafloor rises to form the volcanic islands of Tonga and Fiji. Beyond this island arc, we return to the abyssal plain, now on the Australian Plate, and come to the continental rise, slope, and shelf of eastern Australia, which has a profile similar to that of the east coast of North America.

Main Features of the Deep Seafloor

Away from continental margins and subduction zones, the deep seafloor is constructed primarily by volcanism related to seafloor spreading and secondarily by sedimentation in the open ocean.

MID-OCEAN RIDGES Mid-ocean ridges are the sites of the most intense volcanic and tectonic activity on the deep seafloor. The main rift valley is the center of the action. The valley walls are faulted and intruded by basaltic sills and dikes (see Figure 20.24), and the floor of the valley is covered with flows of basalt and talus from the valley walls, mixed with small amounts of sediment that have settled from surface waters.

Hydrothermal vents form on the rift valley floor as seawater percolates into fractures in the basalt on the flanks of the ridges, is heated as it flows downward to encounter hotter basalt, and finally exits at the valley floor (see Figure 12.22). There, it boils up at temperatures as high as 380°C. Some springs are "black smokers," full of dissolved hydrogen sulfide and metals that the hot water has leached from the basalt (see Figure 11.15). Others are cooler "white smokers" that release various compounds of barium, calcium, and silicon. Both black smokers and white smokers produce mounds of iron-rich clay minerals, iron and manganese oxides, and large deposits of iron–zinc–copper sulfides.

Mid-ocean ridges are offset at many places by transform faults that displace the rift valleys laterally (see Figure 20.22b). Large earthquakes occur on these faults as one plate slips past the other. Rocks collected from the walls of the transform faults often have the olivine-rich compositions typical of the mantle, rather than the basaltic composition of oceanic crust. This observation suggests that the igneous process that creates oceanic crust may operate less efficiently where the spreading center abuts a fault.

ABYSSAL HILLS AND PLAINS The deep seafloor away from the mid-ocean ridges is a landscape of hills, plateaus, sediment-floored basins, and seamounts. Abyssal hills are ubiquitous on the slopes of mid-ocean ridges. They are typically 100 m or so high and lined up parallel to the ridge crest (see Figure 20.24). They are formed primarily by normal faulting of the newly formed oceanic crust as it moves out of the rift valley. Almost all of this faulting occurs during the first million years of the crust's existence, after which the faults bounding the hills become inactive.

As the new oceanic crust drifts away from the spreading center, it cools and contracts, lowering the seafloor. Its hilly, subsiding surface receives a steady rain of sediments from surface waters, gradually becoming covered with deep-sea muds and other deposits. Near the continental margins, terrigenous sediments moving down the continental slopes add to this sediment cover and create the flat, unbroken expanses of the abyssal plains. These plains are the flattest solid surfaces on the planet.

SEAMOUNTS, HOT-SPOT ISLAND CHAINS, AND PLATEAUS The seafloor is littered with tens of thousands of volcanoes. Most are submerged seamounts, but some rise above the ocean surface as volcanic islands. Seamounts and volcanic islands may be isolated or found in clusters or chains. Most seamounts, though not all, are created by eruptions near active spreading centers or where a plate overrides a hot spot.

Some of the larger seamounts, called **guyots,** have flat tops, the result of erosion that occurred when they were volcanic islands. These islands later became submerged as the plate on which they ride cooled, contracted, and subsided as it moved away from the spreading center or hot spot that formed them.

Among the most surprising features of deep ocean basins are large basalt plateaus. Some appear to have formed near triple junctions where three spreading centers meet. Others are associated with massive eruptions at hot spots far from spreading centers. Some scientists believe that deep-sea basalt plateaus of the latter type, like continental flood basalts, can be explained by the mantle plume hypothesis (see Chapter 12).

■ Ocean Sedimentation

Almost everywhere oceanographers search the seafloor, they find a thick blanket of sediments. The ceaseless sedimentation in the world ocean modifies the structures formed by plate tectonic processes and creates its own topography at sites of rapid deposition. The sediments are mainly of two kinds: terrigenous muds and sands eroded from the continents, and biologically precipitated shells and skeletons of marine organisms. In regions of the ocean near subduction zones, sediments derived from volcanic ash and lava flows are abundant. In tropical arms of the sea where evaporation is intense, evaporite sediments are deposited.

Sedimentation on the Continental Shelf

Waves and tides are responsible for terrigenous sedimentation on the continental shelf. The waves produced by intense storms move sediments over the shallow and moderate depths of the shelf, and tidal currents flow over the shelf. These waves and currents distribute the sediments brought in by rivers and eroded at the shoreline into long ribbons of sand and layers of silt and mud. Turbidity currents carry these sediments over the edge of the shelf into the deep sea.

Biological sedimentation on the shelf results from the buildup of the calcium carbonate shells and skeletons of organisms living in shallow waters. Most of these organisms cannot tolerate muddy waters and are found only where terrigenous materials are scarce or absent, such as along the extreme southern coast of Florida or off the coast of the Yucatán Peninsula in Mexico. Here, coral reefs thrive and organisms build up large thicknesses of carbonate sediments (see Chapter 5).

Deep-Sea Sedimentation

Far from the margins of continents, fine-grained terrigenous and biologically precipitated particles suspended in seawater slowly settle from the surface to the bottom. These open-ocean sediments, called **pelagic sediments,** are characterized by their great distance from continental margins, their fine particle size, and their slow-settling mode of deposition. The terrigenous materials are mainly clays, which accumulate on the seafloor at a very slow rate—a few millimeters every 1000 years. A small fraction, about 10 percent, may be blown by winds to the open ocean.

The most abundant biologically precipitated pelagic sediments are the shells of foraminifera. These tiny single-celled animals float in the surface waters of the oceans, and their calcium carbonate shells fall to the bottom after the occupants die. There they accumulate as **foraminiferal**

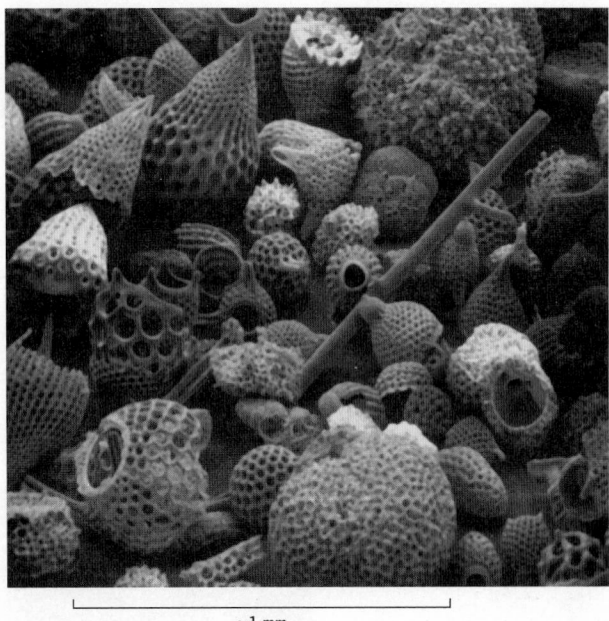

FIGURE 20.26 ■ Scanning electron micrograph of oceanic ooze containing shells of both carbonate-secreting and silica-secreting microorganisms. [Scripps Institution of Oceanography, University of California, San Diego.]

oozes, sandy and silty sediments composed of foraminiferal shells (**Figure 20.26**). Other carbonate oozes are made up of different types of carbonate shells, called *coccoliths.*

Foraminiferal and other carbonate oozes are abundant at depths of less than about 4 km, but they are rare on the deeper parts of the seafloor. This rarity cannot result from a lack of shells, because the surface waters are full of foraminifera everywhere, and the living organisms are unaffected by processes at the bottom far below. The explanation for the absence of carbonate oozes on the deep seafloor is that the shells dissolve below a certain depth, called the **carbonate compensation depth** (**Figure 20.27**).

FIGURE 20.27 ■ The carbonate compensation depth is the level below which calcium carbonate dissolves. As the carbonate shells of dead foraminifera and other organisms settle into deep ocean waters, they enter an environment undersaturated in calcium carbonate and therefore dissolve.

Google Earth Project

The amount of water that covers Earth's surface makes our planet unique in the solar system. Earth's large ocean basins are formed by plate tectonic processes, and the water contained within them forms a tremendous heat sink that has profound effects on global climate. The dissipation of wave, storm, and tidal energy at the ocean surface effectively modifies the edges of large continental land masses. It is at the interface between land and ocean that we see the greatest amount of geologic change visible with GE. In the latest edition of GE, we can view much of the topography of ocean basins as well, but not at the same resolution as most shorelines. To appreciate some of the unique characteristics of coastlines and ocean basins, we will travel to the east coast of North America, which has some of the most striking shoreline sedimentary environments on Earth.

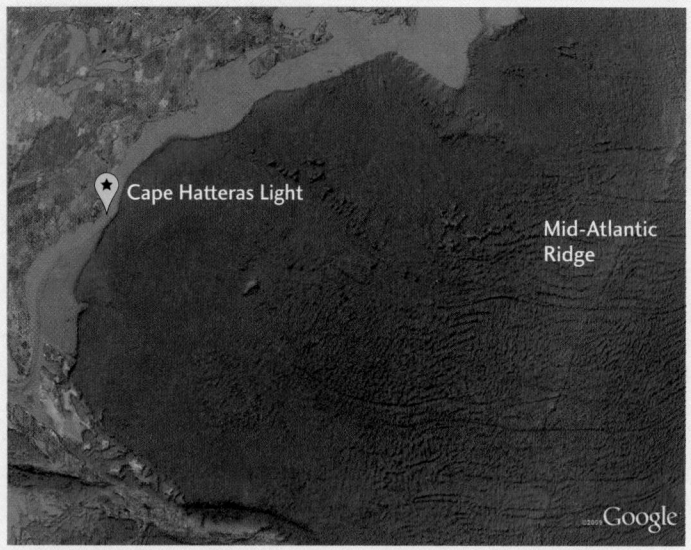

Data SIO, NOAA, U.S. Navy, NGA, GEBCO Image ©2009 TerraMetrics Image USDA Farm Service Agency

LOCATION   Cape Hatteras, North Carolina; Mid-Atlantic Ridge

GOAL   Appreciate the relationship between the coastline, continental shelf, and abyssal plain along a passive continental margin

LINKED   Figures 20.17 and 20.23

Because of thermohaline circulation, deeper waters differ from shallower waters in three ways:

1. Deeper waters are colder. Colder, denser polar waters sink beneath warmer tropical waters and travel toward the equator along the ocean bottom.

2. Deeper waters contain more carbon dioxide. Not only do colder waters absorb more CO_2 than warmer waters, but any organic matter they carry tends to be oxidized to CO_2 in the course of their long circulation.

3. Deeper waters are under higher pressure due to the greater weight of overlying water.

These three factors make calcium carbonate more soluble in deep waters than in shallow ones. As the shells of dead foraminifera fall below the carbonate compensation depth, they enter an environment undersaturated with calcium carbonate, and they dissolve.

Another kind of biologically precipitated sediment, **siliceous ooze,** is produced by sedimentation of the silica

1. Type "Cape Hatteras Light, North Carolina," into the GE search window and zoom out to an eye altitude of 200 km once you arrive there. What kind of landform is the current lighthouse complex constructed on?

   a. A spit
   c. A barrier island
   b. Sea cliffs
   d. A wave-cut terrace

2. Zoom in to an eye altitude of 1 km and focus your view on the swath of cleared land that extends from the current lighthouse location to the northeast. That swath was cut and the lighthouse moved 870 m inland between 1999 and 2000. Investigate the nature of the coastline here and consider the example in Figure 20.17. Why do you think the lighthouse was moved?

   a. The sand beneath the lighthouse was sinking, and the lighthouse no longer stood high enough to warn ships of danger.
   b. The shoreline was steadily eroding, creating the risk of eventual flooding and collapse of the lighthouse.
   c. The lighthouse was at risk of being toppled by an incoming tsunami.
   d. The lighthouse was moved to allow for automobile traffic to better view the ocean.

3. Using the Cape Hatteras lighthouse as a starting point, zoom out to an eye altitude of 2500 km and focus your view on the shoreline of North America here. Consider the source of the currents at work here, and use Figure 20.7 and Figure 15.3a for additional reference if needed. Should the water along this section of the east coast be warmer or cooler than that on the west coast of North America at the same latitude?

   a. Coastal water should be cooler here due to the presence of water circulating from the Arctic.
   b. Coastal water should be warmer here due to the presence of water circulating from equatorial regions.
   c. Coastal water should be warmer here due to the influx of warm air off the southern North American continent.
   d. Coastal water should be cooler here due to the influx of cold air flowing off the Greenland ice cap.

4. Moving eastward from the North American coast near Cape Hatteras, you will notice a relatively flat surface extending offshore to a distance of about 45 km: the continental shelf. It has a maximum depth of only about 50 m over this distance, then drops off steeply along the continental slope that leads to the abyssal plain of the deep sea. Use your cursor to determine the approximate relief of the continental slope. What is that relief?

   a. 3000 m
   c. 70,500 m
   b. 250 m
   d. 55 m

Optional Challenge Question

5. As you continue to move eastward from the continental slope off the Cape Hatteras shore, you will notice that the topography of the abyssal plain remains relatively flat and deep. Eventually you will encounter a gradual decrease in ocean depth where you would not expect it: in the middle of the Atlantic Ocean basin. This decrease in ocean depth and increase in surface relief is associated with the Mid-Atlantic Ridge (see Figure 20.23). In addition to the north-south–oriented central rift valley, there are numerous perpendicular scars along the ridge that trend east-west. What are these features?

   a. These features are small trenches associated with localized subduction.
   b. These features are transform faults that allow lateral movement along the rift.
   c. These features are an artifact of the satellite imagery and are not really present.
   d. These features are dredged shipping lanes to allow large cargo ships to cross the Atlantic Ocean.

shells of *diatoms* (a type of green algae) and *radiolarians* (single-celled protists). Both types of planktonic organisms are abundant in ocean surface waters. After burial on the seafloor, siliceous oozes are cemented into the siliceous rock *chert*.

Some components of pelagic sediments form by chemical reactions of seawater with sediments on the seafloor. The most prominent examples are manganese nodules: black, lumpy accumulations ranging from a few millimeters to many centimeters across. These nodules cover large areas of the deep seafloor—as much as 20 to 50 percent in the Pacific Ocean basin. Rich in nickel and other metals, they are a potential commercial resource if some economical way can be found to mine them and if legal issues regarding their ownership can be resolved. In 1982, the United Nations adopted an agreement known as the Convention on the Law of the Sea, which governs the territorial and economic rights of nations in the world's oceans. At the time of this writing, however, the United States has not signed the agreement.

SUMMARY

How does the geology of ocean basins differ from that of continents? The dominant geologic processes in ocean basins are volcanism and sedimentation. Deformation, weathering, and erosion are much less important in ocean basins than they are on land. Oceanic crust forms at mid-ocean ridges where plates separate, accumulates sediments, and is destroyed by subduction over a few tens of millions of years. Deep-sea sediments, however, provide a nearly continuous record of their relatively brief geologic history.

What coastal processes act on shorelines? Winds blowing over the ocean generate swell; as those waves approach the shore, they are transformed into breakers in the surf zone. Wave refraction results in longshore drift and longshore currents, which transport sand along beaches. Tides, generated by the gravitational pull of the Moon and the Sun on ocean water, can also generate currents that transport sediments.

How do hurricanes affect coastal areas? Hurricanes are intense tropical storms with extremely high winds and very low atmospheric pressures. The low pressure results in the formation of a dome of seawater, known as a storm surge. As the hurricane moves ashore, the storm surge floods low-lying areas, often causing more extensive damage than the storm's high winds.

What processes shape shorelines? Waves and tides, interacting with plate tectonic processes, shape the topography of coastlines, which vary from beaches and tidal flats to uplifted rocky coasts.

What are the major components of a continental margin? A continental margin is made up of a shallow continental shelf; a steeply descending continental slope; and a continental rise, a gently sloping apron of sediment deposited at the lower edges of the continental slope. Active continental margins result from subduction near a continent. Passive continental margins form when seafloor spreading carries a continent far from a plate boundary. Waves and tides affect the continental shelf, but the continental slope is shaped primarily by turbidity currents, which carry large loads of sediment down the slope. Turbidity currents also produce submarine fans and submarine canyons.

What are the main features of the deep seafloor? The deep seafloor is constructed as basalt is extruded in rift valleys along mid-ocean ridges. Abyssal hills are formed by normal faulting as the newly formed oceanic crust moves away from the rift valley. The newly formed crust is soon covered with fine-grained sediments precipitated from surface waters. Near the continental margins, terrigenous sediments add to this sediment cover to create the flat abyssal plains. Volcanic islands, submerged seamounts and guyots, and basalt plateaus are produced on the seafloor by igneous processes.

What kinds of sedimentation occur in and near ocean basins? The two main types of marine sediments are terrigenous sediments and biologically precipitated sediments. Terrigenous sediments are primarily muds and sands eroded from continents and deposited by wave action and tidal currents along the continental shelf. Biological sedimentation on the shelf results from the buildup of calcium carbonate from the shells and skeletons of organisms. Pelagic sediments consist of clay particles and foraminiferal and siliceous oozes composed of the biologically precipitated calcium carbonate and silica shells of planktonic organisms living in the surface waters.

KEY TERMS AND CONCEPTS

abyssal hill (p. 576)

abyssal plain (p. 570)

active margin (p. 570)

barrier island (p. 568)

beach (p. 564)

carbonate compensation
  depth (p. 579)

continental margin
  (p. 570)

continental rise (p. 570)

continental slope (p. 570)

foraminiferal ooze (p. 579)

guyot (p. 578)

hurricane (p. 557)

longshore current (p. 556)

passive margin (p. 570)

pelagic sediment (p. 579)

seamount (p. 576)

shoreline (p. 553)

siliceous ooze (p. 580)

spit (p. 568)

storm surge (p. 561)

tidal flat (p. 557)

tide (p. 556)

turbidity current (p. 571)

wave-cut terrace (p. 567)

PRACTICING GEOLOGY EXERCISE

Does Beach Restoration Work?

Beach erosion is a problem facing many communities that have come to enjoy the scenic beauty of their beaches and depend on them to support tourism and economic de-velopment. Erosion of beaches is often driven by natural processes; in some cases, however, it has been greatly enhanced by failed engineering practices intended to prevent it. In recent years, scientists and engineers have pooled their efforts to create new approaches that have led to greater success in protecting beaches.

The beaches of Monmouth County, New Jersey, on the Atlantic coast of the United States, are among the most intensely studied on Earth. Human modification of these beaches commenced in 1870 with the construction of the New York and Long Branch Railroad. Access by railroad allowed tourism to develop and eventually permitted commuting to New York City by full-time residents, who began to alter the coastline. Concrete seawalls replaced beaches, and sand dunes and rock jetties were built about every quarter mile along the county's 12-mile shoreline. Little by little over the next 100 years, the Monmouth County beaches became very narrow, until miles of shoreline were without a sand beach of any kind. The only bathing beaches were found in tiny pockets tucked into the corners made by the seawall and a jetty. Winter storms in 1991 and 1992 did substantial damage to the entire Monmouth County shoreline, driving the boardwalk back onto streets as splintered debris. Damage to homes occurred as the ocean easily overtopped the almost nonexistent beaches and insufficient seawalls.

By 1994, the state of New Jersey became serious about finding a solution to the beach erosion problem and appealed to the federal government for help. Congress subsequently authorized funding for the nation's largest beach restoration project ever attempted, covering 20 miles of shoreline in Monmouth County, from the township of Sea Bright to Manasquan Inlet. The restoration project involved pumping enough sand from offshore areas to construct a restored beach 100 feet wide with an elevation of 10 feet above mean low water. The project includes periodic nourishment of the restored beaches on a 6-year cycle for 50 years from the start of the initial beach construction in 1994.

Beginning in 1994 and ending in 1997, 57 million cubic meters of sand were pumped from about a mile offshore, at a cost of $210,000,000. This initial placement volume provided a vast supply of new sand to the beaches of 9 out of 12 oceanfront municipalities. The earliest sites restored have responded well, requiring little sand augmentation since the project started.

At the outset, it was not obvious that the Monmouth County beach restoration project would succeed. Some people predicted total loss of the sand within a year or two. Nevertheless, the project has performed far better than all expectations. The results have been tracked by monitoring of changes in sand volume along a 13-km-long segment of the restoration zone.

The accompanying table provides a more quantitative sense of seasonal changes in sand volume along the shoreline due to erosion and deposition by natural processes. Monitoring of sand erosion and deposition on a seasonal basis between 1998 and 2004 has yielded an average value for cubic meters of sand lost (or gained) per meter of shoreline (m³/m) in each season. When this seasonal value is multiplied by the 13-km length of the

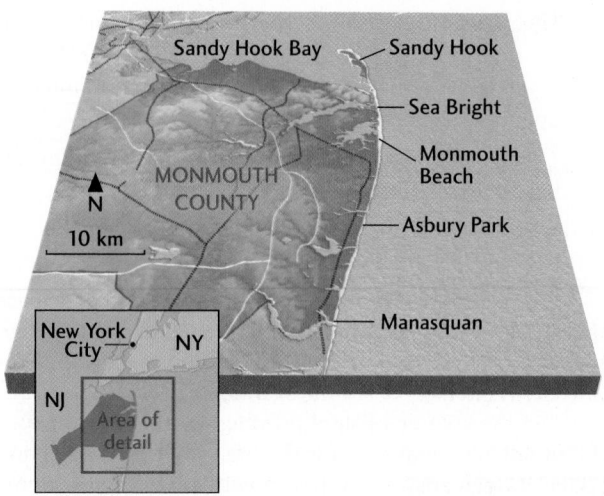

shoreline, the change in the volume of the shoreline (m³) can be calculated. Note that in fall 2002, the shoreline was augmented by an additional, artificially supplied volume of sand. This maintenance fill was designed to offset the expected removal of sand by natural processes.

Changes in Sand Volume for a 13-km Length of Shoreline, Monmouth County, New Jersey, Fall 1998–Fall 2004

Loss (−) or Gain (+) per Meter of Shoreline (m³)	Total Loss or Gain over Shoreline (m³/m)	Period
+1.41	+18,330	Fall 1998
+0.16	+2080	Spring 1999
−22.97	−298,610	Fall 1999
−42.09	−547,170	Spring 2000
−24.7	−321,100	Fall 2000
−29.82	−387,660	Spring 2001
−43.44	−564,720	Fall 2001
−1.02	−13,260	Spring 2002
+522.47	+6,792,110	Fall 2002*
−101.64	−1,321,320	Spring 2003
−77.00	−1,001,000	Fall 2003
−38.84	−504,920	Spring 2004
−79.53	−1,033,890	Fall 2004

*This gain represents the maintenance fill in fall 2002.

From these data we can draw the following conclusions:

1. The shoreline lost an average of 20 m³/m of the initial placement volume per season from the time of initial placement through spring 2002. (This figure is the average of the first column of numbers up until the maintenance fill of fall 2002.)

2. The average seasonal shoreline loss rate increased to 74 m³/m following the maintenance fill of fall 2002. (This figure is the average of the first column of numbers after the maintenance fill of fall 2002.)

3. The shoreline experienced a net loss of 162 m³/m from the time of initial placement through spring 2002. (This figure is the average of the second column of numbers up until the maintenance fill of fall 2002.)

4. The shoreline experienced a net loss of 297 m³/m following the maintenance fill of fall 2002. (This figure is the average of the second column of numbers after the maintenance fill of fall 2002.)

It is not known what factors contributed to the increase in sand loss after 2002, but scientists would want to investigate processes such as increased frequency of storms or increased intensity of storms over that period.

Did the volume of sand provided by the maintenance fill of fall 2002 make up for the losses between 1998 and 2004? We can answer this question by summing the numbers in the second column of the table (5,973,240 m³) and comparing that sum with the volume of sand added in the maintenance fill of fall 2002 (6,792,110 m³). These numbers are close enough that we can conclude that the losses due to natural causes were balanced by the artificial maintenance fill.

BONUS PROBLEM: Given the total cost of the initial restoration project that was started in 1994, and the volume of sand that was pumped to the shoreline at that time,

Sand placement at the southern end of Monmouth Beach, Monmouth County, New Jersey. This erosion control project by the U.S. Army Corps of Engineers includes periodic nourishment of the restored beaches on a 6-year cycle for a period of 50 years. [Country of Army Corps of Engineers, New York District.]

calculate the average cost per cubic meter of sand. Then use this value to estimate the cost of the maintenance fill that was provided in fall 2002. Do you think this continuing cost—every 6 years—is worth it?

EXERCISES

1. How are ocean waves formed?

2. How does wave refraction concentrate erosion at headlands?

3. What is a storm surge?

4. How has human interference affected some beaches?

5. Along what kinds of continental margins do we find broad continental shelves?

6. Where and how is the deep seafloor created by volcanism?

7. What plate tectonic process is responsible for deep-sea trenches?

8. Where do turbidity currents form?

9. What are pelagic sediments?

10. How does the Moon interact with Earth to create tidal currents?

THOUGHT QUESTIONS

1. Why would you want to know the timing of high tide if you wanted to observe a wave-cut terrace?

2. In a 100-year period, the southern tip of a long, narrow, north-south beach has become extended about 200 m to the south by natural processes. What shoreline processes could have caused this extension?

3. After a period of calm along a section of the Gulf Coast of North America, a severe storm with high winds passes over the shore and up the Mississippi Valley. Describe the state of the surf zone before the storm, during it, and after the storm has passed. What would happen to inland rivers?

4. What are the chief differences between the Atlantic and Pacific oceans with regard to topography, plate tectonics, volcanism, and other seafloor processes?

5. How might plate tectonics account for the contrast between the broad continental shelf off the east coast of North America and the narrow, almost nonexistent shelf off the west coast?

6. A major corporation hires you to determine whether New York City's garbage can be dumped at sea within 100 km offshore. What kinds of places would you explore, and what would your concerns be?

7. There is very little sediment on the floor of the central rift valley of the Mid-Atlantic Ridge. Why is this so?

8. A plateau rising from the deep seafloor to within 2000 m of the surface is covered with foraminiferal ooze, whereas the deep seafloor below the plateau, about 5000 m deep, is covered with clay. How can you account for this difference?

9. You are studying a sequence of sedimentary rocks and discover that the beds near the base are shallow marine sandstones and mudstones. Above these beds is an unconformity, overlying which are nonmarine sandstones. Above this group of beds is another unconformity, and overlying it are marine beds similar to those at the base. What could account for this sequence?

- Ice as a Rock   588

- How Glaciers
  Form   591

- How Glaciers
  Move   594

- Glacial
  Landscapes   598

- Glacial Cycles and
  Climate Change   606

Several glaciers flow together near Mt. Waddington in the Coast Mountains of British Columbia, Canada. [Chris Harris/Getty Images, Inc.]

GLACIERS: THE WORK OF ICE

21

THE VIEW OF EARTH from space is painted with the colors of water: vast blue oceans, swirling white clouds, and the frozen whites of solid ice and snow. The Earth system is constantly moving water across the planetary surface in ever-changing patterns. Among the main reservoirs of water, it is the system's icy component—the *cryosphere*—that waxes and wanes most visibly during climate cycles.

The ice sheets of Greenland and Antarctica, as huge as they are, now blanket only about one-tenth of Earth's land surface. But as recently as 20,000 years ago, continental glaciers covered almost three times more land than they do today, extending across all of Canada and deep into the midwestern United States. Within the next century or so, global warming could melt large parts of the existing ice sheets, with worldwide effects on human society. Sea level would rise, submerging low-lying cities. Climate zones would migrate, changing wet zones into deserts and vice versa. Given these threats, there is no doubt that understanding Earth's cryosphere—always an interesting scientific subject—has become an extremely practical goal.

The landscapes of many continents have been sculpted by glaciers now melted away. In mountainous regions, glaciers have eroded steep-walled valleys, scraped their bedrock surfaces smooth, and plucked huge blocks from their rocky floors. During the ice ages of the Pleistocene epoch, glaciers pushed across entire continents, carving far more topography than water and wind. Glacial erosion creates enormous amounts of debris, and glaciers transport huge tonnages of sediment, depositing it at their edges, where it may be carried away by meltwater streams. Glacial processes affect the discharges and sediment loads of river systems, the erosion and sedimentation of coastal areas, and the quantity of sediment delivered to the oceans.

In this chapter, we take a close look at Earth's glaciers, how they form and change over time, and how they leave their marks on Earth's surface by eroding and depositing material as they advance and retreat. We examine the role glaciers play in the climate system and discover what the geologic record of glaciation can tell us about climate change over time.

Ice as a Rock

To a geologist, a block of ice is a rock, a mass of crystalline grains of the mineral ice. Like igneous rock, it is formed by the freezing of a fluid. Like sedimentary rock, it is formed from materials deposited in layers at Earth's surface and can accumulate to great thicknesses. Like metamorphic rock, it is transformed by recrystallization under pressure; that is, glacial ice is formed by the burial and metamorphism of the "sediment" snow. Loosely packed snowflakes—each a single crystal of the mineral ice—age and recrystallize into a solid rock of ice (**Figure 21.1**).

Ice has some unusual properties. Its melting temperature is very low (0°C), hundreds of degrees below the melting temperatures of silicate rock. Most rocks are denser than their melts, which is why magma rises buoyantly through the lithosphere. But ice is less dense than its melt, which is why icebergs float on the ocean. And although ice may seem hard, it is much weaker than most rocks.

Because ice is so weak, it flows readily downhill like a viscous fluid. **Glaciers** are large masses of ice on land that show evidence of being in motion, or of once having moved, under the force of gravity. We divide glaciers, on the basis of their size and shape, into two basic types: *valley glaciers* and *continental glaciers*.

Valley Glaciers

Many skiers and mountain climbers are familiar with **valley glaciers,** sometimes called *alpine glaciers* (**Figure 21.2**). These rivers of ice form in the cold heights of

FIGURE 21.1 ■ A typical mosaic of crystals of glacial ice. The tiny circular and tubular spots are bubbles of air. [Courtesy Joan Fitzpatrick.]

FIGURE 21.2 ■ Herbert Glacier, a valley glacier near Juneau, Alaska. [Greg Dimijian/Science Source.]

mountain ranges, where snow accumulates. They then move downslope, either flowing down an existing stream valley or carving out a new one. A valley glacier usually occupies the complete width of the valley and may bury its floor under hundreds of meters of ice. In warmer, low-latitude climates, valley glaciers are found only at the heads of valleys on the highest mountain peaks. An example is the glacial ice that covers the Mountains of the Moon at elevations over 5000 m along the Uganda-Zaire border in east-central Africa. In colder, high-latitude climates, valley glaciers may descend many kilometers, down the entire length of a valley. Broad lobes of ice may descend into lower lands bordering mountain fronts. Valley glaciers that flow down coastal mountain ranges at high latitudes may terminate at the ocean's edge, where masses of ice break off and form icebergs—a process called **iceberg calving** (Figure 21.3).

FIGURE 21.3 ■ Iceberg calving at Dawes Glacier, Alaska. Calving occurs when huge blocks of ice break off at the edge of a glacier that has moved to a shoreline. [© Paul Souders/Corbis.]

Continental Glaciers

A **continental glacier** is a thick, slow-moving sheet of ice (sometimes called an *ice sheet*) that covers a large part of a continent or other large landmass (**Figure 21.4**). Today, the world's largest continental glaciers overlie much of Greenland and Antarctica, covering about 10 percent of Earth's land surface and storing about 75 percent of the world's fresh water.

In Greenland, 2.6 million cubic kilometers of ice occupy 80 percent of the island's total area of 4.5 million square kilometers (**Figure 21.5**). The upper surface of the ice sheet resembles an extremely wide convex lens. At its highest point, in the middle of the island, the ice is more than 3200 m thick. From this central area, the ice surface slopes to the ocean on all sides. At the mountain-rimmed coast, the ice sheet breaks up into narrow tongues resembling valley glaciers that wind through the mountains to reach the ocean, where icebergs form by calving.

The bowl shape of the bedrock beneath the Greenland ice sheet, evident in the cross section at the bottom of Figure 21.5, is caused by the weight of the ice in the middle of the island. This consequence of isostasy explains why mountains rim the Greenland coast.

Though very large, the Greenland glacier is dwarfed by the Antarctic ice sheet. Ice blankets 90 percent of Antarctica, covering an area of about 13.6 million square kilometers and reaching thicknesses of 4000 m (**Figure 21.6**). The total volume of Antarctic ice—about 30 million cubic kilometers—constitutes over 90 percent of the cryosphere. As in Greenland, the ice forms a dome in the center and slopes down to the margins of the continent.

Parts of Antarctica are rimmed by thinner sheets of ice—**ice shelves**—floating on the ocean and attached to the main glacier on land. The best known of these is the Ross Ice Shelf, a thick layer of ice about the size of Texas that floats on the Ross Sea.

FIGURE 21.4 ■ Sentinel Range, Antarctica. These mountains stick up more than 4000 m through the thick ice of the Antarctic continental glacier. [© Google 2009 Data SIO, NOAA, U.S. Navy, NGA, GEBCO Image U.S. Geological Survey]

(a)

(b)

FIGURE 21.5 ■ Topographic map and cross section of the Greenland continental glacier. (a) The extent and elevation of the Greenland ice sheet. (b) A generalized cross section of south-central Greenland shows the lenslike shape of the glacier. The ice moves down and out from the thickest section. [After R. F. Flint, *Glacial and Quaternary Geology*. New York: Wiley, 1971.]

Ice caps are the masses of ice that blanket Earth's North and South Poles. Most of the Arctic ice cap, located at the highest latitudes of the Northern Hemisphere, lies over water and is not a glacier. Almost all of the Antarctic ice cap lies on the continent of Antarctica and is therefore a continental glacier.

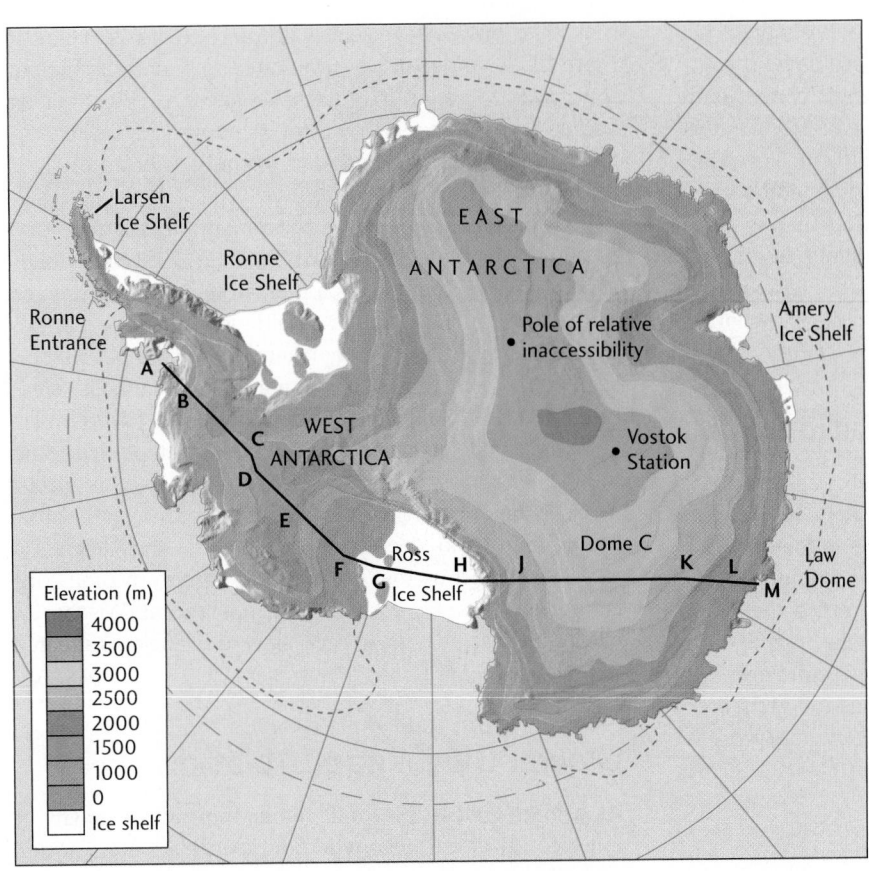

FIGURE 21.6 ■ Topographic map and cross section of the Antarctic continental glacier. (a) The extent and elevation of the Antarctic ice sheet. Ice shelves are shown in white. (b) A generalized cross section of the ice sheet and the land beneath it. [After U. Radok, "The Antarctic Ice," *Scientific American* (August 1985): 100; based on data from the International Antarctic Glaciological Project.]

(a)

(b)

■ How Glaciers Form

A glacier starts with abundant winter snowfall that does not melt away in the summer. The snow is slowly converted into ice, and when the ice is thick enough, it begins to flow.

Basic Ingredients: Freezing Cold and Lots of Snow

For a glacier to form, temperatures must be low enough to keep snow on the ground year-round. These conditions occur at high latitudes, because the Sun's rays strike Earth at low angles there (see Figure 19.1), and at high elevations, because the atmosphere becomes steadily cooler up to altitudes of about 10 km (see Figure 15.2). Therefore, the height of the *snow line*—the elevation above which snow does not completely melt in summer—generally decreases toward the poles, where snow and ice cover the ground year-round even at sea level. Near the equator, glaciers form only on mountains that are higher than about 5000 m.

The precipitation of snow and the formation of glaciers require moisture as well as cold. Moisture-laden winds tend to drop most of their snow on the windward side of a high mountain range, so the leeward side is likely to be dry and unglaciated. Parts of the high Andes of South America, for instance, lie in a belt of prevailing easterly winds. Glaciers form on the moist eastern slopes, but the dry western side has little snow and ice.

The coldest climates are not necessarily the snowiest. Nome, Alaska, has a polar climate with an average annual maximum temperature of 9°C, but it gets only about 4.4 cm of precipitation a year, virtually all of it as snow. In comparison, Caribou, Maine, has a cool temperate climate with an average annual maximum temperature of 25°C, and its average annual snowfall is a whopping 310 cm. Nevertheless, the conditions around Nome, where little of the snow melts, are better for the formation of glaciers than those in Caribou, where all the snow melts in summer. In arid climates, glaciers are unlikely to form unless the temperature is so frigid throughout the year that very little snow melts, as in Antarctica.

Glacial Growth: Accumulation

A fresh snowfall is a fluffy mass of loosely packed snowflakes. As these small, delicate ice crystals age on the ground, they shrink and become grains, and the mass of snowflakes compacts to form a dense, granular snow (**Figure 21.7**). As new snow falls and buries the older snow, that older, granular snow compacts further into an even denser form, called *firn*. Further burial and aging eventually produce solid glacial ice as the grains recrystallize and are cemented together. The whole process

FIGURE 21.7 ■ Stages in the transformation of snow crystals, first into granular ice, then into firn, and finally into glacial ice. A corresponding increase in density accompanies this transformation as air is eliminated from the crystals.

[After H. Bader et al., "Der Schnee und seine Metamorphose," *Beiträge zur Geologie der Schweiz* (1939).]

may take only a few years, although 10 to 20 years is more likely. When ice accumulates to a mass sufficient for gravity to act on it, a glacier is born. A typical glacier adds a layer of ice each winter as snow falls on its surface. The amount of ice added to the glacier annually is its **accumulation.**

As glacial snow and ice accumulate, they entrap and preserve valuable relics of Earth's past. In 1991, mountaineers discovered the body of a prehistoric human preserved for more than 5000 years in Alpine ice on the border between Austria and Italy. In northern Siberia, extinct animals such as the woolly mammoth, a great elephant-like creature that once roamed icy terrains, have been found frozen and preserved in ancient ice. Ancient dust particles and bubbles of atmospheric gases are also preserved in glacial ice (see Figure 21.1). Chemical analyses of air bubbles found in very old, deeply buried Antarctic and Greenland ice have shown us that atmospheric carbon dioxide concentrations were lower during the most recent (Wisconsin) glaciation than they have been since the glaciers retreated (see Figure 15.11).

Glacial Shrinkage: Ablation

As a glacier flows downhill under the pull of gravity, it flows to lower altitudes where temperatures are warmer, and it loses ice. The total amount of ice that a glacier loses annually is called **ablation.** Four mechanisms are responsible for ablation:

1. *Melting.* As the ice melts, the glacier loses material.
2. *Iceberg calving.* Pieces of ice break off and form icebergs when a glacier reaches a shoreline (see Figure 21.3).
3. *Sublimation.* In cold climates, water can be transformed directly from its solid state (ice) into its gaseous state (water vapor).
4. *Wind erosion.* Strong winds can erode glacial ice, primarily by melting and sublimation.

Most ablation takes place at the glacier's leading edge. Therefore, even when a glacier is advancing downward or outward from its center, its leading edge—the *ice front*—may be retreating. The two mechanisms by which glaciers lose the most ice are melting and iceberg calving.

The Glacial Budget: Accumulation Minus Ablation

The relationship between accumulation and ablation, called the *glacial budget,* determines the growth or shrinkage of a glacier (**Figure 21.8**). When accumulation equals ablation over a long period, the glacier remains a constant size, even as it continues to flow downslope from the area where it forms. Such a glacier accumulates snow and ice in its upper reaches as an equal amount is ablated in its

FIGURE 21.8 ■ Accumulation takes place mainly by snowfall over a glacier's colder upper regions. Ablation takes place by melting, iceberg calving, and sublimation in the warmer lower regions. The difference between accumulation and ablation is the glacial budget.

lower reaches. If accumulation exceeds ablation, the glacier grows; if ablation exceeds accumulation, the glacier shrinks.

Glacial budgets vary from year to year. Over the past several thousand years, many glaciers have maintained a constant average size, though some show evidence of growth or shrinkage in response to short-term regional climate variations. In the last century, however, glaciers in many low-latitude regions have been shrinking in response to global warming (**Figure 21.9**). Because glacial shrinkage is a good indicator of climate change, glacial budgets are now carefully monitored.

(a)

(b)

FIGURE 21.9 ■ Photographs of Qori Kalis Glacier in Peru from the same vantage point (a) in July 1978 and (b) in July 2004. Between 1998 and 2001, the ice front retreated an average of 155 m/year, an alarming 32 times faster than the average annual retreat from 1963 to 1978. [Lonnie G. Thompson, Byrd Polar Research Center, the Ohio State University/courtesy NSIDC.]

■ How Glaciers Move

When ice becomes thick enough for gravity to overcome its resistance to movement—normally at least several tens of meters—it starts to flow, and thus becomes a glacier. The ice in a glacier moves downhill in the same kind of laminar flow as a slowly flowing stream of water (see Figure 18.14). Unlike the readily observed flow of a stream, however, glacial movement is so slow that the ice seems not to move at all from day to day, giving rise to the expression "moving at a glacial pace."

Mechanisms of Glacial Flow

The flow of glaciers occurs in two ways: by plastic flow and by basal slip (**Figure 21.10**). In plastic flow, the movement occurs as deformation within the glacier. In basal slip, the glacier slides downslope as a single unit along its base, like a block of ice sliding down a ramp.

MOVEMENT BY PLASTIC FLOW The force of gravity exerted on a glacier causes individual crystals of ice to slip tiny distances relative to each other—on the order of a ten-millionth of a millimeter—over short intervals (Figure 21.10a). The sum of many such movements among the enormous number of ice crystals that make up a glacier deforms the whole mass of ice in a process known as **plastic flow.** To visualize this process, think of a random pile of playing cards; the whole pile can be made to shift by inducing small slips on the many surfaces between cards. As ice crystals grow under pressure deeper in the glacier, their microscopic slip planes become more parallel, increasing plastic flow rates.

Plastic flow is most important in very cold regions, where the ice throughout the glacier, including its base, is well below the freezing point and the basal ice is frozen to the ground (Figure 21.10b). Most of the movement of these cold, dry glaciers takes place above the base by plastic flow. Movements near the frozen base detach and transport pieces of bedrock and soil. Because of this mixing of rock material with ice, the interface between the overlying ice and the underlying ground is usually not a sharp boundary, especially where the ground consists of sediments or weak sedimentary rocks. Instead, the interface becomes a transition between ice loaded with debris and deformed ground containing appreciable ice.

MOVEMENT BY BASAL SLIP The other mechanism of glacial movement is **basal slip,** the sliding of a glacier along the boundary between the ice and the ground (Figure 21.10c). The melting point of ice decreases as pressure increases, so ice at the base of a glacier, where the weight of the overlying ice is greatest, melts at a lower temperature than ice within the glacier. The melted ice lubricates the base of the glacier, causing it to slip downslope. This same effect helps make ice skating possible: the weight of the body on the narrow skate blade provides enough pressure to melt just a little ice under the blade, which lubricates the blade so that it can slide easily along the surface.

In temperate regions, where the air temperature is above freezing during parts of the year, ice may be at the melting point within a glacier as well as at its base. Plastic flow contributes a small amount of internal heating from the friction generated by microscopic slips of ice crystals. In these glaciers, water occurs within the ice as small drops between crystals. Water seeping through cracks in the ice forms pools or streams of meltwater that carve tunnels in the ice. The water throughout the glacier eases internal slip between layers of ice.

Flow in Valley Glaciers

Louis Agassiz, a nineteenth-century Swiss zoologist and geologist, was the first to precisely measure how a valley glacier moves. As a young professor in the 1830s, he pounded stakes into a glacier in the Swiss Alps and measured their positions over several years. He observed that the stakes along the centerline of the glacier moved the fastest, at about 75 m/year, whereas the stakes nearer the valley walls traveled more slowly. Later, the deformation of long vertical tubes pounded deep into the glacier demonstrated that ice near the base flowed more slowly than ice in the center.

This type of deformation, in which the central part of a glacier moves faster than its sides or its base, is diagnostic of plastic flow (Figure 21.10d). Other valley glaciers have been observed to move at more uniform speeds, sliding as a single unit almost entirely by basal slip along a lubricating layer of meltwater next to the ground. Most often, however, valley glaciers flow by a combination of mechanisms—partly by plastic flow within the mass of ice and partly by basal slip.

A sudden period of fast movement of a valley glacier, called a **surge,** sometimes occurs after a long period of little movement. Surges may last several years, and during that time the ice may speed along at more than 6 km/year—a thousand times the normal velocity of a glacier. In many cases, surges follow the buildup of water pressure at or near the base of the glacier. This pressurized water greatly enhances basal slip.

The upper parts of a glacier have little pressure on them. At these low pressures, the ice at the surface of the glacier (shallower than about 50 m) behaves as a rigid, brittle solid, cracking as it is dragged along by the plastic flow of the ice below. The cracks, called **crevasses,** break up the surface ice into many small and large blocks (**Figure 21.11**). Crevasses are most likely to occur in places where glacial deformation is strong—such as where the ice drags against the valley walls at curves in the valley, at irregularities in the valley floor, and where the slope steepens sharply. The movement of brittle surface ice at these places is a "flow" resulting from slipping movements across the surfaces of these irregular blocks, similar in some ways to faulting in crustal rocks.

Antarctica in Motion

Antarctica may appear to be a land frozen in time, but it certainly does not stand still. Glaciers plow downward from

(a)

Ice crystals

Individual ice crystals may stretch and rotate, grow and recrystallize, or slip tiny distances relative to each other.

Overall movement of plastic flow

(b) **PLASTIC FLOW**

Plastic flow dominates in cold regions where the ice at the base of the glacier is frozen to the bedrock or soil.

As a result of frictional forces, the rate of movement decreases toward the base.

(d) Valley glaciers in cold regions move mostly by plastic flow. If one drives a set of stakes deep into the glacier in a line across its flow,…

…after several years, the stakes in the center will have moved farther downhill and will slant forward, indicating faster movement in the center and at the top of the glacier.

(c) **BASAL SLIP**

Basal slip dominates in temperate regions where the pressure of overlying ice melts water at the glacier's base.

Liquid water

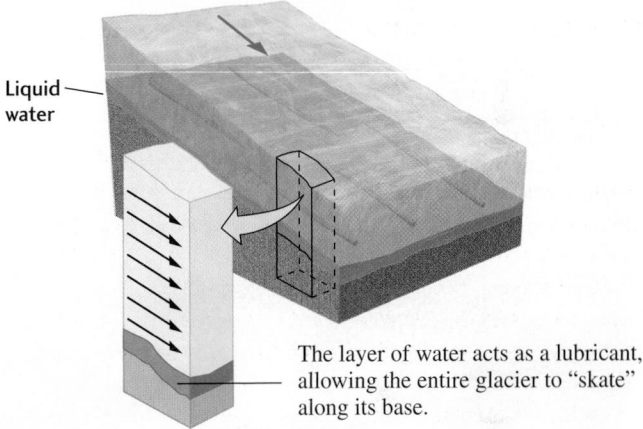

The layer of water acts as a lubricant, allowing the entire glacier to "skate" along its base.

(e) In continental glaciers, the ice moves down and out from the thickest section, like pancake batter poured on a griddle, as shown by the arrows.

FIGURE 21.10 ■ Glaciers move by two primary mechanisms: plastic flow and basal slip. (a) Deformation in plastic flow. (b) Plastic flow. (c) Basal slip. (d) Flow in valley glaciers. (e) Flow in continental glaciers.

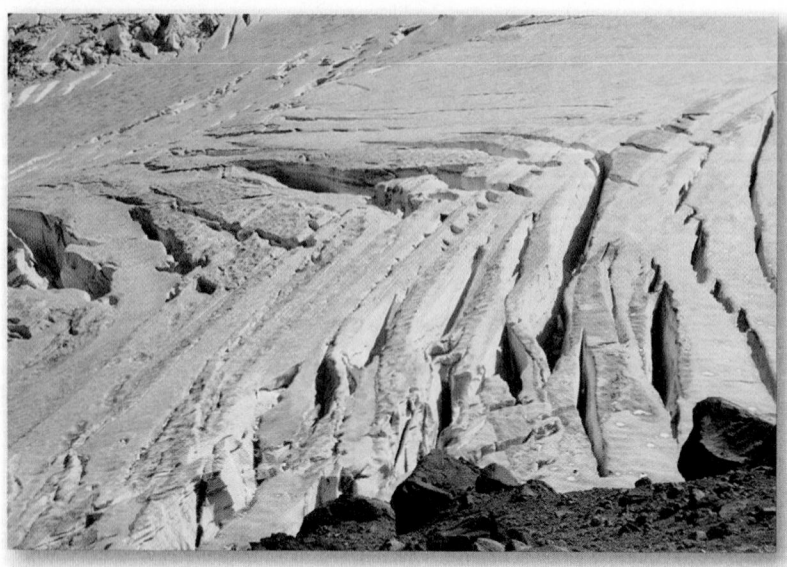

(a)

FIGURE 21.11　■　(a) Crevasses cover the Emmons Glacier on the northeast flank of Mount Rainier, Washington State. [© 2002 Walter Siegmund.] (b) Crevasses in a valley glacier are most likely to occur where ice deformation is strong.

Crevasses form where a glacier moves over steps in bedrock topography...

...and where it curves around topographic features.

(b)

the continent's center to the ocean, icebergs snap off and crash into the ocean, and great rivers of ice snake through the ice sheet. All these movements provide evidence of the dynamic relationship between this remote continent and the global climate system. Continental glaciers in polar climates, where basal slip is minor or absent in most places, have their highest rates of movement in the center of the ice. Pressure there is high, and the main force retarding movement is the friction between layers of ice moving at different speeds (Figure 21.10e).

Geologists use satellites and airborne radar to map the shapes and overall movements of glaciers. These measure-ments show that Antarctic glaciers flow rapidly in **ice streams** 25 to 80 km wide and 300 to 500 km long (**Figure 21.12**). These streams reach velocities of 0.3 to 2.3 m/day, compared with the flow rate of 0.02 m/day in the adjacent ice sheet. Boreholes drilled in the ice reveal that the base of the ice stream is at the melting point and that the meltwater is mixed with soft sediment. One theory is that the rapid movement of ice streams is related to deformation of the water-saturated basal sediment. Ice streams may form during climate warming, leading to ice breakup and rapid deglaciation. In the current period of global warming, they may be contributing to the retreat of glaciers and the instability of the ice sheet in West Antarctica.

(a)

(b)

FIGURE 21.12 ■ (a) Lambert Glacier, Antarctica, showing lineations in the foreground where the ice is flowing more rapidly. (b) Velocity map for Lambert Glacier and its main ice stream. The arrows show the directions in which the ice is flowing. The areas of no movement (yellow) are either exposed land or stationary ice. The smaller tributary glaciers generally have low velocities of 100 to 300 m/year (green), which gradually increase as they flow down the sloping surface of the continent and intersect the upper reaches of Lambert Glacier. Most of Lambert Glacier has velocities of 400 to 800 m/year (blue). As it extends into and across the Amery Ice Shelf and the ice sheet spreads out and thins, velocities increase to 1000 to 1200 m/year (pink/red). The area of the image is about 570 km by 380 km. [(a) Courtesy of Richard Stanaway, Geodynamics Group, Research School of Earth Sciences, Australian National University; (b) RADARSAT imagery from the 2000 Antarctic Mapping Mission, NASA Visible Earth.]

Using high-resolution radar satellite mapping, geologists have observed that several Antarctic glaciers have retreated more than 30 km in just 3 years. Over the past 20 years or so, enormous pieces of ice have snapped off Antarctic glaciers. In March 2000, an iceberg slightly less than 10,000 km² (larger than the state of Delaware) calved from the Ross Ice Shelf. In February and March of 2002, a portion of the Larsen Ice Shelf larger than Rhode Island (about 3250 km²) shattered and separated from the northeastern side of the Antarctic Peninsula (**Figure 21.13**). The fracturing of this piece of the ice sheet produced thousands of icebergs.

Geologists who monitor the Larsen Ice Shelf were able to predict this episode of ice shelf collapse. Field and satellite-based observations showed that the flow rate of the ice stream leading to it had increased dramatically, which was interpreted as a sign of instability. After the collapse of the ice shelf, the flow rate of the ice stream increased further. In general, the destruction of ice shelves tends to destabilize the continental glaciers that feed them, causing the glaciers to flow more rapidly into the oceans.

Ice shelf instabilities are increasing in size and frequency at an alarming rate. The latest example is the Wilkins Ice Shelf, which occupies an area of 14,000 km² on the southwestern side of the Antarctic Peninsula. It began to break apart in early 2008 and, as of 2013, appears to be on the verge of further collapse. Although these collapses are a worrisome symptom of global warming, they will not, by themselves, contribute to sea level rise (see Earth Issues 21.1).

■ Glacial Landscapes

The movement of glaciers is responsible for the immense amount of geologic work done by ice: erosion, transportation, and sedimentation. Just as you cannot see your own footprint in the sand while your foot is covering it, you cannot see the effects of an active glacier at its base or sides. Only when the ice melts is its geologic work revealed. We can infer the physical processes driven by moving ice from the topography of formerly glaciated areas and the distinctive landforms left behind.

Glacial Erosion and Erosional Landforms

Ice is a far more efficient agent of erosion than water or wind. A valley glacier only a few hundred meters wide can tear up and crush millions of tons of bedrock in a single year. The ice carries this heavy load of sediment to the ice front, where it is dropped as the ice melts. The total amount

FIGURE 21.13 ■ Collapse of the Larsen Ice Shelf. This satellite image was taken on March 7, 2002, toward the end of a 2-month period during which a huge piece of the ice shelf separated from land and splintered into thousands of icebergs. The darkest colors on the right-hand side of the image represent open seawater. The white parts are icebergs, the remaining parts of the ice shelf, and glaciers on land. The bright blue area is a mixture of seawater and highly fractured ice. The area of the image is about 150 km by 185 km. [NASA/GSFC/LaRC/JPL, MISR Team.]

Earth Issues

21.1 Isostasy and Sea Level Change

If the ice shelves around Antarctica continue to collapse, will sea level rise? It turns out that even if all of Earth's ice shelves were to break off into the ocean over the next few years, sea level wouldn't change much at all. The reason is related to the principle of isostasy, which we discussed in Chapter 14. Ice shelves, like icebergs, float on ocean waters. When they melt, there is no change in sea level for the same reason that, when ice cubes in your drink melt, the level of the liquid in your glass doesn't change.

The buoyancy of icebergs and ice shelves results from the fact that the volume of the ice *below* the sea surface weighs less than the volume of water it displaces. This buoyancy force counteracts the gravitational force that pulls down on the iceberg.

Bigger icebergs stand higher above the sea surface, but also have a deeper root below the surface that provides greater buoyancy. When an iceberg melts, there is no change in sea level, because the water volume resulting from the melting exactly equals the water volume that had been displaced by the iceberg.

In contrast, when glaciers on land melt, most of the water flows into the ocean, increasing the ocean volume and thereby raising the sea level. And when glaciers on land flow into the ocean, the water in the oceans is displaced by the icebergs they shed, which also raises sea level.

Therefore, the destruction of an ice shelf will raise sea level only if part of the ice shelf that was grounded on land slides into the ocean. In that case, the weight of the ice is no longer supported by the continent, but rather by the seawater it displaces, causing a sea-level rise.

Gravitational force

1 The gravitational force that pulls a floating iceberg downward is counterbalanced by the buoyancy force that pushes it upward. This is the principle of isostasy.

Buoyancy force

2 Larger icebergs have deeper roots and rise higher above sea level.

3 Volume of water displaced upward, to create sea level rise, is equal to the volume of water represented by icebergs.

4 If an iceberg in the ocean melts, sea level does not change.

Glacial ice

Land surface

Sea level low

5 If ice on land melts or slides into the ocean, sea level rises.

Glacial ice melts

Glacial ice ruptures and slides into ocean

Sea level rises

Sea level rises higher

According to the principle of isostasy, ice shelves and icebergs float and thus displace a mass of water equal to their own mass; therefore, their melting does not change sea level. The melting of ice sheets on land, however, injects new water into the oceans, causing sea level to rise.

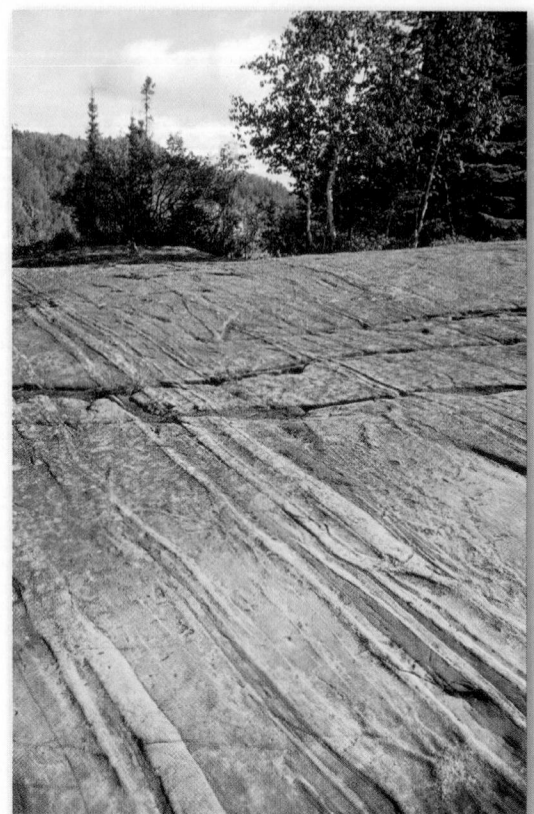

FIGURE 21.14 ■ Glacial striations on bedrock in Quebec, Canada. Striations provide evidence of the direction of ice movement and are especially important clues for reconstructing the movement of continental glaciers. [Michael P. Gadomski/Science Source.]

of sediment deposited in the world ocean per year has been several times larger during recent ice ages than during interglacial periods.

At its base and sides, a glacier engulfs jointed, cracked blocks of rock, breaks off pieces, and grinds them against the bedrock below. This grinding action fragments rocks into a great range of sizes, from boulders as big as houses to fine silt- and clay-sized material called *rock flour*. Rock flour is subject to rapid chemical weathering because of its small size and correspondingly large surface area. Where glacial debris is still encased in ice and the ground is overlain by thick ice, chemical weathering is slower than on ice-free terrain. When rock flour is freed from the ice at the melting edge of a glacier, it dries out and is blown into the air as dust. As we saw in Chapter 19, wind can carry this dust over great distances, ultimately depositing it as the loess that is so common in formerly glaciated regions.

As a glacier drags rocks along its base, those rocks scratch or groove the bedrock beneath it. Such abrasions are termed **striations.** The orientation of striations shows us the direction of ice movement—an especially important factor in the study of continental glaciers, which lack obvious valleys. By mapping striations over wide areas formerly covered by a continental glacier, we can reconstruct the glacier's flow patterns (**Figure 21.14**).

Advancing glacial ice smooths small hills of bedrock—known as *roches moutonnées* ("sheep rocks") for their resemblance to a sheep's back—on their upstream side and plucks them to a rough, steep slope on their downstream side (**Figure 21.15**). These contrasting slopes indicate the direction of ice movement.

Ice polishes the upstream slope to a smooth surface.

Crevasses form as ice moves over an irregularity in the bedrock.

Ice movement

Jointed bedrock

Ice plucks rock fragments from the downstream slope, producing a rough, steep surface.

(a)

(b)

FIGURE 21.15 ■ (a) A *roche moutonnée* is a small hill of bedrock, smoothed by glacial ice on the upstream side and plucked to a steep, rough face on the downstream side as the moving ice pulls rock fragments from joints and cracks. (b) A *roche moutonnée* known as The Beehive rises above Sand Beach at Acadia National Park in Maine. [Jerry and Marcy Monkman/Aurora Photos.]

Rounded mountain profiles

1 **Before glaciation,** a mountain river cuts through a V-shaped valley.

Tributary valley

V-shaped valley

Tributary glaciers

Main valley glacier

2 **During glaciation,** cirques and arêtes form. The glacier moving down from its cirque creates a U-shaped valley.

Horn   Cirques

Arête

3 **After glaciation,** when ice melts and retreats, the tributary valleys are left as hanging valleys.

Hanging valley with waterfall

U-shaped valley

A cirque is the head of a glacial valley, with nearly vertical upper walls and a flat or hollowed-out base.

Arêtes are sharp, jagged crests along an eroded divide.

A glacial valley is U-shaped, with steep sides, hanging valleys, and waterfalls.

A fjord is a U-shaped valley occupied by an arm of the sea.

FIGURE 21.16 ■ Erosion by valley glaciers creates distinctive landforms. [Photos (top right) Marli Miller; (bottom, left to right) © Stephen Matera/Alamy; © Radomir Rezny/Alamy; Philippe Body/Age Fotostock/Robert Harding Picture Library.]

A flowing valley glacier carves a series of erosional landforms as it flows from its origin to its lower edge (**Figure 21.16**). At the head of the valley, the plucking and tearing action of the ice tends to carve out an amphitheater-like hollow called a **cirque,** usually shaped like half

of an inverted cone. With continued erosion, cirques at the heads of adjacent valleys gradually meet to form sharp mountaintops, or *horns,* and jagged crests called *arêtes* along the divide. As the glacier flows down from its cirque, it excavates a new valley or deepens an existing stream valley, creating a characteristic **U-shaped valley.** Glacial valley floors are wide and flat and have steep walls, in contrast to the V-shaped valleys of many mountain streams (see Chapter 18).

Glaciers and streams also differ in how their tributaries form junctions. Although the ice surface is level where a tributary glacier joins a main valley glacier, the floor of the main valley may be carved much more deeply than that of the tributary valley. When the ice melts, the tributary valley is left as a **hanging valley**—one whose floor lies high above the main valley floor (see Figure 21.16). After the ice is gone

and streams occupy the valleys, the junction is marked by a waterfall as the stream in the hanging valley plunges over the steep cliff separating it from the main valley below.

Valley glaciers at coastlines may erode their valley floors far below sea level. When the ice retreats, these steep-walled valleys—which still maintain a U-shaped profile—are flooded with seawater (see Figure 21.16). These arms of the sea carved out by glaciers, called **fjords,** create the spectacular rugged scenery for which the coasts of Alaska, British Columbia, Norway, and New Zealand are renowned.

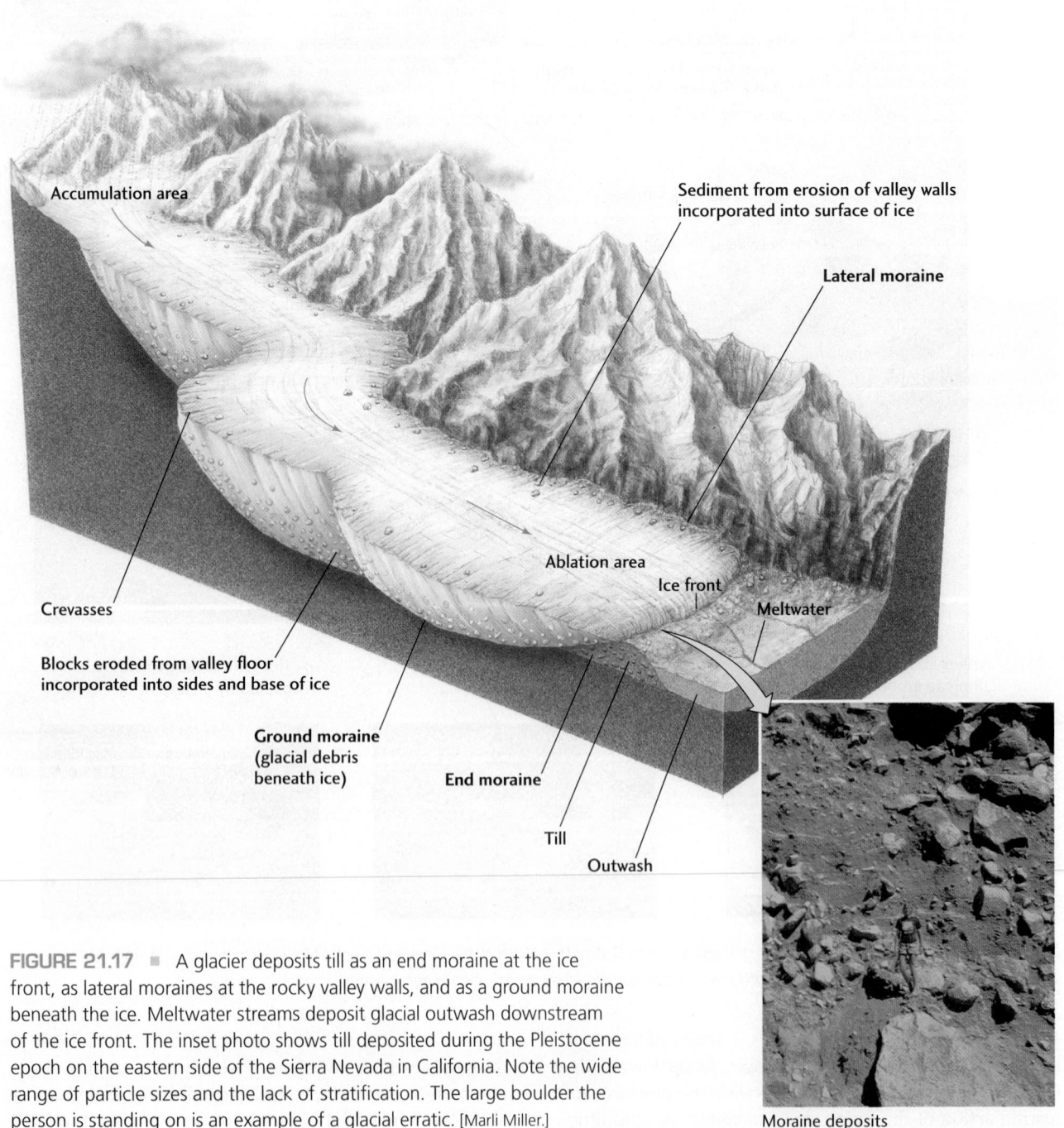

Accumulation area

Sediment from erosion of valley walls incorporated into surface of ice

Lateral moraine

Crevasses

Blocks eroded from valley floor incorporated into sides and base of ice

Ground moraine (glacial debris beneath ice)

End moraine

Till

Outwash

Ablation area

Ice front

Meltwater

Moraine deposits

FIGURE 21.17 ■ A glacier deposits till as an end moraine at the ice front, as lateral moraines at the rocky valley walls, and as a ground moraine beneath the ice. Meltwater streams deposit glacial outwash downstream of the ice front. The inset photo shows till deposited during the Pleistocene epoch on the eastern side of the Sierra Nevada in California. Note the wide range of particle sizes and the lack of stratification. The large boulder the person is standing on is an example of a glacial erratic. [Marli Miller.]

Glacial Sedimentation and Sedimentary Landforms

Glaciers transport eroded rock materials of all kinds and sizes downstream, eventually depositing them where the ice melts. Ice is a very effective transport agent because the material it picks up does not settle out like the load of sediment carried by a stream. Like water and wind currents, flowing ice has both a competence (an ability to carry particles of a certain size) and a capacity (a total amount of sediment that it can transport). Ice has extremely high competence: it can carry huge blocks many meters in diameter that no other transport agent could budge. Ice also has a tremendous capacity. Some glacial ice is so full of rock material that it is dark and looks like sediment cemented with ice.

When glacial ice melts, it deposits a poorly sorted, heterogeneous load of boulders, pebbles, sand, and clay. A wide range of particle sizes is the characteristic that differentiates glacial sediments from the much better sorted material deposited by streams and winds. This heterogeneous material puzzled early geologists, who were not aware of its glacial origins. They called it *drift* because it seemed to have drifted in somehow from other areas. The term **drift** is now used for *all* material of glacial origin found anywhere on land or beneath the ocean.

Some drift is deposited directly by melting ice. This unstratified and poorly sorted sediment is known as **till,** and it may contain all sizes of rock fragments from clay to boulders (**Figure 21.17**). The large boulders often contained in till are called *erratics* because of their seemingly random composition, often very different from that of local rocks.

Other deposits of drift are laid down as the ice melts and releases water and sediment. Meltwater flowing in tunnels within and beneath the ice and in streams at the ice front may pick up, transport, and deposit some of the drift. Deposits of drift may trap some of the meltwater, causing it to pool and form lakes. Like any other waterborne sediment, the drift transported by meltwater is stratified and well sorted and may be cross-bedded. Drift that has been picked up and distributed by meltwater streams is called **outwash,** and it often forms broad sedimentary plains downstream of melting glaciers, known as *outwash plains.* Strong winds can transport fine-grained material from outwash plains over long distances and deposit it as loess.

Glacial sedimentary sequences can be identified by the distinctive textures of interbedded tills, outwash, and loess, as well as by striations and other erosional forms that may be preserved. Mapping of such sequences has allowed geologists to infer many glaciations of past geologic times.

ICE-LAID DEPOSITS A **moraine** is an accumulation of rocky, sandy, and clayey material carried by glacial ice or deposited as till. There are many types of moraines, each named for its position with respect to the glacier that formed it (**Table 21.1**). One of the most prominent in size and appearance is an *end moraine,* formed at the ice front. As the ice flows steadily downstream, it brings more and more sediment to its melting edge. The unsorted material accumulates there as a hilly ridge of till. Regardless of their shape or location, moraines of all kinds consist of till. Figure 21.17 illustrates the processes by which various kinds of moraines form as a glacier works its way through a valley.

TABLE 21-1 Types of Glacial Moraines

Type of Moraine	Location with Respect to Ice Front	Comments
End moraine	At ice front	After glacier melts, seen as ridge parallel to former ice front
Terminal moraine	At ice front marking farthest advance of ice	Type of end moraine
Lateral moraine	Along edge of glacier where it scrapes side walls of valley	Heavy sediment load eroded from valley walls; when ice melts, seen as ridge parallel to valley walls
Medial moraine	Formed as two joined glaciers merge their lateral moraines below junction	Inherits its sediment load from lateral moraines that formed it; forms ridge parallel to valley walls
Ground moraine	Beneath the ice as a layer of glacial debris	Ranges from thin and patchy to a thick blanket of tills

Some continental glacial terrains display prominent landforms called **drumlins:** large, streamlined hills of till and bedrock that parallel the direction of ice movement (**Figure 21.18**). Drumlins, usually found in clusters, are shaped like long, inverted spoons with the gentlest slopes on the downstream sides. They may be 25 to 50 m high and a kilometer long. Drumlins form when the sediment-rich layer at the base of a glacier encounters a knob of bedrock or other obstacle and the excess pressure squeezes out water and drops the sediment.

WATER-LAID DEPOSITS Deposits of outwash by glacial meltwater take a variety of forms. *Kames* are small hills of sand and gravel created when drift fills a hole in a glacier and is left behind when the glacier recedes. Some kames are deltas built into a lake at the ice front. When the lake drains, the deltas are preserved as flat-topped hills. Kames are often exploited as commercial sand and gravel pits.

Eskers are long, narrow, winding ridges of sand and gravel found in the middle of ground moraines (see Figure 21.18). They may run for kilometers in a direction roughly parallel to the direction of ice movement. The origin of eskers is suggested by the well-sorted, water-laid character of their materials and the sinuous, channel-like courses of the ridges themselves: eskers are deposited by meltwater streams flowing in tunnels along the bottom of a melting glacier.

DURING ICE MELTING

A large block of melting ice is isolated from the main ice mass on an outwash plain surrounded by outwash sediment.

Braided meltwater streams

Till

AFTER COMPLETE DEGLACIATION

A kettle remains after the ice block melts; a lake forms if the kettle base is below the water table.

Outwash plain

Lake

Water table

Drumlins

Kettle lake

Eskers

Varved clay

FIGURE 21.18 ■ Ice-laid and water-laid glacial deposits. *Drumlins,* found in Patagonia, Argentina. [© Hauke Steinberg.] Kettle lake, northern Minnesota. [Carlyn Iverson/Getty Images, Inc.] Pleistocene *varved clay,* from an excavation in Stockholm, Sweden. The light layers are the coarse sediments deposited in a lake during warm seasons. The dark layers are the fine clays deposited when the lake was frozen in winter. [University of Washington Libraries, Special Collections, John Shelton Collection, KC4536.] *Esker,* near Whitefish Lake, Northwest Territories, Canada. [All Canada Photos/Alamy.]

Kettles are hollows or undrained depressions that often have steep sides and may be occupied by ponds or lakes. Modern glaciers, which may leave behind huge isolated blocks of ice in outwash plains as they melt, offer the clue to the origin of kettles. A block of ice a kilometer in diameter may take 30 years or more to melt. During that time, the melting block may be partly buried by outwash sand and gravel carried by meltwater streams—usually braided—coursing around it. By the time the block has melted completely, the ice front has retreated so far that little outwash reaches the area. The sand and gravel that formerly surrounded the block of ice now surround a depression. If the bottom of the kettle lies below the groundwater table, a lake will form.

Varves are formed when valley glaciers deposit silts and clays on the bottom of a lake in a series of alternating coarse and fine layers (see Figure 21.18). A **varve** is a pair of layers formed in one year by seasonal freezing of the lake surface. In summer, when the lake is free of ice, coarse silt is laid down by abundant meltwater streams flowing from the glacier into the lake. In winter, when the surface of the lake is frozen, the finest clays settle, depositing a thin layer on top of the coarse summer layer.

Some lakes formed by continental glaciers were huge, many thousands of square kilometers in area. The till dams that created these lakes were sometimes breached and carried away at a later time, causing the lakes to drain rapidly and creating huge floods. In eastern Washington State, an area called the Channeled Scablands (**Figure 21.19**) is covered by broad, dry stream channels, relics of torrential floodwaters draining from Lake Missoula, a large glacial lake, now completely emptied. From the giant ripples, sandbars, and coarse gravels found there, geologists have estimated that this flood discharged 21 million cubic meters of water per second, flowing as fast as 30 m/s. For comparison, the velocities of ordinary river flows are measured in fractions of a meter per second, and the discharge rate of the Mississippi River in full flood is less than 50,000 m^3/s.

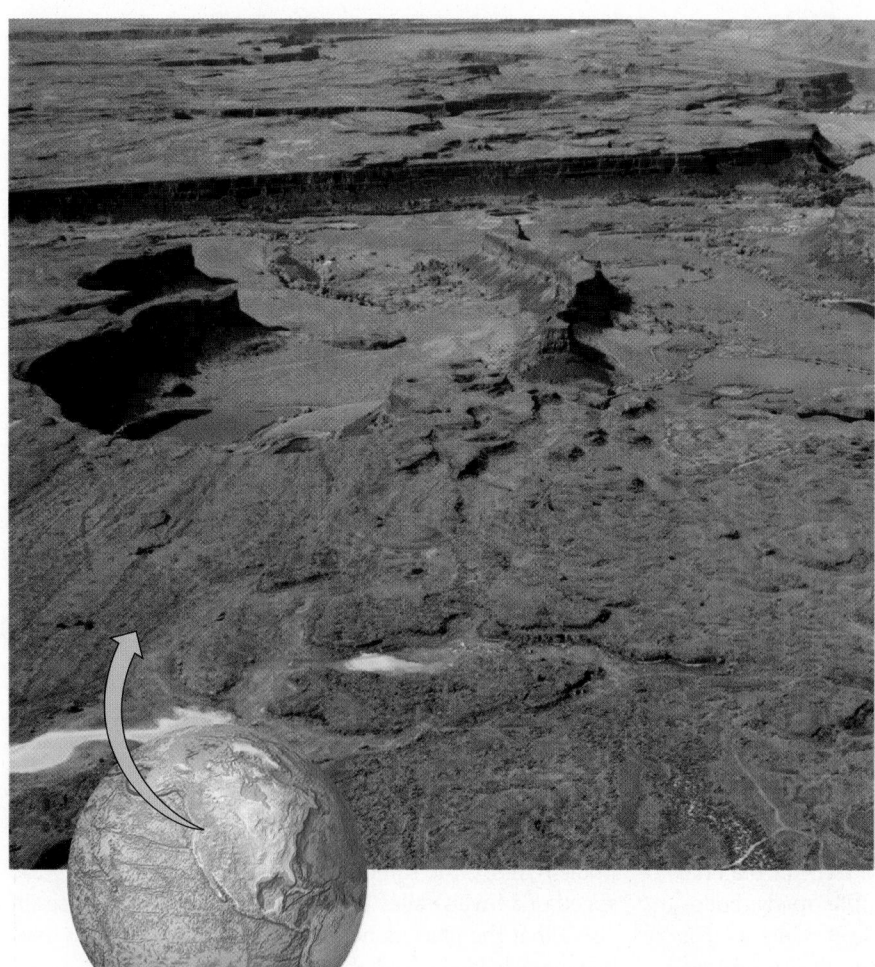

FIGURE 21.19 ■ The Channeled Scablands in eastern Washington State contain unique erosional features formed by catastrophic flooding that resulted from the draining of Lake Missoula, a huge glacial lake. This aerial photograph shows Dry Falls, a 350-foot-high, 3-mile-wide group of scalloped cliffs created by the flood. [Bruce Bjornstad.]

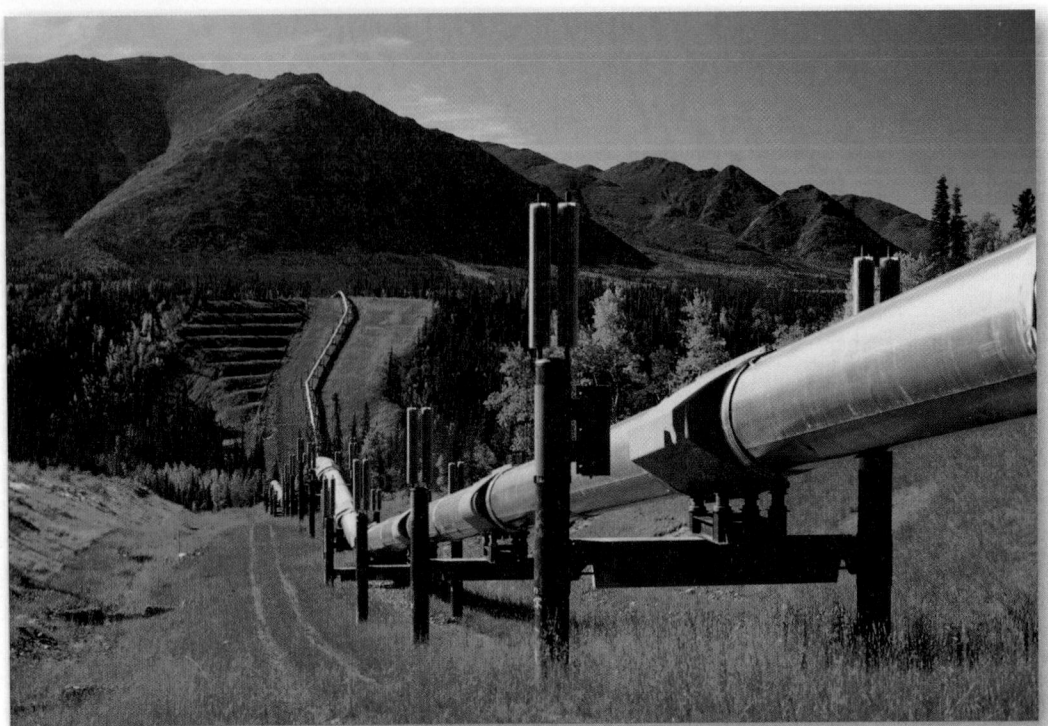

FIGURE 21.20 ■ Permafrost melting could destabilize structures at high latitudes, such as the Trans-Alaska oil pipeline, whose 1300-km (800-mile) route from Prudhoe Bay to Valdez crosses 675 km of permafrost. Where the pipeline crosses permafrost, it is perched on specially designed vertical supports. Because thawing of the permafrost would cause the supports to become unstable, they are outfitted with heat pumps designed to keep the ground around them frozen. The heat pumps contain anhydrous ammonia that vaporizes below ground and rises and condenses above ground, discharging heat through the two aluminum radiators atop each of the vertical supports. [George F. Herben/Getty Images, Inc.]

Permafrost

The ground is always frozen in very cold regions where the summer temperature never gets high enough to melt more than a thin surface layer. Perennially frozen soil, or **permafrost,** today covers as much as 25 percent of Earth's total land area. In addition to the soil itself, permafrost includes aggregates of ice crystals in layers, wedges, and irregular masses. The proportion of ice to soil and the thickness of the permafrost vary from region to region. Permafrost is defined solely by temperature, not by soil moisture content, overlying snow cover, or location: any rock or soil remaining at or below 0°C for 2 or more years is considered permafrost.

In Alaska and northern Canada, permafrost may be as thick as 300 to 500 m. The ground below the permafrost layer, insulated from the bitterly cold temperatures at the surface, remains unfrozen. It is warmed from below by Earth's internal heat. Permafrost is a difficult material to handle in engineering projects—such as roads, building foundations, and pipelines—because it melts as it is excavated. The meltwater cannot infiltrate the still-frozen soil below the excavation, so it stays at the surface, waterlogging the soil and causing it to creep, slide, and slump.

Engineers decided to build part of the Trans-Alaska pipeline above ground when an analysis showed that the pipeline would thaw the permafrost around it in some places and lead to unstable soil conditions (**Figure 21.20**).

Permafrost covers about 82 percent of Alaska and 50 percent of Canada, as well as a large proportion of Siberia (**Figure 21.21**). Outside the polar regions, it is present in high mountainous areas such as the Tibetan Plateau. Permafrost extends downward several hundred meters in shallow marine areas off the Arctic coasts, presenting difficult engineering problems for offshore oil drillers.

■ Glacial Cycles and Climate Change

Louis Agassiz, the same geologist who first measured the speed of a Swiss valley glacier, was the first to propose (in 1837) that the glaciers he observed in the Alps must have been much larger and thicker in the geologically recent past. He suggested that during a past ice age, Switzerland was covered by an extensive continental glacier almost

FIGURE 21.21 ■ A map of the Northern Hemisphere with the North Pole at its center, showing the distribution of permafrost on the northern continents. The large area of high mountain permafrost at the top of the map is on the Tibetan Plateau. [After a map by T. L. Pewe, Arizona State University.]

as thick as its mountains were tall, similar to the one in Greenland today. Among the evidence he cited was the obvious glacial sculpting of high Alpine peaks such as the mighty Matterhorn (**Figure 21.22**). Agassiz's hypothesis was controversial and was not immediately accepted.

Agassiz emigrated to the United States in 1846 and became a professor at Harvard University, where he continued his studies in geology and other sciences. His research took him to many places in the northern parts of Europe and North America, from the mountains of Scandinavia and New England to the rolling hills of the American Midwest. In all these diverse regions, Agassiz saw signs of glacial erosion and sedimentation. In the flat country of the American Great Plains, he observed deposits of glacial drift that reminded him of the end moraines of Swiss valley glaciers (**Figure 21.23**). The heterogeneous material of the drift, including erratic boulders, convinced him of its glacial origin, and the freshness of the soft sediments indicated that they were deposited in the recent past.

The areas covered by this drift were so vast that the ice that deposited them must have been a continental glacier much larger than the ones that now cover Greenland and Antarctica. Agassiz expanded his ice age hypothesis, proposing that a great continental glaciation had extended the polar ice caps far into regions that now enjoy more temperate climates. For the first time, people began to talk seriously about ice ages.

The Wisconsin Glaciation

Geologists have determined the ages of the glacial sediments Agassiz studied by isotopic dating, using carbon-14 in logs buried in the drift. The most recent drift was deposited by ice during the latter part of the Pleistocene epoch. Along the east coast of the United States, the southernmost advance of this ice is recorded by the enormous terminal moraines that form Long Island and Cape Cod. North American geologists named this glaciation after

FIGURE 21.22 ■   The high mountains of the Alps, such as the famous Matterhorn, shown here, were sculpted by a continental glacier nearly as thick as the peaks are tall. These obviously sculpted peaks provide compelling evidence for an ice age in the recent geologic past. [Hubert Stadler/Corbis.]

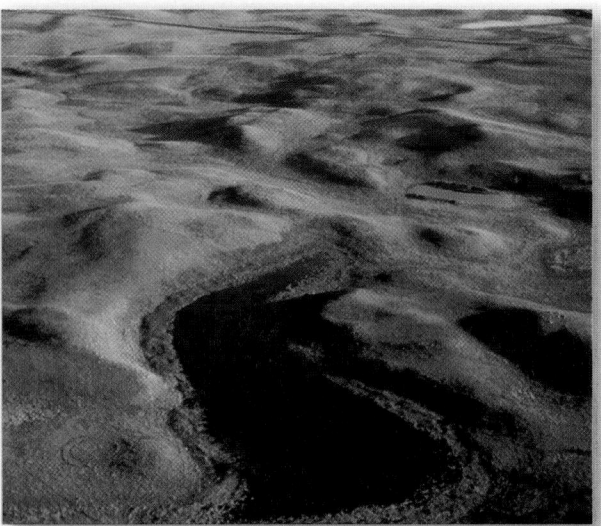

FIGURE 21.23 ■   Irregular hills alternate with lakes in a terrain of glacial till in Coteau des Prairies, South Dakota. Such landscapes provided evidence for the great continental glaciations of the Pleistocene ice ages. [University of Washington Libraries, Special Collections, John Shelton Collection, KC10367.]

Wisconsin because its effects are particularly well manifested in the glacial terrains of that state. The Wisconsin glaciation reached its maximum 21,000 to 18,000 years ago. **Figure 21.24** shows the distribution of ice near the end of the glacial maximum.

The Wisconsin glaciation was a global event, though geologists in various parts of the world have therefore given it their own local names (calling it the Würm glaciation in the Alps, for example). Ice sheets with thicknesses of 2 to 3 km built up over the northern parts of North America, Europe, and Asia. In the Southern Hemisphere, the Antarctic ice sheet expanded, and the southern tips of South America and Africa were covered with ice.

Glaciation and Sea Level Change

At the Wisconsin glacial maximum, the continents were slightly larger than they are today because the continental

FIGURE 21.24 ■   The extent of continental glaciers (white area) and sea ice (gray area) in the Northern Hemisphere near the end of the Wisconsin glacial maximum, around 18,000 years ago. [Mark McCaffrey, National Oceanic and Atmospheric Administration Paleoclimatology Program.]

shelves surrounding them—some more than 100 km wide—were exposed by a drop in sea level of about 130 m (see Figure 15.11). This drop in sea level was due to the enormous amount of water transferred from the hydrosphere to the cryosphere. Rivers extended across the newly emergent continental shelves and began to erode channels in the former seafloor. Early cultures, such as those of prehistoric Egypt, were evolving in the lands beyond the ice sheets, and humans lived in these low coastal plains.

The relationship between sea level change and glacial cycles illustrates the interaction of the hydrosphere and the cryosphere within the climate system (see Chapter 15). As Earth warms or cools, the volume of the cryosphere shrinks or grows. However, as a result of isostasy, only changes in the ice volume on continents directly affect sea level (see Earth Issues 21.1). As continental glaciers grow, the volume of the ocean decreases, and sea level falls; as continental glaciers melt, their volume decreases, and sea level rises. Thus, sea level change is indirectly linked to climate change through changes in temperature and ice volume. Were global warming to melt parts of the remaining ice sheets in Greenland and Antarctica, sea level could rise significantly, posing serious problems for human civilization (see the Practicing Geology exercise at the end of the chapter). We will discuss those problems further in Chapter 23.

The Geologic Record of Pleistocene Glaciations

Soon after Agassiz's ice age hypothesis became widely accepted in the mid-nineteenth century, geologists discovered that there had been multiple ice ages during the Pleistocene epoch, with warmer interglacial periods between them. As they mapped glacial deposits in more detail, they became aware of several distinct layers of drift, the lower ones corresponding to earlier ice ages. Between these older layers of glacial material were well-developed soils containing fossils of warm-climate plants. These fossils provided evidence that the glaciers had retreated as the climate warmed. By the early part of the twentieth century, scientists were convinced that at least four major glaciations had affected North America and Europe during the Pleistocene epoch. In North America, these ice ages, from youngest to oldest, are named after the U.S. states where the evidence of glacial advance is best preserved: Wisconsin, Illinois, Kansas, and Nebraska.

In the late twentieth century, geologists and oceanographers examined marine sediments for evidence of past glaciations, as described in Chapter 15. These sediments, which had accumulated continuously in undisturbed ocean basins, contained a much more complete geologic record of the Pleistocene than did continental glacial deposits, and they showed a much more complex history of glacial advance and retreat. By analyzing oxygen isotope ratios in marine sediments from around the world, geologists have constructed a record of climate history millions of years into the past (see Figure 15.11). More recently, studies of glacial ice cores have provided more detailed information on temperature changes during the most recent ice ages, as well as information on the role of greenhouse gases in glacial cycles (see Figure 15.12).

The Geologic Record of Ancient Glaciations

The Pleistocene glacial cycles were not unique in Earth's history. Since the early part of the twentieth century, we have known from glacial striations and lithified ancient tills, called **tillites,** that glaciers covered parts of the continents several times in the distant geologic past, long before the Pleistocene. Tillites record major continental glaciations during Permian-Carboniferous time, during Ordovician time, and at least twice during Precambrian time (**Figure 21.25**). The Permian-Carboniferous glaciation covered much of southern Gondwana about 300 million years ago, leaving deposits that have been preserved as tillites across much of the Southern Hemisphere (see Figure 21.25a, b). The joining of the southern continents near the South Pole to form Gondwana may have triggered the cooling that led to this glaciation. The Ordovician glaciation was more limited in its distribution and is best preserved in northern Africa.

The oldest confirmed glaciation occurred during the Proterozoic eon, about 2.4 billion years ago. Its glacial deposits are preserved in Wyoming, along the Canadian portion of the Great Lakes, in northern Europe, and in South Africa. Some geologists argue for an even older glaciation in the Archean eon, almost 3 billion years ago, but that interpretation is disputed.

The youngest Proterozoic glaciation, which spanned a period between 750 million and about 600 million years ago, involved several ice ages separated by warm interglacial periods. Glacial deposits of this age have been found on every continent (Figure 21.25c). Curiously, the reconstruction of paleocontinents indicated that the ice sheets in the Northern Hemisphere had extended much farther south than during the Pleistocene glaciations, perhaps all the way to the equator! This evidence has provoked some geologists to speculate that Earth may have been completely covered by ice, from pole to pole—a bold hypothesis called *Snowball Earth* (Figure 21.25d).

According to the Snowball Earth hypothesis, there was ice everywhere—even the oceans were frozen. The average global temperature would have been about −40°C, like that of the Antarctic today. Except for a few warm spots near volcanoes, very little life would have survived. How could such an apocalyptic event have occurred? And how could it have ended, returning us to the climate we know today? The answers may lie in the feedbacks that occur within the climate system (described in Chapter 15).

(a) Evidence of glaciation

PACIFIC
OCEAN

TETHYS
OCEAN

Africa

South
America

India

Australia

Antarctica

1 The Permian-Pennsylvanian glaciation covered southern Gondwana,...

2 ...and is recorded in tillites preserved today across much of the Southern Hemisphere.

ATLANTIC
OCEAN

India

Africa

PACIFIC
OCEAN

South
America

INDIAN
OCEAN

Australia

(b) Permian glacial deposits

Glacial
tillite

Glacial
striations

(d) A ball of ice?

(c) Late Proterozoic glacial deposits

Glacial
dropstones

4000 Ma   3000 Ma   2000 Ma   1000 Ma   0 Ma

HADEAN EON   ARCHEAN EON   PROTEROZOIC EON   PHANEROZOIC EON

FIGURE 21.25 ■ Ancient glaciations. (a) The first map shows the extent of the Permian-Carboniferous glaciation, which occurred more than 300 million years ago. At that time, the southern continents were assembled into the giant continent Gondwana, and the ice cap was situated in the Southern Hemisphere, centered over Antarctica, which is home to a continental glacier today. The second map shows the distribution of Permian-Carboniferous glacial deposits today. (b) Permian glacial deposits from South Africa. (c) Late Proterozoic glacial deposits. (d) The development of a late Proterozoic Snowball Earth. Geologists debate the extent to which ice covered the globe, but some think even the oceans became frozen. [John Grotzinger.]

According to one scenario, as Earth initially cooled, ice sheets at the poles spread outward, their white surfaces reflecting more and more sunlight away from Earth. The increase in Earth's albedo cooled the planet, which further expanded the ice sheets. This self-reinforcing process continued until it reached the tropics, encasing the planet in a layer of ice as much as 1 km thick. This scenario is an example of albedo feedback gone wild.

Earth remained buried in ice for millions of years, but the few volcanoes that poked above the surface slowly pumped carbon dioxide into the atmosphere. When the concentration of carbon dioxide reached a critical level, temperatures rose, the ice melted, and Earth again became a greenhouse.

The Snowball Earth hypothesis is very controversial, and many geologists disagree with the idea that the oceans were completely frozen. Nevertheless, the evidence for glaciation at low latitudes is strong, and the hypothesis serves as an example of the potential of feedbacks in Earth's climate system to produce extreme change. Geologists have their work cut out for them in trying to understand the extremes of Earth's climate system.

SUMMARY

What are the basic types of glaciers? Glaciers are divided into two basic types. A valley glacier is a river of ice that forms in the cold heights of mountain ranges and moves downslope through a valley. A continental glacier is a thick, slow-moving sheet of ice that covers a large part of a continent or other large landmass. Today, continental glaciers cover much of Greenland and Antarctica.

How do glaciers form? Glaciers form where climates are cold enough that snow, instead of melting completely in summer, is transformed into ice by recrystallization. As snow accumulates, either at the tops of valley glaciers or at the domed centers of continental glaciers, the ice thickens. Its thickness increases until it becomes so massive that gravity starts to pull it downhill.

How do glaciers shrink or grow? Glaciers lose ice by melting, sublimation, iceberg calving, and wind erosion. The glacial budget is the relationship between ablation (the amount of ice a glacier loses annually) and accumulation. If ablation is balanced by accumulation of new snow and ice in the glacier's upper reaches, the size of the glacier remains constant. If ablation is greater than accumulation, the glacier shrinks; conversely, if accumulation exceeds ablation, the glacier grows.

How do glaciers move? Glaciers move by a combination of plastic flow and basal slip. Plastic flow dominates in very cold regions, where the glacier's base is frozen to the ground. Basal slip is more important in warmer climates, where meltwater at the glacier's base lubricates the ice.

How do glaciers shape the landscape? Glaciers erode bedrock by scraping, plucking, and grinding it into sizes ranging from boulders to fine rock flour. Valley glaciers erode cirques, horns, and arêtes at their heads; excavate U-shaped and hanging valleys; and create fjords by eroding their valleys below sea level at the coast. Glacial ice has both high competence and high capacity, which enable it to carry abundant sediment particles of all sizes. Glaciers transport huge quantities of sediments to the ice front, where melting releases them. The sediments may be deposited directly by the melting ice as till or picked up by meltwater streams and laid down as outwash. Moraines and drumlins are characteristic landforms deposited by ice. Eskers and kettles are formed by meltwater. Permafrost forms where summer temperatures never rise high enough to melt more than a thin surface layer of soil.

What does the geologic record tell us about past ice ages? Glacial drift of Pleistocene age is widespread over high-latitude regions that now enjoy temperate climates. This widespread drift is evidence that continental glaciers once expanded far beyond the polar regions. Studies of the geologic ages of glacial deposits on land and in marine sediments show that continental ice sheets advanced and retreated many times during the Pleistocene epoch. The most recent glacial advance, known as the Wisconsin glaciation, covered the northern parts of North America, Europe, and Asia with ice and exposed large areas of continental shelves. During interglacial intervals, sea level rose and submerged the shelves.

KEY TERMS AND CONCEPTS

ablation (p. 592)	drumlin (p. 604)	iceberg calving (p. 589)	surge (p. 594)
accumulation (p. 592)	esker (p. 604)	kettle (p. 605)	till (p. 603)
basal slip (p. 594)	fjord (p. 602)	moraine (p. 603)	tillite (p. 609)
cirque (p. 601)	glacier (p. 588)	outwash (p. 603)	U-shaped valley (p. 602)
continental glacier (p. 590)	hanging valley (p. 602)	permafrost (p. 606)	valley glacier (p. 588)
crevasse (p. 594)	ice shelf (p. 590)	plastic flow (p. 594)	varve (p. 605)
drift (p. 603)	ice stream (p. 596)	striation (p. 600)	

Google Earth Project

Glaciers are the most visible features of Earth's cryosphere. Their movements erode the rocks beneath them and deposit huge amounts of sediments. Glaciers have created some spectacular features on Earth's surface in the recent geologic past that can be easily seen using Google Earth.

In this Google Earth Project, you will explore glaciers and glacial landscapes at a number of locations around the world. For this project, you will need to turn on the Terrain Layer and, in the 3D View window of Options, choose "Decimal degrees" in the Show Lat/Long box and "Meters, Kilometers" in the Show Elevation box. You can navigate to the initial geographic position for each exercise by typing the listed coordinates into the "Fly To" search window and clicking on the Search button. You can then navigate by using the Zoom slider to zoom the eye altitude in or out and the Look joystick to rotate the compass azimuth of the view or to tilt the view toward the horizontal. (For these exercises, it's better to turn off "Automatically tilt while zooming" in the Navigation window of Options.) Use the Move joystick to translate your position while maintaining the same Look angles.

LOCATION   Glaciers and glacial landscapes around the world

GOAL   Learn how to identify the types of glaciers and glacial features

LINKED   Figures 21.10, 21.11, and 21.16 and Table 21.1

1. Navigate to 61.385° N, 148.500° W in south central Alaska, zoom to an eye altitude of 4.0 km, rotate the view to look east, and tilt the view to see an expanse of ice: the Knik glacier. Using your cursor, explore the elevation of the ice surface to observe the direction of its slope. Based on this information, which of the following is the best description of the ice mass?

   a. A continental glacier flowing outward from its highest point near the center of the glacier
   b. A continental glacier flowing westward from its highest point on the east side of the glacier
   c. A valley glacier flowing westward
   d. A valley glacier flowing eastward

© 2009 Google Image © 2009 Digital Globe Image IBCAO Image © TerraMetrics

2. Navigate to 64.400° N, 16.800° W on the south side of Iceland, zoom to an eye altitude of 150 km, and examine the large mass of ice below you, which the Icelanders call the Vatnajökull glacier. Using the ruler tool, measure the size of the glacier, and using the cursor, find the region of the glacier with the highest

elevation. Explore the glacier looking for evidence of flow. Based on this information, which of the following is the best description of the ice mass?

   a. A continental glacier flowing outward from its highest point near the center of the glacier
   b. A continental glacier flowing westward from its highest point on the east side of the glacier
   c. A valley glacier flowing westward
   d. A valley glacier flowing eastward

© 2009 Google Image IBCAO © 2009 Cnes/Spot Image

3. Navigate to 37.730° N, 119.580° W in Yosemite National Park, California, zoom to an eye altitude of 3 km, rotate the view to look northeast, and tilt the view to see Yosemite Valley. Observe the shape of the valley perpendicular to its axis and, using your cursor, explore the elevation of the valley floor to observe the direction of its slope. Which of the following is the best description of Yosemite Valley?

a. A V-shaped valley cut by a stream flowing to the southwest

b. A U-shaped valley cut by a glacier flowing to the southwest

c. A V-shaped valley cut by a stream flowing to the northeast

d. A U-shaped valley cut by a glacier flowing to the northeast

to look up the valley; using the Move joystick, move eastward so you can view the position of the lake relative to the Teton mountain front. Which of the following terms best describes the ridge that encircles Jenny Lake?

a. Esker

b. Drumlin

c. Terminal moraine

d. Medial moraine

© 2009 Google Image © 2009 DigitalGlobe Image USDA Farm Service Agency

© 2009 Google Image © Digital Globe Image USDA Farm Service Agency

4. Navigate to 45.100° S, 167.020° E on the west coast of South Island, New Zealand. Zoom to an eye altitude of 1 km, rotate the view to look southeast, and tilt the view to see a water-filled valley in the mountainous terrain. Explore the extent of this water-filled valley. Which of the following terms best describes it?

a. Glacial lake

b. Outwash lake

c. Kettle lake

d. Fjord

© 2009 Google Image © 2009 DigitalGlobe © 2009 Cnes/Spot Image
Data SIO, NOAA, U.S. Navy, NGA, GEBCO

5. Navigate to 43.765° N, 110.730° W in Grand Teton National Park, zoom to an eye altitude of 7 km, and examine Jenny Lake, which lies at the eastern mouth of a large valley. Use your cursor to profile the elevation. You will observe that the lake is rimmed on its eastern side by a narrow ridge, green with trees, that rises up to 30 m above the lake surface. Zoom in to 3.5 km, rotate the view to look west, and tilt the view

Optional Challenge Question

6. Navigate to 46.014° N, 7.616° E in the Swiss Alps, zoom to an eye altitude of 3.5 km, and observe the cracks in the glacial ice. Using your cursor, explore the elevation of the ice surface to observe how the slope changes. Which of the following is the best explanation of the cracks?

a. Crevasses along the side of a valley glacier caused primarily by a bend in flow direction

b. Crevasses across a valley glacier caused primarily by a slope increase in the direction of flow

c. Crevasses across a valley glacier caused primarily by a blockage of flow by an end moraine

d. Crevasses along the side of valley glacier caused primarily by a constriction of the flow by the valley walls

© 2009 Google Image © 2009 Cnes/Spot Image

PRACTICING GEOLOGY EXERCISE

Why Is Sea Level Rising?

Over the twentieth century, sea level rose about 200 mm, and it is currently rising at a rate of about 3 mm/year. Rising sea level is a grave threat to human society because it could inundate deltas, atolls, and other coastal lowlands, and it could erode beaches, increase coastal flooding, and threaten water quality in estuaries and aquifers. Why is sea level rising, and can we predict the rate of its rise in the future?

We know that anthropogenic warming of the polar regions is reducing the amount of sea ice and causing the breakup of large ice shelves (see Figure 21.13). Because of isostasy, however, this decrease in the volume of floating ice does not contribute to sea level rise. Melting ice can cause sea level to change only if the ice is on land, not floating in water (see Earth Issues 21.1).

Most of the world's ice is locked up in the huge continental glaciers that cover Antarctica and Greenland. Is global warming causing these ice sheets to melt faster than they can be regenerated by new snowfall? In the past, it has been difficult to answer this question because scientists had to compute the glacial budget; that is, they

had to figure out the difference between accumulation and ablation, and these are hard quantities to estimate accurately over large regions. But now radar instruments mounted on Earth-orbiting satellites can directly measure changes in the ice volume of a region. The results have been surprising.

First, according to the IPCC's most recent assessment, the East Antarctic ice sheet, the largest ice reservoir on Earth (see Figure 21.6), has been *gaining* ice mass at about 21 Gt/year in the period 1993–2010. Recent climate changes have evidently increased the amount of snowfall in East Antarctica so that accumulation exceeds ablation. This net accumulation is good news because it subtracts from any sea level rise. Unfortunately, the West Antarctic ice sheet is losing mass at a much higher rate, at about 118 Gt/year, and the smaller Greenland ice sheet is losing about 121 Gt/year. Most surprising of all is a net loss of 57 Gt/year in the mass of continental valley glaciers and smaller ice sheets (such as those in Iceland), which together account for less than 1 percent of the total ice volume of the cryosphere. The rates are especially high for valley glaciers in temperate and tropical regions, which are vanishing very quickly.

Summing up these numbers yields 275 Gt/year as the current rate of continental ice loss. Essentially all of this mass goes into the ocean. One gigaton of water occupies one cubic kilometer (its density is 1 g/cm^3), so the increase in ocean volume is about 275 km^3/year. We can convert this volume change into sea level change using the formula

$$\text{sea level rise} = \text{ocean volume increase} \div \text{ocean area}$$

From Appendix 2, we obtain an ocean area of 3.6 × 10^8 km^2, so

$$\text{sea level rise} = 275 \text{ km}^3/\text{year} \div 3.6 \times 10^8 \text{ km}^2 = 7.6 \times 10^{-7} \text{ km/year}$$

or about 0.8 mm/year.

This figure is only a fraction of the current rate of sea level rise. The rest is coming from the warming of the ocean itself. In the twentieth century, the sea surface temperature increased by nearly 1°C, which caused the water in the upper portion of the ocean to expand a tiny fraction, about 0.01 percent. That small increase in volume can account for most of the 200 mm rise in sea level during that period.

We can conclude that melting of continental ice has thus far contributed only a small amount to sea level rise. However, the process of glacial thinning is increasing rapidly, primarily by the acceleration of glacial flow

Sea level change

Sea-surface temperature change

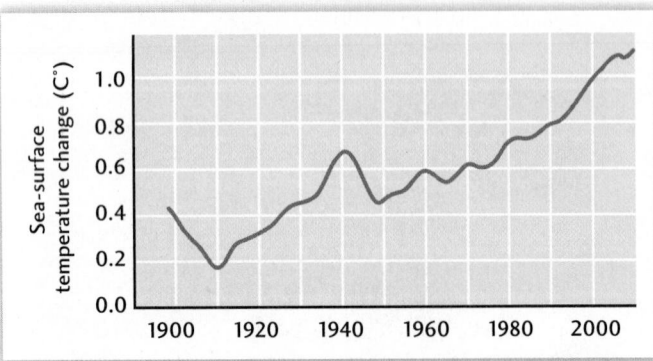

During the twentieth century, sea level rose by about 200 mm (top panel), and the global average sea surface temperature increased by about 1°C (bottom panel). [Sea level change data from B. C. Douglas; sea-surface temperature change data from British Meteorological Office.]

(see Figure 21.12). The satellite observations reveal that flow accelerations of 20 to 100 percent have occurred over the past decade. A key question that concerns scientists is whether these accelerations will increase in the future.

BONUS PROBLEM: If seawater expands by 0.01 percent for each 1°C of temperature increase, how deep is the layer of the ocean that must be heated by 1°C to explain the twentieth-century sea level rise of 200 mm?

EXERCISES

1. How are valley glaciers distinguished from continental glaciers?

2. How is snow transformed into glacial ice?

3. How does glacial growth or shrinkage result from the balance between ablation and accumulation?

4. What are the mechanisms of glacial flow?

5. How do glaciers erode bedrock?

6. What information do striations provide about glaciation?

7. Describe three kinds of glacial sediment.

8. Describe three landforms made by glaciers.

9. Will the melting of ice shelves due to global warming increase sea level? Explain your answer.

10. What type of sedimentary deposit marks the farthest advance of a glacier?

11. Why are kettles described as water-laid sedimentary deposits, rather than ice-laid deposits?

12. Why does sea level drop during ice ages?

THOUGHT QUESTIONS

1. Some parts of a glacier contain a lot of sediment; others contain very little. What accounts for the difference?

2. Contrast the kinds of till that you would expect to find in two glaciated areas: one a terrain of granitic and metamorphic rocks; the other a terrain of soft shales and loosely cemented sands.

3. What geologic evidence would you search for if you wanted to know the direction of ancient glacial movements across the Canadian Shield?

4. You are walking over a winding ridge of glacial drift. What evidence will you look for to discover whether you are on an esker or an end moraine?

5. One of the dangers of exploring glaciers is the possibility of falling into a crevasse. What topographic features of a valley glacier or its surroundings would you use to infer that you were on a part of the glacier that was badly crevassed?

6. You live in New Orleans, not far from the mouth of the Mississippi River. What might be your first indication that Earth is entering a new ice age?

7. Some geologists think that one result of continued global warming could be the shrinkage and collapse of the West Antarctic ice sheet. How might this affect the populations of North America and Europe?

8. Evidence from boreholes drilled into the ice shows that there is liquid water at the base of some glaciers. What kinds of glaciers might have liquid water at the base? What factors might be responsible for the melting of ice at the bottom of these glaciers?

9. The density of ice (0.92 g/cm^3) is less than that of water (1.0 g/cm^3), which is why icebergs float. Using the principle of isostasy, compute what fraction of an iceberg's mass floats above the sea surface.

MEDIA SUPPORT

 21-1 Animation: Glacier Formation

 21-2 Animation: Ice Ages

 21-1 Video: Glacial Deposits: Till, Outwash, Erratics, and Loess

 21-2 Video: Glacial Lakes and Wetlands

 21-3 Video: Landforms Produced by Continental Glaciation

■ Topography, Elevation,
and Relief   618

■ Landforms: Features
Sculpted by Erosion
and Sedimentation 621

■ How Interacting
Geosystems Control
Landscapes   626

■ Models of Landscape
Development   630

Rapidly uplifting mountains form terraces, marking the former position of the river as it cuts through solid bedrock. The terrace in the middle foreground was formed along the Indus River, which cuts through the middle of the Himalaya. [D. W. Burbank.]

LANDSCAPE DEVELOPMENT 22

HAVE YOU EVER GAZED out over the horizon and wondered why Earth's surface has the shape it does, or what forces created that shape? From high snowcapped peaks to broad rolling plains, Earth's landscapes comprise a diverse array of landforms—large and small, rough and smooth. These landscapes develop through slow changes as the processes of tectonic uplift, weathering, erosion, transportation, and sedimentation combine to sculpt the land surface.

In the past, changes in landscapes were imperceptible on a human time scale, but new technologies now permit us to measure the rates of many of these changes directly. These technologies have revitalized the branch of Earth science known as *geomorphology*—the study of landscapes and their development. Understanding the slow processes by which landscapes develop helps us not only to manage our land resources, but also to appreciate the linkages between the plate tectonic and climate systems. Geomorphology represents a great challenge to geologists because it requires them to integrate many branches of Earth science.

In the most basic sense, landscapes can be viewed as the result of competition between processes that raise Earth's surface and those that lower it. Processes driven by the plate tectonic system raise the elevation of Earth's crust to form mountain ranges and high plateaus. The uplifted rocks are exposed to weathering and erosion—processes driven by the climate system. Thus, the landscape is an expression of the interaction between these two geosystems.

The uplifted areas of Earth's crust can be narrow or broad, and uplift rates can be fast or slow. Similarly, weathering and erosion may operate over narrow or broad areas, and their intensity can be high or low. Thus, landscapes themselves depend on balance between these processes. Moreover, tectonic and surface processes influence each other. For example, uplift of a mountain range may drive regional (or even global) climate change and thus change weathering rates, which in turn influence the further uplift of the mountains.

In this chapter, we will look closely at how the plate tectonic and climate systems—and their component processes, including tectonic uplift, weathering, erosion, mass wasting, and sediment transport and deposition—interact in the dynamic process that sculpts Earth's landscapes.

■ Topography, Elevation, and Relief

The term **geomorphology** refers both to the shape of a landscape and to the branch of Earth science concerned with that shape and how it develops. We begin our study of geomorphology here with the elementary observations of any landscape that are obvious as one examines

Earth's surface: the height and ruggedness, or roughness, of mountains and lowlands.

Topography is the general configuration of varying heights that gives shape to Earth's surface (see Figure 1.8). We compare the heights of landscape features with sea level: the average height of the ocean surface throughout the world. We then express the vertical distance above or below sea level as **elevation.** A topographic map shows the distribution of elevations in an area. This distribution is most often represented by **contours,** lines that connect

FIGURE 22.1 ■ The topography of a mountain peak (*left*) and a stream valley (*right*) can be accurately depicted on a flat topographic map by contours, lines that connect points of equal elevation. The more closely spaced the contours, the steeper the slope. [After A. Maltman, *Geological Maps: An Introduction.* New York: Van Nostrand Reinhold, 1990, p. 17. Topographic maps from USGS/DRG.]

points of equal elevation (**Figure 22.1**). The more closely spaced the contours, the steeper the slope.

Centuries ago, geologists learned to survey topography and construct maps to plot and record geologic information. Although land surveying based on age-old methods is still used for some purposes, modern mapmakers rely on satellite photographs, radar imaging, airborne laser range finders, and other technologies that enable them to discern elevation and other topographic properties (**Figure 22.2**).

One of the properties of topography is **relief:** the difference between the highest and lowest elevations in a particular area (**Figure 22.3**). As this definition implies, relief varies with the scale of the area over which it is measured. In studies of geomorphology, it is useful to define three fundamental components of relief: hillslope relief (the decrease in elevation between mountain summits or ridgelines and the point at which stream channels begin), tributary relief (the decrease in elevation along tributary stream channels from their beginning to the main, or *trunk,*

FIGURE 22.2 ▪ Topographic maps for Turkey and the adjacent region. (a) A digital elevation model, or DEM. The elevation values are depicted digitally, with each pixel representing one elevation value. (b) To produce this slope map, elevation values from the DEM were used to calculate the slopes between adjacent pixels. The slopes are represented by angles measured in degrees from the horizontal. A slope map is useful for identifying places where changes in topography are particularly abrupt, such as along mountain fronts or active fault scarps. [Marin Clark.]

Hillslope relief is the decrease in elevation between mountain summits/ridgelines and the point where channels begin.

Tributary relief is the elevation decrease along the tributaries.

Trunk channel relief is the decrease from the highest tributary to the end of the channel.

FIGURE 22.3 ■ Relief is the difference between the highest and lowest elevations in a particular area.

stream with which they merge), and trunk channel relief (the decrease in elevation from the highest tributary to the end of the trunk stream channel).

To estimate relief in an area of interest from contours on a topographic map, we subtract the elevation of the lowest contour, usually at the bottom of a stream valley, from that of the highest contour at the top of the highest hill or mountain. Relief is a measure of the roughness of a terrain: the greater the relief, the more rugged the topography. Mount Everest in the Himalaya—the highest mountain in the world, with an elevation of 8850 m—is located in an area of extremely high relief (**Figure 22.4a**). In general, most regions of high

elevation are also areas of high relief, and most areas of low elevation are areas of low relief. There are exceptions, however. For example, the Dead Sea, between Israel and Jordan, has the lowest land elevation in the world at 392 m below sea level, but it is flanked by impressive mountains that provide significant relief in this small region (Figure 22.4b). Other regions, such as the Tibetan Plateau, lie at high elevations but have relatively low relief (see Figure 22.4a).

If we fly over North America, we can see many kinds of topography. **Figure 22.5** is a computer-processed digital relief map that shows the details of large- and small-scale landforms. This digital map provides an

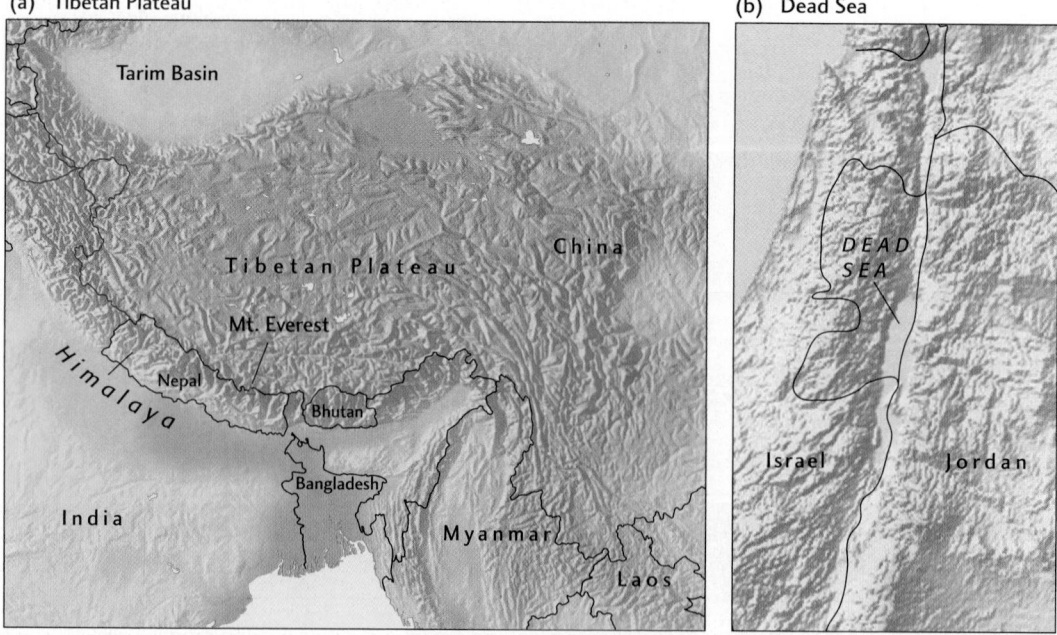

FIGURE 22.4 ■ Areas of high relief are usually, but not always, areas of high elevation. (a) Mount Everest, the highest mountain in the world, is located in an area of high relief. The Tibetan Plateau, to its north, however, is a region of high elevation but relatively low relief. (b) The Dead Sea, with the lowest land elevation in the world, is located in an area of relatively high relief. [Marin Clark and Nathan Niemi.]

FIGURE 22.5 ■ A digital shaded relief map of landforms in the contiguous United States and Canada. [Gail P. Thelin and Richard J. Pike/USGS, 1991.]

overall view of the continent while showing features as small as 2.5 km across. The moderate elevations and relief of the elongate ridges and valleys of the Appalachian Mountains contrast with the low elevations and relief of the midwestern plains. Even more striking is the contrast between the plains and the Rocky Mountains. As we examine these different types of topography more closely, we can characterize them not only by elevation and relief, but also by their landforms: the steepness of their slopes, the shapes of their mountains or hills, and the forms of their valleys.

■ Landforms: Features Sculpted by Erosion and Sedimentation

Streams, glaciers, and wind leave their marks on Earth's surface in a variety of **landforms:** rugged mountain slopes, broad valleys, floodplains, dunes, and many others.

The scale of landforms ranges from regional to very local. At the largest scale (tens of thousands of kilometers), mountain belts form topographic walls along the boundaries of lithospheric plates. At the smallest scale (meters), the topography of an individual outcrop may be shaped by differential weathering of the rocks of differing hardnesses that compose it. This section focuses mostly on the regional-scale features that define the overall topography of Earth's surface.

Mountains and Hills

We have used the word *mountain* many times in this textbook, yet we can define it no more precisely than to say that a mountain is a large mass of rock that projects well above its surroundings. Most mountains are found with others in ranges, where peaks of various heights are easier to distinguish than distinct separate mountains (**Figure 22.6**). Mountains that rise as single peaks above the surrounding lowlands are usually isolated volcanoes or erosional remnants of former mountain ranges.

FIGURE 22.6 ■ Most mountains are found in ranges, not as individual peaks. In this glacially sculpted terrain in southern Argentina, all the peaks are sharp arêtes. [© Renato Granieri/ Alamy.]

We distinguish between mountains and hills only by size and custom. Landforms that would be called mountains in regions of overall lower elevation are called hills in regions of high elevation. In general, however, landforms more than several hundred meters above their surroundings are called mountains.

Mountains are direct or indirect manifestations of plate tectonic activity. The more recent the activity, the more likely the mountains are to be high. The Himalaya, the highest mountains in the world, are also among the youngest. The steepness of the slopes in mountainous and hilly areas generally correlates with elevation and relief.

FIGURE 22.7 ■ Two topographic views of the Tibetan Plateau, the highest and most extensive plateau on Earth.

The steepest slopes are usually found on high mountains in areas of high relief. The slopes of mountains in areas of lower elevation and relief are less steep and rugged. As we will see later in this chapter, the relief of a mountain range depends greatly on how much the bedrock has been incised by glaciers and streams relative to the amount of tectonic uplift.

Plateaus

A **plateau** is a large, broad, flat area of appreciable elevation above the neighboring terrain. Most plateaus have elevations of less than 3000 m, but the Altiplano of Bolivia lies at an elevation of 3600 m, and the extraordinarily high Tibetan Plateau, which extends over an area of 1000 km by 5000 km (well over half the size of the United States), has an average elevation of almost 5000 m (**Figure 22.7**). Plateaus form where tectonic activity produces regional uplift.

Smaller plateau-like features may be called *tablelands*. In the western United States, a small, flat, elevated landform with steep slopes on all sides is called a **mesa** (from the Spanish word for "table") (**Figure 22.8**). Mesas result from differential weathering of bedrock of varying hardness.

Stream Valleys

Observations of stream valleys in various regions led to one of the important early theories of geology: the idea that stream valleys were created through erosion by the streams that flowed in them. Geologists could see that the sedimentary rock formations on one side of a valley matched the formations on the opposite side. Such observations led them to conclude that the formations had once been deposited as continuous beds of sediment, but that the stream had removed enormous quantities of the original formation by breaking up the rock and carrying it away.

How a stream erodes soil and rock depends on its **stream power**—which is the product of its slope and its discharge—balanced by the streambed's ability to resist erosion—which is the product of the volume and the particle size of the sediment in the stream channel (**Figure 22.9**). If stream power is high enough to wash away the sediment, resistance to erosion is mainly a function of bedrock hardness.

As it turns out, rates of bedrock erosion increase dramatically as stream power increases. On most days, a flowing stream accomplishes little erosion because discharge, and thus stream power, is low. However, on the rare days when discharge (and thus stream power) is very high, erosion rates can be dramatically high. This relationship illustrates a fundamental characteristic shared by many of Earth's geosystems: large, rare events often create much more change than small, frequent ones.

Three principal processes erode bedrock in mountainous terrain. The first is abrasion of the bedrock by suspended and saltating sediment particles moving along the bottom and sides of the channel (see Chapter 18).

FIGURE 22.8 ■ A mesa in Monument Valley, Arizona. This flat-topped structure is essentially a stack of erosion-resistant beds. [Raymond Siever.]

(a) Increasing sediment size, sediment volume, and bedrock hardness increase the resistance to erosion.

Increasing slope and stream discharge increase stream power and thus erosion.

(b) In steep, wet terrain, stream power overcomes resistance to erosion. Sediment particles are transported away, and bedrock hardness becomes the principal factor in resistance to erosion.

(c) Where slopes are gentler, stream discharge is lower, and therefore stream power is lower. Thus, sediment begins to be deposited, armoring the streambed and stopping erosion. At this point, stream power and resistance to erosion are in balance.

(d) Where slopes are much flatter, stream power is so decreased that much sediment is deposited and the streambed builds up and fills the valley with sediment.

FIGURE 22.9 ■ (a) Erosion is controlled by a balance between stream power and resistance to erosion. (b) Yellowstone River, Yellowstone National Park; (c) Snake River, Suicide Point, Idaho; (d) Denali National Park, Alaska. [(a) After D. W. Burbank and R. S. Anderson, *Tectonic Geomorphology.* Oxford: Blackwell, 2001; (b) Karl Weatherly/Getty Images, Inc.; (c) Dave G. Houser/Corbis; (d) Dennis Macdonald/ Getty Images, Inc.]

Second, the drag force of the current itself abrades the bedrock as it plucks rock fragments from the channel. Third, at higher elevations, glacial erosion forms valleys that can then be occupied by streams. Determining the relative importance of these three processes in mountainous terrain is one way geologists can distinguish between the influences of climate and of plate tectonic processes on landscape development (see Practicing Geology exercise).

Stream valleys have many names—canyons, gulches, arroyos, gullies—but all have the same general geometry. A vertical cross section through a young mountain stream valley with little or no floodplain has a simple V-shaped profile (Figure 22.9b). A broad, low stream valley with a wide floodplain has a cross section that is more open, but is still distinct from the U-shaped profile of a glacial valley. Regions with different topographies and types of bedrock produce stream valleys of varying shapes and widths (Figure 22.9b–d). Valleys range from the narrow gorges of erosion-resistant mountain belts to the wide, shallow valleys of easily eroded plains. Between these extremes, the width of a valley generally corresponds to the erosional state of the region. Valleys are somewhat broader in mountains that have begun to be lowered and rounded by erosion and are much broader in low-lying hilly topography.

A **badland** is a deeply gullied landscape resulting from the rapid erosion of easily erodible shales and clays (**Figure 22.10**). Virtually the entire area is a proliferation of gullies and valleys, with little flat land between them.

FIGURE 22.10 ■ Gully erosion in the Badlands of South Dakota. Numerous gullies have formed in easily eroded sedimentary rocks. [© Ilene MacDonald/Alamy.]

Structurally Controlled Ridges and Valleys

In young mountain belts, during the early stages of tectonic folding and uplift, the upward folds (anticlines) form ridges and the downward folds (synclines) form valleys (**Figure 22.11**). As weathering and erosion begin to predominate and gullies and valleys bite deeper into the underlying geologic structures, the topography may become inverted, so that anticlines form valleys and synclines form ridges. This happens where the rocks—typically sedimentary rocks such as limestones, sandstones, and shales—exert strong control on the topography by their variable resistance to erosion. If the rocks beneath an anticline are easily erodible, as shales are, the core of the anticline may be eroded to form an anticlinal valley (**Figure 22.12**). In a region that has been eroded for many millions of years, a pattern of linear anticlines and synclines produces a series of ridges and valleys such as those of the Valley and Ridge province of the Appalachian Mountains (**Figure 22.13**).

Structurally Controlled Cliffs

The folds and faults produced by deformation during mountain building leave their marks on Earth's surface in other ways as well. **Cuestas** are asymmetrical ridges that form in a tilted and eroded series of beds with alternating resistance to erosion. One side of a cuesta has a long, gentle slope determined by the dip of an erosion-resistant bed. The other side is a steep cliff formed at the edge of the resistant bed where it is undercut by erosion of a weaker bed beneath it (**Figure 22.14**). Much more steeply dipping or vertical beds of hard strata erode more slowly to form **hogbacks:** steep, narrow, more or less symmetrical ridges (**Figure 22.15**). Fault scarps are steep cliffs produced by nearly vertical faults in which one side rises higher than the other (see Figure 7.9).

■ How Interacting Geosystems Control Landscapes

Broadly speaking, the interaction of Earth's internal and external heat engines controls the development of landscapes. Earth's internal heat engine drives plate tectonic processes, which elevate mountain belts and give rise to volcanoes. Earth's external heat engine, powered by the Sun, drives the climate system, and thus controls the processes at Earth's surface that wear away mountains and fill

FIGURE 22.11 ■ Valley and ridge topography formed on folded sedimentary rock in the Zagros Mountains of Iran. The deformation is so recent (Pliocene) that erosion has not yet significantly modified the original geologic structure of anticlines (ridges) and synclines (valleys). [NASA.]

TIME 1

Harder, erosion-resistant rocks lie over softer, more erodible layers. Ridges are over anticlines and streams flow in valleys formed by synclines. Tributary streams on anticline slopes flow faster and with more power than valley streams. They erode the slopes faster than main streams erode the valleys.

TIME 2

Tributaries over the synclines cut through resistant rock layers and start to quickly carve the softer underlying rock into steep valleys over the anticlines.

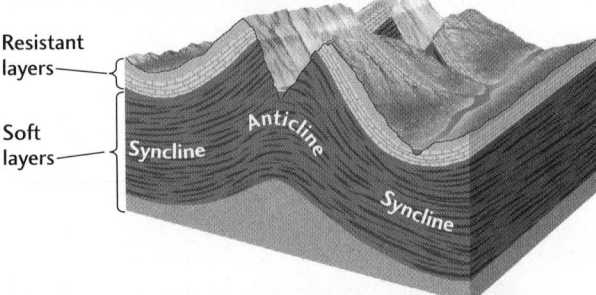

TIME 3

As the process continues, valleys form over the anticlines and ridges capped by resistant strata are left over the synclines.

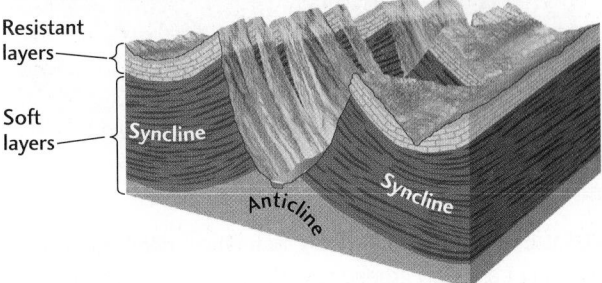

FIGURE 22.12 ■ Stages in the development of ridges and valleys in folded sedimentary rock. In early stages (time 1), ridges are formed by anticlines and valleys by synclines. In later stages (times 2 and 3), the anticlines may be breached. Ridges held up by caps of resistant rock remain as erosion forms valleys in less resistant rock.

FIGURE 22.13 ■ The Valley and Ridge province of the Appalachian Mountains has the structurally controlled topography of linear anticlines and synclines exposed to millions of years of erosion. The prominent ridges, shown in reddish orange, are composed of erosion-resistant sedimentary rock. [Courtesy MDA Information Systems LLC.]

(a)

Erosion of weak shale undercuts sandstone.

Erosion-resistant sandstone

Easily eroded shale

(b)

FIGURE 22.14 ■ (a) Cuestas form where gently dipping beds of erosion-resistant rock, such as sandstone, are undercut by erosion of an easily eroded underlying rock, such as shale. (b) Cuestas formed on structurally tilted sedimentary rocks in Dinosaur National Monument, Colorado. [Marli Miller.]

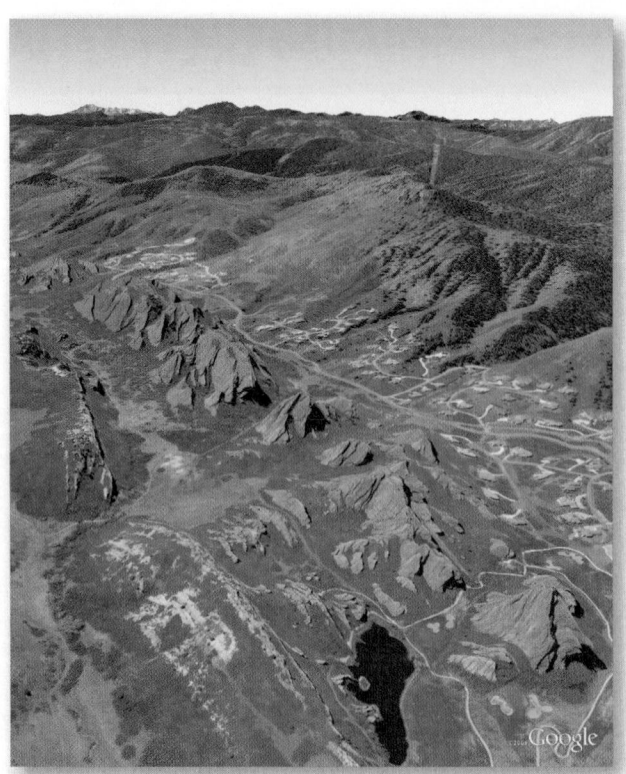

FIGURE 22.15 ■ Hogback ridges in the Rocky Mountains near Roxborough, Colorado. [Image © 2009 DigitalGlobe Image U.S. Geological Survey; Image USDA Farm Service Agency.]

basins with sediment. Solar radiation powers the atmospheric circulation that produces Earth's climates, including its different temperature and precipitation regimes. Thus, landscapes are controlled by the interaction of two global geosystems (Figure 22.16).

Feedback Between Climate and Topography

Many of the forces of weathering and erosion operate at different rates at different elevations. Thus, climate, which changes with elevation, modulates weathering and erosion, and therefore also modulates the uplift of mountain ranges.

Chapter 16 described some of the effects of climate on weathering, erosion, and mass wasting. Climate influences rates of freezing and thawing and of expansion and contraction of rock due to heating and cooling. Climate also affects the rate at which water dissolves minerals. Rainfall and temperature—the principal components of climate—affect weathering and erosion through infiltration and runoff, streamflow, and the formation of glaciers, all of which help to break up rock and mineral particles and carry them downslope.

High elevation and relief enhance the fragmentation and mechanical breakup of rock, partly by promoting freezing and thawing. At high elevations, where the climate is cool, mountain glaciers scour bedrock and carve out deep

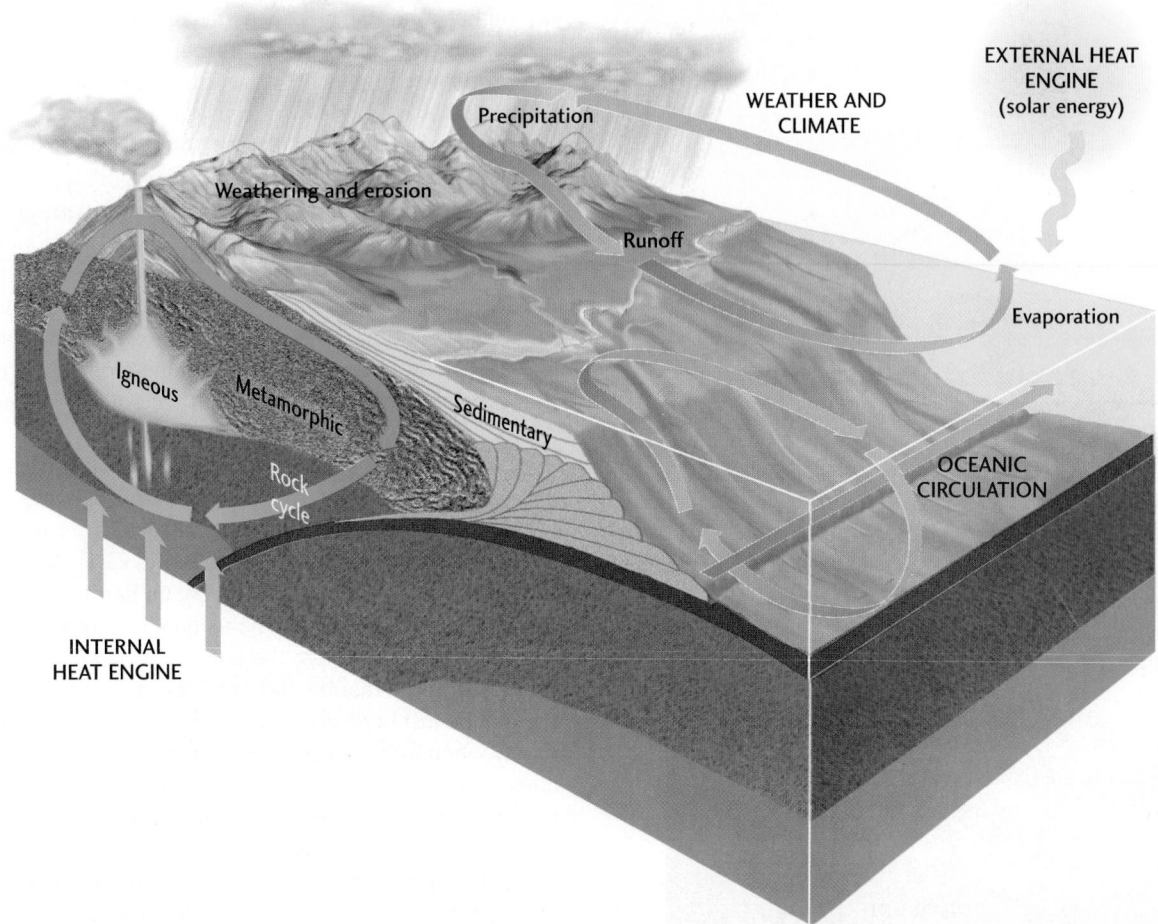

FIGURE 22.16 ■ Landscape development is controlled by interactions between the plate tectonic system and the climate system.

valleys. Rainfall lubricates rock on mountain slopes, which moves downhill quickly in landslides and other mass movements, exposing fresh rock to attack by weathering. Streams run faster in mountains than in lowlands and therefore erode and transport sediment more rapidly. Chemical weathering plays an important role in the erosion of high mountains, but the mechanical breakup of rocks is so rapid that most of the debris appears to be almost unweathered. The products of chemical weathering—dissolved materials and clay minerals—are carried down from steep mountain slopes as soon as they form. The intense erosion that occurs at high elevations produces a topography of steep slopes; deep, narrow stream valleys and narrow floodplains and drainage divides (see Figure 22.9b).

In lowlands, by contrast, weathering and erosion are slower, and the clay mineral products of chemical weathering accumulate as thick soils. Physical weathering occurs, but its effects are small compared with those of chemical weathering. Most streams run over broad floodplains and do little mechanical cutting of bedrock. Glaciers are absent, except in polar regions. Even in lowland deserts, strong winds merely abrade rock fragments and outcrops rather

than breaking them up. A lowland thus tends to have a gentle topography with rounded slopes, rolling hills, and flat plains (see Figure 22.9d).

Just as climate affects topography, topography can affect climate. For example, dry areas called rain shadows may form on the leeward slopes of mountain ranges (see Figure 17.3). Rain shadows result in preferential erosion on the windward side of a mountain belt (**Figure 22.17**). Geologists believe that in places such as New Guinea, where the difference between rainfall on the windward and leeward sides of the mountain belt is extreme, the rate of exhumation of metamorphic rocks buried deep in the crust is influenced by the history of rainfall at Earth's surface.

Feedback Between Uplift and Erosion

The ceaseless competition between plate tectonic processes, which tend to create mountains and build topography, and surface processes, which tend to tear them down, is the focus of intensive study by geomorphologists. Tectonic

FIGURE 22.17 ▪ Easterly view of the escarpment along the Arabian Sea at the Yemen-Oman border. This three-dimensional satellite image illustrates how topography determines local climate, which in turn controls erosion and landscape development. Although the Arabian Peninsula is arid, the steep escarpment of the Qara Mountains wrings moisture from the seasonal rains. That moisture allows natural vegetation to grow (green areas along the mountain fronts and in the canyons) and soil to develop (dark brown areas). In contrast, the light-colored areas are mostly dry desert. This climate focuses erosion on the ocean side of the mountain range. That intense erosion, in turn, has caused the escarpment to retreat landward, from right to left. [NASA.]

uplift provokes an increase in erosion (**Figure 22.18a**), so the higher mountains rise, the faster erosion wears them down. But as long as mountain building continues, their elevations stay high or increase. When mountain building slows, however—perhaps because of a change in the rate of plate movement—the mountains rise more slowly or stop rising entirely. As their growth slows or stops, erosion starts to dominate, and the mountains are worn down to lower elevations. This process explains why old mountains, such as the Appalachians, are relatively low compared with much younger mountains, such as the Rockies. As the mountains continue to be worn down, erosion also slows, and the whole process eventually tapers off. Elevation is thus a balance between the rate of tectonic uplift and the rate of erosion.

Curiously, over shorter time scales of thousands to millions of years, the plate tectonic and climate systems can interact such that mountains rise *higher* as a result of erosion (Figure 22.18b; see also Earth Issues 22.1). As we have seen, continents and mountains float on Earth's mantle because they are less dense than the mantle material. Beneath a mountain range, where the crust is thickest, a deep root projects into the mantle and provides buoyancy. Although the mantle just beneath the crust is solid rock, it flows very slowly when forces are applied to it over thousands to millions of years (see Earth Issues 14.1). The principle of isostasy implies that, at these time scales, the mantle has little strength, and thus behaves like a viscous fluid when it is forced to support the weight of continents and mountains. The principle of isostasy also implies that as a mountain range forms, it slowly sinks under the force of gravity, and the continental crust bends downward. When enough of a root bulges into the mantle to provide buoyancy, the mountain range floats. When the valleys in a mountain range are deepened by erosion, however, the mass of the mountain range decreases, and less root is needed for buoyancy. Thus, as the valleys erode, the mountains float upward. This process, called *isostatic rebound,* results in mountain summits being elevated to new heights (see Figure 22.18b). Over longer time scales, however, erosion will inevitably wear those mountain summits down (see Figure 22.18a).

▪ Models of Landscape Development

The strong contrasts of different landscapes stimulated early geologists to speculate about their causes. Three prominent and influential geologists who did so were William Morris Davis, Walther Penck, and John Hack. Davis believed that an initial burst of tectonic uplift is followed by a long period of erosion, during which landscape morphology depends mainly on geologic age. Davis's view was so dominant during the early 1900s that it

(a) Over long time scales, elevation is a balance between uplift and erosion.

1 Tectonic action elevates mountains. Uplift rate is greater than the rate of erosion.

2 Uplift slows, and erosion is in balance with uplift. Elevations stay high.

3 Uplift slows more, and erosion starts to dominate and elevation begins to decrease.

4 Uplift is almost stopped. Lower elevations—low hills—generate less weather, and erosion slows.

5 Uplift stops and erosion slows further. The elevation decreases further as the landscape evolves into lowlands and plains.

(b) Over short time scales, elevation increases as a result of erosion.

1 Continents collide, uplifting a high plateau.

2 Uplift creates a weather pattern that increases erosion rate.

3 Crust rebounds isostatically and causes mountain summits to be uplifted above pre-erosion elevation.

FIGURE 22.18 ■ Elevation is a balance between the rate of tectonic uplift and the rate of erosion. [After D. W. Burbank and R. S. Anderson, *Tectonic Geomorphology*. Oxford: Blackwell, 2001, p. 9.]

overshadowed that of his contemporary, Penck, who argued that tectonic uplift competes with erosion to control landscape morphology. In the 1960s, another conceptual breakthrough occurred when Hack recognized that uplift could not raise elevation above some critical limit, even if it were sustained for long periods. Rising mountains, in the absence of erosion, will collapse under their own weight.

Modern views of landscape development incorporate parts of these early ideas and acknowledge that there is a natural, time-dependent progression of landscape form. Geologists now understand that the developmental path of landscapes depends strongly on the *time scale* over which geomorphic change occurs. The importance of the different landscape-modifying processes depends on the interval of time over which the landscape is observed to change. For

Earth Issues

22.1 Uplift and Climate Change: A Chicken-and-Egg Dilemma

One of the clearest illustrations of how Earth's climate and plate tectonic systems are coupled is provided by the feedbacks between climate and the elevation of mountain belts. Currently, there is controversy over the directionality of these feedbacks. Some geologists argue that tectonic uplift of mountainous regions leads to climate change, others that climate change may promote tectonic uplift. This type of debate is well characterized by the classic "chicken-and-egg dilemma": which came first?

The debate is fueled by the observation that the cooling of the Northern Hemisphere's climate and the uplift of the Himalaya and Tibetan Plateau may have been synchronous. The Tibetan Plateau is the most imposing topographic feature on the surface of Earth (see Figure 22.7). It is so high and so extensive in area that it may influence patterns of atmospheric circulation in the Northern Hemisphere. If the Tibetan Plateau did not exist, the climate of the Northern Hemisphere would probably be quite different.

Unfortunately, although the timing of the Northern Hemisphere cooling has been well calibrated by the ages of glacial deposits and by the isotopic record of temperature changes in deep-sea sediments, the timing of the uplift of the Tibetan Plateau is less certain. This is where the debate begins. If the uplift preceded the onset of the cooling, then it might be argued that tectonic uplift indirectly caused climate change. On the other hand, if the uplift lagged behind the onset of the cooling, then it might be argued that climate change promoted uplift as an isostatic response to enhanced erosion rates.

Point: Negative Feedback

The possibility that mountain building may have promoted cooling and glaciation in the Northern Hemisphere has been recognized for more than 100 years. Geologists who currently advocate this point of view believe that several important processes occurred in the uplift of the Tibetan Plateau that led to a negative feedback process. According to this scenario, uplift led to a change in atmospheric circulation, which led to cooling in the Northern Hemisphere, and that cooling resulted in an increase in precipitation, glaciation, and stream runoff in the Himalaya and on the plateau. These changes, in turn, produced higher rates of weathering, which led to removal of CO_2—an important greenhouse gas—from the atmosphere. The reduced atmospheric CO_2 concentrations led to further cooling, increased precipitation, and increased weathering and erosion. Over time, erosion will wear down the mountains, and their elevations will decrease. Effectively, an elevation increase—which then modulates climate—results in an elevation decrease—a negative feedback process.

Counterpoint: Positive Feedback

Over the past decade, geologists have discovered that climate change might lead to uplift in mountainous regions. According to this unexpected and counterintuitive scenario, an initial cooling would stimulate an increase in precipitation rates, which in turn would lead to enhanced erosion by glaciers and streams. In the absence of isostatic responses, an increase in erosion would act as a negative feedback to lower the mountain ranges. However, when we consider the influence of isostasy, we see that erosion could result in a decrease in the mass of the mountain range as a whole, so that the mountains would be uplifted and their peaks elevated to new and higher positions (see Figure 22.18b). The rising mountains would act as a positive feedback to modify climate further, thereby increasing precipitation and erosion rates and further enhancing uplift.

---

example, variations in climate have been important factors in landscape development over the past 100,000 years, but they are only a minor factor over time scales of 100 million years. Over these longer intervals of geologic time, the history of tectonic uplift is probably most important.

Davis's Cycle: Uplift Is Followed by Erosion

William Morris Davis, a Harvard geologist in the early 1900s, studied mountains and plains throughout the world. Davis proposed a *cycle of erosion* that progresses from the high, rugged, tectonically uplifted mountains of a young landscape to the rounded hills of maturity and the worn-down plains of old age and tectonic stability

(Figure 22.19a). Davis believed that a strong, rapid pulse of tectonic uplift begins the cycle. All the topography is built during this first stage. Erosion eventually wears down the landscape to a relatively flat surface, leveling all structures and differences in bedrock. Davis observed the flat surfaces of extensive unconformities and saw them as evidence of such flat plains in past geologic times. Here and there, an isolated hill might stand as an uneroded remnant of former heights. Most geologists at that time accepted Davis's assumption that mountains are elevated suddenly over short geologic time scales and then remain tectonically stable as erosion slowly wears them down. Davis's cycle was accepted partly because geologists could find many examples of what seemed to be landscapes in his different stages of youth, maturity, and old age.

(a) Davis's theory

(b) Penck's theory

(c) Hack's theory

FIGURE 22.19 ■ Classic models of landscape development by tectonic uplift and erosion. [After D. W. Burbank and R. S. Anderson, *Tectonic Geomorphology.* Oxford: Blackwell, 2001, p. 5.]

Google Earth Project

Wind, water, ice, and gravity can erode materials and transport them across Earth's surface, influencing the form of landscapes. Many locations in the interior of the western United States have experienced these processes in recent geologic history, which have left clear marks in the form of unique and spectacular landscapes. Because these landscapes are prominently visible at Earth's surface, we can use GE to appreciate their specific characteristics. By traveling to the Teton Range, Death Valley, and the eastern Sierra Nevada, we can appreciate the dramatic impact of elevation and relief in a landscape. Moving along to the Gunnison River in Colorado, we can see for ourselves the amazing landscape that can emerge from the unique interaction of climate change, tectonic uplift, and rocks of various types.

Image © 2009 DigitalGlobe Image USDA Farm Service Agency

LOCATION   Teton Range, Wyoming; Death Valley, California; eastern Sierra Nevada, California; Gunnison River, Colorado; Susquehanna River, Pennsylvania

GOAL   Appreciate the interplay between uplift and erosion in creating unique landforms

LINKED   Figure 22.3, Figure 22.8, and Figure 22.13

1. The Teton Range in Wyoming is famous for its rugged mountains with knife-edged ridges. Its form is the result of uplift and erosion over geologic time scales. To the immediate west is the much lower Snake River Plain, which is flat enough to sustain agricultural production. At an eye altitude of 10 km, locate Grand Teton, the highest peak in the range, at 43°44′28.5″ N, 110°48′08.5″ W and compare its elevation with that of the town of Driggs, Wyoming, on the plain to the west. Note these elevations and measure the horizontal distance between the two locations. What is the slope of the western mountain front here?

Penck's Model: Erosion Competes with Uplift

Davis's view was challenged by his contemporary, Walther Penck, who argued that the magnitude of tectonic deformation and uplift gradually increases to a climax and then slowly wanes (Figure 22.19b). Unfortunately, Davis, with his greater professional stature and prolific publication style, was able to promote his ideas more effectively. Penck's ideas were not given the attention they deserved until the 1950s, more than two decades after Davis's death.

Penck proposed that geomorphic surface processes attack the rising mountains throughout the interval of uplift. Eventually, as the rate of deformation wanes, erosion predominates over uplift, resulting in a gradual decrease in both relief and elevation. This model was a conceptual breakthrough because it recognized that landscape development may result from competition between uplift and erosion. Davis's model, in contrast, emphasized the *a.* 1.025

b. 0.030

c. 0.290

d. 0.095

2. Every July, participants in the Badwater Ultra Marathon run from the lowest point in the United States (Badwater) to the base of the highest mountain in the contiguous United States (Mount Whitney). Hundreds of individuals compete, and the winner typically completes the 135-mile race in approximately 24 hours. Fortunately, the runners do not have to race to the top of Mount Whitney, although the race was originally conceived to do that. To begin to appreciate the unique challenges of this race, locate the peak of Mount Whitney, California, and determine its maximum elevation. Compare it with the elevation of the axis of Badwater Basin in Death Valley, California, to the east. What is the relief between these two points?

a. 4500 m

b. 4000 m

c. 3500 m

d. 3000 m

3. Now travel to Gunnison, Colorado, in the southern Rocky Mountains. Just to the west of town is a long body of water called Morrow Point Reservoir. Immediately to the south of this reservoir, you can see some features trending perpendicular to the shoreline. These features are generated by differential rates of erosion of various rock types. The rock types that make up this landscape are horizontally interbedded sandstones and siltstones. It may help to tilt your frame of view in order to appreciate the relief of these features. Based on your investigation of this area, how would you describe these landscape features?

a. Plateaus

b. Mesas

c. Ridges

d. Anticlines

4. Moving west approximately 30 km from Morrow Point Reservoir, you will encounter the Black Canyon of the Gunnison River. The Gunnison is a tributary of the Colorado River, and its geologic history is related to that of the entire Colorado Plateau. Because rivers often respond to tectonic uplift by incising a canyon, geologists can sometimes use the rate of incision as an approximation for the rate of the uplift that generated it. At an eye altitude of 4 km, navigate to 38°32′29″ N, 107°40′00″ W and measure the elevation at that location. The channels of small streams such as this one are known to contain gravels mixed with volcanic ash. Dating of these minerals yields an estimated deposition time of 640,000 years ago. Both the gravels and the ash have been incised by the Gunnison River since their deposition. By determining the difference between the elevation where this volcanic ash was deposited and the elevation of the modern river channel, you can calculate the average rate of uplift for this part of the Colorado Plateau. What is that rate?

a. 0.0625 cm/year

b. 155 cm/year

c. 0.00045 cm/year

d. 5 cm/year

Optional Challenge Question

5. Now travel to the eastern United States and locate Millersburg, Pennsylvania. At an eye altitude of 110 km, notice the unique trend of the ridge tops that zigzag back and forth across the landscape here. These patterns are characteristic of folds represented by anticlines and synclines that are plunging in and out of the land surface. Look in detail at the path of the Susquehanna River. How do you think the path of this river has been affected by the valley and ridge topography that was created by the folds?

a. The river is forced to curve around ridges.

b. The river ignores the ridges and cuts through them.

c. There is evidence for both of these processes.

d. Each ridge is so resistant that it forms a waterfall along the course of the river.

temporal distinction between these two processes. In his model, landscape age was the primary determinant of form.

Choosing between alternative theories of landscape development required that rates of uplift and erosion in regions of mountain building be determined. Technologies such as the Global Positioning System (GPS) and radar signals from other orbiting satellites have produced spectacular maps of crustal deformation and uplift rates. A number of dating methods (Table 22.1) have helped us determine the ages of geomorphically informative surfaces, such as

stream valley terraces (see Figure 18.27) that date back 1 million years.

A promising new dating scheme is based on the fact that cosmic rays penetrating the uppermost meter of rock or soil exposed at Earth's surface lead to the production of very small quantities of certain radioactive isotopes. One of them is beryllium-10, which accumulates to a greater degree the longer the rock or soil is exposed and to a lesser degree the deeper it is buried. Geologists used beryllium-10 to compare terrace ages along the course of

TABLE 22-1 Methods for Absolute Dating of Landscapes

Method	Useful Range (Years Ago)	Materials Needed
RADIOISOTOPIC		
Carbon-14	35,000	Wood, shell
Uranium/thorium	10,000–350,000	Carbonate (corals)
Thermoluminescence (TL)	30,000–300,000	Quartz silt
Optically stimulated luminescence	0–300,000	Quartz silt
COSMOGENIC		
In situ beryllium-10, aluminum-26	3–4 million	Quartz
Helium, neon	Unlimited	Olivine, quartz
Chlorine-36	0–4 million	Quartz
CHEMICAL		
Tephrochronology	0–several million	Volcanic ash
PALEOMAGNETIC		
Identification of reversals	>700,000	Fine sediments, volcanic lava flows
Secular variations	0–700,000	Fine sediments
BIOLOGICAL		
Dendrochronology (tree rings)	10,000	Wood

Source: D. W. Burbank and R. S. Anderson, *Tectonic Geomorphology.* Oxford: Blackwell, 2001, p. 39.

the Indus River in the Himalaya. They then plotted height changes against time to find average erosion and uplift rates. Stream erosion rates in the Himalaya were found to vary between 2 and 12 mm/year (see Practicing Geology on pages 637–638). In other high mountain ranges, tectonic uplift rates have been measured in the same general range, from 0.8 to 12 mm/year.

Hack's Model: Erosion and Uplift Achieve Equilibrium

John Hack elaborated on the idea that erosion competes with uplift. He believed that when uplift and erosion rates are sustained over long periods, landscape development will achieve a balance, or dynamic equilibrium (Figure 22.19c). During the period of equilibrium, landforms may undergo minor adjustments, but the overall landscape will look more or less the same.

Hack recognized that the height of mountains could not increase forever, even if uplift rates were extremely high. Rocks break if large enough stresses are imposed on them, so it stands to reason that if mountains become

steeper than their angle of repose, they will collapse under their own weight. Thus, with continued uplift beyond some critical limit, slope failures and mass wasting alone will prevent further increases in elevation. Consequently, rates of uplift and rates of erosion come into a long-term balance. Unlike the models of Davis and Penck, Hack's model does not require uplift rates to decrease.

A fascinating implication of Hack's model is that geomorphology does not have to change at all as long as uplift and erosion rates are balanced. Nevertheless, Earth's history teaches us that whatever goes up eventually does come down. Over very long time scales, the models of Davis and Penck are more accurate descriptions of how landscapes ultimately change in form. When erosion exceeds uplift, slopes become lower and more rounded (see Figure 22.18a). Because few areas of the world remain tectonically quiescent for as long as 100 million years, however, the perfectly flat plains that Davis proposed would have formed only rarely in Earth's history. Hack's model of dynamic equilibrium is perhaps most appropriate for landscapes in tectonically active areas where a given uplift rate can be sustained for more than a million years or so.

SUMMARY

What are the principal components of a landscape? A landscape is described in terms of topography, which includes elevation, the vertical distance above or below sea level, and relief, the difference between the highest and the lowest elevations in a region. A landscape comprises the varied landforms produced through erosion and sedimentation by streams, glaciers, mass wasting, and the wind. The most common landforms are mountains and hills, plateaus, and structurally controlled cliffs and ridges—all of which are produced by tectonic activity modified by erosion.

How do the climate and plate tectonic systems interact to control landscapes? Landscapes are shaped by plate tectonic processes, weathering, erosion, and resistance to erosion. Plate tectonic processes lift up mountains and expose rock. Erosion carves rock into valleys and slopes. Climate, in turn, affects rates of weathering and erosion. Variations in climate and bedrock type strongly modify landscape development, making desert and glacial landscapes very different.

How do landscapes develop? The development of landscapes depends strongly on competition between the forces of uplift and the forces of erosion. Landscapes begin their development with tectonic uplift, which in turn stimulates erosion. When tectonic uplift rates are high, erosion rates also tend to be high, and mountains are high and steep. As uplift rates decrease, erosion rates remain high; the land surface is lowered and slopes are rounded. When uplift ends, erosion becomes the dominant process and wears down the former mountains to gentle hills and broad plains.

KEY TERMS AND CONCEPTS

badland (p. 625)

contour (p. 618)

cuesta (p. 626)

elevation (p. 618)

geomorphology (p. 618)

hogback (p. 626)

landform (p. 621)

mesa (p. 623)

plateau (p. 623)

relief (p. 619)

stream power (p. 623)

topography (p. 618)

PRACTICING GEOLOGY EXERCISE

How Fast Do Streams Erode Bedrock?

In mountainous regions, the flowing water in streams cuts downward into bedrock that is uplifted by tectonic processes. It is hard to imagine that something as hard as rock could be cut by flowing water! The purpose of this exercise is to learn how to measure the rate of this erosion, which can be surprisingly fast.

The rate of erosion depends on how quickly the water flows and on how much sediment is suspended in the

Upper strath terrace
Elevation: 250 m
Age: 50,000 years

Middle strath terrace
Elevation: 80 m
Age: 10,000 years

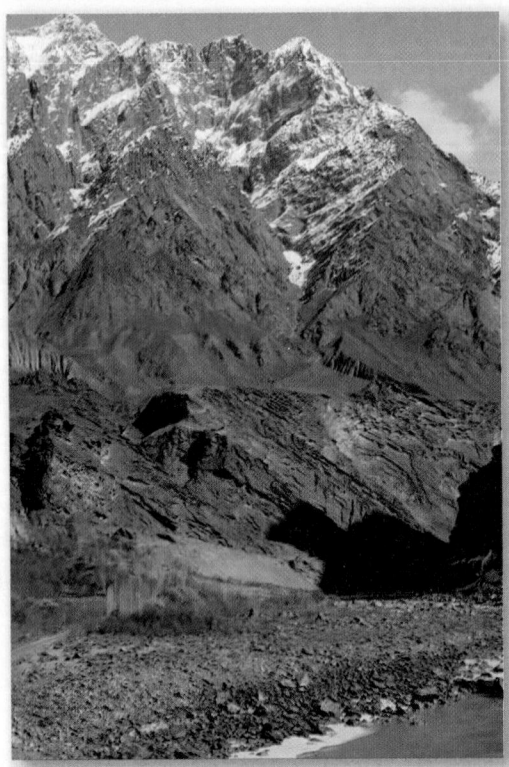

Strath terraces in the Middle Indus basin. [D. W. Burbank.]

flowing water. The erosion of stream channels cut into bedrock is the result of the plucking of large pieces of rock and of abrasion by sediment particles. Plucking of rock occurs most readily when the bedrock is penetrated by tightly spaced sets of fractures or consists of sedimentary rocks with well-defined, closely spaced bedding planes. The zones of weakness inherent in such rocks allow blocks to be broken off the streambed and transported downstream; once loose, the moving blocks can strike other projecting rocks and break them loose as well. Abrasion occurs when sediment particles impact the bedrock and break tiny pieces—smaller than the size of the impacting grain—from the bedrock. While this effect is trivially small on a grain-by-grain basis, it becomes very important when the effects of millions of impacting grains are added up. This is important because the rate of bedrock erosion is what controls the aesthetic beauty of the most dramatic landscapes on Earth, such as the Sierra Nevada mountains, the Grand Tetons, the Alps, and the Himalaya.

Flow velocity is also extremely important. Observations in both the field and the laboratory show that the erosion rate depends on the velocity of the flow raised to the fifth power (erosion rate ~V^5). This means that small increases in current velocity cause dramatic increases in erosion rates. Simply doubling the flow velocity might increase the erosion rate 30 times over what it was initially; double it again and the erosion rate might increase 1000 times over what it was initially. Therefore, it is during rare but powerful flows, such as floods, that the current is capable of eroding bedrock; most of the rest of the time, little erosion occurs.

To understand bedrock erosion, geologists study high-mountain streams such as the Indus River in northern Pakistan. The Indus is one of the world's largest rivers, and it cuts through the heart of Earth's tallest mountain ranges: the Himalaya, Karakoram, and Hindu Kush. Its discharge is 6600 m^3/s—enough to fill an average two-story house *six times every second!* Where the river passes through Nanga Parbat—forming a very narrow and deep gorge—this extremely high discharge results in some of the highest erosion rates on Earth. There, the river is narrow and its slope is steep; thus, current velocities are high—and bedrock erosion rates are high.

Geologists measure bedrock erosion rates using several methods. The most straightforward is to create markers, such as drill holes, on the bottom of a river channel and then monitor the changes in those markers over time. Holes are drilled to a known depth; then, at the end of a specified period—say, after a year—the decrease in the hole depth is measured. This method yields an erosion rate representing an average of all rates during that year, including shorter periods when erosion rates were higher or lower.

Erosion rates are harder to measure over longer periods. Geologists often search for *strath terraces* preserved along the slopes that lead down into river channels. Strath terraces are topographic benches carved in bedrock, and sometimes overlain by gravel deposits, that mark the former positions of the river channel bottom. They are preserved as the river cuts down to new levels. If the age of a strath terrace and its height above the current streambed can both be measured, then the long-term erosion rate can be calculated. Ages of strath terraces are commonly measured using the beryllium-10 method mentioned later in this chapter.

In the accompanying diagram, each of two strath terraces is labeled with an age and a corresponding elevation. For the lower terrace, with an age of 10,000 years and an elevation of 80 m above the streambed, the long-term erosion rate is given by

Erosion rate = elevation terrace streambed/terrace age

= 80 m/10,000 years

= 0.008 m/year (8 mm/year)

BONUS PROBLEM: What is the long-term erosion rate represented by the upper strath terrace?

EXERCISES

1. Give three examples of landforms.

2. What is relief, and how is it related to elevation?

3. Why does relief vary with the scale of the area over which it is measured?

4. How do faulting and uplift control topography?

5. Compare the erosional processes in topographically high and low areas.

6. How do slope and discharge affect stream power?

7. How does climate affect topography, and vice versa?

8. How does the balance between tectonic uplift and erosion affect mountain heights?

9. In what regions of North America do active plate movements currently affect landscapes?

THOUGHT QUESTIONS

1. The summits of two mountain ranges lie at different elevations: range A at about 8 km and range B at about 2 km. Without knowing anything else about these ranges, could you make an intelligent guess about the relative ages of the mountain-building processes that formed them?

2. If you were to climb 1 km from a stream valley to a mountaintop 2 km high in a tectonically active area versus a tectonically inactive area, which would probably be the more rugged climb?

3. A young mountain range of uniform age, rock type, and structure extends from a far northern frigid climate through a temperate zone to a southern tropical rainy climate. How would the topography of the mountain range differ in each of the three climates?

4. Describe the main landforms in a low-lying humid region where the bedrock is limestone.

5. In what landscapes would you expect to find lakes?

6. What changes could you predict for the landscape of the Himalaya and the Tibetan Plateau in the next 10 million years? The next 100 million years?

7. What changes in the landscape of the Rocky Mountains of Colorado might result from a change in the present temperate but somewhat dry climate to a warmer climate with a large increase in rainfall?

8. Over short time scales (thousands of years), isostatic uplift may temporarily raise mountain summits to higher elevations. However, over the longer term (millions of years), continued erosion will reduce those summits to progressively lower elevations. As this happens, what does the principle of isostasy predict about the depth of the base of continental lithosphere beneath the mountains? Should this depth increase or decrease as mountains are worn down?

- Civilization as a Global Geosystem   642
- Fossil-Fuel Resources   646
- Alternative Energy Resources   655
- Global Change   660
- Earth System Engineering and Management   667

North and South America at night, showing the lights of our globalized, energy-intensive civilization. [Image and data processing by NOAA's National Geophysical Data Center, Earth Observation Group (http://www.ngdc.noaa.gov/dmsp). DMSP data collected by US Air Force Weather Agency.]

THE HUMAN IMPACT ON EARTH'S ENVIRONMENT

23

IN THE FOREGOING CHAPTERS, we have seen how a better understanding of the Earth system can improve the human condition by helping us find natural resources, sustain the natural environment, and reduce the risks from natural hazards. But the progress of human civilization cannot be taken for granted. The human population is growing at a phenomenal rate, and Earth's natural resources are necessarily limited. Environmental conditions and overall prosperity are not improving in some parts of the world, and the prospects for detrimental changes to the global environment loom large. Balancing the benefits we reap from our use of natural resources against the costs of that use—such as harmful changes to the geosystems that sustain us—raises new challenges for Earth science and society.

In this final chapter, we survey the energy resources that power our economy and examine how our use of those resources affects our environment. We focus on two of civilization's most pressing problems: the need for more energy resources to power economic development and the potential for climate change that arises from our economic activities.

Our economy depends on the burning of a nonrenewable energy resource (fossil fuels) that produces a potentially dangerous greenhouse gas (carbon dioxide). This stark reality poses some difficult questions: How long will our fossil-fuel resources last? To what extent will the increase in atmospheric carbon dioxide concentrations caused by fossil-fuel burning adversely affect the global climate? How quickly will we need to replace fossil fuels with alternative energy sources? These questions have political and economic dimensions that extend far beyond Earth science, so they do not have strictly scientific answers. Nevertheless, the decisions we will make as a society must be informed by our best and most realistic scientific predictions about how the Earth system will change over the next decades and centuries. Reasonable predictions can be made only if we include human civilization as part of the Earth system.

■ Civilization as a Global Geosystem

The human habitat is a thin interface where Earth meets sky, where the global geosystems—the climate system, the plate tectonic system, and the geodynamo—interact to provide a life-sustaining environment. We have increased our standard of living by discovering clever ways to exploit this environment: to grow food, extract minerals, build structures, transport materials, and manufacture goods of all kinds. One result has been an explosion in the human population.

Early in the Holocene, about 10,000 years ago, when the climate was warming and agriculture first began to flourish, roughly 100 million people were living on the planet. Population grew slowly. The first doubling, to 200 million, was achieved early in the Bronze Age, about 5000 years ago, when humans first learned how to mine ores and refine them into metals such as copper and tin (of which bronze is an alloy). The second doubling, to 400 million, was not achieved until the Middle Ages, about 700 years ago. But once industrialization began, in the early nineteenth century, the global population really took off, climbing to 1 billion in about 1800, 2 billion in 1927, and 4 billion in 1974. By the mid-twentieth century, the doubling time for the human population had dropped to only 47 years—less than a human lifetime. The world population exceeded 7 billion in early 2012, and, although its growth rate is expected to decline, the total will almost surely top 8 billion by 2030 (**Figure 23.1**).

As our population has exploded, our appetites for energy and other natural resources have become voracious. The demand for natural resources is skyrocketing as civilization expands and people around the world strive to improve the quality of their lives. Our energy usage, for example, has risen 1000 percent over the last 70 years and is now increasing twice as fast as the human population. The view of Earth from space in this chapter's opening photograph shows a glowing lattice of highly energized urbanization spreading rapidly across the planet's surface.

Human civilization has altered the environment by deforestation, agriculture, and other land-use changes since it began. But our effects in earlier times were usually restricted to local or regional habitats. Today, energy production on an industrial scale makes it possible for humans to compete with the climate system and the plate tectonic system in modifying Earth's surface environment, as illustrated by some startling observations:

■ Dams and reservoirs built by humans now trap about 30 percent of the sediments transported down the world's rivers.

■ In most developed countries, construction workers move more tons of soil and rock each year than do all natural erosional processes combined.

FIGURE 23.1 ■ The black line charts the global growth of human population since 1800. The colored lines show future population growth in three scenarios used by the IPCC to estimate human impacts on Earth's climate. In scenario A (red line), world population continues to grow into the twenty-second century. In scenario B (green line), it levels off in the late twenty-first century, and in scenario C, it declines after 2070. [IPCC, *Climate Change 2013: The Physical Science Basis.*]

■ Within 50 years after the invention of the artificial coolant freon, enough of it had leaked out of refrigerators and air conditioners and floated into the upper atmosphere to damage Earth's protective ozone layer.

■ Humans have converted about one-third of the world's forested area to other land uses, primarily agriculture, in the last half century.

■ Since the industrial revolution began in the early nineteenth century, deforestation and the burning of fossil fuels have increased the concentration of carbon dioxide in the atmosphere by almost 50 percent.

We are not just part of the Earth system; we are transforming how the Earth system works, perhaps in fundamental ways. In a geologic instant, human civilization has developed into a full-fledged global geosystem.

Natural Resources

The term **natural resources** refers to the energy, water, and raw materials used by human civilization that are available from the natural environment. **Renewable resources** are those natural resources that are continually produced in the environment; for example, if we chop down a forest for wood, it can be regrown and harvested again. **Nonrenewable resources** are those natural resources that we are using up much faster than they are being produced by geologic processes. Organic material must be buried and heated for millions of years, for example, to produce petroleum.

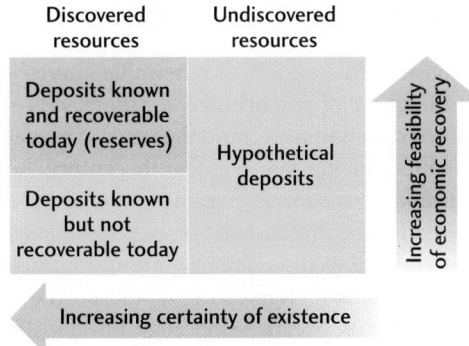

FIGURE 23.2 ■ Resources include reserves, plus known but currently unrecoverable deposits, plus undiscovered deposits that geologists think may eventually be found.

Reserves are considered a dependable measure of supply as long as economic and technological conditions remain constant. As conditions change, some resources become reserves, and vice versa. In many cases, resources that are too poor in quality or quantity to be worth exploiting, or that are too difficult to retrieve, become reserves when new technology is developed or prices rise.

Geologists are experts at discovering new resources. It should be kept in mind, however, that the assessment of resources is much less certain than the assessment of reserves. Any figure cited as representing the resources of a particular material is only an educated guess as to how much will be available in the future. We can better understand how to manage our natural resources by considering the geologic circumstances in which they are found and the problems related to their recovery and use.

Energy Resources

Energy is required to do work, so it is fundamental to all aspects of human civilization. A crisis in the supply of energy can bring a modern society to a halt. Wars have been fought over access to supplies of fuel; economic recession and destructive currency inflation have resulted from fluctuations in the price of oil and other fuels.

The supply of any material we take from Earth's crust is finite. Its availability depends on its distribution in accessible deposits, as well as on how much we are willing to pay to get it out of the ground. Geologists use two measures to describe the supply of these nonrenewable resources. **Reserves** are deposits of a given material that have already been discovered and can be exploited economically and legally at the present time. In contrast, **resources** comprise the entire amount of the material, including the amount that may become available for use in the future. Resources include reserves plus known but currently unrecoverable supplies plus undiscovered supplies that geologists think may eventually be found (**Figure 23.2**).

Our energy use has been increasing over the last two centuries, but our energy sources have also been changing since the industrial revolution began (**Figure 23.3**). A century and a half ago, most of the energy used in the United States came from the burning of wood. A wood fire, in chemical terms, is the combustion of *biomass*—organic matter consisting of carbon and hydrogen compounds.

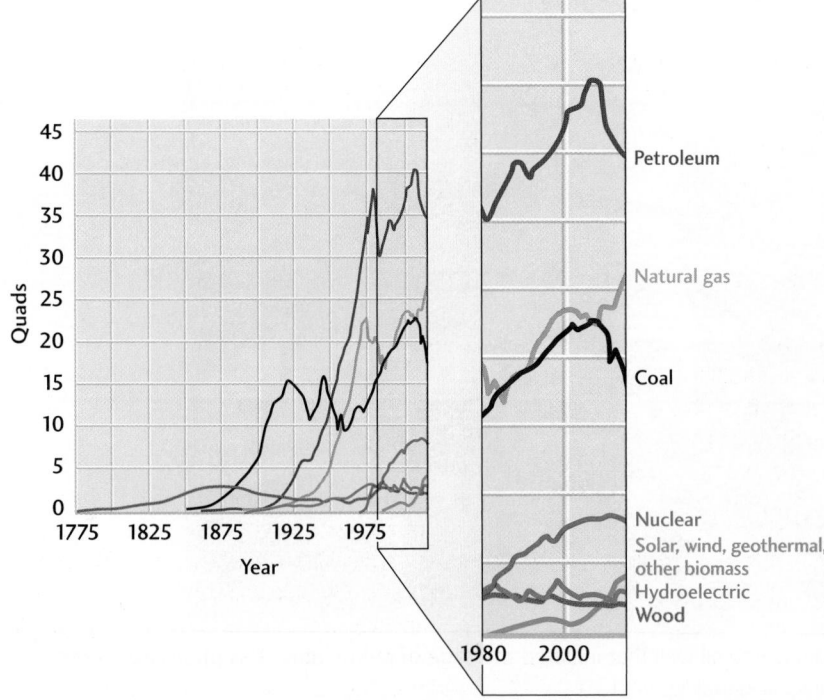

FIGURE 23.3 ■ U.S. energy consumption, shown here for the period 1775–2011, has successively been dominated by the burning of wood, coal, and oil. In just the past few years (magnified in the inset), the energy from petroleum and coal has dropped while that from natural gas and renewable sources such as solar and wind has increased. Here and elsewhere in this chapter, the energy units are quads (1 quad = 10^{15} Btu). [U.S. Energy Information Agency.]

Biomass is produced by plants and animals in a food web that is based on photosynthesis. Thus, the ultimate source of the energy in wood is the sunlight plants use to convert carbon dioxide and water into carbohydrates. Combustion of wood or other biomass produces heat energy and returns carbon dioxide and water to the environment. In this capacity, the biomass acts as a short-term reservoir for storing solar energy. It is a renewable energy resource because the biosphere is constantly producing new biomass. Before the mid-nineteenth century, the burning of wood and other biomass derived from plants and animals (e.g., whale oil, dried buffalo dung) satisfied most of society's need for fuel. Even today, the energy derived from biomass equals the total derived from all other renewable resources.

Some of the biomass that was buried in sedimentary rock formations millions of years ago, particularly during the Carboniferous period, has been transformed into a combustible rock called *coal.* When we burn coal, we are using stored energy from Paleozoic sunlight. Thus, the primary source of this "fossilized" energy is the same solar power that drives the climate system. Our other major fuels, crude oil (petroleum) and natural gas, are also created by diagenesis and metamorphism of dead organic matter. Coal, oil, and natural gas are known collectively as **fossil fuels.** At current rates of use, our reserves of these non-renewable energy resources will be exhausted long before geologic processes can replenish them.

Rise of the Carbon Economy

Humans have used a variety of renewable energy sources to power mills and other machinery for thousands of years, including wind, falling water, and the work of horses, oxen, and elephants. By the late eighteenth century, however, industrialization was increasing the demand for energy beyond what these traditional renewable sources could supply. At about that time, James Watt and others developed coal-fired steam engines that could do the work of hundreds of horses. Steam technology lowered the price of energy dramatically, in part because it made coal mining possible on a large scale. The availability of cheap energy sparked the industrial revolution. By the end of the nineteenth century, coal accounted for more than 60 percent of the U.S. energy supply (see Figure 23.3).

The first oil well was drilled in Pennsylvania by Colonel Edwin L. Drake in 1859. The idea that petroleum could be profitably mined like coal provoked skeptics to call the project "Drake's Folly" (**Figure 23.4**). They were wrong, of course: by the early twentieth century, oil and natural gas were beginning to displace coal as the fuels of choice. Not only did they burn more cleanly than coal, producing no ash, but they could be transported by pipeline as well as by rail and ship. Moreover, gasoline and diesel fuels refined from crude oil were suitable for burning in the newly invented internal combustion engine.

FIGURE 23.4 ■ Edwin L. Drake (*right*) in front of the oil well that initiated the "age of petroleum." This photo was taken by John Mather in 1866 in Titusville, Pennsylvania. [Bettmann/CORBIS.]

Today, the engine of civilization runs primarily on fossil fuels. Taken together, oil, natural gas, and coal account for 85 percent of global energy consumption. We can fairly call the civilization fed by this energy system a **carbon economy.**

Global Energy Consumption

Energy use is often measured in units appropriate to the fuel—for example, barrels of oil, cubic feet of natural gas, and tons of coal. But comparisons are easier if we use a standard unit of energy such as the British thermal unit (Btu). One Btu is the amount of energy needed to raise the temperature of 1 pound of water by 1°F (1054 joules). When we measure large quantities, such as a nation's annual energy use, we use units of 100 quadrillion (10^{15}) Btu, or **quads.**

In 2012 the United States used about 95 quads of energy (**Figure 23.5**), compared with a global total of 530 quads. Thus, the United States, with 4.5 percent of the world's population, consumes about four times more energy

per person than the global average. Fossil fuels provided 82 percent of that total, and renewable biomass accounted for another 4.5 percent. You will notice that the flow of energy through this system is not particularly efficient: about 39 percent of the energy performed useful work, while 61 percent was wasted. You can also see that the system released about 1.4 gigatons (Gt) of carbon into the atmosphere in 2012, primarily as CO_2 (1 Gt = 1 billion tons = 10^{12} kg).

There are promising signs that new energy-efficient technologies and increasing energy conservation are beginning to lessen U.S. energy appetites. In fact, the total annual U.S. consumption of energy of all types dropped by 6 percent between 2007 and 2012, the first multiyear drop in recent history. On a global basis, however, modest reductions in energy consumption by the United States, Japan, and Western Europe have been more than offset by increases in the developing world, led by the world's two most populous countries, China (+8 percent per year) and India (+7 percent per year). In 2007, China's total energy consumption exceeded that of the United States for the

FIGURE 23.5 ▪ Energy consumption in the United States in 2012 (in quads). Energy from primary fuel sources (boxes on left side) is delivered to the residential, commercial, industrial, and transportation sectors (boxes in middle to right side). Not represented are small contributions to electric power generation from geothermal energy (0.2 quad). [After Lawrence Livermore National Laboratory, based on data from the Energy Information Administration.]

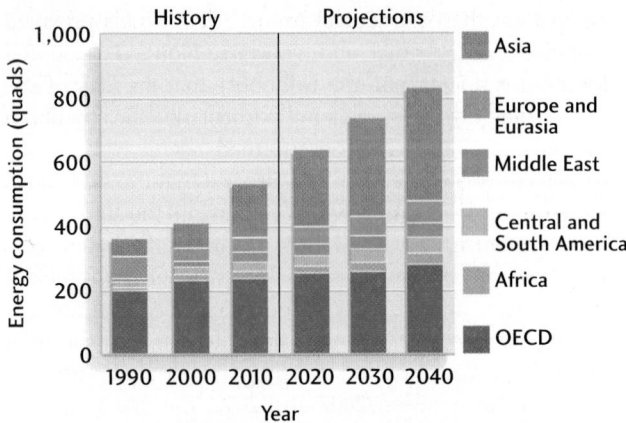

FIGURE 23.6 ■ Historical and projected energy consumption in quads, 1990–2040, by world regional groupings. The Organization for European Economic Cooperation (OECD) includes countries in Western Europe, North America, and Australia. [U.S. Energy Information Agency.]

first time. Still, China's average individual (or *per capita*) energy use remains almost eight times less, owing to its much larger population. As China and other developing economies strive to improve their standards of living, global energy use per capita is bound to rise, accelerating overall energy consumption. Global annual energy consumption is projected to exceed 600 quads by 2020 (**Figure 23.6**).

Energy Resources for the Future

Figure 23.7 gives a rough estimate of the world's remaining fossil-fuel reserves. Simply dividing total reserves (about 53,000 quads) by the current global annual consumption estimate from Figure 23.6 (about 530 quads) might lead one to conclude (mistakenly) that many decades' worth of resources remain before we have to worry about depletion of

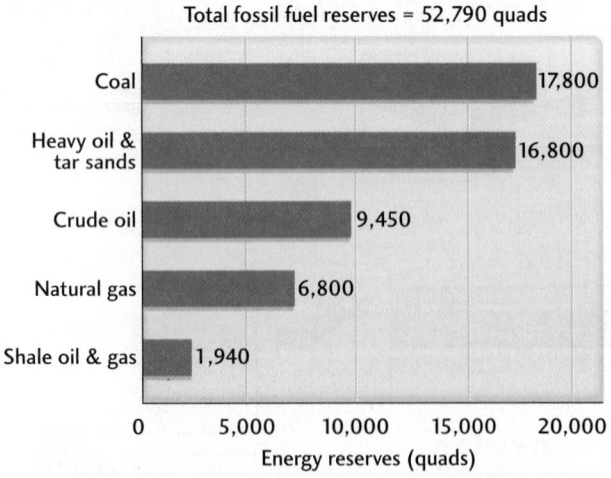

FIGURE 23.7 ■ A rough estimate of total remaining fossil-fuel energy reserves amounts to about 53,000 quads. [World Energy Council.]

our energy supplies. The economics of this issue are much more complicated, however, as we will see in the rest of this chapter. Some energy sources will give out before others, the various sources of energy are not readily interchangeable, and the environmental costs of converting some of them into useful forms of energy may be too great.

Of course, we may begin to meet our energy needs through increased efficiency in the use of fossil fuels and through the development and use of alternatives such as nuclear power and renewable energy sources. Current projections indicate that energy production from renewable sources, such as solar, wind, water, and geothermal power and biofuels, will fall short of our needs for many decades to come, unless there are unanticipated technological breakthroughs. Even so, developing alternative energy sources would reduce the pressure on our fossil-fuel resources as well as their negative environmental effects.

Carbon Flux from Energy Production

One of the most serious environmental costs of using fossil fuels may be climate change caused by the influence of our carbon economy on the global carbon cycle. In the pre-human world, the exchange of carbon between the lithosphere and the other components of the Earth system was regulated by the slow rates at which geologic processes buried and unearthed organic matter. This natural carbon cycle has been disrupted by the rise of the carbon economy, which is now pumping huge amounts of carbon from the lithosphere directly into the atmosphere. As we saw in Chapter 15, the climate system is tightly coupled to the global carbon cycle because carbon dioxide is a greenhouse gas. The concentration of this gas has risen rapidly from its preindustrial level of about 270 ppm, reaching 400 ppm in 2013. If the burning of fossil fuels continues unabated, the amount of CO_2 in the atmosphere will double its preindustrial level by mid-century.

Anthropogenic increases in carbon dioxide and other greenhouse gases have already led to enhancement of the greenhouse effect and global climate warming. It is clear that the future of the climate system and its living component, the biosphere, depends on how our society manages its energy resources, which we will now consider in more detail.

■ Fossil-Fuel Resources

Economically valuable deposits of our most important energy sources, fossil fuels, develop under special environmental and geologic conditions. Fossil fuels come from the organic debris of former life: plants, algae, bacteria, and other microorganisms that have been buried, transformed, and preserved in sediments.

How Do Oil and Gas Form?

Oil and gas begin to form in sedimentary basins where the production of organic matter is high and the supply of oxygen in the sediments is inadequate to decompose all the organic matter they contain. Many offshore thermal subsidence basins on continental margins satisfy both these conditions. In such environments, and to a lesser degree in some river deltas and inland seas, the rate of sedimentation is high, and organic matter is buried and protected from decomposition.

During millions of years of burial, chemical reactions triggered by the elevated temperatures and pressures found deep in the sediments slowly transform some of the organic material in these *source beds* into combustible hydrocarbons. The simplest hydrocarbon is methane gas (CH_4), the compound we call *natural gas*. Raw petroleum, or *crude oil*, includes a diverse class of liquids composed of more complex hydrocarbons.

Crude oil forms at a limited range of pressures and temperatures, known as the **oil window,** usually found at depths between about 2 and 5 km (see the Practicing Geology Exercise in Chapter 5). Above the oil window, temperatures are too low (generally below 50°C) for the maturation of organic material into hydrocarbons, whereas below the oil window, temperatures are so high (greater than 150°C) that the hydrocarbons that form are broken down into methane, producing only natural gas.

As burial progresses, compaction of the source beds forces crude oil and natural gas into adjacent beds of permeable rock (such as sandstones or porous limestones), which act as *hydrocarbon reservoirs*. The relatively low densities of oil and gas cause them to rise, so that they float atop the water that almost always occupies the pores of permeable rock formations.

Where Do We Find Oil and Gas?

The conditions that favor large-scale accumulation of oil and natural gas are combinations of geologic structures and rock types that create an impermeable barrier to upward migration, forming an **oil trap** (**Figure 23.8**). Some oil traps, called *structural traps,* are created by deformation structures. One type of structural trap is formed by an anticline in which an impermeable layer of shale overlies a permeable sandstone formation (Figure 23.8a). The oil and gas accumulate at the crest of the anticline—the gas highest, the oil next—both floating on the groundwater that saturates the sandstone (see Chapter 7, Practicing Geology).

(a) Anticlinal trap
Oil pumps
Gas
Oil
Impermeable shale
Water-saturated permeable reservoir rock

(b) Fault trap
Gas
Oil
Impermeable shale
Water-saturated permeable reservoir rock

(c) Stratigraphic trap
Gas well
Gas
Oil
Impermeable shale
Water-saturated permeable reservoir rock

(d) Salt dome trap
Gas
Impermeable shale
Gas
Water-saturated permeable reservoir rock
Impermeable salt dome
Oil

FIGURE 23.8 ■ Oil and gas accumulate in traps formed by geologic structures. Four types of oil traps are illustrated here.

Similarly, an angular unconformity or displacement at a fault may place a dipping permeable limestone formation opposite an impermeable shale, creating another type of structural trap (Figure 23.8b). Other types of oil traps are created by the original pattern of sedimentation, as when a dipping permeable sandstone formation thins out against an impermeable shale (Figure 23.8c). These structures are called *stratigraphic traps.* Oil can also be trapped against an impermeable mass of salt in a *salt dome trap* (Figure 23.8d).

The hydrocarbon reservoirs that hold oil and natural gas are complex geologic systems. Geologists can map the reservoir rocks in three dimensions using various techniques, such as seismic imaging (see Figure 14.6). The three-dimensional models they obtain show them where the bulk of the oil and gas is located and allow them to predict how it will flow from holes drilled into the reservoir.

In their search for petroleum resources, geologists have seismically mapped thousands of oil traps throughout the world. Only a fraction of them have proven to contain economically valuable amounts of oil or gas, because traps alone are not enough to create a hydrocarbon reservoir. A trap will contain oil only if source beds were present, the necessary chemical reactions took place, and the oil migrated into the trap and stayed there without being disturbed by subsequent heating or deformation. Although oil and gas are not rare, most of the large, easy-to-find deposits have already been located, and the discovery of new resources is becoming more difficult.

Efforts are now under way to find more efficient ways to extract oil and natural gas from deep rock formations. Drilling holes deep into Earth's crust has become a very sophisticated and expensive business (**Figure 23.9**). Petroleum engineers use three-dimensional models to steer drill bits on swooping paths into the richest parts of a reservoir. They inject water into stubborn formations to coax the oil out, a processes known as *hydraulic fracturing* ("fracking"),

and they pump carbon dioxide down strategically positioned drill holes to push the oil into areas where it can be more efficiently recovered through other drill holes. These methods have increased the fraction of oil and gas that can be extracted, increasing reserves.

Distribution of Oil Reserves

In the decade 2002–2012, the world consumed about 0.3 trillion barrels of oil (1 barrel = 42 gallons); yet, the worldwide reserves of oil did not decline; in fact, they increased from about 1.3 trillion barrels to almost 1.7 trillion barrels. Oil exploration is an immensely successful geologic activity!

Oil reserves and their decadal changes are displayed by region in **Figure 23.10**. The oil fields of the Middle East—including Iran, Kuwait, Saudi Arabia, Iraq, and the Baku region of Azerbaijan—account for 48 percent of the world's total. Here, sediments rich in organic material have been folded and faulted by the closure of the ancient Tethys Ocean, forming a nearly ideal environment for oil accumulation. The extensive reservoirs discovered in this vast convergence zone include the world's largest, the Ghawar field in Saudi Arabia. Ghawar has produced more than 70 billion barrels of oil since its opening in 1948 and may produce another 70 billion barrels over its remaining lifetime.

Most of the oil reserves in the Western Hemisphere are located in the highly productive Gulf Coast–Caribbean area, which includes the Louisiana-Texas region, Mexico, Colombia, and Venezuela. The threefold increase in South American oil reserves from 2002 to 2012 (see Figure 23.10) came mainly from improvements in oil-recovery technology, which will allow the heavy oil of Venezuela's Orinoco Basin to be exploited economically, as well as the discovery of huge new oil fields in the Atlantic Ocean off Brazil.

The United States' oil reserves also increased, from 31 billion barrels in 2002 to 35 billion barrels in 2012, placing

FIGURE 23.9 ■ New technologies used aboard offshore platforms in the Gulf of Mexico can recover oil and gas from rock reservoirs below very deep waters. Drilling from a single platform like this one can cost over $100 million. [Larry Lee Photography/Corbis.]

Proved reserves
(in billions of barrels)

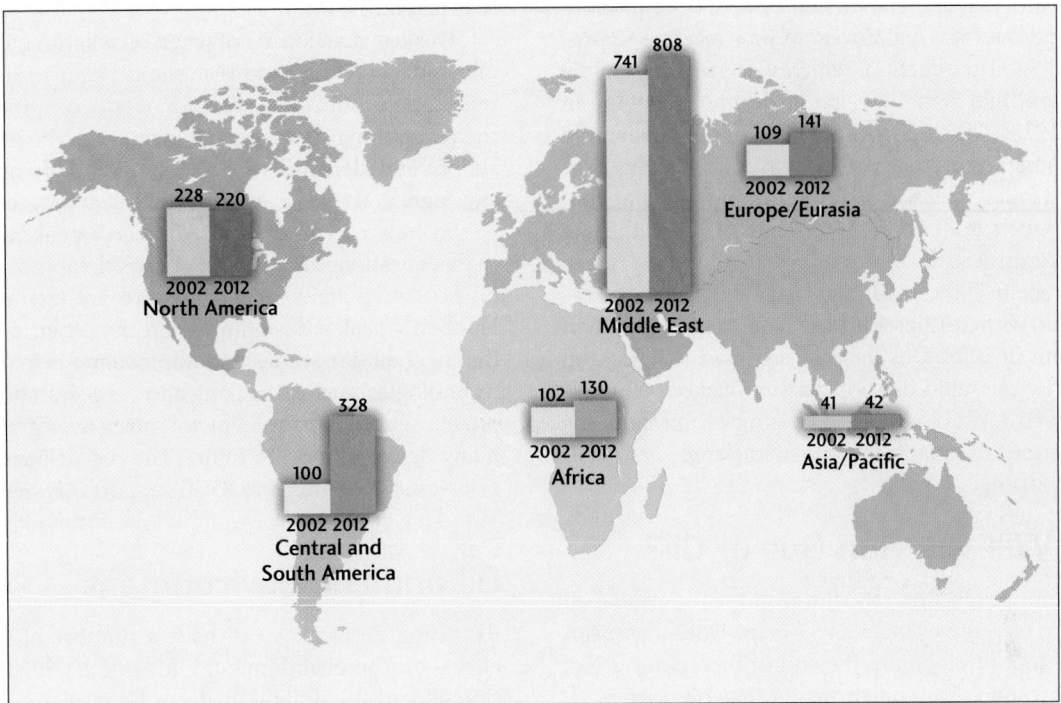

FIGURE 23.10 ▪ ▪ Regional estimates of world oil reserves in 2002 (left bar) and 2012 (right bar) in billions of barrels (bbl). [*British Petroleum Statistical Review of World Energy 2013*, June 2013.]

it tenth worldwide. Thirty-one U.S. states have commercial oil reserves, and small, noncommercial resources can be found in most of the others.

Oil Production and Consumption

Global oil production in 2012 was about 31 billion barrels per year worldwide. The United States produced 3.2 billion barrels, more than any other nation except Saudi Arabia and Russia, but it consumed about 6.8 billion barrels. This gap between U.S. production and consumption, about 3.5 billion barrels, must be filled by importing oil. This imbalance—$327 billion in 2011—has contributed more than any other factor to the massive U.S. foreign trade deficit.

The United States is a "mature" oil producer, in the sense that most of the petroleum reserves within its borders have already been exploited. Production reached a maximum in 1970 and declined, approximately following a bell-shaped curve (**Figure 23.11**). The high point of

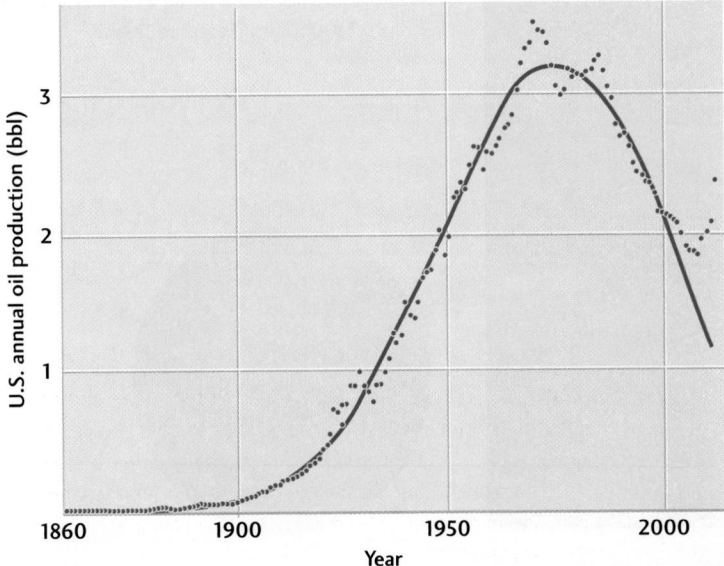

FIGURE 23.11 ▪ U.S. annual oil production in billions of barrels (bbl) from 1860 to 2012. The points show production figures for each year. The solid line is similar to Hubbert's 1959 projection, which predicted the peak in the 1970s and the subsequent decline. However, production reached a minimum in 2008 and has since been increasing. [From K. Deffeyes, *Hubbert's Peak*. Princeton, NJ: Princeton University Press, 2001, modified with data from the U.S. Energy Information Administration.]

the curve is referred to as *Hubbert's peak*, named for petroleum geologist M. King Hubbert. In 1956, Hubbert used a simple mathematical relationship between the production rate and the rate of discovery of new reserves to predict that U.S. oil production, which was growing rapidly at the time, would actually begin to decline sometime in the early 1970s. His arguments were roundly dismissed as overly pessimistic, but history proved him right: production did indeed peak in 1970, beginning a decline that continued much as Hubbert predicted throughout the late twentieth century.

However, in 2009, U.S. oil production began to increase again, with 2012 production 30 percent higher than its minimum in 2008. This increase signals a new U.S. oil boom, fed by the rapid development of offshore oil fields and improved technology for recovering oil on land, including the controversial technique of fracking.

When Will We Run Out of Oil?

At the current production rate, the world will consume all of its known oil reserves in about 55 years. Does that mean we will run out of oil before the end of this century? No, because oil resources are much greater than oil reserves.

In fact, we will never really "run out" of oil. As resources diminish, prices will eventually rise so high that we cannot afford to waste oil by burning it as a fuel. Its main use will then be as a raw material for producing plastics, fertilizers, and a host of other *petrochemical* products. The petrochemical industry is already a very big business, consuming 7 percent of global oil production. As oil geologist Ken Deffeyes has noted, future generations will probably look back

on the Petroleum Age with a certain amount of disbelief: "They burned it? All those lovely organic molecules, and they just burned it?"

The key question is not when oil will run out, but when oil production will stop rising and begin to decline. This milestone—Hubbert's peak for world oil production—is the real tipping point; once it is reached, the gap between supply and demand will grow rapidly, driving oil prices sky-high.

So how close are we to Hubbert's peak? The answer to this question is the subject of considerable debate. Some oil pessimists have argued that we are fast approaching Hubbert's peak. Oil optimists, on the other hand, believe that new oil discoveries and improvements in oil-recovery technologies, such as fracking and deep-water drilling, will provide enough petroleum to satisfy world demand for many decades into the future. The recent increases in oil reserves shown in Figure 23.10 support this view.

Oil and the Environment

Extracting fossil fuels can have a number of detrimental effects on the environment. On April 20, 2010, an explosion aboard the drilling platform *Deepwater Horizon* killed 11 men and injured 17 others. This blowout resulted in the largest marine oil spill in history, releasing 5 million barrels of crude oil into the Gulf of Mexico during the next 3 months (**Figure 23.12**). The oil spill caused significant environmental damage to ecosystems along the Gulf Coast (**Figure 23.13**).

This accident, like earlier spills off the Yucatan coast in 1979 and Santa Barbara in 1969, renewed the long-running

FIGURE 23.12 ▪ Oil slick in the Gulf of Mexico imaged on May 24, 2010 by NASA's Terra satellite, 34 days after the explosion of the *Deepwater Horizon*. [NASA/Goddard/MODIS Rapid Response Team.]

FIGURE 23.13 ■ The oil spilled by the *Deepwater Horizon* blowout harmed wildlife along the Gulf Coast. [AP Photo/Bill Haber.]

debate about whether to allow drilling for oil and natural gas in fragile habitats such as the Arctic National Wildlife Refuge (ANWR) on the coastal plain of northern Alaska (**Figure 23.14**). The total petroleum resource in ANWR has not been fully evaluated, but it could be as much as 40 billion barrels of oil. The USGS estimates that if oil prices were high enough, 6 billion to 16 billion barrels of this oil could be produced economically using current technologies. There is no doubt that these resources would contribute to the national economy. But oil and gas production would require the building of roads, pipelines, and housing in a delicate environment that is an important breeding area for caribou, musk-oxen, snow geese, and other wildlife. Policy makers must weigh the short-term economic benefits of drilling against possible long-term environmental losses in making this decision.

Natural Gas

The world's reserves of natural gas are comparable to its crude oil reserves (see Figure 23.7) and will likely exceed them in the decades ahead. Estimates of natural gas resources have been rising in recent years because exploration for natural gas has increased, and gas reservoirs have been identified in new settings, such as very deep rock formations, overthrust structures, coal beds, tight (less permeable) sandstones, and shales.

As in the case of oil, newer technologies have increased the efficiency of gas production and the types of rock formations from which gas can be extracted. In a method called **hydraulic fracturing** (or "fracking"), large amounts of water are injected through the drilling pipe into a hard rock formation, such as a shale, to create tiny fissures in the rock, which allows gas to flow more readily into the

FIGURE 23.14 ■ Herd of caribou in the Arctic National Wildlife Refuge (ANWR). An intense controversy surrounds proposals to drill for oil and natural gas in this pristine region. [Prisma Bildagentur/Alamy.]

Wellhead

Shale
formation

FIGURE 23.15 ■ Hydraulic fracturing or "fracking" is a technique for withdrawing oil and gas from shale and other tight formations by first pumping water and sand into a borehole at high pressures to create fractures through which the oil and gas can more readily flow. The boreholes are commonly drilled horizontally through nearly flat-lying shale formations.

(a)

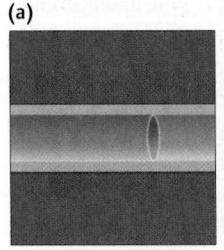

Borehole is cased and surrounded by cement.

(b)

Small holes are blasted through casing and cement.

(c)

Surrounding rocks are hydraulically fractured by pumping water and sand into borehole at high pressure.

(d)

Fracking generates small fissures that are kept open by the sand, allowing oil and gas to flow up the borehole to the wellhead.

pipe (**Figure 23.15**). This technology has created a boom in the extraction of natural gas from shale formations such as the Marcellus shale that underlies the northern Appalachian Mountains and the Allegheny Plateau of the eastern United States (see Chapter 5). The production of "shale gas" has increased tenfold in the last decade and now accounts for almost one-third of U.S. natural gas production (**Figure 23.16**).

The environmental cost associated with fracking can be steep, however, because the process uses huge amounts of water, and wastes from shale gas production can contaminate the local water supply. Moreover, the disposal of waste water and chemicals used in fracking is often done by injecting these fluids into deep wells, which lubricates old faults in Earth's crust, causing earthquakes (see Chapter 13, Practicing Geology). This practice has increased earthquake activity in many regions of the United States where the seismicity has been historically low, such as Oklahoma, Texas, and Ohio.

FIGURE 23.16 ■ Shale gas now accounts for almost one-third of natural gas production in the United States, and it is expected to constitute almost half by 2040.

Natural gas is a premium fuel for a number of reasons. In combustion, methane combines with atmospheric oxygen, releasing energy in the form of heat and producing only carbon dioxide and water. Natural gas therefore burns much more cleanly than oil or coal, which also produces sulfur dioxide (the major cause of acid rain). Moreover, natural gas emits 30 percent less CO_2 per unit of energy than oil and more than 40 percent less than coal. Therefore, substituting natural gas for coal as, say, the fuel for power plants lowers the **carbon intensity** of energy production—that is, carbon emissions per quad of electricity. Natural gas is easily transported across continents through pipelines. Getting it from source to market across oceans has been more difficult. The construction of tankers and ports that can handle liquefied natural gas (LNG) is beginning to solve this problem, although the potential dangers (such as the risk of a large explosion) have made LNG facilities controversial in the communities where they would be located.

Natural gas accounts for about 33 percent of all fossil-fuel consumption in the United States each year (see Figure 23.5). More than half of U.S. homes and a great majority of commercial and industrial buildings are connected to a network of underground pipelines that draw gas from fields in the United States, Canada, and Mexico. The rise of natural gas resources has led some observers to speculate that we are now transitioning from a "petroleum economy" to a "methane economy."

Coal

The abundant plant fossils found in coal beds show that coal is a biological sediment formed from large accumulations of plant material in wetlands. As the luxuriant plant growth of a wetland dies, leaves, twigs, and branches fall to the waterlogged soil. Rapid burial and immersion in water protect this plant material from complete decay because the bacteria that decompose organic matter are cut off from the oxygen they need. The plant material accumulates and gradually turns into *peat*, a porous brown mass of organic matter in which twigs, roots, and other plant parts can still be recognized (**Figure 23.17**). The accumulation of peat in oxygen-poor environments can be seen in modern swamps and peat bogs. When dried, peat burns readily because it is 50 percent carbon.

Over time, with continued burial, the peat is compressed and heated. Chemical transformations increase the peat's already high carbon content, and it becomes *lignite*, a very soft, brownish black, coal-like material containing about 70 percent carbon. The higher temperatures and deformation that accompany greater depths of burial may transform

1 Surface accumulations of plant material in a wetland are buried, partly decayed, and compressed into peat.

2 Shallow burial transforms peat into lignite.

3 Further burial transforms lignite into soft (bituminous) coal.

4 Continued burial and structural deformation, plus heat, metamorphose soft coal into hard coal (anthracite).

FIGURE 23.17 ■ The formation of coal begins with the deposition of vegetation in oxygen-poor environments.

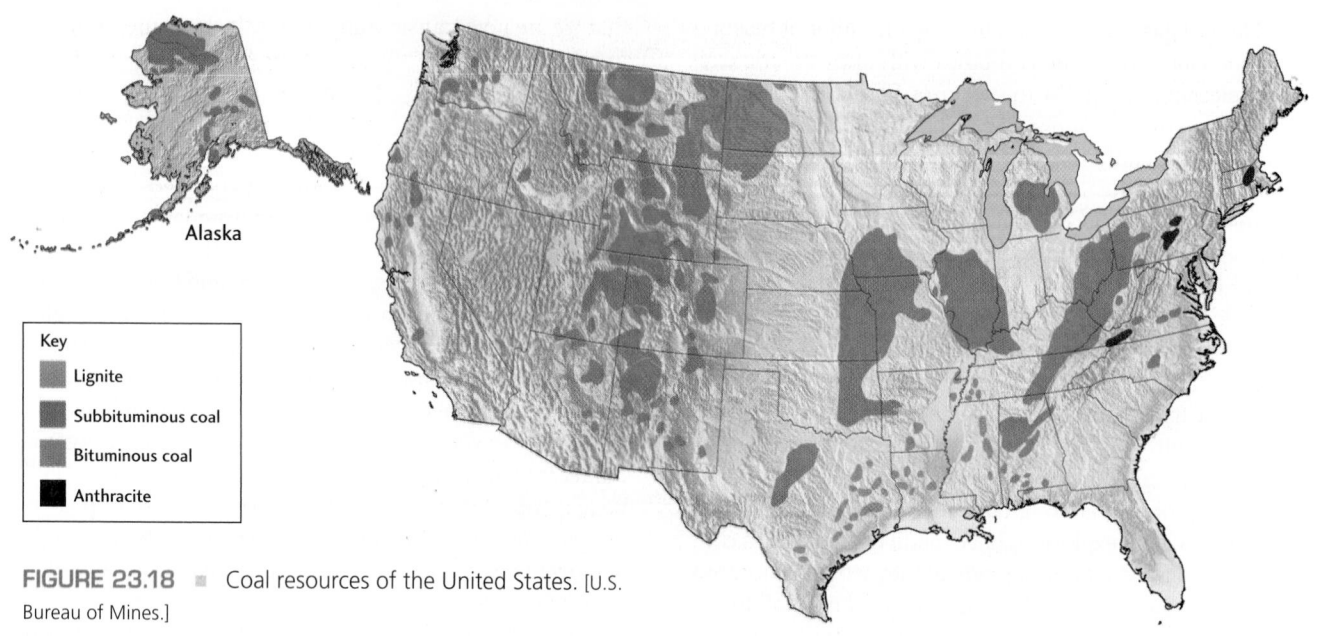

FIGURE 23.18 ■ Coal resources of the United States. [U.S. Bureau of Mines.]

Key
- Lignite
- Subbituminous coal
- Bituminous coal
- Anthracite

Alaska

lignite into *subbituminous* and *bituminous coal,* or soft coal, and ultimately into *anthracite,* or hard coal. The higher the grade of metamorphism, the harder and more vitreous the coal, and the higher its carbon content, and therefore its energy content. Anthracite is more than 90 percent carbon.

COAL RESOURCES There are huge resources of coal in sedimentary rocks. Although coal has been a major energy source since the late nineteenth century, only about a few percent of the world's coal reserves have been consumed. According to the best estimates, these reserves amount to 860 billion metric tons, which are capable of producing 17,800 quads of energy, more than any other fossil fuel (see Figure 23.7). About 85 percent of the world's coal resources are concentrated in the former Soviet Union, China, and the United States; these areas are also the world's largest coal producers. The United States has extensive deposits of coal in many states (**Figure 23.18**)—enough to last for a few hundred years at the nation's current rate of use (about a billion tons per year). From 1975, when the price of oil began its precipitous rise, until 2005, coal supplied an increasing proportion of U.S. energy needs, primarily as fuel for electrical power generation. But coal usage has since declined as natural gas production has increased (see Figure 23.3). Coal currently accounts for about 18 percent of U.S. energy consumption.

THE COSTS OF COAL The extraction and combustion of coal present serious problems that make it a less desirable fuel than oil or natural gas. Underground coal mining is a dangerous profession; more than 2000 miners are killed each year in China alone. Many more coal miners suffer from black lung, a debilitating inflammation of the lungs caused by the inhalation of coal particles. Surface or "strip"

mining—the removal of soil and surface sediments to expose coal beds—is safer for the miners, but it can ravage the countryside if the land is not restored. An especially destructive type of surface mining, now common in the Appalachian Mountains of the eastern United States, is "mountaintop removal," in which up to 300 vertical meters of a mountain crest is blasted away to expose underlying coal beds (**Figure 23.19**). The excess rock and soil are dumped into the surrounding valleys.

Coal is a notoriously dirty fuel. When burned, it produces, on average, 25 percent more CO_2 per unit of energy than oil and 70 percent more than natural gas; in other words, its carbon intensity is high. Most coal also contains appreciable amounts of pyrite, which is released into the atmosphere as noxious sulfur-containing gases when the coal is burned. Acid rain, which forms when these gases combine with rainwater, has been a severe problem in Canada, Scandinavia, and the northeastern United States, and eastern Europe.

An inorganic residue, called coal ash, remains after coal is burned. Coal ash contains all the metals that were present in the coal, some of which, such as mercury, are toxic. Coal ash can amount to several tons for every 100 tons of coal burned, so it poses a significant disposal problem. Ash can also escape from smokestacks, creating a health risk to people downwind.

U.S. government regulations now require industries that burn coal to adopt technologies for "clean" coal combustion, which have reduced emissions of sulfur and toxic chemicals. Federal laws also mandate the restoration of land disrupted by surface mining and the reduction of dangers to miners. These measures are expensive and add to the cost of coal, but it is still a relatively inexpensive fuel compared with petroleum.

FIGURE 23.19 ■ Mountaintop removal mining in the Appalachian Mountains of West Virginia. [Rob Perks, NRDC.]

Unconventional Hydrocarbon Resources

Extensive deposits of hydrocarbons occur in two other forms: source beds that are rich in organic material but never reached the oil window, and formations that once contained oil but have since "dried out," losing many of their volatile components, to form *heavy oil* or a tarlike substance called *natural bitumen* (not to be confused with bituminous coal).

A hydrocarbon resource of the first type is *oil shale*, a fine-grained, clay-rich sedimentary rock containing large amounts of organic matter. In the 1970s, oil producers began trying to commercialize the extensive oil shales of western Colorado and eastern Utah, but those efforts were largely abandoned by the 1980s as oil prices fell, concerns over environmental damage increased, and technical problems persisted. New oil-recovery technologies such as hydraulic fracturing has made energy production from oil shales much more efficient, but the environmental costs per unit of energy remain high. As we have previously noted, the process of extracting oil and gas from shales by fracking requires huge amounts of water, a scarce resource in the western United States, and the reinjection of the waste water into the crust can cause earthquakes, even in seismically inactive areas.

One deposit of the second type, the *tar sands* of Alberta, Canada, is estimated to contain a hydrocarbon reserve equivalent to 170 billion barrels of oil and a total resource perhaps 10 times that amount. More than 600 million barrels of oil are now extracted from the Alberta tar sands each year, and Canadian production is projected to increase fivefold by 2030, providing as much as 5 percent of world demand for fossil fuels. Development of the tar sands, like that of oil shales, raises important environmental concerns, however. It takes 2 tons of mined sand to produce 1 barrel of oil, leaving lots of waste sand, which is an environmental pollutant. Moreover, production of oil from the tar sands is an inefficient process that sucks up about two-thirds of the energy they ultimately render and emits considerably more CO_2 than conventional oil production.

■ Alternative Energy Resources

As we continue to deplete our fossil-fuel resources, alternative energy resources will have to take up more and more of the demand. How quickly will this transition to a postpetroleum economy occur? Which alternative sources of energy have the greatest potential to replace fossil fuels?

Nuclear Energy

The first large-scale use of the radioactive isotope uranium-235 to produce energy was in the atomic bomb in 1944, but the nuclear physicists who first observed the vast energy released when its nucleus split spontaneously (a phenomenon called *fission*) foresaw the possibility of peaceful applications of this new energy source. After World War II, countries around the world built nuclear reactors to produce **nuclear energy.** In these reactors, the fission of uranium-235 releases heat that is used to make steam, which then drives turbines to create electricity.

FIGURE 23.20 ■ Japan's Kashiwazaki-Kariwa facility is the world's largest nuclear power plant, with seven reactors and a total generating capacity exceeding 8200 megawatts. It was damaged by a powerful earthquake (magnitude 6.8) that struck the region on July 16, 2007. [STR/AFP/Getty Images.]

A typical commercial reactor produces about 1000 megawatts of electricity (1 megawatt = 1 million watts). Large nuclear facilities may have multiple reactors (**Figure 23.20**).

Nuclear power supplies a substantial fraction of the electric energy used by some countries, such as France (75 percent in 2012), Slovakia (54 percent), and Sweden (38 percent), but this proportion is much smaller in the United States (19 percent). Overall, the nation's 110 nuclear reactors account for about 8.5 percent of total U.S. energy demand (see Figure 23.14). The early expectation that nuclear fuels would provide a large, low-cost, environmentally safe source of energy has not been realized, primarily because of problems with reactor safety, disposal of radioactive wastes, and nuclear security.

URANIUM RESERVES Uranium is found as a trace element in some granites at an average concentration of only 0.00016 percent of the rock. Moreover, only a small proportion of uranium is uranium-235; its other, much more abundant isotopes (for example, uranium-238) are not radioactive enough to be used as fuel. Uranium is nevertheless the world's largest minable energy resource by far, with a potential energy-generating capacity of at least 240,000 quads, much larger than any fossil fuel. Minable concentrations are typically found as small quantities of uraninite, a uranium oxide mineral (also called pitchblende), in veins in granites and other felsic igneous rocks. Where groundwater is present, uranium in igneous rocks near Earth's surface may oxidize and dissolve, be transported in groundwater, and later be reprecipitated as uraninite in sedimentary rocks.

HAZARDS OF NUCLEAR ENERGY The biggest drawbacks of nuclear energy are concerns about the safety of nuclear reactors, the risk of environmental contamination with radioactive material, and the potential use of radioactive fuels for making nuclear weapons.

In the United States, damage to a reactor at the Three Mile Island reactor in Pennsylvania in 1979 released radioactive debris. Although very little radioactive fuel left the containment building, and no one was harmed, it was a close call. Much more serious was the destruction of a nuclear reactor in the town of Chernobyl, Ukraine, in 1986. The reactor went out of control because of poor design and human error. A plume of radioactive debris was carried by winds over Scandinavia and western Europe. Contamination of buildings and soil has made hundreds of square miles of land surrounding Chernobyl uninhabitable. Food supplies in many countries were contaminated by the radioactive fallout and had to be destroyed. Deaths from cancer caused by exposure to the fallout may be in the thousands.

The most serious disaster occurred when the tsunami from the great Tohoku earthquake of March 11, 2011, inundated the Fukushima Daiichi nuclear power plant on the northeastern coast of Honshu, Japan (see Figure 13.31). The reactors shut down, as designed, but the tsunami destroyed the backup diesel generators, cutting power to the water pumps that were supposed to cool the still-hot reactors. Three of the six reactors suffered complete or partial meltdowns, and explosions of hydrogen gas generated during the meltdowns destroyed the reactor containment buildings, releasing radioactive debris into the atmosphere. Water sprayed to cool the damaged reactors carried radioactive material into the ocean. Radioactive materials from these reactors are still leaking into the environment.

FIGURE 23.21 ■ Aerial view of the north entrance to the Yucca Mountain Nuclear Waste Repository developed at the Nevada Test Site, north of Las Vegas. Yucca Mountain is the high ridge to the right of the entrance. Federal funding for this project was terminated in 2010. [Department of Energy.]

The uranium consumed in nuclear reactors leaves behind dangerous radioactive wastes. A system of safe long-term waste disposal is not yet available, and reactor wastes are being held in temporary storage facilities at reactor sites. (Spent fuel rods stored on site contributed to the radioactive debris released at Fukushima.) Many scientists believe that geologic containment—the burial of nuclear wastes in deep, stable, impermeable rock formations—would provide safe storage of the most dangerous wastes for the hundreds of thousands of years required before they decay. France and Sweden have built underground nuclear waste repositories. A national repository, the Yucca Mountain Nuclear Waste Repository, was being developed in the United States at Yucca Mountain, Nevada (**Figure 23.21**), but local opposition caused the federal government to terminate funding for the site in 2010. At present the United States has no long-term plan for nuclear waste disposal.

Biofuels

Before the coal-fired industrial revolution of the mid-nineteenth century, the burning of wood and other biomass derived from plants and animals satisfied most of society's energy needs. Even today, the energy derived from biomass exceeds the total derived from all other renewable resources.

Biomass is an attractive alternative to fossil fuels because, at least in principle, it is *carbon-neutral;* that is, the CO_2 produced by the combustion of biomass is eventually removed from the atmosphere by plant photosynthesis and used to produce new biomass. In particular, liquid **biofuels** derived from biomass, such as *ethanol* (ethyl alcohol: C_2H_6O), could replace gasoline as our main automobile fuel.

The use of biofuels in transportation is hardly new. The first four-stroke internal combustion engine, invented by Nikolaus Otto in 1876, ran on ethanol, and the original diesel engine, patented by Rudolf Diesel in 1898, ran on vegetable oil. Henry Ford's Model T car, first produced in 1903, was designed to operate on ethanol. But soon thereafter, petroleum from the new reserves discovered in Pennsylvania and Texas became widely available, and cars and trucks were converted almost entirely to petroleum-based gasoline and diesel fuel.

Ethanol can be mixed with gasoline to run most car engines built today. It is produced mainly from corn in the United States and from sugarcane in Brazil. For the last 35 years, the Brazilian government has been pushing to replace imported oil with domestic ethanol; in 2012, about 35 percent of Brazil's automobile fuels came from sugarcane, saving the country about $50 billion in oil imports. In 2012, Brazil and the United States accounted for 68 percent of the global biofuel production.

A promising biomass crop is switchgrass, a perennial plant native to the Great Plains (**Figure 23.22**). Switchgrass has the potential to produce up to 1000 gallons of ethanol per acre per year, compared with 665 gallons for

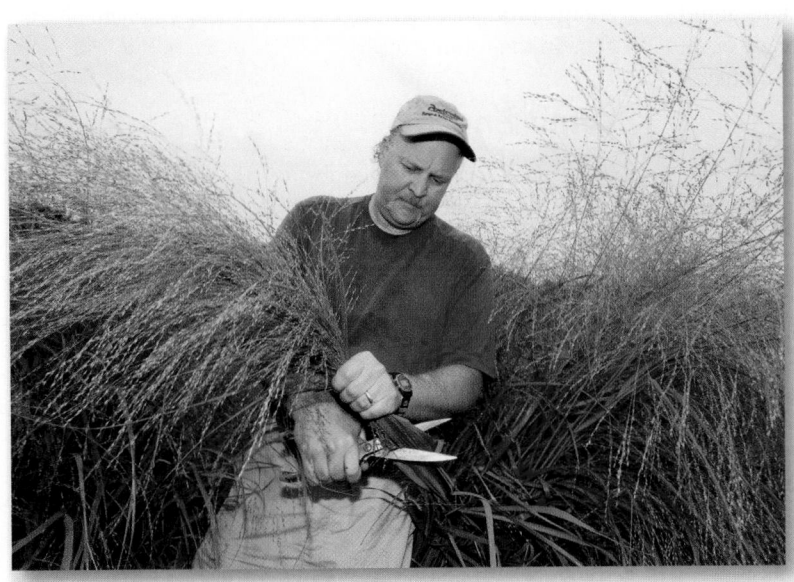

FIGURE 23.22 ■ Switchgrass, a perennial plant native to the Great Plains, is an efficient source of ethanol, the most popular biofuel. Here, geneticist Michael Casler harvests switchgrass seed as part of a breeding program to increase the plant's ethanol yield. [Wolfgang Hoffmann/USDA.]

sugarcane and 400 gallons for corn, and it can be cultivated on grasslands of marginal utility for other types of agriculture. Nevertheless, biofuel production competes with food production, so increasing the former drives up the price of the latter, which reduces the economic benefits of biofuels.

What about the environmental benefits of biofuels? Can they really be carbon-neutral? If the energy used to fertilize plants, transform them into biofuels, and deliver the biofuels to market comes primarily from fossil fuels, then the answer is no. The widespread use of biofuels for transportation would no doubt reduce the pumping of carbon from the lithosphere to the atmosphere, but experts are still arguing about the magnitude of that reduction.

Solar Energy

Solar energy enthusiasts remind us that "every hour Earth receives from sunlight more energy than human civilization uses in one year." **Solar energy** is the prime example of a resource that cannot be depleted by usage: the Sun will continue to shine for at least the next several billion years. Although using solar energy to hear water for homes, industries, and agriculture is economically profitable with existing technology, the methods for the large-scale conversion of solar energy into electricity are still inefficient and expensive. Nevertheless, the solar generation of electricity is increasing rapidly as large power plants are being built in response to voter mandates and government subsidies. The Ivanpah solar electric generating system in California's Mojave Desert, commissioned in 2013, is the world's largest, capable of producing up to 392 megawatts of electricity (**Figure 23.23**).

In the United States, solar energy conversion rose from 0.065 quad in 2004 to almost 0.20 quad in 2012, a threefold increase in just over 8 years. Yet this amounts to only 0.2 percent of U.S. energy consumption. Optimistic projections are that, worldwide, solar conversion could increase to as much as 12 quads per year in a decade or so, which would amount to about 2 percent of total energy production.

Hydroelectric Energy

Hydroelectric energy is derived from water moving under the force of gravity to drive turbines that generate electricity. Artificial reservoirs behind dams usually provide the water. Hydroelectric energy depends on the Sun, whose energy drives the climate system and produces rainfall; thus, like solar energy, it is renewable. It is also relatively clean, risk-free, and cheap to produce.

The Three Gorges Dam on the Yangtze River in China (**Figure 23.24**) is the world's largest hydroelectric facility. It is capable of generating 22,500 megawatts—nearly 5 percent of China's total electricity demand. The project

FIGURE 23.23 ■ The Ivanpah solar electric generating system in California's Mojave Desert, commissioned in 2013, is the world's largest. More than 170,000 mirrors focus sunlight on three towers filled with water, producing steam that spins turbines that can generate up to 392 megawatts of electricity. [Gilles Mingasson/Getty Images for Bechtel.]

has been controversial, however, because the damming of the Yangtze caused flooding that has displaced over a million people.

In the United States, hydroelectric dams deliver about 2.7 quads annually, or a little less than 3 percent of the nation's annual energy consumption. The U.S. Department of Energy has identified more than 5000 sites where new hydroelectric dams could be built and operated economically. Such expansion would be resisted, however, because the dams would drown farmlands and wilderness areas under artificial reservoirs while adding only a small amount of energy to the U.S. supply. For this reason, most energy experts expect that the proportion of the nation's energy produced by hydroelectric power will actually decline in the future.

Wind Energy

Wind power is produced by using windmills to drive electric generators (**Figure 23.25**). Today, the generation of electricity by high-efficiency wind turbines is a fast-growing source of renewable energy. Wind farms containing hundreds of turbines can produce as much electric power as a mid-sized nuclear power plant. Worldwide, the amount of wind-generated electric power grew tenfold between 2000 and 2010. Denmark now produces 21 percent of its electric power by wind; Portugal, 18 percent. In the United States, electricity from wind sources increased threefold from 2005 to 2010, and wind now accounts for about 1.4 percent of all U.S. energy production (see Figure 23.5).

FIGURE 23.24 ■ The Three Gorges Dam on China's Yangtze River is about 2335 m (7660 feet) long and 185 m (616 feet) high. Its 32 generators are capable of producing 22,500 megawatts of hydroelectric power. [AP photo/Xinhua Photo, Xia Lin.]

The U.S. Department of Energy estimates that winds sufficient for power generation blow across 6 percent of the land area of the continental United States, and that those winds have the potential to supply more than one and a half times the nation's current electricity demand. But harvesting this energy would require placing millions of windmills, each 100 m tall, across hundreds of thousands of square kilometers of land. Changes to the landscape required for industrial wind farming, as well as the low-frequency noise generated by the turbines, have made the siting of new facilities a controversial environmental issue in some regions.

FIGURE 23.25 ■ These rows of windmills are part of the Alta Wind Energy Center, located in the Tehachapi Pass of Kern County, California, the largest wind farm in the world, capable of generating up to 1,320 megawatts of electricity. [Lowell Georgia/Science Source.]

Geothermal Energy

Earth's internal heat can be tapped as a source of *geothermal energy,* as we described in Chapter 12. According to one Icelandic estimate, as much as 40 quads of electricity could be generated each year from accessible geothermal energy sources, but so far only a tiny fraction of that amount, about 0.3 quad per year, is actually being generated. Another 0.3 quad of geothermal energy is used for direct heating. At least 46 countries now use some form of geothermal energy.

Although geothermal energy is unlikely to replace petroleum as a major source of power, it may help to meet our future energy needs. Like most of the other energy sources we have looked at, geothermal energy presents some environmental problems. Regional ground subsidence can occur if hot groundwater is withdrawn without being replaced. In addition, hydrothermal waters can contain salts and toxic materials dissolved from the hot rock. As in the case of fracking, the disposal of these waste waters by reinjection into the crust can trigger earthquakes.

■ Global Change

The expression **global change** entered the world's vocabulary when it became clear that emissions from fossil-fuel burning and other human activities were beginning to alter the chemistry of the atmosphere. People are becoming increasingly concerned about the anthropogenic changes that have been observed in every component of the climate system. This section will describe three of the most serious forms of anthropogenic global change:

■ Global warming due to increased concentrations of carbon dioxide and other greenhouse gases in the atmosphere

■ Ocean acidification due to increased carbon dioxide dissolved in the hydrosphere

■ Losses of species diversity due to changes in the biosphere

The consequences of anthropogenic global change are motivating politicians to work together in ways they never have before as we all try to avoid the "tragedy of the commons": the spoiling of our commonly held environmental resources by overexploitation. Neighboring nations are enacting mutually beneficial regulations to address regional environmental problems, and new multinational treaties are being formulated in attempts to manage anthropogenic effects on the global environment. Earth science provides the knowledge needed to make rational choices about global environmental management.

Greenhouse Gases and Global Warming

Since the beginning of the industrial era, fossil-fuel burning, deforestation, land-use changes, and other human activities have caused a significant rise in the concentrations of greenhouse gases in the atmosphere. **Figure 23.26** shows

FIGURE 23.26 ■ Atmospheric concentrations of carbon dioxide, methane, and nitrous oxide over the last 10,000 years (large panels) and for 1750–2000 (inset panels). These measurements, compiled by the Intergovernmental Panel on Climate Change, were derived from ice cores and atmospheric samples. Shaded bands show the uncertainties in the measurements. [IPCC, *Climate Change 2007: The Physical Science Basis.* Figure SPM.1, Cambridge University Press.]

the atmospheric concentrations of three greenhouse gases—carbon dioxide, methane, and nitrous oxide—over the past 10,000 years. In all three cases, the concentrations remained relatively constant through most of the Holocene, but shot upward after the industrial revolution.

The global atmospheric concentration of methane has increased by 150 percent from its preindustrial value, and that of carbon dioxide has increased 48 percent. In both cases, the observed increases can be explained by human activities, predominantly agriculture and fossil-fuel use. Methane's greenhouse effect is weaker than that of carbon dioxide, however, so even though its relative concentration has gone up more, its contribution to greenhouse warming is only about 30 percent as large. The postindustrial increase in nitrous oxide, primarily from agriculture, has been about 20 percent, contributing only a small fraction to greenhouse warming.

These increases in greenhouse gas concentrations have been accompanied by a rise in average temperatures at Earth's surface (see Figure 15.21). The United Nations, recognizing the potential problems this warming trend poses, established an Intergovernmental Panel on Climate Change (IPCC) in 1988 to assess the likelihood of anthropogenic climate change, its potential effects, and possible solutions to those problems (see Earth Issues 15.2). The IPCC provides a continuing forum for hundreds of scientists, economists, and policy experts to work together to understand these issues.

In its major assessment reports, the latest of which was released in 2013, the IPCC drew the following conclusions:

■ From the beginning of the twentieth century until 2012, the average temperature of Earth's surface has risen, on average, by about 0.9°C.

■ Most of this warming has been caused by anthropogenic increases in atmospheric greenhouse gas concentrations.

■ Concentrations of atmospheric greenhouse gases will continue to increase throughout the twenty-first century, primarily because of human activities.

■ The increase in atmospheric greenhouse gas concentrations will cause significant global warming during the twenty-first century.

Predictions of Future Global Warming

The warming trend described in Chapter 15 has continued into the twenty-first century. The decade 2001–2010 was the warmest recorded since accurate temperature measurements began in 1880. The warmest year on record was 2010, followed by 2005. All ten of the warmest years have occurred since 1997.

How much hotter will the planet get, and how will the global warming affect local climates? Answering such questions requires predictions based on complex models of the Earth system. The predictions are uncertain because they depend strongly on how human population and the global economy will evolve, including how energy supplies will be exploited and what political decisions will be made (if any) to limit greenhouse gas emissions. The IPCC has forecast increases in atmospheric CO_2 concentrations under a series of scenarios that sample the possibilities. Each scenario is characterized by a *representative concentration pathway* or RCP that corresponds to a net concentration of greenhouse gases in Earth's atmosphere by the year 2100. Three of these scenarios are depicted in **Figure 23.27**:

■ *Scenario A* assumes a continued reliance on fossil fuels as our major energy source and thus an increasing concentration of greenhouse gases, as given by the red lines in Figure 23.27. In this scenario, called "RPC8.5" by the IPCC, the carbon dioxide concentration would top 900 ppm by 2100, more than three times its preindustrial level.

■ *Scenario B* (green line, "RCP6") assumes that carbon dioxide concentrations will begin to stabilize in the later part of the twenty-first century, reaching just over 600 ppm, more than twice the preindustrial level, by 2100. This scenario would require a big shift toward fossil fuels with less carbon intensity, such as natural gas, as well as an increasing reliance on nuclear energy and renewable energy sources.

■ *Scenario C* (blue line, "RCP2.6") has a peak in carbon dioxide concentrations around 2050, followed by a modest decline to carbon dioxide concentrations near the present level (400 ppm) by the end of the century. This scenario would require a much more rapid conversion from fossil fuels to cleaner alternatives.

Here we can add that the scenarios for global population growth shown in Figure 23.1 were developed by the IPCC to be consistent with the three RCPs listed above. For example, scenario C corresponds to a declining world population in 2100.

The IPCC has used these scenarios to predict average global surface temperatures (**Figure 23.28**). Allowing for uncertainties in the Earth system models, it found that the possible range of global temperature increases during the twenty-first century could range from about 0.5°C to 5.5°C. The lower values can be achieved only through rapid reductions in fossil-fuel burning and conversion to clean and resource-efficient energy technologies. Under the less radical (but still optimistic) scenario B, the temperature rise by 2100 would likely exceed 2°C, over twice the twentieth-century warming. In scenario A, the most pessimistic, the temperature increase would probably exceed 4°C.

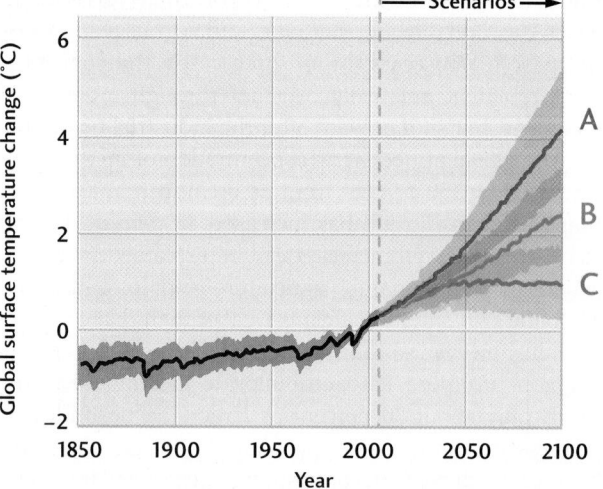

FIGURE 23.28 ■ IPCC projections of average surface temperatures over the twenty-first century derived from scenarios A (red line, RCP8.5), B (green line, RCP6), and C (blue line, RPC2.6). Gray-shaded band gives the uncertainties in past measurements; color-shaded band shows the prediction uncertainties due to incomplete knowledge of the climate system. [IPCC, *Climate Change 2013: The Physical Science Basis.*]

Consequences of Climate Change

It seems clear that human emissions of greenhouse gases will cause further global warming and may result in major changes to the climate system. These changes have the potential to affect civilization in both positive and negative ways. Some regional climates may improve while others may deteriorate. The authors' home city of Los Angeles, for example, is likely to become hotter and drier. Some potential effects of climate change are listed in **Table 23.1.**

CHANGES IN REGIONAL WEATHER PATTERNS

How will the enhanced greenhouse effect, along with other factors, such as deforestation, change temperatures across Earth's surface? **Figure 23.29** maps the regional temperature increases predicted by the three IPCC scenarios. The predicted geographic patterns of temperature change display some similarities to the observed pattern of late-twentieth-century warming in Figure 15.21. In particular, the warming is greater over land than over the oceans, and the temperate and polar regions of the Northern Hemisphere show the most warming. Thus, geographic patterns of climate change in the twenty-first century are likely to be similar to those observed over the past several decades.

The IPCC has documented a number of current trends in regional weather patterns that are likely to continue:

■ The relative humidity and frequency of heavy precipitation events have increased over many land areas in a manner consistent with the observed

FIGURE 23.27 ■ Three scenarios or "representative concentration pathways" (RPCs) projected by the IPCC for carbon dioxide, methane, and nitrous oxide during the twenty-first century. Scenario A (red line) implies continuing high rates of fossil-fuel burning (RCP8.5); scenario B (green line) implies stabilization of emission rates in the latter part of the twenty-first century (RCP6); scenario C (blue line) implies a rapid conversion to nonfossil fuels (RCP2.6). [IPCC, *Climate Change 2013: The Physical Science Basis.*]

TABLE 23-1	Potential Effects of Climate Change on Ecosystems and Resources
System	**Potential Effects**
Forests and other ecosystems	Migration of vegetation; reduction in ecosystem ranges; altered ecosystem composition
Species diversity	Loss of diversity; migration of species; invasion of new species
Coastal wetlands	Inundation of wetlands; migration of wetland vegetation
Aquatic ecosystems	Loss of habitat; migration to new habitats; invasion of new species
Coastal resources	Inundation of coastal structures; increased risk of flooding
Water resources	Changes in water supplies; changes in patterns of drought and flooding; changes in water quality
Agriculture	Changes in crop yields; shifts in relative productivity among regions
Human health	Shifts in ranges of infectious disease organisms; changes in patterns of heat-stress and cold-weather afflictions
Energy	Increase in cooling demand; decrease in heating demand; changes in hydroelectric energy resources

Source: Office of Technology Assessment, U.S. Congress.

Scenario A

Scenario B

Scenario C

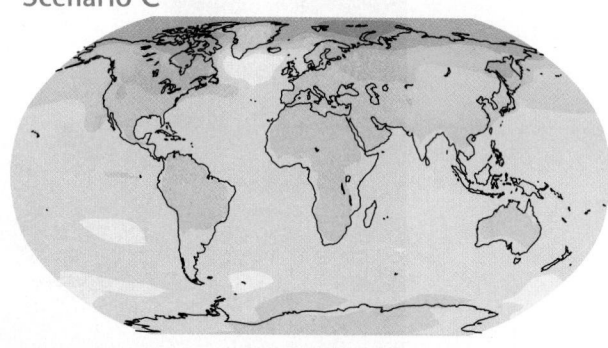

-2 -1.5 -1 -0.5 0 0.5 1 1.5 2 3 4 5 7 9 11

Temperature change (°C)

FIGURE 23.29 ■ Average surface temperatures predicted for 2080–2100 by the three IPCC scenarios of Figure 23.28, expressed as differences from average surface temperatures measured at the same locations during the baseline period 1986–2005. [IPCC, *Climate Change 2013: The Physical Science Basis.*]

temperature increases. Increased precipitation has been observed in eastern parts of North and South America, northern Europe, and northern and central Asia.

■ Drying has been observed in the Sahel, the Mediterranean, southern Africa, and parts of southern Asia. More intense and longer droughts have been observed over wider areas since the 1970s, particularly in the tropics and subtropics.

■ Widespread changes in temperature extremes have been observed over the last 50 years. Cold days, cold nights, and frost have become less frequent, while hot days, hot nights, and heat waves have become more frequent.

■ Intense hurricane activity in the North Atlantic has increased in a manner consistent with increases in tropical sea surface temperatures. Although there is no clear trend in the annual number of hurricanes, the number of very strong hurricanes (category 4 and 5 storms) has almost doubled over the past three decades.

CHANGES IN THE CRYOSPHERE Nowhere are the effects of global warming more evident than in polar regions. The amount of sea ice in the Arctic Ocean is decreasing, and the downward trend seems to be accelerating. The sea ice cover in September 2012 was the lowest for that month since the keeping of satellite records began in 1978: 3.6 million square kilometers, down by a factor of two from the 1979 minimum of 7.2 million square kilometers (Figure 23.30). According to climate models, much of the Arctic Ocean will become ice-free within a few decades. The shrinkage of sea ice is already severely disrupting Arctic ecosystems (Figure 23.31).

Temperatures at the top of the permafrost layer in the Arctic have risen by 3°C since the 1980s, and the melting of permafrost is destabilizing structures such as the Trans-Alaska oil pipeline (see Figure 21.20). The maximum area covered by seasonally frozen ground has decreased by about 7 percent in the Northern Hemisphere since 1900, with a decrease in spring of up to 15 percent. Valley glaciers at lower latitudes retreated during the twentieth-century warming (see Figure 21.9). Fieldwork has demonstrated that rates of glacial retreat and snow cover loss are increasing in both hemispheres. According to a study by the USGS, Glacier National Park in northern Montana will lose its large glaciers by 2030.

SEA LEVEL RISE As we saw in Chapter 21, the melting of sea ice does not affect sea level, but the melting of continental glaciers causes sea level to rise. Sea level also rises as the temperature of ocean water increases, increasing its overall volume by a tiny fraction (see Chapter 21, Practicing Geology). Sea level has risen more than 2 m since

FIGURE 23.30 ■ Global warming is melting the Arctic ice cap. These images of the Arctic, derived from NASA satellite data, compare the minimum extent of the polar ice cap in September 1979 (top) with the minimum extent in 2012 (bottom). One near-term benefit to human society will be the opening of the Northwest Passage and other shorter sea routes between the Atlantic and Pacific Oceans (show on bottom panel as red lines). [NASA/Goddard Space Flight Center Scientific Visualization Studio.]

FIGURE 23.31 ■ Climate change is already disrupting Arctic ecosystems, adversely affecting the habitat of Arctic animals such as polar bears. [Thomas & Pat Leeson/Science Source.]

the industrial revolution, and it is currently rising at about 3 mm per year. Climate models based on the IPCC scenarios indicate that sea level could rise by as much as a meter during the twenty-first century (**Figure 23.32**), creating serious problems for low-lying countries such as Bangladesh (**Figure 23.33**). On the Eastern Seaboard and Gulf Coast of the United States, flooding during coastal storm surges, such as that witnessed in Hurricanes Katrina and Sandy (see Chapter 20), could become much worse.

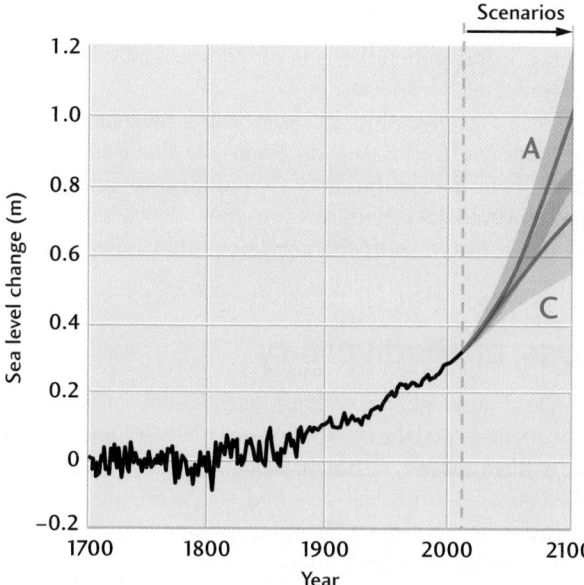

FIGURE 23.32 ■ Sea-level rise 1700–2100. The black line shows the observed value through 2010. The red curve is the IPCC prediction of future sea-level rise during the remainder of the twenty-first century according to scenario A; the blue curve is the prediction according to scenario C.

SPECIES AND ECOSYSTEM MIGRATION

As local and regional climates change, ecosystems will change with them. Many plant and animal species will have difficulty adjusting to rapid climate change or migrating to more suitable climates. Those that cannot cope with the rapid warming could become extinct. Global warming is already being blamed for a variety of adverse ecological effects, such as the disruption of Arctic ecosystems as sea ice and permafrost melt and the spread of tropical diseases like malaria as more of the world experiences a tropical climate.

THE POTENTIAL FOR CATASTROPHIC CHANGES TO THE CLIMATE SYSTEM

The current atmospheric concentrations of carbon dioxide and methane far exceed anything seen in the last 650,000 years. Our climate system is therefore entering unknown territory. Some observers think that the credibility of climate change projections suffers from "the Chicken Little problem": too many people are running around yelling "The sky is falling!" Yet most scientists think that those projections may be too conservative because they do not properly take into account some of the positive feedbacks within the climate system that could greatly enhance global change. Ocean acidification, mentioned earlier in this chapter, is a good example. Here are a few more:

■ *Destabilization of continental glaciers.* The surface melting of the Greenland glacier in 2012 was the largest on record, and there are indications that glacial streams within the ice sheet are accelerating much faster than expected. If the Greenland and Antarctic glaciers begin to shed ice faster than snowfall can generate new glacial ice, sea level could begin to rise much faster than the current IPCC predictions.

■ *Shutoff of thermohaline circulation.* Changes in precipitation and evaporation patterns are decreasing the salinity of seawater at mid- and high latitudes. Some scientists have speculated that this change could substantially reduce global thermohaline circulation (see Figure 15.3b), which is driven by differences in temperature and salinity. Major changes in the Gulf Stream and other aspects of the climate system could result.

■ *Methane release from seafloor sediments and permafrost.* Recall from Chapter 11 that a massive release of methane from shallow seafloor sediments about 55 million years ago might have caused abrupt global warming and led to the mass extinction at the Paleocene-Eocene boundary. Today, there is far more methane stored in shallow seafloor sediments and in permafrost than was released at the end of the Paleocene. If global warming begins to thaw those methane deposits, another runaway cycle of extreme warming could begin.

FIGURE 23.33 ■ Aerial view of a coastal region of Bangladesh flooded by a storm surge in May 2009. This low-lying country would be subject to disastrous flooding if sea level rises due to global warming. [James P. Blair/National Geographic.]

Ocean Acidification

As we saw in Chapter 15, about 30 percent of the carbon dioxide emitted into the atmosphere by fossil-fuel combustion is being absorbed by the oceans (see Figure 15.19). Scientists are concerned that the resulting increase in seawater acidity (see Figure 15.15) could decrease the calcification processes underlying the growth of shellfish and the formation of the hard exterior skeletons of corals.

In January 2009, 155 marine scientists from 26 countries, convened under the auspices of the United Nations, issued the Monaco Declaration, which states, "We are deeply concerned by recent, rapid changes in ocean chemistry and their potential, within decades, to severely affect marine organisms.... Severe damages are imminent." The scientists pointed specifically to observations of acidification-related decreases in shellfish weights and slowed growth of coral reefs.

Ocean acidification is likely to affect many types of marine organisms, not just those with calcium carbonate shells and skeletons. Anemones and jellyfish, for example, appear to be susceptible to even small changes in seawater acidity, and larger increases cause changes in seawater chemistry that can undermine the health of sea urchins and squid. The growing acidity of ocean surface waters is also likely to affect the concentrations of trace metals such as iron, an essential nutrient for the growth of many organisms.

As human activities continue to pump more CO_2 into the atmosphere, the ocean will continue to acidify. Whether marine organisms can adapt to the changes in store remains to be seen, but the effects on human society could be substantial. In the short term, damage to coral reef ecosystems and the fisheries and recreation industries that depend on them could result in economic losses amounting to many billions of dollars per year. In the longer term, changes in the stability of coastal reefs may reduce the protection they offer to coasts, and there may also be direct and indirect effects on commercially important fish and shellfish species.

Ocean acidification is essentially irreversible during our lifetimes. Even if we were magically able to reduce the atmospheric concentration of CO_2 to the level of 200 years ago, it would take tens of thousands of years for ocean chemistry to return to the conditions that existed at that time.

Loss of Biodiversity

In 2003, atmospheric chemist and Nobel laureate Paul Crutzen proposed the recognition of a new geologic epoch—the **Anthropocene,** or Age of Man—beginning about 1780, when James Watt's coal-powered steam engine launched the industrial revolution. The global changes that mark the Holocene-Anthropocene boundary are just now getting under way, so a future scientist with a full record of the next few thousand years may place that boundary at a somewhat later date. Crutzen's main point, however, is that global changes are proceeding so rapidly that such quibbles are likely to be minor. As with many previous geologic boundaries, the main marker will be a mass extinction.

FIGURE 23.34 ■ The Caribbean island country of Haiti is now 98 percent deforested. [Daniel Morel/AP World Wide.]

Between 1850 and 1880, as much as 15 percent of the land surface was deforested, and rates of deforestation have continued to rise. According to the United Nations, over 150,000 km² of tropical rain forests—about 1 percent of the total resource—are being converted each year to other land uses, mostly agricultural ones. In 1950, forests covered approximately 25 percent of Haiti (a Caribbean island country the size of Maryland); its forested area now stands at less than 2 percent (**Figure 23.34**). Other developing nations face similar problems.

Given these rates of habitat loss, it is not surprising that the number of extant species—the most important measure of *biodiversity*—is declining. Biologists estimate that there are over 10 million different species alive on the planet today, although only 1.5 million have been officially classified. Extinction rates are difficult to quantify, but most knowledgeable scientists believe that up to one-fifth of all species will disappear during the next 30 years, and that as many as one-half may go extinct during the twenty-first century. One respected biologist, Peter Raven, has put the problem bluntly:

We are confronting an episode of species extinction greater than anything the world has experienced for the past 65 million years. Of all the global problems that confront us, this is the one that is moving the most rapidly and the one that will have the most serious consequences. And, unlike other global ecological problems, it is completely irreversible.

Some observers, such as sociobiologist E. O. Wilson, have gone so far as to call the current rapid worldwide decline in biodiversity the "Sixth Extinction," placing it in the same rank as the "Big Five" mass extinctions of the Phanerozoic eon (see Figure 11.17). Others consider this extrapolation to be premature, however, because even the rapid losses in biodiversity we see today will not necessarily affect the fossil record as profoundly as, say, the mass extinction at the end of the Cretaceous period, or even the less severe mass extinction associated with global warming at the Paleocene-Eocene boundary.

■ Earth System Engineering and Management

By any measure, the problems we face in confronting global change are daunting. If the human population and its per capita energy use continue to grow at their current rates, our continuing reliance on fossil fuels will cause the rate of carbon emissions to nearly double in 50 years, from 8 Gt per year in 2010 to at least 15 Gt per year in 2060. Under IPCC's extreme scenario A, the CO_2 concentration in the atmosphere could exceed 600 ppm and continue to increase thereafter, with the potentially disastrous consequences we have described. Controlling our carbon emissions—perhaps civilization's most important task—will require an extraordinary collaboration of Earth scientists, policy makers, and the public.

Energy Policy

There is little question that we will need to make changes in the energy sources we use and the ways in which we use them. One set of questions policy makers must tackle is how much money we should spend to curb anthropogenic carbon emissions, and whether the benefits of doing so will justify the costs. Too much spending could depress the economy and cause job losses, yet preventing the most drastic effects of climate change might be less costly than coping with those disasters after they happen.

A partial solution—and certainly the most economical one—is to improve energy use efficiency and reduce waste. In a real sense, using energy more efficiently is like discovering a new source of fuel. Some experts believe that the United States could reduce its emissions of greenhouse gases by as much as 50 percent from current levels by implementing efficiency measures that cost relatively little—for example, insulating buildings, replacing incandescent light bulbs with fluorescent bulbs, increasing the fuel efficiency of motor vehicles, and making greater use of natural gas. The savings in energy costs could amount to hundreds of billions of dollars per year. These modest steps

would offer other fringe benefits as well, including lowered manufacturing costs and improved air quality.

Many observers would say that fossil fuels are simply too cheap in the United States. Carbon emissions are not taxed at a national level, as they are in many other developed nations; consequently, there is little incentive for energy conservation or conversion to new energy sources. The full economic costs of fossil fuels include the costs of cleaning up atmospheric pollution, oil spills, and other environmental damage; the costs of trade deficits; and the military costs of defending oil supplies, as well as the costs of global warming. If these costs were included in energy pricing, alternative energy sources would become much more competitive with fossil fuels. Such *full-cost accounting* has not been politically popular in the United States, however.

We also face the issue of fairness in international politics. The United States, Canada, the European Union, and Japan—with less than one-quarter of the world's population—are responsible for about three-quarters of the global increase in atmospheric greenhouse gas concentrations. These rich industrial nations are better able to pay the costs of reducing their greenhouse gas emissions than the developing countries. China, for example, depends on its huge coal deposits for its rapid economic growth; it became the world's leader in greenhouse gas emissions in 2007 (**Figure 23.35**). Developing nations argue that they will need financial and technological support from the developed countries to help them reduce emissions. Policy

FIGURE 23.35 ■ A large coal-fired power plant near Ordos, a city in northern China. In 2007, China replaced the United States as the nation with the highest rate of greenhouse gas emissions. The carbon economies of China, India, and other developing countries will have a huge influence on future climates. [ZumaWire/Newscom.]

makers have come to agree that the problems of global climate change cannot be solved on a national scale and will have to be addressed through international cooperation.

Use of Alternative Energy Resources

As we have seen, no one alternative energy source will be able to replace fossil fuels quickly. However, some renewable energy resources, such as solar power, wind power, and biofuels, are becoming more important contributors to our energy system. If these technologies were aggressively implemented during the next 50 years, together they could reduce carbon emissions by gigatons per year.

Another step that could be taken is to increase the use of nuclear energy. The capacity of nuclear power plants, which today is approximately 350 gigawatts, could easily be tripled in the next 50 years, but this option is unattractive to many people for the reasons we have discussed. The potential exists for cleaner nuclear technologies, such as *fusion power:* the use of small, controlled thermonuclear explosions to generate energy. But scientific progress toward this goal has been slow, and conceptual breakthroughs will be required.

Engineering the Carbon Cycle

What about the possibility of engineering the carbon cycle to reduce the accumulation of greenhouse gases in the atmosphere? Several promising technologies aim to reduce greenhouse gas emissions by pumping the CO_2 generated by fossil-fuel combustion into reservoirs other than the atmosphere—a procedure known as **carbon sequestration.**

One obvious alternative reservoir for carbon is the biosphere. In Chapter 15, we saw that forests withdraw CO_2 from the atmosphere in surprisingly large amounts. Land-use policies that would not only slow the current high rates of deforestation but also encourage reforestation and other biomass production might help to mitigate anthropogenic climate change.

Biotechnology might provide some ways of increasing the capacity of the biosphere to sequester carbon. One possibility is the engineering of genetically modified bacteria that would be capable of metabolizing methane, sequestering the carbon it contains, and giving off hydrogen. Hydrogen is the ultimate clean fuel; burning it produces only water.

Another controversial possibility is fertilization of the marine biosphere. We know that *phytoplankton* (small photosynthetic marine organisms) take up CO_2 from the atmosphere by photosynthesis. In most regions of the ocean, phytoplankton productivity is limited by the lack of nutrients, such as iron. Preliminary experiments in the 1990s suggested that the growth of phytoplankton could be stimulated by dumping modest amounts of iron into the ocean. Unfortunately, it appears that fertilizing the ocean

in this manner also stimulates the growth of animals that eat the phytoplankton and quickly return the CO_2 to the atmosphere.

One straightforward technology for carbon sequestration—underground storage of CO_2—offers considerable promise. Carbon dioxide captured from oil and gas wells is already being pumped back into the ground as a means of moving oil toward the wells. If capture and underground storage of the CO_2 from coal-fired power plants were economically feasible, the world's abundant coal resources would become much more attractive as a replacement for petroleum.

Stabilizing Carbon Emissions

The strategies and technologies we have just discussed may seem promising, but will they be enough? Under the IPCC's scenario A, carbon emissions are expected to increase by at least 7 Gt per year during the next half century. How can this increase be stopped? In other words, what would it take to *stabilize carbon emissions at current levels?*

Two scientists from Princeton University, Stephen Pacala and Robert Socolow, have provided a simple quantitative framework to address this particular problem. They begin by admitting that there is no single solution to the problem—no "silver bullet." Instead, they break the problem into what they call **stabilization wedges,** each of which offsets the projected growth of carbon emissions by 1 Gt per year in the next 50 years (**Figure 23.36**). Therefore, one wedge roughly corresponds to one-seventh of the solution.

Implementing each stabilization wedge will be a monumental task. To achieve wedge 1, for example, the average gasoline mileage of the world's entire fleet of passenger vehicles, which will grow to 2 billion by mid-century, will have to be steadily increased from 30 miles per gallon (mpg) to 60 mpg. This calculation assumes that a car is driven 10,000 miles per year, the current annual average. An alternative, not shown in the figure, would be to maintain gas mileage at 30 mpg but reduce the average amount of driving by half to 5000 miles per year. Yet another alternative (wedge 2) would be to convert all cars to biofuels. Growing that much biofuel would take up one-sixth of the world's total cropland, so this strategy could adversely affect agricultural productivity and food supplies.

Some of the stabilization wedges involve controversial or expensive technologies, such as expanding nuclear power by a factor of three (wedge 3), increasing the number of large windmills into the millions (wedge 4), or covering large desert areas with solar panels (wedge 5). At least one of the proposed wedges, the capture and storage of carbon emitted from coal-fired power plants (wedge 6), is at the margin of current technological feasibility. The last option, elimination of tropical deforestation and the reforestation of huge additional land areas (wedge 7), is favored by many people in principle, but would be difficult to achieve without imposing severe restrictions on developing countries such as Brazil.

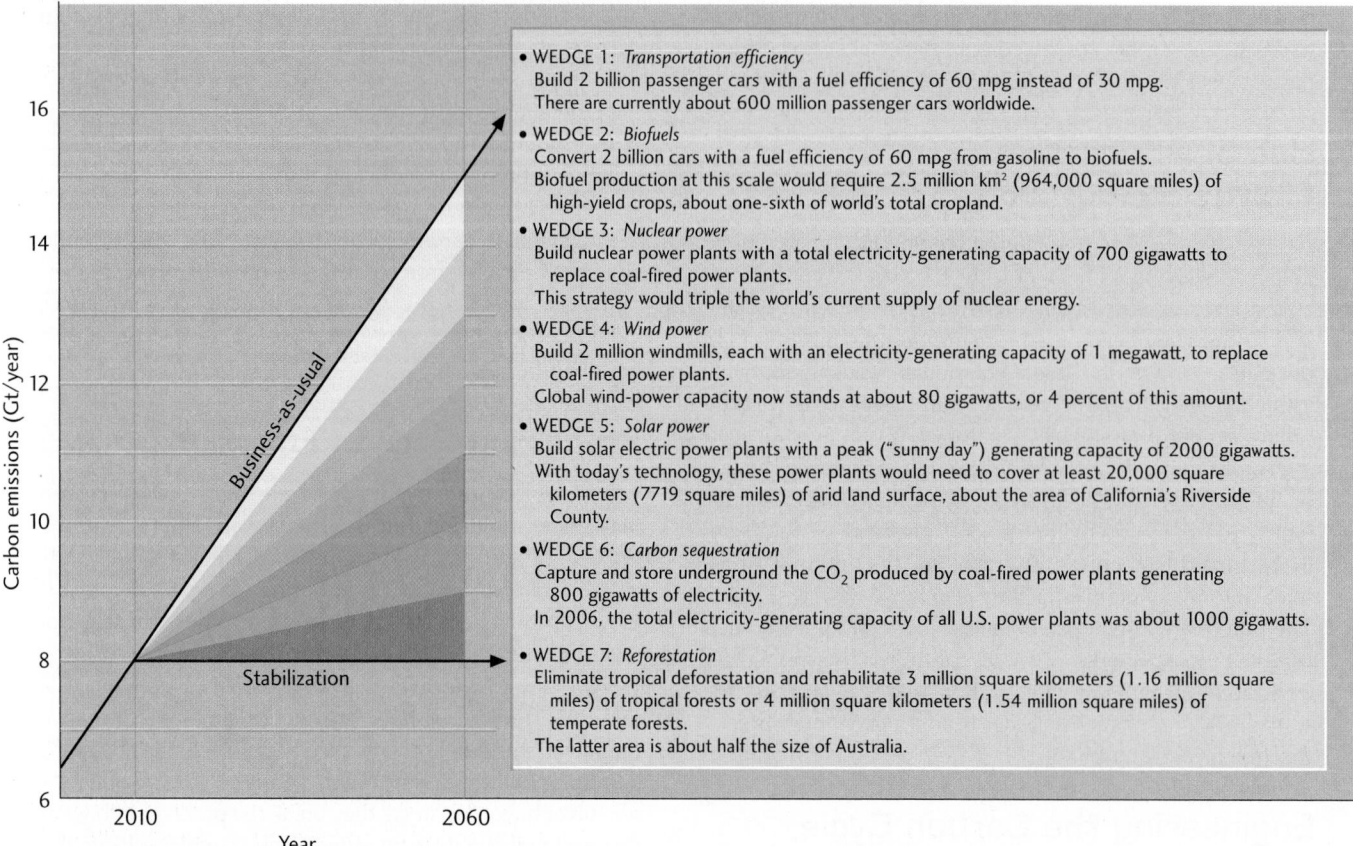

FIGURE 23.36 ■ Under IPCC scenario A, carbon emissions are expected to increase by at least 7 Gt per year during the next 50 years. The problem of stabilizing carbon emissions at their 2010 level of 8 Gt/year can be broken into seven stabilization wedges, each representing a reduction in emissions of 1 Gt per year by 2060. Possible actions that use existing technologies to achieve one-wedge reductions are listed next to each wedge. [Modified from S. Pacala, R. Socolow. "Stabilization Wedges: Solving the Climate Problem for the Next 50 Years with Current Technologies." *Science*, 305: 968–972 (2004).]

The stabilization of carbon emissions at current emission rates would reduce, but not eliminate, the threat of global climate change. The 50-year stabilization scenario (which is intermediate to scenarios B and C in Figure 23.36) would still allow the atmospheric concentration of CO_2 to grow to 500 ppm, almost twice the preindustrial value. Further reductions in carbon emissions during the second half of the twenty-first century would be necessary to maintain atmospheric concentrations below that value. Climate models indicate that such a scenario would still increase the average global temperature by about 2°C, more than three times the total twentieth-century warming.

Nevertheless, the continued rise of atmospheric CO_2 concentrations is not inevitable. The available inventory of stabilization wedges constitutes a technological framework for concerted action by governments. Taking on the stabilization problem involves other difficulties such as developing broad public consensus and creating international agreements. Yet, as the Pacala-Socolow analysis demonstrates, there is still time for actions that can substantially reduce anthropogenic global change. Whether we can grasp this opportunity will depend on our understanding of the problem, its potential solutions, and the consequences of inaction.

Sustainable Development

The term **sustainable development** appears with increasing frequency in newspapers, public debates, classroom discussions, and scholarly journals. The concept was popularized in *Our Common Future*, a 1987 report by the World Commission on Environment and Development (also known as the Brundtland Commission), where it was defined as "development that meets the needs of the present without compromising the ability of future generations to meet their own needs." Sustainable development is difficult to define more precisely, but it offers an appealing, if somewhat utopian, vision: a civilization that carefully manages its interactions with the Earth system to ensure a hospitable environment for future generations.

Sustainability involves many economic and political issues about which many nations do not agree, so forging a global strategy that moves civilization toward this goal will not be easy. As a prerequisite, Earth science will have to provide better knowledge of how geosystems operate, interact, and are perturbed by human activities.

As French novelist Marcel Proust once wrote, "The real voyage of discovery consists not in seeking new lands, but in seeing with new eyes." We hope this textbook has given you new eyes to see the critical issue of global change and the other problems of Earth science that confront your generation.

SUMMARY

In what sense is human civilization a global geosystem? Human society has harnessed the means of energy production on a global scale and now competes with the plate tectonic and climate systems in modifying Earth's surface environment. Most of the energy used by human civilization today comes from carbon-based fuels. The rise of this carbon economy has altered the natural carbon cycle by creating a huge new flux of carbon from the lithosphere to the atmosphere. If that flow continues unabated, CO_2 concentrations in the atmosphere will double by the mid-twenty-first century.

How do we categorize our natural resources? Natural resources can be classified as renewable or nonrenewable, depending on whether they are replenished by geologic processes at rates comparable to the rates at which we are consuming them. Reserves are the known supplies of natural resources that can be exploited economically under current conditions.

What is the origin of oil and natural gas? Oil and natural gas form from organic matter deposited in oxygen-poor sedimentary basins, typically on continental margins. These organic materials are buried as the sedimentary layers grow in thickness. Under elevated temperatures and pressures, the buried organic matter is transformed into liquid and gaseous hydrocarbons. Oil and gas accumulate where geologic structures called oil traps create impermeable barriers to their upward migration.

Why is there concern about the world's oil supply? Oil is a nonrenewable resource: at current rates of use, it is being depleted far faster than geologic processes can replenish it. Therefore, as oil is withdrawn from the hydrocarbon reservoirs of the world, its availability will diminish and its price will rise. The key issue is not when oil will run out, but when global oil production will reach Hubbert's peak—when it will stop rising and begin to decline. The current data supports the oil optimists, who argue that oil resources will meet demand for decades to come.

What is the origin of coal, and what are the consequences of burning it? Coal is formed by the burial, compression, and diagenesis of wetland vegetation. There are huge resources of coal in sedimentary rocks. Coal combustion is a major source of atmospheric CO_2 as well as sulfur-containing gases that contribute to acid rain. Furthermore, coal mining and toxic substances produced by coal burning present risks to human life and to the environment. Because of its abundance and low cost, however, the use of coal is likely to increase on a global scale.

What are the prospects for alternative energy sources? Alternative energy sources include nuclear power, biofuels, and solar, hydroelectric, wind, and geothermal energy. Taken together, these energy sources currently supply only a small percentage of world energy demand. Nuclear energy produced by the fission of uranium, the world's most abundant minable energy resource, could be a major energy source, but only if the public can be assured of its safety and security. With advances in technology and reductions in cost, renewable sources such as solar energy, wind energy, and biomass could become major contributors in the twenty-first century.

How much global warming will there be in the twenty-first century, and what will be its consequences? Atmospheric concentrations of greenhouse gases will continue to rise throughout the twenty-first century, primarily because of fossil-fuel combustion and other human activities. The magnitude of the increase will depend on whether human society takes active steps to limit its greenhouse gas emissions. Projections of climate warming during the twenty-first century are highly uncertain, but the range of likely warming is 0.5°C to 5.5°C. This warming will disrupt ecosystems and increase the rate of species extinctions. The oceans will warm and expand, raising sea level as much as a meter. The Arctic ice cap will continue to shrink rapidly, and much of the Arctic Ocean is expected to become ice-free.

What other types of anthropogenic global change are degrading our environment? Ocean acidification is decreasing the ability of shellfish and corals to calcify their shells and skeletons and may adversely affect many other types of marine organisms, disrupting marine ecosystems. The biodiversity of ecosystems on land is declining through loss of habitat as well as the effects of global warming. The current rapid rate of species extinction may eventually lead to a decline in biodiversity equal to major mass extinctions of the past.

How might we stabilize carbon emissions at their current levels? If human civilization continues to rely on fossil fuels, anthropogenic carbon emissions will increase by at least 7 Gt per year during the next 50 years. This problem could be addressed by implementing seven stabilization wedges, each of them a strategy for reducing the projected growth of carbon emissions by 1 Gt per year.

KEY TERMS AND CONCEPTS

Anthropocene (p. 666)

biofuel (p. 657)

carbon economy (p. 645)

carbon intensity (p. 653)

carbon sequestration
(p. 669)

fossil fuel (p. 644)

global change (p. 660)

hydraulic fracturing
(fracking) (p. 651)

hydroelectric energy
(p. 658)

natural resource (p. 642)

nonrenewable resource
(p. 642)

nuclear energy (p. 655)

oil trap (p. 647)

oil window (p. 647)

quad (p. 645)

renewable resource
(p. 642)

reserve (p. 643)

resource (p. 643)

solar energy (p. 658)

stabilization wedge
(p. 669)

sustainable development
(p. 670)

EXERCISES

1. Describe some of the ways in which human civilization is fundamentally different from the natural geosystems we have studied in this textbook.

2. Which fossil fuel produces the least amount of CO_2 per unit of energy: oil, natural gas, or coal? Which produces the most?

3. What are the prerequisites for oil traps to contain oil?

4. Explain which of the following factors are important in estimating the future supply of oil and natural gas: (a) the rate of oil and gas accumulation, (b) the rate of depletion of known reserves, (c) the rate of discovery of new reserves, (d) the total amount of oil and gas now present on Earth.

5. An aggressive drilling program in the Arctic National Wildlife Refuge could produce as much as 16 billion barrels of oil. At current consumption rates, for how many years would this resource supply U.S. oil demand?

6. Which three countries have the largest coal reserves?

7. If we keep pumping CO_2 into the atmosphere and Earth's climate warms significantly in the next 100 years, how might the global carbon cycle be affected?

8. An economist once wrote: "The predicted change in global temperature due to human activity is less than the difference in winter temperature between New York and Florida, so why worry?" Should he worry? Why or why not?

THOUGHT QUESTIONS

1. In what ways does Earth's internal heat engine contribute to the formation of fossil-fuel resources?

2. Are you an oil optimist or an oil pessimist? Explain why.

3. What issues related to the use of nuclear energy can be addressed by geologists?

4. Contrast the risks and benefits of nuclear fission and coal combustion as energy sources.

5. What do you think will be the major sources of the world's energy in the year 2030? In the year 2100?

6. Do you think we should act now to reduce carbon emissions or delay until the functioning of the climate system is better understood?

7. Is the United States justified in insisting that developing countries that now use much less fossil fuel than developed countries agree to limit their future carbon emissions?

8. Do you think that future scientists and engineers will be able to modify the natural carbon cycle to prevent catastrophic changes in the climate system?

9. Do you think a geologist several thousand years in the future will consider the industrial revolution the beginning of a new geologic epoch?

MEDIA SUPPORT

 23-1 Animation: Oil Field Formation

LENGTH

1 centimeter	0.3937 inch
1 inch	2.5400 centimeters
1 meter	3.2808 feet; 1.0936 yards
1 foot	0.3048 meter
1 yard	0.9144 meter
1 kilometer	0.6214 mile (statute); 3281 feet

LENGTH

1 mile (statute)	1.6093 kilometers
1 mile (nautical)	1.8531 kilometers
1 fathom	6 feet; 1.8288 meters
1 angstrom	10^{-8} centimeter
1 micrometer	0.0001 centimeter

VELOCITY

1 kilometer/hour	27.78 centimeters/second
1 mile/hour	17.60 inches/second

AREA

1 square centimeter	0.1550 square inch
1 square inch	6.452 square centimeters
1 square meter	10.764 square feet; 1.1960 square yards
1 square foot	0.0929 square meter
1 square kilometer	0.3861 square mile
1 square mile	2.590 square kilometers
1 acre (U.S.)	4840 square yards

VOLUME

1 cubic centimeter	0.0610 cubic inch
1 cubic inch	16.3872 cubic centimeters
1 cubic meter	35.314 cubic feet
1 cubic foot	0.02832 cubic meter
1 cubic meter	1.3079 cubic yards
1 cubic yard	0.7646 cubic meter
1 liter	1000 cubic centimeters; 1.0567 quarts (U.S. liquid)
1 gallon (U.S. liquid)	3.7853 liters

MASS

1 gram	0.03527 ounce
1 ounce	28.3495 grams
1 kilogram	2.20462 pounds
1 pound	0.45359 kilogram

PRESSURE

1 kilogram/square centimeter	0.96784 atmosphere; 0.98067 bar; 14.2233 pounds/square inch
1 bar	0.98692 atmosphere; 10^5 pascals

ENERGY

1 joule	0.239 calorie; 9.479×10^{-4} Btu
1 British thermal unit (Btu)	251.9 calories; 1054 joules
1 quad	10^{15} Btu

POWER

1 watt	0.001341 horsepower (U.S.); 3.413 Btu/hour

Equatorial radius	6378 kilometers
Polar radius	6357 kilometers
Radius of sphere with Earth's volume	6371 kilometers
Volume	1.083×10^{27} cubic centimeters
Surface area	5.1×10^{18} square centimeters
Percent surface area of oceans	71
Percent surface area of land	29
Average elevation of land	623 meters
Average depth of oceans	3.8 kilometers
Mass	5.976×10^{27} grams
Density	5.517 grams/cubic centimeters
Gravity at equator	978.032 centimeters/second/second
Mass of atmosphere	5.1×10^{21} grams
Mass of ice	$25–30 \times 10^{21}$ grams
Mass of oceans	1.4×10^{24} grams
Mass of crust	2.5×10^{25} grams
Mass of mantle	4.05×10^{27} grams
Mass of core	1.90×10^{27} grams
Mean distance to Sun	1.496×10^8 kilometers
Mean distance to Moon	3.844×10^5 kilometers
Ratio: Mass of Sun/mass of Earth	3.329×10^5
Ratio: Mass of Earth/mass of Moon	81.303
Total geothermal energy reaching Earth's surface each year	10^{21} joule; 2.39×10^{20} calories; 949 quads
Earth's daily receipt of solar energy	14,137 quads; 1.49×10^{22} joules
U.S. energy consumption, 2012	95 quads

Electron Shells and Ion Stability

Electrons surround the nucleus of an atom of each element in a unique set of concentric spheres called electron shells. Each shell can hold a certain maximum number of electrons. In the chemical reactions of most elements, only the electrons in the outermost shells interact. In the reaction between sodium (Na) and chlorine (Cl) that forms sodium chloride (NaCl), for example, the sodium atom loses an electron from its outer shell of electrons, and the chlorine atom gains an electron in its outer shell (see Figure 3.4).

Before reacting with chlorine, the sodium atom has one electron in its outer shell. When it loses that electron, its outer shell is eliminated and the next shell inward, which has eight electrons (the maximum that shell can hold), becomes the outer shell. The original chlorine atom had seven electrons in its outer shell, with room for a total of eight. By gaining an

electron, it fills its outer shell. Many elements have a strong tendency to acquire a full outer electron shell, some by gaining electrons and some by losing them in the course of a chemical reaction.

Many chemical reactions entail gains and losses of several electrons as two or more elements combine. The element calcium (Ca), for example, becomes a doubly charged cation, Ca^{2+}, as it reacts with two chlorine atoms to form calcium chloride. In the chemical formula for calcium chloride, $CaCl_2$, the presence of two chloride ions is symbolized by the subscript 2. Chemical formulas thus show the relative proportions of atoms or ions in a compound. Common practice is to omit the subscript 1 next to single ions in a formula.

The periodic table organizes the elements (from left to right along a row) in order of atomic number (the number of protons), which also means increasing the numbers of electrons in the outer shell. The third row from the top, for exam-

The periodic table.

ple, starts at the left with sodium (atomic number 11), which has one electron in its outer shell. The next is magnesium (atomic number 12), which has two electrons in its outer shell, followed by aluminum (atomic number 13), with three, and silicon (atomic number 14), with four. Then come phosphorus (atomic number 15), with five; sulfur (atomic number 16), with six; and chlorine (atomic number 17), with seven. The last element in this row is argon (atomic number 18), with eight electrons, the maximum possible, in its outer shell. Each column in the table forms a vertical grouping of elements with similar electron-shell patterns.

Elements That Tend to Lose Electrons

The elements in the leftmost column of the table all have a single electron in their outer shells and have a strong tendency to lose that electron in chemical reactions. Of this group, hydrogen (H), sodium (Na), and potassium (K) are found in major abundance at Earth's surface and in its crust.

The second column from the left includes two more elements that are abundant on Earth: magnesium (Mg) and calcium (Ca). Elements in this column have two electrons in their outer shells and a strong tendency to lose both of them in chemical reactions.

Elements That Tend to Gain Electrons

Toward the right side of the table, the two columns headed by oxygen (O), the most abundant element on Earth, and fluorine (F), a highly reactive toxic gas, contain elements that tend to gain electrons in their outer shells. The elements in the column headed by oxygen have six of the possible eight electrons in their outer shells and tend to gain two electrons. Those in the column headed by fluorine have seven electrons in their outer shells and tend to gain one.

Other Elements

The elements in the columns between those farthest to the left and those farthest to the right have varying tendencies to gain, lose, or share electrons. The column toward the right of the table headed by carbon (C) includes silicon (Si), another of the most abundant elements on Earth. Both silicon and carbon tend to share electrons.

The elements in the last column on the right, headed by helium (He), have full outer shells and thus no tendency either to gain or to lose electrons. As a result, these elements, in contrast with those in other columns, do not react chemically with other elements, except under very special conditions.

	Mineral or Group Name	Structure or Composition	Varieties and Chemical Composition	Form, Diagnostic Characteristics	Cleavage, Fracture		Hardness
LIGHT-COLORED MINERALS, VERY ABUNDANT IN EARTH'S CRUST IN ALL MAJOR ROCK TYPES	FELDSPAR	FRAMEWORK SILICATES	*ORTHOCLASE FELDSPARS* $KAlSi_3O_8$ Sanidine Orthoclase Microcline	Cleavable coarsely crystalline or finely granular masses; isolated crystals or grains in rocks, most commonly not showing crystal faces	Two at right angles, one perfect and one good; pearly luster on perfect cleavage	White to gray, frequently pink or yellowish; some green	6
			PLAGIOCLASE FELDSPARS $NaAlSi_3O_8$ Albite $CaAl_2Si_2O_8$ Anorthite		Two at nearly right angles, one perfect and one good; fine parallel striations on perfect cleavage	White to gray, less commonly greenish or yellowish	
	QUARTZ		SiO_2	Single crystals or masses of 6-sided prismatic crystals; also formless crystals and grains or finely granular or massive	Very poor or nondetectable; conchoidal fracture	Colorless, usually transparent; also slightly colored smoky gray, pink, yellow	7
	MICA	SHEET SILICATES	*MUSCOVITE* $KAl_2(AlSi_3O_{10})(OH)_2$	Thin, disc-shaped crystals, some with hexagonal outlines; dispersed or aggregates	One perfect; splittable into very thin, flexible, transparent sheets	Colorless; slight gray or green to brown in thick pieces	$2-2\frac{1}{2}$
DARK-COLORED MINERALS, ABUNDANT IN MANY KINDS OF IGNEOUS AND METAMORPHIC ROCKS			*BIOTITE* $K(Mg,Fe)_3AlSi_3(OH)_2$	Irregular, foliated masses; scaly aggregates	One perfect; splittable into thin, flexible sheets	Black to dark brown; translucent to opaque	$2\frac{1}{2}-3$
			CHLORITE $(Mg,Fe)_5(Al,Fe)_2Si_3O_{10}(OH)_8$	Foliated masses or aggregates of small scales	One perfect; thin sheets flexible but not elastic	Various shades of green	$2-2\frac{1}{2}$
	AMPHIBOLE	DOUBLE CHAINS	*TREMOLITE-ACTINOLITE* $Ca_2(Mg,Fe)_5Si_8O_{22}(OH)_2$	Long, prismatic crystals, usually 6-sided; commonly in fibrous masses or irregular aggregates	Two perfect cleavage directions at 56° and 124° angles	Pale to deep green Pure tremolite white	5–6
			HORNBLENDE Complex Ca, Na, Mg, Fe, Al silicate				
	PYROXENE	SINGLE CHAINS	*ENSTATITE-HYPERSTHENE* $(Mg,Fe)_2Si_2O_6$	Prismatic crystals, either 4- or 8-sided; granular masses and scattered grains	Two good cleavage directions at about 90°	Green and brown to grayish or greenish white	5–6
			DIOPSIDE $(Ca,Mg)_2Si_2O_6$			Light to dark green	
			AUGITE Complex Ca, Na, Mg, Fe, Al silicate			Very dark green to black	

(Continued)

Mineral or Group Name	Structure or Composition	Varieties and Chemical Composition	Form, Diagnostic Characteristics	Cleavage, Fracture	Color	Hardness
OLIVINE	ISOLATED TETRAHEDRA	$(Mg,Fe)_2SiO_4$	Granular masses and disseminated small grains	Conchoidal fracture	Olive to grayish green and brown	$6\frac{1}{2}$–7
GARNET		Ca, Mg, Fe, Al silicate	Isometric crystals, well formed or rounded; high specific gravity, 3.5–4.3	Conchoidal and irregular fracture	Red and brown, less commonly pale colors	$6\frac{1}{2}$–7
CALCITE	CARBONATES	$CaCO_3$	Coarsely to finely crystalline in beds, veins, and other aggregates; cleavage faces may show in coarser masses; calcite effervesces rapidly in acid, but dolomite effervesces slowly and only if crushed into powder	Three perfect cleavages, at oblique angles; splits to rhombohedral cleavage pieces	Colorless, transparent to translucent; variously colored by impurities	3
DOLOMITE		$CaMg(CO_3)_2$				$3\frac{1}{2}$–4
CLAY MINERALS	HYDROUS ALUMINO-SILICATES	KAOLINITE $Al_2Si_2O_5(OH)_4$; ILLITE Similar to muscovite $+Mg,Fe$; SMECTITE Complex Ca, Na, Mg, Fe, Al silicate $+ H_2O$	Earthy masses in soils; bedded; in association with other clays, iron oxides, or carbonates; plastic when wet; montmorillonite swells when wet	Earthy, irregular	White to light gray and buff; also gray to dark gray, greenish gray, and brownish depending on impurities and associated minerals	$1\frac{1}{2}$–$2\frac{1}{2}$
GYPSUM	SULFATES	$CaSO_4 \cdot 2H_2O$	Granular, earthy, or finely crystalline masses; tabular crystals	One perfect, splitting to fairly thin slabs or sheets; two other good cleavages	Colorless to white; transparent to translucent	2
ANHYDRITE		$CaSO_4$	Massive or crystalline aggregates in beds and veins	One perfect, one nearly perfect, one good; at right angles	Colorless, some tinged with blue	3–$3\frac{1}{2}$
HALITE	HALIDES	NaCl	Granular masses in beds; some cubic crystals; salty taste	Three perfect cleavages at right angles	Colorless, transparent to translucent	$2\frac{1}{2}$
OPAL-CHALCEDONY	SILICA	SiO_2 [Opal is an amorphous variety; chalcedony is a formless microcrystalline quartz.]	Beds in siliceous sediments and chert; in veins or banded aggregates	Conchoidal fracture	Colorless or white when pure, but tinged with various colors by impurities in bands, especially in agates	5–$6\frac{1}{2}$
MAGNETITE	IRON OXIDES	Fe_3O_4	Magnetic; disseminated grains, granular masses; occasional octahedral isometric crystals; high specific gravity, 5.2	Conchoidal or irregular fracture	Black, metallic luster	6

LIGHT-COLORED MINERALS, TYPICALLY AS ABUNDANT CONSTITUENTS OF SEDIMENTS AND SEDIMENTARY ROCKS

DARK-COLORED MINERALS, COMMON IN MANY ROCK TYPES

Mineral or Group Name	Structure or Composition	Varieties and Chemical Composition	Form, Diagnostic Characteristics	Cleavage, Fracture	Color	Hardness
HEMATITE	IRON OXIDES	Fe_2O_3	Earthy to dense masses, some with rounded forms, some granular or foliated; high specific gravity, 4.9–5.3	None; uneven, sometimes splintery fracture	Reddish brown to black	5–6
"LIMONITE"		*GOETHITE* [the major mineral of the mixture called "limonite", a field term] $FeO(OH)$	Earthy masses, massive bodies or encrustations, irregular layers; high specific gravity, 3.3–4.3	One excellent in the rare crystals; usually an early fracture	Yellowish brown to dark brown and black	$5–5\frac{1}{2}$
KYANITE	ALUMINO-SILICATES	Al_2SiO_5	Long, bladed or tabular crystals or aggregates	One perfect and one poor, parallel to length of crystals	White to light-colored or pale blue	5 parallel to crystal length 7 across crystals
SILLIMANITE		Al_2SiO_5	Long, slender crystals or fibrous, felted masses	One perfect parallel to length, not usually seen	Colorless, gray to white	6–7
ANDALUSITE		Al_2SiO_5	Coarse, nearly square prismatic crystals, some with symmetrically arranged impurities	One distinct; irregular fracture	Red, reddish brown olive–green	$7\frac{1}{2}$
FELDSPATHOIDS		*NEPHELINE* $(Na,K)AlSiO_4$	Compact masses or as embedded grains, rarely as small prismatic crystals	One distinct; irregular fracture	Colorless, white, light gray; gray-greenish in masses, with greasy luster	$5\frac{1}{2}–6$
		LEUCITE $KAlSi_2O_6$	Trapezohedral crystals embedded in volcanic rocks	One very imperfect	White to gray	$5\frac{1}{2}–6$
SERPENTINE		$Mg_3Si_4O_{10}(OH)_8$	Fibrous (asbestos) or platy masses	Splintery fracture	Green; some yellowish brownish or gray; waxy or greasy luster in massive habit; silky luster in fibrous habit	4–6
TALC		$Mg_3Si_4O_{10}(OH)_2$ masses or aggregates	Foliated or compact masses or aggregates	One perfect, making thin flakes or scales; soapy feel	White to pale green; pearly or greasy luster	1
CORUNDUM		Al_2O_3	Some rounded barrel-shaped crystals; most often as disseminated grains or granular masses (emery)	Irregular fracture	Usually brown, pink, or blue; emery black. Gemstone varieties: ruby, sapphire	9

LIGHT-COLORED MINERALS, MAINLY IN METAMORPHIC AND IGNEOUS ROCKS AS COMMON OR MINOR CONSTITUENTS

	Mineral or Group Name	Structure or Composition	Varieties and Chemical Composition	Form, Diagnostic Characteristics	Cleavage, Fracture	Color	Hardness
DARK-COLORED MINERALS, COMMON IN METAMORPHIC ROCKS	EPIDOTE	SILICATES	$Ca_2(Al,Fe)Al_2Si_3O_{12}(OH)$	Aggregates of long prismatic crystals, granular or compact masses, embedded grains	One good, one poor at greater than right angles; conchoidal and irregular fracture	Green, yellow-green, gray, some varieties dark brown to black	$6–7$
	STAUROLITE		$Fe_2Al_9Si_4O_{22}(O,OH)_2$	Short prismatic crystals, some cross-shaped, usually coarser than matrix of rock	One poor	Brown, reddish, or dark brown to black	$7–7\frac{1}{2}$
METALLIC LUSTER, COMMON IN MANY ROCK TYPES, ABUNDANT IN VEINS	PYRITE	SULFIDES	FeS_2	Granular masses or well-formed cubic crystals in veins and beds or disseminated; high specific gravity, 4.9–5.2	Uneven fracture	Pale brass-yellow	$6–6\frac{1}{2}$
	GALENA		PbS	Granular masses in veins and disseminated; some cubic crystals; very high specific gravity, 7.3–7.6	Three perfect cleavages at mutual right angles; giving cubic cleavage fragments	Silver-gray	$2\frac{1}{2}$
	SPHALERITE		ZnS	Granular masses or compact crystalline aggregates; high specific gravity, 3.9–4.1	Six perfect cleavages at 60° to one another	White to green, brown and black; resinous to submetallic luster	$3\frac{1}{2}–4$
	CHALCOPYRITE		$CuFeS_2$	Granular or compact masses; disseminated crystals; high specific gravity, 4.1–4.3	Uneven fracture	Brassy to golden-yellow	$3\frac{1}{2}–4$
	CHALCOCITE		Cu_2S	Fine-grained masses; high specific gravity, 5.5–5.8	Conchoidal fracture	Lead-gray to black; may tarnish green or blue	$2\frac{1}{2}–3$
MINERALS FOUND IN MINOR AMOUNTS IN A VARIETY OF ROCK TYPES AND IN VEINS OR PLACERS	RUTILE	TITANIUM OXIDES	TiO_2	Slender to prismatic crystals; granular masses; high specific gravity, 4.25	One distinct, one less distinct; conchoidal fracture	Reddish brown, some yellowish, violet, or black	$6–6\frac{1}{2}$
	ILMENITE		$FeTiO_3$	Compact masses, embedded grains, detrital grains in sand; high specific gravity, 4.79	Conchoidal fracture	Iron-black; metallic to submetallic luster	$5–6$
	ZEOLITES	SILICATES	Complex hydrous silicates; many varieties of minerals, including analcime, natrolite, phillipsite, heulandite, and chabazite	Well-formed radiating crystals in cavities in volcanics, veins, and hot springs; also as fine-grained and earthy bedded deposits	One perfect for most	Colorless, white, some pinkish	$4–5$

Chapter 1 $1.08 \times 10^{12}\,\text{km}^3$

Chapter 2 The distance between the North American and Charleston, South Carolina, continental margin and the African continental margin near Dakar, Senegal, measured using the Google Earth Ruler tool, is about 6300 km. From the isochron map in Figure 2.15, you can estimate that the two continents began to rift apart about 200 million to 180 million years ago (see also Figure 2.16). Assuming that the continents rifted apart 200 million years ago gives

6300 km × 200 million years = 31.5 km/year = 31.5 mm/year

Assuming that the rifting age is 180 million years gives

6300 km × 180 million years = 35 km/year = 35 mm/year

This corresponds to answer (d) of Chapter 2 Google Earth Exercise 3.

Chapter 3 $163,200,000 - 120,000,000 = 43,200,000$ profit. So yes, it is worth it.

Chapter 4 Plagioclase feldspar will settle at a rate of 1.18 cm/hour, which is slower than olivine.

Chapter 5 125°C

Chapter 6 A change in pressure at constant temperature could indicate that the rocks were moved upward or downward in a subduction zone. Movements in subduction zones can be so fast that temperatures don't have time to change even though pressures may be changing quickly.

Chapter 7 The chances of encountering oil in the reservoir rock at point C are poor. From the geologic structure exposed at the surface, you can see that the anticline is plunging to the northeast with a dip of about 30°. Therefore, the depth to the sandstone reservoir rock is increasing to the northeast, and drilling at point C is likely to encounter water, not oil.

Chapter 8 The rock age is given by

$$T = \log(1.0143)/\log(2)$$
$$= 0.00617/0.301$$
$$= 0.0205 \text{ half-lives}$$

Multiplying this result by the half-life of rubidium-87 yields an age of

0.0205 × 49 billion years = 1.00 billion years

Chapter 9 The maximum variation in elevation at a landing site that could be tolerated by a lander with a fuel tank volume of 200 L is 750 m. The maximum acceptable variation if the final descent rate were 1 m/s instead of 2 m/s is 500 m.

Chapter 10 The fraction of relative plate movement taken up by thrust faulting is 20 mm/year ÷ 54 mm/year = 0.37. The remaining movement, about 60 percent of the total, is accommodated by faulting and folding north of the Himalaya, primarily by movement on the Altyn Tagh and other major strike-slip faults as China and Mongolia are pushed eastward (see Figure 10.16).

Chapter 11 Rock A: $R = 5$; Rock A does not record a distinctive signature of biological processes.

Rock C: $R = -50$; this large negative ratio, which is similar to that of the ratio for methane, does show a distinctive signature of biological processes.

Chapter 12 The production rate of the Hawaiian basalts is

$$100{,}000\,\text{km}^3 \times 1 \text{ million years} = 0.1\,\text{km}^3/\text{year}$$

The length of the Nazca-Pacific plate boundary needed to produce this amount is given by the following equation:

$$1.4 \times 10^{-4}\,\text{km/year} \times 7\,\text{km} \times length = 0.1\,\text{km}^3/\text{year}$$

or

$$length = 0.1\,\text{km}^3/\text{year} \div (1.4 \times 10^{-4}\,\text{km/year} \times 7\,\text{km}) = 102\,\text{km}$$

Chapter 13 A magnitude 6 rupture has 100 times the area of a magnitude 4 rupture (because $10^{(6-4)} = 10^2$) and 10 times the slip (because $10^{(6-4)/2} = 10^1$); therefore, it takes 100 × 10 = 1000 magnitude 4 earthquakes to equal a magnitude 6 earthquake.

Chapter 14 The appropriate isostatic equation is

Tibetan Plateau elevation = 0.15 × Tibetan Plateau crustal thickness − 0.12 × 7.0 km − 0.70 × 4.5 km

Solving for crustal thickness, we obtain

$$\text{Tibetan Plateau crustal thickness} = \frac{(\text{Tibetan Plateau elevation} + 0.12 \times 7.0\,\text{km} + 0.70 \times 4.5\,\text{km})}{0.15}$$

For an elevation of 5 km, this formula gives a crustal thickness of 60 km, which agrees with seismic data collected in the Tibetan Plateau.

Chapter 15 Carbon balance requires that

emissions – (atmosphere-ocean flux) –
(atmosphere-land surface flux) = atmospheric accumulation

For the 1990s, the numbers are

6.4 Gt/year – 2.2 Gt/year – 1.0 Gt/year = 3.2 Gt/year

For 2000–2005, the numbers are

7.2 Gt/year – 2.2 Gt/year – 0.9 Gt/year = 4.12 Gt/year

Comparing the two results we see that the rate at which carbon is accumulating in the atmosphere rose by 0.9 Gt/year, which is not good news!

Chapter 16

Table C

SAFETY FACTOR (F$_S$)

	Loose Soil	Slate	Granite
5°	3	10	50
20°	0.6	2	10
30°	0.12	0.4	2

A safety factor value greater than one indicates that a slope is stable enough to build on.

Chapter 17 Silty sand: 0.0015 m³/day
Well-sorted gravel: 18.75 m³/day

Chapter 18 10 feet = about 6000 cubic feet per second
25 feet = about 40,000 cubic feet per second

Chapter 19 Each box is 100 km².
Minimum area susceptible to desertification:
391 × 100 km² = 31,000 km²
Maximum area susceptible to desertification:
513 × 100 km² = 51,300 km²

Chapter 20 The 2002 maintenance fill cost about $25 million.

Chapter 21 200 mm ÷ 0.0001 = 2 × 10⁶ mm = 2 km

Chapter 22 Elevation = 250 m; age = 50,000 years

250 m/50,000 years = 0.005 m/year (5 mm/year)

GLOSSARY

Words in *italics* have separate entries in the Glossary. Specific minerals are defined and described in Appendix 4.

ablation   The total amount of ice that a *glacier* loses each year. (Compare *accumulation.*)

abrasion   The erosive action that occurs when suspended and saltating *sediment* particles move along the bottom and sides of a stream channel.

absolute age   The actual number of years elapsed from a geologic event until now. (Compare *relative age.*)

abyssal hill   A hill on the slope of a *mid-ocean ridge,* typically 100 m or so high and parallel to the ridge crest, formed primarily by normal faulting of newly formed oceanic *crust* as it moves out of a rift valley.

abyssal plain   A wide, flat plain that covers large areas of the ocean floor at depths of about 4000 to 6000 m.

accreted terrain   A piece of continental *crust,* tens to hundreds of kilometers in extent, with common characteristics and a distinct origin, usually transported great distances by plate movements and plastered onto the edge of a continent.

accretion   A process of continental growth in which buoyant fragments of *crust* are attached (accreted) to existing continental masses by horizontal transport during plate movements. (Compare *magmatic addition.*)

accumulation   The amount of snow added to a *glacier* annually. (Compare *ablation.*)

active margin   A *continental margin* where tectonic forces caused by plate movements are actively deforming the continental crust. (Compare *passive margin.*)

aftershock   An *earthquake* that occurs as a consequence of a previous earthquake of larger magnitude. (Compare *foreshock.*)

albedo   The fraction of *solar energy* reflected by a surface. (From the Latin *albus,* meaning "white.")

alluvial fan   A cone- or fan-shaped accumulation of *sediment* deposited where a *stream* widens abruptly as it leaves a mountain front and enters a broad, relatively flat *valley.*

amphibolite   (1) A usually *granoblastic rock* made up mainly of amphibole and plagioclase feldspar, typically formed by medium- to high-grade metamorphism of mafic volcanic rock. Foliated amphibolites can be produced by *deformation.* (2) The metamorphic grade above *greenschist.*

andesite   An *intermediate igneous rock* with a composition between that of *dacite* and that of *basalt;* the extrusive equivalent of *diorite.*

andesitic lava   A lava type of intermediate composition that has a higher silica content than *basalt,* erupts at lower temperatures, and is more viscous.

angle of repose   The maximum angle at which a slope of loose material can lie without sliding downhill.

anion   A negatively charged ion. (Compare *cation.*)

antecedent stream   A *stream* that existed before the present *topography* was created and so maintained its original course despite changes in the structure of the underlying *rocks* and in the topography. (Compare *superposed stream.*)

Anthropocene   The "Age of Man," a geologic epoch beginning about 1780, when the coal-powered steam engine launched the industrial revolution; proposed by atmospheric chemist Paul Crutzen to recognize the speed and magnitude of the changes industrial society is causing in the *Earth system.*

anticline   An archlike fold of layered *rocks* that contains older rock layers in the core of the fold. (Compare *syncline.*)

aquiclude   A relatively impermeable *formation* that bounds an *aquifer* above or below and acts as a barrier to the flow of *groundwater.*

aquifer   A porous *formation* that stores and transmits *groundwater* in sufficient quantity to supply wells.

arkose   A sandstone containing more than 25 percent feldspar.

artesian flow   A spontaneous flow of *groundwater* through a confined *aquifer* to a point where the elevation of the ground surface is lower than that of the *groundwater table.*

ash-flow deposit   An extensive sheet of hard volcanic *tuff* produced by a continental eruption of *pyroclasts.*

asteroid   One of the more than 10,000 small celestial bodies orbiting the Sun, most of them between the orbits of Mars and Jupiter.

asthenosphere   The weak, *ductile* layer of *rock* that constitutes the lower part of the upper *mantle* (below the *lithosphere*) and over which the lithospheric plates slide. (From the Greek *asthenes,* meaning "weak.")

astrobiologist A scientist who searches for the chemical building blocks of life, environments that may have been habitable for life, or even life itself on other worlds.

atomic mass The sum of an atom's protons and neutrons.

atomic number The number of protons in the nucleus of an atom.

autotroph A producer; an organism that makes its own food by manufacturing organic compounds, such as carbohydrates, that it uses as sources of energy. (Compare *heterotroph.*)

badland A deeply gullied landscape resulting from the rapid *erosion* of easily erodible *shales* and *clays.*

banded iron formation A *sedimentary rock* formation composed of alternating thin layers of iron oxide minerals and silica-rich minerals, precipitated from seawater when oxygen was first produced by *cyanobacteria* and reacted with iron dissolved in seawater.

barrier island A long, offshore sandbar that builds up to form a barricade between open ocean waves and the main *shoreline.*

basal slip The sliding of a *glacier* along the boundary between the ice and the ground. (Compare *plastic flow.*)

basalt A dark, fine-grained, *mafic igneous rock* composed largely of plagioclase feldspar and pyroxene; the *extrusive* equivalent of *gabbro.*

basaltic lava A lava type of mafic composition that has a low silica content, erupts at high temperatures, and flows readily.

base level The *elevation* at which a *stream* ends by entering a large standing body of water.

basin A synclinal structure consisting of a bowl-shaped depression of *rock* layers in which the beds *dip* radially toward a central point. (Compare *dome.*)

batholith A great irregular mass of *intrusive igneous rock* that covers at least 100 km²; the largest type of *pluton.*

beach A *shoreline* environment made up of *sand* and pebbles.

bed load The material a *stream* carries along its bed by sliding and rolling. (Compare *suspended load.*)

bedding The formation of parallel layers, or beds, by deposition of *sediment* particles.

bedding sequence A sequence of interbedded and vertically stacked layers of different *sedimentary rock* types.

bioclastic sediment A shallow-water *sediment* made up of fragments of shells or skeletons directly precipitated by marine organisms and consisting primarily of two calcium carbonate *minerals*—calcite and aragonite—in variable proportions.

biofuel A fuel, such as ethanol, derived from biomass.

biogeochemical cycle The pattern of flux of a chemical between the biological ("bio") and environmental ("geo") components of an *ecosystem.*

biological sediment A *sediment* formed near its place of deposition as a result of direct or indirect mineral *precipitation* by organisms. (Compare *chemical sediment.*)

biosphere The component of the *Earth system* that contains all its living organisms.

bioturbation The process by which organisms rework existing *sediments* by burrowing through them.

blueschist A *metamorphic rock* formed under high pressures and moderate temperatures, often containing glaucophane, a blue amphibole.

bomb A *pyroclast* 2 mm or larger, usually consisting of a blob of *lava* that cools in flight and becomes rounded, or a chunk torn loose from previously solidified volcanic rock. (Compare *volcanic ash.*)

bottomset bed A thin, horizontal bed of *mud* deposited seaward of a *delta* and then buried by continued delta growth.

braided stream A *stream* whose *channel* divides into an interlacing network of channels, which then rejoin in a pattern resembling braids of hair.

breccia A volcanic *rock* formed by the *lithification* of large *pyroclasts.* (Compare *tuff.*)

brittle Pertaining to a material that undergoes little *deformation* under increasing stress until it breaks suddenly. (Compare *ductile.*)

building code A set of standards for the design and construction of new buildings that specifies the intensity of shaking a structure must be able to withstand during an *earthquake.*

burial metamorphism Low-grade metamorphism in which buried *sedimentary rocks* are altered by a progressive increase in pressure exerted by growing layers of overlying *sediments* and by the increase in heat associated with increased depth of burial.

caldera A large, steep-walled, basin-shaped depression formed by a violent volcanic eruption in which large volumes of *magma* are discharged rapidly from a large *magma chamber,* causing the overlying volcanic structure to collapse catastrophically through the roof of the emptied chamber.

Cambrian explosion The rapid *evolutionary radiation* of animals during the early Cambrian period, after almost 3 billion years of very slow evolution, in which all the major branches of the animal tree of life originated within about 10 million years.

capacity The total *sediment* load carried by a current. (Compare *competence.*)

carbon cycle The continual movement of carbon among different components of the *Earth system.*

carbon economy The economy of modern industrial civilization, so-called because it runs primarily on fossil fuels.

carbon sequestration The pumping of CO_2 generated by fossil-fuel combustion into reservoirs other than the atmosphere.

carbonates A class of minerals composed of carbon and oxygen—in the form of the carbonate anion (CO_3^{2-})—in combination with calcium and magnesium.

carbonate compensation depth The ocean depth below which the seawater is sufficiently undersaturated with calcium carbonate that calcium carbonate shells and skeletons dissolve.

carbonate rock A *sedimentary rock* formed from *carbonate sediment.*

carbonate sediment A *sediment* formed from the accumulation of carbonate minerals directly or indirectly precipitated by marine organisms.

carbon intensity The amount of carbon released into the atmosphere per amount of energy produced by burning of a fossil fuel. For example, burning methane releases 14.5 Gt of carbon per quad of energy produced, so its carbon intensity is 14.5 Gt/quad.

cation A positively charged ion. (Compare *anion.*)

cementation A diagenetic change in which *minerals* are precipitated in the pores between *sediment* particles and bind them together.

channel A well-defined trough through which the water in a *stream* flows.

chemical sediment A *sediment* formed at or near its place of deposition from dissolved materials that *precipitate* from water. (Compare *biological sediment.*)

chemical stability A measure of a substance's tendency to retain its chemical identity rather than reacting spontaneously to become a different chemical substance.

chemical weathering *Weathering* in which the *minerals* in a *rock* are chemically altered or dissolved. (Compare *physical weathering.*)

chemoautotroph An *autotroph* that derives its energy not from sunlight but from the chemicals produced when minerals are dissolved.

chemofossil The chemical remains of an organic compound made by an organism while it was alive.

chert A *sedimentary rock* made up of chemically or biologically precipitated silica.

cirque An amphitheater-like hollow carved at the head of a glacial valley by the plucking and tearing action of ice.

clay A *siliciclastic sediment* in which most of the particles are less than 0.0039 mm in diameter and which consists largely of clay minerals; the most abundant component of fine-grained *sedimentary rocks.*

claystone A *sedimentary rock* made up exclusively of *clay*-sized particles.

cleavage (1) The tendency of a *crystal* to break along planar surfaces. (2) The geometric pattern produced by such breakage.

climate The average conditions of Earth's surface environment and their variation.

climate model Any representation of the *climate system* constructed to reproduce one or more aspects of its behavior.

climate system The global *geosystem* that includes all the components of the *Earth system* and all the interactions among these components needed to determine climate on a global scale and how it changes over time.

coal A *biological sedimentary rock* composed almost entirely of organic carbon formed by the *diagenesis* of wetland vegetation.

color A property of a *mineral* imparted by transmitted or reflected light.

compaction A diagenetic decrease in the volume and *porosity* of a *sediment* as its particles are squeezed closer together by the weight of overlying sediments.

competence The ability of a current to carry material of a given size. (Compare *capacity.*)

compressional wave A *seismic wave* that propagates by expanding and compressing the material it moves through. (Compare *shear wave.*)

compressive force A force that squeezes or shortens a body. (Compare *shearing force; tensional force.*)

concordant intrusion An igneous intrusion whose boundaries lie parallel to layers of bedded *country rock.* (Compare *discordant intrusion.*)

conduction The mechanical transfer of heat energy by the jostling of thermally agitated atoms and molecules. (Compare *convection.*)

conglomerate A *sedimentary rock* composed of pebbles, cobbles, and boulders; the lithified equivalent of *gravel.*

consolidated material *Sediment* that is compacted and bound together by mineral cements. (Compare *unconsolidated material.*)

contact metamorphism Metamorphism resulting from heat and pressure in a small area, as in rocks in contact with and near an igneous intrusion.

continental drift The large-scale movements of continents across Earth's surface driven by the *plate tectonic system.*

continental glacier A thick, slow-moving sheet of ice that covers a large part of a continent or other large landmass. (Compare *valley glacier.*)

continental margin The *shoreline*, shelf, and slope of a continent.

continental rise An apron of muddy and sandy *sediment* extending from the foot of the *continental slope* to the *abyssal plain.*

continental shelf A broad, flat, submerged platform, consisting of a thick layer of flat-lying shallow-water *sediment*, that extends from the *shoreline* to the edge of the *continental slope.*

continental slope A steep slope that descends from the edge of the *continental shelf* to the *continental rise.*

contour A line that connects points of equal *elevation* on a topographic map.

convection The mechanical transfer of heat energy that occurs as a heated material expands, rises, and displaces cooler material, which is itself heated and rises to continue the cycle.

convergent boundary A boundary between lithospheric plates where the plates move toward each other and one plate is recycled into the *mantle.* (Compare *divergent boundary; transform fault.*)

core The dense central part of Earth below the *core-mantle boundary*, composed principally of iron and nickel. (See also *inner core; outer core.*)

core-mantle boundary The boundary between Earth's *core* and its *mantle*, about 2890 km below Earth's surface.

country rock The rock surrounding an igneous intrusion.

covalent bond A bond between atoms in which electrons are shared. (Compare *ionic bond.*)

crater (1) A bowl-shaped pit found at the summit of most *volcanoes*, centered on the vent. (2) A depression caused by the impact of a *meteorite.*

craton A stable region of ancient continental *crust*, often made up of continental *shields* and platforms.

cratonic keel A mechanically stable and chemically distinct portion of the lithospheric *mantle* that extends some 200 to 300 km beneath a *craton* into the *asthenosphere* like the hull of a boat into water.

creep A slow downhill *mass movement* of *soil* or other debris at a rate ranging from about 1 to 10 mm/year.

crevasse A large vertical crack in the surface of a *glacier*, caused by the cracking of brittle surface ice as it is dragged along by the *plastic flow* of the ice below.

cross-bedding A *sedimentary structure* consisting of beds deposited by currents of wind or water and inclined at angles as much as 35° from the horizontal.

crude oil An organic sediment formed by diagenesis from organic material in the pores of sedimentary rocks; a diverse class of liquids composed of complex hydrocarbons. Also called petroleum.

crust The thin outer layer of Earth, averaging from about 8 km thick under the oceans to about 40 km thick under the continents, consisting of relatively low-density silicates that melt at relatively low temperatures.

crystal An ordered three-dimensional array of atoms in which the basic arrangement is repeated in all directions.

crystal habit The shape in which a mineral's individual *crystals* or aggregates of crystals grow.

crystallization The formation of a solid *mineral* from a gas or liquid whose constituent atoms come together in the proper chemical proportions and ordered three-dimensional arrangement.

cuesta An asymmetrical ridge formed from a tilted and eroded series of beds with alternating weak and strong resistance to *erosion.*

cyanobacteria A group of microorganisms that produce carbohydrates and release oxygen by *photosynthesis* and that probably originated the process early in life's history.

dacite A light-colored, fine-grained *intermediate igneous rock* with a composition between that of *rhyolite* and that of *andesite*; the *extrusive* equivalent of *granodiorite.*

Darcy's law A summary of the relationships among the volume of water flowing through an *aquifer* in a certain time, the vertical drop of the flow, the flow distance, and the *permeability* of the aquifer.

decompression melting The spontaneous melting of rising *mantle* material as it reaches a level where pressure decreases below a critical point, without the introduction of any additional heat. (Compare *fluid-induced melting*.)

deflation   The removal of *clay, silt,* and *sand* from dry *soil* by strong winds, which gradually scoop out shallow depressions in the ground.

deformation   The modification of rocks due to folding, faulting, shearing, compression, or extension by plate tectonic forces.

delta   A large, flat-topped deposit of *sediments* formed where a *river* enters an ocean or lake and its current slows.

dendritic drainage   An irregular *drainage network* that resembles the limbs of a branching tree. (From the Greek *dendron,* meaning "tree.")

density   The mass per unit volume of a substance, commonly expressed in grams per cubic centimeter (g/cm³). (Compare *specific gravity*.)

depositional remanent magnetization   A weak magnetization of *sedimentary rock* created by the parallel alignment of magnetic *sediment* particles in the direction of Earth's *magnetic field* as they settle and preserved when the sediments are lithified.

desert pavement   A coarse, gravelly ground surface left when continued *deflation* removes the smaller *sand* and *silt* particles from desert *soils*.

desert varnish   A distinctive dark brown, sometimes shiny coating found on many desert rock surfaces, consisting of a mixture of clay minerals with smaller amounts of manganese and iron oxides.

desertification   The transformation of semiarid lands into deserts.

diagenesis   The physical and chemical changes, caused by pressure, heat, and chemical reactions, by which buried *sediments* are lithified to form *sedimentary rocks*.

diatreme   A structure formed when a volcanic vent and the feeder channel below it are left full of *breccia* as an explosive eruption wanes.

dike   A sheetlike *discordant igneous intrusion* that cuts across layers of bedded *country rock*. (Compare *sill*.)

diorite   A coarse-grained *intermediate igneous rock* with a composition between that of *granodiorite* and that of *gabbro*; the *intrusive* equivalent of *andesite*.

dip   The amount of tilting of a *rock* layer; the angle at which a rock layer inclines from the horizontal, measured at right angles to the *strike*.

dipole   Pertaining to two oppositely polarized magnetic poles.

dip-slip fault   A *fault* on which the relative movement of opposing blocks of rock has been up or down the *dip* of the fault plane.

discharge   (1) The volume of *groundwater* leaving an *aquifer* in a given time. (Compare *recharge*.) (2) The volume of water that passes a given point in a given time as it flows through a *channel* of a certain width and depth.

discordant intrusion   An igneous intrusion that cuts across the layers of the *country rock* it intrudes. (Compare *concordant intrusion*.)

disseminated deposit   A deposit of ore minerals that is scattered through volumes of rock much larger than a vein.

distributary   A smaller *stream* that receives water and *sediment* from the main *channel* of a *river*, branches off downstream, and thus distributes the water and sediment into many channels; typically found on a *delta*.

divergent boundary   A boundary between lithospheric plates where two plates move apart and new *lithosphere* is created. (Compare *convergent boundary; transform fault*.)

divide   A ridge of high ground along which all rainfall runs off down one side or the other.

dolostone   An abundant *carbonate rock* composed primarily of dolomite and formed by the *diagenesis* of *carbonate sediments* and *limestones*.

dome   An anticlinal structure consisting of a broad circular or oval upward bulge of rock layers in which the beds dip radially away from a central point. (Compare *basin*.)

drainage basin   An area of land, bounded by *divides*, that funnels all its water into the network of *streams* draining the area.

drainage network   The pattern of connections of all the large and small *streams* in a *drainage basin*.

drift   All material of glacial origin found anywhere on land or at sea.

drought   A period of months or years when precipitation is much lower than normal.

drumlin   A large streamlined hill of *till* and bedrock deposited by a *continental glacier* that parallels the direction of ice movement.

dry wash   A desert *valley* that carries water only briefly after a rain. Called a *wadi* in the Middle East.

ductile   Pertaining to a material that undergoes smooth and continuous *deformation* under increasing stress without fracturing and does not spring back to its original shape when the stress is released. (Compare *brittle*.)

dune   An elongated mound or ridge of *sand* formed by a current of wind or water.

dust   Windborne material usually consisting of particles less than 0.01 mm in diameter (including *silt* and *clay*) but often including somewhat larger particles.

dwarf planet   Any of several tiny objects of the outer solar system (including Pluto) that are composed of a frozen mixture of gases, ice, and rock and that orbit the Sun in an unusual pattern that sometimes brings them closer to the Sun than Neptune.

Earth system   The collection of Earth's open, interacting, and often overlapping *geosystems.*

earthquake   The violent motion of the ground that occurs when brittle *rock* under stress suddenly breaks along a *fault.*

eclogite   An ultra-high-pressure *metamorphic rock* formed at the base of the *crust* at moderate to high temperatures, typically containing *minerals* such as coesite (a very dense, high-pressure form of quartz).

ecosystem   An organizational unit at any scale composed of biological and physical components that function in a balanced, interrelated fashion.

El Niño   An anomalous warming of the eastern tropical Pacific Ocean that occurs every 3 to 7 years and lasts for a year or so.

elastic rebound theory   A theory of faulting and *earthquake* generation holding that, as the crustal blocks on either side of a *fault* are deformed by tectonic forces, they remain locked in place by friction, accumulating elastic strain energy, until they fracture and rebound to their undeformed state.

electron sharing   The mechanism by which a *covalent bond* is formed between the elements in a chemical reaction.

electron transfer   The mechanism by which an *ionic bond* is formed between the elements in a chemical reaction.

elevation   The vertical distance above or below sea level.

ENSO (El Niño–Southern Oscillation)   A natural cycle of variation in the exchange of heat between the atmosphere and the tropical Pacific Ocean, of which *El Niño* and a complementary cooling event, known as La Niña, are a part.

eolian   Pertaining to wind.

eon   The largest division of geologic time, including multiple *eras.*

epeirogeny   Gradual downward and upward movements of broad regions of *crust* without significant folding or faulting. (From the Greek *epeiros,* meaning "mainland.")

epicenter   The geographic point on Earth's surface directly above the *focus* of an *earthquake.*

epoch   A division of geologic time representing one subdivision of a *period.*

era   A division of geologic time representing one subdivision of an *eon* and including multiple *periods.*

erosion   The set of processes that loosen *soil* and *rock* and move them downhill or downstream.

esker   A long, narrow, winding ridge of *sand* and *gravel* found in the middle of a ground *moraine,* running roughly parallel to the direction of ice movement, deposited by meltwater streams flowing in tunnels along the bottom of a melting *glacier.*

evaporite rock   A *sedimentary rock* formed from *evaporite sediment.*

evaporite sediment   *Chemical sediment* that is precipitated from evaporating seawater or lake water.

evolution   Systematic change in organisms over time, driven by the process of *natural selection.*

evolutionary radiation   The relatively rapid evolution of many new types of organisms from a common ancestor.

exfoliation   A *physical weathering* process in which large flat or curved sheets of *rock* are detached from an outcrop.

exhumation   The transportation of subducted *metamorphic rocks* back to Earth's surface.

exoplanet   A planet outside the solar system.

extremophile   A *microorganism* that lives in environments that would kill most other organisms.

extrusive igneous rock   A fine-grained or glassy *igneous rock* formed from *magma* that erupts at Earth's surface as lava and cools rapidly. (Compare *intrusive igneous rock.*)

fault   A fracture in *rock* that displaces the rock on either side of it.

fault mechanism   The orientation of the fault rupture and the slip direction of a fault that caused an *earthquake.*

fault slip   The distance of the displacement of the two blocks of *rock* on either side of a *fault* that occurs during an *earthquake.*

faunal succession, principle of   See *principle of faunal succession.*

felsic rock   Light-colored *igneous rock* that is poor in iron and magnesium and rich in high-silica *minerals* such as quartz, orthoclase feldspar, and plagioclase feldspar. (Compare *mafic rock; ultramafic rock.*)

fissure eruption A volcanic eruption emanating from a large, nearly vertical crack in Earth's surface rather than a central vent.

fjord A former glacial valley with steep walls and a U-shaped profile, now flooded with seawater.

flake tectonics The tectonic process of a planet with a vigorously convecting mantle underlying a thin *crust,* which break up into flakes or crumple like a rug; thought to occur on Venus and possibly on early Earth.

flexural basin A type of *sedimentary basin* that develops at a *convergent boundary* where one lithospheric plate pushes up over the other and the weight of the overriding plate causes the underlying plate to bend or flex downward.

flood Inundation that occurs when increased *discharge,* resulting from a short-term imbalance between inflow and outflow, causes a *stream* to overflow its banks.

flood basalt An immense basalt *plateau* formed by *fissure eruptions* of highly fluid *basaltic lava.*

floodplain A flat area about level with the top of a *channel* that lies on either side of the channel; the part of a *valley* that is flooded when a *stream* overflows its banks.

fluid-induced melting Melting of *rock* induced by the presence of water, which lowers its melting point. (Compare *decompression melting.*)

focus The point along a *fault* at which slipping initiates an *earthquake.*

fold A curved deformation structure formed when an originally planar structure, such as a sedimentary sequence, is bent by tectonic forces.

foliated rock *Metamorphic rock* that displays *foliation.* Foliated rocks include *slate, phyllite, schist,* and *gneiss.* (Compare *granoblastic rock.*)

foliation A set of flat or wavy parallel *cleavage* planes produced by *deformation* under directed pressure; typical of regionally metamorphosed rock.

foot wall The block of rock below a dipping *fault* plane. (Compare *hanging wall.*)

foraminifera A group of single-celled planktonic organisms that live in ocean surface waters and whose calcite shells account for most of the *carbonate sediments* of the deep seafloor.

foraminiferal ooze A sandy and silty *sediment* composed of the shells of dead *foraminifera.*

foreset bed A gently inclined deposit of fine-grained *sand* and *silt,* resembling large-scale cross-beds, on the outer front of a *delta.*

foreshock A small *earthquake* that occurs in the vicinity of, but before, a main shock. (Compare *aftershock.*)

formation A distinct set of *rock* layers that can be identified throughout a region by its physical properties and possibly by the assemblage of *fossils* it contains.

fossil A trace of an organism that has been preserved in the *geologic record.*

fossil fuel An energy *resource* formed by the burial and heating of dead organic matter, such as *coal, crude oil,* or *natural gas.*

fractional crystallization The process by which the *crystals* formed in cooling *magma* are segregated from the remaining liquid *rock,* usually by settling to the floor of the *magma chamber.*

fracture The tendency of a *crystal* to break along irregular surfaces other than *cleavage* planes.

frost wedging A *physical weathering* process in which the expansion of freezing water in cracks in *rock* breaks the rock.

gabbro A dark gray, coarse-grained *igneous rock* containing an abundance of mafic *minerals,* particularly pyroxene; the *intrusive* equivalent of *basalt.*

gas See *greenhouse gas; natural gas.*

genes Large molecules within the cells of every organism that encode all the information that determines what the organism will look like, how it will live and reproduce, and how it differs from all other organisms.

geobiology The study of interactions between the *biosphere* and Earth's physical environment.

geochemical cycle The pattern of flux of a chemical from one component of the *Earth system* to another.

geochemical reservoir A component of the *Earth system* where a chemical is stored at some point in its *geochemical cycle.*

geodesy The science of measuring the shape of Earth and locating points on its surface.

geodynamo The global *geosystem* that produces Earth's *magnetic field,* driven by *convection* in the *outer core.*

geologic cross section A diagram showing the geologic features that would be visible if vertical slices were made through part of the *crust.*

geologic map   A two-dimensional map representing the rock formations exposed at Earth's surface.

geologic record   Information about geologic events and processes that has been preserved in rocks as they have formed at various times throughout Earth's history.

geologic time scale   A worldwide history of geologic events that divides Earth's history into intervals, many of which are marked by distinctive sets of *fossils* and bounded by times when those sets of fossils changed abruptly.

geology   The branch of Earth science that studies all aspects of the planet: its history, its composition and internal structure, and its surface features.

geomorphology   (1) The shape of a landscape. (2) The branch of Earth science concerned with the shapes of landscapes and how they develop.

geosystem   A subsystem of the *Earth system* that produces specific types of geologic activity.

geotherm   The curve that describes how Earth's temperature increases with depth.

geothermal energy   Energy produced when underground water is heated as it passes through a subsurface region of hot *rock*.

glacial cycle   A climate cycle alternating between cold glacial periods, or *ice ages,* during which temperatures decline, water is transferred from the hydrosphere to the cryosphere, ice sheets expand into lower latitudes, and sea level falls, and warm *interglacial periods,* during which temperatures rise abruptly, water is transferred from the cryosphere to the hydrosphere, and sea level rises.

glacial rebound   A mechanism of epeirogeny in which continental lithosphere depressed by the weight of a large glacier rebounds upward for tens of millennia after the glacier melts.

glacier   A large mass of ice on land that shows evidence of being in motion, or of once having moved, under the force of gravity. (See also *continental glacier; valley glacier.*)

global change   Change in the *climate system* that has worldwide effects on the *biosphere,* atmosphere, and other components of the *Earth system.*

gneiss   A light-colored, poorly *foliated,* high-grade *metamorphic rock* with coarse bands of segregated light and dark *minerals* throughout.

graded bedding   A bed that shows progressive change in grain size from large sediment particles at the bottom to small particles at the top, usually indicating a weakening of the current that deposited the particles.

graded stream   A *stream* in which the slope, velocity, and *discharge* combine to transport its *sediment* load, with neither net sedimentation nor net *erosion* in the stream or its floodplain.

grain   A crystalline particle of a *mineral.*

granite   A felsic, coarse-grained *igneous rock* composed of quartz, orthoclase feldspar, sodium-rich plagioclase feldspar, and micas; the *intrusive* equivalent of *rhyolite.*

granoblastic rock   A nonfoliated *metamorphic rock* composed mainly of *crystals* that grow in equant shapes, such as cubes and spheres, rather than in platy or elongate shapes. Granoblastic rocks include *hornfels, quartzite, marble, greenstone, amphibolite,* and *granulite.* (Compare *foliated rock.*)

granodiorite   A light-colored, coarse-grained *intermediate igneous rock* that is similar to *granite* in containing abundant quartz, but whose predominant feldspar is plagioclase, not orthoclase; the *intrusive* equivalent of *dacite.*

granulite   (1) A high-grade, medium- to coarse-grained *granoblastic rock.* (2) The highest metamorphic grade.

gravel   The coarsest *siliciclastic sediment,* consisting of particles larger than 2 mm in diameter and including pebbles, cobbles, and boulders.

gravitational differentiation   The transformation of a planet by gravitational forces into a body whose interior is divided into concentric layers that differ from one another both physically and chemically.

graywacke   A sandstone composed of a heterogeneous mixture of *rock* fragments and angular *grains* of quartz and feldspar in which the *sand* grains are surrounded by a fine-grained *clay* matrix.

greenhouse effect   A global warming effect that results when a planet with an atmosphere containing *greenhouse gases* radiates *solar energy* back into space less efficiently than it would without such an atmosphere.

greenhouse gas   A gas that absorbs and reradiates energy when it is present in a planet's atmosphere. Greenhouse gases in Earth's atmosphere include water vapor, carbon dioxide, and methane.

greenschist   (1) A low-grade *metamorphic rock* formed from mafic volcanic rock and containing abundant chlorite. (2) The metamorphic grade above the zeolite grade.

greenstone   A low-grade *granoblastic rock* produced by the metamorphism of mafic volcanic *rock* and containing abundant chlorite, which accounts for its greenish cast.

groundwater   The volume of water that flows beneath Earth's surface.

groundwater table   The boundary between the *unsaturated zone* and the *saturated zone.*

guyot   A large, flat-topped *seamount* resulting from the *erosion* of a volcanic island when it was above sea level.

habitable zone   The distance from a star at which water is stable as a liquid; if a planet's orbit is within this zone, there is a chance that life might have originated there.

half-life   The time required for one-half the original number of parent atoms in a radioactive *isotope* to decay.

hanging valley   A *valley* formed by a tributary glacier that enters a deeper glacial valley high above the main valley floor.

hanging wall   The block of rock above a dipping *fault* plane. (Compare *foot wall.*)

hardness   A measure of the ease with which the surface of a *mineral* can be scratched.

Heavy Bombardment   A time early in the early history of the solar system when planets were subjected to very frequent crater-forming impacts.

hematite   The principal iron *ore;* the most abundant iron oxide at Earth's surface.

heterotroph   A consumer organism that gets its food by feeding directly or indirectly on *autotrophs.* (Compare *autotroph.*)

high-pressure metamorphism   Metamorphism occurring at pressures of 8 to 12 kbar.

hogback   A landscape feature similar to a *cuesta,* consisting of steep, narrow, more or less symmetrical ridges, formed by the *erosion* of steeply dipping or vertical beds of hard strata.

hornfels   A *granoblastic rock* of uniform grain size that has undergone little or no *deformation;* usually formed by *contact metamorphism* at high temperatures.

hot spot   A region of intense, localized volcanism found far from a plate boundary; hypothesized to be the surface expression of a *mantle plume.*

Hubbert's peak   The high point of a bell-shaped curve representing the rate of oil production; the point at which oil production peaks and then begins to decline.

humus   An organic component of *soil* consisting of the remains and waste products of the many organisms living in that soil.

hurricane   A great storm that forms over the warm surface waters of tropical oceans (between 8° and 20° latitude) in areas of high humidity and light winds, producing winds of at least 119 km/hour (74 miles/hour) and large amounts of rainfall.

hydraulic fracturing (fracking)   A technique for withdrawing oil and gas from shale and other tight formations by first pumping water and sand into a borehole at high pressures to create fractures through which the oil and gas can more readily flow.

hydraulic gradient   The ratio between the difference in *elevation* between two points in the *groundwater table* and the flow distance that water travels between the two points.

hydroelectric energy   Energy derived from water moving under the force of gravity driving a turbine that generates electricity.

hydrologic cycle   The cyclical movement of water from the ocean to the atmosphere by evaporation, to the surface by *precipitation,* to *streams* through *runoff* and *groundwater,* and back to the ocean.

hydrology   The science that studies the movements and characteristics of water on and under Earth's surface.

hydrothermal activity   The circulation of water through hot volcanic *rocks* and *magmas.*

hydrothermal solution   A hot water solution formed when circulating *groundwater* or seawater comes into contact with a hot magmatic intrusion, reacts with it, and carries off significant quantities of elements and ions released by the reaction, which may be deposited later as ore minerals.

ice age   The cold period of a *glacial cycle,* during which Earth cools, water is transferred from the hydrosphere to the cryosphere, ice sheets expand, and sea level drops. Also called glacial period. (Compare *interglacial period.*)

ice shelf   A sheet of ice floating on the ocean that is attached to a *continental glacier* on land.

ice stream   A current of ice within a *continental glacier* that flows faster than the surrounding ice.

iceberg calving   The process by which pieces of ice break off a *valley glacier* and form icebergs when the glacier reaches a *shoreline.*

igneous rock   A *rock* formed by the solidification of *magma.* (From the Latin *ignis,* meaning "fire.")

infiltration   The movement of water into *rock* or *soil* through cracks or small pores between particles.

inner core   The central part of Earth below a depth of 5150 km, consisting of a solid sphere, composed of iron and nickel, suspended within the liquid *outer core.*

intensity scale A scale for estimating the intensity of a destructive geologic event, such as an earthquake or a hurricane, directly from the event's destructive effects.

interglacial period The warm period of a *glacial cycle* during which ice sheets melt, water is transferred from the cryosphere to the hydrosphere, and sea level rises. (Compare *ice age.*)

intermediate igneous rock An *igneous rock* midway in composition between mafic and felsic, neither as rich in silica as *felsic rock* nor as poor in it as *mafic rock.*

intrusive igneous rock A coarse-grained *igneous rock* formed from *magma* that intrudes into *country rock* deep in Earth's *crust* and cools slowly. (Compare *extrusive igneous rock.*)

ion An atom or group of atoms that has an electrical charge, either positive or negative, because of the loss or gain of one or more electrons.

ionic bond A bond formed by electrostatic attraction between *ions* of opposite charge when electrons are transferred. (Compare *covalent bond.*)

iron formation A *sedimentary rock* that usually contains more than 15 percent iron in the form of iron oxides and some iron silicates and iron carbonates.

island arc A chain of volcanic islands formed on the overriding plate at a *convergent boundary* by *magma* that rises from the *mantle* as water released from the subducting lithospheric slab causes *fluid-induced melting.*

isochron A *contour* that connects rocks of equal age.

isostasy A principle stating that the buoyancy force that pushes upward a lower-density body (such as a continent or an iceberg) floating in a higher-density medium (such as the *asthenosphere* or seawater) must be balanced by the gravitational force that pulls it downward. (From the Greek for "equal standing.")

isotope One of two or more forms of atoms of the same element that have different numbers of neutrons and therefore different *atomic masses.*

isotopic dating The use of naturally occurring radioactive elements to determine the ages of *rocks.*

joint A crack in a *rock* along which there has been no appreciable movement.

kaolinite A white to cream-colored *clay* produced by the *weathering* of feldspar.

karst topography An irregular, hilly type of terrain characterized by *sinkholes*, caves, and a lack of surface *streams*; formed in regions with humid climates, abundant vegetation, extensively jointed limestone formations, and appreciable *hydraulic gradients.*

kettle A hollow or undrained depression that often has steep sides and may be occupied by a pond or lake; formed in glacial deposits when *outwash* is deposited around a residual block of ice that later melts.

lahar A torrential mudflow of wet volcanic debris.

laminar flow Fluid movement in which straight or gently curved streamlines run parallel to one another without mixing or crossing between layers. (Compare *turbulent flow.*)

landform A characteristic landscape feature on Earth's surface shaped by the processes of *erosion* and sedimentation.

large igneous province (LIP) A voluminous emplacement of predominantly *mafic extrusive* and *intrusive igneous rock* whose origins lie in processes other than normal *seafloor spreading.* LIPs include continental *flood basalts*, oceanic basalt plateaus, and aseismic ridges produced by hot spots.

lava *Magma* that flows out onto Earth's surface.

limestone A *carbonate rock* composed mainly of calcium carbonate in the form of the mineral calcite.

liquefaction The temporary transformation of solid material to a fluid state when it is saturated with water.

lithic sandstone A sandstone containing many particles derived from fine-grained rocks, mostly *shales*, volcanic rocks, and fine-grained *metamorphic rocks.*

lithification The conversion of *sediment* into solid *rock* by *compaction* and *cementation.*

lithosphere The strong, rigid outer shell of Earth that comprises the *crust* and the uppermost part of the *mantle* down to an average depth of about 100 km. (From the Greek *lithos*, meaning "stone.")

loess A blanket of unstratified, wind-deposited, fine-grained *sediment.*

longitudinal profile The smooth, concave-upward curve that represents a cross-sectional view of a *stream*, from notably steep near its head to almost level near its mouth.

longshore current A shallow-water current that runs parallel to the shore.

lower mantle A relatively homogeneous region of the *mantle* about 2200 km thick, extending from the *phase change* at about 660 km in depth to the *core-mantle boundary.*

low-velocity zone   A layer near the base of the *lithosphere*, beginning at a depth of about 100 km, where *S-wave* speed abruptly decreases, marking the top part of the *asthenosphere.*

luster   The way the surface of a *mineral* reflects light. (See Table 3.3.)

mafic rock   Dark-colored *igneous rock* containing *minerals* such as pyroxenes and olivines that are rich in iron and magnesium and relatively poor in silica. (Compare *felsic rock; ultramafic rock.*)

magma   Hot, molten *rock.*

magma chamber   A large pool of *magma* that forms in the *lithosphere* as rising magmas melt and push aside surrounding solid rock.

magmatic addition   A process of continental growth in which low-density, silica-rich *rock* differentiates in the *mantle* and is transported vertically to the *crust.* (Compare *accretion.*)

magmatic differentiation   A process by which *rocks* of varying composition arise from a uniform parent *magma* as various *minerals* are withdrawn from it by *fractional crystallization* as it cools, changing its composition.

magnetic anomaly   One in a pattern of long, narrow bands of high or low magnetic intensity on the seafloor that are parallel to and almost perfectly symmetrical with respect to the crest of a mid-ocean ridge.

magnetic field   The region of influence of a magnetized body or an electric current.

magnetic time scale   The detailed history of Earth's *magnetic field* reversals as determined by measuring the thermoremanent magnetization of *rock* samples whose ages are known.

magnitude scale   A scale for estimating the size of an *earthquake* using the logarithm of the largest ground motion registered by a *seismograph* (Richter magnitude) or the logarithm of the area of the *fault* rupture (moment magnitude).

mantle   The region that forms the main bulk of Earth, between the *crust* and the *core*, containing *rocks* of intermediate *density*, mostly compounds of oxygen with magnesium, iron, and silicon.

mantle plume   A narrow, cylindrical jet of hot, solid material rising from deep within the *mantle*, thought to be responsible for intraplate volcanism.

marble   A *granoblastic rock* produced by the metamorphism of *limestone* or *dolostone.*

mass extinction   A short interval during which a large proportion of the species living at the time disappear from the *geologic record.*

mass movement   A downslope movement of masses of *soil, rock, mud,* or other materials under the force of gravity.

mass wasting   All the processes by which weathered and unweathered Earth materials move downslope in large amounts and in large single events, usually under the influence of gravity.

meander   A curve or bend in a *stream* that develops as the stream erodes the outer bank of a bend and deposits *sediment* against the inner bank.

mélange   A distinct metamorphic assemblage that forms where oceanic *lithosphere* is subducted beneath a plate carrying a continent on its leading edge.

mesa   A small, flat, elevated landform with steep slopes on all sides, created by differential *weathering* of bedrock of varying hardness. (From the Spanish word for "table.")

metabolism   All the processes organisms use to convert inputs (such as sunlight, water, and carbon dioxide) into outputs (such as oxygen and carbohydrates).

metallic bond   A type of *covalent bond* in which freely mobile electrons are shared and dispersed among ions of metallic elements, which have the tendency to lose electrons and pack together as *cations.*

metamorphic facies   Groupings of *metamorphic rocks* of various mineral compositions formed under different grades of metamorphism from different parent rocks.

metamorphic rock   *Rock* formed by high temperatures and pressures that cause changes in the mineralogy, texture, or chemical composition of any kind of preexisting rock while maintaining its solid form. (From the Greek *meta*, meaning "change," and *morphe*, meaning "form.")

metasomatism   Change in the composition of a *rock* by fluid transport of chemical substances into or out of the rock.

meteoric water   Rain, snow, or other forms of water derived from the atmosphere.

meteorite   A chunk of material from outer space that strikes Earth.

microbial mat   A layered microbial community commonly occurring in *tidal flats,* hypersaline lagoons, and thermal springs.

microfossil   A trace of an individual *microorganism* preserved in the *geologic record.*

microorganism   A single-celled organism. Microorganisms include bacteria, some fungi and algae, and most protists.

mid-ocean ridge   An undersea mountain chain at a *divergent boundary*, characterized by earthquakes, volcanism, and rifting, all caused by the tensional forces of mantle convection that are pulling the two plates apart.

migmatite A mixture of *igneous* and *metamorphic rock* produced by incomplete melting, typically badly deformed and contorted and penetrated by many *veins*, small pods, and lenses of melted rock.

Milankovitch cycle A pattern of periodic variations in Earth's movement around the Sun that affects the amount of solar energy received at Earth's surface. Milankovitch cycles include variations in the eccentricity of Earth's orbit, the tilt of Earth's axis of rotation, and precession—Earth's wobble about its axis of rotation.

mineral A naturally occurring, solid crystalline substance, generally inorganic, with a specific chemical composition.

mineralogy (1) The branch of geology that studies the composition, structure, appearance, stability, occurrence, and associations of *minerals*. (2) The relative proportions of a *rock's* constituent minerals.

Mohorovičić discontinuity The boundary between the *crust* and the *mantle*, at a depth of 5 to 45 km, marked by an abrupt increase in *P-wave* velocity to more than 8 km/s. Also called Moho.

Mohs scale of hardness An ascending scale of mineral *hardness* based on the ability of one *mineral* to scratch another. (See Table 3.2.)

moraine An accumulation of rocky, sandy, and clayey material carried by glacial ice and deposited as *till*.

mud A fine-grained *siliciclastic sediment* mixed with water, in which most of the particles are less than 0.062 mm in diameter.

mudstone A blocky, poorly bedded, fine-grained *sedimentary rock* produced by the *lithification* of *mud*.

natural gas Methane gas (CH_4), the simplest hydrocarbon.

natural levee A ridge of coarse material built up by successive *floods* that confines a *stream* within its banks between floods, even when water levels are high.

natural resource A supply of energy, water, or raw material used by human civilization that is available from the natural environment. (See also *resource*.)

natural selection The process by which inherited traits within a population of organisms make it more likely for an organism to survive and successfully reproduce over successive generations.

nebular hypothesis The idea that the solar system originated from a diffuse, slowly rotating cloud of gas and fine dust (a "nebula") that contracted under the force of gravity and eventually evolved into the Sun and planets.

negative feedback A process in which one action produces an effect (the feedback) that tends to counteract the original action and stabilize the system against change. (Compare *positive feedback*.)

nonrenewable resource A *natural resource* that is produced at a rate much slower than the rate at which human civilization is using it up; for example, *fossil fuels*. (Compare *renewable resource*.)

normal fault A *dip-slip fault* in which the *hanging wall* moves downward relative to the *foot wall*, extending the structure horizontally.

nuclear energy Energy produced by the fission of the radioactive isotope uranium-235, which can be used to make steam and drive turbines to create electricity.

obsidian A dense, glassy volcanic *rock*, usually of felsic composition.

ocean acidification A process in which carbon dioxide from the atmosphere dissolves into the ocean and reacts with seawater to form carbonic acid (H_2CO_3), increasing the acidity of the ocean.

oil See *crude oil*.

oil shale A fine-grained, clay-rich *sedimentary rock* containing relatively large amounts of organic matter, from which combustible oil and gas can be extracted.

oil trap An impermeable barrier that blocks the upward migration of *crude oil* or *natural gas*, allowing them to collect beneath the barrier.

oil window The limited range of pressures and temperatures, usually found at depths between about 2 and 5 km, at which *crude oil* forms.

ophiolite suite An assemblage of *rocks*, characteristic of the sea-floor but found on land, consisting of deep-sea *sediments*, submarine *basaltic lavas*, and *mafic igneous* intrusions.

ore A *mineral* deposit from which valuable metals can be recovered profitably.

organic sedimentary rock A *sedimentary rock* that consists entirely or partly of organic carbon-rich deposits formed by the burial and *diagenesis* of once-living material.

original horizontality, principle of See *principle of original horizontality*.

orogen An elongated mountain belt, usually formed by an episode of compressive *deformation*.

orogeny Mountain building by tectonic forces, particularly through the folding and faulting of *rock* layers, often with accompanying volcanism. (From the Greek *oros*, meaning "mountain," and *gen*, meaning "be produced.")

outer core The layer of Earth extending from the *core-mantle boundary* to the *inner core*, at depths of 2890 to 5150 km, composed of molten iron and nickel and minor amounts of lighter elements, such as oxygen or sulfur.

outwash Glacial *drift* that has been picked up and distributed by meltwater *streams*.

oxbow lake A crescent-shaped, water-filled loop created in the former path of a *stream* when it bypasses a *meander* and takes a new, shorter course.

oxides A class of minerals that are compounds of the oxygen anion (O^{2-}) and metallic cations.

P-T path The history of changing temperature (T) and pressure (P) conditions that is reflected in the texture and *mineralogy* of a *metamorphic rock*.

P wave The first type of *seismic wave* to arrive at a *seismograph* from the *focus* of an *earthquake;* a type of *compressional wave*.

paleomagnetism The *geologic record* of ancient magnetization.

Pangaea A supercontinent that coalesced in the late Paleozoic *era* and comprised all present continents, then began to break up in the Mesozoic era.

partial melting Incomplete melting of a *rock* that occurs because the *minerals* that compose it melt at different temperatures.

passive margin A *continental margin* far from a plate boundary. (Compare *active margin*.)

peat A rich organic material, made up of accumulated vegetation preserved from decay in a wetland environment, that contains more than 50 percent carbon.

pediment A broad, gently sloping platform of bedrock left behind as a mountain front erodes and retreats from its *valley*.

pegmatite A *vein* of extremely coarse-grained *granite*, crystallized from a water-rich *magma* in the late stages of solidification, that cuts across much finer grained *country rock* and may contain rich concentrations of rare *minerals*.

pelagic sediment An open-ocean *sediment* composed of small terrigenous and biologically precipitated particles that slowly settle out of suspension in seawater.

peridotite A coarse-grained, dark greenish gray, *ultramafic intrusive igneous rock* composed primarily of olivine with smaller amounts of pyroxene and other minerals such as spinel or garnet; the dominant rock in Earth's *mantle* and the source rock of basaltic magmas.

period A division of geologic time representing one subdivision of an *era*.

permafrost Perennially frozen soil containing aggregates of ice crystals; any *rock* or *soil* remaining at or below 0°C for 2 or more years.

permeability The ability of a solid to allow fluids to pass through it.

phase change A transformation of a rock's *crystal* structure (but probably not its chemical composition) by changing conditions of temperature and pressure, signaled by a change in *seismic wave* velocity.

phosphorite A chemical or biological *sedimentary rock* composed of calcium phosphate precipitated from phosphate-rich seawater and formed diagenetically by the interaction of calcium phosphate with muddy or *carbonate sediments*. Also called phosphate rock.

photosynthesis The process by which organisms such as plants and algae use energy from sunlight to convert water and carbon dioxide into carbohydrates and oxygen.

phyllite A *foliated rock* that is intermediate in metamorphic grade between *slate* and *schist*, containing small *crystals* of mica and chlorite that give it a more or less glossy sheen.

physical weathering *Weathering* in which solid *rock* is fragmented by mechanical processes that do not change its chemical composition. (Compare *chemical weathering*.)

planetesimal Any of the numerous kilometer-sized chunks of material that accreted by gravitational attraction early in the history of the solar system.

plastic flow The deformation of a *glacier* that results from the sum of all the small slips of the ice *crystals* within it. (Compare *basal slip*.)

plate tectonic system The global *geosystem* that includes the convecting *mantle* and its overlying mosaic of lithospheric plates.

plate tectonics The theory that describes and explains the creation and destruction of Earth's lithospheric plates and their movement over Earth's surface. (From the Greek *tekton*, meaning "builder.")

plateau A large, broad, flat area of appreciable *elevation* above the neighboring terrain.

playa A flat bed of *clay* and encrusting precipitated salts, formed by the complete evaporation of a *playa lake*.

playa lake A permanent or temporary lake in an arid mountain *valley* or *basin*, where dissolved *minerals* may be concentrated and precipitated as the water evaporates.

pluton   A large igneous intrusion, ranging in size from a cubic kilometer to hundreds of cubic kilometers, formed deep in the *crust*.

point bar   A curved sandbar deposited along the inside bank of a *stream*, where the current is weakest.

polymorph   One of two or more alternative possible *crystal* structures for a single chemical compound; for example, the *minerals* quartz and cristobalite are polymorphs of silica (SiO_2).

porosity   The percentage of a rock's volume consisting of open pores between particles.

porphyroblast   A large *crystal,* surrounded by a much finer grained matrix of other *minerals,* formed in *metamorphic rock* from a mineral that is stable over a broad range of temperatures and pressures.

porphyry   An *igneous rock* of mixed *texture* in which large *crystals* (phenocrysts) "float" in a predominantly fine-grained matrix.

positive feedback   A process in which one action produces an effect (the feedback) that tends to enhance the original action and amplify change in the system. (Compare *negative feedback.*)

potable   Pertaining to water that tastes agreeable and is not dangerous to human health.

pothole   A hemispherical hole in the bedrock of a streambed, formed by *abrasion* by small pebbles and cobbles rotating in a swirling eddy.

precipitate   (1) (verb) To drop out of a saturated solution as *crystals.* (2) (noun) The crystals that drop out of a saturated solution.

precipitation   (1) A deposit on Earth's surface of condensed atmospheric water vapor in the form of rain, snow, sleet, hail, or mist. (2) The condensation of a solid from a solution during a chemical reaction.

pressure-temperature path   See *P-T path.*

principle of faunal succession   A stratigraphic principle stating that the *sedimentary rock* strata in an outcrop contain distinct *fossils* in a definite sequence.

principle of original horizontality   A stratigraphic principle stating that *sediments* are deposited as essentially horizontal beds.

principle of superposition   A stratigraphic principle stating that each *sedimentary rock* stratum in a tectonically undisturbed sequence is younger than the one beneath it and older than the one above it.

principle of uniformitarianism   A principle stating that the processes we see in action on Earth today have worked in much the same way throughout the geologic past.

pumice   A volcanic *rock,* usually *rhyolitic* in composition, containing numerous cavities (vesicles) that remain after trapped gas has escaped from solidifying *lava.*

pyroclast   A *rock* fragment ejected into the air by a volcanic eruption. (See also *bomb; volcanic ash.*)

pyroclastic flow   A glowing cloud of hot ash, dust, and gases ejected by a volcanic eruption that rolls downhill at high speeds.

quad   A unit consisting of 1 quadrillion (10^{15}) British thermal units (Btu), used to measure large quantities of energy.

quartz arenite   A sandstone made up almost entirely of quartz grains, usually well sorted and rounded.

quartzite   A very hard, white *granoblastic rock* derived from quartz-rich *sandstone.*

radiation   See *evolutionary radiation.*

rain shadow   An area of low rainfall on the leeward slope of a mountain range.

recharge   The *infiltration* of water into any subsurface rock formation.

recurrence interval   The average time between large *earthquakes* at a particular location; according to the elastic rebound theory, the time required to accumulate the strain that will be released by fault slipping in a future earthquake.

red bed   An unusual stream deposit of *sandstones* and *shales* bound together by iron oxide cement, which gives the bed its red color.

reef   A moundlike or ridgelike organic structure constructed of the carbonate skeletons and shells of marine organisms.

regional metamorphism   Metamorphism caused by high pressures and temperatures that extend over large regions; typical of convergent boundaries where two continents collide. (Compare *contact metamorphism.*)

rejuvenation   Renewed uplift in a previously existing mountain chain that returns it to a more youthful stage.

relative age   The age of one geologic event in relation to another. (Compare *absolute age.*)

relative humidity   The amount of water vapor in the air, expressed as a percentage of the total amount of water the air could hold at the same temperature if it were saturated.

relative plate velocity   The velocity at which one lithospheric plate moves relative to another.

relief The difference between the highest and lowest *elevations* in a particular area.

renewable resource A *natural resource* that is produced at a rate rapid enough to match the rate at which human civilization is using it up; for example, wood. (Compare *nonrenewable resource*.)

reserve The supply of a *natural resource* that has already been discovered and can be exploited economically and legally at the present time. (Compare *resource*.)

reservoir See *geochemical reservoir*.

residence time The average time an atom of a particular element spends in a *geochemical reservoir* before leaving it.

resource (1) The entire amount of a given material, including the amount that may become available for use in the future; includes *reserves* plus known but currently unrecoverable supplies plus undiscovered supplies that geologists think may eventually be found. (Compare *reserve*.) (2) A *natural resource*.

respiration The metabolic process by which organisms release the energy stored in carbohydrates; requires oxygen.

reverse fault A *dip-slip fault* in which the *hanging wall* moves upward relative to the *foot wall*, compressing the structure horizontally.

rhyolite A light brown to gray, fine-grained *felsic igneous rock*; the *extrusive* equivalent of *granite*.

rhyolitic lava The lava type that is the richest in silica, erupts at the lowest temperatures, and is the most viscous.

rift basin A *sedimentary basin* that develops at a *divergent boundary* at an early stage of plate separation as the stretching and thinning of the continental crust results in subsidence. (Compare *thermal subsidence basin*.)

ripple A very small ridge of *sand* or *silt* whose long dimension is at right angles to the current that formed it.

river A major branch of a *stream* system.

rock A naturally occurring solid aggregate of minerals or, in some cases, nonmineral solid matter.

rock cycle The set of geologic processes that convert *rocks* of each of the three major types—*igneous, sedimentary,* and *metamorphic*—into the other two types.

Rodinia A supercontinent older than *Pangaea* that formed about 1.1 billion years ago and began to break up about 750 million years ago.

runoff The sum of all *precipitation* that flows over the land surface, including not only streams but also the fraction that temporarily infiltrates near-surface *soil* and *rock* and then flows back to the surface.

S wave The second type of *seismic wave* to arrive at a *seismograph* from the *focus* of an *earthquake;* a type of *shear wave*. S waves cannot travel through liquids or gases.

salinity The total amount of dissolved substances in a given volume of water.

saltation The transportation of *sand* or smaller *sediment* particles by a current in such a manner that the particles move along in a series of short intermittent jumps.

sand A *siliciclastic sediment* consisting of medium-sized particles, ranging from 0.062 to 2 mm in diameter.

sandblasting *Erosion* of a solid surface by *abrasion* caused by the high-speed impact of *sand* grains carried by wind.

sandstone The lithified equivalent of *sand.*

saturated zone The level below the *groundwater table,* in which the pores of *soil* or *rock* are completely filled with water. Also called the phreatic zone. (Compare *unsaturated zone*.)

schist An intermediate-grade *metamorphic rock* characterized by pervasive coarse, wavy *foliation* known as schistosity.

scientific method A general procedure, based on systematic observations and experiments, by which scientists propose and test hypotheses that explain some aspect of how the physical universe works.

seafloor metamorphism A form of *metasomatism* associated with mid-ocean ridges, in which seawater infiltrates hot basaltic lava, is heated, circulates through the newly forming oceanic crust by convection, and reacts with and alters the chemical composition of the basalt.

seafloor spreading The mechanism by which new oceanic *crust* is formed at a *spreading center* on the crest of a mid-ocean ridge. As two plates move apart, *magma* wells up into the rift between them to form new crust, which spreads laterally away from the rift and is replaced continually by newer crust.

seamount A submerged *volcano,* usually extinct, found on the seafloor.

sediment Material deposited on Earth's surface by physical agents (wind, water, and ice), chemical agents (*precipitation* from oceans, lakes, and *rivers*), or biological agents (living and dead organisms).

sedimentary basin A region where the combination of sedimentation and *subsidence* has formed thick accumulations of *sediment* and *sedimentary rock.*

sedimentary environment A geographic location characterized by a particular combination of climate conditions and physical, chemical, and biological processes.

sedimentary rock A *rock* formed by the burial and *diagenesis* of layers of *sediment.*

sedimentary structure Any kind of *bedding* or other feature (such as *cross-bedding, graded bedding,* or *ripples*) formed at the time of *sediment* deposition.

seismic hazard The intensity of shaking and ground disruption by *earthquakes* that can be expected over the long term at some specified location.

seismic ray path The path along which seismic energy propagates. Ray paths are perpendicular to wave fronts.

seismic risk The earthquake damage that can be expected over the long term in a specified region, usually measured in average dollar losses per year.

seismic tomography A technique that uses differences in the travel times of *seismic waves* produced by *earthquakes* and recorded on *seismographs* to construct three-dimensional images of Earth's interior.

seismic wave A ground vibration produced by an *earthquake.* (See also *P wave; S wave; surface wave.*) (From the Greek *seismos,* meaning "earthquake.")

seismograph An instrument that records the *seismic waves* generated by earthquakes.

settling velocity The speed at which particles of various weights suspended in a current settle to the bed.

shadow zone (1) A zone beyond 105° from the *focus* of an *earthquake* where *S waves* are not recorded because they are not transmitted through Earth's liquid *outer core.* (2) A zone at angular distances of 105° to 142° from the focus of an earthquake where *P waves* are not recorded because they are refracted downward into the core and emerge at greater distances after the delay caused by their detour through the core.

shale A fine-grained *sedimentary rock* composed of *silt* plus a significant component of *clay,* which causes it to break readily along bedding planes.

shear wave A seismic wave that propagates by moving the material it travels through from side to side. Shear waves cannot propagate through any fluid—air, water, or the liquid iron in Earth's *outer core.* (Compare *compressional wave.*)

shearing force A force that pushes two sides of a body in opposite directions. (Compare *compressive force; tensional force.*)

shield A large *tectonic province* within a continent that is tectonically stable and where ancient crystalline basement rocks are exposed at the surface.

shield volcano A broad, shield-shaped *volcano* many tens of kilometers in circumference and more than 2 km high, built by successive flows of *basaltic lava* from a central vent.

shock metamorphism Metamorphism that occurs when *minerals* are subjected to high pressures and temperatures by heat and shock waves generated when a *meteorite* collides with Earth.

shoreline The line where the ocean surface meets the land surface.

silicates The most abundant class of minerals in Earth's crust, composed of oxygen (O) and silicon (Si), mostly in combination with cations of other elements.

siliceous ooze A biologically precipitated *pelagic sediment* produced by sedimentation of the silica shells of diatoms and radiolarians.

siliciclastic sediment *Sediment* formed from clastic particles produced by the *weathering* of *rocks* and physically deposited by running water, wind, or ice. (From the Greek *klastos,* meaning "broken.")

sill A sheetlike *concordant igneous intrusion* formed by the injection of *magma* between parallel layers of bedded *country rock.* (Compare *dike.*)

silt A *siliciclastic sediment* in which most of the particles are between 0.0039 and 0.062 mm in diameter.

siltstone A *sedimentary rock* that contains mostly *silt* and looks similar to *mudstone* or very fine grained *sandstone;* the lithified equivalent of silt.

sinkhole A small, steep depression in the land surface formed when the thin roof of a limestone cave collapses suddenly.

slate A fine-grained *foliated rock* that is easily split into thin sheets, formed primarily by low-grade metamorphism of *shale.*

slip face The steep leeward slope of a *dune* on which *sand* is deposited in cross-beds at the *angle of repose.*

slump A slow *mass movement* of *unconsolidated material* that travels as a unit.

soil An intricate combination of weathered *rock* and organic material.

soil profile The composition and appearance of a *soil,* usually characterized by distinct layers.

solar energy Energy derived from the Sun.

solar forcing Cyclical variation in the amount of solar energy received at Earth's surface.

solar nebula According to the *nebular hypothesis,* a disk of gas and dust that surrounded the proto-Sun from which the planets of the solar system formed.

sorting The tendency for variations in current velocity to segregate *sediments* according to size.

specific gravity The weight of a substance divided by the weight of an equal volume of pure water at 4°C. (Compare *density.*)

spit A narrow extension of a *beach* formed by *longshore currents* that carry *sand* to its downcurrent end.

spreading center A *divergent boundary,* marked by a rift at the crest of a mid-ocean ridge, where new oceanic *crust* is formed by *seafloor spreading.*

stabilization wedge A strategy for reducing carbon emissions by 1 gigaton per year in the next 50 years relative to a business-as-usual scenario. About seven stabilization wedges will be necessary to stabilize carbon emissions at current levels.

stock A *pluton* less than 100 km² in area.

storm surge A dome of seawater, formed by a *hurricane,* that rises above the level of the surrounding ocean surface.

stratigraphic succession A chronologically ordered set of *rock* strata.

stratigraphy The description, correlation, and classification of strata in *sedimentary rocks.*

stratosphere The cold, dry layer of the atmosphere above the *troposphere* that extends from about 11 to 50 km in altitude. (Compare *troposphere.*)

stratovolcano A concave-shaped *volcano* formed from alternating layers of lava flows and beds of *pyroclasts.*

streak The *color* of the fine deposit of *mineral* powder left on an abrasive surface when a mineral is scraped across it.

stream Any body of water, large or small, that flows over the land surface.

stream power The product of stream slope and stream *discharge.*

stress The force per unit area acting on any surface within a solid body.

striation A scratch or groove left on bedrock by a *glacier* dragging rocks along its base; may show the direction of glacial movement.

strike The compass direction of a line formed by the intersection of a rock layer's surface or a fault surface with a horizontal surface.

strike-slip fault A *fault* on which the relative movement of the opposing blocks of rock has been horizontal, parallel to the *strike* of the fault plane.

stromatolite A rock with distinctive thin layers, believed to have been formed by ancient *microbial mats;* one of the most ancient *fossil* types on Earth.

subduction The sinking of oceanic *lithosphere* beneath overriding oceanic or continental lithosphere at a convergent plate boundary.

subsidence Depression or sinking of a broad area of *crust* relative to the surrounding crust, induced partly by the weight of *sediments* on the crust but driven mainly by plate tectonic processes.

sulfates A class of *minerals* that are compounds of the sulfate anion (SO_4^{2-}) and metallic *cations.*

sulfides A class of minerals that are compounds of the sulfide anion (S^{2-}) and metallic *cations.*

superposed stream A *stream* that erodes a gorge in a resistant *formation* because its course was established at a higher level on uniform *rock* before downcutting began. (Compare *antecedent stream.*)

superposition, principle of See *principle of superposition.*

surface wave A type of *seismic wave* that travels around Earth's surface from the *focus* of an *earthquake* and arrives at a *seismograph* later than S waves.

surge A sudden period of fast movement of a *valley glacier.*

suspended load All the material temporarily or permanently suspended in the flow of a current. (Compare *bed load.*)

sustainable development Development that meets the needs of the present without compromising the ability of future generations to meet their own needs.

suture A narrow zone where two continental blocks have been juxtaposed by plate convergence and the ocean basin that once separated them has been entirely subducted. Suture zones are often marked by *ophiolite suites.*

syncline A troughlike fold of layered *rocks* that contains younger rock layers in the core of the fold. (Compare *anticline.*)

talus Large blocks of broken *rock* that fall from a steep cliff of *limestone* or hard, cemented sandstone and accumulate in a gentler slope at the foot of the cliff.

tar sands A deposit of *sand* or *sandstone* that once contained oil but has lost many of its volatile components, leaving a tarlike substance called natural bitumen.

tectonic age The time that a *rock* was last subjected to crustal *deformation* intense enough to reset the isotopic clocks within the rock by metamorphism.

tectonic province A large-scale region formed by particular tectonic processes.

tensional force   A force that stretches a body and tends to pull it apart. (Compare *compressive force; shearing force.*)

terrace   A flat, steplike surface in a stream valley that parallels a stream above its *floodplain,* often paired one on each side of the stream, marking a former floodplain that existed at a higher level before regional uplift or an increase in *discharge* caused the stream to erode into the former floodplain.

terrestrial planet   Any of the four inner planets of the solar system (Mercury, Venus, Earth, and Mars) that formed from dense matter close to the Sun, where conditions were so hot that most of their volatile materials boiled away. Also called Earthlike planets.

terrigenous sediment   *Sediment* eroded from the land surface.

texture   The sizes and shapes of a rock's mineral crystals and the way they are put together.

thermal subsidence basin   A *sedimentary basin* that develops in the later stages of plate separation as *lithosphere* that was thinned and heated during the earlier rifting stage cools, becomes more dense, and subsides below sea level. (Compare *rift basin.*)

thermohaline circulation   A global three-dimensional oceanic circulation pattern driven by differences in the temperature and the salinity—and therefore in the density—of ocean waters.

thermoremanent magnetization   Permanent magnetization of magnetizable materials in *igneous rocks* when groups of atoms of the material align themselves in the direction of the *magnetic field* that exists when the material is hot and are then locked into place when the material cools below about 500°C.

thrust fault   A low-angled *reverse fault*—one with a dip of less than 45°.

tidal flat   A muddy or sandy area that is exposed at low *tide* but is flooded at high tide.

tide   The twice-daily rise and fall of the ocean caused by the gravitational attraction between Earth and the Moon.

till   Unstratified and poorly sorted *drift* deposited directly by a melting *glacier,* containing particles of all sizes from *clay* to boulders.

tillite   The lithified equivalent of *till.*

topography   The general configuration of varying heights that gives shape to Earth's surface, which is measured with respect to sea level.

topset bed   A horizontal bed of *sediment*—typically *sand*—deposited on top of a *delta.*

trace element   An element that makes up less than 0.1 percent of a mineral.

transform fault   A plate boundary at which the plates slide horizontally past each other and *lithosphere* is neither created nor destroyed.

transition zone   The portion of the *mantle* bounded by two abrupt *phase changes* at depths of about 410 and 660 km.

tributary   A *stream* that discharges water into a larger stream.

troposphere   The lowest layer of the atmosphere, which has an average thickness of about 11 km, contains about three-fourths of the atmosphere's mass, and convects vigorously due to the uneven heating of Earth's surface by the Sun. (From the Greek *tropos,* meaning "turn" or "mix.") (Compare *stratosphere.*)

tsunami   A fast-moving sea wave, generated by an *earthquake* that lifts the seafloor, that propagates across the ocean and increases in size when it reaches the shore.

tuff   A volcanic *rock* formed by the *lithification* of small *pyroclasts.* (Compare *breccia.*)

turbidity current   A *turbulent flow* of water carrying a *suspended load* of *mud* that flows down the *continental slope* beneath the overlying clear water.

turbulent flow   Fluid movement in which streamlines mix, cross, and form swirls and eddies. (Compare *laminar flow.*)

twentieth-century warming   The rise in Earth's average surface temperature by about 0.6°C between the end of the nineteenth century and the beginning of the twenty-first.

U-shaped valley   A deep *valley* with steep upper walls that grade into a flat floor; the typical shape of a valley eroded by a *glacier.*

ultra-high-pressure metamorphism   Metamorphism occurring at pressures greater than 28 kbar.

ultramafic rock   An *igneous rock* consisting primarily of mafic *minerals* and containing less than 10 percent feldspar. (Compare *felsic rock; mafic rock.*)

unconformity   A surface between two *rock* layers in a *stratigraphic succession* that were laid down with a time gap between them.

unconsolidated material   *Sediment* that is loose and uncemented. (Compare *consolidated material.*)

uniformitarianism, principle of   See *principle of uniformitarianism.*

unsaturated zone   The level above the *groundwater table,* in which the pores of the *soil* or *rock* are not completely filled with water. Also called the vadose zone. (Compare *saturated zone.*)

upper mantle   The portion of the *mantle* that extends from the *Mohorovičić discontinuity* to the base of the *transition zone* at about 660 km in depth.

valley   The entire area between the tops of the slopes on both sides of a *stream.*

valley glacier   A river of ice that forms in the cold heights of a mountain range, where snow accumulates, then moves downslope, either flowing down an existing stream valley or carving out a new valley. (Compare *continental glacier.*)

varve   One pair in a series of alternating coarse and fine *sediment* layers deposited on a lake bottom by a valley glacier, formed in one year by the seasonal freezing of the lake surface.

vein   A sheetlike deposit of *minerals* precipitated in fractures or *joints* in *country rock,* often by a *hydrothermal solution.*

ventifact   A pebble with several curved or almost flat surfaces that meet at sharp ridges, formed by *sandblasting* of the pebble's windward side.

viscosity   A measure of a fluid's resistance to flow.

volcanic ash   *Pyroclasts* less than 2 mm in diameter, usually glass, that form when escaping gases force a fine spray of *magma* from a *volcano.* (Compare *bomb.*)

volcanic geosystem   The total system of *rocks, magmas,* and processes needed to describe the entire sequence of events from melting to the eruption of *lava* from a *volcano* at Earth's surface.

volcano   A hill or mountain constructed from the accumulation of erupted *lava* and *pyroclasts.*

wadi   A desert *valley* that carries water only briefly after a rain. Called a *dry wash* in the western United States.

wave-cut terrace   A level surface formed by wave *erosion* of a rocky *shoreline* beneath the surf zone, which may be visible at low *tide.*

weathering   The processes by which *rocks* are broken down at Earth's surface to produce *sediment* particles. (See also *chemical weathering; physical weathering.*)

Wilson cycle   The sequence of tectonic events on continents caused by the formation and closure of ocean basins. The cycle comprises (1) rifting during the breakup of a continent, (2) *passive margin* cooling and *sediment* accumulation during *seafloor spreading* and ocean opening, (3) *magmatic addition* and *accretion* during *subduction* and ocean closure, and (4) orogeny during continent-continent collision.

zeolite   (1) A class of silicate *minerals* containing water within their *crystal* structure, formed by metamorphism at very low temperatures and pressures. (2) The lowest metamorphic grade.

Aa, 316
Ablation, glacial, 592–593, *593*
Abrasion, 121, *121,* **508**
Absolute ages
   explanation of, **196**
   geological time scale and, *210,*
      210–211
Abyssal hills, **576,** 578
Abyssal plain, **570**
Acadian orogeny, 269
Accreted terrains, **263,** 265
Accretion
   areas of, *264,* 264–265, *265*
   explanation of, **263**–264, *264*
   tectonic processes and, 265–266
Accretionary wedge, 571
Accumulation, in glaciers, **592–593,** *593*
Acidic water, 292, *292*
Acidophiles, 290, 291t
Acid rain, 654
Acids, 439
Acid test, 69, *69*
Active margins, **260,** 570, **570**–571
Adamantine luster, 72, 72t
African Plate, 31, *34,* 38
Aftershocks, **351**–352, *353*
Agassiz, Louis, 606, 607
Age. *See* Dating; Isotopic dating
A-horizon, 447
Alabaster. *See* Gypsum
Alaska earthquake (1964), 452–453, *453*
Albedo
   climate system and, 610
   explanation of, **413**
Albedo feedback, 415
Alberts, Bruce, 5
Allegheny Plateau, 256
Alluvial environments, 124
Alluvial fans, **521,** *522*
Alluvial sediments, in deserts, 543–544,
   544
Alpine glaciers. *See* Valley glaciers
Alpine-Himalayan belt, 267, *267*
Alpine-Himalayan orogeny, 266, 267,
   267, 268
Alps Mountains, 267
Alternative energy
   biofuels as, 657–658
   geothermal energy as, 660
   hydroelectric energy as, 658, *659*
   increasing use of, 669
   nuclear energy as, 655–657, *656, 657*
   solar energy as, 658, *658*
   wind energy as, *659,* 658–659

Aluminum, 62, 67
Amino acids, origin of life and, 298
Amorphous, 58
Amphiboles, **158**
   cleavage in, 71, *71*
   metamorphism and, 153
   in rocks, 76t, 95t
   structure of, *66*
Amphibolites, 160
Anaerobes, 290, 291t
Anaerobic zones, 291, 293, 294
Anatolian Plate, 31
Ancient water, 487
Andean orogeny, 266
Andesitic lava, *317,* 317–318, *318*
Andes Mountains, *35,* 591
Angle of repose
   explanation of, **450**
   of unconsolidated material, *451*
   of weathered rubble, 451
Angular unconformities, 200, *201,* 204
Anhydrite, as sulfate, 68
Anhydrous, 68
Anions
   classification by, 64–65, 65t
   explanation of, **61**
Antarctica
   climate change and, 45
   continental glaciers in, 590, *590, 591*
   glacier movement in, 594, 596,
      597, 598
   ice-core drilling in, 419
   ice sheets of, 587
Antecedent stream, **505,** *505,* 507
Anthracite, 654
Anthropocene, **666**
Anthropogenic emissions, global
   warming and, 16, 646
Anthropogenic global change
   carbon emissions and, 432–433
   fingerprints of, 428–430
Anticlinal theory, 192
Anticlinal trap, 192, *192, 647*
Anticlines, **180**
Apatite, Mohs scale of hardness and, 69t
Apollo missions (spacecraft), 221, 225,
   230
Appalachian fold belt, 254, 256, *257,* 267
Appalachian Mountains, 38, 44, *627,* 652
Appalachian orogeny, 269
Aquicludes, **482**
Aquifers
   ancient microordanisms in, 494
   coastal, 484, *485*

confined and unconfined, 482, *482*
   explanation of, **482**
   recharge and discharge in, 482–484,
      485
   speed of flow and, 485
Arabian Plate, 123
Aragonite, precipitation of, 493
Archaea, 290
Archean eon, *18–19,* 210, 274, *275*
Archimedes, 404
Arctic ice cap, 590, *664,* 665
Arctic National Wildlife Refuge (ANWR),
   651, *651*
Arenal volcano, *319*
Argon, in atmosphere, 409
Aristotle, 8
Arkoses, **134**
Army Corp of Engineers, 501
Arroyos. *See* Valleys
Arsenic, 491, *492*
Artesian flow, **482**
Artesian wells, 482
Asbestos, crystal habit of, 73
Aseismic ridge, 331
Ash, volcanic, **93**
Ash-flow deposits, **326**
Asteroids
   explanation of, **224**
   impact of, 236, 237t
   mass extinctions and, 236, 303–304
Asthenosphere
   explanation of, *14,* **16**
   temperature within, 314
Astrobiologists, **306**
Astrobiology
   explanation of, 306–307
   habitable environments on Mars and,
      308, *308*
   habitable zones around stars and, *307,*
      307–308
Astronomical positioning, 41–42
Asymmetrical fold, 181, *181*
Atlantic Ocean
   profile of, 575–577, *576*
   seafloor spreading in, 257
Atmosphere
   climate system and, 15–16,
      408–409
   composition of, 409
   explanation of, *14*
   formation of, *227,* 227–228
   layers of, 408, *409*
   origin of Earth's, 299–300
   volcanism and, *328,* 328–329

Atmosphere-biosphere gas exchange, 426–427
Atmosphere-ocean gas exchange, 426
Atolls, 139
Atomic mass, **59**
Atomic number, **59**
Atomic structure
    explanation of, 59, *59*
    of minerals, 61–62, *62*
Autotrophs, **285**
Axial plane, of fold, 180

Backshore, 565
Backwash, 555
Bacteria, 290
Badlands, **625,** *625*
Baguios, 558. *See also* Hurricanes
Baikal Lake, 522
Baja California, 44, 53, *53*
Baltica, 267, 269
Banded iron formations, **300,** *300*
Barchans, **539**
Barrier islands, **568,** *569*
Basaltic lavas, *315,* 315–317, *316*
Basalts
    composition of, **96–97**
    as igneous rock, 74, 95
    magmatic differentiation and, 101
    texture of, *92*
Basil slip, **594,** *595*
Basin and Range Province, 185
Basins
    drainage, 503–504, *504*
    explanation of, **182,** *182,* 183
    flexural, 123, 183
    rift, 122, 183
Batholiths, **103**
Bay of Fundy, 556
Bays, wave refraction and, 556
Beaches
    common forms of, 566
    erosion of, 520, 565–567, 582–583
    explanation of, **564**
    preservation of, 566–567, *567*
    restoration of, 567, 582–584
    sand budget of, 565–566, *566*
    sea level change and, 568–570
    structure of, 565
Beagle (ship), 200
Bedding, **77–78**
Bedding sequences, **129**–130, *130*
Bed load, **510**
Bedrock, 173, 181–182, 623, 625, 637–638
B-horizon, 447
Big Bang, 19, 222. *See also* Cambrian explosion
Bioclastic sediments, **119**
Biodiversity, 666–667
Biofilms, 289
Biofuels, *657,* 657–658

Biogeochemical cycles
    explanation of, **288**
    microorganisms and, 294
Biological sedimentary environment, 126–127, *127*
Biological sediments, 77, *77,* **118**–119, *119*
Biology, 283. *See also* Geobiology
Biomass, 643–644
Biosphere
    biogeochemical cycle of phosphorus and, 288, *289*
    climate system and, 412
    explanation of, 13, *14,* **284,** 412, *412*
    magnetic fields and, 402–403, *403*
    as system, *284,* 284–288, *285,* 286t, *287,* 287t
Biotechnology, 669
Biotite, in igneous rocks, 95t
Bioturbation, **128,** *129*
Bituminous coal, 654
Black Hills dome, *182*
Blowout dunes, **539**
Blue Ridge province, 256
Blueschist, 164
Blyde River Canyon Nature Reserve, *508*
Bombardment. *See* Heavy Bombardment; Late Heavy Bombardment; Meteorites
Bombs, **93–94**
Bonds. *See* Chemical bonds; Covalent bonds; Ionic bonds
Bora Bora atoll, 139, *139*
Boreholes. *See* Drilling
Bottomset beds, **513**
Bowen's reaction series, 99, *100*
Bowknot Bend (Utah), *196*
Braided streams, *502,* 502–503
Brakers, 554
Branching networks, 505
Breakwaters, 566
Bressicas, **320**
Brittle material, **176**
Bryce Canyon, 204
Building codes, **372–373**
Buoyant fragment, accretion of, *264*
Burial
    in rock cycle, 117, *117*
    of sediments, 130, 131
Burial metamorphism, *153,* 154
Bushveld deposit (South Africa), 112

Calcite
    acid test for, 69, *69*
    as biological sediment, 77
    as carbonate, 67, 69
    cleavage in, 70, *71*
    as mineral, 58, *58*
    Mohs scale of hardness and, 69t
    precipitation of, 493
    in rocks, 76t
    in sedimentary rocks, 78, 122, 419

Calcium, in seawater, 122, 137, 138
Calcium carbonate. *See also* Calcite
    at hot springs, 493
    indirect precipitation of, *293*
    windblown, 533
Calcium cycle, *425,* 425–426
Calcium ions, 122
Caldera collapse, 335–336
Calderas, *322,* 323
Caledonian orogeny, 269
Cambrian explosion, **301,** *301,* 303
Canada. *See* North America
Canadian Shield
    epeirogeny and, 271
    explanation of, 254–255, *255*
Canyons. *See specific canyons*
Capacity, sediment load, **510**
Cape Cod, 568, *569*
Cape Cod National Seashore, 567
Cape Hatteras, 580, *580*
Carbohydrates, 285
Carbon. *See also* Diamonds; Graphite
    atomic structure of, *59, 60*
    as building block of life, 286, 310
    inorganic, 58
    isotopic dating and, 209
    methane production and, 305
Carbon-12, 59, 209, 310
Carbon-13, 310
Carbon-14, 209, 310, 607
Carbonate compensation depth, **579,** *579*
Carbonate environments, 127
Carbonate ion, 67
Carbonate platforms, 137, *138,* 142
Carbonates
    in seawater, 140
    structure of, 67, *67,* 70
Carbonate sediments/rocks
    direct biological precipitation of, 136–137
    explanation of, 78, **136**
    indirect biological precipitation of, 138, 140
    reefs and evolutionary processes and, 137–138, *138*
    vuggy pore networks in, 479
Carbon cycle
    atmosphere-biosphere gas exchange and, 426–427
    atmosphere-ocean gas exchange and, 426
    engineering of, 669
    explanation of, **423,** 426, *427*
    function of, *426,* 426–427, *427*
    human activities and, 427–428, 432–433
    lithosphere-atmosphere gas exchange and, 427
Carbon dioxide
    in ancient ice, 592
    in atmosphere, 409

climate regulation and, 16
global warming and, 428, 429, *429,*
439, 660, 660, 661, 667–668
photosynthesis and, 310, *311,* 432, 433
in rainwater, 488
weathering and, 439, *440*
Carbon economy, 644–645
Carbon emission stabilization, 669–670,
670
Carbonic acid, 439, *440,* 441, 488
Carboniferous period, 644
Carbon intensity, **653**
Carbon isotopes, 298, 310
Carbon sequestration, **669**
Caribou, Maine, 592
Carlsbad Caverns, 488–489
Cascade Mountains, 473
Cascadia subduction zone, 38, 371
Casita volcano, 449
Casler, Michael, *657*
Cassini-Huygens mission (spacecraft),
246, 248
Cataclastic texture, 184
Cations
explanation of, **61**
in minerals, 61–62, 67
Cation substitution, 62, 67
Caucasus Mountains, 267
Cavendish, Henry, 10
Caves, 488, *488*
Cementation, 77, **131,** 132
Cenozioc era, 206
Central eruptions, 321–324, *322–324*
Central vent eruptions, 315, *315*
Chalcopyrite, 82, *83*
Challenger (ship), 551
Channeled Scablands, 605, *605*
Channelization, 501–502
Channels
in braided streams, *502,* 502–503
explanation of, 500, **500**
origin of, 507–509, *507–509*
patterns of, 500–501, *501*
Chelyabinsk meteor explosion, *15*
Chemical bonds, 61
Chemical compounds, *60,* 60–61
Chemical contaminates, in groundwater,
490
Chemical reactions, *60,* 60–61
Chemical sedimentary environment,
126–127, *127*
Chemical sediments, 77, *77,* **118,** *118*
Chemical stability, **442**
Chemical stratigraphy, 213
Chemical weathering. *See also*
Weathering
carbon dioxide and, 439, *440*
chemical stability and, 442
in deserts, 542, *543*
erosion and, 508
explanation of, **116,** 436

organisms and, 443
oxygen and, 441–442
water and, 438
Chemoautotrophs, **294**
Chemofossils, **299**
Chert, **143**–144
Chicxulub site, 304, *304*
Chile earthquake, 364
Chlorinated solvents, 490
Chomolungma, 279
C-horizon, 447
Chromite, as ore mineral, 83, *84*
Chromium, 67
Chromium ore, 83, *84*
Chrysotile, *73*
Cinder cones, 321
Cinder-cone volcanoes, *322*
Cinnabar, 82, *83*
Circular structures, *182,* 182–183
Circumference, of Earth, 23–24
Cirque, **601**
Cities, on floodplains, 506, *506*
Civilization, as global geosystem, 642
Clark, William, 499
Clastic particles, 117–118, 121
Clay, **135**
Clay minerals
structure of, 67
weathering and, 438
Claystones, **135**
Clean coal technologies, 654
Cleavage
explanation of, **70,** 74t
in metamorphic rocks, 155
number of planes and pattern of,
70–71, *71*
quality of surfaces and, 71
Cliffs, structurally controlled, 626
Climate
effect on streams, 521
explanation of, 15, 408
feedback between topography and,
628–629
hydrology and, 472–473, *473*
plate tectonic processes and, 473
rock cycle and, 79
solar energy and, 13, *14*
Climate change. *See also* Global warming
biodiversity loss and, 666–667
carbon dioxide and, 16
catastrophic changes to climate system
and, 665
cryosphere change and, 664, *665*
desertification and, 541–542
effects on ecosystems and resources
of, 663t
fossil fuel consumption and, 646
movement of tectonic plates and, 45
ocean acidification and, 666
regional weather patterns and,
662–663, *663*

sea level rise and, 664–665, *665*
species and ecosystem migration and,
665
uplift and, 632
Climate models, **415**–416
Climate system
anthropogenic global change and,
428–430, *429, 431*
atmosphere and, 408–409, *409*
biosphere and, 412, *412*
calcium cycle and, 425, 425–426
carbon cycle and, 423, *426,* 426–428
components of, 15, 408–412
cryosphere and, 411, *411*
dating of, 213, *213*
explanation of, *14,* **15**–16, 408
feedbacks in, 414–415
glaciations and, 411, 609
greenhouse effect and, *413–415,*
413–416
hydrosphere and, 409–411, *410*
importance of, 407
lithosphere and, 411–412
plate tectonics and, 79, *80,* 81, 166,
473, 632
rock cycle and, 79, *80*
Climate variation
explanation of, 416–417
long-term global, 417–422, *418,*
420, 421
during most recent glacial cycle,
422–423
short-term regional, 417
Coal
environmental costs of, 654
explanation of, **144,** 644, 653–654
formation of, 58, *653,* 653–654
resources of, 654, *654*
Coal ash, 654
Coal mining, 654, *655*
Coastal aquifers, 484, *485*
Coastal plain, 256–257
Coastal processes
explanation of, 552–553
hurricanes and storm surges and,
557–559, *559,* 560t, 561–564, *562–564*
surf zone and, *554,* 554–555
tides and, 556–557, *557*
wave motion and, 553–554, *554*
wave refraction and, 555–556
Coastlines
beaches and, 564–567, *565, 566*
depositional, 568, *569*
effect of coastal processes on, 564
effect of sea level change on, 568–570
erosion along, 567–569, *568*
erosion and deposition at, 567–568,
569
explanation of, 552–553, *553*
sea level change and, 568–570
shaping of, 564–570, *565–569*

Coastlines (*Contd.*)
  tides near, 557
  variations in, *553*
Coconino Sandstone, 204
Cocos Plate, North American Cordillera
    and, 258, *259*
Code of ethics, for scientists, 5
Coesite, 155
Coleridge, Samuel Taylor, 469
Collision
  continent-continent, 165–166
  paleocontinental, 44
Color
  in felsic rocks, 95, 96
  mineral, *72,* **72**–73, *73,* 74t
Colorado Plateau
  epeirogenic processes in, 272
  explanation of, 259
  stratigraphy of, 204, *205*
Colorado River
  drainage basin of, *504*
  sedimentary deposits in, 214
Columbia (supercontinent), 272
Columbia Plateau, *326,* 333
Columbus, Christopher, 23
Combustion engine, 644
Comets, exploration of, 225
Compaction, 77, **131**–132
Competence, **510**
Compressional waves, 10, 12, 355, 357, 384,
    386. *See also* P waves; Seismic waves
Compressive forces, **172**
Compressive tectonics, **185**
Computer simulations, 5
Concentration factor, 81
Conchoidal fracture, 71
Concordant intrusions, **103,** *104*
Conduction, **391**
Cone of depression, 483, *484*
Confined aquifers, 482, *482*
Conglomerate, **133**
Consolidated materials, **449,** 450
Contact metamorphism, *78,* **79,** *153,* 154
Continental basin, 260
Continental crust. *See also* Crust
  deformational style of, 184–185
  explanation of, 11, *11*
  formation of, 273–274, *275*
  geologic record of, 253
  rifting of, 36, *37*
Continental drift. *See also* Plate tectonics;
    Seafloor spreading
  assembly of Pangaea by, 44
  explanation of, 7, *28,* **28**–29
  reconstruction of history of, 44
Continental glaciers. *See also* Glaciers
  climate change and, 665
  explanation of, **590,** *591*
  history of, 587
  ice distribution and, 608, *608*
  lakes formed by, 605

Continental margins
  continental shelf and, 571
  continental slope and rise and, 571, *572*
  environments in, 126
  explanation of, **570**
  submarine canyons and, 571–572
  types of, *570,* 570–571
Continental rifting
  accretion of, *265*
  explanation of, 36
Continental rise, **570,** 571
Continental sedimentary environments,
    124, *125*
Continental shelf
  environments in, 124, 126
  explanation of, **122,** 571
  of passive margins, 571
  sediment in, 122–123, 579
Continental slope, **570,** 571
Continent-continent collision, 165–166,
    267
Continent-continent convergence, 38
Continents
  age of, 262
  deep structure and, *276,* 276–278
  grand reconstruction of, 44–45
  growth of, 262–266, *263*–*265*
  modification of, 266–267, *266*–*271,*
      269, 271–273, 272t, *273*
  ocean basins vs., 552
  tectonic provinces in, 261, *261*
Contours, **618**–620
Convection
  explanation of, *16,* **16**–17
  in mantle and core, 392
Convention on the Law of the Sea, 581
Convergence
  continent-continent, 38
  ocean-continent, 37–38, 163–165
  ocean-ocean, 36–37
Convergent boundaries
  earthquakes at, 364
  explanation of, **31,** *34*–*35,* 36
  illustration of, *32*–*33*
Copper ores, 82, *83*
Coral reefs. *See also* Reefs
  Darwin and, 139, *139*
  explanation of, 137
Cordillera, North American, 254,
    257–259, *258, 263,* 266
Cordón Caulle volcano, 328
Core. *See also* Inner core;
    Outer core
  composition of, *12,* 12–13, 226, 390
  explanation of, **10**
  inner and outer, 11–12, *14*
  seismic data on, 390
Core-mantle boundary, 10, **390**
Coriolis effect
  explanation of, 530–531
  hurricanes and, 563

Coronae, 235
Corundum, Mohs scale of hardness
    and, 69t
Country rock, **93**
Covalent bonds, **61,** 70
Covelite, 81
Crater counting, 230
Crater Lake Caldera, 323, *323*
Craters
  explanation of, 323
  impact, *236,* 304
  lunar, 230
  on Mars, 238, 241
  on Mercury, 232
  on Venus, 233, 234
Cratonic keels
  age of, 289
  composition of, *276,* 277–278
  explanation of, **276**–277
Cratons
  explanation of, **260**
  origins of, 273–274, *274, 275*
Creep, **456,** *457*
Cretaceous period, 206, 210, 257
Cretaceous-Tertiary mass extinction,
    303, *305*
Crevasses, in valley glaciers, **594,** *596*
Cristobalite, 64
Cross-bedding, **127,** *128*
Cross-cutting relationships
  explanation of, 201–202, *202, 203*
  relative age and, *203*
Cross sections, geologic, *174,* 175, *175,*
    182
Crude oil, **144,** 191–192, 644, 647. *See
    also* Oil; Petroleum
Crust
  composition of, *12,* 13, 226–227, 388
  deformational behavior of, 254
  explanation of, **11,** *11*
  extended, 260
  felsic, 263
  seismic data on, 388
  water in, 492–494, *493, 494*
Cryosphere
  climate change and, 664, *665*
  climate system and, 411, 609
  explanation of, *14,* 15, 411, 587
Crystal faces, 62, *63*
Crystal habit, **73,** 74t
Crystalline, 58
Crystallization
  explanation of, **61**
  of igneous rocks, 92
  laboratory studies of, 92
  of magma, 112–113
  of minerals, 62, 62–63, *63*
Crystals, geologic history in, 167–168
Cuestas, **626,** *628*
Curie, Marie Sklodowska, 208
Curiosity (spacecraft), 238, 240–243, *241*

Currents. *See also* Wind
  ocean, 515
  rip, 556
  strength of, 120, *121*
  as transport agents, 119–121, *120,*
    509–511, *509–511*
  turbidity, 126, 128, 571, *572*
Cyanobacteria, 299
Cyclones, 558
Cynobacteria, **288**

Dacite, **96**
Dams, stream, 520
Darcy, Henri, 485, 496
Darcy's law, **485,** 486, 496–497
Darwin, Charles, 139, 303
Dasuopu Glacier, *431*
Dating, 196. *See also* Isotopic dating
Daughter atoms, 208, 209
Davis, William Morris, 630–632, 634
Debris. *See* Unconsolidated materials
Debris avalanches, *459,* 459–460
Debris flows, 457, *458*
Decompression melting, **98,** 107, *109*
Deep Impact (spacecraft), 245, *246*
Deep mantle, *14*
Deep-sea drilling, 41, *41*
Deep-sea environments, 126
Deepwater Horizon explosion, 650, *651,*
  652
Deflation, *534,* 534–535, *535*
Deformation
  brittle and ductile behavior and,
    175–176, *176*
  circular structures and, *182,*
    182–183
  continental, 184–188, *185–188*
  explanation of, 171, **172**
  faults and, 177–179, *177–179*
  folds and, *179–181,* 179–182
  geologic history and, 188, *188*
  joints and, 183, *183*
  plate tectonics and, 172
  textures in, 183–184, *184*
Delaware Water Gap, *505*
Delta plain, 513
Deltas
  effect of ocean currents, tides and
    plate tectonic processes on, 515
  explanation of, **513,** *513*
  growth of, 513, *514,* 515
  human effects on, 515
Dendritic drainage, *504,* **505**
Density
  of Earth, 10
  of minerals, **64,** 73, **74t**
  of rocks, 10
Deposition
  by glaciers, 603–606, *604, 605*
  in rock cycle, 116, *117*
  sediment transportation and, 119

  at shorelines, 567–568
  by wind, 535–540, *536–540*
Depositional remanent magnetization,
  401
Desertification
  climate change and, 541–542
  effect of humans on, 542, 548
  explanation of, 529
  prediction of, 548–549
Desert landscapes, 544–546, *545*
Desert pavement, **534**
Deserts
  alluvial sediments in, 543–544
  colors of, 543
  eolian sediments in, 544
  evaporite sediments in, 544
  expansion of, 529
  floods in, 543, *544*
  landscape of, 544–546, *545*
  locations of, 540–542, *541*
  nature of, 124, 540
  plate tectonics and, 541
  role of human activities on, 542
  weathering and erosion in, 542–543
Desert varnish, **543,** *543*
Desert Watch project, 541
Development. *See* Sustainable
    development
Diagenesis
  explanation of, **130**
  metamorphism and, 153
  in rock cycle, 117, *117*
  of sediments, 130–132, *131*
Diamonds
  as chemical compound, 60, *60*
  cleavage in, 70
  formation of, 63–64, *64*
  in kimberlite pipes, 277, *277*
  microscopic, 155
  Mohs scale of hardness and, 69t
Diatoms, 581
Diatremes, 155, 323–324, *324*
Diesel, Rudolf, 657
Dietz, Robert, 30
Differentiation, gravitational, 225. *See
    also* Magmatic differentiation
Digital elevation model (DEM), 280, *619*
Dikes, **104**
Diorite, **96**
Dip, **173,** *173*
Dipolar field, 17, *17*
Dipole magnetic field, 396, **396**
Dip-slip faults, **177,** *178,* 179
Discharge
  in aquifer, 482–483, *484*
  calculation of, 517
  explanation of, **480, 516**
  of stream flow, 516–518, *517, 518*
Disconformities, 200
Discordant intrusions, **103**
Disseminated deposits, 82–83, *83*

Dissolution rates, 442
Distributaries, **513**
Divergent boundaries
  continental rifting and, 36
  earthquakes at, 364
  explanation of, **31,** *34,* 36, 330
  illustration of, *32–33*
  oceanic spreading centers and, 36
Divide, **503,** 504, *504*
Dolomite
  as carbonate, 67
  cleavage in, 70
  in rocks, 76t, 78, 136
Dolostone, **136**
Domes, *182,* **182–183**
Double-chain structures
  in amphiboles, 71
  explanation of, 65, *66*
Drainage basins, **503–504,** *504*
Drainage networks
  explanation of, **505**
  geologic history and, 505, 507
Drake, Edwin, 644
Drake's Folly, 644, *644*
Drift, **603.** *See also* Continental drift; Plate
    tectonics; Seafloor spreading
Drilling
  deep-sea, 41, *41*
  oil, 648
Drinking water, 491–492
Droughts
  examples of, 473, *474,* 475
  explanation of, 473, *474*
  severity of, 475
Drumlins, **604,** *604*
Dry washes, **545**
Ductile material, **176**
Dunes. *See* Sand dunes
Dust, **531,** 532
Dust devils, 533
Dust storms
  characteristics of, 532
  on Mars, 238, *238,* 533, *533*
Dwarf planets, **229**

Eagle Nebula, *223*
Earth. *See also* Interior of Earth
  ancient views of, 9
  characteristics of, 228, *228,* 236
  circumference of, 23, 24
  composition of, 10–13, *12, 388,*
    388–390, *389*
  core of, 10–13, *11, 12,* 226, 390
  crust of, 11, *11, 12,* 13, 226–227, 254,
    260, 388, 492–494, *493, 494*
  density of, 10, *12*
  evolution of life on, 19–21
  formation of, 19, 225–227, *227,*
    296–297
  geobiological events in history of,
    296–300, *297–300*

Earth. *See also* Interior of Earth (*Contd.*)
  gravitational differentiation of, 225, *227*
  interactions shaping, 13
  layers of, 9–13, *10*
  magnetic field on, *396*, 398–399, *398–400*
  mantle of, 10, *11*, 12, *12*, *14*, 107, 227, 388–390, *389*, 394, *395*
  orbital variations of, 17, 420
  origin of life on, 296–299
  origin of oxygenated atmosphere on, 299–300
  ray path of seismic waves through, 384–387, *385*, *386*
  rotation of, 530, 556
  shape of, 8–9, 397, *397*
  surface of, 8–9, 230
  temperature of, 12, 273
  as terrestrial planet, 224
  visualizing three-dimensional structure of, 394, *395*, 396
Earthflows, 456–457
Earthquake engineering, 372–373
Earthquakes. *See also* Seismic waves; Tsunamis; *specific earthquakes*
  at convergent boundaries, 34
  at divergent boundaries, *34*
  early warning systems for, 373
  elastic rebound theory and, 349, *350*, *351*, *351*
  explanation of, 9, **348**–349
  faulting and shaking from, *366*, 366–367
  fault patterns and, 362, *363*, 364–365, *365*
  fault rupture during, 348–349, 351
  fires following, 369
  foreshocks and aftershocks from, 351, 352, *352*
  frequency of, 359–360
  hazards and risks of, 365–369, *366–369*
  intraplate, 364
  landslides and ground failures from, 367, *453*
  locating focus of, 357
  magnitude of, 357, *359*, 359–360
  measuring size of, 357, *358–359*, 359–362, 360t, *361*, *362*
  on megathrusts, 185–186, 367, *368*
  methods to study, 353–362
  in the Pacific Ring of Fire, *31*
  plate tectonics and, 348
  prediction of, *375*, 375–378
  reducing risk of, 369–373, *370–373*
  safety steps for, 374
  seismic waves generated by, 343, 355, *356*, 357, *358*
  seismographs to study, 354, *354*
  shaking intensity of, 360–361
  silent, 362

  in subduction zones, 185–186, 347
  tsunamis and, 367–369, *368*, *369*
  volcanic eruptions and, 338
Earth system
  carbon in, 286
  climate system and, *15*, 15–16
  components of, *14*, 14–15
  explanation of, 13, **13**–15, *14*
  geodynamo and, *17*, 17–18
  interactions in, 18, *18*
  as open system, *13*, 14
  plate tectonic system and, *16*, 16–17
East African Rift Valley, 36, 184
Ebb tide, 557
Eclogites, 155
Ecology, 6
Ecosystems
  biogeochemical cycles in, 288
  climate change and, 663t, 665, *665*
  explanation of, *284*, **284**–285
  inputs in, 285–286, 286t
  processes and outputs in, 286, *287*, 287t, 288
Effluent streams, 480
E-horizon, 447
Einstein, Albert, 5
Elastic rebound theory
  explanation of, **349,** *350*, *351*
  forecasting and, 376–377
El Capitan mountain (Texas), *126*
El Chichòn volcano, 412
Electrons, 59, *59*
Electron sharing, **60**
Electron transfer, **60**
Elements, 113
Elevation
  explanation of, **618**
  landscape development and, *630*
  weathering and erosion and, 628–629
El Niño, 417, *417*
El Niño – Southern Oscillation (ENSO), *416*, 417
Emergency management
  for earthquakes, 373, *374*
  for tsunamis, 373
Emmons Glacier, *596*
Endeavour (space shuttle), 279
End moraine, 603t
Energy. *See also* Alternative energy; Fossil fuels
  explanation of, 286
  geothermal, 340–341
  global consumption of, 645–646, *646*
Energy conservation, 645
Energy-efficient technologies, 645
Energy policy, 668–669
Energy resources
  explanation of, *643*, 643–644
  future outlook for, 646
Enhanced greenhouse effect, 428, 430

Environmental impact
  of coal, 654
  of hydraulic fracturing, 652
  of oil, *650*, 650–651, *651*
Eolian processes, **530,** 533
Eolian sediments, 544. *See also* Sand dunes
Eons, **210**–211
Epeirogeny, **272**–273, *273*
Epicenter, **351,** 362
Epochs, **202**
Equator, wind belts and, 530, 531
Eras, **202**
Eratosthenes, 9, 23–24
Ergs, 544
Erosion. *See also* Mass wasting; Weathering
  of beaches, 565–567, 582–583
  of coastlines, 567–569, *568*
  in deserts, 542–543
  elevation and, 628–629
  explanation of, **77,** 116, *117*, 436
  glacial, 587–588
  by groundwater, 488–489, *488–490*
  headward, 508
  precipitation and, 473
  rates of, 637–638
  sediment transport and, 510–511
  at shorelines, 567–568
  by streams, 507–509, *507–509*, 543, 623, *624*, 625
  uplift and, *279*, 279–280, 632, *633*, 634–636
  weathering and, 117–119, 118t, 444–445, *445*, 508
  wind and, *534*, 534–535, *537*
Erosion of the Mother Lode, 85
Erratics, 603
Eruptions. *See also* Volcanoes
  central vent, 315, *315*
  clouds from, 336–337
  fissure, *325*, 325–326, *326*, 333, 334
  phreatic, 318, *318*
Eskers, **604**
Ethanol, 657
Eukaryota, 290
Eukaryotic algae, 300
Eukaryotic fossils, 300
Euphrates River, 506
Eurasian Plate, 31, *34*, *35*, 38, 123, 267
Evaporation
  of seawater, 140–141, *141*
  of water from oceans, 470–471
Evaporite environments, 127
Evaporites
  marine, 140–141, *141*
  nonmarine, 141–142
Evaporite sediments
  marine, 140–141, *141*
  nonmarine, 141–142
  playas and, 544

Evolution
  explanation of, **301**
  geologic record and, 19–21
  mass extinctions and, 21
  microorganisms and, 297
  natural selection and, 301, 303
  reefs and, 137–138, *139*
Evolutionary radiation, **300,** *302, 303*
Ewing, Maurice ("Doc"), 29, *29*
Exfoliation, **444,** *444*
Exhumation, **161,** 164, 166
Exotic terrains, 264–265
Explanets, **248**
Extended crust, 260
External heat. *See* Solar energy
External heat engine, 13
Extinctions, 285. *See also* Mass extinctions
Extremophiles
  explanation of, **290**–291, 291t
  in hot springs, 493
Extrusive igneous rocks, 76, 76t, **93**–95,
  94
Extrusive puroclasts, *93*
Eyjafjallajökull volcano (Iceland), 331,
  336–337

Farallon Plate
  North American Cordillera and, 258,
  259
  subduction of, 262
Fault breccia, 184, *184*
Faulting, as earthquake hazard,
  366–367
Fault mechanisms
  explanation of, 361, *361*
  plate boundaries and, 362, *363*
Fault rupture, during earthquakes, 351
Faults
  earthquakes and, 348–349, 362, *363,*
  364–365, *365*
  explanation of, *172,* 172–173, *173,* **177**
  identification of, 177–179, *178, 179*
  types of, 177, *186,* 361
Fault scarps, 177–178, *179*
Fault slip, **349,** *350, 352*
Fault systems, 364–365, *365*
Fault traces, 178
Fault trap, *647*
Faunal succession, 199, *199*
Feedback, 414–415
Feldspars
  formation of, 63
  in igneous rocks, 76t, 95t
  structure of, *66,* 67
  weathering of, 438, 439, *440,* 441
Felsic crust, 263
Felsic minerals
  explanation of, 94, 95, 95t
  melting temperature for, 97, *97*
Felsic rocks, 95–97, *96*
Fibrous fracture, 71

Fig Tree formation, 299
Fire, from earthquakes, 369
Firn, **592**
Fission, 655
Fissure eruptions, *325,* 325–326,
  333, 334
Fjords, **602**
Flake tectonics, **235**
Flank collapse, 335
Flash floods, in deserts, 543, *544*
Flexural basins, **123,** 183
Flint, 143–144
Flood basalts
  explanation of, **326**
  mass extinction and, 343–344
Flood frequency curve, 518, *519*
Floodplains
  cities on, 506, *506*
  development on, 503, *503*
  explanation of, **500,** *501*
  terraces in former, 521–522
Floods
  in deserts, 543, *544*
  explanation of, **518**
  flash, 543, *544*
  global warming and, 665, *666*
  from hurricanes, 558
  prediction of, 518
  protection from, 506
  sediments deposited by, 503
Flood tide, 557
Florida Peninsula, 257
Flow, 453, 470
Fluid-induced melting, **98,** 109, *110*
Fluids, metamorphic process and, 153
Fluorite, and Mohs scale of hardness, 69t
Flux, 284
Focus (earthquake)
  explanation of, **351**
  method to locate, 357, *358*
Fold and thrust belts, *185,* ¶186–187
Fold axis, 180, *180*
Folds
  explanation of, *172,* 172–173, **179**
  relationships in, 181–182
  size of, *179,* 179–180, *180*
  types of, 180–181, *181*
Foliated rocks, **155**–157, *157*
Foliation, 79, **155,** *156*
"Fool's gold." *See* Pyrite
Foot wall, **177**
Foraminifera, **136**–138
Foraminiferal oozes, **579**
Ford, Henry, 657
Forearc basin, 571
Forecasting
  long-term, 375
  medium-term, 376–378
Foreset beds, **513**
Foreshore, 565
Formations, **173**

Fossil fuels. *See also* Coal; Natural gas;
  Oil
  climate change and, 646
  consumption of, 645, *645*
  explanation of, **644**
  resources of, 646–655
Fossils
  accreted terrain and, 264
  Cambrian, 303
  continental drift and, 28–29, *29*
  eukaryotic, 300
  explanation of, 197, *197*
  geologic time scale and, 297
  origin of life and, 298–299
  Pleistocene, 419
  Precambrian, 301, *302, 303*
  as recorders of geologic time,
  198–200
Fracking, 135, 651–652, *652*
Fractional crystallization, **99**–100, *100*
Fracture, **71,** 74t
Fracture porosity, 478
Franklin, Benjamin, 28
Fresh water, 471
Frost wedging, **443**
Full-cost accounting, 668
Fumaroles, 327, *327*
Fusion power, 669

Gabbro, 95, **96**
Galápagos Islands, 332
Galena, as ore mineral, 82, *83*
Ganges River, 503
Garnet
  chemical composition of, 167–168
  in mantle, 107
  in rocks, 76t
  structure of, 71
Garnierite, 81
Gases, volcanic, 326, 327
Geco Topaz (vessel), *387*
General circulation models, 415
*General Map of Strata in England and
  Wales* (Smith), 199
Genes, **289**
Genetic information, in living organisms,
  289–290
Geobiological events, geologic time scale
  of, 296, *297*
Geobiologists, 288
Geobiology, 6, *6,* **283**
Geochemical cycles
  chemical reactions and, 423–424, *424*
  explanation of, **423**
  residence time and, 423
  transport across interfaces and,
  424–425
Geochemistry, 6, *6*
Geodesy
  explanation of, **8**–9, **41**
  plate movement and, 41–42

Geodynamo
  Earth's magnetic field and, 17
  explanation of, *14*, **17**–18
Geoid, 397, *397*
Geologic cross sections, *174*, 175, *175*
Geologic deduction, intermittent
    eruption of geysers and, 493
Geologic faults, *172*, 172–173. *See also*
    Faults
Geologic folds, *172*, 172–173. *See also*
    Folds
Geologic history, of region, 188, *188*
Geologic maps
  cross sections in, *174*, 175, *175*, *182*
  explanation of, **173**–174, *174*
  oil exploration using, 191–192
  of Siberian Traps, 343
  symbols on, 174, *174*
Geologic records
  absolute age and, 196–197
  cross-cutting relationships and,
    201–202, *202*, *203*
  drainage patterns and, 505, 507
  earth's origin and, 19
  evolution of life and, 19–21
  explanation of, **6**–7, *7*
  fossils as recorders of, 198–200, *199*
  gaps in, 200, *200*, *201*
  glaciations and, 609–611, *610*
  isotopic clocks and, 207–209, *208*, 209t
  overview of, 18–19, 195–196, 253
  time line of events in, *18–19*
Geologic time scale
  eons and, *210*, 210–211
  explanation of, **202**, *206*
  geobiological events and, 296, *297*
  intervals on, 202
  mass extinctions and, 202, 206
  perspectives on, 211, *211*
  petroleum source rocks and, 206–207,
    207
  as relative time scale, 207
  stratigraphy and, 197–198, 204, *205*,
    212, 212–213, *213*
Geology
  explanation of, **4,** 283
  as science, 5–8, *5–8*
Geomorphology, 617, **618,** 636. *See also*
    Landscapes
Geophysics, 6, *6*
Geosystems. *See also specific system*
  civilization as global, 642
  explanation of, **15**
  interactions among, 18
  interactions of landscapes and, 626,
    628–630, *629*, *630*
  interactions of volcanoes and other,
    326–329, *327*, *328*
  soils as, *446*, 446–447
  streams as, 515–522, *516–522*
  volcanic, 314, 315

Geotherm, **392**
Geothermal energy, 340–341, 660
Geothermal gradient
  explanation of, 151, *151*, 392, 394
  in sedimentary basin, 145
Geothermometer, 151
Geysers
  explanation of, 327
  hydrothermal waters and, 493
  intermittent eruption of, 493–494, *495*
  method to study, 494
The Geysers (California), 341, *342*
Giant's Causeway basalts, 96–97
Gilbert, William, 396
Glacial budget, 592–593, *593*
Glacial cycles
  ancient glaciations and, 609–611, *610*
  explanation of, **418**, 606–607
  Milankovitch cycles and, 420–422, *421*
  Pleistocene glaciations and, 609
  sea level change and, 608–609
  variations in most recent, 422–423
Glacial drift, 603
Glacial environments, 124
Glacial erosion
  effects of, 587–588
  process of, 598, 600–602, *600–602*
Glacial rebound, **272**, 393, *393*
Glacial sedimentation
  explanation of, 603
  ice-laid depositions and, 603–604
  water-laid deposits and, *604*, 604–605,
    605
Glaciation. *See also* Ice ages
  ancient, 609–611, *611*
  continental drift and, 29
  sea level change and, 606–607
  Wisconsin, 422, 607–608, *608*
Glacier National Park, *104*
Glaciers
  continental, 587, 590, *591*, 605
  explanation of, **588**
  formation of, 591–593, *592*, *593*
  growth and shrinkage of, *592*,
    592–593, *593*
  landscapes formed by, 598, *600–602*,
    600–606, 603t, *604*, *605*
  movement of, 594, *595–597*, *596*, 598
  valley, 588–589, *589*, 594, *595*, *596*
Global change, **660**. *See also* Global
    warming; Ocean acidification
Global Positioning System (GPS)
  explanation of, 42, *42*
  seismograph positioning and, 354
  silent earthquakes and, 362
Global warming. *See also* Climate change
  carbon dioxide and, 428, 429, *429*
  climate change and, 428, 430
  greenhouse gases and, 16, *660*,
    660–661
  ice sheets and, 587

  mass extinction and, 304–306, *305*
  methane deposits and, 305
  predictions of future, 661, *662*
  public debate over, 16
  sea-level rise and, 484
  twentieth-century, 428–430
Gneiss, **157,** *157*
Gobi Desert, 540
Goddard, Robert H., *223*
Gold
  in native state, 81, *81*
  panning for, 85, *85*
Gondwana, 28, 44, 267, 269, 271, 609
Google Earth Project, 21–22, 50–51,
    142–143, 189–190, 214–215, 247,
    340–341, 463–464, 523–524, 546,
    580, 612–613, 634–635
GPS. *See* Global positioning system
    (GPS)
Grade, of ore minerals, 87, 88
Graded bedding, **127**–128
Graded streams, **521**
Grains, **63**
Grand Canyon Beds, 204
Grand Canyon National Park
  erosion in, 520
  Paradise Marble Canyon in, *479*
  stratigraphy of, *198*
  unconformities in, 200, *201*
  Vishnu schist in, *262*
Granite
  composition of, 95
  cooling process and, 92–93
  density of, 10
  magmatic differentiation and, 101
  texture of, *93*, 93–94
  weathering of, 438, *438*, 443, *444*
Granite-greenstone terrains, **274**
Granoblastic rocks, **157**–158
Granodiorite, **96**
Granulite
  explanation of, **158**
  metamorphic grade and, 160–161
Graphite, 63, 64, *64*
Gravel, **133**
Gravitation
  formation of planets and, 223–224, *224*
  formation of Sun and, 223
  ocean tides and, 556
Gravitational differentiation, **225,** *227*
Graywackes, **134**
Great Barrier Reef, 142
Great Basin (Nevada), 151, 185, 540
Great Pyramid, 42
Great Rift Valley, 36
Great Salt Lake, *141*, 141–142
Great Synthesis (1963-1968), 30–31
Great Trigonometrical Survey of India,
    279–280
Greenhouse effect
  Earth's atmosphere and, 414, *414*

enhanced, 428, 430
explanation of, 15–16, 413–414, **414**
Greenhouse gases
climate models and, *415,* 415–416
climate regulation and, 15–16, 412, 414–415
in developing countries, *668,* 668–669
explanation of, **414**
function of, 413–414
global warming and, 16, *660,* 660–661
habitable zone and, 308
Greenland
continental glaciers in, 590, *590*
ice-core drilling in, 419
ice sheets of, 587
Greenschists, **160**
Greenstones, **158**
Grenville province, 271
Groins, 566, 567, *567*
Gros Ventre landslide, *461,* 461–462
Grotzinger, John, 238
Ground moraine, 603t
Groundwater
aquifers and, *481,* 482, *482*
balanced recharge and discharge and, 482–484, *484*
erosion by, 488–489, *488–490*
explanation of, **470,** 477
flow rates for, 485, 486, 496–497
management of, 485–488, *487*
pollutants in, 490–491, *491*
porosity and permeability and, 478–479, *479,* 480t
potable, 491–492
speed of movement in, 485
Groundwater table, **480,** *481*
Gulches. *See* Valleys
Gulf of California, 36
Gulf of Mexico, as sedimentary basin, 257
Gulf Stream, 410–411, 501
Gully, 507, *507*
Gunflint formation, 299
Gutenberg, Beno, 10
Gypsum
calcium cycle and, 426
as chemical sediment, 78
as evaporite, 140, *141*
Mohs scale of hardness and, 69t
in rocks, 76t
selenite crystals as form of, *63*
as sulfate, 68, *69*

Habitable zones, *307,* **307**–308
Hack, John, 630, 636
Hadean eon, *18,* 210
Hager, Brad, 397
Haiti, 667, *667*
Haiti earthquake, *367*
Half-life, **208,** 209, *209*
Halides, 65, 65t

Halite
as chemical sediment, 77, 78
classification of, 64–65
as evaporite, 140, 442
precipitation of, *64*
in rocks, 76t
Halley's Comet, 225
Hallucigenia, 302
Halophiles, 291, *291,* 291t
Hanging valleys, *601,* **602**
Hanging walls, **177,** 179
Hardness, mineral, **69**–70, 74t
Harrisburg, Pennsylvania, 506
Hawaiian Islands
mantle plume and, 50
seafloor of, 335
tides in, 556
volcanoes in, 331–332, *332*
Hazards. *See also* Emergency management
risks vs., 369–370
seismic, *370,* 370–372
Headward erosion, 508
Health, coal mining and, 654
Heat engines, 13, *13*
Heavy Bombardment, **230,** 231, *231,* 273
Hector Mine earthquake, 365
Heezen, Bruce, 29, 30, *31*
Helium, 223
Hematite
on Mars, 240
as ore mineral, 81
as oxide, 67
Herbert Glacier, *589*
Hercynian orogeny, 269
Hess, Harry, 30
Heterotrophs, **285**
High-grade metamorphic rocks, 150
High-grade metamorphic terrains, **274**
High-pressure metamorphism, *153,* 154–155
Himalaya Mountains
age of, 622
continent-continent collision and, 165–166, 267
as convergent boundary, *35*
orogeny and, 266–267, *267, 268,* 271
topography of, *9, 20*
uplift and erosion of, *279,* 279–280
Holmes, Arthur, 29, 45
Holocene, 422, 642
Homo sapiens. See Humans
Horizontal fold, *180,* 181
Hornfels, **157**–158
Horns, 601
Hot spot island chains, 578
Hot spots, 50, 331–333, **332,** *332*
Hot spot tracks, 333
Hot springs, 493, *495*
Huang Ho river, 506

Hubbert, M. King, 650
Hubbert's peak, 650
Hubble, Edwin Powell, 221, 222
Hubble Space Telescope, 222, *223*
Human impact
on beaches, 566–567
on carbon cycle, 427–428, 432–433
on climate system, 412
on deltas, 515
on deserts, 542, 548
on mass movements, 462
Humans
earthquake impact on, 347–348, 365–369, *366–368*
population growth, 642
volcanism impact on, 313, 334–341, *335–337, 339, 342*
Humidity, relative, 472
Humus, **445**
Hunt, T. Sterry, 192
Hurricane Katrina, 506, 558, *559,* 562–564
Hurricane Mitch, 449
Hurricanes
explanation of, **557**–558
formation of, 558–559
intensity of, 559, 560t
landfall of, 563–564, *564*
storm surges from, *561,* 561–563
tracking of, 563–564
Hurricane Sandy, 562–563
Hutton, James, 7, 8, 93
Hydraulic conductivity, 496
Hydraulic fracturing, 135, 648, **651**–652, *652*
Hydraulic gradient, **485,** 496
Hydrocarbon reservoirs, 647, 648
Hydrocarbon resources, 655
Hydrochloric acid
for acid test, 69, *69*
weathering and, 439
Hydroelectric energy, **658,** *659*
Hydrogen sulfide, microbes and, 293
Hydrogen sulfide gas, 293
Hydrologic cycle, **470,** *472*
Hydrology
climate and, 472–473, *473*
explanation of, 470
of groundwater, 477–488, *479–485,* 480t, *487*
of runoff, 475–477, *476–478*
Hydrolysis, 438
Hydrosphere. *See also* Streams
composition of, 409–410
explanation of, *14,* 15, *409, 412*
oceanic circulation patterns and, *410,* 410–411
volcanism and, *327,* 327–328, *328*
Hydrothermal activity, **327**
Hydrothermal deposits, *82,* 82–83, *83*
Hydrothermal fluids, metamorphism and, 153

Hydrothermal solutions, **82**
Hydrothermal vents
    microorganisms in, 299, *299*
    plumes in, 327
Hydrothermal waters, 492–494, *493, 494*
Hydroxides, 65
Hypotheses, 4, 5

Iapetus, 269
Ice. *See also* Glaciers
    as agent of erosion, 598, 600
    cryosphere and, 411
    properties of, 588
    as rock, 588–590
    transformation of snow crystals to,
        592, *592*
Ice ages
    explanation of, 418
    Paleozoic and Proterozoic, 422
    Pleistocene, 418–421, 587
Iceberg calving, **589,** *589*
Icebergs, 588, 599, *599*
Ice caps, 590
Ice-core drilling, 419
Iceland, as spreading center, 36
Ice sheets, 587. *See also* Continental
    glaciers
Ice shelves
    breakup of, 599, *599*
    explanation of, **590**
Ice streams, **596,** *597*
Igneous deposits, 83–84, *84*
Igneous intrusions
    explanation of, 102
    plutons and, 103, *103*
    sills and dikes and, 103–104, *104*
    veins and, 105, *105*
Igneous rocks
    chemical and mineral composition of,
        94–97, 95t, *96,* 97t
    classification of, 92, *92,* 94, *96*
    common minerals of, 76
    explanation of, **74,** *75*
    extrusive, 76, *76,* 93–95, *94*
    formation of, 76, *76,* 106
    intermediate, 96
    intrusive, 76, *76,* **93,** *94*
    magma formation and, 97–99
    magmatic differentiation and, 99–101,
        100, 102
    magmatic intrusions and, 102–105,
        102–105
    minerals in, 76t, 95t
    plate tectonics and, 105–111, *106–110*
    porosity of, 478
    texture of, 92–94, *92–94*
Iguaçú River falls, *509*
Impact craters, *236*
Incised meanders, 501, *501*
Index minerals, 159, *160*
Indian Plate, *35,* 38, *268*

Indus River, 506
Infiltration, **470**
Influent streams, 480
Inner core. *See also* Core; Outer core
    composition of, 12, *12*
    explanation of, 11–**12,** *14*
Inner planets. *See* Earth; Mars; Mercury
    (planet); Venus
Intensity scales, 360–361, 360t
Interglacial period, **418**
Intergovernmental Panel on Climate
    Change (IPCC), 428, 430, 433, 661,
    662, 667
Intergranular porosity, 478
Interior of Earth. *See also* Earth
    heat flow through, 390–392, *391, 392*
    layering and composition of, 388–390,
        388i, 389
    magnetic field and biosphere and,
        402–403, *403*
    magnetic field and geodynamo of, 396,
        398, 398–399, *399, 400*
    paleomagnetism and, 400–403, *401*
    seismic waves to explore, *384–387,*
        384–388
    temperature of, 390–391, *391,* 394
    three-dimensional structure and, 394,
        395, 396
Intermediate igneous rocks, **96**
Internal heat engines, 13, *13*
Intraplate earthquakes, 364
Intraplate volcanism, 331–334, *332–334*
Intrusions, igneous, 102–105, *102–105*
Intrusive igneous rocks, 76, *76,* **93,** *94*
Ionic bonds, **61,** 70
Ions
    calcium, 122
    carbonate, 67, *67*
    explanation of, **61**
    mineral color and, 72
    in rock-forming minerals, *62*
    silicate, 65, *66*
    sodium, 122
IPCC. *See* Intergovernmental Panel on
    Climate Change (IPCC)
Iridium, 303
Iron, in core, *12*
Iron formations, **144**
Iron oxides, 144, 235, *442*
Island arcs
    accretion and, *264,* 265
    continental growth and, *263*
    explanation of, **37**
Islands, barrier, 568
Isochron map, *43,* 44
Isochrons, **43,** 44
Isograds, 159–160, *160*
Isolated tetrahedra, 65, *66*
Isostasy, 388, 393, *393,* 404–405, *599*
Isostasy principle, 404–405
Isotherm, *151*

Isotopes
    carbon, 298, 310
    explanation of, **59,** 310
    radioactive, 208
Isotopic dating
    absolute ages and, 210–211
    of Cambrian fossils, 303
    explanation of, **208**
    of glaciers, 607
    major elements used in, 209t
    of meteorites, 224
    methods for, 209, 216–218, *217*
    of rock samples from Moon, 230–231
Isotropic clocks
    measuring absolute time with, 207–209
    tectonic age and, 261
Isthmus of Panama, 45
Ivanpah solar energy generating system,
    658, *658*

Jack Hills (Australia), 211, *211*
Jarosite, 244
Jefferson, Thomas, 499
Joekulgikvisl River, *502*
Johnston, David A., 338
JOIDES Resolution, 41, 213
Joints
    explanation of, **183,** *183*
    weathering and, 443
Juan de Fuca Plate
    explanation of, 31
    location of, *31–32*
    North American Cordillera and,
        258, *259*
    subduction of, 371
    volcanoes and, 38
Jupiter, 224, 229
Jurassic period, 206
Juvinas meteorite, 217

Kaibab Limestone, 204
Kalahari Desert, 540
Kames, 604
Kant, Immanuel, 222
Kaolinite, 67, **438,** 439
Karst topography, **490,** *490, 491*
Keels. *See* Cratonic keels
Kenya aquifer, 487
Kerguelen Plateau, 333
Kettles, **605**
Kilauea volcano, *315, 316,* 335
Kimberlite, 155, 324
Kimberlite pipes, 277, *277, 324*
Klarälven River, 516
Kobe earthquake, 365
Komatiite, 273
Krafft, Katia, 320, *320*
Krafft, Maurice, 320
Krakatau volcano, 412
Kudzu, *285*
Kyanite, 76t

Laboratory experiments
  on crystallization, 92
  on deformation, 175–176, *176*
  on fractional crystallization, 99–100
  on melting, 97–98
  at metamorphic pressures, 162–163
  rock magnetism and, 17
Lagoons, 137, 139
Lahars, **335,** *336,* 458
Lake Baikal, 522
Lakes. *See also specific lakes*
  environments in, 124
  formation of, 522
Lambert Glacier, *597*
Lamberton, William, 279
Laminar flow, **509,** 510, *510*
Landers earthquake, 365
Landforms
  explanation of, **621**
  mountains and hills as, *621,* 621–622,
    622
  plateaus as, *622, 623, 624, 625, 625*
  scale of, 621
  stream valleys as, 623, 625
  structurally controlled cliffs as, 626,
    628
  structurally controlled ridges and
    valleys as, 626, *626, 627*
Landscape development models,
    630–632, *633,* 634–636
Landscapes
  absolute dating of, 636t
  desert, 544–546
  geosystems and, 626–630
  glacial, 598, *600–602,* 600–606, 603t,
    604, 605
  overview of, 617
  precipitation and, 473
  topography, elevation, and relief and,
    618–621, *618–621*
Landslides
  from earthquakes, 367, *453*
  from hurricanes, 449
  as mass movements, 453, 462
La Niña, 417
L'Aquila earthquake, *376,* 376–377
Large igneous provinces, *333,* 333–334
Larochelle, A., 39
Larsen Ice Shelf, 598, *598*
Late Heavy Bombardment, 231
Lateral moraine, 603t
Laurasia, 44
Laurentia, 267, 269, 271
Lavas
  andesitic, *317,* 317–318
  basaltic, *315,* 315–317, *316*
  composition of, 315
  explanation of, 92, 93
  layers of, 39
  mantle plumes and, 50
  meanders of, 501

rhyolitic, 318, *318*
  from vents in oceans, 329
Lawrence, George, *348*
Lead pollution, 490
Left-lateral faults, 177
Lehmann, Inge, 12
Levees, natural, 503, *503*
Lewis, Meriwether, 499
Life, evolution of, 19–21
Life Finder (spacecraft), 248
Light waves, 384
Lignite, 653–654
Limbs, of fold, 180
Limestone
  calcite in, *67, 67, 78*
  formation of, 58, *140*
  groundwater in, 488, 492
  porosity of, 478
  as sedimentary rock, 115, **136**
  slope stability of, 451
Liquefaction, from earthquakes, 367
Liquified natural gas (LNG), 653
Listric faults, 185
Lithic sandstones, **134**
Lithification, **77, 132**
Lithosphere
  climate system and, 411–412
  conduction through, 391
  convergence and oceanic vs.
    continental, 37
  cratonic keels and, 276
  explanation of, *14,* **16,** 27
  interactions between hydrosphere
    and, 328
  magma in, 588
  plates of, *34,* 348
  plate tectonics and, 31, 45, 48, *48*
  volcanism in, 412
Lithosphere-atmosphere gas exchange,
    427
Liuzhou, China, *506*
Loess, **538**–540
Loma Prieta earthquake, 365, 367
London, England, windstorm in, 529
Long Island, groundwater on, 488
Longitudinal dunes, **539**
Longitudinal profiles, **519,** *520*
Longshore current, **556,** 565
Longshore drift, 556, 565
Long-term forecasting, earthquake, 375
Lower mantle, **390.** *See also* Mantle
Low-grade metamorphic rock, 150
Low-velocity zone, **389**
Lunar highlands, 230, *230*
Lunar maria, *230,* 230–231
Luray Caverns, Virginia, *488*
Luster, **71**–72, *72,* 74t
Lyell, Charles, 7, 195

Mafic minerals
  explanation of, 94–95, 95t

melting temperature for, 97, *97*
  in ultramafic rocks, 97
Mafic rocks, **96**–97
Magellan (spacecraft), *233,* 233–235
Magma
  from asthenosphere, 314
  composition of, 110–111
  crystallization in, 63, 112–113
  decompression melting and, 108, *109*
  explanation of, 36, **63,** *314,* 314–315
  formation of, 97–99, 105, *329*
  igneous rocks and, 76, 76t
  mantle plumes and, 111
  melting and composition of, 101
  subduction zones and, 108–111, *110*
Magma chambers, **99,** 101, *102,* 110, 112
Magma ocean, 225
Magmatic addition, **262**–263
Magmatic differentiation
  complexities of, 101, *102*
  explanation of, **99,** 262
  fractional crystallization and, 99–101,
    100
Magnesium, 13, 67
Magnetic anomalies
  explanation of, **39**
  patterns of, 39, 41
  seafloor and, 39, *40,* 41
Magnetic chrons, 39
Magnetic fields
  biosphere and, 402–403, *403*
  complexity of, 396, *398,* 398–399, *399,*
    400
  dipole, 396
  explanation of, **17,** *18*
  nondipole, 398
Magnetic reversals, 17–18, 39, 41, 399,
    400
Magnetic subchrons, 39
Magnetic time scale, **39**
Magnetite
  density of, 73
  depositional remanent magnetization
    and, 401
  indirect precipitation of, 293, *293*
Magnitude
  explanation of, 357, *359*
  frequency and, 359–360
  moment, 359
  Richter, 357, 359
Maiandros River, 501
Mainshock, 351
Mammals, evolution of modern, 305
Mammoth Cave, 488
Manganese nodules, 581
Mantle
  composition of, *12, 12,* 227, 388–390,
    389
  deep, *14*
  explanation of, **10,** *11*
  peridotite in, 107

Mantle (*Contd.*)
　seismic data on, 388–390, *389*
　three-dimensional model of, 394, *395*
Mantle convection
　explanation of, *16*, 16–17, 45
　history of, 273
　plate tectonics and, 45, *48*, 48–51, *49*
Mantle plumes
　explanation of, *49*, **49**–50, **332**
　as magma factories, 111
　magma production and, 105
　mass extinctions and, 50–51
Maps, geologic, *173*, **173**–175, *174*
Marble Canyon, *198*
Marbles, 152, **158**
Marcellus Formation, 135, *135*
Mariana Islands, as island arc, *34*
Marianas Trench, as convergent
　　boundary, *34*
Marine evaporites, 140–141, *141*
Mariner missions (spacecraft), 231–232,
　　238–239
Marine sedimentary environments, 124,
　　125, 126
Mars
　atmosphere of, 308
　characteristics of, *228*, 228–229,
　　235–236
　cross-bedding on, 538
　dust storms on, 238, *238*, 533, *533*
　environmental evolution of, 243–245,
　　244
　exploration of, 221, 222, 237–245,
　　238–244, 247, 249–250, 307, 308
　habitable environment on, 308, *308*
　liquid water on, 222
　sulfates on, 68
　surface of, 230
　as terrestrial planet, 224
　topography of, 235, *235*
Mars Exploration Rovers (spacecraft), 237,
　　240–243, *242*, 245, 247, 306–308, 533
Mars Express (spacecraft), 237, 238
Mars Global Surveyor (spacecraft),
　　239–241
Mars Odyssey (spacecraft), 238, 241
Mars Polar Lander (spacecraft), 243
Mars Reconnaisance Orbiter (spacecraft),
　　237–238, 243
Mars Science Laboratory (spacecraft),
　　240–243, 307
Mass, of ore, 87
Mass extinctions
　asteroids and, 236, 303–304
　causes of, 20
　Cretaceous-Tertiary, 303
　of dinosaurs, 303–304
　effects of, 21
　explanation of, 19–20, **202**, 206
　geologic time scale and, 202, 206
　mantle plumes and, 50–51

Paleocene-Eocene, 304–306, *305*
Permian, 306, 344
Phanerozoic eon and, 300–301
Siberian Traps and, 343–344
Mass movements
　classification of, 453, *454*
　explanation of, **448**–449, 453, *454*, 455,
　　455
　factors that influence, 450t
　human activities that promote or
　　trigger, 462, *462*
　natural causes of, *461*, 461–462
　observation of, 463–464
　of rock, 455–456, *456*
　triggers of, 452–453, *453*
　of unconsolidated material, 456–460,
　　457–460
Mass spectrometer, 216, *217*
Mass wasting. *See also* Erosion; Mass
　　movements
　explanation of, **326**
　factors influencing, 449
　in Southern California, 447–448
Mathews, D. H., 39
Matter, structure of, *59*, 59–61, *60*
McPhee, John, 7
Meanders, *501*, **501**–502, *502*
Medial moraine, 603t
Mediterranean Sea, 552
Medium-term forecasting, 376–378
Megathrusts, earthquakes on, 185–186,
　　367, *368*
Mélange, **163,** 164
Melting
　decompression, 98, 107, *109*
　fluid-induced, 98, 109, *110*
　partial, 98
　rock, *97*, 97–98
　water and, 98
Mercalli, Giuseppe, 360
Mercury (planet)
　characteristics of, 228, *228*, 232, *233*
　formation of, 224
　surface of, 230, 232, *232*
Mesas, **623,** *623*
Mesopotamia, 506
Mesopotamian Basin, 123
Mesosaurus (reptile), 28–29, *29*
Mesozoic era, 206, 210
Messenger (spacecraft), 232–233
Metabolism, **286**
Metallic bonds, **61**
Metallic luster, 72, 72t
Metallic ores
　explanation of, 82, *83*
　formation of, 112–113, 153
Metamorphic facies, 161–162, *162*
Metamorphic grade, 150, 160–161
Metamorphic P-T paths, 162–163
Metamorphic rocks
　classification of, 159t

common minerals of, 76t, 79
contact metamorphism and, 79
eclogites as, 155
explanation of, 74, *75*, **76,** 78, 150
formation of, *150–154*, 150–155
high-grade, 150
low-grade, 150
plate tectonics and, 151–152, 162–166,
　164, 165
porosity of, 478
regional metamorphism and, 78–79
texture of, 155–159, *156–158*, 159t
Metamorphism. *See also* Deformation
　causes of, 150–153
　continent-continent collision and,
　　165–166
　as dynamic process, 149
　exhumation and, 166
　explanation of, *78*, 78–79
　intensity of, 155, *157*
　ocean-continent convergence and,
　　163–165, *164*
　plate tectonics and, 162–166, *164, 165*
　pressure-temperature paths and,
　　162–165
　regional, 159–162
　role of fluids in, 153
　role of pressure in, *152*, 152–153,
　　162–163
　role of temperature in, *151*, 151–152,
　　162–163
　subduction-related, *164*, 164–165
　types of, 153–155
Meteor Crater (Arizona), *231*
Meteoric water, **477,** 493
Meteorites
　composition of, 10, 224–225
　Earth's composition and, 10, *11*
　explanation of, **224**
　impact of, 236, 237t
　isotopic dating of, 224
　shock metamorphism and, 155
　types of, *11*
Meteorology, 6
Methane. *See also* Natural gas
　in atmosphere, 409
　deposits of, 306
　global warming and, 305, 660, *660,*
　　661, 665
　natural gas and, 653
Micas
　cleavage in, 70, *70*
　in igneous rocks, 76t, 95t
　in metamorphic rocks, 76t, 153
　structure of, *66*, 67
Michigan Basin
　epeirogeny and, 272
　sediment in, 182, *182*
Microbial decomposers, 293
Microbial mats, *293*, **294**–295
Microbial respiration, 292–293

Microfossils, **290,** 299
Microorganisms
    abundance and diversity of, 289–291, *290, 291,* 291t
    biogeochemical cycles and, 294, *295*
    in deep aquifers, 494
    explanation of, **288**
    in groundwater, 490–491
    mineral interactions with, 292–296, *293–296*
    uses of, 288
Mid-Atlantic Ridge
    basaltic eruptions at, 330
    as divergent plate boundary, 36, *36*
    explanation of, 29, *30, 39*
    magnetic anomalies near, *40*
    profile of, 576, *576*
    spreading center on, 36
    spreading rate at, 44
Mid-ocean ridges. *See also* Spreading centers
    characteristics of, 578
    explanation of, **36**
    magma formation at, 106, *106*
    transform-fault boundaries along, 38
Migmatite, **157,** *157*
Milankovitch cycles
    explanation of, 420
    glacial cycles and, 420–422, *421*
Miller, Stanley, 297, *297,* 298
Mineral dissolution, 294
Mineral dust, 532
Mineralogy, **58,** 74
Mineral precipitation
    examples of, 293, *293, 294*
    explanation of, **63,** *64,* 292–293
Mineral resources
    explanation of, 81, *81*
    in hydrothermal solutions, 82–83
    in igneous rocks, 83–84, *84*
    rock cycle and, 81, *82*
    in sedimentary rocks, 84–85, *85*
Minerals. *See also* Rock-forming minerals; *specific minerals*
    atomic structure of, 61–62
    carbonate, 67, *67*
    characteristics of, 58
    chemical bonds and, 61
    cleavage of, *70,* 70–71, *71*
    color of, *72,* 72–73, *73*
    crystal habit of, *73, 73*
    crystallization of, *62,* 62–63, *63*
    density of, 64, 73
    explanation of, **58**
    felsic, 94–96, 95t
    formation of, 63–64, *64*
    fracture of, 71
    hardness of, 69–70
    in igneous rocks, 76t, 95t, *96*
    index, 159, *160*
    luster of, 71–72, 72t

mafic, 94–95, 95t
    microorganism interactions with, 292–296, *293–296*
    oxide, 67, *68*
    physical properties of, 69–74, *74*
    relative stability of common, 443t
    rock-forming, 64–65, 65t, *66,* 67–68, *69,* 76t, 78, 442, 443t
    silicate, 65, *66, 67*
    structure of matter and, 59–61
    sulfate, 68, *69*
    sulfide, 67–68, *68*
Mines/mining
    coal, 654, 655
    economics of, 87–88
    open-pit, *84*
Miocene epoch, 141
Mississippian period. *See* Carboniferous period
Mississippi River
    delta of, 513, *513,* 515
    as drainage network, *477*
    floodplains of, 506
    loess in valley of, 540
    meanders on, 501
    sedimentary basins and, 119
    sedimentary processes and, 116
    sediment transport in, 523
    streams feeding to, 476, 515
Modified Mercalli intensity scale, 360, 360t
Moenkopi formation, 204
Mohorovicic (Moho) discontinuity, 11, 12, 388
Mohs, Friedrich, 69
Mohs scale of hardness, **69,** 69t, 70
Mojave Desert, 540, 658
Moment magnitude, 359, *359*
Mongolia, compression and, 267
Mont Pelée volcano, 320
Mont-Saint-Michel tidal flat, *557*
Monument Valley, *623*
Moon (Earth)
    characteristics of, 228, *228,* 229
    exploration of, 221
    formation of, 19
    glacial ice on, 589
    surface of, 230, *230*
    tides and, 556
    time scale for, 230–231, *232*
Moraines, **603,** 603t
Morgan, W. Jason, 51
Morley, L., 39
Mountain building. *See* Orogeny
Mountains. *See also specific mountain ranges*
    explanation of, 621–622
    hills vs., 622
    plate tectonics and, 622–623, 629–630
    in ranges, 622
Mount Etna volcano, 323

Mount Everest, Earth's topography and, 9
Mount Fuji volcano, 340
Mount Merapi (Indonesia), 335
Mount Pinatubo volcano (Philippines), 319, 338
Mount Rainer, 337, *337*
Mount St. Helens volcano
    eruption of, 317, *317,* 325
    predicting eruption of, 338
Mount Tambura volcano, 412
Mount Unzen volcano, 321
Mud, **135**
Mudflows, 457–458, *458*
Mudslides, from hurricanes, 449
Mudstones
    erosion of, 174
    explanation of, **135**
Murchison meteorite, 298
Muscovite
    cleavage in, 70, 71
    structure of, 67
Mylonites, 184, *184*

Namacalathus, 302
Namib Desert, *532,* 546, *546*
National Hurricane Center, 564
National Water Quality Assessment Program, *492*
Native elements, 64, 65, 65t
Natural gas. *See also* Methane; Resources
    explanation of, **144,** 647, 651
    exploration for, 145
    extraction of, 651–652, *652*
    formation of, 647, *647*
    as fuel, 653
Natural gas reserves, 648
Natural levees, **503,** *503*
Natural resources. *See also* Resources
    demand for, 642
    explanation of, **642,** 643
    types of, 642, *643*
    from volcanoes, 339–341
Natural selection, **301,** 303
Nazca Plate
    as convergent boundary, *35,* 37
    explanation of, 38–39
Neap tides, 556
Nebulae, 222–224
Nebular hypothesis, **222**–223, *224*
Negative feedback, **415**
Neogene period, 210
Neptune, 224, 229
Neutrons, 59, *59*
Nevado de Huascarán debris avalanche, 459, *459*
Nevado del Ruiz volcano, 37–38
New Horizons (spacecraft), 229
New Madrid earthquake, 360, *361*
New Orleans, Louisiana, 506

Newspaper Rock, Canyonlands, Utah, *543*
Newton, Isaac, 10, 556
Nile River, 503, 506
Nitrogen
  in atmosphere, 409
  microbial metabolism and, 294
Nitrous oxide, 660, *660*, 661
Nome, Alaska, 591–592
Nonconformities, 200
Nondipole magnetic field, **398**
Nonrenewable resources, **642**
Normal fault, **177,** 179, 361
North America
  Appalachian fold belt in, 256, *257*
  coastal plain and continental shelf in, 256–257
  drainage basins in, 504, *505*
  North American cordillera in, 257–259, *258*
  paleogeographic reconstructions of, *270*
  stable interior of, 254–256, *255, 256*
  structure of, 254–259, *254–260*
North American Cordillera, 254, 257–259, *258, 263,* 266
North American Plate, 31, *34, 35, 39,* 258
Northridge earthquake, 365, *366*
Nuclear disasters, 656, *656*
Nuclear energy, 655–657, *656, 657*
Nuclear fusion, 223
Nuclear waste, 657
Nucleus, 59, *59*
Numerical climate models, 415, *415*
Nutrients, 386

Oblique-slip faults, **177,** *178*
Obsidian, **94,** 319
Ocean acidification, 666
Ocean basins
  continents vs., 552
  profile of, 580
  topographical map of, 572, *573*
Ocean-continent convergence, 37–38, 163–165
Ocean currents, effect on deltas, 515
Ocean drilling, 41, *41*
Oceanic crust. *See also* Crust
  explanation of, 11, *11*
  seafloor spreading and, 29–30
  sediments in, 552
Ocean-ocean convergence, 36–37
Oceanography, 6
Oceans. *See also* Seafloor; Seafloor spreading; *specific oceans*
  as chemical mixing vats, 122
  coastal processes and, 552–564
  connection between, 552
  continental margins and, 570–572
  exploration of, 551
  formation of, *227,* 227–228

sediments in, 578–581
shorelines of, 543–570
Offshore, 565
Ogallala aquifer, *486,* 486–487
O-horizon, 447
Oil. *See also* Resources
  environmental impact of, *650,* 650–651, *651*
  exploration for, 145–146, 644, *644,* 648, *648*
  in flexural basin, 123
  formation of, 647, *647*
  geologic maps to find, 191–192
  production and consumption of, *649,* 649–650
Oil reserves
  depletion of, 650
  distribution of, 648–649, *649*
  explanation of, 647–648
  technologies for recovery of, *648*
Oil reservoirs, 647, 648
Oil spills, *650,* 650–651, *651*
Oil traps, *647,* **647**–648
Oil wells, 644
Oil window, **647**
Old Faithful (geyser), 327
Oldham, Robert, 10
Olivine
  composition of, 58, 62
  density of, 73
  formation of, 63, 112
  in mantle, 107
  in rocks, 76t, 95t
  structure of, *66*
Olympus Mons volcano (Mars), 235, *235*
Ontong Java Plateau, 333
Opal, 493
Open-pit mines, *84*
Open system, *13,* 14
Ophiolite suites, **106**–107, *108*
Opportunity (spacecraft). *See Mars Exploration Rovers* (spacecraft)
Orange County, groundwater in, 488
Ordovician period, 269
Ore minerals. *See also* Mines/mining
  deposits of, 81, *81*
  exploration for, 87, *87*
  grade and mass of, 87, 88
  hydrothermal deposits of, *82,* 82–83
  in igneous rocks, 83–84, *85*
Ores
  explanation of, **81**
  metallic, 82, *83*
Organic matter, 58
Organic reefs, environment in, 126
Organic sedimentary rocks, **144**
Organisms. *See also* Microorganisms
  in ecosystem, 284–288
  as producers and consumers, 286t
  weathering and, 443
Original horizontality principle, **197**

Origin time, 357
Orogens, **260**
Orogeny
  Alpine-Himalayan, 266, 267, *267, 268*
  Andean, 266
  early, 269, *270, 271*
  explanation of, **266**
  Paleozoic, 267, 269
Orthoclase
  in igneous rocks, 95
  Mohs scale of hardness and, 69t
Otto, Nikolaus, 657
Outcrop
  explanation of, 173
  on Mars, 244, *244*
Outer core. *See also* Core; Inner core
  composition of, 12, *12,* 13
  explanation of, **12,** *14*
Outer planets. *See* Jupiter; Neptune; Pluto; Saturn; Uranus
Outwash, **603**
Overthrust structures, 186, *186*
Overturned fold, 181, *181*
Owens Valley earthquake, 365
Oxbow lake, **501**
Oxides
  explanation of, 65, 65t
  as ore minerals, 81
  structure of, 67, *68*
Oxygen
  in atmosphere, 409
  in Earth, 13
  evolution of atmospheric, 297, 299–300, *300*
  respiration and, 293
  weathering and, 441–442
Oxygen-16, 419
Oxygen-18, 419
Ozone, 409

Pacific Ocean, profile of, *577,* 577–578
Pacific Plate, *34*
Pacific Ring of Fire, 30, *31,* 329
Pacific Tsunami Warning Center, 373
Pahoehoe, 316, *316*
Paleocene epoch, 304–305
Paleocontinents, 44
Paleomagnetic stratigraphy, 213, 401–402
Paleomagnetic time scale, *402*
Paleomagnetism, **400**–402, *401*
Paleontology, 199–200
Paleosols, 447
Paleozoic era, 210, 422
Paleozoic orogeny, 267, 269
Paleozoic sunlight, 644
Palisades intrusion, 99–100, 112
Pangaea
  assembly of, 43, 44, *46, 48,* 269
  breakup of, 29, 44, *47*
  explanation of, **28,** 42
Papua New Guinea tsunami, 373

Parent atoms, 208, 209, 217, *217*
Partial melting, **98**
Particle size, 121
Passive margins, **260, 570,** *570*
Pathfinder (spacecraft), 238
Patterson, Clair, 208
Peat, **144**
Pediments, *545,* **545**–546
Pegmatites, 84, **105**
Pelagic sediments, **579,** 581
Penck, Walther, 630, 634–636
Perched water tables, 482, *483*
Peridotite, 324
    in kimberlite pipes, 277
    in mantle, 107
Period, wave, 554
Periods (geologic time), **202**
Permafrost, **606,** *606, 607*
Permeability
    in crustal rock, 492, *493*
    explanation of, **479**
Permian-Carboniferous glaciation, 609
Permian Kupferschiefer beds (Germany), 85
Permian period, 333
    mass extinctions in, 306, 344
Perovskite, 73
Peru-Chile subduction zone, 39
Peru-Chile Trench, *35*
Petroleum, 644
Petroleum source rocks, ages of, 206–207, *207*
Phanerozoic eon, *19,* 211, 290
Phanerozoic orogen, 260
Phase change, **389**
Phenocrysts, 94
Philippine Islands, *263*
Philippine Plate, *34*
Phoenix (spacecraft), 238, 243
Phosphorite, **144**
Phosphorus, 288, *289*
Photoautotrophs, 305
Photomicrographs, 92
Photosynthesis
    carbon dioxide and, 310, *311,* 432, 433
    the climate system and, 412
    explanation of, *287,* **287**–288
    respiration vs, 287–288, 287t
Phreatic eruptions, 318, *318*
Phyllites, **156,** *157*
Physical laws, 4
Physical weathering. *See also* Weathering
    in deserts, 542
    erosion and, 444–445, *445,* 508
    explanation of, **116,** 436
    processes of, 443–444
Piedmont, 256
Piers, 566
Pillow lavas, 317, *317*
Placers, 85

Plagioclase
    fractional crystallization and, 101
    in igneous rocks, 95, 96
Planetary science, 6
Planetary time scale, 230–231
Planetesimals, **223**–224
Planets. *See also specific planets*
    characteristics of, *228*
    diversity of, 228–229
    dwarf, 229
    formation of, 223–224, *224*
    giant outer, 224, *225*
    terrestrial, 224, *225,* 230
Plant growth feedback, 415
Plastic flow, **594,** *595*
Plateaus, 578, *622,* **623**
Plate boundaries, 31, *31–38, 36–39.*
        See also Convergent boundaries;
        Divergent boundaries; Transform-
        fault boundaries
Plate tectonics. *See also* Continental drift;
        Seafloor spreading
    Baja California and, 53–54
    climate system and, 79, *80, 81,* 166, 473, 629–630, 632
    continental drift and, 28, 28–29, 44
    deformation and, 172
    deltas and, 515
    deserts and, 541
    discovery of, 28–31
    earthquakes and, 348
    exhumation and, 166
    explanation of, 27, **31**
    grand reconstruction and, 42–45, *43, 46–47*
    Great Synthesis and, 30–31
    igneous processes and, 105–111, *106–110*
    mantle convection and, 45, *48, 48–51, 49*
    metamorphic rocks and, 151–152, 162–166, *164, 165*
    plates and plate boundaries and, 31, *31–38, 36–39*
    rates and history of plate movements and, 39, *40–42,* 41–42
    rock cycle and, 79, *80, 81*
    the scientific method and theory of, 52
    seafloor spreading and, 29–30, *30*
    sedimentary rocks and, 115
    volcanism and, *329,* 329–334, *330, 332*
Plate tectonic system, *14,* 16–**17**
Platforms, 260
Platinum, 83, 112
Platy, 73
Playas, 544, *544*
Pleistocene epoch, *213,* 417–419, *418,* 421–423, 587
Pliocene epoch, *213,* 421
Plumes, 327

Plunging fold, *180,* **181**
Pluto, 229
Plutons, **103**
Point bars, **501,** 503
Pollution
    of beaches, 566, 568
    groundwater, 490–491, *491*
    in lakes, 522
Polymorphs, **63**
Porosity
    in crustal rock, 492, *493*
    explanation of, **131**
    groundwater and, 478–479, *479,* 480t
Porphyritic crystals, *94*
Porphyroblasts, *158,* 158–**159**
Porphyry, **94**
Positive feedback, **415**
Potable water, **491**–492
Potholes, **508,** *509*
Precambrian, 204, 207, 210, 301
Precipitation
    air temperature and water vapor and, 472, 473
    average annual, *476*
    chemical and biological sedimentation and, 122
    ending up as runoff, 475
    explanation of, **470**
    groundwater from atmospheric precipitation, 477
    of snow, 591
    streams and, 515
Precipitation (mineral). *See* Mineral precipitation
Predictions
    of desertification, 548–549
    of earthquakes, *375,* 375–378
    of floods, 518
    of future global warning, 661, *662*
    of hurricanes, 563–564
    of volcano eruptions, 338–339
Press, Frank, *29*
Pressure
    metamorphism and, *152,* 152–153
    rock melting and, 98
Principle of faunal succession, **199,** *199*
Principle of isostasy, 404–405
Principle of original horizontality, **197**
Principle of superposition, **197**
Principle of uniformitarianism, **7,** 8
Probabilities, 518
Prograde paths, **163**
Proterozoic, *19,* 210–211, 422, *609*
Protons, 59, *59*
Proust, Marcel, 671
P-T path, **161**–162, 164–166
Puget Sound, 556
Pumice, **94,** 319
Purple loosestrife, *285*
P waves, **355,** *356,* 357, *362,* 384–386, *385, 386*

Pyrite
    as ore mineral, 82, *83*
    as sulfide, 67–68, *68*
Pyrite precipitation, 293, *294, 295*
Pyroclastic deposits, *319,* 319–320, *320*
Pyroclastic flows, **320,** *321*
Pyroclasts, **93,** *93,* 319
Pyroxenes
    cleavage in, 71, *71*
    in mantle, 107
    in rocks, 76t, 95t
    structure of, *66*

Qori Kalis Glacier, *593*
Quads, **645**
Quartz
    crystal faces in, 62, *63*
    in igneous rocks, 76t, 95t
    as mineral, 58
    Mohs scale of hardness and, 69t
    as polymorph, 64
    structure of, *66, 67, 71*
    weathering of, 442
Quartz arenites, **134**
Quartzites, **158**

Radar, 233
Radial drainage, **504**
Radiation (electromagnetic). *See* Solar
        energy
Radiation balance, 413, 414, *414*
Radioactive damping, 415
Radioactive decay, 208, *209,* 216
Radioactive isotopes, *208,* 208–209, 209t
Radioactive waste, 657
Radioactive wastes, in groundwater, 490
Radioactivity, discovery of, 208
Radiolarians, 581
Radiometric dating. *See* Dating; Isotopic
        dating
Radon, *492*
Rain shadows, 473, *473*
Rainwater
    aquifers and, 482
    carbon dioxide in, 439
    carbonic acid in, 439, 441
    in deserts, 540, 543
    flow of, 500
    stream channels and, 507–509,
        507–509
Ramp morphology, 137
Rapids, 508
Recharge
    artificial, 488
    balance between discharge and, 516
    balancing with discharge, 482–484, *484*
    explanation of, **480**
    of Ogallala aquifer, 486
Rectangular drainage, **504**
Recurrence intervals, **349,** *350*
Red beds, **300**

Red Sea
    deformational style in, 184
    explanation of, 36
Reefs
    Darwin and, 139, *139*
    evolutionary processes and, 137–138, *139*
    explanation of, **137**
Regional metamorphism. *See also*
        Metamorphism
    explanation of, *78,* **78**–79, *153,* 154
    grade and parent rock composition
        and, 160–161, *161*
    metamorphic facies and, 161–162, *162*
    mineral isograds and, 159–160, *160*
Rejuvenation, **258**
Relative age. *See also* Geologic time scale
    cross-cutting relationships and,
        201–202, *203*
    explanation of, **196**
Relative humidity, **472**
Relative plate velocity, **41**
Relief, *620*
    explanation of, **619,** *620*
Renewable resources, **642**
Reserves, **643**
Reservoirs, 470
Resources. *See also* Alternative energy;
        Energy resources; Fossil fuels;
        Natural resources; Ore minerals
    climate change and, 663t
    explanation of, **643**
    groundwater, 485–488
    mineral, 81–85, *81–85*
Respiration
    the carbon cycle and, 432
    explanation of, **288**
    microbial, 292–293
    oxygen for, 293
    photosynthesis vs., 287–288, 287t
Resurgent calderas, 323
Reverse fault, 177, 361
Rhine Valley
    deformational style in, 184
    as divergent boundary, 36
R-horizon, 447
Rhyolite, 95
Rhyolite lavas, 318, *318*
Ribbon Canyon (Wyoming), 189, *189*
Richards, Mark, 397
Richter, Charles, 357
Richter magnitude scale, **357,** 359
Ridges, structurally controlled, 626, *626*
Rift basins, **122,** 183
Rift valleys
    in coastal plain, 256–257
    deformational style in, 184, *186*
    at divergent boundaries, 36, *37*
    seafloor spreading and, 29
Rimmed shelf morphology, 137
Ring of Fire, Pacific, 30, *31,* 329
Rio Tinto (Spain), *291,* 292

Rip current, 556
Ripples, **128,** *128, 129,* **511**–512, *512*
Risk
    hazard vs., 369–370
    seismic, 370
Rivers. *See also* Streams; *specific rivers*
    explanation of, **500**
    exploration of, 499
    recording flow of water in, 525–526
    runoff in, 475, 476
    water flows in, 477t, 510
Roches moutonnées, 600, *600*
Rock avalanches, 456, *456*
Rock cycle
    explanation of, **79,** *80,* 81
    mineral concentrations and, 81, 82
    surface processes in, 116–122, *117–121*
    weathering and, 436
Rockets, 221. *See also* Space exploration
Rockfalls, 455, *455*
Rock flour, 600
Rock-forming minerals. *See also* Minerals
    carbonates as, 67, *67*
    explanation of, 64–65, 65t
    oxides as, 67, *68*
    silicates as, 65, *66,* 67
    stability of, 442, 443t
    sulfates as, 68, *69*
    sulfides, 67–68, *68*
Rocknest basin, 145–146
Rocks. *See also* Deformation; Igneous
        rocks; Metamorphic rocks;
        Sedimentary rocks
    dating methods for, 209
    density of, 10
    evidence of life in, 310–311
    explanation of, **74,** *75*
    failure of, 348
    ice as, 588–590
    isotopic age of, 209
    mafic, 96–97
    mass movements of, 455–456, *456*
    melting of, *97,* 97–98
    minerals in, 64–65, 65t, *66,* 67–68, *69,* 76t
    natural zones of weakness in, 444
    permeability of, 479, 480t, 492, *493*
    porosity of, 478–479, *479,* 480t, 492,
        493
    properties of, 74, *75,* 76
    seafloor, 43
    strength of, 348
    that break, 443–444
    volcanic, 92–94
Rockslides, 455, *455*
Rock slopes, 451–452, *452*
Rock travertine, 493
Rocky Mountains, 258–259
Rodinia, 44, *46,* 267, 271
Ross Ice Shelf, 590
Rubble, 451
Rubidium, 208, *208,* 217, *217*

Runoff
  explanation of, **470**
  hydrology of, 475–477, *478*
Rutherford, Ernest, 208

Safety factor (*Fs*) equation, 466
Saffir-Simpson hurricane intensity scale,
      559, 560t
Saguaro National Park, *544*
Sahara Desert
  climate of, *542*
  droughts in, 473, *474*
  expansion of, 541
  rainfall in, 540
Sahel, 473
Salinity, **122**
Saltation, **511,** 536
Salt dome trap, *647,* 648
Saltwater margin, 484, *485*
San Andreas fault
  deformation textures in, 184
  earthquakes along, 184, 347, 349, 365,
      365
  as fault system, *187,* 187–188, *349*
  housing tracts near, *372*
  maps of, *177, 187*
  plate tectonics and, *53,* 53–54
  as transform-fault boundary, *35, 38,*
      38, 258
Sand
  on beaches, 565–566
  explanation of, **133**
  mass wasting and, 449–450
  mineralogy of, 134
  ripples in, 536, *537*
  sizes and shapes of, 133–134
  windblown, 532–533
Sandbars, 520, 568
Sandblasting, **534**
Sand budget, 565–566
Sand castles, *452*
Sand drift, 537
Sand dunes
  in deserts, 532–533
  explanation of, **511**–512
  formation of, 536–538, *537, 538*
  location of, 535–536
  types of, 538, *539*
Sand rights, 566
Sandstone
  explanation of, **133**
  groundwater and, 482
  mineralogy of, 134, *134*
  slope stability of, 451
  types of, 134, *134*
San Fernando earthquake, 364–365, 372
San Francisco earthquake (1906), *348,*
      349, 351, 369
San Juan River, *501*
Satellites
  historical background of, 221

seafloor charting by, 575
Saturated zone, **480**
Saturation, 63, 442
Saturn
  composition of, 224, 229
  exploration of, *246*
Saudi Arabia, 123
Schistosity, 156
Schists, 79, **156**–157, *157*
Scientific method
  example of, 24
  overview of, **4**–5
  plate tectonics and, 52
Scientific models
  explanation of, 5
  geologic maps as, 175
Scientific research, 4, *4*
Scientists, code of ethics for, 5
SeaBeam (ship), 573
Seafloor. *See also* Oceanic crust; Plate
      tectonics
  across two oceans, 575–578, *576, 577*
  age of, 41, 43
  explanation of, **29,** *30,* 552
  magnetic anomalies on, 39, *40,* 41
  main features of, 578
  studies of, 572, *573*–574, 575
  surface ships to probe, 572, *573, 574,* 575
  topography of, 572, *573*–577, 575–578
  volcanic island and, 139, *139*
Seafloor metamorphism, *153,* 154
Seafloor spreading
  in Atlantic Ocean, 257
  calculating speed of, 54
  effects of, 44, 253
  explanation of, 39, 107, 108
  igneous rock formation and, 106
  measurement of, *40,* 41
  rock cycle and, 79
Sea level
  explanation of, 9, *9*
  regional base levels and, 520
Sea level change
  climate change and, 570, 664–665, *665*
  conclusions related to, 614–615
  effects of, 568–570, *569*
  glaciation and, 606–607
  isostasy and, *599*
Seamounts, **576,** 578
Seawalls, 566, 567
Seawater. *See also* Oceans
  calcium and carbonate minerals in,
      122, 137, 138
  composition of, 140
  evaporation of, 140–141, *141*
  sodium chloride in, 122, 140
Secular variation, 398
Sedimentary basins
  explanation of, 116, **122**
  flexural, 123
  formation of, 122–123, *123*

in North America, 255–256, *257*
oil and gas resources in, 145
rift, 122
thermal subsidence, 122
transportation to, 119–121, *120*
Sedimentary deposits
  explanation of, 84–85, *85*
  floodwaters and, 503, *503*
  glaciers and, 603–606, *604, 605*
Sedimentary environments
  continental, 124, *125*
  explanation of, **124,** *125*
  marine, 124, *125,* 126
  shoreline, 124, *125*
  siliciclastic vs. chemical and biological,
      126–127, *127*
Sedimentary rocks
  carbonate, 136–140, *137, 138*
  chemical and biological, 135–144, 136t,
      137–141
  classification of, 132–142
  evaporite, 140–142, *141*
  explanation of, **74,** *75,* 77, 115
  formation of, *77,* 77–78, 142–143, 171
  granite and, 93
  minerals in, 76t, 78
  organic, 144
  plate tectonics and, 115
  porosity of, 478
  rock cycle and, 116–122, *117–121*
  siliciclastic, 132–135, 132t, *133–135*
Sedimentary sequence, 200
Sedimentary structures
  bedding sequences as, 129–130, *130*
  bioturbation as, 128, *129*
  cross-bedding as, 127, *128*
  explanation of, **127**
  graded bedding as, 127–128
  ripples as, 128, *128*
Sedimentation. *See* Deposition
Sediments
  bioclastic, 119
  biological and chemical, 77, *118,*
      135–144, *119,* 136t, *137–141*
  burial of, 130, 131
  in deserts, 543–544, *544*
  diagenesis of, 130–132, *131*
  eolian, 544
  erosion and, 117–119, 118t
  evaporite, 140–142, *141, 544, 544*
  explanation of, **77,** 115
  iron oxide, 144
  layers of, 77–78
  in oceans, 578–581, *579*
  organic, 144
  pelagic, 579
  phosphorite, 144
  porosity and permeability of, 478–479,
      480t
  precipitation and, 122
  process to solid rock from, 77

Sediments (*Contd.*)
  rock cycle and, 116–122, *117–121*
  siliceous, 143–144
  siliciclastic, 77, 117–118, 132–142, 132t
  stream transport of, 509–512, *510–512*, 515, *516*
  terrigenous, 126
Seine River, 506
Seismic hazard
  explanation of, *370*, **370**–372
  land-use policy and, 372
Seismic hazard map, *371*
Seismicity maps, 362
Seismic profiles, 387
Seismic ray paths, **384**–387, *385, 386*
Seismic risk, **370**
Seismic risk map, *371*
Seismic tomography, **394**, *395, 396*
Seismic waves. *See also* Waves
  earthquakes and, 355
  explanation of, **9,** 355
  to explore near-surface layering, *387*, 387–388
  exploring earth's interior with, *384–387*, 384–388
  focus and, 357
  ground deformation and, *358*
  ray paths of, 384–387, *385, 386*
  reflected, 387–388
  shaking from, 366–367
  travel-time curves for, 357
  types of, 10, 355, *356, 357*, 384
Seismograms, 354
Seismographs, 9–10, **354,** *354*
Seismometers, 9
Seitz, Gordon, *375*
Sequence stratigraphy, *212*, 212–213
Sequoia National Forest (California), *152*
Settling velocity, **511**
Shadow zones, **385,** *385*
Shaking, from seismic waves, 366–367
Shale gas, 652, *652*
Shales
  explanation of, **135**
  exploration for, 145–146
  metamorphism of, 160, *161*
  slope stability of, 451–452
Shark Bay (Australia), 295, *296*
Shearing, tectonics, **185,** *187*, 187–188
Shearing forces, **172**
Shearing tectonics, *185*
Shear strength, 466
Shear stress, 466
Shear waves, 10–12, 357, 384. *See also* Seismic waves; S waves
Sheet structures, 67
Shields, **255,** 260
Shield volcanoes, 234, 315, 321, *322*
Shock metamorphism, *153*, 155
Shoemaker, Eugene, 230, *230*
Shorelines. *See* Coastlines

Shoreline sedimentary environments, 124, *125*
Short-term prediction, earthquake, 375–376
Shuttle Radar Topography Mission (SRTM), 279–280
Siberian Traps, 343–344
Silent earthquakes, 362
Silica. *See also* Quartz
  early uses for, 143–144
  in igneous rocks, 94, 95
  polymorphs of, 64
  structure of, 58
Silicate ions, 62
Silicates
  cleavage in, 71
  composition of, 62, 94
  explanation of, 65, 65t
  framework, *66, 67*
  hardness of, 70
  in igneous rocks, 94
  as ore minerals, 81
  in sedimentary rocks, 78
  structures of, 65, *66, 67*
  weathering of, 438, 439, *440*
Siliceous environments, 127
Siliceous ooze, **580**–581
Siliceous sediments, 143–144
Siliciclastic sedimentary environments, 126
  coarse-grained, 133, *133*
  fine-crained, 132–136
  medium-grained, *133*, 133–134
Siliciclastic sediments
  classification of, 132–142, 132t
  explanation of, 77, *77*, 78, 117–**118**, 132
  streams carrying, 511
Silicon
  in crystal structures, 62
  in Earth, 13
Silicon ion, 65
Sills, **103**–104, *104*
Silt
  explanation of, **135**
  mass wasting and, 449–450
Siltstone, **135**
Single-chain structures
  explanation of, 65, *66*
  in pyroxenes, 71
Sinkhole, **490,** *490*
Sklodowska Curie, Marie, 208
Slates, **156,** *157*
Slip face, **537**
Slope instability, *465*, 465–466
Slope materials, 449–450
Slump, 460, *460*
Smith, William, 198–199
Snow. *See also* Precipitation
  explanation of, 470, 592
  transformation to ice, 592, *592*
Snowball Earth hypothesis, 609, *610*, 611

Snow line, 591
Sodium chloride
  as chemical compound, 60, *60*
  crystallization of, 61
  in seawater, 122, 140
Sodium ions, 122
Soil profiles, **447**
Soils
  ancient, 447
  explanation of, **445**
  as geosystems, *446*, 446–447
  porosity of, 478
  types of, 447, 448t
  weathering and, 441, 445–447, *446*
Sojourner (spacecraft), 238
Solar energy. *See also* Sun
  climate system and, 413, *413*
  explanation of, 13, *13*, 658, *658*
Solar forcing, **408**
Solar nebula, **223**
Solar radiation
  explanation of, 413
  Milankovitch cycles and, *421*
Solar system. *See also* Planets; *specific planets*
  exoplanets and, 248
  origin of, 14, 222–225, *224*
  small bodies of, 224–225
  telescopes for exploration of, 245–246
Solifluction, 456
Solubility, 442
Solutions, 63
Sonar, 572, 575
Sorting, **121**
Sound waves, 384
Source beds, 647
South African Plateau, epeirogeny and, 272–273
South American Plate, *35, 37, 38*
South Atlantic Ocean, 44
Southern Ocean, 44
Southern Oscillation, 417
Space exploration
  historical background of, 221–222
  landing spacecrafts and, 249–250
  questions addressed by, *223*
Specific gravity, **73**
Sphalerite, 82, *83*
Spinels, 67, *68*
Spirit (spacecraft). *See Mars Exploration Rovers* (spacecraft)
Spit, **568**
Splintery fracture, 71
Spreading centers. *See also* Mid-ocean ridges
  explanation of, **36**
  volcanism at, 330
Spring tides, 556
Stabilization wedges, **669**–670, *670*
Stalactites, 488, *489*
Stalagmites, 488, *488*

Staurolite, in rocks, 76t
Steam technology, 644
Steno, Nicolaus, 197
Stillwater deposit (Montana), 112
Stishovite, 155
Stocks, **103**
Stokes' law, 112–113
Storm surges, hurricane, *561,* **561**–563
Storm tide, 561
Strain, 348
Strath terraces, 638
Stratification. *See* Bedding
Stratigraphic succession, **197,** 198
Stratigraphic trap, *647, 648*
Stratigraphy
    chemical, 213
    of Colorado Plateau, 204, *205*
    explanation of, **197**
    of Mars, 244–245
    paleomagnetic, 213
    principles of, 197–198
    sequence, *212,* 212–213
Stratosphere, **408**
Stratovolcanoes, 321, *322*
Streak, **72,** 74t
Streak plate, 72, *72*
Streamgauge data, 525–526
Streamlines, 509, *510*
Stream power, **623,** *624*
Streams. *See also* Rivers
    antecedent, 505, *505,* 507
    base level of, 520, *520, 521*
    braided, *502,* 502–503
    channel patterns of, 500–501, *501*
    current flow and transport sediment
        in, 509–512, *510–512*
    deltas and, 513, *513, 514,* 515
    in deserts, 545
    discharge and, 516–518, *517, 518*
    effect of climate on, 521
    effluent, 480
    erosion by, 507–509, *507–509,* 543, 623,
        624, 637–638
    explanation of, 500, **500**
    exploration of, 499, 500
    floodplains created by, 503, *503*
    forms of, 500–505, *501–505,* 507, *510*
    as geosystems, 515–522, *516–522*
    graded, 521
    influent, 480
    longitudinal profile of, 519, *520*
    meanders by, *501,* 501–502, *502*
    recharge in, 480
    recording flow of water in, 525–526
    runoff in, 475, 476
    sediment transport by, 509–512,
        510–512
    source of, 507–509, *507–509*
    superposed, 507, *507*
    water in, 470
Stream valleys, 500, 623, *624,* 625

Stress, 348
Striations, **600,** *600*
Strike, **173,** *173*
Strike-slip faults
    explanation of, **177,** *178,* 179
    fault mechanisms and, 361
Stromatolites, **295**–296, *296,* 298, *298*
Strontium, 208, *208,* 217, *217*
Structures, sedimentary. *See* Sedimentary
    structures
Subbituminous coal, 654
Subduction, 263–264, 552
Subduction-related metamorphism, *164,*
    164–165
Subduction zones
    earthquakes in, 185–186, 347
    as magma factories, 108–111, *110*
    mantle convection and, 45
    volcanism at, 330–331
Sublimation, 470
Submarine canyons, 571–572
Subsidence
    explanation of, **122**
    from groundwater pumping, 484
Suess, Eduard, 28
Sulfate ion, 68
Sulfates
    explanation of, 65, 65t
    structure of, 68, *69,* 70
Sulfide ion, 67
Sulfides
    acidic water and, 292, *292*
    explanation of, 65, 65t
    as ore minerals, 81–83, *83*
    structure of, 67–68
Sulfur, microbial metabolism and, 294
Sulfur cycle, 294, *295*
Sulfur dioxide, 488
Sumatra earthquake, 347, 364, 373
Sumatra tsunami, 368–369, *369*
Sun. *See also* Solar energy
    Earth's climate and, 13, 15
    formation of, 223
    ocean tides and, 556
    radiation from, 413
    wind belts and, 530
Superposed streams, **507,** *507*
Superposition. *See* Principle of
    superposition
Surface temperature
    climate change and, *663*
    explanation of, 408, 413
Surface tension, 480
Surface waves, **355,** *356,* 357. *See also*
    Seismic waves
Surf zone, 554–555
Surges, in valley glaciers, **594**
Suspended load, **510**
Sustainable development, **670**–671
Suture, **165**
Swash, 555

Swash zones, 565
S waves, **355,** *356,* 357, 384–386, *385, 386*
S-wave velocity, 389, *389*
Switchgrass, *657,* 657–658
Syene (Egypt), 24
Symmetrical fold, 181, *181*

Tablelands, 623
Tadzhikistan earthquake, 458
Tahiti, 331–332
Talc, and Mohs scale of hardness of, 69t
Talus, **451**
Tangshan earthquake, 375
Tapeats Sandstone, 204
Tectonic age, **261**–262
Tectonic provinces
    explanation of, **254**
    of North America, 254, *254*
    types of, 260, *261*
Tectonics. *See also* Plate tectonics
    accretion and, 265–266
    base level of streams and, 520, *520,*
        521
    compressive, 185–187, *186*
    continental modification and, 266
    deformational styles and, 184–188,
        185–187
    explanation of, 28
    flake, 235
    landscapes and, 630–631, *631*
    shearing, *187,* 187–188
    tensional, 184–185, *185*
Tectonic uplift. *See* Uplift
Tektites, 155
Telescopes, 222, 245
Temperature
    in interior of Earth, 12, 392, *392,* 394
    metamorphism and, *150,* 150–152, *151*
    rock melting and, 97, *97,* 98
    seasonal, 408, *408*
    surface, 408, 413
Tensional forces
    deformational style and, 184–185, *185*
    explanation of, **172**
Tensional tectonics, **185**
Terminal moraine, 603t
Terraces, **521**–522, *522*
Terrains
    accreted, 263–265, *264*
    exotic, 264–265
Terrestrial planets, **224**. *See also* Earth;
        Mars; Mercury (planet); Venus
Terrigenous sediments, **126**
Tethys Ocean
    closing of, 267
    narrowing of, 44
Teton Range, 634, *634*
Tetrahedron. *See also* Silica
    carbon atoms bonding to, 61
    explanation of, 65
    isolated, 65

Tetrahedron. *See also* Silica (*Contd.*)
    in sheet structures, 67
    silicate, 65, *66*, 67
Texture
    deformation, 183–184, *184*
    explanation of, **74**
    of igneous rocks, 92–94, *92–94*
    metamorphic, 155–159, *156–158*, 159t
Thames River, 506
Tharp, Marie, 29, 30, *31*
Theories, 4–5
Thermal subsidence basins, **122**
Thermohaline circulation, **411**, 665
Thermophiles, 290–292, 291t
Thermoremanent magnetization, 39, **401**
Thompson, Lonnie, *431*
Three Gorges Dam, 658, *659*
Thrust fault, **177,** 179, 186
Thrust fault slope, 280
Tiber River, 506
Tibetan Plateau
    as convergent boundary, *35*
    formation of, 267
    topographic view of, *622*
Tidal flats, **557,** *557*
Tides
    effect on deltas, 515
    explanation of, **556,** 557
    variations in, 556–557
Tide terrace, 565
Tigris River, 506
Till, **603**
Tillites, **609**
Time scale, 631. *See also* Geologic time
        scale; Magnetic time scale; Planetary
        time scale
Titanium, 67
Tohoku earthquake, 185–186, 353, 355,
        365, 373, 656
Tohoku tsunami, *367,* 367–368, *372*
Tonga Trench, 577–578
Tongue stones, 197
Topaz, and Mohs scale of hardness, 69t
Topographic maps, 619, *619,* 620
Topography
    climate system and, 411, 628–629
    of deep seafloor, 572, *573–577,*
        575–578
    explanation of, **9,** *9,* **618,** *618,* 619
    geology and, 174
    of Mars, 235, *235*
Topset beds, **513**
Trace elements
    explanation of, **72**
    mineral color and, 72–73, *73*
Trade winds, 530
Transformations, 447
Transform fault, accretion of, *265*
Transform-fault boundaries
    earthquakes at, 364
    explanation of, **31,** *32,* 38

plate movement and, 44
tectonics at, *35*
Transition zone, **389**
Translocations, 447
Transpiration, 470
Transport
    in rock cycle, 116, *117*
    to sedimentary basins, 119–121, *120*
    wind as agent of, *531,* 531–533, *532*
Transverse dunes, **539**
Travel-time curves, 357
Travertine, 493, *494*
Trellis drainage, **504**
Triassic age, 204
Tributary, **505**
Trichloroethylene (TCE), 490
Trilobites, 204
Tropical depression, 559
Tropical storm, 559
Troposphere, **408**–409
Trunk, 619–620
Tsunamis
    causes of, 367–368, *368*
    explanation of, **367**
    from megathrust earthquakes,
        367–369, *369*
    sediment deposits by, 335
    warning systems for, 368–369, 373, 375
Tsunami stone (Aneyoshi), 355, *355*
Tuffs, **320**
Turbidites, 128, 571
Turbidity currents, 126, 128, 571, *572*
Turbulent flow, in streams, **509,** 510, *510*
Twentieth-century warming, **428**–430
Typhoons, 558. *See also* Hurricanes

Ultra-high-pressure metamorphism,
        154–155
Ultramafic rocks, **97**
Unconfined acquifers, 482
Unconformities, **200,** *200*
Unconsolidated material
    erosion of, 507–508
    mass movements of, 456–460, *457–460*
Unconsolidated materials
    angle of repose of, 450, *451*
    explanation of, **449**
    slopes of, 449–450
Unconsolidated mixtures, 450
Uniformitarianism principle, 540
United States. *See* North America
United States Geological Survey (USGS),
        525, 651
Universal ancestor, 290, 297
Universal tree of life, 289–290, *290*
Unsaturated zone, **480**
Uplift
    climate change and, 632
    erosion and, *279,* 279–280, 632, *633,*
        634–636
Upper mantle, **388**. *See also* Mantle

Ural Mountains, 44
Ural orogeny, 269
Uranium
    in groundwater, *492*
    isotopic dating and, 209
Uranium reserves, 656
Uranus, composition of, 224, 229
USGS. *See* United States Geological
        Survey (USGS)
U-shaped valleys, **602**

Vaiont Dam slide, 462, *462*
Valles Caldera, 323
Valley and Ridge province, 256
Valley glaciers. *See also* Glaciers
    erosion by, 601, *601*
    explanation of, **588**–589, *589*
    flow in, 594, *595, 596*
Valleys
    creation of, 500
    stream, 500, 507
Vallis Marineris canyon (Mars), 235, *235*
Vapor. *See* Water
Variscan orogeny, 269
Varve, **605**
Veins, **82, 105,** *105*
Ventifacts, **534,** *534*
Venus
    atmosphere of, 222
    characteristics of, 228, *228, 233,*
        233–234, *234*
    surface of, 230
    surface temperature of, 307
    as terrestrial planet, 224
Vesicles, 319, *319*
Viking missions (spacecraft), 238, 239
Vine, F. J., 39
Viscosity
    current flow and, 509–510
    explanation of, **97**
Vishnu schist, 204, *262*
Vittorini, Vincenzo, 377
Volcanic ash, **93,** 331, 532
Volcanic domes, 234, 321, *322*
Volcanic ejecta, 319–320
Volcanic gases, 326, 327
Volcanic geosystems, **314,** 315
Volcanic glass, 94
Volcanic rocks
    composition of, 92–94, 104, 160
    textures of, 319, *319*
    volcanism and, 314
Volcanism
    atmosphere and, *328,* 328–329
    effect on humans, 313, 334–341,
        335–337, 339, 342
    explanation of, 314
    hydrosphere and, *327,* 327–328, *328*
    intraplate, 331–334, *332–334*
    in lithosphere, 412
    patterns of, 329, *330*

plate tectonics, *329, 329–334, 330, 332*
  at spreading centers, 330
  in subduction zones, 330–331
Volcanoes. *See also specific volcanoes*
  at convergent boundaries, *34–35*, 37–38
  at divergent boundaries, *34*, 36
  eruptive styles and landforms of,
    320–321, *322–326, 323–326*
  explanation of, *314,* **314**–315
  gases emitted by, 439
  geosystem interactions and, 326–329,
    327, 328
  as geosystems, *314,* 314–315
  hazards to humans presented by, 313,
    334–339, *335–337*
  kimberlite pipes and, 277
  lava from, 315–318, *316–318*
  on Mars, 235, *235*
  natural resources from, 339–341
  ocean and atmosphere formation and,
    227, 227–228
  in the Pacific Ring of Fire, *31*
  predicting eruptions of, 338–339
  pyroclasts from, *319,* 319–320, *321*
  reducing risk of, 337–339
  seafloor and, 139, *139*
  shield, 315
  on Venus, 233–235
Volcanology, 337
Vostok station, 419, 420, *420*
Vuggy porosity, 478–479

Wadis, **545**
Water. *See also* Glaciers; Groundwater;
    Hydrologic cycle; Ice; Oceans;
    Rainwater
  acidic, 292, *292*
  ancient, 487
  as chemical compound, 60
  chemical weathering and, 438
  distribution of, 470, *471*
  drinking, 491–492
  fresh, 471
  geologic cycling of, 470–471, *471*
  for human use, 471
  hydrothermal, 492–494, *493, 494*
  importance of, 469
  as input for life, 286
  meteoric, 477, 493

potable, 491–492
  rock melting and, 98
  total supply of, 470
  viscosity of, 509–510
Waterfalls, 508–509
Water quality
  drinkability and, 491–492
  explanation of, 490
  water supply contamination and,
    490–491, *491*
Water resources, 474–475
Water rights, 474
Water vapor
  in atmosphere, 409
  explanation of, 470
  precipitation and, 472
Water vapor feedback, 415
Watt, James, 644, 666
Wave-cut terrace, **567,** *568*
Wave motion, 553–554, *554*
Wave refraction, *555,* 555–556
Waves
  height of, 554
  length of, 554
  shorelines and dynamics of, 553–554,
    554
  surf zone and, 554–555
Weather. *See also* Climate
  climate change and regional, 662–663
  explanation of, 15
Weathering. *See also* Chemical
    weathering; Erosion; Physical
    weathering
  chemical, 116, 436, *438,* 438–442, 443t,
    508
  controls on, 436–438, *437*
  in deserts, 542–543
  elevation and, 628–629
  erosion and, 117–119, 118t, 444–445,
    445
  explanation of, **77,** 116, 436
  factors controlling rates of, 436t
  organisms and, 443
  physical, 116, 436, 443–445, *443–445,*
    508
  precipitation and, 473
  rock cycle and, 79
  soil and, 441, 445–447, *446*
Wegener, Alfred, 28, *28*

Wells. *See also* Drilling; Oil
  discharge in, 483, *484*
  production of, 496–497
Westerlies, 530
Wetlands, disappearance of, 477
White Sands National Monument, 533
Wiechert, Emil, 10, 13
Wilson, J. Tuzo, 30, 272
Wilson cycle, *271,* 271–272
Wind. *See also* Currents
  as depositional agent, 535–540,
    536–540
  effects of, 529, 546
  as erosion agent, *534,* 534–535, *537*
  explanation of, 530
  in hurricanes, 559, *559*
  materials carried by, 531–533, *532, 533*
  over surface of oceans, 554
  strength of, 531
  as transport agent, 531–532
Wind belts
  deserts and, *541*
  explanation of, 530, *530*
Wind energy, 659, 658–659
Wind farms, 658, *659*
Wind power, 658, *659*
Windstorms, effects of, 529
Wisconsin glaciation, 422, 607–608, *608*
World ocean, 552
Wrangellia, 265

Xenoliths, 277, 278
Xenon, 487

Yangtze River, 506
Yellowstone National Park
  ash-flow deposits in, 326
  caldera in, 323
  geysers in, 327, *327*
  hot springs in, 493
  volcano in, 313
Yucca Mountain Nuclear Waste
    Repository, 657, *657*

Zagros Mountains, 267
Zebra mussels, *285*
Zeolites, **160**
Zion Canyon, 204
Zircon, 209

Divergent boundaries: plates move apart and create new lithosphere.

Convergent boundaries: plates move together, oceanic lithosphere is recycled back into the mantle, continental plates are deformed.

Transform-fault boundaries: plates slide horizontally past each other.